CALCULUS
WITH AN INTRODUCTION
TO LINEAR ALGEBRA

JOHN G. HOCKING
Michigan State University

HOLT, RINEHART AND WINSTON, INC.
New York Chicago San Francisco Atlanta
Dallas Montreal Toronto London Sydney

CALCULUS

WITH AN INTRODUCTION
TO LINEAR ALGEBRA

Copyright © 1970, by Holt, Rinehart and Winston, Inc.
All Rights Reserved
SBN 03-077910-3
Library of Congress Catalog Number: 77-101084
Manufactured in the United States of America
01234   19   987654321

Text and cover design by The Etheredges
Illustrations by Earl Kvam

PREFACE

This book is designed as a textbook for a first course in the calculus. It obviously reflects the personal convictions and preferences of its author. In my case these convictions have grown during twenty years at a large state university serving a broad cross-section of the college population.

MOTIVATION AND RIGOR

My first goal was to write to a wide variety of students. I deliberately have wed motivational material, and examples and problems from many different disciplines. Surely every student should be aware of the wide utility of the mathematics he studies. At the same time no student of mathematics can or should avoid completely the rigorous proofs that truly constitute the subject. Of course, the degree of rigor required in the elementary calculus varies with the individual student and his curriculum. I have tried to write a book that can be used at several different levels of rigor.

PROOFS

At several important places I use a "two-stage" proof. This is a favorite device of many teachers. The first stage is an heuristic discussion outlining the proof, making the approach natural and isolating the chief difficulties. The second stage presents the rigorous details, the "epsilontics" in the current idiom. There are other deviations from the strictly logical progression. A proof may be delayed to avoid a break in an important discussion, two or more approaches may be made to the same topic, and so on. No apologies need be made for such devices.

EXERCISES AND PROBLEMS

I consider the problem sets to be a vital part of a textbook. There are a large number of exercises and problems here, and I do draw a distinction between "exercise" and "problem." To my mind an exercise is intended to develop manipulative skill and confidence in routine operations. A student should do as many exercises as possible. Less obvious applications of the material, extensions of the theory, and foreshadowing of advanced course material are labelled as "problems." These are often identified further as stemming from such fields as physics and economics. The instructor can tailor his course to the needs of several different groups within a class simply by the appropriate assignment of problems. The exercise lists have been arranged with the harder exercises toward the end of each list. On the other hand, the problem lists are deliberately ungraded, partially reflecting the fact that practical problems do not come to us labelled as to level of difficulty.

INFINITE SEQUENCES AND INFINITE SERIES

If there is novelty in my presentation, it lies partly in the very early appearances of infinite sequences and infinite series. Note however that the two topics appear separately, sequences (as functions) in Chapter 2 and series in Chapter 5. Infinite sequences are used to motivate the concepts of tangent line and rate of change. This approach has always reached my students. Furthermore, the limit of a sequence is surely easier to understand than is the limit of a function of a continuous real variable. I have found that inability to compute the value of a logarithm (without tables) has caused undue concern in many students. This is avoided by having infinite series as an early working tool.

THE COMPUTER

This book is not oriented toward computational techniques, but I have tried to keep the reader aware of the existence and utility of the computer. Several years of teaching numerical analysis courses is surely reflected here in several places. On the other hand I have been prevented by colleagues from using a "flow-diagram" approach. I often use this as a classroom technique, but it does not seem to come across well in the formal exposition of a text.

GEOMETRY

Geometry is used very sparingly in the beginning of this book. A deep-seated confusion occurs in many of my students as they read the older texts. The geometry of the coordinate plane and quantities associated with the graph of a function drawn in that plane are considered to be identical. This confusion becomes critical in many applications of the definite integral, for instance. By working only on the affine plane (ignoring "distance" as long as possible) and by placing early emphasis upon such matters, much of this confusion is avoided. I am indebted to Professor A. W. Tucker for many fruitful discussions on such matters.

ORDER OF TOPICS. OPTIONS

In giving a course based on this book, many options are available. For example, much of *Chapter 1* can be omitted or assigned as remedial reading where needed by the student. The material on real numbers essentially follows Professor R. L. Wilder, whose one-time student I am proud to be. It might be wise to have every student read the first three sections (at least) for uniformity of notation. Sections 4, 5, 6, and 7 can be omitted, but I do think every student should read Sections 8 to 11.

The calculus course begins in *Chapter 2*. Care and time are taken with linear functions (Sections 1, 2, 3) before the introduction of the general concept of function. Sections 4 and 5 on functions and their graphs may be well known to most well-prepared freshmen. Nonetheless I usually ask these students to do more—rather than fewer—of the problems. This seems to prevent a certain complacency that would be ruinous at this stage. In Section 6 infinite sequences and limits are defined. This is the first true analysis in the book. The rate of change of a function (Section 7) and tangent lines (Section 8) are defined tentatively in terms of sequences. This provides motivation and a link to computers. Sections 9 and 10 deal with the concept of a limit of a function and its properties.

Chapter 3 introduces the derivative and puts it to immediate use in locating extreme values of a function (Section 2). It seems more natural that the discussion of continuity follow the introduction to differentiation, as is the case here. Section 4 (on the invariance of connectedness and compactness under continuous functions) need not be studied by all students, but this material is the very basis of all theory to follow. Sections 5, 6, 7, and 8 develop the chief properties of the derivative and Section 9 returns to the subject of the extreme values of a function. Polynomial and algebraic functions occupy the next two sections, which can be used primarily as supplemental material. Chapter 3 ends with the mean value theorem and its most important consequences.

Several application of differentiation are taken up in *Chapter 4*. Newton's iterative method, rectilinear motion, rates and related rates, increments and differentials, some mathematical economics, and indeterminate forms are the chief topics.

In *Chapter 5* infinite series are studied, beginning with Taylor's theorem and including power series and Taylor series. Some topics, such as the integral test for convergence (Chapter 7) and examples involving transcendental functions (Chapters 8 and 9), must be delayed, but this location of infinite series seems advisable. In addition to presenting an important computational tool, this position treats the infinite series as an application of differentiation.

Chapter 6 constitutes the theory of integration as I believe it should be presented to beginners. After a discussion of "area" and of the average value of a function (Sections 1 and 2), and the introduction of the summation symbol (Section 3), the definite integral is defined as the limit of a sequence, or rather, as the common limit of an upper and a lower sequence. This definition simply refines the discussions in Sections 1 and 2. Properties of the definite integral are given in Section 5, and Section 6 presents the precise definition of the integral in order to complete the theory. Note, however, that the discussion is limited to continuous functions. At this stage there seems to be no advantage in being as general as possible.

The fundamental theorem of calculus occupies the first sections of *Chapter 7*. A start is made on techniques of integration, including integration by parts. The applications of integration are largely to physical problems, but there is also a special treatment of economic renewal theory. Many students need not study all these matters, of course.

Exponential and logarithm functions are considered at length in *Chapter 8*. After preminary sections on inverse functions and their derivatives, I treat exponential and logarithm functions in three steps. The first step is the natural follow-up of the usual secondary school method (Sections 3 and 4). Then the functional equation approach is given (Section 5) and finally the integral definition is explored. All three are tied together with the simple proof of the limit result in

Section 6. Differentiation and integration formulas are developed, logarithmic differentiation is mentioned, and power series expansions are given. Lastly, the general first-order linear differential equation is solved and the usual applications of this equation are made.

The study of trigonometric functions in *Chapter 9* begins with a long section on analytic trigonometry. Much of this will be review material for most students. Section 2 on differentiation of sines and cosines cannot be omitted from this chapter, of course, but Section 3 can be shortened to include just the differentiation formulas. Similarly Section 4 on graphs need not be covered by the student with a good trigonometry course in his preparation. The study of power series for sin x, cos x, and tan x makes the trigonometric tables less mysterious. Section 6 gives and discusses the solution of the differential equation $f''(x) + k^2 f(x) = 0$. Inverse trigonometric functions, hyperbolic functions and their inverses then lead to the usual list of differentiation formulas.

Chapter 10 on methods of integration is rather short. It starts with a list of integration formulas, then discusses the integration of trigonometric functions, trigonometric substitutions, integration of rational functions, and a few common rationalizing substitutions. Section 6 on the integration of power series and the lengthy discussion in Section 7 of Simpson's rule for numerical integration complete the Chapter. Notice that it is possible for a student of, say, economics to skip all of Chapter 9 and 10 except perhaps these final two sections.

With Chapter 10 the calculus of functions in one variable is put aside. The student should have the usual computational techniques and an honest, if incomplete, idea of the basic theory. This much material can be covered quite easily in two semesters or hastily in two quarters. Since the pace depends upon the students' preparation and perhaps upon the instructor's augmentation, it is not possible to be more explicit.

FUNCTIONS OF SEVERAL VARIABLES

Chapters 11 and 12 constitute a brief introduction to the calculus of functions of several variables. After a non-vectorial study of lines and planes in space, functions, limits and continuity, partial derivatives, and the chain rules are presented. Taylor's theorem, extreme values, and Lagrange multipliers are introduced. This material does not require lengthy preparation in geometry and should be readily accessible to the non-physical scientist. Very few proofs are included in Chapter 12. It is essentially a preview of advanced work and consists of basic computational techniques with no pretense of theoretical completeness. The late and brief presentation of polar, cylindrical, and spherical coordinates reflects my conviction that students do not require this material in detail until they are well able to study it themselves.

Although it is deliberately placed late in the book, I consider the geometry in *Chapter 13* to be an absolute minimum for physical science majors. The treatment is vectorial; free vectors and vector algebra are used to redefine the basic concepts about lines and planes. Directional derivatives, vector-valued functions and some differential geometry of curves incorporate basic elements of vector analysis as applied to equations of motion. Of the remaining material the brief and unified treatment of conic sections as a one-parameter family deserves the most attention.

LINEAR ALGEBRA

Chapter 14 is a study of those aspects of linear algebra that depend upon the concept of rank. This material is motivated by the basic problem of solving a system of linear equations, and involves a computer-inspired emphasis on pivots, row-echelon form and other fundamental ideas. Matrix algebra is introduced and immediately related to the reduction of systems of linear equations to row-echelon form. The importance of square matrices is both practical and theoretical, but determinants and Cramer's rule are de-emphasized. Following a section on vector spaces and subspaces, the principal theorems concerning the solution set of a system of linear equations are given. The final six sections provide a bridge to physical applications and also foreshadow advanced work in algebra and geometry as well as in analysis.

The last four chapters (11–14) of the book have been arranged so that considerable flexibility is available. For example, with only an explanation in the definition of the product of matrices, Chapter 14 (up to Section 9) can be studied immediately after Sections 11.1 and 11.2. Chapters 12 and 13 (as well as 9 and 10) might be omitted entirely in a course for social science majors. Chapter 13 can readily precede Chapter 12 if course requirements make it necessary. On the other hand, there is every reason to suppose that some instructors will want to cover Chapter 11 right after Chapter 4 or 5. I have tried to make this feasible (with a bare minimum of omissions). In short, several rearrangements of the material have been envisioned and some care has been taken to help the instructor to design his own course.

ACKNOWLEDGEMENTS

Many colleagues, friends, and students contributed to this book during its long gestation period. To all of you I repeat my heartfelt thanks. The consulting editor, Professor R. M. Thrall, is practically a coauthor as well as a cograndfather. The editor, Walter Brownfield of Holt, Rinehart and Winston, truly earned his slave-driver's license, but he remains a friend. The illustrator, Earl Kvam, is surely one of the best in the field. Marvin Camburn did invaluable work in solving the problems.

Lastly, I should like to dedicate this book to my family. It must be wearisome living with a grouchy author who covers every horizontal surface in the house with books and papers, all labelled "Do Not Touch." Thank you, loved ones.

John G. Hocking
East Lansing, Michigan
January, 1970

CONTENTS

1 PRELIMINARIES 1

1.1 The number systems 2
1.2 The order relation 8
1.3 Some set-theoretic notation 16
1.4 Mathematical induction 17
1.5 Geometric representation of real numbers 24
1.6 An important property of real numbers 28
1.7 Factored inequalities 30
1.8 Numerical vectors 33
1.9 Geometric representation of two-dimensional numerical vector 37
1.10 Geometric representation of three-dimensional numerical vectors 41
1.11 The norm of a vector. Distance 44

2 FUNCTIONS, GRAPHS, AND LIMITS 49

2.1 Linear functions and lines 50
2.2 Parallel lines 56
2.3 The slope of a line 59
2.4 Functions and their graphs 61
2.5 Graphing a function (introduction) 69
2.6 Infinite sequences 77
2.7 The rate of change of a function 86
2.8 Tangent lines to a graph 89
2.9 The limit of a function 92
2.10 Properties of limits 100

3 THE DERIVATIVE 105

3.1 The derivative of a function 106
3.2 Extreme values of a function 112
3.3 Continuity 118
3.4 Two important properties of continuous functions 123
3.5 The algebra of functions 129
3.6 Another notation for the derivative 135
3.7 The composition of functions. The chainrule 135
3.8 Higher derivatives 143

3.9	Extremum values	152
3.10	Polynomial functions	156
3.11	Algebraic functions	164
3.12	A mean value theorem for derivatives	168
3.13	Some consequences of the mean value theorem	172

4 APPLICATIONS OF DIFFERENTIATION — 177

4.1	The Newton iterative method for solving equations	178
4.2	Rectilinear motion	183
4.3	Rates and related rates	188
4.4	Increments and differentials	193
4.5	Another notation for the derivative	200
4.6	Some applications to economics	201
4.7	Indeterminate forms	207

5 INFINITE SERIES — 213

5.1	Taylor's theorem	214
5.2	Infinite series	221
5.3	Power series	234
5.4	Taylor series	243

6 THE DEFINITE INTEGRAL — 249

6.1	Area under a curve	250
6.2	The average value of a function	257
6.3	The summation notation Σ	262
6.4	The definite integral	266
6.5	Properties of the definite integral	271
6.6	The existence theorem	276

7 THE INDEFINITE INTEGRAL — 283

7.1	Functions defined by integrals	284
7.2	The fundamental theorem of calculus	288
7.3	The indefinite integral	293
7.4	Area problems	300
7.5	Substitutes of variables	303

7.6	Integration by parts	309
7.7	Improper integrals	315
7.8	Volumes	320
7.9	Arc length	324
7.10	Area of a surface of revolution	327
7.11	Work	330
7.12	Fluid pressure	335
7.13	Renewal theory	338

8 EXPONENTIAL AND LOGARITHM FUNCTIONS 341

8.1	Inverse functions	342
8.2	The derivatives of an inverse function	347
8.3	Exponential and logarithm functions	349
8.4	Graphs of exponential and logarithm functions	354
8.5	A functional equation approach	356
8.6	An integral approach	360
8.7	Differentiation and integration formulas	362
8.8	Logarithmic differentiation	373
8.9	Power series for e^x and $\ln x$	376
8.10	A useful differential equation	382
8.11	Application	386

9 TRIGONOMETRIC FUNCTIONS 397

9.1	Analytic trigonometry	398
9.2	Further properties of cos and sin	406
9.3	The other trigonometric functions	412
9.4	Graphs of trigonometric functions	421
9.5	Power series expansions	427
9.6	Another important differential equation	432
9.7	Inverse trigonometric functions	436
9.8	The hyperbolic functions	445
9.9	Inverse hyperbolic functions	452
9.10	Integration of inverse functions	457
9.11	The differentiation formulas	459

10 METHODS OF INTEGRATION — 461

10.1	Integration formulas	462
10.2	Integration of trigonometric functions	463
10.3	Trigonometric substitutions	471
10.4	Integration of rational functions	479
10.5	Rationalizing substitutions	484
10.6	Integration of power series	489
10.7	Simpson's rule for numerical integration	495

11 THE DIFFERENTIAL CALCULUS OF FUNCTIONS OF SEVERAL VARIABLES — 503

11.1	Lines in space	504
11.2	Linear functions and planes	506
11.3	Regions	511
11.4	Functions of several variables	514
11.5	Limits and continuity	520
11.6	Partial derivatives	524
11.7	Tangent lines and planes	533
11.8	Differentiable functions	538
11.9	The chainrule	541
11.10	Taylor's theorem in two variables	548
11.11	Extreme values	552
11.12	Extreme value with constraints	561
11.13	Exact differentials	568

12 MULTIPLE INTEGRALS — 573

12.1	Integration of a function of two variables	574
12.2	Volume	580
12.3	The double integral	586
12.4	Polar coordinates	590
12.5	First moments and the center of mass	598
12.6	Moments of inertia	601
12.7	The triple integral	605
12.8	Cylindrical coordinates	611
12.9	Spherical coordinates	614
12.10	Heterogeneous solids	618

13 ANALYTIC AND VECTOR GEOMETRY — 623

- 13.1 Free vectors — 624
- 13.2 Vector algebra — 629
- 13.3 Lines and planes — 636
- 13.4 Scalar fields and level sets — 641
- 13.5 Directional derivatives, the gradient — 645
- 13.6 Conics — 651
- 13.7 Quadrics — 660
- 13.8 Vector functions; parametric equations — 667
- 13.9 Differentiation of vector functions — 672
- 13.10 Arc length — 677
- 13.11 Curvature — 683
- 13.12 Velocity and acceleration in polar coordinates — 690

14 LINEAR ALGEBRA — 695

- 14.1 Systems of linear equations — 696
- 14.2 Reduction to row-echelon form — 702
- 14.3 The algebra of matrices — 713
- 14.4 Square matrices — 720
- 14.5 Determinants — 727
- 14.6 Vector spaces and subspaces — 736
- 14.7 The solution set of a system of linear equations — 742
- 14.8 Linear mappings — 749
- 14.9 Transformation of coordinates in $\mathbf{R}^n$ — 758
- 14.10 Symmetric matrices and diagonalization — 765
- 14.11 Quadratic forms — 773

Tables — T-1

Answers to selected exercises and problems — A-1

Index — I-1

CALCULUS
WITH AN INTRODUCTION
TO LINEAR ALGEBRA

1
PRELIMINARIES

This chapter collects a few of the basic mathematical facts needed for a successful study of the calculus. First, we consider again the real number system and its properties. Second, we shall speak about numerical vectors and their geometric representation. This leads toward the very small amount of analytic geometry that we shall need for our early purposes.

1.1 THE NUMBER SYSTEMS

The fundamental theorems of the calculus depend very heavily upon properties of the real number system. We need devote but little time to the familiar algebraic properties of real numbers. Certain nonalgebraic properties (see Section 1 to 7 below) are very important, however, and the definitions given here have been adopted to facilitate precise discussions of these properties.

The first system of numbers we learned is the system of *natural numbers*,

$$1, 2, 3, 4, \cdots$$

It is a profound discovery when one realizes that this system is nonterminating; that is, there is no largest natural number. Our first mathematical experience consisted of learning about the two operations: addition $(+)$ and multiplication $(\times)$ on natural numbers. During this early period our attention was called to certain basic properties of these operations, properties such as the associative laws, the commutative laws, and the distributive laws. We also learned that the natural numbers are closed under these operations, and this is another fundamental fact.

Even the most simple algebraic problems force us out of the system of natural numbers, however. For instance, there is no natural number x satisfying the equation $x + 15 = 7$. In short, the natural numbers are *not* closed under subtraction $(-)$. This shortcoming forces the introduction of the *integers*,

$$\cdots, -3, -2, -1, 0, 1, 2, 3, \cdots$$

The operations $+$ and $\times$ are redefined for these numbers, of course. And the newly defined addition and multiplication are chosen to agree with those for the natural numbers and to satisfy the same basic properties, the associative, commutative, and distributive laws.

Again we find the system of integers is not enough. Such an equation as $3x = 11$ has no integer solution. In order to solve all equations of the form $qx = p$, where p and q are integers, we are forced to introduce the *rational numbers*, all numbers of the form p/q where p and q are integers and $q \neq 0$. In terms of $+$ and $\times$ for integers, the operations of addition and multiplication of rational numbers are defined by

$$\frac{m}{n} + \frac{p}{q} = \frac{mq + np}{nq} \qquad \frac{m}{n} \times \frac{p}{q} = \frac{mp}{nq}$$

(We have used juxtaposition of letters in place of the "times" sign where convenient.)

1.1 THE NUMBER SYSTEMS

By the use of these definitions, it is easy to prove the following six basic algebraic properties:

F1. If x and y are rational numbers, then $x + y$ and $x \times y$ are also rational numbers. (The *closure* properties.)

F2. If x is a rational number, then $x + 0 = x$ and $x \times 1 = x$. (The existence of additive and multiplicative *identity elements*.)

F3. If x, y, and z are rational numbers, then

$$x + (y + z) = (x + y) + z,$$
$$x \times (y \times z) = (x \times y) \times z$$

(The *associative laws*.)

F4. For each rational number x, there is a rational number $-x$ such that $x + (-x) = 0$. If $x \neq 0$, there is a rational number x^{-1} such that $x \times x^{-1} = 1$. (The existence of additive and multiplicative *inverses*.)

F5. If x and y are rational numbers, then

$$x + y = y + x, \qquad x \times y = y \times x$$

(The *commutative laws*.)

F6. If x, y, and z are rational numbers, then

$$x \times (y + z) = (x \times y) + (x \times z)$$

(The *distributive law*.)

These properties F1 to F6 are the axioms of arithmetic. (In technical language these are the axioms for an *algebraic field*.) The reader should be familiar with these properties and with the usual theorems and equation-solving techniques based upon these properties. Two fundamental such theorems are

1. For each rational number x, $x \times 0 = 0$.
2. If $x \times y = 0$, then either $x = 0$ or $y = 0$.

[Assuming (1), we prove (2) as follows: If $x = 0$, then there is nothing to prove. If $x \neq 0$, then F4 says there is a rational number x^{-1} such that $x \times x^{-1} = 1$. Then F5 says that $x^{-1} \times x = 1$. Multiplying both sides of the equation by x^{-1},

$$x^{-1} \times (x \times y) = x^{-1} \times 0 = 0 \text{ [by (1) above]}$$

By F3, $x^{-1} \times (x \times y) = (x^{-1} \times x) \times y = 1 \times y = y$, using F2, whence $y = 0$.]

The extension of the number system does not stop at rational numbers, however. Pythagoras (circa 530 B.C.) is said to have discovered to his dismay that the equation $x^2 = 2$ has no rational solution. In other words, he proved that $\sqrt{2}$ is not a rational number. His proof was *contrapositive*; he assumed that $\sqrt{2}$ is a rational number and, using impeccable logic, produced a contradiction of a known fact.

To understand Pythagoras' proof, we only need to know that the product of two odd integers is an odd integer and the product of even integers is even. From this it follows that the square p^2 of an integer p is even if and only if p is even. (The reader may provide proofs of the statements.) Now suppose that there are integers p and q, $q \neq 0$, such that $\sqrt{2} = p/q$. Suppose further that the fraction p/q is in lowest terms, in the sense that p and q have no common factors. In particular, then, not both p and q are even. This is a permissable assumption, of course, and it is this fact that we now contradict.

If $\sqrt{2} = p/q$, then $2 = p^2/q^2$, or $2q^2 = p^2$. This says that p^2 is an even integer, which implies that p is even, that is, $p = 2m$ for some integer m. Then we have the equation $2q^2 = p^2 = (2m)^2 = 4m^2$ or $q^2 = 2m^2$. Therefore q^2, and hence q, is even. But this contradicts the fact that not both p and q are even. Each step in the proof is logically correct, so that the false conclusion forces us to admit that the assumption $\sqrt{2} = p/q$ is false. ∎

Analogous proofs show that such numbers as $\sqrt{5}$, $\sqrt{7}$, and $\sqrt{11}$ are not rational. Such proofs are based upon the following fact: The prime number p divides the square k^2 of an integer k if and only if p divides k. Several of the problems at the end of this section call for an application of this fact.

Since there are "numbers" such as $\sqrt{2}$ that are not rational, we must extend the number system once again. This time we adopt a definition not based upon the algebraic requirements of solving certain equations.

A *real number* x is a nonterminating decimal (with a sign $+$ or $-$) of the form

$$x = \pm a_n a_{n-1} \cdots a_1 a_0 \cdot d_1 d_2 d_3 \cdots d_k \cdots$$

where n is a nonnegative integer and where each a_i and d_j is a *digit*, 0, 1, 2, $\cdots$, 9. If n is positive, then $a_n \neq 0$.

We are just using the familiar decimal position system, so that

$$x = \pm[a_n \times 10^n + \cdots + a_1 \times 10 + a_0 \times 10^0 + d_1 \times 10^{-1} + d_2 \times 10^{-2} + \cdots + d_k \times 10^{-k} + \cdots].$$

This definition is not yet precise, because certain numbers have two such decimal expansions. For instance, we may write the number $\frac{3}{4}$ as either

$$\tfrac{3}{4} = 0.75000 \cdots 000 \cdots$$

or

$$\tfrac{3}{4} = 0.74999 \cdots 999 \cdots$$

(The latter *is* an expression for $\frac{3}{4}$, as we shall see shortly.) This sort of ambiguity is eliminated simply by selecting one of these forms. For technical reasons we choose the second form. We now stipulate that, with the exception of zero, no real number is written so as eventually to repeat all zeros. The number 0 is written as $0.00000 \cdots 000 \cdots$ and is the only exception to this rule.

1.1 THE NUMBER SYSTEMS

The purpose of this definition of a real number is the desire to give precise proofs, particularly of certain nonalgebraic properties of the real number system. This does *not* mean that we must give up such convenient symbols as 2, $\frac{3}{4}$, $\sqrt{6}$, π, and so forth. Indeed, we shall actually use these nonterminating decimals only a very few times. However, they do lurk behind the scenes, so to speak, ready to supply some precision to our arguments when the need arises.

A definition of the arithmetic operations $+$ and $\times$ on nonterminating decimals would be quite unwieldy. The well-known algorithms for adding and multiplying finite decimals do not apply. Even if we suppose that the numbers

$$+ \; 2.333 \cdots 333 \cdots$$
$$+ \; 1.712712712 \cdots 712 \cdots$$

repeat as they seem to do, just where do we begin to add them together? While it *is* possible to provide rigorous definitions of $+$ and $\times$ as operations on nonterminating decimals, we do not do so here. It is enough to make the following assumption:

The arithmetic operations $+$ and $\times$ are defined on the real numbers and satisfy the axioms F1 to F6.

How do the rational numbers fit into the real number system? Most readers know already.

The real number $x = a_n \cdots a_1 a_0 \cdot d_1 d_2 \cdots d_k \cdots$ is a rational number if and only if there are integers k_1 and k_2 such that, wherever k is larger than k_1, we have $d_{k+k_2} = d_k$.

This means that, after the k_1 decimal place, the decimal repeats in blocks of length k_2. For instance, the number

$$x = 93.17352438624386 \cdots 24386 \cdots$$

starts repeating after the fourth decimal place, $k_1 = 4$, and repeats in blocks of length 5, $k_2 = 5$.

For simple examples it is easy to express such a repeating decimal as a rational number p/q. For instance, consider

$$x = 0.74999 \cdots 999 \cdots \qquad (k_1 = 2, k_2 = 1)$$

Multiplying x by $10^{k_2} = 10$, we may write

$$10x = 7.4999 \cdots 999 \cdots$$

Now we *can* see how to subtract x from $10x$

$$\begin{array}{r} 7.4999 \cdots 999 \cdots \\ -0.7499 \cdots 999 \cdots \\ \hline 10x - x = 9x = 6.7500 \cdots 000 \cdots \end{array}$$

or

$$x = \frac{1}{9}(6.75) = \frac{675}{900} = \frac{3}{4}$$

(We see that $0.7499 \cdots 999 \cdots$ *is* equal to $\frac{3}{4}$, as claimed earlier.)
For the number

$$x = 2.372727 \cdots 72 \cdots \qquad (k_1 = 1, k_2 = 2)$$

we multiply by $10^{k_2} = 10^2 = 100$ to get

$$100x = 237.272727 \cdots 27 \cdots$$
and subtract
$$x = 2.372727 \cdots 27 \cdots$$
$$\overline{99x = 234.900 \cdots 00 \cdots}$$

Thus

$$x = \frac{1}{99}(234.9) = \frac{2349}{990} = \frac{261}{110} \quad \blacksquare$$

An eventually repeating nonterminating decimal is a rational number. Conversely, there must be *irrational numbers;* we need only write a decimal that does *not* repeat. For example,

$$x = 1.101001000100001000001 \cdots$$

does not repeat, because the nth nonzero digit 1 and the $(n+1)$st digit 1 are separated by n zeros. This means that there is no (fixed) integer k_2, and hence this number does not satisfy our definition of a repeating decimal. Of course, there are many other ways to be certain that one writes a nonrepeating decimal. The reader may wish to find some of these other ways.

EXERCISES 1.1 Verify each of the identities 1 to 4, justifying each step with a reference to one of the axioms F1 to F6.

1. $(x \pm y)^2 = x^2 \pm 2xy + y^2$
2. $(x + y)(x - y) = x^2 - y^2$
3. $(x \pm y)^3 = x^3 \pm 3x^2y + 3xy^2 \pm y^3$
4. $(x \pm y)(x^2 \pm xy + y^2) = x^3 \pm y^3$

Using axioms F1 to F6 and the statements (1) and (2) on page 3, prove statements 5 to 8.

5. If $x^2 = 0$, then $x = 0$.
6. If $xyz = 0$, then $x = 0, y = 0$ or $z = 0$.
7. If $a + b = a + c$, then $b = c$.
8. If $ab = ac$ and $a \neq 0$, then $b = c$.
9. Prove that the numbers 1 and -2 are the only roots of the equation

$$x^2 + x - 2 = 0$$

10. Prove that
$$(x + a)^2 + (y + a)^2 - 2(x - a)(y + a)$$
does not depend upon the value a.

11. For any real number x, show that $(-x)^2 = x^2$.

12. Find the real numbers a and b such that
$$\frac{a}{x+1} + \frac{b}{x-2} = \frac{x}{(x+1)(x-2)}$$

13. Is it ever true that
$$\frac{1}{a} + \frac{1}{b} = \frac{1}{a+b}?$$
If so, when?

14. If $x \neq 0$ and $x \neq -1$, show that
$$\frac{2xy + x^2}{x^2 + x} = \frac{x + 2y}{x + 1}$$

15. If a is an irrational number, prove that both $a + 1$ and $2a$ are also irrational.

PROBLEMS 1.1

1. Prove that the rational numbers with the operations $+$ and $\times$ as defined in Section 1.1 satisfy axioms F1 to F6 and therefore constitute an algebraic field.

2. Which of axioms F1 to F6 fail to hold for the addition and multiplication of integers?

3. Prove that each rational number p/q has an eventually repeating decimal expansion.

4. Verify that axioms F1 to F6 hold in the "number system" having just the three elements 0, 1, 2 with "addition" and "multiplication" defined by the following tables:

+	0	1	2
0	0	1	2
1	1	2	0
2	2	0	1

×	0	1	2
0	0	0	0
1	0	1	2
2	0	2	1

5. Prove that there are no integers x and y satisfying the equation $2x + 4y = 9$.

6. Let n be a positive integer. Prove that $\sqrt{n}$ is rational if and only if n is a perfect square.

7. Let λ be a fixed real number not zero or one. Consider the six numbers λ, $1 - \lambda$, $1/\lambda$, $1/(1 - \lambda)$, $(\lambda - 1)/\lambda$, $\lambda/(\lambda - 1)$. Among these six numbers define the binary operation $x \circ y$ to mean "substitute x for λ in the expression for y." Thus
$$(1 - \lambda) \circ \left(\frac{1}{1 - \lambda}\right) = \frac{1}{1 - (1 - \lambda)} = \frac{1}{\lambda}.$$

Verify axioms F1, F2, F3, and F4 for this operation $\circ$ in this collection of six numbers.

8. Determine the real numbers a, b, c, and d such that
$$\frac{x^2}{(x^2+2)(x+1)^2} = \frac{a}{x+1} + \frac{b}{(x+1)^2} + \frac{cx+d}{x^2+2}$$
Can you develop a theory to cover such manipulations?

9. Prove that $\log_{10} 3$ is irrational. (That is, prove that there is no rational number p/q such that $10^{p/q} = 3$.)

10. Use just the axioms F1 to F6 to prove that there is only one additive identity and only one multiplicative identity. (That is, show that, if $x + 0' = x$ for each real number x, then $0' = 0$; and similarly, if $x \cdot 1' = x$ for each x, then $1' = 1$.)

1.2 THE ORDER RELATION

If p and q are integers, then p *is* less than q, and we write $p < q$ (or $q > p$), if there is a natural number n such that $p + n = q$. This definition can be used to give precise proofs of all the well-known facts about inequalities concerning integers. More important to us is the fact that this definition extends easily to rational numbers.

First we define two rational numbers m/n and p/q to be *equal* if and only if $mq = np$. Then we define $m/n < p/q$ if and only if $mq < np$. With this definition we can prove the usual theorems about (rational number) inequalities. In particular, four basic properties can be established easily (Problem 2):

O1. If r and s are rational numbers, then one and only one of the relations $r = s$, $r < s$, or $s < r$ is true. (The *trichotomy property*.)

O2. If r, s, and t are rational numbers and if both $r < s$ and $s < t$, then $r < t$. (The *transitive law*.)

O3. If r, s, and t are rational numbers and $r < s$, then $r + t < s + t$.

O4. If r, s, and t are rational numbers and if both $r < s$ and $0 < t$, then $rt < st$.

Using properties O1 to O4 as axioms, we prove the following results which provide us with some working tools.

■ **THEOREM 1.1.** Let r and s be rational numbers. Then $r < s$ if and only if there is a rational number t, $0 < t$, such that $r + t = s$.

PROOF: The technical details which follow are simply the recognition that we are dealing with the number $t = s - r$. The reader may wish to skip these details or to provide a proof for himself. Notice while reading how heavily we depend upon properties of integer inequalities.

Without the loss of generality we assume that $r = m/n$ has the denominator $n > 0$, and similarly that $s = p/q$ has $q > 0$. Now $m/n < p/q$ tells us that $mq < np$ and hence that $0 < np - mq$. Defining the rational number
$$t = \frac{np - mq}{nq},$$

we easily see that $0 < t$ and $r + t = s$.

Conversely, suppose that $m/n + l/k = p/q$ where $0 < l/k$. Again we may assume that $0 < k$. The equality is written as

$$\frac{mk + nl}{nk} = \frac{p}{q}$$

which is true if and only if $(mk + nl)q = nkp$. Since

$$mkq + nlq = nkp$$

and $0 < nlq$ (Why?), we have $mkq < nkp$, and since $0 < k$, we have $mq < np$ which is equivalent to $m/n < p/q$. ∎

■ **THEOREM 1.2.** Let r be a rational number. Then $r < 0$ if and only if $0 < -r$.

PROOF: If $r < 0$, then by Theorem 1.1 there is a rational number $t > 0$ such that $r + t = 0$. Clearly, $t = -r$. Since $0 < t$, we have $0 < -r$.

Conversely, if $0 < -r$, then by O3 we add r to both sides of the inequality to obtain $r < -r + r = 0$. ∎

■ **THEOREM 1.3.** If $r < s$ and $t < u$, then $r + t < s + u$.

PROOF: By O3, $r + t < s + t$ and also $s + t < s + u$. Then by O2, $r + t < s + t$ and $s + t < s + u$ implies that $r + t < s + u$. ∎

■ **THEOREM 1.4.** If $r < s$ and $t < 0$, then $rt > st$.

PROOF: Theorem 1.2 tells us that $0 < -t$. Then by O4 we have

$$r(-t) < s(-t) \quad \text{or} \quad -rt < -st$$

Using O3 we add $rt + st$ to both sides of this inequality, obtaining

$$rt + st - rt < rt + st - st \quad \text{or} \quad st < rt \quad ∎$$

Our next step is to extend the meaning of the order relation $<$ to apply to real numbers. As was just done for rational numbers, we first give the definition of equal real numbers. Note that the following definition simply says that two equal nonterminating decimals agree at each decimal place: Let

$$x = \pm a_m \cdots a_1 a_0 \cdot d_1 d_2 \cdots d_k \cdots$$

and

$$y = \pm b_n \cdots b_1 b_0 \cdot e_1 e_2 \cdots e_k \cdots$$

be two real numbers. Then $x = y$ if and only if all four of the following conditions are satisfied:

1. the signs of x and y agree,
2. $m = n$,
3. for each i, $a_i = b_i$, and
4. for each j, $d_j = e_j$.

Note that it was necessary to rule out one of the two decimal expansions of such numbers as $\frac{3}{4}$, or else this definition of equality would be invalid.

In most cases it is much easier to ascertain the relative size of two decimals than to determine which of two rational numbers is the smaller. For instance, to show that $\frac{723}{952} < \frac{1813}{2387}$, we must show that $723 \times 2387 < 952 \times 1813$, an easy but tedious job. But presenting these numbers as decimals

$$\frac{723}{952} = +0.75945\cdots$$
$$\frac{1813}{2387} = +0.75953\cdots$$

we see at a glance that the first is less than the second. That is, we simply note that the two decimals agree up to the third place, and in the fourth place the first number has the smaller digit. We refine this observation into a definition.

If the real numbers x and y given above are not equal, then they must differ somewhere. We simply use the several ways in which they can differ to define the order relation $<$. First, assume that

$$x = +a_m \cdots a_1 a_0 \cdot d_1 d_2 \cdots d_k \cdots$$

and

$$y = +b_n \cdots b_1 b_0 \cdot e_1 e_2 \cdots e_k \cdots$$

are positive real numbers. Then $x < y$ if one of the following conditions holds:

(0) $a_m \cdots a_1 a_0 < b_n \cdots b_1 b_0$ (as integers)
(1) $a_m \cdots a_1 a_0 = b_n \cdots b_1 b_0$ but $d_1 < e_1$
(2) $a_m \cdots a_1 a_0 = b_n \cdots b_1 b_0$ and $d_1 = e_1$ but $d_2 < e_2$
$\cdots$
(k) $a_m \cdots a_1 a_0 = b_n \cdots b_1 b_0$, $d_1 = e_1, \cdots, d_{k-1} = e_{k-1}$ but $d_k < e_k$.

The definition of $<$ is completed by agreeing that

(a) $0 < x$ for each positive real number x
(b) if y is a negative number and either $0 = x$ or $0 < x$, then $y < x$
(c) if both x and y are negative, then $x < y$ if and only if $-y < -x$

Using the above explicit definition, it is possible to prove the order properties 01 to 04 and Theorems 1.1 to 1.4 for real numbers. We shall simply accept these as axioms, leaving the proofs to the enterprising reader.

REMARK: In an algebra course one might be asked to show that

$$\sqrt{4} + \sqrt{8} < \sqrt{5} + \sqrt{7}$$

We do not need decimal expansions to do this (although it could be done that way). We simply square both sides

$$4 + 2\sqrt{4 \times 8} + 8 < 5 + 2\sqrt{5 \times 7} + 7$$

1.2 THE ORDER RELATION

(Justify this step.) Subtracting 12, we have
$$2\sqrt{32} < 2\sqrt{35}$$
and then dividing by 2,
$$\sqrt{32} < \sqrt{35}$$
Since $32 < 35$, this last is true. (Again, some justification is required.) ∎

It is very easy, using property 04 and Theorem 1.4, to prove that
$$0 < x^2$$
for any real number $x \neq 0$. Since no real number has a negative square, such an equation as $x^2 + 2 = 0$ has no real number solution. We can certainly write $x = \sqrt{-2}$, but this symbol does not represent a real number. This observation leads to yet another extension of the number system—the *complex numbers*. This book does not require the complex number system, so we do not study this system.

REMARK: An algebraic expression such as $\sqrt{3x + 2}$ represents a real number if and only if $3x + 2$ is not negative. That is, we must have $0 \leq 3x + 2$. Using properties 02 and 03 we easily reduce the inequality $0 \leq 3x + 2$ to $-\frac{2}{3} \leq x$. Thus $\sqrt{3x + 2}$ is a real number if and only if $-\frac{2}{3} \leq x$. ∎

On the other hand, while negative numbers have no (real) square roots, we can state the following:

1. Each positive real number has a square root.
2. Each positive real number has an nth root for any natural number n.
3. The nth root of any real number is again a real number if n is an odd natural number.

We shall give an indication of how these statements are proved in Section 1.6.

A *linear inequality in one variable* is an algebraic statement of the form
$$ax + b < cx + d$$
where a, b, c, and d are constants (fixed real numbers). Such a linear inequality is "solved" by replacing it with an equivalent statement of the simpler forms $m < x$ or $x < m$. This replacement is done via a sequence of applications of properties 01 to 04 and Theorems 1.1 to 1.4.

∎ *Example.* Consider the linear inequality
$$3x + 1 < x + 7$$

Applying property 03, we add -1 to both sides, which gives us the equivalent statement,
$$3x < x + 6$$
Next, add $-x$ to both sides (property 03 again) to obtain
$$2x < 6$$
Last, multiply by $\frac{1}{2}$ (property 04) to obtain the "solution"
$$x < 3$$

The *absolute value* $|x|$ of a real number x is defined by
$$|x| = x \text{ if } x \geq 0$$
$$ = -x \text{ if } x < 0$$
Recalling that $\sqrt{a}$ always denotes a nonnegative number (for $2 > 0$, of course), we have the equivalent definition
$$|x| = \sqrt{x^2}$$
We shall assume that the reader knows (or will prove for himself) the next result.

■ **THEOREM 1.5.** Let x and y be real numbers. Then

1. $|-x| = |x|$
2. $|xy| = |x| \times |y|$
3. $|x + y| \leq |x| + |y|$

■ **THEOREM 1.6.** Suppose $a > 0$ is a real number. Then the inequality $|x| < a$ is equivalent to the inequality $-a < x < a$.
  PROOF: Suppose that $|x| < a$. If $x \geq 0$, then $|x| = x$, so that we have $x < a$. Thus x satisfies $-a < x < a$. If $x < 0$, then $|x| = -x < a$, whence by Theorem 1.4, $-a < x < 0$ and again x satisfies $-a < x < a$. This proves that, if x satisfies $|x| < a$, then x satisfies $-a < x < a$.
  Conversely, suppose that x satisfies $-a < x < a$. If x also satisfies $0 \leq x$, then $|x| = x$ whence $0 \leq x < a$ implies that $|x| < a$. If $x < 0$, then $|x| = -x$. Since $-a < x < 0$, Theorem 1.4 applies to say that $-x < a$ or $|x| < a$. This shows that, if x satisfies $-a < x < a$, then x also satisfies $|x| < a$. ■

■ **COROLLARY 1.7.** Let b and ϵ be real numbers, $\epsilon > 0$. Then the inequalities $|x - b| < \epsilon$ and $b - \epsilon < x < b + \epsilon$ are equivalent.
  PROOF: In view of Theorem 1.6, $|x - b| < \epsilon$ is equivalent to $-\epsilon < x - b < \epsilon$. By property 03 we can add b to both sides, obtaining the equivalent inequality $b - \epsilon < x < b + \epsilon$. ■

EXERCISES 1.2

1. For each of the following pairs of numbers determine which is the larger:
   (a) $\sqrt{2}, \sqrt[3]{3}$
   (b) $\sqrt{3}, \sqrt[3]{5}$
   (c) $\sqrt{6}, \sqrt{2} \times \sqrt{3}$
   (d) $\sqrt{2} + \sqrt{6}, \sqrt{3} + \sqrt{5}$
   (e) $2\sqrt{2}, \sqrt[3]{23}$
   (f) $\sqrt{19} + \sqrt{21}, \sqrt{17} + \sqrt{23}$

2. (a) If $a < b$, show that $a < \frac{1}{2}(a + b) < b$.
   (b) If $0 < a < b$, show that $a < \sqrt{ab} < b$.

3. For what values of x is each of the following expressions a real number?
   (a) $\sqrt{3 + 5x}$
   (b) $\sqrt{1 - 2x}$
   (c) $\sqrt{b^2 - 4ax}$
   (d) $\sqrt{x^2 + 4}$

4. Solve the following equations:
   (a) $|x - 1| = |x - 3|$
   (b) $|3 - x| = |x + 3|$
   (c) $|2x - 1| = |x + 1|$
   (d) $|x - a| = |x - b|$

5. Show the equivalence of each pair of the following inequalities:
   (a) $|x - 3| < 1,\ 6 < x + 4 < 8$
   (b) $|x - 1| < 2,\ 0 \leq |2x - 3| < 5$
   (c) $|x - 3| < 1,\ \dfrac{1}{8} < \dfrac{1}{x + 4} < \dfrac{1}{6}$
   (d) $|x - 4| < 1,\ \dfrac{1}{5} < \dfrac{1}{x - 2} < 1$

6. Solve the following linear inequalities:
   (a) $3x + 1 < x + 5$
   (b) $1 - 2x < 5x - 2$
   (c) $\dfrac{x}{4} + \dfrac{3 - x}{6} > 0$
   (d) $-1 < \dfrac{2x + 3}{5} < 1$
   (e) $-0.02 < \dfrac{3 - 2x}{4} < 0.02$
   (f) $-\epsilon < \dfrac{ax + b}{c} < \epsilon$ $\quad (c > 0)$
   (g) $\dfrac{1}{x + 3} < 0$
   (h) $\dfrac{1}{5 - 2x} > 0$

7. Establish the equalities.
   (a) $|xy| = |x| \times |y|$
   (b) $\left|\dfrac{x}{y}\right| = \dfrac{|x|}{|y|}$

8. Solve the inequalities.
   (a) $|x - 3| < |x - 1|$
   (b) $|x - 3| < 2|x + 5|$
   (c) $|x - 1| + |x + 3| \geq 6$
   (d) $4|x^2 - 1| + |x^2 - 4| \geq 6$

9. Let x and y be positive real numbers. Prove that
$$\dfrac{x}{\sqrt{y}} + \dfrac{y}{\sqrt{x}} \geq \sqrt{x} + \sqrt{y} \geq \sqrt{x + y}$$
   When does equality hold on the left?

10. If x and y are nonnegative real numbers, show that
    (a) $\dfrac{1}{2}(x^6 + y^6) \geq \left(\dfrac{x + y}{2}\right)\left(\dfrac{x^2 + y^2}{2}\right)\left(\dfrac{x^3 + y^3}{2}\right)$
    (b) $\dfrac{1}{2}(x^{11} + y^{11}) \geq \left(\dfrac{x + y}{2}\right)\left(\dfrac{x^3 + y^3}{2}\right)\left(\dfrac{x^7 + y^7}{2}\right)$

PROBLEMS 1.2

1. Show that $|x| = \sqrt{x^2}$ for each real number x.

2. Use the definition of $<$ for rational numbers (together with axioms F1 to F6) to establish 01, 02, and 03 for rational numbers.

3. Let x and y be real numbers satisfying $x < y$. Prove that there is at least one rational number p/q satisfying the inequality

$$\begin{matrix} x < p \\ q < y \end{matrix}$$

(We express this property by saying "The rational numbers are *dense* in the real numbers.")

4. For any rational number $p/q > 0$ and any integer $n > 0$, prove that there is a positive integer $m > 0$ satisfying

$$n < m\left(\frac{p}{q}\right)$$

(This is called the *Archimedean property* of the order relation $<$.)

5. Use the results of Problems 3 and 4 above to prove that the order relation $<$ on the real numbers is also Archimedean.

6. For what positive integers n is $2^n > n^2$? Is $2^n > n^3$?

7. Let x and y be positive real numbers. Prove the inequalities

    (a) $x + \dfrac{1}{x} \geq 2$

    (b) $\dfrac{x}{5y} + \dfrac{5y}{4x} \geq 1$

    (c) $\sqrt{\dfrac{x}{y}} + \sqrt{\dfrac{y}{x}} \geq 2$

    (d) $\dfrac{x + 3y}{3y} \geq \dfrac{4x}{x + 3y}$

8. Let x be a positive real number and prove that

    (a) $\dfrac{x}{2 + x} < \sqrt{1 + x} - 1 < \dfrac{x}{2}$

    (b) $\dfrac{x}{3 + x} < \sqrt[3]{1 + x} - 1 < \dfrac{x}{3}$

9. If the numbers $x_1, x_2, x_3,$ and x_4 satisfy both

$$x_1 \leq x_2 \leq x_3 \leq x_4 \quad \text{and} \quad x_1 = x_4,$$

show that all four numbers are equal.

10. If x and y are not zero, show that

$$\frac{x^2}{y^2} + \frac{y^2}{x^2} + 6 \geq \frac{4x}{y} + \frac{4y}{x}$$

11. If $x, y,$ and z are positive real numbers, show that

    (a) $(x + y + z)\left(\dfrac{1}{x} + \dfrac{1}{y} + \dfrac{1}{z}\right) \geq 9$

    (b) $\dfrac{x}{y} + \dfrac{y}{z} + \dfrac{z}{x} \geq 3$

12. Give very short proofs of the inequalities
    (a) $|x - y| \geq |x| - |y|$
    (b) $|x + y| \geq |x| - |y|$

1.2 THE ORDER RELATION

13. Let k and n be natural numbers satisfying $1 < k < n$. Prove that
$$\sqrt{n-k} + \sqrt{n+k} < \sqrt{n-1} + \sqrt{n+1}$$

14. Let a, b, and c be positive numbers and let e, f, and g be any numbers. For what number x is the expression
$$a|x-e| + b|x-f| + c|x-g|$$
a minimum?

15. Determine the smallest number M such that
$$-19 + 12x - 2x^2 \leq M$$
for all real numbers x.

16. Determine the largest number m such that
$$m \leq x^2 - 4x + 12$$
for all real numbers x.

17. If r is a rational number between 1 and 2 approximating $\sqrt{2}$, prove that the number $(r+2)/(r+1)$ is a better approximation. (This requires that you show that $|(r+2)/(r+1) - \sqrt{2}| < |r - \sqrt{2}|$.) Use this fact to compute $\sqrt{2}$ to four-decimal accuracy.

18. Prove that
$$\begin{aligned}\tfrac{1}{2}(x + |x|) &= x \quad \text{if } x \geq 0 \\ &= 0 \quad \text{if } x < 0\end{aligned}$$

19. If $|h|$ is small compared to n^3, show that $n + h/3n^2$ is a good approximation to $\sqrt[3]{n^3 + h}$. In particular, for $n > 0$, prove that
$$\left| n + \frac{h}{3n^2} - \sqrt[3]{n^3 + h} \right| < \frac{|h^2|}{8n^5}$$
Use this to approximate $\sqrt[3]{1725} = \sqrt[3]{12^3} - 3$. How accurate is the approximation?

20. Show that
$$\begin{aligned}\tfrac{1}{2}(x + y + |x - y|) &= x \quad \text{if } x > y \\ &= y \quad \text{if } x < y\end{aligned}$$

21. Show that
$$\begin{aligned}\tfrac{1}{2}(x + y - |x - y|) &= x \quad \text{if } x < y \\ &= y \quad \text{if } x > y\end{aligned}$$

22. Let a, b, and c be positive real numbers and d, e, and f be any real numbers. Prove that
$$\text{minimum of } \left(\frac{d}{a}, \frac{e}{b}, \frac{f}{c}\right) \leq \frac{d - e + f}{a - b + c} \leq \text{maximum of } \left(\frac{d}{a}, \frac{e}{b}, \frac{f}{c}\right).$$

23. If a, b, and c are positive real numbers, then prove that
$$\tfrac{1}{3}(a + b + c) \geq \sqrt{\tfrac{1}{3}(ab - ac + bc)} \geq \sqrt[3]{abc}$$

24. (Physics) The attraction exerted by a thin spherical shell of radius a upon a particle at distance $b \neq a$ from the center of the sphere is given by the formula

$$A = \frac{k}{b^2}\left(1 - \frac{a-b}{\sqrt{a^2 - 2ab + b^2}}\right)$$

where k is a constant. Show that $A = 2k/b^2$ if $b > a$ and $A = 0$ if $b < a$. Interpret this result in physical language.

25. Prove that the "equality" of rational numbers as defined here is a reflexive, symmetric, and transitive relation.

26. If $x > 0$ and $y > 0$ and if $x^2 < y^2$, show that $x < y$.

1.3 SOME SET-THEORETIC NOTATION

The concept of a *set* of objects is basic to our very thought processes. Beyond remarking that the word is used as a synonym for such words as *collection, class, family,* and so on, we attempt no definition of the word "set." In short, we accept the concept of a set as a primitive, an undefined logical term.

If S is a set and x is an object, then the relationship expressed in the sentence "x is an element of S" is also a primitive term in our logical system. We denote this relationship by

$$x \in S$$

where the symbol $\in$ is read as the phrase "is an element of." Of course,

$$x \notin S$$

means, and is read, "x is not an element of S."

The four sets of numbers discussed in Sections 1.1 and 1.2 find such frequent use that we shall assign names (letters) to them permanently. In this book, we shall always use the boldface letters as follows:

$\mathbf{N}$ = the set of all natural numbers
$\mathbf{Z}$ = the set of all integers
$\mathbf{Q}$ = the set of all rational numbers
$\mathbf{R}$ = the set of all real numbers

Thus we can make such statements as $7 \in \mathbf{N}$ but $-3 \notin \mathbf{N}$, $-3 \in \mathbf{Z}$ but $\frac{9}{5} \notin \mathbf{Z}$, $\frac{9}{5} \in \mathbf{Q}$ but $\sqrt{2} \notin \mathbf{Q}$, and $\sqrt{2} \in \mathbf{R}$ but $\sqrt{-2} \notin \mathbf{R}$.

There is a convenient and widely used notation that symbolizes the general verbal expression "the set of all (variable objects) such that (the rule describing the set under consideration)." This symbol has the form

$$\{\cdots : \cdots\}$$

For example,

$$\{x : x \in \mathbf{R},\ x < 3\}$$

is "the set of all x such that x is a real number and $x < 3$." Acutally, we say "the set of all real numbers less than three" when we read the symbol. Similarly,

$$\{(x, y) : x \in \mathbf{R}, y \in \mathbf{R}, x + y = 1\}$$

is "the set of all ordered pairs of real numbers whose sum is one." We normally omit the statement $x \in \mathbf{R}$, when we deal with real numbers, and write these sets as

$$\{x : x < 3\}, \{(x, y) : x + y = 1\}$$

This notation is used throughout the book and should be familiar to the reader.

A set S *is contained in another set* T, and we write $S \subset T$, if each element of S is also an element of T. Briefly, $S \subset T$ if $x \in S$ implies $x \in T$. Thus we have $\mathbf{N} \subset \mathbf{Z}, \mathbf{Z} \subset \mathbf{Q}$ and $\mathbf{Q} \subset \mathbf{R}$.

Two sets S and T are *equal*, $S = T$, if both $S \subset T$ and $T \subset S$.

Given sets S and T, two new sets are defined by the set equations

$$S \cup T = \{x : x \in S \text{ or } x \in T\}$$ ("or" here is used in the inclusive sense of and/or)

$$S \cap T = \{x : x \in S \text{ and } x \in T\}$$

We call $S \cup T$ the *union* of S and T and $S \cap T$ the *intersection* of S and T. If the sets S and T have no elements in common, they are said to be *disjoint*. In such a case, $S \cap T$ is said to be *empty* and we write

$$S \cap T = \phi$$

where ϕ is our symbol for the empty set.

This brief introduction simply standardizes the notation to be used in the rest of the book. We assume that the reader has an acquaintance with elementary set theory as is found today in most schools.

1.4 MATHEMATICAL INDUCTION

Many theorems involve some such phrase as "for all natural numbers" or "for each natural number n," and so forth. Here are three examples:

1. For any $x \in \mathbf{R}$ and any n numbers, $y_1, y_2, \cdots, y_n \in \mathbf{R}$,

$$x(y_1 + y_2 + \cdots + y_n) = xy_1 + xy_2 + \cdots + xy_n.$$

(A generalized distributive law.)

2. For each $n \in \mathbf{N}$,

$$1 + 2 + \cdots + n = \tfrac{1}{2}n(n + 1)$$

3. For each $n \in \mathbf{N}$, $n(n + 1)(n + 2)$ is divisible by 6.

In essence, such a statement is an unending set of theorems, one for each $n \in \mathbf{N}$, and demands a proof by mathematical induction. We give proofs of statements 2 and 3.

■ *Example 1.* The formula $1 + 2 + \cdots + n = \frac{1}{2}n(n + 1)$ is certainly true if $n = 1$. Suppose it fails to be true for some natural numbers $n > 1$. Surely there is then a smallest number k for which the formula is false and we know that $k > 1$. It follows that the formula is true for the number $k - 1$, that is

$$1 + 2 + \cdots + (k - 1) = \tfrac{1}{2}(k - 1)[(k - 1) + 1] = \tfrac{1}{2}(k - 1)k$$

Now we may add k to both sides of this true equation, obtaining

$$\begin{aligned} 1 + 2 + \cdots + (k - 1) + k &= \tfrac{1}{2}(k - 1)k + k \\ &= k(\tfrac{1}{2}(k - 1) + 1) \quad \text{simple} \\ &= \tfrac{1}{2}k(k - 1 + 2) \quad \text{factoring} \\ &= \tfrac{1}{2}k(k + 1) \end{aligned}$$

Thus, counter to assumption, the formula is true for the number k. This contradiction tells us that there can be no natural number n for which the formula is false.

■ *Example 2.* If $n = 1$, then $n(n + 1)(n + 2) = 6$ and is certainly divisible by 6. Suppose there are natural numbers n for which the number $n(n + 1)(n + 2)$ is not divisible by 6. Let k be the smallest such number. We know that $k > 1$, since we first remarked that the theorem is true for $n = 1$. It follows that the number

$$(k - 1)[(k - 1) + 1][(k - 1) + 2] = (k - 1)k(k + 1)$$

is divisible by 6. But now it is easy to verify that

$$k(k + 1)(k + 2) = (k - 1)k(k + 1) + 3k(k + 1)$$

Either k or $k + 1$ is an even number, so $3k(k + 1)$ is divisible by 6. Knowing that $(k - 1)k(k + 1)$ is divisible by 6, it follows that $k(k + 1)(k + 2)$ is divisible by 6. Again we have achieved a contradiction, which proves that there are no natural numbers n for which $n(n + 1)(n + 2)$ is not divisible by 6.

The reader should have spotted an unjustified assumption in the above "proofs." Why *must* there be a smallest natural number k for which a certain formula is false? We actually require an axiom here.

The well-ordering axiom for the natural numbers:
Let $A \subset \mathbf{N}$ be any subset of the natural numbers. If $A \neq \phi$, then A contains a smallest element.

This is a very believable axiom. For if $A \neq \phi$, we ask first, is $1 \in A$? If $1 \in A$, then obviously 1 is the smallest number in A. If $1 \notin A$, is $2 \in A$? If so, then 2 is the smallest number in A, and if not, we examine 3, and so forth. Doesn't it seem obvious that we shall eventually arrive at the first number in A? Now this line of argument is *not* a proof at all. At best, it is

1.4 MATHEMATICAL INDUCTION

just an indication of why we should be willing to accept the axiom as part of our logical system.

Notice that the set $\mathbf{Z}$ does not have this well-ordering property. The nonempty subset $\{x : x \in \mathbf{Z}, x < 0\}$ of all negative integers has no smallest element. Similarly, neither $\mathbf{Q}$ nor $\mathbf{R}$ has the well-ordering property. For example, the nonempty subset $\{x : x \in \mathbf{Q}, 0 < x < 1\}$ has no smallest element. (To be precise, we should say that the set $\mathbf{N}$ is well-ordered by the usual order relation $<$, while $\mathbf{Z}, \mathbf{Q},$ and $\mathbf{R}$ are not well-ordered by $<$. It is conceivable that $\mathbf{R}$ may be well-ordered by some other order relation.)

If we accept the well-ordering axiom for $\mathbf{N}$, then the proofs of Examples 1 and 2 are indeed proofs. Let us construct another precise proof using this axiom.

■ **Example 3.** For each $n \in \mathbf{N}$,
$$1^2 + 2^2 + \cdots + n^2 = \tfrac{1}{6}n(n+1)(2n+1)$$

PROOF: Let A be the subset of $\mathbf{N}$ consisting of all $n \in N$ for which the formula is false. Note that
$$1^2 = (\tfrac{1}{6})(1)(1+1)(2+1)$$
that is, the formula is true for $n = 1$. Thus $1 \notin A$. Suppose $A \neq \phi$. By the well-ordering axiom, A contains a smallest number, call it k. Because we know $k > 1$, we have $k - 1 \in \mathbf{N}$ while $k - 1 \notin A$. Hence we have the true equation
$$1^2 + 2^2 + \cdots + (k-1)^2$$
$$= \tfrac{1}{6}(k-1)[(k-1)+1][2(k-1)+1]$$
$$= \tfrac{1}{6}(k-1)(k)(2k-1)$$

We add k^2 to both sides to obtain
$$1^2 + 2^2 + \cdots + (k-1)^2 + k^2 = \tfrac{1}{6}(k-1)(k)(2k-1) + k^2$$
$$= k[\tfrac{1}{6}(k-1)(2k-1) + k]$$
$$= \tfrac{1}{6}k[(k-1)(2k-1) + 6k]$$
$$= \tfrac{1}{6}k[2k^2 + 3k + 1]$$
$$= \tfrac{1}{6}k(k+1)(2k+1)$$

This says that the formula is true for k, that is, $k \notin A$. Thus we have arrived at the contradictory statements, $k \in A$ and $k \notin A$. This contradiction is the logically derived consequence of assuming that $A \neq \phi$. Our only alternative is to conclude that $A = \phi$, which is precisely the theorem we wanted to prove. ■

Note that we had to prove that the desired formula *does* hold for some natural number. We must know that the set $\mathbf{N} - A = \{x : x \in \mathbf{N}, x \notin A\}$ is not empty. Otherwise, we might be proving that the formula is always wrong.

■ *Example 4.* For each $n \in \mathbf{N}$,

$$\frac{1}{1 \cdot 2} + \frac{1}{2 \cdot 3} + \cdots + \frac{1}{n(n+1)} = \frac{n}{n+1}$$

PROOF: First note that, for $n = 1$, the formula is true because $1/(1 \cdot 2) = 1/(1 + 1)$. Next let A be the set of all natural numbers for which the formula is false. We know that $1 \notin A$ so that $\mathbf{N} - A \neq \phi$. If the theorem is not true, then A is not empty and hence has a smallest element, say k. Since $k > 1$, it follows that $k - 1 \in \mathbf{N} - A$. Thus we have the true equation

$$\frac{1}{1 \cdot 2} + \frac{1}{2 \cdot 3} + \cdots + \frac{1}{(k-1)(k)} = \frac{k-1}{k}$$

Adding $1/[k(k+1)]$ to both sides, we have

$$\frac{1}{1 \cdot 2} + \frac{1}{2 \cdot 3} + \cdots + \frac{1}{(k-1)(k)} + \frac{1}{k(k+1)}$$

$$= \frac{k-1}{k} + \frac{1}{k(k+1)}$$

$$= \frac{(k-1)(k+1) + 1}{k(k+1)}$$

$$= \frac{k^2}{k(k+1)}$$

$$= \frac{k}{k+1}$$

This tells us that $k \notin A$. The contradictory statements, $k \in A$ and $k \notin A$, then imply that $A = \phi$ and complete the proof. ■

We have been using *mathematical induction* in a slightly disguised form.

The principle of mathematical induction:
Let $a \in \mathbf{N}$ be a fixed natural number and for each natural number $n \geq a$, let p_n denote a mathematical statement.

1. If p_a is true, and
2. the truth of p_k implies the truth of p_{k+1} for each $k \geq a$, then p_n is true for each $n \geq a$.

Using the well-ordering axiom, we can prove this principle as a theorem.
Let B denote the set $\{n : n \in \mathbf{N}, n \geq a\}$. Let A denote the set of all natural numbers n such that $n \geq a$ and p_n is false. By assumption 1, $a \notin A$. Now if $A \neq \phi$, then A has a smallest element, say k. We know that $k > a$

whence $k - 1 \geq a$ and $k \notin A$. That is, the statement p_{k-1} is true. Then by assumption 2, the truth of p_{k-1} implies the truth of p_k, so that p_k is true. But this says that $k \notin A$. The contradictory statements, p_k is false ($k \in A$) and p_k is true ($k \notin A$), have followed logically from assuming that A was nonempty. Thus A must be empty and p_n is true for each natural number $n \geq a$. ∎

The principle of mathematical induction, as it is stated above, gives rise to the induction proofs with which the reader is already familiar. Recall that an induction proof *must* consist of two steps. First, it must be shown that the given statement is true for some $a \in \mathbf{N}$ (often $a = 1$). Secondly, we show that the truth of the statement for any natural number $k \geq a$ implies the truth of the statement for the next number $k + 1$.

■ **Example 5.** For each natural number $n \geq 2$,

$$\left(1 - \frac{1}{4}\right)\left(1 - \frac{1}{9}\right) \cdots \left(1 - \frac{1}{n^2}\right) = \frac{n + 1}{2n}$$

PROOF: 1. If $n = 2$, then

$$1 - \frac{1}{4} = \frac{2 + 1}{2 \cdot 2} = \frac{3}{4}$$

Thus we have the needed start. (Note that the statement is false for $n = 1$.)

2. Suppose that

$$\left(1 - \frac{1}{4}\right)\left(1 - \frac{1}{9}\right) \cdots \left(1 - \frac{1}{k^2}\right) = \frac{k + 1}{2k}$$

Multiplying both sides by $1 - 1/(k + 1)^2$, we obtain

$$\left(1 - \frac{1}{4}\right)\left(1 - \frac{1}{9}\right) \cdots \left(1 - \frac{1}{k^2}\right)\left(1 - \frac{1}{(k + 1)^2}\right)$$

$$= \left(\frac{k + 1}{2k}\right)\left(1 - \frac{1}{(k + 1)^2}\right)$$

$$= \left(\frac{k + 1}{2k}\right)\left(\frac{k^2 + 2k}{(k + 1)^2}\right)$$

$$= \frac{k + 2}{2(k + 1)}$$

$$= \frac{(k + 1) + 1}{2(k + 1)}$$

By the induction principle, the proof is complete. ∎

EXERCISES 1.4

Establish formulas 1 to 12 using mathematical induction.
1. $1 + 3 + 5 + \cdots + (2n - 1) = n^2$
2. $1 + 5 + 9 + \cdots + (4n - 3) = n(2n - 1)$

3. $1^2 + 3^2 + 5^2 + \cdots + (2n - 1)^2 = \tfrac{1}{3}n(2n - 1)(2n + 1)$
4. $1^3 + 2^3 + 3^3 + \cdots + n^3 = \tfrac{1}{4}n^2(n + 1)^2 = (1 + 2 + 3 + \cdots + n)^2$
5. $1 \cdot 2 + 2 \cdot 3 + 3 \cdot 4 + \cdots + n(n + 1) = \tfrac{1}{3}n(n + 1)(n + 2)$
6. $1 \cdot 3 + 3 \cdot 5 + 5 \cdot 7 + \cdots + (2n - 1)(2n + 1) = \tfrac{1}{3}n(4n^2 + 6n - 1)$
7. $3 \cdot 4 + 4 \cdot 7 + 5 \cdot 10 + \cdots + (n + 2)(3n + 1) = n(n + 2)(n + 3)$
8. $\dfrac{1}{1 \cdot 3} + \dfrac{1}{3 \cdot 5} + \dfrac{1}{5 \cdot 7} + \cdots + \dfrac{1}{(2n - 1)(2n + 1)} = \dfrac{n}{2n + 1}$
9. $\dfrac{1}{1 \cdot 4} + \dfrac{1}{4 \cdot 7} + \dfrac{1}{7 \cdot 10} + \cdots + \dfrac{1}{(3n - 2)(3n + 1)} = \dfrac{n}{3n + 1}$
10. $1 + 2 + 4 + 8 + \cdots + 2^n = 2^{n+1} - 1$
11. $2 + 6 + 18 + \cdots + 2 \times 3^n = 3^{n+1} - 1$
12. $(1 - \tfrac{1}{4})(1 - \tfrac{1}{9}) \cdots (1 - 1/n^2) = (n + 1)/2n$ for $n \geq 2$

For Exercises 13 to 16, assume that $n \geq 1$.

13. $4^n - 1$ is always divisible by 3.
14. $3^{2n} - 1$ is divisible by 4.
15. $7^{2n} - 1$ is divisible by 8.
16. If $x \neq 1$, then $x^n - 1$ is divisible by $x - 1$.

PROBLEMS 1.4

1. If $x \neq 1$, show that
$$1 + x + x^2 + \cdots + x^{n-1} = \frac{x^n - 1}{x - 1}$$

2. If $x > 0$, show that
$$\frac{1}{x(x + 1)} + \cdots + \frac{1}{(x + n - 1)(x + n)} = \frac{n}{x(x + n)}$$

3. If neither $x = 1$ nor $x = -1$, show that
$$\frac{1}{1 + x} + \frac{2}{1 + x^2} + \frac{4}{1 + x^4} + \cdots + \frac{2^n}{1 + x^{2^n}} = \frac{1}{x - 1} + \frac{2^{n+1}}{1 - x^{2^{n-1}}}$$

4. Prove that $\tfrac{1}{5}n^5 + \tfrac{1}{3}n^3 + \tfrac{7}{15}n$ is an integer for all $n \geq 1$.

5. Let x and y be integers, $x \neq y$. Show that $x^n - y^n$ is divisible by $x - y$ for all $n \geq 1$.

6. Let x and y be integers, $x + y \neq 0$. Show that $x^{2n} - y^{2n}$ is divisible by $x + y$ for all $n \geq 1$.

7. Let x and y be integers, $x + y \neq 0$. Show that $x^{2n+1} + y^{2n+1}$ is divisible by $x + y$ for all $n \geq 0$.

8. For any integer $n > 1$ and any real number $x > 0$, show that
$$(1 + x)^n > 1 + nx$$

9. If $n > 3$, show that $\sqrt[n]{2} - 1 < 1/(n + 1)$.

10. Let x and y be real numbers satisfying $0 < y < x$. Show that
$$nx^{n-1} > \frac{x^n - y^n}{x - y} > ny^{n-1}$$
for all $n \geq 2$.

11. If $n \geq 1$, show that $2 \leq (1 + 1/n)^n$.

12. If $n \geq 1$, show that $3 > (1 + 1/n)^n$.

13. If $n > 1$, show that $n! < \left(\frac{n+1}{2}\right)^n$.

14. If $n > 1$, show that $(n!)^2 \geq n^n > 1 \times 3 \times 5 \times \cdots \times (2n - 1)$.

15. If $n > 1$, show that $\left(1 + \frac{1}{n}\right)^n < \left(1 + \frac{1}{n+1}\right)^{n+1}$.

16. If $n > 1$, show that $\left(1 - \frac{1}{n^2}\right)^n > 1 - 1/n$.

17. If $n > 1$, show that
$$\left(1 + \frac{1}{6n}\right)^{-n} > \frac{5}{6}.$$

18. Let n and p be any positive integers. Prove that
$$n^p < \frac{(n+1)^{p+1} - n^{p+1}}{p+1} < (n+1)^p.$$

19. Given two positive integers n and b with $b < n$, prove that there exist positive integers q and r, $0 \leq r < b$, such that
$$n = qb + r$$

20. Let $a_1 = 1$ and $a_2 = 2$. Then define recursively
$$a_{n+1} = a_{n-1} + a_n$$
for $n \geq 2$. Prove that
$$a_n < \left(\frac{1 + \sqrt{5}}{2}\right)^n$$
for all $n \geq 1$.

21. Let each of the numbers $a_1, a_2, \cdots, a_n$ be nonnegative. Prove that
$$\frac{1}{n}(a_1 + a_2 + \cdots + a_n) \geq \sqrt[n]{a_1 \times a_2 \times \cdots \times a_n}$$
with equality holding if and only if all of the a_i are equal.

22. Let $a_1, a_2, \cdots, a_n$ and $b_1, b_2, \cdots, b_n$ be real numbers and assume that no b_i is zero. Show that
$$(a_1^2 + \cdots + a_n^2)(b_1^2 + \cdots + b_n^2) \geq (a_1 b_1 + \cdots + a_n b_n)^2$$
with equality holding if and only if $a_1/b_1 = a_2/b_2 = \cdots = a_n/b_n$.

1.5 GEOMETRIC REPRESENTATION OF REAL NUMBERS

The analytic methods to be studied in this text have been and are immensely successful in applications. Much of this success stems from an identification of the set of all real numbers with the set of points on a straight line. This "geometric representation" of real numbers is presented here.

Let l denote a geometric straight line, unending in both directions. Choose any point O on l to be the *origin*. Then choose any other point U on l and call U the *unit point*. These successive choices of two points on l accomplish two important tasks. We have selected a *unit length*, the segment OU, and we have chosen a *positive direction* on l, the direction from O to U. The opposite direction from U to O is then the *negative direction* on l.

A drawing can only show a segment of the line l, of course. An arrowhead is placed on such a segment to indicate the positive direction. This convention is used in Fig. 1.1.

FIGURE 1.1
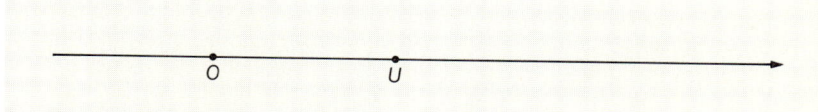

The identification of real numbers with the points on the line l starts by identifying the point O with the number 0 and the point U with the number 1. Laying off the unit length OU in the positive direction, we successively locate the points identified with the numbers 2, 3, 4, $\cdots$. Starting at O and doing the same in the negative direction, those points identified with the negative integers are located. This process convinces us that there is a set of points on l corresponding to the set $\mathbf{Z}$ of all integers. In Fig. 1.2 the numbers are used as labels for the points to which they correspond, a practice we shall follow throughout the book.

FIGURE 1.2
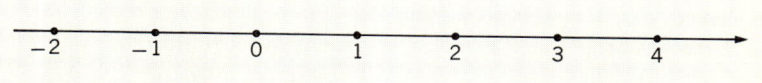

Now consider any real number $x = \pm a_n \cdots a_1 a_0 \cdot d_1 d_2 \cdots d_k \cdots$. If $x \in \mathbf{Z}$, then the corresponding point is located already. If $x \notin \mathbf{Z}$, then there is an integer $p \in \mathbf{Z}$ such that $p < x < p + 1$. Subdividing the interval from p to $(p + 1)$ into 10 equal parts, we locate points to be identified with the rational numbers $p + \frac{1}{10}, p + \frac{2}{10}, \cdots, p + \frac{9}{10}$. If x is one of these numbers, no more need be done. If x is not one of these, then $p + (d_1/10) < x < p + (d_1 + 1)/10$. Then the interval from $p + (d_1/10)$ to $p + (d_1 + 1)/10$ is subdivided into 10 equal parts and the process is repeated. These successive subdivisions locate the desired point on l as accurately as we please.

1.5 GEOMETRIC REPRESENTATION OF REAL NUMBERS

There is a subtle question involved in this "construction." Is there *really* a point on l corresponding to the number x? Just because we can locate the supposed point more and more accurately does not force the point to exist. We must make a basic assumption about the geometric line.

The fundamental assumption of analytic geometry:
To each real number x the process just described provides one and only one point on the line l. Conversely, to each point on l there corresponds one and only one real number.

Our assignment of a real number to each point on the line l imposes a *coordinate system* on l; l becomes a *coordinate line*. The number x assigned to a particular point of l is the *coordinate* of that point. We think of the coordinate line l and the set of real numbers $\mathbf{R}$ as the same set. Thus l is a "geometric representation" of the real numbers.

Just for this paragraph, let P_x denote the point on a coordinate line l having the coordinate $x \in \mathbf{R}$. The distance between the origin O and P_x is defined to be

$$d(O, P_x) = |x|$$

(See Fig. 1.3.) More generally, if P_x and P_y are two points on l with coordinates x and y, respectively, then the distance between P_x and P_y is

$$d(P_x, P_y) = |x - y|$$

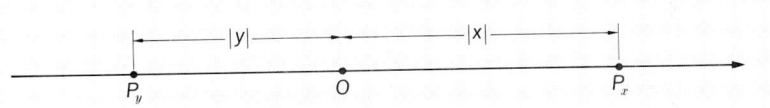

FIGURE 1.3

This can be "proved" if we accept the definition $d(O, P_x) = |x|$. (See Problem 1.5.2.)

Now we drop the P_x notation, identifying a point on line l with its coordinate. Thus we speak of "the distance between two real numbers x and y" and the *distance formula* is written as

$$d(x, y) = |x - y|$$

To each pair of points on l (or to each pair of real numbers) this formula assigns a real number. As shall be discussed in Chapter 2, this assignment constitutes a "function." The four basic properties of this distance function are

D1. $d(x, y) \geq 0$
D2. $d(x, y) = 0$ if and only if $x = y$
D3. $d(x, y) = d(y, x)$
D4. $d(x, y) + d(y, z) \geq d(x, z)$

The proofs of D1 to D4 are easy exercises in the use of absolute values, and we leave them for the reader to provide. We shall discover later that the distance formula for points in a coordinate plane satisfies the same four basic properties. The fact that these four properties are indeed basic will be established by many, many applications.

Three interrelated sets of properties of the real numbers **R** have been discussed. These are the algebraic axioms F1 to F6, the order axioms 01 to 04, and now the distance formula and its properties D1 to D4. Analysis, as we study it, includes the application of all of these properties. It is no wonder, then, that analysis can be complicated.

The basic tools are now at hand to make an analytic study of the geometry on a line. We need only a very few geometric figures on a line, however; these are as follows: Let $a \in \mathbf{R}$ be a fixed real number. Then the set

$$R_a = \{x : x > a\} \text{ is a } \textit{open right ray}$$
$$\overline{R}_a = \{x : x \geq a\} \text{ is a } \textit{closed right ray}$$
$$L_a = \{x : x < a\} \text{ is an } \textit{open left ray}$$

and

$$\overline{L}_a = \{x : x \leq a\} \text{ is a } \textit{closed left ray}$$

If $a, b \in \mathbf{R}$ and $a < b$, then the set $\{x : a < x < b\}$ is an *open interval*, the sets $\{x : a \leq x < b\}$ and $\{x : a < x \leq b\}$ are *half-open intervals* and the set $\{x : a \leq x \leq b\}$ is a *closed interval*.

The union and intersection properties of these rays and intervals are quite easy to work out (see Problems 1.5.4–5). For instance, let a and b be fixed real numbers, $a < b$. Then $R_a \cap L_b = \{x : a < x < b\}$, while $R_a \cup L_b = \mathbf{R}$. Similarly, $\overline{R}_a \cap \overline{L}_b = \{x : a \leq x \leq b\}$.

We shall need two simple results.

■ **THEOREM 1.8.** If x and y are two points on a coordinate line, then the midpoint of the line segment between x and y is the point

$$z = \tfrac{1}{2}(x + y)$$

PROOF: We need only compute that

$$d(x, z) = |x - \tfrac{1}{2}(x + y)| = |\tfrac{1}{2}x - \tfrac{1}{2}y| = \tfrac{1}{2}|x - y| = \tfrac{1}{2}d(x, y)$$

and

$$d(z, y) = |\tfrac{1}{2}(x + y) - y| = |\tfrac{1}{2}x - \tfrac{1}{2}y| = \tfrac{1}{2}d(x, y)$$

Therefore $d(x, z) = \tfrac{1}{2}d(x, y) = d(z, y)$. ■

■ **THEOREM 1.9.** The point $z = (1 - r)x + ry$ divides the segment between two points x and y, so that $d(x, z) = rd(x, y)$ where $0 < r < 1$.

PROOF: We compute that

1.5 GEOMETRIC REPRESENTATION OF REAL NUMBERS

$$d(x, z) = |x - (1 - r)x - ry| = |x - x + rx - ry| = |r(x - y)|$$
$$= r|x - y| = rd(x, y)$$

(Because r is positive, we have $|r| = r$.) ∎

Setting $r = \frac{1}{2}$, Theorem 1.9 becomes Theorem 1.8. That is, Theorem 1.9 is a generalization of Theorem 1.8.

In both the statements and the proofs of these results, points on the line and real numbers were used interchangeably. The exact meaning of each usage should be obvious from context.

EXERCISES 1.5

On a coordinate line, indicate graphically the sets in Exercises 1 to 8.
1. $\{x : x = 1/n, n = 1, 2, 3, 4, 5\}$
2. $\{x : x > 2\}$
3. $\{x : -1 < x < 3\}$
4. $\{x : |x| < 1\}$
5. $\{x : |3x - 2| < 1\}$
6. $\{x : |x - 1| < \frac{1}{4}\}$
7. $\{x : |2x - 1| < \frac{1}{10}\}$
8. $\{x : |x - 4| < \frac{1}{100}\}$

Prove the set equalities in Exercises 9 to 12.

9. $\{x : 0 \leq x \leq 1\} = \{x : x = r, 0 \leq r \leq 1\}$
10. $\{x : 1 \leq x \leq 2\} = \{x : x = 1 + r, 0 \leq r \leq 1\}$
11. $\{x : 2 \leq x \leq 3\} = \{x : x = 2(1 - r) + 3r, 0 \leq r \leq 1\}$
12. $\{x : -1 \leq x \leq 2\} = \{x : x = -(1 - r) + 2r, 0 \leq r \leq 1\}$

In Exercises 13 to 16, determine the point that separates the given segment into two segments the ratio of whose lengths is the number r.

13. $\{x : -1 \leq x \leq 3\}; r = \frac{1}{2}$
14. $\{x : -1 \leq x \leq 3\}; r = \frac{1}{4}$
15. $\{x : |2x - 1| \leq 3\}; r = \frac{1}{2}$
16. $\{x : |3 - 5x| \leq 18\}; r = \frac{1}{5}$
17. Describe in words the set $\{x : 0 < |x - 1| < \frac{1}{2}\}$.

PROBLEMS 1.5

1. Prove the set equality
$$\{x : a \leq x \leq b\} = \{x : x = (1 - r)a + rb, 0 \leq r \leq 1\}$$
2. Using the definition $d(O, P_x) = |x|$, prove that $d(P_x, P_y) = |x - y|$.
3. Establish properties D1 to D4.
4. Let R_a, R_b, L_c, and L_d be open rays. For the several different orderings of the defining numbers $a, b, c,$ and d, determine the intersections and unions

$$R_a \cap R_b \qquad R_a \cap L_c \qquad L_c \cap L_d$$
$$R_a \cup R_b \qquad R_a \cup L_c \qquad L_c \cup L_d$$

Then do the same for the corresponding closed rays.

5. Determine the intersection and union properties of open intervals, of both kinds

of half-open intervals and of closed intervals. For example, show that the intersection of two closed intervals is either empty or is a closed interval.

1.6 AN IMPORTANT PROPERTY OF REAL NUMBERS

Let $X \subset \mathbf{R}$ be a set of real numbers. The set X is *bounded below* if there is a real number $r \in \mathbf{R}$ such that $r \leq x$ for each number $x \in X$. The number r is called a *lower bound* for X. Similarly, the set X is *bounded above* if there is a real number $q \in \mathbf{R}$ such that $x \leq q$ for each $x \in X$. The number q is called an *upper bound* for X. If X is both bounded below and bounded above, then X is said to be *bounded*.

As examples, an interval (open, half-open, or closed) is bounded. The set $\mathbf{N}$ is bounded below but not bounded above. A left ray is bounded above but not bounded below. The set $\mathbf{Z}$ is neither bounded above nor below. For the set $X = \{1/n : n \in \mathbf{N}\}$, the number 0 is a lower bound, while the number 1 is an upper bound.

Let $X \subset \mathbf{R}$ be bounded below. Then the number $m \in \mathbf{R}$ is the *greatest lower bound* of X if (1) m is a lower bound for X but (2) no number greater than m is a lower bound for X. Similarly, if X is bounded above, then the number $M \in \mathbf{R}$ is the *least upper bound* of X if (1) M is an upper bound for X but (2) no number less than M is an upper bound for X.

For the interval $\{x : a < x < b\}$, the number a is the greatest lower bound and the number b is the least upper bound. Also, the closed interval $\{x : a \leq x \leq b\}$ has the same greatest lower bound a and least upper bound b. This tells us that, for instance, the greatest lower bound of a set X may be a member of X, but it need not be. The same observation can be made about the set $X = \{1/n : n \in \mathbf{N}\}$ mentioned above. Here the least upper bound is 1 and is in X, while the greatest lower bound is 0 and is not in X.

The important property we are discussing here is embodied in the following result.

■ **THEOREM 1.10.** If the set $X \subset \mathbf{R}$ has a lower bound, then X has a greatest lower bound. If $X \subset \mathbf{R}$ has an upper bound, then X has a least upper bound.

We do not give a real proof but only an indication of the proof of the first part. First we remark that there must be an integer p such that p is a lower bound for X but $p + 1$ is not a lower bound for X. (A precise proof of this fact can be based upon the well-ordering principle.) Now consider the 10 rational numbers $p + \frac{1}{10}, p + \frac{2}{10}, \cdots, p + \frac{9}{10}, p + 1$. If $p + \frac{1}{10}$ is not a lower bound for X, then the rational number $p + \frac{0}{10}$ is a first decimal approximation to the desired greatest lower bound. In general, if $p + d_1/10$ is a lower bound, but $p + (d_1 + 1)/10$ is not, then $p + d_1/10$ is a first decimal approximation.

Next we examine the ten rational numbers

$$p + \frac{d_1}{10} + \frac{1}{10^2}, p + \frac{d_1}{10} + \frac{2}{10^2}, \cdots, p + \frac{d_1 + 1}{10}$$

For some digit d_2, $p + d_1/10 + d_2/10^2$ is a lower bound, while $p + d_1/10 + (d_2 + 1)/10^2$ is not. This provides us with a second decimal approximation $p + d_1/10 + d_2/10^2$ to the desired greatest lower bound. Continuing in this way, we construct a nonterminating decimal $p + d_1/10 + d_2/10^2 + \cdots + d_k 10^k + \cdots$, which is the desired greatest lower bound.

The property of **R** stated in Theorem 1.10 is known as the *completeness* of **R**. We say that **R** is *complete*. The set **Q** of rational numbers is not complete, that is, the greatest lower bound of a set of rational numbers is not necessarily rational. For example, the set $\{r : r \in \mathbf{Q}, r^2 > 2\}$ has as its greatest lower bound the irrational number $\sqrt{2}$. It is precisely the completeness of **R** that we use in analysis. And because **Q** is not complete, we are forced to use **R**. More about this will be said later.

The completeness of **R** can be used to prove results that are otherwise very difficult.

■ **THEOREM 1.11.** Let x be any positive real number. Then $\sqrt{x}$ is also a real number.

PROOF: The set $Y = \{y : y > 0, y^2 > x\}$ is bounded below; for instance, by the number 0. Hence, in view of Theorem 1.10, Y has a greatest lower bound m. Similarly, the set $Z = \{z : z \in \mathbf{R}, z^2 < x\}$ is bounded above; for instance, by x if $x \geq 1$ or by 1 if $x < 1$. Thus Z has a least upper bound M. It should be obvious that $m = M$ and that $m^2 = x$. ■

■ **THEOREM 1.12.** Let x be a real number. If $x > 0$ and $n \in \mathbf{N}$, then $\sqrt[n]{x}$ is a real number. If $x < 0$ and $n = 2k - 1$ is an odd natural number, then $\sqrt[n]{x}$ is a real number.

The proof of Theorem 1.12 is analogous to that of Theorem 1.11 and need not be given here.

EXERCISES 1.6

1. Give an example of a set of real numbers which
    (a) contains its least upper bound and its greatest lower bound,
    (b) contains its least upper bound but not its greatest lower bound,
    (c) contains neither its least upper bound nor its greatest lower bound.

2. Can a set of irrational numbers have a rational least upper bound? If not, prove it. If so, give an example.

3. Prove that 0 is the greatest lower bound of the set
$$\{\frac{1}{n} : n \in \mathbf{N}\}$$

4. Prove that 1 is the least upper bound of the set
$$\{\frac{n}{n+1} : n \in \mathbf{N}\}$$

5. Suppose that a set X contains a largest number q. Prove that q is the least upper bound of X.

6. Find the least upper bound of the set
$$\left\{\frac{2^n - 1}{2^n} : n \in \mathbf{N}\right\}$$
and prove your claim.

7. Determine a general formula for the members of the set
$$\{\tfrac{1}{3}, \tfrac{4}{9}, \tfrac{13}{27}, \tfrac{40}{81}, \cdots \}$$
Then establish the least upper bound of this set.

8. What is the greatest lower bound of the set $\{p/q : p, q \in \mathbf{N}, p^3 > 6q^3\}$?

PROBLEMS 1.6

1. Define
$$a_1 = \frac{1}{1 \cdot 2}, \quad a_2 = \frac{1}{1 \cdot 2} + \frac{1}{2 \cdot 3},$$
and, in general, define $a_n = a_{n-1} + 1/[n(n+1)]$. Does the least upper bound of the set $\{a_n : n \in \mathbf{N}\}$ exist? If so, what is it?

2. Let x and y be real numbers. Assume that $x \leq y + \epsilon$ for each $\epsilon > 0$ and prove that this implies $x \leq y$. If $x < y + \epsilon$ for each $\epsilon > 0$, what can you conclude?

3. Use Theorem 1.10 to prove the Archimedean property of the order relation $<$.

4. Let $k \in \mathbf{N}$ be a fixed number, not a perfect square. Define $a_1 = \sqrt{k}$, $a_2 = \sqrt{k\sqrt{k}}$, and in general, $a_{n+1} = \sqrt{ka_n}$. Prove that k is an upper bound for the set $\{a_n : n \in \mathbf{N}\}$. If k is the least upper bound, then prove this too.

5. Let $k \in \mathbf{N}$ be a fixed number. Define $a_1 = \sqrt{k}$, $a_2 = \sqrt{k + \sqrt{k}}$, and in general, $a_{n+1} = \sqrt{k + a_n}$. Prove that k is an upper bound for the set $\{a_n : n \in \mathbf{N}\}$. If k is the least upper bound, prove this too.

6. Let X be a set of numbers with greatest lower bound b and least upper bound c. For fixed numbers d and e, define the set $dX + e = \{dx + e : x \in X\}$. Determine and establish the greatest lower bound and least upper bound of this new set.

7. Describe two different sets of rational numbers having the same number $\sqrt{2}$ as least upper bound.

1.7 FACTORED INEQUALITIES

Many problems in mathematics call for the solution of an inequality that can be factored into a product of linear factors such as
$$(ax + b)(cx + d) > 0$$
An easy graphical method of solving such inequalities is based upon the fact that, in a product of several numbers if an even number of factors is negative, then the product is positive; while if an odd number of factors is negative, the product is negative.

The solution of a linear inequality $ax + b > 0$, or its equivalent inequality, $x > -b/a$ (assuming $a > 0$) can be pictured as a number line drawn solid for $x > -b/a$ and dashed for $x < -b/a$. An example is shown in Fig. 1.4.

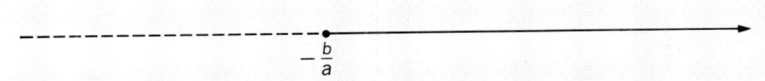

FIGURE 1.4

We draw such a line for each factor $ax + b$ and $cx + d$, as in Fig. 1.5. We easily see that the inequality $(ax + b)(cx + d) > 0$ holds wherever both lines are solid or both are dashed. Similarly, $(ax + b)(cx + d) < 0$ wherever one line is solid and the other dashed.

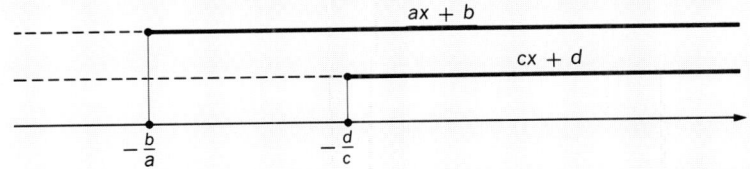

FIGURE 1.5

■ **Example 1.** The solutions of $2x + 3 > 0$ and $3x - 8 > 0$ are shown in Fig. 1.6. It is easy to see that both factors $2x + 3$ and $3x - 8$ are positive if $x > \frac{8}{3}$ and both are negative if $x < -\frac{3}{2}$. Therefore we have $(2x + 3)(3x - 8) > 0$ if $x < -\frac{3}{2}$ or if $x > \frac{8}{3}$. At the same time we see that $(2x + 3)(3x - 8) < 0$ if $-\frac{3}{2} < x < \frac{8}{3}$. A simple inspection of the diagram in Fig. 1.6 has determined the solution sets

$$\{x : (2x + 3)(3x - 8) > 0\} = \{x : x < -\tfrac{3}{2}\} \cup \{x : x > \tfrac{8}{3}\}$$
$$\{x : (2x + 3)(3x - 8) < 0\} = \{x : -\tfrac{3}{2} < x < \tfrac{8}{3}\}$$

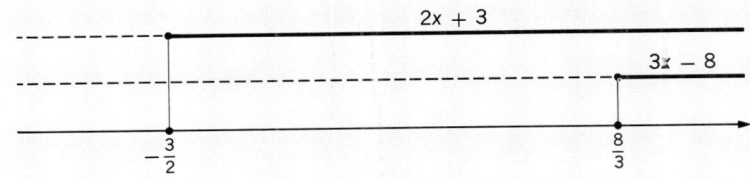

FIGURE 1.6

■ **Example 2.** Given the quadratic inequality

$$3 + 2x - x^2 > 0$$

we factor the quadratic expression to obtain

$$(x + 1)(-x + 3) > 0$$

The linear inequalities $x + 1 > 0$ and $-x + 3 > 0$ are solved and the solutions put into a line chart as shown in Fig. 1.7. Then from this figure we see that $\{x : 3 + 2x - x^2 > 0\} = \{x : -1 < x < 3\}$.

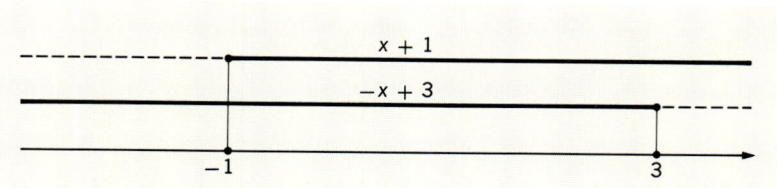

FIGURE 1.7

■ *Example 3.* The line chart method still works easily when there are many factors. Given the factored inequality

$$(x + 2)(x - 3)(2x + 1)(3x - 4) > 0$$

we draw the line chart shown in Fig. 1.8. Then from this picture the solution is quickly seen to be

$$\{x : (x + 2)(x - 3)(2x + 1)(3x - 4) > 0\}$$
$$= \{x : x < -2\} \cup \{x : -\tfrac{1}{2} < x < \tfrac{4}{3}\} \cup \{x : x > 3\}$$

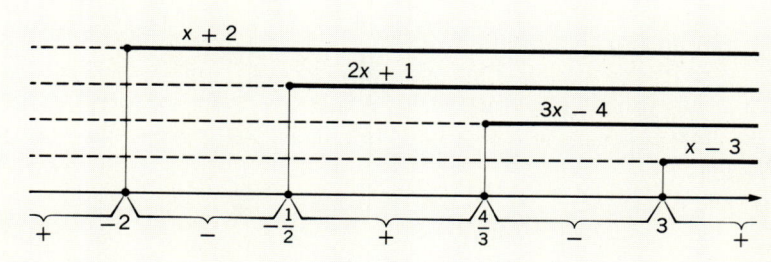

FIGURE 1.8

■ *Example 4.* The line chart can also be used on such factored inequalities as

$$\frac{x - 1}{(x + 2)(2x + 1)} < 0$$

We simply draw the line chart as in Fig. 1.9; an inspection gives us the solution set

$$\left\{x : \frac{x - 1}{(x + 2)(2x + 1)} < 0\right\}$$
$$= \{x : x < -2\} \cup \left\{x : -\tfrac{1}{2} < x < 1\right\}$$

It should be obvious that the solution of such an inequality as

$$(5x + 4)(x - 2) \geq 0,$$

which is not a strict inequality, can be given by solving the strict inequality $(5x + 4)(x - 2) < 0$ with the graphical method. Then the desired solu-

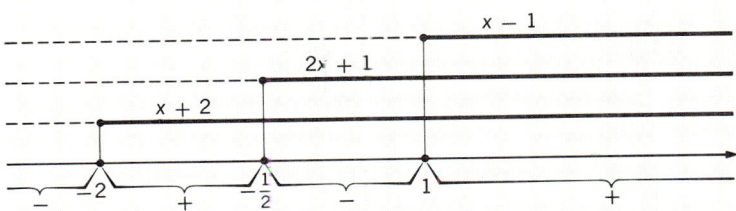

FIGURE 1.9

tion set of $(5x + 4)(x - 2) \geq 0$ is the complement of the solution set of $(5x + 4)(x - 2) < 0$.

We add that the solution set of the equation

$$(ax + b)(cx + d) = 0$$

is the complement of the solution set of both strict inequalities. That is, we have $\{x : (ax + b)(cx + d) = 0\}$ is the complement of

$$\{x : (ax + b)(cx + d) < 0 \text{ or } (ax + b)(cx + d) > 0\}.$$

Of course, this is the technique whereby we locate the latter set in the first place.

EXERCISES 1.7

Use the line chart method to determine the solution set of each of the following inequalities.

1. $(x + 1)(x - 2) > 0$
2. $(4x - 7)(x + 2) < 0$
3. $(x + 1)(2 - x) \geq 0$
4. $(3x - 8)(3x + 8) < 0$
5. $(x - 4)(3x - 5)(4x + 9) > 0$
6. $(x - 3)(2x + 3)(x + 2)(2x - 1) \leq 0$
7. $\dfrac{(x - 1)(x + 2)}{x - 2} > 0$
8. $\dfrac{(x + 1)(x - 3)}{(2x - 1)(x + 4)} < 0$
9. $x^2 - 4 \leq 0$
10. $2x^2 - x - 15 < 0$
11. $-6 + 7x - 2x^2 > 0$
12. $6x^3 - x^2 - 29x + 14 \leq 0$
13. $\dfrac{2x^2 - x - 6}{4 + x} > 0$
14. $\dfrac{2 - x}{x^2 + 3x + 2} \geq 0$
15. $\dfrac{2x^2 - x - 10}{3x^2 - 10x - 8} < 0$

1.8 NUMERICAL VECTORS

Many applications of mathematics involve a simultaneous consideration of two or more numerical quantities. Problems concerning moving bodies involve position and time. Economic problems often deal with the input and output quantities of some manufacturing process. In biology we find a relation between the strength of a muscle and its cross-sectional area.

A situation involving two distinct numerical quantities is described mathematically by an *ordered pair* of real numbers. For instance, a number pair (x_1, x_2) might symbolize the input and output quantities, respectively, of some manufacturing process. Note that (x_1, x_2) is an "ordered" pair in the sense that the first number x_1 measures input, while the second number x_2 measures output. In such a context, obviously, the pair $(3, 5)$ is not the same as the pair $(5, 3)$. In particular, $(a, b) = (c, d)$ if and only if $a = c$ and $b = d$.

Similarly, an application involving three distinct numerical quantities requires an *ordered triple* (x_1, x_2, x_3) of real numbers for its description and, in general, a mathematical description of a problem involving n distinct quantities requires an *ordered n-tuple* $(x_1, x_2, \cdots, x_n)$ of real numbers. Such an ordered n-tuple $(x_1, x_2, \cdots, x_n)$ is also called an *n-dimensional numerical vector*. Each number x_i is a *component* of the vector. Two vectors are equal if and only if corresponding components are equal.

When dealing with two-dimensional numerical vectors, we usually denote an unspecified vector as an ordered pair of letters such as (x, y) or (s, t). Similarly, three-dimensional vectors are denoted by (x, y, z). This practice avoids a tedious use of subscripts.

Now suppose there are two manufacturing processes using the same input material and producing the same output material. Let the first process have an input of a units and an output of b units, while the second process inputs c units and outputs d units. If the two processes are taken together and considered as one process, that is, if we "add" them together, then the combined input is $a + c$ units and the combined output is $b + d$ units. Therefore if the ordered pair (a, b) denotes the input-output quantities of the first process and (c, d) that of the second, then for the combined processes the input-output quantities must be denoted by $(a + c, b + d)$. This observation leads us to a definition.

The *sum of two numerical vectors* (a, b) and (c, d) is given by

$$(a, b) + (c, d) = (a + c, b + d)$$

This process is called the *component-wise addition* of numerical vectors. Two n-dimensional numerical vectors are also added component-wise, the formula being

$$(x_1, x_2, \cdots, x_n) + (y_1, y_2, \cdots, y_n)$$
$$= (x_1 + y_1, x_2 + y_2, \cdots, x_n + y_n)$$

Note that two vectors of differing dimensions cannot be added together.

We shall use boldface letters to denote vectors. Thus,

$$\mathbf{x} = (x_1, x_2, \cdots, x_n)$$
$$\mathbf{y} = (y_1, y_2, \cdots, y_n)$$

and so forth. In particular, the *zero vector* $(0, 0, \cdots, 0)$ is denoted by $\mathbf{0}$.

The following properties of component-wise addition of numerical vectors are easily proved (Problem 1.8.1):

V1. If **x** and **y** are n-dimensional numerical vectors, then $\mathbf{x} + \mathbf{y}$ is an n-dimensional numerical vector.

V2. For any three n-dimensional vectors **x**, **y**, and **z**,
$$\mathbf{x} + (\mathbf{y} + \mathbf{z}) = (\mathbf{x} + \mathbf{y}) + \mathbf{z}$$

V3. For any n-dimensional numerical vector **x**,
$$\mathbf{x} + \mathbf{0} = \mathbf{x}$$

V4. For each n-dimensional numerical vector **x**, there is an n-dimensional numerical vector $-\mathbf{x}$ such that $\mathbf{x} + (-\mathbf{x}) = \mathbf{0}$.

V5. For each pair of n-dimensional numerical vectors **x** and **y**,
$$\mathbf{x} + \mathbf{y} = \mathbf{y} + \mathbf{x}$$

The similarity between V1 to V5 and the properties F1 to F5 of the addition of real numbers is intentional. In technical language V1 to V5 are the axioms for an *abelian group*, a concept to be found in higher algebra courses.

Returning to the manufacturing process with input a and output b, what does it mean to double the process? An obvious meaning would be to create a new process with input $2a$ and output $2b$. We symbolize this sort of "multiplication" by writing
$$2(a, b) = (2a, 2b)$$
A similar rule surely applies if we "multiply" the process by 4 or by $\frac{1}{3}$. This suggests a definition.

The product of a real number r and a numerical vector
$$\mathbf{x} = (x_1, x_2, \cdots, x_n) \text{ is given by the formula}$$
$$r\mathbf{x} = r(x_1, x_2, \cdots, x_n) = (rx_1, rx_2, \cdots, rx_n)$$

This so-called *scalar multiplication* has the following properties (see Problem 1.8.2).

V6. For any real number r and any n-dimensional numerical vector **x**, $r\mathbf{x}$ is also an n-dimensional numerical vector.

V7. For any numerical vector **x**,
$$0\mathbf{x} = \mathbf{0}, \ 1\mathbf{x} = \mathbf{x}$$

V8. Given real numbers r and s and a numerical vector **x**,
$$(r + s)\mathbf{x} = r\mathbf{x} + s\mathbf{x}$$

V9. For any real number r and numerical vectors **x** and **y**,
$$r(\mathbf{x} + \mathbf{y}) = r\mathbf{x} + r\mathbf{y}$$

V10. Given real numbers r and s and a numerical vector $\mathbf{x}$,
$$(rs)\mathbf{x} = r(s\mathbf{x})$$

Properties V1 to V10 are axioms for a *vector space*. We do not need this technical word, but the properties themselves are essential. They will be used very frequently in the sequel. Note their easy use in the following instances:

To prove that two sets are equal, we must show that each set contains the other. For example, we claim that
$$\{(x,y) : x - y = 2\} = \{(1-r)(1,-1) + r(-1,-3) : r \in \mathbf{R}\}$$
A proof consists of two steps. We show first that each vector of the form
$$(1-r)(1,-1) + r(-1,-3)$$
satisfies the equation $x - y = 2$ and hence lies in the first set. This is easy for
$$(1-r)(1,-1) + r(-1,-3) = (1-r, -1+r) + (-r, -3r)$$
$$= (1-2r, -1-2r)$$
and clearly $(1-2r) - (-1-2r) = 2$. Conversely, we show that each vector (x,y) satisfying $x - y = 2$ lies in the second set. That is, we prove that there is a real number r such that
$$(x,y) = (1-r)(1,-1) + r(-1,-3) = (1-2r, -1-2r)$$
By setting $x = 1 - 2r$ or $r = \frac{1}{2}(1 - x)$, the equation $x - y = 2$ yields
$$y = x - 2 = 1 - 2r - 2 = -1 - 2r$$
Therefore, for any vector (x,y) satisfying $x - y = 2$, we have
$$(x,y) = [1 - \tfrac{1}{2}(1-x)](1,-1) + \tfrac{1}{2}(1-x)(-1,-3)$$
$$= (\tfrac{1}{2} + \tfrac{1}{2}x)(1,-1) + \tfrac{1}{2}(1-x)(-1,-3)$$
$$= (\tfrac{1}{2} + \tfrac{1}{2}x, -\tfrac{1}{2} - \tfrac{1}{2}x) + (-\tfrac{1}{2} + \tfrac{1}{2}x, -\tfrac{3}{2} + \tfrac{3}{2}x)$$
$$= (\tfrac{1}{2} + \tfrac{1}{2}x - \tfrac{1}{2} + \tfrac{1}{2}x, -\tfrac{1}{2} - \tfrac{1}{2}x - \tfrac{3}{2} + \tfrac{3}{2}x)$$
$$= (x, x - 2)$$

as required. This proves that the second set contains the first; hence the two sets are equal.

EXERCISES 1.8

1. Consider the two-dimensional numerical vectors $\mathbf{a} = (1, -1)$, $\mathbf{b} = (0, 2)$ and $\mathbf{c} = (3, -2)$. Compute
    (a) $\mathbf{a} + \mathbf{b} - \mathbf{c}$
    (b) $2\mathbf{a} - \mathbf{b} + 2\mathbf{c}$
    (c) $15\mathbf{a} + 3\mathbf{b} - 5\mathbf{c}$
    (d) $6\mathbf{a} + \mathbf{b} - 2\mathbf{c}$
    (e) $4(\mathbf{a} + \mathbf{b}) - 2(\mathbf{b} + \mathbf{c})$
    (f) $3(\mathbf{a} - \mathbf{b} + \mathbf{c}) + 2(\mathbf{a} + \mathbf{b} - \mathbf{c})$

2. Let $\mathbf{a} = (1, -1)$ and $\mathbf{b} = (0, 2)$. Solve the equations below for the real numbers x and y.
    (a) $x\mathbf{a} + y\mathbf{b} = (1, 0)$
    (b) $x\mathbf{a} + y\mathbf{b} = (0, 1)$

1.9 GEOMETRIC REPRESENTATION OF TWO-DIMENSIONAL NUMERICAL VECTORS

3. Let $\mathbf{a} = (1, -1, 0)$, $\mathbf{b} = (1, 2, 1)$, $\mathbf{c} = (0, -1, 2)$ and compute
   (a) $\mathbf{a} + \mathbf{b} - \mathbf{c}$
   (b) $2\mathbf{a} - \mathbf{b} + \mathbf{c}$
   (c) $\frac{5}{7}\mathbf{a} + \frac{2}{7}\mathbf{b} - \frac{1}{7}\mathbf{c}$
   (d) $-\frac{2}{7}\mathbf{a} - \frac{2}{7}\mathbf{b} - \frac{1}{7}\mathbf{c}$

4. With $\mathbf{a}$, $\mathbf{b}$, and $\mathbf{c}$ as in Exercise 3, solve the equation
$$x\mathbf{a} + y\mathbf{b} + z\mathbf{c} = (0, 0, 1)$$

5. Establish the set equalities:
   (a) $\{(x, y) : x + y = 1\} = \{(1 - r)(1, 0) + r(0, 1) : r \in \mathbf{R}\}$
   (b) $\{(x, y) : 3x - y + 4 = 0\} = \{r(1, 7) + (1 - r)(-1, 1) : r \in \mathbf{R}\}$
   (c) $\{(x, y) : x + 3y = 6\} = \{(1 - r)(6, 0) + r(0, 2) : r \in \mathbf{R}\}$
   (d) $\{(x, y, z) : x + y + z = 1\} = \{r(2, -1, 0) + s(0, 2, -1)$
       $+ (1 - r - s)(0, 0, \_) : r, s \in \mathbf{R}\}$
   (e) $\{(x, y, z) : 3x - y + 2z = 6\} = \{r(2, 0, 0) + s(1, -1, 1)$
       $+ (1 - r - s)(0, 2, 4) : r, s \in \mathbf{R}\}$
   (f) $\{(x, y, z) : x + 2y - z = 2\} \cap \{(x, y, z) : 2x - 4y + z = 6\}$
       $= \{r(2, -1, -2) + (1 - r)(3, \frac{1}{2}, 2) : r \in \mathbf{R}\}$

PROBLEMS 1.8

1. Using the known properties of real numbers and the definition of the addition of numerical vectors, verify axioms V1 to V5.

2. Using axioms F1 to F6, the definition of scalar multiplication and axioms V1 to V5 where needed, verify axioms V6 to V10.

3. Let m be a fixed real number. Show that the set of numerical vectors $\{(x, y) : y = mx\}$ also satisfies axioms V1 to V10.

4. Show that the set of numerical vectors
$$\{(x, y, z) : x - y + 3z = 0 = 2x + y - z\}$$
satisfies V1 to V10.

5. Let V^2 denote the set of all two-dimensional numerical vectors. Define a "multiplication" by setting $(a, b) \times (c, d) = (ac - bd, ad + bc)$. Show that this operation and vector addition satisfy axioms F1 to F6.

6. Using only axioms V1 to V5, prove that, if $\mathbf{a} + \mathbf{x} = \mathbf{x}$ for all vectors $\mathbf{x}$, then $\mathbf{a} = \mathbf{0}$ (that is, show that the additive identity is unique). Similarly, if $\mathbf{a} + \mathbf{x} = \mathbf{0}$, show that $\mathbf{x} = -\mathbf{a}$ and hence prove that additive inverses are unique.

7. Using only axioms V1 to V5, show that each equation $\mathbf{a} + \mathbf{x} = \mathbf{b}$ can be solved for $\mathbf{x}$ and that the solution is unique.

1.9 GEOMETRIC REPRESENTATION OF TWO-DIMENSIONAL NUMERICAL VECTORS

There are two equivalent methods of giving a geometric representation of an ordered pair of real numbers. We consider the more obvious one first.

Let Π be a geometric plane and in Π choose two perpendicular lines l and k. Let the intersection of l and k be the origin on each line and construct a coordinate system on each line as in Section 1.5. The result might look like Fig. 1.10. Note that it is conventional but not necessary to choose l to be directed positively to the right and k to be directed positively upward.

A number pair (x, y) is pictured as that point in the plane Π which is at the directed distance x from the line k and at the directed distance y

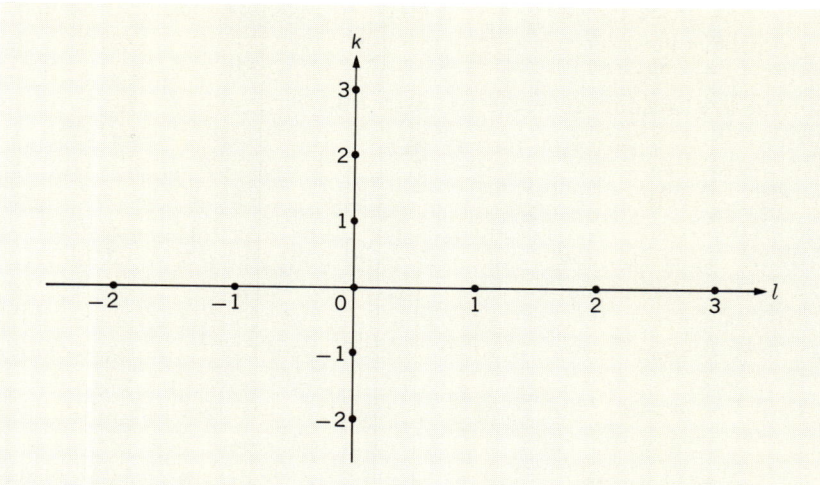

FIGURE 1.10

from the line l. Equivalently, the point corresponding to the pair (x, y) is located by measuring a distance x from the origin along the line l and then measuring a distance y parallel to the line k. In Fig. 1.11, several points are located and labeled with their corresponding number pairs.

The unit distances on the coordinate lines l and k need *not* be equal. This reflects the fact that the two components of a numerical vector (x, y) often refer to quantities which are in no way commensurable. For instance, x may be a length of time and y a geometric distance. In this case the equality of units in our geometric representation would be meaningless and could even be misleading.

This assignment of a point in the plane Π to each ordered pair (x, y)

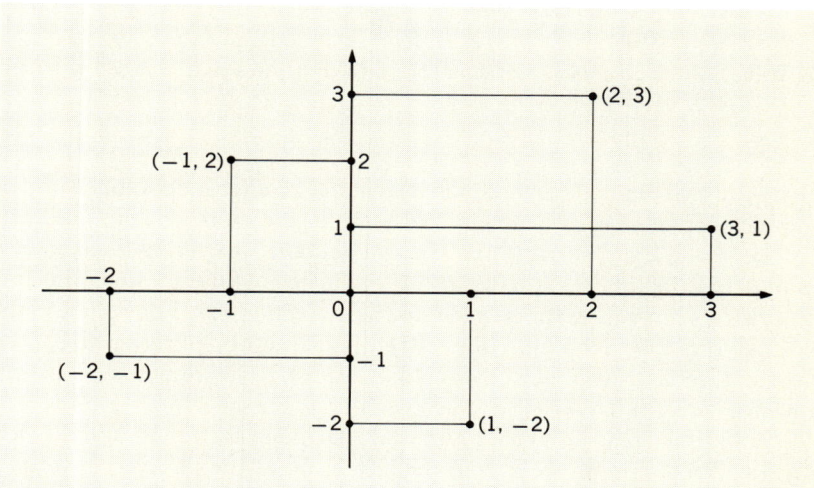

FIGURE 1.11

1.9 GEOMETRIC REPRESENTATION OF TWO-DIMENSIONAL NUMERICAL VECTORS

is a *one-to-one correspondence*. This means that to each ordered pair there corresponds one and only one point in Π and, conversely, to each point in Π there corresponds one and only one ordered pair. (We are using the fundamental assumption of analytic geometry here, of course.) Just as we identified the set of real numbers with the set of points on a line, we now identify the set of two-dimensional numerical vectors with the set of points in a plane.

When the construction above has been made, we call Π a *coordinate plane*, and say that a *coordinate system* has been put upon the plane. If (x, y) is a point on a coordinate plane, then the number x is called the *first coordinate* of the point and the number y is the *second coordinate*. The lines l and k are the *first* and *second coordinate axes*, respectively. We often speak of l as the x *axis* and of k as the y *axis*. [By calling the number pair (x, y) a "point" we are already using the identification mentioned in the preceding paragraph.]

The second method of representing a numerical vector is by means of a "position vector." We use this method often hereafter. Let (x, y) be a point on a coordinate plane. The line segment directed from the origin $(0, 0)$ to the point (x, y) is called the *position vector* of the point. The point (x, y) is called the *tip* of its position vector. (See Fig. 1.12.) We refer to this as the *vector representation* of numerical vectors.

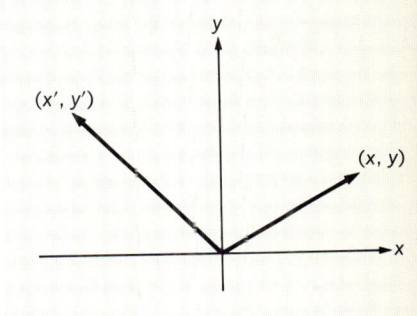

FIGURE 1.12

The unit point $(1, 0)$ on the first coordinate axis is the tip of a position vector **u**. The position vector **v** is that of the unit point $(0, 1)$ on the second coordinate axis. (Note that all position vectors are to be denoted by boldface letters.)

Now let (x, y) be any numerical vector. By properties V1 to V10, we may rewrite this numerical vector as follows:

$$(x, y) = (x, 0) + (0, y) + x(1, 0) + y(0, 1)$$

This leads us to express the position vector **p** of the point (x, y) as

$$\mathbf{p} = x\mathbf{u} + y\mathbf{v}$$

We interpret this vector addition geometrically. First note that $x\mathbf{u} = (x, 0)$ is a vector on the x axis, which is x times as long as $\mathbf{u} = (1, 0)$. Analogously, $y\mathbf{v}$ is on the y axis and is y times as long as **v**. In Fig. 1.13 we picture the position vectors $\mathbf{u}, \mathbf{v}, x\mathbf{u}, y\mathbf{v}$ and $\mathbf{p} = x\mathbf{u} + y\mathbf{v}$. The vector $\mathbf{p} = x\mathbf{u} + y\mathbf{v}$ is the diagonal of the rectangle whose adjacent sides are the segments $x\mathbf{u}$ and $y\mathbf{v}$.

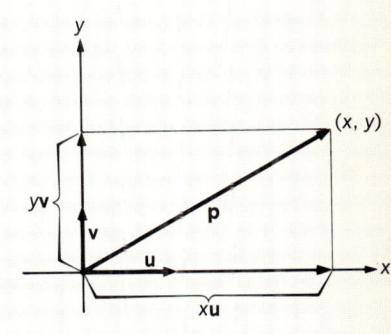

FIGURE 1.13

Every position vector can be expressed as a sum $x\mathbf{u} + y\mathbf{v}$ of multiples of the vectors **u** and **v**. For this reason, **u** and **v** are called the *unit basis vectors*.

Next we give a geometric interpretation of the vector addition

$$(a, b) + (c, d) = (a + c, b + d)$$

Rewriting this equation in terms of position vectors, we obtain

$$(a\mathbf{u} + b\mathbf{v}) + (c\mathbf{u} + d\mathbf{v}) = (a + c)\mathbf{u} + (b + d)\mathbf{v}$$

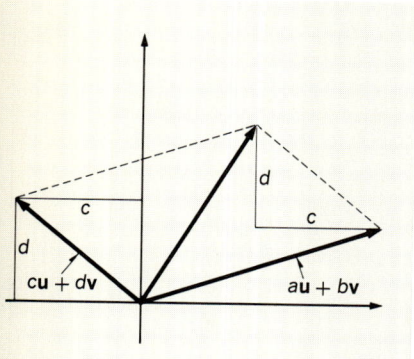

FIGURE 1.14

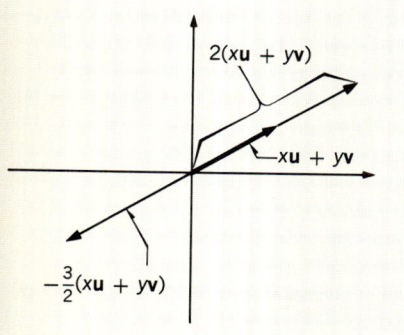

FIGURE 1.15

A typical picture of such vectors is shown in Fig. 1.14. Note that the horizontal directed distance between points (a, b) and $(a + c, b + d)$ is the number c and the vertical directed distance between these points is the number d. This shows that the line segment from (a, b) to $(a + c, b + d)$ is parallel to the position vector $c\mathbf{u} + d\mathbf{v}$. Therefore the vector $(a + c)\mathbf{u} + (b + d)\mathbf{v}$ is the diagonal of the parallelogram whose adjacent sides are the position vectors $a\mathbf{u} + b\mathbf{v}$ and $c\mathbf{u} + d\mathbf{v}$. It follows that *The component-wise addition of numerical vectors corresponds to the addition of position vectors by the parallelogram rule.*

In the physical world, quantities such as force or velocity having both magnitude and direction are called *vector quantities*. Such quantities obey the parallelogram rule of addition. This fact provides strong justification for the component-wise addition of numerical vectors. Indeed, mathematics has simply borrowed this idea from the physical world.

The multiplication of a numerical vector (x, y) by a real number r is defined by

$$r(x, y) = (rx, ry)$$

Using the algebra of numerical vectors, we write

$$(rx, ry) = (rx, 0) + (0, ry) = rx(1, 0) + ry(0, 1)$$

In the notation of position vectors, then,

$$r(x\mathbf{u} + y\mathbf{v}) = (rx)\mathbf{u} + (ry)\mathbf{v}$$

It is an easy matter to show that the point (rx, ry) lies on the straight line through $(0, 0)$ and (x, y), as shown in Fig. 1.15.

The set of all ordered pairs of real numbers

$$\{(x, y) : x, y \in \mathbf{R}\}$$

and the set of all position vectors in a coordinate plane are identified via the one-to-one correspondence

$$(x, y) \leftrightarrow x\mathbf{u} + y\mathbf{v}$$

This means that a subset of the ordered pairs can be pictured geometrically, either as a set of points or a set of position vectors in a coordinate plane. We shall choose the interpretation that seems most useful for a particular purpose.

A warning must be issued here. If a numerical vector (x, y) is represented as a position vector $x\mathbf{u} + y\mathbf{v}$, then the position vector may have geometric properties that have no meaning for the numerical vector. For instance, the position vector has a "length" whereas for the numerical vector the concept of "length" may have no meaning. For example, if x is the input and y the output of some manufacturing process, the economist does not assign meaning to the "length of the vector (x, y)." On the other hand, when the length of a vector has meaning, we shall certainly want to use it.

EXERCISES 1.9

1. On a coordinate plane locate and label the four points in each of the following sets. Draw the position vectors.
   - (a) $\{(3, 0), (-2, 0), (0, 2), (0, -1)\}$
   - (b) $\{(1, 2), (-1, 2), (1, -2), (-1, -2)\}$
   - (c) $\{(3, 2), (-1, -3), (5, -1), (0, -4)\}$
   - (d) $\{(-2, -3), (-1, -3), (0, -3), (3, -3)\}$
   - (e) $\{(1, -2), (1, 1), (1, 3), (1, 5)\}$
   - (f) $\{(0, 4), (1, 3), (2, 2), (3, 1)\}$

2. Locate and label at least four points in each of the following sets. Can you deduce an important property common to all of these sets of points?
   - (a) $\{(x,y) : x = 0\}$
   - (b) $\{(x,y) : y = -3\}$
   - (c) $\{(x,y) : x = 1\}$
   - (d) $\{(x,y) : x + y = 4\}$
   - (e) $\{(x,y) : x - y = 1\}$
   - (f) $\{(x,y) : 2x + y = 0\}$
   - (g) $\{(x,y) : y = x - 2\}$
   - (h) $\{(x,y) : y = 3 - x\}$
   - (i) $\{(x,y) : 3y = x + 2\}$
   - (j) $\{(x,y) : 2y = 4 - 3x\}$

3. For each of the following sets, draw at least four position vectors. Again all of the sets have an important property in common. What is it?
   - (a) $\{r(1, 2) : r \in \mathbf{R}\}$
   - (b) $\{r(-3, 2) : r \in \mathbf{R}\}$
   - (c) $\{(1, 0) + r(1, 2) : r \in \mathbf{R}\}$
   - (d) $\{(0, 1) + r(-3, 2) : r \in \mathbf{R}\}$
   - (e) $\{(-1, 1) + r(1, 1) : r \in \mathbf{R}\}$
   - (f) $\{(-2, 0) + r(-1, -1) : r \in \mathbf{R}\}$
   - (g) $\{(1 - r)(1, 2) + r(-1, 1) : r \in \mathbf{R}\}$
   - (h) $\{(1 - r)(0, 2) + r(1, 3) : r \in \mathbf{R}\}$
   - (i) $\{(1 - r)(1, -1) + r(-2, 2) : r \in \mathbf{R}\}$
   - (j) $\{(1 - r)(3, 4) - r(5, -2) : r \in \mathbf{R}\}$

4. Let $\mathbf{a} = (2, -1)$ and $\mathbf{b} = (1, 2)$. Determine numbers x and y satisfying each of the following equations.
   - (a) $x\mathbf{a} + y\mathbf{b} = (1, 0)$
   - (b) $x\mathbf{a} + y\mathbf{b} = (0, 1)$
   - (c) $x\mathbf{a} + y\mathbf{b} = (-1, 1)$
   - (d) $x\mathbf{a} + y\mathbf{b} = (3, -2)$
   - (e) $x\mathbf{a} + y\mathbf{b} = (5, 6)$
   - (f) $x\mathbf{a} + y\mathbf{b} = (c, d)$

PROBLEMS 1.9

1. Let $\mathbf{u} = (1, 0)$ and $\mathbf{v} = (0, 1)$. Show that each position vector $\mathbf{p}$ can be written in the form $x\mathbf{u} + y\mathbf{v}$.

2. Let $\mathbf{a} = (a_1, a_2) \neq (0, 0)$ and $\mathbf{b} = (b_1, b_2)$. Show that $\mathbf{b} = r\mathbf{a}$ for some $r \in \mathbf{R}$ if and only if $a_1 b_2 = a_2 b_1$.

3. Given $\mathbf{a} = (a_1, a_2)$ and $\mathbf{b} = (b_1, b_2)$, assume that $a_1 b_2 \neq a_2 b_1$. Prove that, for any position vector $\mathbf{p}$, there are real numbers r and s such that $\mathbf{p} = r\mathbf{a} + s\mathbf{b}$. Draw an example to illustrate this situation.

4. Let $x\mathbf{u} + y\mathbf{v}$ be a position vector whose tip is on the line through the points $(-1, 1)$ and $(1, 3)$. Find a relationship between the numbers x and y.

1.10 GEOMETRIC REPRESENTATION OF THREE-DIMENSIONAL NUMERICAL VECTORS

In an exact analogy to the preceding section, consider three mutually perpendicular planes π_1, π_2, and π_3 in space. Each pair of these planes meets in a line; the three lines of intersection are also mutually perpendicular. Let the point of intersection be the origin on each line and construct a coordinate system on each line. The result might appear as in Fig. 1.16.

Each plane π_i, $i = 1, 2, 3$, is called a *coordinate plane* and the line l_i

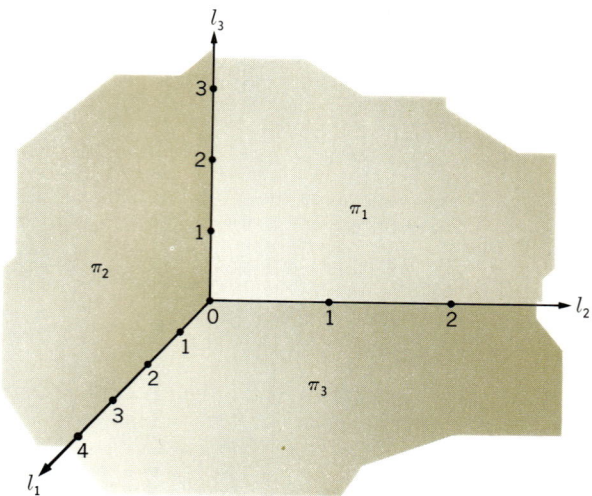

FIGURE 1.16

perpendicular to π_i is a *coordinate axis*. We shall usually call $l_1 = \pi_2 \wedge \pi_3$ the *x axis*, $l_2 = \pi_1 \wedge \pi_3$ the *y axis* and $l_3 = \pi_1 \wedge \pi_2$ the *z axis*.

An ordered number triple (x, y, z) is now pictured as that point in space which is at the (perpendicular directed) distance x from π_1, at a directed distance y from π_2 and at a directed distance z from π_3. In Fig. 1.17 several such numerical vectors are shown as points. Note that the coordinate plane subdivides the space into *octants* and that we have shown points in only three of the eight octants.

If (x, y, z) is a point in the coordinate three-space, then the line seg-

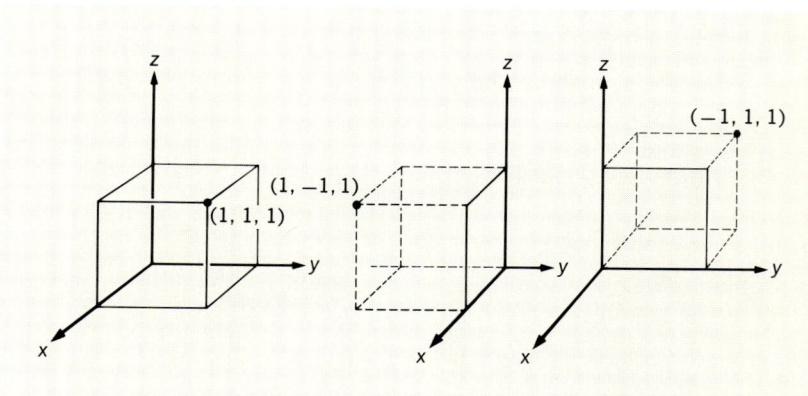

FIGURE 1.17

1.10 GEOMETRIC REPRESENTATION OF THREE-DIMENSIONAL NUMERICAL VECTORS

ment directed from the origin $(0, 0, 0)$ to (x, y, z) is called the *position vector* of that point. Again we speak of the point as the *tip* of its position vector. The unit point $(1, 0, 0)$ on the x axis is the tip of a position vector $\mathbf{u}$, $\mathbf{v}$ is the position vector of the point $(0, 1, 0)$ and $\mathbf{w}$ is that of $(0, 0, 1)$. These are the *unit vectors*.

Given any numerical vector (x, y, z), properties V1 to V10 permit us to write

$$(x, y, z) = x(1, 0, 0) + y(0, 1, 0) + z(0, 0, 1).$$

Therefore the position vector $\mathbf{p}$ of the point (x, y, z) may be expressed as

$$\mathbf{p} = x\mathbf{u} + y\mathbf{v} + z\mathbf{w}$$

a linear combination of the unit vectors. (See Fig. 1.18.)

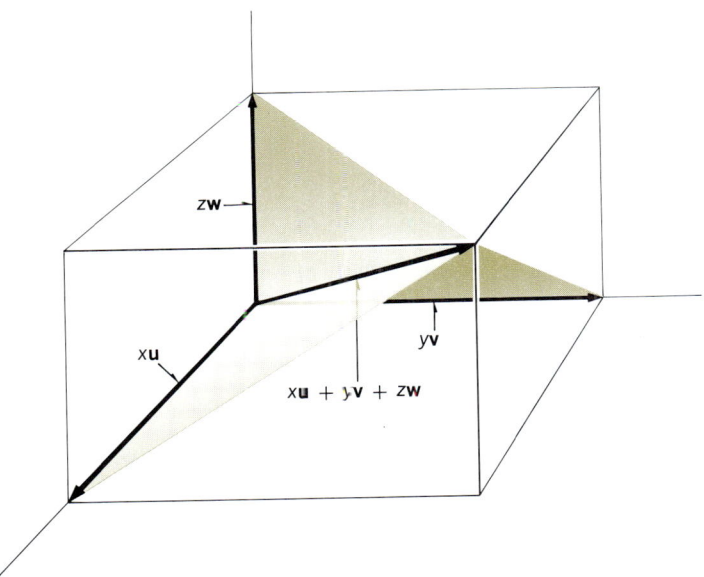

FIGURE 1.18

Two noncollinear position vectors in space determine a unique plane, namely, the plane containing the origin and the tips of the vectors. In this plane we use the parallelogram rule to define the sum of the vectors. Thus the vector sum $\mathbf{p} + \mathbf{q}$ is defined to lie in the plane determined by $\mathbf{p}$ and $\mathbf{q}$, if these vectors are not collinear. If $\mathbf{p}$ and $\mathbf{q}$ are collinear, then $\mathbf{p} + \mathbf{q}$ lies on the same line. In any case the vector sum is defined and coincides with the component-wise addition of numerical vectors. Similarly, the reader may prove that multiplication by a real number has the same geometric meaning in three dimensions as it does in two.

Exercises follow the next section.

1.11 THE NORM OF A VECTOR. DISTANCES

Although we do not put much emphasis upon geometric applications, it should be obvious that a coordinate plane or coordinate space can be used for geometric purposes. When we do so, each line segment must be assigned a length. In particular, then, each position vector must have a length and the unit basis vectors must have the same length, namely, one unit of distance.

We adopt a widely used convention and, *for geometric applications,* denote the unit basis vectors in the plane by

$$\mathbf{i} = (1, 0), \mathbf{j} = (0, 1)$$

and the unit basis vectors in space by

$$\mathbf{i} = (1, 0, 0), \mathbf{j} = (0, 1, 0), \mathbf{k} = (0, 0, 1)$$

Each position vector $\mathbf{p}$ can be written in one and only one way as a linear combination of the basis vectors. Thus, if $\mathbf{p}$ is a position vector in the plane, the coordinates (x, y) of the tip of $\mathbf{p}$ provide the equation

$$\mathbf{p} = x\mathbf{i} + y\mathbf{j}$$

while if $\mathbf{p}$ is in space, the coordinates (x, y, z) of its tip satisfy

$$\mathbf{p} = x\mathbf{i} + y\mathbf{j} + z\mathbf{k}$$

The length of a position vector $\mathbf{p}$ is called its *norm* and will be denoted by

$$\|\mathbf{p}\|$$

From Fig. 1.19 it is easy to see that the Pythagorean theorem applies to prove the following formulas:

$$\|x\mathbf{i} + y\mathbf{j}\| = \sqrt{x^2 + y^2}$$
$$\|x\mathbf{i} + y\mathbf{j} + z\mathbf{k}\| = \sqrt{x^2 + y^2 + z^2}$$

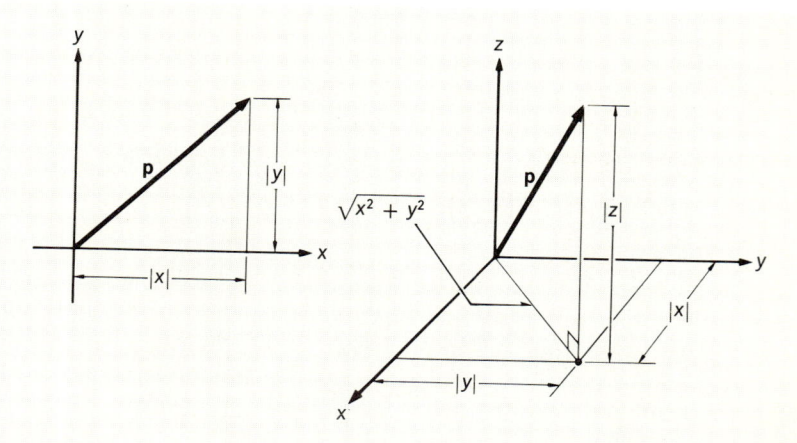

FIGURE 1.19

1.11 THE NORM OF A VECTOR. DISTANCES

■ **THEOREM 1.13.** The norm satisfies the properties:

N1. $\|\mathbf{p}\| \geq 0$ for any position vector $\mathbf{p}$
N2. $\|\mathbf{p}\| = 0$ if and only if $\mathbf{p}$ is the zero vector
N3. $\|r\mathbf{p}\| = |r| \cdot \|\mathbf{p}\|$ for any real number r and any position vector $\mathbf{p}$
N4. $\|\mathbf{p} + \mathbf{q}\| \leq \|\mathbf{p}\| + \|\mathbf{q}\|$ for any two position vectors

Because it is just an exercise in elementary algebra, we leave the proof of Theorem 1.13 to the reader.

Let $\mathbf{p}$ and $\mathbf{q}$ be two position vectors. The line segment joining the tips of these vectors is parallel to the vector position $\mathbf{p} - \mathbf{q}$ and has a length $\|\mathbf{p} - \mathbf{q}\|$. This fact follows from the parallelogram rule of vector addition. Looking at Fig. 1.20, we see that the line segment from the tip of $\mathbf{p}$ to the tip of $\mathbf{q}$ is precisely the side opposite $\mathbf{p} - \mathbf{q}$ in the parallelogram defining $\mathbf{p}$ as the vector sum $\mathbf{q} + (\mathbf{p} - \mathbf{q})$. (Note that Fig. 1.20 could be either on a plane on in space; the parallelogram law of vector addition is the same in either case.) It follows that the distance between the tips of $\mathbf{p}$ and $\mathbf{q}$ is given by the number $\|\mathbf{p} - \mathbf{q}\|$. Thus, in the plane, if $\mathbf{p} = x_1 \mathbf{i} + y_1 \mathbf{j}$ and $\mathbf{q} = x_2 \mathbf{i} + y_2 \mathbf{j}$, the distance from the point (x_1, y_1) to the point (x_2, y_2) is

$$d[(x_1, y_1), (x_2, y_2)] = \sqrt{(x_1 - x_2)^2 + (y_1 - y_2)^2}$$

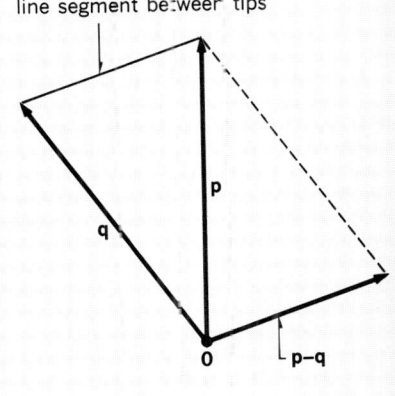

FIGURE 1.20

Similarly, in space

$$d[(x_1, y_1, z_1), (x_2, y_2, z_2)] = \sqrt{(x_1 - x_2)^2 + (y_1 - y_2)^2 + (z_1 - z_2)^2}$$

These are the *distance formulas*.

■ **THEOREM 1.14.** Let P, Q, and R be any three points (either on a plane or in space). Then

D1. $d(P, Q) \geq 0$
D2. $d(P, Q) = 0$ if and only if $P = Q$
D3. $d(P, Q) = d(Q, P)$
D4. $d(P, Q) + d(Q, R) \geq d(P, R)$

Property D4 is the *triangle inequality*. Note that Theorem 1.14 is an immediate consequence of Theorem 1.13.

Many geometric problems can be solved analytically using the distance formulas. We illustrate several techniques in the following examples.

■ *Example 1.* Consider the planar triangle with vertices $(1, 3), (4, -1)$ and $(8, 2)$. The lengths of its sides are

$$d[(1, 3), (4, -1)] = \sqrt{(1 - 4)^2 + (3 - (-1))^2}$$
$$= \sqrt{9 + 16} = 5$$
$$d[(1, 3), (8, 2)] = \sqrt{(1 - 8)^2 + (3 - 2)^2}$$
$$= \sqrt{49 + 1} = 5\sqrt{2}$$
$$d[(4, -1), (8, 2)] = \sqrt{(4 - 8)^2 + (-1 - 2)^2}$$
$$= \sqrt{16 + 9} = 5$$

The equality of two sides proves that this is an isosceles triangle. Also, because $5^2 + 5^2 = (5\sqrt{2})^2$, the Pythagorean theorem tells us that it is a right triangle.

■ *Example 2.* Let the midpoint of the line segment between $(2, 5)$ and $(4, -3)$ be denoted by (x, y). Then we must have

$$d[(x,y), (2, 5)] = d[(x,y), (4, -3)] = \tfrac{1}{2} d[(2, 5), (4, -3)]$$
$$= \tfrac{1}{2}\sqrt{68} = \sqrt{17}$$

We square the equations

$$\sqrt{(x-2)^2 + (y-5)^2} = \sqrt{17}$$
$$\sqrt{(x-4)^2 + (y-3)^2} = \sqrt{(x-2)^2 + (y-5)^2}$$

to obtain

$$(x-2)^2 + (y-5)^2 = 17$$

and $(x-4)^2 + (y+3)^2 = (x-2)^2 + (y-5)^2$. This last equation reduces to

$$x = 4y - 1$$

Substituting this value of x into the first equation, we have

$$(4y-3)^2 + (y-5)^2 = 17$$

or

$$17y^2 - 34y + 17 = 17(y-1)^2 = 0$$

It follows that $y = 1$ and therefore $x = 4(1) - 1 = 3$. The midpoint of the given line segment is therefore $(3, 1)$.

■ *Example 3.* The three points $(2, 1, 1)$, $(0, 3, 0)$ and $(1, 5, 2)$ in space are the vertices of an isosceles right triangle. We prove this simply by computing the lengths of the sides:

$$d[(2, 1, 1), (0, 3, 0)] = \sqrt{(2-0)^2 + (1-3)^2 + (1-0)^2}$$
$$= 3$$
$$d[(2, 1, 1), (1, 5, 2)] = \sqrt{(2-1)^2 + (1-5)^2 + (1-2)^2}$$
$$= 3\sqrt{2}$$
$$d[(0, 3, 0), (1, 5, 2)] = \sqrt{(0-1)^2 + (3-5)^2 + (0-2)^2}$$
$$= 3$$

We also see that the right angle is opposite the longest side and hence is at the vertex $(0, 3, 0)$.

■ **THEOREM 1.15.** The set $\{(x, y) : (x - a)^2 + (y - b)^2 = r^2\}$ in the plane is the circle of the radius $|r|$ and the center (a, b). Similarly, the set

$\{(x, y, z) : (x - a)^2 + (y - b)^2 + (z - c)^2 = r^2\}$ in space is the sphere of the radius $|r|$ and the center (a, b, c).

The proof is just an application of the distance formula and need not be written out here. The reader may wish to do so for himself.

EXERCISES 1.11

1. The vertices of four triangles are given here. Prove that each is an isosceles triangle. Draw the figure.
   (a) $(1, 2), (5, 5), (-2, -2)$
   (b) $(0, 3), (6, 2), (1, -3)$
   (c) $(-1, 3, 2), (4, 2, -1), (3, -1, 3)$
   (d) $(2, -3, -1), (1, -1, 2), (4, 1, 3)$

2. The vertices of four triangles are given here. Prove that each is a right triangle and determine its area. Draw the figure.
   (a) $(-1, 2), (4, -1), (7, 4)$
   (b) $(5, 4), (2, 8), (-2, 5)$
   (c) $(1, -1, 2), (3, 2, -1), (4, 1, 6)$
   (d) $(-1, 2, 4), (2, 1, -3), (3, -1, 2)$

3. Two sets of four points are given below. Prove that each set constitutes the vertices of a parallelogram. Draw the figure.
   (a) $(2, 2), (5, 5), (7, 2), (10, 5)$
   (b) $(1, -1, 2), (0, 3, 4), (3, 3, 0), (2, 7, 2)$

4. Show that the points $(4, 5), (-3, 4)$, and $(-2, 5)$ lie on a circle with a center $(1, 1)$.

5. Show that the points $(1, -2, 4), (-5, 6, 2), (3, -2, 2)$ and $(1, 4, 2\sqrt{7})$ are on a sphere with a center $(-1, 2, 0)$.

6. Find the radius and center of the circle whose equation is written as $x^2 + y^2 + 4x - 6y - 12 = 0$. (Hint: Complete the squares.)

7. Find the radius and center of the sphere whose equation is written as $2x^2 + 2y^2 + 2z^2 - 16x + 2y - 3z + 19 = 0$.

8. (a) Determine the midpoint of the line segment from $(3, 4, -2)$ to $(1, -2, 2)$.

   (b) Find the point (a, b) on the line segment from $(4, -1)$ to $(1, 2)$ such that $d[(a, b), (4, -1)] = \frac{1}{3}d[(4, -1), (1, 2)]$.

PROBLEMS 1.11

In each of Problems 1 to 4 a "distance" formula is defined. Prove that each satisfies properties D1 to D4 of Theorem 1.14.

1. For a pair of points $(x_1, y_1), (x_2, y_2)$ in the plane, define
$$d[(x_1, y_1), (x_2, y_2)] = |x_1 - x_2| + |y_1 - y_2|$$

2. For points $(x_1, y_1), (x_2, y_2)$ in the plane, define
$$d_2[(x_1, y_1), (x_2, y_2)] = \max(|x_1 - x_2|, |y_1 - y_2|)$$

3. For points $(x_1, y_1, z_1), (x_2, y_2, z_2)$ in space define
$$d_3[(x_1, y_1, z_1), (x_2, y_2, z_2)] = |x_1 - x_2| + |y_1 - y_2| + |z_1 - z_2|$$

4. For points $(x_1, y_1, z_1), (x_2, y_2, z_2)$ in space, define
$$d_4[(x_1, y_1, z_1), (x_2, y_2, z_2)] = |x_1 - x_2| + \sqrt{(y_1 - y_2)^2 + (z_1 - z_2)^2}$$

2

FUNCTIONS, GRAPHS, AND LIMITS

We meet many practical situations in which two variable quantities are known to be related. For instance, the output of a manufacturing process surely depends upon the input of raw material; a sensory reaction depends upon the intensity of the stimulus; the amount of current flowing in an electrical circuit depends upon the applied voltage; the strength of a muscle depends upon its cross-sectional area. These relationships are made precise, and hence useful in a quantitative manner, by means of the concept of "function," one of the most important concepts in mathematics.

2.1 LINEAR FUNCTIONS AND LINES

We approach the concept of a function by first studying a special case. In this way the several new ideas involved in the concept, and the new language to be learned, can be absorbed slowly.

Consider a set $\{(x,y)\}$ of numerical vectors. If the y component of each vector in the set is determined from the x component by means of an equation of the form

$$y = ax + b,$$

where a and b are fixed real numbers, then we say that y *is a linear function of* x. Thus a linear function determines or corresponds to a set of numerical vectors of the form

$$\{(x,y) : y = ax + b\}$$

Some restriction may be imposed upon the values of x if the situation requires it.

Considered as a set of points in a coordinate plane, such a set $\{(x,y) : y = ax + b\}$ is always a straight line. (This fact accounts for the name "linear" function, of course.) This statement requires proof for which the following preliminaries are needed.

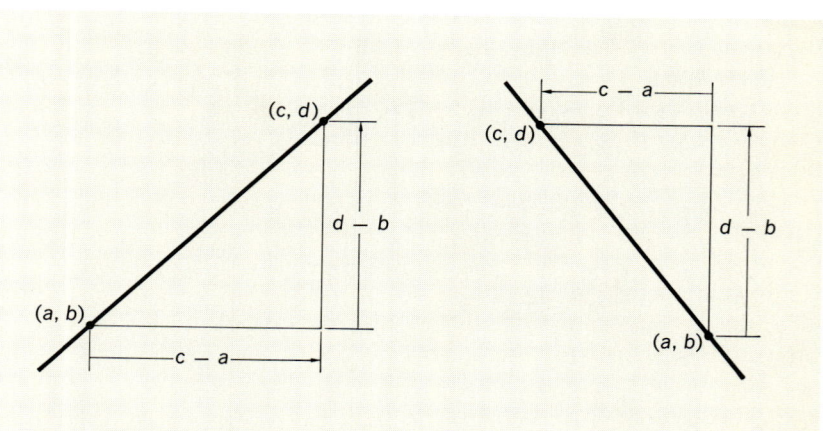

FIGURE 2.1

2.1 LINEAR FUNCTIONS AND LINES

Let (a, b) and (c, d) be any two points in the coordinate plane, $a \neq c$. The ratio of the directed vertical distance $d - b$ to the directed horizontal distance $c - a$,

$$\frac{d - b}{c - a}$$

is called the *slope* of the line segment joining (a, b) and (c, d). Figure 2.1 illustrates this idea.

Any two segments of the same line must necessarily have the same slope. In Fig. 2.2 we show two such segments and the associated right triangles. The fact that the triangles obviously have equal corresponding angles proves that they are similar and this is all the proof we need here.

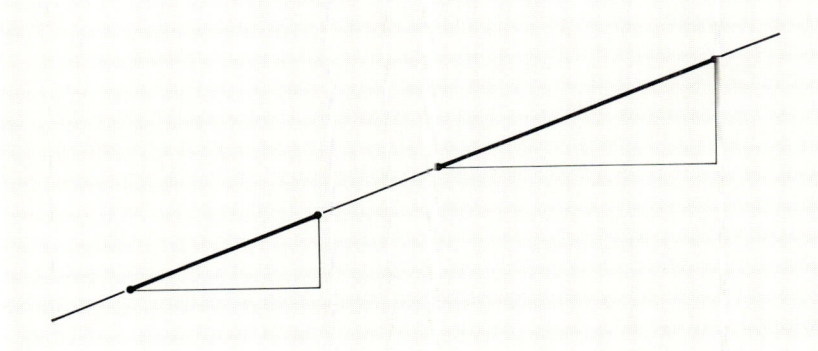

FIGURE 2.2

If two line segments have a common endpoint and have equal slopes, then they must be in the same line. To see this, consider the segments PQ and PR as shown in Fig. 2.3. We note that unless Q lies on the segment PR, the two triangles shown cannot be similar as required by the condition on slopes. This again is sufficient proof.

Now again let (a, b) and (c, d), $a \neq c$, be any two points in the coordinate plane. If (x, y) is any other point on the line through these points, then the slope of the segment from (a, b) to (x, y) must equal the slope of the segment from (a, b) to (c, d). This provides the equation

$$\frac{y - b}{x - a} = \frac{d - b}{c - a}$$

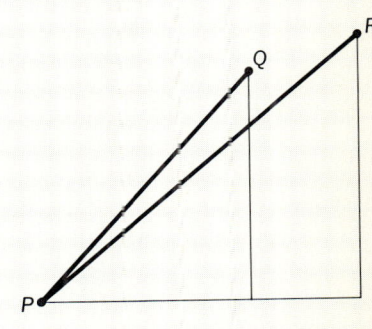

FIGURE 2.3

Conversely, the preceding paragraph says that any point (x, y) which satisfies this equation must be on the line through (a, b) and (c, d). This proves the difficult part of the following result:

■ **THEOREM 2.1.** If (x, y) is on the straight line through the distinct points (a, b) and (c, d), then (x, y) is in the set

$$L = \{(x,y) : (d-b)(x-a) = (c-a)(y-b)\}.$$

Conversely, if (x,y) is in the set L, then (x,y) is on the straight line through the points (a,b) and (c,d).

PROOF: The two paragraphs preceding the theorem prove it for all cases except when $a = c$. For this special case, the proof is as follows: If (x,y) is on the line through distinct points (a,b) and (a,d), then surely $x = a$. Hence (x,y) satisfies

$$x - a = 0$$

and therefore satisfies

$$(d-b)(x-a) = 0 = (a-a)(y-b)$$

i.e. (x,y) is in L.

Conversely, if (x,y) is in L (and $c = a$, of course), then

$$(d-b)(x-a) = 0.$$

Because $(a,b) \neq (a,d)$, we have $d \neq b$ and hence $x - a = 0$. It follows that $x = a$ and hence that (x,y) is on the straight line through (a,b) and (a,d). ∎

We see that the set of points on a straight line corresponds to the set of numerical vectors satisfying a certain (linear) equation. Such an equation is said to be *an equation of the line*. The following theorem presents the *general equation of a line*.

■ **THEOREM 2.2.** Each line in a coordinate plane has an equation of the form $Ax + By + C = 0$, A and B not both zero, and conversely the set of points

$$\{(x,y) : Ax + By + C = 0, A \text{ and } B \text{ not both zero}\}$$

is a straight line.

PROOF: Given a straight line, choose any two distinct points (a,b) and (c,d) on the line. Then by Theorem 2.1, an equation of this line is $(d-b)(x-a) = (c-a)(y-b)$. We simply define

$$A = d - b, \quad B = -(c-a) \quad \text{and} \quad C = -a(d-b) + b(c-a).$$

Because the points (a,b) and (c,d) are distinct, we cannot have both $d - b = 0$ and $c - a = 0$ whence not both A and B are zero. Of course, the equation is then $Ax + By + C = 0$.

Conversely, consider the set

$$\{(x,y) : Ax + By + C = 0\}$$

and suppose first that $B \neq 0$. Then the equation $Ax + By + C = 0$ can be written as

$$\left[\left(A - \frac{C}{B}\right) - \left(-\frac{C}{B}\right)\right](x - 0) = -(B - 0)\left(y - \left(-\frac{C}{B}\right)\right).$$

By Theorem 2.1, this is the equation of a straight line through the points $(0, -C/B)$ and $(B, A - C/B)$. If $B = 0$, then $A \neq 0$ and we can repeat this procedure using A in place of B. ∎

▪ **COROLLARY 2.3.** Each line not parallel to the y axis is a coordinate plane has an equation of the form

$$y = mx + b$$

and conversely each such linear function has a line not parallel to the y axis as its geometric representation.

This is the result we earlier said we could not prove. We state it here and point out that the equation $y = mx + b$ is often called the *slope-intercept form* of the equation of the line.

Many problems in geometry involve finding the point of intersection of two lines. If the equations of the lines are known, this job can be done algebraically. Let the lines be

$$\{(x, y) : A_1x + B_1y + C_1 = 0\}, \quad \{(x, y) : A_2x + B_2y + C_2 = 0\}$$

The intersection of these two sets is precisely

$$\{(x, y) : A_1x + B_1y + C_1 = 0 \quad \text{and} \quad A_2x + B_2y + C_2 = 0\}$$

This means that the desired point (number pair) is the simultaneous solution of the two linear equations

$$A_1x + B_1y + C_1 = 0$$
$$A_2x + B_2y + C_2 = 0$$

These equations can be solved simultaneously for x and y, provided that $A_1B_2 \neq A_2B_1$. The solution is

$$x = \frac{B_1C_2 - B_2C_1}{A_1B_2 - A_2B_1}, \quad y = \frac{A_2C_1 - A_1C_2}{A_1B_2 - A_2B_1},$$

formulas that need not be memorized. We usually use an elimination of variables to solve such systems of linear equations.

▪ **Example 1.** The equation of the line through the points $(-2, 3)$ and $(4, 1)$ is determined simply by putting these numbers into the general equation

$$(d - b)(x - a) = (c - a)(y - b)$$

This results in

$$(1 - 3)(x - (-2)) = (4 + 2)(y - 3)$$

or
$$2(x + 2) = 6(y - 3)$$
or
$$2x + 6y - 14 = 0$$
or
$$x + 3y - 7 = 0$$

■ *Example 2.* We find the point of intersection of the lines
$$\{(x,y) : x - y - 2 = 0\}, \quad \{(x,y) : 3x + 2y + 4 = 0\}$$
by solving simultaneously the equations
$$x - y - 2 = 0$$
$$3x + 2y + 4 = 0$$

The first equation is multiplied by two and added to the second equation to obtain $5x = 0$ or $x = 0$. Putting $x = 0$ into either of the original equations, we obtain $y = -2$. Thus the point of intersection is $(0, -2)$.

■ *Example 3.* The two points $(0, 4)$ and $(-2, 0)$ determine one line and the points $(0, 1)$ and $(3, -2)$ determine another. The equation of the first line is
$$(4 - 0)(x - 0) = (0 - (-2))(y - 4)$$
or
$$2x - y + 4 = 0$$

The equation of the second line is
$$(1 - (-2))(x - 0) = (0 - 3)(y - 1)$$
or
$$x + y - 1 = 0$$

Solving these equations simultaneously, the point of intersection of the lines is found to be $(-1, 2)$.

EXERCISES 2.1 In Exercises 1 to 10 are given pairs of points. Find the equation of the line through each pair. Reduce the equation to general form. Draw each line on a coordinate plane.

1. $(3, 1), (2, -3)$
2. $(0, 4), (1, 3)$
3. $(-1, -1), (4, 2)$
4. $(-3, 1), (2, -2)$
5. $(5, 0), (0, 2)$
6. $(-3, 0), (0, 1)$
7. $(1, 4), (-1, 4)$
8. $(2, -1), (-3, -1)$
9. $(3, 1), (3, 4)$
10. $(-1, 2), (2, -4)$

2.1 LINEAR FUNCTIONS AND LINES

In each of Exercises 11 to 20 determine the point of intersection of the given lines.

11. $\{(x,y) : x - y = 2\}$, $\{(x,y) : 2x + 3y = 4\}$
12. $\{(x,y) : 3x + y = 5\}$, $\{(x,y) : 2x + y = 4\}$
13. $\{(x,y) : y = 3 - x\}$, $\{(x,y) : 3y = x + 13\}$
14. $\{(x,y) : 2x - y - 7 = 0\}$, $\{(x,y) : y = 2x - 1\}$
15. $\{(x,y) : 2y = x + 4\}$, $\{(x,y) : 18x + 6y = 37\}$
16. $\{(x,y) : 3x - 2y = 6\}$, $\{(x,y) : 6x - 4y = -2\}$
17. $\{(x,y) : x - 4y + 4 = 0\}$, $\{(x,y) : x + 3y = 11\}$
18. $\{(x,y) : x + 2y + 20 = 0\}$, $\{(x,y) : y = 2(x - 5)\}$
19. $\{(x,y) : Ax + 2y = 5\}$, $\{(x,y) : Ax + 4y = 10\}$
20. $\{(x,y) : \frac{x}{\frac{3}{2}} + \frac{y}{-3} = 1\}$, $\{(x,y) : \frac{x}{-1} + \frac{y}{-\frac{1}{2}} = 1\}$

21. Why is the set

$$\{(x,y) : \frac{y-1}{x-1} = 1\}$$

not a line through (1, 1)?

PROBLEMS 2.1

1. Suppose that neither a nor b is zero. Prove that the equation of the line through the points $(a, 0)$ and $(0, b)$ can be written in the form

$$\frac{x}{a} + \frac{y}{b} = 1$$

which is called the *intercept form* of the equation of a line.

2. Prove that a line passes through the origin if and only if its equation is of the form $Ax + By = 0$.

3. Two lines are parallel if they do not intersect. Prove that the line through $(0, -3)$ and $(1, 0)$ is parallel to that through $(1, 4)$ and $(-1, -2)$.

4. Suppose you had two job offers for sales work. The first offer is for a straight 15 percent commission, the second offer is for a salary of $60 per week plus 5 percent commission. How much must you sell each week in order that the first job pay as much as the second? Draw diagrams of the income functions as functions of weekly sales volume.

5. (Economics) The demand x for a certain commodity is found to be 7000 lb if the price is 20 cents per lb and is found to be 4000 lb if the price is 40 cents per lb. Assume that x is a linear function of p and determine the demand if the price were 35 cents per lb.

6. (Economics) If we assume conditions of "pure competition," a *supply function* determines the amount of a commodity that producers will make available at any given unit price of the commodity. We say that *market equilibrium* occurs when supply equals demand. Now, for a certain commodity, suppose that the demand function is described by the equation $x = 1000 - 5p$ and the supply function is found to be $x = 6p + 55$. At market equilibrium, what is the unit price p and how many units will be supplied at this price?

7. In the mathematical technique known as *linear programming*, systems of linear inequalities play an important role. Find and shade the area in a coordinate plane where each of the following systems is satisfied.
   (a) $x \geq 0, y \geq 0, y \leq 4, x + y \leq 8$
   (b) $x \geq 0, y \geq 0, -x + y \leq 6, y \leq 5, x \leq 6$
   (c) $x \geq 0, y \geq 0, x + 4y \leq 12, 4x + y \leq 12$
   (d) $x \geq 0, y \geq 0, y \leq 3, 2x + y \leq 7, x - y \geq 2$

2.2 PARALLEL LINES

Two lines are parallel (by definition) if either they coincide or they do not intersect. Assuming that the lines

$$\{(x,y) : A_1x + B_1y + C_1 = 0, A_1^2 + B_1^2 \neq 0\}$$
$$\{(x,y) : A_2x + B_2y + C_2 = 0, A_2^2 + B_2^2 \neq 0\}$$

do not coincide, they are parallel if and only if the system of equations

$$A_1x + B_1y + C_1 = 0$$
$$A_2x + B_2y + C_2 = 0$$

has no simultaneous solution.

■ *Example 1.* The lines

$$\{(x,y) : 3x - y - 2 = 0\} \quad \text{and} \quad \{(x,y) : 12x - 4y + 3 = 0\}$$

are parallel. To see this, we first note that the point $(0, -2)$ is on the first line but not on the second. Hence they do not coincide. Then trying to solve the equations

$$3x - y = 2$$
$$12x - 4y = -3$$

we multiply the first equation by four and subtract the result from the second equation. This results in the false statement $0 = -11$ and tells us that the equations have no simultaneous solution and hence that the lines do not intersect.

There is a purely algebraic criterion for the parallelism of two lines. We need only determine when the equations

$$A_1x + B_1y = -C_1$$
$$A_2x + B_2y = -C_2$$

are either equivalent to each other or have no simultaneous solution. The reader may already know that the determinant of the coefficients of these equations

$$\begin{vmatrix} A_1 & B_1 \\ A_2 & B_2 \end{vmatrix} = A_1B_2 - A_2B_1$$

is equal to zero if and only if one of these two conditions hold. This proves a result.

2.2 PARALLEL LINES

■ **THEOREM 2.4.** The lines $\{(x,y) : A_1x + B_1y + C_1 = 0\}$ and $\{(x,y) : A_2x + B_2y + C_2 = 0\}$ are parallel if and only if $A_1B_2 = A_2B_1$.

When do we have $A_1B_2 = A_2B_1$? If none of these numbers is zero, we may write this equation as

$$\frac{A_1}{A_2} = \frac{B_1}{B_2}$$

If the common ratio is denoted by k, then we see that $A_1 = kA_2$, $B_1 = kB_2$. Conversely, if these equations both held for some constant $k \neq 0$,

$$A_1B_2 - A_2B_1 = (kA_2)B_2 - A_2(kB_2) = k(A_2B_2 - A_2B_2) = 0$$

If none of the numbers A_1, A_2, B_1, B_2 is zero, then $A_1B_2 = A_2B_1$ if and only if there is a number $k \neq 0$ such that $A_1 = kA_2$ and $B_1 = kB_2$.

Suppose that one of these numbers is zero. If $A_1 = 0$, then B_1 is not zero because $A_1^2 + B_1^2 \neq 0$. Hence the equation $A_1B_2 = 0 = A_2B_1$ implies that $A_2 = 0$ and therefore that $B_2 \neq 0$. By setting $k = B_1/B_2$, we can again write $A_1 = kA_2$ and $B_1 = kB_2$. Hence in this case, too, we know that $A_1B_2 = A_2B_1$ if and only if there is a constant $k \neq 0$ such that $A_1 = kA_2$ and $B_1 = kB_2$.

One final observation will complete our discussion. Two linear equations $Ax + By + C = 0$ and $kAx + kBy + kC = 0$, where $k \neq 0$, are equivalent in the sense that a pair (x,y) satisfies $Ax + By + C = 0$ if and only if it also satisfies $kAx + kBy + kC = 0$. This says that we have the equality

$$\{(x,y) : Ax + By + C = 0\} = \{(x,y) : kAx + kBy + kC = 0; k \neq 0\}$$

of the solution sets. Or, in other words, the equations describe the same line. All of the above information can be compiled into one result.

■ **THEOREM 2.5.** The lines

$$\{(x,y) : A_1x + B_1y + C_1 = 0\}$$
$$\{(x,y) : A_2x + B_2y + C_2 = 0\}$$

and

are parallel if and only if there is a constant $k \neq 0$ such that

$$A_1 = kA_2 \quad \text{and} \quad B_1 = kB_2.$$

If we also have $C_1 = kC_2$, then the lines are coincident.

■ **Example 2.** The lines $\{(x,y) : 2x - 3y + 5 = 0\}$ and $\{(x,y) : 4x - 6y + 3 = 0\}$ are parallel because $A_1 = 2 = \frac{1}{2}(4) = \frac{1}{2}A_2$, $B_1 = -3 = \frac{1}{2}(-6) = \frac{1}{2}B_2$. They do not coincide because

$$C_1 = 5 \neq \frac{1}{2}(3) = \frac{1}{2}C_2$$

■ **Example 3.** Consider the line through $(1, 1)$ and $(2, 5)$ and the line through $(-1, 1)$ and $(1, 9)$. The first line has the equation

$$(5 - 1)(x - 1) = (2 - 1)(y - 1)$$

or

$$4x - y - 3 = 0$$

The second line has the equation

$$(9 - 1)(x + 1) = (1 + 1)(y - 1)$$

or

$$8x - 2y + 10 = 0$$

Because $4 = \frac{1}{2}(8)$ and $-1 = \frac{1}{2}(-2)$ but $-3 \neq \frac{1}{2}(10)$, Theorem 2.5 *tells us that these lines are parallel but not coincident.*

EXERCISES 2.2

1. In each of the following cases, determine whether or not the line through the points P_1 and Q_1 is parallel to the line through P_2 and Q_2.
   (a) $P_1 = (1, -1)$, $Q_1 = (2, 1)$; $P_2 = (-1, -1)$, $Q_2 = (1, 3)$
   (b) $P_1 = (0, 3)$, $Q_1 = (-1, 4)$; $P_2 = (2, 0)$, $Q_2 = (4, -2)$
   (c) $P_1 = (3, 14)$, $Q_1 = (-3, 18)$; $P_2 = (6, 4)$, $Q_2 = (-6, 13)$
   (d) $P_1 = (3, 3)$, $Q_1 = (-3, -1)$; $P_2 = (3, 0)$, $Q_2 = (6, 2)$
   (e) $P_1 = (1, -1)$, $Q_1 = (5, 2)$; $P_2 = (-1, 0)$, $Q_2 = (7, 7)$
   (f) $P_1 = (2, 0)$, $Q_1 = (-2, 3)$; $P_2 = (4, -3)$, $Q_2 = (10, -1)$

2. Which of the sets of four points below are the vertices of a parallelogram?
   (a) $\{(1, 1), (7, 3), (3, -5), (9, -3)\}$
   (b) $\{(-2, 0), (1, 4), (5, 2), (8, 6)\}$
   (c) $\{(1, 1), (3, 5), (8, 1), (6, -3)\}$
   (d) $\{(-1, 3), (6, 4), (1, -1), (8, 1)\}$

3. The points $(1, 3), (-2, 1)$ and $(4, -6)$ are vertices of a parallelogram. Find the fourth vertex. (Two solutions.)

4. In each of the following cases, find the equation of the line through the given point and parallel to the given line.
   (a) $(0, 0)$; $\{(x, y) : 3x + y = 4\}$
   (b) $(0, 4)$; $\{(x, y) : x + y = 1\}$
   (c) $(-1, 3)$; $\{(x, y) : 2y = x - 1\}$
   (d) $(3, -2)$; $\{(x, y) : 3x - 2y = -1\}$

PROBLEMS 2.2

1. Prove that the line through (a, b) and parallel to the line $\{(x, y) : Ax + By + C = 0\}$ has equation $A(x - a) + B(y - b) = 0$.

2. Prove that, for any choice of the constants k_1 and k_2 (not both zero), the equation

$$k_1(A_1x + B_1y + C_1) + k_2(A_2x + B_2y + C_2) = 0$$

is that of a line through the point of intersection of $\{(x, y) : A_1x + B_1y + C_1 = 0\}$ and $\{(x, y) : A_2x + B_2y + C_2 = 0\}$, and conversely, every line through this point of intersection has an equation of the above form.

3. Use the result of Problem 2 to determine the line parallel to $\{(x, y) : x - 2y = 3\}$ and passing through the intersection of the lines $\{(x, y) : 3x + 2y = 12\}$ and $\{(x, y) : 2x - 6y + 7 = 0\}$.

4. For what values of t is $\{(x, y) : 3tx + 2y = 4\}$ parallel to $\{(x, y) : 6x - 2y = 3\}$?

2.3 THE SLOPE OF A LINE

Consider a linear function $y = Ax + B$ and its geometric representation, the line $\{(x,y) : y = Ax + B\}$ in some coordinate plane. In Section 2.1 we stated that any two segments of this line have the same slope. Hence there is a unique number, the common value of the slopes of all segments, associated with such a line. Precisely, if $(x_1, Ax_1 + B)$ and $(x_2, Ax_2 + B)$ are two different points on this line, then the slope of the line segment joining them is

$$\frac{Ax_1 + B - Ax_1 - B}{x_2 - x_1} = \frac{A(x_2 - x_1)}{x_2 - x_1} = A$$

Therefore the coefficient A of x in the linear function $y = Ax + B$ is the slope of the corresponding line.

Given the value $x = a$, the corresponding value $Aa + B$ is called the *functional value* of the linear function $y = Ax + B$ at the point $x = a$. Geometrically, this just means that $(a, Aa + B)$ is a point on the line $\{(x, y) : y = Ax + B\}$.

Now suppose we add an amount h to the number a. This yields a new value $a + h$ of x. The corresponding functional value is

$$A(a + h) + B = (Aa + B) + Ah$$

Thus a change in x by the amount h results in a change in the functional value by the amount Ah. (See Fig. 2.4.) For this reason, the number A is also called the *rate of change of the linear function* $y = Ax + B$ with respect to x.

■ **THEOREM 2.6.** Two lines

$$\{(x,y) : A_1x + B_1y + C_1 = 0\}$$

and

$$\{(x,y) : A_2x + B_2y + C_2 = 0\}$$

not parallel to the y axis, are parallel if and only if they have the same slope.

PROOF: The slope of the line $\{(x,y) : Ax + By + C = 0, B \neq 0\}$ is found simply by solving the equation for y in terms of x. This yields

$$y = -\frac{A}{B}x - \frac{C}{B}$$

Hence the slope of the given line is $-(A/B)$.

Now, if the two given lines have the same slope, then

$$-\frac{A_1}{B_1} = -\frac{A_2}{B_2}$$

Multiplying by $-B_1$ and by B_2, we have

$$A_1B_2 = A_2B_1$$

which tells us that the two lines are parallel (or coincident).

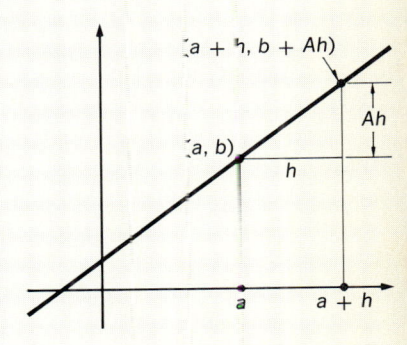

FIGURE 2.4

Conversely, if the lines are parallel, then there is a constant $k \neq 0$ such that $A_1 = kA_2$ and $B_1 = kB_2$. Then

$$-\frac{A_1}{B_1} = -\frac{kA_2}{kB_2} = -\frac{A_2}{B_2}$$

and hence the slopes are equal.

The slope of a line can be used to determine whether or not three or more points are collinear. For example, consider the points $(-1, -6)$, $(2, 0)$ and $(4, 4)$. The slope of the line through $(-1, -6)$ and $(2, 0)$ is $[0 - (-6)]/[2 - (-1)] = 2$. Similarly, the slope of the line through $(-1, -6)$ and $(4, 4)$ is $[4 - (-6)]/[4 + (-1)] = 2$. Therefore the line through $(-1, -6)$ with slope two contains both of the other points and hence the three points lie on one line.

In general, the n points (a_1, b_1), (a_2, b_2), $\cdots$, (a_n, b_n) are collinear if and only if

$$\frac{b_i - b_1}{a_i - a_1} = \frac{b_2 - b_1}{a_2 - a_1}$$

for each $i \in \mathbf{N}$, $2 < i \leqq n$. (See Problem 2.3.1.) ∎

EXERCISES 2.3

1. Determine the slope of the line through the given pair of points in each of the following cases.
    (a) $(3, 1), (-2, 1)$
    (b) $(3, 1), (-2, -1)$
    (c) $(-1, 2), (2, -1)$
    (d) $(0, 2), (4, 0)$
    (e) $(5, -1), (-3, -1)$
    (f) $(2, -1), (2, 5)$
    (g) $(1, 1), (3, -2)$
    (h) $(-1, -1), (1, 1)$

2. Determine the slope of each line below. Locate the intercepts of each line and draw it.
    (a) $\{(x,y) : y = 3x - 6\}$
    (b) $\{(x,y) : x = 4y - 3\}$
    (c) $\{(x,y) : \frac{x}{2} + \frac{y}{3} = 1\}$
    (d) $\{(x,y) : x + 4y = 7\}$
    (e) $\{(x,y) : 3x - 2y = 4\}$
    (f) $\{(x,y) : 2x + 7y = 8\}$
    (g) $\{(x,y) : \frac{x}{a} + \frac{y}{b} = 1\}$
    (h) $\{(x,y) : x = 3t + 1, y = -t + 2; t \in \mathbf{R}\}$

3. Which of the following pairs of lines are parallel?
    (a) $\{(x,y) : 3x - 2y + 4 = 0\}$, $\{(x,y) : 4y = 6x - 5\}$
    (b) $\{(x,y) : 2x + 5y = 0\}$, $\{(x,y) : \frac{x}{5} + \frac{y}{2} = 1\}$
    (c) $\{(x,y) : y = \frac{1}{2}x - 1\}$, $\{(x,y) : x - 2y = 3\}$
    (d) $\{(x,y) : 2x + 6y = 3\}$, $\{(x,y) : x = 3t + 1, y = -t + 2; t \in \mathbf{R}\}$

4. Determine the number c, so that the line through $(c, 5)$ and $(3, -1)$
    (a) has slope 2
    (b) has slope zero
    (c) has no definable slope

5. Which of the following sets of points are collinear?
   (a) $\{(1, 1), (3, 5), (-4, -9)\}$
   (b) $\{(0, 4), (1, -1), (3, -11)\}$
   (c) $\{(-1, 8), (1, 4), (3, 0)\}$
   (d) $\{(2, 0), (4, -1), (-2, 2), (-16, 10)\}$
   (e) $\{(3, -1), (-6, -4), (0, -2), (12, 2)\}$
   (f) $\{(2, 0), (4, 8), (-1, -12), (-3, -18)\}$

PROBLEMS 2.3

1. Prove that the points (a_i, b_i), $i = 1, 2, \ldots, n$, are collinear if and only if $(a_2 - a_1)(b_i - b_1) = (a_i - a_1)(b_2 - b_1)$ for each $i = 3, 4, \ldots, n$.

2. (Physics) The relation between the centigrade and Fahrenheit scales of temperature is linear. Knowing that the freezing point of water is $0°C$ and $32°F$, and that the boiling point of water is $100°C$ and $212°F$, find C as a linear function of F, and conversely. At what temperature, if any, do the two agree?

3. (Economics) If the demand x for a certain commodity is a linear function $x = mp + b$ of the unit price p, then the slope m of the line $\{(p, x) : x = mp + b\}$ is called the *marginal demand*. Interpret this concept in economic terms.

2.4 FUNCTIONS AND THEIR GRAPHS

We have discussed linear functions. Using this and the examples given below we motivate a careful definition of the word "function" as it is used in mathematics.

■ **Example 1.** The area A of a square is a "function" of the length s of its sides; the number A is determined by the number s via the formula

$$A = s^2$$

The formula provides a nonnegative number A for each real number s but, for practical reasons, the length s is not taken to be negative. Hence the pairs of nonnegative numbers $(s, A) = (s, s^2)$ constitute a description of this "function."

■ **Example 2.** In economics one may assume that the consumer demand x for a certain commodity depends only upon its unit price p. For a small range of values of the price p, such a *demand function* might be assumed to be linear

$$x = Ap + B$$

(Here the rate of change A is negative because demand decreases as the unit price increases.) This linear function corresponds to the set of numerical vectors

$$\{(p, x) : x = Ap + B\}$$

■ **Example 3.** A manufacturer finds that the total cost C of producing x units of his product is given by the equation

$$C = x^3 + \frac{15}{x^2} - 75x + 950, \quad x \geq 1$$

Here C is the function of x that is described by the set of numerical vectors

$$\left\{(x, C) : x \geqq 1, C = x^3 + \frac{15}{x^2} - 75x + 950\right\}$$

■ *Example 4.* The volume V of a right circular cylinder is a function of the radius r of its base and its height h, $V = \pi r^2 h$. The relationship under consideration here is described by the set of three-dimensional numerical vectors:

$$\{(r, h, V) : r \geqq 0, h \geqq 0, V = \pi r^2 h\}$$

■ *Example 5.* For each natural number $n \in N$, define the sum

$$s_n = 1 + \frac{1}{2} + \frac{1}{4} + \cdots + \frac{1}{2^{n-1}} = 2 - \frac{1}{2^{n-1}}$$

This defines s_n as a function of n. As a set of vectors we have

$$\left\{(x, y) : x = n \in N, y = s_n = 2 - \frac{1}{2^{n-1}}\right\}$$

In each example just given, we point out that a "function" has three distinct features. Two of these features are sets X and Y, while the third is some rule that assigns an element of Y to each element of X. In Example 1, we have $X = \{s : s \geqq 0\}$ because the length of a side of a square is not negative. Here the set Y must contain all nonnegative numbers because each such number can be the area of a square. Similarly, in Example 2, we assume that $X = \{p : p \geqq 0\}$ because prices are not negative. It then follows that the set Y must contain the set $\{x : x \leqq B\}$. (Explain this statement.) In Example 4, we should note that a value V of the volume is assigned to each pair (r, h) of nonnegative numbers. Hence $X = \{(r, h) : r \geqq 0, h \geqq 0\}$. We see that Y must include all nonnegative numbers because any nonnegative number is the volume of some cylinder. Lastly, concerning Example 5, we remark that our definition assigns a value of the sum S_n to a natural number. Thus we must have $X = N$. Since Y must contain all the values S_n, we can take $Y = \{s : 1 \leqq s \leqq 2\}$, for instance.

Each example above explicitly includes a formula or "rule" that provides a single element of Y for each element of X. This rule is the third feature of a function. Abstracting the essence of these situations, we are led to a tentative definition of the word "function":

A function consists of two sets X and Y and a rule f that assigns to each element of X a single element of Y.

This definition views a function in the active sense as a "transformation" of the elements of X into elements of Y. The set X is called the *domain* of the function; the set Y is called a *codomain* of the function. For each $x \in X$, that element $y \in Y$ which corresponds to x under the functional rule f is called the *image of x* or the *functional value at x* of the function. The functional value at x is denoted by the symbol

$$f(x),$$

which is read "f of x." The subset $\{y : y = f(x)\} \subset Y$ of all functional values is called the *range* or the *image set* of the function. It is denoted by

$$f(X)$$

Any set Y that contains $f(X)$ can be a codomain of the function.

The drawing in Fig. 2.5 is meant only as a mnemonic device. The domain X is transformed, element by element, into the codomain Y by the "machine" f. When an element $x \in X$ is placed into f, the output is the element $f(x) \in Y$. With this in mind, it should be impossible to confuse the "function," which is the entire manufacturing process, either with the rule f (the machine) or with the functional value $f(x)$ (the output of the process).

Mathematicians often use the symbol

$$f : X \to Y$$

for an unspecified function. Note that the symbol explicitly incorporates the three features of a function and the arrow suggests the active nature of a function as a transformation of the domain X into the codomain Y.

Two functions $f : X \to Y$ and $g : X \to Y$ are equal if and only if they have the same domain and $f(x) = g(x)$ for each point x in the domain. This equality of functional values obviously implies that the ranges of f and g coincide.

With this definition of the equality of functions we are led to a second definition of a function. Referring back to our examples again, we point out that each function has an associated set of ordered pairs. In Example 1, the set of ordered pairs is $\{(s, A) : s \geq 0, A = s^2\}$. In Example 2, the set is the "line" $\{(p, x) : x = Ap + B\}$. The set of ordered pairs in Example 4 has elements of the form

$$((r, h), V)$$

in which the first member of the pair is itself an ordered pair. Our second definition of "function" is based upon these ordered pairs. To do so, we need a new concept.

Let X and Y be any two sets (which need not be distinct). The *product set*, denoted by $X \times Y$, consists of all ordered pairs (x, y) with $x \in X$ and $y \in Y$. In symbolic form,

$$X \times Y = \{(x, y) : x \in X, y \in Y\}$$

This idea is not really new to us. The set of all two-dimensional numerical vectors is precisely the set $\mathbf{R} \times \mathbf{R}$, for instance.

Each function we have discussed gives rise to a set of ordered pairs, the pairs of the form $(x, f(x))$. And each such pair $(x, f(x))$ is an element

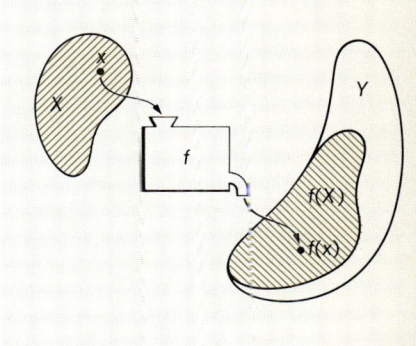

FIGURE 2.5

of the product set $X \times Y$. Thus the set of ordered pairs that corresponds to a function is

$$\{(x,y) : y = f(x)\} \subset X \times Y$$

Knowing this set of ordered pairs, we can reconstruct the rule f. Indeed, the two concepts are equivalent. This leads to our second, and more precise, definition of a function.

A function consists of two sets X, the domain, and Y, the codomain, and a subset f of $X \times Y$ such that each element $x \in X$ appears in one and only one ordered pair in the subset f. (This restriction on f is the equivalent of saying that there is only one functional value for each $x \in X$.)

Under this definition of function, two functions are equal if and only if they are equal as sets of ordered pairs.

There is an unfortunate but seemingly inevitable dichotomy in many mathematical notations; this dichotomy is nowhere better seen than in the discussion of functions. The definition of a function in terms of the three sets X, Y, and $f \subset X \times Y$ lends itself well to general discussions, to proving theorems about arbitrary functions and to understanding the concept completely. But in order to discuss a particular function, we normally revert to less formal notation.

It is common practice to specify a particular function simply by giving a formula for the functional value $f(x)$. For instance, we might write

$$f(x) = x^2 - 3x + 2$$

thereby giving only the rule relating elements of the domain X to those in the range Y. When this is done, the sets X and Y are still implicit in the discussion, of course. Thus in Example 1 the function is usually given just by writing the equation $A = f(s) = s^2$. We are to understand from the geometric situation that $X = \{s : s \geqq 0\}$.

When a specific function has been given merely by means of a formula for $f(x)$, we encounter the problem of determining the domain and range of the function. The following examples provide a few useful techniques.

■ *Example 6.* Suppose that we are given that $f(x) = x - 1$. This formula has no built-in restrictions. Hence if we know nothing more about the situation under discussion, we may assume that the domain X is the entire set **R**. It then follows that the range Y is also **R**. (Why?) A complete description of the function is now: $X = \mathbf{R}$, $Y = \mathbf{R}$, $f = \{(x,y) : y = x - 1\}$.

■ *Example 7.* If we are given the formula $f(x) = x^2 + 1$, then we may again assume that **R** is the domain X. But now the range is $\{y : y \geqq 1\}$, a subset of **R**. This fact is obvious, because $x^2 + 1 \geqq 1$ for any $x \in \mathbf{R}$. So we now know X and $f(X)$ and $f = \{(x,y) : y = x^2 + 1\}$. Note that each value $y \in X$, except $y = 1$, is the image of two values of x, namely, $x = \sqrt{y-1}$ and $x = -\sqrt{y-1}$. For instance, $f(2) = 5 = f(-2)$.

■ **Example 8.** Consider the "function" $f(x) = \sqrt{1-x^2}$. If we insist that the number $y = f(x)$ be real, then the radicand $1 - x^2$ must be non-negative. It follows that $1 - x^2 \geq 0$ or $x^2 \leq 1$ gives the domain of this function as

$$X = \{x : -1 \leq x \leq 1\}$$

Then it is easy to compute that the range is $\{y : 0 \leq y \leq 1\}$. Finally, the set of ordered pairs is

$$f = \{(x,y) : y = \sqrt{1-x^2}\} \subset X \times Y$$

■ **Example 9.** Consider $y = f(x) = 1/\sqrt{x+1}$. Again insisting that y be a real number, we must assume that the radicand $x + 1$ is nonnegative. Also, we must omit $x + 1 = 0$ because division by zero is not defined. Therefore we have $X = \{x : x + 1 > 0\} = \{x : x > -1\}$. Then $f(x) = \{y : y > 0\}$, a fact easily proved by noting that the quantity $1/\sqrt{x+1}$ can take on any positive number y. In particular, if we set $x = (1/y^2) - 1$, then $y = 1/\sqrt{x+1}$.

Until Chapter 13 we shall limit our discussions to *real-valued functions of a real variable*. These are functions having both domain and range in the set **R** of all real numbers.

Graphs. Given some particular real-valued function of a real variable, we find that the set of ordered pairs $(x, f(x))$ may be identified with a subset of a coordinate plane. The resulting geometric figure is called *the graph of the function*. Such a picture is an excellent aid to understanding the function. Some techniques for drawing the graph of a given function will be introduced in the next section, but one basic procedure is illustrated in the next two examples.

■ **Example 10.** Given a linear function $f(x) = Ax + B$, we already know that the set $f = \{(x,y) : y = Ax + B\}$ corresponds to a straight line in a coordinate plane. This line is the graph of the function. We can draw it simply by finding two points on the line. We let x take on any two (different) convenient values, say a and b. The two points $(a, Aa + B)$, $(b, Ab + B)$ determine the graph.

■ **Example 11.** Let $f(x) = x^2 + 1$. In Example 7, we found the set f to be

$$f = \{(x,y) : x \in \mathbf{R}, y = x^2 + 1\}$$

We can find some points on the graph by assigning values to x and computing the corresponding values of $y = f(x)$. For instance, if $x = 0$, then

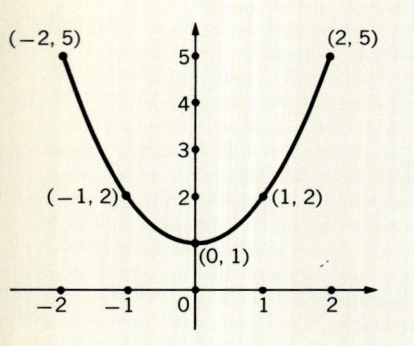

FIGURE 2.6

$y = f(0) = 0^2 + 1 = 1$. Similarly, such points as $(1, 2)$, $(-1, 2)$, $(2, 5)$ and $(-2, 5)$ are on the graph. These points are then located on a coordinate plane. By drawing a smooth curve through these points, we obtain a fairly good picture of the graph. See Fig. 2.6.

The process of computing and plotting points on the graph of a function is always available as an aid to drawing the graph. Despite more sophisticated techniques to come, this one is always used. The more points we locate on the graph, the more accurate our drawing will be.

Let us mention a warning again. Just because the graph of a function is drawn as some curve in a coordinate plane, it need *not* follow that all such geometric quantities as length, area, and so forth, have useful or meaningful interpretations. But it is important to know which geometric quantities *are* meaningful, as in the following examples.

■ *Example 12.* Suppose that $s = f(t) = 3t - 2$ describes a distance s as a function of elapsed time t. In the graph of this function, the slope three of the line would be measured in units with the dimensions

$$\frac{\text{(distance)}}{\text{(time)}} = \text{(velocity)}$$

In this case, then, the geometric quantity "slope" has a physical interpretation as the velocity of a moving body. On the other hand, an "area" in this coordinate plane would be measured in units with the dimensions of (time) × (distance) and would have no recognized physical interpretation.

■ *Example 13.* If an economic demand function $x = f(p)$ is pictured as a graph in a coordinate plane, then in this plane both the geometric quantities of area and slope have a useful meaning. (What are they?) On the other hand, the "distance" between two points in this coordinate plane is not assigned a meaning by the economist.

■ *Example 14.* Suppose the function $F = f(d)$ describes a force F that depends upon a distance d. The coordinate plane on which the graph of this function is drawn has coordinates measured in terms of distance and force. Thus an "area" in this plane has dimensions (distance) × (force) = (work). But the slope of a line would have dimensions (force)/(distance), which is not given a physical interpretation.

Situations such as these just described are found in most applications of mathematics to real-world problems. It is usually very easy to determine which geometric quantities do have useful interpretations, so easy indeed that we often fail to make explicit mention of the fact. The point here is that such a determination must be made if misunderstandings are to be avoided. The reader should do this himself if we do not.

2.4 FUNCTIONS AND THEIR GRAPHS

Another remark should be made here. Most of our functions shall have the functional value $f(x)$ given by a single expression (in terms of the independent variable x). But this restriction is not part of our definition of a function. Well-defined functions can be given by some very peculiar rules.

■ *Example 15.* The rule

$$f(x) = -1 \quad \text{if } x < 0$$
$$f(x) = 0 \quad \text{if } x = 0$$
$$f(x) = 1 \quad \text{if } x > 0$$

defines a perfectly good function $f : \mathbf{R} \to Y$ where $Y = \{-1, 0, 1\}$ consists of just three numbers. The graph of this function appears in Fig. 2.7 where we have encircled the *isolated point* $(0, 0)$, which is on the graph but is near no other point on the graph.

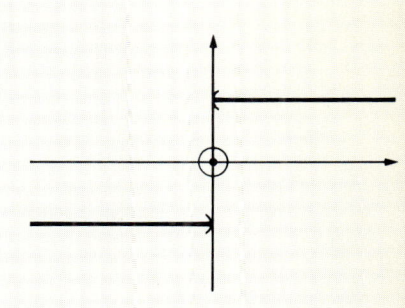

FIGURE 2.7

■ *Example 16.* The function

$$f(x) = \tfrac{1}{2}x, \quad x < 0$$
$$f(x) = 2x - x^2, \quad 0 \leq x \leq 1$$
$$f(x) = 1, \quad x > 1$$

is well-defined. Its graph appears in Fig. 2.8.

■ *Example 17.* A function can be given in a form that does not exhibit the defining rule explicitly. For instance, the set of ordered pairs

$$f = \{(x, y) : y \geq 1, y^2 - 4x - 2y = 3\}$$

is a disguised form of the function

$$y = f(x) = 1 + 2\sqrt{x + 1}$$

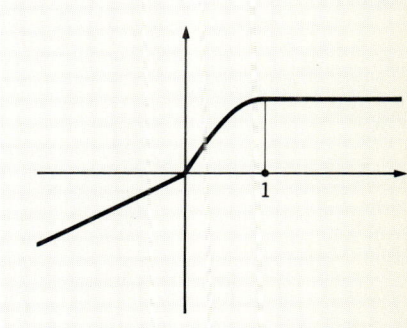

FIGURE 2.8

Even if we do not or cannot determine an explicit expression $f(x)$, however, the point-plotting method can still be used to help draw the graph. For the example here, we would assign values to y and compute corresponding values of x from the equation $x = \tfrac{1}{4}(y^2 - 2y - 3)$. In this way we find that the points $(-1, 1)$, $(-\tfrac{3}{4}, 2)$, $(0, 3)$, $(\tfrac{5}{4}, 4)$ are on the graph, which is shown in Fig. 2.9.

Finally, we have seen that the variables in a function may be denoted by any convenient letters. In the same way we may use any letter to denote a function. Such letters as $g, h, F, G, \varphi, \psi$ are often used in place of the letter f.

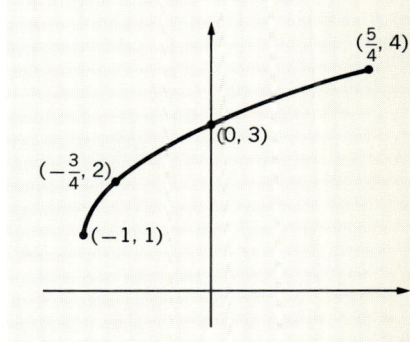

FIGURE 2.9

EXERCISES 2.4

1. For each of the functions whose values are given below determine the range and the implied domain:

   (a) $f(x) = \dfrac{1}{x^2 - 4}$

   (b) $f(x) = \sqrt{4 - x^2}$

   (c) $f(x) = \sqrt{x^2 - 4}$

   (d) $f(x) = (4 - x^2)^{-1/2}$

(e) $g(t) = \dfrac{2t}{t+1}$ (f) $g(t) = \sqrt{\dfrac{2t}{t+1}}$

(g) $h(u) = \dfrac{\sqrt{u+2}}{\sqrt{1-u^2}}$ (h) $h(z) = \sqrt{z^2 - z - 2}$

2. Plot at least five points on the graph of each function whose value and domain are given below. (If no domain is given, then assume that the domain is **R**.) Use the plotted points and any other information to draw the graph of the function.

(a) $f(x) = |x|$; **R**

(b) $f(x) = \sqrt{x}$; $\{x : x \geq 0\}$

(c) $f(x) = \dfrac{1-x}{1+x}$; $\{x : x \neq -1\}$

(d) $f(x) = \sqrt{9 - x^2}$; $\{x : -3 \leq x \leq 3\}$

(e) $g(t) = (4 - t^2)$; $\{t : -3 \leq t \leq 3\}$

(f) $g(t) = \dfrac{t^2}{t+1}$; $\{t : t \geq 0\}$

(g) $g(t) = \dfrac{t}{1 - t^2}$; $\{t : -1 < t < 1\}$

(h) $h(x) = \dfrac{x-1}{x^2}$; $\{x : x \neq 0\}$

(i) $h(x) = \dfrac{4 - x^2}{1 + x^2}$

(j) $y = 2x$ if $x \leq 1$
    $= -x + 3$ if $x \geq 1$

3. If $f(x) = (1 + x)/(1 - x)$, evaluate the expression
$$\dfrac{f(x) - f(y)}{1 + f(x)f(y)}$$

4. For each of the functions whose value is given below, determine (and simplify) the expression $\dfrac{1}{h}[f(x + h) - f(x)]$, $h \neq 0$:

(a) $f(x) = 3x + 2$ (b) $f(x) = 4 - x$
(c) $f(x) = Ax + B$ (d) $f(x) = x^2$
(e) $f(x) = 2x^2 - 3x$ (f) $f(x) = Ax^2 + Bx + C$

(g) $f(x) = x^3$ (h) $f(x) = \dfrac{1}{x}$

(i) $f(x) = \dfrac{1}{x^2}$ (j) $f(x) = \dfrac{x}{x + 1}$

PROBLEMS 2.4

1. Let S be the sphere circumscribed about a regular tetrahedron and let s be the sphere inscribed in the tetrahedron. Determine the volumes of S and s as functions of the volume v of the tetrahedron.

2. Let $f(x) = (1 - x)^{-1}$ and determine $f(f(x)), f(f(f(x)))$, and so forth.

3. Determine the function $F(n)$ that gives the number of zeros at the end of the integer $n!$. For example, $F(4) = 0$ because $4! = 24$ but $F(5) = 1$ because $5! = 120$.

4. *Heaviside's unit function* is defined by
$$H(t) = 1 \text{ if } t \geq 0$$
$$= 0 \text{ if } t < 0$$

Graph $H(t)$ and each of the following functions
$$f(t) = H(t) - H(t-1)$$
$$g(t) = H(t^2 - 4)$$
$$h(t) = t^2[H(t) + H(t+1)]$$

5. (Psychology) A test is to be replaced by an equivalent test n times as long. The reliability of the lengthened test is often taken to be the *Spearman-Brown function*
$$R_n(r) = \frac{nr}{1+(n-1)r}, \qquad 0 \leq r \leq 1$$
where r is the reliability of the original test.
   (a) Show that $0 \leq R_n(r) \leq 1$ for all r in the domain.
   (b) Prove and interpret the equation $R_{1/n}(R_n(r)) = r$.
   (c) Let
$$F_n(r) = n\left(\frac{1}{R_n(r)} - 1\right)$$
   and prove that $F_n(r) = F_m(r)$ for any m and n. Why is the function $F_n(r)$ useful in psychology?

6. Let $f(x) = x^2 + 2$ for all real x and let $g(y) = \sqrt{y-2}$, $y \geq 2$. Prove that $f(g(y)) = y$. For what values of x do we also have the identity $g(f(x)) = x$?

7. What is the implied domain of the function whose values are
$$f(x) = \sqrt{x - \sqrt{|x| - x^2}} \,?$$

2.5 GRAPHING A FUNCTION (INTRODUCTION)

Until Chapter 11 we discuss only real-valued functions of a real variable. Therefore the set f of ordered pairs is a subset of $\mathbf{R} \times \mathbf{R}$, the subset of all two-dimensional numerical vectors. Because we identify $\mathbf{R} \times \mathbf{R}$ with a coordinate plane, the set f is pictured as a subset of the plane. This geometric figure is called the graph of the function.

We stipulated that each element of the domain X of a function appears in one and only one ordered pair in the set f. This means that the function is *single-valued* and, in the case of a real-valued function of a real variable, has the following geometric interpretation:

The graph of a function intersects any vertical line in at most one point. To see why this must be true, we simply determine the intersection of the set $f = \{(x,y) : x \in X, y = f(x)\}$ and a vertical line $\{(x,y) : x = a\}$.

$$\{(x,y) : x \in X, y = f(x)\} \cap \{(x,y) : x = a\}$$
$$= \{(x,y) : x = a, y = f(x)\} \quad \text{if} \quad a \in X$$
$$= \{(a, f(a))\} \quad \text{if} \quad a \in X$$

Thus if $a \in X$, the intersection is a single point. If $a \notin X$, then there is no point of intersection. The curves in Fig. 2.10 are *not* graphs of functions.

A particular function is usually specified just by giving a rule for computing the functional value $f(x)$. Knowing only the relation $y = f(x)$,

70 | FUNCTIONS, GRAPHS, AND LIMITS

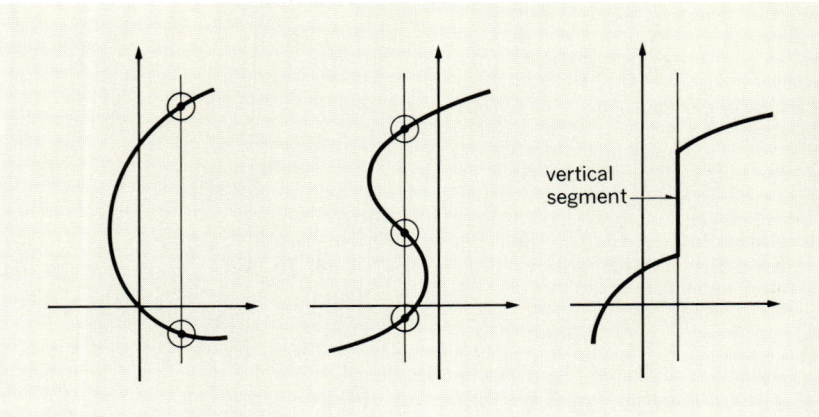

FIGURE 2.10

the problem of drawing the graph $\{(x,y) : y = f(x)\}$ can present some difficulties. A few useful techniques are given here; other methods will be added later on.

Intercepts. In plotting points on a graph $\{(x,y) : y = f(x)\}$, we always locate those points (if any) at which the graph intersects the coordinate axes. The y axis is the set $\{(x,y) : x = 0\}$. This vertical line meets the graph in the single point $(0, f(0))$ if and only if 0 is in the domain of f. The point $(0, f(0))$ is called the y *intercept* of the graph. Note that there can be at most one such y intercept for any graph.

Similarly, the x axis is the line $\{(x,y) : y = 0\}$. Its intersection with the graph is $\{(x,y) : y = f(x)\} \cap \{(x,y) : y = 0\} = \{(x, 0) : f(x) = 0\}$. Therefore the x *intercepts* of the graph are the points of the form $(r, 0)$ where r is a root of the equation $f(x) = 0$. Note that there may be many x intercepts on a given graph.

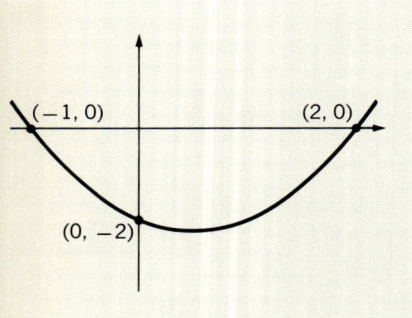

FIGURE 2.11

■ **Example 1.** Let $f(x) = x^2 - x - 2$. Setting $x = 0$, we find that the point $(0, -2)$ is the y intercept. Setting $f(x) = 0$, we have the equation $x^2 - x - 2 = 0$, which has roots $r_1 = 2, r_2 = -1$. Hence the x intercepts are $(2, 0)$ and $(-1, 0)$. These points are located on the graph as pictured in Fig. 2.11.

■ **Example 2.** Suppose that $f(x) = \sqrt{1 - x^2}$. Setting $x = 0$, we see that $f(0) = 1$. Therefore the y intercept of the graph $\{(x,y) : y = \sqrt{1 - x^2}\}$ is the point $(0, 1)$. Setting $y = 0$, we obtain the equation $\sqrt{1 - x^2} = 0$, which has roots $r_1 = 1, r_2 = -1$. Hence the x intercepts of the graph are the points $(1, 0)$ and $(-1, 0)$. The graph of this function is shown in Fig. 2.12.

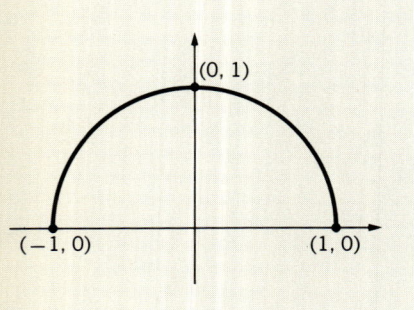

FIGURE 2.12

■ **Example 3.** If $f(x) = 1/\sqrt{x + 1}$, then $f(0) = 1$. Thus the point $(0, 1)$ is the y intercept of the graph $\{(x,y) : y = 1/\sqrt{x + 1}\}$. But setting

$y = 0$, we obtain an equation

$$\frac{1}{\sqrt{x+1}} = 0$$

which has no roots. Therefore this graph has no x intercepts. The graph is shown in Fig. 2.13.

Symmetry. The graph $\{(x, y) : y = f(x)\}$ is *symmetric to the y axis* if, for each point (x_0, y_0) on the graph, the point $(-x_0, y_0)$ is also on the graph. In other words, if x is in the domain of $f(x)$, then $-x$ is also in the domain and furthermore

$$f(-x) = f(x)$$

Figure 2.14 illustrates three examples of such symmetry. Note that each of these graphs can be reflected in the y axis without being changed.

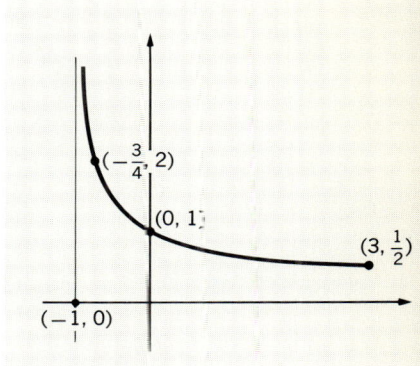

FIGURE 2.13

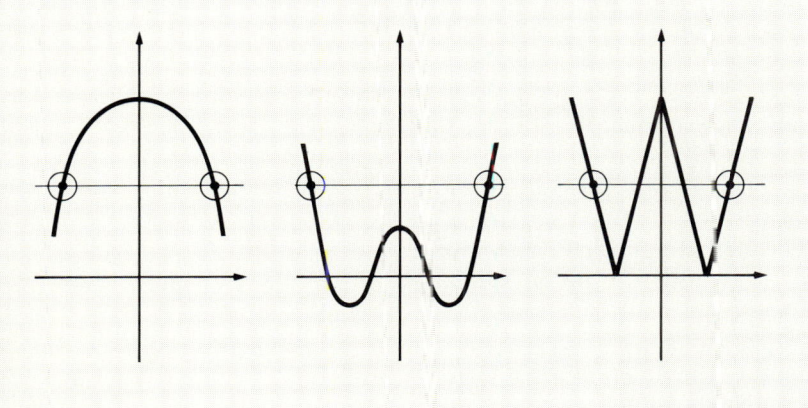

FIGURE 2.14

A function whose values satisfy the equation $f(-x) = f(x)$ is called an *even function*. Thus the graph of an even function is symmetric to the y axis. The name "even function" comes about as follows: If the algebraic expression for $f(x)$ contains only even powers of x, then we always have $f(-x) = f(x)$. For instance, if $f(x) = x^2 + 1$, then $f(-x) = (-x)^2 + 1 = x^2 + 1 = f(x)$. Similarly, if $f(x) = x^4 - 4/x^2$, then

$$f(-x) = (-x)^4 + \frac{4}{(-x)^2} = x^4 + \frac{4}{x^2} = f(x)$$

We can also define symmetry with respect to a vertical line other than the y axis. The graph $\{(x, y) : y = f(x)\}$ is symmetric to the vertical line $\{(x, y) : x = a\}$ if, for each point $(a + h, f(a + h))$ on the graph, the point $(a - h, f(a - h))$ is also on the graph. This occurs if and only if the equation

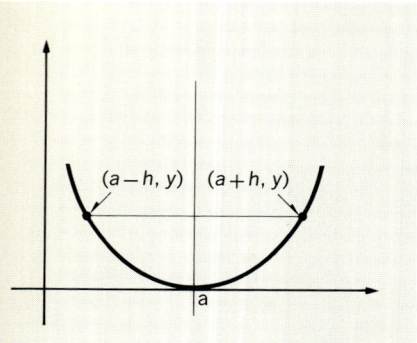

FIGURE 2.15

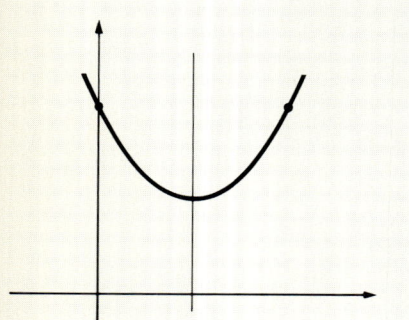

FIGURE 2.16

$$f(2a - x) = f(x)$$

holds. For then, if $x = a + h$, we have

$$f(2a - a - h) = f(a - h) = f(a + h)$$

Note that $\{(x,y) : y = f(x)\}$ will be symmetric to the vertical line $\{(x,y) : x = a\}$ if the algebraic expression for $f(x)$ can be written in even powers of $x - a$. Figure 2.15 illustrates this kind of symmetry.

■ **Example 4.** Consider $f(x) = x^2 - 2x + 2$. Because $x^2 - 2x + 2 = (x - 1)^2 + 1$, the graph $\{(x,y) : y = x^2 - 2x + 2\}$ is symmetric to the vertical line $\{(x,y) : x = 1\}$. (See Fig. 2.16.)

■ **Example 5.** The graph $\{(x,y) : y = Ax^2 + Bx + C, A \neq 0\}$ of a *quadratic function* always has a vertical axis of symmetry. To prove this, we merely complete the square in x as follows:

$$\begin{aligned} f(x) &= Ax^2 + Bx + C \\ &= A\left(x^2 + \frac{B}{A}x\right) + C \\ &= A\left(x^2 + \frac{B}{A}x + \frac{B^2}{4A^2}\right) + C - \frac{B^2}{4A} \\ &= A\left(x + \frac{B}{2A}\right)^2 + C - \frac{B^2}{4A} \end{aligned}$$

Therefore the graph is symmetric to the vertical line $\{(x,y) : x = -B/2A\}$.

One other type of symmetry is easy to recognize. The graph $\{(x,y) : y = f(x)\}$ is *symmetric to the origin* if, for each point (x_0, y_0) on the graph, the point $(-x_0, -y_0)$ is also on the graph. Thus if x is in the domain of the function, then so is $-x$ and

$$f(-x) = -f(x)$$

A function satisfying this equation is called an *odd function*. The name was adopted because an algebraic expression involving x only to odd powers satisfies the equation above. For instance, if $f(x) = 3x$, then $f(-x) = 3(-x) = -3x = -f(x)$. Another odd function is $f(x) = x + 1/x$ because

$$f(-x) = -x + \frac{1}{-x} = -x - \frac{1}{x} = -\left(x + \frac{1}{x}\right) = -f(x).$$

Figure 2.17 shows these two examples and an additional one.

It is easy to construct even and odd functions. Given any function f whose domain is **R,** the function

$$g(x) = f(x) + f(-x)$$

is an even function, while the function defined by

$$h(x) = f(x) - f(-x)$$

2.5 GRAPHING A FUNCTION (INTRODUCTION)

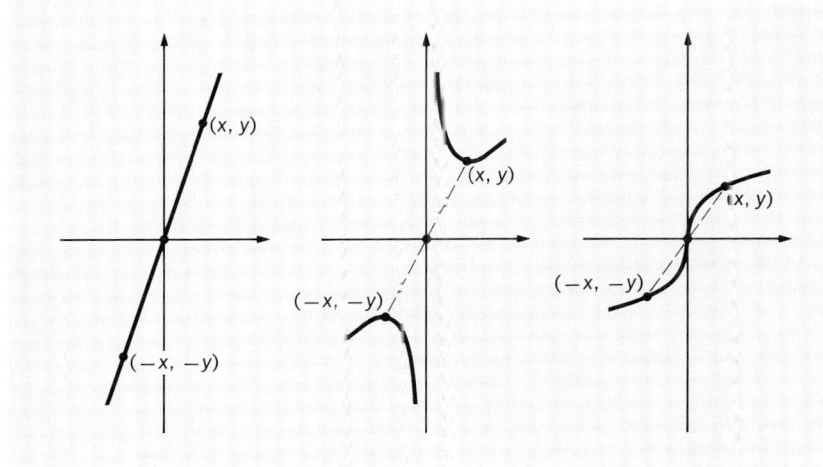

FIGURE 2.17

is an odd function. We prove these statements by computing

$$g(-x) = f(-x) + f(-(-x)) = f(-x) + f(x)$$
$$= f(x) + f(-x) = g(x)$$

and

$$h(-x) = f(-x) - f(-(-x)) = f(-x) - f(x)$$
$$= -[f(x) - f(-x)] = -h(x)$$

■ *Example 6.* Let $f(x) = x^3 + x^2 - 3x + 2$. The function f so defined is neither even nor odd. However

$$f(x) + f(-x) = x^3 - x^2 - 3x + 2 + (-x)^3 + (-x)^2$$
$$- 3(-x) + 2$$
$$= x^3 + x^2 - 3x + 2 - x^3 + x^2 + 3x + 2$$
$$= 2x^2 + 4, \quad \text{defining an even function}$$

Similarly,

$$f(x) - f(-x) = x^3 + x^2 - 3x + 2 - (-x)^3 - (-x)^2$$
$$+ 3(-x) - 2$$
$$= x^3 + x^2 - 3x + 2 + x^4 - x^2 - 3x - 2$$
$$= 2x^6 - 6x, \quad \text{defining an odd function}$$

Vertical asymptotes. The graph $\{(x, y) : y = f(x)\}$ is said to be *asymptotic* to the vertical line $\{(x, y) : x = a\}$ if the functional value $|f(x)|$ grows indefinitely large as x approaches the fixed value a. For instance, the graph $\{(x, y) : y = x + (1/x)\}$ shown as the second curve in Fig. 2.17 exhibits this behavior with respect to the y axis. (A more precise definition of an asymptote must wait until we have discussed "limits" in Chapter 4, but the above intuitive definition will suffice for many simple cases.)

Vertical asymptotes of a graph $\{(x, y) : y = f(x)\}$ are located as follows: We look for values of x for which $f(x)$ is *not* defined. If k is a number

74 | FUNCTIONS, GRAPHS, AND LIMITS

for which $f(x)$ is not defined, while $f(x)$ is defined on an interval having k as an endpoint, then we strongly suspect that the vertical line $\{(x,y) : x = k\}$ is a vertical asymptote. We test for a vertical asymptote simply by finding a few points on the graph for values of x approaching the number k. If $f(x)$, or more precisely, if $|f(x)|$ grows indefinitely large as the value of x approaches k, then we are sure.

■ *Example 7.* Let $f(x) = 1/(x - 1)$. The number 1 is not in the domain of this function, but $f(x)$ *is* defined for any $x \neq 1$. Let h be a number such that $|h|$ is small. Then

$$f(1 + h) = \frac{1}{1 + h - 1} = \frac{1}{h}$$

Hence, the smaller $|h|$ is chosen, the larger $|f(1 + h)|$ will be. In particular, the points $(\frac{1}{2}, -2)$, $(\frac{3}{4}, -4)$, $(\frac{7}{8}, -8)$, $(\frac{3}{2}, 2)$, $(\frac{5}{4}, 4)$ and $(\frac{9}{8}, 8)$ are on the graph. Plotting these gives a good example of the behavior of the graph near the vertical line $\{(x,y) : x = 1\}$. (See Fig. 2.18.)

■ *Example 8.* Let $f(x) = 1/\sqrt{x + 1}$. Then $f(x)$ is defined for $x > -1$ but is not defined at $x = -1$. We suspect then that $\{(x,y) : x = -1\}$ is a vertical asymptote. A test can be made by letting $h^2 > 0$ be a small positive number. Then we have

$$f(-1 + h^2) = \frac{1}{\sqrt{-1 + h^2 + 1}} = \frac{1}{\sqrt{h^2}} = \frac{1}{|h|}$$

Thus as h^2 gets smaller and smaller, $f(-1 + h^2)$ grows larger and larger. Such points as $(0, 1)$, $(-\frac{3}{4}, 2)$, $(-\frac{15}{16}, 4)$ are shown on the graph in Fig. 2.19.

We leave the study of asymptotes other than vertical until Section 13.6 where a precise definition is given and some other methods of locating asymptotes are developed.

We summarize the graphing techniques that we now have on hand. The following steps will usually result in a reasonably accurate picture of the graph of the function under study.

1. Determine the domain and range of the function. This information obviously helps us in drawing the graph.
2. Find the intercepts of the graph and plot them on the axes.
3. Determine the symmetries of the graph. Knowledge of symmetry is useful in drawing the picture.
4. Locate the vertical asymptotes and plot a few points on the graph near each asymptote.
5. Plot additional points as needed.

When these steps are completed an accurate picture of the graph should be easy to draw.

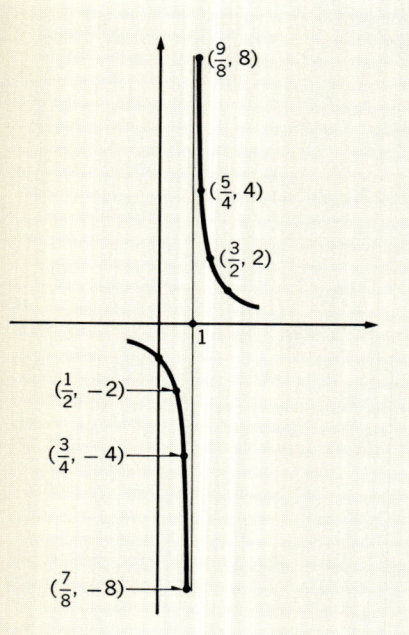

FIGURE 2.18

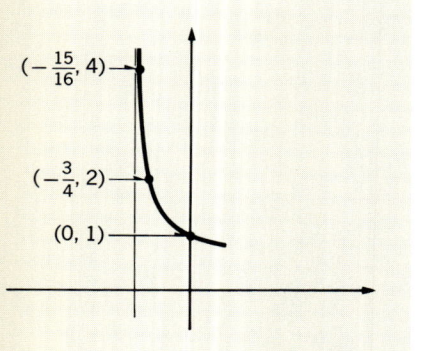

FIGURE 2.19

2.5 GRAPHING A FUNCTION (INTRODUCTION)

■ **Example 9.** Let $f(x) = \sqrt{1 - x^2}$. In Example 7 of Section 2.4, it was stated that the domain of this function is $X = \{x : -1 \leq x \leq 1\}$ and the range $f(x) = \{y : 0 \leq y \leq 1\}$. Therefore the graph $\{(x,y) : y = \sqrt{1 - x^2}\}$ only appears in the shaded rectangle in Fig. 2.20(a).

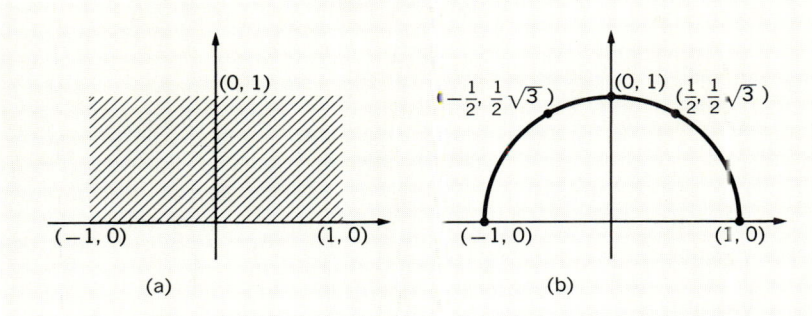

FIGURE 2.20

In Example 2 of this section the intercepts of this graph were determined to be the points $(0, 1)$, $(-1, 0)$, $(1, 0)$.

Because $f(-x) = \sqrt{1 - (-x)^2} = \sqrt{1 - x^2} = f(x)$, the graph of this function is symmetric to the y axis.

There is no vertical asymptote.

Lastly, we decide that we need at least two more points on the graph. Setting $x = \frac{1}{2}$, we have $f(\frac{1}{2}) = \sqrt{\frac{3}{4}} = \frac{1}{2}\sqrt{3}$. Therefore the point $(\frac{1}{2}, \frac{1}{2}\sqrt{3})$ is on the graph. Then by symmetry the point $(-\frac{1}{2}, \frac{1}{2}\sqrt{3})$ is also on the graph. All of our information is plotted on the coordinate plane and the graph is drawn (Fig. 2.20(b)).

■ **Example 10.** Consider $f(x) = x/(x^2 - 1)$. Obviously, $f(x)$ is defined except when $x = \pm 1$. Therefore the domain of this function is $X = \{x : |x| \neq 1\}$. We determine the range of the function as follows: Let r be a real number and ask whether or not the equation $r = x/(x^2 - 1)$ has a solution. If there is a solution, then r is in the range. But the equation $rx^2 - x - r = 0$ actually has two roots. In particular

$$x = \frac{1 \pm \sqrt{1 + 4r^2}}{2r} \quad \text{if } r \neq 0$$

and $x = 0$ if $r = 0$. Hence $f(x)$ can be any real number and so the range Y is $\mathbf{R}$.

The y intercept is $(0, f(0)) = (0, 0)$. This is the only x intercept, too, because the equation $x/(x^2 - 1) = 0$ has only the one root $x = 0$.

In testing for symmetry, we find that

$$f(-x) = \frac{-x}{(-x)^2 - 1} = \frac{-x}{x^2 - 1} = -\frac{x}{x^2 - 1} = -f(x)$$

Therefore the graph is symmetric to the origin.

We should already suspect that the lines $\{(x,y) : x = 1\}$ and $\{(x,y) : x = -1\}$ are vertical asymptotes. Let h be a small number and compute $f(1 + h)$ and $f(-1 + h)$ as follows.

$$f(1+h) = \frac{1+h}{(1+h)^2 - 1} = \frac{1+h}{2h+h^2} = \frac{1}{h}\left(\frac{1+h}{2+h}\right)$$

$$f(-1+h) = \frac{1+h}{(-1+h)^2 - 1} = \frac{-1+h}{-2h+h^2} = \frac{1}{h}\left(\frac{-1+h}{-2+h}\right)$$

From here it is easy to see that $|f(x)|$ grows indefinitely large as x approaches either $+1$ or -1. In computing, we find that the points $(\tfrac{1}{2}, -\tfrac{2}{3})$, $(\tfrac{3}{4}, -\tfrac{12}{7})$, $(\tfrac{7}{8}, -\tfrac{56}{15})$, $(\tfrac{3}{2}, \tfrac{6}{5})$, $(\tfrac{5}{4}, \tfrac{20}{9})$ and $(\tfrac{9}{8}, \tfrac{72}{17})$ are on the graph. Then either by computing or by using symmetry with respect to the origin, the points $(-\tfrac{1}{2}, \tfrac{2}{3})$, $(-\tfrac{3}{4}, \tfrac{12}{7})$, $(-\tfrac{7}{8}, \tfrac{56}{15})$, $(-\tfrac{3}{2}, -\tfrac{6}{5})$, $(-\tfrac{5}{4}, -\tfrac{20}{9})$ and $(-\tfrac{9}{8}, -\tfrac{72}{17})$ are also on the graph. All of this information is used to draw the graph $\{(x,y) : y = x/(x^2 - 1)\}$ as it appears in Fig. 2.21.

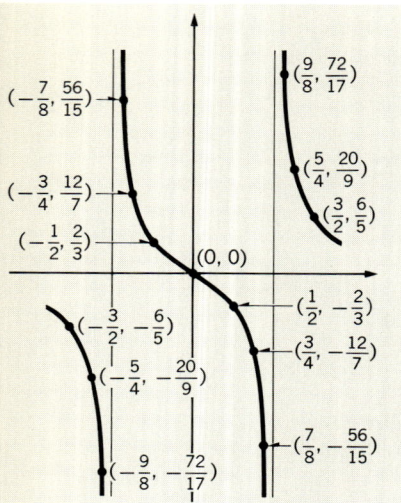

FIGURE 2.21

EXERCISES 2.5

For each function whose value is given below, determine the implied domain and the range of the function. Find the intercepts, symmetries, and asymptotes of its graph and plot at least six points on the graph. Then draw the graph.

1. $f(x) = x^2 - 4$
2. $f(x) = 2 - x^2$
3. $f(x) = x^2 - 3x + 2$
4. $f(x) = 1 - 2x - 2x^2$
5. $f(x) = x^3 - 16x$
6. $f(x) = x^3 - 3x^2 + 4x - 2$
7. $f(x) = \sqrt{x+4}$
8. $f(x) = \dfrac{1}{\sqrt{x+4}}$
9. $f(x) = \sqrt{x^2 - 2x}$
10. $f(x) = \sqrt{4x - x^2}$
11. $f(x) = -\dfrac{1}{x}$
12. $f(x) = \dfrac{1-x}{1+x}$
13. $f(x) = \dfrac{x}{1+x^2}$
14. $f(x) = \dfrac{x}{(1+x)^2}$
15. $f(x) = \dfrac{x-1}{x^2+x}$
16. $f(x) = \dfrac{x^2 - 2x}{1+x}$
17. $f(x) = \dfrac{x^2}{1 - x^2}$
18. $f(x) = \dfrac{x^2 - 1}{x - 1}$
19. $f(x) = \dfrac{1}{x^4 + 4x^3 + 4x^2 + 1}$
20. $f(x) = \dfrac{x^2 - 1}{x^4 + x^2}$

PROBLEMS 2.5

Even and odd functions

1. Can a function be both even and odd?

For Problems 2 to 5, let $E_1(x)$ and $E_2(x)$ be the values of two even functions and let $O_1(x)$ and $O_2(x)$ be the values of two odd functions.

2. Show that the functions with values $E_1(x) + E_2(x)$ and $E_1(x) - E_2(x)$ are also even.

3. Show that the functions with values $O_1(x) + O_2(x)$ and $O_1(x) - O_2(x)$ are also odd.

4. Show that the functions with values $E_1(x) \times E_2(x)$ and $O_1(x) \times O_2(x)$ are both even, while the function with values $E(x) \times O_1(x)$ is odd.

5. Prove that $E_1(x) + O_1(x) = E_2(x) + O_2(x)$ for all x if and only if $E_1(x) = E_2(x)$ and $O_1(x) = O_2(x)$ for all x.

6. Let $f(x)$ denote the values of an arbitrary function defined on $\mathbf{R}$. Define functions with the values
$$E(x) = \tfrac{1}{2}[f(x) + f(-x)], \quad O(x) = \tfrac{1}{2}[f(x) - f(-x)]$$
Prove then that any function can be expressed as the sum of an even function and an odd function. What does Problem 5 tell you about this representation of the function?

2.6 INFINITE SEQUENCES

We define *an infinite sequence of real numbers* to be a function $\varphi: \mathbf{N} \to Y$ whose domain is the set $\mathbf{N}$ of natural numbers and whose range Y is a subset of $\mathbf{R}$. Such a function provides a real number $\varphi(n)$ for each natural number n. The functional value $\varphi(n)$ is called the *nth term* of the sequence φ.

■ *Example 1.* Let $\varphi : \mathbf{N} \to Y$ have functional values
$$\varphi(n) = \frac{1}{n}$$
The range is $Y = \{1, \tfrac{1}{2}, \tfrac{1}{3}, \cdots, 1/n, \cdots\}$. Notice that the terms of this sequence approach zero as n increases.

■ *Example 2.* The equation $\varphi(n) = (\tfrac{1}{2})^n = 1/2^n$ defines an infinite sequence whose terms are $\tfrac{1}{2}, \tfrac{1}{4}, \tfrac{1}{8}, \tfrac{1}{16}, \tfrac{1}{32}, \cdots$. Here again, $\varphi(n)$ approaches zero as n increases.

■ *Example 3.* Let $\varphi(n) = 1 + 1/n$. As n increases, the distance between $\varphi(n)$ and one gets as small as you please, or $\varphi(n)$ gets as close to one as you please. This idea of the terms of an infinite sequence "converging" to a fixed real number is extremely important. We say that $\varphi(n) = 1 + 1/n$ converges to one because a value $\varphi(n)$ can be chosen so as to approximate one as accurately as might be required. In other words, regardless of what limitation is imposed upon the error $|\varphi(n) - 1|$, as long as zero error is not required, we can choose a value $\varphi(n)$ within the desired distance of one. Furthermore, once we have selected a particular number n_0 which makes the error $|\varphi(n_0) - 1|$ as small as was asked, the each subsequent term $\varphi(n)$, $n > n_0$, also has at least the same accuracy.

To be precise, suppose some small positive number ϵ has been given as the error limitation. We must determine n_0 so that

$$|\varphi(n_0) - 1| = \left|1 + \frac{1}{n_0} - 1\right| = \frac{1}{n_0}$$

is less than ϵ. The inequality

$$\frac{1}{n_0} < \epsilon$$

can be solved by taking n_0 to be the first natural number satisfying $n_0 > 1/\epsilon$. For any $n > n_0$, we also have $1/n < \epsilon$, so each term $\varphi(n)$ after $\varphi(n_0)$ is also within distance ϵ of one. This discussion leads to the following precise definition:

The infinite sequence φ *converges* to the real number y if, given any positive number ϵ, there is a natural number n_0 (which depends on ϵ) such that

$$|\varphi(n) - y| < \epsilon \qquad \text{whenever } n > n_0$$

If φ converges to any real number, we say φ is *convergent*.

Crudely speaking, the terms $\varphi(n)$ get closer and closer to y as n increases. Any desired degree of accuracy can be obtained by going out far enough in the sequence.

There is a better way of saying that φ converges to y. We say that *the sequence φ is eventually in the open interval $\{x : a < x < b\}$* if, at most, a finite number of terms fail to lie in this interval. This implies that there is some natural number n_0 such that, for any $n > n_0$, $\varphi(n)$ is in this interval. Then we say that φ converges to y if φ is eventually in any open interval containing y. We shall not use this language here but it is to be found in later courses.

The symbol used to state that the sequence φ converges to y is

$$\lim_{n \to \infty} \varphi(n) = y$$

It is read as "the limit of $\varphi(n)$ as n increases indefinitely is y," or more briefly, as "the limit of $\varphi(n)$ is y."

In Example 1 we claimed that $\lim_{n \to \infty} 1/n = 0$. We now prove this fact by appealing to our definition. Let $\epsilon > 0$ be given. We must show that there is a natural number n_0 such that

$$\left|\frac{1}{n} - 0\right| = \left|\frac{1}{n}\right| = \frac{1}{n} < \epsilon \qquad \text{whenever } n > n_0$$

The desire to have $1/n < \epsilon$ is clearly equivalent to wanting $n > 1/\epsilon$. Thus we may choose n_0 to be any natural number larger than $1/\epsilon$. Then, if $n > n_0 > 1/\epsilon$, we have $1/n < \epsilon$; so the proof is complete.

■ **Example 4.** Let r be any real number satisfying $0 < r < 1$. Then

$$\lim_{n \to \infty} r^n = 0$$

To prove this, we first notice that there is a positive number h such that

$$r = \frac{1}{1+h}$$

and hence
$$r^n = \frac{1}{(1+h)^n}$$

In Section 1.4, Problem 5, it was stated that $(1+h)^n \geq 1 + nh$ for all $n \in \mathbf{N}$ if $h > 0$. Using this, we write
$$r^n = \frac{1}{(1+h)^n} \leq \frac{1}{1+nh} < \frac{1}{nh}$$

Hence we need only show that there is, for any $\epsilon > 0$, an $n_0 \in \mathbf{N}$ such that
$$\frac{1}{nh} < \epsilon \quad \text{whenever } n > n_0$$

It will then follow that
$$|r^n - 0| = r^n < \frac{1}{nh} < \epsilon$$

Of course, we choose n_0 as we did for the first two examples worked out.

- **Example 5.** Let r be any positive real number. Then
$$\lim_{n \to \infty} r^{1/n} = 1$$

We prove this in two steps. First, assume that $r > 1$. We define a new sequence
$$\psi(n) = r^{1/n} - 1$$
Then
$$r^{1/n} = 1 + \psi(n)$$
and
$$r = (1 + \psi(n))^n \geq 1 + n\psi(n)$$
Therefore
$$\psi(n) \leq \frac{r-1}{n}$$

Given $\epsilon > 0$, we must choose $n_0 \in \mathbf{N}$ such that
$$|r^{1/n} - 1| = r^{1/n} - 1 = \psi(n) < \epsilon \quad \text{whenever } n > n_0$$

If we can choose $n_0 \in \mathbf{N}$ such that
$$\frac{r-1}{n} < \epsilon \quad \text{whenever } n > n_0$$

we will be finished. But we can do *this* by taking n_0 to be any natural number larger than $(r-1)/\epsilon$.

Next, assume that $0 < r < 1$. It follows that $r^{1/n} < 1$. Again we define a new sequence
$$\psi(n) = \frac{1}{r^{1/n}} - 1$$

Then
$$\frac{1}{r^{1/n}} = 1 + \psi(n)$$
and
$$\frac{1}{r} = (1 + \psi(n))^n \geqq 1 + n\psi(n)$$
Therefore
$$\psi(n) \leqq \frac{(1/r) - 1}{n}$$

An easy argument then concludes that $\lim_{n\to\infty} \psi(n) = 0$. We conclude that our limit, $\lim_{n\to\infty} r^{1/n} = 1$, is true.

In truth, we need a few theorems about infinite sequences if the "proof" given in Example 4 is to be precise.

■ **THEOREM 2.7.** Suppose that $\lim_{n\to\infty} \varphi(n) = y$, and let ψ be another infinite sequence such that $\psi(n) = \varphi(n)$ for all $n \geqq n_0$. Then
$$\lim_{n\to\infty} \psi(n) = y$$

This result says that, by altering a finite number of terms of a convergent infinite sequence, we do not change its limit. The proof of Theorem 2.7 is extremely easy and is left as an exercise.

■ **THEOREM 2.8.** Let $\varphi(n)$ and $\psi(n)$ be two convergent infinite sequences with $\lim_{n\to\infty} \varphi(n) = y$ and $\lim_{n\to\infty} \psi(n) = z$. We construct the new sequences $\varphi(n) + \psi(n)$, $\varphi(n) \times \psi(n)$ and $\varphi(n)/\psi(n)$. (The latter requires that $\psi(n)$ never be zero.) Then

1. $\lim_{n\to\infty} [\varphi(n) + \psi(n)] = y + z$
2. $\lim_{n\to\infty} [\varphi(n) \times \psi(n)] = yz$, and
3. if $z \neq 0$, $\lim_{n\to\infty} \frac{\varphi(n)}{\psi(n)} = \frac{y}{z}$

We shall not prove this theorem, since it is a special case of later results (Theorems 2.14 to 2.17). Note that the theorem says that the limit of a sum (product) (quotient) is the sum (product) (quotient) of the limits. Also we may point out that, by taking $\psi(n) = k$ to be a constant sequence, the following corollary is immediately proved.

■ **COROLLARY 2.9.** If $\lim_{n\to\infty} \varphi(n) = y$, then
$$\lim_{n\to\infty} [\varphi(n) + k] = y + k \quad \text{and} \quad \lim_{n\to\infty} [k\varphi(n)] = ky$$

For the reader interested in theoretical arguments, we have asked that Corollary 2.9 be proved from the definition (Problem 2.6.6).

Another very useful theorem which allows us to show that a limit exists, requires two definitions.

2.6 INFINITE SEQUENCES

A sequence φ is said to be *bounded* if the set $\{\varphi(n) : n \in \mathbf{N}\}$ of all of its terms is bounded. (See Section 1.6.)

A sequence φ is *monotone increasing* if $\varphi(n) \leq \varphi(n+1)$ for each $n \in \mathbf{N}$. A sequence ψ is *monotone decreasing* if $\varphi(n) \geq \varphi(n+1)$ for each $\mathbf{N} \in \mathbf{N}$. We use the word *monotone* to mean either case.

■ **THEOREM 2.10.** A bounded monotone sequence converges.

PROOF: We prove the theorem when φ is a monotone increasing sequence. The analogous proof for the monotone decreasing case is left to the reader.

Since the set $\{\varphi(n)\}$ is bounded above, Theorem 1.10 says there is a least upper bound y. Given any $\epsilon > 0$, there must be some natural number n_0 such that $\varphi(n_0) > y - \epsilon$. For if no term of φ were bigger than $y - \epsilon$, then $y - \epsilon$ would be an upper bound for the set $\{\varphi(n)\}$, but there can be no upper bound smaller than the *least* upper bound y. If $\varphi(n_0) > y - \epsilon$, then for any $n > n_0$ we also have $\varphi(n) \geq \varphi(n_0)$. Therefore $\varphi(n) > y - \epsilon$ whenever $n > n_0$. It follows that $y - \varphi(n) = |\varphi(n) - y| < \epsilon$. ■

■ *Example 6.* Define $\varphi(n) = 1 + \frac{1}{2} + \frac{1}{6} + \frac{1}{24} + \cdots + 1/r!$. We obtain $\varphi(n+1)$ by adding $1/(n+1)!$ to $\varphi(n)$ so that the sequence is obviously monotone increasing. It is easy to prove that $n! \geq 2^{n-1}$ for all $n \in N$. Hence we have

$$\varphi(n) \leq 1 + \frac{1}{2} + \frac{1}{4} + \cdots + \frac{1}{2^{n-1}} = 2 - \frac{1}{2^{n-1}} < 2$$

Therefore we see that φ is bounded and monotone increasing. By Theorem 2.10 there is a number y such that $\lim_{n \to \infty} \varphi(n) = y$. We shall learn later that $y = 1.71828 \cdots$.

Another useful result permits us to evaluate otherwise difficult limits.

■ **THEOREM 2.11.** Let φ_1, φ_2, and ψ be an infinite sequence and suppose that $\varphi_1(n) \leq \psi(n) \leq \varphi_2(n)$ for each $n \in \mathbf{N}$. If φ_1 and φ_2 have the same limit y, then $\lim_{n \to \infty} \psi(n) = y$, also.

This is also a special case of a later result (Theorem 2.18), so we shall not prove it here. In the two examples presented, note that $\varphi_1(n)$ is a constant sequence in each example.

■ *Example 7.* We prove that $\lim_{n \to \infty} n/2^n = 0$. The key to the proof is to define the sequence

$$\sqrt{\frac{n}{2^n}} = \frac{\sqrt{n}}{(\sqrt{2})^n}$$

By setting $h = \sqrt{2} - 1$, we may write

$$\frac{\sqrt{n}}{(\sqrt{2})^n} = \frac{\sqrt{n}}{(1+h)^n} \leq \frac{\sqrt{n}}{1+nh} < \frac{\sqrt{n}}{hn} = \frac{1}{h\sqrt{n}}$$

It follows that
$$0 < \frac{n}{2^n} < \frac{1}{h^2 n}$$
Since
$$\lim_{n\to\infty} \frac{1}{h^2 n} = \frac{1}{h^2} \lim_{n\to\infty} \frac{1}{n} = 0,$$
Theorem 2.11 completes the proof.

■ **Example 8.** We prove that $\lim_{n\to\infty} \sqrt[n]{n} = 1$. We first remark that $1 \leq \sqrt[n]{n}$ for each $n \in \mathbf{N}$. A new sequence φ is defined by
$$\varphi(n) = \sqrt{\sqrt[n]{n} - 1} = \sqrt{\sqrt[n]{n} - 1}$$
Clearly we have
$$\sqrt[n]{\sqrt{n}} = 1 + \varphi(n)$$
and therefore
$$\sqrt{n} = [1 + \varphi(n)]^n \geq 1 + n\varphi(n).$$
It follows that
$$\varphi(n) \leq \frac{\sqrt{n}-1}{n} < \frac{\sqrt{n}}{n} = \frac{1}{\sqrt{n}}$$
Then
$$\sqrt{\sqrt[n]{n}} = 1 + \varphi(n) < 1 + \frac{1}{\sqrt{n}}$$
or
$$\sqrt[n]{n} < \left(1 + \frac{1}{\sqrt{n}}\right)^2 = 1 + \frac{2}{\sqrt{n}} + \frac{1}{n}$$
The combined inequality
$$1 \leq \sqrt[n]{n} \leq 1 + \frac{2}{\sqrt{n}} + \frac{1}{n}$$
and Theorem 2.11 completes the proof. ■

If the infinite sequence φ does not converge, then we say that it *diverges*. For instance, $\varphi(n) = n^2$ is a divergent sequence. Here the terms simply grow indefinitely large. Divergence of this kind is often denoted by writing the symbol
$$\varphi(n) = +\infty$$
Divergence also occurs if the sequence φ "approaches" more than one number. For instance, the sequence
$$\lim_{n\to\infty} \varphi(n) = (-1)^n \left(1 + \frac{1}{n}\right)$$

has terms $-2, \frac{3}{2}, -\frac{4}{3}, \frac{5}{4}, -\frac{6}{5}, \frac{7}{6}, \cdots$. The even-numbered terms $\varphi(2k)$ "approach" 1, while the odd-numbered terms approach -1. This situation brings up another topic, which we shall need later on.

Let $n_1, n_2, n_3, \cdots, n_k, \cdots$ be a strictly increasing infinite sequence of natural numbers. Given any infinite sequence φ, then

$$\psi(k) = \varphi(n_k), \qquad k \in \mathbf{N}$$

defines a new infinite sequence ψ called a *subsequence* of φ. (We will give a more elegant definition in Section 2.7.) For instance, the even-numbered terms

$$\varphi(2), \varphi(4), \varphi(6), \varphi(8), \cdots, \varphi(2n), \cdots$$

of a sequence constitute the subsequence defined by the strictly increasing sequence of natural numbers $2, 4, 6, 8, \cdots, 2n, \cdots$. In the same way the subsequence

$$\varphi(2), \varphi(4), \varphi(8), \varphi(16), \cdots, \varphi(2^n), \cdots$$

is the subsequence of φ defined by the sequence $2, 4, 8, \cdots, 2^n, \cdots$ of natural numbers.

We only require one result about subsequences.

■ **THEOREM 2.12.** Let the sequence φ converge to y. Then any subsequence of φ also converges to y.

PROOF: By the definition of $\lim_{n \to \infty} \varphi(n) = y$, if we are given $\epsilon > 0$, then we find $n_0 \in \mathbf{N}$ such that $|\varphi(n) - y| < \epsilon$ whenever $n > n_0$. Now suppose the strictly increasing sequence $n_1 < n_2 < \cdots < n_k < \cdots$ defines a subsequence of φ. Then there is some value k_0 such that $n_{k_0} > n_0$. It follows that $|\varphi(n_k) - y| < \epsilon$ whenever $k > k_0$; hence $\lim_{k \to \infty} \varphi(n_k) = y$. ■

EXERCISES 2.6

1. By appealing to the theorems and examples in this section, establish the following limits.

   (a) $\lim_{n \to \infty} \frac{n+4}{n} = 1$

   (b) $\lim_{n \to \infty} \frac{2n+3}{n} = 2$

   (c) $\lim_{n \to \infty} \frac{1-n}{n} = -1$

   (d) $\lim_{n \to \infty} \frac{n+1}{2n} = \frac{1}{2}$

   (e) $\lim_{n \to \infty} \left(\frac{2}{3}\right)^n = 0$

   (f) $\lim_{n \to \infty} \left(\frac{2n}{n+1} - \frac{n+1}{2n}\right) = \frac{3}{2}$

2. Establish the following limits by application of the definition.

   (a) $\lim_{n \to \infty} \frac{n-1}{2n+3} = \frac{1}{2}$

   (b) $\lim_{n \to \infty} \frac{3n+1}{5n-1} = \frac{3}{5}$

   (c) $\lim_{n \to \infty} \frac{1}{\sqrt{n}} = 0$

   (d) $\lim_{n \to \infty} \frac{n^2-2}{n^2+2} = 1$

   (e) $\lim_{n \to \infty} \frac{5n+8}{n^2+1} = 0$

   (f) $\lim_{n \to \infty} \frac{3n^2+2}{4n^2+3n} = \frac{3}{4}$

   (g) $\lim_{n \to \infty} \frac{2n^3-3n}{5n^3+6n^2} = \frac{2}{5}$

   (h) $\lim_{n \to \infty} \left(\frac{n^2}{2n+1} - \frac{n^2}{2n-1}\right) = -\frac{1}{2}$

3. For each of the following sequences, determine the limit by any means you can and establish it by applying the definition.

(a) $\varphi(n) = \dfrac{n+2}{3n-1}$ 　　(b) $\varphi(n) = \dfrac{2n+1}{3-4n}$

(c) $\varphi(n) = \dfrac{n+1}{n^2}$ 　　(d) $\varphi(n) = \dfrac{1}{\sqrt{n^2+1}}$

(e) $\varphi(n) = \dfrac{2n^2-n+2}{n^2}$ 　　(f) $\varphi(n) = \dfrac{n+2}{2n}$

4. Establish the following limits.

(a) $\lim\limits_{n\to\infty} \left(1 + \dfrac{1}{2} + \dfrac{1}{4} + \cdots + \dfrac{1}{2^{n-1}}\right) = 2$

(b) $\lim\limits_{n\to\infty} \left(2 + \dfrac{2}{3} + \dfrac{2}{9} + \cdots + \dfrac{2}{3^{n-1}}\right) = 3$

(c) $\lim\limits_{n\to\infty} \left(1 + \dfrac{1}{10} + \dfrac{1}{100} + \cdots + \dfrac{1}{10^{n-1}}\right) = \dfrac{10}{9}$

(d) $\lim\limits_{n\to\infty} (a + ar + ar^2 + \cdots + ar^{n-1}) = \dfrac{a}{1-r}$ if $0 < r < 1$

5. For each sequence $\varphi(n)$, define $f(n) = \varphi(n+1)/\varphi(n)$ and determine the limit of $f(n)$ in each case.

(a) $\varphi(n) = \dfrac{n}{4^n}$ 　　(b) $\varphi(n) = \dfrac{2n}{1+3n}$

(c) $\varphi(n) = \dfrac{1}{n^2+4}$ 　　(d) $\varphi(n) = n^3 + 2$

PROBLEMS 2.6

1. For each $n \in \mathbf{N}$, prove that $0 < n/3^n < (\tfrac{2}{3})^n$ and then use this fact to prove that $\lim_{n\to\infty} n/3^n = 0$.

2. Determine
$$\lim_{n\to\infty} \left[\frac{1}{(n+1)^2} + \frac{1}{(n+2)^2} + \cdots + \frac{1}{(n+n)^2}\right]$$
If you cannot do so, can you prove that the limit exists?

3. Two banks are competing for savings accounts. One offers 5 percent interest compounded annually. The second offers 5 percent compounded semiannually. The first bank counters with an offer of 5 percent compounded quarterly. The second bank raises the bid to 5 percent compounded monthly, which is surpassed by 5 percent compounded weekly. Compute the value of $1 at the end of one year under each of these offers. What is this value if 5 percent interest is compounded every day? Every hour? Every minute? (We are involved here with the sequence $\varphi(n) = (1 + 1/n)^n$. As we shall see in Chapter 8, this sequence converges to a number, denoted by the letter e, which is very important in mathematics.)

4. A "superball" when dropped rebounds to 90 percent of the height from which it fell. If such a ball is dropped from a height of 6 ft, how far will it travel? (Assume that it bounces infinitely often.)

5. Let b be a positive real number. If $r_1 = \sqrt{b}$, then obviously
$$\frac{b}{r_1} = \sqrt{b}$$

2.6 INFINITE SEQUENCES

If $r_1 < \sqrt{b}$, then $b/r_1 > \sqrt{b}$; while, if $r_1 > \sqrt{b}$, then $b/r_1 < \sqrt{b}$. This suggests that, if r_1 is an approximation of $\sqrt{b}$, then the average of r_1 and b/r_1 will be a better approximation. Let us define

$$r_2 = \frac{1}{2}\left(r_1 + \frac{b}{r_1}\right)$$

Then the number

$$r_3 = \frac{1}{2}\left(r_2 + \frac{b}{r_2}\right)$$

should be a better approximation. This leads to a *recursively defined sequence*

$$f(1) = r_1$$

where r_1 is an initial approximation to $\sqrt{b}$ and

$$f(n) = \frac{1}{2}\left[f(n-1) + \frac{b}{f(n-1)}\right], \quad n > 1$$

Prove that

$$\lim_{n \to \infty} f(n) = \sqrt{b}$$

as we hoped. (Such a sequence can be computed, term by term, very rapidly by an electronic computer and is actually used to compute square roots.)

6. Prove Corollary 2.9 using the definition of a convergent sequence.

7. Prove that the limit of a sequence is unique. This means that, if $\lim_{n \to \infty} \varphi(n) = y$ and $\lim_{n \to \infty} \varphi(n) = z$, then $y = z$.

8. Prove that a convergent sequence is bounded.

9.   (a) If $\lim_{n \to \infty} \varphi(n) = y$, then show that $\lim_{n \to \infty} |\varphi(n)| = |y|$.
     (b) If $\lim_{n \to \infty} |\varphi(n)| = 0$, then show that $\lim_{n \to \infty} \varphi(n) = 0$.
     (c) If $\lim_{n \to \infty} |\varphi(n)| = y \neq 0$, there is no assurance that φ even converges. Find an example such that $|\varphi|$ converges, but φ does not.

10. Prove that, if two subsequences of a given sequence converge to different limits, then the given sequence diverges.

11. Let φ be a monotone increasing sequence. Prove that, if φ has a convergent subsequence, then φ is convergent.

12. If φ does not have a strictly increasing subsequence, prove that it must have a monotone decreasing subsequence.

13. If the even-numbered terms of φ converge to y and the odd-numbered terms converge to y, show that φ converges to y.

14. If $\varphi(n)$ converges to y, show that $[\varphi(n)]^2$ converges to y^2.

15. Prove that the sequence

$$\varphi(n) = 1 + \frac{1}{2^2} + \frac{1}{3^3} + \frac{1}{4^2} + \cdots + \frac{1}{n^2}$$

converges.

16. The number π equals the area of a circle of radius 1. Inscribe regular polygons with n-sides, $n = 3, 4, 5, 6, \cdots$, in the circle, compute their areas, and express π as the limit of a sequence.

2.7 THE RATE OF CHANGE OF A FUNCTION

The rate of change of the linear function $f(x) = Ax + B$ with respect to x is the number A (p. 59). Here we make use of this fact and of infinite sequences to give a temporary definition of the rate of change of more general functions. The temporary definition is actually valid for a wide class of functions (see Theorem 3.9). It also has the dual advantage of leading us to the important concept of a *derivative* and of being a working tool if we have a computer available. (More will be said about this application shortly.)

Suppose that the domain of a function f includes an open interval $\{x : a < x < b\}$. Let c and d be any two points, $c < d$, in the open interval. Then the difference $f(d) - f(c)$ is *the change in the functional value $f(x)$* due to the change $d - c$ in the variable x from c to d. The *average rate of change of the function between c and d* is defined to be the ratio

$$\frac{f(d) - f(c)}{d - c}$$

These quantities are shown in Fig. 2.22.

The ratio $[f(d) - f(c)]/(d - c)$ is precisely the slope of the line through the points $(c, f(c))$ and $(d, f(d))$. Or in other words, that ratio is the rate of change of the linear function

$$y = \left[\frac{f(d) - f(c)}{d - c}\right] x + \frac{df(c) - cf(d)}{d - c}$$

Note that there was no need to assume that $c < d$. The same formula would be valid if $d < c$. Hence, if $c \neq d$, then the average rate of change of $f(x)$ between c and d is the ratio

$$\frac{f(d) - f(c)}{d - c}$$

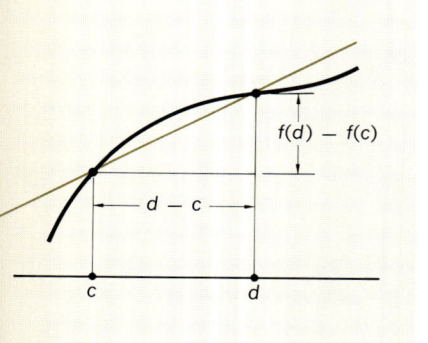

FIGURE 2.22

A crucial process is now set in motion. Holding c fixed, we let d approach c. What happens to the average rate of change of $f(x)$ between c and d as d approaches c? A good indication of the answer can be obtained in the following way: Suppose that d takes on successively the values $c + 1/n, n \in \mathbf{N}$. These values surely approach c. The rate of change of $f(x)$ between c and $c + 1/n$ depends on n and is

$$\varphi(n) = \frac{f(c + 1/n) - f(c)}{c + 1/n - c} = \frac{f(c + 1/n) - f(c)}{1/n}$$

$$= n\left[f\left(c + \frac{1}{n}\right) - f(c)\right]$$

This defines an infinite sequence φ. Now a good answer to our question above comes from finding the limit of the sequence φ.

2.7 THE RATE OF CHANGE OF A FUNCTION

■ **Example 1.** Let $f(x) = x^2$ and take $c = 1$. Then the terms of the sequence of average rates of change are of the form

$$n\left[\left(1 + \frac{1}{n}\right)^2 - 1^2\right] = n\left[1 - \frac{2}{n} + \frac{1}{n^2} - 1\right]$$

$$= n\left[\frac{2}{n} + \frac{1}{n^2}\right] = 2 + \frac{1}{n}$$

Surely the sequence $\varphi(n) = 2 + 1/n$ has limit 2.

■ **Example 2.** Let $f(x) = x^2$, but now let c be arbitrary. In this case, the sequence of average rates of change are

$$n\left[\left(c + \frac{1}{n}\right)^2 - c^2\right] = n\left[c^2 - \frac{2c}{n} + \frac{1}{n^2} - c^2\right]$$

$$= n\left[\frac{2c}{n} + \frac{1}{n^2}\right] = 2c + \frac{1}{n}$$

We see that $\varphi(n) = 2c + 1/n$ has limit $2c$.

■ **Example 3.** Again let $f(x) = x^2$ and let c be arbitrary, but now let d take on the values $c - 1/n$, $n \in \mathbb{N}$, which also approach c, of course. Then the sequence of average rates of change of $f(x)$ becomes

$$\frac{[(c - 1/n)^2 - c^2]}{c - 1/n - c} = \frac{c^2 - 2c/n + 1/n^2 - c^2}{-1/n}$$

$$= -n\left(-\frac{2c}{n} + \frac{1}{n^2}\right) = 2c - \frac{1}{n}$$

Again the sequence $2c - 1/n$ has limit $2c$.

These examples give strong indication (but not proof) that the average rate of change of $f(x) = x^2$ from c to a number $d \neq c$ approaches the value $2c$ as d approaches c. If there is meaning to the phrase

"the rate of change of $f(x) = x^2$ at the point c,"

then this rate must be $2c$. We are led to propose the following temporary definition:

The *rate of change of the function f at the point c* is the limit (if it exists) of the infinite sequence

$$\varphi(n) = n\left[f\left(c + \frac{1}{n}\right) - f(c)\right]$$

We compute several examples using this heuristic definition.

Example 4. Let $f(x) = x^2 - 3x + 2$ and let c be arbitrary. Then

$$\varphi(n) = n\left[\left(c + \frac{1}{n}\right)^2 - 3\left(c + \frac{1}{n}\right) + 2 - c^2 + 3c - 2\right]$$

$$= n\left[c^2 + \frac{2c}{n} + \frac{1}{n^2} - 3c + 2 - c^2 + 3c - 2\right]$$

$$= n\left[\frac{2c}{n} - \frac{3}{n} + \frac{1}{n^2}\right] = 2c - 3 + \frac{1}{n}$$

As n increases indefinitely, $\varphi(n) = 2c - 3 + 1/n$ approaches $2c - 3$. Therefore the rate of change of $f(x) = x^2 - 3x + 2$ at the point c is the number $2c - 3$.

Example 5. Let $f(x) = 1/x$ ($x \neq 0$) and take $c = 2$. The sequence of average rates of change of $1/x$ is

$$\varphi(n) = n\left[\frac{1}{2 + 1/n} - \frac{1}{2}\right] = n\left[\frac{2 - 2 - 1/n}{4 + 2/n}\right]$$

$$= n\left(\frac{-1/n}{4 + 2/n}\right) = -\frac{1}{4 + 2/n}$$

Because the denominator $4 + 2/n$ approaches 4 as n increases, we claim that the rate of change of $1/x$ at the point 2 is the number $-\frac{1}{4}$.

Example 6. Let $f(x) = x^2/(x+1)$ ($x \neq -1$) and let $c \neq -1$ be any fixed value. Then our sequence is

$$n\left[\frac{(c + 1/n)^2}{c + 1/n + 1} - \frac{c^2}{c + 1}\right]$$

$$= n\left[\frac{(c^2 + 2c/n + 1/n^2)(c + 1) - c^2(c + 1 + 1/n)}{(c + 1 + 1/n)(c + 1)}\right]$$

$$= n\left[\frac{c^3 + 2c^2/n + c/n^2 + c^2 + 2c/n + 1/n^2 - c^3 - c^2/n - c^2}{(c + 1)^2 + 1/n(c + 1)}\right]$$

$$= n\left[\frac{(1/n)(c^2 + 2c) + (1/n^2)(c + 1)}{(c + 1)^2 + (1/n)(c + 1)}\right]$$

$$= \left[\frac{c^2 + 2c + (1/n)(c + 1)}{(c + 1)^2 + (1/n)(c + 1)}\right]$$

The terms $(1/n)(c + 1)$ in numerator and denominator approach 0 as n increases. Thus we claim that the rate of change of $x^2/(x + 1)$ at the point c is the number $(c^2 + 2c)/[(c + 1)^2]$.

The terms of the sequence $\varphi(n) = n[f(c + 1/n) - f(c)]$ can be computed for a given number c very rapidly by an electronic computer. The machine need only be programmed to evaluate the function $f(x)$ and to stop computing when two successive terms of the sequence differ by less than some small number chosen by the operator. Then the last term computed is printed out as an accurate approximation of the rate of change of $f(x)$ at c.

There are some serious questions that we have glossed over in the above discussion. These questions will be raised and answered in Chapter 3.

EXERCISES 2.7

For each function and given point c in Exercises 1 to 20, compute and simplify the expression $\varphi(n) = n[f(c + 1/n) - f(c)]$. Then determine the rate of change of $f(x)$ at c by evaluating $\lim_{n \to \infty} \varphi(n)$.

1. $f(x) = 3x^2 - 4x - 2;\ c = 1$
2. $f(x) = 3x^2 - 4x - 2;\ c = -2$
3. $f(x) = 2x^3 - 3x;\ c = \frac{1}{2}$
4. $f(x) = x^3 + x^2 - 3x + 2;\ c = -1$
5. $f(x) = 3x^4 - 8x^3 - 6x^2 + 24x;\ c = 2$
6. $f(x) = \dfrac{x}{x+1};\ c = 2$
7. $f(x) = \dfrac{1}{x-1};\ c = 3$
8. $f(x) = x + 1/x;\ c = 1$
9. $f(x) = \dfrac{1}{x^2};\ c = -1$
10. $f(x) = \dfrac{1}{x^2+1};\ c = 0$
11. $f(x) = \dfrac{x^2-1}{x^2+1};\ c = 1$
12. $f(x) = \sqrt{2x};\ c = 8$
13. $f(x) = \sqrt{x+1};\ c = 3$
14. $f(x) = \sqrt{3-x};\ c = -1$
15. $f(x) = \sqrt{x^2-4};\ c = 4$
16. $f(x) = 1/\sqrt{x};\ c = 4$
17. $f(x) = \dfrac{1}{\sqrt{x+1}};\ c = 0$
18. $f(x) = \dfrac{1}{\sqrt{4-x^2}};\ c = ?$
19. $f(x) = x\sqrt{x};\ c = 4$
20. $f(x) = x\sqrt{x};\ c > 0$

2.8 TANGENT LINES TO A GRAPH

Recall that the tangent line at a point P on a circle and the radius drawn to the same point P are perpendicular. This fact permits us to draw a tangent line to a circle wherever we wish. How do we draw tangent lines to other curves? In fact, how do we even *define* the tangent line at a point of a curve $\{(x, y) : y = f(x)\}$? The answer to this question is another instance in which the limit concept is used to provide a useful definition.

If $(c, f(c))$ is the point on the graph at which we wish to draw the tangent line, then we only have one point on the desired line. This is not enough to locate the line, of course. For the moment, then, let us be satisfied with finding a close approximation to the required tangent line. We can do that. We simply select another point $(d, f(d))$ on the graph and very close to the point $(c, f(c))$. This is illustrated in Fig. 2.23 with the two points separated enough so that we can see them. Because we now have two points, we can draw a line, a so-called *secant line*, which is close to the desired tangent line.

What is more important, the equation of the "approximate" tangent line is

$$(f(d) - f(c))(x - c) = (d - c)(y - f(c))$$

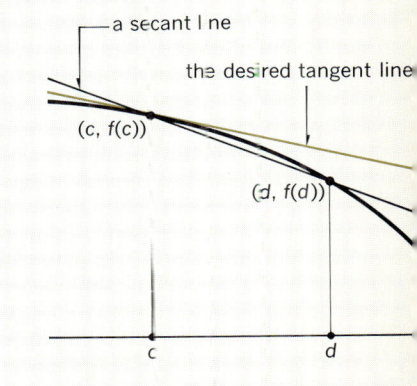

FIGURE 2.23

Then, because $c \neq d$, we divide by $d - c$ to obtain the equation

$$y = \left[\frac{f(d) - f(c)}{d - c}\right](x - c) + f(c)$$

Hence the slope of the secant line is the "rate of change" ratio

$$\frac{f(d) - f(c)}{d - c}$$

If we let d approach c, it seems obvious that the successive secant lines are better and better approximations to the desired tangent line. Also the slope of the secant line approaches the rate of change of $f(x)$ at the point c. We made this discussion to justify the following tentative definition.

Let $\{(x,y) : y = f(x)\}$ be the graph of a function f and let c be a fixed point in the domain of fx. Let m be the rate of change of $f(x)$ at c. Then the tangent line to the graph at $(c, f(c))$ has the equation

$$y = m(x - c) + f(c)$$

■ **Example 1.** From Example 1 of Section 2.7, we know that the rate of change of $f(x) = x^2$ at $c = 1$ is the number 2. Hence the equation of the tangent line to the graph $\{(x,y) : y = x^2\}$ at the point $(1, 1)$ is

$$y = 2(x - 1) + 1 = 2x - 1$$

■ **Example 2.** In Example 2 of Section 2.7 we found that the rate of change of $f(x) = x^2$ at the value c is the number $2c$. It follows from our definition that the equation of the tangent line to $\{(x,y) : y = x^2\}$ at the point (c, c^2) is $y = 2c(x - c) + c^2 = 2cx - c^2$.

■ **Example 3.** Suppose that we want the tangent line to

$$\{(x,y) : y = 2x + \frac{8}{x} ; x \neq 0\}$$

at the point $(2, 8)$. We first compute the rate of change of $2x + 8/x$ at the point 2. Our defining infinite sequence is

$$\varphi(n) = n\left[2\left(2 + \frac{1}{n}\right) + \frac{8}{2 + 1/n} - 4 - \frac{8}{2}\right]$$

$$= n\left[4 + \frac{2}{n} - 4 + \frac{8}{2 + 1/n} - \frac{8}{2}\right]$$

$$= n\left[\frac{2}{n} + \frac{-8/n}{4 + 2/n}\right] = 2 - \frac{8}{4 + 2/n}$$

Now, as n increases, the denominator $4 + 2/n$ approaches 4. Therefore the rate of change of $2x + 8/x$ at 2 is zero. Then the equation of the desired tangent line is $y = 0(x - 2) + 8$ or $y = 8$. Thus the tangent line is horizontal.

2.8 TANGENT LINES TO A GRAPH

The ability to draw a tangent line wherever we wish can be very useful in drawing the graph of a function. By drawing a short segment of the tangent line each time we plot a point on the graph, the entire graph is much easier to envision. With just a few such points and tangent segments, we can "see" the graph very clearly. The procedure outlined in Example 4 is a good pattern to follow.

■ **Example 4.** The problem is to draw the graph $\{(x,y) : x^2/(x-1); x \neq -1\}$. In Example 6 of Section 2.7, we found that the rate of change of $x^2/(x+1)$ at c is the number

$$m = \frac{c^2 + 2c}{(c+1)^2}$$

We tabulate several values of x, $f(x)$, and m.

x	0	1	2	3	4	$-\frac{1}{2}$	$-\frac{3}{4}$
$f(x)$	0	$\frac{1}{2}$	$\frac{4}{3}$	$\frac{9}{4}$	$\frac{16}{5}$	$\frac{1}{2}$	$\frac{9}{4}$
m	0	$\frac{3}{4}$	$\frac{8}{9}$	$\frac{15}{16}$	$\frac{24}{25}$	-3	-15

Each time we plot a point $(x, f(x))$ we also draw a short segment of slope m through the point. This is done in Fig. 2.24(a); the graph is added in Fig. 2.24(b). Notice how easy it is to "see" the graph in the first drawing.

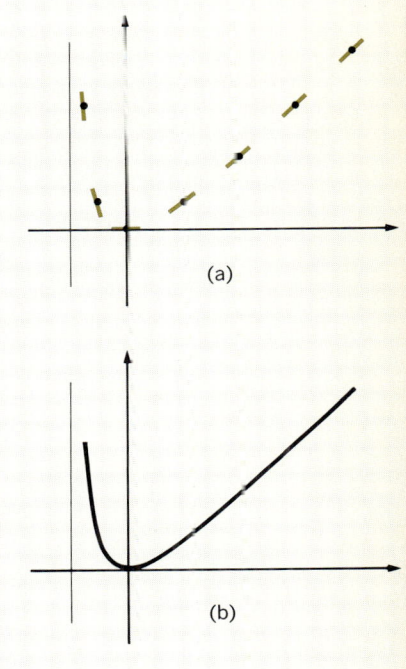

FIGURE 2.24

EXERCISES 2.8

In Exercises 1 to 10, apply the technique of Section 2.8 to obtain the equation of the tangent line to the graph of f at the given point.

1. $f(x) = 3x - 2$ at $(1, 1)$
2. $f(x) = 2x^2$ at $(1, 2)$
3. $f(x) = 2x^2 - 1$ at $(1, 1)$
4. $f(x) = 3x^2 + 2x$ at $(0, 0)$
5. $f(x) = 3x^2 + 2x$ at $(-1, 1)$
6. $f(x) = \frac{1+x}{x}$ at $(1, 2)$
7. $f(x) = \frac{x}{x^2 + 1}$ at $(0, 0)$
8. $f(x) = \sqrt{2x}$ at $(8, 4)$
9. $f(x) = \sqrt{x^2 + 9}$ at $(4, 5)$
10. $f(x) = \sqrt{1 - x^2}$ at $(0, 1)$

In Exercises 11 to 18, apply the technique of Example 4 of Section 2.8 to assist in drawing the graph of f.

11. $f(x) = x^2 - 2x + 2$
12. $f(x) = 4 - 3x - x^2$
13. $f(x) = x^3 - 2x$
14. $f(x) = \frac{1}{x+1}$ $(x \neq -1)$
15. $f(x) = \frac{x}{x+1}$ $(x \neq -1)$

16. $f(x) = \sqrt{4 - x^2}$ $(-2 \leq x \leq 2)$

17. $f(x) = \dfrac{1}{\sqrt{x}}$ $(x > 0)$

18. $f(x) = \dfrac{\sqrt{x + 1}}{x}$ $(x \geq -1, x \neq 0)$

PROBLEMS 2.8

1. Determine the points at which the graph $\{(x, y) : y = Ax^2 + Bx + C\}$ has a horizontal tangent line.

2. Show that there are two points on $\{(x, y) : y = x^2 - 5x + 9\}$ for which the tangent line passes through the origin.

3. Show that there is only one point on $\{(x, y) : y = x^3 - 3x\}$ for which the tangent line passes through the point $(0, 2)$.

4. From a given point (x_0, y_0), how many tangent lines can be drawn to the graph $\{(x, y) : y = Ax^2 + Bx + C\}$?

5. Let (x_1, y_1) be a point on the graph $\{(x, y) : 4py = x^2\}$. Show that the equation of the tangent line to the graph at (x_1, y_1) can be written as $x_1 x = 2p(y + y_1)$.

6. Find the equation of the tangent line to $\{(x, y) : y = \sqrt[3]{x}\}$ at the origin.

7. Show that the graphs $\{(x, y) : y = 4x^2 - 7x\}$, $\{(x, y) : y = x^3 - 3x\}$ have a common tangent line at one of their points of intersection.

8. The point $(1, 1)$ is on both graphs. Also,

$$\{(x, y) : x \geq 0, y = x^2\} \quad \text{and} \quad \{(x, y) : x \geq 0, y = \sqrt{x}\}.$$

What is the relationship between the slopes of the two tangent lines? Generalize this observation to the case of a point (a, a^2) on the first graph and the point (a^2, a) on the second.

2.9 THE LIMIT OF A FUNCTION

In order to give a precise discussion of the rate of change of a function (Section 2.7), we must know the meaning of the phrase "the ratio $[f(d) - f(c)]/(d - c)$ approaches the number m" as d approaches c. This is actually a generalization of the notion of an infinite sequence "approaching" a limit and we shall make this obvious shortly.

Let $\varphi(x)$ denote the functional values of a real-valued function of a real variable and let c be some fixed point in the domain of φ. Then the number L *is the limit of* $\varphi(x)$ *as* x *approaches* c provided that, given any positive number ϵ, there is a positive number δ such that for any value of x [in the domain of $\varphi(x)$] which satisfies the inequality $0 < |x - c| < \delta$, the functional value $\varphi(x)$ will satisfy the inequality $|\varphi(x) - L| < \epsilon$.

This definition is precise but difficult. Its meaning and its implications will become more clear as we work with it. First we recall that the inequality $0 < |x - c| < \delta$ is equivalent to the two inequalities $c - \delta < x < c$ and $c < x < c + \delta$. The inequality $|\varphi(x) - L| < \epsilon$ is equivalent to $L - \epsilon < \varphi(x) < L + \epsilon$.

The given number ϵ determines a (small) open interval of length 2ϵ centered at the number L on the y axis. Hence, ϵ determines a horizontal strip $\{(x, y) : L - \epsilon < y < L + \epsilon\}$ as shown in Fig. 2.25(a). The number δ similarly determines a separated vertical strip of width 2δ, i.e., the set

$$\{(x,y) : c + \delta < x < c \quad \text{or} \quad c < x < c + \delta\}$$

(see Fig. 2.25(b)). The vertical line $\{(x,y) : x = c\}$ is not in the separated strip.

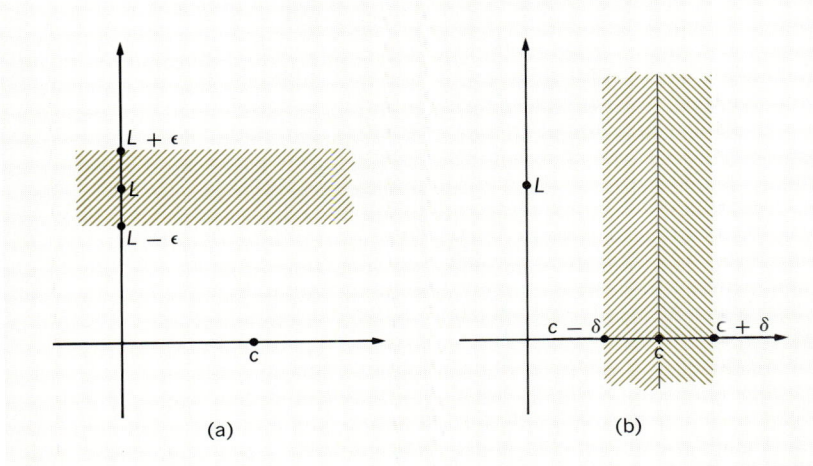

FIGURE 2.25

Now, in order that the definition be satisfied, we must be able to choose δ so that the piece of the graph $\{(x,y) : y = \varphi(x)\}$, which lies in the separated vertical strip, also lies in the given horizontal strip (Fig. 2.26).

In using the definition of "limit of $\varphi(x)$" the number δ, which depends both upon ϵ and upon c, is determined analytically. This requires considerable skill in manipulating inequalities, and is perhaps the most difficult procedure we shall encounter. (Incidentally, this procedure is often called "epsilontics" by the mathematician.) In each of the next four examples, we shall first give an argument involving infinite sequences to make the limit plausible. Then we give the precise use of our definition.

■ **Example 1.** Let $\varphi(x) = 3x - 1$ and take $c = 1$. By letting x take on successively the values $c + 1/n = 1 + 1/n$, we have $\varphi(1 + 1/n) = 3(1 + 1/n) - 1 = 2 + 3/n$. As n increases, $1 + 1/n$ approaches 1 and $\varphi(1 + 1/n)$ approaches 2. Thus we claim that the limit of $3x - 1$ as x approaches 1 is the number 2.

Now we prove this claim by appeal to the definition. Let $\epsilon > 0$ be given. We must find $\delta > 0$ such that whenever x satisfies $0 < |x - 1| < \delta$, we have $|(3x - 1) - 2| < \epsilon$. But the latter inequality is $|3x - 3| < \epsilon$ or $3|x - 1| < \epsilon$. It is easy to see that $3|x - 1| < \epsilon$ will be satisfied if

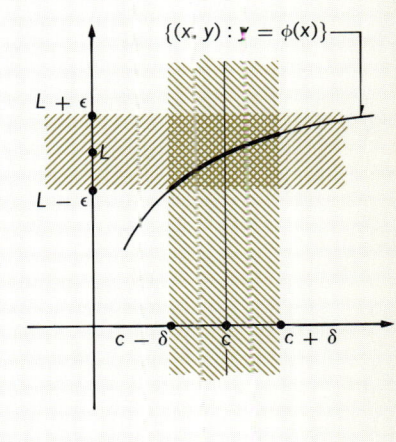

FIGURE 2.26

$|x - 1| < \epsilon/3$. Therefore we choose $\delta = \epsilon/3$. Checking back, if $|x - 1| < \epsilon/3$, then $3|x - 1| < \epsilon$ and hence $|3x - 3| = |(3x - 1) - 2| < \epsilon$. The choice $\delta = \epsilon/3$ satisfies our definition so 2 *is* the limit of $3x - 1$ as x approaches 1.

■ **Example 2.** Let $\varphi(x) = x^2 - 2x$ and let $c = 1$. Then

$$\varphi\left(1 + \frac{1}{n}\right) = \left(1 + \frac{1}{n}\right)^2 - 2\left(1 + \frac{1}{n}\right) = -1 + \frac{1}{n^2}$$

Since $\varphi(1 + 1/n)$ approaches -1 as n increases, we believe that -1 is the limit of $x^2 - 2x$ as x approaches 1.

The proof is as follows: Let $\epsilon > 0$ be given. We must choose $\delta > 0$ such that $|\varphi(x) - L| = |(x^2 - 2x) - (-1)| < \epsilon$ whenever x satisfies $0 < |x - 1| < \delta$. But $(x^2 - 2x) - (-1) = x^2 - 2x + 1 = (x - 1)^2$. Therefore $|(x^2 - 2x) - (-1)| = (x - 1)^2$. That is, we want $(x - 1)^2 < \epsilon$ whenever $0 < |x - 1| < \delta$. Clearly, $(x - 1)^2 < \epsilon$ is equivalent to $|x - 1| < \sqrt{\epsilon}$. Hence, if we choose $\delta = \sqrt{\epsilon}$, the inequality $|(x^2 - 2x) - (-1)| < \epsilon$ will hold whenever $0 < |x - 1| < \sqrt{\epsilon}$ and our definition is satisfied.

In Examples 1 and 2 it turned out that "the limit of $\varphi(x)$ as x approaches c" was precisely the number $\varphi(c)$. However, nothing in the definition of limit implies that $\varphi(x)$ is even defined at $x = c$. The following example shows that the "limit of $\varphi(x)$ as x approaches c" can exist when $\varphi(c)$ is not defined.

■ **Example 3.** Let $\varphi(x) = (x^2 - 1)/(x - 1)$, $x \neq 1$, and let $c = 1$. Obviously, $\varphi(1) = 0/0$ is not defined. However, by letting x take on successively the values $1 + 1/n$, $n \in \mathbf{N}$, we have the sequence

$$\varphi\left(1 + \frac{1}{n}\right) = \frac{(1 + 1/n)^2 - 1}{1 + 1/n - 1} = \frac{1 + 2/n + 1/n^2 - 1}{1/n}$$

$$= n\left[\frac{2}{n} + \frac{1}{n^2}\right] = 2 + \frac{1}{n}$$

Hence we surmise that the limit of $\varphi(x) = (x^2 - 1)/(x - 1)$ as x approaches 1 is the number 2.

PROOF: Given $\epsilon > 0$, we must choose δ so that

$$\left|\frac{x^2 - 1}{x - 1} - 2\right| < \epsilon$$

whenever $0 < |x - 1| < \delta$. Since we have $0 < |x - 1|$, we may assume that $x \neq 1$, whence

$$\left|\frac{x^2 - 1}{x - 1} - 2\right| = \left|\frac{x^2 - 1 - 2(x - 1)}{x - 1}\right| = \left|\frac{x^2 - 2x + 1}{x - 1}\right|$$

$$= |x - 1|$$

2.9 THE LIMIT OF A FUNCTION

Therefore, $|\varphi(x) - 2| = |x - 1| < \epsilon$ whenever $0 < |x - 1| < \epsilon$, and we may simply choose $\delta = \epsilon$. ∎

■ **Example 4.** Let $f(x) = x^2$ and let c be any fixed value of x. Form the new function of x defined by

$$\varphi(x) = \frac{f(x) - f(c)}{x - c} = \frac{x^2 - c^2}{x - c}$$

Notice that $\varphi(x)$ is *not* defined when $x = c$. In Example 2 of Section 2.7, however, we showed that the sequence $\varphi(c + 1/n)$ approaches $2c$ as n increases. This leads us to believe that the limit of $(x^2 - c^2)/(x - c)$ as x approaches c is the number $2c$.

To prove that $2c$ is the limit, let $\epsilon > 0$ be given. When $x \neq c$, we have

$$\left| \frac{x^2 - c^2}{x - c} - 2c \right| = \left| \frac{x^2 - c^2 - 2c(x - c)}{x - c} \right| = \left| \frac{x^2 - 2cx + c^2}{x - c} \right|$$
$$= |x - c|$$

Therefore we choose $\delta = \epsilon$ and the inequality $0 < |x - c| < \epsilon$ implies that

$$\left| \frac{x^2 - c^2}{x - c} - 2c \right| < \epsilon$$

That is, our definition is satisfied and the proof is complete.

The sentence "L is the limit of $\varphi(x)$ as x approaches c" is written in symbolic form as

$$\lim_{x \to c} \varphi(x) = L$$

The four examples just given are written symbolically as

1. $\lim_{x \to 1} (3x - 1) = 2$
2. $\lim_{x \to 1} (x^2 - 2x) = -1$
3. $\lim_{x \to 1} \left(\frac{x^2 - 1}{x - 1} \right) = 2$
4. $\lim_{x \to c} \left(\frac{x^2 - c^2}{x - c} \right) = 2c$

REMARK: If L is a real number and we use the symbol $\lim_{x \to c} \varphi(x) = L$, then we imply that the limit of $\varphi(x)$ as x approaches c both exists and equals L.

■ **THEOREM 2.13.** If the limit of $\varphi(x)$ as x approaches c exists, then the limit is unique.

PROOF: We assume that $\lim_{x \to c} \varphi(x) = L$ and that $\lim_{x \to c} \varphi(x) = M$. We prove that $L = M$. To do so, we use a trick that is quite commonly used. We write

$$|L - M| = |\varphi(x) - M - (\varphi(x) - L)| \leq |\varphi(x) - M| + |\varphi(x) - L|$$

making use of properties of absolute values. Now given any $\epsilon > 0$, there is a value of x so close to c that we have both $|\varphi(x) - M| < \epsilon$ and $|\varphi(x) - L| < \epsilon$. Thus for any $\epsilon > 0$, $|L - M| < 2\epsilon$. Finally, two numbers whose difference is less than any given positive number are necessarily equal. Hence $L = M$. ∎

■ *Example 5.* Let $\varphi(x) = (x + 1)/x$ and take $c = 1$. We arrive at a guess about $\lim_{x \to 1} [(x + 1)/x]$ simply by setting $x = 1$ in the expression $(x + 1)/x$. This yields the number 2. A somewhat more convincing argument is to let x take on values $1 + 1/n$ and note that

$$\varphi\left(1 + \frac{1}{n}\right) = \frac{1 + 1/n + 1}{1 + 1/n} = \frac{2 + 1/n}{1 + 1/n}$$

Since the numerator approaches 2 and the denominator approaches 1 as n increases, our limit should be 2.

PROOF: Let $\epsilon > 0$ be given. We must choose δ so that

$$\left|\frac{x + 1}{x} - 2\right| < \epsilon$$

whenever $0 < |x - 1| < \delta$. First we remark that

$$\left|\frac{x + 1}{x} - 2\right| = \left|\frac{x + 1 - 2x}{x}\right| = \left|\frac{1 - x}{x}\right| = \frac{|x - 1|}{|x|}$$

Eventually x will be close to 1 anyway, so we do not lose generality by first assuming that x satisfies $\frac{1}{2} < x < \frac{3}{2}$. (This only means that we shall have to choose $\delta \leq \frac{1}{2}$.) For values of $x > \frac{1}{2}$, we surely have

$$\frac{|x - 1|}{|x|} < \frac{|x - 1|}{1/2} = 2|x - 1|$$

Therefore, given any value of x satisfying $|x - 1| < \frac{1}{2}$, we have

$$\frac{|x - 1|}{|x|} < 2|x - 1|$$

Then $2|x - 1|$ will be less than ϵ if we choose $|x - 1| < \epsilon/2$. Therefore we choose δ to be the smaller of the numbers $\frac{1}{2}$ and $\epsilon/2$ and our definition is satisfied. ∎

■ *Example 6.* We should call this a counterexample. Let $\varphi(x) = 1/(x - 1)$ and take $c = 1$. Consider the infinite sequence

$$\varphi\left(1 + \frac{1}{n}\right) = \frac{1}{1 + 1/n - 1} = \frac{1}{1/n} = n$$

As n increases indefinitely, so does $\varphi(1 + 1/n)$. This should convince us that $\varphi(x)$ approaches no number at all when x approaches 1.

We prove this claim as follows: Let L be any real positive number, however large. Then there is a natural number n such that $2L < n$.

Therefore

$$\frac{1}{1 + 1/n - 1} - L = n - L > 2L - L = L$$

Obviously, then, we cannot make the quantity

$$\left|\frac{1}{x - 1} - L\right|$$

arbitrarily small as x approaches 1.

The situation in Example 6, where $\varphi(x)$ grows indefinitely large as x approaches c, can be described by another use of the limit symbol. The functional value $\varphi(x)$ is said to approach positive infinity as x approaches c, and we write

$$\lim_{x \to c} \varphi(x) = +\infty$$

if for any $n \in \mathbf{N}$, there exists $\delta > 0$ such that $\varphi(x) > n$ whenever $0 < |x - c| < \delta$.

▪ **Example 7.** Let $\varphi(x) = 1/x^2$. Then $\lim_{x \to 0} (1/x^2) = +\infty$. To prove this, let n be any integer > 0. We want to have $1/x^2 > n$ or $x^2 < 1/n$. Obviously, by choosing $\delta = 1/\sqrt{n}$, we shall have $1/x^2 > n$ whenever $0 < |x - 0| < 1/\sqrt{n}$.

▪ **Example 8.** Let $\varphi(x) = 1/x$. Now we cannot say that $\lim_{x \to 0} 1/x = +\infty$ because negative values of x yield negative values of $1/x$. Thus for any $n \in \mathbf{N}$, $1/x < n$ if x is negative, no matter how close to zero x may be.

The functional value $\varphi(x)$ approaches minus infinity as x approaches c, or

$$\lim_{x \to c} \varphi(x) = -\infty$$

if, given $n \in \mathbf{N}$, there exists $\delta > 0$ such that $\varphi(x) < -n$ whenever $0 < |x - c| < \delta$.

▪ **Example 9.** If $\varphi(x) = 2 - 1/x^2$, then $\lim_{x \to 0} \varphi(x) = -\infty$. The proof consists in noting that for any $n \in \mathbf{N}$, we want $2 - 1/x^2 < -n$ or $2 + n < 1/x^2$ or $x^2 < 1/(2 + n)$. By choosing $\delta = 1/\sqrt{2 + n}$, we see that, if $0 < |x - 0| = |x| < 1/\sqrt{2 + n}$, then $2 - 1/x^2 < -n$.

Returning to Example 8, we see that $\lim_{x \to 0} 1/x \ne -\infty$ either. Yet, in a sense, $1/x$ does grow "large" as x approaches zero. The following definition provides just the answer to this situation: The functional value $\varphi(x)$ approaches infinity as x approaches c, and we write

$$\lim_{x \to c} \varphi(x) = \infty$$

if $\lim_{x \to c} |\varphi(x)| = +\infty$. Clearly, then $\lim_{x \to 0} 1/x = \infty$ in this new sense.

Still another use of the limit symbol is the following: The functional values $\varphi(x)$ approach the limit L as x approaches positive infinity,

$$\lim_{x \to +\infty} \varphi(x) = L$$

if, given $\epsilon > 0$, there exists $n \in \mathbf{N}$ such that $|\varphi(x) - L| < \epsilon$ whenever $x > n$. Similarly, we write

$$\lim_{x \to -\infty} \varphi(x) = L$$

if, given $\epsilon > 0$, there exists $n \in \mathbf{N}$ such that $|\varphi(x) - L| < \epsilon$ whenever $x < -n$.

■ **Example 10.** Let $f(x) = x/(x + 1)$. We can establish that $\lim_{x \to +\infty} x/(x + 1) = 1$. The proof starts with the manipulation

$$\left|\frac{x}{x+1} - 1\right| = \left|\frac{x - x - 1}{x + 1}\right| = \left|\frac{-1}{x + 1}\right| = \frac{1}{|x + 1|}$$

if $x > 0$, then $|x + 1| = x + 1$. Hence if we want $1/(x + 1) < \epsilon$, we need only choose $x > 1/\epsilon - 1$. Letting n be the first natural number larger than $1/\epsilon - 1$, we see that $x > n$ implies $x > 1/\epsilon - 1$, whence $1/(x + 1) < \epsilon$.

EXERCISES 2.9

1. $\lim_{x \to 0} (x^2 + 1) = 1$
2. $\lim_{x \to 1} \dfrac{3x + 1}{x + 1} = 2$
3. $\lim_{x \to 3} \dfrac{3x - 2}{x^2} = \dfrac{7}{9}$
4. $\lim_{x \to -1} \dfrac{x^2 - x - 2}{x + 1} = -3$
5. $\lim_{x \to 2} \dfrac{x^2 - 4}{x - 2} = 4$
6. $\lim_{x \to 1} \dfrac{x^3 - 1}{x - 1} = 3$
7. $\lim_{x \to a} \dfrac{x^3 - ax^2}{x - a} = a^2$
8. $\lim_{x \to a} \dfrac{1/x - 1/a}{x - a} = -\dfrac{1}{a^2}$
9. $\lim_{x \to -1} \sqrt{x^2 - x - 1} = 1$
10. $\lim_{x \to a} \dfrac{\sqrt{x} - \sqrt{a}}{x - a} = \dfrac{1}{2\sqrt{a}}$
11. $\lim_{x \to a} \dfrac{\sqrt{x^2 + 1} - \sqrt{a^2 + 1}}{x - a} = \dfrac{a}{\sqrt{a^2 + 1}}$
12. $\lim_{x \to a} \dfrac{\dfrac{x}{x+1} - \dfrac{a}{a+1}}{x - a} = \dfrac{1}{(a + 1)^2}$

For each of Exercises 13 to 24, determine and establish the value of the given limit.

13. $\lim_{x \to 0} (x^3 - 3x + 4)$
14. $\lim_{x \to 1} (x^2 - 2x + 1)$
15. $\lim_{x \to 1} \dfrac{x^2 - 1}{x^2 - 2x + 1}$
16. $\lim_{x \to 6} \dfrac{x^2 - 5x - 6}{x^2 - 2x - 24}$
17. $\lim_{x \to 1} \dfrac{x^3 + x^2 - 2x}{x^3 - 2x^2 - x + 2}$
18. $\lim_{x \to -2} \dfrac{(x + 2)(x^2 - x + 3)}{x^2 + 3x + 2}$

2.9 THE LIMIT OF A FUNCTION

19. $\lim_{x \to a} \dfrac{x^4 - a^4}{x^3 - a^3}$

20. $\lim_{x \to a} \dfrac{x^2 - a^2}{x^3 - a^3}$

21. $\lim_{x \to a} \dfrac{x^2 - a^2}{x^3 - ax^2}$

22. $\lim_{x \to a} \dfrac{\sqrt[3]{x} - \sqrt[3]{a}}{x - a}$

23. $\lim_{x \to a} \dfrac{1/x^2 - 1/a^2}{x - a}$

24. $\lim_{x \to a} \dfrac{\sqrt{x} - |a|}{x - a^2}$

In Exercises 25 to 32, establish the limit equations given.

25. $\lim_{x \to 2} (x - 2)^{-2} = +\infty$

26. $\lim_{x \to 0} (3 + x^{-2}) = +\infty$

27. $\lim_{x \to 1} \dfrac{1 - 2x}{(x - 1)^2} = -\infty$

28. $\lim_{x \to -1} \dfrac{x}{(1 + x)^2} = -\infty$

29. $\lim_{x \to +\infty} 1/x = 0$

30. $\lim_{x \to +\infty} \dfrac{x^2 + 1}{x^2 - 1} = 1$

31. $\lim_{x \to -\infty} \dfrac{2x}{x + 1} = 2$

32. $\lim_{x \to -\infty} \dfrac{x^2}{3x + 1} = -\infty$

For each of the functions listed in Exercises 33 to 40, form the new function

$$\varphi(h) = \dfrac{f(x + h) - f(x)}{h}$$

and then determine $\lim_{h \to 0} \varphi(h)$.

33. $f(x) = 3x + 2$

34. $f(x) = x^2 - 2x$

35. $f(x) = 1 - 2x - x^2$

36. $f(x) = 3 - 6x - x^3$

37. $f(x) = x + 1/x$

38. $f(x) = \sqrt{x}$

39. $f(x) = 1/\sqrt{x}$

40. $f(x) = \dfrac{x}{\sqrt{x + 1}}$

The one-sided limits of $f(x)$ at $x = a$ are defined as follows. The number R is the *right limit* of $f(x)$ at $x = a$, and we write

$$\lim_{x \to a^+} f(x) = R$$

if for each $\epsilon > 0$, there is a $\delta > 0$ such that $|f(x) - R| < \epsilon$ whenever $a < x < a + \delta$. Similarly, the *left limit* of $f(x)$ at $x = a$ equals L,

$$\lim_{x \to a^-} f(x) = L$$

if for each $\epsilon > 0$, there is a $\delta > 0$ such that $|f(x) - L| < \epsilon$ whenever $a - \delta < x < a$.

PROBLEMS 2.9
One-sided limits

1. Evaluate

   (a) $\lim_{x \to 1^+} |x - 1|$

   (b) $\lim_{x \to 1^-} |x - 1|$

   (c) $\lim_{x \to 0^+} \dfrac{x}{|x|}$

   (d) $\lim_{x \to 0^-} \dfrac{x}{|x|}$

   What do these results tell you about absolute values?

2. Find a function f that satisfies all three of the following conditions:

   (a) $\lim_{x \to 0} |f(x)| = 1$;

   (b) $\lim_{x \to 0^-} f(x) = -1$;

   (c) $\lim_{x \to 0^+} f(x)$ does not exist.

3. Prove that $\lim_{x \to a} f(x)$ exists if and only if the one-sided limits $\lim_{x \to a^+} f(x)$ and $\lim_{x \to a^-} f(x)$ both exist and are equal.

4. Consider the function $f(x) = [x]$, where $[x]$ is defined to be the largest integer $\leq x$. (Thus, $[\frac{7}{2}] = 3$, $[\pi] = 3$, $[-\frac{3}{2}] = -2$, and so on.) Prove that $\lim_{x \to a^+} [x] = [a]$. Where does the left limit fail to equal the right limit?

2.10 PROPERTIES OF LIMITS

This final section of the chapters contains proofs of the basic theorems about limits. We emphasize that the proofs here are no more than exercises in the careful use of definitions. However, a complete understanding will require careful study. The reader may decide for himself the depth of understanding he must acquire.

Our first theorem can be remembered as "the limit of a sum is the sum of the limits." In other words, the operation of taking a limit distributes over the addition of functions.

■ **THEOREM 2.14.** If $\lim_{x \to a} f(x) = L$ and $\lim_{x \to a} g(x) = K$, then

$$\lim_{x \to a} [f(x) \pm g(x)] = L \pm K$$

PROOF: Let $\epsilon > 0$ be given. Our proof makes precise the following statement: If $f(x)$ is within the distance $\epsilon/2$ of L and $g(x)$ is within the distance $\epsilon/2$ of K, then $f(x) + g(x)$ must be within the distance $\epsilon/2 + \epsilon/2 = \epsilon$ of $L + K$.

Because $\lim_{x \to a} f(x) = L$, there exists $\delta' > 0$, so that $|f(x) - L| < \epsilon/2$ whenever $0 < |x - a| < \delta'$. Similarly, since $\lim_{x \to a} g(x) = K$, there exists $\delta'' > 0$, so that $|g(x) - K| < \epsilon/2$ whenever $0 < |x - a| < \delta''$. Now take δ to be the smaller of δ' and δ''. Then, if $0 < |x - a| < \delta$, we have

$$|[f(x) + g(x)] - (L + K)| = |(f(x) - L) + (g(x) - K)|$$
$$\leq |f(x) - L| + |g(x) - K|$$
$$< \frac{\epsilon}{2} + \frac{\epsilon}{2} = \epsilon$$

Similarly, we have

$$|[f(x) - g(x)] - (L - K)| = |(f(x) - L) - (g(x) - K)|$$
$$= |f(x) - L| + |g(x) - K|$$
$$< \frac{\epsilon}{2} + \frac{\epsilon}{2} = \epsilon$$

This completes the proof. ■

■ **THEOREM 2.15.** If $\lim_{x \to a} f(x) = L$ and $\lim_{x \to a} g(x) = K$, then

$$\lim_{x \to a} [f(x)g(x)] = LK$$

PROOF: By assumption we can make $|f(x) - L|$ and $|g(x) - K|$ as small as we please for x in a neighborhood of a. We must use this control to

2.10 PROPERTIES OF LIMITS

make $|f(x)g(x) - LK|$ as small as is asked. This is done via the following: By adding zero in the special form $-f(x)K + f(x)K$, we have
$$f(x)g(x) - LK = f(x)g(x) - f(x)K + f(x)K - LK$$
$$= f(x)(g(x) - K) + K(f(x) - L)$$

Since we can make $|g(x) - K|$ and $|f(x) - L|$ small, we can also do the same for $|f(x)g(x) - LK|$. The argument below makes this precise.

Given $\epsilon > 0$, we must chose δ so that
$$|f(x)(g(x) - K) + K(f(x) - L)| < \epsilon \quad \text{whenever} \quad 0 < |x - a| < \delta$$

We first consider the case where $K \neq 0$. Since $\lim_{x \to a} f(x) = L$, there exist $\delta' > 0$ such that
$$|f(x) - L| < \frac{\epsilon}{2|K|} \quad \text{whenever } 0 < |x - a| < \delta'$$

Hence we have
$$|K| \cdot |f(x) - L| < \frac{\epsilon}{2} \quad \text{whenever } 0 < |x - a| < \delta'$$

Now we know that
$$L - \frac{\epsilon}{2|K|} < f(x) < L + \frac{\epsilon}{2|K|}, \text{ if } 0 < |x - a| < \delta'$$

If we let k be the larger of the numbers $|L - \epsilon/2|K||$ and $|L + \epsilon/2|K||$, we see that $f(x)$ satisfies
$$|f(x)| < k, \quad \text{whenever } 0 < |x - a| < \delta'$$

Now because $\lim_{x \to a} g(x) = K$, there is a number $\delta'' > 0$ such that
$$|g(x) - K| < \frac{\epsilon}{2k}, \quad \text{whenever } 0 < |x - a| < \delta''$$

Choosing δ to be the smaller of δ' and δ'', we simultaneously have all three inequalities
$$|f(x)| < k$$
$$|g(x) - K| < \frac{\epsilon}{2k}$$
and
$$|f(x) - L| < \frac{\epsilon}{2|K|}$$
whenever $0 < |x - a| < \delta$. Then
$$|f(x)g(x) - LK| = |f(x)(g(x) - K) + K(f(x) - L)|$$
$$\leq |f(x)| \cdot |g(x) - K| + |K| \cdot |f(x) - L|$$
$$< k \cdot \frac{\epsilon}{2k} + |K| \cdot \frac{\epsilon}{2|K|} = \epsilon$$

This completes the argument for the case $K \neq 0$.

Suppose that $K = 0$. We first impose a bound upon $f(x)$ by letting $\epsilon = 1$ in the definition of $\lim_{x \to a} f(x) = L$. Then we find $\delta' > 0$, so that $|f(x) - L| < 1$ whenever $0 < |x - a| < \delta'$. Thus we have

$$L - 1 < f(x) < L + 1 \qquad \text{if } 0 < |x - a| < \delta'$$

Setting k equal to the larger of the numbers $|L - 1|$ and $|L + 1|$, we have $|f(x)| < k$ whenever $0 < |A - a| < \delta'$. Now, since $\lim_{x \to a} g(x) = 0$, for any given $\epsilon > 0$, there is a number $\delta'' > 0$ such that $|g(x)| < \epsilon/k$ whenever $0 < |x - a| < \delta''$. It follows that by taking δ as the smaller of δ' and δ'', we have

$$|f(x)g(x)| = |f(x)| \cdot |g(x)| < k \cdot \frac{\epsilon}{k} = \epsilon$$

whenever $0 < |x - a| < \delta$. ∎

■ **THEOREM 2.16.** If $\lim_{x \to a} g(x) = K$ and $K \neq 0$, then

$$\lim_{x \to a} \frac{1}{g(x)} = \frac{1}{K}$$

PROOF: By writing

$$\left| \frac{1}{g(x)} - \frac{1}{K} \right| = \left| \frac{K - g(x)}{g(x)K} \right| = \frac{1}{|K| \cdot |g(x)|} \cdot |g(x) - K|$$

it is readily seen that we can control the size of $|1/g(x) - 1/K|$ by controlling the size of $|g(x) - K|$. We do this in detail.

We first make sure that $|g(x)|$ cannot be near zero as follows: Use $\frac{1}{2}|K|$ as ϵ in the definition of $\lim_{x \to a} g(x) = K$ to find $\delta' > 0$ such that $|g(x) - K| < \frac{1}{2}|K|$ whenever $0 < |x - a| < \delta'$. It will follow that

$$K - \tfrac{1}{2}|K| < g(x) < K + \tfrac{1}{2}|K| \qquad \text{if } 0 < |x - a| < \delta'$$

Letting k be the smaller of the numbers $|K - \tfrac{1}{2}|K||$, $|K + \tfrac{1}{2}|K||$ we see that $|g(x)| > k$, or $1/|g(x)| < k$, when $0 < |x - a| < \delta'$.

Then, given any $\epsilon > 0$, there is a $\delta'' > 0$ such that

$$|g(x) - K| < \epsilon \cdot k|K| \qquad \text{whenever } 0 < |x - a| < \delta''$$

If δ is the smaller of δ' and δ'', then we have

$$\left| \frac{1}{g(x)} - \frac{1}{K} \right| = \frac{1}{|K| \cdot |g(x)|} \cdot |g(x) - K|$$

$$< \frac{1}{|K| \cdot k} \cdot \epsilon \cdot k|K| = \epsilon, \qquad \text{if } 0 < |x - a| < \delta$$

and this completes the proof. ∎

■ **THEOREM 2.17.** If $\lim_{x \to a} f(x) = L$ and $\lim_{x \to a} g(x) = K$, $K \neq 0$, then

$$\lim_{x \to a} \frac{f(x)}{g(x)} = \frac{L}{K}$$

PROOF: We simply write $f(x)/g(x)$ as $f(x) \cdot 1/g(x)$ and apply Theorems 2.16 and 2.15 above. ∎

Our last result here is a very useful method of determining an unknown limit. We shall not use this result immediately, but will refer back to it several times in later chapters.

■ **THEOREM 2.18.** If $f(x) \leq g(x) \leq h(x)$ for all values of x in an interval containing the point a, and if $\lim_{x \to a} f(x) = L = \lim_{x \to a} h(x)$, then $\lim_{x \to a} g(x)$ also equals L.

PROOF: By assumption there exist numbers δ' and δ'' such that

$$|f(x) - L| < \epsilon \quad \text{wherever } 0 < |x - a| < \delta'$$

and

$$|h(x) - L| < \epsilon \quad \text{whenever } 0 < |x - a| < \delta''$$

Choosing δ to be the smaller of δ' and δ'', we have both of these inequalities simultaneously if $0 < |x - a| < \delta$. Writing these inequalities as follows

$$L - \epsilon < f(x) < L + \epsilon, \quad L - \epsilon < h(x) < L + \epsilon$$

we see from the assumed inequality on functional values that

$$L - \epsilon < f(x) \leq g(x) \leq h(x) < L + \epsilon$$
$$L - \epsilon < g(x) < L + \epsilon$$

This is equivalent to saying that $|g(x) - L| < \epsilon$ whenever $0 < |x - a| < \delta$, which completes the proof. ∎

Many problems in the next chapter involve limits. For this reason, we include just a few problems here.

PROBLEMS 2.10

1. If $f(x) = g(x)$ except for $x = a$, then prove that $\lim_{x \to a} f(x) = \lim_{x \to a} g(x)$, if either limit exists.

2. (a) If $\lim_{x \to a} f(x) = L$ and $L > 0$, then for any natural number n,
   $$\lim_{x \to a} \sqrt[n]{f(x)} = \sqrt[n]{L}$$
   (b) If $\lim_{x \to a} f(x) = L$ and $L < 0$, then for any odd natural number n,
   $$\lim_{x \to a} \sqrt[n]{f(x)} = \sqrt[n]{L}$$

3. If $\lim_{x \to a} f(x) = L$ and $L > 0$, then there is an open interval containing the point $x = a$ such that $f(x) > 0$ for all points x in the interval except (possibly) $x = a$.

4. State and prove theorems corresponding to those in this section for one-sided limits.

5. State and prove theorems corresponding to those in this section for limits as x approaches infinity.

3

THE DERIVATIVE

With the derivative we begin the study of the calculus proper. This concept is one of the most important ideas in mathematics. It has become an invaluable tool in both the physical and social sciences. In this chapter we introduce the derivative and develop its basic properties. Typical applications of the derivative will be considered in Chapter 4.

3.1 THE DERIVATIVE OF A FUNCTION

The limit concept of Section 2.10 permits a precise definition of the "rate of change of a function," as was first mentioned in Section 2.7. Suppose that the domain of a function f contains an open interval and let x be some fixed point in this open interval. We then add an amount h to x, where h may be positive or negative but not zero. Assume that $|h|$ is small enough so that $x + h$ also lies in the given open interval.

The average rate of change of f from x to $x + h$ is then

$$\frac{f(x+h) - f(x)}{x + h - x} = \frac{f(x+h) - f(x)}{h}$$

We define *the rate of change of f at the point x* to be

$$\lim_{h \to 0} \frac{f(x+h) - f(x)}{h}$$

if this limit exists as a finite number. If this limit does not exist, then the rate of change of f at x is not defined.

For the fixed value x, the ratio

$$\frac{f(x+h) - f(x)}{h}$$

depends only upon h. But this ratio does depend upon x, of course. Let us denote the ratio as a function of h and of x by writing

$$\varphi_x(h) = \frac{f(x+h) - f(x)}{h}$$

Then, for each fixed x, $\varphi_x(0)$ is undefined. However, the limit $\lim_{h \to 0} \varphi_x(h)$ may still exist as we saw in Section 2.9. If $\lim_{h \to 0} \varphi_x(h)$ exists, then in general the limit depends upon x. Therefore this limit is a new function of x called the *derived function* of f, or the *derivative* of f.

We denote the derivative of f by f'. The prime indicates that a new function has been constructed by the process of *differentiation*. By definition, then, we have

$$f'(x) = \lim_{h \to 0} \frac{f(x+h) - f(x)}{h}$$

■ *Example 1.* Let $f(x) = Ax^2 + Bx + C$. Then

$$\begin{aligned} f(x+h) - f(x) &= A(x+h)^2 + B(x+h) + C \\ &\quad - (Ax^2 + Bx + C) \\ &= Ax^2 + 2Axh + Ah^2 + Bx + Bh + C \\ &\quad - Ax^2 - Bx - C \\ &= 2Axh + Bh + Ah^2 \end{aligned}$$

Therefore
$$\frac{f(x+h)-f(x)}{h} = 2Ax + B + Ah$$

It follows easily that
$$f'(x) = \lim_{h \to 0} (2Ax + B + Ah) = 2Ax + B$$

- **Example 2.** Let $f(x) = 1/x$; $x \neq 0$. Then
$$f(x+h) - f(x) = \frac{1}{x+h} - \frac{1}{x} = \frac{x-(x+h)}{x(x+h)} = \frac{-h}{x(x+h)}$$

Thus
$$\frac{f(x+h)-f(x)}{h} = \frac{-1}{x(x+h)}$$

and
$$f'(x) = \lim_{h \to 0} \frac{-1}{x(x+h)}$$

It is easy to prove that this limit equals $-1/x^2$ by appeal to the theorems in Section 2.10.

- **Example 3.** Let $f(x) = x/(x+1)$, $x \neq -1$. Then
$$f(x+h) - f(x) = \frac{x+h}{x+h+1} - \frac{x}{x+h}$$
$$= \frac{(x+h)(x+1) - x(x+h+1)}{(x+1)(x+h+1)}$$

or
$$\frac{f(x+h)-f(x)}{h} = \frac{1}{(x+1)(x+h+1)}$$

A guess at the value of $f'(x)$ is obtained by setting $h = 0$ in this reduced form of the ratio. This gives us $f'(x) = 1/[(x+1)^2]$.

The accuracy of our guess is verified by appealing to the theorems of Section 2.10. We leave the details to the reader.

- **Example 4.** Let $f(x) = (x^2 + 1)/x$; $x \neq 0$. Then we compute
$$f(x+h) - f(x) = \frac{(x+h)^2 + 1}{x+h} - \frac{x^2 + 1}{x}$$
$$= \frac{x[(x+h)^2 + 1] - (x+h)(x^2+1)}{x(x+h)}$$
$$= \frac{hx^2 - h + h^2 x}{x(x+h)}$$

Then, for any $h \neq 0$, the ratio reduces to

$$\frac{f(x+h)-f(x)}{h} = \frac{x^2 - 1 + hx}{x(x+h)}$$

Setting $h = 0$ in this reduced ratio gives us the value $(x^2 - 1)/x^2$. Hence we expect that

$$f'(x) = \frac{x^2+1}{x^2}$$

Note that the procedure of setting $h = 0$ can only be done *after* the ratio has been reduced, that is, after h has been canceled as a factor of the denominator. Setting $h = 0$ in the reduced ratio is *not* a proof, of course. And we shall see instances later in which we cannot use the procedure at all.

If the derivative f' exists at x, that is, if

$$\lim_{h \to 0} \frac{f(x+h)-f(x)}{h}$$

exists, then we say that the function f is *differentiable at* x. Then if $f'(x)$ exists at each point of an open interval, we say that f is *differentiable on the interval*. Our interest is primarily in differentiable functions, but the reader should be warned that there are nondifferentiable functions.

■ **Example 5.** Let $f(x) = x^{2/3} = \sqrt[3]{x^2}$. Then $f(x)$ is defined at each point x but has no derivative at $x = 0$. We determine this fact as follows: For $h \neq 0$,

$$\frac{f(0+h)-f(0)}{h} = \frac{\sqrt[3]{h^2}-0}{h} = \frac{1}{\sqrt[3]{h}}$$

Clearly, we have no limit of $1/\sqrt[3]{h}$ as h approaches zero. Indeed, $\lim_{h \to 0} h^{-1/3} = \infty$. Therefore $f(x) = x^{2/3}$ has no derivative at $x = 0$.

The first application of the derivative was seen in Section 2.8. There we introduced a technique for drawing tangent lines to the graph of a function. The reader will easily see that we were really using the derivative in that discussion. A direct restatement of the tangent line technique of Section 2.8 is as follows.

Let f be differentiable at the point x_0. Then the line

$$\{(x,y) : y = f'(x_0)(x - x_0) + f(x_0)\}$$

is the tangent line at $(x_0, f(x_0))$ to the graph $\{(x,y) : y = f(x)\}$.

More generally, two differentiable functions f and g are said to be *tangent at the point* x_0 if (1) $f(x_0) = g(x_0)$ and (2) $f'(x_0) = g'(x_0)$. This simply means that the graphs $\{(x,y) : y = f(x)\}$ and $\{(x,y) : y = g(x)\}$ share the same tangent line at their common point $(x_0, f(x_0)) = (x_0, g(x_0))$.

The definition of the tangency of two functions agrees with our definition of a tangent line. In other words, the differentiable function f and the linear function $l(x) = f'(x_0)(x - x_0) + f(x_0)$ are tangent at x_0, because (1) $f(x_0) = l(x_0)$ and (2) $f'(x_0) = l'(x_0)$. (This latter fact can be verified by the reader.)

■ **Example 6.** The functions $f(x) = x^2$ and $g(x) = -x^2 + 4x - 2$ are tangent at $x = 1$.

PROOF:
1. $f(1) = 1^2 = 1$ and $g(1) = -(1)^2 + 4(1) - 2 = 1$
2. From Example 1, $f'(x) = 2x$ and $g'(x) = -2x + 4$. Then

$$f'(1) = 2(1) = 2 \quad \text{and} \quad g'(1) = -2(1) + 4 = 2$$

The graphs of these functions are shown in Fig. 3.1. ■

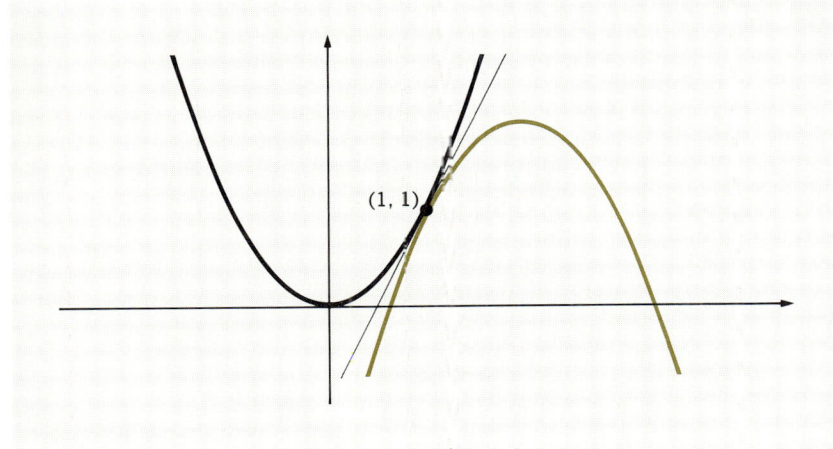

FIGURE 3.1

■ **Example 7.** Let $f(x) = x^2$ and $g(x) = -x^2 + ax + b$. Suppose we want to choose the coefficients a and b, so that f and g are tangent at $x = 2$. Then we want (1) $f(2) = g(2)$ and (2) $f'(2) = g'(2)$. From (1),

$$f(2) = 4 = g(2) = -4 + 2a + b \quad \text{or} \quad 2a + b = 8$$

Knowing from Example 1 that $f'(x) = 2x$ and $g'(x) = -2x + a$, condition (2) provides the equation

$$f'(2) = 4 = g'(2) = -4 + a \quad \text{or} \quad a = 8$$

The equations $a = 8$, $2a + b = 8$ tell us that $b = -8$. Therefore the function $f(x) = -x^2 + 8x - 8$ is tangent to $f(x) = x^2$ at $x = 2$.

EXERCISES 3.1

1. Find the derivative of each of the following functions and then determine the domain in which the function is differentiable.
    (a) $f(x) = 3x - 1$
    (b) $f(x) = 4 - 2x$
    (c) $f(x) = x^2 - 2x + 3$
    (d) $f(x) = 6 - 4x - x^2$

(e) $f(x) = x^3 + 2x + 1$ (f) $f(x) = x^3 - 4x^2 + 4x - 1$

(g) $f(x) = \dfrac{x - 2}{x}$ (h) $f(x) = \dfrac{2x + 1}{x^2}$

(i) $f(x) = \sqrt{2x + 3}$ (j) $f(x) = \sqrt{x^2 + 2x + 2}$

2. For each of the following functions, draw both of the graphs $\{(x,y) : y = f(x)\}$ and $\{(x,y) : y = f'(x)\}$ on the same coordinate system.
   (a) $f(x) = 2x + 3$ (b) $f(x) = x^2 - 3x + 1$
   (c) $f(x) = \frac{1}{3}x^3 + 2x$ (d) $f(x) = x^4 - 4x^2 + 4$

3. Prove that the given functional equations are satisfied by the corresponding functions.
   (a) If $f(x) = x^2 + 2$, then $xf'(x) + 4 = 2f(x)$.
   (b) If $f(x) = 1/x$, then $xf'(x) + f(x) = 0$.
   (c) If $f(x) = \dfrac{x}{x + 2}$, then $2f^2(x) = x^2 f'(x)$.
   (d) If $f(x) = \sqrt{2x - 1}$, then $f(x)f'(x) = 1$.

4. Prove that the graphs of each of the following pairs of functions are tangent at the given point.
   (a) $f(x) = 2 - 3x$, $g(x) = x^2 - 5x + 3$; $(1, -1)$
   (b) $f(x) = x^2 + 1$, $g(x) = -1 - 4x - x^2$; $(-1, 2)$
   (c) $f(x) = x^2$, $g(x) = x^3 - 2x^2 - 4$; $(2, 4)$
   (d) $f(x) = x^3 - 3x^2 + x$, $g(x) = \dfrac{x}{x + 1}$; $(0, 0)$

PROBLEMS 3.1
Mathematics

The *right derivative* and *left derivative* of $f(x)$ at a point $x = a$ are defined by

$$f'_+(a) = \lim_{h \to 0^+} \frac{f(a + h) - f(a)}{h}$$

$$f'_-(a) = \lim_{h \to 0^-} \frac{f(a + h) - f(a)}{h}$$

1. Prove that f is differentiable at $x = a$ if and only if both $f'_+(a)$ and $f'_-(a)$ exist and $f'_+(a) = f'_-(a)$.

2. Show that $f(x) = |x|$ has right and left derivatives at $x = 0$, but is not differentiable at $x = 0$.

3. Show that $f(x) = x|x|$ is differentiable everywhere.

4. Let f be defined by setting

$$f(x) = x \quad \text{if} \quad x = \frac{1}{n}, n \in \mathbf{N}$$

$$= 0 \quad \text{if} \quad x \text{ is not of the form } \frac{1}{n}$$

Is this function differentiable at $x = 0$?

5. Suppose that a function f is defined for all $x \in \mathbf{R}$. Assume that this function has the three properties
   1. $f(a + b) = f(a) \cdot f(b)$ for any pair $a, b \in \mathbf{R}$
   2. $f(0) = 1$, and
   3. $f'(0)$ exists

Prove that f is differentiable everywhere by showing that
$$f'(x) = f'(0) \cdot f(x)$$

The *total cost* of producing x units of a commodity is often considered to be a function $C(x)$ of x alone. The *average unit cost* would then be given by $C(x)/x$. If the number of units produced is changed to $x + h$, then the average rate of change of cost is

$$\frac{C(x+h) - C(x)}{x + h - x}$$

The limit of this ratio as h approaches zero is called the *marginal cost*. That is, the marginal cost is the derivative $C'(x)$, which we think of as the rate of change of the cost per unit change in output.

PROBLEMS 3.1
Economics

1. A manufacturer finds that, if he makes x units per day of his product, his costs consist of (1) a fixed cost of \$200 per day, (2) a cost of 10 cents per unit, and (3) additional costs of maintenance, and so forth, of $2x^2$ cents per day. What is his average cost per unit and his marginal cost? What level of production makes the average unit cost a minimum?

2. A wholesale dealer has discovered that his fixed warehouse costs amount to \$60,000 per year, and that if his inventory averages x units, then his storage costs are $20x + 8x^3$ cents per year. What is his average unit cost and his marginal cost? What inventory x minimizes the average unit cost?

   Let x denote the number of units of a certain commodity demanded of a producer by the market and let $p(x)$ denote the unit price corresponding to the demand x. [We assume that the price $p(x)$ is a function of the demand, as is obvious.] Then the *total revenue* to the producer (from this commodity) is

   $$R(x) = xp(x)$$

   and the rate of change $R'(x)$ of total revenue with respect to demand is called the *marginal revenue*. We consider it as the rate of change in total revenue due to a unit increase in demand. (See Section 6.6.)

3. Suppose that the price of a commodity with demand x is

   $$p(x) = 15 - 2x + \frac{1}{x}$$

   Determine total and marginal revenue and interpret the results.

4. In Problem 1, suppose that $p(x) = 1650 - 5x$ cents. Determine the total and marginal revenue and the *net profit* $= R(x) - C(x)$. What level of production maximizes net profit?

PROBLEMS 3.1
Physical Science

1. The force attracting two unlike magnetic poles at a distance x apart is given by
   $$F = \frac{k}{x^2}$$
   where k is a positive constant. At what rate does F change with respect to x when $x = 4$?

2. A 16-ft ladder leans against a vertical wall. The lower end rests on a horizontal floor. Now the lower end is pulled away from the wall at the rate of 3 ft per sec. At what rate does the top end slide down the wall when the lower end is (a) 4 ft and (b) 8 ft from the wall? Over what time interval does the top end move faster than the bottom?

3. Two ships are sailing along lines that intersect at an angle of 60°. The first ship, sailing at 15 knots, passes the point of intersection of their paths at 11:00 A.M. The second ship sails at 12 knots and passes this point at 12:00 noon. At t hours past noon, what is the rate of change of the distance between the ships? (There should be two answers. Why?)

4. Flour is being poured onto a flat surface at the rate of 2 cu ft per min. It forms a conical pile whose base diameter is always $2\frac{1}{2}$ times its height. How fast is the height increasing 10 min after the flour starts to pour? How long does it take to make a pile 2 ft high and how fast is the height of the pile increasing when it is 2 ft high?

3.2 EXTREME VALUES OF A FUNCTION

The derivative finds many applications to practical problems. One of the simplest applications consists in locating the extreme values of a function. We start with some examples.

■ *Example 1.* A movie theater has 480 seats. The manager has found by experience that the number x of tickets sold for each performance is related to the cost p in dollars of each ticket by the formula

$$p = \left(\frac{3}{2} - \frac{x}{800}\right)^2$$

If x tickets are sold each at the price p, then the total revenue r is

$$r = xp = x\left(\frac{3}{2} - \frac{x}{800}\right)^2 = \frac{9}{4}x - \frac{3}{800}x^2 + \frac{1}{640000}x^3$$

In Fig. 3.2 the graph $\{(x, r) : r = xp\}$ is sketched. The curve is dashed after $x = 480$, indicating that the curve is not meaningful to us for $x > 480$.

By finding the highest point on this curve in the closed interval $\{x : 0 \leq x \leq 480\}$ the maximum total revenue, the theater manager knows the number x of tickets he should sell and hence the price p that should be charged for each ticket.

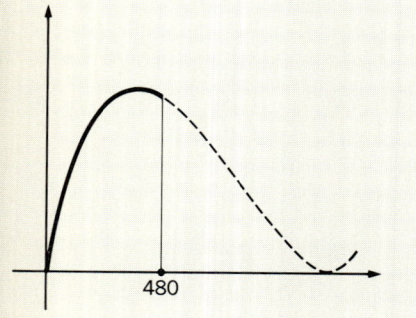

FIGURE 3.2

■ *Example 2.* The strength S of a wooden beam of given length is known to vary directly as the product of its width x and the square of its depth y. That is, the formula is

$$S = kxy^2$$

where k is a factor of proportionality depending upon the length of the beam, the kind of wood, the orientation of the grain of the wood, and so forth.

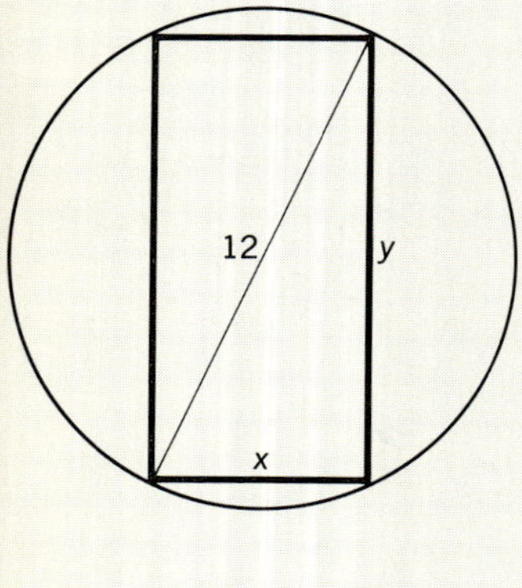

Suppose there is a 10-ft log 12 in. in diameter available. What are the dimensions of the strongest beam (10 ft long) that can be cut from this log? Assuming that the log has circular cross section, we should cut a beam shown in cross section in Fig. 3.3. In view of the Pythagorean theorem, the relation $x^2 + y^2 = 144$ holds. Hence the strength S of the beam can be expressed as a function of x

FIGURE 3.3

$$S = kxy^2 = kx(144 - x^2)$$

3.2 EXTREME VALUES OF A FUNCTION

The practical problem has been reduced to finding the maximum value of the function $kx(144 - x^2)$ over the interval $\{x : 0 \leq x \leq 12\}$. (For values of x not in this interval, the function has no practical meaning to us.)

This section, then, introduces some basic tools for solving such problems.

Increasing and decreasing functions. A function f is said to be *increasing over an interval* if, for any two points x_1, x_2 of the interval such that $x_1 < x_2$, we have $f(x_1) < f(x_2)$. Similarly, f is *decreasing over the interval* if $x_1 < x_2$ implies $f(x_1) > f(x_2)$.

The following result shows that the derivative can help in determining whether a function is increasing or decreasing.

■ **THEOREM 3.1.** Let f be a differentiable function. Then f is increasing over any interval on which $f'(x)$ is always positive and f is decreasing over any interval on which $f'(x)$ is always negative.

The proof of Theorem 3.1 will be given in Section 3.13. (We include the theorem here to allow us to get to some applications rapidly.) Intuitively, however, this theorem says that, if the tangent line to $\{(x,y) : y = f(x)\}$ always has positive slope, then the function is increasing. At the very least, this sounds plausible.

Let f be a differentiable function. A point x_0 for which

$$f'(x_0) = 0$$

is called a *critical value* in the domain of f. The function is said to be *stationary* at such a critical value and the point $(x_0, f(x_0))$ is called a *critical point* on the graph of f. By our definition, the tangent line to the graph $\{(x,y) : y = f(x)\}$ at a critical point $(x_0, f(x_0))$ has slope zero, hence this tangent line is parallel to the x axis. A graph can exhibit this behavior in several ways. The function whose graph appears in Fig. 3.4 has critical values x_1, x_2, x_3.

A function f has a *relative maximum value* at x_0 if there is an interval $\{x : a < x < b\}$ containing x_0 such that $f(x_0) - f(x) \geq 0$ for all points x in the interval. Similarly, f has a *relative minimum value* at x_1 if there is an interval $\{x : c < x < d\}$ containing x_1 such that $f(x_0) - f(x) \leq 0$ for all x in the interval. The term *relative extremum* covers both situations.

The adjective "relative" as used in the definition just given means only that $f(x_0)$ is an extremum value with respect to functional values at nearby points. For instance, the function shown in Fig. 3.5(a) has three relative maximum values, only one of which is the absolute maximum value of $f(x)$. The function pictured in Fig. 3.5(b) has a relative maximum but has no absolute maximum at all.

Suppose that $f(x_0)$ is a relative maximum value of f and that $x < x_0$ is a point in the interval $\{x : a < x < b\}$. Then the line through the two points $(x_0, f(x_0))$ and $(x, f(x))$ has slope

$$\frac{f(x_0) - f(x)}{x_0 - x}$$

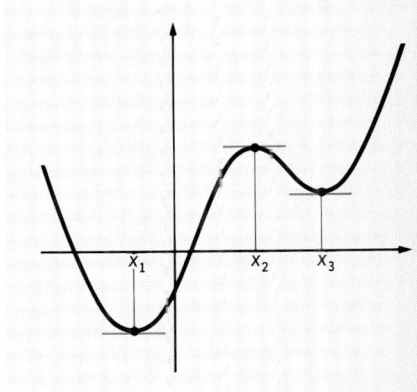

FIGURE 3.4

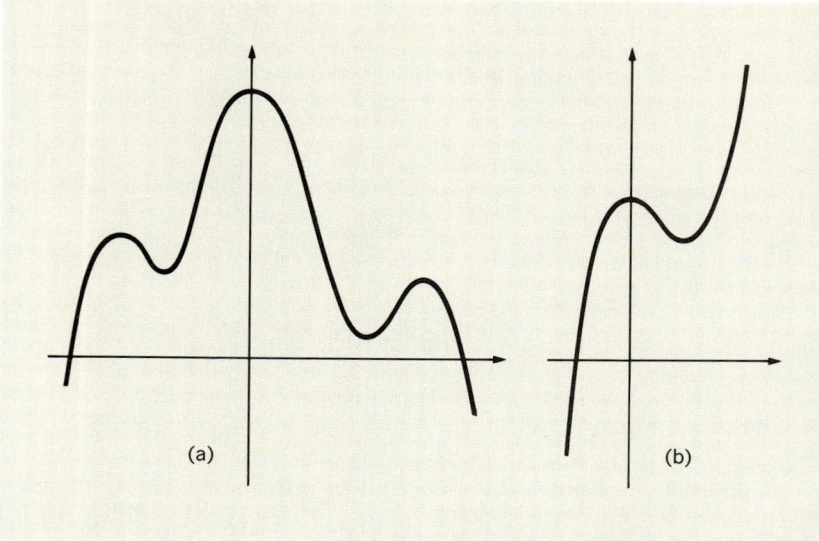

FIGURE 3.5   (a)   (b)

Because $x < x_0$ and $f(x) < f(x_0)$, this slope is positive. On the other side of x_0, if $x_0 < x < b$, then $x_0 < x$, while $f(x) < f(x_0)$. Hence the slope is negative.

■ **THEOREM 3.2.** The function f has a relative maximum value at x_0 if there is an interval $\{x : a < x < b\}$ containing x_0 such that the line through $(x_0, f(x_0))$ and $(x, f(x))$ has positive slope if $a < x < x_0$, and has negative slope if $x_0 < x < b$. Similarly, f has a relative minimum value at x_0 if the slope is negative when $a < x < x_0$ and is positive when $x_0 < x < b$.

PROOF: The slope of the line in question is

$$\frac{f(x_0) - f(x)}{x_0 - x}$$

If this ratio is positive for $a < x < x_0$, then because the denominator is positive, so is the numerator, that is, $x < x_0$ implies $f(x_0) - f(x) > 0$. If this ratio is negative when $x_0 < x < b$, then, because the denominator is negative, the numerator must be positive and again $f(x_0) - f(x) > 0$. Thus if $x \neq x_0$, $f(x_0) - f(x) > 0$. Analogous arguments prove the second part of the theorem. ■

Note that Theorem 3.2 does not *locate* extremum values, although it does decide whether or not a suspected value *is* extremum and hence is useful. Our next result yields a method for locating extremum values.

■ **THEOREM 3.3.** Let the function f have a relative extremum value at x_0. If f is differentiable at x_0, then $f'(x_0) = 0$.

3.2 EXTREME VALUES OF A FUNCTION

PROOF: First suppose that $f(x_0)$ is a relative maximum value and consider the ratio

$$\frac{f(x_0 + h) - f(x_0)}{h}$$

We have assumed that the limit

$$\lim_{h \to 0} \frac{f(x_0 + h) - f(x_0)}{h} = f'(x_0)$$

exists. Now $f'(x_0)$ cannot be positive because $f(x_0 + h) - f(x_0)$ is always negative, whereas if h is positive, this ratio is negative. A set of negative numbers cannot have a positive limit, so $f'(x_0) \leq 0$. Similarly, $f'(x_0)$ cannot be negative because for negative values of h the ratio $[f(x_0 + h) - f(x_0)]/h$ is positive. A set of positive numbers cannot have a negative limit hence $f'(x_0) \geq 0$. It follows that $f'(x_0)$ must be zero.

A precise argument is given by contradiction. Suppose that $f'(x_0) \neq 0$ and assume first that $f'(x_0) > 0$. Define $\epsilon = \frac{1}{2}f'(x_0) > 0$. By the definition of a limit, there is a positive number δ such that, whenever $0 < |h| < \delta$, we have

$$\left| \frac{f(x_0 + h) - f(x_0)}{h} - f'(x_0) \right| < \frac{1}{2}f'(x_0)$$

or

$$-\frac{1}{2}f'(x_0) < \frac{f(x_0 + h) - f(x_0)}{h} - f'(x_0) < \frac{1}{2}f'(x_0)$$

or

$$\frac{1}{2}f'(x_0) < \frac{f(x_0 + h) - f(x_0)}{h} < \frac{3}{2}f'(x_0)$$

Thus for any h satisfying $0 < |h| < \delta$, the ratio $[f(x_0 + h) - f(x_0)]/h$ is positive. But this is false, because any positive value of h makes this ratio negative. We have here the contradiction proving that $f'(x_0)$ cannot be positive. A similar argument to show that $f'(x_0)$ cannot be negative should be supplied by the reader. Then it will follow that $f'(x_0) = 0$. The case of a relative minimum value is treated in the same way with some obvious changes in the direction of the inequalities used. The details are also left to the reader as an exercise.

To use Example 1 again, the total revenue r from a certain theater was given as a function of the number x of tickets sold,

$$r(x) = \frac{9}{4}x - \frac{3}{800}x^2 + \frac{1}{640{,}000}x^3; \qquad 0 \leq x \leq 480$$

The reader should verify that

$$r'(x) = \frac{9}{4} - \frac{3}{400}x + \frac{3}{640,000}x^2$$

$$= \frac{3}{640,000}(x^2 - 1600x + 480,000)$$

$$= \frac{3}{640,000}(x - 400)(x - 1200)$$

Therefore $r'(x) = 0$ if and only if $x = 400$ or $x = 1200$. The value $x = 1200$ is not in the domain of $r(x)$, so we ignore it. We suspect that $r(x)$ has a maximum value at $x = 400$.

A proof that $r(400)$ is a maximum value entails proving that

$$r(400) - r(x) > 0$$

for any point $x \neq 400$. Because $r(400) = 400$, we must show that

$$400 - \frac{9}{4}x + \frac{3}{800}x^2 - \frac{1}{640,000}x^3 > 0$$

for any x near 400. Multiplying by 640,000, this inequality becomes

$$256,000,000 - 1,440,000x + 2400x^2 - x^3 > 0$$

or

$$(400 - x)^2(1600 - x) > 0$$

Now this expression $(400 - x)^2(1600 - x)$ *is* positive for any point in $\{x : 0 \leqq x \leqq 480\}$. Hence $x_0 = 400$ is a critical value.

Lastly, the price p for each ticket should be

$$p = \left(\frac{3}{2} - \frac{x_0}{800}\right)^2 = \left(\frac{3}{2} - \frac{400}{800}\right)^2 = 1$$

That is, the theater manager sets the ticket price at \$1 if he wishes to maximize his total revenue.

To use Example 2 again, the strength S of a beam cut from a circular log of diameter 12 in. was expressed as a function of its width x,

$$S(x) = kx(144 - x^2) \qquad 0 \leqq x \leqq 12$$

The reader should verify that

$$S'(x) = k(144 - 3x^2)$$

Then $S'(x) = 0$ if and only if $3x^2 = 144$ or $x = \pm 4\sqrt{3}$. The number $-4\sqrt{3}$ is not in the domain of our function S, so we suspect that S has a relative maximum value at $x_0 = 4\sqrt{3}$. Now

$$S(4\sqrt{3}) = 384k\sqrt{3}$$

Then $384k\sqrt{3} - kx(144 - x^2) = k(x - 4\sqrt{3})^2(x + 8\sqrt{3})$. For values of x near $4\sqrt{3}$, we do have $S(4\sqrt{3}) - S(x) > 0$ and hence we have found the desired maximum value. ∎

3.2 EXTREME VALUES OF A FUNCTION

EXERCISES 3.2

1. Determine those intervals and rays over which each of the following functions is (1) increasing and (2) decreasing. In each case find all of the stationary values and draw the graph.
   (a) $f(x) = x^2 - 3x - 4$
   (b) $f(x) = x^2 - x + 2$
   (c) $f(x) = 3 - 2x - x^2$
   (d) $f(x) = 5 - 6x - x^2$
   (e) $f(x) = 2x^3 - 3x^2 - 12x + 4$
   (f) $f(x) = x^3 - x^2 + 4x$
   (g) $f(x) = 3x^4 - 8x^3 - 6x^2 + 24x - 10$
   (h) $f(x) = 3x^5 - 25x^3 + 60x$
   (i) $f(x) = \dfrac{x-3}{2x+1}$
   (j) $f(x) = \dfrac{x^2}{x+1}$

2. Locate and identify the relative maximum and minimum values of each of the following functions. Draw the graph in each case.
   (a) $f(x) = 2x^2 - 3x$
   (b) $f(x) = x^2 - 4x + 4$
   (c) $f(x) = 6x - x^2$
   (d) $f(x) = 1 - 4x - x^2$
   (e) $f(x) = x^3 - 4x^2 + 2x + 1$
   (f) $f(x) = 4x - \frac{1}{3}x^3$
   (g) $f(x) = 4x^4 + 8x^3 + 6x^2 + 2x + \frac{1}{4}$
   (h) $f(x) = x^5 - 5x$
   (i) $f(x) = \dfrac{x}{x+1}$
   (j) $f(x) = \sqrt{2x^2 + 3}$

3. Prove that $f(1)$ is a relative maximum value for each of the following functions.
   (a) $f(x) = 3 + 2x - x^2$
   (b) $f(x) = 3x - x^3$
   (c) $f(x) = x^4 - 4x^3 + 4x^2$
   (d) $f(x) = \dfrac{1 - 2x}{2x^2}$

4. Prove that $f(-1)$ is a relative minimum value for each of the following functions.
   (a) $f(x) = x^2 + 2x - 3$
   (b) $f(x) = 3x - x^3$
   (c) $f(x) = 3x^4 + 4x^3 - 6$
   (d) $f(x) = \sqrt{x^2 + 2x + 4}$

5. Determine constants a and b so that the given function shall have a relative maximum value at the given point:
   (a) $f(x) = a + bx - x^2$; $(-1, 2)$
   (b) $f(x) = a + bx - x^3$; $(-1, 0)$
   (c) $f(x) = a + bx^3 - 3x^4$; $(1, 3)$
   (d) $f(x) = \sqrt{a + bx + x^2}$; $(1, 2)$

6. Prove that
$$f(x) = 8x^3 - 36x^2 + 54x - 25$$
has no relative extremum values.

7. Find the maximum *slope* of the graph of $f(x) = 24x^2 - x^4$; $x \geq 0$.

8. Find both the maximum and minimum slopes of the graph of
$$f(x) = \dfrac{x^2 - 1}{x^2 + 1}$$

PROBLEMS 3.2

1. Prove that the rectangle having maximum area for a given fixed perimeter is a square.

2. Find the rectangle of maximum area that can be inscribed in a given semicircle.

3. Consider the graph $\{(x, y) : y = ax^3 + bx^2 + cx + d, a > 0\}$. What conditions must the coefficients a, b, c, and d satisfy in order that the graph have (a) two distinct critical points, (b) one critical point and (c) no critical points? In case (b) what can you say about the critical point?

4. The sum of three numbers is 40. The first number plus three times the second plus four times the third is 80. Choose the numbers so that their product is a maximum.

5. An open-topped rectangular box is to be made from a piece of sheet metal 8 in. wide and 15 in. long by cutting a square from each corner and bending up the sides. What are the dimensions of the box of largest volume that can be made in this way?

6. A building in the form of a rectangle 108 by 256 ft is located at the intersection of two perpendicular roads. Find the length of the shortest straight path from one road to the other passing behind the building.

7. A rectangular lot is to be fenced off against the wall of a building. There is to be a 24-ft opening in the front of the fence. If side fencing costs $1 per ft and front fencing costs $1.50 per ft, what are the dimensions of the largest lot that can be so fenced for $300?

8. Two posts are placed 15 ft apart. One post is 8 ft high and the other is 12 ft high. A guy wire is attached to the top of each post and to a stake at ground level between them. Where should the stake be placed in order to use the least length of wire?

9. A billboard is to display 80 sq ft of advertising material with borders of 5 ft top and bottom and 4 ft on the sides. Find the dimensions if the total area of the billboard is to be minimized.

10. Postal regulations require that a package sent parcel post be such that the combined length and girth shall not exceed 60 in. What dimensions should be used to construct a box with a square cross section which will comply with this regulation and have maximum volume?

More applications of extremums are found after Section 3.9.

3.3 CONTINUITY

In the limit definition

$$f'(x) = \lim_{h \to 0} \frac{f(x+h) - f(x)}{h}$$

we pointed out that the ratio $[f(x+h) - f(x)]/h$ is not defined when $h = 0$. Now we emphasize that the limit $f'(x_0)$ can only exist if the numerator $f(x+h) - f(x)$ approaches zero as the denominator h approaches zero.

■ **THEOREM 3.4.** If f is differentiable at x_0, then

$$\lim_{h \to 0} [f(x_0 + h) - f(x_0)] = 0$$

PROOF: Intuitively, we argue that, if the numerator of the ratio $[f(x_0 + h) - f(x_0)]/h$ were to approach a number L other than zero, then the entire ratio would approach $L/0$ and hence would not have a limit.

The precise proof is made as follows: Given $\epsilon > 0$, we must find $\delta > 0$ such that

$$|f(x_0 + h) - f(x_0)| < \epsilon \qquad \text{whenever } 0 < |h| < \delta$$

Now since $\lim_{h\to 0} [f(x_0 + h) - f(x_0)]/h = f'(x_0)$, there is a $\delta' > 0$ such that

$$\left|\frac{f(x_0 + h) - f(x_0)}{h} - f'(x_0)\right| < \epsilon \quad \text{whenever } 0 < |h| < \delta'$$

This inequality is equivalent to

$$-\epsilon < \frac{f(x_0 + h) - f(x_0)}{h} - f'(x_0) < \epsilon$$

or

$$f'(x_0) - \epsilon < \frac{f(x_0 + h) - f(x_0)}{h} < f'(x_0) + \epsilon$$

If $h > 0$, multiply by h to obtain

$$h[f'(x_0) - \epsilon] < f(x_0 + h) - f(x_0) < h[f'(x_0) + \epsilon]$$

and if $h < 0$, multiplication by h results in

$$h[f'(x_0) - \epsilon] > f(x_0 + h) - f(x_0) > h[f'(x_0) + \epsilon]$$

In either case, let k be the larger of the two numbers $|f'(x_0) - \epsilon|$, $|f'(x_0) + \epsilon|$. Lastly, choose δ to be the smaller of the numbers δ' and ϵ/k. Then for any h satisfying $0 < |h| < \delta \leq \epsilon/k$, we have

$$|f(x_0 + h) - f(x_0)| < hk < \frac{\epsilon}{k} \cdot k = \epsilon \quad \blacksquare$$

The property of a differentiable function just established in Theorem 3.4 has a name and is an important concept in analysis. A function f is said to be *continuous at a point* x_0 if

$$\lim_{h\to 0} [f(x_0 + h) - f(x_0)] = 0$$

or, equivalently, if

$$\lim_{x\to x_0} f(x) = f(x_0)$$

The second defining equation actually implies that three conditions are satisfied

1. $f(x_0)$ exists, that is, f is defined at x
2. $\lim_{x\to x_0} f(x)$ exists
   and
3. the two numbers are equal

Because most of the functions we have considered are continuous, we look at three counterexamples first.

The function $f(x) = 1/x$ cannot be continuous at $x = 0$, because it is not defined at $x = 0$. However, this function is continuous at each point $x \neq 0$.

The function
$$f(x) = 0 \quad \text{if } x < 0$$
$$f(x) = 1 \quad \text{if } x \geq 0$$
is defined at $x = 0$, but $\lim_{x \to 0} f(x)$ does not exist. (Give a proof of this.) Hence this function is not continuous at $x = 0$.

The function with values
$$f(x) = \frac{x^2 - 1}{x - 1} \quad \text{if } x \neq 1$$
$$f(1) = 1$$
has $f(1) = 1$ while
$$\lim_{x \to 1} \frac{x^2 - 1}{x - 1} = 2$$
Hence this function is not continuous at $x = 1$.

Each of the examples above has a *discontinuity*, a point at which it is not continuous.

Theorem 3.4 may be restated as follows: *If f is differentiable at x_0, then f is continuous at x_0.* Briefly, "differentiable" implies "continuous." The converse implication is false. For example, we showed in Example 5 of Section 3.1 that $f(x) = x^{2/3}$ is *not* differentiable at $x = 0$. But it is easy to prove that $\lim_{x \to 0} x^{2/3} = 0 = 0^{2/3}$ so this function *is* continuous at $x = 0$.

A function is *continuous on an open interval* if it is continuous at each point of the interval.

THEOREM 3.5. A constant function $f(x) = k$ is continuous everywhere.

PROOF: For any point x, we have
$$\lim_{h \to 0} [f(x + h) - f(x)] = \lim_{h \to 0} (k - k) = \lim_{h \to 0} 0 = 0$$

The graph of a constant function $f(x) = k$ is the horizontal line $\{(x, y) : y = k\}$, which has slope zero at each point. Hence the next result should be expected. ∎

THEOREM 3.6. A constant function $f(x) = k$ has a derivative $f'(x) = 0$ for all x.

PROOF: The ratio whose limit is $f'(x)$ is
$$\frac{f(x + h) - f(x)}{h} = \frac{k - k}{h} = \frac{0}{h} = 0$$
and again $\lim_{h \to 0} 0 = 0$. ∎

In a later theorem we prove the converse of 3.6, namely, if the derivative of a function is zero everywhere, then that function is constant.

A constant function $f(x) = k$ can also be denoted just by the letter k. Thus such symbols as $0, 2, \sqrt{5}, \pi$, and so forth, may denote either numbers or constant functions. The proper interpretation should be obvious from the context in most cases. We shall say "the number 2" or "the function 2," as the case may be, if there seems to be a possibility of misinterpretation.

The *identity function* is that function with values $f(x) = x$. Most often we denote this function simply by the letter x, following the convention of denoting a function by its functional values alone. However, in certain contexts we need a special symbol for the identity function and we have chosen the capital letter E. Thus we define

$$E(x) = x$$

■ **THEOREM 3.7.** The identity function E is continuous everywhere and its derivative is the constant function 1.

PROOF: In view of Theorem 3.4, we need only prove the second half of the statement. If x is any real number, then we have

$$\frac{E(x+h) - E(x)}{h} = \frac{x+h-x}{h} = \frac{h}{h} = 1$$

Then $\lim_{h \to 0} 1 = 1$, which completes the proof. ■

Our next result of this section has important theoretical value, which will be cited later. Its proof is an exercise in the use of the precise definition of continuity.

■ **THEOREM 3.8.** Let f be continuous on the interval $\{x : a < x < b\}$ and suppose that $f(x_0) > 0$ for some point x_0 in the interval. Then there is an open interval $\{x : x_0 - \delta < x < x_0 + \delta\}$ at each point of which $f(x)$ is positive.

PROOF: In the definition of the limit $\lim_{x \to x_0} f(x) = f(x_0)$ for continuity, we choose $\epsilon = f(x_0) > 0$. By assumption, there exists $\delta > 0$ such that

$$|f(x) - f(x_0)| < f(x_0) \quad \text{whenever } |x - x_0| < \delta$$

Rewriting this, we have

$$-f(x_0) < f(x) - f(x_0) < f(x_0) \quad \text{whenever } -\delta < x - x_0 < \delta$$

or

$$0 < f(x) < 2f(x_0) \quad \text{whenever } x_0 - \delta < x < x_0 + \delta$$

The first part of this inequality proves the theorem. ■

Our last theorem here constitutes some justification for using sequences to compute limits as we did in Section 2.9, for instance.

■ **THEOREM 3.9.** If f is continuous at x_0, then $\lim_{n \to \infty} f(x_0 + 1/n) = f(x_0)$.

PROOF: Given $\epsilon > 0$, we must find $n_0 \in \mathbf{N}$ such that

$$\left| f\left(x_0 + \frac{1}{n}\right) - f(x_0) \right| < \epsilon, \quad \text{whenever } n \geq n_0$$

But by continuity at x_0, there is a number $\delta > 0$ such that

$$\left| f\left(x_0 + \frac{1}{n}\right) - f(x_0) \right| < \epsilon, \quad \text{whenever } \left| x_0 + \frac{1}{n} - x_0 \right|$$
$$= \frac{1}{n} < \delta$$

Hence if we choose n_0 to be any natural number larger than $1/\delta$, we shall have $1/n \leq 1/n_0 < \delta$ whenever $n \geq n_0$.

The converse of Theorem 3.9 is *not* true. That is, if $\lim_{n \to \infty} f(x_0 + 1/n) = f(x_0)$, it need not follow that f is continuous at x_0. A simple counterexample is the function

$$f(0) = 0$$
$$f\left(\frac{1}{n}\right) = 0, \, n \in \mathbf{N}$$
$$f(x) = 1 \quad \text{otherwise}$$

Clearly, $\lim_{n \to \infty} f(0 + 1/n) = 0 = f(0)$ but f is not continuous at x_0. (In an advanced course in analysis, the reader may see a kind of converse to Theorem 3.9: If $\lim_{n \to \infty} f(x_n) = f(x_0)$ for each sequence $\varphi(n) = x_n$ approaching x_0, then f is continuous at x_0.) ■

EXERCISES 3.3

1. Prove that each of the following functions is continuous at $x = 1$ by an explicit application of the definition. That is, given any $\epsilon > 0$, find $\delta > 0$ such that $|f(x) - f(1)| < \epsilon$ whenever $|x - 1| < \delta$.
   (a) $f(x) = 3x + 4$
   (b) $f(x) = 1 - 2x$
   (c) $f(x) = x^2$
   (d) $f(x) = 3x - x^2$
   (e) $f(x) = x^3$
   (f) $f(x) = -2 + 3x + x^2 - x^3$
   (g) $f(x) = \dfrac{x}{x+1}$
   (h) $f(x) = \dfrac{x^2}{x+1}$
   (i) $f(x) = \sqrt{x}$
   (j) $f(x) = \sqrt{x^2 + 2x}$

2. Show that each of the following functions is continuous everywhere by proving that $\lim_{h \to 0} [f(x + h) - f(x)] = 0$ for any value of x.
   (a) $f(x) = 5x - 3$
   (b) $f(x) = 3 - \frac{1}{2}x$
   (c) $f(x) = x^2 - 2$
   (d) $f(x) = 4 - 2x - x^2$
   (e) $f(x) = x^3 - 3x$
   (f) $f(x) = 4 - 4x^2 - x^4$
   (g) $f(x) = \dfrac{x}{x^2 + 1}$
   (h) $f(x) = \sqrt{x^2 + 1}$
   (i) $f(x) = \dfrac{1}{\sqrt{x^2 + 1}}$
   (j) $f(x) = \sqrt[3]{x}$

3. For what values of x is each of the following functions discontinuous?
   (a) $f(x) = \dfrac{x-1}{x}$
   (b) $f(x) = \dfrac{3}{x+2}$
   (c) $f(x) = \dfrac{1}{x^2 - 2x - 3}$
   (d) $f(x) = \dfrac{x^2 - 4}{x - 2}$
   (e) $f(x) = \dfrac{x^2 + 3x - 1}{x^2 - 3x}$
   (f) $f(x) = \dfrac{x^2 + 1}{x^3 - x}$
   (g) $f(x) = \sqrt{x+1}$
   (h) $f(x) = \sqrt{x^2 - 1}$
   (i) $f(x) = \dfrac{1}{\sqrt{1-x^2}}$
   (j) $f(x) = \dfrac{\sqrt{x^2 - 3x - 4}}{x+2}$

4. What value must be assigned to $f(1)$ so that $f(x) = (x^2 - 3x + 2)/(x^2 - 1)$, $x \neq \pm 1$, will be continuous at $x = 1$? Can $f(-1)$ be defined so that this function is continuous at $x = -1$?

Discuss the continuity of each of the functions 1 to 10 and sketch its graph.

PROBLEMS 3.3
Mathematics

1. $f(x) = |x|$
2. $f(x) = \dfrac{x}{|x|}$
3. $f(x) = \dfrac{x^3 + 8}{x^2 - 4}$
4. $f(x) = \dfrac{|x^2 - 1|}{x+1}$
5. $f(x) = \dfrac{2x}{x^2 - 4}, \ 0 < x < 2$
   $= 3x - 5, \ 2 \leq x \leq 5$
   $= x^2 + 6, \ 5 < x < 7$
6. $f(x) = \dfrac{x^2}{a}, \ x < a$
   $= 2a - \dfrac{a^3}{x^2}, \ x \geq a$
7. $f(x) = 0, \ x \neq \dfrac{1}{n}$; $= \dfrac{1}{n}, \ x = \dfrac{1}{n}$, $n \in \mathbf{N}$
8. $f(x) = x$ if x is irrational
   $= \sqrt{(1 + p^2)/(1 + q^2)}$ if $x = p/q$
   where p and q have no common factors.
9. $f(x) = 1 - x + [x] - [1 - x]$
10. $f(x) = [x + \tfrac{1}{2}] - 2[x] + [x - \tfrac{1}{2}]$
11. A function f has a *removable discontinuity* at $x = a$ if $f(a)$ is not defined but $\lim_{x \to a} f(x)$ exists. In such a case, show that the function defined by
    $$g(x) = f(x) \quad x \neq a$$
    $$= \lim_{x \to a} f(x), \quad x = a$$
    is continuous at $x = a$.
12. Using one-sided limits, define the meaning of "The function f is continuous on the closed interval $\{x : a \leq x \leq b\}$."

3.4 TWO IMPORTANT PROPERTIES OF CONTINUOUS FUNCTIONS

The first of the two properties to which the section title refers can be stated as follows: *If the graph of a continuous function appears both below and above the x axis, then it must cross the x axis.* Since there are no breaks or jumps in such a graph, this is not a very surprising result. However, the subtlety required in the proof we give below may be startling.

■ **THEOREM 3.10.** Let f be continuous on an interval containing points a and b, $a < b$, and suppose that $f(a) \cdot f(b) < 0$. Then there is at least one point c, $a < c < b$, such that $f(c) = 0$.

PROOF: To be definite, assume that $f(a) > 0$ while $f(b) < 0$. Define a set X of real numbers by the condition $x_0 \in X$ if and only if $f(x) > 0$ for all values of x in the interval $\{x : a \leq x \leq x_0\}$. This set X is not empty, because by Theorem 3.8 there is a positive number δ such that $f(x) > 0$ on $\{x : a \leq x \leq a + \delta\}$ and hence X contains this interval.

The point b is obviously an upper bound for X because $f(b) < 0$. Therefore X is bounded above. Now Theorem 1.10 states that X has a least upper bound c. We claim that $f(c) = 0$. If we suppose that $f(c) > 0$, then Theorem 3.8 applies to provide an interval $\{x : c - \delta < x < c + \delta\}$ over which $f(x) > 0$. But then the number $c + \delta/2$ would be in X. Since $c + \delta/2 > c$, this contradicts the fact that c is an upper bound for X. Next suppose that $f(c) < 0$. Again Theorem 3.8 applies to give us an interval $\{x : c - \delta < x < c + \delta\}$ over which $f(x) < 0$. But this tells us that $c - \delta$ would be an upper bound for X and this contradicts the fact that c is the *least* upper bound for X. Since $f(c)$ exists and is neither positive nor negative, $f(c) = 0$.

The above result is the basis of a well-known technique for solving equations by successive approximation. The technique is known as *binary splitting*. It requires many computations and so is most often used in conjunction with a high speed electronic computer.

Let f be continuous and suppose that $f(a) \cdot f(b) < 0$. Theorem 3.10 says that there is a root of the equation $f(x) = 0$ somewhere between a and b. Next we compute $f[(a + b)/2]$, the value of f at the midpoint of the interval $\{x : a \leq x \leq b\}$. If $f(a) \cdot f[(a + b)/2] < 0$, then there is a root between a and $(a + b)/2$, that is, we have located a root more accurately. (Of course, if $f[(a + b)/2] = 0$, we have found the root and if $f[(a + b)/2] \cdot f(b) < 0$, then the root lies between $(a + b)/2$ and b.) For a closer approximation we compute the value of f at the midpoint

$$\frac{1}{2}\left(a + \frac{1}{2}(a + b)\right) = \frac{3a + b}{4}$$

of the interval $\{x : a \leq x \leq \frac{1}{2}(a + b)\}$. The value $f[(3a + b)/4]$ is used to locate the desired root in some interval of length $\frac{1}{4}(b - a)$. Repeating this procedure, we can locate the root in an interval of length $1/2^n(b - a)$ after the nth repetition. Obviously, we need either a large amount of patience or a fast computer for this kind of procedure.

■ *Example 1.* We compute $\sqrt{2}$ approximately by solving $x^2 - 2 = 0$ with binary splitting. Because $f(1) = -1$ and $f(2) = 2$, there is a root between 1 and 2. Because $f[(1 + 2)/2] = f(\frac{3}{2}) = \frac{1}{4} > 0$, there is a root between 1 and $\frac{3}{2}$. Next, $f[(1 + \frac{3}{2})/2] = f(\frac{5}{4}) = -\frac{9}{16}$. Therefore there is a root between $\frac{5}{4}$ and $\frac{3}{2}$. Then $f[(\frac{5}{4} + \frac{3}{2})/2] = f(\frac{11}{8}) = -\frac{7}{64}$, so that the root is between $\frac{11}{8}$ and $\frac{3}{2}$. The value $f(\frac{1}{2}(\frac{11}{8} + \frac{3}{2})) = f(\frac{23}{16}) = \frac{17}{256}$, and hence the root lies between $\frac{11}{8}$ and $\frac{23}{16}$. For our final step

$$f\left(\frac{1}{2}\left(\frac{11}{8} + \frac{23}{16}\right)\right) = f\left(\frac{45}{32}\right) = -\frac{3}{1024}$$

This tells us that the root is between $\frac{45}{32}$ and $\frac{23}{16}$, an interval of length $\frac{1}{32}$. The midpoint of this interval is $\frac{1}{2}(\frac{45}{32} + \frac{23}{16}) = \frac{91}{64} = 1.42 \cdots$, a rough approximation of $\sqrt{2}$. It is just such repetitive computation that the electronic computer does so well.

The following result is usually called the *intermediate value theorem*.

■ **THEOREM 3.11.** Let f be continuous in an interval containing the points a and b. Then $f(x)$ takes on every value between $f(a)$ and $f(b)$.

PROOF: To be definite, suppose that $f(a) < f(b)$. Let k be any number satisfying $f(a) < k < f(b)$. Define the function $g(x) = f(x) - k$. Clearly, g is continuous wherever f is continuous. Then $g(a) = f(a) - k < 0$, while $g(b) = f(b) - k > 0$. Therefore Theorem 3.10 says that there is a point c, $a < c < b$, such that $g(c) = f(c) - k = 0$. Thus $f(c) = k$, which is just what we wanted to prove. The other case in which $f(a) > f(b)$ is proved similarly. Finally, if $f(a) = f(b)$, then there is nothing to prove. ■

A preliminary result is needed before we can establish the second important property of continuous functions.

■ **THEOREM 3.12.** Let f be continuous on an interval containing the points a and b, $a < b$. Then the set $\{y : y = f(x), a \leq x \leq b\}$ of functional values is bounded.

We give only an indication of the proof as follows: In the definition of the statement $\lim_{x \to a} f(x) = f(a)$, take $\epsilon = 1$. Then there exists $\delta > 0$ such that

$$|f(x) - f(a)| < 1, \quad \text{whenever } |x - a| < \delta$$

In particular, then, on the interval $\{x : a \leq x \leq a + \delta/2\}$, we have

$$f(a) - 1 < f(x) < f(a) + 1$$

Thus $f(x)$ is bounded on this interval.

Now define a set X of real numbers by agreeing that $x_0 \leq b$ is in X if and only if $f(x)$ is bounded on the interval $\{x : a \leq x \leq x_0\}$. The set X is not empty, because we found that $a + (\delta/2) \in X$. Note that if $x_0 \in X$, then the entire interval $\{x : a \leq x \leq x_0\}$ is contained in X. Also, by definition, the set X is bounded and hence has a least upper bound c. If $c = b$, then $X = \{x : a \leq x \leq b\}$ and our proof will be complete.

Suppose, to the contrary, that $c < b$. Because f is continuous at c, we take $\epsilon = 1$ and find $\delta > 0$ such that $|f(x) - f(c)| < 1$ whenever $|x - c| < \delta$. Arguing as before, $f(x)$ is bounded on the interval

$$\left\{x : c - \frac{\delta}{2} \leq x \leq c + \frac{\delta}{2}\right\}.$$

By definition of c, any number smaller than c must lie in X. Therefore $f(x)$ is bounded on $\{x : a \leq x \leq c - \delta/2\}$. Because $f(x)$ is bounded on both intervals $\{x : a \leq x \leq c - \delta/2\}$ and $\{x : c - \delta/2 \leq x \leq c + \delta/2\}$, $f(x)$ must also be bounded on the interval $\{x : a \leq x \leq c + \delta/2\}$. Thus, $c + \delta/2$ is also in X. This contradicts the assumption that c is an upper bound for X. Therefore the assumption that $c < b$ is untenable and hence $c = b$.

The second important property of continuous functions is contained in the next result.

■ **THEOREM 3.13.** Let f be continuous on an interval containing the points a and b, $a < b$. Then there exist two points x_0, x_1 in the closed interval $\{x : a \leq x \leq b\}$ such that, for each point x in this closed interval, we have $f(x_0) \leq f(x) \leq f(x_1)$.

We discuss this result before proving it. First of all, the theorem says that on a closed interval a continuous function actually attains both a minimum value $f(x_0)$ and a maximum value $f(x_1)$. There are two major hypotheses here, the continuity of the function and the fact that the interval is closed. Both are necessary if the conclusion is to hold. If the function is not continuous, then it need not attain a maximum or a minimum. The function

$$f(x) = \frac{1}{x}, \quad \text{if } x \neq 0$$
$$= 0, \quad \text{if } x = 0$$

is well defined and fails to be continuous at just the one point $x = 0$. However, on the closed interval $\{x : -1 \leq x \leq 1\}$, this function attains no maximum and no minimum. The graph of this function appears in Fig. 3.6.

The conclusion of Theorem 3.13 need not hold if the interval over which we consider the function is not closed. For instance, the function $f(x) = 1/x$ is continuous on the half-open interval $\{x : 0 < x \leq 1\}$ but it does not attain a maximum value in this interval. These examples show that both hypotheses are required.

PROOF OF THEOREM 3.13: By Theorem 3.12, the set of functional values $\{y : y = f(x), a \leq x \leq b\}$ is bounded. Then Theorem 1.10 indicates this set has both a least upper bound M and a greatest lower bound m. We prove by contradiction that there exists at least one point x_1, $a \leq x_1 \leq b$, such that $f(x_1) = M$.

If this were not true then for every point x in $\{x : a \leq x \leq b\}$, we should have $f(x) < M$ or $M - f(x) > 0$. Hence the function

$$g(x) = \frac{1}{M - f(x)}$$

is also continuous because the denominator is never zero. Applying Theorem 3.12 to the function g, there is an upper bound M' for the values of g.

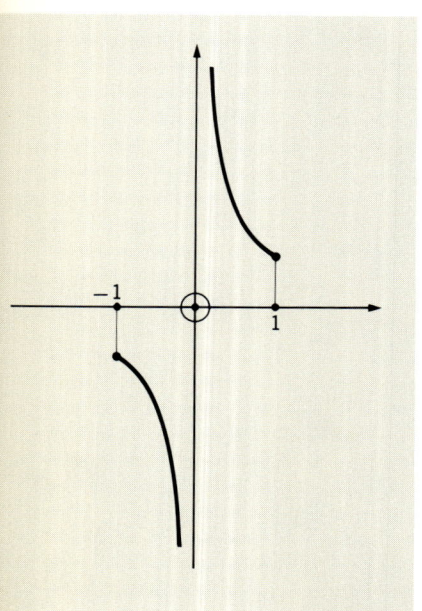

FIGURE 3.6

3.4 TWO IMPORTANT PROPERTIES OF CONTINUOUS FUNCTIONS

Obviously, M' must be positive because $g(x) > 0$ for all x in the interval. That is, we have

$$g(x) = \frac{1}{M - f(x)} < M' \qquad \text{for all } x, \ a \leq x \leq b$$

But then

$$1 < M'(M - f(x))$$

or

$$\frac{1}{M'} < M - f(x)$$

or

$$f(x) < M - \frac{1}{M'}, \qquad a \leq x \leq b$$

This says that the number $M - 1/M' < M$ is also an upper bound for $f(x)$, which contradicts the statement that M is the *least* upper bound. The assumption that $f(x) < M$ for all x is seen to be untenable; hence there must be at least one point x_1 in $\{x : a \leq x \leq b\}$ for which $f(x_1) = M$. The proof of the existence of a point x_0 in this interval such that $f(x_0) = m$ is completely analogous. ■

Combining Theorems 3.11 and 3.13, we obtain our final result.

■ **THEOREM 3.14.** If f is continuous on an interval containing the points a and b, $a < b$, then the functional values $\{y : y = f(x), a \leq x \leq b\}$ form a closed interval.

EXERCISES 3.4

1. Locate the real roots of each equation between successive integer points:
   (a) $x^2 - 3x - 5 = 0$
   (b) $4x^2 + 7x - 1 = 0$
   (c) $x^3 + 2x - 4 = 0$
   (d) $x^3 - 4x^2 - 2x + 7 = 0$
   (e) $x^4 - 3x + 3 = 0$
   (f) $2x^4 + 20x^3 - 17x - 5 = 0$
   (g) $\dfrac{1}{x+2} + \dfrac{1}{x+1} = 1$
   (h) $\dfrac{4}{x+4} + \dfrac{5}{x+1} = 1$
   (i) $\dfrac{1}{\sqrt{x+2}} + \dfrac{1}{x+2} = 1$
   (j) $\dfrac{2}{\sqrt{x+3}} + \dfrac{4}{\sqrt{x+5}} = 3$

2. Use binary splitting to locate the positive roots of each equation within intervals of length $\frac{1}{64}$.
   (a) $x^2 = 3$
   (b) $x^2 = 7$
   (c) $x^2 - 3x - 5 = 0$
   (d) $x^2 + 4x - 6 = 0$
   (e) $x^3 - 3x - 4 = 0$
   (f) $x^3 - 2x^2 - 4x - 6 = 0$
   (g) $x^4 - 6 = 0$
   (h) $\dfrac{1}{x+3} + \dfrac{1}{x-1} = 2$
   (i) $\dfrac{1}{\sqrt{x+3}} + \dfrac{1}{\sqrt{x+1}} = 1$
   (j) $\dfrac{1}{\sqrt{x+3}} + \sqrt{x+3} = 3$

PROBLEMS 3.4
Mathematics

1. Prove that each polynomial function of odd degree has at least one real zero.

2. Let f be defined and continuous on the closed interval $\{x : a \leq x \leq b\}$ (see Problem 3.3.12). Show that, if $f(r) = 0$ for each rational number r in $\{x : a \leq x \leq b\}$, then f is identically zero on this interval.

3. Apply the intermediate value theorem to prove that $\sqrt[n]{x}$, $x > 0$, exists for any natural number n.

4. Let a_1, a_2, and a_3 be positive constants and let b_1, b_2, b_3 be constants such that $b_1 > b_2 > b_3$. Prove that the equation
$$\frac{a_1}{b_1^2 + x} + \frac{a_2}{b_2^2 + x} + \frac{a_3}{b_3^2 + x} = 1$$
has precisely three real roots x_1, x_2, x_3 satisfying
$$-b_1^2 < x_1 < -b_2^2 < x_2 < -b_3^2 < x_3$$

3.5 THE ALGEBRA OF FUNCTIONS

The algebraic operations of addition, multiplication, and division can be performed on functions, too. The following definitions make implicit use of our earlier definition of the equality of functions.

Let f and g be two real-valued functions having domains X_f and X_g, respectively. We define three new functions as follows: The *sum function* $f + g$, having the values
$$(f + g)(x) = f(x) + g(x)$$
the *product function* fg with values
$$(fg)(x) = f(x) \cdot g(x)$$
and the *quotient function* f/g, which is defined to have values
$$\left(\frac{f}{g}\right)(x) = \frac{f(x)}{g(x)}$$
whenever $g(x) \neq 0$. Of course, these functions are only defined when both $f(x)$ and $g(x)$ are defined. Hence their domains are
$$X_{f+g} = X_f \cap X_g$$
$$X_{fg} = X_f \cap X_g$$
$$X_{f/g} = X_f \cap X_g \cap \{x : g(x) \neq 0\}$$

■ **Example 1.** If $f(x) = \sqrt{1 - x^2}$ and $g(x) = x^2 - 3x$, then
$$(f + g)(x) = \sqrt{1 - x^2} + x^2 - 3x$$
$$(fg)(x) = \sqrt{1 - x^2}\,(x^2 - 3x)$$
$$\left(\frac{f}{g}\right)(x) = \frac{\sqrt{1 - x^2}}{x^2 - 3x}$$

Clearly, we have $X_f = \{x : -1 \leq x \leq 1\}$, while $X_g = R$. Thus X_{f+g} and X_{fg} equal $X_f \cap X_g = X_f \cap R = X_f$. To get $X_{f/g}$, we omit those values

where $g(x) = 0$ whence
$$X_{f/g} = \{x : -1 \leq x < 0\} \cup \{x : 0 < x \leq 1\}$$

■ **Example 2.** Recalling that $E(x) = x$ is the identity function, we see that $(EE)(x) = E^2(x) = E(x) \cdot E(x) = x^2$ and, in general, that
$$E^n(x) = x^n$$
It follows that we may write a polynominal function $f(x) = 4x^2 - 3x + 1$ as
$$f = 4E^2 - 3E + 1$$
thinking here of 1 as the constant function.

■ **THEOREM 3.15.** Let the functions f and g be continuous at the point $x = a$. Then $f + g$ and fg are continuous at $x = a$. If $g(a) \neq 0, f/g$ is also continuous at $x = a$.

PROOF: This result is an easy consequence of the limit theorems in Section 2.10. We are assuming that $\lim_{x \to a} f(x) = f(a)$ and $\lim_{x \to a} g(x) = g(a)$. In particular, $f(a)$ and $g(a)$ exist. Hence $(f + g)(a) = f(a) + g(a)$, $(fg)(a) = f(a)g(a)$ and, if $g(a) \neq 0$, $(f/g)(a) = f(a)/g(a)$ all exist. We need only show that
$$\lim_{x \to a} (f + g)(x) = f(a) + g(a) = \lim_{x \to a} f(x) + \lim_{x \to a} g(x)$$
$$\lim_{x \to a} (fg)(x) = f(a) \cdot g(a) = \lim_{x \to a} f(x) \cdot \lim_{x \to a} g(x)$$
and
$$\lim_{x \to a} \left(\frac{f}{g}\right)(x) = \frac{f(a)}{g(a)} = \frac{\lim_{x \to a} f(x)}{\lim_{x \to a} g(x)}, \quad \text{if } g(a) \neq 0$$
These are just the basic limit theorems in Section 2.10. ■

While the preceding result is primarily of theoretical interest, the next forms the basis of three fundamental differentiation formulas.

■ **THEOREM 3.16.** Let f and g be differentiable at the point $x = a$. Then $f + g, fg$, and, if $g(a) \neq 0, f/g$ are all differentiable at $x = a$. Furthermore, we have the formulas
$$(f + g)'(a) = f'(a) + g'(a)$$
$$(fg)'(a) = f'(a)g(a) + f(a)g'(a)$$
$$\left(\frac{f}{g}\right)'(a) = \frac{f'(a)g(a) - f(a)g'(a)}{g^2(a)}$$

PROOF: This result is also a consequence of the basic limit theorems. We prove each formula separately: For the first, we simply write out

$$\lim_{h \to 0} \frac{(f+g)(a+h) - (f+g)(a)}{h}$$

$$= \lim_{h \to 0} \left[\frac{f(a+h) + g(a+h) - f(a) - g(a)}{h} \right]$$

$$= \lim_{h \to 0} \left[\frac{f(a+h) - f(a)}{h} + \frac{g(a+h) - g(a)}{h} \right]$$

$$= \lim_{h \to 0} \frac{f(a+h) - f(a)}{h} + \lim_{h \to 0} \frac{g(a+h) - g(a)}{h}$$

$$= f'(a) + g'(a)$$

To prove the second formula, we consider

$$(fg)(a+h) - (fg)(a) = f(a+h)g(a+h) - f(a)g(a)$$

By adding and subtracting $f(a)g(a+h)$ here, we get

$$f(a+h)g(a+h) - f(a)g(a+h) + f(a)g(a+h) - f(a)g(a)$$
$$= [f(a+h) - f(a)]g(a+h) + f(a)[g(a+h) - g(a)]$$

Divide by h and take the limit as follows:

$$\lim_{h \to 0} \left[\frac{f(a+h) - f(a)}{h} \cdot g(a+h) + f(a) \cdot \frac{g(a+h) - g(a)}{h} \right]$$

$$= \lim_{h \to 0} \frac{f(a+h) - f(a)}{h} \cdot \lim_{h \to 0} g(a+h)$$

$$+ f(a) \cdot \lim_{h \to 0} \frac{g(a+h) - g(a)}{h}$$

where we have used the limit theorems to simplify the expression. Knowing that g is differentiable at $x = a$, it is necessarily continuous there, so that $\lim_{h \to 0} g(a+h) = g(a)$. Therefore the above limit is precisely

$$f'(a)g(a) + f(a)g'(a)$$

The third formula is proved by showing that

$$\lim_{h \to 0} \frac{\frac{f(a+h)}{g(a+h)} - \frac{f(a)}{g(a)}}{h} = \frac{f'(a)g(a) - f(a)g'(a)}{g^2(a)}$$

We first write

$$\frac{f(a+h)}{g(a+h)} - \frac{f(a)}{g(a)} = \frac{f(a+h)g(a) - f(a)g(a+h)}{g(a+h)g(a)}$$

By adding and subtracting $f(a)g(a)$ in the numerator, we obtain

$$\frac{f(a+h)g(a) - f(a)g(a) - f(a)g(a+h) + f(a)g(a)}{g(a+h)g(a)}$$

$$= \frac{[f(a+h) - f(a)]g(a) - f(a)[g(a+h) - g(a)]}{g(a+h)g(a)}$$

3.5 THE ALGEBRA OF FUNCTIONS

Next we multiply by $1/h$ and take the limit as h goes to zero:

$$\lim_{h \to 0} \frac{\left[\frac{f(a+h) - f(a)}{h}\right] g(a) - f(a) \left[\frac{g(a+h) - g(a)}{h}\right]}{g(a+h)g(a)}$$

The limit formulas in Section 2.10 now complete the proof, because this limit can be rewritten as

$$\frac{\lim_{h \to 0} \left[\frac{f(a+h) - f(a)}{h}\right] g(a) - f(a) \cdot \lim_{h \to 0} \left[\frac{g(a+h) - g(a)}{h}\right]}{\lim_{h \to 0} g(a+h)g(a)}$$

$$= \frac{f'(a)g(a) - f(a)g'(a)}{g^2(a)} \quad \blacksquare$$

■ **COROLLARY 3.17.** If f and g are differentiable over an interval $\{x : a < x < b\}$, then $f + g, fg$, and, if $g(x)$ is never zero, f/g are differentiable over this interval. Furthermore, we have the differentiation formulas:

$$(f + g)' = f' + g'$$
$$(fg)' = f'g + fg'$$
$$\left(\frac{f}{g}\right)' = \frac{f'g - fg'}{g^2}$$

■ **Example 3.** Let $f(x) = x^2 - 3x$ and $g(x) = (x^2 + 1)/x$. In examples 1 and 4 of Section 3.1, we found $f'(x) = 2x - 3$ and $g'(x) = (x^2 - 1)/x^2$. Hence, from Corollary 3.17,

$$(f + g)'(x) = 2x - 3 + \frac{x^2 - 1}{x^2}$$

$$(fg)'(x) = (2x - 3)\left(\frac{x^2 + 1}{x}\right) - (x^2 - 3x)\left(\frac{x^2 - 1}{x^2}\right)$$

$$\left(\frac{f}{g}\right)'(x) = \frac{(2x - 3)[(x^2 - 1)/x] - (x^2 - 3x)[(x^2 - 1)/x^2]}{[(x^2 + 1)/x]^2}$$

■ **Example 4.** We know that the identity function $E(x) = x$ has the constant derivative $E'(x) = 1$, while the constant function 1 has the derivative 0. Hence the function $(1/E)(x) = 1/x$ has a derivative

$$\left(\frac{1' \cdot E - 1 \cdot E'}{E^2}\right)(x) = \frac{0 \cdot x - 1 \cdot 1}{x^2} = -\frac{1}{x^2}$$

■ **THEOREM 3.18.** For each integer $p \in Z$, the function $E^p(x) = x^p$ has the derivative

$$(E^p)'(x) = (x^p)' = px^{p-1}$$

(Note that, if $p < 0$, then x^p is not defined at $x = 0$.)

PROOF: We first prove the result for positive integers by induction. If $n = 1$, then $(x)' = 1 \cdot x^0 = 1$, so that the formula is true in this case. Suppose that the formula is true for all positive integers $n \leq k$. If $n = k + 1$, then we use Corollary 3.17 to write

$$(x^{k+1})' = (x^k \cdot x)' = (x^k)' \cdot x + (x^k)(x')$$
$$= (kx^{k-1})(x) = (x^k)(1) \text{ by the induction hypothesis}$$
$$= kx^k + x^k = (k + 1)x^k$$

This completes the induction, so that the formula is true for all positive integers. The case where $p = 0$ is trivial; so we next consider the case in which $p < 0$.

If $p < 0$, then $-p > 0$ and $x^p = 1/x^{-p}$. By the quotient rule of differentiation, we have

$$(x^p)' = \left(\frac{1}{x^{-p}}\right)' = \frac{1' \cdot x^{-p} - 1 \cdot (x^{-p})'}{(x^{-p})^2}$$
$$= \frac{0 \cdot x^{-p} - (-p)(x^{-p-1})}{x^{-2p}}$$
$$= \frac{px^{-p-1}}{x^{-2p}}$$
$$= px^{-p-1}x^{2p} = px^{p-1}$$

Hence the formula is true for all integers. ∎

As another easy corollary of Corollary 3.17, we have the next result.

■ COROLLARY 3.19. If f is differentiable and k is a constant, then kf is differentiable and

$$(kf)' = kf'$$

PROOF: By Corollary 3.17, $(kf)' = k'f + kf'$, but $k' = 0$ by Theorem 3.6. ∎

The following result makes it possible to differentiate a polynomial function by inspection alone.

■ **THEOREM 3.20.**

$$(a_0 + a_1x + a_2x^2 + \cdots + a_nx^n)' = a_1 + 2a_2x + \cdots + na_nx^{n-1}$$

PROOF: Since by 3.10 the derivative of a sum is the sum of the derivatives, we have

$$(a_0 + a_1x + a_2x^2 + \cdots + a_nx^n)' = (a_0)' + (a_1x)' + (a_2x^2)' + \cdots + (a_nx^n)'$$

From 3.19 we have $(a_ix^i)' = a_i(x^i)'$ and Theorem 3.18 tells us that $(x^i)' = ix^{i-1}$. Hence the result is obvious. ∎

3.5 THE ALGEBRA OF FUNCTIONS

EXERCISES 3.5

1. For each pair f and g, determine the domains of $f + g, fg, f/g$.
   (a) $f(x) = \sqrt{x}$, $g(x) = x - 2$
   (b) $f(x) = x - 1$, $g(x) = \sqrt{x^2 - 1}$
   (c) $f(x) = \dfrac{x-1}{x+1}$, $g(x) = \sqrt{x}$
   (d) $f(x) = \sqrt[3]{x+1}$, $g(x) = \sqrt[4]{x-1}$

2. Verify the continuity of each of the following functions by appealing to known theorems.
   (a) $f(x) = x - 3x^2$
   (b) $f(x) = x^2(1-x)$
   (c) $f(x) = \dfrac{3x}{x^2+1}$
   (d) $f(x) = \dfrac{(2x-1)^4}{x^2+x+1}$

For each of Exercises 3 to 20, use the formulas to compute f'.

3. $f(x) = 2 - 3x + x^2$
4. $f(x) = 3 + 4x - 2x^2$
5. $f(x) = 4x + x^3$
6. $f(x) = 2 - x + x^2 + 3x^3$
7. $f(x) = 3 + 2x^2 = x^4$
8. $f(x) = 4 - 2x - 6x^2 + 2x^3 + x^4$
9. $f(x) = (3 + 2x)^3$
10. $f(x) = (1 + x)(2 + x^2)$
11. $f(x) = \dfrac{1-x}{x}$
12. $f(x) = \dfrac{1+x}{x}$
13. $f(x) = \dfrac{1}{1+x}$
14. $f(x) = \dfrac{1}{1+x^2}$
15. $f(x) = \dfrac{3-4x}{2+3x}$
16. $f(x) = \dfrac{3-4x+x^2}{2+3x}$
17. $f(x) = \dfrac{1}{1-x^2}$
18. $f(x) = \dfrac{x}{1-x^2}$
19. $f(x) = \dfrac{1-x^3}{1+x^3}$
20. $f(x) = \dfrac{(1+x)(2+x^2)}{x(1-x)}$

Find and identify the critical points of the functions in Exercises 21 to 30.

21. $f(x) = 4x - x^2$
22. $f(x) = (3-x)(4+3x)$
23. $f(x) = (1-x)(1+x^2)$
24. $f(x) = 12x - 4x^2 - 4x^3 + 2x^4$
25. $f(x) = \dfrac{4}{1+x^2}$
26. $f(x) = \dfrac{4x}{1+x^2}$
27. $f(x) = \dfrac{3+x^2}{x-1}$
28. $f(x) = \dfrac{x^2-4}{x^2+4}$
29. $f(x) = \dfrac{2+3x-x^2}{2+x}$
30. $f(x) = \dfrac{x^5}{x^2-6}$

Use the derivative to aid in sketching the graph of each function in Exercises 31 to 38.

31. $f(x) = -1 + 4x - x^2$
32. $f(x) = 2 - 3x + x^2$
33. $f(x) = (2+3x)(-1+2x)$
34. $f(x) = (1+x)(1-x^2)$
35. $f(x) = \dfrac{1}{1+x}$
36. $f(x) = \dfrac{1+x}{x}$

37. $f(x) = \dfrac{1-x^2}{x}$ 38. $f(x) = \dfrac{1-x^2}{1+x^2}$

For each function in Exercises 39 to 46 determine those points on the graph where the tangent line has slope 1. Sketch your result in each case.

39. $f(x) = 2 + x^2$ 40. $f(x) = 3 - 5x - 2x^2$

41. $f(x) = x^2(3-x)$ 42. $f(x) = 2x(3-x^2)$

43. $f(x) = \dfrac{1}{1+x}$ 44. $f(x) = \dfrac{x}{1+x}$

45. $f(x) = \dfrac{2x^2}{x-1}$ 46. $f(x) = \dfrac{2x}{1+x^2}$

For each function in Exercises 47 to 50, determine the point at which the tangent line to the graph $\{(x,y) : y = f(x)\}$ at $(a, f(a))$ crosses the x axis.

47. $f(x) = 4 - 2x - 3x^2$ 48. $f(x) = -1 - 3x + x^3$

49. $f(x) = \dfrac{1}{x}$ 50. $f(x) = \dfrac{1-x^2}{x}$

PROBLEMS 3.5
Mathematics

1. In Problem 3.3.12, the reader was asked to define the meaning of "continuous on a closed interval." Now we let $C[a, b]$ denote the set of all functions f continuous on the closed interval $\{x : a \leq x \leq b\}$.
   (a) Prove that $C[a, b]$ is a vector space.
   (b) For two functions f and g in $C[a, b]$, define the "distance" between f and g by the formula
   $$p(f, g) = \text{maximum } |f(x) - g(x)|$$
   Assume that $\varphi(f, g)$ exists for each pair and prove that φ is a metric (see Section 1.5).

2. Suppose the functions f and g satisfy the equation $f(x)g(x) = k$ (a constant). Prove that, whenever $g(x) \neq 0$ and $g'(x) \neq 0$, we have
$$\frac{f'(x)}{g'(x)} = -\frac{f(x)}{g(x)}$$

3. If f is any real-valued function, then the *negative function* $-f$ is defined by $(-f)(x) = -f(x)$. Prove that
$$(-f)' = -f'$$
by direct application of the definitions.

4. In Problem 1, $C[a, b]$ is shown to be a vector space. Now verify the formulas
   (a) $f \times 1 = f$ (b) $fg = gf$
   (c) $f(gh) = (fg)h$ (d) $f(g+h) = fg + fh$
   [These show that $C[a, b]$ is a *commutative ring with identity*.]

5. Prove that the sum of even functions is an even function and that the sum of odd functions is an odd function. Find and prove similar results about products and quotients of even and odd functions.

6. Let the function f be differentiable and positive everywhere. Define a new function g by the equation $g^2(x) = f(x)$. Prove that
$$g'(x) = \frac{f'(x)}{\pm 2\sqrt{f(x)}}$$

3.6 ANOTHER NOTATION FOR THE DERIVATIVE

There are times when the use of f' for the derivative of a function f causes notational problems. Several other symbols for the derivative are in common use—we introduce one of them here.

If f is a differentiable function, then its derivative is also denoted by

$$Df$$

and hence the values of the derivative by

$$Df(x)$$

The rules of differentiation obtained so far are listed below in both the "prime" and the "D" notation.

1. The derivative of a constant is zero.

$$k' = 0 \quad \text{or} \quad Dk = 0$$

2. The derivative of the identity function x is 1.

$$(x)' = 1 \quad \text{or} \quad Dx = 1$$

3. $(f + g)' = f' + g'$ or $D(f + g) = Df + Dg$
4. $(fg)' = f'g + fg'$ or $D(fg) = g\,Df + f\,Dg$
5. $\left(\dfrac{f}{g}\right)' = \dfrac{f'g - fg'}{g^2}$ or $D\left(\dfrac{f}{g}\right) = \dfrac{g\,Df - f\,Dg}{g^2}$

The other differentiation rules stated so far are consequences of these basic five. For instance, we used rules 4 and 5 to derive the next rule.

6. For each $p \in Z$,

$$(x^p)' = px^{p-1} \quad \text{or} \quad Dx^p = px^{p-1}$$

Similarly, with rules 1 and 4 we proved

7. If k is a constant, then

$$(kf)' = kf' \quad \text{or} \quad D(kf) = k\,Df$$

3.7 THE COMPOSITION OF FUNCTIONS. THE CHAIN RULE

Consider the function with values $(3 - 2x + x^2)^3$. We can construct this function as a "composition of functions" in the following way: Define $t = f(x) = 3 - 2x + x^2$ and let $g(t) = t^3$. Then we can write

$$g(t) = g(f(x)) = g(3 - 2x + x^2) = (3 - 2x + x^2)^3$$

Therefore the given function has been expressed by substituting the functional value of the function f for the variable in the function g. This situation is generalized in this section.

Let f be a real-valued function with domain X_f and g be a real-valued function with domain T_g. (The variable in f is to be denoted by x and the variable in g is to be denoted by t.) Then the *composite function* $g \circ f$ has values

$$(g \circ f)(x) = g(f(x))$$

The domain of $g \circ f$ is the set

$$\{x : x \in X_f \text{ and } f(x) \in T_g\}$$

This means that $g \circ f$ is defined at the point x if and only if f is defined at x and $f(x)$ lies in the domain of g.

■ **Example 1.** The function with values $\sqrt{x^2 - 3x}$ can be expressed as a composite function by setting $t = f(x) = x^2 - 3x$ and $g(t) = \sqrt{t}$. Then

$$(g \circ f)(x) = g(f(x)) = g(x^2 - 3x) = \sqrt{x^2 - 3x}$$

Because $X_f = \mathbf{R}$ and $T_g = \{t : t \geqq 0\}$, we have

$$\begin{aligned} X_{g \circ f} &= \{x : x \in \mathbf{R} \quad \text{and} \quad x^2 - 3x \geqq 0\} \\ &= \{x : x \leqq 0\} \cup \{x : x \geqq 3\} \end{aligned}$$

■ **Example 2.** If $t = f(x) = (x-1)/x$ and $g(t) = \sqrt{t}$ again, then

$$(g \circ f)(x) = g\left(\frac{x-1}{x}\right) = \sqrt{\frac{x-1}{x}}$$

Clearly, $X_f = \{x : x \neq 0\}$ and $T_g = \{t : t \geqq 0\}$. We compute

$$\begin{aligned} X_{g \circ f} &= \{x : x \neq 0 \quad \text{and} \quad \frac{x-1}{x} \geqq 0\} \\ &= \{x : x < 0\} \cup \{x : x \geqq 1\} \end{aligned}$$

The reader should verify these computations of $X_{g \circ f}$.

The composition of functions provides the proper way to define a subsequence of a sequence (See Section 2.6). Let φ be a sequence and let $f : N \to N$ be a strictly increasing sequence of natural numbers. Then the composite function $\psi = \varphi \circ f$ is a subsequence of φ. For if we denote $f(k)$ by n_k, we have assumed that $n_1 < n_2 < \cdots < n_k < \cdots$. Also we have $\psi(k) = (\varphi \circ f)(k) = \varphi(f(k)) = \varphi(n_k)$.

Our first result here is the customary theoretical one.

■ **THEOREM 3.21.** Let f be continuous at $x = a$ and g be continuous at $t = f(a)$. Then $g \circ f$ is continuous at $x = a$.

PROOF: By hypothesis we have the limits

$$\lim_{x \to a} (f)x = f(a) \quad \text{and} \quad \lim_{t \to f(a)} g(t) = g(f(a))$$

We must prove that

$$\lim_{x \to a} g(f(x)) = g(f(a))$$

The argument simply refines the intuitive feeling that, as x gets close to a, $f(x)$ gets close to $f(a)$ and hence $g(f(x))$ gets close to $g(f(a))$.

Let $\epsilon > 0$ be given. Because g is continuous at $f(a)$, there is a number $\eta > 0$ such that

$$|g(t) - g(f(a))| < \epsilon \quad \text{whenever } |t - f(a)| < \eta$$

3.7 THE COMPOSITION OF FUNCTIONS. THE CHAIN RULE

In turn, the continuity of f at $x = a$ applied to the number $\eta > 0$ implies that there is a number $\delta > 0$ such that

$$|f(x) - f(a)| < \eta \quad \text{whenever } |x - a| < \delta$$

Therefore we have

$$|g(f(x)) - g(f(a))| < \epsilon \quad \text{whenever } |x - a| < \delta \quad \blacksquare$$

■ **COROLLARY 3.22.** If f and g are both continuous on their domains, then $g \circ f$ is continuous on its domain.

The *chain rule* developed next is one of the most frequently used rules of differentiation.

■ **THEOREM 3.23.** If f is differentiable at a and g is differentiable at $f(a)$, then $g \circ f$ is differentiable at a and

$$D(g \circ f)(a) = Dg(f(a)) \cdot Df(a)$$

PROOF: We have two limits given,

$$\lim_{h \to 0} \frac{f(a + h) - f(a)}{h} = DF(a)$$

and

$$\lim_{k \to 0} \frac{g(f(a) + k) - g(f(a))}{k} = D(f(a))$$

From the existence of these limits, we are to prove that

$$\lim_{h \to 0} \frac{g(F(a + h)) - g(f(a))}{h} = D(f(a)) \cdot Df(a)$$

The proof is in two steps. First, we set $k = F(a + h) - F(a)$ and define a new function $\varphi(k)$ by

$$\varphi(k) = \frac{g(f(a) + k) - g(f(a))}{k} \quad \text{if } k \neq 0$$

$$= D(f(a)) \quad \text{if } k = 0$$

We know that φ is continuous at $k = 0$, because we assumed that

$$\lim_{k \to 0} \varphi(k) = \lim_{k \to 0} \frac{g(f(a) + k) - g(f(a))}{k} = Dg(f(a)) = \varphi(0)$$

Furthermore, if $f(a + h) - f(a) \neq 0$, we have

$$\varphi[F(a + h) - F(a)] = \frac{g[f(a) + f(a + h) - f(a)] - g(f(a))}{f(a + h) - f(a)}$$

$$= \frac{g[f(a + h)] - g(f(a))}{f(a + h) - f(a)}$$

while, if $f(a + h) - f(a) = 0$, then

$$\varphi[f(a + h) - f(a)] = Dg(f(a))$$

Because f is differentiable at $x = a$, it is continuous there; hence
$$\lim_{h \to 0} [f(a + h) - f(a)] = 0$$

Then φ being continuous at $k = 0$, we have
$$\lim_{h \to 0} \varphi[f(a + h) - f(a)] = \varphi(0) = Dg(f(a))$$

Next we write
$$\frac{g[f(a + h)] - g(f(a))}{h} = \frac{g[f(a + h)] - g(f(a))}{f(a + h) - f(a)} \cdot \frac{f(a + h) - f(a)}{h},$$
$$\text{if } f(a + h) \neq f(a)$$
$$= 0, \quad \text{if } f(a + h) = f(a)$$

In either case, we have
$$\frac{g[f(a + h)] - g(f(a))}{h} = \varphi[f(a + h) - f(a)]$$
$$\cdot \frac{f(a + h) - f(a)}{h}$$

Taking the limit as h approaches zero on both sides of this equation, we find that
$$\frac{g[f(a + h)] - g(f(a))}{h} = \lim_{h \to 0} \varphi[f(a + h) - f(a)]$$
$$\cdot \lim_{h \to 0} \frac{f(a + h) - f(a)}{h}$$

or
$$D(g \circ f)(a) = Dg(f(a)) \cdot Df(a) \quad \blacksquare$$

We note that $Dg(f(a))$ is precisely $(Dg \circ f)(a)$. This explains the way in which we write our eighth basic differentiation rule, the so-called *chain rule* of differentiation.

8. $D(g \circ f) = (Dg \circ f) \cdot Df$

■ **Example 3.** To determine the derivative $D(x^2 - 3x)^3$, we may set $f(x) = x^2 - 3x$ and $g(t) = t^3$ as in the introductory remarks at the beginning of this section. Then
$$Df(x) = 2x - 3, \quad Dg(t) = 3t^2$$
Therefore $Dg(f(x)) = 3(f(x))^2 = 3(x^2 - 3x)^2$ and
$$D(x^2 - 3x)^3 = Dg(f(x)) \cdot Df(x) = 3(x^2 - 3x)^2(2x - 3)$$

3.7 THE COMPOSITION OF FUNCTIONS. THE CHAIN RULE

This particular example can also be handled as follows: Use the binomial theorem to write

$$(x^2 - 3x)^3 = (x^2)^3 + 3(x^2)^2(-3x) + 3(x^2)(-3x)^2 + (-3x)^3$$
$$= x^6 - 9x^5 + 27x^4 - 27x^3$$

Then in view of Theorem 3.20,

$$D(x^2 - 3x)^3 = 6x^5 - 45x^4 + 108x^3 - 81x^2$$
$$= 3(2x^5 - 15x^4 + 36x^3 - 27x^2)$$
$$= 3(2x - 3)(x^4 - 6x + 9x^2)$$
$$= 3(2x - 3)(x^2 - 3x)^2$$

Clearly, using the chain rule is more efficient.

■ **Example 4.** To determine $D[1/(3x - 1)^2]$, we may set $t = f(x) = 3x - 1$ and then $g(t) = t^{-2}$. Then $Df(x) = 3$ and $Dg(t) = 2t^{-3}$. Therefore $Dg(f(x)) = -2[f(x)]^{-3} = -2(3x - 1)^{-3}$ and hence

$$D\frac{1}{(3x - 1)^2} = Dg(f(x)) \cdot Df(x)$$
$$= -2(3x - 1)^{-3} \cdot 3$$
$$= -6(3x - 1)^{-3}$$

Let f be a real-valued function, and let $p \in \mathbf{Z}$. Then the function with values $[f(x)]^p$ is denoted by f^p. Of course, we may write f^p as a composite function by setting $t = f(x)$ and $g(t) = t^p$. It follows that $Dg(f(x)) = p[f(x)]^{p-1} = pf^{p-1}(x)$. Hence by the chain rule,

$$Df^p(x) = Dg(f(x)) \cdot Df(x) = pf^{p-1}(x) \cdot Df(x)$$

We have thus derived a general formula differentiating a power of a function,

$$Df^p = pf^{p-1} \cdot Df, \quad p \in \mathbf{Z}$$

This formula can be generalized still further, however. Let $f(x) = x^{p/q} = \sqrt[q]{x^p}$ where $p, q \in \mathbf{Z}, q \neq 0$. Raising both sides of the defining equation to the qth power, we have

$$f^q(x) = x^p$$

Using the previous formula, we differentiate both sides of this equation, obtaining

$$Df^q(x) = qf^{q-1}(x) \cdot Df(x) = px^{p-1}$$

But now

$$f^{q-1}(x) = \frac{f^q(x)}{f(x)} = \frac{x^p}{x^{p/q}}$$

and hence

$$qf^{q-1}(x) \cdot Df(x) = q\frac{x^p}{x^{p/q}} \cdot Df(x) = px^{p-1}$$

Therefore

$$Df(x) = \frac{p}{q} \cdot \frac{x^{p/q}}{x^p} \cdot x^{p-1} = \frac{p}{q} \cdot x^{p/q} \cdot x^{-1} = \frac{p}{q} x^{p/(q-1)}$$

This argument has proved the following result.

■ **THEOREM 3.24.** For each rational number $r \in \mathbf{Q}$,

$$Dx^r = rx^{r-1}$$

■ **THEOREM 3.25.** If f is a differentiable function and $r \in \mathbf{Q}$, then f^r is differentiable and

$$Df^r(x) = rf^{r-1}(x) \cdot Df(x)$$

PROOF: Set $t = f(x)$ and $g(t) = t^r$. By Theorem 3.24, we have $Dg(t) = rt^{r-1}$ and hence $Dg(f(x)) = rf^{r-1}(x)$. Then by the chain rule,

$$Df^r(x) = Dg(f(x)) \cdot Df(x)$$
$$= rf^{r-1}(x) \cdot Df(x) \quad \blacksquare$$

This is a generalization of differentiation rule 6. We shall replace the old rule with this one:

6. $Df^r = rf^{r-1} \cdot Df \qquad$ for any $r \in Q$

■ **Example 5.** If $f(x) = x^{2/3}$, then Theorem 3.24 yields

$$Dx^{2/3} = \tfrac{2}{3}x^{-1/3}$$

Note that $x^{2/3}$ is continuous at $x = 0$, but its derivative $\tfrac{2}{3}x^{-1/3}$ is not defined at $x = 0$.

■ **Example 6.** We differentiate $\sqrt{x^2 + a^2} = (x^2 + a^2)^{1/2}$ as follows: Letting $f(x) = x^2 + a^2$, we are required to find $Df^{1/2}$ which by rule 5 is

$$\tfrac{1}{2} F^{-1/2} Df = \tfrac{1}{2}(x^2 + a^2)^{-1/2} \cdot 2x = \frac{x}{\sqrt{x^2 + a^2}}$$

■ **Example 7.** In Fig. 3.7, the horizontal line represents a straight road through a heavy forest. A man at point P wishes to go to a point Q on the road. Suppose that he can walk through the forest at 3 ft per sec and can walk along the road at 6 ft per sec. What is the shortest length of time in which he can get to point Q?

We express this problem mathematically by introducing a coordinate system with the origin at the point 0 and the road as the x axis. The point

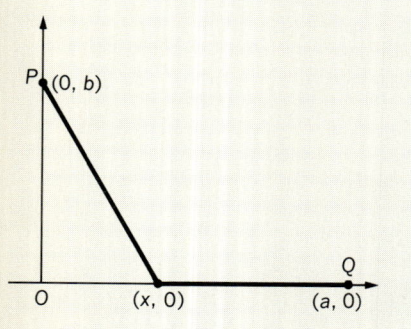

FIGURE 3.7

3.7 THE COMPOSITION OF FUNCTIONS. THE CHAIN RULE

P has coordinates $(0, b)$ and the point Q has coordinates $(a, 0)$. Assume that the man walks along a straight line from $(0, b)$ to a point $(x, 0)$ somewhere between $(0, 0)$ and $(a, 0)$ and then walks along the road. By the Pythagorean theorem, the distance from $(0, b)$ to $(x, 0)$ is $\sqrt{x^2 + b^2}$. At 3 ft per sec, this distance will require $\frac{1}{3}\sqrt{x^2 + b^2}$ sec of walking. The distance from $(x, 0)$ to $(a, 0)$ is $a - x$ and, at 6 ft per sec, requires $\frac{1}{6}(a - x)$ sec. The total number of seconds for the entire walk is then

$$T = \tfrac{1}{3}\sqrt{x^2 + b^2} + \tfrac{1}{6}(a - x)$$

Obviously, T is a function of x.

We determine the critical values of the function T by setting $D_x T$ equal to zero and solving the resulting equation for x. From Example 6 we have

$$D_x T(x) = \frac{1}{3} \frac{x}{\sqrt{a^2 + b^2}} - \frac{1}{6}$$

Hence we solve the equation

$$\frac{1}{3} \frac{x}{\sqrt{x^2 + b^2}} - \frac{1}{6} = 0$$

The solution is $x = b/\sqrt{3}$.

Finally, if $a > b/\sqrt{3}$, we obtain a minimum value of T by choosing $x = b\sqrt{3}$. This minimum value of T is

$$\frac{1}{3}\sqrt{b^2/3 + b^2} + \frac{1}{6}\left(a - \frac{b}{\sqrt{3}}\right) = \frac{b}{2\sqrt{3}} + \frac{a}{6}$$

On the other hand, if $a < b/\sqrt{3}$, our walker should walk in a straight line from $(0, b)$ to $(a, 0)$, a distance of $\sqrt{a^2 - b^2}$ ft. The required time will be $\frac{1}{3}\sqrt{a^2 + b^2}$ in this case. Notice that, if $a = b/\sqrt{3}$, both formulas give the same solution, $2b/3\sqrt{3}$ sec.

■ **Example 8.** A balloon is 500 ft above the ground and rising at 20 ft per sec at the instant that an automobile passes beneath the balloon. If the automobile is traveling at 88 ft per sec [60 miles per hour (mph)], how fast is the distance between the balloon and the automobile increasing 5 sec later?

We first express the distance between the two as a function of the number t of seconds that have elapsed since the automobile passed under the balloon. At the end of t sec, the balloon is $500 + 20t$ ft above the ground. The automobile is $88t$ ft from the point directly beneath the balloon. Hence, by the Pythagorean theorem, the straight line distance between the two is

$$d = \sqrt{(88t)^2 + (500 + 20t)^2}$$
$$= \sqrt{8144t^2 + 20000t + 250000}$$

The rate of change of d with respect to t is the derivative

$$Dd = \frac{8144t + 10000}{\sqrt{8144t^2 + 20000t + 250000}}$$

Finally, we set $t = 5$ to solve the problem. At $t = 5$, the rate of change is

$$\frac{1268}{\sqrt{346}} \cong 68.2 \text{ ft per sec}$$

EXERCISES 3.7

1. Write each of the following as a composite function.
   (a) $(x + 1)^3$
   (b) $(2 - 3x)^{1/2}$
   (c) $\sqrt{\dfrac{x+1}{x-1}}$
   (d) $(x^2 - 3x + 2)^4$
   (e) $\sqrt[3]{x^2 - 3x}$
   (f) $\dfrac{1}{(x+2)^2}$
   (g) $(3x^2 - 4x + 2)^{-2/3}$
   (h) $(x^4 - 1)^{-1/2}$

2. Determine the domain and range of $g \circ f$ for the following cases.
   (a) $f(x) = 3x - 1$, $g(t) = t^2$
   (b) $f(x) = x + 2$, $g(t) = \sqrt{t}$
   (c) $f(x) = \sqrt{1-x}$, $g(t) = t^3$
   (d) $f(x) = \dfrac{x}{1+x}$, $g(t) = \sqrt{t}$

3. Differentiate each of the following functions in two different ways and compare the results.
   (a) $f(x) = (x + 1)^2$
   (b) $f(x) = (2x^2 - 3x)^2$
   (c) $f(x) = \left(x + \dfrac{1}{x}\right)^3$
   (d) $f(x) = (3x - 1)^{-3}$

4. Make full use of the derivative to aid in drawing the graph of each of the following functions.
   (a) $f(x) = (x + 1)^2$
   (b) $f(x) = \dfrac{1}{(x+1)^2}$
   (c) $f(x) = (x^2 + 1)^2$
   (d) $f(x) = (x^2 + 1)^{1/2}$
   (e) $f(x) = x\sqrt{3 - x}$
   (f) $f(x) = \sqrt{\dfrac{x-1}{x+1}}$
   (g) $f(x) = x(x^2 + 1)^{1/2}$
   (h) $f(x) = x\sqrt{4 - x^2}$

5. Locate and identify the critical values of each of the following functions.
   (a) $f(x) = (3x - 1)^2$
   (b) $f(x) = (3x - 1)^3$
   (c) $f(x) = x\sqrt{3x - 1}$
   (d) $f(x) = x^2\sqrt{3x - 1}$
   (e) $f(x) = \dfrac{x}{\sqrt{x+1}}$
   (f) $f(x) = (x^2 - 1)^2$
   (g) $f(x) = \dfrac{2x}{x^2 + 1}$
   (h) $f(x) = \dfrac{3x - 2}{2x + 3}$
   (i) $f(x) = x^2\sqrt{5 - x}$
   (j) $f(x) = x^{2/3}(x + 6)^{1/3}$

6. Show that each of the following functions f satisfies the accompanying equation involving its derivative.
   (a) $f(x) = \sqrt{\frac{1}{2}x^2 - 1}$; $xf(x)f'(x) = 1 + f^2(x)$

(b) $f(x) = \dfrac{-2}{x^2 + k}$; $f'(x) = xf^2(x)$

(c) $f(x) = kx^{p/q}$; $qxf'(x) = pf(x)$

(d) $f(x) = \sqrt{kx^4 - x^2}$; $xf(x)f'(x) = x^2 + 2f^2(x)$

(e) $f(x) = \dfrac{x}{x+1}$; $x^2 f'(x) = f^2(x)$

PROBLEMS 3.7
Physical Science

1. A spherical snowball is melting at the rate of 1 cu in. per min. At what rate is the radius decreasing when the snowball is 4 in. in diameter?

2. A swimming pool has the shape of a rectangle 15 yd wide and 25 yd long. The bottom is a slanting plane, the pool is 12 ft deep at one end and 3 ft deep at the other. Water is being pumped into this pool at the rate of 5000 cu ft per hr. At what rate is the surface rising when the water is 8 ft deep at the deeper end?

3. A street lamp is mounted on a post 24 ft high. A man 6 ft tall walks directly away from the post at a rate of 5 ft per sec. How rapidly is the length of his shadow increasing when he is 36 ft away from the post?

4. A bucket is suspended on a rope that passes over a pulley 20 ft above the ground. A man holds the end of the rope and walks directly away from the point under the pulley at 4 ft per sec. How fast is the bucket rising when it is 8 ft above the ground?

5. A bridge over a canal has its surface 20 ft above the surface of the water. A boat moving at 15 ft per sec passes under the bridge at the same moment that a car passes over the bridge at 30 ft per sec. How rapidly is the distance between the two increasing 10 sec later?

6. During adiabatic expansion of air, the pressure p and the volume v are related by the equation

$$pv^{7/5} = k \text{ (a constant)}$$

At a given moment the pressure in a certain container is 50 lb per sq in. and the volume of the container is 32 cu in. If the volume is decreasing at the rate of 4 cu in. per min at this moment, how fast is the pressure changing?

3.8 HIGHER DERIVATIVES

If the function f is differentiable on the interval $\{x : a < x < b\}$, then its derivative f' is a function of x on the same interval. As a function, f' may also have a derivative $(f')'$, which is called the *second derivative* of the function f. We usually denote the second derivative by f'' or in the "D" notation, by $D(Df) = D^2 f$.

The new function f'' may itself be differentiable and we could obtain the *third derivative* of f, denoted by $(f'')' = f'''$, or by $D(D^2 f) = D^3 f$. Similarly, we repeat the process to obtain the fourth, fifth, and so forth, derivatives of f, any one of which is a *higher derivative* of f. Since we run out of space to keep adding primes to f, we usually write

$$(f''')' = f^{(4)}$$

or

$$D(D^3 f) = D^4 f$$

for the fourth derivative and, in general, the *nth derivative* of f is defined inductively by

$$(f^{(n-1)})' = f^{(n)}$$

or

$$D(D^{n-1}f) = D^n f$$

■ **Example 1.** If $f(x) = x^3 - 3x^2 + 4$, then $f'(x) = 3x^2 - 6x$, $f''(x) = 6x - 6$ and $f'''(x) = 6$, whence $f^{(n)}(x) = 0$ for all $n \geq 4$.

■ **Example 2.** If $f(x) = (x-1)/x$, then $f'(x) = x^{-2}$, $f''(x) = -2x^{-3}$, $f'''(x) = 6x^{-4}$, $f^{(4)}(x) = -24x^{-5}$, and so forth. Note that none of these higher derivatives is the zero function.

The value $f'(x_0)$ of the derivative of a function may be interpreted as the slope of the tangent line to the graph $\{(x,y) : y = f(x)\}$ at the point $(x_0, f(x_0))$. We shall give a geometric interpretation of the *sign* of a value $f''(x_0)$, not of $f''(x_0)$ itself.

A function f is said to be *smooth* on an interval $\{x : a < x < b\}$ if both f' and f'' exist for all points x in the interval. The graph

$$\{(x, y) : y = f(x), a < x < b\}$$

is called a *smooth curve*. Note that the existence of f' implies that f is continuous and, similarly, the existence of $f'' = (f')'$ implies that f' is continuous (Theorem 4.5). Because f' is continuous, we often say that the graph has a *continuously turning tangent line*.

A smooth curve $\{(x, y) : y = f(x)\}$ is *convex* over an interval $\{x : a < x < b\}$ if the following condition holds: For any choice of two points $(x_1, f(x_1))$ and $(x_2, f(x_2))$, $a < x_1 < x_2 < b$, on the curve, the secant line between these points is above the curves between these points. Thus, analytically,

$$f(x) \leq f(x_1) + \frac{f(x_2) - f(x_1)}{x_2 - x_1}(x - x_1)$$

whenever $x_1 \leq x \leq x_2$

Similarly, the curve is *concave* over the interval if, for any choice of x_1 and x_2, $a < x_1 < x_2 < b$, we have

$$f(x) \geq f(x_1) + \frac{f(x_2) - f(x_1)}{x_2 - x_1}(x - x_1)$$

whenever $x_1 \leq x \leq x_2$

Figure 3.8 illustrates these properties and includes typical secant lines.

■ **THEOREM 3.26.** The smooth curve $\{(x, y) : y = f(x)\}$ is convex over any interval on which f' is strictly increasing and is concave over any interval on which f' is strictly decreasing.

3.8 HIGHER DERIVATIVES

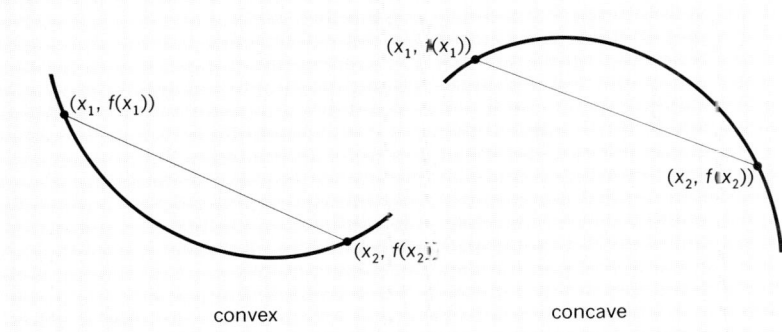

convex   concave   FIGURE 3.8

A proof of this theorem will be called for in a problem after we have studied the theorem of the mean (Section 3.13).

The real working tool in determining convexity is the following.

■ **THEOREM 3.27.** The graph of a smooth function f is convex over any interval on which $f''(x) > 0$ and is concave over any interval on which $f''(x) < 0$.

PROOF: We note that, if $f''(x) > 0$ on an interval, then Theorem 3.1 concludes that f' is an increasing function on this interval. Therefore Theorem 3.26 can be applied. The second part has an analogous proof. ■

The smooth curve $\{(x, y) : y = f(x)\}$ is *convex at a point* $(x_0, f(x_0))$ if there is an open interval containing x_0 over which the curve is convex; similarly, the curve is *concave at a point* $(x_1, f(x_1))$ if x_1 lies in an open interval over which the curve is concave.

■ **THEOREM 3.28.** Assume that the second derivative f'' of a function f is continuous. Then the curve $\{(x, y) : y = f(x)\}$ is convex at any point $(x, f(x))$ for which $f''(x) > 0$ and is concave at any point $(x, f(x))$ where $f''(x) < 0$.

PROOF: Theorem 3.8 tells us that the continuous function f'' is positive over an entire interval containing x if it is positive at x. Then Theorem 3.27 applies. For the second part of the theorem, we consider the function $-f''$ and apply the same argument. ■

Note that Theorem 3.28 provides no information about points $(x_0, f(x_0))$ on the graph for which $f''(x_0) = 0$. Indeed, if $f''(x_0) = 0$, the curve may be convex, concave, or neither at $(x_0, f(x_0))$. Consider the function $f(x) = x^4$. Then $f'(x) = 4x^3$ and $f''(x) = 12x^2$. We have $f''(0) = 0$, but surely $f'(x) = 4x^3$ is increasing over the entire set R. Hence the graph $\{(x, y) : y = x^4\}$ is convex at every point. On the other hand, if $f(x) = -x^4$, we again have $f''(0) = 0$ but the graph $\{(x, y) : y = -x^4\}$ is concave at every point. Finally, if $f(x) = x^3$, then $f'(x) = 3x^2$ and $f''(x) = 6x$. Again $f''(0) = 0$. However, $f''(x) < 0$ if $x < 0$ and $f''(x) > 0$ if $x > 0$. It follows that the graph $\{(x,y) : y = x^3\}$ is convex if $x > 0$ and is concave if $x < 0$. At the point $(0, 0)$, the graph is neither convex nor concave.

A point $(x_0, f(x_0))$ is a *point of inflection* of the graph $\{(x,y) : y = f(x)\}$ if the graph is convex on one side of x_0 and is concave on the other side of x_0. More precisely, there are intervals $\{x : a < x < x_0\}$ and $\{x : x_0 < x < b\}$ over one of which the graph is convex and over the other of which the graph is concave. (See Fig. 3.9.)

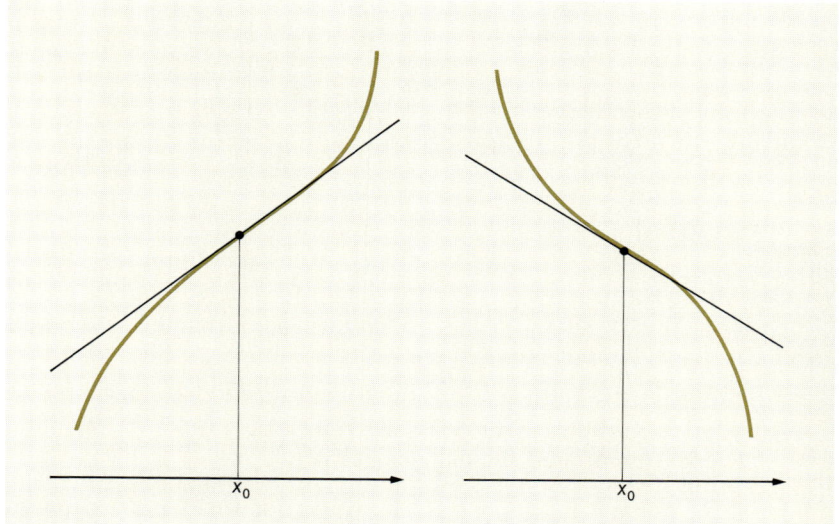

FIGURE 3.9

Assuming that f'' is continuous, then $f''(x)$ has opposite signs for $x > x_0$ and $x < x_0$. It follows that $f''(x_0) = 0$, that is, *the second derivative is zero at a point of inflection*. The example $f(x) = x^4$ tells us that the converse of this statement is not true. That is, the fact that $f''(x_0) = 0$ does not necessarily imply that $(x_0, f(x_0))$ is a point of inflection.

Another aid in drawing graphs can now be introduced. Each time we compute a point $(x_0, f(x_0))$ on a graph $\{(x,y) : y = f(x)\}$, we not only obtain the slope $f'(x_0)$ at that point, but we also compute $f''(x_0)$ to tell us whether the curve is convex or concave at the point $(x_0, f(x_0))$. This knowledge will surely be of use in drawing the graph.

One of our techniques is to locate all possible extremal points on the graph by equating $f'(x)$ to zero and solving the resulting equation for x. Now we see that we can *locate all possible points of inflection of the graph by equating $f''(x)$ to zero*. The step-by-step procedure in the examples that follow is quite efficient and the reader should follow it in his own work.

■ **Example 3.** Let $f(x) = 5 - 12x - 3x^2 + 2x^3$. Then $f'(x) = -12 + 6x + 6x^2$ and $f''(x) = 6 + 12x$.

Step 1. Set $f'(x)$ equal to zero to find the critical values:

$$-12 + 6x + 6x^2 = 6(x + 2)(x - 1) = 0$$

The critical values are $x = 1$ and $x = -2$ and hence the stationary points on the graph $\{(x, y) : y = 5 - 12x + 3x^2 + 2x^3\}$ are $(1, -2)$ and $(2, 25)$.

Step 2. Set $f''(x)$ equal to zero to find points of inflection

$$6 + 12x = 6(1 + 2x) = 0$$

The solution $x = -\frac{1}{2}$ tells us that $(-\frac{1}{2}, \frac{23}{2})$ is the only possible point of inflection on the graph.

Step 3. Determine where the curve is convex by solving the inequality $f''(x) > 0$ and where the curve is concave by solving $f''(x) < 0$.

In the present case, $6 + 12x > 0$ whenever $x > -\frac{1}{2}$ and $6 + 12x < 0$ whenever $x < -\frac{1}{2}$. Hence the curve is convex over the set $\{x : x > -\frac{1}{2}\}$ and is concave over the set $\{x : x < -\frac{1}{2}\}$. Because the curve is convex on one side of $x = -\frac{1}{2}$ and concave on the other, we now know that $(-\frac{1}{2}, \frac{23}{2})$ is indeed a point of inflection.

Step 4. Prepare a table of values of x, $f(x)$, and $f'(x)$, including the stationary points and the points of inflection: For the example under discussion, we have computed the following table.

x	-3	-2	-1	$-\frac{1}{2}$	0	1	2
$f(x)$	14	25	18	$\frac{23}{2}$	5	-2	9
$f'(x)$	24	0	-12	$-\frac{27}{2}$	-12	0	24

Step 5. Plot the points computed in step 4 and draw a short segment of the tangent line at each point.

Step 6. Draw the curve making use of the knowledge of convexity and continuity.

These last two steps are illustrated in Fig. 3.10, which illustrates the graph

$$\{(x, y) : y = 5 - 12x + 3x^2 + 2x^3\}$$

■ **Example 4.** Let $f(x) = x/(x^2 + 1)$. The reader may verify that

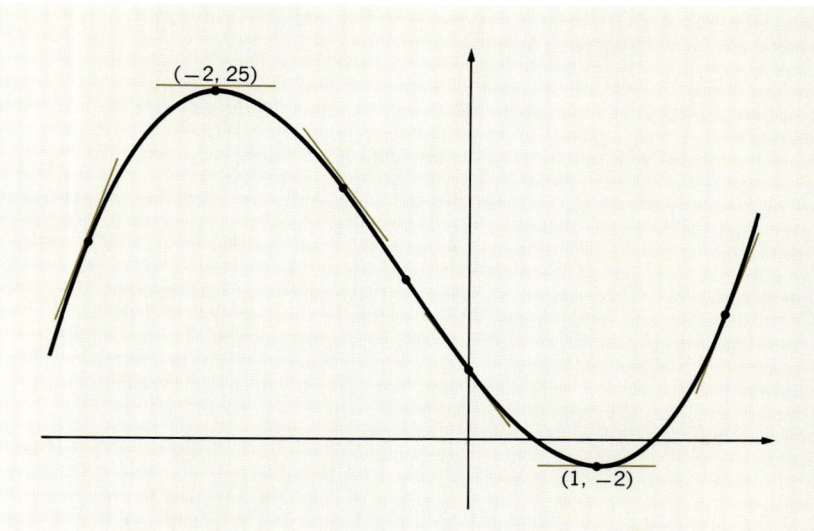

FIGURE 3.10

$$f'(x) = \frac{1 - x^2}{(x^2 + 1)^2}, \quad f''(x) = \frac{2x(x^2 - 3)}{(x^2 + 1)^3}$$

Step 1. The equation

$$f'(x) = \frac{1 - x^2}{(x^2 + 1)^2} = 0$$

has solutions $x = 1$, $x = -1$, whence the stationary points are $(1, \frac{1}{2})$ and $(-1, -\frac{1}{2})$.

Step 2. The equation

$$f''(x) = \frac{2x(x^2 - 3)}{(x^2 + 1)^3} = 0$$

has roots $x = 0$, $x = \sqrt{3}$, and $x = -\sqrt{3}$. Therefore the possible points of inflection are $(0, 0)$, $(\sqrt{3}, \frac{1}{4}\sqrt{3})$ and $(-\sqrt{3}, -\frac{1}{4}\sqrt{3})$.

Step 3. The inequality

$$f''(x) = \frac{2x(x^2 - 3)}{(x^2 + 1)^3} > 0$$

has solution sets $\{x : x > \sqrt{3}\}$ and $\{x : -\sqrt{3} < x < 0\}$, so that the curve is convex over these intervals. Similarly,

$$f''(x) = \frac{2x(x^2 - 3)}{(x^2 + 1)^3} < 0$$

3.8 HIGHER DERIVATIVES

is solved to prove that the curve is concave over $\{x : x < -\sqrt{3}\}$ and $\{x : 0 < x < \sqrt{3}\}$. Because f'' changes sign at each of the values 0, $\sqrt{3}$, and $-\sqrt{3}$, it follows that each point determined in step 2 actually is a point of inflection.

Step 4. A table for this function follows.

x	-3	-2	$-\sqrt{3}$	-1	0	1	$\sqrt{3}$	2	3
$f(x)$	$-\frac{3}{10}$	$-\frac{2}{5}$	$-\frac{1}{4}\sqrt{3}$	$-\frac{1}{2}$	0	$\frac{1}{2}$	$\frac{1}{4}\sqrt{3}$	$\frac{2}{5}$	$\frac{3}{10}$
$f''(x)$	$-\frac{8}{100}$	$-\frac{3}{25}$	$-\frac{1}{4}$	0	1	0	$-\frac{1}{4}$	$-\frac{3}{25}$	$-\frac{8}{100}$

Adding the facts that the curve is symmetric to the origin and that its only intercept is the origin, we picture this graph in Fig. 3.11.

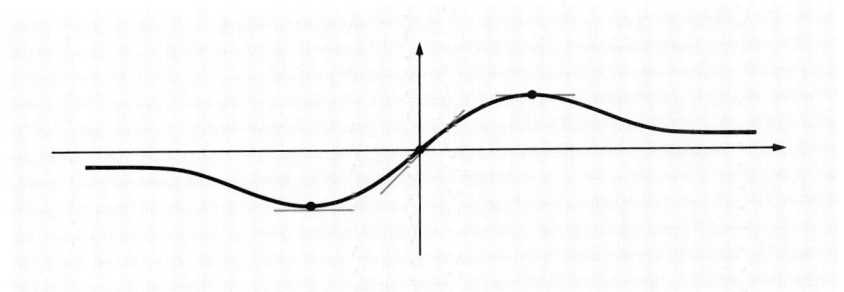

FIGURE 3.11

Two functions f and g are *tangent of order n* at a point x_0 if the $n+1$ equations

$$f(x_0) = g(x_0)$$
$$f'(x_0) = g'(x_0)$$
$$\vdots$$
$$f^{(n)}(x_0) = g^{(n)}(x_0)$$

are all satisfied. (Of course, we are assuming that these derivatives all exist.) This generalizes the definition of Section 3.1.

■ **Example 5.** Let $f(x) = x^2$ and $g(x) = -1 + 3x - 2x^2 + x^3$. Then we have

$$f(1) = 1 = g(1)$$
$$f'(1) = 2 = g'(1)$$
$$f''(1) = 2 = g'(2)$$

Hence these functions are tangent of order 2 at $x = 1$.

It is difficult to draw functions with higher order tangency because they are so very much alike near the point of tangency. We shall not try.

■ **Example 6.** Let $f(x) = 1 - 3x + 2x^2$ and let $g(x) = a_0 + a_1 x + a_2 x^2 + x^3$. We can determine the coefficients a_0, a_1, and a_2 so that f and g are tangent of order 2 at $x = 0$. The defining conditions are

$$f(0) = 1 = g(0) = a_0$$
$$f'(0) = -3 = g'(0) = a_1$$
$$f''(0) = 4 = g''(0) = 2a_2$$

Therefore, choosing $a_0 = 1, a_1 = -3$, and $a_2 = 2$, we have obtained the function $g(x) = 1 - 3x + 2x^2 + x^3$ of the given form that is tangent of order 2 at $x = 0$ to the function $f(x) = 1 - 3x + 2x^2$.

EXERCISES 3.8

1. Determine both the second and third derivatives of each of the following functions.
    (a) $f(x) = 2 - 4x - x^2$
    (b) $f(x) = 4 - 6x + 2x^2$
    (c) $f(x) = 6 + 2x + x^3$
    (d) $f(x) = 2 - 3x + 3x^2 + x^3$
    (e) $f(x) = \dfrac{x}{x-1}$
    (f) $f(x) = \dfrac{1}{x^2 + 4}$
    (g) $f(x) = \sqrt{x^2 + 1}$
    (h) $f(x) = x\sqrt{x+1}$
    (i) $f(x) = \dfrac{x^2 - 1}{x^2 + 1}$
    (j) $f(x) = \dfrac{x}{\sqrt{x+1}}$

2. For each of the following functions determine where the graph is (1) convex and (2) concave. Find the points of inflection. Draw the graph.
    (a) $f(x) = 3 - 6x + 8x^2$
    (b) $f(x) = 4 + 2x - 3x^2$
    (c) $f(x) = 1 - 3x + x^3$
    (d) $f(x) = -1 + 2x - 6x^2 + x^3$
    (e) $f(x) = \dfrac{x+2}{x-1}$
    (f) $f(x) = \dfrac{x^2}{x-1}$
    (g) $f(x) = \dfrac{2x}{x^2 + 4}$
    (h) $f(x) = \sqrt{\dfrac{x}{x+1}}$
    (i) $f(x) = \dfrac{x}{\sqrt{x^2 + 1}}$
    (j) $f(x) = \dfrac{x^2 - 1}{\sqrt{x^2 + 1}}$

3. Sketch a graph of a function satisfying each of the following sets of conditions:
    (a) $f(0) = 4, f'(0) = 1, f''(0) > 0$
    (b) $f(-1) = 1, f'(-1) = -1, f''(-1) < 0$
    (c) $f(1) = 2, f'(1) = -1, f''(1) > 0$
    (d) $f(2) = -3, f'(2) = \frac{1}{2}, f''(2) < 0$

4. Determine constants a, b, and c for each function g, so that it and the given function f shall be tangent of order 2 at the point $(1, 0)$.
    (a) $f(x) = x^2 - 1; g(x) = a + bx + cx^2 + x^3$
    (b) $f(x) = 2 - 3x + x^3; g(x) = a + bx + cx^2 - x^3$
    (c) $f(x) = \dfrac{4(x-1)}{x+1}; g(x) = a + bx + cx^2$
    (d) $f(x) = \dfrac{x-1}{x^2+1}; g(x) = -2 + ax + bx^2 + cx^3$

3.8 HIGHER DERIVATIVES

5. Determine the constants a, b, and c so that the graph of the function $f(x) = a + bx + cx^2 + x^3$ has a relative maximum point at $(-1, 7)$ and a relative minimum point on the vertical line $\{(x,y) : x = 1\}$.

6. Determine the constants a, b, c, and d so that the graph of the function $f(x) = a + bx + cx^2 + dx^3$ passes through the points $(1, 8)$ and $(3, 0)$, has a point of inflection at $(3, 0)$, and has a tangent line of slope -6 at the point of inflection.

7. Determine the constants a, b, c, d, and e so that the graph of the function $f(x) = a + bx + cx^2 + dx^3 + ex^4$ passes through the points $(1, 2)$, $(0, 5)$ and $(-7, 2)$, has a point of inflection at $(1, 2)$, and has a horizontal tangent line at this point of inflection.

8. Prove that each function below satisfies the accompanying equation
   (a) $f(x) = a + bx^2$; $xf'' = f'$

   (b) $f(x) = a + bx^2 + \dfrac{1}{x}$; $x^3 f'' - x^2 f' = 3$

   (c) $f(x) = a + \frac{1}{3}bx^3 + \frac{1}{4}x^4$; $xf'' - 2f' = x^3$

   (d) $f(x) = \sqrt{2x - x^2}$; $ff'' + (f')^2 = -1$

PROBLEMS 3.8
Mathematics

1. Provide formulas, in terms of the derivatives of f and g, for
   (a) $D^2(fg)$ (b) $D^3(fg)$
   (c) $D^2(g \circ f)$ (d) $D^3(g \circ f)$

2. Provide a formula for $f^{(n)}(x)$ for each of the following functions:
   (a) $f(x) = \dfrac{1}{x}$ (b) $f(x) = \dfrac{1}{1-x}$

   (c) $f(x) = \dfrac{1}{(2 - 3x)^2}$ (d) $f(x) = \dfrac{1}{\sqrt{x+1}}$

3. Assume that the function f has continuous second derivative at every point x. Furthermore, suppose that $f(\pm 1/n) = 1/n$ if n is an even natural number, and that $f(\pm 1/n) = 1/n^2$ if n is an odd natural number. What are the values of $f(0)$, $f'(0)$, and $f''(0)$? Can you say anything about the convexity of such a function at $x = 0$?

4. Derive formulas for f' and f'' if:
   (a) $f(x) = u(x)v(x)w(x)$ (b) $f(x) = \dfrac{u(x)v(x)}{w(x)}$

   (c) $f(x) = [u(x)]^m[v(x)]^n$ (d) $f(x) = [u(x)]^m[v(x)]^n[w(x)]^p$

PROBLEMS 3.8
Velocity and acceleration

If an object moves along a line, then its distance from a fixed point on the line is a function $s(t)$ of the number t of units of time that have elapsed since some fixed instant. The time rate of change of the distance $s(t)$ is the *velocity* v of the moving object,

$$v = Ds$$

The time rate of change of the velocity is the *acceleration* a experienced by the object,

$$a = Dv = D^2s$$

An object falling in air is subjected to the acceleration of gravity. If air resistance is neglected, then this is the only acceleration acting on the object. Its height in feet above the surface of the earth is given by $s(t) = s_0 + v_0 t - 16t^2$, where s_0 is its height above the surface at $t = 0$ and v_0 is its (vertical) velocity at $t = 0$. Then its velocity at time t is $Ds = v_0 - 32t$ and its acceleration is the constant -32 (ft per sec per sec). If one knows these facts, the following problems are easily solved.

1. From a point 56 ft above the ground, with what velocity must a stone be thrown down in order to reach the ground in 1 sec?

2. A stone is thrown vertically upward from the top of a tower. Its initial velocity is 100 ft per sec; it reaches the ground with a velocity of 140 ft per sec. Find the height of the tower.

3. A stone is thrown vertically upward from the top of a tower. At the end of two sec, it is still moving up at a velocity of 10 ft per sec and is 380 ft above the ground. Find the height of the tower.

4. From what height should an automobile be dropped in order to produce the same impact as a 60-mph crash into a solid wall?

   Other problems concerning velocity and acceleration occur in later sections. At the present, the following is enough to suggest other applications.

5. A rocket leaves the surface of the earth. Its distance s in feet from the center of the earth is given by

$$s = R\left(1 + \frac{12t}{\sqrt{R}}\right)^{2/3}$$

where R is the radius of the earth in feet and t is the number of seconds elapsed since its flight began. Show that

$$v = \frac{8R}{\sqrt{s}}$$

$$a = -\frac{32R^2}{s^2}$$

Discuss the motion and find the initial velocity (called the *escape velocity*).

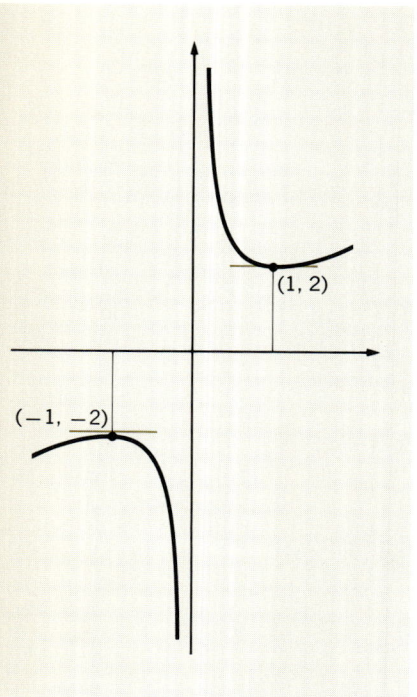

FIGURE 3.12

3.9 EXTREMUM VALUES

The second derivative can be used to identify the relative extremum values of a function. In most instances, the *second derivative test* given in the following theorem is much easier than the methods given in Section 3.2.

■ **THEOREM 3.29.** Let the function f have a continuous second derivative. If $f'(x_0) = 0$, then $f(x_0)$ is a relative maximum value if $f''(x_0) < 0$ and $f(x_0)$ is a relative minimum value if $f''(x_0) > 0$.

PROOF: Theorem 3.28 applies to show that $\{(x,y) : y = f(x)\}$ is concave at $(x_0, f(x_0))$ if $f''(x_0) < 0$. That is, f' is a decreasing function on some interval containing x_0. Because f' is continuous and $f'(x_0) = 0$, Theorem 3.2 tells us that $f(x_0)$ is a relative maximum value. The proof of the second part is analogous. ■

■ **Example 1.** The quadratic function $f(x) = a_0 + a_1 x + a_2 x^2$ has $f'(x) = a_1 + 2a_2 x$ and $f''(x) = 2a_2$. The critical value is $x = -a_1/2a_2$ and the stationary point on the graph is $(-a_1/2a_2, a_0 - a_1^2/4a_2)$. If $a_2 <$

0, then the graph is concave everywhere, so that the value $a_0 - a_1^2/4a_2$ is a maximum. On the other hand, if $a_2 > 0$, the graph is convex everywhere and $a_0 - a_1^2/4a_2$ is a minimum value.

■ **Example 2.** If $f(x) = (x^2 + 1)/x$, then the reader should verify that

$$f'(x) = \frac{x^2 - 1}{x^2} \quad \text{and} \quad f''(x) = \frac{2}{x^3}$$

By setting $f'(x)$ equal to zero, we determine the stationary points of the graph $\{(x,y) : y = (x^2 + 1)/x\}$ to be $(1, 2)$ and $(-1, -2)$. Because $f''(1) = 2 > 0$, $f(1) = 2$ is a relative minimum value and because $f''(-1) = -2 < 0, f(-1) = -2$ is a relative maximum value.

It may seem peculiar to have a relative maximum smaller than a relative minimum, as we found here, but Fig. 3.12 (on p. 152) shows how this happens in our present case.

EXERCISES 3.9

1. Find and identify the stationary points of each of the following functions.
   (a) $f(x) = 3 - 4x - 2x^2$
   (b) $f(x) = 24 + 48x - 7x^2 - 4x^3$
   (c) $f(x) = x^4 + 6x^2$
   (d) $f(x) = 8 + x - x^2 + 4x^4$
   (e) $f(x) = \frac{1}{x^2}$
   (f) $f(x) = \frac{x^2}{1 + x}$
   (g) $f(x) = \frac{x}{1 - x^2}$
   (h) $f(x) = \sqrt{8 - 12x + 3x^2}$
   (i) $f(x) = \frac{2x}{\sqrt{1 + x}}$
   (j) $f(x) = x^2\sqrt{-2 - 4x + 3x^2}$

2. Determine the value of the constant c so that the function
   $$f(x) = x^2 + \frac{c}{x}$$
   has
   (a) a relative minimum at $x = 2$,
   (b) a relative minimum at $x = -3$,
   (c) a point of inflection at $x = 1$, and
   (d) any relative maximum

3. Sketch the graph of a function f, $x > 0$, that has $f(1) = 0$ and $f'(x) = 1/x$. Can you determine the convexity properties of such a function?

4. Show that $f(x) = a + bx + cx^2 + x^3$ has an extremum value if and only if $c^2 > 3b$. Discuss the situation if $c^2 = 3b$ and if $c^2 < 3b$. Draw a graph illustrating each of the three situations.

PROBLEMS 3.9
Mathematics

1. Consider the set of all right circular cones that can be inscribed in a given sphere. What is the maximum of the volumes of these cones?

2. Consider the set of all right circular cones that can be circumscribed about a given sphere. What is the minimum of the volumes of these cones?

3. Find the dimensions of the rectangle of perimeter 36 that will sweep out the maximum volume when revolved about the following:
   (a) a vertical line through the center of the rectangle,
   (b) one of its sides, and
   (c) a line parallel to a side and one unit away from the side

4. A right circular cylinder of radius x is inscribed in a right circular cone of base radius r and altitude h. Under what conditions is there such a cylinder of maximum surface area? Prove that the inscribed cylinder with maximum volume has $\frac{4}{9}$ the volume of the cone.

5. One corner of a sheet of paper $8\frac{1}{2}$ inches wide is folded over to just touch the opposite edge of the paper. Find the width of the folded part when (a) the crease has a minimal length, and (b) the area of the folded-over triangle is a minimum.

6. Consider $f(x) = \sqrt{r^2 + a^2 - 2ax}$ where $0 < a < r$. This function is defined on $\{x : -r \leq x \leq r\}$ and the derivative $f'(x)$ is never zero in this interval. Nevertheless, the function attains both a maximum value and a minimum value in this interval. Draw a graph of the function and explain this phenomenon.

7. Show that the equation $k + x(1 - x^3) = 0$ has two real roots for each constant $k > -\frac{3}{8}\sqrt[3]{2}$.

PROBLEMS 3.9
Physical Sciences

1. An open storage box with a square base and vertical sides is to be made from a given fixed amount of lumber. Neglecting waste and the thickness of the lumber, find the dimensions of the box of maximum volume that can be so constructed.

2. A long strip of galvanized sheet metal 12 in. wide is to be formed into an open gutter by bending up the edges to form a trough with rectangular cross section. What dimensions provide the gutter of maximum volume?

3. A Norman window consists of a rectangle surmounted by a semicircle. What proportions yield the window of maximum area for a given perimeter?

4. A building site is in the form of a right triangle with sides 120 and 160 ft long. What is the floor area of the largest rectangular (one-story) building facing the hypotenuse that can be built on this site?

5. Two towns are located along a straight lake shore at perpendicular distances of 1 and 6 miles from the water. The straight line distance between the towns is 13 miles. Where on the lake should a water pump be located so as to service both towns with a minimum total length of pipe?

6. An oil can is designed to be a cylinder surmounted by a cone. The height of the cone is to be $\frac{6}{5}$ of its diameter. Determine the most economical proportions for the oil can.

7. The stiffness of a rectangular beam is proportional to its width and to the cube of its thickness. Find the proportions of the stiffest beam that can be cut from a round log of given diameter.

8. If a wire is bent into the shape of a square of side $2a$, then the attraction on a particle on the line perpendicular to the plane of the square and through the center of the square is proportional to

$$\frac{x}{\sqrt{x^2 + a^2}\sqrt{x^2 + 2a^2}}$$

where x is the distance from the particle to the center of the square. Find the point of maximum attraction.

9. The amount of water flowing over the spillway of a certain dam is given by
$$V = bd\sqrt{2g(h-d)}$$
where V is the number of cubic feet of water flowing per second, b is the width of the spillway, d is the depth of the water as it flows over the spillway, $g = 32$ ft per sec and h is the head of water. If b and h are fixed, what value of d maximizes V?

10. The intensity of illumination at any point is proportional to the strength of the light source and inversely proportional to the square of the distance from the light source. Suppose two sources of light with relative strength s and t are a distance d apart. At what point between these light sources is the intensity of illumination a maximum?

11. A beam of length L is cantilevered from a wall at one end and is simply supported at the other end. The deflection y at a distance x from the wall is given by
$$48Ely = w(2x^4 - 5Lx^3 + 3L^2x^2)$$
where E, l, and w are constants. Where is the deflection a maximum?

12. Find the length of the longest thin rod that can be carried horizontally around a corner from a hall 8 ft wide into a hall 4 ft wide.

13. Repeat Problem 12 if the halls have 8-ft ceilings and the rod may be tilted in negotiating the corner.

14. A cylindrical hole of radius x is bored through a ball of radius R with the axis of the hole passing through the center of the ball. Determine x so that the complete surface area of the remaining solid is a maximum and show that this maximum area is precisely $\frac{3}{4}\sqrt{3}$ times the surface area of the ball.

15. A circular ring of radius r is uniformly charged with electricity, the total charge being Q. A unit particle on the axis of the ring and at a distance x from the center of the ring experiences a force due to the charge of
$$F = Qx(x^2 + r^2)^{-3/2}$$
Find the maximum value of F.

16. In a particular electrostatic field the electric intensity E at a given point on the x axis is given by
$$E(x) = \left(x^2 + \frac{c^4}{x^2}\right)^{-1/2}, \quad c > 0$$
For what values of x is $E'(x) > 0$? At what points is E maximum?

PROBLEMS 3.9
Economics and management

1. An oil field now has 40 oil wells averaging 150 barrels of oil per day each. If new wells are drilled, it is estimated that the average yield will drop 3 barrels per day for each new well added to the field. What number of wells will provide the maximum daily yield?

2. An apartment complex has 150 units. When the rent of each unit is $150 per month, all units are occupied. For each $5 increase in rent, it is found that one unit is vacated and remains empty. If each occupied unit requires $15 per month of services and repairs, what rent should be charged in order to obtain a maximum profit?

3. Show that a manufacturer maximizes (or minimizes) his profit if his production level makes his marginal revenue equal his marginal cost.

4. It costs d dollars to manufacture and distribute a certain item. If each item is priced at x dollars, then total sales are estimated to be

$$y = \frac{a}{x-d} + b(100 - x)$$

where a and b are positive constants. At what price should these items be sold in order to maximize profits?

5. Suppose a manufacturer can sell x units of his product per week if the unit price is $p = 360 - 0.02x$ cents. If it costs $80x + 30{,}000$ cents to make x items, what production level provides maximum profit?

6. In Problem 5, suppose that the government imposes a tax of 12 cents per item on each item sold. How much of this tax should the manufacturer absorb and how much should he pass on to the customer? Explain your response.

7. An oil refinery produces $(25 - 2x)/(15 - x)$ thousand gallons per day of a premium grade gasoline in conjunction with x thousand gallons of regular gasoline. The refinery must produce at least 1000 gallons per day of regular gasoline; its total capacity is 12,500 gallons of regular gasoline per day. If the premium gasoline brings in $\frac{9}{5}$ as much per gallon as does the regular gasoline, what output of regular gasoline maximizes the total revenue?

8. The fuel cost for a river boat running at x miles per hr in still water is $\$x^3/32$ per hr. Other operational costs total $\$160$ per hr. What is the most economical speed this boat can run against a 4-mph current in making an upstream trip?

9. A highway truck has a minimum speed in high gear of 10 miles per hr and a maximum speed of 65 mph. When traveling at v miles per hr, the truck uses $4/v + v/256$ gallons of fuel per mile. Suppose that the fuel costs 22 cents per gallon. (a) What steady speed minimizes the total cost of fuel for a given trip? (b) What steady speed minimizes total costs if the driver is paid $\$2.50$ per hr?

10. A wholesale dealer is faced with the following problem: If he carries a large inventory, then carrying costs (such as storage fees, damages, insurance, and so forth) could be too high. On the other hand, if his inventory is low, then reorder costs could be excessive. Suppose that he orders x units at a time and that the total cost of holding one unit for a year is p dollars. Since we may assume that his average inventory is $\frac{1}{2}x$ (Why?), his total carrying cost will be $\frac{1}{2}xp$ dollars per year.

If the dealer expects to sell q units per year and to receive x units per shipment, then he must get q/x shipments. If b represents fixed shipping costs and c is the cost of shipping x units, then the total cost of a shipment is $b + cx$. Hence the annual reorder cost is $(q/x)(b + cx)$ and the total inventory cost is

$$C(x) = \frac{1}{2}px + \frac{q}{x}(b + cx)$$

Minimize C and prove that the dealer should increase his inventory in proportion to the square root of expected sales.

3.10 POLYNOMIAL FUNCTIONS

In this and the following section, we must draw upon the reader's knowledge of algebra. Several algebraic theorems are used without explicit statement.

A *polynomial function* p has values

$$p(x) = a_0 + a_1x + a_2x^2 + \cdots + a_nx^n$$

where each coefficient a_i, $i = 0, 1, \cdots, n$, is a real number ($a_n \neq 0$) and where n is a nonnegative integer called the *degree* of the polynomial.

A polynomial function of degree zero is a constant function and a polynomial function of degree one is a linear function. A polynomial function $a_0 + a_1x + a_2x^2$ of degree two is called a *quadratic function* and a polynomial function $a_0 + a_1x + a_2x^2 + a_3x^3$ is a *cubic function*. We shall not give special names to polynomial functions of degree $n > 3$.

The graph $\{(x,y) : y = p(x)\}$ of a polynomial function is called a *polynomial curve*.

■ **THEOREM 3.30.** A straight line intersects a polynomial curve of degree n ($n > 1$) in at most n points.

PROOF: A vertical line $\{(x,y) : x = c\}$ meets a curve $\{(x,y) : y = p(x)\}$ in exactly one point $(c, p(c))$, because p is a function.

Any nonvertical line has the form $\{(x,y) : y = mx + b\}$. The intersection of such a line with the graph of $p(x) = a_0 + a_1x + a_2x^2 + \cdots + a_nx^n$ is the set

$$\begin{aligned}\{(x,y) : y = mx + b\} &\cap \{(x,y) : y = p(x)\} \\ &= \{(x,y) : y = mx + b \quad \text{and} \quad y = p(x)\} \\ &= \{(x,y) : y = mx + b = p(x)\}\end{aligned}$$

This means that we find the x coordinates of the points of intersection to be the *real* roots of the equation

$$mx + b = p(x) = a_0 + a_1x + a_2x^2 + \cdots + a_nx^n$$

or

$$(a_0 - b) + (a_1 - m)x + a_2x^2 + \cdots + a_nx^n = 0$$

We know from our algebra courses that such a polynomial equation of degree n has precisely n roots, counting the complex roots and multiplicity. Hence it cannot have more than n real roots and the theorem follows.

We know that the polynomial function p has the derivative

$$p'(x) = (a_0 + a_1x + a_2x^2 + \cdots + a_nx^n)' = a_1 + 2a_2x + \cdots + na_nx^{n-1}$$

which we recognize as a polynomial function of degree $n - 1$. Now the domain X_p of a polynomial function is the set $\mathbf{R}$ of all real numbers. Hence we say that *a polynomial function is differentiable everywhere*. But because p' is again a polynomial function, it follows that p'' exists. Indeed,

$$\begin{aligned}p''(x) &= (a_1 + 2a_2x + 3a_3x^2 + \cdots + na_nx^{n-1})' \\ &= 2a_2 + 6a_3x + \cdots + (n - 1)na_nx^{n-2}\end{aligned}$$

Hence *every polynomial function is smooth*. ■

■ **THEOREM 3.31.** Let $p(x) = a_0 + a_1x + a_2x^2 + \cdots + a_nx^n$ be a polynomial function of degree n. Then the nth derivative of p is the constant function

$$p^{(n)}(x) = n!a_n$$

where $n! = 1 \cdot 2 \cdot 3 \cdots n$, and hence all higher derivatives $p^{(n+k)}$ are zero.

PROOF: We easily prove the theorem for $n = 1$. If $p(x) = a_0 + a_1 x$, then $p'(x) = a_1 = 1!a_1$. As the second step in a proof by mathematical induction, assume the theorem to be true for all $n \leq k$ and let

$$p(x) = a_0 + a_1 x + a_2 x^2 + \cdots + a_{k+1} x^{k+1}$$

be a polynomial function of degree $k + 1$. We then write

$$D^{k+1} p = D^k(Dp) = D^k(a_1 + 2a_2 x + \cdots + (k+1)a_{k+1} x^k)$$

By the induction assumption, we have

$$D_x^k(a_1 + 2a_2 x + \cdots + (k+1)a_{k+1} x^k) = k!(k+1)a_{k+1}$$
$$= (k+1)!a_{k+1}$$

which completes the proof. ∎

We interrupt the general development here to consider the relations between quadratic functions and quadratic equations. The reader is probably accustomed to seeing a quadratic equation in the form $Ax^2 + Bx + C = 0$. Hence we shall write the general quadratic function as $Ax^2 + Bx + C$. Obviously,

$$(Ax^2 + Bx + C)' = 2Ax + B$$
$$(Ax^2 + Bx + C)'' = 2A$$

Of course, we assume $A \neq 0$ or else the function would not be *quadratic*.

By Theorem 3.28, the graph $\{(x,y) : y = Ax^2 + Bx + C\}$ is convex everywhere if $A > 0$ and is concave everywhere if $A < 0$. Setting the derivative $2Ax + B$ equal to zero, we find that

$$\left(-\frac{B}{2A}, -\frac{B^2 - 4AC}{4A}\right)$$

is the only stationary point on the graph. Thus, if $A > 0$, $-(B^2 - 4AC)/4A$ is the minimum value of the function and if $A < 0$, this is the maximum value. The two cases are shown in Fig. 3.13.

Suppose for a moment that $A > 0$, whence the function has a minimum value. If the minimum value $-(B^2 - 4AC)/4A$ is positive, then the quantity $B^2 - 4AC$ must be negative. Conversely, if $B^2 - 4AC < 0$, then $-(B^2 - 4AC)/4A > 0$. If the minimum value is positive, then the graph can never cross the x axis. That is, the intersection

$$\{(x,y) : y = Ax^2 + Bx + C\} \cap \{(x,y) : y = 0\}$$
$$= \{(x, 0) : Ax^2 + Bx + C = 0\}$$

is the empty set. Hence the equation $Ax^2 + Bx + C = 0$ has no real roots. This agrees exactly with the algebraic fact that $Ax^2 + Bx + C = 0$ has two complex conjugate roots (no real roots) if the discriminant $B^2 - 4AC$ is negative.

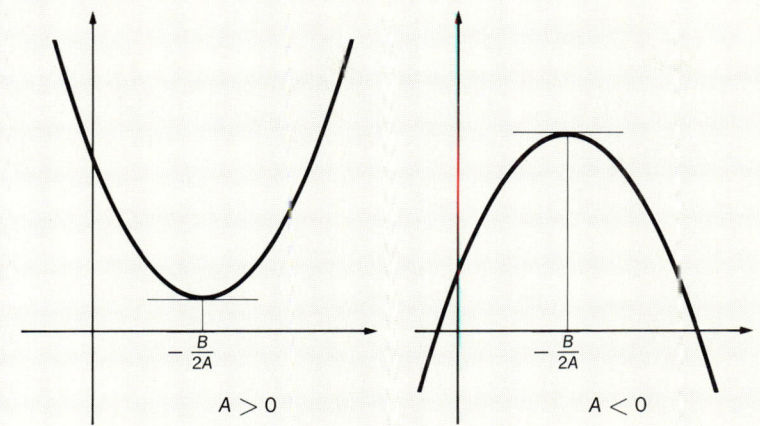

FIGURE 3.13

Now consider the case where $A < 0$. If the maximum value $-(B^2 - 4AC)/4A$ is negative, then again the graph does not cross the x axis and again $B^2 - 4AC$ must be negative. Here, too, the geometric interpretation of a negative discriminant is obvious. That is, the graph $\{(x,y) : y = Ax^2 + Bx + C\}$ fails to touch the x axis if and only if $B^2 - 4AC < 0$.

If either $A > 0$ or $A < 0$, then the extremum value $-(B^2 - 4AC)/4A$ is zero if and only if $B^2 - 4AC = 0$. This tells us that the graph meets the x axis only in the stationary point (see Fig. 3.14). Here we see the geometric equivalent to the fact that $Ax^2 + Bx + C = 0$ has one real (double) root if and only if $B^2 - 4AC = 0$.

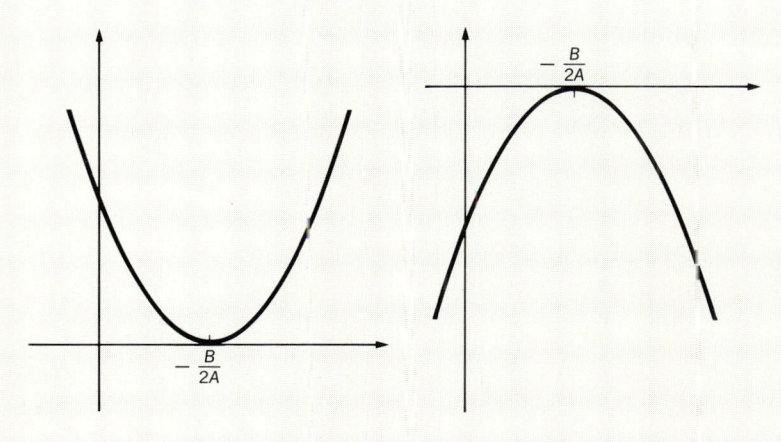

FIGURE 3.14

Finally, assume that $B^2 - 4AC > 0$. If $A > 0$, then the minimum value $-(B^2 - 4AC)/4A$ is negative. Hence the quadratic curve crosses the x axis in two different points. Similarly, if $A < 0$, then the maximum value $-(B^2 - 4AC)/4A$ is positive and again the curve crosses the x axis in two points. This geometric conclusion agrees precisely with the fact that $Ax^2 + Bx + C = 0$ has two distinct real roots if $B^2 - 4AC > 0$. Typical curves appear in Fig. 3.15.

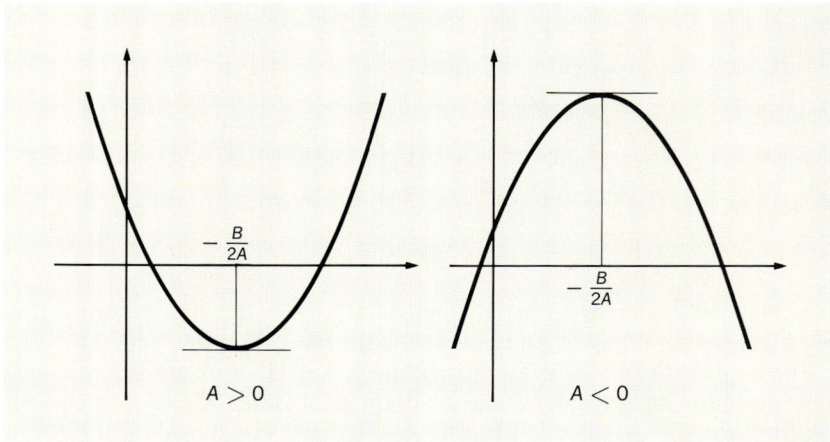

FIGURE 3.15

The quadratic curve $\{(x, y) : y = Ax^2 + Bx + C\}$ is also called a *parabola* (see Chapter 13). Because we can write

$$Ax^2 + Bx + C = A\left(x + \frac{B}{2A}\right)^2 - \frac{B^2 - 4AC}{4A}$$

the above parabola is symmetric to the vertical line $\{(x, y) : x = -B/2A\}$. This line of symmetry is called the *axis* of the parabola; it is the vertical line through the stationary point.

Returning to the general polynomial function, we need another algebraic fact. *If two polynomials of degree n have equal values at more than n distinct points, then the polynomials are identical.* This is an easy consequence of the fact that a polynomial of degree n has exactly n zeros, counting multiplicity.

■ **THEOREM 3.32.** If $(x_0, y_0), (x_1, y_1), \cdots, (x_n, y_n)$ are $n + 1$ points in a coordinate plane with $x_0 < x_1 < \cdots < x_n$, then there is one and only one polynomial function of degree $\leqq n$ whose graph passes through these points.

PROOF: If there is one such polynomial function, then the algebraic fact stated above tells us that it is the only such function. Therefore we need show only that there is such a polynomial function.

3.10 POLYNOMIAL FUNCTIONS

We write a polynomial function of degree n in the following form:
$$p(x) = a_0 + a_1(x - x_0) + a_2(x - x_0)(x - x_1) + \cdots \\ + a_n(x - x_0)(x - x_1) \cdots (x - x_{n-1})$$

The coefficients a_i are now chosen so that $p(x_i) = y_i$ for $i = 0, 1, \cdots, n$. Note that

$$p(x_0) = a_0 + a_1(x_0 - x_0) + a_2(x_0 - x_0)(x_0 - x_1) + \cdots \\ + a_n(x_0 - x_0)(x_0 - x_1) \cdots (x_0 - x_{n-1})$$
$$= a_0$$

Therefore we must choose $a_0 = y_0$.

Similarly,
$$p(x_1) = a_0 + a_1(x_1 - x_0) = y_0 + a_1(x_1 - x_0) = y_1$$

or
$$a_1 = \frac{y_1 - y_0}{x_1 - x_0}$$

Continuing this process,
$$p(x_2) = a_0 + a_1(x_2 - x_0) + a_2(x_2 - x_0)(x_2 - x_1)$$
$$= y_0 + \left(\frac{y_1 - y_0}{x_1 - x_0}\right)(x_2 - x_0) + a_2(x_2 - x_0)(x_2 - x_1) = y_2$$

or
$$a_2 = \frac{y_2 - y_0}{(x_2 - x_0)(x_2 - x_1)} - \frac{y_1 - y_0}{(x_1 - x_0)(x_2 - x_1)}$$

It is now obvious that we may continue this process to obtain all of the coefficients a_i. ∎

A special case of Theorem 3.32 is used in Section 10.7 to develop a formula for approximate integration. Consider three points (x_0, y_0), $(x_0 + h, y_1)$ and $(x_0 + 2h, y_2)$ whose x coordinates are equally spaced. Then there is a polynomial function of degree ≤ 2, whose graph passes through these three points. We first write
$$p(x) = y_0 + a_1(x - x_0) - a_2(x - x_0)(x - x_0 - h)$$

Then
$$p(x_0 + h) = y_0 + a_1(x_0 + h - x_0) = y_1$$

so that
$$a_1 = \frac{1}{h}(y_1 - y_0)$$

Lastly,
$$p(x_0 + 2h) = y_0 + \frac{1}{h}(y_1 - y_0)(x_0 + 2h - x_0) \\ + a_2(x_0 + 2h - x_0)(x_0 + 2h - x_0 - h) = y_2$$

or
$$y_2 = y_0 + \frac{1}{h}(y_1 - y_0)(2h) + a_2(2h)(h)$$
so that
$$a_2 = \frac{1}{2h^2}(y_2 - 2y_1 + y_0)$$

Our polynomial is therefore
$$p(x) = y_0 + \frac{1}{h}(y_1 - y_0)(x - x_0)$$
$$+ \frac{1}{2h^2}(y_2 - 2y_1 + y_0)(x - x_0)(x - x_0 - h)$$

It is to this formula we shall refer in Section 10.7.

The formula just derived is *Newton's forward interpolation formula* for $n = 2$. It gives considerably more accuracy than the familiar linear interpolation. The reader can derive the general form of Newton's formula by considering equally spaced values of x, $x_i = x_0 + ih$, $i = 0, 1, 2, \cdots, n$ and using the method in the proof of Theorem 3.32. Such formulas find many applications in *numerical analysis*.

The customary methods of evaluating a polynomial are not easy to reduce to a simple repetitive routine suitable for an electronic computer. Computer programmers use the following method: Suppose that the polynomial
$$a_0 + a_1 x + a_2 x^2 + a_3 x^3 + a_4 x^4 + a_5 x^5$$
is to be evaluated for some fixed x. We write the polynomial as
$$a_0 + a_1 x + a_2 x^2 + a_3 x^3 + x^4(a_4 + x a_5)$$
$$= a_0 + a_1 x + a_2 x^2 + x^3(a_3 + x(a_4 + x a_5))$$
$$= a_0 + a_1 x + x^2(a_2 + x(a_3 + x(a_4 + x a_5)))$$
$$= a_0 + x(a_1 + x(a_2 + x(a_3 + x(a_4 + x a_5))))$$

Now the evaluation is a simple repetition of the two steps: (1) multiply a number by x, and (2) add a number. Thus we cause the machine to first form

(1) $x \cdot a_5$

and then

(2) $a_4 + x \cdot a_5$

The first repetition computes

(1) $x(a_n + x \cdot a_5)$

and then

(2) $a_3 + x(a_4 + x \cdot a_5)$

The second repetition yields

$$(1) \quad x(a_3 + x(a_4 + xa_5))$$

and then

$$(2) \quad a_2 + x(a_3 + x(a_4 + xa_5))$$

Having instructed the machine to stop computing after the second step of the fourth repetition, the value of the polynomial has been formed. This method is quite suitable for hand computation, too.

PROBLEMS 3.10

1. For what values of the constant c does
   (a) $p(x) = 6 - 2cx + x^2$ have two distinct roots?
   (b) $p(x) = 4c + 3x + x^2$ have no real roots?
   (c) $p(x) = 3 - 4x + cx^2$ have one double root?

2. Obtain a formula for $f^{(k)}(x)$ if $f(x) = (1 + nx^m)(1 + mx^n)$, $m, n \in \mathbf{N}$.

3. Show that each polynomial function of odd degree has the entire set $\mathbf{R}$ as its range (and hence has at least one real zero).

4. Given the formula

$$1 + x + x^2 + \cdots + x^n = \frac{x^{n+1} - 1}{x - 1}, \quad x \neq 1$$

use differentiation to obtain a formula for $x + 4x^2 + 9x^3 + \cdots + n^2 x^n$.

5. (a) Find a cubic polynomial $f(x)$ such that, for all $x \in \mathbf{R}$,

$$f(x + 1) - f(x) = 3x^2$$

   (b) Find a fourth-degree polynomial such that, for all $x \in \mathbf{R}$,

$$f(x + 2) - 2f(x + 1) + f(x) = 12x^2$$

6. Let $p(x) = a_0 + a_1 x + \cdots + a_n x^n$ be a polynomial function and let c be any point. Prove that there exist unique constants $b_0, b_1, \cdots, b_n$ such that $p(x) = b_0 + b_1(x - c) + \cdots + b_n(x - c)^n$. (Hint: Consider the higher derivatives evaluated at $x = c$.)

7. For what polynomials $p(x)$ is
   (a) $p(nx) = np(x)$?
   (b) $p(x + h) = p(x) + p(h)$?

8. Prove the converse of Theorem 3.31 by showing that $D^n f(x) = 0$ only if f is a polynomial of degree less than n. (What other assumption would make this an easy problem?)

9. A short table of square roots follows. Fit a polynomial of degree 3 to the first four points and use the result to obtain an interpolated value of $\sqrt{4.5}$.

x	4	5	6	7	8
$\sqrt{x}$	2	2.2361	2.4495	2.6458	2.8284

10. Prove that the quadratic function $f(x) = Ax^2 + 2Bx + C$ is always nonnegative only if $B^2 \leq AC$. Apply this result to the function

$$g(x) = (a_1 x + b_1)^2 + (a_2 x + b_2)^2 + \cdots + (a_n x + b_n)^2$$

to deduce *Schwartz' inequality*

$$(a_1 b_1 + a_2 b_2 + \cdots + a_n b_n)^2$$
$$\leq (a_1^2 + a_2^2 + \cdots + a_n^2)(b_1^2 + b_2^2 + \cdots + b_n^2)$$

11. If $f(x)$ is a polynomial function, what relationships exist between the number of real roots of the two equations $f(x) = 0$ and $f'(x) = 0$? Can $f(x) = 0$ have three distinct real roots, while $f'(x) = 0$ has no real root? Can $f(x) = 0$ have no roots, while $f'(x) = 0$ has two real roots? Develop a general theorem.

12. Prove that the set $P[x]$ of all polynomials has precisely the same properties under addition and multiplication as does the set **Z**.

13. The numbers $a_1, a_2, \cdots, a_n$ were recorded as measurements of a quantity x in n repetitions of a certain experiment. Determine the value of x that minimizes the quantity $(a_1 - x)^2 + (a_2 - x)^2 + \cdots + (a_n - x)^2$. Discuss your result.

14. Let $r > 1$ be a rational number and prove that the function

$$f(x) = (1 + x)^r - (1 + rx), \qquad x \geq -1$$

has (a) an absolute minimum value of zero occurring only at $x = 0$ and (b) if $x > -1$ and $x \neq 0$, then

$$(1 + x)^r > 1 + rx$$

15. Let $m \geq 2$ and $n \geq 2$ be integers and define

$$f(x) = (x - 1)^m (x + 1)^n$$

Show that f has three critical points at $x = 1$, $x = -1$, and $x = \dfrac{m - n}{m + n}$.

Find the extremum values of $f(x)$ when
(a) m and n are both even
(b) m and n are both odd
(c) m is even and n is odd
(d) m is odd and n is even

16. Let $q(x)$ be a polynomial such that $q(c) \neq 0$. Define the new polynomial $p(x) = (x - c)^k q(x)$. Show that

$$p(c) = p'(c) = \cdots = p^{(k-1)}(c) = 0$$

but that

$$p^{(k)}(c) = n!\, q(c)$$

3.11 ALGEBRAIC FUNCTIONS

The polynomial functions are the simplest type of algebraic functions. Problem 3.10.12 claims that the set of all polynomial functions has properties very similar to those of the set of integers. This analogy between numbers and functions is carried on here.

A *rational function* f is the quotient function of two polynomial functions, that is,

$$f = \frac{p}{q}$$

where p and q are polynomial functions. Since p and q are differentiable everywhere, the rational function is differentiable at each point x for which $q(x) \neq 0$. Note that the derivative

$$\left(\frac{p}{q}\right)' = \frac{p'q - pq'}{q^2}$$

is again a rational function.

All of the techniques developed so far are used in drawing the graph of a rational function, of course. But we can add another useful technique here. In addition to the vertical asymptotes, we can locate three other types:

1. If the degree of the numerator $p(x)$ is less than the degree of the denominator $q(x)$, then the x axis is a horizontal asymptote to the curve $\left\{(x,y) : y = \frac{p(x)}{q(x)}\right\}$.

2. If the degrees of $p(x)$ and $q(x)$ are equal, then the horizontal line $\{(x,y) : y = a_n/b_n\}$ is an asymptote to $\left\{(x,y) : y = \frac{p(x)}{q(x)}\right\}$, where a_n and b_n are the coefficients of the highest power of x in $p(x)$ and $q(x)$, respectively.

3. If the degree of $p(x)$ is larger by one than that of $q(x)$, and if the degree of $q(x)$ is not zero, then the line

$$\left\{(x,y) : b_{n-1}y = a_n x + a_{n-1} - \frac{a_n b_{n-2}}{b_{n-1}}\right\}$$

is an asymptote (a_n and b_{n-1} as in case 2.)

We shall not prove these rules. They can be established with the aid of the following definition: A line $\{(x,y) : y = mx + b\}$ is asymptotic to the curve $\{(x,y) : y = f(x)\}$ if

$$\lim_{x \to \infty} |f(x) - mx - b| = 0$$

■ **Example 1.** Let

$$f(x) = \frac{x}{x^2 + 1}$$

According to rule 1 above, the x axis is a horizontal asymptote. This curve was discussed in Example 4 of Section 3.8 and is shown in Fig. 3.11.

■ **Example 2.** Let

$$f(x) = \frac{x^2 + 1}{2x^2 - 3x}$$

Because the denominator has zeros at $x = 0$ and $x = \frac{3}{2}$, the vertical lines $\{(x,y) : x = 0\}$ and $\{(x,y) : x = \frac{3}{2}\}$ are asymptotes. In addition, rule 2 above says that the line $\{(x,y) : y = \frac{1}{2}\}$ is a horizontal asymptote.

■ *Example 3.* Let
$$f(x) = \frac{x^2 - 1}{2x}$$

In addition to the vertical asymptote $\{(x,y) : x = 0\}$, rule 3 tells us that the line $\{(x,y) : y = \frac{1}{2}x\}$ is an asymptote. We readily prove that there is no y intercept and that $(1, 0)$ and $(-1, 0)$ are the x intercepts. Also, we see that f is an odd function, so that its graph is symmetric to the origin. The derivative is $f'(x) = (x^2 + 1)/2x^2$ and is never zero. Hence the graph has no stationary points. The second derivative is $f''(x) = -1/x^3$. Hence the curve is convex when $x < 0$ and is concave when $x > 0$. Since $f''(x)$ is never zero, there are no points of inflection. See Fig. 3.16.

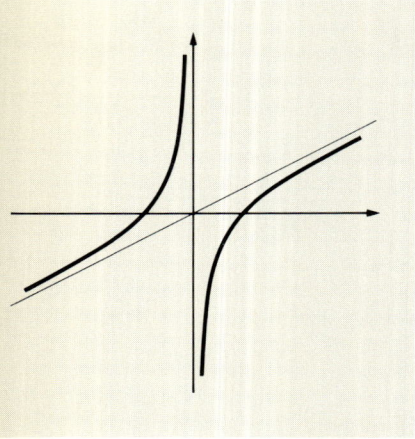

FIGURE 3.16

A number x is said to be algebraic if it satisfies a polynomial equation
$$a_0 + a_1 x + \cdots + a_n x^n = 0$$
where each $a_i \in \mathbf{Z}$. In precise analogy, the function f is *algebraic* if it satisfies an equation
$$p_0 + p_1 f + p_2 f^2 + \cdots + p_n f^n = 0 \qquad \text{(the zero function)}$$
where each coefficient p_i, $i = 0, 1, \cdots, n$, is a polynomial function.

■ *Example 4.* The function $f(x) = x^{1/2}$ is algebraic because it satisfies the equation
$$x - f^2 = 0$$

■ *Example 5.* The function $f(x) = x(1 + x^2)^{-1/2}$ is algebraic because it satisfies the equation
$$x^2 - (1 + x^2)f^2 = 0$$
where the coefficients x^2, an unwritten 0, and $-(1 + x^2)$ are polynomials.

The set of all algebraic functions is very large, and two algebraic functions can exhibit widely differing properties. It would be much too time-consuming to attempt a systematic study here of the general algebraic function. However, all of the general theorems about continuity and differentiation apply to these functions. We need only use special care in determining the domain of such a function.

A student often asks "In what order shall I apply the rules of differentiation to this algebraic function?" We describe here a mnemonic device to help the beginner to differentiate algebraic functions.

■ *Example 6.* Let $f(x) = x(1 - x^2)^{1/2}$. To differentiate this, we first apply the product rule of differentiation to obtain

$$Df(x) = (Dx)(1 - x^2)^{1/2} + xD(1 - x^2)^{1/2}$$
$$= (1 - x^2)^{1/2} + xD(1 - x^2)^{1/2}$$

The problem has been reduced to finding $D(1 - x^2)^{1/2}$, which we shall do with the power rule of differentiation.

The point here is that, if we were to evaluate this function $f(x)$ for some fixed number x, then the last arithmetic operation would be multiplication. For instance, to compute $f(\frac{1}{2})$ we first evaluate $1 - x^2$, then take the square root $\sqrt{1 - \frac{1}{4}}$, and finally multiply $\frac{1}{2} \times \sqrt{1 - \frac{1}{4}}$.

■ **Example 7.** If

$$f(x) = \frac{(4 + x)^{2/3}}{(2 + x^2)^{1/2}}$$

then we first apply the quotient rule to obtain

$$Df(x) = \frac{(2 + x^2)^{1/2}D(4 + x)^{2/3} - (4 + x)^{2/3}D(2 + x^2)^{1/2}}{(2 + x^2)}$$

Then the power rule applies to compute $D(4 + x)^{2/3}$ and $D(2 + x^2)^{1/2}$. Again we point out that in evaluating $f(2)$, for instance, we compute $(4 + 2)^{2/3}$ and $(2 + 2)^{1/2}$ and our final step is the division of $(4 + 2)^{2/3}$ by $(2 + 2^2)^{1/2}$.

■ **Example 8.** If $f(x) = \sqrt{(x + 1)/(x - 1)}$, then the power rule is applied first,

$$D\left(\frac{x + 1}{x - 1}\right)^{1/2} = \frac{1}{2}\left(\frac{x + 1}{x - 1}\right)^{-1/2} D\left(\frac{x + 1}{x - 1}\right)$$

Then the quotient rule applies to compute $D[(x + 1)/(x - 1)]$. Once more, the last step in computing, say $f(4)$, would be to extract the square root ($\frac{1}{2}$ power) of $(4 + 1)/(4 - 1)$.

This is our mnemonic device: *The first rule of differentiation to be applied to a function corresponds to the last arithmetic operation that would be performed in evaluating the function.* This device works equally well in making any subsidiary calculations, of course, and hence provides a systematic procedure for differentiating algebraic functions. We do not try to justify this device. We simply claim that it works.

EXERCISES 3.11

For each function 1 to 10, find f' and f''.

1. $f(x) = \dfrac{x + 1}{(x - 2)^2}$

2. $f(x) = \dfrac{(x + 2)^2}{x - 2}$

3. $f(x) = \dfrac{x^2}{x^2 - 4}$

4. $f(x) = \dfrac{2 + x + 2x^2}{1 + x + x^2}$

5. $f(x) = x\sqrt{x + 1}$

6. $f(x) = \dfrac{\sqrt{4 - x^2}}{x}$

7. $f(x) = \dfrac{x + 5}{\sqrt{x^2 + 1}}$

8. $f(x) = \dfrac{\sqrt[3]{2x + 5}}{\sqrt{4 - x^2}}$

9. $f(x) = x(x - 1)^{3/4}(x + 2)^{-1/3}$

10. $f(x) = \dfrac{(x^2 + \sqrt{x + 1})^3}{(x^2 + 3)^2}$

PROBLEMS 3.11

For each of Exercises 1 to 6, apply all of the methods you know to draw a careful sketch of the graph of the function.

1. For what rational functions $r(x)$ is $r(x) = r(2x)$ for each $x \in \mathbf{R}$?

2. Let $x_0 = x$ and define x_{n+1} inductively by
$$x_{n+1} = \frac{1 - x_n}{1 + x_n}$$
Find a formula for x_{n+1} as a function of x.

3. Show that a rational function is algebraic.

4. Is the function $f(x) = |x|$ algebraic?

5. Show that the following functions are algebraic:

   (a) $f(x) = \dfrac{x + \sqrt{x^2 + c^2}}{\sqrt{x^2 + c^2}}$ (b) $f(x) = \dfrac{\sqrt{x^2 - 1} - \sqrt[3]{x^2 + 1}}{\sqrt{x^3 - 2x}}$

6. Determine the relationship among the constants a, b, and c, so that the function
$$f(x) = \frac{(x - a)(x - b)}{(x - c)}$$
has the entire set $\mathbf{R}$ as its range.

7. Let $p(x) = a_0 + a_1 x + \cdots + a_m x^m$ and $q(x) = b_0 + b_1 x + \cdots + b_n x^n$ be polynomials. Show that

   (a) If $n > m$, then $\lim\limits_{x \to \infty} \dfrac{p(x)}{q(x)} = 0$

   (b) If $n = m$, then $\lim\limits_{x \to \infty} \dfrac{p(x)}{q(x)} = \dfrac{a_n}{b_n}$

   (c) If $n = m - 1$, then $\lim\limits_{x \to \infty} \dfrac{p(x)}{xq(x)} = \dfrac{a_m}{b_{m-1}}$

8. Prove *Leibnitz'* rule:
$$D^n(fg) = (D^n f)g + n(D^{n-1}f)(Dg) + \cdots$$
$$+ \frac{n(n - 1) \cdots (n - k + 1)}{k!}(D^{n-k}f)(D^k g) + \cdots + f(D^n g)$$

3.12 A MEAN VALUE THEOREM FOR DERIVATIVES

The "mean value theorem" established here is one of the most useful theoretical results in our theory. Its importance is illustrated in the following section by giving several applications.

The first theorem is a special case of our main result. It is known as *Rolle's theorem*.

3.12 A MEAN VALUE THEOREM FOR DERIVATIVES

■ **THEOREM 3.33.** Let f be continuous on an interval containing the points a and b, $a < b$. Assume that $f'(x)$ exists at each point of $\{x : a < x < b\}$. If $f(a) = f(b)$, there is at least one point x_0 in $\{x : a < x < b\}$ such that $f'(x_0) = 0$.

PROOF: In view of Theorem 3.13, there exist points x_1 and x_2 in $\{x : a \leq x \leq b\}$ such that $f(x_1) \leq f(x) \leq f(x_2)$ for each point x in the interval. Let $k = f(a) = f(b)$. If $f(x_1) = k = f(x_2)$, then f is the constant function k whence $f' = 0$, the constant function. In this case any point x_0, $a < x_0 < b$, satisfies the conclusion of the theorem.

If $f(x_1) < f(x_2)$, then we must have $f(x_1) \leq k$ and $f(x_2) \geq k$ with equality holding in at most one case. If $f(x_1) < k$, then because $f(x_1)$ is a minimum value of f and $a < x_1 < b_1$, Theorem 3.3 concludes that $f'(x_1) = 0$. Similarly, if $f(x_2) > k$, then $f(x_2)$ is a maximum value of f and $a < x_2 < b$ and again Theorem 3.3 tells us that $f'(x_2) = 0$. ■

Figure 3.17 illustrates three functions satisfying the hypotheses, and hence the conclusion, of Rolle's theorem.

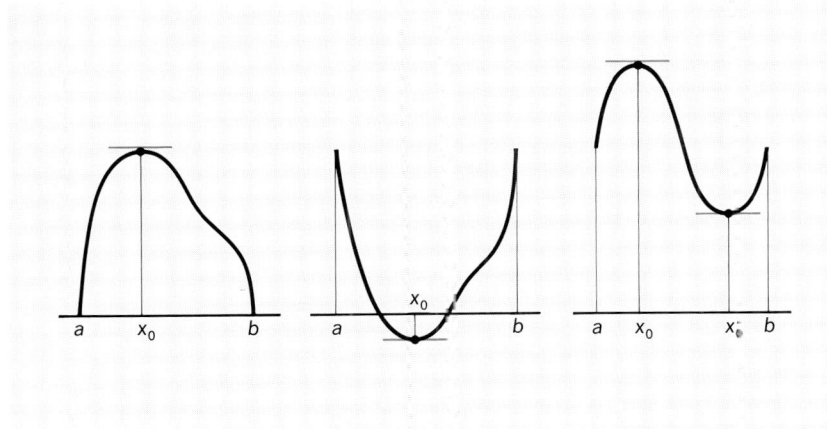

FIGURE 3.17

We do need each of the hypotheses of Rolle's theorem. To prove this, we shall negate the hypotheses one at a time and construct examples that fail to satisfy the conclusion of the theorem. First, if the function fails to be continuous at one of the endpoints of the interval $\{x : a \leq x \leq b\}$, then the conclusion need not follow. (Note that f must be continuous on $\{x : a < x < b\}$, the open interval, if we assume that $f'(x)$ exists at each point.) Our example here is the function

$$f(x) = \frac{1}{x}, \quad x \neq 0$$
$$= 1, \quad \text{if } x = 0$$

Then $f(0) = 1 = f(1)$ and $f'(x) = -1/x^2$ exists on $\{x : 0 < x < 1\}$. But $f'(x)$ is never zero on the interval $\{x : 0 < x < 1\}$.

If the derivative $f'(x)$ fails to exist at just one point of the open interval $\{x : a < x < b\}$, then the conclusion of Rolle's theorem may fail to hold. Consider the function $f(x) = x^{2/3}$ which is continuous everywhere. Note that $f(-1) = 1 = f(1)$. The derivative $f'(x) = \frac{2}{3}x^{-1/3}$ fails to exist at $x = 0$, so the hypotheses of Rolle's theorem are satisfied except for this one little omission. But note that $f'(x)$ is never zero, so that the conclusion of Rolle's theorem does not apply. Fig. 3.18 shows why this failure occurs. Note that the minimum value of the functions occurs exactly where the derivative fails to exist.

The main result here is the *mean value theorem* for derivatives, as follows.

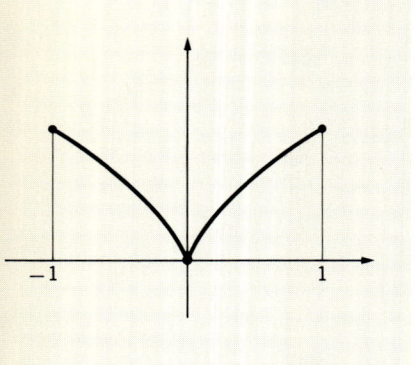

FIGURE 3.18

■ **THEOREM 3.34.** Let f be continuous on an interval containing the points a and b, $a < b$, and let f be differentiable on the open interval $\{x : a < x < b\}$. Then there is at least one point x_0 in $\{x : a < x < b\}$ such that

$$f'(x_0) = \frac{f(b) - f(a)}{b - a}$$

PROOF: We construct a new function

$$g(x) = f(x) + Ax + B$$

in such a way that g satisfies the hypotheses of Rolle's theorem. Because g has the same continuity and differentiability properties as does f, we need only select the constants A and B, so that $g(a) = g(b) = 0$. That is, A and B are chosen to satisfy

$$g(a) = f(a) + Aa + B = 0$$
$$g(b) = f(b) + Ab + B = 0$$

Solving these for A and B, we obtain

$$A = -\frac{f(b) - f(a)}{b - a}, \quad B = \frac{af(b) - bf(a)}{b - a}$$

With these choices of A and B, the function g satisfies all three hypotheses of Rolle's theorem. Hence there is a point x_0 in $\{x : a < x < b\}$ such that $g'(x_0) = 0$. But obviously $g'(x) = f'(x) + A$, so that we must have

$$g'(x_0) = f'(x_0) + A = f'(x_0) - \frac{f(b) - f(a)}{b - a} = 0$$

and the proof is complete. ■

In case $f(a) = k = f(b)$, the constant $[f(b) - f(a)]/(b - a)$ equals zero. It follows that Rolle's theorem is a special case of the mean value theorem. This is one of many instances in mathematics where a theorem is proved by first proving a special case of that theorem.

A geometric interpretation of the mean value theorem is easy to understand. First note that the quantity $[f(b) - f(a)]/(b - a)$ is the slope of the line through the points $[a, f(a)]$ and $[b, f(b)]$ on the graph of the function. We call this the secant line. Next recall that $f'(x)$ is the slope of the tangent line to the graph at the point $[x, f(x)]$. Therefore the mean value theorem says that there is at least one point $[x_0, f(x_0)]$ on the graph where the tangent line is parallel to the secant line (see Fig. 3.19).

The conclusion of the mean value theorem is often written in slightly different forms. By solving the equation $f'(x_0) = [f(b) - f(a)]/(b - a)$ for the quantity $f(b)$, we can state the conclusion as follows: There exists a point x_0, $a < x_0 < b$, such that

$$f(b) = f(a) + (b - a)f'(x_0)$$

Then by simply setting $b - a = h$, we have $b = a + h$ and may write the equation as

$$f(a + h) = f(a) + hf'(x_0)$$

for some x_0 between a and $a + h$. We may interpret this equation as an expression for the functional value $f(a - h)$ in terms of the value $f(a)$, the interval length h, and some intermediate value $f'(x_0)$ of the derivative. In this form the mean value theorem is often used to obtain close approximations to the value $f(a + h)$. We shall do precisely this at several places later.

Another form of the mean value theorem is obtained as follows: Because $a < x_0 < a + h$, we define the number $\theta = (x_0 - a)/h$. Then, clearly, we have $0 < \theta < 1$ and $x_0 = a + \theta h$. Therefore the mean value theorem may be stated as follows: There exists a number θ, $0 < \theta < 1$, such that

$$f(a + h) = f(a) + hf'(a + \theta h)$$

This form is also used occasionally.

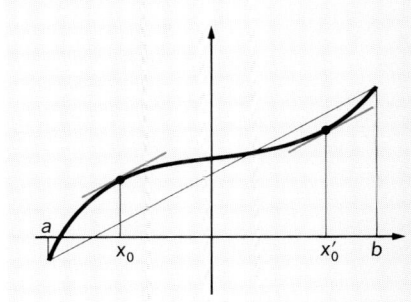

FIGURE 3.19

EXERCISES 3.12

1. Each of the following functions satisfies Rolle's theorem on the given interval. Determine a value x_0 such that $f'(x_0) = 0$
   (a) $f(x) = 60x - x^2$; $\{x : 0 \leq x \leq 60\}$
   (b) $f(x) = x^3 - 3x$; $\{x : 0 \leq x \leq \sqrt{3}\}$
   (c) $f(x) = \dfrac{x^2 - 1}{x + 2}$; $\{x : -1 \leq x \leq 1\}$
   (d) $f(x) = \dfrac{x^2 + 2x}{x^2 + 1}$; $\{x : -2 \leq x \leq 0\}$

2. Each of the following functions satisfies the mean value theorem on the given interval. Find an x_0 such that $(b - a)f'(x_0) = f(b) - f(a)$.
   (a) $f(x) = 4 - x^2$; $\{x : 1 \leq x \leq 2\}$
   (b) $f(x) = x^3 + 4x$; $\{x : -1 \leq x \leq 1\}$
   (c) $f(x) = \dfrac{x}{x+1}$; $\{x : 1 \leq x \leq 2\}$
   (d) $f(x) = \dfrac{x^2}{x-1}$; $\{x : -3 \leq x \leq -1\}$

3. As in Exercise 2 above, determine θ, $0 < \theta < 1$, such that
$$hf'(a + \theta h) = f(a + h) - f(a)$$
   (a) $f(x) = 1 + 3x - x^2$; $a = 2$, $h = 3$
   (b) $f(x) = 4x^2 - x^3$; $a = 2$, $h = 1$
   (c) $f(x) = \dfrac{x-1}{x}$; $a = 2$, $h = 2$
   (d) $f(x) = \sqrt{2x + 3}$; $a = 3$, $h = 8$

4. Use the mean value theorem to establish the following inequalities:
   (a) $\sqrt{1 + x} < 4 + \frac{1}{8}(x - 15)$ if $x > 15$
   (b) $1 + \dfrac{x}{2\sqrt{1+x}} < \sqrt{1+x} < 1 + \dfrac{1}{2}x$ if $-1 < x < 0$ or if $x > 0$
   (c) $1 - \dfrac{1}{2}x < \dfrac{1}{\sqrt{1+x}} < 1 - \dfrac{x}{2(1+x)^{3/2}}$ if $-1 < x < 0$ or if $x > 0$
   (d) $p(x - 1) < x^p - 1 < px^{p-1}(x - 1)$ if $p > 1$ and $x > 1$

3.13 SOME CONSEQUENCES OF THE MEAN VALUE THEOREM

Theorem 3.6 states that the derivative of a constant function is the zero function. A very important consequence of the mean value theorem is the converse statement.

■ **THEOREM 3.35.** If f is defined on an interval $\{x : a < x < b\}$ and if $f'(x) = 0$ for each point x in the interval, then $f(x)$ is constant over this interval.

PROOF: Because $f'(x)$ exists at each point, f is continuous at each point. Let x_1 and x_2 be any two points in the interval. Then the mean value theorem applies to provide a point x_0 between x_1 and x_2 such that
$$f(x_2) = f(x_1) + (x_2 - x_1)f'(x_0)$$
But by assumption, $f'(x_0) = 0$. Therefore $f(x_2) = f(x_1)$. Because the points x_1 and x_2 were selected arbitrarily, it follows that $f(x)$ is constant over the interval. ■

The next result is a corollary to Theorem 3.35 but it is extremely important in the theory of integration (see Chapter 7).

3.13 SOME CONSEQUENCES OF THE MEAN VALUE THEOREM

■ **THEOREM 3.36.** Let f and g be differentiable on an interval $\{x : a < x < b\}$. If $f'(x) = g'(x)$ for each point x in the interval, then there is a constant c such that $g(x) = f(x) + c$ for each point x.

PROOF: The function $g - f$ has derivative $(g - f)' = g' - f'$. By assumption $g' - f'$ is the zero function and hence by Theorem 3.35, $g - f$ is a constant function. That is, there exists a real number c such that $g(x) - f(x) = c$ for each point x in the interval.

Back in Section 3-2 we omitted the proof of Theorem 3.1. This result states that "A function f is increasing [decreasing] on any interval over which $f'(x)$ is always positive [negative]." This theorem is also a consequence of the mean value theorem. ■

PROOF OF THEOREM 3.1: Suppose that $f'(x) > 0$ at each point x in the interval $\{x : a < x < b\}$. Let x_1 and x_2, $x_1 < x_2$, be any two points in this interval. Then the mean value theorem applies to conclude that there exists a point x_0, $x_1 < x_0 < x_2$, such that

$$f(x_2) = f(x_1) + (x_2 - x_1)f'(x_0)$$

Because $x_2 - x_1 > 0$ and $f'(x_0) > 0$, it follows that $f(x_2) > f(x_1)$. This proves that for any two points satisfying $x_1 < x_2$, we have $f(x_1) < f(x_2)$, which is the definition of an increasing function. An analogous argument proves the second part of the theorem. ■

Our next result is an *extended mean value theorem*. We shall find use for it several times, the first of which is in the first section of the next chapter.

■ **THEOREM 3.37.** Assume that the function f and its derivative f' are continuous on an interval containing the points a and b, $a < b$, and assume that $f''(x)$ exists at each point x in $\{x : a < x < b\}$. Then there exists at least one point x_0, $a < x_0 < b$, such that

$$f(b) = f(a) + (b - a)f'(a) + \tfrac{1}{2}(b - a)^2 f''(x_0)$$

PROOF: We define a new function g of the form

$$g(x) = f(x) - f(a) - f'(a)(x - a) - B(x - a)^2$$

By construction, $g(a) = 0$ and we select the constant B so that $g(b) = 0$. That is, we choose B to satisfy the equation

$$g(b) = 0 = f(b) - f(a) - f'(a)(b - a) - B(b - a)^2$$

from which we obtain

$$B = \frac{f(b) - f(a)}{(b - a)^2} - \frac{f'(a)}{b - a}$$

With this choice of B, the function g satisfies Rolle's theorem. Hence

there exists a point x_1, $a < x_1 < b$, such that $g'(x_1) = 0$. We easily compute that
$$g'(x) = f'(x) - f'(a) - 2B(x - a)$$
and hence
$$g'(x_1) = f'(x_1) - f'(a) - 2B(x_1 - a) = 0$$
Also we have
$$g'(a) = f'(a) - f'(a) - 2B(a - a) = 0$$

Our assumptions on f imply that the function g' is differentiable on the interval $\{x : a < x < x_1\}$. Therefore g' satisfies Rolle's theorem on the interval $\{x : a \leq x \leq x\}$ and so there is a point x_0, $a < x_0 < x_1$, such that $(g')'(x_0) = 0$. But $(g')' = g''$, so $g''(x_0) = 0$. By computation
$$g''(x) = f''(x) - 2B$$

We have proved that there is a point x_0, $a < x_0 < b$, for which $f''(x_0) = 2B$ or $B = \tfrac{1}{2} f''(x_0)$. Recalling the value of B, we have
$$\frac{f(b) - f(a)}{(b - a)^2} - \frac{f'(a)}{b - a} = \frac{1}{2} f''(x_0)$$
and the theorem follows immediately. ∎

Using Theorem 3.37, we can provide a more sophisticated test for identifying relative extremum values of a function. Remember the function $f(x) = x^4$ with $f'(x) = 4x^3$ and $f''(x) = 12x^2$. The critical point of this function is $x = 0$. But $f''(x) = 0$, so our second derivative test fails. Indeed, we have no theorem that gives us an easy test for such functions. The following result provides a test that applies to many such cases.

■ **THEOREM 3.38.** Let f and its derivatives f' and f'' be defined on an interval $\{x : a < x < b\}$ and suppose that $f'(x_0) = 0$ for some point x_0 in this interval. If there is an interval $\{x : x_0 - \delta < x < x_0 + \delta\}$ containing x_0 over which f'' is never positive [negative], then $f(x_0)$ is a relative maximum [minimum] value.

PROOF: Let x be any point in the interval $\{x : x_0 - \delta < x < x_0 + \delta\}$. Applying Theorem 3.37 to the interval between x_0 and x, there is a point x_1 between x_0 and x such that
$$f(x) = f(x_0) + f'(x_0)(x - x_0) + \tfrac{1}{2} f''(x_1)(x - x_0)^2$$
Because $f'(x_0) = 0$, this equation reduces to
$$f(x) = f(x_0) + \tfrac{1}{2} f''(x_1)(x - x_0)^2$$
Now if $f''(x) \leq 0$, we have $f(x) \leq f(x_0)$ whence $f(x_0)$ is a relative maximum value. If $f''(x) \geq 0$, then $f(x) \geq f(x_0)$ so $f(x_0)$ is a relative minimum value. ∎

3.13 SOME CONSEQUENCES OF THE MEAN VALUE THEOREM

■ **Example 1.** Let $f(x) = x - 4x^2 + 6x^3 - 4x^4 + x^5$. Then

$$f'(x) = 1 - 8x + 18x^2 - 16x^3 + 5x^4$$
$$f''(x) = -8 + 36x - 48x^2 + 20x^3$$

By inspection we can see that $x = 1$ is a zero of both f' and f''. From there it is easy to factor these expressions, obtaining

$$f'(x) = (5x - 1)(x - 1)^3$$
$$f''(x) = 4(5x - 2)(x - 1)^2$$

The critical points are $x = \frac{1}{5}$ and $x = 1$. The interval $\{x : \frac{3}{4} < x < \frac{5}{4}\}$, for instance, contains $x = 1$ and $f''(x) \geq 0$ on this entire interval. Therefore $f(1) = 0$ is a relative minimum value. On the other hand, $f''(x) \leq 0$ on the interval $\{x : 0 < x < \frac{2}{5}\}$ containing $x = \frac{1}{5}$, so $f(\frac{1}{5}) = \frac{256}{3125}$ is a relative maximum value.

PROBLEMS 3.13

1. The mean value theorem states that

    $$f(b) = f(a) + (b - a)f'(x)$$

    for some point x between a and b. If c and b are close together, then $f'(x)$ and $f'(a)$ are almost equal. Thus $f(b)$ can be approximated by the quantity $f(a) + (b - a)f'(a)$. Apply this idea to approximate
    (a) $\sqrt{15}$ by taking $f(x) = \sqrt{x}$, $a = 16$ and $b = 15$
    (b) $(2.002)^2$ by taking $f(x) = x^2$, $a = 2$ and $b = 2.002$
    (c) $\sqrt[3]{30}$ by taking $f(x) = x^{1/3}$, $a = 27$ and $b = 30$
    (d) $\frac{1}{998}$ by taking $f(x) = 1/x$, $a = 1000$ and $b = 998$

2. Establish the following inequality for all $x \in \mathbf{R}$:

    $$\frac{1}{5} \leq \frac{x^2 - 4x + 9}{x^2 + 4x + 9} \leq 5$$

3. Let f be differentiable for all $x \in \mathbf{R}$ and suppose that $x = c$ is the largest critical value of f. Prove that
    (a) $f(x) \neq f(c)$ for every $x > c$
    (b) If $f(x_1) > f(c)$ for some one point $x_1 > c$, then f is increasing over the entire ray $\{x : c < x\}$.
    (c) If $f(x_1) < f(c)$ for some one point $x_1 > c$, then f is decreasing over the entire ray $\{x : c < x\}$.

4. Suppose that f is continuous on the closed interval $\{x : 0 \leq x \leq 1\}$ and that $0 \leq f(x) \leq 1$ for each point x in this interval. Prove that there is at least one point x_0 in the interval such that $f(x_0) = x_0$.

5. Let f be a polynomial function and suppose that x_1 and x_2 are real roots of the equation $f(x) = 0$. Show that the equation $f'(x) = 0$ has a real root between x_1 and x_2. State a more general theorem.

6. Suppose that $f''(x) > 0$ for each point in the interval $\{x : a < x < b\}$. How many real roots of the equations $f(x) = 0$ and $f'(x) = 0$ can exist in this interval?

7. Suppose that f and g are differentiable in the interval $\{x : a < x < b\}$ and that, for each point x in this interval,

    $$f'(x)g(x) \neq f(x)g'(x)$$

Show that there is a root of the equation $g(x) = 0$ between each two real roots of $f(x) = 0$, and conversely.

8. Let f and g be differentiable in an interval containing points a and b. Suppose $g(b) \neq g(a)$. Apply Rolle's theorem to an appropriately defined function to prove that there is a point x_0 between a and b such that

$$\frac{f(b) - f(a)}{g(b) - g(a)} = \frac{f'(x_0)}{g'(x_0)}$$

9. Theorem 3.26 assumes that f' is strictly increasing on an interval. Given two points x_1, and x_2, $x_1 < x_2$, on this interval, let x be any point between them. The theorem of the mean applies to give points d and e such that

$$f(x) - f(x_1) = (x - x_1)f'(d), \quad x_1 < d < x$$
$$f(x_2) - f(x) = (x_2 - x)f'(e), \quad x < e < x_2$$

Clearly $f'(d) < f'(e)$, so we have

$$f(x_2) - f(x) > (x_2 - x)f'(d)$$

Adding $f(x) - f(x_1) = (x - x_1)f'(d)$ to this inequality, we obtain

$$f(x_2) - f(x_1) > (x_2 - x_1)f'(d)$$

Follow this line of reasoning to a proof of Theorem 3.26.

10. Suppose that f is differentiable on an interval containing the points $a < b$ and that $f(a) = a$ and $f(b) = b$. Prove that there are two points x_1 and x_2, $a < x_1 < x_2 < b$, such that

$$\frac{1}{f'(x_1)} + \frac{1}{f'(x_2)} = 2$$

11. Prove the formulas:

   (a) $f''(x) = \lim_{h \to 0} \dfrac{f(x + h) - 2f(x) + f(x - h)}{h^2}$

   (b) $f''(x) = \lim_{h \to 0} \dfrac{f(x + 2h) - 2f(x + h) + f(x)}{h^2}$

12. Derive expressions for $f'''(x)$ analogous to those in Problem 11.

4

APPLICATIONS OF DIFFERENTIATION

4.1 THE NEWTON ITERATIVE METHOD FOR SOLVING EQUATIONS

This chapter consists of several more or less unrelated topics, each depending upon the derivative in some essential way. These topics will appear in exercises and problems in subsequent chapters as we study new functions. Our work here should be considered introductory.

A brief mention of "binary splitting" was made in Section 3.4. This is an iterative method for finding approximations to the real roots of an equation. It has the advantage of being simple, but it does require very many computations of functional values. Indeed, we probably would only use this method if we had an electronic computer to do the work for us. Even with a computer, however, it is economical to use an iterative process that gives good approximations quickly. Sir Isaac Newton developed such a method.

Let f be a differentiable function and suppose we want the real roots of the equation

$$f(x) = 0$$

In other words, we want to determine the x intercepts of the graph $\{(x,y) : y = f(x)\}$. Suppose that α is one of the desired real roots and that we have some first approximation x_0 to the root α. Our purpose is to obtain a better approximation x_1.

Think of replacing the graph $\{(x,y) : y = f(x)\}$ by its tangent line at the point $(x_0, f(x_0))$. If x_0 is close to α, then the point x_1 at which this tangent line crosses the x axis will, in general, be closer to α than is x_0. The geometry of this situation is illustrated in Fig. 4.1.

Now the tangent line to the graph $\{(x,y) : y = f(x)\}$ at the point $(x_0, f(x_0))$ is $\{(x,y) : y - f(x_0) = (x - x_0) f'(x_0)\}$. Its intersection with the x axis is the set $\{(x, 0) : 0 - f(x_0) = (x - x_0) f'(x_0)\}$. Thus we obtain the better approximation x_1 simply by solving the equation $-f(x_0) = (x - x_0) f'(x_0)$ to get

$$x_1 = x_0 - \frac{f(x_0)}{f'(x_0)}$$

Having the approximate root x_1, we can proceed to an even better approximation x_2 by repeating this process. That is, we locate x_2 by determining the point of intersection of the x axis and the tangent line at the point $(x_1, f(x_1))$ on the graph. But this is done by means of the same formula, so we have

$$x_2 = x_1 - \frac{f(x_1)}{f'(x_1)}$$

In general, then, we have an *iterative procedure* in which we obtain successively better approximations to the root α by means of the formula

$$x_{n+1} = x_n - \frac{f(x_n)}{f'(x_n)}$$

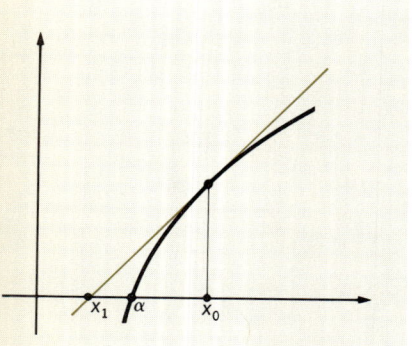

FIGURE 4.1

4.1 THE NEWTON ITERATIVE METHOD FOR SOLVING EQUATIONS

■ **Example 1.** To determine the positive root of the equation
$$f(x) = x^2 - 3x - 2 = 0$$
we first note that $f(3) = -2$ while $f(4) = 2$, so that there is a root α between 3 and 4. We might try $x_0 = 3.5$ as a first approximation to this root.
In this special case, the iteration formula above becomes
$$x_{n+1} = x_n - \frac{f(x_n)}{f'(x_n)} = x_n - \frac{x_n^2 - 3x_n - 2}{2x_n - 3} = \frac{x_n^2 + 2}{2x_n - 3}$$
Therefore, using $x_0 = 3.5$, we have
$$x_1 = \frac{(3.5)^2 + 2}{2(3.5) - 3} = 3.5625$$
Since we want a bit more accuracy, we should compute x_2, which is
$$x_2 = \frac{(3.5625)^2 + 2}{3(3.5625) - 2} \cong 3.56155$$
The positive root of $x^2 - 3x - 2 = 0$ is actually
$$\alpha = \tfrac{1}{3}(3 + \sqrt{17}) = 3.561555 \cdots$$
so two iterations have given quite good accuracy.

■ **Example 2.** To obtain approximations to a square root $\sqrt{r} = \alpha$, we may apply Newton's method to the equation $f(x) = x^2 - r = 0$. In this special case, the iterative formula becomes
$$x_{n+1} = x_n - \frac{f(x_n)}{f'(x_n)} = x_n - \frac{x_n^2 - r}{2x_n} = \frac{x_n^2 + r}{2x_n}$$
$$= \frac{1}{2}\left(x_n + \frac{r}{x_n}\right)$$

For instance, suppose we want to determine $\sqrt{2.1}$. Knowing that $\sqrt{2}$ is approximately 1.4 and that $\sqrt{2.25} = 1.5$, we would perhaps choose $x_0 = 1.45$ as our first approximation. Then
$$x_1 = \frac{1}{2}\left(1.45 + \frac{2.1}{1.45}\right) \cong 1.4491$$
and this is accurate to five significant figures as an approximation of $\sqrt{2.1}$.

So far, then, we seem to have rather good luck with our examples. This is not altogether without cause, of course. Examining the formula
$$x_{n+1} = x_n - \frac{f(x_n)}{f'(x_n)}$$
we can see that we may run into trouble if the derivative f' ever takes on the value zero (or even comes close to zero) during our iterative process. And there can be other difficulties, too. Look at the situation in Fig. 4.2.

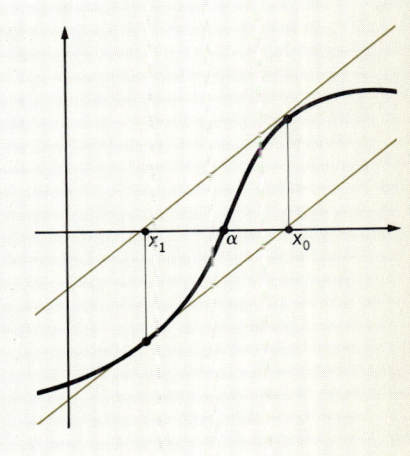

FIGURE 4.2

Here we see that x_1 is *not* a better approximation to the root α than is x_0; what is even worse, we shall have $x_2 = x_0$, $x_3 = x_1$, $x_4 = x_0$, and so forth. This could be discouraging.

We want to be certain that the sequence of iterates $x_0, x_1, x_2, \cdots, x_n, \cdots$ actually converges to the root α. This requires that we impose some conditions upon the function f and upon the choice of the initial approximation x_0. These conditions are as follows.

Let f be twice differentiable and let α be a real root of the equation $f(x) = 0$. Assume that α is a point of an interval over which $f'(x)$ is bounded away from zero and $f''(x)$ is bounded. Then for any initial approximation x_0 close enough to α, we have $\lim_{n \to \infty} x_n = \alpha$.

Note that we are assuming that there are positive numbers m and M such that $0 < m \leq |f'(x)|$ and $|f''(x)| \leq M$ for all points of an interval containing α. Choose a point x_0 of this interval such that $|\alpha - x_0| < 2m\alpha/M$.

We apply the extended mean value theorem (3.34) to the function f on the interval from x_0 to α. This provides a point c between x_0 and α such that

$$0 = f(\alpha) = f(x_0) + (\alpha - x_0)f'(x_0) + \tfrac{1}{2}(\alpha - x_0)^2 f''(c)$$

Now the next approximation x_1 is defined by the equation

$$0 = f(x_0) + (x_1 - x_0)f'(x_0)$$

Subtracting the second equation from the first, we obtain

$$(\alpha - x_1)f'(x_0) + \tfrac{1}{2}(\alpha - x_0)^2 f''(c) = 0$$

or

$$\alpha - x_1 = -(\alpha - x_0)^2 \frac{f''(c)}{2f'(x_0)}$$

Having assumed that $|f'(x_0)| \geq m$ and $|f''(c)| \leq M$, we have the inequality

$$|\alpha - x_1| \leq (\alpha - x_0)^2 \frac{M}{2m}$$

Now suppose we had chosen x_0 so that

$$|\alpha - x_0| \cdot \frac{M}{2m} = r < 1$$

Then we would have

$$|\alpha - x_1| \leq |\alpha - x_0| \cdot r$$

which means that x_1 is closer to α than is x_0.

By exactly the same argument, we next show that

$$|\alpha - x_2| \leq (\alpha - x_1)^2 \frac{M}{2m} \leq \left[(\alpha - x_0)^2 \frac{M}{2m}\right]^2 \cdot \frac{M}{2m}$$

4.1 THE NEWTON ITERATIVE METHOD FOR SOLVING EQUATIONS

or
$$|\alpha - x_2| \leq |\alpha - x_0| \cdot r^3$$

Continuing in the same way, we obtain
$$|\alpha - x_n| \leq |\alpha - x_0| \cdot r^{2n-1}$$

Because $r < 1$, it follows immediately that $\lim_{n \to \infty} x_n = \alpha$. ∎

The fact that the error $|\alpha - x_1|$ is bounded by $(\alpha - x_0)^2 \dfrac{M}{2m}$ and, in general,

$$|\alpha - x_{n+1}| \leq (\alpha - x_n)^2 \frac{M}{2m}$$

gives us a good idea of the rate of improvement in our successive approximations. For suppose that x_n is accurate to n decimals. This means that

$$|\alpha - x_n| \leq \frac{5}{10^{n+1}}$$

Then from the above formula

$$|\alpha - x_{n+1}| \leq \frac{25}{10^{2n+2}} \cdot \frac{M}{2m}$$

Therefore x_{n+1} is accurate to about $2n$ decimals. That is, we double the decimal accuracy with each pass through the Newton method.

■ **Example 3.** Consider the equation $x^2 - 3x - 2 = 0$ again. Since $f(x) = x^2 - 3x - 2$, we have $f'(x) = 2x - 3$ and $f''(x) = 2$. Recall that the positive root α lies between 3 and 4. On the interval $\{x : 3 \leq x \leq 4\}$, the minimum value of $f'(x)$ is 3, so $m = 3$, and the maximum value of $f''(x) = 2$, so $M = 2$. Hence we choose x_0 so that

$$|\alpha - x_0| \cdot \frac{2}{(2)(3)} = |\alpha - x_0| \cdot \frac{1}{3} < 1$$

or
$$|\alpha - x_0| < 3$$

The choice $x_0 = 3.5$ we made before was a good one, because $|\alpha - 3.5| < \frac{1}{2}$.

■ **Example 4.** We apply the Newton method to the equation $f(x) = x^q - r = 0$ to compute the qth root of r. We have $f(x) = x^q - r$, whence $f'(x) = qx^{q-1}$ and $f''(x) = (q-1)qx^{q-2}$. We locate the desired root α in an interval not containing zero. Then $|f'(x)|$ attains its minimum value at one end of this interval, while $|f''(x)|$ attains its maximum value at the other end. This tells us how to choose x_0. Of course, the iterative formula is

$$x_{n+1} = x_n - \frac{x_n^q - r}{qx_n^{q-1}} = \frac{(q-1)x_n^q + r}{qx_n^{q-1}}$$

For instance, we compute $\sqrt[3]{4}$ by solving $f(x) = x^3 - 4 = 0$, whence $f'(x) = 3x^2$ and $f''(x) = 6x$. Because $f(1) = -3$ and $f(2) = 4$, the root α lies between 1 and 2. Clearly, on this interval $\{x : 1 \leq x \leq 2\}$, $M = |\max f''(x)| = 12$ and $m = \min |f'(x)| = 3$. Therefore we shall choose x_0, so that

$$|\alpha - x_0| \cdot \frac{12}{2 \times 3} = |\alpha - x_0| \cdot 2 < 1$$

or

$$|\alpha - x_0| < \tfrac{1}{2}$$

We do not know α, of course, but choosing $x_0 = 1.5$ cannot be wrong. The iterative formula is

$$x_{n+1} = \frac{2x_n^3 + 4}{3x_n^2}$$

The iterates are $x_0 = 1.5$, $x_1 = 1.592592 \cdots$, $x_2 = 1.5874 \cdots$, and x_2 is already accurate to five significant figures.

EXERCISES 4.1

Use the Newton iterative method to compute the real roots of each of the following equations to four significant figure accuracy.

1. $x^2 - 3x - 1 = 0$
2. $x^2 + x - 1 = 0$
3. $x^2 - 15 = 0$
4. $x^2 + 4x - 2 = 0$
5. $x^3 + x - 1 = 0$
6. $3x^3 - x^2 - 4 = 0$
7. $x^3 = 5.32$
8. $5x^3 + 6x - 10 = 0$
9. $x^3 - 2x - 2 = 0$
10. $x^4 - 2 = 0$
11. $x^4 + x - 1 = 0$
12. $x^4 - 6x^2 + 3x - 7 = 0$
13. $\dfrac{1}{x^2 + 1} = 3x - 1$
14. $\sqrt{x^2 + 4} = \dfrac{1}{x - 6}$

15. What would happen if our first guess x_0 were lucky and we chose $x_0 = \alpha$, where α is the desired root?

PROBLEMS 4.1

1. Construct an example in which the successive "approximations" by the Newton method grow worse and worse in the sense that $|\alpha - x_{n+1}|$ exceeds $|\alpha - x_n|$ at each step. [Consider a function f with $f(\alpha) = 0$ but such that $f'(\alpha)$ fails to exist.]

2. In most computers the process of division is much slower than are addition and multiplication. Give an iterative method for computing the reciprocal $1/r$ of a number r that does not involve division. (A computer programmer often causes his machine to compute a/b by first computing $1/b$ and then multiplying $a \times 1/b$.)

3. To compute $\alpha = r^{p/q}$, we apply Newton's method to the equation $x^{q/p} - r = 0$. Show that x_{n+1} is a "weighted average" of the two numbers x_n and $rx_n/x_n^{q/p}$.

4. In computing $\alpha = \sqrt[p]{r}$, the *Bailey root extraction method* is even faster than Newton's method. It uses the iterative formula:

$$x_{n+1} = \frac{x_n[(p-1)x_n^p + (p+1)r]}{(p+1)x_n^p + (p-1)r}$$

For extracting square roots, this formula becomes

$$x_{n+1} = \frac{x_n(x_n^2 + 3r)}{3x_n^2 + r}$$

Determine $\sqrt{2}$ starting with $x_0 = 1$ and stopping with x_2.

If we set

$$g(x) = \frac{x[(p-1)x^p + (p+1)r]}{(p+1)x^p + (p-1)r}$$

show that

(1) $g(\sqrt[p]{r}) = \sqrt[p]{r}$
(2) $g'(\sqrt[p]{r}) = 0 = g''(\sqrt[p]{r})$

What does this say about the function $g(x)$ and what does it mean in regard to the rapidity of convergence of the sequence of iterates x_n?

5. The equation $ax^3 - x^2 + x - 2 = 0$ has a root $\alpha = -1$ if $a = 0$. We should expect it to have a root near -1 if the constant a is small. Show that $-1 + a/3(a+1)$ is a good approximation to this root.

6. (Physics) If cork has a specific gravity of 0.24, how far will a cork sphere of diameter 1 ft sink into water? (The solution is a root of a cubic equation, which can be "solved" by Newton's method.)

4.2 RECTILINEAR MOTION

When an object moves along a straight line we say that the motion is *rectilinear*. The motion of the object is described mathematically in the following way: We introduce a coordinate system on the line of motion. Then the directed distance x from the origin locates the object at any instant. A particular instant is chosen from which to measure elapsed time t. Then the distance s of the object from the origin is a function

$$s = f(t)$$

of the elapsed time t. This is called an *equation of motion*.

■ **Example 1.** Suppose a fireworks rocket is launched straight up and reaches a height of 200 ft above ground with a velocity of 100 ft per sec at the instant its fuel is spent. We wish to describe its subsequent unpowered flight. It is customary to select the origin at ground level and to measure distance positively upward. If we also start measuring time (in seconds) at the instant the fuel is burned out, the height of the rocket above ground is expressed in feet as (approximately)

$$s = 200 + 100t - 16t^2$$

We often draw the graph $\{(t, s) : s = f(t)\}$ on a (t, s)-coordinate plane to help envision the motion. Note, however, that this is one of those situations in which certain geometric properties of the coordinate plane are not usually given a physical interpretation. For instance, "area" in the (t, s) coordinate plane has the dimensions of (time) × (distance), which is not a standard physical unit.

One geometric property does have an important physical interpretation in the time-distance coordinate plane. If (t_1, s_1) and (t_2, s_2) are points in the (t, s)-plane, then the slope $(s_2 - s_1)/(t_2 - t_1)$ of the line through these points has the dimensions of (distance) ÷ (time) or *velocity*. We explore this fact.

Let an object move rectilinearly with equation of motion $s = f(t)$. For two values of t, say t_1 and t_2, the quantity of $f(t_2) - f(t_1)$ is the displacement of the object during the time interval from t_1 to t_2. Thus the ratio

$$\frac{f(t_2) - f(t_1)}{t_2 - t_1}$$

is the *average velocity* of the object during this time interval. This leads us to define the *instantaneous velocity at time t* to be

$$Ds(t) = \lim_{h \to 0} \frac{f(t + h) - f(t)}{h}$$

(The quantity shown by an automobile speedometer is instantaneous velocity.)

In Example 1 the velocity of the rocket is

$$v = Ds(t) = 100 - 32t$$

Setting Ds equal to zero, we obtain the value $t = \frac{25}{8}$ at which the distance s is the greatest, $s(\frac{25}{8}) = 356\frac{1}{4}$ ft.

The rate of change of velocity with respect to time is called the *instantaneous acceleration* of time t of the moving body. That is, we define

$$a = Dv = D^2s$$

The time rate of change of acceleration $Da = D^3s$ is sometimes used in problems in mechanics.

■ **Example 2.** Suppose an object moves rectilinearly with equation of motion $s = 12t - t^3$. Then both its velocity and its acceleration are functions of time

$$v = 12 - 3t^2, \qquad a = -6t$$

Falling bodies. A moving object subject only to the acceleration of gravity is called a *free-falling body*. When such a body is close to the surface of the earth, the acceleration of gravity may be taken to be a constant $-g$.

It is chosen to be negative because the acceleration is directed opposite to the increasing distance above the surface of the earth. It follows that

$$a = D^2s = -g$$

regardless of the choice of origin used in constructing an equation of motion.

The function $\varphi(t) = -gt$ has a derivative $D\varphi = -g$. Thus Theorem 3.36 tells us that we may write

$$v = Df(t) = b - gt$$

for some constant b. Similarly, the function $\psi(t) = bt - \frac{1}{2}gt^2$ has a derivative $D\psi = b - gt$. If we again apply Theorem 3.36, it follows that

$$s = f(t) = a + bt - \frac{1}{2}gt^2$$

for some constant a. Thus the equation of motion of a free-falling body always has the above form.

The constants a and b in this equation of motion are easy to identify. For any choice of time origin whatsoever, we may set t equal to zero and see that the constant a is precisely the distance of the body from the origin at the instant we start measuring elapsed time. This distance is usually denoted by s_0 and is called the *initial distance*. Similarly, since $v = b - gt$, the constant b is the velocity of the falling body at $t = 0$. This is called the *initial velocity* and is usually denoted by v_0. That is, the equation of motion of a falling body is written as

$$s = s_0 + v_0 t - \frac{1}{2}gt^2$$

where s_0 and v_0 are the initial distance and initial velocity, respectively.

The above discussion constitutes the solution of a *differential equation*. We started out knowing only the second derivative $f''(t)$ of some unknown function and then determined the general form that the function $f(t)$ must have. More such problems occur later on.

■ **Example 3.** Suppose a stone is dropped from the top of a cliff 100 ft high. Take the distance s of the stone above the base of the cliff as measured positively upward, and let elapsed time be measured from the instant the stone is released. We then have $s_0 = 100$ ft and because the stone is dropped (not thrown), its initial velocity is zero. Therefore the equation of motion of the stone may be written as

$$s = 100 - \frac{1}{2}gt^2$$

The constant g is approximately 32 ft per sec per sec (or 980 centimeters (cm) per sec per sec). Using this approximate value, we get

$$s = 100 - 16t^2, \qquad t \geq 0$$

as a complete description of the motion of the stone.

Now, to find how long it takes for the stone to reach the base of the cliff, we simply set s equal to zero. The equation $0 = 100 - 16t^2$ and the condition $t \geq 0$ imply that $t = \frac{5}{2}$, that is, it takes $2\frac{1}{2}$ sec for the stone to fall to the ground. By differentiation, we have

$$v = -32t$$

and at $t = 2\frac{1}{2}$, we get $v = -32(\frac{5}{2}) = -80$. Thus the stone has a velocity of 80 ft per sec when it strikes the ground. (The minus sign merely means that the stone is moving down when it strikes bottom.)

■ *Example 4.* Suppose that the stone in Example 3 is not dropped but is thrown upward with a velocity of 60 ft per sec. We still have $s_0 = 100$, but now $v_0 = 60$ (positive because the initial velocity is directed upward). The equation of motion this time is

$$s = 100 + 60t - 16t^2, \qquad t \geq 0$$

whence

$$v = 60 - 32t$$

Again this equation tells us all we need to know about the motion.

For instance, we may set v equal to zero to find that instant when the stone reaches the highest point in its flight. (Why *is* it the highest point?) This gives us $t = \frac{60}{32} = \frac{15}{8}$ sec. At this instant, $s = 100 + 60(\frac{15}{8}) - 16(\frac{15}{8})^2 = 156\frac{1}{4}$ ft, which is the highest distance (above the base of the cliff) that the stone reaches. Again we can set s equal to zero to determine the time at which the stone hits the ground at the base of the cliff. The equation

$$100 + 60t - 16t^2 = 0$$

has only one positive root, $t = 5$. Hence the stone hits the ground 5 sec after it was thrown and it hits with a velocity $v = -32(5) + 60 = -100$ ft per sec.

In the discussion of falling bodies here, the effect of air resistance on the motion is completely ignored. For slowly moving bodies, the air resistance is quite small and can be omitted from our consideration. Dealing with more rapidly moving objects, air resistance becomes an important factor. In such cases another differential equation

$$a = Dv = -32 - k^2 v$$

comes into play. We cannot solve this yet, but will do so in Chapter VIII.

EXERCISES 4.2

1. Draw graphs of the equations of motion given below over the indicated domain:
   (a) $s(t) = 200 - 10t - 16t^2$; $0 \leq t \leq 3$
   (b) $s(t) = 80t - 16t^2$; $0 \leq t \leq \frac{5}{2}$
   (c) $s(t) = 3 - 4t + t^3$; $-1 \leq t \leq 3$
   (d) $s(t) = 5 - 2t^2 - 3t^3$; $-2 \leq t \leq 1$

2. For each graph in Exercise 1, draw the graphs of $v(t)$ and $a(t)$.

3. For each equation of motion below, determine the velocity function $v(t)$ and the acceleration function $a(t)$. Solve the equation $v(t) = 0$ and the inequalities $v(t) < 0$ and $v(t) > 0$. Describe the meaning of your results in words.
   (a) $s(t) = 2 + 12t - 6t^2 + t^3$
   (b) $s(t) = 4 - 7t + 2t^2 + t^3$
   (c) $s(t) = 7 - 6t^2 + t^4$
   (d) $s(t) = -125t + 30t^2 + 10t^3 + 5t^4 + 3t^5$

4. A diving platform is 10 meters (m) above the surface of the water. How long does it take to fall this distance and with what velocity does a diver enter the water? (Use the metric system; it's easier.)

5. A stone is thrown vertically downward from the top of a cliff 144 ft high with an initial velocity of 40 ft per sec. With what velocity does it strike the ground at the base of the cliff?

6. Given that the acceleration function of a body moving rectilinearly is
$$a(t) = 24 - 6t$$
and that $v_0 = 21$ and $s_0 = 0$, determine $v(t)$ and $s(t)$.

PROBLEMS 4.2

1. According to Newtonian theory of gravitation, the force acting on a body of mass m at a distance r from the center of the earth is
$$F = ma = -\frac{kmM}{r^2}$$
where a is the acceleration, k is a constant, and M is the mass of the earth. (At the surface of the earth, we have $mg = kmM/R^2$ where R is the radius of the earth.) Prove that the minimum initial velocity imparted to a body at the surface in order that the body never return is $\sqrt{2gR}$.

2. A rocket is fired vertically upward from the surface of the earth with the "escape velocity" $v_0 = \sqrt{2gR}$ of Problem 1. Show that its equation of motion can be written as
$$r(t) = R\left(1 + \frac{3v_0 t}{2R}\right)^{2/3}$$
where r is the distance from the center of the earth. Show that its velocity decreases toward zero with increasing time. Knowing that $F = ma$, show that the force acting on the rocket at any instant is inversely proportional to r^2.

3. If in Problem 2 the initial velocity is $v_0 = q\sqrt{2gR}$, where $0 < q < 1$, what height will the rocket reach?

4. A stone is dropped from the top of a cliff. One second later a second stone is thrown down the cliff with an initial velocity of 40 ft per sec. The two stones reach the base of the cliff at the same moment. How high is the cliff?

5. Neglecting air resistance, how long does it take an object to reach the ground falling from an altitude of 25,000 ft?

6. How fast must a ball be thrown vertically upward in order that it return to the surface in $4\tfrac{1}{2}$ sec?

7. A dragster covers a quarter mile from a standing start in 8 sec. Assuming constant acceleration, what fraction of the acceleration g of gravity has the machine experienced?

8. A man runs the 100-yd dash by maintaining a constant acceleration for the first 24 yd and then running at a constant velocity. What are the constant acceleration and constant velocity if he runs the 100 yd in 10 sec?

9. In the device known as *Atwood's machine*, two masses $M_1 > M_2$ are suspended by a rope over a pulley. Neglecting the mass of the rope and the pulley, the total mass being moved is $M_1 + M_2$. Neglecting frictional and other such factors, the force producing the motion is $M_1 g - M_2 g$. Hence from the basic equation $F = ma$, we have

$$(M_1 + M_2)\frac{dv}{dt} = (M_1 - M_2)g$$

or

$$\frac{dv}{dt} = \frac{M_1 - M_2}{M_1 + M_2} \cdot g$$

In a particular experiment, $M_1 = 51$ grams, $M_2 = 50$ grams and the heavier weight falls from rest a distance of 10 cm. In several trials the *time for the fall averaged 1.435 sec. What value of g does this yield?*

4.3 RATES AND RELATED RATES

This section considers time rates of change that are not necessarily velocities. First, however, suppose that y is a function of x and x is a function of t. Then there is a rate of change of y with respect to x, and a rate change of y with respect to t. We cannot use the same symbol for each of these, so we put a subscript on the D notation to distinguish between the two rates. Thus $D_x y$ denotes the rate of change of y with respect to x and $D_t y$ is the rate of change of y with respect to t. Of course, the chain rule relates these functions via the equation

$$D_t y = D_x y \cdot D_t x$$

As an example, suppose that a spherical balloon is being inflated, its radius r increasing. The volume $V = \frac{4}{3}\pi r^3$ and the area $A = 4\pi r^2$ have $D_r V = 4\pi r^2$ and $D_r A = 8\pi r$ as rates of change with respect to r. Also we can write $r = (3/4\pi)^{1/3} V^{1/3}$ and $A = (36\pi)^{1/3} V^{2/3}$, whence

$$D_V r = \frac{1}{3}\left(\frac{3}{4\pi}\right)^{1/3} V^{-2/3}, \qquad D_V A = \frac{2}{3}(36\pi)^{1/3} V^{-1/3}$$

■ **Example 1.** Two straight highways meet at right angles. A car traveling 45 mph on one highway passes through the intersection. Three minutes later another car traveling 60 mph on the other highway reaches the intersection. How fast is the distance between the cars increasing 5 min after the second car reaches the intersection?

We first express the distance s between the cars as a function of the number t of minutes elapsed since the second car reached the intersection.

Then the rate of change $D_t s$ is evaluated at $t = 5$ to give the answer to the question.

At 45 mph the first car travels $\frac{3}{4}$ miles per min and therefore travels $\frac{3}{4}t$ miles in t min. Having a 3-min lead, the first car is $\frac{9}{4}$ miles away from the intersection at the time $t = 0$, the moment the second car reaches the intersection. Hence the first car is $\frac{9}{4} + \frac{3}{4}t$ miles away from the intersection t min later. The second car, moving at 60 mph, travels t miles in t min. Looking at Fig. 4.3, we see that the Pythagorean theorem applies to yield

$$s = \sqrt{t^2 + (\tfrac{9}{4} + \tfrac{3}{4}t)^2}$$
$$= \tfrac{1}{4}\sqrt{81 + 54t + 25t^2}$$

The rate of change of s with respect to t is

$$D_t s = \tfrac{1}{4} D_t \sqrt{81 + 54t + 25t^2}$$
$$= \frac{27 + 25t}{4\sqrt{81 + 54t + 25t^2}} \text{ miles per min}$$

The question is answered then by evaluating $D_t s(5)$;

$$D_t s(5) = \frac{27 + 25(5)}{4\sqrt{81 + 54(5) + 25(5)^2}} = 1.216 \text{ miles per min}$$

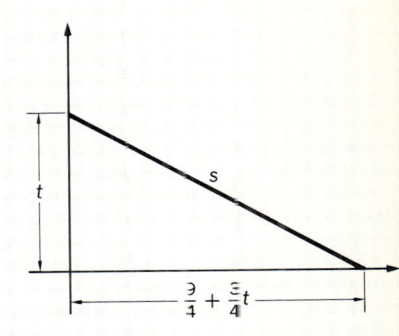

FIGURE 4.3

or 72.98 mph. Although this rate of change has the dimensions of velocity, it is actually a *relative velocity* of one car with respect to the other.

■ *Example 2.* A stone thrown into calm water produces concentric circular waves. Suppose the radius r of such a wave increases at the rate of 2 ft per sec. The area enclosed in the circle is

$$A = \pi r^2$$

and hence

$$D_r A = 2\pi r$$

Applying the chain rule to find the rate of change of A with respect to time t, we get

$$D_t A = D_r A \cdot D_t r = 2\pi r \cdot D_t r$$

Knowing that $D_t r = 2$ ft per sec, we find that

$$D_t A = 4\pi r \text{ sq ft per sec}$$

Many problems involve two or more *related rates*. Our next two examples illustrate both the nature of such problems and the method of solution.

■ *Example 3.* A 20-ft ladder leans against a vertical wall. The lower end is being pulled out along the horizontal floor at the constant rate of

3 ft per sec. How fast is the upper end sliding down the wall when the lower end is 12 ft from the wall?

Let x denote the distance of the foot of the ladder from the wall and let y be the distance of the top of the ladder from the floor. (See Fig. 4.4.) By the Pythagorean theorem, $x^2 + y^2 = 20^2$ or

$$y = \sqrt{400 - x^2}$$

We are given the rate of change $D_t x = 3$ ft per sec and are asked for a particular value of $D_t y$.

Using the chain rule of differentiation, we find that

$$D_t y = D_x y \cdot D_t x = D_x y \cdot 3$$

Now

$$D_x y = \tfrac{1}{2}(400 - x^2)^{-1/2} D_x(400 - x^2) = -x(400 - x^2)^{-1/2}$$

and hence

$$D_t y = -\frac{3x}{\sqrt{400 - x^2}}$$

The answer to the question is then the number $D_t y(12)$,

$$D_t y(12) = \frac{-3(12)}{\sqrt{400 - (12)^2}} = \frac{-36}{\sqrt{256}} = -\frac{9}{4}$$

That is, at the instant when the foot of the ladder is 12 ft out from the wall, the top is sliding down at the rate of $-\tfrac{9}{4}$ ft per sec, where the minus sign indicates that y is decreasing.

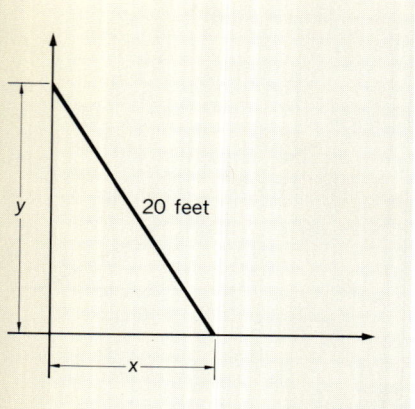

FIGURE 4.4

■ **Example 4.** Water is being poured into a hemispherical tank at the rate of 25 cu ft per min. The tank is 20 ft in diameter. How fast is the water rising when it is 5 ft deep at the center?

When the water is x ft deep, the volume of water in the tank (see Fig. 4.5) is given by

$$V = \pi(10x^2 - \tfrac{1}{3}x^3)$$

(This formula is derived in Section 7.8.) Therefore *the rate of change of V with respect to x* is

$$D_x V = \pi(20x - x^2)$$

We are given that $D_t V = 25$ cu ft per min. By the chain rule

$$D_t V = D_x V \cdot D_t x$$

or

$$25 = \pi(20x - x^2) \cdot D_t x$$

This gives the expression

$$D_t x = \frac{25}{\pi(20x - x^2)}$$

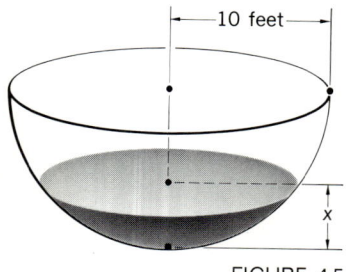

FIGURE 4.5

which holds for any depth x of the water. The answer to our question, finally, is

$$D_t x(5) = \frac{25}{\pi(100 - 25)} = \frac{1}{3\pi} \text{ ft per min}$$

Related rate problems all have the same form. Two variables x and y are given. They are functions of time, but the explicit functions are not given. The variables x and y are related to each other by some equation usually stemming from a physical situation. Both the value of one variable and its rate of change D_t is given at a certain instant; it is required to find the rate of change of the other variable at the same instant.

The method of solution is always the same. If x and y are related by the equation

$$y = f(x)$$

then the chain rule of differentiation provides the relation

$$D_t y = D_x y \cdot D_t x = D_x f \cdot D_t x$$

between the rates of change. Whichever rate is given, we can find the other from this equation.

After this has been done in general, then and not until then do we insert the data giving the desired answer. That is, we obtain a function whose values provide us with the answer for any data; we then evaluate this function at a particular point to obtain an answer.

PROBLEMS 4.3
Physical Sciences

1. A trough 20 ft long has as its cross section an isosceles triangle (base upward) with a base of 2 ft and depth 2 ft. Water is pumped into this trough at the rate of 2 cu ft per min. How fast is the depth of the water increasing when the water is 6 in. deep?

2. A trough with a trapezoidal cross section is 10 ft long, 3 ft wide at the top, 1 ft wide at the bottom and 2 ft deep. If water is pumped into this trough at 3 cu ft per min, how fast is the level of the water rising when the water is 1 ft deep?

3. Water is draining from a conical container 48 in. deep and 24 in. across the top. If the volume changes at 6 cu in. per min, how fast is the level of the water falling when the water is 1 ft deep?

4. Water is poured into a hemispherical bowl of radius 8 in. at the rate of 10 cu in. per min. Find the rate at which the level of the water is rising when the water is 4 in. deep. (This requires knowing the volume of a spherical segment.)

5. If l_0 is the length of a piece of copper wire at $0°C$, its length at $T°$ is (approximately)

$$l(T) = l_0\left(1 + \frac{16T}{10^6} + \frac{T^2}{10^8}\right)$$

What is the rate of change of l with respect to time if $T = 50°$ and is changing at $3°$ per min.

6. A constant voltage of 24 volts is applied to a certain electrical circuit. As the circuit heats up, the resistance increases at the rate of $\frac{1}{10}$ ohm per sec. Find the

rate of change of current when the resistance is 4 ohms. (Use Ohm's law, $E = RI$.)

7. A load of hay is lifted by a rope running in a pulley in the eaves of a barn 40 ft above the load. A tractor pulls the other end of the rope away from the barn at 6 ft per sec. How fast is the load rising when the tractor is 30 ft away from the barn?

8. A rocket is fired vertically upward from a launching pad on the ground 3000 ft from an observer. If it climbs at 2000 ft per sec for the first 4 sec, how fast is it moving away from the observer at the end of 2 sec?

9. A kite is 100 ft high with 140 ft of string paid out. If the wind is blowing at 10 miles per hr, how fast must the string be paid out at this moment so that the kite maintains its height?

10. A man 6-ft tall walks away from a lamppost 15-ft tall. If he walks at 4 miles per hr, how fast does the tip of his shadow move?

11. Assume that a drop of water is a perfect shape. By evaporation it loses moisture at a rate proportional to its surface area. Show that the radius decreases at a constant rate.

12. A balloon is 200 ft above the ground and is rising vertically at 10 ft per sec. At this moment a car passes beneath the balloon moving on a straight road at 60 ft per sec. How fast are the two separating 4 sec later?

13. An airplane is in level flight at an altitude of 6 miles. Its flight path will carry it directly over a radar installation. How fast is the airplane moving if the distance from the airplane to the radar set is decreasing at 4 miles per minute when this distance is 8 miles?

14. Two airplanes fly on parallel courses 8 miles apart in the same direction. One flies at 300 mph and the other at 540 mph. How fast is the distance between them decreasing when the slower one is still 6 miles ahead of the faster?

15. A truck moving at 40 ft per sec crosses a bridge at the same moment that a boat moving at 20 ft per sec passes under the bridge. The course of the boat is at right angles to the road, and the water level is 30 ft below the road. How fast are the two separating 1 sec later?

16. Two roads meet at right angles. A car passes through the intersection at 60 mph. At the moment another car is 660 ft from the intersection approaching the intersection on the other road at 40 mph. How fast is the distance between the cars changing when the second car is still 330 ft from the intersection? When the second car is at the intersection?

17. A ship is steaming due north at 10 knots. At noon it is 40 miles north and 50 miles east of a port. At 12:15 P.M. a second ship leaves the port and steams due east at 12 knots. When will the two be nearest each other?

18. A lamppost is 15 ft high. A stone is thrown vertically upward with an initial velocity of 25 ft per sec from a point 30 ft from the base of the lamppost. At what rate is the shadow of the stone moving along the level ground when the stone is 7 ft above the ground?

19. When air expands adiabatically (without change of heat content) the pressure p and volume v satisfy the equation $pv^{1.4} = k$ (a constant). Suppose that at a certain instant $p = 50$ lb per sq in. and $v = 243$ cu in. If v is decreasing at 4 cu in. per sec, how rapidly is the pressure changing at that instant?

20. Regardless of the shape of the container, the rate in cu ft per min at which water escapes through a small circular hole at the bottom of the container is (approximately)

$$r = 4.8a\sqrt{h}$$

where a is the area of the hole in sq ft and h is the height of the water level above the hole. If the container is a right circular cylinder of diameter 4 ft, if the original depth of water was 9 ft and if the hole of diameter 1 in. is opened at time $t = 0$, express the height h of the water as a function of t and thus determine how long it will take to drain the tank. Note that

$$\frac{1}{\sqrt{h}} D_t h = 2 D_t \sqrt{h}.$$

4.4 INCREMENTS AND DIFFERENTIALS

The familiar quantity

$$f(x + h) - f(x)$$

is called the *increment* in the functional value $f(x)$ due to the increment h in the variable x. For a particular function f, the increment $f(x + h) - f(x)$ obviously depends upon both the numbers x and h. That is, the increment is a function of ordered pairs (x, h). This new function is denoted by Δf and has values

$$\Delta f(x, h) = f(x + h) - f(x)$$

Figure 4.6 illustrates the quantity $\Delta f(x, h)$.
For example, if $f(x) = x^2 - 2$, then

$$\Delta f(x, h) = (x + h)^2 - 2 - (x^2 - 2)$$
$$= 2hx + h^2$$

In particular,

$$\Delta f(2, \tfrac{1}{10}) = 2(\tfrac{1}{10})(2) + (\tfrac{1}{10})^2 = \tfrac{41}{100} \blacksquare$$

If we construct the increment function ΔE for the identity function $E(x) = x$, we obtain $\Delta E(x, h) = x + h - x = h$. Thus the increment function of the identity function is itself the identity function of the increment h. Now we may write the derivative f' as

$$f'(x) = \lim_{h \to 0} \frac{f(x + h) - f(x)}{h} = \lim_{h \to 0} \frac{\Delta f(x, h)}{\Delta x(x, h)}$$

or, more briefly,

$$f'(x) = \lim_{\Delta x \to 0} \frac{\Delta f}{\Delta x}$$

A problem involving an error or an estimate is the natural place to use increments. For example, suppose a carpenter is building a cubical box that is to have a capacity of 8 cu ft. However, his ruler is in error and reads 12 in. when the true measurement is $12 \tfrac{1}{16}$ inches. What error will there be in the volume of the box?

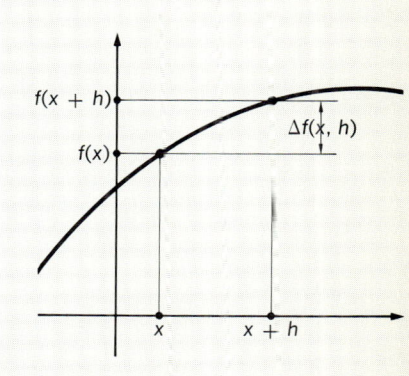

FIGURE 4.6

Obviously, each side of the box will be $2(12\frac{1}{16}) = 24\frac{1}{8}$ in. long instead of 24 in. long. Hence its volume will be $(24\frac{1}{8})^3$ cubic inches when it should have been $(24)^3$ cubic inches. The difference,

$$(24\tfrac{1}{8})^3 - (24)^3 = 217\tfrac{65}{512} \text{ cubic inches}$$

or approximately $\frac{1}{8}$ cubic foot, is the error.

In the notation of increments, the function here is $V = s^3$; hence the increment is

$$\Delta V(s, h) = (s + h)^3 - s^3$$
$$= 3s^2 h + 3sh^2 + h^3$$

Evaluating this increment at the pair $(24, \frac{1}{8})$ gives the error.

A similar problem often arises in a design situation. An engineer is designing a spherical tank whose volume is to be 4500π cu ft, that is, the radius is to be 15 ft. He can allow no more than an error of 20π cu ft in the volume of the completed tank. What error can be allowed in the radius of the tank during construction?

The volume of the tank must lie between the numbers $4500\pi - 20\pi = 4480\pi$ and $4500\pi + 20\pi = 4520\pi$. Therefore the volume $\frac{4}{3}\pi r^3$ must satisfy the inequality

$$4480\pi \leqq \tfrac{4}{3}\pi r^3 \leqq 4520\pi$$

Dividing by $\frac{4}{3}\pi$, we have

$$3360 \leqq r^3 \leqq 3390$$

or

$$\sqrt[3]{3360} \leqq r \leqq \sqrt[3]{3390}$$

The radius of the completed tank must satisfy

$$14.977 \leqq r \leqq 15.023$$

(The engineer would write $r = 15 \pm .023$.) That is, in order to obtain the specified accuracy in volume, the engineer must specify that the error in the radius of the tank be no more that 0.023 ft or 0.276 in.

In increment notation, the accuracy of the volume condition is

$$|\Delta V(15, h)| \leqq 20\pi$$

where h is the permissable error in the 15 ft radius. Since $V = \frac{4}{3}\pi r^3$, we have $\Delta V(r, h) = \frac{4}{3}\pi(r + h)^3 - \frac{4}{3}\pi r^3 = 4\pi(r^2 h + rh^2 + \frac{1}{3}h^3)$. Hence we are led to the problem of solving

$$|4\pi(15^2 h + 15h^2 + \tfrac{1}{3}h^3)| \leqq 20\pi$$

or

$$-20\pi \leqq 4\pi(225h + 15h^2 + \tfrac{1}{3}h^3) \leqq 20\pi$$

or

$$-5 \leqq 225h + 15h^2 + \tfrac{1}{3}h^3 \leqq 5$$

4.4 INCREMENTS AND DIFFERENTIALS

Such a cubic inequality is difficult to solve, but we can play a simplifying game. The quantity h is obviously going to be very small; hence we shall just ignore the terms $15h^2 + \frac{1}{3}h^3$. This gives us an estimate

$$-5 \leq 225h \leq 5$$

or

$$-\tfrac{1}{45} \leq h \leq \tfrac{1}{45}$$

The specification of the accuracy would then be $r = 15 \pm h = 15 \pm \frac{1}{45} = 15 \pm 0.022$, which agrees well with the value obtained previously.

The computational difficulty we just encountered in using the increment function Δf can be circumvented frequently. We use a close approximation to the increment. The *differential* of a function f is the new function df of the variables x and h whose values are

$$df(x, h) = f'(x)h,$$

the product of $f'(x)$ and the increment h.

The quantity $df(x, h)$ is readily identified. At a particular point $(x_0, f(x_0))$, the tangent line to the graph $\{(x, y) : y = f(x)\}$ is the graph of the linear function

$$g(x) = f(x_0) + f'(x_0)(x - x_0)$$

Then the increment of g is

$$\begin{aligned}\Delta g(x_0, h) &= f(x_0) + f'(x_0)(x_0 + h - x_0) \\ &\quad - [f(x_0) + f'(x_0)(x - x_0)] \\ &= f'(x_0)h \\ &= df(x_0, h)\end{aligned}$$

Therefore the differential $df(x_0, h)$ is precisely the increment of the linear function $g(x)$ whose graph is the tangent line at $(x_0, f(x_0))$ to the graph of f. We picture this situation in Fig. 4.7. Note that, even for a large increment h as in the figure, the quantities $\Delta f(x, h)$ and $df(x, h)$ are almost the same.

An analytic relation between $\Delta f(x, h)$ and $df(x, h)$ can be given using the extended mean value theorem (Theorem 3.37). Assume that f is twice differentiable in an interval containing x and $x + h$. Then Theorem 3.37 applies to say that there is a point c between x and $x + h$ such that

$$f(x, h) = f(x) + f'(x)h + \tfrac{1}{2}f''(c)h^2$$

or

$$f(x + h) - f(x) = f'(x)h + \tfrac{1}{2}f''(c)h^2$$

or

$$\Delta f(x, h) = df(x, h) + \tfrac{1}{2}f''(c)h^2$$

This formula tells us that, for small values of h, the increment and the differential are close together. In using the differential in place of the in-

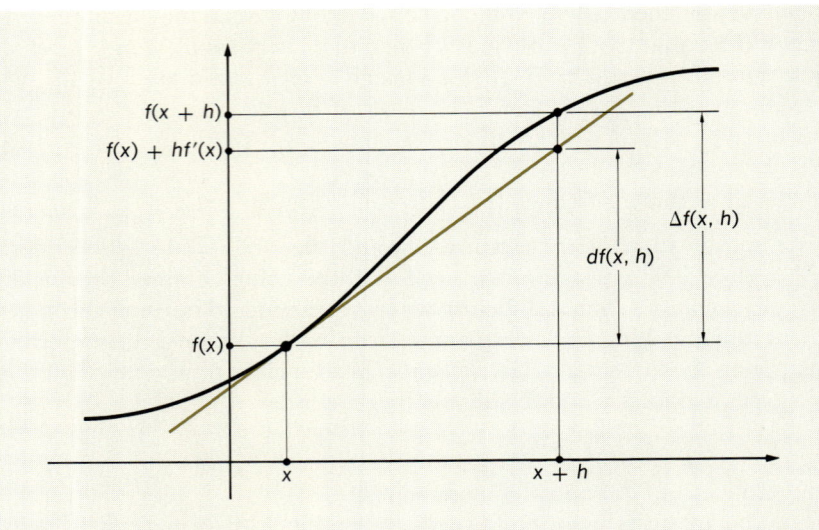

FIGURE 4.7

crement, we commit an error

$$\Delta f(x, h) - df(x, h) = \tfrac{1}{2} f''(c) h^2$$

If we can estimate $f''(c)$ accurately, then we can estimate this error accurately.

■ *Example 1.* If $f(x) = 1 - 2x + 3x^2$, then $f'(x) = -2 + 6x$ and

$$df(x, h) = (-2 + 6x)h$$

Thus, if $x = 2$ and $h = \tfrac{1}{10}$, we have

$$df(2, \tfrac{1}{10}) = (-2 + 6(2))(\tfrac{1}{10}) = 1$$

On the other hand,

$$\begin{aligned}\Delta f(x, h) &= 1 - 2(x + h) + 3(x + h)^2 - [1 - 2x + 3x^2] \\ &= -2h + 6xh + 3h^2 \\ &= df(x, h) + 3h^2\end{aligned}$$

And at $x = 2$, $h = \tfrac{1}{10}$, we have

$$\Delta f(2, \tfrac{1}{10}) = 1.03$$

so that

$$\Delta f(2, \tfrac{1}{10}) - df(2, \tfrac{1}{10}) = 0.03$$

■ *Example 2.* If $f(x) = (4 - x^2)^{1/2}$, then $f'(x) = -x(4 - x^2)^{-1/2}$ and

$$df(x, h) = -xh(4 - x^2)^{-1/2}$$

Also we have $f''(x) = -4(4 - x^2)^{-3/2}$. From the formula above

$$\Delta f(x, h) - df(x, h) = \frac{1}{2} \frac{-4}{(4 - c^2)^{3/2}} \cdot h^2$$

$$= \frac{-2h^2}{(4 - c^2)^{3/2}}$$

Now suppose that $x = 1$ and $h = \frac{1}{5}$. The unknown number c then lies between 1 and $1\frac{1}{5}$. Clearly, f'' is increasing over the interval $\{x : 1 \leq x \leq 1\frac{1}{5}\}$; hence its maximum value is at $1\frac{1}{5}$. Therefore,

$$\left|\Delta f\left(1, \frac{1}{5}\right) - df\left(1, \frac{1}{5}\right)\right| = \left|\frac{-2h^2}{(4 \cdot c^2)^{3/2}}\right| \leq \frac{2(1/5)^2}{(4 - 36/25)^{3/2}}$$

$$\cong 0.019$$

We can use $df(1, \frac{1}{5})$ in place of $\Delta f(x, h)$, knowing that we make an error no larger than 0.019. ■

The differential is useful in computing approximate values. The reasoning here is this: By definition

$$f(x + h) = f(x) + \Delta f(x, h)$$

Because $\Delta f(x, h)$ is approximately equal to $df(x, h)$, we can write

$$f(x + h) \cong f(x) + df(x, h)$$

■ **Example 3.** The number $\sqrt{25.5}$ is simply the value of $f(x) = x^{1/2}$ at $x = 25.5$. If we write $25.5 = 25 + 0.5$, then we can say that

$$f(25.5) = f(25 + 0.5) = f(25) + \Delta f(25, 0.5)$$
$$\cong f(25) + df(25, 0.5)$$

Now

$$f(25) = 5 \quad \text{and} \quad f'(x) = \frac{1}{2}x^{-1/2} = \frac{1}{2\sqrt{x}}$$

Hence $df(x, h) = h/2\sqrt{x}$. Therefore

$$\sqrt{25.5} \cong \sqrt{25} + \frac{0.5}{2\sqrt{25}} = 5.05$$

(To five decimals, $\sqrt{25.5} = 5.04975$.) ■

As we have pointed out, the increment $\Delta f(x, h)$ is often considered as the *error* in the functional value $f(x)$ due to an error h in the variable x. By replacing the increment by the differential, we obtain an approximate value of the error and the approximate error is more easily computed.

■ **Example 4.** In a time-honored experiment, the physics student is required to measure the period T of a pendulum of length s and then compute the acceleration of gravity by means of the formula

$$g = \frac{4\pi^2 s}{T^2}$$

Suppose that the error in the length s is negligible. Then the error $\Delta g(T, h)$ due to an error h in measuring the period T is approximated by

$$dg(T, h) = -\frac{8\pi^2 sh}{T^3} \quad \blacksquare$$

Two more ideas are seen often in dealing with errors. These are the *relative error,* defined by

$$\frac{\Delta f(x, h)}{f(x)}$$

and the *percentage error,* which is the relative error times 100 percent. Here again we use the differential in place of the increment to obtain approximations.

■ *Example 5.* In Example 4 above, we computed that

$$dg(T, h) = -\frac{8\pi^2 sh}{T^3}$$

Therefore a good approximation to the relative error is

$$\frac{dg(T, h)}{g(T)} = \frac{-8\pi^2 sh/T^3}{4\pi^2 s/T^2} = -\frac{2h}{T}$$

The relative error in T is

$$\frac{\Delta T(T, h)}{T} = \frac{h}{T}$$

Therefore the relative error in g is (approximately) twice the relative error in measuring T. The negative sign simply says that a positive error in T results in a negative error in g.

EXERCISES 4.4

1. Determine the new functions Δf and df for the following:

   (a) $f(x) = -3 + 2x + x^2$
   (b) $f(x) = \dfrac{2x}{1 + x^2}$
   (c) $f(x) = \sqrt{1 - x^2}$
   (d) $f(x) = x\sqrt{1 - x^2}$
   (e) $f(x) = x^2(1 - x^2)^{1/2}$
   (f) $f(x) = x(1 - x^2)^{-1/2}$
   (g) $f(x) = x^{-2}(1 - x^2)^{1/2}$
   (h) $f(x) = x^{-1}(25 - x^2)^{1/2}$
   (i) $f(x) = (2 + 3x)(4 + x)^{1/2}$
   (j) $f(x) = (x + x^{1/2})^{1/2}$

2. Evaluate both $\Delta f(x_0, h)$ and $df(x_0, h)$ for each of the following cases.

   (a) $f(x) = 2 + x - x^2$; $x_0 = 1$, $h = \dfrac{1}{5}$

4.4 INCREMENTS AND DIFFERENTIALS

(b) $f(x) = \dfrac{2-x}{4-x}$; $x_0 = -2$, $h = \dfrac{1}{2}$

(c) $f(x) = \dfrac{x^3}{2+x}$; $x_0 = -1$, $h = -\dfrac{1}{4}$

(d) $f(x) = \dfrac{1}{x^2} - x^2$; $x_0 = 2$, $h = -\dfrac{1}{8}$

(e) $f(x) = \sqrt{9+x^2}$; $x_0 = 4$, $h = \dfrac{1}{10}$

(f) $f(x) = (19+x^3)^{1/3}$; $x_0 = 2$, $h = 1$

(g) $f(x) = \sqrt[3]{4x+3x^2}$; $x_0 = 4$, $h = \dfrac{1}{5}$

(h) $f(x) = \dfrac{\sqrt{3+x^2}}{4-x}$; $x_0 = 1$, $h = \dfrac{1}{10}$

3. Without computing Δf, give an estimate of $|\Delta f - df| = \tfrac{1}{2}|f''(x_0)|h^2$ in each of the following cases.
   (a) $f(x) = -4 + 2x + 3x^2$; $x_0 = 2$, $h = \tfrac{1}{2}$
   (b) $f(x) = 4x + x^3$; $x_0 = 1$, $h = -\tfrac{1}{5}$
   (c) $f(x) = (1+x)^{-1}$; $x_0 = 4$, $h = \tfrac{1}{10}$
   (d) $f(x) = x\sqrt{9+x^2}$; $x_0 = -4$, $h = -\tfrac{1}{10}$

4. Use the method of Example 3 to give approximate values of
   (a) $\sqrt{101}$ (b) $\sqrt{16.4}$
   (c) $\sqrt{49.2}$ (d) $\sqrt{224}$
   (e) $\sqrt[3]{8.3}$ (f) $\sqrt[3]{26.9}$
   (g) $\sqrt[4]{80.6}$ (h) $\dfrac{1}{1002}$
   (i) $\dfrac{1}{(2.1)^2}$ (j) $\dfrac{1}{(3.9)^3}$

PROBLEMS 4.4

1. Compute the approximate and relative errors in the area of a square for which one edge is measured as 24 in. with a possible error of $\pm \tfrac{1}{16}$ in.

2. If the diameter of a circle is measured as 12 in. with a maximum error of $\pm \tfrac{1}{32}$ in., what are the approximate and relative errors in computing the area of the circle?

3. If the diameter of a spherical balloon is measured as 3 ft with a possible error of 1 in., what approximate error is made in computing (a) the volume, and (b) the area of the balloon?

4. In constructing a cubical box the length of the edges can be held to at best an error of 2 percent. What percentage error must be permitted in the volume of the box?

5. Two concentric circles have radii 6 and $6\tfrac{1}{2}$ in. Give an approximate value of the area between the two circles.

6. A pipe 30 ft long has an inner diameter of 3 in. and a wall thickness of $\tfrac{1}{4}$ in. If the material weighs 250 lb per cu ft, give an approximate weight of the pipe.

7. A tin can is 5 in. in diameter and 10 in. high. If the thickness of the metal is 0.05 in., how much metal is required to make 10,000 such cans?

8. What relative error is permissable in the sides of a square if the permissable relative error in area is $\frac{1}{100}$?

9. Bernoulli's equation in fluid dynamics is
$$P = \tfrac{1}{2} p v^2 = H$$
where P is the pressure, p is density (here a constant), v is the velocity, and H is a constant. Find an approximate formula for the change in pressure due to a small change in velocity. What relative change in velocity is permissable if this formula is to be accurate to within 1 percent?

10. The cylinder bore of an automobile engine is 4.250 in. and the stake is 4.000 in. If each cylinder is bored out to 4.355 inches, what is the new stroke volume of each cylinder? Conversely, what should be the new bore if each cylinder is to have its stroke volume increased by 2 cu in.?

4.5 ANOTHER NOTATION FOR THE DERIVATIVE

The differential function $df(x, h)$ obeys the same rules of operation as does the derivative. Thus we have

1. $dk = 0$ for any constant function k
2. $dx = h = \Delta x(x, h)$
3. $d(f + g) = df + dg$
4. $d(fg) = (df)g + f(dg)$
5. $d\left(\dfrac{f}{g}\right) = \dfrac{(df)g - f(dg)}{g^2}$
6. $df^r = rf^{r-1}\, df,\ r \in \mathbf{Q}$
7. $d(kf) = k\, df$
8. $d(g \circ f) = (D_f g)\, df$

The differential also leads us to a widely used symbol for the derivative. Recalling that $dx(x, h) = h$, we may write
$$df(x, h) = f'(x)h = f'(x)\, dx(x, h)$$
and hence
$$f'(x) = \frac{df(x, h)}{dx(x, h)}$$
Briefly, we write
$$f' = \frac{df}{dx}$$
and this is the new symbol for a derivative.

When we use this symbol we need not think of df/dx as a ratio, of course, but the derivative is the limit of a ratio, however. Precisely, we may write
$$f' = \frac{df}{dx} = \lim_{\Delta x \to 0} \frac{\Delta f}{\Delta x}$$

recalling that $h = \Delta x = dx$. We also write the definition of the differential in the easily remembered formula

$$df = \frac{df}{dx} dx = \frac{df}{dx} \Delta x$$

In this new notation our rules of differentiation are

1. $\frac{dk}{dx} = 0$ if k is a constant function

2. $\frac{dx}{dx} = 1$

3. $\frac{d}{dx}(f+g) = \frac{df}{dx} + \frac{dg}{dx}$

4. $\frac{d}{dx}(fg) = g\frac{df}{dx} + f\frac{dg}{dx}$

5. $\frac{d}{dx}\left(\frac{f}{g}\right) = \frac{g(df/dx) - f(dg/dx)}{g^2}$

6. $\frac{d}{dx}(x^p) = px^{p-1}, p \in \mathbf{Z}$

7. $\frac{d}{dx}(kf) = k\frac{df}{dx}$

8. $\frac{d}{dx}(g \circ f) = \frac{dg}{df} \cdot \frac{df}{dx}$

9. $\frac{d}{dx}x^r = rx^{r-1}, r \in \mathbf{Q}$

The corresponding symbol for a higher derivative is given by the formula

$$D_x^n f = \frac{d}{dx}\left(\frac{d^{n-1}f}{dx^{n-1}}\right) = \frac{d^n f}{dx^n}$$

The reader should use this notation occasionally in order to become familiar with it.

4.6 SOME APPLICATIONS TO ECONOMICS

The mathematical study of economics has recently become subject of increasing importance in business management and in economics itself. It has also been the source of some very interesting developments in mathematics. This brief section collects a few of the concepts introduced earlier and presents some new ideas.

The *demand* for a particular commodity depends upon its price, the price of similar competing commodities and other such variables as the time of the year, the state of the general economy, and so forth. This complex situation is often simplified (in order to obtain either a first approximation or perhaps short-term information) by assuming that the *unit price p* alone determines the demand *x*. Thus we may write

$$x = f(p)$$

The seller will not tolerate negative prices, of course, and a negative demand is unreal. Hence we assume that both $p \geqq 0$ and $x \geqq 0$. We usually assume that demand will decrease if the price is increased, so that $f(p)$ is taken to be a decreasing function. Also, by raising the price too high, you can price your product out of the market. Thus $f(p)$ is zero if p exceeds a certain value. Finally, even if the price were zero, the demand would remain finite. Thus demand functions look a lot like Fig. 4.8.

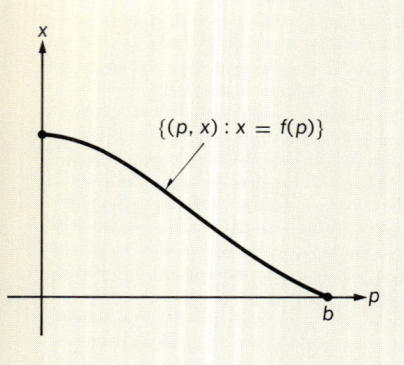

FIGURE 4.8

In an analogous manner, the *supply* of a certain commodity may be taken to be a function of the price at which it can be sold. Again, both the supply *x* and the unit price *p* are nonnegative. More suppliers will get into the market as the price goes up and makes the activity look more profitable. Thus the supply function $x = g(p)$ is generally an increasing function. Also, if the price falls below a certain value, no supplier will bother producing the commodity. Hence a typical *supply curve* looks like that in Fig. 4.9.

In economics the rate of change of a function is usually called a *marginal* of that function. The rate of change of a demand function $x = f(p)$ with respect to p is the *marginal demand*. In mathematical terms, this is simply the derivative

$$Dx = Df(p)$$

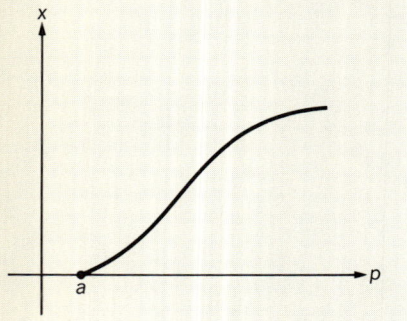

FIGURE 4.9

which is measured in terms of units of the commodity, per unit price.

There is no fixed unit in economic activity. Various commodities are measured in such diverse units as pounds, quarts, acres, board feet, and so forth. Hence a marginal demand has no uniform interpretation. To free our mathematical work from a dependence upon such units, we often speak of relative or percentage changes in a quantity. Thus the *average relative change in a demand function* $f(p)$ due to a change h in the price p is given by

$$\frac{f(p+h) - f(p)}{f(p)}$$

This ratio is a pure number, that is, it has no unit of measurement associated with it. When this number is small, the price change h may not affect the buyer very much. For instance, increasing the price of an automobile by $1 is a relatively small change, whereas a $1 increase in the price of a haircut is a relatively large change. The latter change would be more of an influence on the customer's buying habits.

It seems obvious that the demand $f(p)$ responds to price changes, according to the relative change in price. This response is called the *elasticity of demand* and is defined as follows: The *average elasticity of demand* is

4.6 SOME APPLICATIONS TO ECONOMICS

the ratio of the relative change in demand
$$\frac{f(p+h) - f(p)}{f(p)}$$
to the relative change in price $[(p+h) - p]/p = h/p$ and hence is
$$\frac{[f(p+h) - f(p)]/f(p)}{h/p} = \frac{p}{f(p)}\left[\frac{f(p+h) - f(p)}{h}\right]$$

Then the *elasticity of demand* with respect to price is the function
$$E(p) = \lim_{h \to 0} \frac{p}{f(p)}\left[\frac{f(p+h) - f(p)}{h}\right] = \frac{p}{f(p)} Df(p)$$

Since $E(p)$ is the limit of pure numbers, it is itself a pure number. It measures the response in demand due to a small change away from a given price p.

■ **Example 1.** Consider the demand function $x = 100 - 5p - 2p^2$. The elasticity of demand is
$$E(p) = \frac{p}{100 - 5p - 2p^2}(-5 - 4p) = -\frac{5p + 4p^2}{100 - 5p - 2p^2}$$

In particular, when $p = 4$, we have
$$E(4) = \frac{-5(4) - 4(4)^2}{100 - 5(4) - 2(4)^2} = -\frac{84}{48} = -1.75$$

This has the following meaning: When the price is four, an increase of 1 percent will result in a decrease of $1\frac{3}{4}$ percent in the demand. (Recall that this is an approximation, that we are in essence using a differential.)

■ **Example 2.** A demand function of the form $x = ap^{-r}$, where a is some constant, has the constant elasticity of demand $E(p) = -r$. Since $D_p x = -rap^{-r-1}$, we have
$$E(p) = \frac{p}{x} Dx = \frac{p}{ap^{-r}}(-rap^{-r-1}) = -r\frac{ap^{-r}}{ap^{-r}} = -r$$

Conversely, this is the only form of a demand function that does have a constant elasticity of demand. That is, if $E(p) = p/f(p) Df(p) = -r$, then $f(p) = ap^{-r}$. This is another differential equation. It can be written as
$$p \, Dx = -rx$$
or
$$\frac{Dx}{x} = -\frac{r}{p}$$

We shall solve such equations in Section 7.10.

The *total cost* of producing x units of a certain commodity is often taken to be a function of the number x alone. If this function is denoted by $C(x)$, then the average cost per unit is

$$\frac{C(x)}{x}, \quad \text{in dollars per unit, for instance}$$

The *marginal cost* is the derivative $DC(x)$ and represents (approximately) the change in the total cost for each addition unit of production. That is, the increment $\Delta C(x, 1)$ is approximately equal to $DC(x)$. Similarly, we may speak of the *marginal average cost*

$$D\left(\frac{C(x)}{x}\right) = \frac{x\,DC(x) - C(x)}{x^2}$$

Two frequently used cost functions are the *linear cost function*

$$C(x) = a_0 + a_1 x$$

and the *quadratic cost function*

$$C(x) = a_0 + a_1 x + a_2 x^2$$

The linear cost function has the average cost function $a_0 x^{-1} + a_1$ and hence has marginals $C'(x) = a_1$ and $D(a_0 x^{-1} + a_1) = -a_0 x^{-2}$. For the quadratic cost function, the average cost is $a_0 x^{-1} + a_1 + a_2 x$ and the corresponding marginals are $C'(x) = a_1 + 2a_2 x$ and $D(a_0 x^{-1} + a_1 + a_2 x) = -a_0 x^{-2} + a_2$.

In order to achieve a most economical situation, one often wishes to find the *minimum average cost* of producing a commodity. This can be done by the methods of Sections 3.2 and 3.9.

■ *Example 3.* A contractor has fixed overhead expenses of $30,000 per year. His other costs amount to $7500 for each house he builds plus $100x^2$ dollars, where x is the number of houses built per year. (This latter cost may be due to the fact that the extensive use of equipment increases replacement costs, for instance.) His total yearly cost is

$$C(x) = 30{,}000 + 7500x + 100x^2$$

and the average cost per house is

$$\frac{30{,}000}{x} + 7500 + 200x$$

For profit reasons, assume that he wishes to keep his average cost per house below $12,500. That is, the number x of houses built must satisfy

$$\frac{30{,}000}{x} + 7500 + 100x < 12{,}500$$

or

$$\frac{300}{x} + x < 50$$

or
$$300 + x^2 < 50x$$
or
$$x^2 - 50x + 300 < 0$$
This can be solved to yield
$$25 - 5\sqrt{13} < x < 25 + 5\sqrt{13}$$
or, approximately,
$$6.97 < x < 43.03$$

Thus he must build between 7 and 43 houses per year if he is to keep his average cost below $12,500.

On the other hand, his average cost $30{,}000/x + 7500 + 100x$ will be a minimum if its derivative is zero, that is, if x satisfies
$$-\frac{30{,}000}{x^2} + 100 = 0$$
or
$$x^2 = 300$$
or
$$x = 10\sqrt{3}$$

Of course, such a number of houses cannot be built. But $17 < 10\sqrt{3} < 18$, so the contractor computes his two average costs,
$$\frac{30{,}000}{17} + 7500 + 100(17) = 10{,}964.70$$
$$\frac{30{,}000}{18} + 7500 + 100(18) = 10{,}966.66$$

Thus by building 17 houses per year he has a minimum average cost. ∎

Given a demand function $x = f(p)$, the *total revenue* R obtained by producing x units of the commodity is simply px. That is, the total revenue function is
$$R(p) = px = pf(p)$$
The demand function $x = f(p)$ is often solved for p in terms of x, so that we obtain the price p as a function of the demand x,
$$p = g(x)$$
Then the total revenue is a function of the demand x,
$$R(x) = px = xg(x)$$

A business monopoly can fix its output x, so that the *total profit*
$$P(x) = R(x) - C(x)$$

is a maximum. Clearly, this maximum will occur when the *marginal profit*
$$DP(x) = DR(x) - DC(x)$$
is zero, or when the *marginal revenue* equals the *marginal cost*
$$DR(x) = DC(x)$$
Thus this equation is a necessary condition for maximum profit.

■ *Example 4.* In a monopolistic situation, suppose that we find the linear demand function
$$x = 5000 - 10p$$
or
$$p = 500 - \frac{x}{10}$$
If the total cost function is
$$C(x) = 6000 + 20x + 4x^2$$
then the total profit is
$$\begin{aligned}P(x) &= R(x) - C(x)\\ &= xg(x) - C(x)\\ &= x\left(500 - \frac{x}{10}\right) - (6000 + 20x + 4x^2)\\ &= -6000 + 480x - \frac{41}{10}x^2\end{aligned}$$

Then $P'(x) = 480 - \frac{41}{5}x$, while $P''(x) = -\frac{41}{5} < 0$. Therefore the critical value $x = \frac{5}{41}(480) \cong 58.5$ yields a maximum value for $P(x)$. If the units of our commodity are indivisible, it is impossible to produce $58\frac{1}{2}$ units. Hence we compute the two profit values
$$P(58) = -6000 + 480(58) - \frac{41}{10}(58)^2 = \$8047.60$$
$$P(59) = -6000 + 480(59) - \frac{41}{40}(59)^2 = \$8047.90$$

The larger occurs by taking $x = 59$ (although the 30 cents is a very small relative difference). The actual maximum value is
$$P\left(\frac{2400}{41}\right) = -6000 + 480\left(\frac{2400}{41}\right) - \frac{41}{10}\left(\frac{2400}{41}\right)^2 \cong 8078.78$$
not significantly different from the previous amounts.

PROBLEMS 4.6

1. In a certain soil it has been found that, if 30 apple trees are planted per acre, each tree yields 400 apples (on the average); but for each tree in excess of 30 per acre, the average yield is reduced by 10 apples per tree. What number of trees per acre provides the highest total yield?

2. The manufacturer of a certain article finds that, if he produces x articles per day, $0 \leq x \leq 4000$, he has fixed costs of $900 per day, a unit production cost of $0.90, and other costs amounting to $(0.0001)x^2$ dollars per day. What number of articles produced minimizes average cost?

3. It costs a supermarket $15 + .03x + .000025x^2$ per can to buy and distribute x cans of peas per day. Assuming a linear demand law $x = 400(23 - p)$ and monopoly conditions, how many cans of peas should be bought and at what price should each can be sold?

4. A steel plant can produce up to 8 tons of low grade steel per day. If x tons, $0 \leq x \leq 8$, of low grade steel are produced, then $(40 - 5x)/(10 - x)$ tons of high grade steel can be produced also. It costs $1\frac{1}{2}$ times as much to produce a ton of the high grade steel as it does to produce a ton of low grade steel, but high grade steel sells for twice as much as low grade steel. If fixed costs are the same for any combination of the two grades produced, how much of each grade should be produced in order to maximize the profit?

5. Determine the extremum values of marginal cost if the total cost of producing x units is $45 + 4x + \frac{1}{2}x^2 + \frac{1}{45}x^3 - \frac{1}{120}x^4$ and if $0 \leq x \leq 6$.

6. In the text, we defined elasticity of demand with *respect to price*. If the demand law $x = f(p)$ is solved for p in terms of x,

$$p = g(x)$$

what form should be taken for the *elasticity of demand with respect to demand?*

7. Suppose that the supply of a certain commodity is a function $S(t)$ of some other variable t (such as average rainfall, the prevailing interest rate, and so on). Define the elasticity of supply with respect to t and give it economic meaning.

8. What is the elasticity of demand if the demand function $x = f(p)$ is constant? Explain your answer in economic terms.

9. Prove that the total revenue is constant over an entire interval of prices if and only if $E(p) = -1$ over the entire interval.

10. Let $x = f(p)$ be a demand law. If $E(p) < -1$ on some interval $\{p : a \leq p \leq b\}$, we say that $f(p)$ is *inelastic*. What effect does a price increase have in such a case?

4.7 INDETERMINATE FORMS

If both $\lim_{x \to a} f(x) = 0$ and $\lim_{x \to a} g(x) = 0$, then the limit

$$\lim_{x \to a} \frac{f(x)}{g(x)}$$

is called an *indeterminate form* of type $0/0$. The derivative

$$f'(x) = \lim_{h \to 0} \frac{f(x+h) - f(x)}{h}$$

is just such an indeterminate form, for example. And just as the derivative can exist, so can the limit called an indeterminate form. Yet another generalization of the mean value theorem helps us to evaluate indeterminate forms.

■ **THEOREM 4.1.** Let the functions f and g be continuously differentiable in an open interval containing the point c. Assume that $g'(c) \neq 0$ and that $g'(x)$ does not change sign in the open interval. Then, for any point $x \neq c$ in the interval, there is a point x_0 between c and x such that

$$\frac{f(x) - f(c)}{g(x) - g(c)} = \frac{f'(x_0)}{g'(x_0)}$$

REMARK: If g is the identity function $g(x) = x$, then the above theorem is the mean value theorem. ∎

PROOF: Again we apply Rolle's theorem to a suitably chosen function. We construct a function $\varphi(t)$ defined on the interval from c to x with the properties that $\varphi(c) = \varphi(x)$ and φ' exists. The existence of a point x_0 between c and x such that $\varphi'(x_0) = 0$ is to imply that

$$\frac{f(x) - f(c)}{g(x) - g(c)} = \frac{f'(x_0)}{g'(x_0)}$$

or

$$[f(x) - f(a)]g'(x_0) = [g(x) - g(c)]f'(x_0)$$

This leads us to consider the function

$$\varphi(t) = [f(x) - f(c)]g(t) - [g(x) - g(c)]f(t)$$

whose derivative is

$$D\varphi(t) = [f(x) - f(c)]g'(t) - [g(x) - g(c)]f'(t)$$

We easily compute that

$$\varphi(c) = [f(x) - f(c)]g(c) - [g(x) - g(c)]f(c)$$
$$= f(x)g(c) - g(x)f(c)$$

and

$$\varphi(x) = [f(x) - f(c)]g(x) - [g(x) - g(c)]f(x)$$
$$= -f(c)g(x) + g(c)f(x)$$
$$= \varphi(c)$$

Therefore Rolle's theorem applies and the desired point x_0 exists. [We omitted one detail here, which we leave to the reader to supply. We did not show that $g(x) \neq g(c)$.] ∎

We apply Theorem 4.1 to prove *L'Hospital's rule* for evaluating indeterminate forms of type 0/0.

■ **THEOREM 4.2.** Let f and g be continuously differentiable in an open interval containing the point c. Suppose that $g'(c) \neq 0$ and that $g'(x)$ does not change sign in the interval. If $f(c) = 0 = g(c)$, then

$$\lim_{x \to c} \frac{f(x)}{g(x)} = \lim_{x \to c} \frac{f'(x)}{g'(x)}$$

4.7 INDETERMINATE FORMS

PROOF: The functions f and g satisfy the conditions of Theorem 4.1. Hence for any point $x \neq c$ in the interval, there is a point x_0 between c and x such that

$$\frac{f(x) - f(c)}{g(x) - g(c)} = \frac{f(x)}{g(x)} = \frac{f'(x_0)}{g'(x_0)}$$

As x approaches c, we see that x_0 also approaches c, so that our limit formula should hold.

The precise details of the proof follow. Suppose that $\lim_{x \to a} f'(x)/g'(x) = L$. Then by definition, given $\epsilon > 0$ there exists $\delta > 0$ such that

$$\left| \frac{f'(x)}{g'(x)} - L \right| < \epsilon \quad \text{whenever } 0 < |x - c| < \delta$$

But then

$$\left| \frac{f(x)}{g(x)} - L \right| = \left| \frac{f'(x_0)}{g'(x_0)} - L \right|$$

for some point x_0 between c and x, which implies that $0 < |x_0 - c| < \delta$ whenever $0 < |x - c| < \delta$. Thus

$$\left| \frac{f(x)}{g(x)} - L \right| < \epsilon \quad \text{whenever } 0 < |x - c| < \delta. \quad \blacksquare$$

- **Example 1.** Let $f(x) = x^3 - 1$ and $g(x) = 1 - x$. Since $f(1) = 0 = g(1)$, these functions satisfy Theorem 4.2. Therefore

$$\lim_{x \to 1} \frac{x^3 - 1}{1 - x} = \lim_{x \to 1} \frac{D(x^3 - 1)}{D(1 - x)} = \lim_{x \to 1} \frac{3x^2}{-1} = -3$$

- **Example 2.** Let $f(x) = x^3 - 1$ and $g(x) = x^2 - 1$. Again $f(1) = 0 = g(1)$ but now

$$\lim_{x \to 1} \frac{x^3 - 1}{x^2 - 1} = \lim_{x \to 1} \frac{3x^2}{2x} = \frac{3}{2}$$

- **Example 3.** A second application of L'Hospital's rule is sometimes needed. If it happens that both

$$f(c) = g(c) = 0$$

and

$$f'(c) = g'(c) = 0$$

but that $g''(c) \neq 0$ and $g''(x)$ is continuous in some interval containing the point c, then

$$\lim_{x \to c} \frac{f(x)}{g(x)} = \lim_{x \to c} \frac{f''(x)}{g''(x)}$$

(See Problem 4.7.4.)

For instance, let $f(x) = x^3 - x^2 - x + 1$ and $g(x) = 2x^3 - x^2 - 4x + 3$. Then $f(1) = 0 = g(1)$ and $f'(1) = 0 = g'(1)$. Therefore

$$\lim_{x \to 1} \frac{x^3 - x^2 - x + 1}{2x^3 - x^2 - 4x + 3} = \lim_{x \to 1} \frac{3x^2 - 2x - 1}{6x^2 - 2x - 4}$$

$$= \lim_{x \to 1} \frac{6x - 2}{12x - 2} = \frac{2}{5} \blacksquare$$

Another type of indeterminate form occurs when $\lim_{x \to c} f(x) = \infty$ and $\lim_{x \to c} g(x) = \infty$. In this case, the limit

$$\lim_{x \to c} \frac{f(x)}{g(x)}$$

is called an *indeterminate form of type* ∞/∞. L'Hospital's rule applies, however, and we still apply the equation

$$\lim_{x \to c} \frac{f(x)}{g(x)} = \lim_{x \to c} \frac{f'(x)}{g'(x)}$$

■ **Example 4.** Let $f(x) = 2 + 3x$ and $g(x) = 1 - 2x$. Then $\lim_{x \to \infty} (2 + 3x) = \infty$ and $\lim_{x \to \infty} (1 - 2x) = \infty$. Then

$$\lim_{x \to \infty} \frac{2 + 3x}{1 + 2x} = \lim_{x \to \infty} \frac{3}{-2} = -\frac{3}{2}$$

■ **Example 5.** If $f(x) = 2 + 3x$ and $g(x) = 1 + x^2$, then

$$\lim_{x \to \infty} \frac{2 + 3x}{1 + x^2} = \lim_{x \to \infty} \frac{3}{2x} = 0$$

■ **Example 6.** An indeterminate form of type ∞/∞ may also need a second application of L'Hospital's rule as in Example 3. Thus if $f(x) = 1 + x - x^2$ and $g(x) = 2 + x^2$, then $f'(x) = 1 - 2x$ and $g'(x) = 2x$. The limit

$$\lim_{x \to \infty} \frac{f'(x)}{g'(x)} = \lim_{x \to \infty} \frac{1 - 2x}{2x}$$

is again an indeterminate form of type ∞/∞. Then we can write

$$\lim_{x \to \infty} \frac{f(x)}{g(x)} = \lim_{x \to \infty} \frac{f''(x)}{g''(x)} = \lim_{x \to \infty} \frac{-2}{2} = -1$$

■ **Example 7.** There are indeterminate forms with no value. For instance, let $f(x) = x - 1$ and $g(x) = x^2 - 2x + 1$. Then $f(1) = 0 = g(1)$, but now $g'(1) = 0$. We can write

$$\lim_{x \to 1} \frac{x - 1}{x^2 - 2x + 1} = \lim_{x \to 1} \frac{1}{2x - 2}$$

but the second limit does not exist. This means that the original indeterminate form has no value.

■ **Example 8.** It may be necessary to use higher derivatives to show that an indeterminate form has no value. Let $f(x) = x^3 - 3x + 2$ and $g(x) = x^3 - 3x^2 + 3x - 1$. Then both $f(1) = 0 = g(1)$ and $f'(1) = 0 = g'(1)$. We may still write

$$\lim_{x \to 1} \frac{f(x)}{g(x)} = \lim_{x \to 1} \frac{f''(x)}{g''(x)} = \lim_{x \to 1} \frac{6x - 3}{6x - 6}$$

but the latter limit does not exist.

EXERCISES 4.7

Apply L'Hospital's rule to evaluate the following indeterminate forms.

1. $\lim_{x \to -1} \dfrac{x^2 - x - 2}{x^2 + 3x + 2}$

2. $\lim_{x \to 1} \dfrac{x^2 - 4x + 3}{2x^2 - x - 1}$

3. $\lim_{x \to 2} \dfrac{x^3 - 9x^2 + 26x - 24}{x^2 - 5x + 6}$

4. $\lim_{x \to -2} \dfrac{2x^2 + 3x - 2}{x^2 + 3x + 2}$

5. $\lim_{x \to -1} \dfrac{x^2 - x - 2}{x^2 + 2x + 1}$

6. $\lim_{x \to 3} \dfrac{x^3 - x^2 - 7x + 3}{x^3 - 8x - 3}$

7. $\lim_{x \to 0} \dfrac{\sqrt{x + 4} - 2}{x}$

8. $\lim_{x \to 0} \dfrac{\sqrt[3]{8 - x} - 2}{x}$

9. $\lim_{x \to 0} \dfrac{x}{\sqrt{1 + x} - \sqrt{1 - x}}$

10. $\lim_{x \to 0} \dfrac{(1 + x)^{-1/2} - 1}{x}$

11. $\lim_{x \to \infty} \dfrac{x - 1}{3x + 2}$

12. $\lim_{x \to \infty} \dfrac{x^2 - 3x + 1}{2x^2 + 4x - 7}$

13. $\lim_{x \to \infty} \dfrac{2x^3 - 4x + 1}{x^3 + 2x^2 + 3}$

14. $\lim_{x \to \infty} \dfrac{x^2 + 3x - 4}{x^3 + x - 1}$

15. $\lim_{x \to \infty} \dfrac{3x^4 - x + 1}{2x^4 + x^3 - 6}$

16. $\lim_{x \to \infty} \dfrac{x^3 - 3x + 2}{2x^2 + 4x + 9}$

17. $\lim_{x \to \infty} \dfrac{x^n + 1}{x^m + 1}$ if $m > n$

18. $\lim_{x \to \infty} \dfrac{x^n + 1}{x^m + 1}$ if $m < n$

19. $\lim_{x \to \infty} \dfrac{x}{\sqrt{x^2 + 1}}$

20. $\lim_{x \to \infty} \dfrac{\sqrt{x^2 + 1}}{\sqrt{4x^2 + 5}}$

PROBLEMS 4.7

1. Theorem 6.18 can be proved quickly by showing that the function

$$\Phi(x) = \begin{vmatrix} f(a) & g(a) & 1 \\ f(b) & g(b) & 1 \\ f(x) & g(x) & 1 \end{vmatrix}$$

satisfies Rolle's theorem and continuing. Do so.

2. Using the method of Problem 1 prove the following result.

Theorem: If f, g, and h are continuous on $\{x : a \leq x \leq b\}$ and are differentiable on $\{x : a < x < b\}$, then there is a point x_0, $a < x_0 < b$, such that

$$\begin{vmatrix} f(a) & g(a) & h(a) \\ f(b) & g(b) & h(b) \\ f'(x_0) & g'(x_0) & h'(x_0) \end{vmatrix} = 0$$

3. State and prove L'Hospital's rule for indeterminate forms of the type ∞/∞.

4. Prove the following generalization of L'Hospital's rule.

Theorem: Let f and g be differentiable of order n on some open interval containing the point c. Assume that $g^{(n)}(c) \neq 0$ and that $g^{(n)}(x)$ does not change sign in this interval. If $f(c) = f'(c) = \cdots = f^{(n-1)}(c) = 0$ and $g(c) = g'(c) = \cdots = g^{(n-1)}(c) = 0$, then

$$\lim_{x \to c} \frac{f(x)}{g(x)} = \lim_{x \to c} \frac{f^{(n)}(x)}{g^{(n)}(x)}$$

provided that the latter limit exists.

INFINITE
SERIES

Infinite series find many applications throughout the calculus. Many otherwise troublesome problems can be successfully approached via some infinite series. Computation with truncated infinite series is the way to evaluate a function on a digital computer, for instance. Furthermore, the theory of infinite series is quite simple and yet very attractive as a mathematical discipline. We shall motivate the study of infinite series with a first section on a generalization of the mean value theorem.

5.1 TAYLOR'S THEOREM

The result known as Taylor's theorem is yet another generalization of the mean value theorem. It leads to the study of infinite series in a very natural way. Also, the material here is of fundamental importance in the science of computation with electronic computers.

Let f be a function defined on an interval containing the point c and suppose that the derivatives $f', f'', \cdots, f^{(n)}$ of f each exist at the point c. The polynomial P whose values are

$$P(x) = f(c) + f'(c)(x-c) + \frac{f''(c)}{2!}(x-c)^2 + \cdots$$
$$+ \frac{f^{(n)}(c)}{n!}(x-c)^n$$

is called the *Taylor polynomial of degree n* for the function f at the point c.

For instance, let $f(x) = x^{1/2}$. Then

$$f'(x) = \tfrac{1}{2}x^{-1/2}, f''(x) = -\tfrac{1}{4}x^{-3/2} \text{ and } f'''(x) = \tfrac{3}{8}x^{-5/2}$$

Thus the Taylor polynomial of degree three at the point one is

$$P(x) = f(1) + f'(1)(x-1) + \tfrac{1}{2}f''(1)(x-1)^2$$
$$+ \tfrac{1}{6}f'''(1)(x-1)^3$$
$$= 1 + \tfrac{1}{2}(x-1) - \tfrac{1}{8}(x-1)^2 + \tfrac{1}{16}(x-1)^3$$

Similarly, the Taylor polynomial for this function at the point four is

$$P(x) = f(4) + f'(4)(x-4) + \tfrac{1}{2}f''(4)(x-4)^2$$
$$+ \tfrac{1}{6}f'''(4)(x-4)^3$$
$$= 2 + \tfrac{1}{4}(x-4) - \tfrac{1}{32}(x-4)^2 + \tfrac{1}{512}(x-4)^3$$

The derivatives of the Taylor polynomial P are easy to compute:

$$P'(x) = f'(c) + f''(c)(x-c) + \frac{3f'''(c)}{3!}(x-c)^2 + \cdots$$
$$+ \frac{nf^{(n)}(c)}{n!}(x-c)^{n-1}$$

$$P''(x) = f''(c) + \frac{2 \cdot 3f'''(c)}{3!}(x-c) + \cdots$$
$$+ \frac{(n-1)nf^{(n)}(c)}{n!}(x-c)^{n-2}$$

$$P'''(x) = f'''(c) + \cdots + \frac{(n-2)(n-1)nf^{(n)}(c)}{n!}(x-c)^{n-3}$$

$$\vdots$$

$$P^{(n)}(x) = f^{(n)}(c)$$

Obviously, then, we have the $n+1$ equations

$$P(c) = f(c)$$
$$P'(c) = f'(c)$$
$$P''(c) = f''(c)$$
$$\vdots$$
$$P^{(n)}(c) = f^{(n)}(c)$$

These equations say that the function f and its Taylor polynomial P are tangent of order n at the point c. Therefore, if x is close to c, $P(x)$ is a very good approximation to the value $f(x)$. The next result gives the accuracy of this approximation.

■ **THEOREM 5.1.** Let f be a function such that $f^{(n+1)}(x)$ exists in an interval $\{x : a < x < b\}$ and let c be a point of this interval. Let P_n be the Taylor polynomial of degree n of f at the point c. Then, for any point x in $\{x : a < x < b\}$, there exists a point x_0 between c and x such that

$$f(x) = P_n(x) + \frac{f^{(n+1)}(x_0)}{(n+1)!}(x-c)^{n+1}$$

PROOF: We apply Rolle's theorem (Theorem 3.33) to an appropriately constructed function.

Assume that $x \neq c$ throughout this discussion. Define a number K by the equation

$$f(x) = P_n(x) + \frac{(x-c)^{n+1}}{(n+1)!}K$$

Explicitly,

$$K = \frac{[f(x) - P_n(x)](n+1)!}{(x-a)^{n+1}}$$

Construct the function

$$\varphi(t) = -f(x) + f(t) + f'(t)(x-t) + \frac{f''(t)}{2}(x-t)^2 + \cdots$$
$$+ \frac{f^{(n)}(t)}{n!}(x-t)^n + \frac{(x-t)^{n+1}}{(n+1)!}K$$

By construction, we have $\varphi(x) = 0$. Furthermore, by choice of the number K, we also have $\varphi(c) = 0$. Because $f^{(n+1)}$ exists at all points in $\{x : a < x < b\}$, it follows that f and all of its derivatives up to and including $f^{(n)}$ are continuous. Therefore Rolle's theorem applies to say that there exists a point x_0 between c and x such that $D_t\varphi(x_0) = 0$.

Now

$$D_t\varphi(t) = -0 + f'(t) + [f''(t)(x-t) - f'(t)]$$
$$+ \left[\frac{f'''(t)}{2}(x-t)^2 - f''(t)(x-t)\right] + \cdots$$
$$+ \left[\frac{f^{(n+1)}(t)}{n!}(x-t)^n - \frac{f^{(n)}(t)}{(n-1)!}(x-t)^{n-1}\right] - \frac{(x-t)^n}{n!}K$$
$$= \frac{(x-t)^n}{n!}[f^{(n+1)}(t) - K]$$

because all of the other terms cancel. Therefore at $t = x_0$, we have

$$D_t\varphi(x_0) = 0 = \frac{(x-x_0)^n}{n!}[f^{(n+1)}(x_0) - K]$$

Recalling that x_0 is between c and x, we know that $x_0 \neq x$, and hence $(x - x_0)^n/n! \neq 0$. Therefore $K = f^{(n+1)}(x_0)$, and the proof is complete. ∎

One of the many uses of Taylor's theorem lies in the art of computation. We can use it to compute approximate functional values and at the same time obtain a bound on the error in the approximation. The following examples illustrate the techniques employed in this kind of work.

■ **Example 1.** Suppose we wish to obtain $\sqrt{26}$ accurate to five decimals. We shall consider the function $f(x) = x^{1/2}$ and take $c = 25$. Then we have

$$f'(x) = \tfrac{1}{2}x^{-1/2}, \quad f''(x) = -\tfrac{1}{4}x^{-3/2},$$
$$f'''(x) = \tfrac{3}{8}x^{-5/2}, \quad f^{(4)}(x) = -\tfrac{15}{16}x^{-7/2}, \quad \text{etc.}$$

By Taylor's theorem, we may conclude that

$$f(26) = f(25) + f'(25)(26-25) + \frac{f''(25)}{2}(26-25)^2$$
$$+ \frac{f'''(25)}{6}(26-25)^3 + \frac{f^{(4)}(x_0)}{24}(26-25)^4$$

for some point x_0, $25 < x_0 < 26$. That is,

$$26^{1/2} = 25^{1/2} + \frac{1}{2}(25)^{-1/2}(1) - \frac{1/4(25)^{-3/2}}{2}(1)^2$$
$$+ \frac{3/8(25)^{-5/2}}{6}(1)^3 - \frac{15/16(x_0)^{-7/2}}{24}(1)^4$$
$$= 5 + \frac{1}{10} - \frac{1}{1000} + \frac{1}{50,000} - \frac{5}{128x_0^{7/2}}$$
$$= 5.09902 - \frac{5}{128x_0^{7/2}}$$

5.1 TAYLOR'S THEOREM

Now, because $x_0 > 25$, we have $x_0^{7/2} > 25^{7/2} = 5^7$. Therefore

$$\frac{5}{128c^{7/2}} < \frac{5}{128(5)^7} = 0.0000005$$

and hence

$$|\sqrt{26} - 5.09902| < 0.0000005$$

Since 5.09902 is in error by an amount that cannot affect the fifth decimal place, we know that $\sqrt{26} = 5.09902$ is accurate to five decimals.

■ **Example 2.** If we are asked to determine $\sqrt[3]{1003}$ accurate to four decimals, then we choose $f(x) = x^{1/3}$ and $c = 1000$. A priori we do not know the degree n of the Taylor polynomial needed to give four-decimal accuracy. We just know that we want the error in our approximation to be less than 0.00005, so that it cannot affect the fourth decimal.

The "error term"

$$\frac{f^{(n+1)}(x_0)}{(n+1)!}(1003 - 1000)^{n+1}$$

is to be chosen so that

$$|f(1003) - P_n(1003)| = \left|\frac{f^{(n+1)}(x_0)}{(n+1)!}(1003 - 1000)^{n+1}\right|$$

$$< 0.00005$$

An appropriate choice of n can be made just by trying successive values of n. For $n = 1$, the error term is

$$\frac{f''(x_0)}{2}(1003 - 1000)^2 = -\frac{\frac{2}{9}x_0^{-5/3}}{2}(3)^2 = -x_0^{-5/3}$$

Therefore

$$|\sqrt[3]{1003} - P_1(1003)| = x_0^{-5/3}$$

where $1000 < x_0 < 1003$. This means that $x_0^{5/3} > 1000^{5/3} = 100{,}000$, so that

$$|\sqrt[3]{1003} - P_1(1003)| < \frac{1}{100{,}000} = 0.00001$$

It follows that $P_1(1003)$ provides four-decimal accuracy.

$$P_1(x) = f(1000) + f'(1000)(x - 1000)$$
$$= 1000^{1/3} + \tfrac{1}{3}(1000)^{-2/3}(x - 1000)$$
$$= 10 + \tfrac{1}{300}(x - 1000)$$

and hence

$$P_1(1003) = 10 + \tfrac{1}{300}(1003 - 1000) = 10.01$$

($\sqrt[3]{1003} = 10.009990$ to six decimals.)

■ **Example 3.** To compute $\frac{1}{98}$, we may take $f(x) = x^{-1}$ and $c = 100$. Having $f'(x) = -x^{-2}, f''(x) = 2x^{-3}, f'''(x) = -6x^{-4}$, and so on, we can write the Taylor polynomial equation

$$\frac{1}{x} = \frac{1}{c} - \frac{c^{-2}}{1}(x-c) + \frac{2c^{-3}}{2}(x-c)^2 - \frac{6x_0^{-4}}{6}(x-c)^3$$

$$= \frac{1}{c} - \frac{x-c}{c^2} + \frac{(x-c)^2}{c^3} - \frac{(x-c)^3}{x_0^4}$$

where x_0 lies between c and x. In particular, for $c = 100$ and $x = 98$,

$$\frac{1}{98} = \frac{1}{100} - \frac{(98-100)}{100^2} + \frac{(98-100)^2}{100^2} - \frac{(98-100)^3}{x_0^4},$$

$$98 < x_0 < 100$$

$$= \frac{1}{100} + \frac{2}{100^2} + \frac{4}{100^3} + \frac{8}{x_0^4}$$

Since $x_0 > 98$, we have $8/x_0^4 < 8/98^4$. Therefore the error in using the value $P_2(98) = 1/100 + 2/100^2 + 4/100^3 = 0.010204$ instead of $1/98$ is less than $8/98^4$, which does not affect the sixth-decimal place.

■ **Example 4.** Let $f(x) = (1+x)^{1/2}$. Then $f'(x) = \frac{1}{2}(1+x)^{-1/2}$, $f''(x) = -\frac{1}{4}(1+x)^{-3/2}$, $f'''(x) = \frac{3}{8}(1+x)^{-5/2}$, and so forth. Letting $c = 0$, we find that Taylor's theorem yields

$$\sqrt{1+x} = (1+0)^{1/2} + \frac{1}{2}(1+0)^{-1/2}(x-0)$$

$$- \frac{1/4(1+0)^{-3/2}}{2} + \frac{3/8(1+x_0)^{-5/2}}{6}(x-0)^3$$

$$= 1 + \frac{1}{2}x - \frac{1}{8}x^2 + \frac{x^3}{16}(1+x_0)^{-5/2}$$

for some point x_0 between 0 and x.

For any $x_0 > 0$, the term $x^3/16(1+x_0)^{-5/2}$ is positive. Therefore we have

$$\sqrt{1+x} > 1 + \tfrac{1}{2}x - \tfrac{1}{8}x^2 \quad \text{if } x > 0$$

Suppose that we want to use the value $1 + \tfrac{1}{2}x - \tfrac{1}{8}x^2$ as an approximation to $\sqrt{1+x}$. If we need, say, four-decimal accuracy, then the error term must satisfy

$$\tfrac{1}{16}(1+x_0)^{-5/2}x^3 < 0.00005$$

The presence of that unknown number x_0 between 0 and x makes it impossible to solve this inequality as it stands. However, we can obtain some information. If $x > 0$, then $0 < x_0 < x$ whence $1 + x_0 > 1$ and $(1 + x_0)^{-5/2} < 1$. This means that

$$\tfrac{1}{16}(1+x_0)^{-5/2}x^3 < \tfrac{1}{16}x^3$$

5.1 TAYLOR'S THEOREM

If we now choose x so that
$$\tfrac{1}{16}x^3 < 0.00005$$
or
$$x^3 < 0.0008$$
or
$$0 < x < \sqrt[3]{0.0008} \cong 0.0928$$
then we shall be certain that
$$|\sqrt{1+x} - 1 - \tfrac{1}{2}x + \tfrac{1}{8}x^2| < 0.00005.$$

This means that we obtain the desired four-decimal accuracy, in using $1 + \tfrac{1}{2}x - \tfrac{1}{8}x^2$ instead of $\sqrt{1+x}$, whenever x satisfies $0 < x < 0.0928$.

The error analyses carried out in these examples is an important part of the work of a numerical analyst. Before a computation is programmed for an electronic computer, the operator must know what degree of accuracy can be expected from the method of computation. Then the machine can be instructed when to stop computing, as well as how to do the arithmetic.

EXERCISES 5.1

Use Taylor's theorem to compute the following numbers to four-decimal accuracy. Use Examples 1 to 3 as guides.

1. $\sqrt{1.02}$
2. $\sqrt{50}$
3. $\sqrt{82}$
4. $\sqrt{99}$
5. $\sqrt{35}$
6. $\sqrt{65}$
7. $\sqrt{142}$
8. $\sqrt{511}$
9. $\sqrt[3]{103}$
10. $\sqrt[3]{28}$
11. $\sqrt[4]{80}$
12. $\sqrt[5]{33}$
13. $\tfrac{1}{49}$
14. $\tfrac{1}{102}$
15. $\tfrac{1}{255}$
16. $\tfrac{1}{1002}$

As in Example 4, use Taylor's theorem to establish the following inequalities.

17. $1 + \tfrac{1}{3}x - \tfrac{1}{9}x^2 < \sqrt[3]{1+x} < 1 + \tfrac{1}{3}x$ if $x > 0$

18. $\dfrac{1}{5} - \dfrac{x}{100} < \dfrac{1}{x} < \dfrac{3}{10} - \dfrac{3x}{100} + \dfrac{x^2}{1000}$ if $x > 10$

19. Show that, if $0 \leq x \leq \tfrac{1}{4}$, then
$$\dfrac{1}{1+x} = 1 - x + x^2 - x^3 + x^4 - x^5 + x^6 + R$$
where $|R| < 2^{-14}$.

20. Show that, if $0 \leq x \leq \tfrac{1}{4}$, then

$$\frac{1}{4+x} = \frac{1}{4} - \frac{1}{16}x + \frac{1}{64}x^2 + R$$

where $|R| < 2^{-14}$.

Determine the Taylor polynomial of the indicated degree at the given point for each of the functions 21 to 30.

21. $f(x) = 2 - 3x + x^2$; $n = 2$ at $c = -1$
22. $f(x) = 1 - 6x + x^3$; $n = 3$ at $c = 2$
23. $f(x) = 4 - 3x + x^2 - 5x^3 + x^4$; $n = 4$ at $c = -3$
24. $f(x) = (1 - 3x + x^2)^3$; $n = 6$ at $c = 1$
25. $f(x) = \dfrac{1}{1+x}$; $n = 3$ at $c = 2^{-1}$
26. $f(x) = \dfrac{x}{x-1}$; $n = 4$ at $c = -1$
27. $f(x) = \dfrac{1}{1+x^2}$; $n = 4$ at $c = 0$
28. $f(x) = \dfrac{x}{x^2 - 4}$; $n = 5$ at $c = 0$
29. $f(x) = \dfrac{1+x}{\sqrt{1-x}}$; $n = 4$ at $c = 0$
30. $f(x) = \dfrac{1 + x + x^2}{(1-x)^2(2-x)}$; $n = 4$ at $c = 0$

PROBLEMS 5.1

1. Establish the inequality

$$1 + rx + \frac{r(r-1)}{2}x^2 \leq (1+x)^r$$

if $x \geq 0$ and $r \geq 2$.

2. Define y as a function of x via the equation

$$\sqrt{x+1} = \frac{6 + 5x - x^4 y}{6 + 2x - \frac{1}{4}x^2}$$

and prove that $\lim_{x \to 0} y = \frac{5}{64}$.

3. Let $f(x) = x^5|x|$. If $c = -1$, $x = +1$ and $n = 3$ in Taylor's theorem, then what is the value of x_0?

4. For some $n \in \mathbf{N}$, suppose that $f^{(n)} = g^{(n)}$. What can you say about the functions f and g?

5. Suppose that $f'(c) = f''(c) = \cdots = f^{(n)}(c) = 0$, but that $f^{(n+1)}(x) > 0$ for all points x in some interval containing the point c. Then, for any such point, we have

$$f(x) - f(c) = f^{(n+1)}(x_0) \frac{(x-c)^{n+1}}{(n+1)!}$$

for some point x_0 between x and c. Thus, if n is odd, we must have $f(c)$ as a minimum value. Now state and prove a general theorem.

5.2 INFINITE SERIES

Let φ be an infinite sequence. There is an important way in which a new sequence Ψ can be constructed from the terms $\varphi(n) = a_n$ of the given sequence. We define

$$\Psi(n) = s_n = a_1 + a_2 + a_3 + \cdots + a_n$$

and call $\Psi(n)$ the nth *partial sum* of the sequence $a_1, a_2, a_3, \cdots, a_n, \cdots$.

If the sequence of partial sums

$$s_1 = a_1, \; s_2 = a_1 + a_2, \; s_3 = a_1 + a_2 + a_3, \cdots,$$
$$s_n = a_1 + a_2 + \cdots + a_n, \cdots$$

converges to a number L, then L is called the *sum* of the terms of the sequence $a_1, a_2, \cdots, a_n, \cdots$ and we write an infinite summation

$$a_1 + a_2 + a_3 + \cdots + a_n + \cdots = L$$

Such an indicated sum of the terms of an infinite sequence is called an *infinite series*. Note that the plus signs in the symbol for an infinite series

$$a_1 + a_2 + a_3 + \cdots + a_n + \cdots$$

not only indicate addition but also suggest the limiting process involved. Similarly, by using the previous equation, we simply mean that

$$\lim_{n \to \infty} (a_1 + a_2 + a_3 + \cdots + a_n) = L$$

In general, then, the infinite series

$$a_1 + a_2 + a_3 + \cdots + a_n + \cdots$$

converges or *diverges* as does the sequence of its partial sums,

$$a_1, a_1 + a_2, \cdots, a_1 + a_2 + \cdots + a_n, \cdots$$

■ **Example 1.** The infinite series

$$1 + \frac{1}{2} + \frac{1}{4} + \frac{1}{8} + \cdots + \frac{1}{2^{n-1}} + \cdots = 2$$

This is proved by recalling that the finite geometric series,

$$\Psi(n) = 1 + \frac{1}{2} + \frac{1}{4} + \cdots - \frac{1}{2^{n-1}}$$

can be evaluated as

$$1 + \frac{1}{2} + \frac{1}{4} + \cdots + \frac{1}{2^{n-1}} = \frac{1 - 1/2^n}{1/2} = 2 - \frac{1}{2^{n-1}}$$

It is now easily shown that

$$\lim_{n \to \infty} \Psi(n) = 2$$

as claimed.

■ **Example 2.** Let $r \in \mathbf{R}$ satisfy $0 < r < 1$. Then

$$1 + r + r^2 + \cdots + r^{n-1} + \cdots = \frac{1}{1-r}$$

Again we evaluate the partial sum $\Psi(n)$ as a finite geometric series

$$\Psi(n) = 1 + r + r^2 + \cdots + r^{n-1} = \frac{1 - r^n}{1 - r}$$

$$= \frac{1}{1-r} - \frac{r^n}{1-r}$$

Then from Theorems 3.2 and 3.3 and Example 3 of Section 3.3, we have

$$\lim_{n \to \infty} \Psi(n) = \frac{1}{1-r}$$

■ **Example 3.** The *harmonic series*

$$1 + \frac{1}{2} + \frac{1}{3} + \cdots + \frac{1}{n} + \cdots$$

is divergent. We prove this by showing that the partial sums grow indefinitely large. We note that

$$\Psi(1) = 1$$
$$\Psi(2) = 1 + \tfrac{1}{2}$$
$$\Psi(3) = 1 + \tfrac{1}{2} + (\tfrac{1}{3} + \tfrac{1}{4}) > 1 + \tfrac{1}{2} + \tfrac{1}{2}$$
$$\Psi(8) = 1 + \tfrac{1}{2} + (\tfrac{1}{3} + \tfrac{1}{4}) + (\tfrac{1}{5} + \tfrac{1}{6} + \tfrac{1}{7} + \tfrac{1}{8})$$
$$> 1 + \tfrac{1}{2} + \tfrac{1}{2} + \tfrac{1}{2} = 1 + 3(\tfrac{1}{2})$$
$$\Psi(16) = 1 + \tfrac{1}{2} + (\tfrac{1}{3} + \tfrac{1}{4}) + (\tfrac{1}{5} + \cdots + \tfrac{1}{8})$$
$$+ (\tfrac{1}{9} + \cdots + \tfrac{1}{16}) > 1 + 4(\tfrac{1}{2})$$

because $\tfrac{1}{9} + \tfrac{1}{10} + \cdots + \tfrac{1}{16} > \tfrac{1}{16} + \tfrac{1}{16} + \cdots + \tfrac{1}{16} = \tfrac{1}{2}$. In general, then,

$$\Psi(2^n) > 1 + n(\tfrac{1}{2})$$

and therefore $\Psi(n)$ grows indefinitely large and cannot converge.

■ **THEOREM 5.2.** Let $a_1 + a_2 + \cdots + a_n + \cdots = L$ be a convergent infinite series. If a finite number of terms are deleted from this infinite series, then the resulting infinite series also converges. Also, if a finite number of terms is added to this infinite series, then the resulting infinite series converges.

PROOF: This result says that the convergence of a series does not depend upon the beginning terms but only upon the "tail" of the series. For if the term a_1 is deleted, then the new sequence of partial sums

$$\Psi(1) = a_2, \ \Psi(2) = a_2 + a_3, \ \cdots,$$
$$\Psi(n) = a_2 + a_3 + \cdots + a_{n+1}$$

obviously converges to $L - a_1$. Similarly, if we add a new term a_0, then the new partial sums

$$\Psi(1) = a_0 + a_1, \quad \Psi(2) = a_0 + a_1 + a_2, \cdots,$$
$$\Psi(n) = a_0 + a_1 + a_2 + \cdots + a_n$$

converges to $a_0 + L$. The precise proofs of these statements are trivial. A finite induction argument then completes the proof. ∎

■ **THEOREM 5.3.** Let $a_1 + a_2 + \cdots + a_n + \cdots = L$ be a convergent infinite series. If a finite number of terms a_i are altered, then the new infinite series also converges.

PROOF: Suppose a_1 is changed to b_1. Then it is easy to show that

$$b_1 + a_2 + a_3 + \cdots + a_n + \cdots = L + (b_1 - a_1)$$

A finite induction completes the proof again. ∎

Our next result gives us a necessary condition for the convergence of an infinite series.

■ **THEOREM 5.4.** If the infinite series $a_1 + a_2 + \cdots + a_n + \cdots$ converges, then

$$\lim_{n \to \infty} a_n = 0$$

PROOF: We are assuming that

$$\lim_{n \to \infty} \Psi(n) = \lim_{n \to \infty} (a_1 + a_2 + \cdots + a_n) = L$$

Therefore, by definition, if we are given $\epsilon > 0$, there exists $n_0 \in \mathbf{N}$ such that

$$|\Psi(n) - L| < \frac{\epsilon}{2} \quad \text{whenever } n > n_0$$

Then we have

$$a_n = \Psi(n) - \Psi(n-1)$$
$$= \Psi(n) - L + L - \Psi(n-1)$$

Hence

$$|a_n| = |\Psi(n) - L + L - \Psi(n-1)|$$
$$\leq |\Psi(n) - L| + |\Psi(n-1) - L|$$

and if $n > n_0 + 1$, we have

$$|a_n| \leq |\Psi(n) - L| + |\Psi(n-1) - L| < \frac{\epsilon}{2} + \frac{\epsilon}{2} = \epsilon$$

This proves that $\lim_{n \to \infty} a_n = 0$. ∎

We point out that Theorem 5.4 gives a necessary but not sufficient condition for the convergence of an infinite series. In other words, if $\lim_{n \to \infty} a_n = 0$, it does *not* follow that the series $a_1 + a_2 + \cdots a_n + \cdots$ must converge. A counterexample is the divergent harmonic series in Example 3.

■ **THEOREM 5.5.** If $a_1 + a_2 + \cdots + a_n + \cdots = L$ is a convergent infinite series, then the infinite series $ka_1 + ka_2 + \cdots + ka_n + \cdots = kL$.

PROOF: The partial sums of the second series are

$$\Psi(n) = (ka_1 + \cdots + ka_n) = k(a_1 + \cdots + a_n)$$

and hence are multiples of the terms of a convergent sequence. Therefore, this result follows from Corollary 2.9. ■

In essence, Theorem 5.5 is an infinite distributive law in that it says that multiplication distributes over an infinite sum *if the infinite sum converges*. Of course, if the infinite sum diverges, then the equation $ka_1 + ka_2 + \cdots + ka_n + \cdots = kL$ has no meaning.

■ **THEOREM 5.6.** If $a_1 + a_2 + \cdots + a_n + \cdots = L$, and $b_1 + b_2 + \cdots + b_n + \cdots = M$ are convergent infinite series, then the infinite series

$$(a_1 + b_1) + (a_2 + b_2) + \cdots + (a_n + b_n) + \cdots = L + M$$

is also convergent.

PROOF: Since

$$\lim_{n \to \infty} (a_1 + \cdots + a_n) = L$$

and

$$\lim_{n \to \infty} (b_1 + b_2 + \cdots + b_n) = M$$

Theorem 2.8 applies to show that

$$\lim_{n \to \infty} [(a_1 + b_1) + \cdots + (a_n + b_n)]$$
$$= \lim_{n \to \infty} [(a_1 + \cdots + a_n) + (b_1 + \cdots + b_n)] = L + M \quad ■$$

We now limit our discussion temporarily to infinite series of nonnegative numbers. Assuming that each term of the defining infinite sequence φ is nonnegative, we are also assuming that the partial sums of the infinite series satisfy

$$0 \leq \Psi(n) = a_1 + a_2 + \cdots + a_n$$
$$\leq a_1 + a_2 + \cdots + a_n + a_{n+1} = \Psi(n + 1)$$

We say that Ψ is a *monotone increasing sequence*.

Our next result is an existence theorem. It tells us that an infinite series converges without our knowing the number to which it converges.

Such information can be useful, both theoretically and practically. Note that the proof of this result, as is true for most of our existence theorems, relies upon basic properties of bounded sets of real numbers.

■ **THEOREM 5.7.** Suppose that each $a_n \geqq 0$ and that each partial sum satisfies
$$a_1 + a_2 + \cdots + a_n \leqq M$$
for some positive number M. Then the infinite series $a_1 + a_2 + \cdots + a_n + \cdots$ is convergent.

This result is an immediate consequence of Theorem 3.10

■ *Example 4.* Consider the infinite sequence $\varphi(n) = 1/n!$ and the associated infinite series
$$1 + \frac{1}{2!} + \frac{1}{3!} + \cdots + \frac{1}{n!} + \cdots$$

It is easy to prove that the partial sums of this series satisfy
$$1 + \frac{1}{2!} + \frac{1}{3!} + \cdots + \frac{1}{n!} \leqq 1 + \frac{1}{2} + \frac{1}{4} + \cdots$$
$$+ \frac{1}{2^{n-1}} = \frac{1 - 1/2^n}{1 - 1/2} = 2 - \frac{1}{2^{n-1}} < 2$$

Therefore by Theorem 5.7, the given series converges, even though we do not know its limit. In fact, the limit of the augmented series
$$1 + 1 + \frac{1}{2!} + \frac{1}{3!} + \frac{1}{4!} + \cdots + \frac{1}{n!} + \cdots$$
is the very important number $e = 2.71328 \cdots$.

In Example 4, we already made use of a "comparison" between two infinite series. The next theorem gives a general *comparison test* for the convergence or divergence of an infinite series.

■ **THEOREM 5.8.** (a) If each $a_n \geqq 0$, if $a_1 + a_2 + \cdots + a_n + \cdots$ converges, and if $0 \leqq b_n \leqq a_n$ for each $n \in \mathbf{N}$, then the infinite series $b_1 + b_2 + \cdots + b_n + \cdots$ also converges.

(b) If each $a_n \geqq 0$, if $a_1 + a_2 + \cdots + a_n + \cdots$ diverges and if $b_n \geqq a_n$ for each $n \in \mathbf{N}$, then the infinite series $b_1 + b_2 + \cdots + b_n + \cdots$ also diverges.

PROOF: (a) Since the sequence of partial sums $\Psi(n) = a_1 + a_2 + \cdots + a_n$ is monotone increasing, the assumption that $\lim_{n \to \infty} \Psi(n) = L$ implies that
$$\Psi(n) = a_1 + a_2 + \cdots + a_n \leqq L \quad \text{for all } n \in \mathbf{N}$$

Then the inequality $0 \leqq b_n \leqq a_n$ (for all n) implies that

$$b_1 + b_2 + \cdots + b_n \leq a_1 + a_2 + \cdots + a_n \leq L$$

whence the partial sums of the second series are bounded and Theorem 5.7 applies.

(b) If $b_1 + b_2 + \cdots + b_n + \cdots$ were convergent, then by part (a) the series $a_1 + a_2 + \cdots + a_n + \cdots$ would also be convergent. Having assumed that $a_1 + a_2 + \cdots + a_n + \cdots$ is divergent, we find that the larger series cannot converge. ∎

■ *Example 5.* We know from Example 2 that the series

$$1 + \frac{1}{3} + \frac{1}{9} + \frac{1}{27} + \cdots + \frac{1}{3^{n-1}} + \cdots$$

is convergent. Because $\dfrac{1}{n \cdot 3^n} \leq 1/3^n$ for all $n \in \mathbf{N}$, it follows from 5.8 that the series

$$1 + \frac{1}{1 \cdot 3} + \frac{1}{2 \cdot 3^2} + \frac{1}{3 \cdot 3^3} + \frac{1}{4 \cdot 3^4} + \cdots + \frac{1}{n \cdot 3^n} n + \cdots$$

also converges.

■ *Example 6.* From Example 3 we have a divergent series

$$1 + \frac{1}{2} + \frac{1}{3} + \cdots + \frac{1}{n} + \cdots$$

Then because $1/n \leq 1/\sqrt{n}$ for all $n \in \mathbf{N}$, Theorem 5.8 tells us that the series

$$1 + \frac{1}{\sqrt{2}} + \frac{1}{\sqrt{3}} + \cdots + \frac{1}{\sqrt{n}} + \cdots$$

diverges, too. ∎

Now we return to an infinite series of arbitrary numbers, that is, we remove the restriction that the terms of the defining sequence be nonnegative.

■ **THEOREM 5.9.** Let $a_1 + a_2 + \cdots + a_n + \cdots$ be an infinite series. If the infinite series of absolute values $|a_1| + |a_2| + \cdots + |a_n| + \cdots$ converges, then the given series converges.

PROOF: We define two new sequences. Let

$$\begin{aligned} b_n &= a_n && \text{if } a_n \geq 0 \\ &= 0 && \text{if } a_n < 0 \end{aligned}$$

and let

$$\begin{aligned} c_n &= -a_n && \text{if } a_n < 0 \\ &= 0 && \text{if } a_n \geq 0 \end{aligned}$$

5.2 INFINITE SERIES

Note that we have both $b_n \geq 0$ and $c_n \geq 0$ for all $n \in \mathbf{N}$. Furthermore, we have

$$b_n - c_n = a_n - 0 \quad \text{if } a_n \geq 0$$
$$= -(-a_n) = a_n \quad \text{if } a_n < 0$$

or

$$b_n - c_n = a_n$$

Similarly,

$$b_n + c_n = a_n - 0 = a_n \quad \text{if } a_n \geq 0$$
$$= 0 + (-a_n) = -c_n \quad \text{if } a_n < 0$$

Hence

$$b_n + c_n = |a_n|$$

Obviously, too, we have

$$0 \leq b_n \leq |a_n|, \; 0 \leq c_n \leq |a_n|$$

for all $n \in \mathbf{N}$. Hence both series $b_1 - b_2 + \cdots + b_n + \cdots$ and $c_1 + c_2 + \cdots + c_n + \cdots$ converge by Theorem 5.8. From Theorem 5.6, then, the series

$$(b_1 - c_1) + (b_2 - c_2) + \cdots + (b_n - c_n) + \cdots$$
$$= a_1 + a_2 + \cdots + a_n + \cdots$$

is convergent. ∎

The next theorem is often called the *ratio test* for convergence. It will be very useful to us in later applications.

■ **THEOREM 5.10.** Let $a_1 + a_2 + \cdots + a_n + \cdots$ be an infinite series and suppose that

$$\lim_{n \to \infty} \left| \frac{a_{n+1}}{a_n} \right| = r$$

If $r < 1$, then the series converges. If $r > 1$, the series diverges. If $r = 1$, no information results from this test.

PROOF: First assume that $r < 1$. Choose any number r_0 such that $r < r_0 < 1$. Then for the positive number $\epsilon = r_0 - r$, there is an $n_0 \in \mathbf{N}$ such that

$$\left| \left| \frac{a_{n+1}}{a_n} \right| - r \right| < r_0 - r \quad \text{whenever } n \geq n_0$$

Therefore,

$$-r_0 + r < \left| \frac{a_{n+1}}{a_n} \right| - r < r_0 - r$$

and in particular,

$$\left|\frac{a_{n+1}}{a_n}\right| < r_0 < 1 \quad \text{whenever } n \geq n_0$$

It follows that

$$|a_{n_0+1}| < r_0|a_{n_0}|$$
$$|a_{n_0+2}| < r_0|a_{n_0+1}| < r_0{}^2|a_{n_0}|$$

and, in general,

$$|a_{n_0+k}| < r_0{}^k|a_{n_0}|$$

In view of Example 2 and Theorem 5.5, $|a_{n_0}| + r_0|a_{n_0}| + \cdots + r_0{}^k|a_{n_0}| + \cdots$ is convergent. Then Theorem 5.2 tells us that the series

$$|a_1| + |a_2| + \cdots + |a_{n_0}| + |a_{n_0+1}| + \cdots$$

converges and, finally, Theorem 5.9 completes the proof.

If $r > 1$, we choose r_0 so that $1 < r < r_0$, repeat the argument above to compare the given series with a divergent series $|a_{n_0}| + r_0|a_{n_0}| + \cdots + r_0{}^k|a_{n_0}| + \cdots$ and the proof is complete. The details are left as an exercise.

Lastly, if $r = 1$, the series may either diverge or converge. For instance, the divergent harmonic series $1 + 1/2 + 1/3 + \cdots + 1/n + \cdots$ has

$$\lim_{n \to \infty} \frac{1/n + 1}{1/n} = \lim_{n \to \infty} \frac{n}{n + 1} = 1$$

On the other hand, the series

$$1 + \frac{1}{2^2} + \frac{1}{3^2} + \cdots + \frac{1}{n^2} + \cdots$$

converges (proof in Section 7.7). However, we again find that

$$\lim_{n \to \infty} \frac{1/(n + 1)^2}{1/n^2} = \lim_{n \to \infty} \frac{n^2}{(n + 1)^2} = 1 \quad \blacksquare$$

An infinite series is said to be *alternating* if $a_n a_{n+1} < 0$ for each $n \in \mathbf{N}$. We shall normally write

$$a_1 - a_2 + a_3 - a_4 + \cdots + (-1)^{n-1}a_n + \cdots,$$

thinking of each a_n as positive, whence the term $(-1)^{n-1}a_n$ is positive if n is odd and is negative if n is even.

■ **THEOREM 5.11.** Let $a_1 - a_2 + a_3 - a_4 + \cdots + (-1)^{n-1}a_n + \cdots$ be an alternating series. If $a_n > a_{n+1}$ for each $n \in \mathbf{N}$ and if $\lim_{n \to \infty} a_n = 0$, then the series converges.

PROOF: Consider the even-numbered partial sums

$$\Psi(2) = a_1 - a_2$$
$$\Psi(4) = (a_1 - a_2) + (a_3 - a_4) > \Psi(2)$$
$$\Psi(6) = \Psi(4) + (a_5 - a_6) > \Psi(4)$$

and, in general,

$$\Psi(2n) = \Psi(2k - 2) + (a_{2k-1} - a_{2k}) > \Psi(2k - 2)$$

Having assumed that $a_{2k-1} > a_{2k}$, we have

$$0 \leq \Psi(2) < \Psi(4) < \cdots < \Psi(2n) < \cdots$$

However, we may write

$$\Psi(2n) = a_1 - (a_2 - a_3) - (a_4 - a_5) - \cdots$$
$$- (a_{2n-2} - a_{2n-1}) - a_{2n}$$

which immediately shows that $\Psi(2n) < a_1$ for all $n \in \mathbf{N}$. From Theorem 5.7, we know that $\lim_{n \to \infty} \Psi(2n)$ exists and does not exceed a_1. Let

$$\lim_{n \to \infty} \Psi(2n) = L$$

Hence, given $\epsilon > 0$, there is an integer $n_0 \in \mathbf{N}$ such that

$$|\Psi(2n) - L| < \frac{\epsilon}{2} \qquad \text{whenever } n \geq n_0$$

Also because $\lim_{n \to \infty} a_n = 0$, there exists $n_1 \in \mathbf{N}$ such that

$$|a_n - 0| = a_n < \frac{\epsilon}{2} \qquad \text{whenever } n \geq n_1$$

If n_2 is the larger of n_0 and n_1, then if $n > n_2$ we have

$$|\Psi(2n) - L| < \frac{\epsilon}{2}$$

and

$$|\Psi(2n + 1) - L| = |\Psi(2n) - L + a_{2n+1}|$$
$$\leq |\Psi(2n) - L| + a_{2n+1} < \frac{\epsilon}{2} + \frac{\epsilon}{2} = \epsilon$$

Thus in every case,

$$|\Psi(n) - L| < \epsilon \qquad \text{whenever } n \geq 2n_2$$

This completes the proof. ∎

■ *Example 7.* The alternating harmonic series

$$1 - \frac{1}{2} + \frac{1}{3} - \frac{1}{4} + \cdots + (-1)^{n-1}\frac{1}{n} + \cdots$$

converges because $1/n > 1/(n + 1)$ and $\lim_{n \to \infty} 1/n = 0$. The series converges very slowly, however.

- **COROLLARY 5.12.** Let $a_1 - a_2 + a_3 - a_n + \cdots + (-1)^{n-1}a_n + \cdots$ be an alternating series and let $\lim_{n \to \infty} \Psi(n) = L$. Then for any $n \in \mathbf{N}$,

$$|L - \Psi(n)| \leq a_{n+1}$$

The proof is actually part of the proof of Theorem 5.11.

EXERCISES 5.2

1. The general terms a_n of certain infinite series are given below. Write out the fifth partial sum $a_1 + a_2 + a_3 + a_4 + a_5$ in each case. Guess whether or not the series converges.

   (a) $a_n = \dfrac{1}{100n}$
   (b) $a_n = \dfrac{1}{2n^2 - n}$

   (c) $a_n = \dfrac{1000}{n^2 + n}$
   (d) $a_n = \dfrac{n^3}{3n}$

   (e) $a_n = \dfrac{1}{(2n+1)(n+1)}$
   (f) $a_n = \dfrac{n!}{100^n}$

   (g) $a_n = \dfrac{1}{n2^n}$
   (h) $a_n = \dfrac{1}{n^n}$

   (i) $a_n = (-1)^{n-1} \dfrac{1}{\sqrt{n^2 + 1}}$
   (j) $a_n = \dfrac{n}{(n+1)2^n}$

2. The nth partial sums $\Psi(n) = a_1 + a_2 + \cdots + a_n$ of certain infinite series are given below. Determine if the series is convergent in each case. Can you find a formula for the general term a_n in each case?

   (a) $\Psi(n) = \dfrac{n+1}{3n}$
   (b) $\Psi(n) = 1 - \dfrac{1}{2^n}$

   (c) $\Psi(n) = \dfrac{3n}{5n+2}$
   (d) $\Psi(n) = \dfrac{n^2}{n+3}$

   (e) $\Psi(n) = 3^n$
   (f) $\Psi(n) = \dfrac{n^2 + 1}{n^2}$

   (g) $\Psi(n) = \left(\dfrac{5}{4}\right)^n - 1$
   (h) $\Psi(n) = (-1)^{n-1}$

3. Show that each of the infinite series below is divergent.

   (a) $1 - 1 + 1 - 1 + \cdots + (-1)^{n-1} + \cdots$

   (b) $1 + \sqrt{2} + 2 + 2\sqrt{2} + 4 + \cdots$

   (c) $\dfrac{1}{3} + \dfrac{2}{4} + \dfrac{3}{5} + \cdots + \dfrac{n}{n+2} + \cdots$

   (d) $\dfrac{11}{10} + \left(\dfrac{11}{10}\right)^2 + \cdots + \left(\dfrac{11}{10}\right)^n + \cdots$

   (e) $\dfrac{1}{1000} + \dfrac{2}{1000^2} + \dfrac{6}{1000^3} + \cdots + \dfrac{n!}{1000^n} + \cdots$

4. For each of the following series, there is a positive number M such that $a_1 + a_2 + \cdots + a_n < M$ for each $n \in \mathbf{N}$. Hence by Theorem 5.7, each series converges. Determine such a number M in each case.

   (a) $4 + 2 + 1 + \dfrac{1}{2} + \dfrac{1}{4} + \cdots + 2^{3-n} + \cdots$

(b) $\dfrac{1}{9} + \dfrac{1}{27} + \dfrac{1}{81} + \cdots + \left(\dfrac{1}{3}\right)^{n+1} + \cdots$

(c) $1 - \dfrac{1}{2} + \dfrac{1}{3} - \dfrac{1}{4} + \dfrac{1}{5} - \dfrac{1}{6} + \cdots + (-1)^{n-1}\dfrac{1}{n} + \cdots$

(d) $1 + \dfrac{1}{4} + \dfrac{1}{9} + \dfrac{1}{16} + \cdots + \dfrac{1}{n^2} + \cdots$

(e) $\dfrac{1}{2} + \dfrac{2}{9} + \dfrac{3}{28} + \cdots + \dfrac{n}{n^3 + 1} + \cdots$

5. The comparison test (Theorem 5.8) is valuable only to the extent that one can find an appropriate comparison series. Do so to determine whether or not each of the following series converges.

(a) $\dfrac{1}{4} + \dfrac{1}{24} + \cdots + \dfrac{1}{4 \cdot 6^{n-1}} + \cdots$

(b) $\dfrac{1}{2 \cdot 1} + \dfrac{1}{4 \cdot 2} + \dfrac{1}{6 \cdot 2^2} + \cdots + \dfrac{1}{2n \cdot 2^n} + \cdots$

(c) $1 + \dfrac{1}{\sqrt{2}} + \dfrac{1}{\sqrt{3}} + \cdots + \dfrac{1}{\sqrt{r}} + \cdots$

(d) $\dfrac{1}{2} + \dfrac{1}{6} + \dfrac{1}{12} + \cdots + \dfrac{1}{n(n+1)} + \cdots$

(e) $\dfrac{1}{2} + \dfrac{1}{2\sqrt[3]{2}} + \dfrac{1}{2\sqrt[3]{3}} + \cdots + \dfrac{1}{2\sqrt[3]{n}} + \cdots$

(f) $\dfrac{1}{2} + \dfrac{2}{5} + \dfrac{3}{10} + \cdots + \dfrac{n}{n^2 + 1} + \cdots$

(g) $1 + \dfrac{1}{2^2} + \dfrac{1}{3^3} + \cdots + \dfrac{1}{n^n} + \cdots$

(h) $\dfrac{2}{1} + \dfrac{4}{2!} + \dfrac{8}{3!} + \cdots + \dfrac{2^n}{n!} + \cdots$

(i) $\dfrac{1}{2} + \dfrac{2}{9} + \dfrac{3}{28} + \cdots + \dfrac{n}{n^3 + 1} + \cdots$

(j) $\dfrac{1}{2} + \dfrac{4}{24} + \dfrac{36}{720} + \cdots + \dfrac{(n!)^2}{(2n)!} + \cdots$

6. The ratio test (Theorem 5.10) applies best when the general term of the series under test contains a factor such as $n!$ or p^n. Apply the ratio test to the following series.

(a) $1 + \dfrac{1}{2} + \dfrac{1}{6} + \dfrac{1}{24} + \cdots + \dfrac{1}{n!} + \cdots$

(b) $2 + \dfrac{3}{2} + \dfrac{4}{6} + \dfrac{5}{24} + \cdots + \dfrac{n+1}{n!} + \cdots$

(c) $5 + \dfrac{11}{2} + \dfrac{16}{6} + \dfrac{21}{24} + \cdots + \dfrac{5n + 1}{n!} + \cdots$

(d) $4 + \dfrac{9}{2} + \dfrac{16}{6} + \dfrac{25}{24} + \cdots + \dfrac{(n+1)^2}{n!} + \cdots$

(e) $2 + \dfrac{4}{2} + \dfrac{8}{6} + \dfrac{16}{24} + \cdots + \dfrac{2^n}{n!} + \cdots$

(f) $\dfrac{3}{4} + \dfrac{9}{16} + \dfrac{27}{64} + \cdots + \left(\dfrac{3}{4}\right)^n + \cdots$

(g) $\dfrac{3}{4} + \dfrac{18}{16} + \dfrac{81}{64} + \cdots + n\left(\dfrac{3}{4}\right)^n + \cdots$

(h) $\dfrac{3}{4} + \dfrac{36}{16} + \dfrac{729}{64} + \cdots + n^2\left(\dfrac{3}{4}\right)^n + \cdots$

(i) $1 + \dfrac{2}{3} + \dfrac{6}{15} + \dfrac{24}{105} + \cdots + \dfrac{n!}{1 \cdot 3 \cdot 5 \cdot \cdots (2n-1)} + \cdots$

(j) $1 + \dfrac{3}{4} + \dfrac{15}{28} + \dfrac{105}{280} + \cdots + \dfrac{1 \cdot 3 \cdot 5 \cdots (2n-1)}{1 \cdot 4 \cdot 7 \cdots (3n-2)} + \cdots$

(k) $1 + \dfrac{2^m}{2!} + \dfrac{3^m}{3!} + \cdots + \dfrac{n^m}{n!} + \cdots$

(l) $m + \dfrac{m^2}{2!} + \dfrac{m^3}{3!} + \cdots + \dfrac{m^n}{n!} + \cdots$

(m) $1 + \dfrac{2^2}{2!} + \dfrac{3^3}{3!} + \cdots + \dfrac{n^n}{n!} + \cdots$

7. A real mark of skill is to apply knowledge without being told how or which facts to use. Here is a mixed selection of series to test for convergence in any way you can.

(a) $\dfrac{1}{10} + \dfrac{2}{19} + \dfrac{3}{28} + \cdots + \dfrac{n}{9n+1} + \cdots$

(b) $\dfrac{1}{3} + \dfrac{1}{4} + \dfrac{1}{5} + \cdots + \dfrac{1}{n+2} + \cdots$

(c) $\dfrac{1}{3} - \dfrac{1}{4} + \dfrac{1}{5} - \dfrac{1}{6} + \cdots + (-1)^{n-1}\dfrac{1}{n+2} + \cdots$

(d) $\dfrac{1}{2 \cdot 4} + \dfrac{1}{3 \cdot 5} + \dfrac{1}{4 \cdot 6} + \cdots + \dfrac{1}{(n+1)(n+3)} + \cdots$

(e) $\dfrac{1}{\sqrt{2}} + \dfrac{1}{\sqrt{5}} + \dfrac{1}{\sqrt{10}} + \cdots + \dfrac{1}{\sqrt{n^2+1}} + \cdots$

(f) $\dfrac{1}{3} + \dfrac{1}{8} + \dfrac{1}{17} + \cdots + \dfrac{1}{2^n + n^2} + \cdots$

(g) $\dfrac{1}{3} + \dfrac{1}{15} + \dfrac{1}{35} + \cdots + \dfrac{1}{4n^2 - 1} + \cdots$

(h) $\dfrac{1}{3} + \dfrac{1}{18} + \dfrac{1}{81} + \cdots + \dfrac{1}{n \cdot 3^n} + \cdots$

(i) $\dfrac{1}{2} + \dfrac{1}{3} + \dfrac{1}{8} + \dfrac{1}{30} + \cdots + \dfrac{n}{(n+1)!} + \cdots$

(j) $\dfrac{1}{6} + \dfrac{1}{24} + \dfrac{1}{60} + \cdots + \dfrac{1}{n(n+1)(n+2)} + \cdots$

(k) $\dfrac{1}{5} + \dfrac{1}{10} + \dfrac{1}{15} + \cdots + \dfrac{1}{5n} + \cdots$

(l) $\dfrac{1}{5} - \dfrac{1}{10} + \dfrac{1}{15} - \dfrac{1}{20} + \cdots + (-1)^{n-1}\dfrac{1}{5n} + \cdots$

(m) $\dfrac{1}{5} + \dfrac{\sqrt{2}}{8} + \dfrac{\sqrt{3}}{13} + \cdots + \dfrac{\sqrt{n}}{n^2+4} + \cdots$

(n) $\dfrac{1}{\sqrt{2}} + \dfrac{1}{\sqrt{8+1}} + \dfrac{1}{\sqrt{27+1}} + \cdots + \dfrac{1}{\sqrt{n^3+1}} + \cdots$

8. Here are several series whose nth term is not given. From the first five terms, deduce the form of the general term a_n. Which are convergent?

(a) $\frac{1}{2} + \frac{1}{5} + \frac{1}{10} + \frac{1}{17} + \frac{1}{26} + \cdots$

(b) $\frac{1}{2} + \frac{1}{5} + \frac{1}{11} + \frac{1}{29} + \frac{1}{83} + \cdots$

(c) $\frac{1}{4} + \frac{1}{7} + \frac{1}{10} + \frac{1}{13} + \frac{1}{16} + \cdots$

(d) $\frac{1}{2} - \frac{2}{5} + \frac{3}{10} - \frac{4}{17} + \frac{5}{26} - \cdots$

(e) $\frac{1}{2} + 1 + \frac{9}{8} + 1 + \frac{25}{32} + \cdots$

(f) $\frac{1}{4} - \frac{1}{10} + \frac{1}{18} - \frac{1}{28} + \frac{1}{40} - \cdots$

(g) $1 + \frac{4}{3} + \frac{4}{3} + \frac{32}{27} + \frac{80}{81} + \cdots$

(h) $1 + \frac{1}{3} + \frac{4}{27} + \frac{2}{27} + \frac{16}{405} + \cdots$

(i) $1 + \frac{1}{2} + \frac{3}{14} + \frac{3}{35} + \frac{3}{91} + \cdots$

(j) $1 + \frac{3}{5} + \frac{1}{3} + \frac{7}{39} + \frac{21}{221} + \cdots$

PROBLEMS 5.2

1. Let $b_n = \varphi(n)$ be a positive sequence and suppose that $\lim_{n \to \infty} b_n = \infty$. Show that the series whose general term is $a_n = b_{n+1} - b_n$ diverges, but that the series whose general term is $c_n = 1/b_n - 1/b_{n+1}$ converges.

2. Suppose that the series of positive terms $a_1 + a_2 + \cdots + a_n + \cdots$ is convergent and has the sum M. Prove that the partial sums $s_n = a_1 + \cdots + a_n$ satisfy

$$\lim_{n \to \infty} \frac{1}{n}(s_1 + s_2 + \cdots + s_n) = M$$

3. Let $a_1 + a_2 + \cdots + a_n + \cdots$ be a convergent series of positive terms. Show that the series $a_1^2 + a_2^2 + \cdots + a_n^2 + \cdots$ is also convergent.

4. Let $a_1, a_2, \ldots, a_n, \ldots$ be a sequence of positive terms such that the series $a_1^2 + a_2^2 + \cdots + a_n^2 + \cdots$ is convergent. Prove that the series

$$a_1 + \frac{1}{2}a_2 + \frac{1}{3}a_3 + \cdots + \frac{1}{n}a_n + \cdots$$

is also convergent.

5. Show that, if the series $a_1 + 2a_2 + 3a_3 + \cdots + na_n + \cdots$ converges, then so does $a_1 + a_2 + a_3 + \cdots + a_n + \cdots$.

6. Let the series of positive terms $a_1 + a_2 + \cdots + a_n + \cdots$ be divergent. Prove that the series

$$\frac{a_1}{1 + a_1} + \frac{a_2}{(1 + a_1)(1 + a_2)} + \cdots + \frac{a_n}{(1 + a_1)(1 + a_2) \cdots (1 + a_n)} + \cdots$$

is convergent and find its sum.

7. Let P and Q be polynomials. If the degree of Q exceeds that of P by at least 2, then prove that the series

$$\frac{P(1)}{Q(1)} + \frac{P(2)}{Q(2)} + \frac{P(3)}{Q(3)} + \cdots + \frac{P(n)}{Q(n)} + \cdots$$

is convergent.

8. Determine
$$\lim_{n\to\infty} \frac{1}{\sqrt{n}}\left(\frac{1}{\sqrt{1}} + \frac{1}{\sqrt{2}} + \cdots + \frac{1}{\sqrt{n}}\right)$$

9. Determine the sum of each of the following series:

   (a) $\dfrac{1}{3} + \dfrac{1}{15} + \cdots + \dfrac{1}{(2n-1)(2n+1)} + \cdots$

   (b) $\dfrac{1}{2} + \dfrac{1}{3} + \cdots + \dfrac{n}{(n+1)!} + \cdots$

   (c) $\dfrac{3}{4} + \dfrac{5}{36} + \cdots + \dfrac{2n+1}{n^2(n+1)^2} + \cdots$

10. Establish the following theorem, often called the *root test*:

    Theorem. Let $a_1 + a_2 + \cdots + a_n + \cdots$ be a series of positive constants. If $\lim_{n\to\infty} \sqrt[n]{a_n} = k < 1$, then the series converges. If $\lim_{n\to\infty} \sqrt[n]{a_n} = k > 1$, then the series diverges. If $\lim_{n\to\infty} \sqrt[n]{a_n} = 1$, then no conclusion can be drawn from this test. (That is, there are series having this property that converge and others that diverge.)

5.3 POWER SERIES

An infinite series of the form

$$a_0 + a_1(x - c) + a_2(x - c)^2 + \cdots + a_n(x - c)^n + \cdots$$

is called a *power series in* $x - c$. For some values of x such a power series is convergent; for other values it may be divergent. In particular, we see that

$$a_0 + a_1(c - c) + a_2(c - c)^2 + \cdots + a_n(c - c)^n + \cdots = a_0$$

so the power series in $x - c$ always converges when $x = c$.

Suppose that there is a number $x \neq c$ for which the power series

$$|a_0| + |a_1| \cdot |x - c| + |a_2| \cdot |x - c|^2 + \cdots + |a_n| \cdot |x - c|^n + \cdots$$

is convergent. If $|x - c| = \epsilon > 0$, then any number y satisfying

$$c - \epsilon < y < c + \epsilon$$

also satisfies $|y - c| < |x - c|$, and hence

$$|a_n| \cdot |y - c|^n < |a_n| \cdot |x - c|^n$$

Therefore Theorem 5.6 tells us that the series

$$|a_0| + |a_1| \cdot |y - c| + |a_2| \cdot |y - c|^2 + \cdots + |a_n| \cdot |y - c|^n + \cdots$$

converges and, by Theorem 5.9, the power series

$$a_0 + a_1(y - c) + a_2(y - c)^2 + \cdots + a_n(y - c)^n + \cdots$$

is also convergent. These remarks lead us to the following definition:

The *radius of convergence* r of the power series

$$a_0 + a_1(x - c) + a_2(x - c)^2 + \cdots + a_n(x - c)^n + \cdots$$

is the least upper bound of all numbers $|x - c|$ for which the series

$$|a_0| + |a_1| \cdot |x - c| + |a_2| \cdot |x - c|^2 + \cdots + |a_n| \cdot |x - c|^n + \cdots$$

is convergent. If the series of absolute values converges for all $x \in \mathbf{R}$, then we say that $r = \infty$.

Because the power series $a_0 + a_1(x - c) + \cdots + a_n(x - c)^n + \cdots$ converges (to a_0) if $x = c$, the number r is always nonnegative. If $r = 0$, then the series only converges for the value $x = c$. Hence our interest lies chiefly in the cases where $r > 0$.

■ **THEOREM 5.13.** Let $r > 0$ be the radius of convergence of the power series $a_0 + a_1(x - c) + a_2(x - c)^2 + \cdots + a_n(x - c)^n + \cdots$. Then the series converges for every point x in the open interval $\{x : c - r < x < c + r\}$ and diverges for every point x not in the closed interval $\{x : c - r \leq x \leq c + r\}$.

PROOF: In view of the discussion preceding the definition above, we only need to prove that the series diverges if $|x - c| > r$. This is done by showing the converse, namely, that if the series converges, then we must have $|x - c| \leq r$. If the series converges, then by Theorem 5.4,

$$\lim_{n \to \infty} |a_n(x - c)^n| = 0$$

Therefore there exists a number $m > 0$ such that

$$|a_n(x - c)^n| < m \qquad \text{for all } n \in \mathbf{N}$$

Now choose any number z such that $|z - c| < |x - c|$ and define $t = |z - c|/|x - c| < 1$. Then we have

$$|a_n(z - c)^n| = |a_n(x - c)^n t^n| < m t^n$$

In view of Theorem 5.8, the series

$$|a_0| + |a_1| \cdot |z - c| + \cdots + |a_n| \cdot |z - c|^n + \cdots$$

is term-by-term less than the convergent series $m + mt + \cdots + mt^n + \cdots$ and therefore converges. It follows that $|x - c|$ cannot exceed the least upper bound r. ■

We note that Theorem 5.13 says nothing about the convergence of the power series $a_0 + a_1(x - c) + \cdots + a_n(x - c)^n + \cdots$ when $|x - c| = r$, the radius of convergence. The two numbers $x = c - r$ and $x = c - r$ are special cases. A given series may converge at either or both of these points, or it may diverge at both. In short, there are four possibilities. A given series may converge for each point in the open interval $\{x : c - r < x < c + r\}$ and diverge elsewhere; it may converge in either the half-open interval $\{x : c - r < x \leq c + r\}$ or $\{x : c - r \leq x < c + r\}$

and diverge elsewhere; or it may converge in the closed interval $\{x : c - r \leq x \leq c + r\}$ and diverge elsewhere. In any case the interval in which the series converges is called the *interval of convergence* of the power series.

■ **Example 1.** Consider the series
$$1 + (x - 2) + (x - 2)^2 + \cdots + (x - 2)^n + \cdots$$
We apply the limit test (Theorem 5.10) as follows:
$$\lim_{n \to \infty} \left| \frac{a_{n+1}}{a_n} \right| = \lim_{n \to \infty} \left| \frac{(x - 2)^{n+1}}{(x - 2)^n} \right| = \lim_{n \to \infty} |x - 2| = |x - 2|$$
Therefore the given series converges if $|x - 2| < 1$ and diverges if $|x - 2| > 1$. That is, the series converges in the interval $\{x : 1 < x < 3\}$, and diverges in the open rays $\{x : x < 1\}, \{x : x > 3\}$. At the points $x = 1$ and $x = 3$, the ratio test fails so we have to look at these cases separately.

When $x = 1$, the series reduces to
$$1 - 1 + 1 - 1 + \cdots + (-1)^n + \cdots$$
and when $x = 3$, we have
$$1 + 1 + 1 + 1 + \cdots + 1 + \cdots$$
Neither of these converge because $\lim_{n \to \infty} a_n \neq 0$. Hence the complete interval of convergence of this series is $\{x : 1 < x < 3\}$.

■ **Example 2.** Consider the series
$$1 + (x - 1) + \frac{1}{2}(x - 1)^2 + \frac{1}{3}(x - 1)^3 + \cdots$$
$$+ \frac{1}{n}(x - 1)^n + \cdots$$
Again the limit test supplies the answer
$$\lim_{n \to \infty} \left| \frac{a_{n+1}}{a_n} \right| = \lim_{n \to \infty} \left| \frac{(x - 1)^{n+1}/(n + 1)}{(x - 1)^n/n} \right|$$
$$= \lim_{n \to \infty} \left| \frac{n}{n + 1}(x - 1) \right| = |x - 1|$$
Therefore the series converges if $|x - 1| < 1$ and diverges if $|x - 1| > 1$. At the point $x = 2$, we have the divergent harmonic series
$$1 + 1 + \frac{1}{2} + \frac{1}{3} + \cdots + \frac{1}{n} + \cdots$$
but at $x = 0$, we have the convergent alternating harmonic series
$$1 - 1 + \frac{1}{2} - \frac{1}{3} + \cdots + (-1)^n \frac{1}{n} + \cdots$$

Therefore the complete interval of convergence of the given series is the half-open interval $\{x : 0 \leq x < 2\}$.

- **Example 3.** The power series (with $c = 0$)

$$1 + x + \frac{1}{4}x^2 + \frac{1}{9}x^3 + \cdots + \frac{1}{n^2}x^n + \cdots$$

has a closed interval of convergence $\{x : -1 \leq x \leq 1\}$. It is easy to apply the limit test to prove that this series converges if $|x - 0| = |x| < 1$ and diverges if $|x| > 1$.

Note that $\dfrac{a_{n+1}}{a_n} = \dfrac{x^{n+1}}{(n+1)^2} \bigg/ \dfrac{x^n}{n^2} = \dfrac{n^2}{(n+1)^2} x$. Hence

$$\lim_{n \to \infty} \left| \frac{a_{n+1}}{a_n} \right| = \lim_{n \to \infty} \left| \frac{n^2}{(n+1)^2} x \right| = |x|$$

which proves our claim. But now at both endpoints $x = -1$ and $x = 1$, we have the convergent series

$$1 - 1 + \frac{1}{4} - \frac{1}{9} + \cdots + (-1)^n \frac{1}{n^2} + \cdots$$

and

$$1 + 1 + \frac{1}{4} + \frac{1}{9} + \cdots + \frac{1}{n^2} + \cdots$$

The first converges by Theorem 5.11 and the second converges, too, as we shall show in Section 7.7.

- **Example 4.** Consider the power series

$$1 + (x - c) + \frac{1}{2!}(x - c)^2 + \cdots + \frac{1}{n!}(x - c)^n + \cdots$$

The ratio of successive terms is

$$\frac{a_{n+1}}{a_n} = \frac{1}{(n+1)!}(x - c)^{n-1} \bigg/ \frac{1}{n!}(x - c)^n$$

$$= \frac{n!}{(n+1)!}(x - c) = \frac{x - c}{n + 1}$$

Thus, for any fixed number x, we have

$$\lim_{n \to \infty} \left| \frac{a_{n+1}}{a_n} \right| = \lim_{n \to \infty} \left| \frac{x - c}{n + 1} \right| = 0$$

This means that the series converges for *any* value of x or, equivalently, that its radius of convergence is infinite. ∎

Let $a_0 + a_1(x - c) + a_2(x - c)^2 + \cdots + a_n(x - c)^n + \cdots$ be a power series with radius of convergence $r > 0$. For any point x in the interval of convergence, the partial sums $\Psi(n) = a_0 + a_1(x - c) + \cdots +$

$a_n(x-c)^n$ have a limit $\lim_{n\to\infty} \Psi(n) = L$ which obviously depends upon the number x. That is, the power series defines a function

$$f(x) = a_0 + a_1(x-c) + a_2(x-c)^2 + \cdots + a_n(x-c)^n + \cdots$$

whose domain is the interval of convergence of the series. We show that such a function is differentiable.

■ **THEOREM 5.14.** Let $a_0 + a_1(x-c) + a_2(x-c)^2 + \cdots + a_n(x-c)^n + \cdots$ be a power series with radius of convergence $r > 0$. Then the function

$$f(x) = a_0 + a_1(x-c) + a_2(x-c)^2 + \cdots + a_n(x-c)^n + \cdots$$

is differentiable in the open interval $\{x : c - r < x < c + r\}$ and its derivative is

$$f'(x) = a_1 + 2a_2(x-c) + 3a_3(x-c)^2 + \cdots \\ + na_n(x-c)^{n-1} + \cdots$$

PROOF: We first show that the series $a_1 + 2a_2(x-c) + \cdots + a_n(x-c)^{n-1}$ converges on the interval $\{x : c - r < x < c + r\}$. Note that the ratio of two successive terms is

$$\frac{(n+1)a_{n+1}(x-c)^n}{na_n(x-c)^{n-1}} = \frac{n+1}{n} \cdot \frac{a_{n+1}}{a_n} \cdot (x-c)$$

Therefore the limit is

$$\lim_{n\to\infty} \left| \frac{n+1}{n} \cdot \frac{a_{n+1}}{a_n} (x-c) \right| = \lim_{n\to\infty} \left| \frac{a_{n+1}}{a_n} \right| \cdot |x-c|$$

This implies that the radius of convergence of the derived series is the same as that of the original series.

Next we show that

$$f'(x) = \lim_{h\to 0} \frac{f(x+h) - f(x)}{h}$$

is the series $a_1 + 2a_2(x-c) + \cdots + na_n(x-c)^{n-1} + \cdots$. Let x be any point in $\{x : c - r < x < c + r\}$ and choose h such that $x + h$ is also in this interval. Then

$$\frac{f(x+h) - f(x)}{h}$$
$$= \frac{1}{h}[a_0 + a_1(x+h-c) + a_2(x+h-c)^2 + \cdots$$
$$+ a_n(x+h-c)^n + \cdots] - \frac{1}{h}[a_0 + a_1(x-c)$$
$$+ a_2(x-c)^2 + \cdots + a_n(x-c)^n + \cdots]$$

$$= \frac{1}{h}[a_1 h + a_2[(x + h - c)^2 - (x - c)^2] + \cdots$$
$$+ a_n[(x + h - c)^n - (x - c)^n] + \cdots]$$
$$= \frac{1}{h}[a_1 h + 2a_2(x - c)h + h^2] + \cdots$$
$$+ a_n[(x + h - c)^n - (x - c)^n] + \cdots]$$

In order to evaluate the term $(x + h - c)^n - (x - c)^n$, we apply the extended mean value theorem (p. 173) to the function $(x - c)^n$ to write

$$(x + h - c)^n = (x - c)^n + n(x - c)^{n-1} h +$$
$$\frac{n(n - 1)}{2}(x_n - c)^{n-2} h^2$$

where x_n is some point between $x - c$ and $x + h - c$. Thus

$$\frac{1}{h} a_n[(x + h - c)^n - (x - c)^n]$$
$$= n a_n (x - c)^{n-1} + \frac{n(n - 1)}{2} a_n (x_n - c)^{n-2} h$$

and we have

$$\frac{f(x + h) - f(x)}{h} = a_1 + 2a_2(x - c) + h + \cdots$$
$$+ n a_n (x - c)^{n-1} + \frac{n(n - 1)}{2} a_n (x_n - c)^{n-2} h + \cdots$$

It is now easy to believe that

$$\lim_{h \to 0} \frac{f(x + h) - f(x)}{h} = f'(x)$$
$$= a_1 + 2a_2(x - c) + \cdots$$
$$+ n a_n (x - c)^{n-1} - \cdots \blacksquare$$

A complete proof can be found in an advanced calculus book such as Devinatz: *Advanced Calculus* (Holt, Rinehart and Winston, Inc.).

■ **COROLLARY 5.15.** A function $f(x) = a_0 + a_1(x - c) + a_2(x - c)^2 + \cdots + a_n(x - c)^n + \cdots$ defined by a power series with radius of convergence $r > 0$ is infinitely differentiable. (That is, $f^{(n)}(x)$ exist for each $n \in \mathbf{N}$ and for each point x in the open interval $\{x : c - r < x < c + r\}$.

PROOF: The argument proving Theorem 5.14 applies equally well to the function $f'(x) = a_1 + 2a_2(x - c) + \cdots + n a_n (x - c)^{n-1} + \cdots$ and hence to $f''(x)$, and so forth. ∎

EXERCISES 5.3

For each of the power series 1 to 40 determine the radius of convergence. By testing endpoints (where feasible) give the entire interval of convergence.

1. $1 + \dfrac{x}{2} + \dfrac{x^2}{3} + \cdots + \dfrac{x^n}{n+1} + \cdots$

2. $x + \dfrac{x^2}{\sqrt{2}} + \dfrac{x^3}{\sqrt{3}} + \cdots + \dfrac{x^n}{\sqrt{n}} + \cdots$

3. $x + \dfrac{x^2}{2^2} + \dfrac{x^3}{3^2} + \cdots + \dfrac{x^n}{n^2} + \cdots$

4. $x + \dfrac{x^2}{2^3} + \dfrac{x^3}{3^3} + \cdots + \dfrac{x^n}{n^3} + \cdots$

5. $1 + \dfrac{x}{6} + \dfrac{x^2}{36} + \cdots + \left(\dfrac{x}{6}\right)^n + \cdots$

6. $1 + \dfrac{4x}{2} + \dfrac{16x^2}{3} + \cdots + \dfrac{(4x)^n}{n+1} + \cdots$

7. $1 + \dfrac{2x}{1!} + \dfrac{3x^2}{2!} + \cdots + \dfrac{(n+1)x^n}{n!} + \cdots$

8. $1 - \dfrac{x^2}{2} + \dfrac{x^3}{6} - \cdots + (-1)^n \dfrac{x^{2n}}{(2n)!} + \cdots$

9. $x - \dfrac{x^3}{6} + \dfrac{x^5}{120} - \cdots + (-1)^{n+1} \dfrac{x^{2n-1}}{(2n-1)!} + \cdots$

10. $1 - \dfrac{x}{2} + \dfrac{x^2}{5} - \cdots + (-1)^n \dfrac{x^n}{1+n^2} + \cdots$

11. $\dfrac{x}{24} - \dfrac{x^2}{210} + \cdots + (-1)^{n+1} \dfrac{x^n}{(3n-1)(3n)(3n+1)} + \cdots$

12. $1 + x + 2x^2 + 6x^3 + \cdots + n!x^n + \cdots$

13. $1 + x + \dfrac{2}{4}x^2 + \dfrac{6}{27}x^3 + \cdots + \dfrac{n!}{n^n}x^n + \cdots$

14. $1 + x + 2x^2 + \dfrac{9}{2}x^3 + \cdots + \dfrac{n^n}{n!}x^n + \cdots$

15. $-\dfrac{x^3}{8} + \dfrac{x^5}{64} - \cdots + (-1)^n \dfrac{x^n}{2^{2n+1} \cdot n!} + \cdots$

16. $1 + \dfrac{1}{2}x^2 + \dfrac{(1\cdot 2)^2}{1\cdot 2\cdot 3\cdot 4}x^4 + \cdots + \dfrac{(n!)^2}{(2n)!}x^{2n} + \cdots$

17. $-\dfrac{x}{3} + \dfrac{x^2}{81} - \cdots + (-1)^n \dfrac{x^n}{(2n-1)3^{2n-1}} + \cdots$

18. $1 + \dfrac{1}{3}x^3 + \dfrac{(1\cdot 2)^3 x^6}{1\cdot 2\cdot 3\cdot 4\cdot 5\cdot 6} + \cdots + \dfrac{(n!)^3}{(3n)!}x^{3n} + \cdots$

19. $\dfrac{1}{4} + \dfrac{1}{5}(x-2) + \dfrac{1}{6}(x-2)^2 + \cdots + \dfrac{1}{n+4}(x-2)^n + \cdots$

20. $1 + 1(x-2) + \dfrac{1}{4}(x-2)^2 + \cdots + \dfrac{1}{n^2}(x-2)^n + \cdots$

21. $1 + \frac{1}{2}(x - 2) + \frac{1}{6}(x - 2)^2 + \cdots + \frac{1}{n(n + 1)}(x - 2)^n + \cdots$

22. $1 + \frac{1}{4}(x - 2) + \frac{2}{8}(x - 2)^2 + \cdots + \frac{n}{4^n}(x - 2)^n + \cdots$

23. $\frac{1}{25}(x - 2) + \frac{4}{125}(x - 2)^2 + \cdots + \frac{n^2}{5^{n+1}}(x - 2)^n + \cdots$

24. $(x + 3) + \frac{(x + 3)^2}{\sqrt{2}} + \frac{(x + 3)^3}{\sqrt{3}} + \cdots + \frac{(x + 3)^n}{\sqrt{n}} + \cdots$

25. $1 + \frac{x + 4}{3} + \left(\frac{x + 4}{3}\right)^2 + \cdots + \left(\frac{x + 4}{3}\right)^n + \cdots$

26. $\frac{1}{3} + \frac{2x + 9}{3(3 + 1)} + \frac{(2x + 9)^2}{9(6 + 1)} + \cdots + \frac{(2x + 9)^n}{3^n(3n + 1)} + \cdots$

27. $\frac{2x - 1}{1 \cdot 2 \cdot 3} + \frac{(2x - 1)^2}{2 \cdot 3 \cdot 4} + \cdots + \frac{(2x - 1)^n}{n(n + 1)(n + 2)} + \cdots$

28. $-\frac{(3x + 4)}{10 \cdot 1} + \frac{(3x + 4)^2}{10^2 \cdot 2^{3/2}} - \cdots + (-1)^n \frac{(3x + 4)^n}{10^n \cdot n^{3/2}} + \cdots$

29. $1 + \frac{6}{16}x + \frac{17}{108}x^2 + \cdots + \frac{3n^2 + 2n + 1}{(n + 1)^3}\left(\frac{x}{2}\right)^n + \cdots$

30. $-\frac{3}{1}x^2 + \frac{3 \cdot 4}{1 \cdot 2}x^4 - \frac{3 \cdot 4 \cdot 5}{1 \cdot 2 \cdot 3}x^6 + \cdots + (-1)^n \frac{3 \cdot 4 \cdot 5 \cdots (n + 2)}{n!}x^{2n} + \cdots$

31. $\frac{1}{2}x^3 + \frac{1 \cdot 3}{2 \cdot 4}x^5 + \frac{1 \cdot 3 \cdot 5}{2 \cdot 4 \cdot 6}x^7 + \cdots + \frac{1 \cdot 3 \cdot 5 \cdots (2n - 1)}{2 \cdot 4 \cdot 6 \cdots (2n)}x^{2n+1} + \cdots$

32. $-\frac{1}{2}x + \frac{1 \cdot 3}{2 \cdot 4}x^2 - \frac{1 \cdot 3 \cdot 5}{2 \cdot 4 \cdot 6}x^3 + \cdots + (-1)^n \frac{1 \cdot 3 \cdot 5 \cdots (2n - 1)}{2 \cdot 4 \cdot 6 \cdots (2n)}x^n + \cdots$

33. $\frac{1}{4}\left(\frac{x + 1}{3}\right) + \left[\frac{1 \cdot 3}{2 \cdot 4}\right]^2\left(\frac{x + 1}{3}\right)^2 + \cdots + \left[\frac{1 \cdot 3 \cdot 5 \cdots (2n - 1)}{2 \cdot 4 \cdot 6 \cdots (2n)}\right]^2\left(\frac{x + 1}{3}\right)^n + \cdots$

34. $-x + 5x^2 - 7x^3 + 17x^4 - 31x^5 + \cdots + [1 + (-2)^n]x^n + \cdots$

35. $(3 - 2\sqrt{2})x^2 + (\sqrt[3]{3} - 1)^3 x^3 + \cdots + (\sqrt[n]{n} - 1)^n x^n + \cdots$

36. $x + \frac{x^2}{2\sqrt{2}} + \frac{x^3}{3\sqrt[3]{3}} + \cdots + \frac{x^n}{n\sqrt[n]{n}} + \cdots$

37. $8(x + 2)^3 + 32(x + 2)^5 + \cdots + [2^n + (-2)^n](x + 2)^{2n-1} + \cdots$

38. $x + x^2 + 2x^4 + 6x^8 + 24x^{16} + \cdots + n!x^{2n} + \cdots$

39. $\frac{x}{a + b} + \frac{x^2}{a^2 + b^2} + \frac{x^3}{a^3 + b^3} + \cdots + \frac{x^n}{a^n + b^n} + \cdots; a$ and $b > 0$

40. $2x + 24x^2 + 720x^6 + \cdots + (2n)!x^{n!} + \cdots$

PROBLEMS 5.3

1. If $|a_n| \leq |b_n|$ for each $n \in \mathbf{N}$, then prove that the radius of convergence of
$a_1(x - c) + a_2(x - c)^2 + \cdots + a_n(x - c)^n + \cdots$ is at least as large as that of
$b_1(x - c) + b_2(x - c)^2 + \cdots + b_n(x - c)^n + \cdots$.

2. Below are twelve peculiar series. Determine the values of x in each case for which the series converges.

(a) $-\dfrac{1}{x} + \dfrac{1}{2x^2} - \dfrac{1}{3x^3} + \cdots + \dfrac{(-1)^n}{nx^n} + \cdots$; $x \neq 0$

(b) $\dfrac{1}{x} + \dfrac{2}{x^2} + \dfrac{3}{x^3} + \cdots + \dfrac{n}{x^n} + \cdots$; $x \neq 0$

(c) $\dfrac{2}{2x} + \dfrac{3}{4x^2} + \dfrac{4}{8x^3} + \cdots + \dfrac{n+1}{(2x)^n} + \cdots$; $x \neq 0$

(d) $\dfrac{-5}{4x} + \dfrac{25}{7x^2} - \dfrac{125}{10x^3} + \cdots + \dfrac{(-5)^n}{(3n+1)x^n} + \cdots$; $x \neq 0$

(e) $\dfrac{1}{x-2} + \dfrac{1}{2(x-2)^2} + \dfrac{1}{3(x-2)^3} + \cdots + \dfrac{1}{n(x-2)^n} + \cdots$; $x \neq 2$

(f) $\dfrac{1}{x} + \dfrac{1}{2x^2} + \dfrac{1}{6x^3} + \cdots + \dfrac{1}{n!x^n} + \cdots$; $x \neq 0$

(g) $\dfrac{1}{x} + \dfrac{2}{x^2} + \dfrac{6}{x^3} + \cdots + \dfrac{n!}{x^n} + \cdots$; $x \neq 0$

(h) $\dfrac{x-1}{2x+5} + \left(\dfrac{x-1}{2x+5}\right)^2 + \cdots + \left(\dfrac{x-1}{2x+5}\right)^n + \cdots$; $x \neq -\dfrac{5}{2}$

(i) $\dfrac{3x-14}{x+3} + 2\left(\dfrac{3x-14}{x+3}\right)^2 + \cdots + n\left(\dfrac{3x-14}{x+3}\right)^n + \cdots$; $x \neq -3$

(j) $\dfrac{3x+4}{4x+5} + \dfrac{1}{\sqrt{2}}\left(\dfrac{3x+4}{4x+5}\right)^2 + \cdots + \dfrac{1}{\sqrt{n}}\left(\dfrac{3x+4}{4x+5}\right)^n + \cdots$; $x \neq -\dfrac{5}{4}$

(k) $\dfrac{x}{1+x^2} + \dfrac{x^2}{2+x^4} + \cdots + \dfrac{x^n}{n+x^{2n}} + \cdots$

(l) $\dfrac{1}{3}(x^2-5) + \dfrac{1}{9}(x^2-5)^2 + \cdots + \dfrac{1}{3^n}(x^2-5)^n + \cdots$

3. Prove that the series

$$\dfrac{x}{(x+1)(0 \cdot x + 1)} + \dfrac{x}{(2x+1)(x+1)} + \cdots$$
$$+ \dfrac{x}{(nx+1)[(n-1)x+1]} + \cdots$$

has sum zero if $x = 0$ and has sum 1 if $x > 0$.

4. If $a_0 + a_1 x + a_2 x^2 + \cdots + a_n x^n + \cdots$ has a radius of convergence r, prove that the radius of convergence of $a_0 + a_1 x^2 + a_2 x^4 + \cdots + a_n x^{2n} + \cdots$ is $\sqrt{r}$.

5. The functions

$$J_0(x) = 1 - \dfrac{x^2}{4} + \dfrac{x^4}{64} - \dfrac{x^6}{2304} + \cdots + (-1)^n \dfrac{x^{2n}}{(n!)^2 2^{2n}} + \cdots$$

and

$$J_1(x) = \dfrac{x}{2} - \dfrac{x^3}{8} + \dfrac{x^5}{384} - \cdots + (-1)^n \dfrac{x^{2n+1}}{n!(n+1)! 2^{2n+1}} + \cdots$$

are *Bessel functions of the first kind* of orders zero and one, respectively. Prove that
(a) both series converge for each $x \in \mathbf{R}$

(b) $J_0'(x) = -J_1(x)$
(c) $xJ_1'(x) = xJ_0(x) - J_1(x)$
(d) with $n = 0, 1, J_n$ satisfies *Bessel's equation*
$$x^2 f''(x) + xf'(x) + (x^2 - n^2)f(x) = 0$$

6. In computing $J_0(.3)$, how many terms of the series should be used in order to be assured of 10-decimal accuracy?

5.4 TAYLOR SERIES

Corollary 5.15 states that a function defined by a power series is infinitely differentiable. Is the converse true? That is, if the function $f(x)$ is infinitely differentiable, can $f(x)$ be expressed as a power series? The answer is no, not always. However, the functions we shall discuss in this book do have such a *power series expansion*.

Suppose $f(x)$ can be expressed as a power series. We mean that
$$f(x) = a_0 + a_1(x - c) + a_2(x - c)^2 + \cdots + a_n(x - c)^n + \cdots$$
for each value of x in some interval containing the point c. We then know that
$$f'(x) = a_1 + 2a_2(x - c) + 3a_3(x - 2)^2 + \cdots + na_n(x - c)^n + \cdots$$
$$f''(x) = 2a_2 + 6a_3(x - c) + 12a_n(x - c)^2 + \cdots + (n - 1)na_n(x - c)^{n-2} + \cdots$$
$$\vdots$$
$$f^{(n)}(x) = n!a_n + 2 \cdot 3 \cdots (n)(n + 1)a_{n+1}(x - c) + \cdots$$
$$\vdots$$

From these equations we easily see that
$$f'(c) = a_1$$
$$f''(c) = 2a_2$$
$$\vdots$$
$$f^{(n)}(c) = n!a_n$$
$$\vdots$$

A simple induction argument proves that we must have
$$a_n = \frac{f^{(n)}(c)}{n!}$$
for each $n \in \mathbf{N}$. It follows that the power series for $f(x)$ is
$$f(x) = f(c) + f'(c)(x - c) + \frac{f^{(2)}(c)}{2!}(x - c)^2 + \cdots + \frac{f^{(n)}(c)}{n!}(x - c)^n + \cdots$$

This is called the *Taylor series for $f(x)$ at the point c*. Of course, the nth partial sum $a_0 + a_1(x - c) + a_2(x - c)^2 + \cdots + a_n(x - c)^n = P_n(x)$ is a Taylor polynomial for $f(x)$ at the point c. By Taylor's theorem (p. 214)

$$f(x) = P_n(x) + \frac{f^{(n+1)}(x_0)}{(n+1)!}(x - c)^{n+1}$$

where x_0 is some point between c and x. We call the term

$$\frac{f^{(n+1)}(x_0)}{(n+1)!}(x - c)^{n+1}$$

a *remainder term* and denote it by $R_n(x)$, so that

$$f(x) = P_n(x) + R_n(x)$$

We shall use this expression to discuss the power series expansion of functions. The basic result about such expansion is the following.

■ **THEOREM 5.16.** If the function $f(x)$ is infinitely differentiable in an interval containing the point c, and if $\lim_{n \to \infty} R_n(x) = 0$ for each point x in the interval, then $f(x)$ may be expanded into a Taylor series at the point c.

PROOF: We merely write $P_n(x) = f(x) - R_n(x)$ and then

$$\lim_{n \to \infty} P_n(x) = \lim_{n \to \infty} [f(x) - R_n(x)] = f(x) - \lim_{n \to \infty} R_n(x) = f(x)$$

This tells us that the power series has the value $f(x)$ at each point. ■

■ **Example 1.** The function $f(x) = 1/x$ is infinitely differentiable in its entire domain. We have $f'(x) = -x^{-2}$, $f''(x) = 2x^{-3}$, $f'''(x) = -6x^{-4}$ and, in general, $f^{(n)}(x) = (-1)^n n! x^{-n-1}$. If we choose $c = 2$, then we have

$$f(2) = \frac{1}{2}, f'(2) = -\frac{1}{4}, f''(2) = \frac{2}{8},$$

$$f'''(2) = -\frac{6}{16}, \cdots, f^{(n)}(2) = (-1)^n \frac{n!}{2^{n+1}}, \cdots$$

Therefore the coefficients of the Taylor series for $f(x) = 1/x$ at the point 2 are

$$a_0 = f(2) = \tfrac{1}{2},\ a_1 = f'(2) = -\tfrac{1}{4},$$
$$a_2 = \tfrac{1}{2}f''(2) = \tfrac{1}{8},\ a_3 = \tfrac{1}{6}f'''(2) = -\tfrac{1}{16},$$

and in general,

$$a_n = \frac{1}{n!}f^{(n)}(2) = (-1)^n \frac{1}{2^{n+1}}.$$

The Taylor series is

$$\frac{1}{x} = \frac{1}{2} - \frac{1}{4}(x - 2) + \frac{1}{8}(x - 2)^2 - \frac{1}{16}(x - 2)^3 + \cdots$$
$$+ \frac{(-1)^n}{2^{n+1}}(x - 2)^n + \cdots$$

It is easily shown that the complete interval of convergence of this series is $\{x : 0 < x < 4\}$.

■ **Example 2.** Consider the polynomial $f(x) = 2 - 3x + x^2 - x^3$. It is infinitely differentiable everywhere and

$$f'(x) = -3 + 2x - 3x^2, \quad f''(x) = 2 - 6x,$$
$$f'''(x) = -6 \quad \text{and} \quad f^{(n)}(x) = 0 \quad \text{if } n > 3$$

Thus for any point c we may write

$$f(x) = f(c) + f'(c)(x - c) + \tfrac{1}{2}f''(c)(x - c)^2 + \tfrac{1}{6}f'''(c)(x - c)^3$$
$$= (2 - 3c + c^2 - c^3) + (-3 + 2c - 3c^2)(x - c)$$
$$+ (1 - 3c)(x - c)^2 - (x - c)^3$$

For $c = 2$, $f(x) = -8$, $f'(2) = -11$, $f''(2) = -10$, and $f'''(2) = -6$, so that we may write

$$2 - 3x + x^2 - x^3 = -8 - 11(x - 2) - 5(x - 2)^2 - (x - 2)^3$$

We say that the polynomial has been expanded in powers of $x - 2$. We shall make frequent use of this technique later.

If the point $c = 0$ is used in a Taylor series, then the resulting series is often called a *Maclaurin series* for $f(x)$. That is, the Maclaurin series for $f(x)$ is

$$f(x) = f(0) + f'(0)x + \frac{f''(0)}{2!}x^2 + \cdots + \frac{f^{(n)}(0)}{n!}x^n + \cdots$$

Taylor series expansions for a function are often used to obtain approximate values of $f(x)$. Such a series usually converges rapidly for points x near the midpoint c of the interval of convergence. Let us illustrate this.

■ **Example 3.** If we choose $x = 2.1$ in the series of Example 1, then we have

$$\frac{1}{2.1} = \frac{1}{2} - \frac{1}{4}(2.1 - 2) + \frac{1}{8}(2.1 - 2)^2$$
$$- \frac{1}{16}(2.1 - 2)^3 + \cdots$$
$$= \frac{1}{2} - \frac{1}{4}(.1) + \frac{1}{8}(.1)^2 - \frac{1}{16}(.1)^3 + \frac{1}{32}(.1)^4$$
$$- \frac{1}{64}(.1)^5 + \cdots$$

By omitting all terms after the fifth in this alternating series, the error is less than the sixth term $\frac{1}{64}(.1)^5$ which is less than 0.0000002. This means that just the first five terms yield accuracy to at least six decimals. In fact, the sum of the first five terms is 0.476190625, whereas $1/2.1 = 0.476190476190 \cdots$.

■ *Example 4.* The function $f(x) = \sqrt{1+x}$ is infinitely differentiable on the open ray $\{x : -1 < x\}$. The first five terms of its Maclaurin series are easily found to be

$$\sqrt{1+x} = 1 + \tfrac{1}{2}x - \tfrac{1}{8}x^2 + \tfrac{1}{16}x^3 - \tfrac{5}{128}x^4 + \cdots$$

Again, for points close to zero, this series converges rapidly. Taking $x = \tfrac{1}{10}$, for instance, we have

$$\sqrt{1+\tfrac{1}{10}} = \sqrt{1.1} = 1 + \tfrac{1}{2}(\tfrac{1}{10}) - \tfrac{1}{8}(\tfrac{1}{10})^2$$
$$+ \tfrac{1}{16}(\tfrac{1}{10})^3 - \tfrac{5}{128}(\tfrac{1}{10})^4 + \cdots$$
$$= 1 + \tfrac{1}{20} - \tfrac{1}{800} + \tfrac{1}{16,000} - \tfrac{5}{1,280,000} + \cdots$$

Using just the first four terms, we obtain the approximate value 1.0488125 for $\sqrt{1.1}$. By our theory this value is in error by no more than the first term omitted, namely, $\tfrac{5}{1,280,000}$. The true value is $\sqrt{1.1} = 1.0488088 \cdots$, so that once again theory and practice agree. ■

If $p \in \mathbf{N}$, then the expression $(1+x)^p$ is "expanded" by the binomial theorem to become

$$(1+x)^p = 1 + px + \frac{p(p-1)}{2!}x^2 + \frac{p(p-1)(p-2)}{3!}x^2 + \cdots$$
$$+ \frac{p(p-1) \cdots (p-k+1)}{k!}x^k + \cdots$$

This "series" terminates with the $p+1$st term, of course.

It is easy to show that the same expansion applies even when the exponent p is not a positive integer. But if p is not a positive integer, then the factor $p - k + 1$ never becomes zero; hence we obtain an infinite *binomial series*. It is a difficult proof (which we do not present) to show that this binomial series actually represents the function $(1+x)^p$. However, it is easy to prove that the radius of convergence is one.

■ *Example 5.* We duplicate the series in Example 4 by writing

$$(1+x)^{1/2} = 1 + \frac{1}{2}x + \frac{1/2(1/2-1)}{2!}x^2$$
$$+ \frac{1/2(1/2-1)(1/2-2)}{3!}x^3$$
$$+ \frac{1/2(1/2-1)(1/2-2)(1/2-3)}{4!}x^4 + \cdots$$
$$= 1 + \frac{1}{2}x - \frac{1}{8}x^2 + \frac{1}{16}x^3 - \frac{5}{128}x^4 + \cdots$$

■ *Example 6.* Simple substitutions render the binomial series more widely useful in obtaining Maclaurin series expansions. We write

$$(1-x^2)^{-1} = 1 + (-1)(-x^2) + \frac{(-1)(-1-1)}{2!}(-x^2)^2$$
$$+ \frac{(-1)(-1-1)(-1-2)}{3!}(-x^2)^3 + \cdots$$
$$= 1 + x^2 + x^4 + x^6 + \cdots$$

EXERCISES 5.4

1. Expand each polynomial in power of $x - c$ for the given value of c:
   (a) $2 - x + 2x^2$; $c = 3$
   (b) $1 + 2x - 3x^2$; $c = -1$
   (c) $4 - 4x + x^2$; $c = 2$
   (d) $2 + 4x - x^2$; $c = 2$
   (e) $1 - x + 2x^2 + x^3$; $c = 1$
   (f) $2 + x - x^2 - 2x^3$; $c = -1$
   (g) $-2 + x - x^2 + 2x^3 + x^4$; $c = 2$
   (h) $3 - 2x + x^2 - 2x^3 - x^4$; $c = 1$

2. Use the binomial series expansion to give the Maclaurin series for each of the following functions:
   (a) $\sqrt{1-x}$
   (b) $\dfrac{1}{\sqrt{1-x}}$
   (c) $\sqrt{1+x^2}$
   (d) $(1-x^2)^{-1/2}$
   (e) $(1-x^3)^{-1/3}$
   (f) $(4-x)^{-1} = \dfrac{1}{4}\left(1 - \dfrac{x}{4}\right)^{-1}$
   (g) $x\sqrt{1+x}$
   (h) $\dfrac{x}{\sqrt{1+x}}$
   (i) $x^2\sqrt{1-x}$
   (j) $x^2(1+x)^{1/3}$

3. Expand each function in a Taylor series at the given point c:
   (a) $f(x) = (4 + x^2)^{-1}$; $c = 1$
   (b) $f(x) = (4 + x^2)^{-1}$; $c = -1$
   (c) $f(x) = \dfrac{1+x}{3-x}$; $c = -1$
   (d) $f(x) = \dfrac{2x}{1+x^2}$; $c = 0$
   (e) $f(x) = \sqrt{2x - x^2}$; $c = 0$
   (f) $f(x) = \dfrac{x}{\sqrt{2x-x^2}}$; $x = 1$
   (g) $f(x) = (1 + x + x^2)^{-2}$; $c = 0$
   (h) $f(x) = (1 + x + x^2)^{-2}$; $c = -1$
   (i) $f(x) = \dfrac{1-x^2}{1+x^2}$; $c = 1$
   (j) $f(x) = \sqrt{\dfrac{1-x}{1+x}}$; $c = 0$

4. Determine the radius of convergence of each series in Exercise 3 for which you obtained a general term.

5. Use an appropriate Taylor series to compute each of the following numbers to at least five decimal accuracy:
   (a) $\sqrt{0.95}$
   (b) $\sqrt{1.2}$
   (c) $\sqrt{4.1}$
   (d) $\sqrt[3]{8.3}$
   (e) $\dfrac{1}{4.2}$
   (f) $\dfrac{1}{\sqrt{11}}$
   (g) $\dfrac{1}{\sqrt{101}}$
   (h) $\sqrt[4]{82}$

6. For each pair of functions below, differentiate the Maclaurin series for the first function to obtain a series for the second. Verify your results by an independent method.

(a) $(1-x)^{-2}$; $(1-x)^{-3}$

(b) $(1+x)^{1/2}$; $(1+x)^{-1/2}$

(c) $(1-x^2)^{1/2}$; $\dfrac{x}{\sqrt{1-x^2}}$

(d) $x(1-x^2)^{1/2}$; $\dfrac{1-2x^2}{\sqrt{1-x^2}}$

PROBLEMS 5.4

1. Suppose that the function $f(x) = 1 + c_1 x + c_2 x^2 + \cdots + c_4 x^n + \cdots$ satisfies the differential equation $f'(x) = f(x)$. Show that $c_n = 1/n!$ for each $n \in \mathbf{N}$.

2. Prove that the series
$$\frac{1}{1-x} + \frac{x^2}{1-x^3} + \frac{x^4}{1-x^5} + \cdots + \frac{x^{2n}}{1-x^{2n+1}} + \cdots$$
converges if $|x| < 1$.

3. From Example 4, we have
$$\sqrt{1+x} = 1 + \tfrac{1}{2}x - \tfrac{1}{8}x^2 + \tfrac{1}{16}x^3 - \tfrac{5}{128}x^4 + \cdots$$
Verify this by computing the first five terms of
$$(1 + \tfrac{1}{2}x - \tfrac{1}{8}x^2 + \tfrac{1}{16}x^3 - \tfrac{5}{128}x^4 + \cdots)^2$$

4. If $|x| > 1$, show that
$$\sqrt{1+x} = \sqrt{x}\left(1 + \frac{1}{2}x^{-1} + \frac{1}{2 \cdot 4}x^{-2} + \cdots\right)$$

5. For each of the following functions verify that the given formula provides the general term of its Maclaurin series

(a) $f(x) = \dfrac{x(1+x)}{(1-x)^3}$; $a_n = n^2$

(b) $f(x) = \dfrac{x}{1+x-2x^2}$; $a_n = \dfrac{1}{3}[1-(-2)^n]$

(c) $f(x) = \dfrac{12-5x}{6-5x-x^2}$; $a_n = 1 + \dfrac{(-1)^n}{6^n}$

(d) $f(x) = \dfrac{x}{(1-x)^2(1+x)}$; $a_n = \dfrac{1}{2}\left[n + \dfrac{1-(-1)^n}{2}\right]$

6

THE DEFINITE INTEGRAL

THE DEFINITE INTEGRAL

The derivative and the integral are the two chief tools of the mathematical analyst. Like the derivative, the integral finds many widely differing applications. Furthermore, the relationship between the derivative and the integral constitutes one of the most important and beautiful theorems in the calculus.

We shall define the definite integral temporarily as the limit of a sequence of successively better approximations. By doing this we are unable to give easy proofs of several key results. Thus in the final section of the chapter we adopt a seemingly more general definition of the integral. This definition permits the desired easy proofs and is not actually more general.

6.1 AREA UNDER A CURVE

We have always spoken confidently about the "area" of a region in the plane. For polygonal figures it is easy to make good sense out of the concept of area. Is it really obvious, however, that a circle of radius r inches encloses an area of πr^2 square inches? We mean that the very word "area" requires careful definition if it is to be used for figures with curvilinear boundaries. Our discussion here is not complete and precise; it merely points out the direction in which the desired precision lies. Advanced courses in the theory of integration eventually will provide the student with full details.

The mathematician considers "area" to be a function A that assigns a real number to each of a certain class of regions in the plane. If R is such a region, we interpret the value $A(R)$ as a number of square units. Guided by geometric experience, we require that the function A have several properties. The four most important properties are as follows:

1. If R is a rectangular region of a length a and a height b, then $A(R) = ab$.
2. If the region R_1 is contained in the region R_2, then $A(R_1) \leqq A(R_2)$.
3. If regions R_1 and R_2 have at most boundary points in common, then $A(R_1 \cup R_2) = A(R_1) + A(R_2)$.
4. If R_1 and R_2 are congruent regions, then $A(R_1) = A(R_2)$.

This first section shows how a limit process can be used to provide the value $A(R)$ to be assigned to a region R with a curvilinear boundary.

We impose a coordinate system on a geometric plane. We ask for the area A of the region R bounded by the x axis, the vertical line $\{(x, y) : x = 1\}$, and the curve $\{(x, y) : y = x^2\}$. (See Fig. 6.1.) Our attack on the problem is by means of a sequence of successively more accurate approximations.

The region R is contained in a unit square. This gives us the inequality

$$0 < A < 1$$

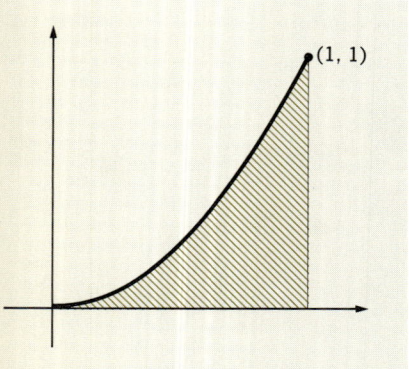

FIGURE 6.1

6.1 AREA UNDER A CURVE

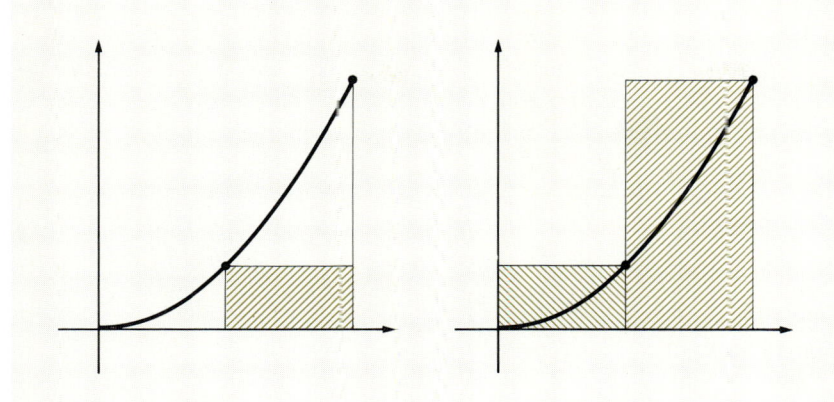

FIGURE 6.2

Next consider the point $(\frac{1}{2}, \frac{1}{4})$ on the curve. Looking at the rectangles in Fig. 6.2 and their relation to the given region, we obtain the inequality

$$0(\tfrac{1}{2}) + \tfrac{1}{4}(\tfrac{1}{2}) < A < \tfrac{1}{4}(\tfrac{1}{2}) + 1(\tfrac{1}{2})$$

or

$$\tfrac{1}{8} < A < \tfrac{5}{8}$$

As a third step we consider the points $(\frac{1}{4}, \frac{1}{16})$, $(\frac{1}{2}, \frac{1}{4})$, $(\frac{3}{4}, \frac{9}{16})$ on the curve and construct two sets of four rectangles as shown in Fig. 6.3.

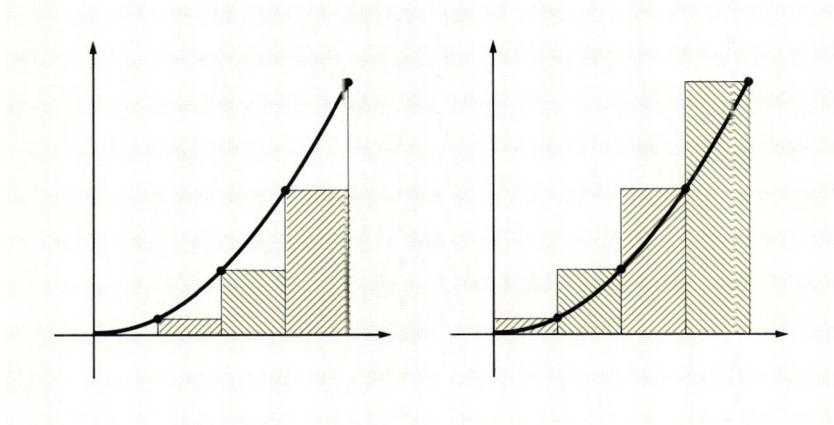

FIGURE 6.3

From this figure we can see that

$$0(\tfrac{1}{4}) + \tfrac{1}{16}(\tfrac{1}{4}) + \tfrac{1}{4}(\tfrac{1}{4}) + \tfrac{9}{16}(\tfrac{1}{4}) < A < \tfrac{1}{16}(\tfrac{1}{4})$$
$$+ \tfrac{1}{4}(\tfrac{1}{4}) + \tfrac{9}{16}(\tfrac{1}{4}) + 1(\tfrac{1}{4})$$

or

$$\tfrac{7}{32} < A < \tfrac{15}{32}$$

Our plan should now be obvious. For each $n \in \mathbf{N}$, we consider the points $(1/2^n, 1/2^{2n})$, $(2/2^n, 4/2^{2n})$, $(3/2^n, 9/2^{2n})$, $\cdots$, $(2^n - 1/2^n, (2^n - 1)^2/2^{2n})$ on the curve. We construct two sets of 2^n rectangles, the first set having its union contained in R and the second set having its union containing R. The sum of the areas in the first set of rectangles is

$$s_n = 0\left(\frac{1}{2^n}\right) + \frac{1}{2^{2n}}\left(\frac{1}{2^n}\right) + \frac{4}{2^{2n}}\left(\frac{1}{2^n}\right) + \frac{9}{2^{2n}}\left(\frac{1}{2^n}\right) + \cdots$$
$$+ \frac{(2^n - 1)^2}{2^{2n}}\left(\frac{1}{2^n}\right)$$

and the sum of the areas of the second set of rectangles is

$$S_n = \frac{1}{2^{2n}}\left(\frac{1}{2^n}\right) + \frac{4}{2^{2n}}\left(\frac{1}{2^n}\right) + \frac{9}{2^{2n}}\left(\frac{1}{2^n}\right) + \cdots$$
$$+ \frac{(2^n - 1)^2}{2^{2n}}\left(\frac{1}{2^n}\right) + 1\left(\frac{1}{2^n}\right)$$

By the construction, we have

$$s_n < A < S_n$$

and by looking at s_n and S_n as given above, we can see that they differ by the amount $1/2^n$. As n increases indefinitely, we surely have $\lim_{n \to \infty} s_n = \lim_{n \to \infty} S_n$ and this limit must also be the number A.
Now

$$s_n = \frac{1}{2^{3n}}[1^2 + 2^2 + 3^2 + \cdots + (2^n - 1)^2]$$

and from Section 1.4, we know that

$$1^2 + 2^2 + 3^2 + \cdots + m^2 = \tfrac{1}{6}m(m + 1)(2m + 1)$$

for each $m \in \mathbf{N}$. Therefore

$$1^2 + 2^2 + 3^2 + \cdots + (2^n - 1)^2 = \tfrac{1}{6}(2^n - 1)(2^n)(2^{n+1} - 1)$$

and so

$$s_n = \frac{1}{6}\frac{(2^n - 1)(2^n)(2^{n+1} - 1)}{2^{3n}}$$
$$= \frac{1}{6}\frac{(2^n - 1)(2^{n+1} - 1)}{2^{2n}} = \frac{1}{6}\frac{2^{2n+1} - 2^{n+1} - 2^n + 1}{2^{2n}}$$
$$= \frac{1}{6}\left(2 - \frac{1}{2^{n-1}} - \frac{1}{2^n} + \frac{1}{2^{2n}}\right)$$
$$= \frac{1}{3} - \frac{1}{6}\left(\frac{1}{2^{n-1}} + \frac{1}{2^n} - \frac{1}{2^{2n}}\right)$$

Clearly then, we have solved our problem, because

$$A = \lim_{n \to \infty} s_n = \lim_{n \to \infty} \left[\frac{1}{3} - \frac{1}{6}\left(\frac{1}{2^{n-1}} + \frac{1}{2^n} - \frac{1}{2^{2n}} \right) \right] = \frac{1}{3}$$

Let us examine the procedure we used to compute A; at the same time we generalize the procedure. We subdivided the interval $\{x : 0 \leq x \leq 1\}$ into m subintervals by choosing points $x_1, x_2, \cdots, x_{m-1}$ satisfying

$$0 = x_0 < x_1 < x_2 < \cdots < x_{m-1} < x_m = 1$$

Each interval of length $x_i - x_{i-1}$, $i = 1, 2, \cdots, m$, is used as the base of two different rectangles, one in each of our two sets of rectangles.

In each interval $\{x : x_{i-1} \leq x \leq x_i\}$, we choose two values of $f(x) = x^2$. The first of these is x_{i-1}^2, which gives the minimum value of $f(x)$ over this subinterval; the second value is x_i^2, which is the maximum value of $f(x) = x^2$ over this subinterval. Using the ith subinterval as the base, we constructed the rectangle of area $x_{i-1}^2(x_i - x_{i-1})$ and the rectangle of area $x_i^2(x_i - x_{i-1})$. The region under the curve and above this subinterval obviously contains the first rectangle and is contained in the second. (See Fig. 6.4.)

Then we formed the two sums of areas

$$s_n = x_0^2(x_1 - x_0) + x_1^2(x_2 - x_1) + \cdots + x_{m-1}^2(x_m - x_{m-1})$$

and

$$S_n = x_1^2(x_1 - x_0) + x_2^2(x_2 - x_1) + \cdots + x_m^2(x_m - x_{m-1})$$

From the geometry of the situation, we asserted that

$$s_n < A < S_n$$

In practice, we selected our subintervals so that the values s_n and S_n were easy to compute and thereby arrived at a value for A as a limit.

The procedure used above can be generalized and used to compute the area of the region under any graph, at least theoretically. Suppose the function f is defined on $\{x : a \leq x \leq b\}$ and that $f(x) \geq 0$ in this interval. We want to compute the area of the region bounded by the x axis, the vertical lines $\{(x,y) : x = a\}$ and $\{(x,y) : x = b\}$ and the curve $\{(x,y) : y = f(x)\}$. (See Fig. 6.5).

Our approach is just the same as before. We select $n - 1$ points $x_1, x_2, \cdots, x_{n-1}$ satisfying

$$a = x_0 < x_1 < x_2 < \cdots < x_{n-1} < x_m = b$$

On the ith subinterval $\{x : x_{i-1} \leq x \leq x_i\}$, we set

m_i = minimum value of $f(x)$
M_i = maximum value of $f(x)$

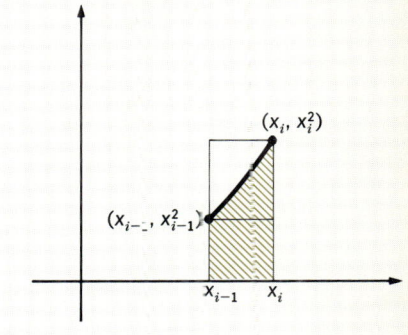

FIGURE 6.4

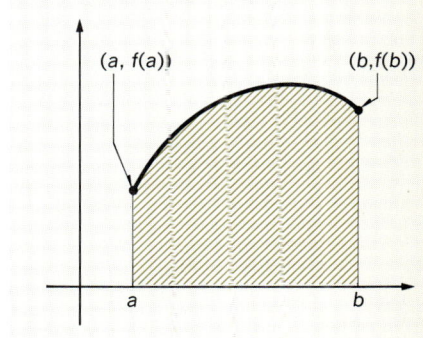

FIGURE 6.5

Letting A_i denote the area under the curve and above the ith subinterval, we have the inequality

$$m_i(x_i - x_{i-1}) \leq A_i \leq M_i(x_i - x_{i-1})$$

for each $i = 1, 2, \cdots, n$. It follows that the desired area $A = A_1 + A_2 + \cdots + A_n$ must satisfy

$$s_n = m_1(x_1 - x_0) + m_2(x_2 - x_1) + \cdots + m_n(x_n - x_{n-1}) \leq A$$

and

$$S_n = M_1(x_1 - x_0) + M_2(x_2 - x_1) + \cdots + M_n(x_n - x_{n-1}) \geq A$$

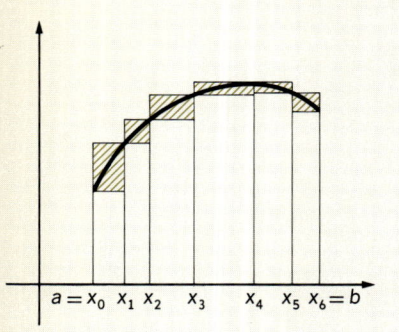

FIGURE 6.6

A particular example is shown in Fig. 6.6 in which the shaded region has an area equal to the difference $S_n - s_n$.

The inequalities $s_n \leq A \leq S_n$ certainly are valid for any choice of points subdividing the interval $\{x : a \leq x \leq b\}$. If we can select a sequence of such subdivisions that allows us to compute s_n and S_n, then the area A can be "squeezed" into successively narrower bounds. The final evaluation process is by a limit, of course.

■ **Example 1.** Let $f(x) = 1 + x^3$ and let the interval be $\{x : 0 \leq x \leq 1\}$. For any selection of $n - 1$ points x_i such that

$$0 = x_0 < x_1 < x_2 < \cdots < x_{n-1} < x_n = 1$$

we have that

$$m_i = \text{minimum of } f(x) \text{ on } \{x : x_{i-1} \leq x \leq x_i\} = 1 + x_{i-1}^3$$

and

$$M_i = \text{maximum of } f(x) \text{ on } \{x : x_{i-1} \leq x \leq x_i\} = 1 + x_i^3$$

for each $i = 1, 2, \cdots, n$. This follows because f is an increasing function on the entire interval $\{x : 0 \leq x \leq 1\}$. Thus we find that

$$(1 + x_0^3)(x_1 - x_0) + (1 + x_1^3)(x_2 - x_1) + \cdots + (1 + x_{n-1}^3)(x_n - x_{n-1}) \leq A$$

and

$$(1 + x_1^3)(x_1 - x_0) + (1 + x_2^3)(x_2 - x_1) + \cdots + (1 + x_n^3)(x_n - x_{n-1}) \geq A$$

For the actual computation we select $x_i = i/n$, $i = 0, 1, 2, \cdots, n$. Then $m_i = 1 + \dfrac{(i-1)^3}{n^3}$, $M_i = 1 + \dfrac{i^3}{n^3}$ and $x_i - x_{i-1} = 1/n$ for all $i = 1, 2, \cdots, n$. The approximating sums are

$$s_n = (1 + 0^3)\left(\frac{1}{n}\right) + \left(1 + \frac{1^3}{n^3}\right)\left(\frac{1}{n}\right) + \cdots + \left(1 + \frac{(n-1)^3}{n^3}\right)\left(\frac{1}{n}\right) \leq A$$

and
$$S_n = \left(1 + \frac{1^3}{n^3}\right)\left(\frac{1}{n}\right) + \left(1 + \frac{2^3}{n^3}\right)\left(\frac{1}{n}\right) + \cdots$$
$$+ \left(1 + \frac{n^3}{n^3}\right)\left(\frac{1}{n}\right) \geq A$$

We rewrite s_n and S_n as follows:
$$s_n = \frac{1}{n}\left[\underbrace{1 + 1 + \cdots + 1}_{n \text{ terms}} + \frac{0^3 + 1^3 + 2^3 + \cdots + (n-1)^3}{n^3}\right]$$
$$= \frac{1}{n}\left[n + \frac{(n-1)^2 n^2}{4n^3}\right] = 1 + \frac{(n-1)^2}{4n^2} = \frac{5}{4} - \frac{1}{2n} + \frac{1}{4n^2}$$

$$S_n = \frac{1}{n}\left[\underbrace{1 + 1 + \cdots + 1}_{n \text{ terms}} + \frac{1^3 + 2^3 + \cdots + n^3}{n^3}\right]$$
$$= \frac{1}{n}\left[n + \frac{n^2(n+1)^2}{4n^3}\right] = 1 + \frac{(n+1)^2}{4n^2} = \frac{5}{4} + \frac{1}{2n} + \frac{1}{4n^2}$$

(We have used a formula for the sum of the first n cubes of integers. See Section 1.4.) From these formulas, we readily see that
$$\lim_{n \to \infty} s_n = \frac{5}{4} = \lim_{n \to \infty} S_n$$

and we conclude that $A = \frac{5}{4}$.

Even for the simple function of our example this procedure is quite laborious. However, if we really must compute such an area, there seems to be little else we can do. Our only recourse is to study this procedure and find means of making it easier. In the end we obtain a very simple method for solving problems of this kind. Our theory will also lead us to many other applications as well.

EXERCISES 6.1

In each of Exercises 1 to 8, use the technique of this section to compute the area of the region under the given line and above the given interval. Note that the region so described is either a triangle or a trapezoid; hence the result is easily verified in each case.

1. $\{(x,y) : y = x\}; \{x : 0 \leq x \leq 1\}$
2. $\{(x,y) : y = x\}; \{x : 0 \leq x \leq 4\}$
3. $\{(x,y) : y = x\}; \{x : 1 \leq x \leq 2\}$
4. $\{(x,y) : y = 2 + x\}; \{x : -2 \leq x \leq 0\}$
5. $\{(x,y) : y = 2 + x\}; \{x : 0 \leq x \leq 2\}$
6. $\{(x,y) : y = -1 + 2x\}; \{x : 1 \leq x \leq 2\}$

7. $\{(x,y) : 2x - 3y = 6\}$; $\{x : 4 \leq x \leq 6\}$

8. $\{(x,y) : 2x + 3y = 6\}$; $\{x : 0 \leq x \leq 2\}$

For Exercises 9 to 16 compute the area of the region under the given quadratic curve and above the given interval. Sketch the region under discussion.

9. $\{(x,y) : y = x^2\}$; $\{x : -1 \leq x \leq 0\}$
10. $\{(x,y) : y = x^2\}$; $\{x : -1 \leq x \leq 1\}$
11. $\{(x,y) : y = x^2\}$; $\{x : 1 \leq x \leq 2\}$
12. $\{(x,y) : y = -1 + x^2\}$; $\{x : 1 \leq x \leq 2\}$
13. $\{(x,y) : y = 4 - x^2\}$; $\{x : -2 \leq x \leq 2\}$
14. $\{(x,y) : y = 4x - x^2\}$; $\{x : 0 \leq x \leq 4\}$
15. $\{(x,y) : y = 1 + 4x - x^2\}$; $\{x : 0 \leq x \leq 4\}$
16. $\{(x,y) : y = 2 + x - x^2\}$; $\{x : -1 \leq x \leq 2\}$

By separate computations verify that the sum of the areas of the regions under the curves $\{(x,y) : y = f(x)\}$ and $\{(x,y) : y = g(x)\}$ equals the area of the region under the curve $\{(x,y) : y = f(x) + g(x)\}$, all regions over the given interval. Sketch the three regions involved.

17. $f(x) = x^2$, $g(x) = x$; $\{x : 1 \leq x \leq 2\}$
18. $f(x) = x^2$, $g(x) = 1 - x$; $\{x : 0 \leq x \leq 1\}$
19. $f(x) = x^2$, $g(x) = 1 - x^3$; $\{x : 0 \leq x \leq 1\}$
20. $f(x) = x - x^2$, $g(x) = 1 - x^3$; $\{x : -1 \leq x \leq 1\}$

By separate computations verify that the area of the region under the curve $\{(x,y : y = f(x)\}$ over $\{x : a \leq x \leq b\}$, plus the area of the region under the same curve and over the interval $\{x : b \leq x \leq c\}$, equals the area of the region over the interval $\{x : a \leq x \leq c\}$.

21. $\{(x,y) : y = 1 + x\}$; $a = -1$, $b = 0$, $c = 1$
22. $\{(x,y) : y = -1 + 2x\}$; $a = 1$, $b = 2$, $c = 4$
23. $\{(x,y) : y = x^2\}$; $a = -1$, $b = 0$, $c = 1$
24. $\{(x,y) : y = 4 - x^2\}$; $a = -2$, $b = 0$, $c = 2$

25. Give an example to show that the product of the area of the region under $\{(x,y) : y = f(x)\}$ and the area of the region under $\{(x,y) : y = g(x)\}$ need not equal the area of the region under $\{(x,y) : y = f(x)g(x)\}$, all over the same interval.

PROBLEMS 6.1

1. Verify that the area of the region under the curve $\{(x,y) : y = x^n\}$ and over the interval $\{x : a \leq x \leq b\}$ is $1/(n+1)(b^{n+1} - a^{n+1})$ for the three cases:
   (a) $n = 1$,   (b) $n = 2$,   (c) $n = 3$.

2. By choosing n sufficiently large (and this determination is the gist of the problem), compute the area of the region under $\{(x,y) : xy = 1\}$ and over $\{x : 1 \leq x \leq 2\}$ to two-decimal accuracy.

3. In each subinterval $\{x : x_{i-1} \leq x \leq x_i\}$, choose any point x_i'. Since $m_i \leq f(x_i') \leq M_i$ for each subinterval, it follows that

$$f(x_1')(x_1 - x_0) + f(x_2')(x_2 - x_1) + \cdots + f(x_n')(x_n - x_{n-1})$$

is also an approximation of the area of the region under $\{(x, y) : y = f(x)\}$ and over the interval $\{x : x_0 \leq x \leq x_n\}$. Try this approximation in the following cases:

(a) $f(x) = x$; $\{x : 0 \leq x \leq 2\}$; $x_i' = \frac{1}{2}(x_{i-1} + x_i)$
(b) $f(x) = x$; $\{x : 0 \leq x \leq b\}$; $x_i' = \frac{1}{2}(x_{i-1} + x_i)$
(c) $f(x) = x^2$; $\{x : 0 \leq x \leq 1\}$; $x_i' = \frac{1}{3}\sqrt{3(x_{i-1}^2 + x_{i-1}x_i + x_i^2)}$
(d) $f(x) = x^2$; $\{x : 0 \leq x \leq b\}$; $x_i' = \frac{1}{3}\sqrt{3(x_{i-1}^2 + x_{i-1}x_i + x_i^2)}$

4. Instead of bounding the region under $\{(x, y) : y = f(x)\}$ and over the ith subinterval $\{x : x_{i-1} \leq x \leq x_i\}$ between two rectangles, we can approximate this region by the trapezoid with vertices $(x_{i-1}, 0)$, $(x_{i-1}, f(x_{i-1}))$, $(x_i, 0)$ and $(x_i, f(x_i))$. Draw this trapezoid and give its area. Then give a formula for the approximate area over the entire interval. Finally apply this approximation to an example of your own choice.

6.2 THE AVERAGE VALUE OF A FUNCTION

If a motorist drives at 40 mph for 3 hr and then at 50 mph for 4 hr, then his average velocity for the seven hours is the weighted average

$$40(\tfrac{3}{7}) + 50(\tfrac{4}{7}) = 45\tfrac{5}{7} \text{ mph}$$

If there were several changes in velocity, then the weighted average method would still apply. But what could we do if the velocity changes constantly?

Our illustrative problem is this: A particle moving rectilinearly has velocity

$$v = f(t) = 3 + 2t$$

What is its *average velocity* during the time interval from $t = 1$ to $t = 3$? The answer to this question is also obtained by a sequence of approximations. First, we choose $n - 1$ points $t_1, t_2, \cdots, t_{n-1}$ in the interval $\{t : 1 \leq t \leq 3\}$ such that

$$1 = t_0 < t_1 < t_2 < \cdots < t_{n-1} < t_n = 3$$

In the ith subinterval $\{t : t_{i-1} \leq t \leq t_i\}$, the particle has an average velocity which we denote by V_i. Then the average velocity V over the entire interval is the weighted average of the quantities V_i;

$$V = V_1\left(\frac{t_1 - t_0}{t_n - t_0}\right) + V_2\left(\frac{t_2 - t_1}{t_n - t_0}\right) + \cdots + V_n\left(\frac{t_n - t_{n-1}}{t_n - t_0}\right)$$

But during the ith subinterval of time, the average velocity V_i must be between the minimum velocity and the maximum velocity. Thus if we set

$$m_i = \text{minimum of } f(t) \text{ on } \{t : t_{i-1} \leq t \leq t_i\}$$

and

$$M_i = \text{maximum of } f(t) \text{ on } \{t : t_{i-1} \leq t \leq t_i\}$$

then we have

$$m_i \leq V_i \leq M_i$$

for each $i = 1, 2, \cdots, n$. By our choice of the numbers t_i, each number $(t_i - t_{i-1})/(t_n - t_0)$ is positive. Hence we may write

$$m_i\left(\frac{t_i - t_{i-1}}{t_n - t_0}\right) \leq V_i\left(\frac{t_i - t_{i-1}}{t_n - t_0}\right) \leq M_i\left(\frac{t_i - t_{i-1}}{t_n - t_0}\right),$$
$$i = 1, 2, \cdots, n$$

Adding these inequalities, we obtain

$$\frac{1}{t_n - t_0}[m_1(t_1 - t_0) + m_2(t_2 - t_1) + \cdots + m_n(t_n - t_{n-1})] \leq V$$

and

$$V \leq \frac{1}{t_n - t_0}[M_1(t_1 - t_0) + M_2(t_2 - t_1) + \cdots + M_n(t_n - t_{n-1})]$$

For computational purposes we choose the points $t_i = 1 + i/n$, $i = 0, 1, 2, \cdots, 2n$. Then each subinterval has a length $1/n$ and there are $2n$ subintervals. Also for each $i = 1, 2, \cdots, 2n$, we easily see that

$$m_i = \text{min of } f(t) \text{ on } \{t : t_{i-1} \leq t \leq t_i\} = 3 + 2t_{i-1}$$

and

$$M_i = \text{max of } f(t) \text{ on } \{t : t_{i-1} \leq t \leq t_i\} = 3 + 2t_i$$

Hence we have

$$\frac{1}{3-1}\left\{[3 + 2(1)]\left(\frac{1}{n}\right) + \left[3 + 2\left(1 + \frac{1}{n}\right)\right]\left(\frac{1}{n}\right) + \cdots + \left[3 + 2\left(1 + \frac{2n-1}{n}\right)\right]\left(\frac{1}{n}\right)\right\} \leq V$$

and

$$\frac{1}{3-1}\left\{\left[3 + 2\left(1 + \frac{1}{n}\right)\right]\left(\frac{1}{n}\right) + \left[3 + 2\left(1 + \frac{2}{n}\right)\right]\left(\frac{1}{n}\right) + \cdots + \left[3 + 2\left(1 + \frac{2n}{n}\right)\right]\left(\frac{1}{n}\right)\right\} \geq V$$

6.2 THE AVERAGE VALUE OF A FUNCTION

We factor out the common factor $1/n$ and combine it with $1/(3-1) = \frac{1}{2}$ to write

$$\frac{1}{2n}\left\{[3 + 2(1)] + \left[3 + 2\left(1 + \frac{1}{n}\right)\right] + \cdots \right.$$
$$\left. + \left[3 + 2\left(1 + \frac{2n-1}{n}\right)\right]\right\} \leq V$$

and

$$\frac{1}{2n}\left\{\left[3 + 2\left(1 + \frac{1}{n}\right)\right] + \left[3 + 2\left(1 + \frac{2}{n}\right)\right] + \cdots \right.$$
$$\left. + \left[3 + 2\left(1 + \frac{2n}{n}\right)\right]\right\} \geq V$$

The sums within the outer brackets are arithmetic series. We recall that the formula for the sum of an arithmetic series is

$$a_0 + (a_0 + d) + \cdots + [a_0 + (m-1)d]$$
$$= \frac{m}{2}[2a_0 + (m-1)d]$$

Therefore

$$[3 + 2(1)] + \left[3 + 2\left(1 + \frac{1}{n}\right)\right] + \cdots + \left[3 + 2\left(1 + \frac{2n-1}{n}\right)\right]$$
$$= \frac{2n}{2}\left[3 + 2(1) + 3 + 2\left(1 + \frac{2n-1}{n}\right)\right]$$
$$= n\left(14 - \frac{2}{n}\right)$$

and

$$\left[3 + 2\left(1 + \frac{1}{n}\right)\right] + \left[3 + 2\left(1 + \frac{2}{n}\right)\right] + \cdots$$
$$+ \left[3 + 2\left(1 + \frac{2n}{n}\right)\right]$$
$$= \frac{2n}{n}\left[3 + 2\left(1 + \frac{1}{n}\right) + 3\left(1 + \frac{2n}{n}\right)\right]$$
$$= n\left(14 + \frac{2}{n}\right)$$

It follows that

$$\frac{1}{2n} \cdot n\left(14 - \frac{2}{n}\right) \leq V \leq \frac{1}{2n} \cdot n\left(14 + \frac{2}{n}\right)$$

or

$$7 - \frac{1}{n} \leq V \leq 7 + \frac{1}{n}$$

Because this inequality is true for each $n \in \mathbf{N}$, we conclude that $V = 7$.

There is a way to verify the result just obtained so laboriously. In Section 6.2 we showed that, if a particle in rectilinear motion has velocity

$$v = 3 + 2t$$

then the distance s traveled by the particle is given by

$$s = s_0 + 3t + t^2$$

When $t = 1$, we have $s = s_0 + 3(1) + 1^2 = s_0 + 4$; when $t = 3$, we have $s = s_0 + 3(3) + (3)^2 = s_0 + 18$. Therefore the distance traveled during the time interval $\{t : 1 \leq t \leq 3\}$ is $18 + s_0 - (4 + s_0) = 14$, and the average velocity is $\frac{14}{2} = 7$.

The simple solution used as a check may raise objections to the difficult work with sums in the previous paragraphs. But it turns out that the two methods are equivalent. Indeed, this is the gist of our theory. Many theoretical results will be needed before we get to this equivalence, however, and we use the summation method as the basic one.

■ *Example 1.* Water weighs approximately 62.4 per cu ft. Therefore a column of water h feet high over an area of A sq ft will contain hA cu ft of water and will weigh $(62.4)hA$ lb. That is, this column of water exerts a force of $(62.4)hA$ lb over the area A. Then the pressure on this area is

$$p = \frac{(62.4)hA}{A} = (62.4)h \text{ lb per sq ft.}$$

Now consider a dam 20 ft high shown in cross section in Fig. 6.7. Our problem is to define the average pressure on the dam when the reservoir is full. Again the definition is obtained by means of successively more accurate approximations.

Since the height of the dam is 20 ft, we subdivide the interval $\{y : 0 \leq y \leq 20\}$ into subintervals by choosing points satisfying

$$0 = y_0 < y_1 < y_2 < \cdots < y_{n-1} < y_n = 20$$

Let P_i denote the average pressure over the ith subinterval $\{y : y_{i-1} \leq y \leq i\}$. Then the average pressure P on the dam is the weighted average

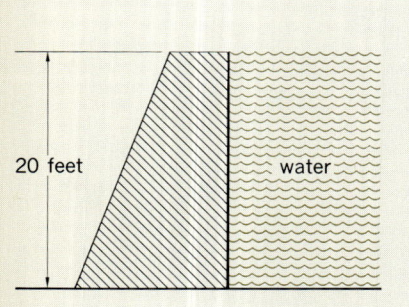

FIGURE 6.7

6.2 THE AVERAGE VALUE OF A FUNCTION

$$P = P_1\left(\frac{y_1 - y_0}{y_n - y_0}\right) + P_2\left(\frac{y_2 - y_1}{y_n - y_0}\right) + \cdots + P_n\left(\frac{y_n - y_{n-1}}{v_n - v_0}\right)$$

$$= \frac{1}{20}[P_1(y_1 - y_0) + P_2(y_2 - y_1) + \cdots + P_n(y_n - y_{n-1})]$$

For each $i = 1, 2, \cdots, n$, the average pressure P_i lies somewhere between the minimum pressure and the maximum pressure over the ith interval. Because the pressure at depth y is $(62.4)y$, the minimum pressure over $\{y : y_{i-1} \leq y \leq y_i\}$ occurs at the upper left-hand endpoint, while the maximum pressure occurs at the lower right-hand endpoint. That is, we set

$$m_i = (62.4)y_{i-1}$$

and

$$M_i = (62.4)y_i$$

Thus we have

$$(62.4)y_{i-1} \leq P_i \leq (62.4)y_i$$

Taking the weighted averages, we obtain

$$(62.4)\left[y_0\left(\frac{y_1 - y_0}{20}\right) + \cdots + y_{n-1}\left(\frac{y_n - y_{n-1}}{20}\right)\right] \leq P$$

$$\leq (62.4)\left[y_i\left(\frac{y_1 - y_0}{20}\right) + \cdots + y_n\left(\frac{y_n - y_{n-1}}{20}\right)\right]$$

Now choose the n subintervals to be of equal length $20/n$. Each term $(y_i - y_{i-1})/20 = (20/n)/20 = 1/n$ and the sums in the square brackets are arithmetic series. Therefore

$$y_0(y_1 - y_0) + \cdots + y_{n-1}(y_n - y_0) = \frac{20}{n}(y_0 + \cdots + y_{n-1})$$

$$= \frac{20}{n} \cdot \frac{n}{2}(y_0 + y_{n-1})$$

But $y_0 = 0$ and $y_{n-1} = 20 - 20/n$, so we have

$$\frac{62.4}{20}[y_0(y_1 - y_0) + \cdots + y_{n-1}(y_n - y_{n-1})]$$

$$= \frac{62.4}{20} \cdot \frac{20}{n} \cdot \frac{n}{2}\left(0 + 20 - \frac{20}{n}\right)$$

$$= \frac{62.4}{2}\left(20 - \frac{20}{n}\right)$$

$$= 624\left(1 - \frac{1}{n}\right)$$

Similarly,

$$\frac{62.4}{20}[y_1(y_1 - y_0) + \cdots + y_n(y_n - y_{n-1})]$$

$$= \frac{62.4}{20} \cdot \frac{20}{n} \cdot \frac{n}{2}\left(\frac{20}{n} + 20\right)$$

$$= 624\left(1 + \frac{1}{n}\right)$$

Because $624(1 - 1/n) \leq P \leq 624(1 + 1/n)$ for each $n \in \mathbf{N}$, we have $P = 624$ lb per sq ft. If the dam is 100 ft long, then its total area is $20 \times 100 = 2000$ sq ft; hence the total force acting on the dam is 624×2000 lb or 624 tons.

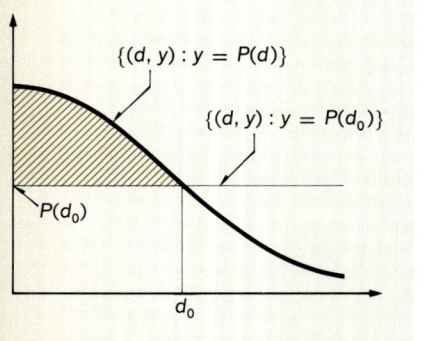

FIGURE 6.8

■ *Example 2.* Let $P(d)$ be the price of a certain commodity, expressed as a function of the demand d. Suppose that market conditions determine a fixed demand d_0 and hence the corresponding price $P(d_0)$. As the demand increases, the price is assumed to decrease, so that $P(d)$ is a monotone decreasing function of d. Now suppose the demand is $d < d_0$ whence the price is $P(d) > P(d_0)$. The consumer has gained because, while he is willing to pay $P(d)$, he need only pay the lower price $P(d_0)$. Under these circumstances, *consumer demand* is defined to be the area of the region shown in Fig. 6.8. Or in other words, consumer demand related to the demand d_0 is the product of the demand d_0 and the average value of the function $P(d) - P(d_0)$ over the interval $\{d : 0 \leq d \leq d_0\}$. Obviously, consumer demand can be calculated by the methods we are developing.

We shall not give exercises for this section. Applications will be much more easily computed after we have developed the theory.

6.3 THE SUMMATION NOTATION Σ

The unwieldy summations used in the two preceding sections have forced a shorthand notation. We use the Greek letter Σ to indicate summation as follows: Let $\varphi : \mathbf{N} \to Y$ be an infinite sequence. We define

$$\sum_{i=1}^{m} \varphi(i) = \varphi(1) + \varphi(2) + \cdots + \varphi(m)$$

The symbol is read as "the sum of the terms $\varphi(i)$ from $i = 1$ to $i = m$", or simply as "sigma $\varphi(i)$ from 1 to m." In the same way, if $m < n$, we define

$$\sum_{i=m}^{n} \varphi(i) = \varphi(m) + \varphi(m+1) + \cdots + \varphi(n)$$

and read this as "sigma $\varphi(i)$ from m to n."

■ **Example 1.** Let $\varphi(n) = n$. Then

$$\sum_{i=1}^{n} \varphi(i) = \sum_{i=1}^{n} i = 1 + 2 + 3 + \cdots + n = \frac{1}{2}n(n+1)$$

Similarly,

$$\sum_{i=1}^{n} i^2 = 1^2 + 2^2 + 3^2 + \cdots + n^2 = \frac{1}{6}n(n+1)(2n+1)$$

and

$$\sum_{i=1}^{n} i^3 = 1^3 + 2^3 + 3^3 + \cdots + n^3 = \frac{1}{4}n^2(n+1)^2$$

Using this notation, the sums in the preceding sections can be written in a very compact form. Recall that each term of the sums was of the form

$$f(\xi_i)(x_i - x_{i-1})$$

where the point ξ_i satisfied $x_{i-1} \leq \xi_i \leq x_i$. Therefore the sum may be written as

$$\sum_{i=1}^{n} f(\xi_i)(x_i - x_{i-1})$$

We shall see that this summation symbol not only saves space, but permits smoother and easier manipulation of such sums.

The reader should become familiar with the sigma symbol as soon as possible. To this end, we leave the proofs of the following basic properties as exercises.

1. Let φ_1 and φ_2 be sequences, and let c and d be real numbers. Then

$$\sum_{i=1}^{n} [c\varphi_1(i) + b\varphi_2(i)] = c \sum_{i=1}^{n} \varphi_1(i) + d \sum_{i=1}^{n} \varphi_2(i)$$

2. If φ is a constant sequence and $\varphi(n) = c$ for each $n \in \mathbf{N}$, then

$$\sum_{i=1}^{m} \varphi(i) = \sum_{i=1}^{m} c = cm$$

and

$$\sum_{i=m}^{n} \varphi(i) = \sum_{i=m}^{n} c = c(n - m + 1)$$

The letter i used in the sigma symbol is a "dummy index." It does not appear in the actual sum when written out. Such a dummy index may be changed to suit any purpose. For instance, given a particular sequence φ, we may write

$$\sum_{i=m}^{n} \varphi(i) = \sum_{j=m}^{n} \varphi(j) = \sum_{k=m}^{n} \varphi(k)$$

and equality holds by definition. Similarly, the index of summation can be changed to accommodate a change in the limits of summation. For example, we have

$$\sum_{i=1}^{n} i^2 = \sum_{i=2}^{n+1} (i-1)^2$$
$$= \sum_{i=m}^{n+m-1} (i-m+1)^2 = 1^2 + 2^2 + \cdots + n^2$$

■ **Example 2.** Let $\varphi(n) = -3 + 2n + 4n^2$. Then

$$\sum_{i=1}^{50} \varphi(i) = \sum_{i=1}^{50} (-3 + 2i + 4i^2)$$
$$= \sum_{i=1}^{50} (-3) + 2 \sum_{i=1}^{50} i + 4 \sum_{i=1}^{50} i^2$$
$$= (-3)(50) + 2\left[\frac{1}{2}(50)(51)\right] + 4\left[\frac{1}{6}(50)(51)(101)\right]$$
$$= 174{,}000$$
(Example 1)

■ **Example 3.** Let $(x_1, x_2, \cdots, x_n)$ and $(y_1, y_2, \cdots, y_n)$ be two n-tuples of real numbers. For each real number t, the inequality

$$\sum_{i=1}^{n} (tx_i + y_i)^2 \geqq 0$$

holds because this is a sum of squares of real numbers. Applying our basic properties, we expand this sum as follows:

$$\sum_{i=1}^{n} (tx_i + y_i)^2 = \sum_{i=1}^{n} (t^2 x_i^2 + 2tx_i y_i + y_i^2)$$
$$= t^2 \sum_{i=1}^{n} x_i^2 + 2t \sum_{i=1}^{n} x_i y_i + \Sigma y_i^2$$

Now set $\Sigma_{i=1}^{n} x_i^2 = A$, $\Sigma_{i=1}^{n} x_i y_i = B$ and $\Sigma_{i=1}^{n} y_i^2 = C$. Then we have a quadratic function of t

$$At^2 + 2Bt + C \geqq 0$$

which is nonnegative for all values of t. The reader should know that this implies that the quantity $4B^2 - 4AC$, the discriminant, must be nonpositive. Since $4B^2 - 4AC \leqq 0$, or $B^2 - AC \leqq 0$, we have $B^2 \leqq AC$ or

$$\left(\sum_{i=1}^{n} x_i y_i\right)^2 \leqq \left(\sum_{i=1}^{n} x_i^2\right)\left(\sum_{i=1}^{n} y_i^2\right)$$

6.3 THE SUMMATION NOTATION Σ

This is known as *Schwartz' Inequality*. By taking square roots, we may also conclude that

$$-1 \leq \frac{\sum_{i=1}^{n} x_i y_i}{\sqrt{\sum_{i=1}^{n} x_i^2} \sqrt{\sum_{i=1}^{n} y_i^2}} \leq 1,$$

a fact to be of use later.

EXERCISES 6.3

1. Prove properties 1 and 2 on p. 263.
2. Prove the equality

$$\sum_{i=1}^{n} \varphi(i) = \sum_{j=2}^{n+1} \varphi(j-1) = \sum_{k=0}^{n-1} \varphi(n-k)$$

3. Prove that

$$\sum_{i=1}^{n} [\varphi(i+1) - \varphi(i)] = \varphi(n-1) - \varphi(1)$$

Then let $\varphi(i) = i^2$ and use the equation

$$\sum_{i=1}^{n} [i^2 - (i-1)^2] = \sum_{i=1}^{n} (2i - 1)$$

to obtain the formula $\sum_{i=1}^{n} i = \tfrac{1}{2} n(n+1)$.

For Exercises 4 to 11, use the known formulas

$$\Sigma i = \tfrac{1}{2} n(n+1), \quad \Sigma i^2 = \tfrac{1}{6} n(n+1)(2n+1), \quad \Sigma i^3 = \tfrac{1}{4} n^2(n+1)^2$$

to evaluate the given sum.

4. $\sum_{i=1}^{n} (i + 1)$

5. $\sum_{i=1}^{n} (3 + 2i)$

6. $\sum_{i=1}^{n} (2i - i^2)$

7. $\sum_{i=1}^{n} (1 + i)^2$

8. $\sum_{i=1}^{n} i(i - 1)$

9. $\sum_{i=1}^{n} (2i + i^3)$

10. $\sum_{i=1}^{n} i(-1 + i^2)$

11. $\sum_{i=1}^{n} i(i + 1)(i + 2)$

12. Let $f(x) = 1 + x$. Find a formula for $\sum_{i=1}^{n} f(1 + i/n) \cdot 1/n$ and then evaluate

$$\lim_{n \to \infty} \sum_{i=1}^{n} f\left(1 + \frac{i}{n}\right) \cdot \frac{1}{n}$$

13. Prove that

$$\lim_{n \to \infty} \sum_{i=1}^{n} \left(1 + \frac{i}{n}\right)^2 \cdot \frac{1}{n} = \frac{7}{3}$$

14. Prove that
$$\lim_{n\to\infty} \sum_{i=1}^{n}\left(c + \frac{i}{n}\right)^2 \cdot \frac{1}{n} = \frac{1}{3}(1 + 3c + 3c^2)$$

15. Use the identity $1/[i(i+1)] = 1/i - 1/(i+1)$ as an aid in evaluating
$$\sum_{i=1}^{n} \frac{1}{i(i+1)}$$

16. Evaluate
$$\sum_{i=1}^{n} \frac{1}{i(i+1)(i+2)}$$

17. In the formula in Exercise 3, let $\varphi(i) = i^3$; obtain the formula $\sum_{i=1}^{n} i^2 = \frac{1}{6}n(n+1)(2n+1)$. Then, successively, let $\varphi(i) = i^4$ and $\varphi(i) = i^5$ to obtain a formula for $\sum_{i=1}^{n} i^4$.

18. Prove that
$$\lim_{n\to\infty} \sum_{i=1}^{n} \frac{i^4}{4n+1} = \frac{1}{5}$$

Then use this fact to show that the region under the curve $\{(x,y) : y = x^4\}$ and over the interval $\{x : a \leqq x \leqq b\}$ has the area $\frac{1}{5}(b^5 - a^5)$.

6.4 THE DEFINITE INTEGRAL

We shall define the definite integral to be the common limit of two infinite sequences. As these sequences are being defined, the reader may keep in mind the approximations of Sections 6.1 and 6.2. Note, however, that the procedure is purely analytic and does not depend upon a geometric interpretation of the operations used. Also, we point out that several important results are left unproved in this and the next section. The mathematical details are left until Section 6.6, so that we can begin to use the facts before attempting the more difficult proofs.

Let the function f be continuous on the closed interval $\{x : a \leqq x \leqq b\}$. For each $n \in \mathbf{N}$, we subdivide this interval into n equal subintervals each of length

$$\Delta x = \frac{b-a}{n}$$

We denote the endpoints of these intervals by

$$a = x_0, \ x_1 = a + \Delta x, \ x_2 = a + 2\,\Delta x, \cdots,$$
$$x_i = a + i\,\Delta x, \cdots, x_n = b$$

For each subinterval $\{x : x_{i-1} \leqq x \leqq x_i\}$, $i = 1, 2, 3, \cdots, n$, we define the two numbers

6.4 THE DEFINITE INTEGRAL

$$m_i = \text{minimum of } f(x) \text{ over } \{x : x_{i-1} \leq x \leq x_i\}$$

and

$$M_i = \text{maximum of } f(x) \text{ over } \{x : x_{i-1} \leq x \leq x_i\}$$

By definition, we have $m_i \leq M_i$ for each i and, since $\Delta x = (b-a)/n > 0$, we also have $m_i \Delta x \leq M_i \Delta x$ for each i.

Finally, we form the two sums

$$s(n) = \sum_{i=1}^{n} m_i \Delta x \quad \text{and} \quad S(n) = \sum_{i=1}^{n} M_i \Delta x$$

We define $s(n)$ to be the nth term of the *lower sequence* and $S(n)$ to be the nth term of the *upper sequence* for the function f over the interval $\{x : a \leq x \leq b\}$. From the inequality $m_i \Delta x \leq M_i \Delta x$ for each i, it follows immediately that we have the inequality

$$s(n) \leq S(n)$$

Furthermore, let m and M be the minimum and maximum, respectively, of $f(x)$ over the entire interval $\{x : a \leq x \leq b\}$. We obviously have the inequality $m \leq m_i \leq M_i \leq M$ for each i, and hence also the inequality

$$\sum_{i=1}^{n} m \Delta x \leq \sum_{i=1}^{n} m_i \Delta x \leq \sum_{i=1}^{n} M_i \Delta x \leq \sum_{i=1}^{n} M \Delta x$$

Now

$$\sum_{i=1}^{n} m \Delta x = m \sum_{i=1}^{n} \Delta x = m(b-a)$$

and similarly

$$\sum_{i=1}^{n} M_i \Delta x = M(b-a)$$

Therefore we have the inequality

$$m(b-a) \leq s(n) \leq S(n) \leq M(b-a)$$

telling us that both sequences are bounded.

We shall show in Section 6.6 that both $s(n)$ and $S(n)$ converge, but for now we ask that the reader accept the following result.

■ **THEOREM 6.1.** If the function f is continuous on the closed interval $\{x : a \leq x \leq b\}$, then the corresponding lower and upper sequences both converge and $\lim_{n \to \infty} s(n) = \lim_{n \to \infty} S(n)$. ■

We assume that f is continuous on $\{x : a \leq x \leq b\}$. The *definite integral* of f from $x = a$ to $x = b$ is defined to be the common limit

$$\lim_{n\to\infty} s(n) = \int_a^b f = \lim_{n\to\infty} (n)$$

of the lower and upper sequences.

The symbol is read as "the integral of f from a to b." It is actually an elongated letter s, serving to remind us of the fact that the integral is the limit of certain sums.

We take one more step here. Again let f be a continuous function on the closed interval $\{x : a \leqq x \leqq b\}$ and consider the nth subdivision of this interval into equal subintervals, each of length $\Delta x = 1/n(b - a)$. In each subinterval we choose some arbitrary point x_i'. Thus we make n choices of points x_i', each satisfying the inequality $x_{i-1} \leqq x_i' \leqq x_i$. By definition of the numbers m_i and M_i, we have the n inequalities

$$m_i = f(x_i') \leqq M_i, \qquad i = 1, 2, \cdots, n$$

Since $\Delta x > 0$, it follows that the inequality

$$s(n) = \sum_{i=1}^n m_i \, \Delta x \leqq \sum_{i=1}^n f(x_i') \, \Delta x \leqq \sum_{i=1}^n M_i \, \Delta x = S(n)$$

holds for any choice whatsoever of the n points x_i'.

A sum of the form

$$\sum_{i=1}^n f(x_i') \, \Delta x$$

is called a *Riemann sum*. The inequality

$$s(n) \leqq \sum_{i=1}^n f(x_i') \, \Delta x = S(n),$$

the fact that $\lim_{n\to\infty} s(n) = \lim_{n\to\infty} S(n)$, and Theorem 2.11 combine to prove the following result.

■ **THEOREM 6.2.** Let f be continuous on $\{x : a \leqq x \leqq b\}$. For each $n \in \mathbf{N}$, choose n points x_i', $i = 1, 2, \cdots, n$, each satisfying the corresponding inequality

$$a + (i - 1)\left(\frac{b - a}{n}\right) \leqq x_i' \leqq a + i\left(\frac{b - a}{n}\right)$$

Then

$$\lim_{n\to\infty} \sum_{i=1}^n f(x_i')\left(\frac{b - a}{n}\right) = \int_a^b f$$

We do not often use a sequence to evaluate a definite integral, despite the definition. However, in some simple instances we can use Riemann sums as the terms of the sequence and obtain easy results. Note that, in view of Theorem 2.12, we can use a subsequence of Riemann sums if this will further simplify matters.

6.4 THE DEFINITE INTEGRAL

■ **Example 1.** Let $E(x) = x$ be the identity function that we take on the interval $\{x : a \leq x \leq b\}$. For each $n \in \mathbf{N}$, we select the midpoint

$$x'_i = \tfrac{1}{2}(x_{i-1} + x_i)$$

in each interval $\{x : x_{i-1} \leq x \leq x_i\}$, $i = 1, 2, \cdots, n$. Using these particular points, we form a Riemann sum for the function E. This yields

$$\sum_{i=1}^n E(x'_i) \Delta x = \sum_{i=1}^n x'_i \Delta x = \sum_{i=1}^n \tfrac{1}{2}(x_{i-1} + x_i)(x_i - x_{i-1})$$

$$= \tfrac{1}{2} \sum_{i=1}^n (x_i^2 - x_{i-1}^2)$$

We have here a *telescoping sum*, that is, we have

$$x_1^2 - x_0^2 + x_2^2 - x_1^2 + x_3^2 - x_2^2 + \cdots$$
$$+ x_{n-1}^2 - x_{n-2}^2 + x_n^2 - x_{n-1}^2 = -x_0^2 + x_n^2$$

because all other terms appear both positively and negatively. Now, since $x_0 = a$ and $x_n = b$ for each $n \in \mathbf{N}$, this particular choice of the points x_i at each stage has given us a constant Riemann sum:

$$\sum_{i=1}^n E(x'_i) \Delta x = \tfrac{1}{2}(b^2 - a^2)$$

This proves that

$$\int_a^b E = \tfrac{1}{2}(b^2 - a^2)$$

■ **Example 2.** Given a continuous function f on a closed interval $\{x : a \leq x \leq b\}$ and given some fixed natural number $n_0 \in \mathbf{N}$, we can readily prove the inequalities

$$s(n_0) \leq s(2n_0), \quad S(2n_0) \leq S(n_0)$$

The proof for the lower sequence is as follows: Consider the ith subinterval of the n_0th subdivision of the interval. In the $2n_0$th subdivision, this subinterval is further subdivided into two equal subintervals of length $(b - a)/2n_0$ by its midpoint $y_i = \tfrac{1}{2}(x_{i-1} + x_i)$. Define

$$m'_i = \text{minimum of } f(x) \text{ on } \{x : x_{i-1} \leq x \leq y_i\}$$

and

$$m''_i = \text{minimum of } f(x) \text{ on } \{x : y_i \leq x \leq x_i\}$$

It is obvious that both $m_i \leq m'_i$, $m_i \leq m''_i$ must be satisfied. Thus

$$m'_i(y_i - x_{i-1}) + m''_i(x_i - y_i) \geq m_i(y_i - x_{i-1}) + m_i(x_i - y_i)$$
$$= m_i(x_i - x_{i-1})$$

Since
$$\sum_{i=1}^{n_0} [m'_i(y_i - x_{i-1}) + m''_i(x_i - y_i)] = s(2n_0)$$
the above inequalities prove that
$$s(n_0) \leqq s(2n_0)$$
An induction argument then proves that, starting with any n_0, the subsequence
$$s(n_0 \cdot 2^k)$$
of the lower sequence is monotone increasing. Theorem 2.12 then tells us that such a sequence must converge. ∎

Finally, the discussions in Sections 6.1 and 6.2 are augmented by the following definitions:

1. Let f be continuous on $\{x : a \leqq x \leqq b\}$ and suppose that $f(x) \geqq 0$ for all x in this interval. Then the area of the region $R = \{(x,y) : a \leqq x \leqq b, 0 \leqq y \leqq f(x)\}$ is defined to be
$$A(R) = \int_a^b f$$

2. Let f be continuous on $\{x : a \leqq x \leqq b\}$. Then the *average value* of f over this interval is defined to be
$$\frac{1}{b-a} \int_a^b f$$

EXERCISES 6.4 For each of Exercises 1 to 10 determine the constants m and M satisfying
$$m \leqq \int_a^b f \leqq M$$

1. $f(x) = \frac{1}{3}x;\ a = 3,\ b = 6$
2. $f(x) = x^2;\ a = 1,\ b = 5$
3. $f(x) = x^4;\ a = -1,\ b = 1$
4. $f(x) = \frac{1}{1+x};\ a = 0,\ b = 2$
5. $f(x) = \frac{1}{1+x^2};\ a = 0,\ b = 2$
6. $f(x) = \frac{x}{1+x^2};\ a = 0,\ b = 4$
7. $f(x) = \sqrt{1+x};\ a = 0,\ b = 3$
8. $f(x) = \sqrt{1+x^2};\ a = 0,\ b = \frac{4}{5}$
9. $f(x) = \frac{1+x}{\sqrt{9+x^2}};\ a = 0,\ b = 4$
10. $f(x) = \frac{4\sqrt{x}}{2+x};\ a = 1,\ b = 4$

For each function and interval in Exercises 11 to 20, evaluate both $s(10)$ and $S(10)$ and write out the inequality
$$s(10) \leqq \int_a^b f \leqq S(10)$$

11. $f(x) = \dfrac{x}{2}$; $a = 2$, $b = 4$

12. $f(x) = x^2$; $a = 2$, $b = 0$

13. $f(x) = x^4$; $a = 0$, $b = 2$

14. $f(x) = \dfrac{1}{x}$; $a = 1$, $b = \dfrac{3}{2}$

15. $f(x) = \dfrac{x}{1+x}$; $a = 1$, $b = 2$

16. $f(x) = \sqrt{3+x}$; $a = 1$, $b = 6$

17. $f(x) = \sqrt{1+x^2}$; $a = 0$, $b = \tfrac{1}{2}$

18. $f(x) = \dfrac{x}{1+x^2}$; $a = 0$, $b = 1$

19. $f(x) = \dfrac{\sqrt{x}}{1+x}$; $a = 1$, $b = 2$

20. $f(x) = \sqrt{1+x^3}$; $a = 0$, $b = 1$

Express each integral in Exercises 21 to 30 as the limit of a sequence. Determine the limit where possible and find an approximate value accurate to two significant figures if you cannot determine the limit.

21. $\int_0^2 (1+x)$

22. $\int_1^3 (2-x)$

23. $\int_1^2 x^2$

24. $\int_{-1}^3 (1+x)^2$

25. $\int_{-1}^0 (1+x-x^2)$

26. $\int_1^3 x^3$

27. $\int_1^2 x^2(x-1)$

28. $\int_0^3 \sqrt{1+x}$

29. $\int_0^{1/2} \sqrt{1+x^2}$

30. $\int_0^1 \sqrt{1+x^3}$

Express each limit in Exercises 31 to 35 as a definite integral of the form $\int_0^1 f$.

31. $\lim\limits_{n \to \infty} \dfrac{1}{n^3} \sum\limits_{i=1}^{n} i^2$

32. $\lim\limits_{n \to \infty} \dfrac{1}{n^4} \sum\limits_{i=1}^{n} i^3$

33. $\lim\limits_{n \to \infty} \dfrac{1}{n^3} \sum\limits_{i=1}^{n} (n^2 + i^2)$

34. $\lim\limits_{n \to \infty} \dfrac{1}{n^5} \sum\limits_{i=1}^{n} i^2(n^2 - i^2)$

35. $\lim\limits_{n \to \infty} \sum\limits_{i=n+1}^{2n} \dfrac{1}{i}$ (Hint: Write $\dfrac{1}{n+1} = \dfrac{1}{1+i/n} \cdot \dfrac{1}{n}$.)

6.5 PROPERTIES OF THE DEFINITE INTEGRAL

The theorems of this section provide the fundamental tools used in dealing with definite integrals. One result (Theorem 6.5) is left unproved. Its proof is to be found in Section 6.6, following the practice of leaving the more complicated mathematical analysis to be studied separately.

■ **THEOREM 6.3.** Let f be continuous on $\{x : a \leq x \leq b\}$ and let k be any constant. Then

$$\int_a^b (kf) = k \int_a^b f$$

PROOF: This is just a special case of the fact that we proved in Corollary 2.9, namely, $\lim_{n \to \infty} k\varphi(n) = k \lim_{n \to \infty} \varphi(n)$. First note that kf

is continuous, so that the integral on the left does exist. Then for any $n \in \mathbf{N}$ and any choice of the n points x_i, $i = 1, 2, \cdots, n$, in the subintervals, we surely have

$$\sum_{i=1}^{n} kf(x_i') \Delta x = k \sum_{i=1}^{n} f(x_i') \Delta x$$

Corollary 2.9 states that

$$\lim_{n \to \infty} k \sum_{i=1}^{n} f(x_i') \Delta x = k \lim_{n \to \infty} \sum_{i=1}^{n} f(x_i') \Delta x = k \int_{a}^{b} f$$

Since

$$\lim_{n \to \infty} \sum_{i=1}^{n} kf(x_i') \Delta x = \int_{a}^{b} (kf),$$

the proof is complete. ∎

■ **THEOREM 6.4.** If f and g are continuous on $\{x : a \leq x \leq b\}$, then

$$\int_{a}^{b} (f + g) = \int_{a}^{b} f + \int_{a}^{b} g$$

PROOF: This follows directly from Theorem 2.8, which tells us that the limit of a sum is the sum of the limits. In particular, since $f + g$ is continuous, all three integrals exist. For any $n \in \mathbf{N}$ and any choice of the points x_i in the nth stage, we have the obvious equation:

$$\sum_{i=1}^{n} (f + g)(x_i') \Delta x = \sum_{i=1}^{n} f(x_i') \Delta x + \sum_{i=1}^{n} g(x_i') \Delta x$$

Then, from Theorem 2.8, we have

$$\lim_{n \to \infty} \left(\sum_{i=1}^{n} f(x_i') \Delta x + \sum_{i=1}^{n} g(x_i') \Delta x \right)$$
$$= \lim_{n \to \infty} \sum_{i=1}^{n} f(x_i') \Delta x + \lim_{n \to \infty} \sum_{i=1}^{n} g(x_i') \Delta x \quad ∎$$

Using our simple definition of the definite integral, the following theorem is rather hard to prove. With a more complicated, but equivalent, definition as given in Section 6.6, the proof becomes much easier. We need this result now, however.

■ **THEOREM 6.5.** Let f be continuous on the closed interval from the smallest to the largest of three numbers a, b, and c. Then

$$\int_{a}^{b} f = \int_{a}^{c} f + \int_{c}^{b} f$$

It is vital to recognize that we need not have the inequality $a < c < b$ in order to apply Theorem 6.5. Furthermore, the three numbers need not be distinct.

- **COROLLARY 6.6.** If $f(a)$ exists, then $\int_a^a f = 0$.
  PROOF: Take $a = b = c$ in Theorem 6.5 to get
  $$\int_a^a f = \int_a^a f + \int_a^a f = 2\int_a^a f$$
  which implies that the integral is zero. ∎

- **COROLLARY 6.7.** If f is continuous on $\{x : a \leq x \leq b\}$, then
  $$\int_b^a f = -\int_a^b f$$
  PROOF: Take $b = a$ in Theorem 6.5 to get
  $$0 = \int_a^a f = \int_a^c f + \int_c^a f$$
  from which the corollary follows. ∎

- **THEOREM 6.8.** If f is continuous on $\{x : a \leq x \leq b\}$ and if $f(x) \geq 0$ for each point x in the interval, then $\int_a^b f \geq 0$.
  PROOF: Surely the minimum value of $f(x)$ cannot be negative. Hence we have
  $$0 \leq m(b - a) \leq s(n)$$
  for all $n \in \mathbf{N}$. Since a nonnegative sequence cannot have a negative limit, the theorem is immediate. ∎

- **THEOREM 6.9.** If f and g are continuous on $\{x : a \leq x \leq b\}$ and if $f(x) \leq g(x)$ for each point x in the interval, then
  $$\int_a^b f \leq \int_a^b g$$
  PROOF: The function $g - f$ is continuous and nonnegative on the interval, that is, $g(x) - f(x) \geq 0$ for each x. Thus Theorem 6.8 says that
  $$\int_a^b (g - f) \geq 0$$
  Then, using Theorems 6.3 and 6.4, we have
  $$\int_a^b (g - f) = \int_a^b g - \int_a^b f \geq 0 \quad \blacksquare$$

■ **COROLLARY 6.10.** If f is continuous on $\{x : a \leq x \leq b\}$, then

$$\left| \int_a^b f \right| \leq \int_a^b |f|$$

PROOF: Since $|f|$ is continuous (see p. 270), the integral on the right exists. From the obvious inequality

$$-|f(x)| \leq f(x) \leq |f(x)|$$

Theorem 6.9 yields

$$-\int_a^b |f| \leq \int_a^b f \leq \int_a^b |f|$$

which is equivalent to the inequality of the corollary. ■

The last result of this section is the *mean value theorem for integrals*.

■ **THEOREM 6.11.** If f is continuous on $\{x : a \leq x \leq b\}$, then there is a point x_0 between a and b such that

$$\int_a^b f = (b - a) f(x_0)$$

PROOF: If m and M are the minimum and maximum values of $f(x)$, respectively, on the entire interval, then we obviously have the inequality $m(b - a) \leq s(n) \leq M(b - a)$, as remarked before. It immediately follows that $m(b - a) \leq \lim_{n \to \infty} s(n) \leq M(b - a)$, or

$$m \leq \frac{1}{b - a} \int_a^b f \leq M$$

Now Theorem 3.11 tells us that $f(x)$ takes on every value between its minimum m and its maximum M so that, in particular, there is a point x_0 between a and b such that

$$f(x_0) = \frac{1}{b - a} \int_a^b f \quad ■$$

This theorem has an easy interpretation: if f is continuous on a closed interval, then f must take on its average value at least once in the interval. As an everyday example we can claim that, if a motorist averages 50 mph on a certain trip, then there must be at least one instant when his average speed is exactly 50 mph.

A geometric interpretation of the mean value theorem for integrals is also of interest. In view of definition 1 on p. 270, the definite integral $\int_a^b f$ is the area of the region bounded by the x axis, the vertical lines $\{(x,y) : x = a\}$, $\{(x,y) : x = b\}$, and the curve $\{(x,y) : y = f(x)\}$. An example is shown in Fig. 6.9. Note that $m(b - a)$ and $M(b - a)$ are areas of rectangles, the first of which is contained in the region under

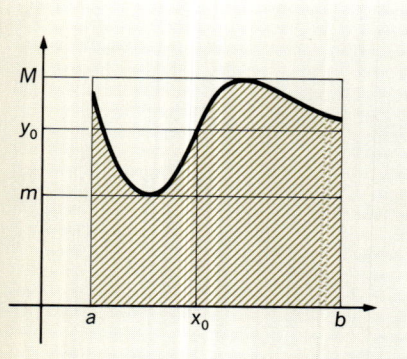

FIGURE 6.9

the curve and the second of which contains this region. Therefore we have

$$m(b-a) \leq \int_a^b f \leq M(b-a)$$

Then it is intuitively obvious that there is a horizontal line $\{(x,y) : y = y_0\}$ with $y_0(b-a) = \int_a^b f$. Since such a line crosses the curve, there is at least one point x_0 such that $f(x_0) = y_0$.

- **Example.** Let f and g be continuous on $\{x : a \leq x \leq b\}$. Then for each $t \in \mathbf{R}$, the function $(tf + g)^2$ is also continuous on the interval. Furthermore,

$$(tf + g)^2(x) = t^2 f^2(x) + 2tf(x)g(x) + g^2(x) \geq 0$$

for each x in the interval. Using the properties stated above, we have

$$\int_a^b (t^2 f^2 + 2tfg + g^2) = t^2 \int_a^b f^2 + 2t \int_a^b fg + \int_a^b g^2 \geq 0$$

Because this quadratic form in t is never negative, its discriminant $B^2 - 4AC$ cannot be positive. Therefore

$$(2\int_a^b fg)^2 - 4(\int_a^b f^2)(\int_a^b g^2) \leq 0$$

or

$$(\int_a^b fg)^2 \leq (\int_a^b f^2)(\int_a^b g^2)$$

This is the *Schwartz inequality for integrals*. It is put to use in Problem 6.5.4.

REMARK: The definite integral should be viewed as an effective means of defining certain quantities, such as area, which are otherwise quite elusive. This viewpoint may change perhaps as computational experience grows, but it should not be forgotten. Furthermore, the properties given in this section remain true in any application of the definite integral. ■

1. Let $f(x) = k$ be a constant function and prove that

$$\int_a^b k = k(b-a)$$

2. (a) If f is an odd function, show that

$$\int_{-a}^a f = 0$$

PROBLEMS 6.5
Mathematics

(b) If f is an even function, show that

$$\int_{-a}^{a} f = 2\int_{0}^{a} f$$

3. Let f be continuous on $\{x : a \leq x \leq b\}$ and let c be a constant. Prove that $\int_{a}^{b} (f - c)^2$ is a minimum when $c = 1/(b - a)\int_{a}^{b} f$. Use this fact to establish the identity

$$\left[\int_{a}^{b} f\right]^2 \leq (b - a)\int_{a}^{b} f^2$$

4. Use the Schwartz' inequality (p. 275) to prove the *Minkowski inequality*. Let f and g be continuous on $\{x : a \leq x \leq b\}$. Then

$$\left[\int_{a}^{b} (f + g)^2\right]^{1/2} \leq \left[\int_{a}^{b} f^2\right]^{1/2} + \left[\int_{a}^{b} g^2\right]^{1/2}$$

5. If f is continuous on $\{x : a \leq x \leq b\}$, if $f(x) \geq 0$ at each point, and if $f(x)$ is not identically zero, prove that $\int_{a}^{b} f > 0$.

6. Let f and g be continuous on $\{x : a \leq x \leq b\}$ and suppose that $g(x)$ does not change sign in this interval. Show that there is a point x_0, $a < x_0 < b$, such that $\int_{a}^{b} fg = f(x_0)\int_{a}^{b} g$.

7. Express the following limit as the definite integral of a function over $\{x : 0 \leq x \leq 1\}$.

$$\lim_{n \to \infty} n\left(\frac{1}{1 + n^2} + \frac{1}{4 + n^2} + \frac{1}{9 + n^2} + \cdots + \frac{1}{2n^2}\right)$$

6.6 THE EXISTENCE THEOREM

Theorems 6.1 and 6.5 have yet to be proved. As we shall mention again later, the difficulties with Theorem 6.5 stem from our restrictive use of subdivisions into equal subintervals. By permitting more general subdivisions, we not only make it easy to prove Theorem 6.5 but we do not increase trouble with Theorem 6.1. However, we must return to the beginning.

We consider a continuous function f on an interval $\{x : a \leq x \leq b\}$. Subdivide the interval into n subintervals by choosing any $n - 1$ points satisfying

$$a < x_1 < x_2 < \cdots < x_{n-1} < b$$

For uniformity of notation, let $a = x_0$ and $b = x_n$. Such a selection of points is called a *net* on the interval and the resulting set of n subintervals

$$\{x : x_{i-1} \leq x \leq x_i\}, \quad i = 1, 2, 3, \cdots, n,$$

is a *partition* of the given interval. We use the generic symbol Δ to denote a partition. The *norm* $|\Delta|$ of a partition Δ is defined to be the maximum length $x_i - x_{i-1}$ of its subintervals.

Given a partition Δ, we define the two numbers

$$m_i = \text{minimum value of } f(x)$$
$$M_i = \text{maximum value of } f(x)$$

over $\{x : x_{i-1} \leq x \leq x_i\}$ for each $i = 1, 2, \cdots, n$. Then we form the *lower* and *upper sums*,

$$s(\Delta) = \sum_{i=1}^{n} m_i(x_i - x_{i-1})$$

$$S(\Delta) = \sum_{i=1}^{n} M_i(x_i - x_{i-1})$$

Because $m_i \leq M_i$ for each i, it follows that

$$s(\Delta) \leq S(\Delta)$$

Furthermore, as in Section 6.4, we have the inequality

$$m(b - a) \leq s(\Delta) \leq S(\Delta) \leq M(b - a)$$

for any partition Δ. [Recall that m and M are the minimum and the maximum, respectively, of $f(x)$ over the entire interval.]

We need two easy preliminary results: Lemmas 6.12 and 6.13.

■ **LEMMA 6.12.** Let Δ be a partition of $\{x : a \leq x \leq b\}$ and let Δ' be another partition obtained by adding a finite number of points to the net defining Δ. Then we have

$$s(\Delta) \leq s(\Delta') \leq S(\Delta') \leq S(\Delta)$$

PROOF: Let $a = x_0 < x_1 < x_2 < \cdots < x_{n-1} < x_n = b$ be the net defining Δ and add one point $\bar{x}$ to this net. Suppose $\bar{x}$ lies in the kth subinterval of Δ, that is, $x_{k-1} < \bar{x} < x_k$. To obtain $s(\Delta')$, we only replace the kth term of $s(\Delta)$. Let

$$m'_k = \text{minimum value of } f(x) \text{ on } \{x : x_{k-1} \leq x \leq \bar{x}\}$$
$$m''_k = \text{minimum value of } f(x) \text{ on } \{x : \bar{x} \leq x \leq x_k\}$$

Clearly, we have both $m_k \leq m'_k$ and $m_k \leq m''_k$. Therefore

$$m_k(x_k - x_{k-1}) = m_k(\bar{x} - x_{k-1}) + m_k(x_k - \bar{x})$$
$$\leq m'_k(\bar{x} - x_{k-1}) + m''_k(x_k - \bar{x})$$

Because we have

$$s(\Delta) = \sum_{i=1}^{k-1} m_i(x_i - x_{i-1}) + m_k(x_k - x_{k-1}) + \sum_{i=k+1}^{n} m_i(x_i - x_{i-1})$$

and

$$s(\Delta') = \sum_{i=1}^{k-1} m_i(x_i - x_{i-1}) + m'_k(\bar{x} - x_{k-1}) + m''_k(x_k - \bar{x})$$
$$+ \sum_{i=k+1}^{n} m_i(x_i - x_{i-1})$$

it follows immediately that $s(\Delta) \leq s(\Delta')$. Now a finite induction argument completes the proof. An analogous argument then shows that $S(\Delta') \leq S(\Delta)$. ■

■ **LEMMA 6.13.** Let Δ and Δ' be any two partitions of the interval $\{x : a \leqq x \leqq b\}$. Then
$$s(\Delta) \leqq S(\Delta')$$

PROOF: Combine the net defining Δ and the net defining Δ' into one net, and let Δ'' be the partition defined by the combined net. Since the combined net is obtained by adding a finite number of points to either of the given nets, Lemma 6.12 provides both
$$s(\Delta) \leqq s(\Delta'') \quad \text{and} \quad S(\Delta'') \leqq S(\Delta')$$
But then $s(\Delta'') \leqq S(\Delta'')$ and the proof is complete. ■

One more fact about continuous functions is needed, but we shall not prove it.

■ **THEOREM 6.14.** Let f be continuous on the closed interval $\{x : a \leqq x \leqq b\}$. Then, given any $\epsilon > 0$, there exists $\delta > 0$ such that
$$|f(x') - f(x'')| < \epsilon, \quad \text{whenever } |x' - x''| < \delta$$

This is actually stronger than the statement of continuity which it resembles. In the definition of "f is continuous as $x = a$," the δ that is postulated depends upon the given $\epsilon > 0$ *and upon the point a*. In Theorem 6.14, the δ depends only upon ϵ, not upon a choice of a point. We say that f is *uniformly continuous* if it satisfies the conclusion of Theorem 6.14. A proof of this theorem can be found in advanced calculus books. (For instance, see p. 80 of *Advanced Calculus* by A. Devinatz, New York: Holt, Rinehart and Winston, Inc., 1968.)

■ **THEOREM 6.15.** Let f be continuous on $\{x : a \leqq x \leqq b\}$ and let $\Delta_1, \Delta_2, \cdots, \Delta_n, \cdots$ be any sequence of partitions such that $\lim_{n \to \infty} |\Delta_n| = 0$. Then
$$\lim_{n \to \infty} [S(\Delta_n) - s(\Delta_n)] = 0$$

PROOF: Given $\epsilon > 0$, use Theorem 6.14 to select $\delta > 0$ such that
$$|f(x') - f(x'')| < \frac{\epsilon}{2(b-a)} \quad \text{whenever } |x' - x''| < \delta$$

By the limit assumption on the norms of the partitions, there is some n_0 so that $|\Delta_n| < \delta$ whenever $n > n_0$. On each subinterval of the nth partition Δ_n, $n > n_0$, we clearly have
$$M_i - m_i < \frac{\epsilon}{2(b-a)}$$

6.6 THE EXISTENCE THEOREM

Therefore we have

$$S(\Delta_n) - s(\Delta_n) = \sum_{i=1}^{r} (M_i - m_i)(x_i - x_{i-1})$$

$$\leq \sum_{i=1}^{r} \frac{\epsilon}{(b-a)}(x_i - x_{i-1})$$

$$= \frac{\epsilon}{2(b-a)} \sum_{i=1}^{n} (x_i - x_{i-1})$$

$$= \frac{\epsilon}{2(b-a)} \cdot (b-a) = \frac{1}{2}\epsilon < \epsilon \;\blacksquare$$

■ **LEMMA 6.16.** Let f be continuous on $\{x : a \leq x \leq b\}$ and consider the partition defined by the net $x_i = a + i(b-a)/2^n, i = 1, 2, \cdots, 2^n$. Let $s(2^n)$ and $S(2^n)$ be the corresponding lower and upper sums. Then both $s(2^n)$ and $S(2^n)$ converge. Note that if they do converge, then Theorem 6.15 says that they have the same limit.

PROOF: It is only necessary to point out that the $(n+1)$st net is obtained from the nth net simply by adding the midpoint of each subinterval in the nth partition. Thus Lemma 6.12 shows that $s(2^n)$ is monotone increasing and $S(2^n)$ is monotone decreasing. Since both sequences are bounded, Theorem 2.10 concludes that both converge. ■

The unproved Theorem 6.1 is just a particular case of the following existence theorem.

■ **THEOREM 6.17.** Let f be continuous on $\{x : a \leq x \leq b\}$ and let $\Delta_1, \Delta_2, \cdots, \Delta_n, \cdots$ be any sequence of partitions such that $\lim_{n \to \infty} |\Delta_n| = 0$. Then both $s(\Delta_n)$ and $S(\Delta_n)$ are terms of convergent sequences and $\lim_{n \to \infty} s(\Delta_n) = \lim_{n \to \infty} S(\Delta_n)$.

PROOF: In view of Lemma 6.16,

$$\lim_{n \to \infty} s(2^n) = \lim_{n \to \infty} S(2^n).$$

We shall denote this common limit by L.

Now suppose $\epsilon > 0$ is given. In view of this remark, there is a natural number k sufficiently large so that both

$$|s(2^k) - L| < \frac{\epsilon}{2}, \quad |S(2^k) - L| < \frac{\epsilon}{2}$$

are true. Since $s(2^n)$ is monotone increasing and $S(2^n)$ is monotone decreasing, we have $|s(2^k) - L| = L - s(2^k)$ and $|S(2^k) - L| = S(2^k) - L$. Thus these inequalities above may be rewritten as

$$L - \frac{\epsilon}{2} < s(2^k), \quad S(2^k) < L + \frac{\epsilon}{2}$$

Next we use the limit condition established in Theorem 6.15 to find an integer n such that $S(\Delta_n) - s(\Delta_n) < \epsilon/2$. From Lemma 6.13 we have the inequalities

$$s(2^k) \leqq S(\Delta_n) \quad \text{and} \quad s(\Delta_n) \leqq S(2^k)$$

Hence we obtain

$$L - \frac{\epsilon}{2} < s(2^k) \leqq S(\Delta_n) < s(\Delta_n) + \frac{\epsilon}{2}$$

$$\leqq S(2^k) + \frac{\epsilon}{2} < L + \frac{\epsilon}{2} + \frac{\epsilon}{2}$$

which says that

$$|S(\Delta_n) - L| < \epsilon$$

This proves that $S(\Delta_n)$ converges to L and, of course, Theorem 6.15 completes the proof. ∎

Using these more general subdivisions, we find that the proof of Theorem 6.5 becomes quite routine. Recall that we are to prove that

$$\int_a^b f = \int_a^c f + \int_c^b f$$

for any three numbers a, b, and c, provided only that f be continuous on the interval from the smallest to the largest. First consider the case in which $a < c < b$. Consider sequences

$$\Delta_1', \Delta_2', \ldots, \Delta_n', \ldots \quad \text{with } \lim_{n \to \infty} |\Delta_n'| = 0$$
$$\Delta_1'', \Delta_2'', \ldots, \Delta_n'', \ldots \quad \text{with } \lim_{n \to \infty} |\Delta_n''| = 0,$$

each Δ_n' a partition of $\{x : a \leqq x \leqq c\}$ and each Δ_n'' a partition of $\{x : c \leqq x \leqq b\}$. By simply combining the partitions Δ_n' and Δ_n'', we obtain a partition Δ_n of the combined interval $\{x : a \leqq x \leqq b\}$. [It is here that we would get into difficulty using equal subintervals. For the lengths $(c - a)/n$ and $(b - c)/n$ are not equal in general, of course.] It is quite obvious now that

$$s(\Delta_n) = s(\Delta_n') + s(\Delta_n'')$$

Therefore

$$\int_a^b f = \lim_{n \to \infty} s(\Delta_n) = \lim_{n \to \infty} s(\Delta_n') + \lim_{n \to \infty} s(\Delta_n'') = \int_a^b f + \int_a^b f$$

The key to the remaining cases of Theorem 6.5 obviously lies in the equation

6.6 THE EXISTENCE THEOREM

$$\int_a^b f = -\int_b^a f$$

if $a > b$. This is proved as follows: Consider a partition Δ of the interval $\{x : b \leq x \leq a\}$. Let the defining net be numbered as if a were less than b. Thus

$$b = x_n < x_{n-1} < \cdots < x_1 < x_0 = a$$

Then we have the lower sum approximating $\int_a^b f$:

$$\sum_{i=1}^n m_i(x_i - x_{i-1}) = -\sum_{i=1}^n m_i(x_{i-1} - x_i)$$

Because $x_i < x_{i-1}$, the sum on the right is precisely a lower sum approximating $\int_b^a f$. This shows that integrating in the negative direction reverses the sign of the integral as claimed. Theorem 6.5 now follows in its entirety. ∎

7
THE INDEFINITE INTEGRAL

In this chapter we study a practical method of evaluating definite integrals. The method involves "antidifferentiation," a process that can be defined as follows: Given a function f, find a function F such that $f = DF$. This leads to many useful applications.

7.1 FUNCTIONS DEFINED BY INTEGRALS

Again we consider a function f continuous on a closed interval $\{x : a \leqq x \leqq b\}$. For any fixed point x_0, $a \leqq x_0 \leqq b$, the function f is also continuous on the subinterval $\{x : a \leqq x \leqq x_0\}$. Therefore the integral

$$\int_a^{x_0} f$$

exists. Obviously, this number depends upon the choice of x_0, that is the integral $\int_a^{x_0} f$ is a function of x_0. We may now omit the subscript on the letter x and define a new function F by setting

$$F(x) = \int_a^x f$$

Note that F is also defined on the closed interval $\{x : a \leqq x \leqq b\}$. In particular, we have $F(a) = \int_a^a f = 0$ and $F(b) = \int_a^b f$. It is this function F that has the property that $DF = f$.

■ **THEOREM 7.1.** Let f be continuous on $\{x : a \leqq x \leqq b\}$. Then the function F,

$$F(x) = \int_a^x f$$

is also continuous on $\{x : a \leqq x \leqq b\}$.

PROOF: Let x' and x'', $x' < x''$, be any two points in the interval $\{x : a \leqq x \leqq b\}$. By definition

$$F(x'') - F(x') = \int_a^{x''} f - \int_a^{x'} f$$

Then by Theorem 6.5 we may write

$$\int_a^{x''} f = \int_a^{x'} f + \int_{x'}^{x''} f$$

and hence

$$F(x'') - F(x') = \int_{x'}^{x''} f$$

Now let M denote the maximum value of $|f(x)|$ on the interval $\{x : a \leqq x \leqq b\}$. We assume $M > 0$ for otherwise $f(x) = 0$ for all x. Corollary 6.10 applies to tell us that

7.1 FUNCTIONS DEFINED BY INTEGRALS

$$|F(x'') - F(x')| = \left|\int_{x'}^{x''} f\right| \leq \int_{x'}^{x''} |f| \leq M(x'' - x')$$

We verify the definition of continuity as follows: Let x_0 be any fixed point in the interval and assume that $\epsilon > 0$ has been given. For any point x of the interval satisfying $|x - x_0| < \epsilon/M$, we have

$$|F(x) - F(x_0)| = \left|\int_{x_0}^{x} f\right| \leq M \cdot |x - x_0| < M \cdot \frac{\epsilon}{M} = \epsilon$$

Therefore choosing $\delta = \epsilon/M$, we have shown that

$$|F(x) - F(x_0)| < \epsilon \quad \text{whenever } |x - x_0| < \delta \quad \blacksquare$$

The next result is the important one in our theory. It lies at the foundation of the relationship between differentiation and integration.

■ **THEOREM 7.2.** Let f be continuous on $\{x : a \leq x \leq b\}$. Then the function F with values $F(x) = \int_c^x f$ is differentiable in the open interval $\{x : a < x < b\}$ and

$$DF = f$$

PROOF: Let x_0 be any point, $a < x_0 < b$. We must show that

$$\lim_{h \to 0} \frac{F(x_0 + h) - F(x_0)}{h} = f(x_0)$$

Because

$$F(x_0 + h) - F(x_0) = \int_{x_0}^{x_0+h} f$$

we are trying to show that

$$\lim_{h \to 0} \frac{1}{h} \int_{x_0}^{x_0+h} f = f(x_0)$$

Now we apply Theorem 6.11 which states that there is a point x' between x_0 and $x_0 + h$ such that

$$\int_{x_0}^{x_0+h} f = f(x')(x_0 + h - x_0) = hf(x')$$

Therefore

$$\frac{1}{h} \int_{x_0}^{x_0+h} f = f(x')$$

for some point x' between x_0 and $x_0 + h$. As h approaches zero, the point x' must approach x_0. Hence

$$\lim_{h \to 0} \frac{1}{h} \int_{x_0}^{x_0+h} f = \lim_{x' \to x_0} f(x') = f(x_0)$$

where the latter equality holds because f is continuous. ■

A function defined by an integral will usually be quite new to us. For instance, it is doubtful that the function

$$F(x) = \int_0^x \frac{(4 + 3t^2)^{1/2}(3 + t^3)^{1/3}}{(5 + t + 2t^2)^2}$$

has ever been studied. Nor would there be much purpose in studying it unless it had arisen in some application. Nevertheless, we know as much about this function F as we do about an algebraic function. Obviously, we can determine any derivative $F^{(n)}$. And individual values such as $F(1)$ can be obtained with any desired degree of accuracy. In short, such a function need not worry us.

If we recognize the function f as the derivative of some other function G, then we can very easily determine the function $F(x) = \int_a^x f$. Knowing that $F'(x) = f(x) = G'(x)$, we conclude that

$$F(x) = G(x) + c$$

for some constant c (Theorem 3.36). Furthermore, we know that

$$F(a) = \int_a^a f = 0$$

It follows immediately that $c = -G(a)$. Thus, if $G'(x) = f(x)$, we have

$$F(x) = \int_a^x f = G(x) - G(a)$$

■ **Example 1.** Let $F(x) = \int_1^x (2 - 2t + 3t^2)$. Recognizing that the function $G(t) = 2t - t^2 + t^3$ satisfies $G'(t) = 2 - 2t + 3t^2 = f(t)$, we may write

$$F(x) = G(x) - G(1)$$

Since $G(1) = 2$, we have $F(x) = -2 + 2x - x^2 + x^3$.

■ **Example 2.** Let

$$F(x) = \int_3^x \frac{1}{\sqrt{1 + t}}$$

We know that, if $G(t) = 2(1 + t)^{1/2}$, then $G'(t) = (1 + t)^{-1/2} = f(t)$. Hence

$$F(x) = 2(1 + x)^{1/2} - 2(1 + 3)^{1/2}$$
$$= 2\sqrt{1 + x} - 4$$

EXERCISES 7.1   For the functions 1 to 10, determine Df.

1. $f(x) = \int_2^x (t - 2)^2$ 2. $f(x) = \int_4^x (t - 2)^2$

3. $f(x) = \int_1^x \sqrt{t}$

4. $f(x) = \int_4^x \sqrt{5+t}$

5. $f(x) = \int_1^x (t^3 - 3t + 2)$

6. $f(x) = \int_1^3 (t^3 - 3t + 2)$

7. $f(x) = \int_x^3 t^2$

8. $f(x) = \int_x^4 \sqrt{t}$

9. $f(x) = \int_x^{2x} t^2$

10. $f(x) = \int_x^{3x} t^3$

11. Determine $D^2 f$ if $f(x) = \int_0^x \left[\int_0^t z^2 \right]$.

12. If $f(x) = (x^2 + 1)^2$, then $f'(x) = 4x(x^2 + 1)$. Use this fact and Theorem 3.36 to evaluate $\int_0^3 4x(x^2 + 1)$.

13. If $f(x) = \sqrt{x^2 + 2}$, then $f'(x) = x/\sqrt{x^2 + 2}$. Use this fact and Theorem 3.36 to evaluate $\int_{\sqrt{2}}^{\sqrt{7}} x(x^2 + 2)^{-1/2}$.

14. Assume that f is continuous and that $g'(x)$ exists. Use the chain rule of differentiation to establish the formula

$$D \int_a^{g(x)} f = f(g(x)) \cdot Dg(x)$$

For functions 15 to 19, use the formula of Exercise 14 to find Df.

15. $f(x) = \int_x^{2x} (t - 2)^2$

16. $f(x) = \int_2^{1+3x} (t - 2)^2$

17. $f(x) = \int_1^{x^2 - 1} \sqrt{t^2 - 1}$

18. $f(x) = \int_1^{x+1} \sqrt{t}$

19. $f(x) = \int_{-1}^{x - x^2} \sqrt{t + 1}$

20. Determine all values $x \in \mathbf{R}$ that satisfy

$$\int_0^x (t^3 - t) = \frac{1}{3} \int_{\sqrt{2}}^x (t - t^3)$$

Determine the functions in Exercises 21 to 30.

21. $\int_0^x (1 + 3t^2)$

22. $\int_1^x (1 + 3t^2)$

23. $\int_0^x 3(1 + t)^2$

24. $\int_{-1}^x (1 + t)^2$

25. $\int_{-1}^x (t + t^2)$

26. $\int_x^{2x} (t + t^2)$

27. $\int_0^x \sqrt{t}$

28. $\int_1^x \dfrac{1}{\sqrt{t}}$

29. $\int_x^{x^2} \sqrt{t}$
30. $\int_1^{2x} \left(\sqrt{t} - \dfrac{1}{\sqrt{t}}\right)$

PROBLEMS 7.1

1. Prove that $D\int_x^b f = -f(x)$.

2. If f is continuous and $u(x)$ and $v(x)$ are differentiable, prove that
$$D\int_{v(x)}^{u(x)} f = f(u(x)) \cdot Du(x) - f(v(x)) \cdot Dv(x)$$

3. Let f be a strictly increasing function and define $F(x) = \int_a^x f$. Prove that
$$F\left(\dfrac{x+y}{2}\right) \leqq \dfrac{1}{2}(F(x) + F(y))$$

4. Not every function $F(x)$ satisfying $F'(x) = f(x)$ can be written in the form $F(x) = \int_a^x f$. Why not?

5. (Management) The operator of a trucking concern wants to determine the optimal period T between overhauls of a truck. Let $f(t)$ be the rate of depreciation of a vehicle and let C denote the fixed cost of an overhaul. Explain why he should minimize the function
$$g(T) = \dfrac{i}{T}\left[C + \int_0^T f(t)\right]$$

7.2 THE FUNDAMENTAL THEOREM OF CALCULUS

Earlier mention was made of an effective method for evaluating definite integrals. The so-called fundamental theorem of calculus that follows is the basis for that method.

■ **THEOREM 7.3.** If f is continuous on $\{x : a \leqq x \leqq b\}$ and if F is any function such that $DF = f$, then
$$\int_a^b f = F(b) - F(a)$$

PROOF: We consider the function
$$G(x) = \int_a^x f$$

From the remarks following Theorem 7.2, for any function F with $F' = f$ we have
$$G(x) = \int_a^x f = F(x) - F(a)$$

(This formula appears on p. 286 with the roles of letters F and G interchanged.) The conclusion of the theorem is now immediately obvious. ■

7.2 THE FUNDAMENTAL THEOREM OF CALCULUS

It is Theorem 7.3 that provides an evaluation method. To evaluate $\int_a^b f$, we need only find some function F such that $DF = f$. Then $F(b) - F(a)$ is the value of the definite integral. While there are some difficulties in finding F as required, this method is the standard one for evaluating definite integrals.

■ **Example 1.** Recalling that $D(x^3) = 3x^2$ and hence $D(\frac{1}{3}x^3) = x^2$, we have

$$\int_a^b f = \int_a^b x^2 = \frac{1}{3}b^3 - \frac{1}{3}a^3 = \frac{1}{3}(b^3 - a^3) = F(b) - F(a)$$

We have $f(x) = x^2$ and $F(x) = \frac{1}{3}x^3$, of course. Notice that, for any constant k, we also have $D(\frac{1}{3}x^3 + k) = x^2$. If we had chosen $F(x) = \frac{1}{3}x^3 + k$, then

$$F(b) - F(a) = (\frac{1}{3}b^3 + k) - (\frac{1}{3}a^3 + k) = \frac{1}{3}b^3 - \frac{1}{3}a^3$$

Hence it does not matter which of the functions F we might use. We always arrive at the same value.

■ **Example 2.** In order to evaluate $\int_a^b (3 - 4x)$, we simply note that $D(3x - 2x^2) = 3 - 4x$. Therefore

$$\int_a^b (3 - 4x) = (3b - 2b^2) - (3a - 2a^2)$$

For instance,

$$\int_0^1 (3 - 4x) = (3 - 2) - (0 - 0) = 1$$

■ **Example 3.** Because $D(-x^{-2}) = 2x^{-3}$, Theorem 7.3 tells us that

$$\int_1^3 2x^{-3} = (-3^{-2}) - (-1^{-2}) = -\frac{1}{3^2} + 1 = \frac{8}{9}$$

More generally, on an interval $\{x : a \leq x \leq b\}$ not containing zero, we have

$$\int_a^b 2x^{-3} = (-b^{-2}) - (-a^{-2}) = \frac{1}{a^2} - \frac{1}{b^2}$$

■ **Example 4.** Knowing that

$$x^n = D\left(\frac{x^{n+1}}{n+1}\right),$$

Theorem 7.3 shows that

$$\int_1^5 x^n = \frac{5^{n+1}}{n+1} - \frac{1^{n+1}}{n+1} = \frac{1}{n+1}(5^{n+1} - 1)$$

But notice that this formula is *not* valid for $n = -1$ because the factor $1/(n + 1)$ would be meaningless.

■ *Example 5.* The definition of a function F by means of an integral
$$F(x) = \int_a^x f$$
often leads to new functions that cannot be expressed in terms of known functions. For instance, if $f(x) = \sqrt{1 - x^3}$, then it is impossible to express the new function
$$F(x) = \int_0^x \sqrt{1 - t^3}, \qquad 0 \leq x \leq 1$$
in terms of any functions we know. Nevertheless, this function is well defined, has a derivative that we know, and so forth. ■

A commonly used symbol for the expression $F(b) - F(a)$ is
$$F(b) - F(a) = F(x)\Big]_a^b$$

For instance,
$$3x^2 - 1\Big]_1^2 = [3(2)^2 - 1] - [3(1)^2 - 1] = 11 - 2 = 9$$
and
$$\frac{x-1}{x+1}\Big]_1^5 = \frac{5-1}{5+1} - \frac{1-1}{1+1} = \frac{4}{6} - \frac{0}{2} = \frac{2}{3}$$

The method for evaluating a definite integral $\int_a^b f$ is to find a function F such that $F'(x) = f(x)$ and then to apply the formula
$$\int_a^b f = F(x)\Big]_a^b = F(b) - F(a)$$

In Section 6.4 the definite integral was shown to be a limit of Riemann sums,
$$\int_a^b f = \lim_{n \to \infty} \sum_{i=1}^n f(x_i')(x_i - x_{i-1})$$
where, for each $i = 1, 2, \cdots, n$, the point x_i' is any point in the ith subinterval $\{x : x_{i-1} \leq x \leq x_i\}$. We remarked that each term of a Riemann sum is of the form
$$f(x_i') \, dx_i$$
where $dx_i = \Delta x_i = x_i - x_{i-1}$ is a differential in the variable x. Now if F is any function such that $F'(x) = f(x)$, we have
$$f(x_i') \, dx_i = DF(x_i') \, dx_i$$

Hence the Riemann sum is a sum of differentials of F and *the definite integral is a limit of sums of differentials.*

The elongated "s" in the integral symbol $\int_a^b f$ reminds us that the definite integral is a limit of sums. But it is most useful to retain the differential $DF(x)\,dx = f(x)\,dx$ as part of the integral symbol. Hence the symbol becomes

$$\int_a^b f(x)\,dx$$

reminding us that the integral is a limit of sums *of differentials.* This becomes extremely important in the manipulation of integrals.

The variable x appearing in this symbol is a *dummy variable,* much like the dummy index of summation. While we call x the *variable of integration,* it does not appear in the value of the definite integral. In short, we may change the variable of integration at our own convenience. For instance, we may write

$$\int_a^b f = \int_a^b f(x)\,dx = \int_a^b f(t)\,dt = \int_a^b f(u)\,du$$

and so forth. We really used this fact already. In Example 5 we wrote

$$F(x) = \int_0^x \sqrt{1 - t^3}$$

Of course, it would make no difference at all if we had written

$$F(x) = \int_0^x \sqrt{1 - u^3}\,du$$

and so on.

Use the fundamental theorem (7.3) to evaluate the definite integrals 1 to 32. Interpret each integral as an area and explain any peculiar results obtained.

EXERCISES 7.2

1. $\int_1^4 5\,dx$
2. $\int_2^3 2x\,dx$
3. $\int_1^2 (4 + 2x)\,dx$
4. $\int_4^5 (3 - x)\,dx$
5. $\int_{-1}^0 3x^2\,dx$
6. $\int_1^2 (x + x^2)\,dx$
7. $\int_0^3 (2 + x)^2\,dx$
8. $\int_{-1}^2 (1 + x)(3 - x)\,dx$
9. $\int_{-1}^2 x^3\,dx$
10. $\int_{-2}^2 x^3\,dx$
11. $\int_0^1 x^4\,dx$
12. $\int_{-1}^1 x^4\,dx$

13. $\int_{-2}^{2} (2x + x^3 + x^5) \, dx$ 14. $\int_{2}^{3} (2 - x + x^2)^2 \, dx$

15. $\int_{1}^{2} \left(\frac{1}{x^2}\right) dx$ 16. $\int_{1}^{2} \left(\frac{1}{x^3}\right) dx$

17. $\int_{1}^{2} (x^{-2} + x^{-3}) \, dx$ 18. $\int_{1}^{3} \frac{3 + 2x}{x^4} \, dx$

19. $\int_{0}^{4} \sqrt{x} \, dx$ 20. $\int_{0}^{2} \sqrt{x} \, dx + \int_{2}^{4} \sqrt{x} \, dx$

21. $\int_{0}^{8} \sqrt{1 + x} \, dx$ 22. $\int_{1}^{9} \left(\frac{1}{\sqrt{x}}\right) dx$

23. $\int_{0}^{8} \left(\frac{1}{\sqrt{1+x}}\right) dx$ 24. $\int_{0}^{2} \frac{x}{\sqrt{4+x^2}} \, dx$

25. $\int_{0}^{8} x^{4/3} \, dx$ 26. $\int_{1}^{8} (x^{-2/3} + x^{1/3}) \, dx$

27. $\int_{1}^{2} x^{p/q} \, dx$ 28. $\int_{0}^{1} (a + bx)^{p/q} \, dx$

29. $\int_{2}^{3} \frac{x}{(x^2+1)^2} \, dx$ 30. $\int_{3}^{2} \frac{x}{(x^2+1)^2} \, dx$

31. $\int_{1}^{8} \frac{1 + x^2}{x^{2/3}}$ 32. $\int_{0}^{2} x(a + bx^2)^{p/q} \, dx$

Express each limit 33 to 36 as a definite integral $\int_{0}^{1} f(x) \, dx$ and then evaluate the integral.

33. $\lim_{n \to \infty} \sum_{i=1}^{n} \frac{1}{n^3} (n + i)^2$ 34. $\lim_{n \to \infty} \frac{1}{n} \sum_{i=1}^{n} \sqrt{\frac{n+i}{n}}$

35. $\lim_{n \to \infty} \sum_{i=1}^{n} \frac{n}{(n+i)^2}$ 36. $\lim_{n \to \infty} \frac{3}{n^{3/2}} \sum_{i=1}^{n} \sqrt{2n + 3i}$

PROBLEMS 7.2

1. Let $F(x) = \int_{a}^{x} f(t) \, dt$, where f is continuous. If an interval $\{x : a \leq x \leq b\}$ is subdivided by means of a net $a = x_0 < x_1 < \cdots < x_n = b$, then the mean value theorem concludes that, for each subinterval, there is a point $\bar{x}_i$, $x_{i-1} < \bar{x}_i < x_i$, such that $F(x_i) - F(x_{i-1}) = f(\bar{x}_i)(x_i - x_{i-1})$. Show that there exist Riemann sums approximating $\int_{a}^{b} f(x) \, dx$ arbitrarily close and having value $F(b) - F(a)$ whence this is the value of the integral.

2. The variables x and y are related by the equation
$$x = \int_{0}^{y} \frac{1}{\sqrt{1 + 4t^2}} \, dt$$
Prove that $D^2 y = 4y$.

3. In each of the following cases, the acceleration of a body in rectilinear motion is given, together with some initial conditions. Find both the velocity and the distance from the origin as functions of time.
   (a) $a = 1 + t$; $v = 0$ and $s = 0$ when $t = 1$
   (b) $a = 1 - 1/t^3$; $v = 2$ and $s = 0$ when $t = 1$
   (c) $a = \sqrt{t}$; $v = -2$ and $s = 3$ when $t = 0$
   (d) $a = \sqrt{t} - 1/\sqrt{t}$; $v = 1$ and $s = -4$ when $t = 1$

4. What percent of the acceleration of gravity must the brakes of an automobile generate in order that a car traveling 60 mph can be brought to a halt 200 ft from the point at which the brakes are applied? Redo the problem when the velocity is 90 mph.

7.3 THE INDEFINITE INTEGRAL

Given a function f, any function F having the property that

$$DF(x) = f(x)$$

is called an *antiderivative* of the function f. Theorem 3.36 provides the fundamental property of antiderivatives: *If F and G are two antiderivatives of the same function, then the difference function $G - F$ is constant.*

Theorem 7.2 gives us an explicit antiderivative of a given function F, namely, the function F defined by

$$F(x) = \int_a^x f$$

If G is any other antiderivative of f, then $G - F$ is constant. Hence

$$G(x) = \int_a^x f + C$$

for some constant C. More generally, if F is any fixed antiderivative of the function f, then the set of all antiderivatives is expressible as

$$\{F(x) + C : C \in R\}$$

The set of all antiderivatives of the function f is denoted by the integral sign without limits of integration. We write

$$\int f \quad \text{or} \quad \int f(x)\,dx$$

for this set and call it the *indefinite integral* of f. It is customary to write

$$\int f(x)\,dx = F(x) + C$$

if $F' = f$ to indicate the general form of the antiderivatives of f. This abuse of the symbols has the weight of tradition behind it and should not cause confusion.

Example 1. If $f(x) = 2x$, then $\int 2x\,dx = x^2 + C$, because we have
$$d(x^2 + C) = (2x + 0)\,dx = 2x\,dx$$
This proves the statement.

Example 2. If $f(x) = 3x^2 + 5$, then the indefinite integral is
$$\int (3x^2 + 5)\,dx = x^3 + 5x + C$$
Again we just compute the differential of $F(x) = x^3 + 5x + C$, noting that
$$d(x^3 + 5x + C) = D(x^3 + 5x + C)\,dx = (3x^2 + 5)\,dx \quad \blacksquare$$

Corresponding to each differentiation formula, there is a corresponding formula for indefinite integration. The formulas that follow are the first of many such integration formulas to be found in this book.

1. $\int dx = x + C$
2. $\int kf(x)\,dx = k \int f(x)\,dx$
3. $\int [f(x) + g(x)]\,dx = \int f(x)\,dx + \int g(x)\,dx$
4. If $r \in \mathbf{Q}$ and $r \neq -1$, then
$$\int x^r\,dx = \frac{x^{r+1}}{r+1} + C$$

These are very easily proved, of course, simply by taking differentials. These four also combine to give the formula for the indefinite integral of a polynomial function.

5. $\int (a_0 + a_1 x + \cdots + a_n x^n)\,dx$
$$= C + a_0 x + \frac{a_2}{2} x^2 + \cdots + \frac{a_n}{n+1} x^{n+1}$$

If we know the derivative f' of a function f, then we can determine f "up to an additive constant." That is, we can find a *family* of functions, each differing from f only by an additive constant and each having f' as its derivative. If the graph of one member of this family is drawn, then the graph of any other member may be obtained simply by translating the given curve parallel to the y axis. For example, suppose that $f'(x) = x^2$. Then the family of antiderivatives consists of all functions of the form $f(x) = \frac{1}{3}x^3 + C$ (Formula 4 above). The graphs of $f(x) = \frac{1}{3}x^3, f(x) = \frac{1}{3}x^3 + 1, f(x) = \frac{1}{3}x^3 + 2$ and $f(x) = \frac{1}{3}x^3 - 2$ are pictured in Fig. 7.1. These curves are "parallel" in the sense that, for any point x, the tangent lines at $(x, f_1(x)), (x, f_2(x))$ are parallel for any two functions f_1 and f_2 in the family.

Given any point (x_0, y_0) in the coordinate plane, there is one and only one of these curves that passes through the given point. For if

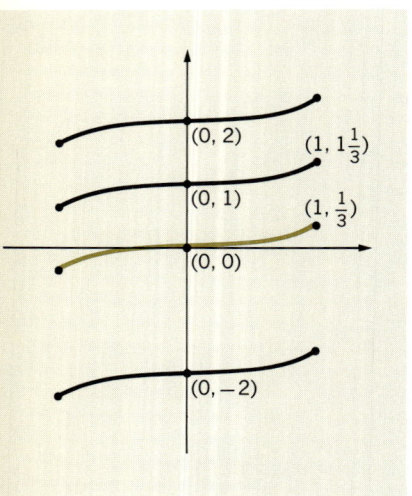

FIGURE 7.1

7.3 THE INDEFINITE INTEGRAL

$f(x_0) = \frac{1}{3}x_0^3 + C$ is to be equal to y_0, then

$$y_0 = \frac{1}{3}x_0^3 + C \quad \text{or} \quad C = y_0 - \frac{1}{3}x_0^3$$

Therefore the particular member of the family of functions,

$$f(x) = \frac{1}{3}x^3 + y_0 - \frac{1}{3}x_0^3$$

is the only one whose graph passes through (x_0, y_0).

The situation above generalizes completely. If we have the family of functions of the form

$$f(x) = \int f'(x)\,dx + C$$

then a particular member is selected by requiring its graph to pass through a given point. In particular, let $F(x)$ be any fixed member of the family. Then the function

$$f(x) = F(x) - F(x_0) + y_0$$

has its graph passing through the point (x_0, y_0).

An equation of the form

$$f'(x) = g(x)$$

is called a *differential equation*. (This is a particularly simple kind of differential equation.) From our discussion we recognize that any function

$$\int g(x)\,dx$$

satisfies the differential equation, which means that the equation does not have a unique solution. The equation *does* determine a family of functions, however. And in order to specify a unique solution, another condition must be given. A condition stating that $f(x_0) = y_0$ is called an *initial condition* and imposing it upon the function f does determine a unique solution of the equation $f'(x) = g(x)$.

■ **Example 3.** A body moving rectilinearly has a velocity

$$v = at$$

where a is a constant and t is the number of seconds of elapsed time. Assume that the distance s is measured from the location of the body at time $t = 0$. Therefore $s = 0$ when $t = 0$. We have the differential equation

$$v = Ds = at$$

together with the initial condition $s(0) = 0$. Then

$$s = \tfrac{1}{2}at^2 + C$$

and, because $s(0) = \frac{1}{2}a(0)^2 + C = C = 0$, the particular function

$$s = \frac{1}{2}at^2$$

satisfies both the differential equation and the initial condition.

A more efficient technique is to write the entire solution in one step

$$s(t) = \int_0^t ax\,dx$$

$$= a\int_0^t x\,dx = a\left(\frac{1}{2}x^2\right)\Big]_0^t = \frac{1}{2}at^2 - \frac{1}{2}a(0)^2 = \frac{1}{2}at^2$$

■ **Example 4.** We derive again the equation of motion of a falling body under the assumption that air resistance is negligible. That is, we assume that a falling body is subjected only to the force of gravity and that this force provides a constant acceleration g. This gives us the differential equation

$$a = Dv = g$$

The general solution of this equation is the family of all functions of the form

$$v(t) = gt + v_0$$

where v_0 is a constant. In particular, v_0 is the velocity of the body at the time $t = 0$.

The solution $v(t)$ above leads us to a second equation

$$Ds = v(t) = gt + v_0$$

which has the general solution

$$s(t) = \frac{1}{2}gt^2 + v_0 t + s_0$$

where s_0 is the distance of the falling body from the origin at time $t = 0$. In a particular application, of course, a coordinate system must be selected before the values of v_0 and s_0 can be assigned. This procedure was exemplified in Section 4.2.

■ **Example 5.** Find the curve whose slope at a point (x, y) is given by $1/x^2$ and which passes through the point $(2, 1)$.

This problem simply requires the particular solution of the differential equation

$$Dy = \frac{1}{x^2}$$

with the initial condition $y(2) = 1$. The solution, by formula 4, is of the form

$$y = -\frac{1}{x} + C$$

Knowing that $y(2) = -\frac{1}{2} + C = 1$, we find that $C = \frac{3}{2}$. Therefore the desired curve has the equation

$$y = -\frac{1}{x} + \frac{3}{2}$$

Again we can write this particular function in one step as

$$y - 1 = \int_2^x \frac{1}{t^2} dt$$

$$= -\frac{1}{t}\Big]_2^x = -\frac{1}{x} - \left(-\frac{1}{2}\right) = -\frac{1}{x} + \frac{1}{2}$$

whence

$$y = -\frac{1}{x} + \frac{3}{2}$$

We shall return to this technique later and then it will be our standard. ■

In the symbol

$$\int f(x)\, dx$$

the function $f(x)$ is the derivative of some function $F(x)$. Therefore the quantity $f(x)\, dx = F'(x)\, dx$ is precisely the differential $dF(x)$ of the function $F(x)$. Our basic equation then becomes

$$\int f(x)\, dx = \int dF(x) = F(x) + C$$

That is, the integration of the differential of a function yields that function plus an arbitrary constant. From this point of view, indefinite integration is the inverse operation of taking differentials.

■ **Example 6.** Suppose we wish to solve the differential equation

$$\frac{dy}{dx} = xy^2, \quad y \neq 0$$

Then we may pass to the differential form

$$dy = xy^2\, dx$$

and hence to

$$\frac{dy}{y^2} = x\, dx$$

Now we have

$$\int \frac{dy}{y^2} = -\frac{1}{y} + C$$

and

$$\int x\, dx = \frac{1}{2}x^2 + D$$

Therefore

$$-\frac{1}{y} + C = \frac{1}{2}x^2 + D$$

$$-\frac{1}{y} = \frac{x^2}{2} + B, \qquad \text{where } B = D - C$$

and thus

$$y = -\frac{1}{\frac{1}{2}x^2 + B}$$

This is the *general solution* of the differential equation.

EXERCISES 7.3 Use formulas 1 to 5 to work out each of the indefinite integrals in Exercises 1 to 24.

1. $\int 3\, dx$
2. $\int x\, dx$
3. $\int (4 - x)\, dx$
4. $\int x^2\, dx$
5. $\int (8 - 2x + x^2)\, dx$
6. $\int (x + x^3)\, dx$
7. $\int (x^2 + x^4)\, dx$
8. $\int (a^4 - 2a^2x^2 - x^4)\, dx$
9. $\int \sqrt{x}\, dx$
10. $\int \frac{dx}{\sqrt{x}}$
11. $\int x^{-2/3}\, dx$
12. $\int (2x^{1/3} + 6x^{-1/3})\, dx$
13. $\int \frac{x+1}{x^3}\, dx$
14. $\int \frac{x^4 + 2}{x^2}\, dx$
15. $\int \frac{x^2 - 1}{\sqrt{x}}\, dx$
16. $\int \sqrt[3]{x}\,(1 - \sqrt[3]{x})\, dx$
17. $\int \frac{x - 1}{\sqrt{x} + 1}\, dx$
18. $\int \frac{x - 1}{\sqrt[3]{x} + 1}\, dx$
19. $\int x\sqrt{1 + x^2}\, dx$
20. $\int x\sqrt{1 + 2x^2}\, dx$

7.3 THE INDEFINITE INTEGRAL

21. $\int \dfrac{x\, dx}{\sqrt{1 + 3x^2}}$

22. $\int x^2(1 + 4x^3)^{-1/3}\, dx$

23. $\int (x + 1)\sqrt{x^2 + 2x - 1}\, dx$

24. $\int \dfrac{(x + 2)\, dx}{\sqrt{3x^2 + 12x + 4}}$

Find the general solution of differential equations 25 to 34.

25. $y' = 2 - x^2$

26. $y' = 2x + x^2$

27. $y' = x^2 + x^{-2}$

28. $y' = \dfrac{3x^2}{y},\ y > 0$

29. $y' = \dfrac{2x}{y^3},\ y > 0$

30. $y' = 4\sqrt{y},\ y > 0$

31. $y' = x\sqrt{y},\ y > 0$

32. $y' = \sqrt{xy},\ x > 0,\ y > 0$

33. $y' = \left(\dfrac{y}{x}\right)^{1/3},\ x > 0,\ y > 0$

34. $y' = (2y + 3)^2,\ y \neq -\tfrac{3}{2}$

Find the particular solution of each differential equation 35 to 40 that satisfies the given initial conditions.

35. $y' = 1 + 2x;\ y(1) = 1$

36. $y' = 2x + x^{-2};\ y(1) = -1$

37. $y' = \dfrac{2x}{3y^2},\ y(0) = 2$

38. $y' = (xy)^{1/3},\ y(0) = 0$

39. $y' = (1 + y)^2,\ y(1) = \tfrac{1}{2}$

40. $y' = xy^2(1 + y^2)^{-1},\ y(0) = 1$ (Do not solve for y in terms of x.)

41. Find a polynomial $P(x)$ such that $P'(x) = x^3 + P(x)$.

42. Find a polynomial $Q(x)$ such that $Q''(x) + cQ'(x) + dQ(x) = x^2$.

PROBLEMS 7.3

1. If $P(x)$ is a quadratic polynomial, then show that
$$\int_a^b P(x)\, dx = \dfrac{1}{6}(b - a)\left[P(a) + 4P\!\left(\dfrac{a + b}{2}\right) + P(b)\right]$$
Show that the same formula holds if $P(x)$ is a cubic polynomial.

2. For a certain structural beam, the deflection y at a distance x from one end satisfies
$$\dfrac{d^2y}{dx^2} = \dfrac{-24Wx}{48E1}$$
where E, 1, and W are constants. Assuming that $y = 0$ when $x = 0$ and that $y = Wl^3/48E1$ when $x = \tfrac{1}{2}l$ (l is the length of the beam), find y as a function of x.

3. The *root mean square* of a function $f(x)$ over an interval $\{x : a \leq x \leq l\}$ is the square root of the average value of $f^2(x)$ over this interval. What is this num-

ber for a linear function $f(x) = mx + b$ over the interval $\{x : 0 \leq x \leq c\}$; over $\{x : -c \leq x \leq c\}$?

4. The attraction exerted by the earth on a particle of mass m at a distance r from the center of the earth is

$$F = -mgR^2 r^{-2}$$

where R is the radius of the earth. (This force is negative because it acts in the direction opposite to increasing r.) Using the basic equation $F = ma$, the acceleration acting on the particle is $a = -gR^2 r^{-2}$. Find r as a function of t if the particle is projected upward from the surface with the initial velocity $v_0 = \sqrt{2gR}$. (Hint: Write $a = v(dv/dr)$ first.)

5. If several volumes $V_1, V_2, \cdots, V_n$ of water at temperatures $T_1, T_2, \cdots, T_n$ are mixed, the temperature of the mixture is the weighted average

$$\frac{\Sigma T_i V_i}{\Sigma V_i} \quad \text{(assuming no heat loss)}$$

Now suppose that water is filling a container and assume that the temperature of the incoming water is a function $T(V)$ of the volume of water already in the container. Let U denote the temperature of the mixture in the container when the volume is V and show that

$$UV = \int_0^V T(u) \, du$$

6. Suppose a body is moving rectilinearly with an acceleration $a = f(s)$, which is a function of the distance s. Prove that

$$v^2 = 2 \int a \, ds$$

7.4 AREA PROBLEMS

Recall that "area" is a function that assigns a nonnegative real number to each of certain regions in the plane. At the end of Section 6.4, we gave the integral definition:

$$A(R) = \int_a^b f(x) \, dx$$

for the area of the region $R = \{(x, y) : a \leq x \leq b, 0 \leq y \leq f(x)\}$. Here we extend this definition to more general regions.

Let f and g be continuous on $\{x : a \leq x \leq b\}$ and suppose that $f(x) \leq g(x)$ for each point x in the interval. Then the area of the region $R = \{(x, y) : a \leq x \leq b, f(x) \leq y \leq g(x)\}$ is defined to be

$$A(R) = \int_a^b [g(x) - f(x)] \, dx$$

Since $g - f$ is nonnegative, Theorem 6.8 tells us that $A(R) \geq 0$, as required. Figure 7.2 shows a typical situation covered by this definition. Note that the region R is bounded above and below by the curves

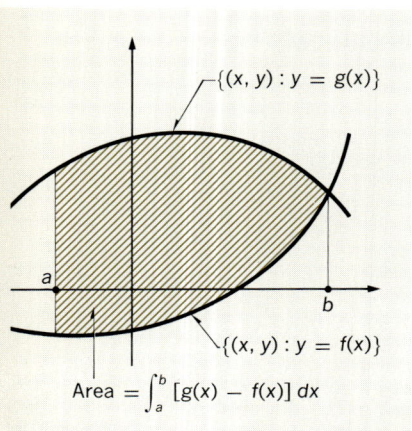

Area $= \int_a^b [g(x) - f(x)] \, dx$

FIGURE 7.2

$\{(x,y) : y = g(x)\}$ and $\{(x,y) : y = f(x)\}$

respectively.

- **Example 1.** The region bounded by the line $\{(x,y) : y = 1 - 2x\}$, the x axis and the two vertical lines $\{(x,y) : x = 1\}$ and $\{(x,y) : x = 3\}$ is shaded in Fig. 7.3. From the figure we see that the region is bounded above by the x axis $\{(x,y) : y = 0\}$ and is bounded below by the line $\{(x,y) : y = 1 - 2x\}$. That is, we have $g(x) = 0$ and $f(x) = 1 - 2x$. Therefore, by definition, the area is given by the definite integral

$$\int_1^3 [0 - (1 - 2x)] \, dx = \int_1^3 (-1 + 2x) \, dx$$

$$= -x + x^2 \Big]_1^3 \quad \text{by formula 5}$$

$$= (-3 + 3^2) - (-1 + 1^2)$$

$$= 6$$

Our result can be checked easily in this simple case. The figure in question is a trapezoid with bases of length 1 and 5 and with height 2. Its area is $\frac{1}{2}h(b_1 + b_2) = \frac{1}{2} \cdot 2(1 + 5) = 6$. The agreement here is a justification of the definition of area as a definite integral.

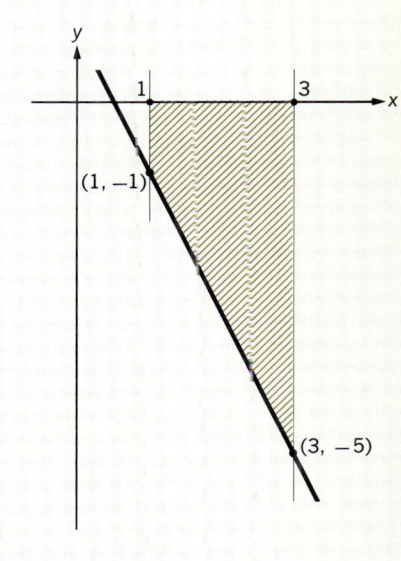

FIGURE 7.3

- **Example 2.** The region bounded by the line $\{(x,y) : y = 2 - x\}$ and the curve $\{(x,y) : y = x^2\}$ is shaded in Fig. 7.4. We note that the region is bounded above by the line $\{(x,y) : y = 2 + x\}$, is bounded below by the curve $\{(x,y) : y = x^2\}$, is bounded on the left by the vertical line $\{(x,y) : x = -1\}$, and is bounded on the right by the vertical line $\{(x,y) : x = 2\}$. The fact that these vertical lines meet the region each in a single point does not violate our definition. It follows that

$$g(x) = 2 + x, \, f(x) = x^2, \qquad a = -1 \text{ and } b = 2$$

Hence the area is

$$\int_{-1}^2 [(2 + x) - x^2] \, dx = \int_{-1}^2 (2 + x - x^2) \, dx$$

$$= 2x + \tfrac{1}{2}x^2 - \tfrac{1}{3}x^3 \Big]_{-1}^2 \quad \text{by formula 5}$$

$$= 2(2) + \tfrac{1}{2}(2)^2 - \tfrac{1}{3}(2)^3$$

$$\quad - [2(-1) + \tfrac{1}{2}(-1)^2 - \tfrac{1}{3}(-1)^3]$$

$$= \tfrac{9}{2}$$

- **Example 3.** The region under the curve $\{(x,y) : y = 1/x^2\}$ above the x axis and to the right of the vertical line $\{(x,y) : x = 1\}$ (see Fig. 7.5) has a finite area even though the region is unbounded in one direction. This claim is substantiated as follows: Let t be any point

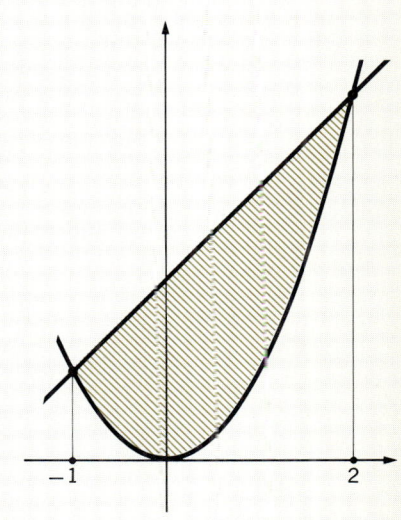

FIGURE 7.4

greater than 1. The region under the curve from 1 to t has an area given by

$$A(t) = \int_1^t \frac{dx}{x^2}$$

Knowing that $D(-1/x) = 1/x^2$, we have

$$A(t) = -\frac{1}{x}\bigg]_1^t = -\frac{1}{t} - \left(-\frac{1}{1}\right) = 1 - \frac{1}{t}$$

Obviously, the area $A(t)$ increases as t increases, but we always have $A(t) < 1$. The only meaningful definition of the total area under the curve is

$$A = \lim_{t \to \infty} A(t) = \lim_{t \to \infty} \left(1 - \frac{1}{t}\right) = 1$$

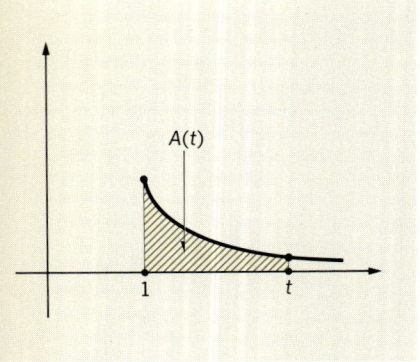

FIGURE 7.5

■ *Example 4.* Even a seemingly easy area problem can lead to an integral that is not easily evaluated. Suppose we consider the region under the curve $\{(x,y) : y = \sqrt{1 + x^3}\}$, above the x axis and between the vertical lines $\{(x,y) : x = 0\}$ and $\{(x,y) : x = 1\}$. Surely the integral

$$\int_0^1 \sqrt{1 + x^3}\, dx$$

exists because the area does, but we cannot evaluate by our antiderivative technique. The function

$$F(x) = \int_0^x \sqrt{1 + t^3}\, dt$$

cannot be expressed in elementary terms. Of course, we do have approximate methods of evaluation that will obtain values of such an integral as accurately as we wish.

We shall study some other geometric applications of the definite integral after looking at two important techniques in evaluating integrals.

EXERCISES 7.4 Find the area of the region bounded by each set of curves 1 to 20. Draw the area.

1. The x axis and $\{(x,y) : y = 4x - x^2\}$
2. The x axis and $\{(x,y) : y = x^2 - x^3\}$
3. $\{(x,y) : y = -3\}$ and $\{(x,y) : y = 2x - x^2\}$
4. $\{(x,y) : y = 2\}$ and $\{(x,y) : y = 11 - x^2\}$
5. $\{(x,y) : y = x\}$ and $\{(x,y) : y = x^2\}$
6. $\{(x,y) : y = \tfrac{1}{2}x\}$ and $\{(x,y) : y = \sqrt{x}\}$
7. $\{(x,y) : x + y = 3\}$ and $\{(x,y) : y = 3x - x^2\}$

8. $\{(x,y) : x - 2y + 4 = 0\}$ and $\{(x,y) : 4y = x^2\}$
9. $\{(x,y) : y = 2(x + 1)\}$ and $\{(x,y) : y = -3 - 2x + x^2\}$
10. $\{(x,y) : y = 5(x - 2)\}$ and $\{(x,y) : y = 8 + 2x - x^2\}$
11. $\{(x,y) : x - y + 7 = 0\}$ and $\{(x,y) : y = 9 - x^2\}$
12. $\{(x,y) : y = 2x^2\}$ and $\{(x,y) : y = x^4 - 2x^2\}$
13. $\{(x,y) : x + 3y = 5\}$ and $\{(x,y) : y = 2(x + 2)^{-1/2}\}$
14. $\{(x,y) : x = -1\}$, $\{(x,y) : y = 1\}$ and $\{(x,y) : y = (2x - 1)^{-2}\}$
15. The x axis, $\{(x,y) : y = (2 - x)^2\}$ and $\{(x,y) : y = x^3\}$
16. $\{(x,y) : x = 4\}$, $\{(x,y) : y = 1\}$ and $\{(x,y) : x^2 y = 1\}$
17. The y axis, $\{(x,y) : x = 2\}$, $\{(x,y) : y = -3\}$ and $\{(x,y) : y = x^3 - 1\}$
18. $\{(x,y) : y = x + 1\}$, $\{(x,y) : y = 2x\}$ and $\{(x,y) : x + y + 3 = 0\}$
19. $\{(x,y) : y = x + 1\}$, $\{(x,y) : x + y = 3\}$ and $\{(x,y) : x - 7y = 11\}$
20. The arc of $\{(x,y) : y = (2x - 3)^2\}$ from $(0, 9)$ to $(\frac{3}{2}, 0)$, the arc of $\{(x,y) : y = 5(2x - 3)\}$ from $(\frac{3}{2}, 0)$ to $(2, 5)$ and the arc of $\{(x,y) : y = 9 - x^2\}$ from $(2, 5)$ to $(0, 9)$.

PROBLEMS 7.4

1. The region bounded by $\{(x,y) : y = 4\}$ and $\{(x,y) : y = x^2\}$ is divided into equal parts by the line $\{(x,y) : y = c\}$. What is c?

2. The rectangle with vertices $(0, 0)$, $(0, b)$, (a, b) and $(a, 0)$ is divided into two parts by the curve $\{(x,y) : a^n y = bx^n; n \in \mathbf{N}\}$. What is the ratio of the areas of these parts?

3. For each $n \in \mathbf{N}$, let $f(n)$ denote the area of the region bounded by the curves $\{(x,y) : y = x^n\}$ and $\{(x,y) : y = x^{n+1}\}$. Determine the sum

$$\sum_{i=1}^{n} f(i)$$

What does this sum approximate when n is large? Give a geometric interpretation of your answer.

4. Assume that $f(x)$ is continuous on $\{x : a \leq x \leq b\}$. What is the area of the region

$$\{(x,y) : a \leq x \leq b, \tfrac{1}{2}[f(x) - |f(x)|] \leq y \leq \tfrac{1}{2}[f(x) + |f(x)|]\}$$

7.5 SUBSTITUTION OF VARIABLES

If the function $F(x)$ is differentiable, then we have the indefinite integral

$$\int F'(x)\, dx = F(x) + C$$

The differential $F'(x)\, dx$ can often be expressed in terms of another variable as an aid in finding the indefinite integral. The theory here is as follows:

Suppose that $F(x)$ is expressed as a composite function
$$F(x) = f(u(x))$$
By the chain rule of differentiation, we have
$$DF(x) = Df(u(x)) \cdot Du(x)$$
Therefore the differential $F'(x)\,dx$ may be written as $Df(u) \cdot Du \cdot dx$, or because $Du \cdot dx = du$, we have
$$F'(x)\,dx = Df(u)\,du = f'(u)\,du$$
It follows that we may write
$$\int F'(x)\,dx = \int f'(u)\,du = f(u) + C = f(u(x)) + C = F(x) + C$$

If we *can* determine the indefinite integral $\int f'(u)\,du$, then $f(u(x)) + C$ is the indefinite integral we wanted. This procedure of integration by means of a *substitution of variables* is best introduced by some examples.

■ **Example 1.** Consider the indefinite integral
$$\int \sqrt{x+6}\,dx$$
Think of the function $u = x + 6$ with $du = dx$. Then the differential $\sqrt{x+6}\,dx$ can be written as $u^{1/2}\,du$. From formula 4 we have
$$\int u^{1/2}\,du = \frac{2}{3}u^{3/2} + C$$
Resubstituting $x + 6$ for u, we obtain
$$\int \sqrt{x+6}\,dx = \frac{2}{3}(x+6)^{3/2} + C$$

■ **Example 2.** Consider the indefinite integral
$$\int (2 + 3x)^{3/2}\,dx$$
If we let $u = 2 + 3x$, then $3\,dx = du$ or $dx = \tfrac{1}{3}\,du$. Making these substitutions, we obtain
$$\int (2 + 3x)^{3/2}\,dx = \int u^{3/2}\left(\frac{1}{3}\,du\right)$$
$$= \frac{1}{3}\int u^{3/2}\,du = \frac{1}{3}\left(\frac{2}{5}u^{5/2}\right) + C$$
$$= \frac{2}{15}u^{5/2} + C$$

Resubstituting $2 + 3x$ for u, we find that

$$\int (2 + 3x)^{3/2}\, dx = \frac{2}{15}(2 + 3x)^{5/2} + C$$

- **Example 3.** If $p \neq -1$, then

$$\int (a + bx)^p\, dx = \frac{1}{b(p + 1)}(a + bx)^{p+1} + C$$

The proof of this statement is made by setting $u = a + bx$ whence $du = b\, dx$ or $dx = 1/b\, du$. Then

$$\int (a + bx)^p\, dx = \int u^p\left(\frac{1}{b}\, du\right) = \frac{1}{b}\int u^p\, du$$

$$= \frac{1}{b}\left(\frac{1}{p - 1} u^{p+1}\right) + C$$

$$= \frac{1}{b(p + 1)}(a + bx)^{p+1} + C$$

- **Example 4.** If $p \neq -1$, then

$$\int x(a + bx^2)^p\, dx = \frac{1}{2b(p + 1)}(a + bx^2)^{p+1} + C$$

Again we set $u = a + bx^2$ whence $du = 2bx\, dx$ or $x\, dx = \frac{1}{2b}\, du$. The integral is

$$\int x(a + bx^2)^p\, dx = \int (a + bx^2)^p (x\, dx)$$

$$= \int u^p\left(\frac{1}{2b}\, du\right)$$

$$= \frac{1}{2b}\int u^p\, du = \frac{1}{2b}\left(\frac{1}{p + 1} u^{p+1}\right) + C$$

$$= \frac{1}{2b(p + 1)}(a + bx^2)^{p+1} + C$$

- **Example 5.** We show that

$$\int (x + 2)\sqrt[3]{1 + 4x + x^2}\, dx = \frac{3}{8}(1 + 4x + x^2)^{4/3} + C$$

Set $1 + 4x + x^2 = u$. Then $(4 + 2x)\, dx = du$ or $(x + 2)\, dx = \frac{1}{2}\, du$. Using these substitutions, we obtain

$$\int (x+2)\sqrt[3]{1+4x+x^2}\,dx = \int (1+4x+x^2)^{1/3}(x+2)\,dx$$

$$= \int u^{1/3}\left(\frac{1}{2}\,du\right)$$

$$= \frac{1}{2}\left(\frac{3}{4}u^{4/3}\right) + C$$

$$= \frac{3}{8}(1+4x+x^2)^{4/3} + C$$

■ **Example 6.** Consider the indefinite integral

$$\int \frac{1+3\sqrt{t}}{\sqrt{t}}\,dt$$

By setting $1 + 3\sqrt{t} = u$, we have

$$3\left(\frac{1}{2t^{-1/2}}\right)dt = \frac{3}{2}\frac{dt}{\sqrt{t}} = du \quad \text{or} \quad \frac{dt}{\sqrt{t}} = \frac{2}{3}\,du$$

With these substitutions, we may write the integral as

$$\int \frac{1+3\sqrt{t}}{\sqrt{t}}\,dt = \int (1+3\sqrt{t})\frac{dt}{\sqrt{t}}$$

$$= \int u\left(\frac{2}{3}\,du\right)$$

$$= \frac{2}{3}\int u\,du$$

$$= \frac{2}{3}\left(\frac{1}{2}u^2\right) + C$$

$$= \frac{1}{3}(1+3\sqrt{t})^2 + C$$

(Note: This integration can be done in another way. The reader should do so and then compare his result with that above.) ■

According to our theory,

$$\int_a^b f(x)\,dx = F(b) - F(a)$$

where F is any function such that $F'(x) = f(x)$. If we use a substitution of variable to find $F(x)$, it is not necessary to resubstitute the variable x in order to evaluate the definite integral. The next example is worked out in two ways, the second of which avoids resubstitution.

■ **Example 7.** Suppose we want to evaluate

$$\int_1^2 (2 + 3x)^{3/2} \, dx$$

By Example 2, we have

$$\int (2 + 3x)^{3/2} \, dx = \frac{2}{15}(2 + 3x)^{5/2} + C$$

This means that we may take $F(x) = \frac{2}{15}(2 + 3x)^{5/2}$ and then

$$\int_1^2 (2 + 3x)^{3/2} \, dx = \frac{2}{15}(2 + 3x)^{5/2} \Big]_1^2$$

$$= \frac{2}{15}(2 + 6)^{5/2} - \frac{2}{15}(2 + 3)^{5/2}$$

$$= \frac{2}{15}(8^{5/2} - 5^{5/2})$$

But let us return to the substitution $u = 2 + 3x$, $\frac{1}{3} du = dx$, used in Example 2. Then when $x = 2$, we have $u = 2 + 3(2) = 8$ and when $x = 1$, $u = 2 + 3(1) = 5$. Then we write

$$\int_1^2 (2 + 3x)^{3/2} \, dx = \int_5^8 u^{3/2} \left(\frac{1}{3} du\right) = \frac{1}{3} \int_5^8 u^{3/2} \, du = \frac{2}{15} u^{5/2} \Big]_5^8$$

$$= \frac{2}{15}(8^{5/2} - 5^{5/2})$$

We justify the second method in Example 7 with the following results.

■ **THEOREM 7.4.** Let f and u be functions satisfying the following conditions:

1. f is continuous on a domain including the closed interval $\{x : a \leq x \leq b\}$,
2. for each point t in the closed interval $\{t : \alpha \leq t \leq \beta\}$, the value $u(t)$ is a point in $\{x : a \leq x \leq b\}$,
3. $u(\alpha) = a$ and $u(\beta) = b$,
4. u' is continuous on $\{t : \alpha \leq t \leq \beta\}$.

Then

$$\int_a^b f(x) \, dx = \int_\alpha^\beta f(u(t)) \cdot u'(t) \, dt$$

PROOF: Define the functions

$$F(z) = \int_a^z f(x) \, dx, \quad G(w) = \int_\alpha^w f(u(t)) \cdot u'(t) \, dt$$

Obviously $F(a) = 0$ and $G(\alpha) = 0$. We show that $F(b) = G(\beta)$. Our conditions surely imply that these integrals exist and also ensure the differentiability of F and G. Indeed,

$$F'(z) = f(z) \quad \text{and} \quad G'(w) = f(u(w)) \cdot u'(w)$$

By the chain rule of differentiation, $F'(u(w)) = f(u(w)) \cdot u'(w)$ from which we conclude that

$$G'(w) = F'(u(w))$$

This tells us that G and F differ by at most a constant, but because they have the same value $G(\alpha) = 0 = F(u(\alpha)) = F(a) = 0$, the two functions agree everywhere. Therefore $G(\beta) = F(u(\beta)) = F(b)$. ∎

EXERCISES 7.5 Evaluate the indefinite integrals 1 to 20 by using the appropriate substitution.

1. $\int 2\sqrt{2x + 1}\, dx$

2. $\int \dfrac{dx}{\sqrt{2x + 1}}$

3. $\int \sqrt{1 - 3x}\, dx$

4. $\int \dfrac{dx}{\sqrt{4x + 1}}$

5. $\int 2x(1 + x^2)^2\, dx$

6. $\int \dfrac{x\, dx}{(2 + x^2)^2}$

7. $\int \dfrac{x\, dx}{\sqrt{1 + x^2}}$

8. $\int x\sqrt[3]{1 + x^2}\, dx$

9. $\int \dfrac{2x\, dx}{(4 + x^2)^{3/2}}$

10. $\int \dfrac{x\, dx}{\sqrt{1 + 8x^2}}$

11. $\int \dfrac{dx}{(4x - 1)^2}$

12. $\int (1 - x)\sqrt{2x - x^2}\, dx$

13. $\int \dfrac{(x + 2)\, dx}{\sqrt{x^2 + 4x + 1}}$

14. $\int (x + 1)(x^2 + 2x - 3)^{-1/3}\, dx$

15. $\int x(a + bx^2)^{p/q}\, dx$

16. $\int x^{2/3}(1 + x^{5/3})^{1/3}\, dx$

17. $\int \dfrac{x\, dx}{\sqrt{a^2 - x^2}}$

18. $\int x^{-2}\sqrt{1 - 1/x}\, dx$

19. $\int \dfrac{\sqrt{1 + \sqrt{x}}}{\sqrt{x}}$

20. $\int \sqrt{x}\sqrt{1 + x\sqrt{x}}\, dx$

Evaluate the definite integrals 21 to 30 both by resubstitution and by avoiding resubstitution (as in Example 7).

21. $\int_0^3 \sqrt{x + 1}\, dx$

22. $\int_0^4 \sqrt{2x + 1}\, dx$

23. $\int_{-1}^0 \sqrt{1 - 8x}\, dx$

24. $\int_4^{12} \dfrac{dx}{\sqrt{2x + 1}}$

25. $\int_0^{1/2} x(1-x^2)^{-2}\,dx$

26. $\int_0^{\sqrt{3}} \dfrac{x\,dx}{\sqrt{1+x^2}}$

27. $\int_1^2 \dfrac{x\,dx}{\sqrt{1+8x^2}}$

28. $\int_0^2 (2-x)\sqrt{4x-x^2}\,dx$

29. $\int_0^1 x^{1/3}(1+x^{4/3})^{-1/2}\,dx$

30. $\int_1^{5/3} \dfrac{\sqrt{1-1/x^2}}{x^3}\,dx$

7.6 INTEGRATION BY PARTS

The product law of differentiation leads to a very useful technique for integration. The fundamental theorem of calculus stated for indefinite integrals is

$$\int F'(x)\,dx = F(x) - C$$

Suppose that $F(x) = u(x)v(x)$ is a product of differentiable functions $u(x)$ and $v(x)$. Then we use the product law of differentiation to write

$$F'(x) = [u(x)v(x)]' = u(x)v'(x) + u'(x)v(x)$$

The indefinite integral is then

$$\int [u(x)v'(x) + v(x)u'(x)]\,dx = u(x)v(x) + C$$

or

$$\int u(x)v'(x)\,dx + \int v(x)u'(x)\,dx = u(x)v(x) + C$$

or

$$\int u(x)v'(x)\,dx = u(x)v(x) - \int v(x)u'(x)\,dx + C$$

This last is our formula for *integration by parts*.

Suppose that we are faced with an integral $\int f(x)\,dx$ which does not respond to a known substitution of variables method. We attempt to express the differential $f'(x)\,dx$ in the form $u(x)v'(x)\,dx$ and to do so in a way that makes the integral $\int v(x)u'(x)\,dx$ easier. Then the integration by parts formula applies to express the difficult integral $\int u(x)v'(x)\,dx$ as $u(x)v(x) - \int v(x)u'(x)\,dx + C$. If we have chosen $u(x)$ and $v(x)$ cleverly, then we will be done.

■ **Example 1.** Faced with the integral

$$\int x\sqrt{1+x}\,dx$$

there are just two choices we could make. Either we take $u(x) = x$ and $v'(x) = (1+x)^{1/2}$ or we take $u(x) = \sqrt{1+x}$ and $v'(x) = x$. Let us make the first choice first, $u(x) = x$ and $v'(x) = (1+x)^{1/2}$. Then $u'(x) = 1$

and $v(x) = \frac{2}{3}(1 + x)^{3/2}$, as was shown in Example 3 of Section 7.5. With these choices, we may write

$$\int x\sqrt{x + 1}\, dx = x\left[\frac{2}{3}(1 + x)^{3/2}\right] - \int \frac{2}{3}(1 + x)^{3/2}(1)\, dx + C$$

$$= \frac{2}{3}x(1 + x)^{3/2} - \frac{2}{3}\int (1 + x)^{3/2}\, dx + C$$

$$= \frac{2}{3}x(1 + x)^{3/2} - \frac{2}{3}\left[\frac{2}{5}(1 + x)^{5/2}\right] + C$$

$$= \frac{2}{3}x(1 + x)^{3/2} - \frac{4}{15}(1 + x)^{5/2} + C$$

On the other hand, suppose we had used the second possibility, $u(x) = \sqrt{1 + x}$ and $v'(x) = x$. Then $u'(x) = \frac{1}{2}(1 + x)^{-1/2}$ and $v(x) = \frac{1}{2}x^2$. Using the integration by parts formula, we obtain

$$\int x\sqrt{1 + x}\, dx = (1 + x)^{1/2}\left(\frac{1}{2}x^2\right)$$

$$- \int \left(\frac{1}{2}x^2\right)\left(\frac{1}{2}(1 + x)^{-1/2}\right) dx + C$$

$$= \frac{1}{2}x^2(1 + x)^{1/2} - \frac{1}{4}\int x^2(1 + x)^{-1/2}\, dx + C$$

But now we have an even harder integral to determine. That is, this particular choice of u and v' only complicates the problem. The moral here is to try again if the first choice does not help.

■ **Example 2.** Consider the integral

$$\int \frac{x^2}{\sqrt{1 + x}}\, dx$$

There are several possible choices for $u(x)$ and $v'(x)$. One that works is to set $u(x) = x^2$ and $v'(x) = (1 + x)^{-1/2}$. Then $u'(x) = 2x$ and $v(x) = 2(1 + x)^{1/2}$ (again Example 3 of Section 7.5 was used to obtain $v(x)$). Integrating by parts, we get

$$\int \frac{x^2}{\sqrt{1 + x}}\, dx = x^2(2(1 + x)^{1/2}) - \int 2(1 + x)^{1/2}(2x)\, dx + C$$

$$= 2x^2(1 + x)^{1/2} - 4\int x(1 + x)^{1/2}\, dx + C$$

From Example 1 above, the integral

$$\int x(1+x)^{1/2}\,dx = \frac{2}{3}x(1+x)^{3/2} - \frac{4}{15}(1+x)^{5/2} + C$$

Therefore

$$\int \frac{x^2}{\sqrt{1+x}}\,dx = 2x^2(1+x)^{1/2} - \frac{8}{3}x(1+x)^{3/2}$$
$$+ \frac{16}{15}(1+x)^{5/2} + C \; \blacksquare$$

From these examples, we see that integration by parts replaces the problem of determining

$$\int u(x)v'(x)\,dx$$

by the two integrations

$$\int v'(x)\,dx, \quad \int v(x)u'(x)\,dx$$

(It is no *problem* to find $u'(x)$.) In applying this technique, then, we must choose $u(x)$ and $v'(x)$ so that we can determine these two integrals. It is often best to make the simplest possible (nontrivial) choice for $u(x)$.

Integration by parts can be applied directly to definite integrals. The theory is as follows: From the fundamental theorem of calculus,

$$\int_a^b [u(x)v(x)]'\,dx = u(b)v(b) - u(a)v(a)$$

Again $[u(x)v(x)]' = u(x)v'(x) + v(x)u'(x)$, and hence we may write

$$\int_a^b [u(x)v'(x) + v(x)u'(x)]\,dx = \int_a^b u(x)v'(x)\,dx$$
$$+ \int_a^b v(x)u'(x)\,dx$$
$$= u(b)v(b) - u(a)v(a)$$

or

$$\int_a^b u(x)v'(x)\,dx = u(b)v(b) - u(a)v(a) - \int_a^b v(x)u'(x)\,dx$$

■ *Example 3.* We evaluate the definite integral

$$\int_0^1 x\sqrt{1+x}\,dx$$

First, from Example 1 above we know that $\frac{2}{3}x(1+x)^{3/2} - \frac{4}{15}(1-x)^{5/2}$ is an antiderivative of $x(1+x)^{1/2}$. Hence

$$\int_0^1 x(1+x)^{1/2}\,dx = \frac{2}{3}x(1+x)^{3/2} - \frac{4}{15}(1+x)^{5/2}\Big]_0^1$$
$$= \frac{2}{3}\cdot 1(1+1)^{3/2} - \frac{4}{15}(1+1)^{5/2}$$
$$- \left(\frac{2}{3}\cdot 0(1+0)^{3/2} - \frac{4}{15}(1+0)^{5/2}\right)$$
$$= \frac{4}{15}(1+\sqrt{2})$$

We can also write

$$\int_0^1 x(1+x)^{1/2}\,dx = \frac{2}{3}x(1+x)^{3/2}\Big]_0^1 - \frac{2}{3}\int_0^1 (1+x)^{3/2}\,dx$$
$$= \frac{1}{3}\cdot 2^{5/2} - \frac{2}{3}\int_0^1 (1+x)^{3/2}\,dx$$

Then the remaining integral can be evaluated to obtain the same result.

■ *Example 4.* Let p and q be positive integers and consider

$$\int_0^1 x^p(1-x)^q\,dx$$

Let $u(x) = x^p$ and $v'(x) = (1-x)^q$. It follows that $u'(x) = px^{p-1}$ and that $v(x) = -(1-x)^{q+1}/(q+1)$ (to do the latter, use a substitution as in Section 7.5). According to our integration by parts formula,

$$\int_0^1 x^p(1-x)^q\,dx = -\frac{x^p(1-x)^{q+1}}{q+1}\Big]_0^1$$
$$- \int_0^1 -\frac{(1-x)^{q-1}}{q+1}(px^{p-1})\,dx$$
$$= 0 + \frac{p}{q+1}\int_0^1 x^{p-1}(1-x)^{q+1}\,dx$$

The identity

$$\int_0^1 x^p(1-x)^q\,dx = \frac{p}{q+1}\int_0^1 x^{p-1}(1-x)^{q+1}\,dx$$

can be used to evaluate the integral on the left. (See Problem 7.6.2.)

■ *Example 5.* Consider

$$\int_0^3 x\sqrt{25-3x}\,dx$$

7.6 INTEGRATION BY PARTS

The successful choice of $u(x)$ and $v'(x)$ is to take $u(x) = x$ and $v'(x) = (25 - 3x)^{1/2}$. Then $u'(x) = 1$ and $v(x) = -\frac{2}{9}(25 - 3x)^{3/2}$ (Example 3 of Section 7.5 again).

Our integration by parts formula yields

$$\int_0^3 x(25 - 3x)^{1/2} \, dx = x\left[-\frac{2}{9}(25 - 3x)^{3/2}\right]_0^3$$

$$- \int_0^3 -\frac{2}{9}(25 - 3x)^{3/2} \, dx$$

$$= -\frac{2}{9}x(25 - 3x)^{3/2}\Big]_0^3$$

$$+ \frac{2}{9}\int_0^3 (25 - 3x)^{3/2} \, dx$$

$$= -\frac{128}{3} + \frac{2}{9}\int_0^3 (25 - 3x)^{3/2} \, dx$$

To evaluate the remaining integral, let $w(x) = 25 - 3x$ whence $dx = -\frac{1}{3}\, dw$. When $x = 0$, $w = 25$; when $x = 3$, $w = 16$. Therefore

$$\int_0^3 (25 - 3x)^{3/2} \, dx = \int_{25}^{16} w^{3/2}\left(-\frac{1}{3}\, dw\right) = -\frac{1}{3}\int_{25}^{16} w^{3/2}\, dw$$

$$= -\frac{1}{3}\left(\frac{2}{15}w^{5/2}\right)\Big]_{25}^{16}$$

$$= -\frac{2}{15}(16)^{5/2} - \left(-\frac{2}{15}(25)^{5/2}\right)$$

$$= -\frac{2043}{15} + \frac{6250}{15} = \frac{4202}{15}$$

Therefore our integral has the value

$$-\frac{128}{3} + \frac{2}{9}\left(\frac{4202}{15}\right) = \frac{2644}{135}$$

Only a few exercises and problems are given here, but there will be many more in Chapters 8 and 10.

EXERCISES 7.6

Use integration by parts to evaluate integrals 1 to 10.

1. $\int \dfrac{x \, dx}{\sqrt{1 - x}}$

2. $\int x(x + 2)^5 \, dx$

3. $\int x\sqrt{x - 1} \, dx$

4. $\int \dfrac{x \, dx}{\sqrt{3x + 1}}$

5. $\int (x - 1)^2(x - 2)^8 \, dx$

6. $\int x^3\sqrt{x^2 + 2} \, dx$

7. $\displaystyle\int_2^4 \frac{x\,dx}{\sqrt{x-1}}$

8. $\displaystyle\int_0^3 x^2\sqrt{1+x}\,dx$

9. $\displaystyle\int_0^4 \frac{x^2\,dx}{(1+x)^{1/2}}$

10. $\displaystyle\int_0^a x\sqrt{a-x}\,dx$

PROBLEMS 7.6

1. Find the quadratic polynomial $P(x)$ such that $P(0) = 1$, $P'(0) = 0$ and the indefinite integral
$$\int \frac{P(x)\,dx}{x^3(1-x)^2}$$
is a rational function.

2. Use the reduction formula of Example 4 to evaluate
$$\int_0^1 x^4(1-x)^7\,dx.$$

3. Assuming that $f''(x) = -af(x)$ and that $g''(x) = bg(x)$ where a and b are constants, determine
$$\int f(x)g(x)\,dx.$$

4. Assuming that $\displaystyle\int_0^a \sqrt{a^2-t^2}\,dt = \frac{1}{4}\pi a^2$, evaluate
$$\int_0^a (a^2-t^2)^{3/2}\,dt.$$

5. Establish the formulas and tell when they are valid:

   (a) $\displaystyle\int_0^a x^2 f'''(x)\,dx = a^2 f''(a) - 2af'(a) + 2f(a) - 2f(0)$

   (b) $\displaystyle\int_a^b f(x)g''(x)\,dx = f(x)g'(x) - f'(x)g(x)\Big|_a^b + \int_a^b f''(x)g(x)\,dx$

6. Assume that $f(x)$, $g(x)$, and $g'(x)$ are continuous on $\{x : a \leq x \leq b\}$ and that $g'(x)$ does not change sign in this interval. Prove that there is a point x_0, $a < x_0 < b$, such that
$$\int_a^b f(x)g(x)\,dx = g(a)\int_a^{x_0} f(x)\,dx + g(b)\int_{x_0}^b f(x)\,dx.$$

(Hint: This is easy enough if you have another form of the mean value theorem for integrals, namely,
$$\int_a^b f(x)g(x)\,dx = f(x_0)\int_a^b g(x)\,dx$$
for some point x_0, $a < x_0 < b$.)

7. Prove the formula
$$\int_0^x \left(\int_0^u f(t)\,dt\right) du = \int_0^x f(u)(x-u)\,du.$$

and state the conditions under which you are sure it holds.

8. Establish the reduction formula, valid for $n > 1$;

$$\int \frac{dx}{(a^2 + x^2)^n} = \frac{x}{2a^2(n-1)(a^2 + x^2)^{n-1}} + \frac{2n - 3}{2a^2(n-1)} \int \frac{dx}{(a^2 + x^2)^{n-1}}$$

7.7 IMPROPER INTEGRALS

Let f be continuous on a closed ray $\{x : a \leq x\}$. Then, for each $x \geq a$, the function

$$F(x) = \int_a^x f(t)\, dt$$

is defined. We define *an improper integral of the first kind* by

$$\int_a^\infty f(x)\, dx = \lim_{x \to \infty} \int_a^x f(t)\, dt$$

if this limit exists (is finite). If the limit does exist, we say that the improper integral is *convergent*. If the limit fails to exist for any reason, then the integral is said to be divergent.

As an illustration, suppose that $f(x) > 0$ for all $x \geq a$. Then the quantity

$$F(b) = \int_a^b f(t)\, dt$$

is the area under the curve $\{(x,y) : y = f(x)\}$, above the x axis and between the vertical lines $\{(x,y) : x = a\}$, $\{(x,y) : x = b\}$. In this case, then, the improper integral has the interpretation of an area of an unbounded region. A typical situation is shown in Fig. 7.6 (see also Fig. 7.5).

The next result provides us with a family of functions, each of which has a convergent improper integral of the first kind.

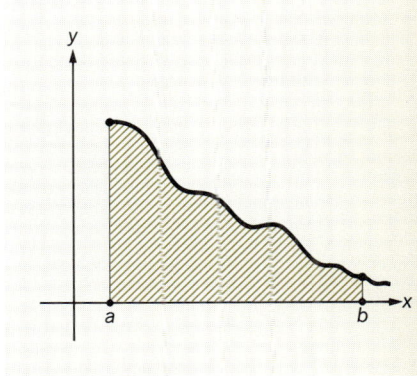

FIGURE 7.6

■ **THEOREM 7.5.** Let $f(x) = x^{-p}$, $p \in \mathbb{Q}$. Then the improper integral

$$\int_1^\infty x^{-p}\, dx$$

converges if $p > 1$ and diverges if $p \leq 1$.

PROOF: If $p \neq 1$, then an antiderivative of x^{-p} is $1/(1-p)x^{1-p}$. Therefore

$$F(x) = \int_1^x t^{-p}\, dt = \frac{t^{1-p}}{1-p} \bigg]_1^x = \frac{x^{1-p}}{1-p} - \frac{1}{1-p}$$

and

$$\lim_{x \to \infty} F(x) = \lim_{x \to \infty} \left(\frac{x^{1-p}}{1-p} - \frac{1}{1-p} \right) = \frac{1}{1-p} \lim_{x \to \infty} x^{1-p} + \frac{1}{p-1}$$

Now if $p > 1$, then $1 - p < 0$ and
$$\lim_{x \to \infty} x^{1-p} = 0$$
Thus, if $p > 1$,
$$\int_1^\infty x^{-p}\, dx = \frac{1}{p-1}$$
On the other hand, if $p < 1$, then $1 - p > 0$ and $\lim_{x \to \infty} x^{1-p} = \infty$, that is, the limit fails to exist whence the integral is divergent.

Because we cannot yet integrate $\int_1^x dt/t$, we shall have to omit the case where $p = 1$. It will be treated in Section 8.7. ∎

Just as for infinite series, we have a *comparison test for improper integrals.*

■ **THEOREM 7.6.** Let f and g be continuous and nonnegative on the closed ray $\{x : a \leq x\}$ and suppose that $f(x) \leq g(x)$ for all $x \geq a$. If the improper integral $\int_a^\infty g(x)\, dx$ converges, so does $\int_a^\infty f(x)\, dx$. Conversely, if $\int_a^\infty f(x)\, dx$ diverges, so does $\int_a^\infty g(x)\, dx$.

PROOF: Because $f(x) \leq g(x)$, Theorem 6.9 tells us that
$$0 \leq F(x) = \int_a^x f(x)\, dx \leq G(x) = \int_a^x g(x)\, dx$$
It follows that, if $\lim_{x \to \infty} G(x) < \infty$, then $\lim_{x \to \infty} F(x) < \infty$, while if $\lim_{x \to \infty} F(x) = \infty$, then $\lim_{x \to \infty} G(x) = \infty$. ∎

■ *Example 1.* Consider the improper integral
$$\int_2^\infty \frac{dx}{\sqrt{x-1}}$$
The function $1/\sqrt{x-1}$ is very much like the function $x^{-1/2}$ when x is large. So we suspect that the integral is divergent. The proof consists in noting that
$$0 \leq \frac{1}{x^{1/2}} < \frac{1}{\sqrt{x-1}}$$
for all $x \geq 2$. We know from Theorem 7.5 that $\int_2^\infty x^{-1/2}\, dx$ is divergent; hence so is the given integral. ∎

Another type of improper integral occurs when the function f is continuous on a half-open interval $\{x : a \leq x < b\}$ and $\lim_{h \to 0} f(b - h^2) = \infty$. For any point x in the half-open interval, the integral
$$F(x) = \int_a^x f(t)\, dt$$

is defined. Then an *improper integral of the second kind* is defined to be

$$\int_a^b f(x)\,dx = \lim_{h \to 0} F(b - h^2) = \lim_{h \to 0} \int_a^{b-h^2} f(x)\,dx$$

Again the improper integral is convergent if this limit is finite and is divergent if the limit fails to exist.

- **Example 2.** Let $f(x) = 1/\sqrt{1-x}$. Then f is continuous on $\{x : 0 \leq x < 1\}$ and

$$\lim_{h \to 0} f(1 - h^2) = \lim_{h \to 0} \frac{1}{\sqrt{1 - (1 - h^2)}} = \lim_{h \to 0} \frac{1}{|h|} = \infty$$

Therefore

$$\int_0^1 \frac{dx}{\sqrt{1-x}} = \lim_{h \to 0} \int_a^{b-h^2} \frac{dt}{\sqrt{1-t}}$$

is an improper integral of the second kind. We easily verify that $D(-2\sqrt{1-x}) = 1/\sqrt{1-x}$. Hence

$$\int_0^x \frac{dt}{\sqrt{1-t}} = -2\sqrt{1-t}\Big]_0^x = -2\sqrt{1-x} - (-2\sqrt{1-0})$$

$$= 2 - 2\sqrt{1-x}$$

The given integral is convergent because

$$\lim_{h \to 0} \int_0^{1-h^2} \frac{dt}{\sqrt{1-t}} = \lim_{h \to 0} [2 - 2\sqrt{1 - (1 - h^2)}]$$

$$= \lim_{h \to 0} [2 - 2|h|] = 2$$

- **Example 3.** The integral $\int_{-1}^1 x^{-2/3}\,dx$ is also improper. Note that the function $x^{-2/3}$ does not exist at $x = 0$. We use this symbol to denote the following sums of one-sided limits

$$\int_{-1}^1 x^{-2/3}\,dx = \lim_{x \to 0^-} \int_{-1}^x x^{-2/3}\,dx + \lim_{x \to 0^+} \int_x^1 x^{-2/3}\,dx$$

Since $\int x^{-2/3}\,dx = 3x^{1/3} + C$, we see that

$$\int_{-1}^x x^{-2/3}\,dx = 3x^{1/3} + 3, \quad \int_x^1 x^{-2/3}\,dx = 3 - 3x^{1/3}$$

Therefore

$$\lim_{x \to 0^-} (3x^{1/3} + 3) = 3 \quad \text{and} \quad \lim_{x \to 0^+} (3 - 3x^{1/3}) = 3$$

and our improper integral has the value 6.

One of the easier applications of an improper integral is the following *integral test* for the convergence of infinite series of positive terms.

■ **THEOREM 7.7.** Let f be a continuous, positive, nonincreasing function on the closed ray $\{x : 1 \leq x\}$. The infinite series $\sum_{n=1}^{\infty} f(n)$ converges if and only if the improper integral $\int_1^{\infty} f(x)\, dx$ converges.

PROOF: Whenever $i \leq x \leq i + 1$, we have $f(i+1) \leq f(x) \leq f(i)$; this implies that $f(i+1) \leq \int_i^{i+1} f(x)\, dx \leq f(i)$ for each $i \in \mathbf{N}$. Thus it follows that

$$\sum_{i=1}^{n-1} f(i+1) \leq \sum_{i=1}^{n-1} \int_i^{i+1} f(x)\, dx \leq \sum_{i=1}^{n-1} f(i)$$

or, if S_n denotes the nth partial sum of the series, we have

$$S_n - f(1) \leq \int_1^n f(x)\, dx \leq S_n - f(n) < S_n$$

From this we also have

$$\int_1^n f(x)\, dx \leq S_n \leq f(1) + \int_1^n f(x)\, dx$$

From these two inequalities we see that, if the partial sums S_n are bounded, then the integral exists and conversely. ■

A very useful comparison series is the so-called *p-series*

$$1 + \frac{1}{2^p} + \frac{1}{3^p} + \cdots + \frac{1}{n^p} + \cdots$$

■ **COROLLARY 7.8.** The *p*-series

$$\sum_{n=1}^{\infty} \frac{1}{n^p}$$

converges if $p > 1$ and diverges if $p \leq 1$.

PROOF: If $p = 1$, this is the divergent harmonic series. All other cases follow immediately from Theorems 7.5 and 7.7.

■ *Example 4.* Consider the infinite series $\sum_{n=1}^{\infty} 1/n\sqrt{n+1}$. It is clear that

$$\frac{1}{n\sqrt{n+1}} < \frac{1}{n\sqrt{n}} = \frac{1}{n^{3/2}}$$

Hence the given series may be compared with the *p*-series $\sum_{n=1}^{\infty} 1/n^{3/2}$ which converges and shows that the given series so converges.

EXERCISES 7.7 Evaluate each of the following improper integrals (if convergent).

1. $\int_1^{\infty} x^{-4/3}\, dx$

2. $\int_{-\infty}^{4} \frac{dx}{(5-x)^2}$

7.7 IMPROPER INTEGRALS

3. $\int_1^\infty \dfrac{dx}{x^{1.01}}$

4. $\int_0^1 \dfrac{dx}{\sqrt{x}}$

5. $\int_0^1 \dfrac{dx}{\sqrt[3]{x}}$

6. $\int_{-1}^1 x^{-1/3}\,dx$

7. $\int_0^4 \dfrac{dx}{x\sqrt{x}}$

8. $\int_2^\infty \dfrac{dx}{(x-1)^2}$

9. $\int_0^2 \dfrac{dx}{(x-1)^2}$

10. $\int_3^6 \dfrac{dx}{\sqrt{5-x}}$

Use Theorem 7.6 to determine whether or not each of the following improper integrals converges.

11. $\int_1^\infty \dfrac{dx}{(2x)^2}$

12. $\int_1^\infty \dfrac{dx}{x(x+1)}$

13. $\int_2^\infty \dfrac{dx}{x\sqrt{x-1}}$

14. $\int_0^\infty \dfrac{(x+1)\,dx}{(x+2)\sqrt{x^3+1}}$

Apply Theorem 7.7 to determine whether or not each of the following infinite series converges.

15. $\displaystyle\sum_{n=1}^\infty \dfrac{1}{(n+3)^2}$

16. $\displaystyle\sum_{n=1}^\infty \dfrac{n}{n^2-4}$

17. $\displaystyle\sum_{n=1}^\infty \dfrac{n+1}{(n^2+2n+3)^2}$

18. $\displaystyle\sum \dfrac{1}{(n+2)^{3/2}}$

19. $\displaystyle\sum_{n=2}^\infty \dfrac{1}{n^2-n}$

20. $\displaystyle\sum_{n=3}^\infty \dfrac{\sqrt{n}}{n^2-4}$

PROBLEMS 7.7

1. For which values of p does $\int_0^1 dx/x^p$ converge?

2. Give an upper bound for the value of $\int_2^\infty dx/(1+x^8)$.

3. It is a fact that $\int_{-\infty}^\infty dx/(x^2+1) = \pi$; on the other hand, the integral
$$\int_{-\infty}^\infty \dfrac{dx}{x^2+2x+1}$$
does not exist. Explain this.

4. Evaluate the limit
$$\lim_{n\to\infty} \dfrac{1}{\sqrt{n}} \sum_{k=1}^n \dfrac{1}{\sqrt{k}}$$

5. If $p > 1$, show that
$$\dfrac{1}{(p-1)(n+1)^{p-1}} < \sum_{k=1}^\infty \dfrac{1}{k^p} - \sum_{k=1}^n \dfrac{1}{k^p} < \dfrac{1}{(p-1)n^{p-1}}$$

Use this to tell the accuracy obtained in approximating $\displaystyle\sum_{n=1}^\infty 1/n^2$ by its tenth partial sum; its 100th partial sum. Then do the same for $\displaystyle\sum_{n=1}^\infty 1/n^3$.

7.8 VOLUMES

The volume of a solid can often be defined by integration. More precisely, consider a solid in which a particular axis has been selected, as shown in Fig. 7.7. For discussion purposes we refer to the axis as the x axis and the interval cut off by the solid as $\{x : a \leq x \leq b\}$. A final stipulation is that the cross-sectional area A at the point x in this interval should be a function of the number x.

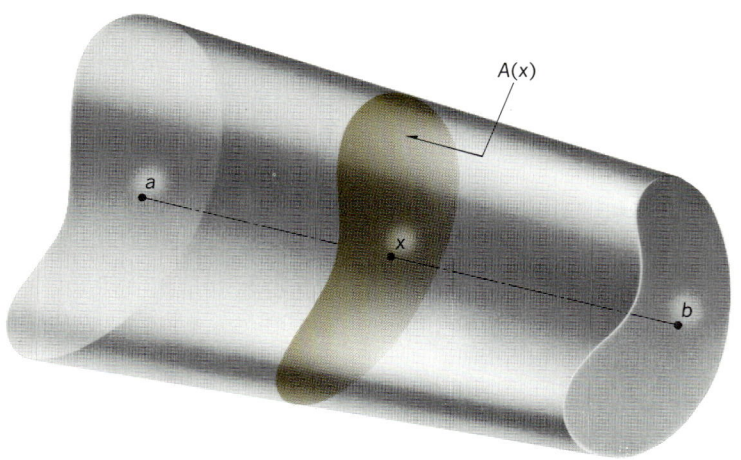

FIGURE 7.7

Now for any net $a = x_0 < x_1 < \cdots < x_n = b$ on the interval, the planes perpendicular to the axis at the points x_i cut the solid into parallel slabs. Each such slab has a volume ΔV_i satisfying

$$A'_i(x_i - x_{i-1}) \leq \Delta V_i \leq A''_i(x_i - x_{i-1})$$

where A'_i is the minimum of $A(x)$ and A''_i is the maximum of $A(x)$ over $\{x : x_{i-1} \leq x \leq x_i\}$. We then have the familiar approximation

$$\sum_{i=1}^{n} A'_i(x_i - x_{i-1}) \leq V \leq \sum_{i=1}^{n} A''_i(x_i - x_{i-1})$$

for the volume V. This leads immediately to the definition

$$V = \int_a^b A(x)\, dx$$

This formula is applicable whenever the solid is divided into parallel slabs by planes parallel to the x axis. (Of course, we are free to choose the x axis to suit our own convenience in any particular instance.)

■ *Example 1.* Consider the tetrahedron in Fig. 7.8. Its base is an equilateral triangle with sides 2 in. long and its height is 3 in. The fourth vertex is directly above one of the base vertices. We shall use the vertical edge as the axis and choose the origin at the base of the tetrahedron. It is an easy matter of considering similar triangles to see that

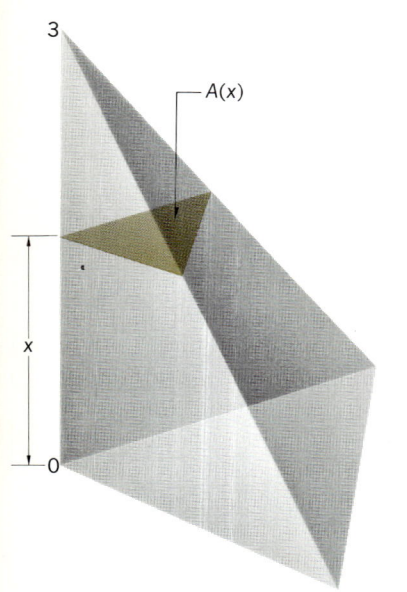

FIGURE 7.8

the area $A(x)$ of the cross section at height x above the base is that of an equilateral triangle with sides $\frac{2}{3}(3 - x)$. Hence

$$A(x) = \frac{\sqrt{3}}{9}(3 - x)^2$$

Therefore

$$V = \int_0^3 \frac{\sqrt{3}}{9}(3 - x)^2 \, dx = \frac{\sqrt{3}}{9}\int_0^3 (3 - x)^2 \, dx$$

$$= \frac{\sqrt{3}}{9}\left(-\frac{1}{3}(3 - x)^3\right)\Big]_0^3 = \sqrt{3}$$

This agrees with the known formula $V = \frac{1}{3}Bh = \frac{1}{3}(\sqrt{3})(3)$, where B is the area of the base and h is the height of the tetrahedron.

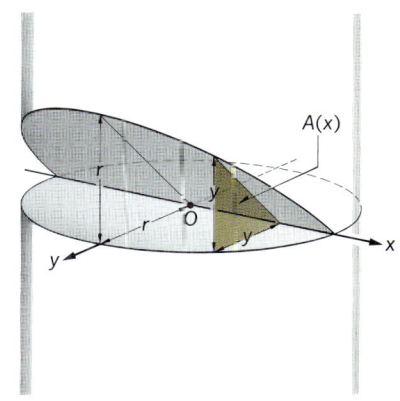

FIGURE 7.9

■ **Example 2.** In felling a tree the lumberjack has cut a wedge as illustrated in Fig. 7.9. What volume of wood has he removed?

By using the diameter of the tree (assumed to be a right circular cylinder of radius r) as an axis and choosing the origin 0 at the center of the tree, we see that each cross section perpendicular to the axis is a triangle with an area $A(x)$. Each such triangle has equal base and height, since each is similar to the one at the center. Now the base of this wedge is bounded by the half circle $\{(x, y) : y = \sqrt{r^2 - x^2}\}$. Hence the area $A(x)$ is given by

$$A(x) = \tfrac{1}{2}y^2 = \tfrac{1}{2}(r^2 - x^2)$$

The desired volume is therefore

$$V = \int_{-r}^r \frac{1}{2}(r^2 - x^2) \, dx = \frac{1}{2}\int_{-r}^r (r^2 - x^2) \, dx$$

$$= \frac{1}{2}\left(r^2 x - \frac{1}{3}x^3\right)\Big]_{-r}^r = \frac{2}{3}r^3$$

Solids of revolution A solid of revolution is generated by rotating a plane region about an axis in its plane. The axis of rotation clearly becomes an axis of symmetry of the solid. In Fig. 7.10, the region bounded by the x axis, the line $\{(x, y) : x = 1\}$ and the curve $\{(x, y) : y = x^2\}$ is rotated (a) about the x axis and (b) about the y axis. Only a half of the resulting solid is shown in each instance. In each case it should be noted that a cross-sectional region perpendicular to the axis of rotation is bounded by one or more circles. Hence the area $A(x)$ is easily computed. In Fig. 7.10(a) we readily see that

$$A(x) = \pi y^2 = \pi x^4$$

whence

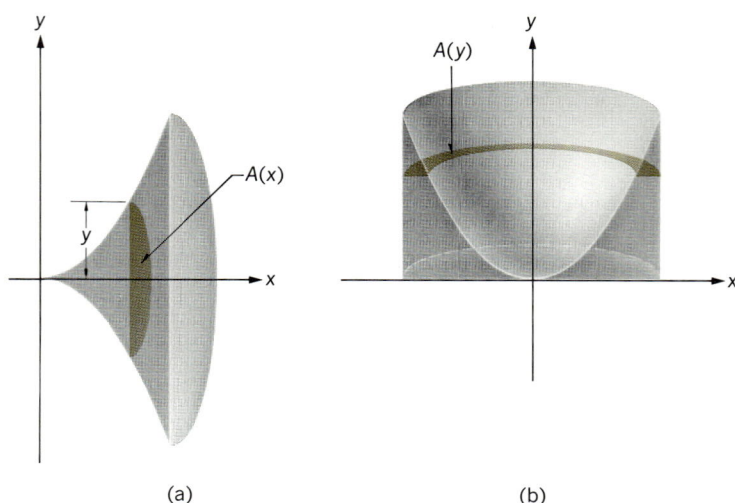

FIGURE 7.10

(a)   (b)

$$V = \int_0^1 \pi x^4 \, dx = \pi \frac{x^5}{5} \Big]_0^1 = \frac{1}{5}\pi$$

In Fig. 7.10(b), the outer circle bounding the cross-sectional area has a radius 1, while the inner circle has a radius $x = \sqrt{y}$. Thus the cross-sectional area $A(y)$ is given by

$$A(y) = \pi - \pi x^2 = \pi(1 - y)$$

It follows that

$$V = \int_0^1 \pi(1 - y) \, dy = \pi\left(y - \frac{1}{2}y^2\right)\Big]_0^1 = \frac{1}{2}\pi$$

As another example, let the area under the half circle $\{(x,y) : y = \sqrt{r^2 - x^2}\}$ be rotated about the x axis to generate a solid ball of radius r. A representative cross-sectional area is then that of a circle of radius $y = \sqrt{r^2 - x^2}$ or

$$A(x) = \pi y^2 = \pi(r^2 - x^2)$$

See Fig. 7.11.

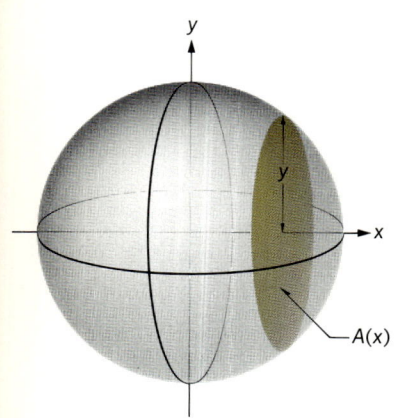

FIGURE 7.11

It then follows that

$$V = \int_{-r}^r \pi(r^2 - x^2) \, dx = \pi\left(r^2 x - \frac{1}{3}x^3\right)\Big]_{-r}^r = \frac{4}{3}\pi r^3$$

EXERCISES 7.8   In each of Exercises 1 to 10 compute the volume of the solid generated by rotating the given region about the given axis. Draw the solid with a representative cross section in each case.

1. The region bounded by the x axis, the line $\{(x,y) : x = 4\}$ and the line $\{(x,y) : y = 2x\}$ rotated about the x axis.

2. The region of Exercise 1 rotated about the y axis.

3. The region of Exercise 1 rotated about the line $\{(x,y) : x = 4\}$.

4. The region of Exercise 1 rotated about the line $\{(x,y) : x = 6\}$.

5. The region bounded by the x axis and the curve $\{(x,y) : y = 4x - x^2\}$ rotated about the x axis.

6. The region of Exercise 5 rotated about the line $\{(x,y) : y = 4\}$.

7. The region of Exercise 5 rotated about the line $\{(x,y) : y = -1\}$.

8. The region bounded by the line $\{(x,y) : y = x\}$ and the curve $\{(x,y) : y = 4x - x^2\}$ rotated about the x axis.

9. The region of Exercise 8 rotated about the line $\{(x,y) : y = -2\}$.

10. The region of Exercise 8 rotated about the line $\{(x,y) : y = 5\}$.

11. The cross sections perpendicular to the x axis of a certain solid are squares whose diagonals are the segments from the curve $\{(x,y) : y = x^2\}$ to the line $\{(x,y) : y = x\}$. What is the volume of this solid?

12. The cross sections perpendicular to the x axis of a certain solid are circles whose diameters are the segments from the line $\{(x,y) : y = 2x\}$ to the curve $\{(x,y) : y = 8 - x^2\}$. What is the volume of this solid?

13. Rotate the region of Exercise 5 about the y axis and compute the volume of the solid so generated. (Hint: Be careful about the description of the representative cross-sectional area.)

14. Rotate the region of Exercise 8 about the y axis and compute the volume of the solid so generated. (Hint: This will require two steps because the description of the cross-sectional area changes form when $y = 3$.)

15. Determine the volume of the frustrum of a right circular cone if its upper base has a radius r, its lower base has a radius R, and its height is h.

PROBLEMS 7.8

1. Two right circular cylinders of radius r intersect, their axes meeting at right angles. What is the volume of the solid of intersection?

2. Rotate the region bounded by the circle $\{(x,y) : x^2 + y^2 = r^2\}$ about the line $\{(x,y) : x = R\}$ where $r < R$. Find the volume of the solid anchor ring so generated. (Hint: The integral $\int_{-r}^{r} \sqrt{r^2 - x^2}\, dx$ must equal $\tfrac{1}{2}\pi r^2$ because it measures the area of a semicircle of radius r.)

3. Prove the following theorem called *Cavalieri's theorem*: Given two solids of equal height, assume that the cross sections made by planes parallel to and at equal distance from their respective bases have equal area. Then the two solids have equal volume.

4. A *prismatoid* is defined to be a solid such that the area $A(h)$ of a cross section parallel to and at a distance h from some fixed plane can be expressed as a quadratic function of h,

$$A(h) = a_0 + a_1h + a_2h^2$$

For any such solid prove that the volume is given by

$$V = \tfrac{1}{6}(B_1 + 4B + B_2)H$$

where B_1 and B_2 are the areas of the bases, where B is the area of the cross section parallel to the bases and halfway between them, and where H is the height of the solid. Conclude by showing that (a) the frustrum of a cone, (b) the pyramid over any base, and (c) a segment of a sphere are prismatoids and compute their volumes.

5. Show that the formula in Problem 4 also applies to the volume of a solid whose cross-sectional area at height h is a cubic function

$$A(h) = a_0 + a_1h + a_2h^2 + a_3h^3$$

6. A cylindrical drinking glass has a radius r and height h. It is tipped to pour the water out. How much water remains when exactly half of the bottom of the glass is exposed?

7. A solid of revolution can also be cut into "shells" by concentric cylinders about the axis of symmetry. This is illustrated in Fig. 7.12. Derive the formula

$$V = 2\pi \int_b^a xf(x)\, dx$$

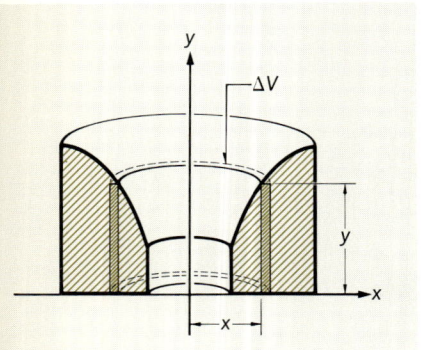

FIGURE 7.12

7.9 ARC LENGTH

Let f be a differentiable function with continuous derivative f' on $\{x : a \leq x \leq b\}$. We shall use a definite integral to define the length of the curve $\{(x,y) : a \leq x \leq b, y = f(x)\}$.

For any $n \in \mathbf{N}$, consider the net

$$x_i = a + i\left(\frac{b-a}{n}\right), \quad i = 1, 2, \cdots, n$$

We approximate the graph of f by a polygonal arc having the $n+1$ vertices $(x_i, f(x_i))$. (See Fig. 7.13 with $N = 4$.)

Using the distance formula, we easily compute the length of this polygonal approximation to be

$$\sum_{i=1}^{n} \sqrt{(\Delta x)^2 + [f(x_i) - f(x_{i-1})]^2}$$

where $\Delta x = (b-a)/n$, as usual. Using the mean value theorem, we find a point $\bar{x}_i$, $x_{i-1} < \bar{x}_i < x_i$, for each $i = 1, 2, \cdots, n$, such that

$$f(x_i) - f(x_{i-1}) = \Delta x\, f'(\bar{x}_i)$$

Substituting these values into the summation above, we can write the length of the polygonal approximation as

$$\sum_{i=1}^{n} \sqrt{(\Delta x)^2 + (\Delta x)^2[f'(\bar{x}_i)]^2}$$

$$= \sum_{i=1}^{n} \sqrt{1 + [f'(\bar{x}_i)]^2}\, \Delta x$$

7.9 ARC LENGTH

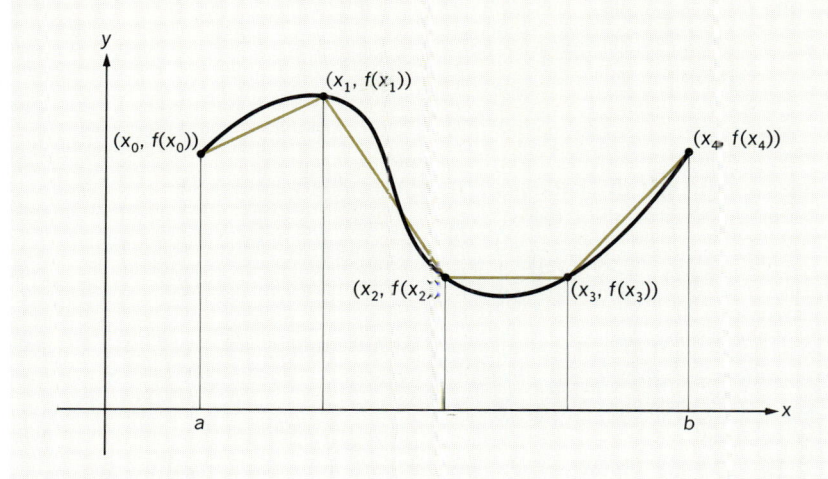

FIGURE 7.13

We see that this has the form of a Riemann sum for the function $\sqrt{1 + [f'(x)]^2}$. Having assumed that f' is continuous, we find that $\sqrt{1 + [f'(x)]^2}$ is also continuous. Therefore the integral

$$\int_a^b \sqrt{1 + [f'(x)]^2}\, dx$$

exists. This number is *defined* to be the length of the graph $\{(x,y) : y = f(x)\}$ between the points $(a, f(a))$ and $(b, f(b))$.

- **Example 1.** The length of the graph $\{(x,y) : y = x^{3/2}\}$ from $(0,0)$ to $(4, 8)$ is obtained as follows: Since $f(x) = x^{3/2}$, $f'(x) = \frac{3}{2}x^{1/2}$ and $1 + [f'(x)]^2 = 1 + \frac{9}{4}x$. Hence the length is

$$\int_0^4 \sqrt{1 + \frac{9}{4}x}\, dx$$

Substitute $u = 1 + \frac{9}{4}x$, whence $dx = \frac{4}{9}\, du$ and $u = 1$ when $x = 0$, $u = 10$ when $x = 4$. Then we have

$$\int_1^{10} \sqrt{u} \cdot \frac{4}{9}\, du = \frac{4}{9} \int_1^{10} u^{1/2}\, du = \frac{8}{27} u^{3/2} \Big]_1^{10} = \frac{8}{27}(10^{3/2} - 1)$$

- **Example 2.** Consider the function $f(x) = \frac{1}{3}(x^2 + 2)^{3/2}$ over the interval $\{x : 0 \le x \le 3\}$. Because

$$f'(x) = \tfrac{1}{3} \cdot \tfrac{3}{2}(x^2 + 2)^{1/2}(2x) = x(x^2 + 2)^{1/2}$$

we have
$$1 + [f'(x)]^2 = 1 + x^2(x^2 + 2) = 1 + 2x^2 + x^4 = (1 + x^2)^2$$
and hence
$$\sqrt{1 + [f'(x)]^2} = 1 + x^2$$
Therefore the length of the graph from $(0, f(0))$ to $(3, f(3))$ is
$$\int_0^3 \sqrt{1 + [f'(x)]^2}\, dx = \int_0^3 (1 + x^2)\, dx = x + \frac{1}{3}x^3 \Big]_0^3 = 12 \quad \blacksquare$$

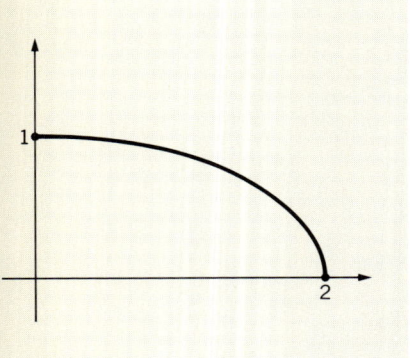

FIGURE 7.14

It is only rarely that the indefinite integral $\int \sqrt{1 + [f'(x)]^2}\, dx$ can be evaluated explicitly by elementary methods. That is, we must usually resort to approximate integration.

■ **Example 3.** The function $f(x) = \frac{1}{2}\sqrt{4 - x^2}$ on $\{x : 0 \leq x \leq 2\}$ has for its graph an arc of an ellipse, as seen in Fig. 7.14. Now $f'(x) = -x/2(4 - x^2)^{-1/2}$ and hence

$$1 + [f'(x)]^2 = 1 + \frac{x^2}{4}(4 - x^2)^{-1} = 1 + \frac{x^2}{16 - 4x^2} = \frac{16 - 3x^2}{16 - 4x^2}$$

The length of this curve is then
$$\int_0^2 \sqrt{\frac{16 - 3x^2}{16 - 4x^2}}\, dx$$

But we have a problem here: The integral is an improper integral and none of our methods serves to determine the indefinite integral
$$\int \sqrt{\frac{16 - 3x^2}{16 - 4x^2}}\, dx$$

EXERCISES 7.9 Set up an integral expressing the length of each of the arcs given in Exercises 1 to 10. Evaluate the integral when possible. (All eventually will be possible.)

1. $\{(x, y) : 0 \leq x \leq 3, y = \frac{2}{3}x^{3/2}\}$
2. $\{(x, y) : 0 \leq x \leq 1, y = 2x^{3/2} - 1\}$
3. $\{(x, y) : 1 \leq x \leq 2, y = (9 - x^{2/3})^{3/2}\}$
4. $\{(x, y) : 1 \leq x \leq 4, y = \frac{1}{3}x^3 + \frac{1}{4}x^{-1}\}$
5. $\{(x, y) : 0 \leq x \leq 4, y = \frac{2}{3}x^{3/2} - \frac{1}{2}x^{1/2}\}$
6. $\{(x, y) : 1 \leq x \leq 2, y = \frac{1}{4}x^4 + \frac{1}{8}x^{-2}\}$
7. $\{(x, y) : -2 \leq x \leq 2, y = 4 - x^2\}$
8. $\{(x, y) : 0 \leq x \leq 3, y = \sqrt{49 - x^2}\}$

9. $\{(x,y) : 0 \leq x \leq \frac{1}{2}\sqrt{3}, y = \sqrt{1-x^2}\}$
10. $\{(x,y) : 2 \leq x \leq 3, y = \sqrt{x^2-1}\}$
11. Find the entire length of the curve $\{(x,y) : x^{2/3} + y^{2/3} = a^{2/3}\}$.
12. Find the length of the curve $\{(x,y) : (x/a)^{2/3} + (y/b)^{2/3} = 1, x \geq 0, y \geq 0\}$.
13. Find the perimeter of the loop in the curve $\{(x,y) : 9y^2 = x(x-3)^2\}$.
14. Find the perimeter of the loop in the curve $\{(x,y) : 9y^2 = x^2(2x+3)\}$.
15. For any $C > 0$, show that the curve

$$\left\{(x,y) : a \leq x \leq b, y = \frac{(2+Cx^2)^{3/2}}{3\sqrt{C}}\right\}$$

has a length $(b-a)[1 + \tfrac{1}{3}C(a^2 - ab + b^2)]$.

16. For any constants C and D such that $CD = \tfrac{1}{12}$, show that the length of the curve $\{(x,y) : a \leq x \leq b, y = Cx^3 + Dx^{-1}\}$ is $(b-a)[C(a^2 + ab + b^2) + D/ab$.

7.10 AREA OF A SURFACE OF REVOLUTION

If a plane curve is rotated about a line in the same plane, the rotating curve sweeps out a surface of revolution. Our problem here is to define the area of such a surface. In addition to the standard tools of integration, we need a fact from elementary geometry.

The frustrum of a cone (see Fig. 7.15) having base radii r_1 and r_2 and slant height s has a lateral surface area

$$\pi s(r_1 + r_2)$$

We limit our consideration to the following situation: Let $f(x)$ be a smooth function that is nonnegative over the interval $\{x : a \leq x \leq b\}$. We rotate the graph $\{(x,y) : a \leq x \leq b, y = f(x)\}$ about the x axis and assign an area to the resulting surface of revolution, which might look like Fig. 7.16.

An approximation to the surface swept out by rotating the curve $\{(x,y) : y = f(x)\}$ can be obtained by rotating a polygonal arc that approximates the curve. We proceed as in the preceding section by taking a net $a = x_0 < x_1 < \cdots < x_n = b$ on the interval. Then the $n+1$ points $(x_i, f(x_i))$ on the curve are used as the vertices of an approximating polygonal arc. (See Fig. 7.17.)

Each line segment in the approximating polygonal arc sweeps out the frustrum of a cone. In particular, the segment with endpoints $(x_{i-1}, f(x_{i-1}))$ and $(x_i, f(x_i))$ sweeps out a frustrum with base radii $f(x_i), f(x_{i-1})$ and slant height

$$\sqrt{(x_i - x_{i-1})^2 + [f(x_i) - f(x_{i-1})]^2}$$

By the formula quoted at the outset, this frustrum has a lateral area

$$\pi[f(x_{i-1}) + f(x_i)]\sqrt{(x_i - x_{i-1})^2 + [f(x_i) - f(x_{i-1})]^2}$$

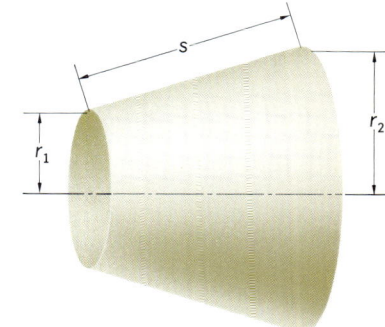

FIGURE 7.15

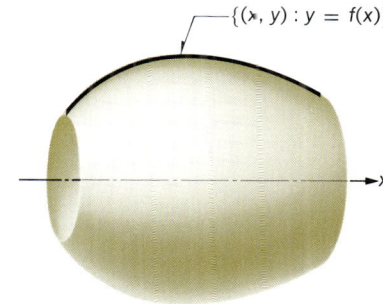

FIGURE 7.16

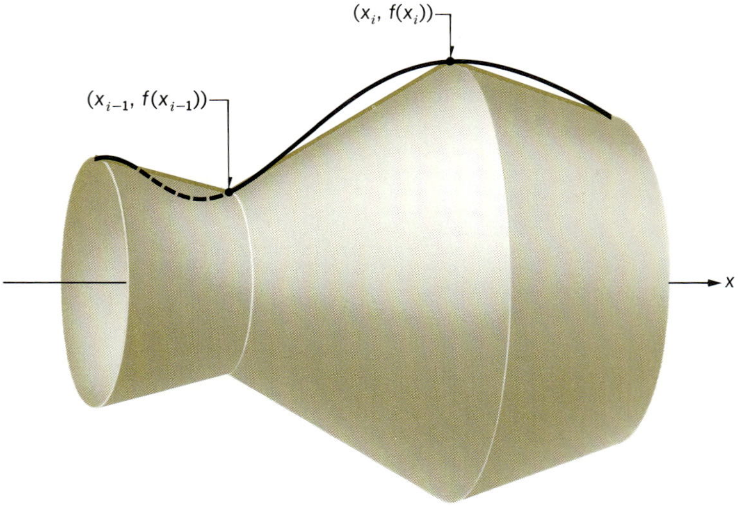

FIGURE 7.17

Again, by the mean value theorem there is a point $\bar{x}_i$, $x_{i-1} < \bar{x}_i < x_i$ such that $f(x_i) - f(x_{i-1}) = (x_i - x_{i-1})f'(\bar{x}_i)$. Therefore the area of this frustrum is

$$\pi[f(x_{i-1}) + f(x_i)]\sqrt{1 + [f'(\bar{x}_i)]^2}(x_i - x_{i-1})$$

One further approximation is made by replacing the sum $f(x_{i-1}) + f(x_i)$ with $2f(\bar{x}_i)$. That is, we say that the quantity

$$2\pi f(\bar{x}_i)\sqrt{1 + [f'(\bar{x}_i)]^2}(x_i - x_{i-1})$$

is an approximation to the area of the surface swept out by the curve $\{(x,y) : x_{i-1} \leq x \leq i, y = f(x)\}$. The entire surface is then approximated by the sum

$$2\pi \sum_{i=1}^{n} f(\bar{x}_i)\sqrt{1 + [f'(\bar{x}_i)]^2}(x_i - x_{i-1})$$

We recognize this as a Riemann sum for the function

$$2\pi f(x)\sqrt{1 + [f'(x)]^2}$$

Since $f'(x)$ is assumed to be continuous, the function

$$2\pi f(x)\sqrt{1 + [f'(x)]^2}$$

is continuous. Thus the integral

$$2\pi \int_a^b f(x)\sqrt{1 + [f'(x)]^2}\, dx$$

exists. This is defined to be the area of the surface of revolution.

7.10 AREA OF A SURFACE OF REVOLUTION

■ **Example 1.** Let $f(x) = \sqrt{r^2 - x^2}$ over the interval $\{x : -r \leq x \leq r\}$. (This function does not quite satisfy the conditions we imposed.) Then $f'(x) = -x/\sqrt{r^2 - x^2}$, so

$$1 + [f'(x)]^2 = 1 + \frac{x^2}{r^2 - x^2} = \frac{r^2}{r^2 - x^2}$$

It follows that

$$f(x)\sqrt{1 + [f'(x)]^2} = \sqrt{r^2 - x^2}\left(\frac{r}{\sqrt{r^2 - x^2}}\right) = r$$

Therefore, if the upper semicircle $\{(x, y) : y = \sqrt{r^2 - x^2}\}$ is rotated about the x axis to sweep out a sphere of radius r, then the area of the sphere is

$$2\pi \int_{-r}^{r} f(x)\sqrt{1 + [f'(x)]^2}\, dx = 2\pi \int_{-r}^{r} r\, dx = 4\pi r^2$$

■ **Example 2.** Determine the area swept out by rotating the curve $\{(x, y) : 0 \leq x \leq 2, y = \sqrt{x}\}$ about the x axis. Here we have

$$f(x) = \sqrt{x}, \quad f'(x) = \tfrac{1}{2}\sqrt{x}$$

whence

$$f(x)\sqrt{1 + [f'(x)]^2} = \sqrt{x}\sqrt{1 + \frac{1}{4x}} = \frac{1}{2}\sqrt{4x + 1}$$

Therefore the desired area is

$$2\pi \int_0^2 \frac{1}{2}\sqrt{4x + 1}\, dx = \frac{13\pi}{3}$$

■ **Example 3.** We compute the area swept out by rotating the curve $\{(x, y) : 1 \leq x \leq 2, y = \tfrac{1}{3}x^3 + \tfrac{1}{4}x^{-1}\}$. Here $f(x) = \tfrac{1}{3}x^3 + \tfrac{1}{4}x^{-1}$ whence $f'(x) = x^2 - \tfrac{1}{4}x^{-2}$. Computing, we obtain

$$\begin{aligned}1 + [f'(x)]^2 &= 1 + (x^2 - \tfrac{1}{4}x^{-2})^2 \\ &= 1 + x^4 - \tfrac{1}{2} + \tfrac{1}{16}x^{-4} \\ &= x^4 + \tfrac{1}{2} + \tfrac{1}{16}x^{-4} \\ &= (x^2 + \tfrac{1}{4}x^{-2})^2\end{aligned}$$

Thus

$$\begin{aligned}f(x)\sqrt{1 + [f'(x)]^2} &= (\tfrac{1}{3}x^3 + \tfrac{1}{4}x^{-1})(x^2 + \tfrac{1}{4}x^{-2}) \\ &= \tfrac{1}{3}x^5 + \tfrac{1}{3}x + \tfrac{1}{16}x^{-3}\end{aligned}$$

It follows that the desired area is

$$2\pi \int_1^2 \left(\frac{1}{3}x^5 + \frac{1}{3}x + \frac{1}{16}x^{-3}\right) dx = \frac{515\pi}{64}$$

EXERCISES 7.10 In each of Exercises 1 to 6 determine the area of the surface formed by rotating the given curve about the x axis.

1. $\{(x,y) : 0 \leq x \leq 2, y = x^3\}$
2. $\{(x,y) : 0 \leq x \leq a, y^2 = ax\}$
3. $\{(x,y) : 0 \leq x \leq \sqrt[4]{24}, y = \frac{1}{3}x^3\}$
4. $\{(x,y) : 5 \leq x \leq 8, y = 4\sqrt{x}\}$
5. $\{(x,y) : 1 \leq x \leq 2, y = \frac{1}{4}x^4 + \frac{1}{8}x^{-2}\}$
6. $\{(x,y) : x^{2/3} + y^{2/3} = a^{2/3}\}$

PROBLEMS 7.10

1. Develop integral formulas for areas of surfaces formed by rotating a curve about
   (a) the y axis
   (b) a horizontal line other than the x axis
   (c) a vertical line other than the y axis
   (d) a line $\{(x,y) : ax + by + c = 0\}$

2. Compute the area of the surface formed by rotating about the x axis the loop of the curve $\{(x,y) : 9y^2 = x(x-3)^2\}$.

3. A *spherical zone* is the area on a sphere between two parallel planes. If the distance between the parallel planes is h and the radius of the sphere is r, prove that the area is $2\pi rh$.

4. If the curve $\{(x,y) : x \geq 1, y = 1/x\}$ is rotated about the x axis, an infinitely long funnel is formed. Prove that the funnel can be filled with paint but its surface cannot be painted. More precisely, show that the infinite volume exists, while the infinite area does not.

7.11 WORK Consider a body in rectilinear motion and a force acting on the body along its line of motion. If the force is a constant F lb and acts while the body moves d ft, then the work done is in foot-pounds (ft-lb):

$$W = Fd \text{ ft-lb}$$

We say that the work is done *by* the force or *against* the force, depending on whether the force acts in the direction of motion or opposite the direction of motion.

If the force F is not constant, then we must use an integral to define the concept of work. Assume that the line of motion of the moving body is the x axis and that a variable force $F(x)$ acts on the body. Let $F(x)$ be a continuous function of the position x of the body. We shall use the now-familiar approximation technique to define the work done by (or against) the force $F(x)$ as the body moves over the interval $\{x : a \leq x \leq b\}$.

7.11 WORK

Let $a = x_0 < x_1 < \cdots < x_n = b$ be a net on the interval and let W_i be the work done as the object moves from the point x_{i-1} to the point x_i. If F_i' and F_i'' are the minimum and maximum, respectively, of the force function $F(x)$ on $\{x : x_{i-1} \leq x \leq x_i\}$, then it seems obvious that, for each $i = 1, 2, \cdots, n$,

$$F_i'(x_i - x_{i-1}) \leq W_i \leq F_i''(x_i - x_{i-1})$$

Adding these n inequalities, we have

$$\sum_{i=1}^{n} F_i'(x_i - x_{i-1}) \leq W \leq \sum_{i=1}^{n} F_i''(x_i - x_{i-1})$$

Since W must lie between the upper and lower sums for any net, it follows that we must accept the definition

$$W = \int_a^b F(x)\, dx$$

of the work done by or against the variable force $F(x)$ acting on the body as it moves from $x = a$ to $x = b$.

■ **Example 1.** A force of 30 lb is required to hold a certain spring extended to the length of 15 in. from its rest length of 10 in. How much work is done in stretching this spring from 12 in. to 18 in. long? We assume the *spring law*, namely, the force $F(x)$ required to hold the spring extended to the length $10 + x$ is proportional to the displacement x. Thus $F(x) = kx$ for some constant k.

Knowing that $F(5) = k(5) = 30$, we have $k = 6$ and hence, for this particular spring, $F(x) = 6x$. Then to change the length of the spring from $10 + 2$ to $10 + 8$ in., the work done is in inch-pounds (in.-lb):

$$W = \int_2^8 F(x)\, dx = \int_2^8 6x\, dx = 3x^2 \Big]_2^8 = 180$$

■ **Example 2.** A standard problem is to determine the work done in emptying a tank by lifting the water up over the top rim of the tank. Suppose that the tank is a vertical cylinder with a diameter of 8 ft and a height of 10 ft. We seek the integral expressing the work done in emptying this tank. Consider a net $0 = x_0 < x_1 < \cdots < x_n = 10$ on the interval $\{x : 0 \leq x \leq 10\}$ determined by the height of the tank. Now the slab of water $x_i - x_{i-1}$ ft thick (see Fig. 7.18) has a volume $16\pi(x_i - x_{i-1})$ cu ft and, because water weighs 62.5 lb per cu ft, this slab weighs $1000\pi(x_i - x_{i-1})$ lb. It is now clear that the work W_i done in lifting this slab of the water to the top of the tank must satisfy

$$1000\pi(x_i - x_{i-1}) \cdot (10 - x_i) \leq W_i$$
$$\leq 1000\pi(x_i - x_{i-1}) \cdot (10 - x_{i-1})$$

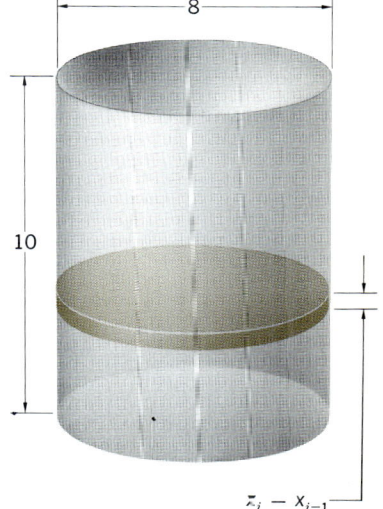

FIGURE 7.18

Thus we see that
$$W_i \approx 1000\pi(10 - x_i)(x_i - x_{i-1})$$
or
$$W_i \approx 1000\pi(10 - x)\,\Delta x$$

From this we immediately set up the required integral
$$W = \int_0^{10} 1000\pi(10 - x)\,dx$$
$$= 1000\pi\left(10x - \frac{1}{2}x^2\right)\Big]_0^{10}$$
$$= 50{,}000 \text{ ft lb}$$

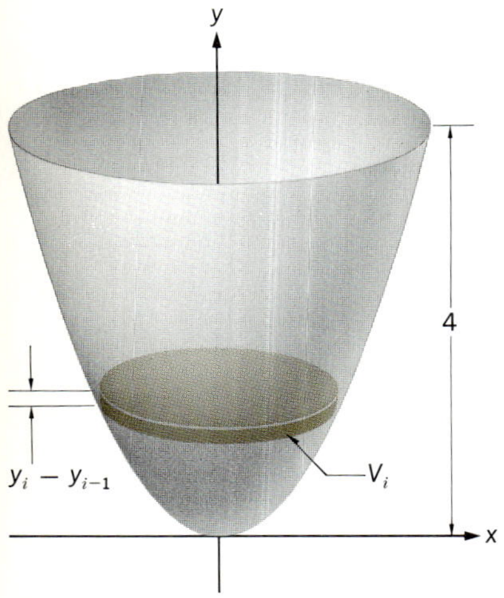

■ **Example 3.** The problem of emptying a tank can be made more complicated by considering a tank with nonconstant cross section and by lifting the water some distance above the top of the tank. Consider a tank, as pictured in Fig. 7.19, which is the volume obtained by rotating about the y axis that area bounded by the y axis, the line $\{(x,y) : y = 4\}$ and the curve $\{(x,y) : y = x^2; x \geq 0\}$.

Again we take an arbitrary net over the interval $\{y : 0 \leq y \leq 4\}$ of the height of the tank. The volume V_i of the slab of water between heights y_{i-1} and y_i surely satisfies
$$\pi y_i(y_i - y_{i-1}) = \pi x_{i-1}^2(y_i - y_{i-1}) \leq V_i$$
$$\leq \pi x_i^2(y_i - y_{i-1}) = \pi y_i(y_i - y_{i-1})$$
and hence the weight of this slab satisfies
$$62.5 y_i(y_i - y_{i-1}) \leq V_i \leq 62.5 y_i(y_i - y_{i-1})$$

The water in this slab is to be lifted a distance between $4 - y_i$ and $4 - y_{i-1}$ to the top of the tank. It follows that the amount of work W_i done in lifting this part of the water to the top of the tank satisfies
$$62.5 y_{i-1}(y_i - y_{i-1}) \cdot (4 - y_{i-1}) \leq W_i$$
$$\leq 62.5 y_i(y_i - y_{i-1}) \cdot (4 - y_{i-1})$$

Thus we write
$$W_i \approx 62.5 y_i(4 - y_i) \cdot (y_i - y_{i-1})$$
and from this the integral follows immediately:
$$W = \int_0^4 62.5\pi y(4 - y)\,dy$$
$$= 62.5\pi\left(2y^2 - \frac{1}{3}y^3\right)\Big]_0^4$$
$$= \frac{2000\pi}{3} \text{ ft lb}$$

FIGURE 7.19

- **Example 4.** Suppose the water in the tank of Example 3 is to be lifted a distance of 6 ft above the top surface of the tank. Then the approximate expression for the work W_i would be

$$W_i \approx 62.5 y_i (10 - y_i)(y_i - y_{i-1})$$

The work done is now

$$\begin{aligned} W &= \int_0^4 62.5\pi y(10-y)\,dy \\ &= 62.5\pi \left(5y^2 - \frac{1}{3}y^3\right)\bigg|_0^4 \\ &= \frac{4750\pi}{3} \text{ ft lb} \end{aligned}$$

- **Example 5.** Two electrons repel each other with a force inversely proportional to the square of the distance between them. By choosing the constants appropriately we may write the force as

$$F(r) = \frac{1}{r^2}$$

where r is the distance between the electrons. How much work must be done against this repelling force in moving an electron from 10 cm away from a fixed other electron to a distance 10^{-1} cm $(= 1$ millimeter) away? It is easy to show that

$$W = \int_{1/10}^{10} \frac{dr}{r^2} = -\frac{1}{r}\bigg]_{1/10}^{10} = -\frac{1}{10} + \frac{1}{10^{-1}} = 9.9$$

(The units here would depend upon the units of force.)

PROBLEMS 7.11

1. A spring has a rest length of 12 in. If a force of 40 lb stretches this spring a distance of 1 in., how much work is done in stretching the spring from 15 to 20 in.?

2. A spring has a rest length of 10 in. If a 10-lb force stretches it to 12 in., how much work is done in stretching it from 10 to 12 in.?

3. An automobile coil spring 8 in. long is compressed 1 in. by a force of 500 lb. How much work is done against the spring force by the compression?

4. If the cylindrical tank of Example 1 had been only half full, how much work would have been required to empty it?

5. A tank is in the form of an inverted cone (point down) of the height 10 ft and base radius 5 ft. How much work is required to empty this tank if the original depth of water is 8 ft?

6. A tank is in the form of an inverted cone of base radius r and height h_1 surmounted by a cylinder of base radius r and height h_2. If the tank is full of

water, how much work must be done to pump the water out of the top of the tank. If the tank is filled to a ditsance $d < h_2$ of the top, how much work is required in pumping it empty?

7. A swimming pool is 25 yd long, 15 yd wide, 5 ft deep at the shallow end and 12 ft deep at the deep end. The bottom is a slanting plane. If the pool is full of water to within 6 in. of the top, find the work needed to pump it empty.

8. A cylinder 1 ft long and 2 ft in diameter is $\frac{2}{3}$ as dense as water and is floating with its axis vertical. How much work must be done to submerge the cylinder?

9. A cube 10 in. on an edge is floating in water. Archimedes' law states that a floating body is buoyed up by a force equal to the weight of the water displaced. How much work is done in forcing this cube down a distance of 2 in. below its equilibrium position?

10. A service station has a cylindrical tank buried on its side with the top of the tank 5 ft below the surface of the driveway. The tank is 6 ft in diameter and 10 ft long. The depth of the gasoline in the tank is 4 ft and gasoline weighs 45 lb per cubic ft. Assuming that the filler cap of each automobile gasoline tank is 2 ft above the ground, how much work is done in emptying the tank into automobiles?

11. A cable 100 ft long and weighing 10 lb per linear foot is to be wound up on a windlass 100 ft above the ground. How much work will be done?

12. An anchor is 50 ft below the capstan on which the anchor chain is to be wound. The anchor weighs 500 lb and the chain weighs 30 lb per ft. How much work is done in raising the anchor?

13. A bucket of water originally contains 50 lb of water and is lifted at the constant rate of 5 ft per min. The water leaks out of this bucket at the rate of 1 lb per min. Find the work done in lifting the water 35 ft.

14. The force of attraction between a body of mass m and the earth (which has mass M) is given by

$$F = \frac{kmM}{r^2}$$

where k is a constant and r is the distance of the body from the center of the earth. If a body weighs 100 lb at the surface (that is, when $r = 4000$ miles), how much work must be done against gravity to lift this body to 1000 miles above the surface?

15. Determine the work done against gravity in rocketing an artificial satellite of mass m from the surface to a height h above the surface. Show that we may express W as $mgRh/(R + h)$. If $h = 150$ miles, how much does this expression differ from the work calculated by assuming gravity to be constant g?

16. If a hole could be bored straight down to the center of the earth, then a particle of mass m in this hole would be attracted toward the center of the earth by a force $F = mgr/R$, where r is the distance from the center, R is the radius of the earth, and g is the acceleration of gravity at the surface. How much work is done by gravity as the particle falls from the surface to the center?

17. How much work is done by gravity on a particle of mass m which falls from infinity to the surface of the earth?

18. A metal bar of length L and cross-sectional area A obeys *Hookes' Law* when stretched. If it is stretched x units, then the force required to hold it elongated is

$$F = \frac{EA}{L} x$$

where E is Young's modulus of elasticity and depends only upon the particular metal used. If $L = 18$ in. and the bar has cross-sectional area 2 sq in., express in terms of E the amount of work done in stretching this bar $\frac{1}{4}$ in.

19. The pressure and volume of a certain gas obey the law

$$pv^{1.4} = 180$$

where p is the pressure in lb per sq in. and v is the volume in cu in. Determine the work done when the gas expands from 32 cu in. to 243 cu in.

20. In the study of a certain steam engine it is found that the pressure p (in lb per sq ft) and the volume v in cu ft satisfy the equation

$$pv^\eta = c$$

where η and c are constants. If $\eta \neq 1$, show that the work done against the pressure in changing the volume from v_1 to v_2 is

$$W = \int_{v_1}^{v_2} p \, dv = \frac{p_2 v_2 - p_1 v_1}{1 - \eta}$$

21. Denote the quantity of electricity in a condenser by Q and the capacity of the condenser by C. Let the potential difference across the condenser be V. Then we have the relation $Q = CV$. The work done in moving a charge Q through a potential difference V equals QV. Since the potential difference V increases with Q, the work done in charging the condenser must be expressed as an integral. Show that the work done is

$$V \, dQ = \frac{1}{C} \int_0^{Q_1} Q \, dQ = \frac{1}{2} Q_1 V_1$$

where Q_1 is the total amount of electricity passed into the condenser and V_1 is the difference in potential across the condenser at the end of the charging process.

7.12 FLUID PRESSURE

If a flat plate lies horizontally at a depth of h feet below the surface of a liquid, then the weight of the liquid above the plate exerts a force on the plate. This force per unit area is called the *pressure*. It is really a remarkable fact that the pressure depends only upon the weight w of the liquid (per cubic foot) and the depth h below the surface, indeed,

$$p = wh$$

If the horizontal plate has an area A sq ft, then the total force exerted on the plate is

$$F = pA = whA$$

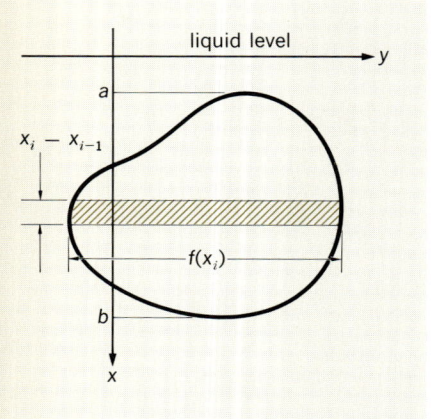

FIGURE 7.20

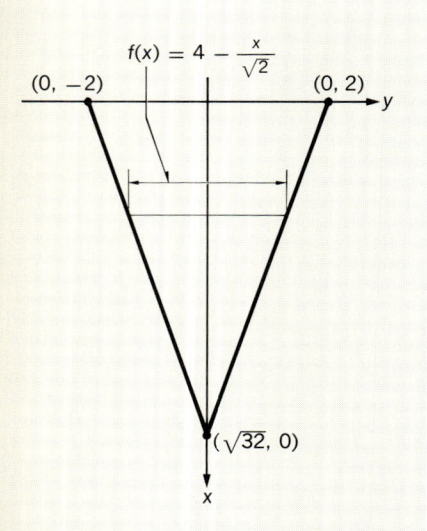

FIGURE 7.21

Now if we consider the flat plate submerged vertically in the liquid, then the pressure p is not constant and it requires an integral to define the total force acting on the plate due to fluid pressure. To be definite, we shall choose our coordinate system with the y axis along the surface of the liquid and the x axis directed positively downward. Then the area of the vertically submerged plate can be given as the integral

$$A = \int_a^b f(x)\,dx$$

where the constants a and b and the function $f(x)$ are as defined by Fig. 7.20.

Choose any net $a = x_0 < x_1 < \cdots < x_n = b$ on the interval $\{x : a \leq x \leq b\}$. Let F_i denote the total force acting on the area from x_{i-1} to x_i (see Fig. 7.20). Obviously, we have the inequality

$$wx_{i-1} \cdot A_i \leq F_i \leq wx_i \cdot A_i$$

where A_i is the area of the ith strip, that is

$$A_i = \int_{x_{i-1}}^{x_i} f(x)\,dx$$

If we let f'_i and f''_i denote the minimum and maximum, respectively, of the width function $f(x)$ on the ith subinterval, then

$$f'_i \cdot (x_i - x_{i-1}) \leq A_i \leq f''_i \cdot (x_i - x_{i-1})$$

and therefore we have

$$wx_{i-1} \cdot f'_i \cdot (x_i - x_{i-1}) \leq F_i \leq wx_i \cdot f''_i \cdot (x_i - x_{i-1})$$

Clearly, too, $x_{i-1} \cdot f'_i$ is the minimum, and $x_i \cdot f''_i$ is the maximum, of the function $xf(x)$ over the ith subinterval. Thus by adding up these inequalities we have the total force F pinned between an upper and a lower sum

$$w \sum_{i=1}^{n} m_i(x_i - x_{i-1}) \leq F \leq w \sum_{i=1}^{n} M_i(x_i - x_{i-1})$$

where m_i and M_i denote the minimum and maximum, respectively, of $xf(x)$ on $\{x : x_{i-1} \leq x \leq x_i\}$. It follows that, since this inequality holds for any net whatsoever, we must define the total force to be

$$F = w \int_a^b xf(x)\,dx$$

■ **Example 1.** One end of a tank is in the form of an inverted isoceles triangle with sides 6 ft and top 4 ft long. We compute the force on this end when the tank is full of water.

Figure 7.21 illustrates the coordinate system discussed in the previous paragraphs. It is relatively easy to find that $f(x) = 4 - x/\sqrt{2}$

and that $a = 0$ and $b = \sqrt{32} = 4\sqrt{2}$. Assuming that water weighs 62.5 lb per cu ft, the total force on this end of the tank is

$$F = 62.5 \int_0^{\sqrt{32}} x\left(4 - \frac{x}{\sqrt{2}}\right) dx$$

$$= 62.5\left(2x^2 - \frac{x^3}{3\sqrt{2}}\right)\Big]_0^{\sqrt{32}}$$

$$= \frac{4000}{3} \text{ lb, or } \frac{2}{3} \text{ of a ton}$$

■ **Example 2.** The vertical face of a dam next to the water has the form of a trapezoid 300 ft long at the top, 180 ft long at the bottom and 50 ft high. We assume that the reservoir behind the dam is full and compute the total force exerted on the dam by the water pressure.

There is no loss of generality in assuming that the face of the dam looks like Fig. 7.22. We compute the width function to be

$$f(x) = 300 - \frac{12}{5}x$$

Hence the integral expressing the total force is

$$F = 62.5 \int_0^{50} x\left(300 - \frac{12}{5}x\right) dx$$

$$= 62.5\left(150x^2 - \frac{4}{5}x^3\right)\Big]_0^{50}$$

$$= 17{,}187{,}500 \text{ lb or about 8600 tons}$$

We shall give several more applications of the definite integral as the material of the subject is developed. For instance, several more geometric and mechanical applications are to be found in Chapter 11. One final application in this chapter follows next.

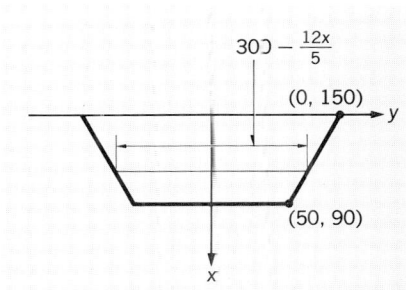

FIGURE 7.22

PROBLEMS 7.12

1. A vertical floodgate in the form of an 8-ft square has its upper edge at the surface of the water. What is the total force on the gate?

2. If the floodgate of Problem 1 had its top edge 8 ft below the surface, what would be the total force?

3. Divide the floodgate of Problem 1 into two triangular parts by means of a diagonal. What is the force of each triangle?

4. Suppose that the 8-ft square floodgate were installed with a diagonal vertical. If the upper corner were 10 ft below the surface of the water, what would the total force be?

5. A cylindrical tank 8 ft in diameter in a horizontal position is full of gasoline weighing 45 lb per cu ft. Express the force on the end of the tank as an integral.

6. A circular valve closing a water main has a radius of 2 ft and its center is 40 ft below the surface of the water. Express the force exerted on the valve.

7. The flat plate enclosed by $\{(x,y):y=8\}$ and $\{(x,y):y=2x^2\}$ is pushed down (along the y axis) into water. What is the total force on this plate when it is half submerged? (Two possible interpretations can be made.)

8. If a surface is submerged as shown in Fig. 7.23, prove that the total force is

$$F = \frac{62.5}{2} \int_a^b f^2(x)\, dx$$

7.13 RENEWAL THEORY

Consider a population, which changes with time, some members dying or becoming inoperative and other members being added. We seek a function $f(t)$ that will express the total number of members of the population at time t. We assume that we know this number $f(0)$ at time $t = 0$.

A *survival function* $s(t)$, perhaps empirically determined, is known. The number $s(t)$ is the fraction of the original population that survives to time t. Thus $f(0) \cdot s(t)$ is the number of original members still operative at time t. This function $s(t)$ is normally a decreasing function and has $\lim_{t \to \infty} s(t) = 0$. Clearly we also have $s(0) = 1$ and $0 \leq s(t) \leq 1$.

Lastly we are given a *renewal function* $r(t)$ telling us the rate at which new members are added to the population. Specifically, consider some fixed time $T > 0$, and let $0 = t_0 < t_1 < \cdots < t_n = T$ be any net on the interval $\{t : 0 \leq t \leq T\}$. We suppose that, at the time t_i, precisely $r(t_i) \cdot (t_i - t_{i-1})$ new members are added to the population. We suppose that these new members are still subject to the same survival function $s(t)$, of course. Thus when we are at time T, exactly $T - t_i$ time units have passed and hence

$$r(t_i) \cdot (t_i - t_{i-1}) \cdot s(T - t_i)$$

of the members added at time t_i survive to time T. Hence our total population at time T is approximated by the sum

$$f(0)s(T) + \sum_{i=1}^{n} s(T - t_i) r(t_i)(t_i - t_{i-1})$$

It follows immediately that the desired function $f(t)$ is

$$f(t) = f(0)s(t) + \int_0^t s(t - x) r(x)\, dx$$

■ *Example.* Suppose that our original population has P units and that the survival function is $s(t) = (1 + t)^{-2}$. If the renewal function is the constant $r(t) = 250$, then at time t the population is

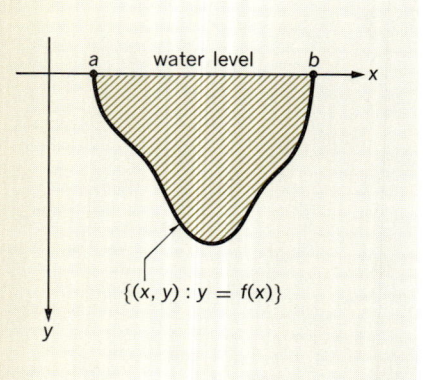

FIGURE 7.23

7.13 RENEWAL THEORY

$$f(t) = P(1 + t)^{-2} + \int_0^t \frac{250}{(1 + t - x)^2} dx$$

$$= \frac{P}{(1 + t)^2} + 250\left(1 - \frac{1}{1 + t}\right)$$

Other and more meaningful examples of this theory will be found after we have studied the transcendental functions of Chapter 8. For this reason we also do not include problems in this section.

8

EXPONENTIAL
AND LOGARITHM
FUNCTIONS

In the first seven chapters we considered only algebraic functions. Here we meet our first nonalgebraic, or *transcendental*, functions. The definition of an algebraic function (Section 3.11) implies that a transcendental function f is one that satisfies *no* functional equation

$$p_0 + p_1 f + p_2 f^2 + \cdots + p_n f^n = 0$$

where the functions p_i are polynomials. (A proof that the exponential and logarithm functions are transcendental is beyond the scope of this book.) We begin with some general theory which will be of use here and also in Chapter 9.

8.1 INVERSE FUNCTIONS

Let the function f be continuous and strictly increasing on an interval $\{x : a \leq x \leq b\}$. That is, if $a \leq x_1 < x_2 \leq b$, then $f(x_1) < f(x_2)$. We set

$$y = f(x), \qquad a \leq x \leq b$$

The claim is that for each $y_0, f(a) \leq y_0 \leq f(b)$, there is one and only one point x_0 in the interval such that $y_0 = f(x_0)$.

That such a point x_0 exists follows from Theorem 3.11 and the uniqueness of the point x_0 is proved by noting that if $x_1 \neq x_0$, then either $x_1 < x_0$ or $x_1 > x_0$. Therefore $f(x_1) < f(x_0)$ or $f(x_1) > f(x_0)$. In any case, $f(x_1) \neq y_0$.

We have shown that, if f is continuous and strictly increasing on $\{x : a \leq x \leq b\}$, and if $y = f(x)$, then x may be considered as a function of y. Let us temporarily denote this function by g so that $x = g(y)$ is defined on the interval $\{y : f(a) \leq y \leq f(b)\}$. It turns out that g is also continuous and strictly increasing. (See Fig. 8.1.)

Consider the composite functions $g \circ f$ and $f \circ g$. By definition,

$$(g \circ f)(x) = g(f(x)) = g(y) = x$$
$$(f \circ g)(y) = f(g(y)) = f(x) = y$$

That is, each of these composite functions is an identity function. These facts lead to the following definition.

If the two functions f and g satisfy the conditions

$$(g \circ f)(x) = g(f(x)) = x$$
$$(f \circ g)(x) = f(g(y)) = y$$

then g is called the *inverse function* of f (and f is the inverse of g). The inverse of the function f is usually denoted by

$$f^{-1}$$

(Note that the superscript -1 in this usage is *not* an exponent.) Therefore the inverse f^{-1} of the function f is defined by the two conditions $(f^{-1} \circ f)(x) = x$ and $(f \circ f^{-1})(y) = y$.

The graphs $\{(x,y) : y = f(x)\}$ and $\{(x,y) : x = f^{-1}(y)\}$ coincide. They are the same set of points. But it is not really natural to study the

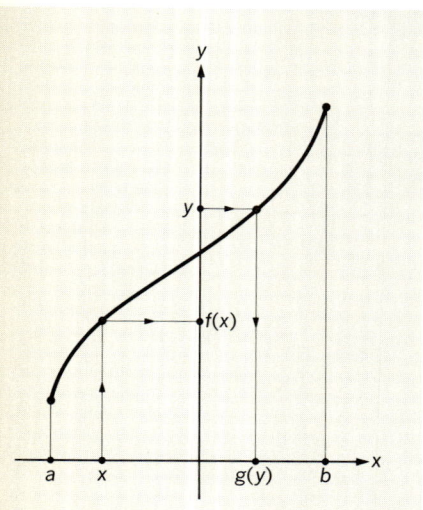

FIGURE 8.1

8.1 INVERSE FUNCTIONS

inverse function f^{-1} by means of the graph of f. The natural way to look at the graph of a function is to consider $\{(x,y) : y = f^{-1}(x)\}$. That is, we interchange the roles of x and y.

The graphs $\{(x,y) : y = f(x)\}$ and $\{(x,y) : y = f^{-1}(x)\}$ are symmetrically located with respect to the line $\{(x,y) : x = y\}$. (See Fig. 8.2.) This just means that if (a, b) is a point on one graph, then (b, a) is a point on the other. Another way to look at this relationship is to note that the two graphs $\{(x,y) : y = f^{-1}(x)\}$ and $\{(x,y) : x = f(y)\}$ coincide. So by drawing $\{(x,y) : x = f(y)\}$, we have the graph of the inverse function.

- **Example 1.** Consider the function $f(x) = x^2$ on $\{x : 1 \leq x \leq 2\}$. We claim that $f^{-1}(y) = \sqrt{y}$ and that f^{-1} is defined on $\{y : 1 \leq y \leq 4\}$. The proof of this claim consists in noting that

$$(f^{-1} \circ f)(x) = f^{-1}(f(x)) = f^{-1}(x^2) = \sqrt{x^2} = |x| = x$$

(because $1 \leq x \leq 2$)

$$(f \circ f^{-1})(y) = f(f^{-1}(y)) = f(\sqrt{y}) = (\sqrt{y})^2 = y$$

(because $1 \leq y \leq 4$) ■

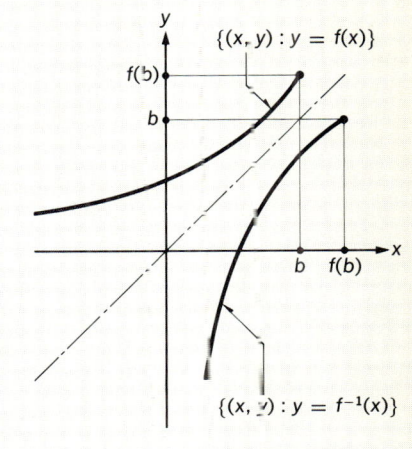

FIGURE 8.2

In general, finding the inverse function f^{-1} entails solving the equation $y = f(x)$ for x in terms of y. We cannot always find an explicit solution, of course, and even if we do find a solution it may not be unique. For instance, consider the function $f(x) = x^2$, but now on the interval $\{x : -1 \leq x \leq 1\}$. Solving the equation

$$y = x^2$$

for x in terms of y, we get two solutions

$$x = \sqrt{y}, \qquad x = -\sqrt{y}, \qquad 0 \leq y \leq 1$$

Neither of these is an inverse function, however, because neither satisfies the defining condition. Let $g_1(y) = \sqrt{y}$ and $g_2(y) = -\sqrt{y}$. Then

$$(g_1 \circ f)(x) = g_1(f(x)) = g_1(x^2) = \sqrt{x^2} = |x|$$

But $x \neq |x|$ when $x < 0$, so that $(g_1 \circ f)(x) \neq x$ on $\{x : -1 \leq x \leq 1\}$. Similarly,

$$(g_2 \circ f)(x) = g_2(f(x)) = g_2(x^2) = -\sqrt{x^2} = -|x|$$

and $x \neq -|x|$ if $x > 0$ so $(g_2 \circ f)(x) \neq x$ on $\{x : -1 \leq x \leq 1\}$. The reason there is no inverse function for $f(x) = x^2$ on this interval is that each value of y determines *two* values of x. Such a situation does not give rise to a function (which is always defined to be single-valued).

In our opening discussion we assumed that the function f was strictly increasing. By just changing the direction of the appropriate inequalities, we can prove the same statements for a strictly decreasing function. A function f is *strictly monotone* on an interval $\{x : a \leq x \leq b\}$ if it is either strictly increasing or strictly decreasing on the interval. The terms *strictly monotone increasing* and *strictly monotone decreasing* are used also to distinguish between the two cases when need be.

▪ **THEOREM 8.1.** Let f be continuous and strictly monotone on $\{x : a \leq x \leq b\}$. Then the inverse function f^{-1} is continuous and strictly monotone on the closed interval between $f(a)$ and $f(b)$. In particular, if f is monotone increasing, then f^{-1} is monotone increasing on $\{y : f(a) \leq y \leq f(b)\}$ and if f is monotone decreasing, then f^{-1} is monotone decreasing on $\{y : f(b) \leq y \leq f(a)\}$.

The only difficulty in the proof of Theorem 8.1 lies in showing that f^{-1} is continuous. We sketch such a proof (for the case f increasing) as follows: Let $f(x_0)$ be a point between $f(a)$ and $f(b)$. Given $\epsilon > 0$, we must find $\delta > 0$ such that

$$|f^{-1}(y) - x_0| < \epsilon \quad \text{whenever } |y - f(x_0)| < \delta$$

Assume that ϵ is small enough so that $x_0 - \epsilon$ and $x_0 + \epsilon$ both lie in $\{x : a \leq x \leq b\}$. Then we shall have

$$x_0 - \epsilon < f^{-1}(y) < x_0 + \epsilon$$

if and only if

$$f(x_0 - \epsilon) < f(f^{-1}(y)) = y < f(x_0 + \epsilon)$$

Now choose δ to be the smaller of the two numbers $|f(x_0) - f(x_0 - \epsilon)|$ and $|f(x_0 + \epsilon) - f(x_0)|$. Then we have

$$f(x_0 - \epsilon) \leq f(x_0) - \delta \quad \text{and} \quad f(x_0) + \delta \leq f(x_0 + \epsilon)$$

Therefore every point y satisfying

$$f(x_0) - \delta < y < f(x_0) + \delta$$

will also satisfy

$$f(x_0 - \epsilon) < y < f(x_0 + \epsilon)$$

Then, because f^{-1} is monotone increasing, we have

$$f^{-1}(f(x_0 - \epsilon)) = x_0 - \epsilon < f^{-1}(y) < f^{-1}(f(x_0 + \epsilon)) = x_0 + \epsilon$$

or

$$|f^{-1}(y) = x_0| < \epsilon$$

This completes the proof for a point $f(x_0)$ such that $a < x_0 < b$. An easy modification, using right-hand and left-hand limits, establishes the continuity of f^{-1} at $f(a)$ and $f(b)$. ▪

One more example will suffice for now.

▪ **Example 2.** Consider the function $f(x) = x^2 - 3x - 2$ on the interval $\{x : -1 \leq x \leq 1\}$. Because $f'(x) = 2x - 3$ is always negative on this interval, f is strictly monotone decreasing on the interval. Therefore Theorem 8.1 applies to tell us that f^{-1} is defined, is continuous, and is monotone decreasing on $\{y: -4 \leq y \leq 2\}$.

Setting $y = x^2 - 3x - 2$ and solving for x in terms of y, we obtain

$$x = \frac{3 \pm \sqrt{17 + 4y}}{2}$$

Thus we have two possible choices

$$g_1(y) = \tfrac{1}{2}(3 + \sqrt{17 + 4y}), \qquad g_2(y) = \tfrac{1}{2}(3 - \sqrt{17 + 4y})$$

for the inverse function. Not both can be valid, of course. To determine which is the desired inverse, apply the defining conditions:

$$\begin{aligned}(g_1 \circ f)(x) = g_1(x^2 - 3x - 2) &= \tfrac{1}{2}(3 + \sqrt{17 + 4x^2 - 12x - 8}) \\ &= \tfrac{3}{2} + \tfrac{1}{2}\sqrt{4x^2 - 12x + 9} \\ &= \tfrac{3}{2} + \tfrac{1}{2}\sqrt{(2x-3)^2} \\ &= \tfrac{3}{2} + \tfrac{1}{2}|2x - 3|\end{aligned}$$

But $\tfrac{3}{2} + \tfrac{1}{2}|2x - 3| \neq x$ if $x < 0$ because $\tfrac{3}{2} + \tfrac{1}{2}|2x - 3| > 0$ always. Therefore g_1 cannot be the inverse. On the other hand,

$$(g_2 \circ f)(x) = g_2(x^2 - 3x - 2) = \tfrac{3}{2} - \tfrac{1}{2}|2x - 3|$$

In the range $-1 \leq x \leq 1$, the quantity $2x - 3$ is always negative and hence $|2x - 3| = 3 - 2x$. Hence, $\tfrac{3}{2} - \tfrac{1}{2}|2x - 3| = \tfrac{3}{2} - \tfrac{1}{2}(3 - 2x) = \tfrac{3}{2} - \tfrac{3}{2} + x = x$ on the interval $\{x: -1 \leq x \leq 1\}$.

The inverse f^{-1} of $f(x) = x^2 - 3x - 2$ on the interval $\{x: -1 \leq x \leq 1\}$ has values

$$f^{-1}(y) = \tfrac{3}{2} - \tfrac{1}{2}\sqrt{17 + 4y}$$

Figure 8.3 pictures this function f. ∎

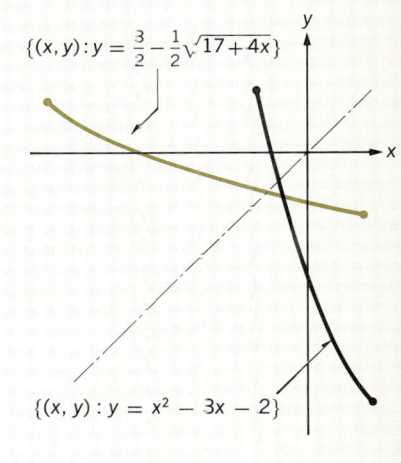

FIGURE 8.3

EXERCISES 8.1

For Exercises 1 to 10, determine whether or not the function has an inverse over the given domain. Determine the domain of the inverse function if it exists.

1. $f(x) = 2x + 3$, $\{x: -\infty < x < \infty\}$
2. $f(x) = x^2 + 2x - 3$, $\{x: 0 \leq x < \infty\}$
3. $f(x) = x^2 + 2x - 3$, $\{x: -3 \leq x \leq 0\}$
4. $f(x) = x^3 + 2x - 3$, $\{x: -\infty < x < \infty\}$
5. $f(x) = \dfrac{x}{x+1}$, $\{x: -1 < x < \infty\}$
6. $f(x) = \dfrac{2+x}{2-x}$, $\{x: 0 \leq x < 2\}$
7. $f(x) = \dfrac{2x+3}{5x-2}$, $\{x: -2 \leq x \leq 0\}$

8. $f(x) = \dfrac{2x}{1+x^2}$, $\{x : 1 \leq x < \infty\}$

9. $f(x) = \dfrac{x^2}{(1+x)^2}$, $\{x : -2 \leq x \leq 2\}$

10. $f(x) = x\sqrt{1+x^2}$, $\{x : -2 \leq x \leq 2\}$

For Exercises 11 to 20, determine those domains of f over which f^{-1} exists.

11. $f(x) = x^2 + 4x - 1$
12. $f(x) = 5x - x^2$

13. $f(x) = 4 + 3x + 2x^2$
14. $f(x) = x + \dfrac{1}{x}$

15. $f(x) = 7 - 9x + 3x^2 + x^3$
16. $f(x) = x^3 + 3x^2 + 6x - 3$

17. $f(x) = x^4 - 2x^2 - 1$
18. $f(x) = \dfrac{2}{3} - 4x^2 + \dfrac{4}{3}x^3 + x^4$

19. $f(x) = \dfrac{2x}{1+x^2}$
20. $f(x) = x$ if $x < 1$
    $= x^2$ if $x \geq 1$

For Exercises 21 to 25, draw the graph of the function f over the given domain and then reflect this graph across the line $\{(x,y) : x = y\}$ to obtain the graph of the inverse function f^{-1}. Compute the slope of $\{(x,y) : y = f(x)\}$ at the point $(1, 2)$ and from this determine the slope of $\{(x,y) : y = f^{-1}(x)\}$ at the point $(2, 1)$.

21. $f(x) = 1 + x$, $\{x : 0 \leq x \leq 3\}$

22. $f(x) = 3x - 1$, $\{x : 0 \leq x \leq 2\}$

23. $f(x) = 1 + x^2$, $\{x : 0 \leq x \leq 2\}$

24. $f(x) = x^3 + 3x - 2$, $\{x : 0 \leq x \leq 2\}$

25. $f(x) = \dfrac{4x}{1+x^2}$, $\{x : -1 \leq x \leq 2\}$

PROBLEMS 8.1

1. Over what domain does the function $f(x) = x/(1 + |x|)$ have an inverse and what is it?

2. Prove that $f(x) = \sqrt[n]{x}$, $n \in \mathbf{N}$, is continuous and strictly increasing on the closed ray $\{x : 0 \leq x\}$.

3. Given any constant b, determine the constant c such that the function $f(x) = (x + b)/(x + c)$ is its own inverse.

4. Determine constants b and c such that the function $f(x) = (bx + 1)/(cx - 1)$ is its own inverse.

5. What conditions on the constants a, b, c, and d ensure that $f(x) = a + bx + cx^2 + dx^3$ have an inverse function over all of $\mathbf{R}$?

6. If $f(x) = |x|$ and $g(y) = y$, then $(g \circ f)(x) = x$, for $x \geq 0$. Is g then equal to f^{-1}?

7. Prove that the inverse function of an algebraic function is again an algebraic function.

8. If f and g are strictly increasing [decreasing], show that $g \circ f$ is also strictly increasing [decreasing] and that $(g \circ f)^{-1} = f^{-1} \circ g^{-1}$.

9. Let f be continuous and strictly increasing on $\mathbf{R}$ and suppose that f satisfies the functional equation
$$f(x+y) = f(x) \cdot f(y)$$
Prove that f^{-1} satisfies the functional equation
$$f^{-1}(x \cdot y) = f^{-1}(x) + f^{-1}(y)$$

8.2 THE DERIVATIVE OF AN INVERSE FUNCTION

Even in such a simple case as the function $f(y) = 2 - 4y + y^3$, it is difficult to obtain the inverse function f^{-1} explicitly. In this instance we must solve the cubic equation $x = 2 - 4y + y^3$ for y in terms of x. Despite such difficulty, however, we can determine the derivative of f^{-1} in an explicit form.

■ **THEOREM 8.2.** Let f be a differentiable monotonic function of a variable y. Then its inverse function f^{-1} is a differentiable function of the variable $x = f(y)$ and

$$Df^{-1}(x) = \frac{1}{Df(y)}$$

PROOF: Apply the chain rule of differentiation to the defining condition

$$(f^{-1} \circ f)(y) = y$$

We obtain

$$D[f^{-1} \circ f](y) = (Df^{-1})(f(y)) \cdot Df(y) = Dy = 1$$

and because $x = f(y)$, we have

$$Df^{-1}(x) \cdot Df(y) = 1$$

or

$$Df^{-1}(x) = \frac{1}{Df(y)} \quad \blacksquare$$

■ **Example 1.** If $f(y) = 2 - 4y + y^3$, then $Df(y) = -4 + 3y^2$. Thus if $x = 2 - 4y + y^3$, we have

$$Df^{-1}(x) = \frac{1}{-4 + 3y^2}$$

Note that this formula yields $Df^{-1}(x)$ in terms of y, not in terms of x as is more natural perhaps. Nevertheless the formula is still valuable. For suppose that the function $f(y)$ is given and we must study its inverse $f^{-1}(x)$. We draw the graph $\{(x,y): y = f^{-1}(x)\}$ simply by drawing the set $\{(x,y): x = f(y)\}$ and then constructing its symmetric image with respect to the line $\{(x,y): x = y\}$.

A less sophisticated procedure for drawing the graph $\{(x,y) : y = f^{-1}(x)\}$ is to assign values of y and compute the corresponding values of $x = f(y)$. Thus we can find as many points (x,y) on $\{(x,y) : y = f^{-1}(x)\}$ as we wish. At the points so located, the slope Df^{-1} of the curve is then given by Theorem 8.2. It does not matter that Df^{-1} is given in terms of y; we can still compute its value wherever it is needed.

■ **Example 2.** Let $f(y) = 2 - 4y + y^3$ as in Example 1. Over the interval $\{y : -1 \leq y \leq 1\}$ we draw the graph $\{(x,y) : y = f^{-1}(x)\}$, recalling from Example 1 that $Df^{-1}(x) = 1/(-4 + 3y^2)$. In the usual way we prepare a table as given below by choosing values of y and computing.

x	$f(y)$	$2 - 4y + y^3$	5	2	-1
y	y	y	-1	0	1
Df^{-1}	$\dfrac{1}{Df(y)}$	$\dfrac{1}{-4 + 3y^2}$	-1	$-\tfrac{1}{4}$	-1

In Fig. 8.4(a), these points are plotted and segments of the tangent lines at these points are drawn. The completed graph appears in Fig. 8.4(b).

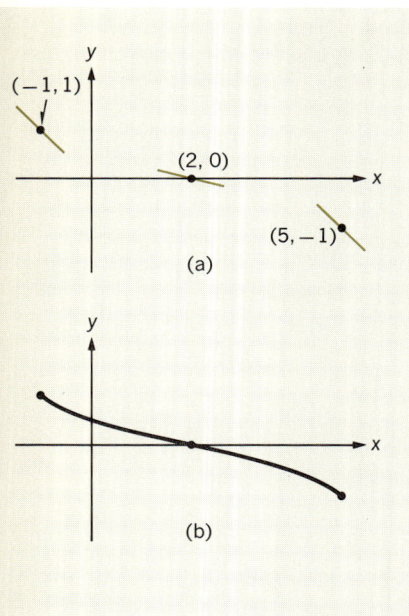

FIGURE 8.4

EXERCISES 8.2

For Exercises 1 to 10, compute Df^{-1}; express the result in terms of x.

1. $f(y) = 3 - 2y$
2. $f(y) = 1 - 2y + y^2, y \geq 1$
3. $f(y) = 4 - 5y + y^2, y \geq \tfrac{5}{2}$
4. $f(y) = \dfrac{y}{1+y}, y \neq 1$
5. $f(y) = \dfrac{3y-2}{2+y}, y \neq -2$
6. $f(y) = \dfrac{1+y}{y^2}, y > 0$
7. $f(y) = \sqrt{16 + y^2}, y \geq 0$
8. $f(y) = \sqrt{9 - y^2}, y \geq 0$
9. $f(y) = y^{2/3}$
10. $f(y) = (4 - y^2)^{-3/2}, y \geq 0$
11. Use Theorem 8.2 to derive the formula
$$Dx^{1/n} = \frac{1}{n} x^{-1+1/n}$$
12. Suppose that a function f satisfies the equation $f'(y) = f(y)$ for all points $y \in \mathbf{R}$. Show that $Df^{-1}(x) = 1/x$.

PROBLEMS 8.2

1. If $s = f(t)$, $v = ds/dt$ and $a = dv/dt$, show that $a = v\, dv/ds$.
2. Derive formulas for $D^2 f^{-1}$ and $D^3 f^{-1}$ in terms of Df, $D^2 f$, and $D^3 f$.
3. Suppose that f^{-1} exists and that we define $F(x) = \int_0^x f(t)\, dt$. Prove that
$$D(xf^{-1}(x) - F(f^{-1}(x))) = f^{-1}(x)$$

4. If $F(x) = \int_0^x f(t)\,dt$ and if $y = F^{-1}(x)$, show that $y' = 1/f(y)$.

5. Define
$$f(x) = \int_0^x \frac{dt}{\sqrt{1+t^3}}, \qquad x \geq 0$$
and prove that $D_y^2 f^{-1}(y) = \tfrac{3}{2}[f^{-1}(y)]^2$.

6. Suppose that $f'(x) = f(x)$ for each x. Prove that
$$\frac{df^{-1}(y)}{dy} = \frac{1}{y}$$

7. Define
$$x = \int_1^y \frac{dt}{t}$$
and prove that $D_x y = y$.

8. Define
$$x = \int_0^y \frac{dt}{\sqrt{1+4t^2}}$$
and prove that $D_x^2 y$ is proportional to y. Determine the constant of proportionality.

8.3 EXPONENTIAL AND LOGARITHM FUNCTIONS

The exponential function a^x and the logarithm function $\log_a x$ are inverses of each other. That is,
$$a^{\log_a x} = x \quad \text{and} \quad \log_a(a^x) = x$$

In theory, therefore, it does not matter which is defined first because the other will follow from Section 8.1. We shall present three different approaches to these functions in this and the following sections. All three approaches contribute to complete understanding. Our first approach is similar to that found in the usual secondary-school mathematics course.

Let $a > 0$ be a positive real number. Then for any rational number p/q, $q > 0$, the number
$$a^{p/q} = \sqrt[q]{a^p}$$
is well defined (see Section 1.6). But what does the symbol a^x mean if x is an irrational number? For instance, how do we define $2^{\sqrt{2}}$? A natural answer is to take the limit of such a sequence as
$$2,\ 2^{1.4},\ 2^{1.41},\ 2^{1.414},\ 2^{1.4142},\ \ldots$$
to be the definition of $2^{\sqrt{2}}$. But such a definition has many difficulties in practice. We adopt the following.

If $a > 1$, then the number a^x is the least upper bound of the set
$$\{a^r : r \in \mathbf{Q}, r < x\}$$

If $0 < a < 1$, then a^x is the least upper bound of the set
$$\{a^r : r \in \mathbf{Q}, r > x\}$$

Given $x \in \mathbf{R}$, let n be the largest integer less than or equal to x. If $a > 1$ and $r < x$, then $r < n + 1$. Then one easily shows that $a^r < a^{n+1}$; hence the set $\{a^r : r \in \mathbf{Q}, r < x\}$ is bounded above by a^{n+1}. If $0 < a < 1$ and $r > x$, then $r > n$. It is again easy to prove that $a^r < a^n$ so the set $\{a^r : r \in \mathbf{Q}, r > x\}$ is bounded above by a^n. Hence, in view of Theorem 1.10, a^x exists and is unique. Furthermore, because $\sqrt[q]{a^p} > 0$ for each rational number p/q, it follows that $a^x \geq 0$ for all $x \in \mathbf{R}$. Indeed, $a^x > 0$ for all $x \in \mathbf{R}$, because there is a rational number $r < x$ and we know that $a^r > 0$.

A function f with values
$$f(x) = a^x, \quad a > 0$$
is called an *exponential function*. By the preceding discussion, such a function is defined and positive at each point $x \in \mathbf{R}$.

The *logarithm function* $\log_a x$, $a > 0$, is defined to be the inverse function of the exponential function a^y. Therefore we have the identities
$$a^{\log_a x} = x \quad \text{and} \quad \log_a(a^y) = y$$

We note that the identity $a^{\log_a x} = x$ plus the fact that $a^y > 0$ for all $y \in \mathbf{R}$ implies that $\log_a x$ is defined only for positive real numbers.

In particular, we know that $a^0 = 1$. Therefore $\log_a(a^0) = \log_a(1) = 0$ by the second identity above. This tells us that the point $(0, 1)$ is the y intercept on each graph $\{(x, y) : y = a^x, a > 0\}$ and that $(1, 0)$ is the x intercept on each graph $\{(x, y) : y = \log_a x, a > 0\}$.

The proof of our first result is not given here but it will be a consequence of a later theorem (8.9).

■ **THEOREM 8.3.** Let a^x, $a > 0$ be an exponential function. Then for any real numbers x and y,

1. $a^x \cdot a^y = a^{x+y}$
2. $(a^x)^y = a^{xy}$

■ **COROLLARY 8.4.** Let $\log_a x$, $a > 0$, be a logarithm function. Then for any positive real numbers x and y,

1. $\log_a(xy) = \log_a x + \log_a y$
2. $\log_a(x^y) = y \log_a x$

The properties of logarithms given in Corollary 8.4 are merely restatements of the properties of exponents given in Theorem 8.3.

■ **THEOREM 8.5.** Let $h > -1$ be any real number. Then for any natural number $n \in \mathbf{N}$,

$$(1 + h)^n \geq 1 + nh$$

PROOF: We use mathematical induction. The result is obviously true for the case $n = 1$. Suppose it is true for $k > 1$. Then $(1 + h)^k \geq 1 + kh$ and hence

$$(1 + h)^{k+1} = (1 + h)(1 + h)^k \geq (1 + h)(1 + kh)$$

because $1 + h > 0$. But then

$$(1 + h)(1 + kh) = 1 + (k + 1)h + kh^2 \geq 1 + (k + 1)h$$

because $kh^2 \geq 0$. Therefore $(1 + h)^n \geq 1 + nh$ for all $n \in \mathbf{N}$. Indeed, if $h \neq 0$, then strict inequality holds for $n > 1$. ∎

■ **THEOREM 8.6.** If $a > 1$, then the exponential function a^x is strictly increasing over the entire set $\mathbf{R}$ and we have

1. $\lim\limits_{x \to +\infty} a^x = +\infty$ 2. $\lim\limits_{x \to -\infty} a^x = 0$

If $0 < a < 1$, then a^x is strictly decreasing over $\mathbf{R}$ and

1. $\lim\limits_{x \to +\infty} a^x = 0$ 2. $\lim\limits_{x \to -\infty} a^x = +\infty$

PROOF: If $a > 1$, set $a = 1 + h$ where $h > 0$. We first show that

$$a^x = (1 + h)^x > 1 \quad \text{whenever } x > 0$$

If $x > 0$, then there is a rational number p/q such that $0 < p/q < x$. We know that $a^x \geq a^{p/q}$, so we need show only that $a^{p/q} > 1$. But the latter inequality is equivalent to $a^p > 1^q = 1$. Also from Theorem 8.5 $a^p = (1 + h)^p > 1 + ph > 1$. This completes the first part.

Now let $x_1 < x_2$ be any two real numbers. Then $x_2 - x_1 > 0$ and hence

$$a^{x_2 - x_1} = a^{x_2} \cdot a^{-x_1} = a^{x_2}/a^{x_1} > a^0 = 1$$

or $a^{x_1} < a^{x_2}$. This proves that a^x, $a > 1$, is strictly increasing.

Next let M be any (large) positive number. From the Archimedean property of the order relation, there exists $n \in \mathbf{N}$ such that $nh \geq M$. Therefore

$$a^n = (1 + h)^n > 1 + nh > M$$

and it follows immediately that $\lim_{x \to +\infty} a^x = +\infty$. On the other hand, let $\epsilon > 0$ be any (small) positive number. Then there exists $n \in \mathbf{N}$ such that $1/nh < \epsilon$. Then

$$a^{-n} = \frac{1}{a^n} = \frac{1}{(1 + h)^n} < \frac{1}{1 + nh} < \epsilon$$

which implies that $\lim_{x \to -\infty} a^x = 0$.

The corresponding statements for the case $0 < a < 1$ are proved analogously by setting $1/a = 1 + h$, $h > 0$. ∎

■ **COROLLARY 8.7.** If $a > 1$, then the logarithm function is strictly increasing over the entire open ray $\{x : 0 < x\}$ and we have

1. $\lim\limits_{x \to +\infty} \log_a x = +\infty$ 2. $\lim\limits_{x \to 0} \log_a x = -\infty$

If $0 < a < 1$, then $\log_a x$ is strictly decreasing over $\{x : 0 < x\}$ and

1. $\lim\limits_{x \to +\infty} \log_a x = -\infty$ 2. $\lim\limits_{x \to 0} \log_a x = +\infty$

PROOF: The statements concerning monotonicity follow from Theorem 8.1. The limit statements follow from the corresponding limits in Theorem 8.6. ∎

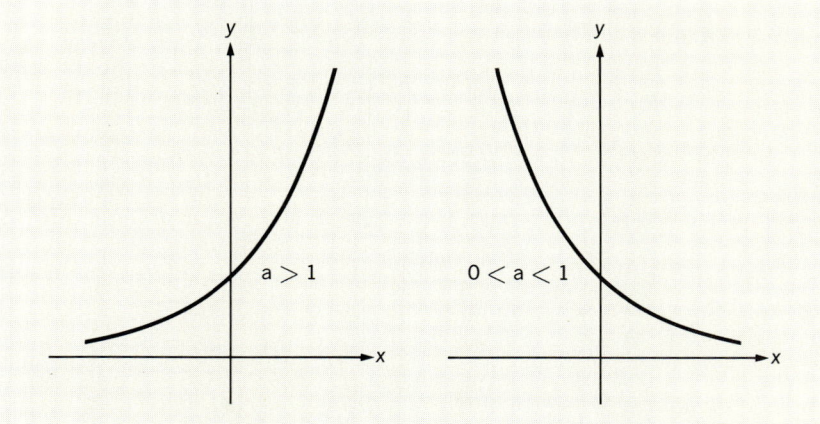

FIGURE 8.5

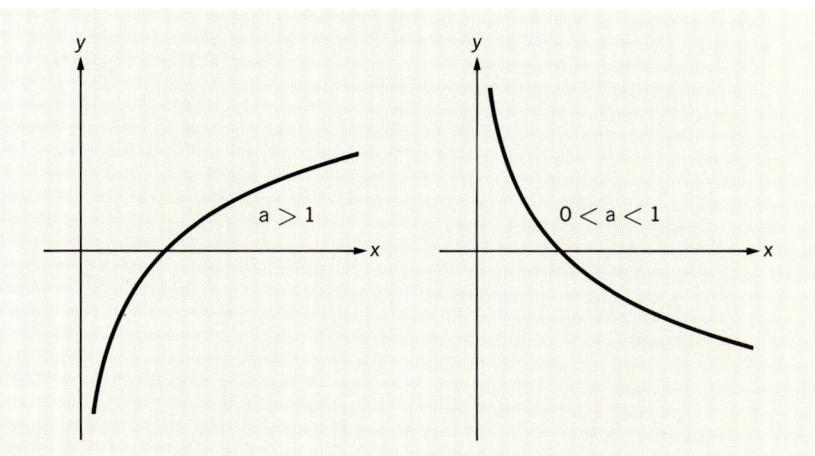

FIGURE 8.6

8.3 EXPONENTIAL AND LOGARITHM FUNCTIONS

The graphs $\{(x, y) : y = a^x\}$ in Fig. 8.5 illustrate the behavior of the exponential functions when $a > 1$ and when $0 < a < 1$. Similarly, Fig. 8.6 pictures two cases, $a > 1$ and $0 < a < 1$, of logarithm functions.

EXERCISES 8.3

Solve the Eqs in Exercises 1 to 16. Check your solutions in order to eliminate extraneous roots.

1. $\log_a x = 1$
2. $\log_a(x - 1) = 2$
3. $4^{x-1} = 2^{x+2}$
4. $2^x + 2^{2-x} = 4$
5. $2^x - 5 = 6(2^{-x})$
6. $2(3^x) - 3^{-x} + 1 = 0$
7. $1 - 3(10^x) = 4(10^{-x})(10^{-x} + 1)$
8. $3^x = 2^{x+1}$
9. $4^{\log_4 5} + 2^{-\log_2 3} = 3^{\log_3 x}$
10. $\log_2(x - 1) + \log_2(x - 2) = \log_2 6$
11. $\log_3(x + 1) + \log_3(x - 2) = \log_3 5$
12. $\log_2(x + 1) = 1 + \log_2(1 - x)$
13. $\log_{10} x + \log_{10}(x - 21) = 2$
14. $\log_3(\sqrt{x} + \sqrt{x + 1}) = 1$
15. $\log_a(2x - 1) = \log_a(x + 2)$
16. $2^x = x^2$

Solve the inequalities in Exercises 17 to 20. Check the solutions obtained.

17. $\log_a|x| < 1$
18. $\log_3(x + 3) < 0$
19. $\log_a x < -1$
20. $\log_a(x - 2) > 2$

For the functions in Exercises 21 to 28, find the inverse function and its domain.

21. $f(y) = 3^{1-y}$
22. $f(y) = 2^{1+y/2} + 1$
23. $f(y) = \frac{1}{2}(a^y - a^{-y})$
24. $f(y) = \frac{1}{4}(a^y + 4a^{-y})$
25. $f(y) = \log_2(y + 1) + \log_2(y - 1)$
26. $f(y) = 2\log_3 y - \log_3(y + 1)$
27. $f(y) = \frac{1}{2}\log_3(3^{1+y/2} - 2)$
28. $f(y) = \log_a(1 - \sqrt{4 - y^2})$

29. If $2x - \log_a(3 + 6a^x + 3a^{2x}) = b - 2\log_a(1 + a^x)$, what is b?
30. Prove that $\log_a(x + \sqrt{x^2 - 1}) = -\log_a(x - \sqrt{x^2 - 1})$.
31. If $h \neq 0$, show that $1/h[\log_a(x + h) - \log_a x] = \log_a(1 + h/x)^{1/h}$.
32. Solve
$$\log_a \frac{1 + x}{1 - x} = \log_a \frac{1 + b}{1 - b} + \log_a \frac{1 + c}{1 - c}$$
given that $b > 0, c > 0$.

PROBLEMS 8.3

1. Prove that there are unique solutions to the equations
   (a) $x + a^x = 0$, $a > 0$
   (b) $x + \log_a x = 0$, $a > 1$

2. Let $0 < a < b$ and show that

$$a^x > b^x \quad \text{if } x < 0$$
$$a^x < b^x \quad \text{if } 0 < x$$

3. Let $1 < a < b$ and show that

$$\log_a x > \log_b x \quad \text{if } x > 1$$
$$\log_a x < \log_b x \quad \text{if } 0 < x < 1$$

4. If $a = b^c$, show that $\log_d(\log_d a) = \log_d c + \log_d(\log_d b)$. For what values of a, b, and c is this equation valid?

8.4 GRAPHS OF EXPONENTIAL AND LOGARITHM FUNCTIONS

We first draw the graphs of 2^x and $\log_2 x$ on the same coordinate plane for comparison purposes. In the following tables notice that the point (a, b) is in one table if and only if (b, a) is in the other. This is just another statement of the fact that one curve is the reflection of the other across the line $\{(x, y) : x = y\}$.

x	-3	-2	-1	0	1	2	3
2^x	$\frac{1}{8}$	$\frac{1}{4}$	$\frac{1}{2}$	1	2	4	8

x	$\frac{1}{8}$	$\frac{1}{4}$	$\frac{1}{2}$	1	2	4	8
$\log_2 x$	-3	-2	-1	0	1	2	3

The diagrams in Fig. 8.7 are typical but in Fig. 8.8 we compare exponential and logarithm functions for the three bases $a = 2, 4$, and 10.

Logarithms to the base 10, the so-called *common logarithms,* are normally used in manual computation. The internal arithmetic of an electronic computer is binary (base 2), so that logarithms to the base 2 are most often used in computer work. In analysis, theoretical reasons make it convenient to use a number denoted by e as the base for logarithms. We shall meet e first in the next section.

It is easy to change the base of logarithms. Suppose we know logarithms to the base a and want a logarithm to a different base b. Now by definition

$$b^{\log_b x} = x$$

We take the logarithm to the base a of both sides of this equation, obtaining

$$\log_a(b^{\log_b x}) = \log_a x$$

Therefore

$$\log_b x \cdot \log_a b = \log_a x$$

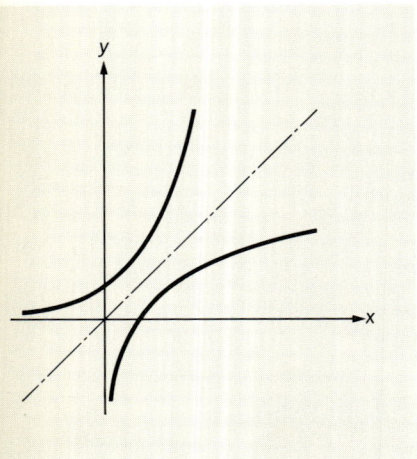

FIGURE 8.7

or

$$\log_b x = \frac{\log_a x}{\log_a b}$$

Note in particular that

$$\log_b a = \frac{1}{\log_a b}$$

As an example suppose we need $\log_2 7$ but only have a table of common logarithms. Then

$$\log_2 7 = \frac{\log_{10} 7}{\log_{10} 2} \cong \frac{0.84510}{0.30103} \cong 2.80736$$

EXERCISES 8.4

For Exercises 1 to 5, sketch graphs of the given pairs of functions, each pair on the same coordinate plane.

1. $f(x) = 2^x$; $g(x) = \log_2 x$
2. $f(x) = 2^{-x}$; $g(x) = -\log_2 x$
3. $f(x) = 2^{1-x}$; $g(x) = 1 - \log_2 x$
4. $f(x) = 2^x + 2^{-x}$; $g(x) = \log_2[\frac{1}{2}(x + \sqrt{x^2 - 4})]$
5. $f(x) = 2^{x^2}$; $g(x) = \sqrt{\log_2 x}$

For Exercises 6 to 20, sketch the graph of f over the given interval.

6. $f(x) = 2^{-x^2}$, $\{x : -2 \leq x \leq 2\}$
7. $f(x) = x 2^x$, $\{x : -1 \leq x \leq 2\}$
8. $f(x) = x^2 2^{-x}$, $\{x : 0 \leq x \leq 4\}$
9. $f(x) = 2^x - 2^{-x}$, $\{x : -2 \leq x \leq 2\}$
10. $f(x) = \frac{2^x - 2^{-x}}{2^x + 2^{-x}}$, $\{x : -2 \leq x \leq 2\}$
11. $f(x) = 2^{2x} - 6(2^x) + 9x$, $\{x : -1 \leq x \leq 2\}$
12. $f(x) = 2^{2x}$, $\{x : -2 \leq x \leq 1\}$
13. $f(x) = \log_2 x^2$, $\{x : -4 \leq x < 0\}$
14. $f(x) = \log_4 |4 - 3x|$, $\{x : -4 \leq x \leq 0\}$
15. $f(x) = x - x \log_2 x$, $\{x : 1 \leq x \leq 4\}$
16. $f(x) = \frac{1}{x} \log_2 x^2$, $\{x : 1 \leq x \leq 4\}$
17. $f(x) = x \log_2(1 + x^2)$, $\{x : 0 \leq x \leq \sqrt{7}\}$
18. $f(x) = \log_2 \frac{1 + x^2}{1 - x^2}$, $\{x : -1 < x < 1\}$
19. $f(x) = \log_2(x + \sqrt{x^2 - 1})$, $\{x : 1 \leq x \leq \frac{5}{3}\}$

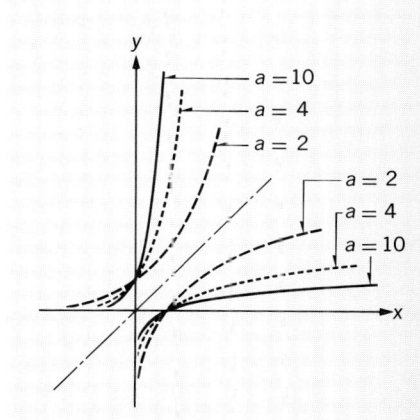

FIGURE 8.8

20. $f(x) = \log_2 2^{2x}$, $\{x : -2 \leq x \leq 2\}$
21. Let $0 < a < 1$ and prove that $\log_a x = -\log_{1/a} x$.
22. Given that $\log_2 10 = 3.3219$, determine
    (a) $\log_2 20$ (b) $\log_2 400$ (c) $\log_2 0.08$
    (d) $\log_{10} 20$ (e) $\log_{10} 400$ (f) $\log_{10}(1.6)^{1.3}$

8.5 A FUNCTIONAL EQUATION APPROACH

The exponential function a^x, $a > 1$, satisfies the functional equation

$$f(x + y) = f(x) \cdot f(y)$$

in view of Theorem 8.3. A converse theorem is also true.

■ **THEOREM 8.8.** If f is a nonconstant continuous function satisfying the functional equation $f(x + y) = f(x) \cdot f(y)$, then $f(x) = a^x$ for some constant a, that is, f is an exponential function.

PROOF: First we note that $f(0) = f(0 + 0) = f(0) \cdot f(0)$, or $f(0) = [f(0)]^2$. Hence $f(0) = 1$. (If $f(0) = 0$, then f would be constant.) An easy induction argument now shows that

$$f(n) = [f(1)]^n, \quad n \in \mathbf{N}$$

This is obviously true for $n = 1$ and if it is true for $n = k$, then

$$f(k + 1) = f(1) \cdot f(k) = f(1) \cdot [f(1)]^k = [f(1)]^{k+1}$$

Also because

$$f(0) = f(n - n) = f(n)f(-n) = 1$$

we have

$$f(-n) = \frac{1}{[f(1)]^n} = [f(1)]^{-n}$$

Thus for each $p \in \mathbf{Z}$,

$$f(p) = [f(1)]^p$$

Another simple induction argument proves that for any $x \in \mathbf{R}$,

$$f(px) = [f(x)]^p$$

for each $p \in \mathbf{Z}$. In particular, then,

$$f(1) = f\left(n\left(\frac{1}{n}\right)\right) = \left[f\left(\frac{1}{n}\right)\right]^n$$

and hence

$$f\left(\frac{1}{n}\right) = [f(1)]^{1/n}$$

It follows that for any rational number p/q, $q > 0$, we have

$$f\left(\frac{p}{q}\right) = [f(1)]^{p/q}$$

If we now define $a = f(1)$, then
$$f(r) = a^r$$
for each $r \in \mathbf{Q}$. From the assumed continuity of f and from the definition of a^x, it follows that
$$f(x) = a^x$$
for each $x \in \mathbf{R}$. ∎

The rather remarkable fact that the equation $f(x+y) = f(x) \cdot f(y)$ characterizes the exponential functions among all continuous functions finds its counterpart in the theory of logarithms.

■ **THEOREM 8.9.** If f is a continuous function satisfying the functional equation
$$f(xy) = f(x) + f(y)$$
for all positive real numbers x and y, then $f(x) = \log_a x$ for some $a > 0$.

The proof of Theorem 8.9 will be left as a problem for the reader.

The defining functional equations can even be used to determine the forms of the derivatives of our functions. Consider the exponential function first. We know that
$$f(x+h) = f(x) \cdot f(h)$$
Therefore $f(x+h) - f(x) = f(x)f(h) - f(x) = f(x)[f(h) - f(1)]$ and the derivative is
$$f'(x) = \lim_{h \to 0} f(x)\left[\frac{f(h)-1}{h}\right] = f(x) \lim_{h \to 0} \frac{f(h) - f(0)}{h}$$
Now, if the limit, $\lim_{h \to 0} [f(h) - f(0)]/h$, exists, then it is $f'(0)$. Hence we have
$$f'(x) = f'(0)f(x)$$
for all $x \in \mathbf{R}$. It follows that
$$f^{(n)}(x) = [f'(0)]^n f(x)$$
for all $n \in \mathbf{N}$. We also have the indefinite integral formula
$$\int f(x)\, dx = \frac{f(x)}{f'(0)} + C$$
In proper form, then
$$Da^x = a^x f'(0)$$
and
$$\int a^x\, dx = \frac{a^x}{f'(0)} + C$$

We can give a Maclaurin series for a^x, too, but this will be done in Section 8.9 for both a^x and $\log_a x$.

The corresponding arguments for a continuous function satisfying $f(xy) = f(x) + f(y)$ are given next. Because

$$f(x + h) - f(x) = f\left(\frac{x + h}{x}\right) = f\left(1 + \frac{h}{x}\right)$$

we have the derivative

$$\lim_{h \to 0} \frac{1}{h} f\left(1 + \frac{h}{x}\right) = f'(x)$$

By setting $h/x = t$, hence $h = xt$, we have

$$\frac{1}{h} f\left(1 + \frac{h}{x}\right) = \frac{1}{xt} f(1 + t) = \frac{1}{x} f[(1 + t)^{1/t}]$$

Clearly, if h approaches zero, so must t. Therefore

$$f'(x) = \lim_{t \to 0} \frac{1}{xt} f(1 + t) = \frac{1}{x} \lim_{t \to 0} f[(1 + t)^{1/t}]$$

We have assumed that f is continuous and therefore

$$\lim_{t \to 0} f[(1 + t)^{1/t}] = f[\lim_{t \to 0} (1 + t)^{1/t}]$$

Our problem is reduced to finding

$$\lim_{t \to 0} (1 + t)^{1/t}$$

or at least to proving that this limit exists.

Two facts are quoted without proof:

1. If $0 < s < t$, then $(1 + t)^{1/t} < (1 + s)^{1/s}$.
2. For any $t > 0$, $(1 + t)^{1/t} < 4$.

If we accept these facts, then it is easy to show that $\lim_{t \to 0} (1 + t)^{1/t}$ is the least upper bound of the set $\{(1 + t)^{1/t} : t > 0\}$ which exists by Theorem 1.10.

The number

$$e = \lim_{t \to 0} (1 + t)^{1/t}$$

is very important in analysis and is always denoted by the letter e. To 20-decimal accuracy

$$e = 2.71828\ 18284\ 59045\ 23536 \cdots$$

It turns out that e is a transcendental number but the proof of this fact is beyond the scope of this book. We shall prove later that e is irrational.

We see now that the derivative of $\log_a x$ can be written as

$$D \log_a x = \frac{1}{x} \log_a e, \qquad x > 0$$

8.5 A FUNCTIONAL EQUATION APPROACH

That inconvenient constant factor $\log_2 e$ in this formula is one of the many reasons why the analyst uses logarithms to the base e. The *natural logarithm function* $\ln x$ is the inverse of the exponential function $x = e^y$. Because $e^1 = e$, we have $\ln e = 1$. Therefore the derivative of the natural logarithm function is

$$D \ln x = \frac{1}{x}$$

It follows from Section 7.1 that

$$\ln x = \int_1^x \frac{dt}{t}$$

a formulation we see again in the next section.

PROBLEMS 8.5

1. Prove that $\lim_{n \to \infty} (1 + 1/n)^n = e$.
2. Prove that $\lim_{n \to \infty} (1 - 1/n)^n = e^{-1}$.
3. Prove that $\lim_{n \to \infty} (1 + 2/n)^n = e^2$.
4. Prove that $\lim_{n \to \infty} (1 + x/n)^n = e^x$.
5. What is the following expression?

$$\lim_{n \to \infty} \frac{1}{n}(e^{1/n} + e^{2/n} + \cdots + e^{1-1/n} + e)$$

6. What is the following expression?

$$\lim_{n \to \infty} \left(1 + \frac{1}{n^2}\right)^n$$

7. Define $a_n = \left(1 + \frac{1}{n}\right)^n$ and $b_n = \left(1 + \frac{1}{n}\right)^{n+1}$ for each $n \in \mathbf{N}$. Prove that

   (1) $a_n < a_{n+1}$ for all $n \in \mathbf{N}$
   (2) $b_{n+1} < b_n$ for all $n \in \mathbf{N}$
   (3) $a_m < b_n$ for any choice of $m, n \in \mathbf{N}$
   (4) $a_n \leqq 1 + 1 + \frac{1}{2} + \frac{1}{6} + \cdots + \frac{1}{n!} < 3$ for all $n \in \mathbf{N}$
   (5) $\lim_{n \to \infty} b_n = e$

8. Let $a_1, a_2, \ldots, a_n, \ldots$ be any sequence of positive numbers such that $\lim_{n \to \infty} a_n = 0$. Prove that $\lim_{n \to \infty} [1 + (a_n x/n)]^n = 1$ for any choice of $x \in \mathbf{R}$.

9. If $\sum_{i=1}^n 1/x_i'(x_i - x_{i-1})$ is a Riemann sum approximating the integral $\int_1^b dt/t$, show that $\sum_{i=1}^n 1/ax_i'(ax_2 - ax_{i-1})$ approximates $\int_a^{ab} dt/t$. What conclusions can you draw from this?

10. Prove by mathematical induction that

$$D^n \ln x = (-1)^{n-1}(n-1)! x^{-n}$$

8.6 AN INTEGRAL APPROACH

In Section 7.3 we give the integration formula

$$\int x^r \, dx = \frac{x^{r+1}}{r+1} + C, \qquad r \neq -1$$

The gap left by the omission of the value $r = -1$ in this formula can be filled now. Indeed, we gave the answer at the end of the preceding section.

The *natural logarithm function* is defined by the integral

$$\ln x = \int_1^x \frac{dt}{t}, \qquad x > 0$$

■ **THEOREM 8.10.** The function $\ln x$ is differentiable (and therefore continuous) on the entire open ray $\{x : 0 < x\}$ and

$$D \ln x = \frac{1}{x}$$

PROOF: This is an immediate consequence of Theorem 7.2. ■

It follows, too, that by applying the chain rule of differentiation, we have

$$D \ln f(x) = \frac{1}{f(x)} Df(x)$$

for any function f that is both differentiable and positive.

Let us unite this definition with our previous definition of a logarithm function. First consider the integral

$$\int_x^{xy} \frac{dt}{t}$$

Apply the substitution $t = xu$, $dt = x \, du$. Then $u = 1$ when $t = x$ and $u = y$ when $t = xy$. Therefore we have

$$\int_x^{xy} \frac{dt}{t} = \int_1^y \frac{x \, du}{xu} = \int_1^y \frac{du}{u} = \int_1^y \frac{dt}{t}$$

(In the last step here we use the fact that the variable of integration in a definite integral is a dummy index.)

By definition, we know that

$$\ln(xy) = \int_1^{xy} \frac{dt}{t}$$

In view of Theorem 6.5, we may write

$$\int_1^{xy} \frac{dt}{t} = \int_1^x \frac{dt}{t} + \int_x^{xy} \frac{dt}{t} = \int_1^x \frac{dt}{t} + \int_1^y \frac{dt}{t}$$

Therefore we have proved that

$$\ln xy = \ln x + \ln y$$

8.6 AN INTEGRAL APPROACH

From Theorem 8.9, then, we know that $\ln x$ *is the inverse of the exponential function* e^x where e is the number defined by the equation

$$\ln e = \int_1^e \frac{dt}{t} = 1$$

A question remains here. What does this number e such that $\ln e = 1$ have to do with the number $e = \lim_{t \to 0} (1 + t)^{1/t}$? We indicate that the two are the same as follows: Consider the limit

$$\lim_{n \to \infty} \int_1^{(1+1/n)^n} \frac{dt}{t}$$

We first show that, for any $a > 0$,

$$\int_1^{a^n} \frac{dt}{t} = n \int_1^a \frac{dt}{t}$$

The proof is by induction, starting with the fact that the formula is obviously true for $n = 1$. If it is true for $n = k$, then

$$\int_1^{a^{k+1}} \frac{dt}{t} = \int_1^{a \cdot a^k} \frac{dt}{t} = \int_1^a \frac{dt}{t} + \int_a^{a^k} \frac{dt}{t} = \int_1^a \frac{dt}{t} + k \int_1^a \frac{dt}{t}$$

$$= (k + 1) \int_1^a \frac{dt}{t}$$

This proves the formula for all $n \in \mathbf{N}$.

Now we know that

$$\int_1^{(1+1/n)^n} \frac{dt}{t} = n \int_1^{1+1/n} \frac{dt}{t}$$

But on the interval from 1 to $1 + 1/n$, the maximum value of $1/t$ is 1 and the minimum value is

$$\frac{1}{1 + 1/n} = \frac{n}{n + 1}$$

The length of this interval is $1/n$. Therefore we have the inequality

$$\left(\frac{n}{n+1}\right)\left(\frac{1}{n}\right) < \int_1^{1+1/n} \frac{dt}{t} < (1)\left(\frac{1}{n}\right)$$

or

$$\frac{n}{n+1} < n \int_1^{1+1/n} \frac{dt}{t} < 1$$

It follows that

$$\lim_{n \to \infty} \int_1^{(1+1/n)^n} \frac{dt}{t} = \lim_{n \to \infty} n \int_1^{(1+1/n)^n} \frac{dt}{t} = 1$$

Having defined $e = \lim_{n \to \infty} (1 + 1/n)^n$ and knowing that $\ln x$ is continuous, we have shown that

$$\ln e = \ln\left(\lim_{n\to\infty}\left(1 + \frac{1}{n}\right)^n\right) = \lim_{n\to\infty} \ln\left(1 + \frac{1}{n}\right)^n = 1$$

This tells us that the two very different definitions of e coincide. ∎

In view of Theorem 8.1, the particular exponential function e^x is continuous because it is the inverse of the continuous function $\ln x$. We can then prove the general result.

■ **THEOREM 8.11.** Let $a > 0$. Then the exponential function a^x is differentiable (and hence continuous) over the entire set $\mathbf{R}$ and

$$Da^x = a^x \cdot \ln a$$

PROOF: For any constant k, the composite function e^{kx} is continuous for all x. If $a > 0$, then $\ln a$ is a constant and $a = e^{\ln a}$. Therefore

$$a^x = (e^{\ln a})^x = e^{x \ln a}$$

is continuous for all x.

Let $x = \ln y = f(y)$. Then $y = e^x$ is the inverse $f^{-1}(x)$. By Theorem 8.2,

$$Df^{-1}(x) = \frac{1}{D \ln y} = \frac{1}{1/y} = y = e^x$$

that is,

$$De^x = e^x$$

Then by the chain rule it follows that

$$Da^x = De^{x \ln a} = e^{x \ln a} D(x \ln a) = e^{x \ln a} \cdot \ln a$$

or

$$Da^x = a^x \ln a$$

Recalling that we had found the formula

$$Da^x = f'(0)a^x$$

where

$$f'(0) = \lim_{h\to 0} \frac{f(h) - 1}{h}$$

we have again closed up a circle by finding that $f'(0) = \ln a$. ∎

Exercises appear after the next section.

8.7 DIFFERENTIATION AND INTEGRATION FORMULAS

We add to the list of our differentiation and integration formulas, starting with the newly derived formulas

$$D \ln x = \frac{1}{x}$$

8.7 DIFFERENTIATION AND INTEGRATION FORMULAS

$$D \log_a x = \frac{1}{x} \log_a e$$

$$De^x = e^x$$

$$Da^x = a^x \ln a$$

First, we note that $|x| > 0$ for all $x \neq 0$. Hence $\log_a|x|$ is defined if $x \neq 0$. If $x > 0$, then $|x| = x$ and

$$D \log_a|x| = D \log_a x = \frac{1}{x} \log_a e$$

On the other hand, if $x < 0$, then $|x| = -x$ and

$$D \log_a|x| = D \log_a(-x)$$
$$= \frac{1}{-x} D(-x) \log_e e \text{ (by the chain rule)}$$
$$= \frac{1}{-x} (-1) \log_a e$$
$$= \frac{1}{x} \log_a e$$

Therefore, in any case, we have

$$D \log_a|x| = \frac{1}{x} \log_a e \qquad x \neq 0$$

and, of course,

$$D \ln|x| = \frac{1}{x}$$

whence

$$\int \frac{dx}{x} = \ln|x| + C$$

More generally yet, the function

$$\log_a|f(x)|$$

is defined whenever $f(x) \neq 0$. Applying the chain rule, we obtain

$$D \log_a|f(x)| = D \log_a|f(x)| \cdot Df(x)$$
$$= \frac{1}{f(x)} \log_a e \cdot f'(x)$$
$$= \frac{f'(x)}{f(x)} \log_a e$$

■ **Example 1.** Consider the function $\log_a \sqrt{1-x^2}$. Clearly, $\sqrt{1-x^2} > 0$ if and only if $1 - x^2 > 0$, or $-1 < x < 1$. Therefore, $\log_a \sqrt{1 - x^2}$ is defined only on the open interval $\{x : -1 < x < 1\}$. Furthermore

$$\log_a \sqrt{1 - x^2} = \log_a(1 - x^2)^{1/2} = \tfrac{1}{2} \log_a(1 - x^2)$$

and hence

$$D \log_a \sqrt{1 - x^2} = \frac{1}{2} D \log_a(1 - x^2)$$

$$= \frac{1}{2}\left(\frac{1}{1 - x^2}\right) D_x(1 - x^2) \cdot \log_a e$$

$$= \frac{1}{2}\left(\frac{1}{1 - x^2}\right)(-2x) \log_a e$$

$$= \left(-\frac{x}{1 - x^2}\right) \log_a e$$

■ **Example 2.** The function $\ln|x/(1 - x)|$ is defined, except when $x = 0$ and $x = 1$. Its derivative is

$$D \ln\left|\frac{x}{1 - x}\right| = D(\ln|x| - \ln|1 - x|)$$

$$= D \ln|x| - D \ln|1 - x|$$

$$= \frac{1}{x} - \frac{1}{1 - x} D(1 - x)$$

$$= \frac{1}{x} - \frac{1}{1 - x}(-1)$$

$$= \frac{1}{x} + \frac{1}{1 - x} = \frac{1}{x(1 - x)}$$

■ **Example 3.** The indefinite integral

$$\int \frac{x\,dx}{1 + x^2}$$

responds neatly to the substitution $u = 1 + x^2$, $du = 2x\,dx$. Then

$$\int \frac{x\,dx}{1 + x^2} = \int \frac{\frac{1}{2}\,du}{u} = \frac{1}{2}\int \frac{du}{u} = \frac{1}{2}\ln|u| + C$$

But $u = 1 + x^2$ so $|u| = |1 + x^2| = 1 + x^2$ and therefore

$$\int \frac{x\,dx}{1 + x^2} = \frac{1}{2}\ln(1 + x^2) + C$$

Next consider the function

$$a^{f(x)}$$

By the chain rule again we obtain

$$Da^{f(x)} = D_f a^{f(x)}\, Df(x)$$
$$= a^{f(x)} \ln a \cdot Df(x)$$

Of course, when $a = e$, we have

$$De^{f(x)} = e^{f(x)}\, Df(x)$$

The integration formulas follow immediately:

$$\int a^{f(x)} Df(x)\, dx = \frac{a^{f(x)}}{\log_a e} + C$$

$$\int e^{f(x)} Df(x)\, dx = e^{f(x)} + C$$

■ **Example 4.** Let $g(x) = 2^{x^2-1}$. Then
$$D2^{x^2-1} = 2^{x^2-1} \cdot \ln 2 \cdot D(x^2 - 1)$$
$$= 2^{x^2-1} \cdot 2x \cdot \ln 2$$
$$= x(2 \cdot 2^{x^2-1}) \ln 2$$
$$= x2^{x^2} \ln 2$$

■ **Example 5.** An important function in probability theory is
$$f(x) = ae^{-k^2 x^2}$$
where a and k are constants. We compute that
$$Dae^{-k^2 x^2} = ae^{-k^2 x^2} D(-k^2 x^2) = -2k^2 a x e^{-k^2 x^2}$$
and
$$D^2 ae^{-k^2 x^2} = -2k^2 ae^{-k^2 x^2} - 2k^2 ax(-2k^2 x e^{-k^2 x^2})$$
$$= 2k^2 a(2k^2 x^2 - 1)e^{-k^2 x^2}$$

Obviously, $f'(x) = 0$ only if $x = 0$ and then $f''(0) < 0$. Therefore $f(0) = a$ is the only extremum value and it is a maximum. Also $f''(x) = 0$ only if $2k^2 x^2 - 1 = 0$ or $x = \pm 1/\sqrt{2k^2}$. These values of x locate the points of inflection of the graph $\{(x, y) : y = ae^{-k^2 x^2}\}$.

■ **Example 6.** Faced with the integral
$$\int xe^{1-x^2}\, dx$$
we make the substitution $u = 1 - x^2$, whence $du = -2x\, dx$ or $x\, dx = -\frac{1}{2} du$. Then the integral becomes
$$\int e^{1-x^2}(x\, dx) = \int e^u \left(-\frac{1}{2} du\right) = -\frac{1}{2} \int e^u\, du = -\frac{1}{2} e^u + C$$
or
$$\int xe^{1-x^2}\, dx = -\frac{1}{2} e^{1-x^2} + C$$

■ **Example 7.** At first glance it would seem that $\int \ln x\, dx$ might give us some trouble. Actually, it is easily done by means of the integration of parts. We set $\ln x = u$ and $dx = dv$. Then $du = dx/x$ and $v = x$. Therefore
$$\int \ln x\, dx = x \ln x - \int x \cdot \frac{dx}{x} = x \ln x - \int dx = x \ln x - x + C$$

The entire list of differentiation and integration formulas encountered so far follows.

1. $Dk = 0$
2. $Dx = 1$ $\qquad \int dx = x + C$
3. $D(f + g) = Df + Dg$ $\qquad \int [f(x) + g(x)]\, dx$
$\qquad\qquad\qquad\qquad\qquad = \int f(x)\, dx + \int g(x)\, dx$
4. $D(fg) = g\, Df + f\, Dg$ $\qquad \int f(x)\, dg(x) = f(x)g(x) - \int g(x)\, df(x)$
$\qquad\qquad\qquad\qquad\qquad$ (integration-by-parts formula)
5. $D\left(\dfrac{f}{g}\right) = \dfrac{g\, Df - f\, Dg}{g^2}$
6. $D(kf) = k\, Df$ $\qquad \int kf(x)\, dx = k \int f(x)\, dx$
7. $D(g \circ f) = Dg \cdot Df$
8. $D(x^r) = rx^{r-1}, r \in \mathbf{Q}$ $\qquad \int x^r\, dr = \dfrac{x^{r+1}}{r+1} + C, r \neq -1$
9. $D \ln x = \dfrac{1}{x}$ $\qquad \int \dfrac{dx}{x} = \ln|x| + C$
10. $D \log_a x = \dfrac{1}{x} \log_a e$
11. $De^x = e^x$ $\qquad \int e^x\, dx = e^x + C$
12. $Da^x = a^x \ln a$

A more general list will appear in Section 9.11.

EXERCISES 8.7

For Exercises 1 to 42, find $f'(x)$.

1. $f(x) = \ln(4x + 8)$
2. $f(x) = \ln x^2$
3. $f(x) = \ln(x - 1)^3$
4. $f(x) = \ln(x^3 + 3x - 2)$
5. $f(x) = \log_a \sqrt{4 - 6x + 3x^2}$
6. $f(x) = (\ln x)^2$
7. $f(x) = -x + x \ln x$
8. $f(x) = -2x + x \ln x^2$
9. $f(x) = x \ln x$
10. $f(x) = \sqrt{\ln x}$
11. $f(x) = \ln \left| \dfrac{x}{2 + 3x} \right|$
12. $f(x) = \ln \left| \dfrac{x}{a + bx} \right|$
13. $f(x) = \ln \dfrac{1 + x^2}{1 - x^2}$
14. $f(x) = \ln \sqrt{\dfrac{1 + x}{1 - x}}$
15. $f(x) = \dfrac{1}{2a} \ln \left| \dfrac{a + x}{a - x} \right|$
16. $f(x) = \dfrac{1}{2\sqrt{ab}} \ln \dfrac{\sqrt{a} + x\sqrt{b}}{\sqrt{a} - x\sqrt{b}}$
17. $f(x) = \ln|x + \sqrt{x^2 + a^2}|$
18. $f(x) = \ln \dfrac{1}{a}(x + \sqrt{x^2 - a^2})$
19. $f(x) = \ln \dfrac{\sqrt{x^2 + a^2} + x}{\sqrt{x^2 + a^2} - x}$
20. $f(x) = x\sqrt{x^2 - 4} - 4 \ln(x + \sqrt{x^2 - 4})$
21. $f(x) = \dfrac{1}{x} \ln x$
22. $f(x) = \dfrac{x^2}{\ln x}$

23. $f(x) = \ln(\ln(x^2 + 1))$
24. $f(x) = e^{-3x}$
25. $f(x) = 3^{-3x}$
26. $f(x) = 10^{-x/5}$
27. $f(x) = e^{2x-(1/2)x^2}$
28. $f(x) = e^{1/x^2}$
29. $f(x) = \frac{1}{2}(e^x + e^{-x})$
30. $f(x) = \frac{1}{2}(e^x - e^{-x})$
31. $f(x) = 2x - \frac{1}{2}e^{2x}$
32. $f(x) = xe^x$
33. $f(x) = e^x \ln x$
34. $f(x) = xe^{-x^2}$
35. $f(x) = \left(\frac{ax-1}{a^2}\right)e^{ax}$
36. $f(x) = x^2 e^{-1/x}$
37. $f(x) = \frac{x^2}{e^x + x}$
38. $f(z) = \frac{e^z}{\sqrt{e^{2z} - 1}}$
39. $f(x) = e^{-2\ln x}$
40. $f(x) = e^{(\ln x)^2}$
41. $f(x) = \ln\left(\frac{1+e^x}{1-e^x}\right)$
42. $f(x) = \ln(1 + 5e^x)$

Work out the indefinite integrals in Exercises 43 to 60. Some will require the use of integration by parts.

43. $\int \frac{dx}{4+x}$
44. $\int \frac{dx}{1+4x}$
45. $\int \frac{x\,dx}{x^2+2}$
46. $\int \frac{x^2+2}{x^3+6x+3}\,dx$
47. $\int \frac{x\,dx}{x+1}$
48. $\int \frac{3x+4}{x}\,dx$
49. $\int \frac{x^3\,dx}{a+bx}$
50. $\int \frac{\ln x}{x}\,dx$
51. $\int e^{x/3}\,dx$
52. $\int 4^x\,dx$
53. $\int xe^{-x^2}\,dx$
54. $\int \frac{e^{\sqrt{x}}}{\sqrt{x}}\,dx$
55. $\int (e^x + e^{-x})\,dx$
56. $\int \frac{e^x + e^{-x}}{e^x - e^{-x}}\,dx$
57. $\int x \ln x\,dx$
58. $\int xe^{-x}\,dx$
59. $\int x^2 \ln x\,dx$
60. $\int x^2 e^{-x}\,dx$

Evaluate the definite integrals in Exercises 61 to 72.

61. $\int_0^2 e^{-4x}\,dx$
62. $\int_0^1 xe^{-x^2}\,dx$
63. $\int_0^2 x^2 e^{-x^3}\,dx$
64. $\int_0^1 \frac{4\,dx}{e^{3x}}$
65. $\int_0^1 \frac{dx}{x+1}$
66. $\int_1^5 \frac{x\,dx}{x^2+1}$
67. $\int_0^3 \frac{x\,dx}{2x^2+5}$
68. $\int_0^1 \frac{e^x\,dx}{1+e^x}$
69. $\int_2^4 \frac{dx}{x \ln x}$
70. $\int_{-1}^2 10^{-x}\,dx$

71. $\int_0^1 5^{x-1}\, dx$
72. $\int_0^2 x \cdot 2^{x^2}\, dx$

Exercises 73 to 84, locate and identify the critical points and find the points of inflection of the graph of the given functions.

73. $f(x) = x + \ln x - 2$
74. $f(x) = 2x - 3 - \ln x$
75. $f(x) = x \ln x$
76. $f(x) = \dfrac{x}{\ln x}$
77. $f(x) = xe^x$
78. $f(x) = x - e^x$
79. $f(x) = e^x + e^{-x}$
80. $f(x) = e^x - e^{-x}$
81. $f(x) = x^2 e^{-x}$
82. $f(x) = e^{1/x}$
83. $f(x) = e^{-x} \ln x$
84. $f(x) = e^{-k^2 x^2}$

For Exercises 85 to 90, use the given function and its derivatives to evaluate the accompanying expression (the numbers a and b are constants).

85. $f(x) = ae^{cx} + be^{-cx};\ f'' - c^2 f$
86. $f(x) = ae^x + be^{-2x};\ f'' + f' - 2f$
87. $f(x) = ae^{cx} + be^{3cx};\ f'' - 4cf' + 3c^2 f$
88. $f(x) = (a + bx)e^{-2x};\ f'' + 4f' + 4f$
89. $f(x) = a + be^{-x} + \tfrac{1}{2}e^x - 3x + \tfrac{3}{2}x^2;\ f'' + f'$
90. $f(x) = ae^{-x} + be^{2x} - 2x^2 + 2x - 3;\ f'' - f' - 2f$

91. Prove that the function $f(x) = \dfrac{1}{x} \log_a x$, $x > 0$, has a maximum value when $x = e$ for any $a > 1$.

For Exercises 92 to 100, locate and identify the critical points of the given function. Find the points of inflection, too. Then sketch the graphs.

92. $f(x) = x^2 - 2 \ln x$
93. $f(x) = 4 - 3x^2 + 12 \ln x$
94. $f(x) = x^2 \ln x$
95. $f(x) = \dfrac{x^2}{\ln x}$
96. $f(x) = x^2 e^x$
97. $f(x) = (ax + b)e^{-x}$
98. $f(x) = xe^{-x^2}$
99. $f(x) = e^{-1/x}$
100. $f(x) = e^x(1 - \ln x)$

Use Newton's approximate method (Section 4.1) to determine the root of each equation in Exercises 101 to 106 to three-decimal accuracy.

101. $\ln x = -x$
102. $e^{-x} = x$
103. $2x - \ln x = 7$
104. $x + e^x = 2$
105. $x \ln x = 2$
106. $xe^{-x} = x^2 + 1$

Use L'Hospital's rule to evaluate the indeterminate forms in Exercises 107 to 118.

107. $\lim\limits_{x \to 1} \dfrac{\ln x}{x - 1}$
108. $\lim\limits_{x \to 0} x \ln x = \lim\limits_{x \to 0} \dfrac{\ln x}{1/x}$

8.7 DIFFERENTIATION AND INTEGRATION FORMULAS

109. $\lim\limits_{x \to 0} x^2 \ln x$

110. $\lim\limits_{x \to \infty} \dfrac{\ln x}{x}$

111. $\lim\limits_{x \to \infty} \dfrac{(\ln x)^n}{x}$

112. $\lim\limits_{x \to 0} x(\log x)^n$

113. $\lim\limits_{x \to 0} \dfrac{4^x - 2^x}{x}$

114. $\lim\limits_{x \to 0} xe^{1/x} = \dfrac{e^{1/x}}{1/x}$

115. $\lim\limits_{x \to \infty} \dfrac{x^3 + x^2}{e^x + 1}$

116. $\lim\limits_{x \to \infty} \dfrac{\ln(1 + e^{3x})}{x}$

117. $\lim\limits_{x \to \infty} x^2 e^{-x}$

118. $\lim\limits_{x \to \infty} x^n e^{-x};\ n = 1, 2, 3, \cdots$

Determine the inverse function of each function in Exercises 119 to 124.

119. $f(x) = \tfrac{1}{2}(e^{3x} + e^{-3x})$

120. $f(x) = \tfrac{1}{4}e^x - 2e^{-x}$

121. $f(x) = \ln\left(\dfrac{x + 1}{x - 1}\right)$

122. $f(x) = 2\ln x - \ln(x + 1)$

123. $f(x) = \ln(e^{2x} - 1)$

124. $f(x) = \ln\sqrt{3e^{x/2} - 2}$

125. Determine the area of the region under the curve $\{(x, y) : y = 2x(x^2 + 3)^{-1}\}$ and over the interval $\{x : 1 \leq x \leq 7\}$.

126. Determine the area of the region bounded by the line through the points $(0, 1)$ and $(1, e)$ and the curve $\{(x, y) : y = e^x\}$.

127. Find the area of the region under the curve $\left\{(x, y) : y = \dfrac{a}{2}(e^{x/a} + e^{-x/a})\right\}$ and over the interval $\{x : -a \leq x \leq a\}$.

128. Find the volume generated by revolving the region of Exercise 127 about the x axis.

129. Determine the volume generated by revolving about the x axis that region bounded by the x axis, the line $\{(x, y) : x = 1\}$ and the curve $\{(x, y) : xy = x - 3\}$.

130. The region under $\{(x, y) : y = 1/x\}$ and over the ray $\{x : 1 \leq x\}$ has infinite area. Show, however, that the volume of the solid generated by rotating this region about the x axis *is* finite.

131. How does each of the following functions relate to $\ln x$?

(a) $\displaystyle\int_2^x \dfrac{dt}{t}$

(b) $\displaystyle\int_2^x \dfrac{dt}{t - 1}$

(c) $\displaystyle\int_{x^{-1}}^1 \dfrac{dt}{t}$

(d) $\displaystyle\int_{x^2}^{x^3} \dfrac{dt}{t}$

PROBLEMS 8.7

1. Show that

$$\ln(1 + x) = x - \dfrac{1}{2}x^2 + \dfrac{1}{3}x^3 - \dfrac{1}{4}x^4 + \dfrac{1}{5}\dfrac{x^5}{(1 + x_0)^5},$$

$$0 < x_0 < x < 1$$

Use this to evaluate (a) $\ln(1.1)$, and (b) $\ln(1.2)$. Give the accuracy of your approximations. Use these values (a) and (b) to compute

(c) $\ln(1.21)$ (d) $\ln(1.44)$ (e) $\ln(1.32)$

2. Show that
$$e^x = 1 + x + \tfrac{1}{2}x^2 + \tfrac{1}{6}x^3 + \tfrac{1}{24}x^4 + \tfrac{1}{120}e^{x_0}x^5$$
where x_0 lies between 0 and x. Use this to evaluate (a) $e^{0.1}$, and (b) $e^{0.2}$. What accuracy can you claim for the values given? Use (a) and (b) to compute

(c) $e^{-0.1}$ (d) $e^{0.3}$ (e) $e^{1.1} = exe^{0.1}$

3. Establish the inequality
$$x - \frac{x^2}{2} + \frac{x^3}{3} - \frac{x^4}{4} < \ln(1 + x) < x - \frac{x^2}{2} + \frac{x^3}{3}; \quad 0 < x < 1$$

4. Establish the inequality
$$2\left(x + \frac{x^3}{3}\right) < \ln\frac{1 + x}{1 - x} < 2\left(x + \frac{x^3}{3} + \frac{x^4}{4}\right), \quad 0 < x \leq \frac{1}{2}$$

5. Define $f_1(x) = \int_0^x te^{-t}\, dt$, $f_2(x) = \int_0^x t^2 e^{-t}\, dt$ and, in general, $f_n(x) = \int_0^x t^n e^{-t}\, dt$. Determine $f_n(x)$ and evaluate $\lim_{n \to \infty} f_n(x)$, $|x| < 1$.

6. Let k be a fixed natural number larger than one. Prove that
$$\lim_{n \to \infty} \left(\frac{1}{n+1} + \frac{1}{n+2} + \cdots + \frac{1}{kn}\right) = \ln k$$

7. For any $n \in \mathbf{N}$ and $x > 0$, prove that
$$\left(1 + \frac{x}{n}\right)^n < e^x < \left(1 - \frac{x}{n}\right)^{-n}$$

8. Establish the formulas
   (a) $D^n \ln x = (-1)^{n-1}(n - 1)! x^{-n}$
   (b) $D^n \ln(1 - x) = -(n - 1)!(1 - x)^{-n}$
   (c) $D^n x e^{-x} = (-1)^n(x - n)e^{-x}$
   (d) $D^{n+1}(x^n \ln x) = \dfrac{n!}{x}$

9. Consider the function
$$f(x) = \frac{x}{1 + e^{1/x}}, \quad x \neq 0$$
$$f(0) = 0$$
Is f continuous at $x = 0$? Is it differentiable at $x = 0$?

10. Consider the function
$$f(x) = x \ln|x|, \quad x \neq 0$$
$$f(0) = 0$$
Is f continuous at $x = 0$? Is it differentiable at $x = 0$?

11. Find the area and the sides of the rectangle of largest area that has one side on the x axis and the other two vertices on $\{(x, y) : y = e^{-x^2}\}$.

8.7 DIFFERENTIATION AND INTEGRATION FORMULAS

12. Atmospheric pressure P in lb per sq in and the altitude h in miles above sea level are related (approximately) by an equation of the form
$$P = ae^{kh}$$
If $P = 14.7$ at sea level and $P = 4.6$ when $h = 4$ miles, determine the constants a and k.

13. A rope hanging over a cylinder of radius r is just barely held in place by friction if one end of the rope is level with the axis of the cylinder, while the other end hangs down a distance d below the axis. The distance d is related to the coefficient of friction u by the equation
$$(1 + u^2)d = 2ru(1 + e^{\pi u})$$
If $d = 2\pi r$, determine u to three significant figures.

14. Two wires of an electrical transmission line have a radius r. If they are x units apart, then magnetic flux per unit length is given by $k \ln[(x - r)/r]$, where k and m are constants. Graph this function and describe what you find about the flux as a function of x.

15. Prove that, for $0 < x < 1$,
$$\frac{x}{1+x} < \ln(1+x) < x$$

16. Prove that
$$\ln \frac{1}{2}(n+1) < \frac{1}{2} + \frac{1}{3} + \cdots + \frac{1}{n} < \ln n$$
for each $n \in \mathbf{N}$.

17. Prove that
$$\lim_{h \to \infty} n(\sqrt[n]{x} - 1) = \ln x$$

18. Determine
$$\lim_{h \to \infty} \frac{1}{h}(e^{1/n} + e^{2/n} + \cdots + e^{1-1/n} + e)$$

19. The demand law for a certain commodity is $p = 16e^{-1/2x}$. Find the maximum total revenue.

20. If the demand law $x(p)$ has a constant elasticity of demand, then determine $x(p)$.

21. Psychologists often use the function
$$f(t) = 1 + t \ln t, \quad 0 < t \leq u$$
$$= 1, \quad t = 0 \ (t \text{ measured in years})$$
to approximate the normal ability of a young child to memorize facts. Determine the extremum values of this function.

22. Let p and q be continuous and define the functions
$$f(x) = \int_0^x p(t) \, dt, \quad g(x) = \int_0^x e^{f(t)} q(t) \, dt$$
Show that, for any constant c, the function $(c + g(x))e^{-f(x)}$ satisfies the differential equation $y' + py = q$.

23. Consider the integrals $\int_1^x (t^{-1-1/n})\, dt$ and $\int_1^x (t^{-1+1/n})\, dt$ to prove that, for any $x > 0$ and $n \in \mathbf{N}$,

$$n(1 - x^{-1/n}) \leqq \ln x \leqq n(x^{1/n} - 1)$$

PROBLEMS 8.7
Economics

One of the important considerations of the economist is the present value of future money. How much is a dollar x years from now worth today? If a dollar is deposited today at an interest rate r and interest is compounded annually, the dollar is worth $(1 + r)^x$ dollars x years from now. If interest is compounded p times per year, then the dollar is worth $(1 + r/p)^{px}$ dollars in x years.

1. Prove that, if interest is compounded continuously, then the dollar is worth e^{rx} dollars in x years. Hence the *present value* of a dollar x years from now is e^{-rx} dollars.

   Suppose a certain business activity produces a profit at the rate of $f(t)$ dollars per year. The profit in a small time interval is then $f(t)\, dt$. Hence the total future profit from now ($t = 0$) to some future time $t = T$ is

   $$P(T) = \int_0^T f(t)\, dt$$

   However, this is not the present value of the future profit. The present value of the profit in the time interval dt is $e^{-rt} f(t)\, dt$. So the present value of all future profit to time T is

   $$\int_0^T e^{-rt} f(t)\, dt$$

   and, for all time to come,

   $$P(r) = \int_0^\infty e^{-rt} f(t)\, dt$$

   Of course, this is idealization.

2. If $f(t)$ is a constant, prove that the present value of all future profit is inversely proportional to the interest rate. And determine $P(r)$.

3. If $f(t) = kt$, show that $P(r)$ is inversely proportional to r^2.

4. At what interest rate r will the present value of a dollar 15 years from now equal the present value of a dollar 10 years from now at 6 percent interest?

   The cost of a piece of equipment is C. Its salvage value t years from now is $f(t)$. Let $p(t)$ be the net revenue from this equipment t years from now. Suppose the interest rate is a constant r. This information is applicable to Problems 5 and 6.

5. In order to determine the length of time T the equipment should be used, prove that we should maximize the function

   $$M(T) = -C + f(T)e^{-rT} + \int_0^T p(t)e^{-rt}\, dt$$

   Also prove that, when $M(T)$ is a maximum, we have $p(T) = rf(T) - f'(T)$.

6. The cost of a gasoline engine is $\$A$ and the cost of a diesel engine is $\$B$, $B > A$. A gasoline engine costs $\$C$ per year to operate, while a diesel engine costs $\$D$ per year to operate, $D < C$. Assuming a fixed interest rate r, show that the diesel engine is more economical if $e^r(B - A) > C - D$.

8.8 LOGARITHMIC DIFFERENTIATION

A function that is expressed as a product of powers of simpler functions may often be differentiated easily by using the properties of logarithms. A few examples should suffice to introduce the technique.

■ **Example 1.** Let $f(x) = \sqrt{(1+x)/(1-x)}$. Taking natural logarithms, we have

$$\ln f(x) = \ln \sqrt{\frac{1+x}{1-x}}$$

$$= \ln\left(\frac{1+x}{1-x}\right)^{1/2}$$

$$= \frac{1}{2}\ln\left(\frac{1+x}{1-x}\right)$$

$$= \frac{1}{2}\ln(1+x) - \frac{1}{2}\ln(1-x)$$

Differentiating the equation $\ln f(x) = \frac{1}{2}\ln(1+x) - \frac{1}{2}\ln(1-x)$, we obtain

$$\frac{f'(x)}{f(x)} = \frac{1}{2}\left(\frac{1}{1+x}\right)(1) - \frac{1}{2}\left(\frac{1}{1-x}\right)(-1)$$

$$= \frac{1}{2}\left(\frac{1}{1+x} - \frac{1}{1-x}\right) = \frac{1}{1-x^2}$$

Therefore

$$f'(x) = \frac{1}{1-x^2}f(x)$$

$$= \frac{1}{1-x^2}\sqrt{\frac{1+x}{1-x}} = \frac{1}{(1-x)\sqrt{1-x^2}}$$

■ **Example 2.** Let $f(x) = \sqrt[3]{1+x}/(1-x)^3$. Then

$$\ln f(x) = \ln\frac{(1+x)^{1/3}}{(1-x)^3} = \frac{1}{3}\ln(1+x) - 3\ln(1-x)$$

Hence

$$\frac{f'(x)}{f(x)} = \frac{1}{3} \cdot \frac{1}{1+x}(1) - 3 \cdot \frac{1}{1-x}(-1) = \frac{10+8x}{3(1-x^2)}$$

and

$$f'(x) = \frac{10+8x}{3(1-x^2)} \cdot \frac{\sqrt[3]{1+x}}{(1-x)^3} = \frac{10+8x}{3(1-x)^4(1+x)^{2/3}}$$

■ **Example 3.** A function that we probably could not differentiate in any other way is $f(x) = x^x$, $x > 0$. But it succumbs to logarithmic differentiation like this:

$$\ln f(x) = \ln x^x = x \ln x$$

Therefore

$$\frac{f'(x)}{f(x)} = (1) \ln x + x\left(\frac{1}{x}\right) = \ln x + 1$$

and hence

$$f'(x) = x^x(1 + \ln x)$$

■ *Example 4.* By using the differential form of the derivative of a logarithm, we can greatly simplify the calculation of relative errors (see Section 4.4). For instance, the volume of a sphere is $V = \frac{4}{3}\pi r^3$. Hence

$$\ln V = \ln \tfrac{4}{3}\pi + \ln r^3$$
$$= \ln \tfrac{4}{3}\pi + 3 \ln r$$

Differentiating with respect to r,

$$\frac{D_r V}{V} = 3\left(\frac{1}{r}\right)$$

Multiplying by dr, we have

$$\frac{dV}{V} = 3\left(\frac{dr}{r}\right)$$

Therefore the relative error in the volume of a sphere is three times the relative error in its radius. ■

The complicated function $(g(x))^{f(x)}$ presents difficult problems just to determine its domain. However, there is no difficulty in differentiating it. Let

$$\varphi(x) = (g(x))^{f(x)}$$

Then

$$\ln \varphi(x) = f(x) \ln g(x)$$

Differentiating both sides of this equation, we obtain

$$\frac{\varphi'(x)}{\varphi(x)} = f'(x) \ln g(x) + f(x)\left(\frac{g'(x)}{f(x)}\right)$$

or

$$\varphi'(x) = \varphi(x)\left[f'(x) \ln g(x) + f(x)\left(\frac{g'(x)}{g(x)}\right)\right]$$
$$= (g(x))^{f(x)}\left[f'(x) \ln g(x) + \frac{f(x)g'(x)}{g(x)}\right]$$
$$= (g(x))^{f(x)}f'(x) \ln g(x) + (g(x))^{f(x)-1}f(x)g'(x)$$

8.8 LOGARITHMIC DIFFERENTIATION

This is the general *power-exponential formula* for differentiation.

$$D(g(x))^{f(x)} = (g(x))^{f(x)} \cdot Df(x) \cdot \ln g(x)$$
$$+ f(x)(g(x))^{f(x)-1} \cdot Dg(x)$$

Note that, if $g(x) = a$ is a constant function, then $g'(x) = 0$ and the above formula reduces to that for an exponential function. On the other hand, if $f(x) = k$ is a constant function, then $f'(x) = 0$. This special case then proves the power rule:

$$D(g(x))^k = k(g(x))^{k-1} Dg(x) \text{ for any real number } k.$$

EXERCISES 8.8

Use logarithmic differentiation (and not formula 18) to find f' for the functions in Exercises 1 to 24.

1. $f(x) = \sqrt{x^2 + 4}$
2. $f(x) = \sqrt[3]{6x - 1}$
3. $f(x) = \dfrac{\sqrt{x+2}}{3x}$
4. $f(x) = \dfrac{2x}{\sqrt{x^2+1}}$
5. $f(x) = \sqrt{\dfrac{x+1}{x-1}}$
6. $f(x) = \sqrt[3]{\dfrac{x+1}{x-1}}$
7. $f(x) = \dfrac{\sqrt[3]{x+1}}{\sqrt{3x+1}}$
8. $f(x) = \dfrac{\sqrt{x+2}}{\sqrt[3]{x+3}}$
9. $f(x) = (x+1)^{1/2}(x-1)^{1/3}$
10. $f(x) = (x-1)^{1/2}(x+1)^{1/3}(x+2)^{1/4}$
11. $f(x) = \dfrac{x\sqrt{x^2-1}}{(x^2+1)^{1/3}}$
12. $f(x) = \sqrt{\dfrac{(x+1)(x+2)}{(x^2+1)(x^2+2)}}$
13. $f(x) = x^{2x}$
14. $f(x) = x^{x+1}$
15. $f(x) = x^{3/x}$
16. $f(x) = x^{\ln x}$
17. $f(x) = (1+x)^{1/x}$
18. $f(x) = (x^2+1)^x$
19. $f(x) = (\ln x)^x$
20. $f(x) = (x^2+1)^{\ln x}$
21. $f(x) = 2^{x^2}$
22. $f(x) = x^{x^2}$
23. $f(x) = x^{e^x}$
24. $f(x) = e^{x^x}$

25. A wooden block is sanded smooth. In the process the size of the cube is reduced. Determine the relations between the relative decreases in volume and the length of an edge, and beween the volume and the area of a side.

26. In a classical experiment in physics the student determines the acceleration of gravity g by timing the period of a pendulum. The formula is

$$g = \frac{4\pi^2 l}{t^2}$$

where l is the length of the pendulum and t is the period measured. Find the relative error in g in terms of the relative error in t.

8.9 POWER SERIES FOR e^x AND $\ln x$

Because $De^x = e^x$, it follows that $D^n e^x = e^x$ for each $n \in \mathbf{N}$. Therefore $D^n e^0 = 1$ and the Maclaurin series for e^x is

$$e^x = 1 + x + \frac{1}{2}x^2 + \frac{1}{6}x^3 + \cdots + \frac{1}{n!}x^n + \cdots$$

The remainder R_n after the nth term is

$$R_n = \frac{e^{x_1}}{n!} x^n$$

where x_1 is some point between 0 and x (see p. 244). It is then easily seen that, for each fixed $x \in \mathbf{R}$,

$$\lim_{n \to \infty} |R_n| = 0$$

Thus Theorem 5.16 tells us that the above Maclaurin series represents e^x for all values of x.

If we choose $x = 1$ in the Maclaurin series for e^x, we get

$$e = 1 + 1 + \frac{1}{2} + \frac{1}{6} + \cdots + \frac{1}{n!} + \cdots$$

This series converges quite rapidly. For instance, the sum of the first nine terms is $2.718278787\ldots$ which is accurate to six significant figures.

We shall use the series for e to show that e is an irrational number. Suppose that e were equal to p/q. Then we would have

$$\frac{p}{q} = 1 + 1 + \frac{1}{2} + \frac{1}{6} + \cdots + \frac{1}{n!} + \cdots$$

Multiplying this equation by $q!$, we obtain

$$\frac{pq!}{q} = p(q-1)! = q! + q! + \frac{q!}{2!} + \frac{q!}{3!} + \cdots$$

$$+ \frac{q!}{q!} + \frac{q!}{(q+1)!} + \cdots$$

or

$$p(q-1)! = (q! + q! + (3 \cdot 4 \cdots q) + (4 \cdot 5 \cdots q) + \cdots + 1)$$

$$+ \frac{1}{q+1} + \frac{1}{(q+1)(q+2)} + \cdots$$

Obviously, the quantity

$$p(q-1)! - q! - q! - (3 \cdot 4 \cdots q) - (4 \cdot 5 \cdots q) - \cdots - 1$$

is an integer. If we prove that the quantity

$$\frac{1}{q+1} + \frac{1}{(q+1)(q+2)} + \cdots + \frac{q!}{n!} + \cdots$$

is *not* an integer, then the equation $p/q = 1 + \sum_{n=1}^{\infty} 1/n!$ is impossible. But it is easy to see that

$$\frac{1}{q+1}\left(1 + \frac{1}{q+2} + \frac{1}{(q+2)(q+3)} + \cdots\right)$$
$$< \frac{1}{q+1}\left(1 + \frac{1}{2} + \frac{1}{4} + \cdots + \frac{1}{2^k} + \cdots\right)$$
$$= \frac{2}{q+1} < 1$$

Hence the equation is impossible and e must be irrational.

The function $\ln x$ is not defined at $x = 0$, so we cannot expand it in a Maclaurin series. Instead we shall give a Taylor series expansion about a point $a > 0$. It is an easy exercise in mathematical induction to show that

$$D^n \ln x = (-1)^{n-1}(n-1)!x^{-n}$$

It follows that $D^n \ln a = (-1)^{n-1}(n-1)!a^{-n}$ for each $n \in \mathbf{N}$. The Taylor series

$$f(x) = f(a) + f'(a)(x-a) + \frac{f''(a)}{2!}(x-a)^2 + \cdots$$
$$+ \frac{f^{(n)}(a)}{n!}(x-a)^n + \cdots$$

for $\ln x$ is

$$\ln x = \ln a + \sum_{n=1}^{\infty} (-1)^{n-1}(n-1)!a^{-n}\frac{(x-a)^n}{n!}$$
$$= \ln a + \sum_{n=1}^{\infty} \frac{(-1)^{n-1}}{na^n}(x-a)^n$$

Applying the ratio test, we compute that

$$\lim_{n\to\infty} \left|\frac{\frac{(-1)^n(x-a)^{n+1}}{(n+1)a^{n+1}}}{\frac{(-1)^{n-1}(x-a)^n}{na^n}}\right| = \lim_{n\to\infty} \frac{n}{n+1} \cdot \frac{|x-a|}{a} = \frac{|x-a|}{a}$$

It follows that the series converges wherever $|x - a| < a$ and diverges if $|x - a| > a$. At $x = 0$, the series is the divergent negative harmonic series $\ln a - 1 - \frac{1}{2} - \frac{1}{3} - \frac{1}{4} - \cdots$ and at $x = 2a$, the series is the very slowly convergent series $\ln a + 1 - \frac{1}{2} + \frac{1}{3} - \frac{1}{4} + \cdots$. Therefore the complete interval of convergence is $\{x : 0 < x \leq 2a\}$.

The series

$$\ln x = \ln a + \frac{x-a}{a} - \frac{(x-a)^2}{2a^2} + \frac{(x-a)^3}{3a^3} - \frac{(x-a)^4}{4a^4} + \cdots$$

converges rapidly enough for computational purposes if x is chosen close to a. Consider the special case $a = 1$, where the series is

$$\ln x = (x - 1) - \tfrac{1}{2}(x - 1)^2 + \tfrac{1}{3}(x - 1)^3 - \tfrac{1}{4}(x - 1)^4 + \cdots$$

(Recall that $\ln 1 = 0$.) Then we have

$$\ln 1.1 = 0.1 - \tfrac{1}{2}(0.1)^2 + \tfrac{1}{3}(0.1)^3 - \tfrac{1}{4}(0.1)^4 + \cdots$$
$$= 0.1 - \tfrac{1}{2}(0.01) + \tfrac{1}{3}(0.001) - \tfrac{1}{4}(0.0001) + \cdots$$
$$= 0.09531$$

using just the first four terms. This value is accurate to five decimals.

There is a clever method that can be used to compute natural logarithms. This is based on the fact that the function $(1 + x)/(1 - x)$ takes on every positive real number over the interval $\{x : -1 < x < 1\}$. We consider

$$\ln\left(\frac{1+x}{1-x}\right) = \ln(1 + x) - \ln(1 - x)$$

In the Taylor series $\ln x = (x - 1) - \tfrac{1}{2}(x - 1)^2 + \cdots$, we replace x by $(1 + x)$ to obtain

$$\ln(1 + x) = (1 + x - 1) - \tfrac{1}{2}(1 + x - 1)^2$$
$$+ \tfrac{1}{3}(1 + x - x)^3 - \tfrac{1}{4}(1 + x - 1)^4 + \cdots$$
$$= x - \tfrac{1}{2}x^2 + \tfrac{1}{3}x^3 - \tfrac{1}{4}x^4 + \cdots$$

Similarly,

$$\ln(1 - x) = (1 - x - 1) - \tfrac{1}{2}(1 - x - 1)^2$$
$$+ \tfrac{1}{3}(1 - x - 1)^3 - \tfrac{1}{4}(1 - x - 1)^4 + \cdots$$
$$= -x - \tfrac{1}{2}x^2 - \tfrac{1}{3}x^3 - \tfrac{1}{4}x^4 - \cdots$$

Then

$$\ln(1 + x) - \ln(1 - x) = (x - \tfrac{1}{2}x^2 + \tfrac{1}{3}x^3 - \tfrac{1}{4}x^4 + \cdots)$$
$$- (-x - \tfrac{1}{2}x^2 - \tfrac{1}{3}x^3 - \tfrac{1}{4}x^4 - \cdots)$$
$$= 2x + \tfrac{2}{3}x^3 + \tfrac{2}{5}x^5 + \cdots$$

or

$$\ln\left(\frac{1+x}{1-x}\right) = 2\left(x + \frac{1}{3}x^3 + \frac{1}{5}x^5 + \cdots \right.$$
$$\left. + \frac{1}{2n+1}x^{2n+1} + \cdots\right)$$

This series converges for each point in the interval $\{x : -1 < x < 1\}$ and can be used to compute the logarithm of any number. For instance, suppose we want ln 3. Set $(1 + x)/(1 - x) = 3$ and solve for x to obtain $x = \frac{1}{2}$. Then

$$\ln 3 = 2[\tfrac{1}{2} + \tfrac{1}{3}(\tfrac{1}{2})^3 + \tfrac{1}{5}(\tfrac{1}{2})^5 + \tfrac{1}{7}(\tfrac{1}{2})^7 + \tfrac{1}{9}(\tfrac{1}{2})^9 + \cdots]$$
$$= 2(\tfrac{1}{2} + \tfrac{1}{24} + \tfrac{1}{160} + \tfrac{1}{896} + \tfrac{1}{4608} + \cdots)$$
$$\cong 1.0985$$

just using the given terms. This is within 0.00015 of the true value.

We use the convention that $0! = 1$ to write the Maclaurin series for e^x as

$$e^x = \sum_{n=0}^{\infty} \frac{x^n}{n!}$$

By Theorem 5.14 this series can be differentiated term by term to obtain De^x. Thus

$$De^x = \sum_{n=0}^{\infty} D_x\left(\frac{x^n}{n!}\right) = \sum_{n=0}^{\infty} \frac{nx^{n-1}}{n!} = \sum_{n=0}^{\infty} \frac{x^{n-1}}{(n-1)!}$$

This last is equivalent to

$$De^x = \sum_{n=1}^{\infty} \frac{x^{n-1}}{(n-1)!} = \sum_{n=0}^{\infty} \frac{x^n}{n!}$$

which is the known result that $De^x = e^x$.

Similarly, we may differentiate the series

$$\ln(1 + x) = x - \frac{1}{2}x^2 + \frac{1}{3}x^3 - \frac{1}{4}x^4 + \cdots$$
$$+ (-1)^{n-1}\frac{x^n}{n} + \cdots$$

term by term to obtain

$$D \ln(1 + x) = \frac{1}{1 + x} = 1 - x + x^2 - x^3 + x^4 - x^5 + \cdots$$

This last can be "verified" simply by multiplying

$$(1 + x)(1 - x + x^2 - x^3 + x^4 - x^5 + \cdots)$$
$$= 1 - x + x^2 - x^3 + x^4 - x^5 + \cdots$$
$$\quad\quad + (x - x^2 + x^3 - x^4 + x^5 - x^6 + \cdots)$$
$$= 1 - x + x + x^2 - x^2 - x^3 + x^3$$
$$\quad\quad + x^4 - x^4 - x^5 + x^5 + \cdots$$
$$= 1$$

More precisely, we could write

$$(1+x)\sum_{n=0}^{\infty}(-1)^n x^n = \sum_{n=0}^{\infty}(-1)^n x^n + x\sum_{n=0}^{\infty}(-1)^n x^n$$

$$= \sum_{n=0}^{\infty}(-1)^n x^n + \sum_{n=0}^{\infty}(-1)^n x^{n+1}$$

$$= \sum_{n=0}^{\infty}(-1)^n x^n - \sum_{n=0}^{\infty}(-1)^{n+1} x^{n+1}$$

$$= 1 + \sum_{n=1}^{\infty}(-1)^n x^n - \sum_{n=0}^{\infty}(-1)^{n+1} x^{n+1}$$

$$= 1 + \sum_{n=1}^{\infty}(-1)^n x^n - \sum_{n=1}^{\infty}(-1)^n x^n$$

$$= 1$$

EXERCISES 8.9 Determine the Maclaurin series for functions in Exercises 1 to 16.

1. $f(x) = e^{-x}$
2. $f(x) = e^{3x}$
3. $f(x) = xe^x$
4. $f(x) = x^2 e^{-x}$
5. $f(x) = \frac{1}{2}(e^x + e^{-x})$
6. $f(x) = \frac{1}{2}(e^x - e^{-x})$
7. $f(x) = e^{-x^2}$
8. $f(x) = e^{x-x^2}$ (first three nonzero terms)
9. $\ln\left(\frac{1}{1-x}\right)$
10. $f(x) = \ln(xe^x)$
11. $\ln(x^2 + 1)$
12. $f(x) = x\ln(1+x)$
13. $\ln(1 - x + x^2)$
14. $f(x) = \ln(1 + 2x + 3x^2)$
15. $\ln(1 + x - 2x^2)$
16. $f(x) = \ln(1 + x + x^2)$

Determine the Taylor series at the given point c for each of the functions in Exercises 17 to 26.

17. $f(x) = e^x$, $c = 1$
18. $f(x) = e^x$, $c = -1$
19. $f(x) = e^{-2x}$, $c = -1$
20. $f(x) = xe^x$, $c = 1$
21. $f(x) = xe^{-x}$, $c = 1$
22. $f(x) = e^{-x^2}$, $c = -1$
23. $f(x) = e^{x^2-1}$, $c = 1$
24. $f(x) = \ln(x^2 + 1)$, $c = 1$
25. $f(x) = x\ln x$, $c = 1$
26. $f(x) = [\ln(2+x)]^2$, $c = -1$

27. Show that

$$\ln(x + \sqrt{1+x^2}) = x - \tfrac{1}{6}x^3 + \tfrac{3}{40}x^5 - \cdots$$

28. Show that

$$[\ln(1-x)]^2 = x^2 + x^3 + \tfrac{11}{12}x^4 + \tfrac{5}{6}x^5 + \cdots$$

8.9 POWER SERIES FOR e^x AND ln x

PROBLEMS 8.9

1. Prove that, for each $n \in \mathbf{N}$,
$$\ln\left(\frac{n+1}{n}\right) = 2\left[\frac{1}{2n+1} + \frac{1}{3(2n+1)^3} + \frac{1}{5(2n+1)^5} + \cdots\right]$$
Use the iterative formula $\ln(n+1) + \ln n + \ln[(n+1)/n]$ and this series to compute ln 2, ln 3, ln 4 and ln 5.

2. The function
$$f(x) = \frac{e^x - 1}{x}, \quad x \neq 0, \quad f(0) = 1$$
is differentiable at $x = 0$. Indeed, $f^{(n)}(0)$ exists for every $n \in \mathbf{N}$. Determine the Maclaurin series for $f(x)$.

3. As in Problem 2 above, the function
$$f(x) = \frac{x}{e^x - 1}, \quad x \neq 0, \quad f(0) = 1$$
has a Maclaurin series. If we write
$$f(x) = \sum_{n=0}^{\infty} \frac{B_n}{n!} x^n$$
then the numbers B_n are called *Bernoulli numbers*. Determine $B_0, B_1, \cdots, B_n$.

4. Formally multiply the series
$$\frac{x}{e^x - 1} = \sum_{n=0}^{\infty} \frac{B_n}{n!} x^n$$
by the series $e^x - 1 = x + x^2/2! + x^3/3! + \cdots$ and equate the result to the function x. Show therefore that the Bernoulli numbers satisfy the equation
$$\binom{n}{0} B_0 + \binom{n}{1} B_1 + \binom{n}{2} B_2 + \cdots + \binom{n}{n-1} B_{n-1} = 0$$
where $\binom{n}{k}$ is a binomial coefficient. Use this recursive formula to determine $B_5, \cdots, B_{10}$.

5. Still considering the Bernoulli numbers B_n, show that

   (a) $\quad x\left(\dfrac{e^x + e^{-x}}{e^x - e^{-x}}\right) = \sum\limits_{n=0}^{\infty} \dfrac{B_{2n}}{(2n)!} (2x)^{2n}$

   (b) $\quad \dfrac{2x}{e^x - e^{-x}} = \sum\limits_{n=0}^{\infty} \dfrac{B_{2n}}{(2n)!} (2 - 2^{2n}) x^{2n}$

6. By a formal term-by-term integration of a Maclaurin series, prove that
$$\int_0^x \frac{\ln(1-t)}{t} dt = -\sum_{n=1}^{\infty} \frac{x^n}{n^2}$$

7. The function
$$f(x) = e^{-1/x^2}, \quad x \neq 0$$
$$f(0) = 0$$

is infinitely differentiable at $x = 0$ (that is, $f^{(n)}(0)$ exists for all $n \in \mathbf{N}$). Nevertheless, $f(x)$ is *not* represented by its Maclaurin series. Prove these facts and explain this phenomenon.

8. Show that the remainder term in Taylor's theorem can be written in this special case as

$$e^x = 1 + x + \frac{1}{2}x^2 + \cdots + \frac{1}{n!}x^n + \frac{1}{n!}\int_0^x t^n e^{x-t}\, dt$$

9. Integrate the Maclaurin series for xe^x term by term between 0 and 1 to prove that

$$\sum_{n=1}^{\infty} \frac{1}{(n+2)n!} = \frac{1}{2}$$

10. Use the series for e^{-1} to help prove that the integer closest to $n!/e$ is divisible by $n-1$.

11. Prove that the infinite sequence defined by

$$a_{2n-1} = \frac{1}{n}$$

$$a_{2n} = \int_n^{n+1} \frac{dx}{x}$$

is monotone decreasing and that $\lim_{n \to \infty} a_n = 0$.

12. If a_n is as defined in Problem 11, then the alternating series $\sum_{n=1}^{\infty}(-1)^{n-1}a_n$ converges to a number γ (called *Euler's constant*). Show that

$$\lim_{n \to \infty}\left(1 + \frac{1}{2} + \cdots + \frac{1}{n} - \log n\right) = \gamma$$

(Note: To 10 decimals, $\gamma = 0.57721\ 56649\cdots$. An unsolved problem of mathematics is to show whether γ is rational or irrational.)

8.10 A USEFUL DIFFERENTIAL EQUATION

The fact that $De^x = e^x$ leads to the solution of many practical problems. Such problems often give rise to an equation of the form

$$f'(x) + a(x)f(x) = b(x)$$

where $a(x)$ and $b(x)$ are known functions, while the function $f(x)$ is to be determined. This is known as a *first-order linear differential equation*. It is a first-order differential equation because it involves both the function and its first derivative but no higher derivatives appear. Of course, it is linear in f and f'.

If the function $b(x)$ is identically zero, then the differential equation above is said to be *homogeneous*. Thus, the homogeneous linear first-order differential equation is

$$f'(x) + a(x)f(x) = 0$$

If we divide by $f(x)$ and use differentials, we obtain the equivalent equation

$$\frac{df(x)}{f(x)} + a(x)\, dx = 0$$

This can be integrated easily and yields

$$\ln y + \int a(x)\, dx = k \quad \text{where } y = f(x)$$

Then we can write

$$e^{\ln y + \int a(x)\, dx} = e^k$$

or

$$e^{\ln y} \cdot e^{\int a(x)\, dx} = e^k$$

or

$$f(x) e^{\int a(x)\, dx} = C$$

From here the solution of the homogeneous equation is easy.

Next we use the solution just derived to help solve the nonhomogeneous case. It is just a matter of computation to check that

$$\frac{d}{dx}[f(x) e^{\int a(x)\, dx}] = [f'(x) + a(x)f(x)] e^{\int a(x)\, dx} = \frac{dC}{dx} = 0$$

Therefore, multiplying the nonhomogeneous equation

$$f'(x) + a(x)f(x) = b(x)$$

by the *integrating factor*

$$e^{\int a(x)\, dx}$$

we obtain

$$[f'(x) + a(x)f(x)] e^{\int a(x)\, dx} = \frac{d}{dx}[f(x) e^{\int a(x)\, dx}] = b(x) e^{\int a(x)\, dx}$$

This equation can be integrated directly, yielding

$$f(x) e^{\int a(x)\, dx} = \int [b(x) e^{\int a(x)\, dx}]\, dx + C$$

From here, $f(x)$ is obtained by division.

- **Example 1.** If $f'(x) - 2f(x) = 3$ is the equation under consideration, we may note by inspection that the function $g(x) = Ce^{2x}$ has the property that $g'(x) - 2g(x) = 0$. Thus if we find any particular function $f_0(x)$ that satisfies the differential equation $f' - 2f = 3$, the function $f_0 + g$ will also solve the equation. A very easily seen particular function solving the equation is $f_0(x) = -\frac{3}{2}$. Therefore a general solution is

$$f(x) = -\frac{3}{2} + Ce^{2x}$$

We can obtain the same solution using the formula involving the integrating factor

$$e^{\int -2\,dx} = e^{-2x}$$

We write

$$[f'(x) - 2f(x)]e^{-2x} = \frac{d}{dx}[f(x)e^{-2x}] = 3e^{-2x}$$

whence

$$f(x)e^{-2x} = -\tfrac{3}{2}e^{-2x} + C$$

or

$$f(x) = -\tfrac{3}{2} + Ce^{2x}$$

■ **Example 2.** We apply the integrating factor approach to

$$f'(x) - xf(x) = x$$

Multiplying by the integrating factor

$$e^{\int -x\,dx} = e^{-x^2/2}$$

we obtain

$$[f'(x) - xf(x)]e^{-x^2/2} = \frac{d}{dx}[f(x)e^{-x^2/2}] = xe^{-x^2/2}$$

This can be integrated directly to give

$$f(x)e^{-x^2/2} = -e^{-x^2/2} + C$$

or

$$f(x) = -1 + Ce^{x^2/2}$$

■ **Example 3.** Consider the equation

$$xf'(x) + f(x) = x^2$$

and note that the left-hand side of the equation is already the derivative of the product $xf(x)$. Thus

$$xf'(x) + f(x) = \frac{d}{dx}[xf(x)] = x^2$$

8.10 A USEFUL DIFFERENTIAL EQUATION

It follows that we have

$$xf(x) = \frac{1}{3}x^3 + C$$

or

$$f(x) = \frac{1}{3}x^2 + \frac{C}{x}$$

The same result comes about by first writing the equation as

$$f'(x) + \frac{1}{x}f(x) = x$$

Then the integrating factor is

$$e^{\int dx/x} = e^{\ln x} = x$$

Multiplying $f'(x) + (1/x)f(x) = x$ by the integrating factor x, we return to the original equation. Therefore, we just saved ourselves a single step.

EXERCISES 8.10

Find the general solution $f(x)$ for each equation in Exercises 1 to 20.

1. $f'(x) = -3f(x)$
2. $f'(x) + 2f(x) = 1$
3. $4f'(x) + f(x) = 8$
4. $f'(x) = 6 - 3f(x)$
5. $f(x) - 4f(x) = 4$
6. $f'(x) + f(x) = e^{-x}$
7. $f'(x) + 3f(x) = x$
8. $xf'(x) + f(x) = x^2$
9. $xf'(x) + f(x) = 6x^3$
10. $xf'(x) + 2f(x) = 6x^4$
11. $f'(x) + f(x) = x + 1$
12. $xf'(x) - 3f(x) = x^4$
13. $xf'(x) - 3f(x) = x + 1$
14. $(1 + x^2)f'(x) - xf(x) = x(1 + x^2)$
15. $(x + 1)f'(x) - 2f(x) = (x + 1)^4$
16. $(x + 1)^2 f'(x) + 3(x + 1)f(x) = 4$
17. $xf'(x) - (1 + x)f(x) = e^x$
18. $xf'(x) + (1 + x)f(x) = e^x$
19. $x^2 f'(x) = 1 + 2xf(x)$
20. $x^2 f'(x) + (1 - 2x)f(x) = x^2$

For Exercises 21 to 30, determine the explicit function that satisfies both the differential equation and the added condition.

21. $f'(x) = -2f(x); f(0) = 2$
22. $f'(x) + 2f(x) = 1; f(1) = 2$
23. $f'(x) - f(x) = 4; f(-1) = 1$
24. $f'(x) - 2f(x) = 3; f(0) = 0$
25. $f'(x) + f(x) = 2; f(x_0) = y_0$
26. $(1 + x)f'(x) + f(x) = 3x^2; f(0) = 2$

27. $x^2 f'(x) + 2xf(x) = -e^x; f(2) = 0$

28. $(1 + x^2)f'(x) + 2xf(x) = 4 + 2x; f(0) = 4$

29. $xf'(x) = 1 + x + f(x); f(e) = -1$

30. $xf(x) + (1 + x)f(x) = 1; f(1) = 2$

PROBLEMS 8.10

1. Determine $f(x)$ if $f'(x) = f^2(x)e^{-x}$ and $f(0) = 2$. (Note: This is *not* a linear differential equation but you can solve it.)

2. A particle moving rectilinearly satisfies the equation of motion
$$v = ks$$
If $s = 2$ when $t = 0$ and $s = 4$ when $t = 10$, find s when $t = 6$.

3. A curve $\{(x,y) : y = f(x)\}$ passes through the point $(-1, 2)$ and at each point $(x, f(x))$ has the slope $\ln|x|$. Find $f(x)$.

4. A curve $\{(x,y) : y = f(x)\}$ has the property that the segment of the tangent line at a point $(x, f(x))$ from the point of tangency to the x axis is always bisected by the y axis. If $f(1) = 2$, find $f(x)$.

5. A curve $\{(x,y) : y = f(x)\}$ passes through the point $(0, 0)$. Given any point $(a, f(a))$ on the curve, the curve separates the rectangle $\{(x, y) : 0 \leq x \leq a, 0 \leq y \leq f(a)\}$ into two regions, one above the curve and the other below. If the upper area is always twice the lower for each point $(a, f(a))$ on the curve, determine $f(x)$.

More problems involving differential equations follow the next section.

8.11 APPLICATIONS

We present several real-world situations that give rise to linear differential equations.

Growth and decay laws. It is a statistical fact that the rate of growth of a population of living organisms, with no restrictions on food and living space, is proportional to the size of the population. Thus if $P(t)$ is the number of organisms present at time t, then

$$DP(t) = kP(t) \quad \text{or} \quad DP(t) - kP(t) = 0$$

where k is the constant of proportionality. In view of Section 8.10, we have

$$P(t) = Ce^{kt}$$

If the size of the population is known at some instant; then we may start measuring elapsed time at that instant. We then know that $P(0) = Ce^0 = C$ and can write

1. $P(t) = P(0)e^{kt}$

In practice we determine the constant k by having another measurement, say $P(t_1)$, of the population. For then

$$P(t_1) = P(0)e^{kt_1}$$

or

$$e^{kt_1} = \frac{P(t_1)}{P(0)}$$

or

$$kt_1 = \ln\left[\frac{P(t_1)}{P(0)}\right]$$

whence

$$k = \frac{1}{t_1}\ln\left[\frac{P(t_1)}{P(0)}\right]$$

Then

$$P(t) = P(0)e^{(t/t_1)\ln[P(t_1)/P(0)]}$$
$$= P(0)(e^{\ln[P(t_1)/P(0)]})^{t/t_1}$$

or

2. $$P(t) = P(0)\left(\frac{P(t_1)}{P(0)}\right)^{t/t_1}$$

Either Eq. 1 or 2 is called the *law of exponential growth*. This holds true in a statistical sense for populations of any kind. Thus any population of living organisms grows according to this law until some external condition alters the situation. It is such exponential growth that is the basis of the current worry about the "population explosion." (See Problem 8.11.2.)

A similar situation occurs in radioactive decay. The number of atoms of a radioactive material that decompose during a given (short) interval dt of time is proportional to the total number of atoms present times the length of time. If $x(t)$ is the number of atoms present at time t, then

$$Dx(t)\,dt = -kx(t)\,dt \qquad \text{or} \qquad Dx(t) + kx(t) = 0$$

where k is the constant of proportionality and Dx is negative because the number x is decreasing.

Again, the formula of Section 8.10 provides a general solution

$$x(t) = Ce^{-kt}$$

If the amount of material present is known at some given instant $t = 0$, then the particular solution can be written as

$$x(t) = x(0)e^{-kt}$$

The *half-life* of a radioactive material is the length of time required for a given amount $x(0)$ to be reduced to the amount $\tfrac{1}{2}x(0)$. We relate this length of time T to the constant k by solving the equation

$$x(0)e^{-kt} = \tfrac{1}{2}x(0)$$

or
$$e^{-kT} = \tfrac{1}{2}$$

This tells us that $-kT = \ln \tfrac{1}{2} = -\ln \tfrac{1}{2}$ or that

$$kT = \ln 2$$

In practice, of course, the physicist measures the half-life T rather than the constant k. Then he obtains

$$k = \frac{\ln 2}{T}$$

and hence

$$\begin{aligned} x(t) &= x(0)e^{-(\ln 2/T)t} \\ &= x(0)e^{(\ln 2)x(-t/T)} \\ &= x(0)(e^{\ln 2})^{-t/T} \end{aligned}$$

Therefore

$$x(t) = x(0) \cdot 2^{-t/T}$$

Exponential decay is the basis of *radiocarbon dating*, a useful technique in archaeology and geology. Carbon 14, C^{14}, has a half-life of about 5570 years. While alive, an organism has an equilibrium concentration of C^{14} atoms. Further accumulation stops at death and the number of atoms begins to diminish in accordance with the law of exponential decay. By comparing the concentration c_0 of C^{14} atoms in a present-day specimen with the concentration c in a long-dead specimen, the time since the death of the old specimen can be determined. The mathematics is as follows:

$$c(t) = c_0 2^{-t/T} = c_0 2^{-t/5570}$$

where t is the number of years since death. Therefore

$$2^{-t/5570} = \frac{c}{c_0}$$

or

$$\left(-\frac{t}{5570}\right) \ln 2 = \ln\left(\frac{c}{c_0}\right)$$

Then

$$t = -\frac{5570}{\ln 2} \ln\left(\frac{c}{c_0}\right) = \frac{5570}{\ln 2} \ln\left(\frac{c_0}{c}\right)$$

Since $\ln 2 \cong 0.693$, the number of years since the death of the specimen is

$$t \cong 8040 \ln\left(\frac{c_0}{c}\right)$$

Due to experimental errors, radiocarbon dating is not really dependable beyond an age of about 65,000 years

Continuous dilution problems. We consider just one example of a dilution problem. Equivalent problems occur in many different areas, some of which are illustrated in the problem set at the end of this section.

Consider the "smoke-filled room" of political fame. At the moment the air conditioner is turned on, suppose that the room contains a 0.25 percent concentration of carbon dioxide, CO_2. The air conditioner brings in fresh air with only a 0.05 percent concentration of CO_2 at the rate of 500 cu ft per min. The fan mixes the air in the room thoroughly, and the mixture is exhausted from the room also at the rate of 500 cu ft per min. If the room has a volume of 8000 cu ft, then describe the concentration of CO_2 as a function of time. In particular, how long does it take to reduce the concentration to 0.10 percent?

We solve such a problem by concentrating upon the *amount* x of CO_2 in the room. Obviously, x is a function of time and $x(0) = 8000 \times 0.25$ percent $= 20$ cu ft. The time rate of change of x equals the rate of input of CO_2 minus the rate of withdrawal of CO_2.

$$\text{rate of input} = (500 \text{ cu ft of air}) \times (0.05)$$
$$= 0.25 \text{ cu ft of } CO_2 \text{ per min}$$

$$\text{rate of output} = (500 \text{ cu ft of air})$$
$$(\text{the fraction of air that is } CO_2)$$
$$= 500 \times \frac{x}{8000} = \frac{x}{16} \text{ cu ft of } CO_2 \text{ per min}$$

Therefore

$$\frac{dx}{dt} = \frac{1}{4} - \frac{x}{16}$$

or

$$\frac{dx}{dt} + \frac{x}{16} = \frac{1}{4}$$

By the formula in Section 8.10

$$x(t) = \frac{1/4}{1/16} + Ce^{-t/16}$$
$$= 4 + Ce^{-t/16}$$

Knowing that $x(0) = 20$, we have the equation

$$20 = 4 + Ce^0$$

or

$$C = 16$$

Thus our function is

$$x(t) = 4 + 16e^{-t/16}$$

where t is measured in minutes.

We wanted to know when the concentration was reduced to 0.10 percent. Now 8000×0.10 percent $= 8$ cu ft of CO_2 in the room Therefore we are to solve for t in the equation

$$8 = 4 + 16e^{-t/16}$$

or

$$e^{-t/16} = \tfrac{1}{4}$$

Thus

$$-\frac{t}{16} = \ln\left(\frac{1}{4}\right) = -\ln 4$$

or

$$t = 16 \ln 4 \cong 24.181$$

That is, it takes just over 24 min to reduce the CO_2 concentration to 0.10 percent.

Motion in a resisting medium. An object falling through the air is acted upon by the force of gravity and by air resistance. If the mass of the body is m, then the force due to gravity is mg. We assume that air resistance is proportional to the velocity v of the body. Hence the total force acting on the object is

$$mg - kv$$

By Newton's third law of motion, force $= ma$, where a is the acceleration. Thus we have the differential equation

$$ma = m\frac{dv}{dt} = mg - kv$$

or

$$\frac{dv}{dt} + \frac{k}{m}v = g$$

The general solution is

$$v = \frac{mg}{k} + Ce^{-kt/m}$$

Now if the object had been dropped from rest, $v(0) = 0$. Therefore the constant $C = -mg/k$ and we have

$$v = \frac{mg}{k}(1 - e^{-kt/m})$$

Note that, as t increases, v increases but never exceeds mg/k. This is the *limiting velocity* of the object.

We also know that $v = ds/dt$, so we may go on to determine the equation of motion:

$$\frac{ds}{dt} = \frac{mg}{k} - \frac{mg}{k}e^{-kt/m}$$

and integrating,

$$s = \frac{mg}{k}t + \frac{m^2g}{k^2}e^{-kt/m} + C$$

If the initial position $s(0)$ is known, then

$$C = s(0) - \frac{m^2g}{k^2}$$

and our solution is complete.

Newton's law of cooling. This law states that the time rate of change of the temperature of a body is proportional to the difference between its temperature and that of the ambient medium. Thus the temperature T of the body satisfies the differential equation

$$\frac{dT}{dt} = -k(T - f(t)),$$

where k is positive and $f(t)$ is a function of time describing the temperature of the ambient medium. Of course, dT/dt is opposite in sign to the quantity $T - f(t)$ because, if the body is warmer than its surroundings, it will cool down, and vice versa. The general solution is obtained from Section 8.10 to be

$$T(t) = e^{-kt}\left[k\int f(t)e^{kt}\,dt + C\right]$$

In particular, suppose that a body at the initial temperature of $100°C$ is placed in a medium with a fixed temperature of $20°C$. Then we have

$$\frac{dT}{dt} = -k(T - 20)$$

We write this as the differential equation

$$\frac{dT}{T-20} = -k\,dt$$

and upon integration we get

$$\ln(T - 20) = -kt + C$$

When $t = 0$, $T = 100$; therefore $\ln(100 - 20) = \ln 80 = C$. This yields
$$\ln(T - 20) - \ln 80 = -kt$$
or
$$\ln\left(\frac{T - 20}{80}\right) = -kt$$

From here it follows that
$$\frac{T - 20}{80} = e^{-kt}$$
or
$$T = 20 + 80e^{-kt}$$

We may determine the constant k if we know the temperature at some other time. For instance, suppose we know that $T(30) = 60°C$. That is, we find that the body cools to $60°C$ in 30 min. Then
$$\ln\left(\frac{60 - 20}{80}\right) = -30k$$
or
$$k = -\frac{1}{30}\ln\left(\frac{1}{2}\right) = \frac{1}{30}\ln 2$$

Putting this value of k into the formula for T, we obtain
$$T = 20 + 80e^{-[(1/30)\ln 2]t}$$
$$= 20 + 80(e^{\ln 2})^{-t/30}$$
$$= 20 + 80 \cdot 2^{-t/30}$$

One hour after the experiment starts we shall expect the temperature
$$T = 20 + 80 \cdot 2^{-60/30} = 20 + 80 \cdot 2^{-2} = 20 + \tfrac{80}{4} = 40°C$$

A simple electrical circuit. Figure 8.9 is a diagram of an electrical circuit where R is a resistance, E is the electromotive force, and L is the inductance. If the switch is closed at time $t = 0$, then the current $I(t)$ flowing in the circuit is a function of the elapsed time t. The quantity $RI(t)$ is the voltage drop across the resistance and the quantity $L\dfrac{d}{dt}I(t)$ is the voltage drop across the inductance. The sum of these voltage drops equals the (constant) electromotive force E. That is,
$$L\frac{d}{dt}I(t) + RI(t) = E$$

Also we have the initial current $I(0) = 0$.

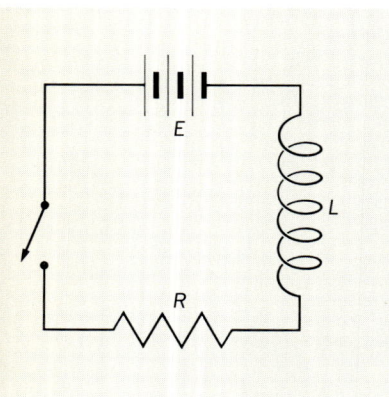

FIGURE 8.9

By the formula, the current function has the form

$$I(t) = \frac{E}{R} + Ce^{-Rt/L}$$

and because $I(0) = 0$, we find that $0 = E/R + C$ or $C = -E/R$. Therefore

$$I(t) = \frac{E}{R}(1 - e^{-Rt/L})$$

Note that $I(t)$ approaches the value E/R as t increases. The value E/R is called the *steady state current* and satisfies Ohm's law, $IR = E$.

PROBLEMS 8.11

1. In 1960 a certain town had 25,000 inhabitants; by 1965 the population had increased to 29,000. What population can the city planning commission expect by 1980?

2. El Salvador has an area of 8260 sq. miles. Its population in 1961 was 2,511,000 and in 1966 was 3,166,000. When can the country expect a population density of 1000 per sq. mile?

3. In a certain culture of bacteria, the number of bacteria increased six-fold in 10 hours. How long would it take to double the number?

4. As a population P grows, the rate of growth dP/dt usually decreases due to some inhibiting factor. For instance, assume that P cannot exceed some fixed value Q. (Perhaps living space is limited.) It is natural then to assume that dP/dt is proportional both to P and to $Q - P$. This gives us the differential equation

$$\frac{dP}{dt} = kP(Q - P)$$

which we rewrite as

$$\frac{dP}{P(Q - P)} = k\, dt$$

and then as

$$\frac{dP}{P} + \frac{dP}{Q - P} = kQ\, dt$$

Determine P as a function of t. (This function very closely approximates such natural phenomena as the growth of bacteria in a fixed container, the spread of an epidemic in a fixed population, and the rate of certain chemical reactions.)

5. The half-life of radium is approximately 1690 years. What percent of an original amount of m grams will remain unaltered after 500 years?

6. A body is cooling in air at $10°C$. If the temperature of the body falls from $75°C$ to $60°C$ in 3 min, what will its temperature be 7 min later? How long will it take to cool to $35°C$?

7. A cake is removed from an oven at $420°F$ and left to cool at a room temperature of $70°F$. In 30 min the cake is at $210°F$. When will its temperature be $100°F$?

8. A thermometer reading 75°F is taken outside. In 5 min it reads 45°F and in another 5 min the reading is 30°F. What is the outdoor temperature?

9. A certain body cools at a rate in minutes equal to $\frac{1}{3}$ of the difference between its temperature and that of the surrounding air. If it is originally at 175°C and the air is at 20°C, find the temperature after 30 min. How long will it take to reach 21°C?

10. A spherical moth ball sublimates at a rate that reduces it by $\frac{1}{3}$ of its volume in 4 months. When will $\frac{2}{3}$ of its volume be gone? (The rate of sublimation is assumed to be proportional to the area of the sphere.)

11. A tank initially contains 1000 gallons of fresh water. A saline solution containing 4 lb of salt per gallon flows into the tank at the rate of 2 gallons per min. The mixture is kept uniform by stirring and runs out at the same rate (so that the volume is constant). How long before there are 100 lb of salt in the tank? How long does it take to reach 90 percent of the upper bound?

12. A tank holds 1000 gallons of a saline solution. Fresh water is pumped in at the rate of 10 gallons per min. The mixture is kept uniform by stirring, and the mixture is removed at the same rate. How long does it take to reduce the salinity to $\frac{1}{3}$ of its original value?

13. The change in the relative length of a certain metal rod due to a change in the temperature is the constant 10^{-5}. If the rod is 75 cm long at 0°C, what is its length at 100°C?

14. What interest rate compounded continuously will double capital in 20 years?

15. Suppose that capital is invested whenever the total capital C is below a value C_0 and is withdrawn if C is above C_0. The rate of investment dC/dt is proportional to $C - C_0$ at time t. With this assumption there is a negative number $-k^2$ such that

$$\frac{dC}{dt} = -k^2(C - C_0)$$

Determine C as a function of t. Show that C approaches C_0 as t increases.

16. A mass of m lb is being dropped across a rough surface by a constant force of F lb. Frictional forces retarding the motion amount to a constant cm where c is the coefficient of friction. Assume that air resistance proportional to velocity also retards the motion. If the body starts from rest at $t = 0$, determine both velocity and displacement from the initial position as functions of t.

17. A parachutist jumps from an altitude of 12,000 ft and falls free for the first 2000 ft, opening his parachute at 10,000 feet. Assume that the air resistance is $\frac{3}{4}v$ lb prior to the opening of the parachute and $12v$ lb thereafter. Assuming that the acceleration of gravity is constant, find the time for him to reach the ground. (This will require Newton's method at two places.)

18. If a voltage V is applied to a nerve, then a certain ionization takes place, causing excitation of the nerve. If x denotes the ion concentration causing nerve excitation, it is assumed that $dx/dt = aV - bx$, where a and b are positive constants. Assume that V is constant and determine x as a function of t. Then assume that $V = ke^{-t}$ and redetermine x.

19. By differentiating both sides of the equation $f'(x) + a(x)f(x) = b(x)$, solve the *second-order differential equation*
$$f''(x) + (a'(x) - a^2(x))f(x) = b'(x) - a(x)b(x)$$

20. A particle moves rectilinearly. Its average velocity over any interval $\{t : c \leq t \leq d\}$ always equals $\frac{1}{2}[v(c) + v(d)]$. Prove that $v(t)$ is a linear function.

9

TRIGONOMETRIC FUNCTIONS

The use of trigonometric functions in geometry should be known to the reader. However, these transcendental functions also find an important application in describing periodic phenomena of all kinds. It is this analytic aspect of the trigonometric functions that concerns us here.

9.1 ANALYTIC TRIGONOMETRY

We shall use the letter pair (c, s) to denote the generic point on a coordinate plane instead of the usual (x, y). Then the circle of radius 1 centered at the origin in the (c, s) plane is the set

$$\{(c, s) : c^2 + s^2 = 1\}$$

Given any real number x, we start at the point $(1, 0)$ on this circle and measure off the distance x *along the circle*. If x is positive, then the distance is laid off in the counterclockwise direction. This is a convention, of course. Then negative distances are laid off in the clockwise direction. In this way, each $x \in \mathbf{R}$ determines a point $P(x)$ on the circle. In particular, $P(0) = (1, 0)$. Several points are so located in Fig. 9.1.

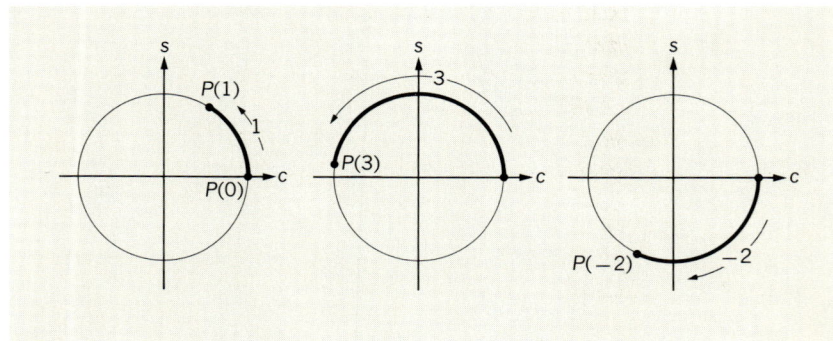

FIGURE 9.1

Because the radius of the circle is 1, its circumference has length 2π. Thus, if we measure off the length 2π around the circle, we return to the point $(1, 0)$. In fact, for any integer $p \in \mathbf{Z}$ located off the length, $p(2\pi)$ brings us back to the point $(1, 0)$. Therefore,

$$P(2p\pi) = (1, 0)$$

for each integer $p \in \mathbf{Z}$.

More important, the numbers x and $(x + 2\pi)$ locate the same point on the circle. Laying off the distance x and then, from the point $P(x)$, laying off another length 2π, we pass once around the circle returning to $P(x)$. It also follows that, for any $x \in \mathbf{R}$

$$P(x + 2p\pi) = P(x)$$

for each $p \in \mathbf{Z}$. This is a basic fact that is important in our development.

9.1 ANALYTIC TRIGONOMETRY

The coordinates of the point $P(x)$ are obviously functions of the real number x. If $P(x) = (c, s)$, then c is called the *cosine* of x and s is called the *sine* of x. The functions are denoted by cos x and sin x, respectively. Therefore, by definition,

$$P(x) = (\cos x, \sin x)$$

Because the point $P(x)$ is on the unit circle its coordinates, cos x and sin x, must satisfy the equation $(\cos x)^2 + (\sin x)^2 = 1$. We usually write this equation as

$$\cos^2 x + \sin^2 x = 1$$

This is a *fundamental identity* relating the two functions cos and sin.

Other properties about these two functions are easily deduced. First, cos x and sin x are real numbers satisfying $\cos^2 x + \sin^2 x = 1$. Therefore the inequalities

$$-1 \leq \cos x \leq 1$$
$$-1 \leq \sin x \leq 1$$

must hold for any $x \in \mathbf{R}$. Geometrically, these inequalities reflect the fact that the unit circle lies entirely between the vertical line $\{(c, s) : c = -1\}$ and $\{(c, s) : c = +1\}$ and lies entirely between the horizontal lines $\{(c, s) : s = -1\}$ and $\{(c, s) : s = 1\}$.

For any $x \in \mathbf{R}$, the points $P(x)$ and $P(-x)$ lie on the same vertical line and are symmetrically located with respect to the c axis. (This situation is illustrated in Fig. 9.2.) One may either accept this as being intuitively obvious or give a "proof" as follows: Moving along the circle a distance $-x$ from $(1, 0)$ is the same as moving a distance x from $(1, 0)$ and then reflecting the circle in the c axis.

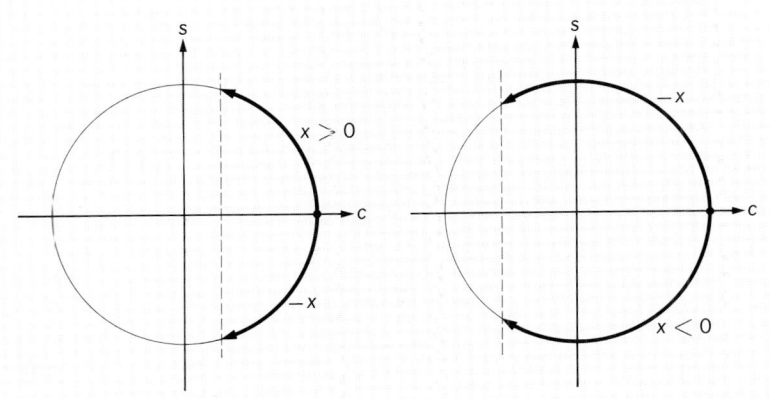

FIGURE 9.2

The fact that $P(x) = (\cos x, \sin x)$ and $P(-x) = (\cos(-x), \sin(-x))$ are symmetrically located with respect to the c axis implies that, for any $x \in \mathbf{R}$,

$$\cos(-x) = \cos x$$
$$\sin(-x) = -\sin x$$

That is,

cos x is an even function of x,
sin x is an odd function of (x)

(See Section 2.5.)

The property that $P(x + 2p\pi) = P(x)$, which we noted above, has an important consequence: For any $x \in \mathbf{R}$,

$$\cos(x + 2p\pi) = \cos x$$
$$\sin(x + 2p\pi) = \sin x$$

for each $p \in \mathbf{Z}$. We say that *cos and sin are periodic functions with period* 2π. Periodicity is precisely what makes these functions so useful in describing waves (sound waves, electromagnetic vibrations, and so forth), alternating electric current, rotary motion, and other period phenomena.

Let us relate the above discussion to trigonometry. Consider a point (c, s) on the unit circle with both c and s positive. Let α be the angle between the position vector of the point $(1, 0)$ and the position vector of (c, s) as shown in Fig. 9.3. From the heavy-lined triangle we see that

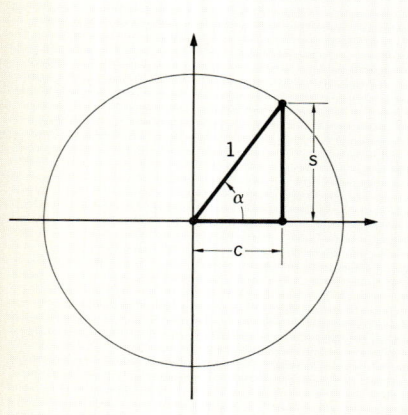

FIGURE 9.3

$$\cos \alpha = \frac{c}{1} = c, \sin \alpha = \frac{s}{1} = s$$

This indicates that our definition coincides with the definition of cosine and sine in terms of the sides and hypotenuse of a right triangle. But in order to let a real number x correspond to the angle α in the most natural way, we need a new unit of measurement for angles.

The real number 1 determines a point $P(1)$ on the unit circle and hence determines the angle from the position vector of $(1, 0)$ to the position vector of the point $P(1)$. It is natural to take this angle as the unit of measurement for angles, for then the real number 1 will correspond to the unit angle. The new unit angle is called a *radian* and is shown in Fig. 9.4 Now to each real number x there corresponds an angle of x radians.

In degree measurement a right angle has $90°$ and a straight angle has $180°$. The total circumference of the unit circle is 2π. Hence, measured along the circle, the length from $(1, 0)$ to $(-1, 0)$ is π. That is, $P(\pi) = (-1, 0)$. Therefore the angle π radians is a straight angle or

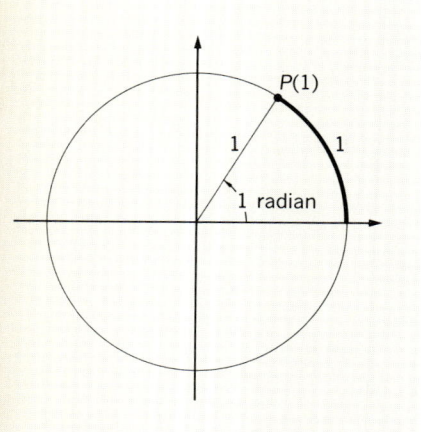

FIGURE 9.4

$$\pi \text{ radians} = 180°$$

or

$$1 \text{ radian} = \frac{180°}{\pi} \cong 57°18'$$

Addition formulas. Let A, B, and C be the sides of a triangle and let θ be the angle opposite the side C. Then the *law of cosines* states that

$$C^2 = A^2 + B^2 - 2AB \cos \theta$$

We digress to prove this formula, using plane trigonometry. In Fig. 9.5 the triangle is shown with several parts labeled. Using plane trigonometry, the height $h = B \sin \theta$. The base of the right triangle having B as the hypotenuse is of length $B \cos \theta$. Therefore the base of the right triangle with the hypotenuse C is $A - B \cos \theta$. The proof is completed by applying the Pythagorean theorem to write

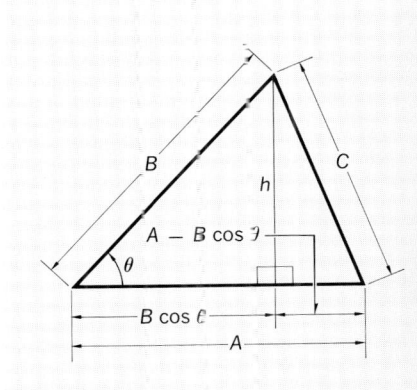

FIGURE 9.5

$$\begin{aligned} C^2 &= h^2 + (A - B \cos \theta)^2 \\ &= (B \sin \theta)^2 + (A - B \cos \theta)^2 \\ &= B^2 \sin^2 \theta + A^2 - 2AB \cos \theta + B^2 \cos^2 \theta \\ &= A^2 + B^2(\cos^2 \theta + \sin^2 \theta) - 2AB \cos \theta \\ &= A^2 + B^2 - 2AB \cos \theta \end{aligned}$$

Now consider the real numbers x and y as angles in our standard position. In Fig. 9.6 we have shown the case in which $0 < x < y < \pi$, but this is just for convenience. The analysis makes no use of this restriction. The triangle with vertices at the origin and the points $P(x)$, $P(y)$ has two sides of length 1, and the third side is the line segment from $P(x)$ to $P(y)$. Furthermore this third side is opposite the angle $y - x$. Therefore the law of cosines tells us that

$$C^2 = 1^2 + 1^2 - 2(1)(1) \cos(y - x) = 2 - 2 \cos(y - x)$$

But $P(x) = (\cos x, \sin x)$ and $P(y) = (\cos y, \sin y)$; hence by the distance formula we also have

$$\begin{aligned} C^2 &= (\cos y - \cos x)^2 + (\sin y - \sin x)^2 \\ &= \cos^2 y - 2 \cos y \cos x + \cos^2 x + \sin^2 y - 2 \sin y \sin x \\ &\qquad + \sin^2 x \\ &= (\cos^2 y + \sin^2 y) + (\cos^2 x + \sin^2 x) \\ &\qquad - 2(\cos y \cos x + \sin y \sin x) \\ &= 2 - 2(\cos y \cos x + \sin y \sin x) \end{aligned}$$

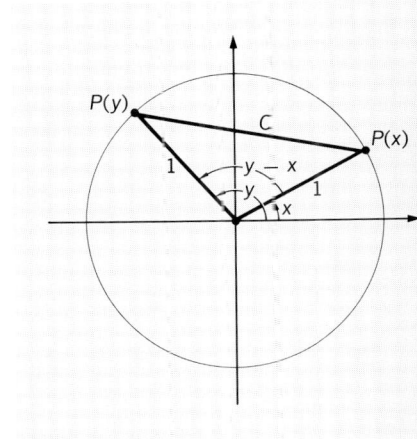

FIGURE 9.6

Comparing the two expressions for C^2, we see that

$$\cos(y - x) = \cos y \cos x + \sin y \sin x$$

This formula is the starting point from which we derive a number of useful trigonometric identities

By writing $y + x = y - (-x)$, the formula above yields
$$\cos(y + x) = \cos(y - (-x)) = \cos y \cos(-x) + \sin y \sin(-x)$$
We know that $\cos(-x) = \cos x$ and $\sin(-x) = -\sin x$. Making these substitutions,
$$\cos(y + x) = \cos y \cos x - \sin y \sin x$$
The two formulas can be written as one *addition formula:*
$$\cos(y \pm x) = \cos y \cos x \mp \sin y \sin x$$
Recalling that $\cos^2(y + x) + \sin^2(y + x) = 1$, we write
$$\begin{aligned}\sin^2(y + x) &= 1 - \cos^2(y + x) \\ &= 1 - (\cos y \cos x - \sin y \sin x)^2 \\ &= 1 - \cos^2 y \cos^2 x + 2 \cos y \cos x \sin y \sin x \\ &\quad - \sin^2 y \sin^2 x\end{aligned}$$

Next we replace $\cos^2 y$ by $1 - \sin^2 y$ and $\sin^2 y$ by $1 - \cos^2 y$ and write
$$\begin{aligned}\sin^2(y + x) &= 1 - (1 - \sin^2 y) \cos^2 x \\ &\quad + \cos y \cos x \sin y \sin x - (1 - \cos^2 y) \sin^2 x \\ &= 1 - \cos^2 x + \sin^2 y \cos^2 x + 2(\sin y \cos x) \\ &\quad (\cos y \sin x) - \sin^2 x + \cos^2 y \sin^2 x \\ &= 1 - (\cos^2 x + \sin^2 x) \\ &\quad + (\sin y \cos x + \cos y \sin x)^2 \\ &= (\sin y \cos x + \cos y \sin x)^2\end{aligned}$$

It easily follows that the *addition formula for the sine function* is
$$\sin(y \pm x) = \sin y \cos x \pm \cos y \sin x$$

The *double-angle formulas*
$$\cos 2x = \cos^2 x - \sin^2 x$$
$$\sin 2x = 2 \sin x \cos x$$
are obtained simply by setting $y = x$ in the addition formulas. Because $\cos^2 x = 1 - \sin^2 x$, we may write
$$\begin{aligned}\cos 2x &= (1 - \sin^2 x) - \sin^2 x \\ &= 1 - 2 \sin^2 x\end{aligned}$$
Similarly, $\sin^2 x = 1 - \cos^2 x$ and hence
$$\begin{aligned}\cos 2x &= \cos^2 x - (1 - \cos^2 x) \\ &= 2 \cos^2 x - 1\end{aligned}$$

We set $2x = y$ and $x = \tfrac{1}{2} y$ in the latter two expressions for $\cos 2x$. This yields the identities
$$\cos y = 1 - 2 \sin^2 \tfrac{1}{2} y$$
$$\cos y = 2 \cos^2 \tfrac{1}{2} y - 1$$

9.1 ANALYTIC TRIGONOMETRY

The following *half-angle formulas* are obtained immediately:

$$\cos \tfrac{1}{2}y = \pm\sqrt{\tfrac{1}{2}(1 + \cos y)}$$
$$\sin \tfrac{1}{2}y = \pm\sqrt{\tfrac{1}{2}(1 - \cos y)}$$

We shall need one more formula shortly. Let $y + x = u$ and $y - x = v$. Then $x = \tfrac{1}{2}(u - v)$ and $y = \tfrac{1}{2}(u + v)$. Using the addition formula for the sine function, we find that

$$\sin u = \sin(y + x)$$
$$= \sin\left[\tfrac{1}{2}(u + v) + \tfrac{1}{2}(u - v)\right]$$
$$= \sin\left(\tfrac{u+v}{2}\right)\cos\left(\tfrac{u-v}{2}\right) + \cos\left(\tfrac{u+v}{2}\right)\sin\left(\tfrac{u-v}{2}\right)$$

Similarly, by the subtraction formula,

$$\sin v = \sin(y - x)$$
$$= \sin\left[\tfrac{1}{2}(u + v) - \tfrac{1}{2}(u - v)\right]$$
$$= \sin\left(\tfrac{u+v}{2}\right)\cos\left(\tfrac{u-v}{2}\right) - \cos\left(\tfrac{u+v}{2}\right)\sin\left(\tfrac{u-v}{2}\right)$$

The *difference formula* for the sine function

$$\sin u - \sin v = 2\cos\left(\tfrac{u+v}{2}\right)\sin\left(\tfrac{u-v}{2}\right)$$

is obtained immediately.

Another property of the cosine and sine functions can also be deduced from the subtraction formulas. The point $P(\pi/2)$ obviously has coordinates $(0, 1)$. Therefore $\cos \pi/2 = 0$ and $\sin \pi/2 = 1$. Then

$$\cos\left(\tfrac{\pi}{2} - x\right) = \cos\tfrac{\pi}{2}\cos x + \sin\tfrac{\pi}{2}\sin x$$
$$= 0 \cdot \cos x + 1 \cdot \sin x$$

or

$$\cos\left(\tfrac{\pi}{2} - x\right) = \sin x$$

Similarly, we have

$$\sin\left(\tfrac{\pi}{2} - x\right) = \sin\tfrac{\pi}{2}\cos x - \cos\tfrac{\pi}{2}\sin x$$
$$= 1 \cdot \cos x - 0 \cdot \sin x$$

or

$$\sin\left(\tfrac{\pi}{2} - x\right) = \cos x$$

A short table of the values of $\sin x$ and $\cos x$ follows.

x	0	$\frac{\pi}{6}$	$\frac{\pi}{4}$	$\frac{\pi}{3}$	$\frac{\pi}{2}$	$\frac{2\pi}{3}$	$\frac{3\pi}{4}$	$\frac{5\pi}{6}$	π	$\frac{3\pi}{2}$	2π
$\sin x$	0	$\frac{1}{2}$	$\sqrt{\frac{2}{2}}$	$\sqrt{\frac{3}{2}}$	1	$\sqrt{\frac{3}{2}}$	$\sqrt{\frac{2}{2}}$	$\frac{1}{2}$	0	-1	0
$\cos x$	1	$\sqrt{\frac{3}{2}}$	$\sqrt{\frac{2}{2}}$	$\frac{1}{2}$	0	$-\frac{1}{2}$	$-\sqrt{\frac{2}{2}}$	$-\sqrt{\frac{3}{2}}$	-1	0	1

EXERCISES 9.1   In Exercises 1 to 10, use the addition formulas, half-angle formulas, and so forth, to compute the given values.

1. $\sin\left(\dfrac{\pi}{4} + \dfrac{\pi}{3}\right)$ 2. $\cos\left(\dfrac{\pi}{4} + \dfrac{\pi}{3}\right)$

3. $\sin\left(\dfrac{\pi}{4} + \dfrac{2\pi}{3}\right)$ 4. $\cos\left(\dfrac{\pi}{4} + \dfrac{5\pi}{6}\right)$

5. $\sin\left(\dfrac{3\pi}{2} - \dfrac{\pi}{4}\right)$ 6. $\cos\left(\dfrac{3\pi}{2} - \dfrac{2\pi}{3}\right)$

7. $\sin\dfrac{\pi}{8}$ 8. $\cos\dfrac{\pi}{8}$

9. $\sin\dfrac{\pi}{12}$ 10. $\cos\dfrac{\pi}{12}$

In Exercises 11 to 20, express the given number in terms of $\sin x$ and $\cos x$.

11. $\sin\left(x + \dfrac{\pi}{4}\right)$ 12. $\cos\left(x + \dfrac{\pi}{3}\right)$

13. $\sin\left(\dfrac{3\pi}{2} - x\right)$ 14. $\cos(\pi - x)$

15. $\sin\left(\dfrac{5\pi}{6} - x\right)$ 16. $\cos\left(\dfrac{\pi}{12} + x\right)$

17. $\sin\left(\dfrac{\pi}{4} - 2x\right)$ 18. $\cos\left(\dfrac{\pi}{3} + 2x\right)$

19. $\sin(x + 2x)$ 20. $\cos(x + 2x)$

PROBLEMS 9.1   1. Establish the following identities:
     (a) $\sin x + \sin y = 2\sin\tfrac{1}{2}(x+y)\cos\tfrac{1}{2}(x-y)$
     (b) $\cos x + \cos y = 2\cos\tfrac{1}{2}(x+y)\cos\tfrac{1}{2}(x-y)$
     (c) $\cos x - \cos y = -2\sin\tfrac{1}{2}(x+y)\sin\tfrac{1}{2}(x-y)$
     (d) $\cos^4 x - \sin^4 x = \cos 2x$
     (e) $\sin 3x = 3\sin x - 4\sin^3 x$
     (f) $\cos 3x = 4\cos^3 x - 3\cos x$
     (g) $\sin 4x = 4\sin x \cos^3 x - 4\sin^3 x \cos x$
     (h) $\cos 4x = 8\cos^4 x - 8\cos^2 x + 1$

(i) $\dfrac{1 + \cos x}{\sin x} + \dfrac{\sin x}{1 + \cos x} = \dfrac{2}{\sin x}$

(j) $\sin^3 x \pm \cos^3 x = (\sin x \pm \cos x)(1 \mp \sin x \cos x)$

(k) $\dfrac{\sin(x+y) + \sin(x-y)}{\cos(x+y) + \cos(x-y)} = \dfrac{\sin x}{\cos x}$

(l) $\dfrac{\sin 3x}{\sin x} - \dfrac{\cos 3x}{\cos x} = 2$

(m) $\dfrac{\cos 6x + \cos 4x}{\sin 6x - \sin 4x} = \dfrac{\cos x}{\sin x}$

(n) $\dfrac{\cos 3x - \cos 7x}{\cos x \sin x} = 4 \sin 5x$

2. Let $f(x) = a \cos(bx + c)$. Determine the constants a, b, and c if this function satisfies the functional equation
$$f(x + 1) = f(x) + \sin x \qquad \text{for all } x \in \mathbf{R}$$

3. Let $f(x) = a \sin x + b \cos x$ and then express $f(x)$ in the form
$$f(x) = A \sin(x + B)$$
Apply the formula just derived to the following special cases:

(a) $\sqrt{3} \cos x - \sin x$
(b) $\sqrt{2}(\sin x + \cos x)$
(c) $5 \cos 2x + 12 \sin 2x$
(d) $\sqrt{3} \sin \dfrac{x}{2} + \cos \dfrac{x}{2}$
(e) $3 \cos x - 4 \sin x$

4. Derive formulas for $\sin x/4$ and $\cos x/4$ in terms of $\sin x$ and $\cos x$.

5. Given that $\sin x/2 \neq 0$, show that
$$\sin x + \sin 2x + \cdots + \sin nx = \dfrac{\sin(nx/2) \sin[(n+1)x]/2}{\sin(x/2)}$$

6. Prove that no algebraic function other than a constant is periodic, thus proving that the sine and cosine functions are transcendental.

7. If the function f has period p, then show that
$$\int_x^{x+p} f(t)\, dt$$
is independent of x.

8. If the functions f and g have rational periods, show that $f + g$ is periodic.

9. If f is periodic and monotone, then show that f is constant.

10. Let f be continuous. If $f(x + 1/n) = f(x)$ for each $n \in \mathbf{N}$ and each $x \in \mathbf{R}$, show that f is constant.

9.2 FURTHER PROPERTIES OF COS AND SIN

We now study the cosine and sine functions for such properties as continuity and differentiability.

■ **LEMMA 9.1.** Let $x \in \mathbf{R}$ satisfy the inequality $-\pi/2 < x < \pi/2$. Then

$$|\sin x| \leq |x|$$

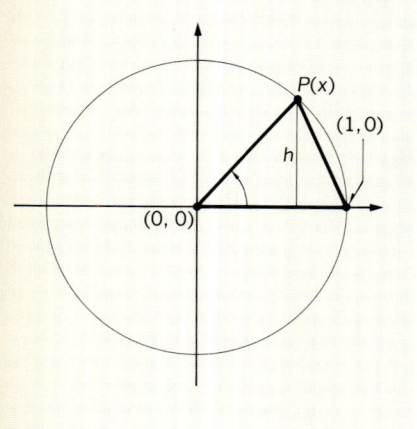

FIGURE 9.7

PROOF: We know that $\sin(-x) = -\sin x$ and hence that $|\sin(-x)| = |\sin x|$. Thus it suffices to prove the lemma only for positive values of x. Our proof is based upon an obvious geometric fact (see Fig. 9.7): The area of the triangle with vertices $(0, 0)$, $(1, 0)$ and $P(x) = (\cos x, \sin x)$ is less than the area of the circular sector in which the triangle is inscribed.

The area of the unit circle is $\pi r^2 = \pi(1)^2 = \pi$. The angle x, measured in radians, is $x/2\pi$ parts of the entire circumference. Therefore the area of the sector with central angle x is $(x/2\pi)\pi = x/2$. The inscribed triangle has a base of length 1 and height $h = \sin x$. Hence its area is $\frac{1}{2}(1)(\sin x) = \frac{1}{2}\sin x$. Our geometric statement about areas concludes that $\frac{1}{2}\sin x < \frac{1}{2}x$ or $\sin x < x$. This is true if x satisfies $0 < x < \pi/2$. If $x = 0$, then $\sin(0) = 0$. Therefore in any case $|\sin x| \leq |x|$. ■

■ **THEOREM 9.2.** The sine function is continuous for all $x \in \mathbf{R}$.

PROOF: This result is obvious from the geometric definition of $\sin x$. For if h is a small real number, the points $P(x)$ and $P(x + h)$ are surely close together. However, we can give a precise proof.

Let $x_0 \in \mathbf{R}$ be fixed and suppose that $\epsilon > 0$ is given. We are required to determine $\delta > 0$ such that

$$|\sin x - \sin x_0| < \epsilon \quad \text{whenever } |x - x_0| < \delta$$

Using the difference formula for the sine function (p. 403), we write

$$|\sin x - \sin x_0| = 2\left|\cos\left(\frac{x + x_0}{2}\right)\right| \cdot \left|\sin\left(\frac{x - x_0}{2}\right)\right|$$

Because $|\cos[(x + x_0)/2]| \leq 1$, we have the inequality

$$|\sin x - \sin x_0| \leq 2\left|\sin\left(\frac{x - x_0}{2}\right)\right|$$

By Lemma 9.1,

$$\left|\sin\left(\frac{x - x_0}{2}\right)\right| \leq \left|\frac{x - x_0}{2}\right| = \frac{1}{2}|x - x_0|$$

Therefore

$$|\sin x - \sin x_0| \leq 2 \times \frac{1}{2}|x - x_0| = |x - x_0|$$

9.2 FURTHER PROPERTIES OF COS AND SIN

It follows that we may choose $\delta = \epsilon$, that is,

$$|\sin x - \sin x_0| < \epsilon \quad \text{whenever } |x - x_0| < \epsilon \quad \blacksquare$$

■ **THEOREM 9.3.** The cosine function is continuous for all $x \in \mathbf{R}$.
PROOF: The linear function $f(x) = \pi/2 - x$ is continuous for all $x \in \mathbf{R}$. Hence by Theorem 9.2, the composite function $\sin(f(x)) = \sin(\pi/2 - x) = \cos x$ is continuous for all $x \in \mathbf{R}$. ■

Now we set out to compute the derivative $D \sin x$. We must evaluate the limit

$$D \sin x = \lim_{h \to 0} \frac{\sin(x + h) - \sin x}{h}$$

The ratio

$$\frac{\sin(x + h) - \sin x}{h}$$

is reduced to one we can handle by means of the difference formula (p. 403)

$$\sin(x + h) - \sin x = 2 \cos\left(\frac{x + h + x}{2}\right) \sin\left(\frac{x + h - x}{2}\right)$$

$$= 2 \cos\left(x + \frac{h}{2}\right) \sin\left(\frac{h}{2}\right)$$

Therefore we have

$$D \sin x = \lim_{h \to 0} \frac{2 \cos(x + h/2) \sin(h/2)}{h}$$

$$= \lim_{h \to 0} \frac{\cos(x + h/2) \sin(h/2)}{h/2}$$

$$= \lim_{h \to 0} \cos\left(x + \frac{h}{2}\right) \cdot \lim_{h \to 0} \frac{\sin(h/2)}{h/2}$$

Since the cosine function is continuous, we know that

$$\lim_{h \to 0} \cos\left(x + \frac{h}{2}\right) = \cos x$$

It follows that

$$D \sin x = \cos x \cdot \lim_{h \to 0} \frac{\sin(h/2)}{h/2}$$

Thus our problem has been reduced to evaluating the limit

$$\lim_{h \to 0} \frac{\sin(h/2)}{h/2}$$

■ **LEMMA 9.4.** Let

$$\lim_{x \to 0} \frac{\sin x}{x} = 1$$

PROOF: From Lemma 9.1, for any x such that $-\pi/2 < x < 0$ or $0 < x < \pi/2$, we have

$$|\sin x| < |x|$$

whence

$$\frac{|\sin x|}{|x|} < 1$$

We next show that, for the same restrictions on x,

$$|\cos x| < \frac{|\sin x|}{|x|}$$

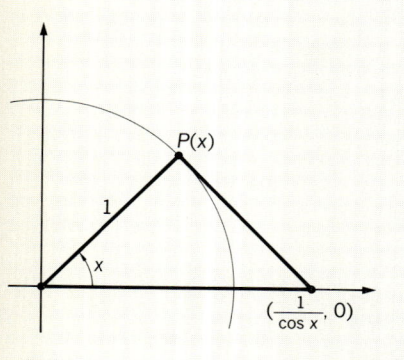

FIGURE 9.8

To do so, consider the right triangle with vertices $(0, 0)$ and $P(x)$, $x > 0$, and having its hypotenuse on the x axis (see Fig. 9.8). Our proof is based upon the geometric fact that the area of this triangle is larger than the area $x/2$ of the circular sector with central angle x. The height of this triangle is $\sin x$. Its base surely satisfies the equation $1/b = \cos x$ or $b = 1/\cos x$. (Consider the right triangle having base b and side 1 adjacent to the angle x.) Therefore the area of this triangle is $\frac{1}{2}(1/\cos x)(\sin x) = \frac{1}{2}\sin x/\cos x$. Our geometric assumption about areas now tells us that

$$\frac{1}{2}\left|\frac{\sin x}{\cos x}\right| > \frac{1}{2}|x|$$

or

$$|\sin x| > |x| \cdot |\cos x|$$

or

$$\frac{|\sin x|}{|x|} > |\cos x|$$

Combining these inequalities,

$$|\cos x| < \frac{|\sin x|}{|x|} < 1 \qquad \text{whenever } 0 < |x| < \frac{\pi}{2}$$

It follows that

$$\lim_{x \to 0} |\cos x| \leq \lim_{x \to 0} \frac{|\sin x|}{|x|} \leq 1$$

9.2 FURTHER PROPERTIES OF COS AND SIN

But $\cos x$ is continuous and so is $|\cos x|$. Therefore $\lim_{x \to 0} |\cos x| = |\cos 0| = 1$ and we have

$$\lim_{x \to 0} \frac{|\sin x|}{|x|} = 1$$

Finally, we note that

$$\frac{|\sin x|}{|x|} = \left|\frac{\sin x}{x}\right| = \frac{\sin x}{x} \qquad \text{if } 0 < |x| < \frac{\pi}{2}$$

because in this interval $\sin x$ and x have the same algebraic sign. ∎

In view of Lemma 9.4, then, we have deduced the formula

$$D \sin x = \cos x$$

Note that we must use x as the radian measure of an angle if we interpret x in this formula as an angle measurement.

Consider a composite function $\sin[f(x)]$. The chain rule of differentiation immediately yields the formula

$$D \sin[f(x)] = \cos[f(x)] \cdot Df(x)$$

In this formula we set $f(x) = \pi/2 - x$. Then we use the identities

$$\sin\left(\frac{\pi}{2} - x\right) = \cos x, \qquad \cos\left(\frac{\pi}{2} - x\right) = \sin x$$

to compute

$$D \cos x = D \sin\left(\frac{\pi}{2} - x\right) = \cos\left(\frac{\pi}{2} - x\right) \cdot D\left(\frac{\pi}{2} - x\right)$$
$$= \sin x \cdot (-1) = -\sin x$$

Thus we have the differentiation formula

$$D \cos x = -\sin x$$

Applying the chain rule again, we obtain

$$D \cos[f(x)] = -\sin[f(x)] \cdot Df(x)$$

■ **Example 1.** Let $f(x) = \sin 2x + \cos^2 x$. Then

$$Df(x) = D \sin 2x + D \cos^2 x$$
$$= \cos 2x \cdot D2x + 2 \cos x \, D \cos x$$
$$= 2 \cos 2x - 2 \cos x \sin x$$
$$= 2 \cos 2x - 2 \sin 2x$$

because $2 \sin x \cos x = \sin 2x$.

■ **Example 2.** If $f(x) = \ln(\cos x)$, then

$$Df(x) = \frac{1}{\cos x} \cdot D \cos x = \frac{1}{\cos x}(-\sin x) = -\frac{\sin x}{\cos x}$$

■ **Example 3.** We apply Newton's iterative method (Section 4.1) to the equation

$$x = \cos x$$

That is, we want to solve $f(x) = x - \cos x = 0$. Because $f'(x) = 1 + \sin x$, the iterative formula is

$$x_{n+1} = x_n - \frac{f(x_n)}{f'(x_n)} = x_n - \frac{x_n - \cos x_n}{1 + \sin x_n} = \frac{x_n \sin x_n + \cos x_n}{1 + \sin x_n}$$

We determine an initial approximation x_0 by looking at a table of the values of $\cos x$ (x in radian measure, of course). We find, for instance,

$$\cos(0.74) = 0.7385$$

so $x_0 = 0.74$ is a good starting point. Then

$$x_1 = \frac{(0.74)\sin(0.74) + \cos(0.74)}{1 + \sin(0.74)} = \frac{(0.74)(0.6743) + (0.7385)}{1 + 0.6743}$$

$$\cong 0.7396$$

Further iterations require more and more accurate values of both the sine and cosine functions. Given any set of tables, we can obtain a solution as accurate as the tables. However, we are not at a loss if we have no tables. Section 9.5 gives computational methods.

EXERCISES 9.2 Determine Df in Exercises 1 to 40.

1. $f(x) = 3 \sin 2x$
2. $f(x) = 2 \cos(3x + 1)$
3. $f(x) = 2 \sin^2 3x$
4. $f(x) = \sqrt{\sin x}$
5. $f(x) = \cos \sqrt{x}$
6. $f(x) = \frac{1}{2} \sin x^2$
7. $f(x) = \cos 2x - 2 \cos x$
8. $f(x) = \frac{1}{2}x + \frac{1}{4} \sin 2x$
9. $f(x) = \frac{1}{2}x - \frac{1}{4} \sin 2x$
10. $f(x) = x \sin 2x + \cos 2x$
11. $f(x) = 2 \sin \frac{x}{2} - x \cos \frac{x}{2}$
12. $f(x) = \sin^3 x - 3 \sin x$
13. $f(x) = \sin 2x - \frac{1}{3} \sin^3 2x$
14. $f(x) = (x^2 - 1) \sin x + x \cos x$
15. $f(x) = (2 - x^2) \cos x + 2x \sin x$
16. $f(x) = 2 \sin x \cos x$
17. $f(x) = \sin^3 x \cos^2 x$
18. $f(x) = \sin^4 x \cos^4 x$
19. $f(x) = \dfrac{\sin 2x}{1 + \cos 2x}$
20. $f(x) = \left(\dfrac{1 - \cos x}{1 + \cos x}\right)^3$

21. $f(x) = \sqrt{1 + \sin x}$

22. $f(x) = \sqrt{1 + 3\cos^2 4x}$

23. $f(x) = \dfrac{\cos x}{x}$

24. $f(x) = \dfrac{x}{\sin 3x}$

25. $f(x) = \dfrac{x}{\sin^2 x}$

26. $f(x) = \ln \sin 2x$

27. $f(x) = \ln \sin^2 x$

28. $f(x) = \ln \cos^3 2x$

29. $f(x) = \ln \sqrt{\dfrac{1 - \cos x}{1 + \cos x}}$

30. $f(x) = \cos(\ln x) - \sin(\ln x)$

31. $f(x) = e^x \cos 2x$

32. $f(x) = e^{-2x} \sin 3x$

33. $f(x) = e^{ax}(a \cos bx - b \sin bx)$

34. $f(x) = e^{\sin x}$

35. $f(x) = \cos(\sin x)$

36. $f(x) = \sin^2(\cos x)$

37. $f(x) = x^{\sin x}$

38. $f(x) = (\sin x)^x$

39. $f(x) = \ln \ln(1 - \cos x)$

40. $f(x) = x^{e^{\sin x}}$

Evaluate the indeterminate forms in Exercises 41 to 50.

41. $\lim\limits_{x \to 0} \dfrac{\sin 2x}{x}$

42. $\lim\limits_{x \to 0} \dfrac{x - \sin x}{x}$

43. $\lim\limits_{x \to 0} \dfrac{x(e^{2x} - 1)}{1 - \cos 2x}$

44. $\lim\limits_{x \to 0} \dfrac{x^2}{1 - \cos x}$

45. $\lim\limits_{x \to 0} \dfrac{\sin^2 x}{1 - \cos x}$

46. $\lim\limits_{z \to 0} \dfrac{1 - \cos x}{x \cos x}$

47. $\lim\limits_{x \to 0} \dfrac{x^2}{\ln \cos x}$

48. $\lim\limits_{x \to 0} \left(\dfrac{1}{\sin^2 x} - \dfrac{\sin x}{x^3}\right)$

49. $\lim\limits_{x \to 0} (1 + \sin^2 x)^{1/x^2}$

50. $\lim\limits_{x \to 0} \left(\dfrac{\sin x}{x}\right)^{1/x^2}$

Locate and identify the critical points of the functions in Exercises 51 to 58.

51. $f(x) = \cos x + \sin x$

52. $f(x) = 3 \cos x + 4 \sin x$

53. $f(x) = 2 \sin x + \sin 2x$

54. $f(x) = \sin^2 x + \cos x$

55. $f(x) = \sin^3 x \cos x$

56. $f(x) = (1 + \cos x) \sin x$

57. $f(x) = \cos x \sqrt{1 + \sin x}$

58. $f(x) = \sin x(1 - \cos x)^2$

59. Establish the differentiation formulas:

(a) $D\left(\dfrac{1}{2}x + \dfrac{1}{4a}\sin 2ax\right) = \cos^2 ax$

(b) $D\left(\dfrac{1}{2}x - \dfrac{1}{4a}\sin 2ax\right) = \sin^2 ax$

(c) $D\left(\dfrac{1}{a}\sin ax - \dfrac{1}{3a}\sin^3 ax\right) = \cos^3 ax$

60. Determine the equation of the tangent line to $\{(x, y) : y = x - \sin x\}$ at the point $(\pi/2, \pi/2 - 1)$ and find where the line crosses the x axis.

In Exercises 61 to 64, find $Df^{-1}(x)$ for the given function.

61. $x = f(y) = \sin y$
62. $x = f(y) = \cos \sqrt{y}$
63. $x = f(y) = \cos^2 y + \sin 2y$
64. $x = f(y) = \sqrt{y - \sin y}$

65. Show that each of the following pairs of functions is tangent of order two at the origin:
   (a) $f(x) = \sin x \cos x$; $g(x) = \frac{1}{2}x + \frac{1}{4}\sin 2x$
   (b) $f(x) = x \sin x$; $g(x) = \dfrac{x \sin x}{\cos x}$
   (c) $f(x) = \dfrac{x \sin x}{\cos x}$; $g(x) = \dfrac{1}{\cos x} - \cos x$
   (d) $f(x) = \frac{1}{2}x^2$; $g(x) = -\ln \cos x$

66. For each function f, determine the value of the expression at the right (a and b denote arbitrary constants):
   (a) $f(x) = a \cos 3x + b \sin 3x$; $f'' + 9f$
   (b) $f(x) = a \cos x + b \sin x + 6 - 2 \cos 2x$; $f'' + f$
   (c) $f(x) = ae^{-x} \cos 2x + be^{-x} \sin 2x$; $f'' + 2f' + 5f$
   (d) $f(x) = ae^x + be^{2x} - e^{2x} \ln \cos(e^{-x})$; $f'' - 3f' + 2f$

67. Discuss the rectilinear motion of a particle whose law of motion is:
   (a) $s = 4 \sin 3t$ (b) $s = 2 \cos 2\pi t$
   (c) $s = 4 \cos t - 6 \sin t$ (d) $s = \sin t + t \cos t$

68. Apply Taylor's theorem to obtain the inequalities:
   (a) $x - \frac{1}{6}x^3 < \sin x < x - \frac{1}{6}x^3 + \frac{1}{120}x^5$
   (b) $1 - \frac{1}{2}x^2 + \frac{1}{24}x^4 > \cos x > 1 - \frac{1}{2}x^2$

69. Use the inequalities in Exercise 68 to approximate $\sin(1.5)$ and $\cos(0.1)$.

70. Obtain two-decimal accuracy in the roots of the following equations, using Newton's iterative method:
   (a) $\sin x + x = 1$ (b) $\cos x = x^2$
   (c) $\sin x = 2x \cos x$ (d) $\sin x + \sin 2x = 1.2$
   (smallest positive root)
   (e) $e^x \sin x = 1$ (f) $e^x = 5 \sin x$
   (smallest positive root)

9.3 THE OTHER TRIGONOMETRIC FUNCTIONS

Four more trigonometric functions are defined in terms of $\cos x$ and $\sin x$. These are:

the *tangent* of x, $\tan x = \dfrac{\sin x}{\cos x}$

the *secant* of x, $\sec x = \dfrac{1}{\cos x}$

the *cosecant* of x, $\csc x = \dfrac{1}{\sin x}$

9.3 THE OTHER TRIGONOMETRIC FUNCTIONS

the *cotangent* of x, $\cot x = \dfrac{1}{\tan x} = \dfrac{\cos x}{\sin x}$

These quotient functions are defined, continuous, and differentiable whenever their denominators are not zero. We know that $P(\pi/2) = (0, 1)$ and hence that $\cos \pi/2 = 0$ and $\sin \pi/2 = 1$. Furthermore, for any even integer $2n$, $P(2n\pi) = (1, 0)$ and for any odd integer $2n + 1$, $P((2n + 1)\pi) = (-1, 0)$. Therefore, $\sin(p\pi) = 0$ for any $p \in \mathbf{Z}$. Using the addition formula for the cosine function,

$$\cos\left(\frac{\pi}{2} + p\pi\right) = \cos\left(\frac{\pi}{2}\right)\cos p\pi - \sin\left(\frac{\pi}{2}\right)\sin(p\pi)$$
$$= 0 \cdot \cos p\pi - \sin\left(\frac{\pi}{2}\right) \cdot 0 = 0$$

Also it is easy to see that we have determined *all* values of x for which $\cos x$ and $\sin x$ vanish. Therefore

$\tan x$ and $\sec x$ are defined except when $x = \dfrac{\pi}{2} + p\pi$, $p \in \mathbf{Z}$

and

$\csc x$ and $\cot x$ are defined except when $x = p\pi$, $p \in \mathbf{Z}$

Starting with the identity $\cos^2 x + \sin^2 x = 1$, we can write

$$\frac{\cos^2 x}{\cos^2 x} + \frac{\sin^2 x}{\cos^2 x} = \frac{1}{\cos^2 x} \quad \text{whenever } \cos x \neq 0$$

This provides us with the identity

$$1 + \tan^2 x = \sec^2 x$$

which holds whenever both $\tan x$ and $\sec x$ are defined. Similarly, whenever $\sin x \neq 0$, we have

$$\frac{\cos^2 x}{\sin^2 x} + \frac{\sin^2 x}{\sin^2 x} = \frac{1}{\sin^2 x}$$

or

$$\cot^2 x + 1 = \csc^2 x$$

We shall have later use for the *subtraction formula for the tangent function*. We write

$$\tan(y - x) = \frac{\sin(y - x)}{\cos(y - x)} = \frac{\sin y \cos x - \cos y \sin x}{\cos y \cos x + \sin y \sin x}$$

Now if both $\cos y \neq 0$ and $\cos x \neq 0$, we can write

$$\frac{\sin y \cos x - \cos y \sin x}{\cos y \cos x + \sin y \sin x} = \frac{\dfrac{\sin y \cos x}{\cos y \cos x} - \dfrac{\cos y \sin x}{\cos y \cos x}}{\dfrac{\cos y \cos x}{\cos y \cos x} + \dfrac{\sin y \sin x}{\cos y \cos x}}$$

$$= \frac{\dfrac{\sin y}{\cos y} - \dfrac{\sin x}{\cos x}}{1 + \left(\dfrac{\sin y}{\cos y}\right)\left(\dfrac{\sin x}{\cos x}\right)}$$

$$= \frac{\tan y - \tan x}{1 + \tan y \tan x}$$

Therefore, if both $\tan x$ and $\tan y$ are defined and if $1 + \tan y \tan x \neq 0$,

$$\tan(y - x) = \frac{\tan y - \tan x}{1 + \tan y \tan x}$$

Deducing differentiation formulas for these trigonometric functions is quite easy. For example,

$$D \tan x = D\left(\frac{\sin x}{\cos x}\right) = \frac{\cos x \, D \sin x - \sin x \, D \cos x}{\cos^2 x}$$

$$= \frac{(\cos x)(\cos x) - (\sin x)(-\sin x)}{\cos^2 x}$$

$$= \frac{1}{\cos^2 x}$$

Therefore,

$$D \tan x = \sec^2 x$$

and, using the chain rule,

$$D \tan[f(x)] = \sec^2[f(x)] \cdot Df(x)$$

Similarly, we have

$$D \sec x = D\left(\frac{1}{\cos x}\right) = \frac{\cos x \, D1 - 1 \, D \cos x}{\cos^2 x}$$

$$= \frac{0 - 1(-\sin x)}{\cos^2 x}$$

$$= \frac{\sin x}{\cos^2 x} = \left(\frac{\sin x}{\cos x}\right)\left(\frac{1}{\cos x}\right)$$

and therefore

$$D \sec x = \tan x \sec x$$

More generally,

$$D \sec[f(x)] = \tan[f(x)] \sec[f(x)] \cdot Df(x)$$

9.3 THE OTHER TRIGONOMETRIC FUNCTIONS

The derivation of the following formulas is left to the reader as exercises.

$$D \cot x = -\csc^2 x$$
$$D \cot[f(x)] = -\csc^2[f(x)] \cdot Df(x)$$
$$D \csc x = -\cot x \csc x$$
$$D \csc[f(x)] = -\cot[f(x)] \csc[f(x)] \cdot Df(x)$$

- **Example 1.** Let $f(x) = \tan x + \sec x$. Then
$$Df(x) = D \tan x + D \sec x$$
$$= \sec^2 x + \tan x \sec x = \sec x(\sec x + \tan x)$$

- **Example 2.** Let us consider
$$D \ln \cos x = \frac{1}{\cos x} D(\cos x) = \frac{1}{\cos x}(-\sin x) = -\tan x \;\blacksquare$$

The corresponding integration formulas are collected below.

$$\int \cos x \, dx = \sin x + C$$
$$\int \cos[f(x)] \cdot df(x) = \sin[f(x)] + C$$
$$\int \sin x \, dx = -\cos x + C$$
$$\int \sin[f(x)] \, df(x) = -\cos[f(x)] + C$$
$$\int \sec^2 x \, dx = \tan x + C$$
$$\int \sec^2[f(x)] \, df(x) = \tan[f(x)] + C$$
$$\int \tan x \sec x \, dx = \sec x + C$$
$$\int \tan[f(x)] \sec[f(x)] \, df(x) = \sec[f(x)] + C$$
$$\int \csc^2 x \, dx = -\cot x + C$$
$$\int \csc^2[f(x)] \, df(x) = -\cot[f(x)] + C$$
$$\int \cot x \csc x \, dx = -\csc x + C$$
$$\int \cot[f(x)] \csc[f(x)] \, df(x) = -\csc[f(x)] + C$$

EXERCISES 9.3

In Exercises 1 to 30, find both Df and D^2f. Use the trigonometric identities to simplify the results as much as you can.

1. $f(x) = 3 \tan 2x$
2. $f(x) = \tan(\pi/4 - x/2)$
3. $f(x) = 6 \sec 2x$
4. $f(x) = 2 \sec(x + 3)$
5. $f(x) = \cot(2x + 3)$
6. $f(x) = 2 \csc(1 - 3x)$
7. $f(x) = \tan 3x^2$
8. $f(x) = \tan^2 3x$
9. $f(x) = \sec^2 x - \tan^2 x$
10. $f(x) = (\csc x + \cot x)^2$
11. $f(x) = \sec^3 x/3$
12. $f(x) = \tan^3 2x$
13. $f(x) = \csc^4 2x$
14. $f(x) = (1 + \sec 3x)^{2/3}$
15. $f(x) = \sec^4 x - \tan^4 x$
16. $f(x) = \sec 3x - 3 \tan x$

17. $f(x) = \cot\left(\dfrac{8}{x}\right)$

18. $f(x) = x + \tan x$

19. $f(x) = \cos x \cot x$

20. $f(x) = \dfrac{\tan 2x}{1 - \cot 2x}$

21. $f(x) = [2 \tan^3 (2x) + 1]^{1/3}$

22. $f(x) = \ln \sec^2 x$

23. $f(x) = \ln(1 - 4 \tan x)$

24. $f(x) = \ln \tan\left(\dfrac{\pi}{4} + \dfrac{x}{2}\right)$

25. $f(x) = \ln\left(\dfrac{\sec x + \tan x}{\sec x - \tan x}\right)$

26. $f(x) = \cot(\ln x)$

27. $f(x) = e^{\tan x}$

28. $f(x) = \tan(x \sin x)$

29. $f(x) = (\csc x)^x$

30. $f(x) = (\sin x)^{\sec x}$

In Exercises 31 to 40, evaluate the indeterminate forms.

31. $\lim\limits_{x \to 0} \dfrac{x + \tan x}{\sin 4x}$

32. $\lim\limits_{x \to 2} \dfrac{\sin \pi x}{x - 2}$

33. $\lim\limits_{x \to 0} \dfrac{x - \tan x}{x \sin^2 x}$

34. $\lim\limits_{x \to 0} \dfrac{\sec x - 1}{x \sin x}$

35. $\lim\limits_{x \to 0} \dfrac{\tan 2x - 2 \sin x}{x^3}$

36. $\lim\limits_{x \to \pi/2} \left(x - \dfrac{\pi}{2}\right) \sec x$

37. $\lim\limits_{x \to 0} e^{x \cot x}$

38. $\lim\limits_{x \to 0} (1 + x)^{\cot 2x}$

39. $\lim\limits_{x \to 0} \left(\dfrac{1}{x^2} - \cot^2 x\right)$

40. $\lim\limits_{x \to 0} (\csc x)^{\sin x}$

Establish the differentiation formulas given in Exercises 41 to 44.

41. $D(-x + \tan x) = \tan^2 x$
42. $D(x + \cot x) = -\cot^2 x$
43. $D \ln|\cos x| = -\tan x$
44. $D \ln|\tan x| = 2 \csc 2x$

Work out the integrals in Exercises 45 to 70.

45. $\int \sin 3x \, dx$

46. $\int \cos 4x \, dx$

47. $\int \sec^2 \dfrac{x}{2} \, dx$

48. $\int x^2 \csc^2 x^3 \, dx$

49. $\int \sin x \sec^3 x \, dx$

50. $\int \dfrac{\cos x \, dx}{1 + \sin x}$

51. $\int \dfrac{\sec^2 x \, dx}{\tan x}$

52. $\int \dfrac{\csc^2 x \, dx}{1 - \cot x}$

53. $\int \cos x e^{\sin x} \, dx$

54. $\int \cot x \, dx$

55. $\int \sec^3 x \tan x \, dx$

56. $\int \dfrac{\sec^2 x \, dx}{\sqrt{3 + \tan x}}$

57. $\int \cot^n x \csc^2 x \, dx, \; n \neq -1$

58. $\int \sin ax \cos^n ax \, dx, \; n \neq -1$

9.3 THE OTHER TRIGONOMETRIC FUNCTIONS

59. $\int \csc^4 ax \cot ax \, dx$

60. $\int \sec^2 ax \tan^3 ax \, dx$

61. $\int_0^{\pi/4} \sin 2x \, dx$

62. $\int_0^{\pi/6} \cos 3x \, dx$

63. $\int_0^{\pi/4} \sec^2 x \, dx$

64. $\int_0^{\pi/4} \sec x \tan x \, dx$

65. $\int_0^{\pi/2} \sin^2 x \cos x \, dx$

66. $\int_0^{\pi/2} \frac{\sin x \, dx}{1 + \cos x}$

67. $\int_{-\pi/4}^{\pi/4} \sec^2 x \tan x \, dx$

68. $\int_{\pi/4}^{\pi/2} \frac{\csc^2 x \, dx}{(1 + \cot x)^3}$

69. $\int_{\pi/3}^{2\pi/3} \sec^2 x \, dx$

70. $\int_0^{\pi/2} \tan^2 x \, dx$

Find the areas of the regions described in Exercises 71 to 76.

71. $\{(x, y) : 0 \leq x \leq \frac{\pi}{3}, 0 \leq y \leq \sin x\}$

72. $\{(x, y) : 0 \leq x \leq \frac{\pi}{4}, 0 \leq y \leq \tan x\}$

73. $\{(x, y) : 0 \leq x \leq \frac{\pi}{3}, \sec^2 x \leq y \leq 2\}$

74. $\{(x, y) : 0 \leq x \leq \frac{\pi}{2}, \cos^2 x \leq y \leq \cos x\}$

75. $\{(x, y) : \frac{\pi}{4} \leq x \leq \frac{5\pi}{4}, \cos x \leq y \leq \sin x\}$

76. The area in the first quadrant bounded by the x axis, the line $\{(x, y) : x = \pi/2\}$ and the curve $\{(x, y) : y = \frac{\sin x}{1 + \cos x}\}$.

77. Revolve the area under one arch of the sine curve about the x axis and determine the resulting volume.

78. Revolve the area bounded by the x axis, the vertical line $\{(x, y) : y = \pi/4\}$, and the curve $\{(x, y) : y = \tan x\}$ about the x axis and determine the resulting volume.

PROBLEMS 9.3

1. Find the minimum value of $f(x) = \tan x + \cot x$ over $\{x : 0 < x < \pi/2\}$.

2. Prove that $f(\frac{1}{2} + n)$ is a relative maximum or minimum value of $f(x) = \pi x - \tan \pi x/2$ according as n is an even or an odd integer.

3. Determine the smallest positive root of $\tan x = x$ to two-decimal accuracy.

4. Consider the function $f(x) = \sin 1/x$, $x \neq 0$. Is it possible to assign a value $f(0)$ to make this function continuous at $x = 0$?

5. Show that the function
$$f(x) = x \sin \frac{1}{x}, \quad x \neq 0$$
$$= 0, \quad x = 0$$
is continuous but not differentiable at $x = 0$.

6. Show that the function

$$g(x) = x^2 \sin \frac{1}{x}, \qquad x \neq 0$$
$$= 0, \qquad x = 0$$

   is differentiable but not twice differentiable at $x = 0$. Generalize this to the functions $h_n(x) = x^n \sin 1/x$, $x \neq 0$, $h_n(0) = 0$.

7. Consider the set of all triangles having one side 16 in. long and the angle opposite this side equal to $7\pi/18$ radians. Find the other angles of that triangle in this set which has maximum area.

8. Two corridors intersect at right angles. One is 125 in. wide and the other is 216 in. wide. Find the length of the longest thin rod that can be carried horizontally around the corner.

9. Two factories A and B are at distances a and b, respectively, from a straight power line. A transformer T is to be located along the power line to serve both factories. If the cost of a cable from T to A is \$$c$ per ft and the cost of a cable from T to B is \$$d$ per ft, determine the form of the cost function.

10. A sphere of radius r rests on a table. Consider the set of right circular cones that cover the sphere, resting with the base on the table. What is the height of the cone of minimum volume in this set?

11. A light is to be hung above the center of a circular table so as to give maximum illumination at the edge of the table. The illumination at a point on the table is inversely proportional to the square of the distance from the point to the light, and is proportional to the sine of the angle between the plane of the table and the line from the point to the light. At what height should the light be hung?

12. A mass of w lb rests on a level surface. A force F acts on the mass at an angle θ with the surface. To overcome friction, the force must satisfy the inequality

$$F \geq \frac{\mu w}{\cos \theta + \mu \sin \theta}$$

   where μ is a constant. Find the minimum force required to move the mass.

13. A man walks along a diameter of a circular courtyard of radius 20 ft. A light at one end of the perpendicular diameter casts his shadow on the circular wall. If he moves 5 ft per sec, find his shadow's speed on the wall when he is 8 ft from the center.

14. (Physics) Let the velocity of light in air and in water be v_1 and v_2, respectively. Prove that a photon from a light source S in the air will arrive in the shortest time at a point O in the water if it travels along the path shown in the accompanying figure (Fig. 9.9) where

$$\frac{\sin \alpha_1}{v_1} = \frac{\sin \alpha_2}{v_2}$$

   (This is a special case of Snell's law of refraction.)

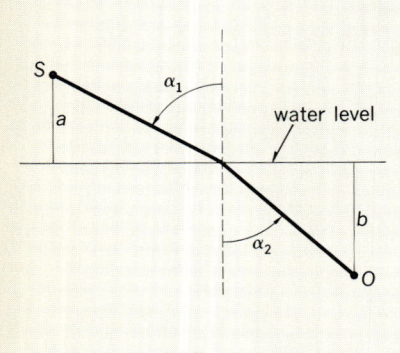

FIGURE 9.9

15. (Engineering) Determine the average speed of a piston in a given gasoline engine as a function of the number of revolutions per minute of the crankshaft.

16. Find the central angle subtended by a chord that cuts off a segment of a circle equal in area to $\frac{1}{8}$ of the area of the circle.

17. A quadrilateral is inscribed in a circle with one side equal to the diameter of the circle, while the other sides have lengths 3, 4, and 5. Find the radius of the circle.

18. Two poles are braced by guy wires 40 ft and 30 ft long as shown in the accompanying figure (Fig. 9.10). If the wires cross at a point 10 ft above the ground, how far are the poles apart?

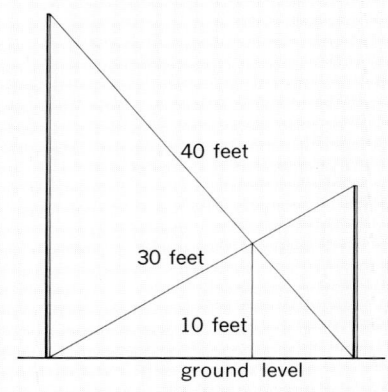

FIGURE 9.10

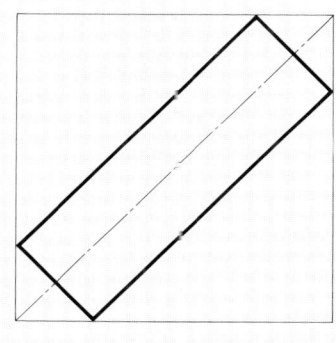

FIGURE 9.11

19. Find the area of the largest rectangle that can be inscribed in a given square with its sides parallel to the diagonals of the square as illustrated in the accompanying figure (Fig. 9.11).

20. Find the area of the largest isosceles triangle that can be inscribed in a given semicircle with a vertex at the center of the semicircle as illustrated in the accompanying figure (Fig. 9.12).

21. Find the area of the largest rectangle that can be inscribed in a given circular sector as illustrated in the accompanying figure (Fig. 9.13).

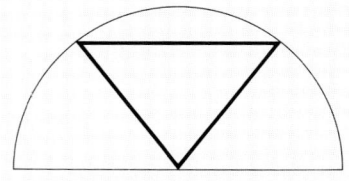

FIGURE 9.12

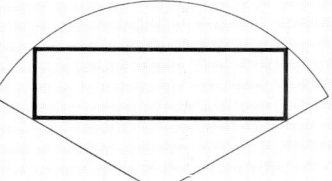

FIGURE 9.13

22. Two equal isosceles triangles are inscribed in a circle with vertices at opposite ends of a fixed diameter, as illustrated in the accompanying figure (Fig. 9.14). What is the area of the largest such six-pointed figure that can be so formed?

23. Two equal rectangles are inscribed in a circle with their sides parallel to perpendicular diameters, as illustrated in the accompanying figure (Fig. 9.15). What is the area of the largest cruciform figure that can be so formed?

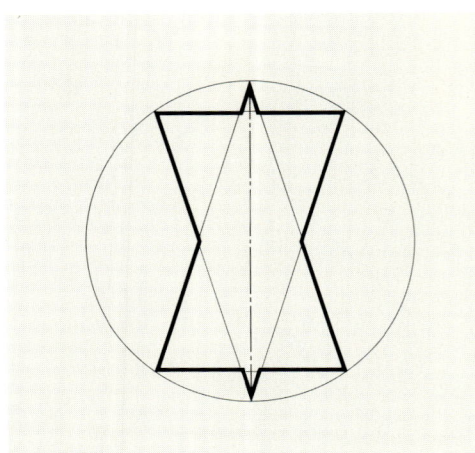

FIGURE 9.14

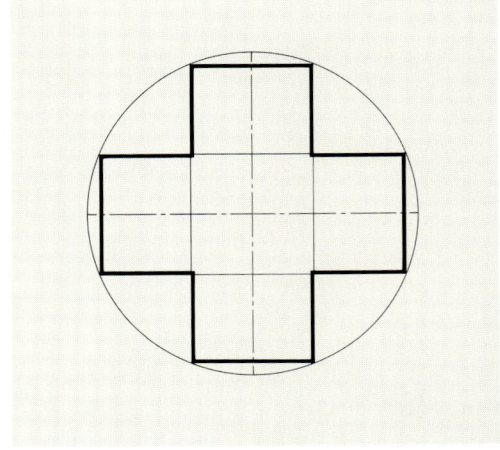

FIGURE 9.15

24. A projectile is fired with initial velocity v_0 at an angle θ with the horizontal. Neglecting air resistance, it follows the curve

$$\left\{(x,y) : y = x \tan \theta - \frac{16}{v_0^2} x^2 \sec^2 \theta \right\}$$

For what value of θ is horizontal range of the projectile the greatest?

25. If a flat surface is inclined at an angle θ with the horizontal and then is moved horizontally, the lift due to air pressure is proportional to $\sin^2 \theta \cos \theta$. What angle maximizes the lift?

26. A circular sector is cut out of a circular piece of sheet metal 12 in. in diameter. The remaining piece is rolled up until the cut edges meet to form a cone. What is the volume of the largest cone that can be made in this way?

27. If the functions f and g satisfy the conditions

$$\begin{aligned} f' &= g, & f(0) &= 0 \\ g' &= -f, & g(0) &= 1 \end{aligned}$$

show that $f(x) = \sin x$ and $g(x) = \cos x$. (Hint: Prove that the function $F(x) = [f(x) - \sin x]^2 + [g(x) - \cos x]^2$ is identically zero.)

28. (a) If $f'' = -f, f(0) = 0$ and $f'(0) = 1$, prove that $f(x) = \sin x$.
    (b) If $g'' = -g, g(0) = 1$ and $g'(0) = 0$, prove that $g(x) = \cos x$.

9.4 GRAPHS OF TRIGONOMETRIC FUNCTIONS

As we have pointed out before, a chief use of the sine and cosine functions is to describe periodic phenomena. A function f is *periodic* if there is a number p such that

$$f(x + p) = f(x)$$

for all values of x such that both x and $x + p$ are in the domain of f. The smallest positive number p having this property is the *period* of f.

The behavior of the periodic function f is completely determined by its behavior over any interval $\{x : x_0 \leq x \leq x_0 + p\}$ of length p. The portion of the graph, $\{(x,y) : y = f(x), x_0 \leq x \leq x_0 + p\}$, is duplicated over the intervals

$$\{x : x_0 + p \leq x \leq x_0 + 2p\}, \{x : x_0 + 2p \leq x \leq x_0 + 3p\}$$

and so on. Figure 9.16 pictures the graph of a function with period 1.

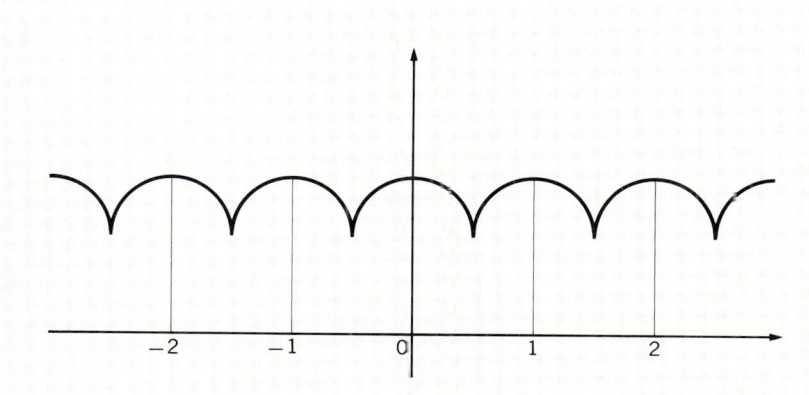

FIGURE 9.16

The sine and cosine functions are both periodic of period 2π (p. 400). Hence we need draw just one segment over the interval $\{x : 0 \leq x \leq 2\pi\}$ in order to know the entire graph. The values of the sine and cosine function for some frequently used values of x are given in the following table.

x	0	$\dfrac{\pi}{6}$	$\dfrac{\pi}{4}$	$\dfrac{\pi}{3}$	$\dfrac{\pi}{2}$	$\dfrac{3\pi}{4}$	π	$\dfrac{5\pi}{4}$	$\dfrac{3\pi}{2}$	$\dfrac{7\pi}{4}$	2π
$\sin x$	0	$\tfrac{1}{2}$	$\sqrt{\tfrac{2}{2}}$	$\sqrt{\tfrac{3}{2}}$	1	$\sqrt{\tfrac{2}{2}}$	0	$-\sqrt{\tfrac{2}{2}}$	-1	$-\sqrt{\tfrac{2}{2}}$	0
$\cos x$	1	$\sqrt{\tfrac{3}{2}}$	$\sqrt{\tfrac{2}{2}}$	$\tfrac{1}{2}$	0	$-\sqrt{\tfrac{2}{2}}$	-1	$-\sqrt{\tfrac{2}{2}}$	0	$\sqrt{\tfrac{2}{2}}$	1

We know that $D \sin x = \cos x$ and hence $D^2 \sin x = D \cos x = -\sin x$. From this we see that the graph $(x,y) : y = \sin x$ has extremum points at $x = \pi/2$ and $x = 3\pi/2$, the first being a maximum point and the second a minimum point. Also, there are points of inflection at $x = 0$, $x = \pi$, and $x = 2\pi$. This information is plotted as in Fig. 9.17(a) and the complete segment of the graph is shown in Fig. 9.17(b). Note that we are talking about functions here, not geometry. Therefore the unit lengths on the axes need no longer be equal.

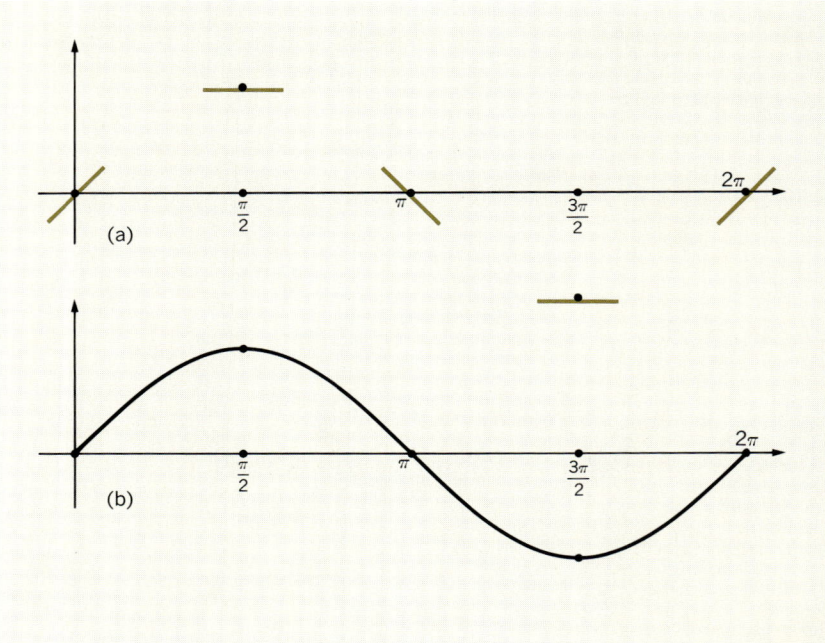

FIGURE 9.17

Knowing this segment, the periodicity of the sine function permits us to draw the graph of $\sin x$ over any interval we wish. Five periods of $\sin x$ are shown in Fig. 9.18. Note that the symmetry with respect to the origin is apparent, as should be the case for an odd function. The resulting graph is called a *sine wave*.

The graph of the cosine function is very much the same as a sine wave. In fact, the identity $\cos x = \sin(\pi/2 - x)$ implies that the graph of the cosine function can be obtained simply by translating the sine wave a distance $-\pi/2$ along the x axis (see Fig. 9.19). As expected for the even function $\cos x$, this graph is symmetric to the y axis.

For describing periodic phenomena it is important to be able to graph the functions

$$a \sin(bx + c), \ a \cos(bx + c)$$

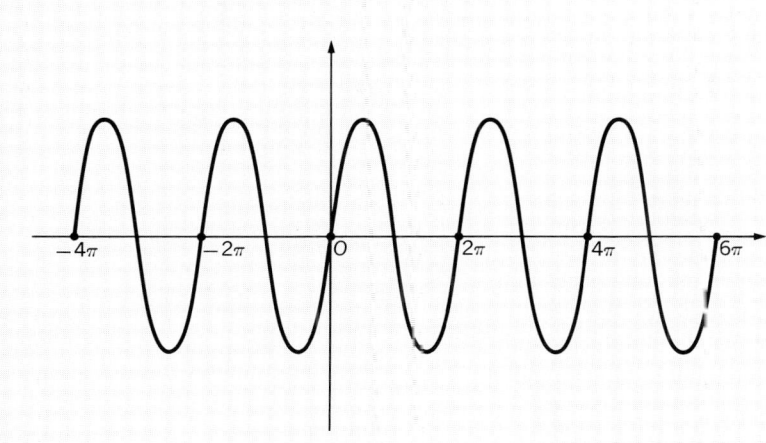

FIGURE 9.18

where a, b, and c are constants and $a > 0$. It is obvious that

$$-a \leqq a \sin(bx + c) \leqq a$$
$$-a \leqq a \cos(bx + c) \leqq a$$

The number a is called the *amplitude* of the sine wave.

Because $b(x + 2\pi/b) + c = bx + c + 2\pi$, we have

$$\sin\left[b\left(x + \frac{2\pi}{b}\right) + c\right] = \sin(bx + c)$$

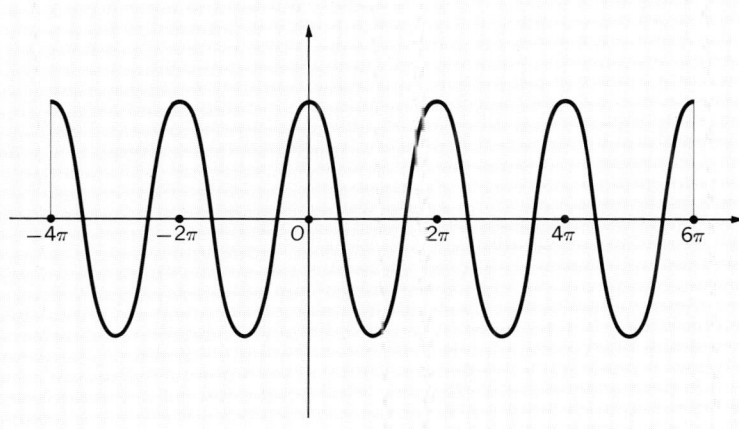

FIGURE 9.19

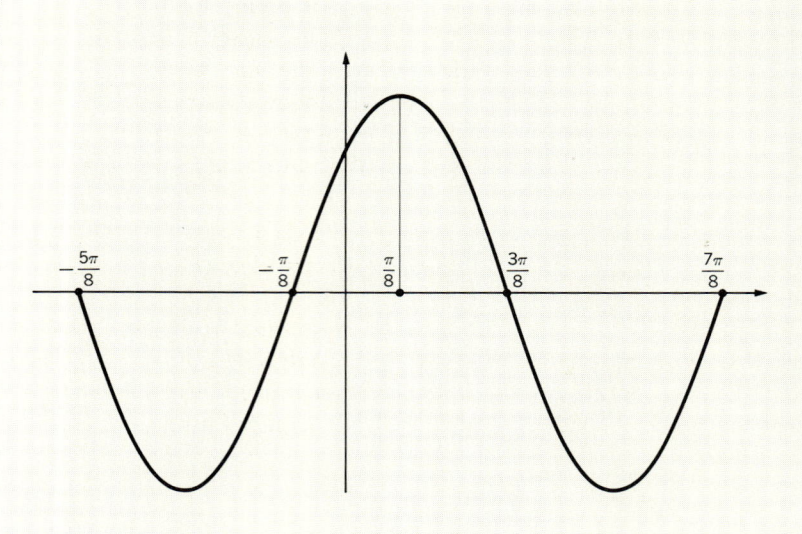

FIGURE 9.20

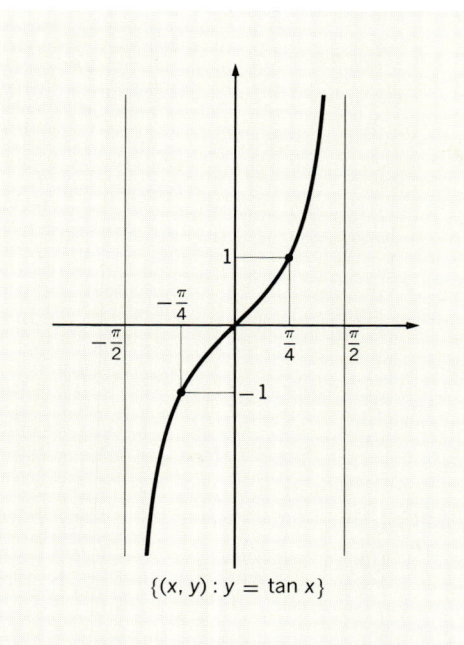

$\{(x, y) : y = \tan x\}$

FIGURE 9.21

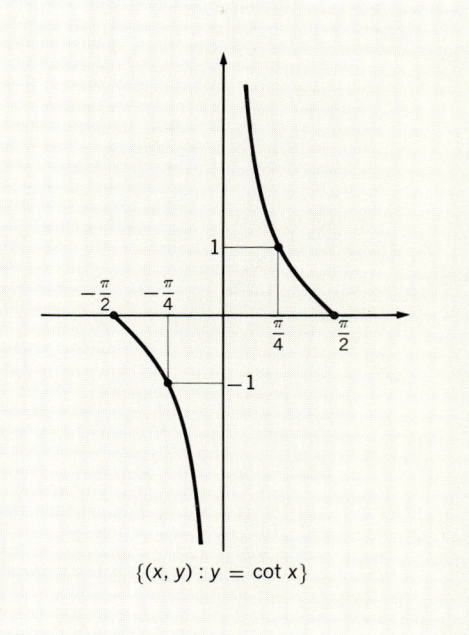

$\{(x, y) : y = \cot x\}$

FIGURE 9.22

It follows that the function $a \sin(bx + c)$ has period $2\pi/b$. Of course, so does $a \cos(bx + c)$. The number b is called the *frequency* of the sine wave.

Lastly, the constant c indicates that the sine wave is translated along the x axis by an amount $-c/b$. This follows immediately from the fact that $\sin(b \cdot 0 + c) = \sin c$ and $\sin(b(-c/b) + c) = \sin 0 = 0$. The number c/b is called the *phase angle* in alternating circuit theory.

■ *Example.* The graph of $3 \sin(2x + \pi/4)$ is a sine wave with amplitude 3, period $2\pi/2 = \pi$, and phase angle $\frac{1}{2}(\pi/4) = \pi/8$. It appears in Fig. 9.20.

The graphs of the remaining four trigonometric functions all possess vertical asymptotes. Figures 9.21 through 9.24 show the graphs of $\tan x$, $\cot x$, $\sec x$, and $\csc x$. Note that $\tan x$ and $\cot x$ are periodic of period π (not 2π). The reader may prove this fact for himself.

The function

$$f(x) = e^{-kx} \sin ax$$

occurs in several applications. We compute that

$$f'(x) = -ke^{-kx} \sin ax + ae^{-kx} \cos ax$$
$$= e^{-kx}(-k \sin ax + a \cos ax)$$

and

$$f''(x) = e^{-kx}[(k^2 - a^2) \sin ax - 2ak \cos ax]$$

Therefore this function has extremum values when

$$a \cos ax = k \sin ax$$

or

$$\tan ax = \frac{a}{k}$$

Also the graph $\{(x,y) : y = e^{-kx} \sin ax\}$ has points of inflection where

$$(k^2 - a^2) \sin ax = 2ak \cos ax$$

or

$$\tan ax = \frac{2ak}{k^2 - a^2}$$

Note that, while $e^{-kx} \sin ax$ is not periodic, it does take on the value zero periodically, and it has both periodic extremum points and periodic points of inflection. From the inequality $-1 \leq \sin ax \leq 1$, we have

$$-e^{-kx} \leq e^{-kx} \sin ax \leq e^{-kx}$$

This tells us that the graph $\{(x,y) : y = e^{-kx} \sin ax\}$ lies entirely be-

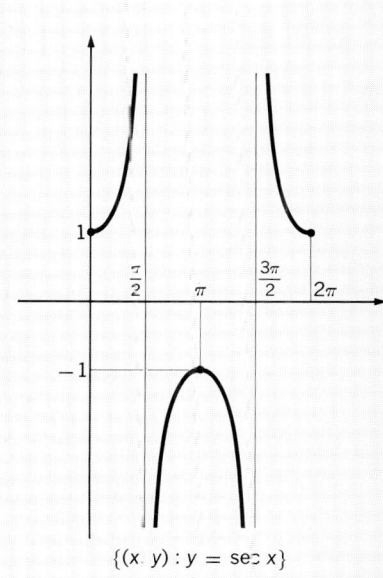

$\{(x, y) : y = \sec x\}$

FIGURE 9.23

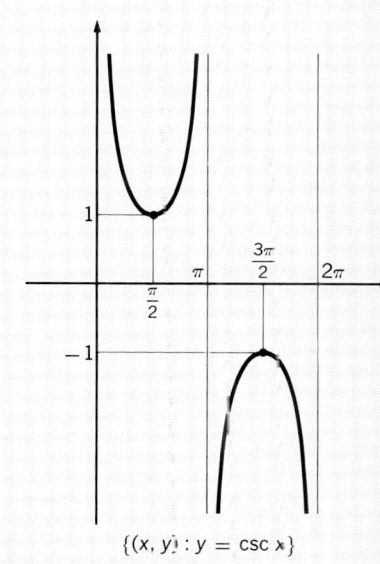

$\{(x, y) : y = \csc x\}$

FIGURE 9.24

tween the two graphs $\{(x,y) : y = -e^{-kx}\}$ below and $\{(x,y) : y = e^{-kx}\}$ above.

Figure 9.25 shows the graph $\{(x,y) : y = e^{-x/5} \sin 2x\}$ over the interval $\{x : 0 \leq x \leq 2\pi\}$. The critical values occur where $\tan 2x = 10$, whence $x \cong 0.73$ radians, and so forth. We have also shown the graphs of $e^{-x/5}$ and $-e^{-x/5}$. The figure exhibits a typical instance of a *damped vibration*.

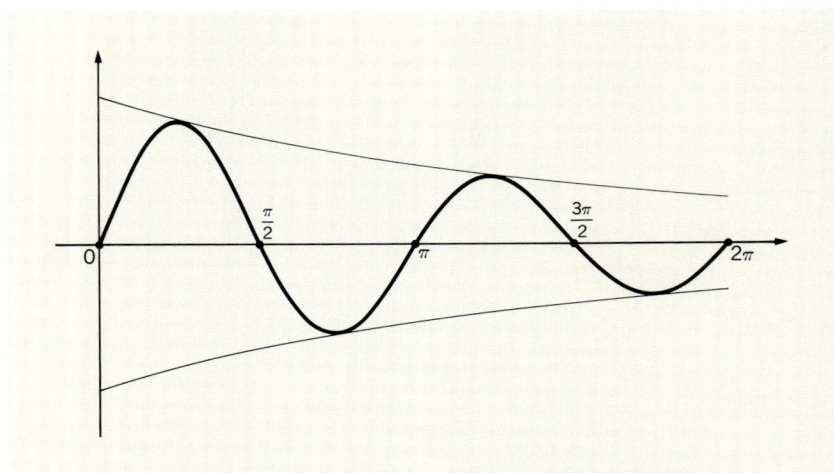

FIGURE 9.25

EXERCISES 9.4   Determine the period of each function in Exercises 1 to 10.

1. $f(x) = 2 \sin 2x$
2. $f(x) = 3 \sin \pi x$
3. $f(x) = 6 \cos \dfrac{\pi x}{2}$
4. $f(x) = 2 \cos \dfrac{x}{3}$
5. $f(x) = 4 \sin\left(2\pi x - \dfrac{\pi}{2}\right)$
6. $f(x) = \cos\left(3x - \dfrac{\pi}{4}\right)$
7. $f(x) = \tfrac{1}{2} \tan 2x$
8. $f(x) = 3 \tan\left(\pi x - \dfrac{\pi}{6}\right)$
9. $f(x) = \sec\left(2x - \dfrac{\pi}{2}\right)$
10. $f(x) = 2 \sin 3x - 3 \cos 2x$

Making full use of the first and second derivatives, draw the graphs in Exercises 11 to 35.

11. $\left\{(x,y) : y = 4 \sin \dfrac{\pi x}{2} ; \ 0 \leq x \leq 4\right\}$
12. $\left\{(x,y) : y = \cos \dfrac{x}{3} ; \ -3\pi \leq x \leq 6\pi\right\}$

13. $\{(x,y) : y = \frac{1}{2} \tan 2x; \ 0 \leq x \leq \frac{\pi}{4}\}$

14. $\{(x,y) : y = \sec \frac{x}{2}; \ 0 \leq x \leq \pi\}$

15. $\{(x,y) : y = -2 \sin \frac{\pi x}{2}; \ -2 \leq x \leq 2\}$

16. $\{(x,y) : y = \frac{1}{2} + \frac{1}{2} \cos 2x; \ 0 \leq x \leq 2\pi\}$

17. $\{(x,y) : 2 \sin\left(x + \frac{\pi}{4}\right); \ 0 \leq x \leq 2\pi\}$

18. $\{(x,y) : \cos\left(2x + \frac{\pi}{3}\right); \ -\pi \leq x \leq \pi\}$

19. $\{(x,y) : \tan\left(2x - \frac{\pi}{4}\right); \ \frac{-\pi}{8} < x \leq \frac{\pi}{4}\}$

20. $\{(x,y) : y = \sin x + \cos x; \ -\pi \leq x \leq 2\pi\}$

21. $\{(x,y) : y = 2 \sin x - \cos x; \ 0 \leq x \leq 2\pi\}$

22. $\{(x,y) : y = \cos 2x - \sqrt{3} \sin 2x; \ -\pi \leq x \leq \pi\}$

23. $\{(x,y) : y = \sin 2x + \cos x; \ 0 \leq x \leq 2\pi\}$

24. $\{(x,y) : y = \sin^2 x, \ 0 \leq x \leq 2\pi\}$

25. $\{(x,y) : y = 2 \cos 3x - 3 \sin 4x; \ 0 \leq x \leq \pi\}$

26. $\{(x,y) : y = x - 2 \sin x; \ 0 \leq x \leq 2\pi\}$

27. $\{(x,y) : y = \cos 2x - x \sin x; \ 0 \leq x \leq 2\pi\}$

28. $\{(x,y) : y = 3 \sec x + 4 \cos x; \ 0 \leq x < \frac{\pi}{2}\}$

29. $\{(x,y) : y = 2 \csc x - \cot x; \ 0 < x \leq \frac{\pi}{2}\}$

30. $\{(x,y) : y = \frac{\sin^2 x}{1 + \cos^2 x}, \ 0 \leq x \leq 2\pi\}$

31. $\{(x,y) : y = e^{-x} \cos x; \ 0 \leq x < \frac{\pi}{2}\}$

33. $\{(x,y) : y = e^{-x/2} \sin \frac{x}{2}; \ 0 \leq x \leq 4\pi\}$

34. $\{(x,y) : y = e^{-x/10} \sin \pi x; \ 0 \leq x \leq 4\}$

35. $\{(x,y) : y = 4e^{-3x} \cos 3\pi x; \ 0 \leq x \leq 2\}$

9.5 POWER SERIES EXPANSIONS

The functions sin, cos, and tan z can each be expanded in a Maclaurin series. We provide these expansions here because they are useful both for computational and for theoretical purposes.

Let $f(x) = \sin x$. Then $f'(x) = \cos x$, $f''(x) = -\sin x$, $f'''(x) = -\cos x$, $f^{(4)}(x) = \sin x$. It then follows that, for any $k \in \mathbf{N}$,

$$f^{(4k)}(x) = \sin x, \qquad f^{(4k+1)}(x) = \cos x$$
$$f^{(4k+2)}(x) = -\sin x, \qquad f^{(4k+3)}(x) = -\cos x$$

Then we have

$$f^{(4k)}(0) = 0 = f^{(4k+2)}(0), \qquad f^{(4k+1)}(0) = 1, \qquad f^{(4k+3)}(0) = -1$$

The general Maclaurin series

$$f(x) = f(0) + f'(0)x + \frac{f''(0)}{2!} x^2 + \cdots + \frac{f^{(n)}(0)}{n!} x^n + \cdots$$

specializes to

$$\sin x = 0 + 1 \cdot x + \frac{0}{2!} x^3 + \frac{-1}{3!} x^3 + \frac{0}{4!} x^4 + \frac{1}{5!} x^5$$
$$+ \frac{0}{6!} x^6 + \frac{-1}{7!} x^7 + \cdots$$
$$= x - \frac{1}{6} x^3 + \frac{1}{120} x^5 - \frac{1}{5040} x^7 + \cdots + \frac{(-1)^k}{(2k+1)!} x^{2k+1} + \cdots$$

Using the ratio test we can easily prove that this series converges for all $x \in \mathbf{R}$ and, applying Taylor's theorem, it is just as easy to show that the series converges to $\sin x$ for each $x \in \mathbf{R}$. For values of x close to zero, the series converges very rapidly. For instance, if $x = \frac{1}{10}$, then

$$\sin \frac{1}{10} = \frac{1}{10} - \frac{1}{6}\left(\frac{1}{10}\right)^3 + \frac{1}{120}\left(\frac{1}{10}\right)^5 - \frac{1}{5040}\left(\frac{1}{10}\right)^7 + \cdots$$
$$= 0.1 - \frac{1}{6}(0.001) + \frac{1}{120}(0.00001)$$
$$- \frac{1}{5040}(0.0000001) + \cdots$$
$$\cong 0.0998334\ldots$$

which was computed using just the first three terms and is accurate as far as it is written. In fact, the error is less than the first term omitted, namely, $\frac{1}{5040}(0.0000001) \cong 2 \times 10^{-11}$ so we could have kept *ten* decimals.

An easy application of the Maclaurin series provides some useful inequalities concerning the sine function. For positive values of x, we have

$$x - \frac{1}{6} x^3 < \sin x < x$$

and
$$x - \frac{1}{6}x^3 < \sin x < x - \frac{1}{6}x^3 + \frac{1}{120}x^5$$
and
$$x - \frac{1}{6}x^3 + \frac{1}{120}x^5 - \frac{1}{5040}x^7 < \sin x < x - \frac{1}{6}x^3 + \frac{1}{120}x^5$$

and so forth.

Next let $f(x) = \cos x$. Then $f'(x) = -\sin x$, $f''(x) = -\cos x$, $f'''(x) = \sin x$ and $f^{(4)}(x) = \cos x$. It follows that, for any $k \in \mathbf{N}$,

$$f^{(4k)}(x) = \cos x, \quad f^{(4k+1)}(x) = -\sin x,$$
$$f^{(4k+2)}(x) = -\cos x, \quad f^{(4k+3)}(x) = \sin x$$

Therefore $f^{(4k)}(0) = 1$, $f^{(4k+2)}(0) = -1$ and $f^{(4k+1)}(0) = 0 = f^{(4k+3)}(0)$. The Maclaurin series for $\cos x$ is then

$$\cos x = 1 + 0 \cdot x + \frac{-1}{2!}x^2 + 0 \cdot x^3 + \frac{1}{4!}x^4 + 0 \cdot x^5 + \frac{-1}{6!}x^6 + \cdots$$
$$= 1 - \frac{1}{2}x^2 + \frac{1}{24}x^4 - \frac{1}{720}x^6 + \cdots + \frac{(-1)^k}{(2k)!}x^{2k} + \cdots$$

This series also converges very rapidly for values of x close to zero.

Note that the series for $\sin x$ contains only odd powers of x and the series for $\cos x$ contains only even powers of x. This is a reflection of the fact that sin is an odd function and cos is an even function.

The Maclaurin series for $\sin x$ and $\cos x$ are true computational tools. With the aid of a computer, tables of values of $\sin x$ and $\cos x$ can be compiled to any desired degree of accuracy.

A comparable Maclaurin series for $\tan x$ is somewhat more difficult to determine. The computation of the first three terms shows why. Let $f(x) = \tan x$. Then

$$f'(x) = \sec^2 x$$
$$f''(x) = 2 \sec x(\sec x \tan x) = 2 \sec^2 x \tan x$$
$$f'''(x) = 4 \sec x(\sec x \tan x) \tan x + 2 \sec^2 x(\sec^2 x)$$
$$= 4 \sec^2 x \tan^2 x + 2 \sec^4 x$$
$$f^{(4)}(x) = 8 \sec x(\sec x \tan x) \tan^2 x + 8 \sec^2 x \tan x(\sec^2 x)$$
$$\qquad + 8 \sec^3 x(\sec x \tan x)$$
$$= 8 \sec^2 x \tan^3 x + 16 \sec^4 x \tan x$$
$$f^{(5)}(x) = 16 \sec x(\sec x \tan x) \tan^3 x + 24 \sec^2 x \tan^2 x(\sec^2 x)$$
$$\qquad + 64 \sec^3 x(\sec x \tan x) \tan x + 16 \sec^4 x(\sec^2 x)$$
$$= 16 \sec^2 x \tan^4 x + 88 \sec^4 x \tan^2 x + 16 \sec^6 x$$

The values $\sec(0) = 1$, $\tan(0) = 0$, imply that

$$f(0) = 0, \quad f'(0) = 1, \quad f''(0) = 0, \quad f'''(0) = 2$$
$$f^{(4)}(0) = 0, \quad f^{(5)}(0) = 16$$

Hence the first three (nonzero) terms of the Maclaurin series are

$$\tan x = x + \tfrac{1}{3}x^3 + \tfrac{2}{15}x^5 + \cdots$$

We know that $\tan x$ is not defined when $x = \pi/2$. Therefore this series cannot converge beyond the interval $\{x : -\pi/2 < x < \pi/2\}$.

■ **Example.** Let $f(x) = e^{-x}\sin x$. Then obviously $f(0) = 0$. Also

$$f'(x) = e^{-x}(\cos x - \sin x) \quad \text{so } f'(0) = 1$$
$$f''(x) = -2e^{-x}\cos x \quad \text{so } f''(0) = -2$$
$$f'''(x) = 2e^{-x}(\cos x + \sin x) \quad \text{so } f'''(0) = 2$$
$$f^{(4)}(x) = -4e^{-x}\sin x \quad \text{so } f^{(4)}(0) = 0$$
$$f^{(5)}(x) = -4e^{-x}(\cos x - \sin x) \quad \text{so } f^{(5)}(0) = -4$$

Therefore the first four nonzero terms of the Maclaurin series are

$$e^{-x}\sin x = x - x^2 + \tfrac{1}{3}x^3 - \tfrac{1}{30}x^5 + \cdots$$

The Maclaurin series

$$\sin x = x - \frac{x^3}{3!} + \frac{x^5}{5!} - \frac{x^7}{7!} + \cdots$$

can be differentiated term by term to give a Maclaurin series for $D \sin x$ (see Section 5.4). Doing so, we see that

$$D \sin x = 1 - \frac{3x^2}{3!} + \frac{5x^5}{5!} - \frac{7x^6}{7!} + \cdots$$

$$= 1 - \frac{x^2}{2!} + \frac{x^4}{4!} - \frac{x^6}{6!} + \cdots = \cos x$$

Similarly,

$$D \cos x = D\left(1 - \frac{x^2}{2!} + \frac{x^4}{4!} - \frac{x^6}{6!} + \frac{x^8}{8!} - \cdots\right)$$

$$= -\frac{2x}{2!} + \frac{4x^3}{4!} - \frac{6x^5}{6!} + \frac{8x^7}{8!} - \cdots$$

$$= -x + \frac{x^3}{3!} - \frac{x^5}{5!} + \frac{x^7}{7!} - \cdots$$

$$= -\left(x - \frac{x^3}{3!} + \frac{x^5}{5!} - \frac{x^7}{7!} + \cdots\right) = -\sin x$$

These calculations tend more to check the validity of term-by-term differentiation of series than to verify our derivative formulas, of course.

9.5 POWER SERIES EXPANSIONS

Verify the Maclaurin series expansions in Exercises 1 to 10.

EXERCISES 9.5

1. $x \sin x = x^2 - \frac{1}{6}x^4 + \frac{1}{120}x^6 - \frac{1}{5040}x^8 + \cdots$
2. $\sin x^2 = x^2 - \frac{1}{6}x^6 + \frac{1}{120}x^{10} - \cdots$
3. $x^2 \tan x = x^2 - \frac{1}{3}x^4 + \frac{2}{45}x^6 - \cdots$
4. $\sin^2 x = x^2 - \frac{1}{3}x^4 + \frac{2}{45}x^6 - \cdots$
5. $\sin^2 x \cos x = x^2 - \frac{5}{6}x^4 + \frac{91}{360}x^6 - \cdots$
6. $e^{-x} \sin x = x - x^2 + \frac{1}{3}x^3 - \frac{1}{30}x^5 + \cdots$
7. $e^x \cos x = 1 + x - \frac{1}{3}x^3 - \frac{1}{6}x^4 - \cdots$
8. $\dfrac{\cos x}{1 - \frac{1}{2}x^2} = 1 + \dfrac{1}{24}x^4 + \dfrac{7}{360}x^6 + \cdots$
9. $\ln \cos x = -\frac{1}{2}x^2 - \frac{1}{12}x^4 - \cdots$
10. $\ln \sec x = \frac{1}{2}x^2 + \frac{1}{12}x^4 + \cdots$

Use the series given in the section and those in Exercises 1 to 10 to evaluate the expressions in Exercises 11 to 20.

11. $\sin(0.2)$
12. $\cos(-0.1)$
13. $\tan(\frac{1}{5})$
14. $\frac{1}{4} \tan \frac{1}{2}$
15. $\frac{1}{5} \sin \frac{1}{5}$
16. $\sin 5° = \sin \dfrac{\pi}{36}$
17. $\tan \dfrac{\pi}{18}$
18. $e^{-1/10} \sin(\frac{1}{10})$
19. $\ln \cos(\frac{1}{4})$
20. $\dfrac{\cos(\pi/4)}{1 - \pi^2/32}$

PROBLEMS 9.5

1. Expand $\sin x$ in a Taylor series at $x = \pi/4$ and use the series to evaluate $\sin 40°$.
2. Expand $\cos x$ in a Taylor series at $x = \pi/3$ and use the series to evaluate $\cos 65°$.
3. Expand $\sin x$ in a Taylor series at $x = \pi/2$ and compare the result with the Maclaurin series for $\cos x$.
4. Obtain the Maclaurin series for $x^2 \cos x$ and show that it is the same as the term-by-term product

$$x^2 \sum_{n=0}^{\infty} (-1)^n \frac{x^{2n}}{(2n)!} = \sum_{n=0}^{\infty} (-1)^n \frac{x^{2n+2}}{(2n)!}$$

5. Use the identity $\cos 2x = 1 - 2\sin^2 x = 2\cos^2 x - 1$ to obtain the Maclaurin series for $\sin^2 x$ and $\cos^2 x$.
6. Use known series to obtain Maclaurin series expansions of the following:
   (a) $f(x) = x \cos x - \sin x$
   (b) $f(x) = x \tan x$

(c) $f(x) = \dfrac{\sin x}{x}, \quad x \neq 0$

$ = 1, \quad x = 0$

(d) $f(x) = \dfrac{1 - \cos x}{x}, \quad x \neq 0$

$ = 0, \quad x = 0$

7. Verify the series expansion

$$\sin^3 x = \dfrac{3}{4} \sum_{n=1}^{\infty} (-1)^{n+1} \dfrac{3^{2n} - 1}{(2n + 1)!} x^{2n+1}$$

8. Suppose that $f(x) = \sum_{n=1}^{\infty} a_n x^n$ satisfies $f'' + f = 0$. Equate coefficients of like powers of x in the equation

$$f''(x) = \sum_{n=2}^{\infty} (n - 1)n a_n x^{n-2} = -f(x) = -\sum_{n=0}^{\infty} a_n x^n$$

to show that f must have the form $f(x) = a_0 \cos x + a_1 \sin x$.

9.6 ANOTHER IMPORTANT DIFFERENTIAL EQUATION

Any function of the form

$$f(x) = a \cos kx + b \sin kx$$

where a and b are arbitrary constants, satisfies the differential equation

$$f''(x) = -k^2 f(x)$$

or

$$f''(x) + k^2 f(x) = 0$$

For by direct calculation,

$$D(a \cos kx + b \sin kx) = a(-k \sin kx) + b(k \cos kx)$$
$$= k(-a \sin kx + b \cos kx)$$
$$D^2(a \cos kx + b \sin kx) = ka(-k \cos kx) + kb(-k \sin kx)$$
$$= -k^2(a \cos kx + b \sin kx)$$

or $f''(x) = -k^2 f(x)$. In a more advanced course in analysis it is also proved that any function satisfying $f'' + k^2 f = 0$ must be of the given form.

A useful trick can be played here to alter the form of the function $f(x) = a \cos kx + b \sin kx$. We choose a number x_0 satisfying the equation

$$\sin kx_0 = -\dfrac{a}{\sqrt{a^2 + b^2}}$$

or

$$\sin(-kx_0) = \dfrac{a}{\sqrt{a^2 + b^2}}$$

9.6 ANOTHER IMPORTANT DIFFERENTIAL EQUATION

Then

$$\cos^2(-kx_0) = 1 - \sin^2(-kx_0) = 1 - \frac{a^2}{a^2 + b^2} = \frac{b^2}{a^2 + b^2}$$

Hence $\cos(-kx_0) = b/\sqrt{a^2 + b^2}$. The function $f(x)$ can now be expressed as

$$\begin{aligned}
f(x) &= a \cos kx + b \sin kx \\
&= \sqrt{a^2 + b^2} \sin(-kx_0) \cos kx \\
&\qquad + \sqrt{a^2 + b^2} \cos(-kx_0) \sin kx \\
&= \sqrt{a^2 + b^2}(\sin kx \cos kx_0 - \cos kx \sin kx_0) \\
&= \sqrt{a^2 + b^2} \sin k(x - x_0)
\end{aligned}$$

Therefore the general solution of the differential equation $f''(x) = -k^2 f(x)$ may be written as

$$f(x) = c \sin k(x - x_0)$$

We recognize that the graph of this function is a *sine wave* (Fig. 9.20).

Simple harmonic motion. If a body moves rectilinearly with the equation of motion

$$s = f(t) = c \sin k(t - t_0)$$

the motion is called a simple harmonic motion. The *amplitude* of the motion is $|c|$, its *period* is $2\pi/|k|$ and its phase constant is t_0. (See Section 9.4.)

■ **Example 1.** Let a mass of weight m be suspended from a spring as shown schematically in Fig. 9.26. If the mass is displaced from its equilibrium position a distance s, then Hooke's law of spring says that the restoring force due to the spring is proportional to the displacement s and is directed opposite to the displacement. If we neglect the effects of gravity and internal friction, then Newton's second law of motion provides the differential equation

$$D_t(mv) = -cs$$

where c is the "spring constant," and v is velocity as a function of t. Setting $c/m = k^2$, we have the equation

$$D_t v = D_t^2 s = -k^2 s$$

Therefore the displacement $s(t)$ has the form

$$s(t) = C \sin k(t - t_0)$$

or the mass will move in a simple harmonic motion.

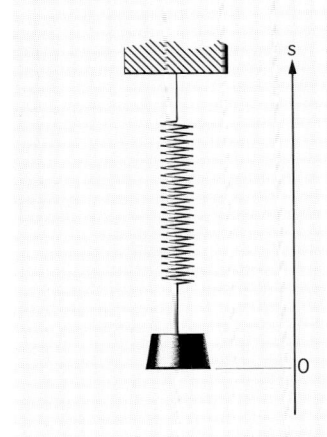

FIGURE 9.26

(Note: In practice, the oscillations of the mass will decrease in amplitude in what is known as a "damped vibration." Our unrealistic result stems from neglecting gravity and friction.)

Suppose the mass is displaced 3 in. and released from rest ($v = 0$) at time $t = 0$. Then we have

$$s(0) = 3 = C \sin k(0 - t_0) = C \sin(-kt_0)$$

and

$$v(0) = 0 = D_t s(0) = Ck \cos k(0 - t_0) = Ck \cos(-kt_0)$$

The only way we can have $Ck \cos(-kt_0) = 0$ is to have $-kt_0 = \pi/2 + p\pi$, for some $p \in \mathbf{Z}$. We choose $t_0 = -\pi/2k$. It follows that

$$3 = C \sin(-kt_0) = C \sin\left(\frac{\pi}{2}\right) = C$$

Therefore, in this particular case, the equation of motion is

$$s(t) = 3 \sin k\left(t - \frac{\pi}{2k}\right)$$

$$= 3 \sin\left(kt - \frac{\pi}{2}\right)$$

Without further knowledge, of course, we cannot determine the spring constant and hence the period $2/|k|$. (Note that k is not the spring constant.)

If a body moves with simple harmonic motion according to the equation

$$s(t) = c \sin k(t - t_0)$$

then its velocity is

$$v(t) = ck \cos k(t - t_0)$$

Therefore

$$\left(\frac{s}{c}\right)^2 + \left(\frac{v}{ck}\right)^2 = \sin^2 k(t - t_0) + \cos^2 k(t - t_0) = 1$$

which reduces to

$$k^2 s^2 + v^2 = c^2 k^2$$

This equation determines the velocity

$$v = k\sqrt{c^2 - s^2}$$

as a function of distance. It also can be used to determine the constants c and k from measurements of distance and velocity.

9.6 ANOTHER IMPORTANT DIFFERENTIAL EQUATION

■ **Example 2.** Suppose a body moving in simple harmonic motion has a speed of 8 ft per sec when it is 2 ft from its mean position and has a speed of 4 ft per sec when it is 4 ft from its mean position. Then we have the two equations

$$k^2(2)^2 + 8^2 = c^2k^2$$
$$k^2(4)^2 + 4^2 = c^2k^2$$

or

$$4k^2 + 64 = 16k^2 + 16$$
$$48 = 12k^2$$
$$k^2 = 4 \quad \text{or} \quad k = 2$$

Then $4k^2 + 64 = 4(4) + 64 = c^2k^2 = 4c^2$ or $c^2 = 20$ and $c = 2\sqrt{5}$. The simple harmonic motion can be expressed by the equation

$$s = 2\sqrt{5} \sin 2(t - t_0)$$

The phase constant t_0 depends upon when we start measuring time and cannot be determined by this method.

PROBLEMS 9.6

1. A particle is in simple harmonic motion. The amplitude of its motion is 6 cm and when the particle is halfway between the center and the end of its interval of motion, its speed is $12\sqrt{3}$ cm per sec. What is the period of vibration?

2. A body in harmonic motion has speed 4 when its displacement is 3 and has speed 6 when its displacement is 2. Determine the equation of motion.

3. A torsion spring obeys the law $ID_t^2\theta = -k\theta$, where I is the moment of inertia of the spring weight, k is the torsion coefficient of the spring, and θ is the angle through which the spring has been twisted from its rest position. What is the period of the spring?

4. In what fraction of the total period of a particle in simple harmonic motion does its velocity fall from its maximum value to one-half of its maximum value? At what fraction of the amplitude does this occur?

5. A pendulum of length l swings from a frictionless bearing. The angle θ between the pendulum arm and a vertical line through the bearing satisfies the differential equation $lD_t^2\theta = -g \sin \theta$, where g is the acceleration of gravity. This equation has no elementary solution. However, for small oscillations, $\sin \theta \cong \theta$. Use this simplifying assumption to determine the period of the pendulum.

6. An off-center circular cam rotates on a shaft whose center is k units from the center of the arm. Prove that a point cam follower constrained to move rectilinearly will experience simple harmonic motion.

7. Let f be a solution of the differential equation $f'' + k^2f = 0$. Define a function u by the equation

$$f(x) = u(x) \cos kx$$

When this form of f is substituted into the differential equation, the result is a first-order linear differential equation in the function u'. Obtain u from this equation and thereby show that f must have the form $f(x) = a \sin kx + b \cos kx$.

8. In the Maclaurin series for e^x, let x be a pure imaginary number iy (where $i = \sqrt{-1}$). By separating the real and imaginary terms of this series, derive *Euler's formula*
$$e^{iy} = \cos y + i \sin y$$

9. For what values of k does the function $f(x) = e^{kx}$ satisfy the differential equation $f'' + af' + bf = 0$? Can you carry out a program using this result and that of Problem 8 to solve all such differential equations?

10. A particle whose equation of motion is of the form
$$s'' + 2cs' + k^2 s = 0, \qquad c \neq 0, k > 0$$
is said to experience a *damped vibration*. Verify the following three cases:
   (a) if $c = k$, then $s = (a + bt)e^{-kt}$ (critical damping)
   (b) if $c > k$, then $s = ae^{r_1 t} + be^{r_2 t}$, where $r_1 = -c + \sqrt{c^2 - k^2}$ and $r_2 = -c - \sqrt{c^2 - k^2}$ (over-critical damping)
   (c) if $c < k$, then $s = ae^{-kt} \sin(\alpha t + \beta)$, where $\alpha = \sqrt{k^2 + c^2}$ (under-critical damping).

9.7 INVERSE TRIGONOMETRIC FUNCTIONS

The function sin is not a monotone function; so, in general, it does not have a (unique) inverse. However, if we restrict the domain of sin to be an interval over which sin *is* monotonic, then there will be a unique inverse function (Theorem 8.1). By convention, we select the interval $\{y : -\pi/2 \leqq y \leqq \pi/2\}$. In this interval, sin is a strictly increasing function. Therefore the inverse function of $x = \sin y$,
$$y = \sin^{-1} x$$
is defined, is continuous and differentiable, and is strictly increasing on the interval from $\sin(-\pi/2) = -1$ to $\sin(\pi/2) = 1$. Note that
$$\frac{-\pi}{2} \leqq \sin^{-1} x \leqq \frac{\pi}{2}, \qquad -1 \leqq x \leqq 1$$

The graph of $\sin^{-1} x$ may be constructed by the methods of Section 8.1. Note that the *graph* $\{(x, y) : y = \sin^{-1} x\}$ in Fig. 9.27(a) is just a portion of the *set* $\{(x, y) : x = \sin y\}$ which is not the graph of a function (Fig. 9.27(b)).

The cosine function is treated similarly. We know that $\cos y$ is a strictly decreasing function on the interval $\{y : 0 \leqq y \leqq \pi\}$. Therefore, over this interval, its inverse function $\cos^{-1} x$ is defined, is continuous and differentiable, and is strictly decreasing. The domain of the function $\cos^{-1}$ is the interval between $\cos(0) = 1$ and $\cos(\pi) = -1$, and its range is given by the inequality
$$0 \leqq \cos^{-1} x \leqq \pi$$

9.7 INVERSE TRIGONOMETRIC FUNCTIONS

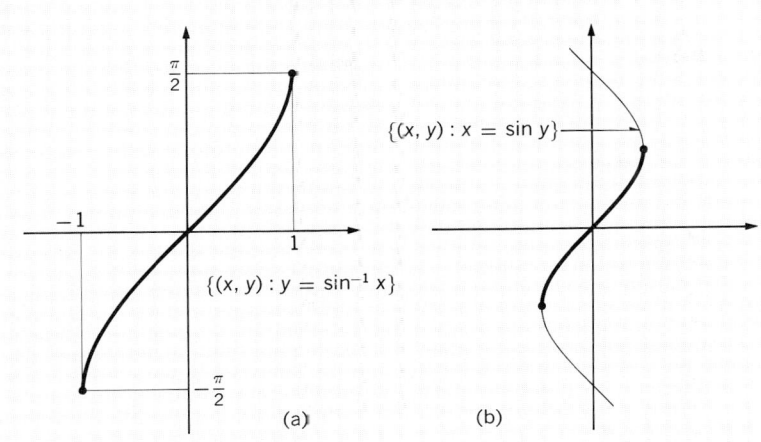

FIGURE 9.27

The graph of $\cos^{-1}$ and the position it occupies in the set $\{(x, y) : x = \cos y\}$ are shown in Fig. 9.28.

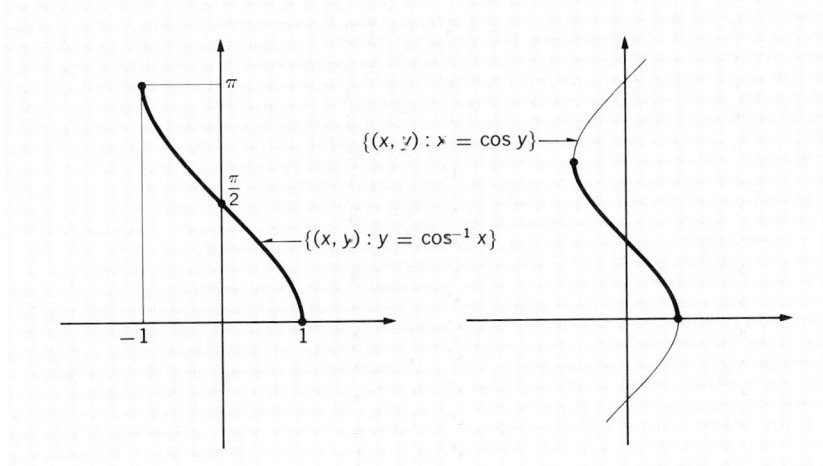

FIGURE 9.28

The other four inverse trigonometric functions have domains and ranges as follows:

$\tan^{-1} x$ is defined for all $x \in \mathbf{R}$ and $-\dfrac{\pi}{2} < \tan^{-1} x < \dfrac{\pi}{2}$

$\cot^{-1} x$ is defined for all $x \in \mathbf{R}$ and $0 < \cot^{-1} x < \pi$

$\sec^{-1} x$ is defined for $|x| \geq 1$ and $0 \leq \sec^{-1} x \leq \pi$

$\csc^{-1} x$ is defined for $|x| \geq 1$ and $-\dfrac{\pi}{2} \leq \csc^{-1} x \leq \dfrac{\pi}{2}$

Graphs of these functions appear in Fig. 9.29.

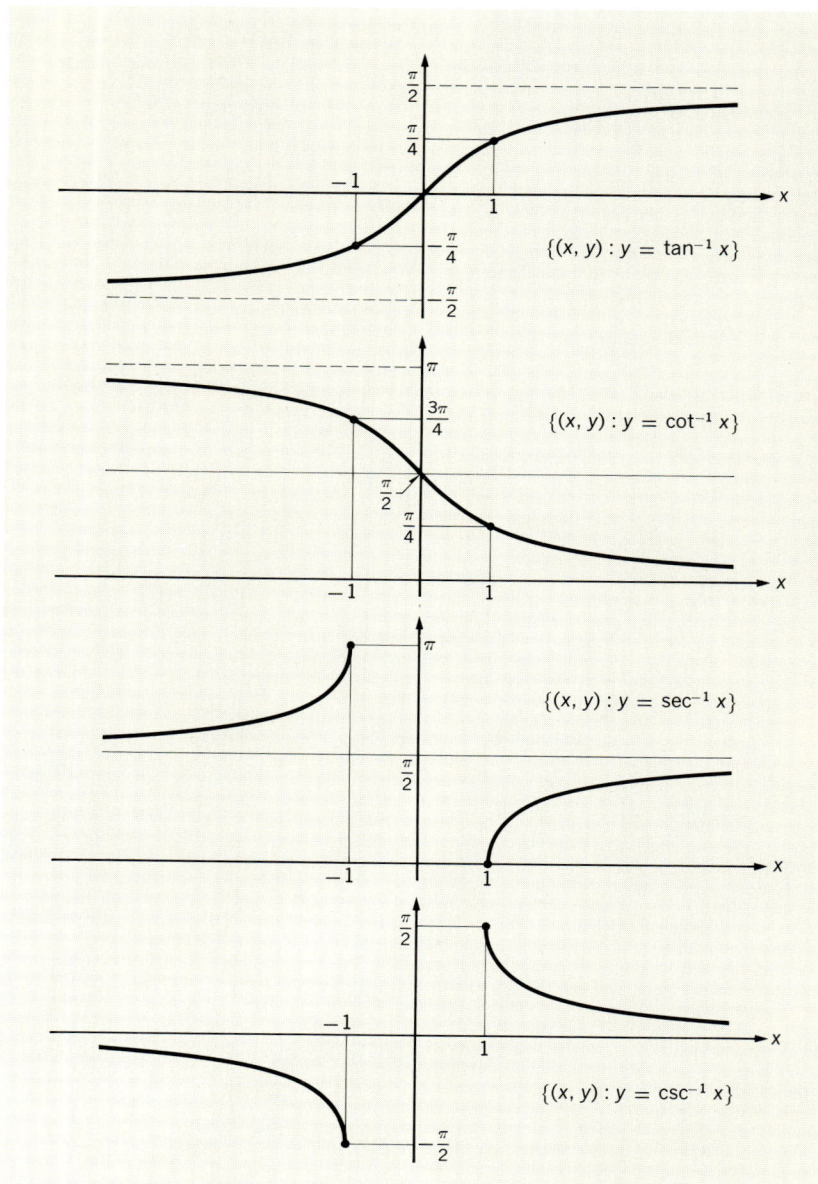

FIGURE 9.29

9.7 INVERSE TRIGONOMETRIC FUNCTIONS

We shall compute the derivatives $\sin^{-1} x$ and $\tan^{-1} x$ by the method of Section 8.2 and leave the remaining four formulas to be verified by the reader.

If $y = \sin^{-1} x$, then we have

$$x = \sin y$$

whence

$$D_y x = \cos y$$

By Theorem 8.2, then,

$$Dy = D \sin^{-1} x = \frac{1}{\cos y}$$

But now $\cos^2 y = 1 - \sin^2 y = 1 - x^2$. Therefore we must have either $\cos y = \sqrt{1 - x^2}$ or $\cos y = -\sqrt{1 - x^2}$. However, on the domain $-\pi/2 \leq y \leq \pi/2$, $\cos y$ is not negative. Hence $\cos y = \sqrt{1 - x^2}$ and we have the formula

$$D \sin^{-1} x = \frac{1}{\sqrt{1 - x^2}}, \qquad -1 < x < 1$$

Note that $D \sin^{-1} x$ does not exist at $x = \pm 1$.

The composite function $\sin^{-1} f(x)$ is defined for all values of x such that $-1 \leq f(x) \leq 1$. Applying the chain rule of differentiation, we obtain the general formula

$$D \sin^{-1} f(x) = \frac{1}{\sqrt{1 - f^2(x)}} Df(x), \qquad -1 < f(x) < 1$$

This formula provides us with the useful integration formula

$$\int \frac{df(x)}{\sqrt{1 - f^2(x)}} = \sin^{-1} f(x) + C$$

■ **Example 1.** Let $f(x) = \sin^{-1} \sqrt{2x - x^2}$. Then

$$Df(x) = \frac{1}{\sqrt{1 - (\sqrt{2x - x^2})^2}} \cdot D\sqrt{2x - x^2}$$

$$= \frac{1}{\sqrt{1 - 2x + x^2}} \cdot \tfrac{1}{2}(2x - x^2)^{-1/2}(2 - 2x)$$

$$= \frac{1 - x}{\sqrt{(1 - x)^2}} \cdot \frac{1}{\sqrt{2x - x^2}}$$

The formula is only valid for the range $-1 < x < 1$, and in this range, $1 - x > 0$. Therefore

$$Df(x) = \frac{1}{\sqrt{2x - x^2}}$$

■ **Example 2.** We may evaluate the integral

$$\int \frac{dx}{\sqrt{6x - 9x^2}}$$

by first writing $6x - 9x^2 = 1 - 1 + 6x - 9x^2 = 1 - (1 - 6x + 9x^2) = 1 - (1 - 3x)^2$. Then we make the substitution $u = 1 - 3x$ whence $du = -3\,dx$ or $dx = -\frac{1}{3}\,du$. The integral becomes

$$\int \frac{-\frac{1}{3}\,du}{\sqrt{1 - u^2}} = -\frac{1}{3} \int \frac{du}{\sqrt{1 - u^2}} = -\frac{1}{3} \sin^{-1} u + C$$

Therefore

$$\int \frac{dx}{\sqrt{6x - 9x^2}} = -\frac{1}{3} \sin^{-1}(1 - 3x) + C \ \blacksquare$$

For the inverse tangent function we write $y = \tan^{-1} x$, and then

$$x = \tan y$$

It follows that $D_y x = D_y \tan y = \sec^2 y = 1 + \tan^2 y = 1 + x^2$. Therefore

$$Dy = D \tan^{-1} x = \frac{1}{1 + x^2}$$

Again, using the chain law, we deduce the general formula

$$D \tan^{-1} f(x) = \frac{1}{1 + f^2(x)} \cdot Df(x)$$

This also provides a useful integration formula, of course. We have

$$\int \frac{df(x)}{1 + f^2(x)} = \tan^{-1} f(x) + C$$

The differentiation formulas for the remaining four inverse trigonometric functions are simply listed here. The reader should verify them as exercises.

$$D \cos^{-1} f(x) = -\frac{1}{\sqrt{1 - f^2(x)}} \cdot Df(x)$$

$$D \cot^{-1} f(x) = -\frac{1}{1 + f^2(x)} Df(x)$$

$$D_x \sec^{-1} f(x) = \frac{1}{|f(x)| \sqrt{f^2(x) - 1}} D_x f(x)$$

$$D_x \csc^{-1} f(x) = -\frac{1}{|f(x)| \sqrt{f^2(x) - 1}} D_x f(x)$$

9.7 INVERSE TRIGONOMETRIC FUNCTIONS

Evaluate each of the expressions in Exercises 1 to 24.

EXERCISES 9.7

1. $\sin^{-1}\left(\frac{\sqrt{2}}{2}\right)$
2. $\sin^{-1}(1)$
3. $\sin^{-1}\left(\frac{-\sqrt{3}}{2}\right)$
4. $\cos^{-1}(-1)$
5. $\cos^{-1}\left(\frac{-\sqrt{2}}{2}\right)$
6. $\cos^{-1}\left(\frac{\sqrt{3}}{2}\right)$
7. $\tan^{-1}(0)$
8. $\tan^{-1}\sqrt{3}$
9. $\tan^{-1}\left(\frac{-1}{\sqrt{3}}\right)$
10. $\sin^{-1}\left(\sin\frac{5\pi}{6}\right)$
11. $\cos(\sin^{-1}\frac{1}{2})$
12. $\tan(\sin^{-1}\frac{1}{2})$
13. $\sin^{-1}\left[\cos\left(\frac{-\pi}{3}\right)\right]$
14. $\sin(\cos^{-1}\frac{1}{2})$
15. $\cos\left(\tan^{-1}\frac{\sqrt{3}}{3}\right)$
16. $\tan^{-1}\left[\cot\left(\frac{-\pi}{3}\right)\right]$
17. $\sin^{-1}\left(\sin\frac{3\pi}{2}\right)$
18. $\sec^{-1}(\cos \pi)$
19. $\cos(\sin^{-1} t)$
20. $\tan(\cos^{-1} t)$
21. $\sin(\tan^{-1} t)$
22. $\cos(\tan^{-1} t)$
23. $\sin\left[\tan^{-1}\left(\frac{t-1}{t+1}\right)\right]$
24. $\cos\left[\tan^{-1}\left(\frac{t-1}{t+1}\right)\right]$

Verify each of the identities in Exercises 25 to 40.

25. $\tan^{-1}(-x) = -\tan^{-1} x$
26. $\cos^{-1}(-x) = \pi - \cos^{-1} x$
27. $\cos^{-1} x = \frac{\pi}{2} - \sin^{-1} x$
28. $\tan^{-1}\frac{1}{x} = \cot^{-1} x$
29. $\sin^{-1}\frac{1}{x} = \csc^{-1} x$
30. $\cos^{-1}\frac{1}{x} = \sec^{-1} x$
31. $\sin^{-1}\frac{x}{\sqrt{1+x^2}} = \tan^{-1} x$
32. $\tan^{-1}\frac{x}{\sqrt{1-x^2}} = \sin^{-1} x$
33. $\sin(\cos^{-1} x) = \sqrt{1-x^2}$
34. $\cos(\tan^{-1} x) = \frac{1}{\sqrt{1+x^2}}$
35. $\tan(2 \tan^{-1} x) = \frac{2x}{1-x^2}$
36. $\sin(2 \cos^{-1} x) = 2x\sqrt{1-x^2}$
37. $\cos(2 \sin^{-1} x) = 1 - 2x^2$
38. $\tan(2 \sin^{-1} x) = \frac{2x\sqrt{1-x^2}}{1-2x^2}$
39. $\cos(2 \tan^{-1} x) = \frac{1-x^2}{1+x^2}$
40. $\tan(2 \cos^{-1} x) = \frac{2x\sqrt{1-x^2}}{2x^2-1}$

Differentiate each of the functions in Exercises 41 to 66.

41. $\sin^{-1}\frac{x}{2}$
42. $\cos^{-1} 2x$

43. $\tan^{-1}\dfrac{1}{x}$

44. $\dfrac{1}{3}\tan^{-1}\dfrac{x}{3}$

45. $\sec^{-1} 3x$

46. $\cos^{-1}\sqrt{x}$

47. $\tan^{-1}(2x + 1)$

48. $\tan^{-1}\left(\dfrac{x-1}{x+1}\right)$

49. $\cot^{-1}\left(\dfrac{1-x}{1+x}\right)$

50. $\tan^{-1}\dfrac{x}{\sqrt{1-x^2}}$

51. $\sin^{-1}\dfrac{x}{\sqrt{1+x^2}}$

52. $\sec^{-1}\dfrac{\sqrt{1+x^2}}{x}$

53. $\tan^{-1}\left(\dfrac{3\sin x}{4+5\cos x}\right)$

54. $\tan^{-1}(2\tan x)$

55. $\dfrac{1}{2}\sqrt{4-x^2} + \sin^{-1}\dfrac{x}{2}$

56. $\sqrt{x^2-9} - 3\sec^{-1}\dfrac{x}{3}$

57. $\dfrac{1}{1+4x^2} + \tan^{-1} 2x$

58. $\dfrac{1}{4}\tan^{-1} x^2 + \dfrac{1}{8}\ln\left(\dfrac{1+x^2}{1-x^2}\right)$

59. $x\sin^{-1} x + \sqrt{1-x^2}$

60. $x\cos^{-1} 2x - \tfrac{1}{2}\sqrt{1-4x^2}$

61. $x\tan^{-1} x - \tfrac{1}{2}\ln(x^2 + 1)$

62. $\ln\dfrac{x}{\sqrt{1+x^2}} - \dfrac{1}{x}\tan^{-1} x$

63. $x\sec^{-1} x - \ln(x + \sqrt{x^2-1})$

64. $x\ln(a^2 + x^2) - 2x + 2a\tan^{-1}\dfrac{x}{a}$

65. $2x^3 \tan^{-1} x - x^2 + \ln(1 + x^2)$

66. $x(\sin^{-1} x)^2 - 2x + 2\sqrt{1-x^2}\sin^{-1} x$

Determine the values of the indeterminate forms in Exercises 67 to 74.

67. $\lim\limits_{x\to 0}\dfrac{x}{\sin^{-1} x}$

68. $\lim\limits_{x\to 0}\dfrac{\tan^{-1} x}{x}$

69. $\lim\limits_{x\to 1}\dfrac{\tan^{-1}\sqrt{x^2-1}}{\ln(1+\sqrt{x^2-1})}$

70. $\lim\limits_{x\to 0}\dfrac{\sin^{-1} x - \tan^{-1} x}{\tan x \sin^2 x}$

71. $\lim\limits_{x\to 0}\left[\dfrac{1}{\ln(1+x)} - \dfrac{1}{\tan^{-1} x}\right]$

72. $\lim\limits_{x\to 0}\left[\dfrac{1}{(\tan^{-1} x)^2} - \dfrac{1}{x^2}\right]$

73. $\lim\limits_{x\to 0}\dfrac{\sin 2x(1-\cos 2x)}{\sin^{-1} x - \sin x}$

74. $\lim\limits_{x\to 0^+}(\sin^{-1} x)\ln x$

Verify the differentiation formulas in Exercises 75 to 78.

75. $D\sin^{-1}\dfrac{x}{a} = \dfrac{1}{\sqrt{a^2-x^2}}$

76. $D\dfrac{1}{a}\tan^{-1}\dfrac{x}{a} = \dfrac{1}{a^2+x^2}$

77. $D\cos^{-1}\left(\dfrac{a-x}{a}\right) = \dfrac{1}{\sqrt{2ax-x^2}}$

78. $D\left(\dfrac{1}{2}x\sqrt{a^2-x^2} + a^2\sin^{-1}\dfrac{x}{a}\right) = \sqrt{a^2-x^2}$

9.7 INVERSE TRIGONOMETRIC FUNCTIONS

In Exercises 79 to 85 solve for x.

79. $\tan^{-1} x + \tan^{-1} \frac{1}{2} = \frac{\pi}{4}$

80. $\sin^{-1} x + \sin^{-1} \frac{4}{5} = \frac{\pi}{2}$

81. $\sec^{-1} x + \tan^{-1} 2 = \frac{\pi}{2}$

82. $4 \tan^{-1} x - \tan^{-1} \frac{1}{239} = \frac{\pi}{4}$

83. $\sin^{-1} x + \sin^{-1} 2x = \frac{\pi}{2}$

84. $\sin^{-1} 3x - \sin^{-1} x = \frac{\pi}{3}$

85. $\tan^{-1} 3x + \tan^{-1} 2x = \frac{\pi}{4}$

Work out the integrals in Exercises 86 to 105.

86. $\int \dfrac{dx}{\sqrt{1 - 4x^2}}$

87. $\int \dfrac{dx}{x^2 + 16}$

88. $\int \dfrac{dx}{1 + 9x^2}$

89. $\int \dfrac{dx}{\sqrt{1 - (x + 2)^2}}$

90. $\int \dfrac{dx}{\sqrt{4 - 9x^2}}$

91. $\int \dfrac{dx}{16 + 9x^2}$

92. $\int \dfrac{dx}{\sqrt{a^2 - b^2 x^2}}$

93. $\int \dfrac{dx}{a^2 + b^2 x^2}$

94. $\int \dfrac{x\,dx}{16 + x^4}$

95. $\int \dfrac{dx}{x\sqrt{1 - \ln^2 x}}$

96. $\int \dfrac{\sin x\,dx}{\sqrt{4 - \cos^2 x}}$

97. $\int \dfrac{\cos x\,dx}{1 + \sin^2 x}$

98. $\int_0^{1/2} \dfrac{dx}{\sqrt{1 - x^2}}$

99. $\int_{-1}^{1} \dfrac{dx}{1 + x^2}$

100. $\int_0^3 \dfrac{dx}{x^2 + 9}$

101. $\int_0^1 \dfrac{dx}{\sqrt{25 - 9x^2}}$

102. $\int_{\sin x}^{\cos x} \dfrac{dt}{\sqrt{1 - t^2}}$

103. $\int_{\cot x}^{\tan x} \dfrac{dt}{1 + t^2}$

104. $\int \dfrac{dx}{x\sqrt{4x^2 - 1}}$

105. $\int_{3\sqrt{2}}^{6} \dfrac{dx}{x\sqrt{x^2 - 9}}$

PROBLEMS 9.7

1. Express each of the following functions in another way and explain why it should be possible.

   (a) $\csc^{-1}\left(\dfrac{1}{x}\right)$ (b) $\cot^{-1}\left(\dfrac{1}{x}\right)$

   (c) $\cos^{-1}(-x)$ (d) $\tan^{-1}\left(\dfrac{x - 1}{x + 1}\right)$

2. Determine the constants c and d in terms of the constants a and b, so that each of the following equations (page 444) is true.

(a) $Dc \sin^{-1} dx = \dfrac{1}{\sqrt{b - ax^2}}$

(b) $Dc \tan^{-1} dx = \dfrac{1}{ax^2 + b}$

(c) $Dc \sec^{-1} dx = \dfrac{1}{x\sqrt{ax^2 - b}}$

(d) $Dc \tan^{-1}\left(\dfrac{2ax + b}{d}\right) = \dfrac{k}{ax^2 + bx + c}$

3. A drive-in movie screen is 80 ft high and its base is 20 ft above eye level. How far back from the base of the support should a car park in order that the figures on the screen appear the largest?

4. An aircraft flies directly over a radar antenna at an altitude of 1 mile and a ground speed of 7 miles per min. How fast must the antenna rotate (in a vertical plane) to keep pointed at the aircraft when it is directly above the antenna? Reduce the altitude successively to $\tfrac{1}{2}$, $\tfrac{1}{4}$, and $\tfrac{1}{8}$ miles and notice what happens.

5. Suppose that a particular rocket climbs to $\tfrac{1}{25} t^{5/2}$ miles in t seconds. If the rocket is launched vertically from a pad 3 miles from a tracking telescope, how fast is the angle of elevation of the telescope increasing when $t = 9$ sec?

6. A wall 8 ft high is 10 ft away from a building. What is the shortest ladder that will lean against the building with one end outside the wall?

7. Of all circular sectors with a fixed perimeter c, which has the largest area?

8. A piece of sheet metal 12 in. wide is to be bent into a gutter whose cross section is a circular arc. Find the radius of the arc which will provide that gutter with the most capacity.

9. Assume that the earth is 8000 miles in diameter. If a tunnel could be dug in a straight line between two points that are 400 miles apart along the surface, how much shorter than 400 miles would the tunnel be?

10. One ship has a mast 60 ft high and another has a mast 80 ft high. How far is each masthead visible from the other?

11. Let $f(x)$ be defined by

$$f(x) = \int_0^x \dfrac{dt}{1 + t^2}$$

Without using the fact that $f(x) = \tan^{-1} x$, prove that for any constants $a > 0$ and b such that $ab < 1$,

$$f(a) + f(b) = f\left(\dfrac{a + b}{1 - ab}\right)$$

12. Use the formula in Problem 11 and the known continuity of the function $f(x)$ defined there to show that

$$\lim_{x \to \infty} f(x) = 2f(1)$$

13. Follow the development begun in Problems 11 and 12 to a complete theory of trigonometric functions based on that one integral.

14. Compute the volume removed from a right circular cylinder of radius R by boring a hole of radius r through the cylinder, the axis of the hole perpendicular to that of the cylinder.

15. Compute the volume removed by boring a hole of radius r through a sphere of radius R, the axis of the hole passing through the center of the sphere.

16. Rotate the circle $\{(x,y) : x^2 + (y - b)^2 = a^2, a < b\}$ about the x axis and determine the area of the torus so formed.

9.8 THE HYPERBOLIC FUNCTIONS

Any function of the form

$$f(x) = ae^{kx} + be^{-kx}$$

where a and b are arbitrary constants satisfies the differential equation

$$f''(x) = k^2 f(x),$$

for $Df(x) = ake^{kx} - bke^{-kx}$ and $D^2 f(x) = ak^2 e^{kx} + bk^2 e^{-kx} = k^2 f(x)$. In an advanced course in analysis, the converse of this statement is proved, that is, if f satisfies $f''(x) = k^2 f(x)$, then $f(x) = ae^{kx} + be^{-kx}$ for some choice of constants a and b.

Linear combinations of the functions e^x and e^{-x} often occur in the solutions of differential equations. Two particular such linear combinations have been given names. The *hyperbolic cosine* cosh x and the *hyperbolic sine* sinh x are defined by

$$\cosh x = \tfrac{1}{2}(e^x + e^{-x})$$
$$\sinh x = \tfrac{1}{2}(e^x - e^{-x})$$

The names hyperbolic "cosine" and "sine" were chosen because these functions have properties very similar to those of the trigonometric functions. The adjective "hyperbolic" stems from the relation between these functions and the hyperbola. This relationship is the subject of a problem in Chapter 13.

Note that

$$\cosh x + \sinh x = \tfrac{1}{2}e^x + \tfrac{1}{2}e^{-x} + \tfrac{1}{2}e^x - \tfrac{1}{2}e^{-x} = e^x$$
$$\cosh x - \sinh x = \tfrac{1}{2}e^x + \tfrac{1}{2}e^{-x} - \tfrac{1}{2}e^x + \tfrac{1}{2}e^{-x} = e^{-x}$$

It follows that any function of the form $ae^x + be^{-x}$ can also be written as $A \cosh x + B \sinh x$. In particular, $A = a + b$ and $B = a - b$.

The fundamental relationship between the functions cosh and sinh is the identity

$$(\cosh x)^2 - (\sinh x)^2 = 1$$

or, as it is usually written,

$$\cosh^2 x - \sinh^2 x = 1$$

To prove this, we simply compute:

$$[\tfrac{1}{2}e^x + \tfrac{1}{2}e^{-x}]^2 - [\tfrac{1}{2}e^x - \tfrac{1}{2}e^{-x}]^2$$
$$= \tfrac{1}{4}(e^{2x} + 2e^x e^{-x} + e^{-2x}) - \tfrac{1}{4}(e^{2x} - 2e^x e^{-x} + e^{-2x})$$
$$= \tfrac{1}{4}(e^{2x} + 2 + e^{-2x} - e^{2x} + 2 - e^{-2x}) = \tfrac{1}{4}(4) = 1$$

The similar identities, $\cosh^2 x - \sinh^2 x = 1$ and $\cos^2 x + \sin^2 x = 1$, should lead us to expect that the hyperbolic functions are analogous to $\cos x$ and $\sin x$. We explore this analogy.

The *hyperbolic cosine is an even function,* and the *hyperbolic sine is an odd function.* That is,

$$\cosh(-x) = \cosh x$$
$$\sinh(-x) = -\sinh x$$

Again the proof is a matter of computation:

$$\cosh(-x) = \tfrac{1}{2}(e^{-x} + e^{-(-x)}) = \tfrac{1}{2}(e^{-x} + e^x) = \cosh x$$
$$\sinh(-x) = \tfrac{1}{2}(e^{-x} - e^{-(-x)}) = \tfrac{1}{2}(e^{-x} - e^x) = -\sinh x$$

Addition formulas for these hyperbolic functions are analogous to those for the trigonometric functions:

$$\cosh(x + y) = \cosh x \cosh y + \sinh x \sinh y$$
$$\sinh(x + y) = \sinh x \cosh y + \cosh x \sinh y$$

We prove the first of these and leave the proof of the second to the reader.

$$\cosh(x + y) = \tfrac{1}{2}(e^{x+y} + e^{-x-y}) = \tfrac{1}{2}(e^x e^y + e^{-x} e^{-y})$$

and

$$\cosh x \cosh y + \sinh x \sinh y$$
$$= \tfrac{1}{2}(e^x + e^{-x}) \cdot \tfrac{1}{2}(e^y + e^{-y}) + \tfrac{1}{2}(e^x - e^{-x})\tfrac{1}{2}(e^y - e^{-y})$$
$$= \tfrac{1}{4}(e^x e^y + e^x e^{-y} + e^{-x} e^y + e^{-x} e^{-y} + e^x e^y - e^x e^{-y} - e^{-x} e^y + e^{-x} e^{-y})$$
$$= \tfrac{1}{4}(2e^x e^y + 2e^{-x} e^{-y}) = \tfrac{1}{2}(e^x e^y + e^{-x} e^{-y})$$

By setting $x - y = x + (-y)$, we also obtain the *subtraction formulas*

$$\cosh(x - y) = \cosh x \cosh(-y) + \sinh x \sinh(-y)$$
$$= \cosh x \cosh y - \sinh x \sinh y$$
$$\sinh(x - y) = \sinh x \cosh(-y) + \cosh x \sinh(-y)$$
$$= \sinh x \cosh y - \cosh x \sinh y$$

Also by setting $y = x$, we have the analogues of the double-angle formulas

$$\cosh 2x = \cosh^2 x + \sinh^2 x$$
$$\sinh 2x = 2 \sinh x \cosh x$$

9.8 THE HYPERBOLIC FUNCTIONS

Then the identity $\cosh^2 x - \sinh^2 x = 1$ implies that

$$\cosh 2x = -1 + 2\cosh^2 x = 1 + 2\sinh^2 x$$

Four more hyperbolic functions are defined in terms of $\cosh x$ and $\sinh x$:

the *hyperbolic tangent* of x, $\tanh x = \dfrac{\sinh x}{\cosh x} = \dfrac{e^x - e^{-x}}{e^x + e^{-x}}$

the *hyperbolic secant* of x, $\operatorname{sech} x = \dfrac{1}{\cosh x} = \dfrac{2}{e^x + e^{-x}}$

the *hyperbolic cosecant* of x, $\operatorname{csch} x = \dfrac{1}{\sinh x} = \dfrac{2}{e^x - e^{-x}}$

the *hyperbolic contangent* of x,

$$\coth x = \dfrac{1}{\tanh x} = \dfrac{\cosh x}{\sinh x} = \dfrac{e^x + e^{-x}}{e^x - e^{-x}}$$

Note that tanh, coth, and csch are odd functions, while sech is an even function.

For large positive values of x, the quantity e^{-x} is very small while e^x is very large. Thus, if x is large positively, $\cosh x$ and $\sinh x$ are very nearly equal to $\tfrac{1}{2}e^x$. More precisely, we write

$$\lim_{x \to +\infty} \frac{\cosh x}{\tfrac{1}{2}e^x} = 1, \qquad \lim_{x \to +\infty} \frac{\sinh x}{\tfrac{1}{2}e^x} = 1$$

Similarly,

$$\lim_{x \to -\infty} \frac{\cosh x}{\tfrac{1}{2}e^{-x}} = 1, \qquad \lim_{x \to -\infty} \frac{\sinh x}{\tfrac{1}{2}e^{-x}} = -1$$

Also, it is easy to show that

$$\lim_{x \to \infty} \tanh x = 1, \qquad \lim_{x \to \infty} \coth x = 1$$
$$\lim_{x \to \infty} \operatorname{sech} x = 0, \qquad \lim_{x \to \infty} \operatorname{csch} x = 0$$

We shall utilize these facts in drawing graphs of these functions. First, however, we deduce the derivative formulas.

It is easy to see that $D\tfrac{1}{2}(e^x + e^{-x}) = \tfrac{1}{2}(e^x - e^{-x})$ and that

$$D\tfrac{1}{2}(e^x - e^{-x}) = \tfrac{1}{2}(e^x + e^{-x}).$$

Therefore we have

$$D \cosh x = \sinh x$$
$$D \sinh x = \cosh x$$

The chain rule of differentiation then provides the general formulas

$$D \cosh[f(x)] = \sinh[f(x)] \cdot Df(x)$$
$$D \sinh[f(x)] = \cosh[f(x)] \cdot Df(x)$$

For the hyperbolic tangent, we compute

$$D \tanh x = D\left(\frac{\sinh x}{\cosh x}\right) = \frac{\cosh x \, D \sinh x - \sinh x \, D \cosh x}{\cosh^2 x}$$

$$= \frac{\cosh^2 x - \sinh^2 x}{\cosh^2 x}$$

$$= \frac{1}{\cosh^2 x} = \text{sech}^2 x$$

Again using the chain rule, we obtain the general formula

$$D \tanh[f(x)] = \text{sech}^2[f(x)] \cdot Df(x)$$

The reader may deduce the following formulas

$$D \coth[f(x)] = -\text{csch}^2[f(x)] \cdot Df(x)$$
$$D \text{sech}[f(x)] = -\text{sech}[f(x)]\tanh[f(x)] \cdot Df(x)$$
$$D \text{csch}[f(x)] = -\text{csch}[f(x)]\coth[f(x)] \cdot Df(x)$$

The corresponding integration formulas are often useful. We list them below for reference.

$$\int \sinh[f(x)] \, df(x) = \cosh[f(x)] + C$$

$$\int \cosh[f(x)] \, df(x) = \sinh[f(x)] + C$$

$$\int \text{sech}^2[f(x)] \, df(x) = \tanh[f(x)] + C$$

$$\int \text{csch}^2[f(x)] \, df(x) = -\coth[f(x)] + C$$

$$\int \text{sech}[f(x)] \tanh[f(x)] \, df(x) = -\text{sech}[f(x)] + C$$

$$\int \text{csch}[f(x)] \coth[f(x)] \, df(x) = -\text{csch}[f(x)] + C$$

Graphs of the six hyperbolic functions appear in Fig. 9.30. The reader will find it instructive to verify that these diagrams do indeed represent the functions. Thus by studying symmetry, intercepts, critical points, points of inflection, asymptotes, and so forth, the reader should draw the graphs for himself.

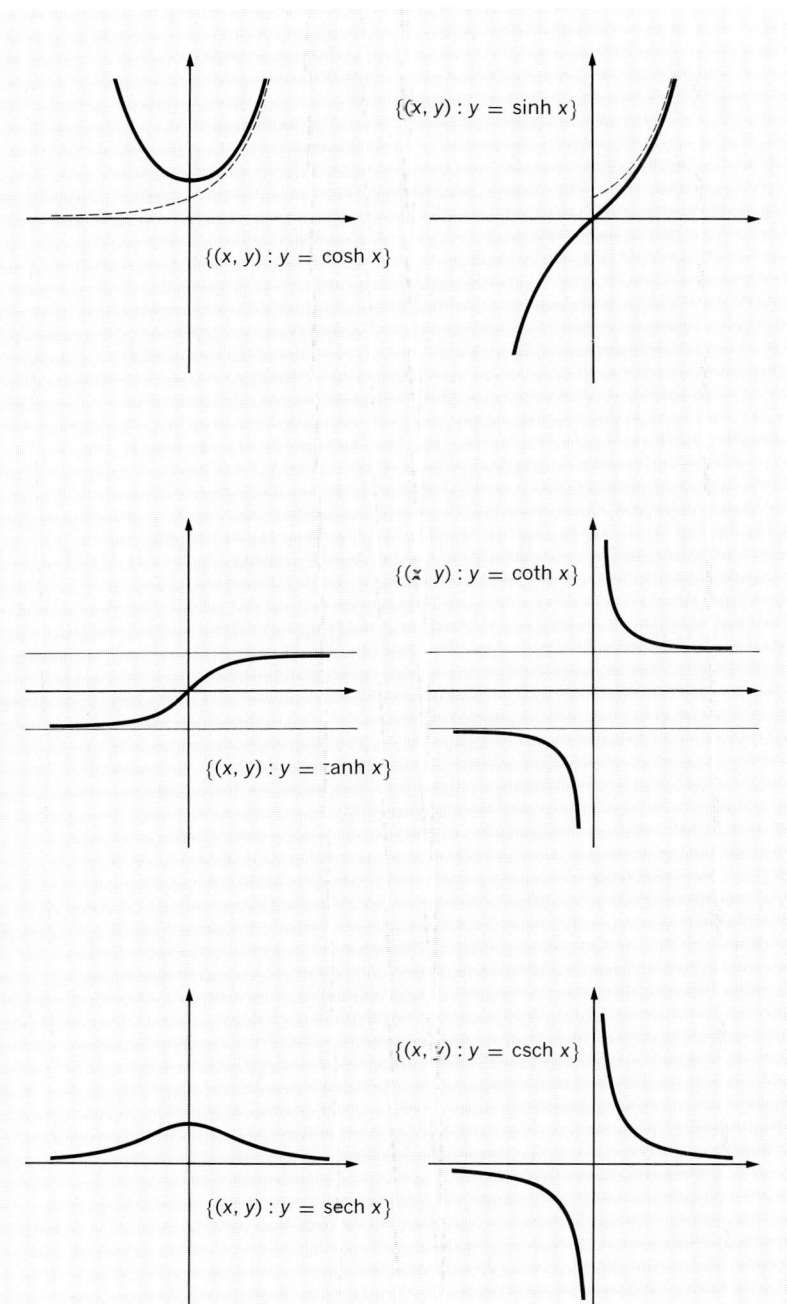

FIGURE 9.30

EXERCISES 9.8 Establish the identities in Exercises 1 to 18.

1. $\cosh^2 x + \sinh^2 x = \cosh 2x$
2. $\cosh 2x = 2\cosh^2 x - 1$
3. $\cosh^2 x = 2\sinh^2 x + 1$
4. $\sinh 2x = 2\sinh x \cosh x$
5. $\cosh \dfrac{x}{2} = \sqrt{\dfrac{1}{2}(\cosh x - 1)}$
6. $\sinh \dfrac{x}{2} = \sqrt{\dfrac{1}{2}(\cosh x + 1)}$
7. $\tanh \dfrac{x}{2} = \dfrac{\cosh x - 1}{\sinh x}$
8. $\dfrac{1 + \tanh x}{1 - \tanh x} = e^{2x}$
9. $\tanh^2 x + \mathrm{sech}^2 x = 1$
10. $\coth^2 x - \mathrm{csch}^2 x = 1$
11. $\tanh x + \coth x = 2\coth 2x$
12. $\tanh 2x = \dfrac{2\tanh x}{1 + \tanh^2 x}$
13. $\sinh x + \sinh y = 2\sinh \tfrac{1}{2}(x+y) \cosh \tfrac{1}{2}(x-y)$
14. $(\cosh x \pm \sinh x)^n = \cosh nx \pm \sinh nx$
15. $\tanh x + \tanh y = \dfrac{\sinh(x+y)}{\cosh x \cosh y}$
16. $\cosh 4x - 4\cosh 2x + 3 = 8\sinh^4 x$
17. $\sin^{-1}(\tanh x) = \tan^{-1}(\sinh x)$
18. $\sin^{-1}(\tanh x) = \cos^{-1}(\mathrm{sech}\, x),\ x \geqq 0$

Find Df for each function in Exercises 19 to 42.

19. $\sinh 2x$
20. $\cosh(3x - 1)$
21. $\tanh(1 - 2x)$
22. $\mathrm{sech}\, 3x$
23. $\sinh^2 3x$
24. $\cosh^2(2x + 3)$
25. $\tanh^2(x - 1)$
26. $x^2 \cosh 2x$
27. $\sinh^2 x - \sinh x^2$
28. $\cosh 2x - \sinh 2x$
29. $(2x - 1)^2 \cosh 3x$
30. $e^{-x} \sinh x$
31. $e^x \cosh x$
32. $\dfrac{2\cosh 3x}{1 + 2\sinh 3x}$
33. $\sqrt{1 + \sinh^2 x}$
34. $\mathrm{csch}\, \ln x - \mathrm{sech}\, \ln x$
35. $\ln \mathrm{csch}\, x - \ln \mathrm{sech}\, x$
36. $\ln(\sinh 2x)$
37. $\ln\!\left(\coth \dfrac{x}{2}\right)$
38. $\ln(\tanh^2 x)$
39. $\tan^{-1}(\sinh x)$
40. $\tanh(x \cot x)$
41. $\sin^{-1}(\tanh x)$
42. $\cosh(\ln \sec x)$

Determine Df^{-1} for each given function in Exercises 43 to 46.

43. $x = f(y) = \sinh y$
44. $x = f(y) = \tanh 3y$

9.8 THE HYPERBOLIC FUNCTIONS

45. $x = f(y) = \coth y - \tanh y$
46. $x = f(y) = \ln(\cosh y)$
47. The graph $\{(x, y) : y = a \cosh x/a\}$ is the curve called a *catenary*. (A flexible cable supported at both ends assumes this shape. Telephone and power lines are common examples.) Draw this graph over the interval $\{x : |x| \leq a\}$.

Evaluate the indeterminate forms given in Exercises 48 to 50.

48. $\lim\limits_{x \to 0} (\csc x - \operatorname{csch} x)$
49. $\lim\limits_{x \to 0} \dfrac{x - \sinh x}{(1 - \cosh x)^2}$
50. $\lim\limits_{x \to 0} \dfrac{\tanh 2x - 2x}{3x - \sinh 3x}$

Work out the integrals in Exercises 51 to 56.

51. $\int \cosh 3x \, dx$
52. $\int \tanh^2 x \, dx$
53. $\int \operatorname{csch}^2(2x - 1) \, dx$
54. $\int \tanh x \, dx$
55. $\int \dfrac{dx}{(e^x + e^{-x})^2}$
56. $\int \dfrac{\cosh \sqrt{x} \, dx}{\sqrt{x}}$

PROBLEMS 9.8

1. If $\sinh y = x$, show that $y = \ln(x + \sqrt{x^2 + 1})$

2. The *gudermannian* $gd(x)$ is defined to be
$$gd(x) = \tan^{-1}(\sinh x)$$
Prove that
   (a) $D gd(x) = \operatorname{sech} x = \cos(gd(x))$
   (b) $(gd)^{-1}(y) = \ln(\sec y + \tan y)$

3. Prove that any function f satisfying the differential equation $f'' = k^2 f$ must have the form $f(x) = ae^{kx} + be^{-kx}$.

4. Prove that $f(x) = e^{ax} \sinh bx$ satisfies the differential equation $f'' - 2af' + (a^2 - b^2)f = 0$. Use this information to deduce the general form of the solution of this equation.

5. Apply Newton's iterative method to obtain the roots of the equations
   (a) $\cosh x = 3x$ (b) $\sinh x = 2x$

6. If $f(x) = \cosh x$, prove that
$$\dfrac{f''(x)}{[1 + (f'(x))^2]^{3/2}} = \operatorname{sech}^2 x$$

7. Draw an accurate graph of the function $f(x) = \ln[(1 + \tanh x)/(1 - \tanh x)]^{1/2}$.

8. Prove that
$$\int_a^b \cosh x \, dx = 2 \cosh \tfrac{1}{2}(b + a) \sinh \tfrac{1}{2}(b - a)$$

9.9 INVERSE HYPERBOLIC FUNCTIONS

The function $\cosh y$ is strictly increasing on the closed ray $\{y : 0 \leq y\}$. Therefore its inverse function $\cosh^{-1} x$ is defined, differentiable, and strictly increasing. Since $\cosh y \geq 1$ for all $y \geq 0$, the domain of $\cosh^{-1} x$ is the closed ray $\{x : 1 \leq x\}$. All of this follows from Section 8.1 and 8.2. However, we can determine an explicit form of the $\cosh^{-1} x$.

Given that $x = \cosh y$, or

$$x = \tfrac{1}{2}(e^y + e^{-y})$$

we simply solve for y in terms of x. (The job can be done here, although it could not be done for $\cos x$, for instance.) Multiply both sides of this equation by $2e^y$ to get

$$2xe^y = (e^y)^2 + 1$$

or

$$(e^y)^2 - 2xe^y + 1 = 0$$

Apply the quadratic formula and obtain

$$e^y = x \pm \sqrt{x^2 - 1}$$

Surely x must be large when y is large, so the negative sign cannot hold. Thus

$$e^y = x + \sqrt{x^2 - 1}$$

and

$$y = \ln(x + \sqrt{x^2 - 1})$$

We have derived the formula

$$\cosh^{-1} x = \ln(x + \sqrt{x^2 - 1}), \qquad x \geq 1$$

The inverse hyperbolic tangent $\tanh^{-1} x$ is obtained similarly. We first note that, if $x = \tanh y$, then $-1 < x < 1$. Therefore the domain of $\tanh^{-1} x$ is $\{x : -1 < x < 1\}$. Also, $\tanh y$ is strictly increasing over $\mathbf{R}$, so that $\tanh^{-1} x$ is strictly increasing. Now set

$$x = \frac{e^y - e^{-y}}{e^y + e^{-y}}$$

or

$$xe^y + xe^{-y} = e^y - e^{-y}$$

Hence

$$(1 + x)e^{-y} = (1 - x)e^y$$

or

$$\frac{1 + x}{1 - x} = e^{2y}$$

9.9 INVERSE HYPERBOLIC FUNCTIONS

Therefore

$$2y = \ln\left(\frac{1+x}{1-x}\right)$$

and

$$\tanh^{-1} x = \frac{1}{2}\ln\left(\frac{1+x}{1-x}\right), \quad -1 < x < 1$$

Using similar techniques, the reader may derive the following formulas:

$$\sinh^{-1} x = \ln(x + \sqrt{x^2 + 1}), \quad x \in \mathbf{R}$$

$$\operatorname{sech}^{-1} x = \ln\left(\frac{1 + \sqrt{1-x^2}}{x}\right) = \cosh^{-1}\left(\frac{1}{x}\right), \quad 0 < x \leq 1$$

$$\operatorname{csch}^{-1} x = \ln\left(\frac{1}{x} + \frac{\sqrt{1+x^2}}{|x|}\right) = \sinh^{-1}\left(\frac{1}{x}\right), \quad x \neq 0$$

$$\coth^{-1} x = \frac{1}{2}\ln\left(\frac{x+1}{x-1}\right) = \tanh^{-1}\left(\frac{1}{x}\right), \quad |x| > 1$$

Differentiation formulas for these functions can be computed from the above expressions or we may use the technique of Section 8.2. For example, let $y = \cosh^{-1} x$. Then $x = \cosh y$ and $D_y x = D_y \cosh y = \sinh y$. Therefore

$$Dy = D \cosh^{-1} x = \frac{1}{\sinh y}$$

But because $\cosh^2 y - \sinh^2 y = 1$, we have $\sinh^2 y = \cosh^2 y - 1 = x^2 - 1$. Thus

$$D \cosh^{-1} x = \frac{1}{\sqrt{x^2 - 1}}, \quad x > 1$$

By direct calculation

$$D \cosh^{-1} x = D \ln(x + \sqrt{x^2 - 1})$$

$$= \frac{1}{x + \sqrt{x^2 - 1}} D(x + \sqrt{x^2 - 1})$$

$$= \frac{1}{x + \sqrt{x^2 - 1}} \cdot \left(1 + \frac{x}{\sqrt{x^2 - 1}}\right)$$

$$= \frac{1}{x + \sqrt{x^2 - 1}} \left(\frac{\sqrt{x^2 - 1} + x}{\sqrt{x^2 - 1}}\right)$$

$$= \frac{1}{\sqrt{x^2 - 1}}, \quad x > 1$$

The chain rule then provides the general formula

$$D \cosh^{-1}[f(x)] = \frac{1}{\sqrt{[f(x)]^2 - 1}} Df(x), \qquad f(x) > 1$$

By using either or both of these methods, the reader should verify the differentiation formulas for the remaining inverse hyperbolic functions:

$$D \sinh^{-1}[f(x)] = \frac{1}{\sqrt{1 + [f(x)]^2}} Df(x)$$

$$D \tanh^{-1}[f(x)] = \frac{1}{1 - [f(x)]^2} Df(x) \quad |f(x)| < 1$$

$$D \coth^{-1}[f(x)] = \frac{1}{1 - [f(x)]^2} Df(x) \quad |f(x)| > 1$$

$$D \operatorname{sech}^{-1}[f(x)] = \frac{-1}{f(x)\sqrt{1 - [f(x)]^2}} Df(x), \qquad 0 < f(x) < 1$$

$$D \operatorname{csch}^{-1}[f(x)] = \frac{-1}{|f(x)|\sqrt{1 + [f(x)]^2}} \cdot Df(x), \qquad f(x) \neq 0$$

The corresponding integration formulas are perhaps more important than the original differentiation formulas.

$$\int \frac{df(x)}{\sqrt{f^2(x) - 1}} = \cosh^{-1}[f(x)] + C$$
$$= \ln[f(x) + \sqrt{f^2(x) - 1}] + C, \qquad f(x) \geq 1$$

$$\int \frac{df(x)}{\sqrt{1 + f^2(x)}} = \sinh^{-1}[f(x)] + C$$
$$= \ln[f(x) + \sqrt{f^2(x) + 1}] + C$$

$$\int \frac{df(x)}{1 - f^2(x)} = \tanh^{-1}[f(x)] + C = \frac{1}{2} \ln\left(\frac{1 + f(x)}{1 - f(x)}\right) + C,$$
$$-1 < f(x) < 1$$

$$\int \frac{df(x)}{f(x)\sqrt{1 - f^2(x)}} = -\operatorname{sech}^{-1}[f(x)] + C$$
$$= -\ln\left[\frac{1 + \sqrt{1 - f^2(x)}}{f(x)}\right] + C,$$
$$0 < f(x) \leq 1$$

$$\int \frac{df(x)}{|f(x)|\sqrt{1 + f^2(x)}} = -\operatorname{csch}^{-1}[f(x)] + C$$
$$= -\ln\left[\frac{1}{f(x)} + \frac{\sqrt{1 + f^2(x)}}{|f(x)|}\right] + C,$$
$$f(x) \neq 0$$

9.9 INVERSE HYPERBOLIC FUNCTIONS

■ **Example.** If a body of mass m falls in water, it encounters resistance which we assume to be proportional to the square of its velocity. The force acting on the body is therefore

$$F = mg - kv^2$$

where k is a constant of proportionality. Newton's second law of motion states that $F = D_t(mv) = mD_t v$. Hence we have

$$mD_t v = mg - kv^2$$

or

$$D_t v = g - \frac{k}{m}v^2 = g\left(1 - \frac{k}{gm}v^2\right)$$

Setting $k/gm = c^2$, whence $c = \sqrt{k/gm}$, we have

$$D_t v = g[1 - c(v)^2]$$

or

$$\frac{D_t v}{1 - (cv)^2} = g$$

Multiplying both sides by cdt, we obtain

$$\frac{cdv}{1 - (cv)^2} = \frac{d(cv)}{1 - (cv)^2} = cg\,dt$$

Now we integrate both sides, using the third integration formula on page 453, we get

$$\tanh^{-1}(cv) = cgt + C$$

or

$$\frac{1}{2}\ln\left(\frac{1 + cv}{1 - cv}\right) = cgt + C$$

If we assume that the body falls from rest, then we have $x = 0$ when $t = 0$. Therefore $\tanh^{-1}(0) = cg(0) + C = 0$, or $C = 0$.

In this case, then, we obtain

$$cv = \tanh(cgt)$$

or

$$v = \frac{1}{c}\tanh(cgt)$$

Recalling that $c = \sqrt{k/gm}$, we have $1/c = \sqrt{gm/k}$ and $cg = \sqrt{gk/m}$. Therefore the velocity function of the body falling in water is

$$v(t) = \sqrt{\frac{gm}{k}}\tanh\left(t\sqrt{\frac{gk}{m}}\right)$$

We recall that $\lim_{x \to +\infty} \tanh x = 1$. This implies that the velocity v approaches the constant $\sqrt{gm/k}$ as t increases. The constant $\sqrt{gm/k}$ is called the *terminal velocity*. The terminal velocity here should be compared with that found in Section 8.11 ("Motion in a resisting medium") where we assumed resistance is proportional to v instead of v^2.

EXERCISES 9.9

1. Prove the identities:
   (a) $\tanh^{-1}(-x) = -\tanh^{-1} x$
   (b) $\sinh^{-1}(-x) = -\sinh^{-1} x$

Derive the logarithmic form of each function in Exercises 2 to 5.

2. $\sinh^{-1} x$
3. $\text{sech}^{-1} x$
4. $\text{csch}^{-1} x$
5. $\coth^{-1} x$

Differentiate each function in Exercises 6 to 20.

6. $\sinh^{-1} 2x$
7. $\cosh^{-1}(x^2 - 1)$
8. $x^2 \cosh^{-1} 3x$
9. $\dfrac{1}{x} \tanh^{-1} x^2$
10. $\dfrac{1}{a} \tanh^{-1} \dfrac{x}{a}$
11. $\coth^{-1}(\sec x)$
12. $\cosh^{-1}(\sec x)$
13. $\tanh^{-1}(\cos x)$
14. $\sinh^{-1}(\tan x)$
15. $\text{sech}^{-1}(\sin 2x)$
16. $x \tanh^{-1} x + \tfrac{1}{2} \ln(1 - x^2)$
17. $x \sinh^{-1} x - \sqrt{1 + x^2}$
18. $x \cosh^{-1} x - \sqrt{x^2 - 1}$
19. $(1 + x^2)^{1/2} \sinh^{-1} x - x$
20. $(x^2 - 1)^{1/2} \cosh^{-1} x - x$

21. The *tractrix* is the curve
$$\left\{ (x, y) : a \, \text{sech}^{-1} \dfrac{y}{a} - \sqrt{a^2 - y^2} = x \right\}$$

Show that the slope of this curve at (x, y) is $Dy = -y/\sqrt{a^2 - y^2}$. Use this information to sketch the curve.

22. If $b < 0$, then
$$D \dfrac{2}{\sqrt{-b}} \tan^{-1} \sqrt{\dfrac{ax + b}{-b}} = \dfrac{1}{x\sqrt{ax + b}}$$

but if $b > 0$, then
$$D \dfrac{-2}{\sqrt{b}} \tanh^{-1} \sqrt{\dfrac{ax + b}{b}} = \dfrac{1}{x\sqrt{ax + b}}$$

Establish these formulas and explain the phenomenon.

Work out the integrals in Exercises 23 to 36.

23. $\displaystyle\int \frac{dx}{\sqrt{x^2-16}}$ 24. $\displaystyle\int \frac{dx}{\sqrt{x^2-2x+26}}$

25. $\displaystyle\int \frac{dx}{\sqrt{x^2-x}}$ 26. $\displaystyle\int \frac{dx}{\sqrt{x^2+4}}$

27. $\displaystyle\int \frac{dx}{4-x^2}$ 28. $\displaystyle\int \frac{dx}{x\sqrt{9-4x^2}}$

29. $\displaystyle\int \frac{\sin x \, dx}{1-\cos^2 x}$ 30. $\displaystyle\int \frac{dx}{3+2x-x^2}$

31. $\displaystyle\int_{\sqrt{2}}^{2} \frac{dx}{\sqrt{x^2-1}}$ 32. $\displaystyle\int_{0}^{3} \frac{dx}{\sqrt{x^2+9}}$

33. $\displaystyle\int_{-1/2}^{1/2} \frac{dx}{1-x^2}$ 34. $\displaystyle\int_{3}^{4} \frac{dx}{\sqrt{4x^2-9}}$

35. $\displaystyle\int_{-1}^{1} \frac{e^t \, dt}{\sqrt{1+e^{2t}}}$ 36. $\displaystyle\int_{1/4}^{1/2} \frac{dx}{x\sqrt{1-4x^2}}$

9.10 INTEGRATION OF INVERSE FUNCTIONS

The inverse functions considered in this chapter can all be integrated by parts. We shall do four of them. The reader is asked to complete the rest as exercises.

1. $\displaystyle\int \sin^{-1} x \, dx = x \sin^{-1} x + \sqrt{1-x^2} + C$

We prove this formula by setting $\sin^{-1} x = u$ and $dx = dv$. Then $du = dx/\sqrt{1-x^2}$ and $v = x$. Therefore

$$\int \sin^{-1} x \, dx = x \sin^{-1} x - \int x \frac{dx}{\sqrt{1-x^2}}$$

The integral

$$\int \frac{x \, dx}{\sqrt{1-x^2}}$$

responds easily to the substitution $w = 1 - x^2$, whence $dw = -2x \, dx$ or $x \, dx = -\frac{1}{2} dw$. Then we have

$$\int \frac{x \, dx}{\sqrt{1-x^2}} = \int \frac{-1/2 \, dw}{w^{1/2}} = -\frac{1}{2} \int \frac{dw}{w^{1/2}} = -\frac{1}{2} \frac{w^{1/2}}{1/2}$$
$$= -w^{1/2} = -\sqrt{1-x^2}$$

Thus

$$\int \sin^{-1} x \, dx = x \sin^{-1} x + \sqrt{1-x^2} + C$$

This may be verified by differentiating the expression on the right-hand side of the equation

2. $\int \tan^{-1} x \, dx = x \tan^{-1} x - \tfrac{1}{2} \ln(1 + x^2) + C$

Again we take $\tan^{-1} x = u$ and $dv = dx$. Then $du = dx/(1 + x^2)$ and $v = x$. Thus

$$\int \tan^{-1} x \, dx = x \tan^{-1} x - \int x \cdot \frac{dx}{1 + x^2}$$

It is easy to see that

$$\int \frac{x \, dx}{1 + x^2} = \frac{1}{2} \int \frac{2x \, dx}{1 + x^2} = \frac{1}{2} \ln(1 + x^2) + C$$

and our formula has been established.

3. $\int \sec^{-1} x \, dx = x \sec^{-1} x - \cosh^{-1} x + C$

This, too, responds to the substitution $u = \sec^{-1} x$, $dv = dx$, whence $du = dx/x\sqrt{x^2 - 1}$ and $v = x$. Thus we have

$$\int \sec^{-1} x \, dx = x \sec^{-1} x - \int x \cdot \frac{dx}{x\sqrt{x^2 - 1}}$$

$$= x \sec^{-1} x - \int \frac{dx}{\sqrt{x^2 - 1}}$$

$$= x \sec^{-1} x - \cosh^{-1} x + C$$

4. $\int \cosh^{-1} x \, dx = x \cosh^{-1} x - \sqrt{x^2 - 1} + C$

Letting $\cosh^{-1} x = u$ and $dx = dv$, we have $du = dx/\sqrt{x^2 - 1}$, $v = x$. Hence

$$\int \cosh^{-1} x \, dx = x \cosh^{-1} x - \int x \frac{dx}{\sqrt{x^2 - 1}}$$

The integral $\int \frac{x \, dx}{\sqrt{x^2 - 1}} = \frac{1}{2} \int \frac{2x \, dx}{\sqrt{x^2 - 1}}$ is evaluated by setting $w = x^2 - 1$, $dw = 2x \, dx$. It then becomes

$$\frac{1}{2} \int \frac{dw}{w^{1/2}} = \frac{1}{2} \frac{w^{1/2}}{1/2} + C = w^{1/2} + C = \sqrt{x^2 - 1} + C$$

and our formula follows immediately.

The remaining inverse functions are all treated in the same way. The reader should work out their integrals as an exercise in the use of integration by parts.

9.11 THE DIFFERENTIATION FORMULAS

We list here all of the formulas we have developed for differentiation.

1. $Dk = 0$
2. $Dx = 1$
3. $D(f + g) = Df + Dg$
4. $D(fg) = gDf + fDg$
5. $D\left(\dfrac{f}{g}\right) = \dfrac{gDf - fDg}{g^2}$
6. $D(g \circ f) = (Dg \circ f) \cdot Df$
7. $Df^r = rf^{r-1} \cdot Df$
8. $D \ln[f(x)] = \dfrac{1}{f(x)} Df(x)$
9. $D \log_a[f(x)] = \dfrac{Df(x)}{f(x)} \log_a e$
10. $De^{f(x)} = e^{f(x)} Df(x)$
11. $Da^{f(x)} = a^{f(x)} Df(x) \cdot \ln a$
12. $D[g(x)]^{f(x)} = [g(x)]^{f(x)} \ln[g(x)] Df(x) + f(x)[g(x)]^{f(x)-1} Dg(x)$
13. $D \sin[f(x)] = \cos[f(x)] Df(x)$
14. $D \cos[f(x)] = -\sin[f(x)] \cdot Df(x)$
15. $D \tan[f(x)] = \sec^2[f(x)] Df(x)$
16. $D \cot[f(x)] = -\csc^2[f(x)] \cdot Df(x)$
17. $D \sec[f(x)] = \sec[f(x)] \tan[f(x)] Df(x)$
18. $D \csc[f(x)] = -\csc[f(x)] \cot[f(x)] Df(x)$
19. $D \sin^{-1}[f(x)] = \dfrac{Df(x)}{\sqrt{1 - f^2(x)}}$
20. $D \cos^{-1}[f(x)] = -\dfrac{Df(x)}{\sqrt{1 - f^2(x)}}$
21. $D \tan^{-1}[f(x)] = \dfrac{Df}{1 + f^2(x)}$
22. $D \cot^{-1}[f(x)] = -\dfrac{Df(x)}{1 + f^2(x)}$
23. $D \sec^{-1}[f(x)] = \dfrac{Df(x)}{|f(x)| \sqrt{f^2(x) - 1}}$
24. $D \csc^{-1}[f(x)] = -\dfrac{Df(x)}{|f(x)| \sqrt{f^2(x) - 1}}$
25. $D \sinh[f(x)] = \cosh[f(x)] \cdot Df(x)$
26. $D \cosh[f(x)] = \sinh[f(x)] \cdot Df(x)$

27. $D \tanh[f(x)] = \text{sech}^2[f(x)] \cdot Df(x)$
28. $D \coth[f(x)] = -\text{csch}^2[f(x)] \cdot Df(x)$
29. $D \,\text{sech}[f(x)] = -\text{sech}[f(x)] \tanh[f(x)] \cdot Df(x)$
30. $D \,\text{csch}[f(x)] = -\text{csch}[f(x)] \coth[f(x)] \cdot Df(x)$
31. $D \sinh^{-1}[f(x)] = \dfrac{Df(x)}{\sqrt{1 + f^2(x)}}$
32. $D \cosh^{-1}[f(x)] = \dfrac{Df(x)}{\sqrt{f^2(x) - 1}}$
33. $D \tanh^{-1}[f(x)] = \dfrac{Df(x)}{1 - f^2(x)}$
34. $D \coth^{-1}[f(x)] = \dfrac{Df(x)}{1 - f^2(x)}$

10

METHODS OF INTEGRATION

This chapter is devoted to some standard integration techniques. When combined with a few standard substitutions, the integration formulas developed in Chapter 9 permit us to integrate a large class of functions.

10.1 INTEGRATION FORMULAS

All of our integration formulas are listed here for easy reference.

1. $\int df(x) = f(x) + C$

2. $\int [f(x) + g(x)] \, dx = \int f(x) \, dx + \int g(x) \, dx$

3. $\int kf(x) \, dx = k \int f(x) \, dx$

4. $\int f(x) \, dg(x) = f(x) g(x) - \int g(x) \, df(x)$ (Integration by parts)

5. $\int [f(x)]^r \, df(x) = \dfrac{1}{r+1} [f(x)]^{r+1} + C, \qquad r \neq -1$

6. $\int \dfrac{df(x)}{f(x)} = \ln |f(x)| + C$

7. $\int e^{f(x)} \, df(x) = e^{f(x)} + C; \qquad \int a^{f(x)} \, df(x) = \dfrac{1}{\ln a} a^{f(x)} + C$

8. $\int \cos[f(x)] \, df(x) = \sin[f(x)] + C$

9. $\int \sin[f(x)] \, df(x) = -\cos[f(x)] + C$

10. $\int \sec^2[f(x)] \, df(x) = \tan[f(x)] + C$

11. $\int \csc^2[f(x)] \, df(x) = -\cot[f(x)] + C$

12. $\int \sec[f(x)] \tan[f(x)] \, df(x) = \sec[f(x)] + C$

13. $\int \csc[f(x)] \cot[f(x)] \, df(x) = -\csc[f(x)] + C$

14. $\int \dfrac{df(x)}{\sqrt{1 - f^2(x)}} = \sin^{-1}[f(x)] + C, \qquad -1 < f(x) < 1$

15. $\int \dfrac{df(x)}{1 + f^2(x)} = \tan^{-1}[f(x)] + C$

16. $\int \dfrac{df(x)}{|f(x)| \sqrt{f^2(x) - 1}} = \sec^{-1}[f(x)] + C, \qquad |f(x)| > 1$

17. $\int \sinh[f(x)] \, df(x) = \cosh[f(x)] + C$

18. $\int \cosh[f(x)]\, df(x) = \sinh[f(x)] + C$

19. $\int \operatorname{sech}^2[f(x)]\, df(x) = \tanh[f(x)] + C$

20. $\int \operatorname{csch}^2[f(x)]\, df(x) = -\coth[f(x)] + C$

21. $\int \operatorname{sech}[f(x)] \tanh[f(x)]\, df(x) = -\operatorname{sech}[f(x)] + C$

22. $\int \operatorname{csch}[f(x)] \coth[f(x)]\, df(x) = -\operatorname{csch}[f(x)] + C$

23. $\int \dfrac{df(x)}{\sqrt{f^2(x) - 1}} = \cosh^{-1}[f(x)] + C$
$= \ln[f(x) + \sqrt{f^2(x) - 1}] + C, \quad f(x) \geqq 1$

24. $\int \dfrac{df(x)}{\sqrt{1 + f^2(x)}} = \sinh^{-1}[f(x)] + C = \ln[f(x) + \sqrt{f^2(x) + 1}] + C$

25. $\int \dfrac{df(x)}{1 - f^2(x)} = \tanh^{-1}[f(x)] + C = \dfrac{1}{2} \ln\left(\dfrac{1 + f(x)}{1 - f(x)}\right) + C,$
$-1 < f(x) < 1$

26. $\int \dfrac{df(x)}{f(x)\sqrt{1 - f^2(x)}} = -\operatorname{sech}^{-1}[f(x)] + C$
$= -\ln\left[\dfrac{1 + \sqrt{1 - f^2(x)}}{f(x)}\right] + C$
$= \ln\left[\dfrac{f(x)}{1 + \sqrt{1 - f^2(x)}}\right] + C,$
$0 < f(x) \leqq 1$

27. $\int \dfrac{df(x)}{|f(x)|\sqrt{1 + f^2(x)}} = -\operatorname{csch}^{-1}[f(x)] + C$
$= -\ln\left[\dfrac{1}{f(x)} + \dfrac{\sqrt{1 + f^2(x)}}{|f(x)|}\right] + C,$
$f(x) \neq 0$

10.2 INTEGRATION OF TRIGONOMETRIC FUNCTIONS

We already have the two formulas

1. $\int \sin[f(x)]\, df(x) = -\cos[f(x)] + C$

2. $\int \cos[f(x)]\, df(x) = \sin[f(x)] + C$

From the fact that

$$\tan[f(x)]\, df(x) = \dfrac{\sin[f(x)]\, df(x)}{\cos[f(x)]} = -\dfrac{d \cos[f(x)]}{\cos[f(x)]}$$

we immediately obtain the following formula:

3. $\int \tan[f(x)]\,df(x) = -\ln|\cos[f(x)]| + C$

Similarly,

$$\cot[f(x)]\,df(x) = \frac{\cos[f(x)]\,df(x)}{\sin[f(x)]} = \frac{d\sin[f(x)]}{\sin[f(x)]}$$

and hence

4. $\int \cot[f(x)]\,df(x) = \ln|\sin[f(x)]| + C$

A simple but clever device yields the integral of $\sec[f(x)]$ and $\csc[f(x)]$. We write the identity

$$\sec[f(x)] = \sec[f(x)]\left(\frac{\sec[f(x)] + \tan[f(x)]}{\sec[f(x)] + \tan[f(x)]}\right)$$
$$= \frac{\sec^2[f(x)] + \sec[f(x)]\tan[f(x)]}{\sec[f(x)] + \tan[f(x)]}$$

It follows that

$$\sec[f(x)]\,df(x) = \frac{\sec[f(x)]\tan[f(x)]\,df(x) + \sec^2[f(x)]\,df(x)}{\sec[f(x)] + \tan[f(x)]}$$
$$= \frac{d(\sec[f(x)] + \tan[f(x)])}{\sec[f(x)] + \tan[f(x)]}$$

Therefore

5. $\int \sec[f(x)]\,df(x) = \ln|\sec[f(x)] + \tan[f(x)]| + C$

In an analogous way, the reader should verify that

6. $\int \csc[f(x)]\,df(x) = \ln|\csc[f(x)] - \cot[f(x)]| + C$

The "double-angle formulas"

$$\cos 2A = 2\cos^2 A - 1 = 1 - 2\sin^2 A$$
$$\sin 2A = 2\sin A \cos A$$

come to our aid in the integration of squares of sine and cosine functions. From the first equation we find that $\sin^2 A = \tfrac{1}{2} - \tfrac{1}{2}\cos 2A$ and $\cos^2 A = \tfrac{1}{2} + \tfrac{1}{2}\cos 2A$. Then we have

$$\int \sin^2[f(x)]\,df(x) = \int \left(\frac{1}{2} - \frac{1}{2}\cos[2f(x)]\right)df(x)$$

10.2 INTEGRATION OF TRIGONOMETRIC FUNCTIONS

$$= \frac{1}{2}\int df(x) - \frac{1}{2}\int \cos[2f(x)] \, df(x)$$

$$= \frac{1}{2}f(x) - \frac{1}{4}\int \cos[2f(x)] \cdot 2 \, df(x)$$

$$= \frac{1}{2}f(x) - \frac{1}{4}\sin[2f(x)] + C$$

$$= \frac{1}{2}f(x) - \frac{1}{4}(2\sin[f(x)]\cos[f(x)]) + C$$

or

7. $\int \sin^2[f(x)] \, df(x) = \frac{1}{2}f(x) - \frac{1}{2}\sin[f(x)]\cos[f(x)] + C$

Analogously, we have

8. $\int \cos^2[f(x)] \, df(x) = \frac{1}{2}f(x) + \frac{1}{2}\sin[f(x)]\cos[f(x)] + C$

Next we have the identities $\tan^2 A = \sec^2 A - 1$, $\cot^2 A = \csc^2 A - 1$, which imply that

$$\int \tan^2[f(x)] \, df(x) = \int (\sec^2[f(x)] - 1) \, df(x)$$

$$= \int \sec^2[f(x)] \, df(x) - \int df(x)$$

or

9. $\int \tan^2[f(x)] \, df(x) = \tan[f(x)] - f(x) + C$

and similarly,

10. $\int \cot^2[f(x)] \, df(x) = -\cot[f(x)] - f(x) + C$

Of course, we have used the formulas

11. $\int \sec^2[f(x)] \, df(x) = \tan[f(x)] + C$

and

12. $\int \csc^2[f(x)] \, df(x) = -\cot[f(x)] + C$

Higher powers of sines and cosines are integrated by similar use of the double-angle formulas, together with the identity $\cos^2 A + \sin^2 A = 1$. We illustrate the technique with two examples.

■ *Example 1*

$$\int \sin^3 x \, dx = \int \sin^2 x \cdot \sin x \, dx$$

$$= \int (1 - \cos^2 x) \cdot \sin x \, dx$$

$$= \int \sin x \, dx - \int \cos^2 x \cdot \sin x \, dx$$

$$= -\cos x - \int \cos^2 x (-d \cos x)$$

$$= -\cos x + \tfrac{1}{3} \cos^3 x + C$$

Any positive odd power of $\sin x$ or $\cos x$ can be treated in the same way. For instance, $\cos^5 x \, dx = (1 - \sin^2 x)^2 \cdot \cos x \, dx = (1 - \sin^2 x)^2 \, d \sin x$. Therefore

$$\int \cos^5 x \, dx = \int (1 - 2 \sin^2 x + \sin^4 x) \, d \sin x$$

$$= \sin x - \tfrac{2}{3} \sin^3 x + \tfrac{1}{5} \sin^5 x + C$$

■ *Example 2*

$$\int \sin^4 x \, dx = \int (\sin^2 x)^2 \, dx = \int \left(\tfrac{1}{2} - \tfrac{1}{2} \cos 2x\right)^2 dx$$

$$= \int \left(\tfrac{1}{4} - \tfrac{1}{2} \cos 2x + \tfrac{1}{4} \cos^2 2x\right) dx$$

$$= \int \left(\tfrac{1}{4} - \tfrac{1}{2} \cos 2x + \tfrac{1}{4}\left(\tfrac{1}{2} + \tfrac{1}{2} \cos 4x\right)\right) dx$$

$$= \int \left(\tfrac{3}{8} - \tfrac{1}{2} \cos 2x + \tfrac{1}{8} \cos 4x\right) dx$$

$$= \tfrac{3}{8} x - \tfrac{1}{4} \sin 2x + \tfrac{1}{32} \sin 4x + C$$

$$= \tfrac{3}{8} x - \tfrac{1}{4} (2 \sin x \cos x)$$

$$ + \tfrac{1}{32} (2 \sin 2x \cos 2x) + C$$

$$= \tfrac{3}{8} x - \tfrac{1}{2} \sin x \cos x$$

$$ + \tfrac{1}{16} (2 \sin x \cos x)(1 - 2 \sin^2 x) + C$$

$$= \tfrac{3}{8} x - \tfrac{3}{8} \sin x \cos x - \tfrac{1}{4} \sin^3 x \cos x + C$$

10.2 INTEGRATION OF TRIGONOMETRIC FUNCTIONS

- **Example 3.** The integral $\int \sec^3 x \, dx$ can be evaluated using integration by parts in a "lift-yourself-by-your-bootstraps" way. We choose $u = \sec x$ and $dv = \sec^2 x \, dx$. Then $du = \sec x \tan x \, dx$ and $v = \tan x$. Putting these into the formula $\int u \, dv = uv - \int v \, du$, we have

$$\int \sec x \cdot \sec^2 x \, dx = \sec x \tan x - \int \tan x \cdot \sec x \tan x \, dx$$

$$= \sec x \tan x - \int \sec x \tan^2 x \, dx$$

$$= \sec x \tan x - \int \sec x (\sec^2 x - 1) \, dx$$

or

$$\int \sec^3 x \, dx = \sec x \tan x - \int \sec^3 x \, dx + \int \sec x \, dx$$

Therefore

$$2 \int \sec^3 x \, dx = \sec x \tan x + \int \sec x \, dx$$

or

$$\int \sec^3 x \, dx = \tfrac{1}{2} \sec x \tan x + \tfrac{1}{2} \ln|\sec x + \tan x| + C$$

Products of powers of sines and cosines also respond to the use of trigonometric identities. We illustrate the methods employed.

- **Example 4**

$$\int \cos^2 x \sin^3 x \, dx = \int \cos^2 x \cdot (1 - \cos^2 x) \sin x \, dx$$

$$= \int (\cos^2 x - \cos^4 x)(-d \cos x)$$

$$= -\tfrac{1}{3} \cos^3 x + \tfrac{1}{5} \cos^5 x + C$$

- **Example 5**

$$\int \cos^2 x \sin^2 x \, dx = \int (1 - \sin^2 x) \sin^2 x \, dx$$

$$= \int \sin^2 x \, dx - \int \sin^4 x \, dx$$

Now we apply formula 7 and Example 2 to write

$$\int \cos^2 x \sin^2 x \, dx = \tfrac{1}{2} x - \tfrac{1}{2} \sin x \cos x$$
$$- (\tfrac{3}{8} x - \tfrac{3}{8} \sin x \cos x - \tfrac{1}{4} \sin^3 x \cos x) + C$$
$$= \tfrac{1}{8} x - \tfrac{1}{8} \sin x \cos x + \tfrac{1}{4} \sin^3 x \cos x + C$$

The reader may verify this by differentiation.

■ **Example 6.** Let $n \in \mathbb{N}$ and consider the integral $\int \tan^n x \, dx$. By writing this as

$$\int \tan^n x \, dx = \int \tan^{n-2} x \cdot \tan^2 x \, dx$$

$$= \int \tan^{n-2} x (\sec^2 x - 1) \, dx$$

$$= \int \tan^{n-2} x \cdot \sec^2 x \, dx - \int \tan^{n-2} x \, dx$$

$$= \int \tan^{n-2} x \, d \tan x - \int \tan^{n-2} x \, dx$$

we obtain the *reduction formula*

$$\int \tan^n x \, dx = \frac{1}{n-1} \tan^{n-1} x - \int \tan^{n-2} x \, dx, \qquad n \neq 1$$

EXERCISES 10.2 Work out the integrals in Exercises 1 to 40.

1. $\int \cos^3 x \, dx$

2. $\int \tan^2 x \sec^2 x \, dx$

3. $\int \cot x \csc^2 x \, dx$

4. $\int \sin x (\cos x)^{3/2} \, dx$

5. $\int \sec^5 x \sin x \, dx$

6. $\int (\tan 2x)^{3/2} \sec^4 2x \, dx$

7. $\int \tan^2 x \cos^3 x \, dx$

8. $\int \sec^3 4x \tan 4x \, dx$

9. $\int \csc^5 x \cot x \, dx$

10. $\int \frac{\cos x \, dx}{(1 + \sin x)^2}$

11. $\int \frac{\cos x \, dx}{1 + \sin^2 x}$

12. $\int \frac{\sin^3 x \, dx}{\cos^2 x}$

13. $\int \frac{\cos^3 x \, dx}{\sin^2 x}$

14. $\int \frac{\cos^3 x \, dx}{\sqrt[3]{\sin x}}$

15. $\int \cot^3 3x \, dx$

16. $\int \sin^4 2x \, dx$

17. $\int \csc^4 2x \, dx$

18. $\int \sin^2 x \cos^2 x \, dx$

19. $\int \frac{\sin^2 x \, dx}{\cos^4 x}$

20. $\int \sin^6 x \, dx$

21. $\int \sec^6 2x \, dx$

22. $\int \cot x \cdot \ln \sin x \, dx$

23. $\int \frac{\cos x \, dx}{9 - \sin^2 x}$

24. $\int \frac{\cos x \, dx}{2 - \cos^2 x}$

10.2 INTEGRATION OF TRIGONOMETRIC FUNCTIONS

25. $\int \dfrac{dx}{1 + \cos x}$

26. $\int \dfrac{dx}{1 + \sec x}$

27. $\int (\sin 2x - \sin x)^2 \, dx$

28. $\int \sin 3x \cos 5x \, dx$

29. $\int \cos 4x \cos 3x \, dx$

30. $\int \sin 2x \sin 3x \sin 5x \, dx$

31. $\int_{\pi/6}^{\pi/4} \cot^3 x \, dx$

32. $\int_0^{\pi/4} \tan^4 x \, dx$

33. $\int_0^{\pi/4} \cos x \cos 2x \, dx$

34. $\int_0^{\pi/4} (\sin 2x + \cos 2x)^2 \sin 2x \, dx$

35. $\int_{-1}^{1} \sin^2 \pi x \cos^2 \pi x \, dx$

36. $\int_0^{\pi/2} \cos^3 x \, dx$

37. $\int_0^{\pi/3} \dfrac{\tan^3 x \, dx}{\sec x}$

38. $\int_0^{\pi/2} \sin^4 x \cos^5 x \, dx$

39. $\int_{\pi/4}^{\pi/2} \dfrac{\cos^4 x \, dx}{\sin^6 x}$

40. $\int_0^{\pi/4} \cos x \cos 3x \, dx$

In Exercises 41 to 56, use integration by parts to evaluate the given integral.

41. $\int x \sin x \, dx$

42. $\int x \cos 2x \, dx$

43. $\int x^2 \sin x \, dx$

44. $\int x \sec^2 2x \, dx$

45. $\int x \csc^2 x \, dx$

46. $\int x^2 \cos x \, dx$

47. $\int x^3 \sin x \, dx$

48. $\int x \tan^{-1} x \, dx$

49. $\int x^2 \sin^{-1} x \, dx$

50. $\int e^x \cos x \, dx$

51. $\int e^{-x} \sin 2x \, dx$

52. $\int e^{3x} \cos 5x \, dx$

53. $\int x \sin^{-1} ax \, dx$

54. $\int \text{sech}^{-1} x \, dx$

55. $\int x^2 \tan^{-1} 2x \, dx$

56. $\int \dfrac{\tan^{-1} x \, dx}{x^2}$

PROBLEMS 10.2

1. (a) Determine the area of the region bounded by the x axis and the graph of $f(x) = \sin^2 x$ between two successive zeros of the function.
   (b) Determine the area bounded by the y axis and the two curves $\{(x, y) : y = \sec^2 x\}$ and $\{(x, y) : y = 2 \tan^2 x\}$.

2. Rotate each of the following areas about the indicated axis and find the volume so generated.
   (a) $\{(x, y) : 0 \leq x \leq \pi, 0 \leq y \leq \sin^2 x\}$; about the x axis

(b) $\{(x,y) : 0 \leq x \leq \pi, 0 \leq y \leq \sin^2 x\}$; about $\{(x,y) : y = -2\}$
(c) $\{(x,y) : 0 \leq x \leq \pi/4, 0 \leq y \leq \sec^2 x\}$; about the x axis

3. Establish the following reduction formulas:

   (a) $\int \sin^n x \, dx = -\dfrac{\sin^{n-1} x \cos x}{n} + \dfrac{n-1}{n} \int \sin^{n-2} x \, dx$

   (b) $\int \cos^n x \, dx = \dfrac{\cos^{n-1} x \sin x}{n} + \dfrac{n-1}{n} \int \cos^{n-2} x \, dx$

   (c) $\int \cot^n x \, dx = -\dfrac{\cot^{n-1} x}{n-1} - \int \cot^{n-2} x \, dx, \quad n \neq 1$

   (d) $\int \sec^n x \, dx = \dfrac{\sec^{n-2} \tan x}{n-1} + \dfrac{n-2}{n-1} \int \sec^{n-2} x \, dx, \quad n \neq 1$

   (e) $\int \csc^n x \, dx = -\dfrac{\csc^{n-2} x \cot x}{n-1} + \dfrac{n-2}{n-1} \int \csc^{n-2} x \, dx, \quad n \neq 1$

4. Determine the constants that make the following reduction formulas valid.

   (a) $\int \sin^m x \cos^n x \, dx = A \sin^{m+1} x \cos^{n-1} x + B \int \sin^m x \cos^{n-2} x \, dx$

   (b) $\int \dfrac{\sin^m x}{\cos^n x} \, dx = C \dfrac{\sin^{m+1} x}{\cos^{n-1} x} + D \int \dfrac{\sin^m x}{\cos^{n-2} x} \, dx \cdot m \neq 1$

5. Evaluate the improper integrals:

   (a) $\int_0^{\pi/2} \sqrt{\dfrac{\sin^2 x}{\cos x}} \, dx$ (b) $\int_0^{\pi/4} \tan^2 2x \, dx$

6. By writing each limit as a definite integral, evaluate:

   (a) $\lim\limits_{n \to \infty} \sum\limits_{k=1}^{n} \dfrac{\pi}{n} \sin \dfrac{k\pi}{n}$

   (b) $\lim\limits_{n \to \infty} \sum\limits_{k=1}^{n} \dfrac{3\pi}{4n} \sin\left(\dfrac{\pi}{4} + \dfrac{3k\pi}{4n}\right)$

7. By substituting $1 - \cos 2x = 2 \sin^2 x$, we may write

   $$\int_0^{2\pi} (1 - \cos 2x)^{3/2} \, dx = 2\sqrt{2} \int_0^{2\pi} \sin^3 x \, dx$$

   However, the integral on the right has value zero, while that on the left is not zero. What is the difficulty here?

8. Establish the following values (which are used in computations with *Fouier series*): For all $m, n \in \mathbf{N}$,

   (a) $\int_{-1}^{1} \cos m\pi x \cos n\pi x \, dx = 0 \quad$ if $m \neq n$
   $\phantom{(a) \int_{-1}^{1} \cos m\pi x \cos n\pi x \, dx} = 1 \quad$ if $m = n$

   (b) $\int_{-1}^{1} \sin \pi m x \cos n\pi x \, dx = 0$

   (c) $\int_{-1}^{1} \sin m\pi x \sin n\pi x \, dx = 0 \quad$ if $m \neq n$
   $\phantom{(c) \int_{-1}^{1} \sin m\pi x \sin n\pi x \, dx} = 1 \quad$ if $m = n$

9. Evaluate the following limits:

(a) $\displaystyle\lim_{n\to\infty}\left(\frac{1}{\sqrt{n^2}}+\frac{1}{\sqrt{n^2+n}}+\cdots+\frac{1}{\sqrt{n^2+(n-1)n}}\right)$

(b) $\displaystyle\lim_{n\to\infty}\left(\frac{n}{n^2+0^2}+\frac{n}{n^2+1^2}+\cdots+\frac{n}{n^2+(n-1)^2}\right)$

10. Validate the following reduction formulas:

(a) $\displaystyle\int x^n \sin ax\, dx = -\frac{1}{a}x^n \cos ax + \frac{n}{a}\int x^{n-1}\cos ax\, dx$

(b) $\displaystyle\int x^n \cos ax\, dx = \frac{1}{a}x^n \sin ax - \frac{n}{a}\int x^{n-1}\sin ax\, dx$

(c) $\displaystyle\int x^n \sin^{-1} x\, dx = \frac{x^{n+1}}{n+1}\sin^{-1} x - \frac{1}{n+1}\int \frac{x^{n+1}\, dx}{\sqrt{1-x^2}}$

(d) $\displaystyle\int x^n \tan^{-1} x\, dx = \frac{x^{n+1}}{n+1}\tan^{-1} x - \frac{1}{n+1}\int \frac{x^{n+1}\, dx}{1+x^2}$

11. Develop reduction formulas for

(a) $\displaystyle\int x^n \sin^2 x\, dx$ (b) $\displaystyle\int x^n \cos^2 x\, dx$

12. Prove that

$$\int_0^{\pi/2} \sin^{2m-1} x \cos^{2n-1} x\, dx = \frac{(m-1)!(n-1)!}{(m+n-1)!}$$

13. Prove that

$$\lim_{x\to\infty}\frac{1}{x}\int_0^x \sin at \cos bt\, dt = 0$$

14. Find the length of each curve given below:

(a) $\left\{(x,y): 0 \leq x \leq \frac{\pi}{3}, y = \ln \cos x\right\}$

(b) $\left\{(x,y): 0 \leq x \leq \frac{\pi}{4}, y = \ln \sec x\right\}$

10.3 TRIGONOMETRIC SUBSTITUTIONS

The theory here is based upon three simple facts. Let $a > 0$ be a positive real number and let $f(x)$ be any number.

1. If $|f(x)| < a$, then there is an angle of u radians, $-\pi/2 \leq u \leq \pi/2$, such that

$$f(x) = a \sin u$$

2. For any $f(x)$, there exists u, $-\pi/2 < u < \pi/2$, such that

$$f(x) = a \tan u$$

3. If $|f(x)| \geq a$, then there exists u, either $0 \leq u < \pi/2$ or $-\pi \leq u < -\pi/2$, such that

$$f(x) = a \sec u$$

The identities $1 - \sin^2 u = \cos^2 u$, $1 + \tan^2 u = \sec^2 u$ and $\sec^2 u - 1 = \tan^2 u$ also enter our manipulations.

We use the three trigonometric substitutions:

1. $f(x) = a \sin u,\quad df(x) = a \cos u\, du$
2. $f(x) = a \tan u,\quad df(x) = a \sec^2 u\, du$
3. $f(x) = a \sec u,\quad df(x) = a \sec u \tan u\, du$

They serve to reduce integrals involving the factors, $\sqrt{a^2 - f^2(x)}$, $\sqrt{a^2 + f^2(x)}$, $\sqrt{f^2(x) - a^2}$, $a^2 + f^2(x)$ and $a^2 - f^2(x)$, to trigonometric integrals.

1. If the integral involves the factor $\sqrt{a^2 - f^2(x)}$, then we set

$$f(x) = a \sin u$$

It follows that $df(x) = a \cos u\, du$ and $\sqrt{a^2 - f^2(x)} = \sqrt{a^2 - a^2 \sin^2 u} = a\sqrt{1 - \sin^2 u} = a \cos u$. For example, consider

$$\int \frac{dx}{\sqrt{a^2 - x^2}}$$

We set $x = a \sin u$, whence $dx = a \cos u\, du$ and

$$\sqrt{a^2 - x^2} = \sqrt{a^2 - a^2 \sin^2 u} = a \cos u$$

The integral becomes

$$\int \frac{a \cos u\, du}{a \cos u} = \int du = u + C$$

Knowing that $\sin u = x/a$, we have $u = \sin^{-1}(x/a)$. Thus

$$\int \frac{dx}{\sqrt{a^2 - x^2}} = \sin^{-1}\left(\frac{x}{a}\right) + C$$

As another example, consider

$$\int \frac{\sqrt{a^2 - x^2}}{x}\, dx$$

If we make the same substitutions, the integral becomes

$$\int \frac{a \cos u}{a \sin u} \cdot a \cos u\, du = a \int \frac{\cos^2 u\, du}{\sin u}$$

$$= a \int \frac{(1 - \sin^2 u)}{\sin u}\, du$$

$$= a \int \left(\frac{1}{\sin u} - \sin u\right) du$$

$$= a \int \csc u\, du - a \int \sin u\, du$$

$$= a \ln|\csc u - \cot u| + a \cos u + C$$

10.3 TRIGONOMETRIC SUBSTITUTIONS

Now we know that $\sin u = x/a$ and that

$$\cos u = \sqrt{1 - \sin^2 u} = \sqrt{\frac{1 - x^2}{a^2}} = \frac{1}{a}\sqrt{a^2 - x^2}$$

Therefore

$$\csc u = \frac{1}{\sin u} = \frac{a}{x}$$

$$\cot u = \frac{\cos u}{\sin u} = \frac{(1/a)\sqrt{a^2 - x^2}}{x/a} = \frac{\sqrt{a^2 - x^2}}{x}$$

and $a \cos u = \sqrt{a^2 - x^2}$. We have shown that

$$\int \frac{\sqrt{a^2 - x^2}}{x} dx = \sqrt{a^2 - x^2} + a \ln\left|\frac{a}{x} - \frac{\sqrt{a^2 - x^2}}{x}\right| + C$$

2. If the integral involves the factor $a^2 + f^2(x)$, then we use the substitution

$$f(x) = a \tan u$$

whence $df(x) = a \sec^2 u\, du$ and

$$a^2 + f^2(x) = a^2 + a^2 \tan^2 u = a^2(1 + \tan^2 u) = a^2 \sec^2 u$$

This reduces the integral to a trigonometric integral which we can hope to solve.

■ *Example 1.* The integral

$$\int \frac{dx}{4 + 9x^2}$$

responds by setting $3x = 2 \tan u$, $dx = \frac{2}{3} \sec^2 u\, du$ and $4 + 9x^2 = 4 + 4 \tan^2 u = 4 \sec^2 u$. For using these substitutions, we have the trigonometric integral

$$\int \frac{\frac{2}{3} \sec^2 u\, du}{4 \sec^2 u} = \frac{2}{12} \int du = \frac{1}{6} u + C$$

Now $\tan u = 3x/2$, so that $u = \tan^{-1}(3x/2)$ and we have

$$\int \frac{dx}{4 + 9x^2} = \frac{1}{6} \tan^{-1}\left(\frac{3x}{2}\right) + C$$

As another example, consider

$$\int \frac{x^2\, dx}{\sqrt{4 + x^2}}$$

This time we set $x = 2 \tan u$, $dx = 2 \sec^2 u\, du$, and

$$\sqrt{4 + x^2} = \sqrt{4 + 4 \tan^2 u} = 2\sqrt{1 + \tan^2 u} = 2 \sec u$$

The integral becomes

$$\int \frac{(2\tan u)^2(2\sec^2 u\, du)}{2\sec u} = 4\int \tan^2 u \sec u\, du$$

$$= 4\int (\sec^2 u - 1)\sec u\, du$$

$$= 4\int \sec^3 u\, du - 4\int \sec u\, du$$

We have worked these out in Section 10.2 to be

$$\int \sec^3 u\, du = \frac{1}{2}\sec u \tan u + \frac{1}{2}\ln|\sec u + \tan u|$$

$$\int \sec u\, du = \ln|\sec u + \tan u|$$

Therefore

$$4\int \sec^3 u\, du - 4\int \sec u\, du$$
$$= 2\sec u \tan u - 2\ln|\sec u + \tan u| + C$$

To return to the variable x, we note that $\tan u = x/2$ and hence $\sec u = \sqrt{1 + \tan^2 u} = \sqrt{1 + x^2/4} = \frac{1}{2}\sqrt{4 + x^2}$. This shows that

$$\int \frac{x^2\, dx}{\sqrt{4 + x^2}} = 2\left(\frac{1}{2}\sqrt{4 + x^2}\right)\left(\frac{x}{2}\right)$$

$$- 2\ln\left|\frac{1}{2}(x + \sqrt{4 + x^2})\right| + C$$

$$= \frac{x}{2}\sqrt{4 + x^2} - 2\ln(x + \sqrt{4 + x^2}) + C'$$

(The term $-2\ln\frac{1}{2}$ has been absorbed into the constant. Since $x + \sqrt{4 + x^2} > 0$ for all x, we omitted the absolute value symbols.)

3. If the integral involves $f^2(x) - a^2$, then we use the substitution

$$f(x) = a\sec u$$

We have $df(x) = a\sec u \tan u\, du$ and $f^2(x) - a^2 = a^2\sec^2 u - a^2 = a^2(\sec^2 u - 1) = a^2\tan^2 u$.

■ **Example 2.** The integral

$$\int \frac{dx}{\sqrt{4x^2 - 9}}$$

is quickly integrated by substituting $2x = 3\sec u$, $dx = \frac{3}{2}\sec u \tan u\, du$ and $\sqrt{4x^2 - 9} = \sqrt{9\sec^2 u - 9} = 3\sqrt{\sec^2 u - 1} = 3\tan u$. Then

10.3 TRIGONOMETRIC SUBSTITUTIONS

$$\int \frac{\frac{3}{2} \sec u \tan u \, du}{3 \tan u} = \frac{1}{2} \int \sec u \, du = \frac{1}{2} \ln|\sec u + \tan u| + C$$

But $\sec u = 2x/3$ and $\tan u = \frac{1}{3}\sqrt{4x^2 - 9}$, so that

$$\int \frac{dx}{\sqrt{4x^2 - 9}} = \frac{1}{2} \ln \left| \frac{2x}{3} + \frac{1}{3}\sqrt{4x^2 - 9} \right| + C$$

$$= \frac{1}{2} \ln \left| 2x + \sqrt{4x^2 - 9} \right| - \frac{1}{2} \ln 3 + C$$

$$= \frac{1}{2} \ln |2x + \sqrt{4x^2 - 9}| + C'$$

As another example, consider

$$\int \frac{dx}{x\sqrt{x^2 - 4}}$$

We set $x = 2 \sec u$, $dx = 2 \sec u \tan u \, du$ and $\sqrt{x^2 - 4} = \sqrt{4 \sec^2 u - 4} = 2 \tan u$. Then

$$\int \frac{dx}{x\sqrt{x^2 - 4}} = \int \frac{2 \sec u \tan u \, du}{2 \sec u \cdot 2 \tan u} = \frac{1}{2} \int du = \frac{1}{2} u + C$$

$$= \frac{1}{2} \sec^{-1}\left(\frac{x}{2}\right) + C$$

The techniques used above can be applied to integrals containing a factor of the form

$$\sqrt{a + bx + cx^2}$$

simply by making a substitution of the form $y = x + k$, $dx = dy$. That is, we want the quadratic function $a + bx + cx^2$ to appear in the form

$$c(y^2 + A)$$

This is done by writing

$$a + bx + cx^2 = cx^2 + bx + a$$

$$= c\left(x^2 + \frac{b}{c}x\right) + a$$

$$= c\left(x^2 + \frac{b}{c}x + \frac{b^2}{4c^2}\right) + a - \frac{b^2}{4c}$$

$$= c\left(x + \frac{b}{2c}\right)^2 + \frac{4ac - b^2}{4c^2}$$

$$= c\left[\left(x + \frac{b}{2c}\right)^2 + \frac{4ac - b^2}{4c}\right]$$

Setting $y = x + b/2c$ and $A = (4ac - b^2)/4c^2$, we have the desired form.

Note that we cannot have both c negative and A positive. For in this case, $c(y^2 + A) = a + bx + cx^2$ would be negative and $\sqrt{a + bx + cx^2}$ would not be real. There we consider only the two cases: (1) $c > 0$, and (2) both $c < 0$ and $A < 0$. Now if $c > 0$, then $\sqrt{a + bx + cx^2} = \sqrt{c(y^2 + A)} = \sqrt{c}\sqrt{y^2 + A}$. Regardless of the sign A, our methods apply. If $c < 0$, then $-c = d > 0$ and similarly, $-A = B > 0$. Then $C(y^2 + A) = -d(y^2 - B) = d(B - y^2)$ so that $\sqrt{a + bx + cx^2} = \sqrt{d}\sqrt{B - y^2}$ and again our method applies.

■ **Example 3.** Consider the integral

$$\int \frac{dx}{x^2 - 4x + 8}$$

We write $x^2 - 4x + 8 = x^2 - 4x + 4 + 4 = (x - 2)^2 + 4$. Setting $y = x - 2$ and hence $dy = dx$, the integral becomes

$$\int \frac{dy}{y^2 + 4}$$

Now use the trigonometric substitution $y = 2 \tan u$, $dy = 2 \sec^2 u \, du$ and $4 + y^2 = 4 + 4 \tan^2 u = 4(1 + \tan^2 u) = 4 \sec^2 u$. We get

$$\int \frac{2 \sec^2 u \, du}{4 \sec^2 u} = \frac{1}{2} \int du = \frac{1}{2} u + C$$

But $u = \tan^{-1}(y/2)$ and $y = x - 2$, so that

$$\int \frac{dx}{x^2 - 4x + 8} = \frac{1}{2} \tan^{-1}\left(\frac{x - 2}{2}\right) + C$$

■ **Example 4.** The integral

$$\int \frac{dx}{\sqrt{-3 + 4x - x^2}}$$

is written as

$$\int \frac{dx}{\sqrt{1 - 4 + 4x - x^2}} = \int \frac{dx}{\sqrt{1 - (x - 2)^2}}$$

Then set $y = x - 2$, $dy = dx$, to get

$$\int \frac{dy}{\sqrt{1 - y^2}} = \sin^{-1} y + C$$

Since $y = x - 2$, we have

$$\int \frac{dx}{\sqrt{-3 + 4x - x^2}} = \sin^{-1}(x - 2) + C$$

10.3 TRIGONOMETRIC SUBSTITUTIONS

Work out the integrals in Exercises 1 to 30.

EXERCISES 10.3

1. $\int \sqrt{16 - x^2}\, dx$
2. $\int \dfrac{dx}{\sqrt{9 - x^2}}$
3. $\int \sqrt{x^2 - 25}\, dx$
4. $\int \sqrt{ax^2 - 16}\, dx$
5. $\int x\sqrt{9x^2 - 16}\, dx$
6. $\int \dfrac{\sqrt{16 - x^2}}{x}\, dx$
7. $\int \dfrac{dx}{(4x^2 - 9)^2}$
8. $\int \dfrac{x\, dx}{(4x^2 - 9)^3}$
9. $\int \dfrac{(x^2 - 9)^{3/2}\, dx}{x}$
10. $\int \dfrac{x^2\, dx}{\sqrt{x^2 - 4}}$
11. $\int \dfrac{dx}{9x^2 + 25}$
12. $\int \dfrac{dx}{(x^2 + 4)^2}$
13. $\int \dfrac{dx}{x\sqrt{x^2 + 25}}$
14. $\int (16 - 9x^2)^{3/2}\, dx$
15. $\int \dfrac{x^2\, dx}{\sqrt{4x^2 + 9}}$
16. $\int \dfrac{dx}{x^2\sqrt{x^2 - 6}}$
17. $\int x^2\sqrt{8 - x^2}\, dx$
18. $\int \dfrac{dx}{(2x^2 - 5)^{3/2}}$
19. $\int \dfrac{\sec^2 x\, dx}{(2 - 5\tan^2 x)^{1/2}}$
20. $\int \dfrac{(4 + \ln^2 x)^{3/2}\, dx}{x}$
21. $\int_0^1 \dfrac{dx}{16x^2 + 9}$
22. $\int_0^1 \sqrt{25 - 9x^2}\, dx$
23. $\int_0^5 \sqrt{4x^2 + 25}\, dx$
24. $\int_1^{\sqrt{2}} \sqrt{25x^2 - 16}\, dx$
25. $\int_3^5 \dfrac{(x^2 - 9)^{1/2}}{x^2}\, dx$
26. $\int_4^8 \dfrac{dx}{(x - 4)^{3/2}}$
27. $\int_5^6 x\sqrt{8x^2 + 7}\, dx$
28. $\int_1^4 \dfrac{dx}{x^2\sqrt{x^2 + 9}}$
29. $\int_0^1 \dfrac{dx}{(x^2 + 2)^2}$
30. $\int_2^3 \dfrac{x\, dx}{(x^2 - 1)^3}$

Derive a formula for each of the integrals in Exercises 31 to 40.

31. $\int \sqrt{x^2 + a^2}\, dx$
32. $\int (x^2 - a^2)^{3/2}\, dx$
33. $\int x^2\sqrt{a^2 - x^2}\, dx$
34. $\int \dfrac{\sqrt{a^2 - x^2}}{x^2}\, dx$
35. $\int \dfrac{x^2\, dx}{\sqrt{a^2 - x^2}}$
36. $\int \dfrac{\sqrt{x^2 + a^2}}{x^2}\, dx$
37. $\int x^2\sqrt{x^2 + a^2}\, dx$
38. $\int \dfrac{\sqrt{x^2 - a^2}\, dx}{x}$
39. $\int \dfrac{\sqrt{x^2 - a^2}\, dx}{x^2}$
40. $\int \dfrac{x^2\, dx}{\sqrt{x^2 - a^2}}$

By completing the square first and then using the appropriate trigonometric substitution, work out each of the integrals in Exercises 4 to 58.

41. $\displaystyle\int \frac{dx}{\sqrt{x^2 - 4x + 20}}$

42. $\displaystyle\int \frac{dx}{x^2 + 4x + 13}$

43. $\displaystyle\int \frac{dx}{\sqrt{3 + 2x - x^2}}$

44. $\displaystyle\int \frac{dx}{\sqrt{4x + x^2}}$

45. $\displaystyle\int \frac{dx}{\sqrt{4x - x^2}}$

46. $\displaystyle\int \frac{x^2\,dx}{\sqrt{4x - x^2}}$

47. $\displaystyle\int \frac{dx}{(x^2 - 4x + 20)^2}$

48. $\displaystyle\int \frac{dx}{(3 - 4x - x^2)^{3/2}}$

49. $\displaystyle\int \frac{x\,dx}{\sqrt{3 - 2x - x^2}}$

50. $\displaystyle\int \frac{x\,dx}{\sqrt{x^2 - 4x + 5}}$

51. $\displaystyle\int \frac{(x + 2)\,dx}{\sqrt{4x^2 + 4x + 17}}$

52. $\displaystyle\int \frac{(x + 1)\,dx}{\sqrt{x^2 - 2x}}$

53. $\displaystyle\int \frac{(x - 1)\,dx}{\sqrt{15 + 2x - x^2}}$

54. $\displaystyle\int \frac{\sqrt{x^2 + 2x - 3}}{x + 1}\,dx$

55. $\displaystyle\int (x + 2)^2 \sqrt{x^2 + 4x + 5}\,dx$

56. $\displaystyle\int \frac{(3x + 5)\,dx}{9x^2 + 24x + 25}$

57. $\displaystyle\int \frac{(3x - 1)\,dx}{\sqrt{x^2 + 4x + 29}}$

58. $\displaystyle\int \frac{\sqrt{x^2 + 2x - 3}}{x + 1}\,dx$

PROBLEMS 10.3

1. Develop the theory of hyperbolic substitutions. For example, for an integrand involving $a^2 + f^2(x)$, we let $f(x) = a \sinh u$. Then obviously $a^2 + f^2(x) = a^2 \cosh^2 u$, and so forth.

2. In each of the following integrals, consider all possible cases in deriving a general formula:

   (a) $\displaystyle\int \frac{dx}{ax^2 + bx + c}$

   (b) $\displaystyle\int \sqrt{ax^2 + bx + c}\,dx$

   (c) $\displaystyle\int \frac{x\,dx}{(ax^2 + bx + c)}$

   (d) $\displaystyle\int \frac{dx}{(ax^2 + bx + c)^2}$

   (e) $\displaystyle\int \frac{x^2\,dx}{\sqrt{ax^2 + bx + c}}$

   (f) $\displaystyle\int \frac{\sqrt{ax^2 + bx + c}\,dx}{x^3}$

3. Work out the following integrals:

   (a) $\displaystyle\int \sin(ax + b) \sin(cx + d)\,dx$

   (b) $\displaystyle\int \sin(ax + b) \cos(cx + d)\,dx$

   (c) $\displaystyle\int \cos(ax + b) \cos(cx + d)\,dx$

   (d) $\displaystyle\int x \sin x \cos 2x\,dx$

   (e) $\displaystyle\int x^2 \sin 2x \cos 3x\,dx$

4. Determine the relationship between each pair of integrals:

(a) $\displaystyle\int \frac{dx}{x\sqrt{a^2+x^2}}$ and $\displaystyle\int \frac{dx}{\sqrt{a^2+x^2}}$

(b) $\displaystyle\int \frac{x^2\,dx}{\sqrt{a^2-x^2}}$ and $\displaystyle\int \frac{dx}{x^3\sqrt{x^2-a^2}}$

5. Determine the area of each of the following regions:

(a) $\{(x,y) : 3 \leq x \leq 5,\ -\tfrac{4}{3}\sqrt{x^2-9} \leq y \leq \tfrac{4}{3}\sqrt{x^2-9}\}$

(b) $\{(x,y) : 2 \leq x \leq 4,\ -\sqrt{x^2+4x-5} \leq y \leq \sqrt{x^2+4x-5}\}$

6. Prove that a circle of radius r encloses an area of πr^2.

7. Obtain a formula for each of the following integrals:

(a) $\displaystyle\int \frac{dx}{x\sqrt{2ax-x^2}}$ (b) $\displaystyle\int \frac{dx}{x\sqrt{2ax+x^2}}$

(c) $\displaystyle\int \frac{dx}{x\sqrt{x^2-2ax}}$

8. Do each of the examples in Problem 7 by a different method.

10.4 INTEGRATION OF RATIONAL FUNCTIONS

Our goal here is to outline a procedure for evaluating integrals of the form

$$\int \frac{p(x)}{q(x)}\,dx$$

where $p(x)$ and $q(x)$ are polynomials. The theory is based upon three algebraic facts, which are stated without proof.

1. Each polynomial $p(x) = a_0 + c_1 x + \cdots + a_n x^n$ has a unique factorization into a product of linear and quadratic factors:

$$p(x) = a_n(x-r_1)^{s_1}(x-r_2)^{s_2}\cdots(x-r_j)^{s_j}(x^2+b_1 x+c_1)^{t_1}$$
$$\cdots (x^2 + b_k x + c_k)^{t_k}$$

The numbers r_i, $i = 1, 2, \cdots, j$, are the real zeros of $p(x)$ with their corresponding multiplicities s_i. Each factor $x^2 + b_i x + c_i$, $i = 1, 2, \cdots, k$, corresponds to a pair of complex conjugate zeros with their multiplicities t_i. We do not actually do much factorization, however.

2. The polynomials $p(x) = a_0 + a_1 x + \cdots + a_m x^m$ and $q(x) = b_0 + b_1 x + \cdots + b_n x^n$ are identical (that is $p(x) = q(x)$ for all $x \in \mathbf{R}$) if and only if $a_i = b_i$ for each i. In particular, then, the polynomials must have the same degree.

The third algebraic fact we shall use involves the use of *partial fractions*. In a sense, the procedure here might be considered as undoing an addition of fractions. For example, we might want to be able to determine coefficients A, B, C, and D such that

$$\frac{5x^2+4x+6}{x^2(x^2+2)} = \frac{A}{x} + \frac{B}{x^2} + \frac{Cx+D}{x^2+2}$$

We can add the fractions on the right-hand side by putting them over a common denominator. This gives us

$$\frac{5x^2 + 4x + 6}{x^2(x^2 + 2)} = \frac{Ax(x^2 + 2) + B(x^2 + 2) + Cx^3 + Dx^2}{x^2(x^2 + 2)}$$

$$= \frac{(A + C)x^3 + (B + D)x^2 + (2A)x + 2B}{x^2(x^2 + 2)}$$

These fractions can only be identical if their numerators are identical. Thus

$$5x^2 + 4x + 6 \equiv (A + C)x^3 + (B + D)x^2 + 2Ax + 2B$$

By the second fact we must have equality of the coefficients of like powers so

$$A + C = 0, \quad B + D = 5, \quad 2A = 4, \quad 2B = 6$$

One easily computes that $A = 2, B = 3, C = -2$, and $D = 2$. Therefore

$$\frac{5x^2 + 4x + 6}{x^2(x^2 + 2)} = \frac{2}{x} + \frac{3}{x^2} + \frac{-2x + 2}{x^2 + 2}$$

This illustration leads to our third algebraic fact:

3. Let $p(x) = a_0 + a_1 x + \cdots + a_m x^m$ and $q(x) = b_0 + b_1 x + \cdots + b_n x^n$ be polynomials with $m < n$. Denote the factorization of $q(x)$ by

$$q(x) = b_n(x - r_1)^{s_1} \cdots (x - r_j)^{s_j}(x^2 + b_1 x + c_1)^{t_1} \cdots (x^2 + b_k x + c_k)^{t_k}$$

Then the rational function $p(x)/q(x)$ can be uniquely decomposed into partial fractions

$$\frac{p(x)}{q(x)} = \frac{A_{11}}{x - r_1} + \frac{A_{12}}{(x - r_1)^2} + \cdots + \frac{A_{1s_1}}{(x - r_1)^{s_1}} + \cdots$$

$$+ \frac{A_{j1}}{x - r_j} + \frac{A_{j2}}{(x - r_j)^2} + \cdots + \frac{A_{js_j}}{(x - r_j)^{s_j}}$$

$$+ \frac{B_{11}x - C_{11}}{x^2 + b_1 x + c_1} + \frac{B_{12}x + C_{12}}{(x^2 + b_1 x + c_1)^2} + \cdots$$

$$+ \frac{B_{1t_1}x + C_{1t_1}}{(x^2 + b_1 x + c_1)^{t_1}}$$

$$+ \cdots + \frac{B_{k1}x + C_{k1}}{x^2 + b_k x + c_k} + \cdots + \frac{B_{kt_k}x + C_{kt_k}}{(x^2 + b_k x + c_k)^{t_k}}$$

This formula states that there must be a partial fraction for each power of each factor of the denominator. Also note that the numerator must have a smaller degree than the denominator in order to apply our third algebraic fact.

■ **Example 1.** For the rational function $(-x+2)/[x(x+1)(x^2-x+1)]$ we write

$$\frac{-x+2}{x(x+1)(x^2-x+1)}$$
$$= \frac{A}{x} + \frac{B}{x+1} + \frac{Cx+D}{x^2-x+1}$$
$$= \frac{A(x+1)(x^2-x+1) + Bx(x^2-x+1) + x(x+1)(Cx+D)}{x(x+1)(x^2-x+1)}$$
$$= \frac{(A+B+C)x^3 + (-B+C+D)x^2 + (B+D)x + A}{x(x+1)(x^2-x+1)}$$

Equating coefficients in the two numerators, we have

$$A+B+C=0, \quad -B+C+D=0, \quad B+D=-1, \quad A=2$$

This gives three equations in the three undetermined coefficients:

$$B+D=-1$$
$$-B+C+D=0$$
$$B+C=-2$$

Standard methods provide the solution $B=-1$, $C=-1$, $D=0$. Therefore

$$\frac{-x+2}{x(x+1)(x^2-x+1)} = \frac{2}{x} - \frac{1}{x+1} - \frac{x}{x^2-x+1}$$

■ **Example 2.** Let

$$\frac{2x^3+x+1}{x^2(x^2+1)^2} = \frac{A}{x} + \frac{B}{x^2} + \frac{Cx+D}{x^2+1} + \frac{Ex+F}{(x^2+1)^2}$$

Adding the partial fractions on the right, we obtain the numerator

$$Ax(x^2+1)^2 + B(x^2+1)^2 + (Cx+D)x^2(x^2+1) + (Ex+F)x^2$$
$$= (A+C)x^5 + (B+D)x^4 + (2A+C+E)x^3 + (2B+D+F)x^2$$
$$+ Ax + B$$

Equating coefficients, we have

$$A+C=0, \quad B+D=0, \quad 2A+C+E=2$$
$$2B+D+F=0, \quad A=1, \quad B=1$$

It follows quickly that $C=-1$, $D=-1$, $E=1$, and $F=-1$. Thus,

$$\frac{2x^3+x+1}{x^2(x^2+1)^2} = \frac{1}{x} - \frac{1}{x^2} - \frac{x+1}{x^2+1} + \frac{x-1}{(x^2+1)^2} \quad \blacksquare$$

We put this to use in our integration theory as follows. Confronted with a rational function $p(x)/q(x)$, we first perform a long division if needed to obtain a rational function $c_0 + c_1 x + \cdots + c_k x^k + r(x)/q(x)$, where the degree of $r(x)$ is less than that of $q(x)$. We can integrate the polynomial part with ease. Then by the method of partial fractions the function $r(x)/q(x)$ is split into a sum of terms, each of which we can integrate with the techniques we have on hand.

- **Example 3.** We illustrate our method with the integral

$$\int \frac{x^5 + x^4 - 3x^2 - x + 2}{x^2(x^2 - x + 1)}$$

By long division we obtain

$$\frac{x^5 + x^4 - 3x^2 - x + 2}{x^2(x^2 - x + 1)} = x + 2 + \frac{3x^3 - 5x^2 - x + 2}{x^2(x^2 - x + 1)}$$

Applying the partial fraction technique, we obtain

$$\frac{x^5 + x^4 - 3x^2 - x + 2}{x^2(x^2 - x + 1)} = x + 2 + \frac{1}{x} + \frac{2}{x^2} - \frac{6}{x^2 - x + 1}$$

Each term can be integrated easily now. We leave the details to the reader, just writing the result

$$\frac{1}{2} x^2 + 2x + \ln|x| - \frac{2}{x} - 4\sqrt{3} \tan^{-1}\left(\frac{2x - 1}{\sqrt{3}}\right) + C$$

- **Example 4.** We work out the integral

$$\int \frac{x \, dx}{x^4 - 1}$$

First, we note that the denominator factors as follows; $x^4 - 1 = (x - 1)(x + 1)(x^2 + 1)$. Thus we must find coefficients A, B, C, and D such that

$$\frac{x}{x^4 - 1} = \frac{A}{x - 1} + \frac{B}{x + 1} + \frac{Cx + D}{x^2 + 1}$$

Adding the fractions, we obtain the numerator

$$A(x + 1)(x^2 + 1) + B(x - 1)(x^2 + 1) + (Cx + D)(x^2 - 1)$$
$$= (A + B + C)x^3 + (A - B + D)x^2$$
$$+ (A + B - C)x + A - B - D$$

The four equations

$$A + B + C = 0$$
$$A - B + D = 0$$
$$A + B - C = 1$$
$$A - B - D = 0$$

determine these coefficients to be $A = \frac{1}{4}$, $B = \frac{1}{4}$, $C = -\frac{1}{2}$, and $D = 0$. Thus

$$\frac{x}{x^4 - 1} = \frac{1/4}{x - 1} + \frac{1/4}{x + 1} + \frac{-1/2x}{x^2 + 1}$$

and we have

$$\int \frac{x\,dx}{x^4 - 1} = \frac{1}{4}\int \frac{dx}{x - 1} + \frac{1}{4}\int \frac{dx}{x + 1} - \frac{1}{2}\int \frac{x\,dx}{x^2 + 1}$$

$$= \frac{1}{4}\ln|x - 1| + \frac{1}{4}\ln|x + 1| - \frac{1}{4}\ln|x^2 + 1| + C$$

$$= \frac{1}{4}\ln\left|\frac{x^2 - 1}{x^2 + 1}\right| + C$$

The reader should verify this result by differentiation.

Evaluate the integrals in Exercises 1 to 20.

EXERCISES 10.4

1. $\int \frac{2x^2 + x - 5}{x^2 + x - 2}\,dx$
2. $\int \frac{(x - 2)\,dx}{x^2 - 6x + 8}$
3. $\int \frac{5x^3\,dx}{x^2 - x - 6}$
4. $\int \frac{x^2\,dx}{x(x - 1)(x + 1)}$
5. $\int \frac{x^2 - 12x - 16}{x(x - 3)(x + 2)}\,dx$
6. $\int \frac{(x + 6)\,dx}{x(x + 2)(x + 3)}$
7. $\int \frac{dx}{x(x - 1)(x + 1)}$
8. $\int \frac{(3x + 1)\,dx}{x^3 + 2x^2 - 5x - 6}$
9. $\int \frac{(3x - 2)\,dx}{x^3 - 3x - 2}$
10. $\int \frac{2x^2 + 5x - 4}{x^3 + x^2 - 2x}\,dx$
11. $\int \frac{x^4 - 1}{x^2(x + 1)^2}\,dx$
12. $\int \frac{3x^3 + 1}{x^4 + 2x^3 + x^2}\,dx$
13. $\int \frac{(2x + 3)\,dx}{x^4 + x^3 - 2x^2}$
14. $\int \frac{x^2\,dx}{x^4 - 1}$
15. $\int \frac{x\,dx}{(x^2 - 1)^3}$
16. $\int \frac{(3x^2 - 1)x\,dx}{x^4 - 16}$
17. $\int \frac{dx}{(4 - x)\sqrt{x}}$
18. $\int \frac{dx}{\sin x \cos^2 x}$
19. $\int \frac{dx}{x(x^3 + 2)}$
20. $\int \frac{dx}{x(1 - x^5)}$

1. Determine the areas of the regions described below and at the top of page 484. Sketch each region.

PROBLEMS 10.4

(a) $\left\{(x,y) : -\frac{\sqrt{3}}{3}a \leq x \leq \frac{\sqrt{3}}{3}a, -\frac{a^2}{(a^2 + x^2)^{3/2}} \leq y \leq \frac{a^2}{(a^2 + x^2)^{3/2}}\right\}$

(b) $\{(x,y) : 0 \leq y \leq (x^2 + 4)^{-2}\}$

(c) $\left\{(x,y) : 0 \leq x \leq 2a, -\sqrt{\dfrac{x^3}{2a - x}} \leq y \leq \sqrt{\dfrac{x^3}{2a - x}}\right\}$

(d) $\left\{(x,y) : 0 \leq x \leq 4, -\dfrac{1}{\sqrt{4x - x^2}} \leq y \leq \dfrac{1}{\sqrt{4x - x^2}}\right\}$

2. In the following integrals, first use integration by parts and then a trigonometric substitution:

(a) $\displaystyle\int \dfrac{\sin^{-1} x/2 \, dx}{x^3}$
(b) $\displaystyle\int x^2 \tan^{-1} \dfrac{x}{3} \, dx$

3. Obtain formulas for the four integrals:

(a) $\displaystyle\int \dfrac{dx}{(ax + b)(cx + d)}$
(b) $\displaystyle\int \dfrac{x \, dx}{(ax + b)(cx + d)}$

(c) $\displaystyle\int \dfrac{x \, dx}{(ax + b)^2 (cx + d)}$
(d) $\displaystyle\int \dfrac{dx}{(ax + b)^2 (cx + d)}$

4. Work out the integrals:

(a) $\displaystyle\int \dfrac{dx}{(x - a)(x - b)(x - c)}$
(b) $\displaystyle\int \dfrac{x \, dx}{(x - a)^2 (x - b)^2}$

(c) $\displaystyle\int \dfrac{dx}{(x - a)^2 (x - b)^2}$

5. Determine the indefinite integral

$$\int \dfrac{dx}{x(x + 1)(x + 2) \cdots (x + n)}$$

6. Prove that

$$\int_{-1}^{1} (1 - x^2)^n \, dx = \dfrac{2^{2n+1}(n!)^2}{(2n + 1)!}$$

10.5 RATIONALIZING SUBSTITUTIONS

The method of partial fractions suffices to reduce a rational function to a sum of terms, each of which can be integrated. There is no hope of reducing the general algebraic function to a rational function by some useful substitution. However, there are three particular kinds of non-rational functions for which a useful substitution can be chosen.

Integrals of the form

$$\int r(x)(ax + b)^{p/q} \, dx$$

where p and q are integers and $r(x)$ is a rational function, become rational when we make the substitution

$$u = (ax + b)^{1/q}, \quad x = \dfrac{1}{a}(u^q - b), \quad dx = \dfrac{q}{a} u^{q-1} \, du$$

10.5 RATIONALIZING SUBSTITUTIONS

Obviously, the integrand becomes

$$r\left[\frac{1}{a}(u^q - b)\right] \cdot u \cdot \frac{q}{a} u^{q-1} \, du$$

- **Example 1.** To work out the integral

$$\int \frac{x^2 - 1}{\sqrt{2x + 3}} \, dx$$

we set $u = \sqrt{2x + 3}$, $x = \frac{1}{2}(u^2 - 3)$, $dx = u \, du$. Then the integral becomes

$$\int \frac{[\frac{1}{4}(u^2 - 3)]^2 - 1}{u} \cdot u \, du = \int \left(\frac{1}{4} u^4 - \frac{3}{2} u^2 + \frac{5}{4}\right) du$$

$$= \frac{1}{20} u^5 - \frac{1}{2} u^3 + \frac{5}{4} u + C$$

This can be put into terms of x by writing it as

$$\tfrac{1}{20}(u^4 - 10u^2 + 25)u + C$$

Since $u^2 = 2x + 3$, we have $u^4 = 4x^2 + 12x + 9$ and hence

$$u^4 - 10u^2 + 5 = 4x^2 - 8x + 5$$

Therefore

$$\int \frac{x^2 - 1}{\sqrt{2x + 3}} \, dx = \frac{1}{20}(4x^2 - 8x + 5)\sqrt{2x + 3} + C$$

- **Example 2.** Consider the integral

$$\int \frac{\sqrt{x + 3}}{x + 2} \, dx$$

We set $u = \sqrt{x + 3}$, whence $x = u^2 - 3$ and $dx = 2u \, du$. Then we have the integral

$$\int \frac{u \cdot 2u \, du}{u^2 - 3 + 2} = 2 \int \frac{u^2 \, du}{u^2 - 1} = 2 \int \left(1 + \frac{1}{u^2 - 1}\right) du$$

$$= 2u + \ln \left|\frac{u - 1}{u + 1}\right| + C$$

Therefore

$$\int \frac{\sqrt{x + 3}}{x + 2} \, dx = 2\sqrt{x + 3} + \ln \left|\frac{\sqrt{2x + 3} - 1}{\sqrt{2x + 3} + 1}\right| + C \quad \blacksquare$$

A second special substitution can be applied to any integral involving a single irrational factor of the form either $\sqrt{a^2 - x^2}$ or $\sqrt{x^2 \pm a^2}$ and an odd power x^{2k+1} of x as a factor. The substitution consists merely of choosing the radical as a new variable u. Thus we take either

$$u = \sqrt{a^2 - x^2} \quad \text{or} \quad u = \sqrt{x^2 \pm a^2}$$

But we *must* have x^{2k+1} as a factor of the integrand if this substitution is used.

■ **Example 3.** Consider the integral

$$\int \frac{x^3}{\sqrt{4 - x^2}} dx$$

We set $u = \sqrt{4 - x^2}$, whence $u^2 = 4 - x^2$, $x^2 = 4 - u^2$, and $x\,dx = -u\,du$. Then by writing the integral as

$$\int \frac{x^3}{\sqrt{4 - x^2}} dx = \int \frac{x^2}{\sqrt{4 - x^2}} \cdot x\,dx$$

the substitutions result in the integral

$$\int \frac{4 - u^2}{u} \cdot (-u\,du) = \int (u^2 - 4)\,du$$

$$= \frac{1}{3} u^3 - 4u + C$$

This is most easily simplified by writing it as

$$\tfrac{1}{3}(u^2 - 12)u = \tfrac{1}{3}(4 - x^2 - 12)\sqrt{4 - x^2} + C$$
$$= -\tfrac{1}{3}(x^2 + 8)\sqrt{4 - x^2} + C$$

■ **Example 4.** The integral

$$\int \frac{x + x^3}{\sqrt{x^2 + 9}} dx$$

is rationalized by choosing $u = \sqrt{x^2 + 9}$ or $u^2 - 9 = x^2$ and $u\,du = x\,dx$. Writing

$$\int \frac{1 + x^2}{\sqrt{x^2 + 9}} x\,dx = \int \frac{1 + u^2 - 9}{u} \cdot u\,du = \int (u^2 - 8)\,du$$

$$= \frac{1}{3} u^3 - 8u + C$$

$$= \frac{1}{3}(u^2 - 24)u + C$$

we quickly obtain

$$\int \frac{x + x^3}{\sqrt{x^2 + 9}}\, dx = \frac{1}{3}(x^2 - 15)\sqrt{x^2 + 9} + C$$

These examples are admittedly among the simple ones. More involved instances are found in the exercises. ∎

A rational function of a single trigonometric function, such as

$$\frac{\sin^2 x + \sin x - 2}{\sin^3 x - 2 \sin x}$$

can be changed to an ordinary rational function by means of the substitution

$$u = \tan \frac{x}{2}, \qquad x = 2 \tan^{-1} u, \qquad dx = \frac{2\, du}{1 + u^2}$$

We note that, if $u = \tan x/2$, then $1 + u^2 = 1 + \tan^2 x/2 = \sec^2 x/2$. Therefore $\cos^2 x/2 = 1/(1 + u^2)$ and $\sin^2 x/2 = 1 - \cos^2 x/2 = 1 - 1/(1 + u^2) = u^2/(1 + u^2)$. From here it follows that

$$\cos x = \cos^2 \frac{x}{2} - \sin^2 \frac{x}{2} = \frac{1}{1 + u^2} - \frac{u^2}{1 + u^2} = \frac{1 - u^2}{1 + u^2}$$

and

$$\sin x = \sqrt{1 - \cos^2 x} = \sqrt{1 - [(1 - u^2)/(1 + u^2)]^2}$$
$$= \sqrt{[4u^2/(1 + u^2)]^2} = \frac{2u}{1 + u^2}$$

Then the example just given becomes

$$\frac{[2u/(1 + u^2)]^2 + [2u/(1 + u^2)] - 2}{[2u/(1 + u^2)]^3 - 2[2u/(1 + u^2)]}$$

a rational function of u.

∎ **Example 5.** We determine

$$\int \frac{dx}{3 - 2 \cos x}$$

Setting $u = \tan x/2$, $x = 2 \tan^{-1} u$ and $dx = 2\, du/(1 + u^2)$, we obtain

$$\int \frac{1}{3 - 2\left(\frac{1 - u^2}{1 + u^2}\right)} \cdot \frac{2\, du}{1 + u^2} = \int \frac{2\, du}{3(1 + u^2) - 2(1 - u^2)}$$

$$= \int \frac{2\, du}{5u^2 + 1}$$

$$= \frac{2}{\sqrt{5}} \tan^{-1}(u\sqrt{5}) + C$$

Therefore,
$$\int \frac{dx}{3 - 2\cos x} = \frac{2}{\sqrt{5}} \tan^{-1}\left(\sqrt{5} \tan \frac{x}{2}\right) + C$$

EXERCISES 10.5 Work out the integrals in Exercises 1 to 56. Some may require the use of several of the integration techniques so far developed.

1. $\int x\sqrt{x + 5}\, dx$
2. $\int x\sqrt{2x - 1}\, dx$
3. $\int \frac{\sqrt{x + 1}}{x + 2}\, dx$
4. $\int x^2 \sqrt{x + 2}\, dx$
5. $\int \frac{x\, dx}{\sqrt{1 - x}}$
6. $\int \frac{x + 4}{\sqrt{x + 9}}\, dx$
7. $\int \frac{x\, dx}{(1 + x)^{3/4}}$
8. $\int \frac{x + 2}{x\sqrt{x + 1}}\, dx$
9. $\int \sqrt{\frac{1 + x}{1 - x}}\, dx$
10. $\int \frac{dx}{1 + \sqrt{x}}$
11. $\int \frac{x\, dx}{1 + \sqrt{x}}$
12. $\int \frac{dx}{\sqrt{1 + \sqrt{x}}}$
13. $\int \frac{dx}{a + b\sqrt{x}}$
14. $\int \frac{dx}{a + \sqrt{x + b}}$
15. $\int \frac{\sqrt{ax + b} - 1}{\sqrt{ax + b} + 1}\, dx$
16. $\int \frac{dx}{x + x^{2/3}}$
17. $\int \frac{dx}{x - x^{5/4}}$
18. $\int \frac{dx}{x\sqrt{4 + x^2}}$
19. $\int \frac{\sqrt{x^2 - 4}}{x}\, dx$
20. $\int \frac{x^5\, dx}{\sqrt{x^2 + 1}}$
21. $\int \frac{x^2\, dx}{(x^2 + 4)^{3/2}}$
22. $\int \frac{2x - x^3}{\sqrt{a^2 + x^2}}\, dx$
23. $\int x^3 \sqrt{7 + 4x^2}\, dx$
24. $\int \frac{x^3\, dx}{(x^2 - 9)^{2/3}}$
25. $\int \frac{dx}{x^3 \sqrt{x^2 - 4}}$
26. $\int \frac{x\, dx}{(1 - x^2)\sqrt{1 + x^2}}$
27. $\int \frac{\sqrt{x^4 - 1}}{x}\, dx$
28. $\int \frac{dx}{x(ax^4 + b)}$
29. $\int \cos \sqrt{x}\, dx$
30. $\int \sin \sqrt{1 - x}\, dx$
31. $\int \sin^{-1} \sqrt{x}\, dx$
32. $\int \tan^{-1} \sqrt{x + 1}\, dx$
33. $\int \frac{dx}{1 - \sin x}$
34. $\int \frac{dx}{1 + 2\sin x}$

35. $\displaystyle\int \frac{dx}{5 + 4\cos 2x}$
36. $\displaystyle\int \frac{dx}{6 + 5\sec x}$
37. $\displaystyle\int \frac{dx}{\sin x + \tan x}$
38. $\displaystyle\int \frac{dx}{4\cos x - \sin x}$
39. $\displaystyle\int \frac{dx}{\cot x - \tan x}$
40. $\displaystyle\int \frac{dx}{2\csc x - \sin x}$
41. $\displaystyle\int \frac{\sin x\, dx}{\sec x + \cos x}$
42. $\displaystyle\int \frac{\tan x\, dx}{\tan x + \sec x}$
43. $\displaystyle\int \frac{\cos 2x - 1}{\cos 2x + 1}\, dx$
44. $\displaystyle\int \frac{dx}{3 + 2\sin x - \cos x}$
45. $\displaystyle\int \frac{dx}{x - \sqrt{1 - x^2}}$
46. $\displaystyle\int_0^3 \frac{x\, dx}{\sqrt{1 + 5x}}$
47. $\displaystyle\int_{1/4}^1 x\sqrt{4x - 1}\, dx$
48. $\displaystyle\int_2^6 \frac{dx}{x^2\sqrt{2x - 3}}$
49. $\displaystyle\int_3^7 x^2\sqrt{x - 3}\, dx$
50. $\displaystyle\int_0^3 x^2(3x - 1)^{1/3}\, dx$
51. $\displaystyle\int_0^{15} \frac{x\, dx}{\sqrt[4]{1 + x}}$
52. $\displaystyle\int_0^1 \frac{x\, dx}{1 + \sqrt{x} + x}$
53. $\displaystyle\int_0^4 \frac{dx}{(x^2 + 9)^2}$
54. $\displaystyle\int_1^2 \frac{\sqrt{4 - x^2}}{x}\, dx$
55. $\displaystyle\int_0^{\pi/2} \frac{dx}{a + b\sin x}$
56. $\displaystyle\int_0^{\pi/2} \frac{dx}{(a + b\sin x)^2}$

10.6 INTEGRATION OF POWER SERIES

Theorem 5.14 says that a power series may be differentiated term by term. That is, if

$$f(x) = \sum_{n=0}^{\infty} a_n(x - c)^n$$

has the open interval of convergence $\{x : c - k < x < c + k\}$, then so does its derived series

$$f'(x) = D_x\left(\sum_{n=0}^{\infty} a_n(x - c)^n\right) = \sum_{n=0}^{\infty} D_x(a_n(x - c)^n) = \sum_{n=0}^{\infty} na_n(x - c)^{n-1}$$

As an exact analogue we can also integrate a power series term by term.

■ **THEOREM 10.1.** Assume that the power series $\sum_{n=0}^{\infty} a_n(x - c)^n$ has the open interval of convergence $\{x : c - k < x < c + k\}$ where

$$\frac{1}{k} = \lim_{n \to \infty}\left|\frac{a_{n+1}}{a_n}\right|$$

Then the integrated series

$$\int_c^x \left[\sum_{n=0}^\infty a_n(x-c)^n\right] dx = \sum_{n=0}^\infty \left[a_n \int_c^x (x-c)^n \, dx\right] = \sum_{n=0}^\infty \frac{a_n}{n+1}(x-c)^{n+1}$$

also converges in the same interval.

PROOF: We need only apply the limit test (Section 5.9) to the series. We get

$$\lim_{n\to\infty} \frac{a_{n+1}/(n+2)}{a_n/(n+1)} = \lim_{n\to\infty} \left|\frac{n+1}{n+2} \cdot \frac{a_{n+1}}{a_n}\right| = \lim_{n\to\infty} \left|\frac{n+1}{n+2}\right| \cdot \lim_{n\to\infty} \left|\frac{a_{n+1}}{a_n}\right|$$

$$= 1 \cdot \frac{1}{k} = \frac{1}{k}$$

It follows immediately that the integrated series converges whenever $|x - c| < 1/k$. ∎

■ **COROLLARY 10.2.** If the function $f(x)$ is represented by its Taylor series,

$$f(x) = f(c) + f'(c)(x-c) + \frac{f''(c)}{2!}(x-c)^2 + \cdots$$

$$+ \frac{f^{(n)}(c)}{n!}(x-c)^n + \cdots$$

then the integral of $f(x)$ is represented by the integrated Taylor series. That is,

$$\int_c^x f(t) \, dt = f(c)(x-c) + \frac{f'(c)}{2!}(x-c)^2$$

$$+ \frac{f''(c)}{3!}(x-c)^3 + \cdots + \frac{f^{(n)}(c)}{(n+1)!}(x-c)^{n+1} + \cdots$$

The proof of this corollary is left as a problem for the reader.

■ **Example 1.** From the known Maclaurin series

$$e^x = 1 + x + \frac{x^2}{2} + \frac{x^3}{3!} + \cdots + \frac{x^n}{n!} + \cdots$$

we obtain

$$\int_0^x e^t \, dt = \int_0^x dt + \int_0^x t \, dt + \int_0^x \frac{t^2}{2} \, dt + \cdots + \int_0^x \frac{t^n}{n!} \, dt + \cdots$$

$$= x + \frac{1}{2}x^2 + \frac{1}{6}x^3 + \cdots + \frac{1}{(n+1)!}x^{n+1} + \cdots$$

We also know that

$$\int_0^x e^t \, dt = e^x - e^0 = e^x - 1$$

10.6 INTEGRATION OF POWER SERIES

The equation then becomes

$$e^x - 1 = x + \frac{1}{2}x^2 + \frac{1}{6}x^3 + \cdots + \frac{1}{n!}x^n + \cdots$$

and this constitutes verification of the term-by-term integration of this power series.

■ **Example 2.** Knowing that

$$\cos x = 1 - \frac{x^2}{2!} + \frac{x^4}{4!} - \frac{x^6}{6!} + \cdots$$

we find from Corollary 10.2 that

$$\int_0^x \cos t \, dt = \sin x = \int_0^x \left[1 - \frac{t^2}{2!} + \frac{t^4}{4!} - \frac{t^6}{6!} + \cdots \right] dt$$

$$= x - \frac{x^3}{3!} + \frac{x^5}{5!} - \frac{x^7}{7!} + \cdots$$

which is another verification of the corollary.

■ **Example 3.** From

$$e^x = 1 + x + \frac{x^2}{2!} + \frac{x^3}{3!} + \cdots + \frac{x^n}{n!} + \cdots$$

it follows that

$$e^{-x} = 1 - x + \frac{x^2}{2!} - \frac{x^3}{3!} + \cdots + \frac{(-1)^n x^n}{n!} + \cdots$$

By adding the two series, we get

$$e^x + e^{-x} = \left(1 + x + \frac{x^2}{2!} + \frac{x^3}{3!} + \cdots \right)$$

$$+ \left(1 - x + \frac{x^2}{2!} - \frac{x^3}{3!} + \cdots \right)$$

$$= 2\left(1 + \frac{x^2}{2!} + \frac{x^4}{4!} + \cdots \right)$$

Therefore we have the Maclaurin series

$$\cosh x = 1 + \frac{x^2}{2!} + \frac{x^4}{4!} + \cdots + \frac{x^{2n}}{(2n)!} + \cdots$$

Similarly, one can obtain

$$\sinh x = x + \frac{x^3}{3!} + \frac{x^5}{5!} + \cdots + \frac{x^{2n-1}}{(2n-1)!} + \cdots$$

Now we verify Corollary 10.2 again by noting that

$$\int_0^x \cosh t \, dt = \sinh x = \int_0^x \left[1 + \frac{t^2}{2!} + \frac{t^4}{4!} + \cdots \right] dt$$

$$= x + \frac{x^3}{3!} + \frac{x^5}{5!} + \cdots$$

and

$$\int_0^x \sinh t \, dt = \cosh x - \cosh 0 = \cosh x - 1$$

$$= \int_0^x \left[t + \frac{t^3}{3!} + \frac{t^5}{5!} + \cdots \right] dt$$

$$= \frac{1}{2} x^2 + \frac{1}{4!} x^4 + \frac{1}{6!} x^6 + \cdots$$

■ **Example 4.** The function $f(t) = \frac{1}{t} \sin t$, $0 < t$, $f(0) = 1$ is continuous at $t = 0$. (The reader should prove this.) It follows that the integral

$$\int_0^x \frac{\sin t}{t} \, dt$$

exists. However, we have no method of determining an indefinite integral of $f(t)$. The integration of power series applies, however. First, set

$$\frac{\sin t}{t} = \frac{1}{t} \left(t - \frac{t^3}{3!} + \frac{t^5}{5!} - \frac{t^7}{7!} + \cdots \right)$$

$$= 1 - \frac{t^2}{3!} + \frac{t^4}{5!} - \frac{t^6}{7!} + \cdots$$

Therefore,

$$\int_0^x \frac{\sin t}{t} \, dt = \int_0^x \left(1 - \frac{t^2}{3!} + \frac{t^4}{5!} - \frac{t^6}{7!} + \cdots \right) dt$$

$$= x - \frac{x^3}{3(3!)} + \frac{x^5}{5(5!)} - \frac{x^7}{7(7!)} + \cdots$$

This series can be used to obtain values of the function

$$g(x) = \int_0^x \frac{\sin t}{t} \, dt$$

to any desired degree of accuracy.

■ **Example 5.** By long division we can establish that

$$\frac{1}{1+u} = 1 - u + u^2 - u^3 + u^4 - u^5 + \cdots$$

10.6 INTEGRATION OF POWER SERIES

and the ratio test tells us that this series converges whenever $|u| < 1$. Now it follows that

$$\frac{1}{1+t^2} = 1 - t^2 + t^4 - t^6 - t^8 - t^{10} + \cdots \qquad |t| < 1$$

Therefore

$$\int_0^x \frac{dt}{1+t^2} = \tan^{-1} x = \int_0^x (1 - t^2 + t^4 - t^6 + \cdots) \, dt$$

$$= x - \frac{1}{3}x^3 + \frac{1}{5}x^5 - \frac{1}{7}x^7 + \cdots$$

or

$$\tan^{-1} x = x - \frac{1}{3}x^3 + \frac{1}{5}x^5 - \frac{1}{7}x^7 + \cdots$$

The integration of power series is a valuable computational tool. A definite integral $\int_c^x f(t) \, dt$ is first evaluated as a power series and then the power series is evaluated approximately by computing enough of its terms. For example, the previous series for $\tan^{-1} x$ implies that

$$\tan^{-1} 1 = \frac{\pi}{4} = 1 - \frac{1}{3} + \frac{1}{5} - \frac{1}{7} + \cdots$$

which is a remarkable connection between the number π and the odd natural numbers. Of course, this series converges very slowly and is not an efficient means of computing the value of π. A much faster method for computing π uses the identity

$$\tan^{-1} 1 = 4 \tan^{-1}(\tfrac{1}{5}) - \tan^{-1}(\tfrac{1}{239})$$

It follows that

$$\frac{\pi}{4} = 4 \left[\frac{1}{5} - \frac{1}{3}\left(\frac{1}{5}\right)^3 + \frac{1}{5}\left(\frac{1}{5}\right)^5 - \frac{1}{7}\left(\frac{1}{5}\right)^7 + \cdots \right]$$

$$- \left[\frac{1}{239} - \frac{1}{3}\left(\frac{1}{239}\right)^3 + \frac{1}{5}\left(\frac{1}{239}\right)^5 - \frac{1}{7}\left(\frac{1}{239}\right)^7 + \cdots \right]$$

Both of these series converge rapidly, the second one very rapidly. A computer was programmed to compute π to 10,000 decimal places, using these series. (It was done just as an exercise.) Reputedly it took about 4 hr of computer time to complete the arithmetic.

EXERCISES 10.6

In exercises 1 to 10, express each function as a Maclaurin series.

1. $\cos^{-1} x$
2. $\sin^{-1} x$
3. $\tan^{-1} x$
4. $\tan^{-1} x^2$

5. $\ln(x + \sqrt{1 + x^2})$
6. $(\sin^{-1} x)^2$
7. $\int_0^x e^{-t^2}\, dt$
8. $\int_0^x \sqrt{1 + t^3}\, dt$
9. $\int_0^x \sin t^2\, dt$
10. $\int_0^x \dfrac{e^{-t} - 1}{t}\, dt$

Evaluate each integral in Exercises 11 to 24 to three-decimal accuracy.

11. $\int_0^{1/2} \dfrac{\sin x}{x}\, dx$
12. $\int_0^{1/2} \sin x^3\, dx$
13. $\int_{1/4}^{1/2} \dfrac{\cos x - 1}{x}\, dx$
14. $\int_0^{1/2} \dfrac{e^{-x} - 1}{x}\, dx$
15. $\int_0^1 \dfrac{e^x - e^{-x}}{x}\, dx$
16. $\int_0^{1/2} \sqrt{1 - x^3}\, dx$
17. $\int_0^{1/10} (1 + x^2)^{4/5}\, dx$
18. $\int_0^{2/5} x\sqrt{1 - x^3}\, dx$
19. $\int_0^{1/2} \dfrac{dx}{\sqrt{16 + x^4}}$
20. $\int_0^{1/10} \dfrac{\ln(1 + x)}{x}\, dx$
21. $\int_0^{1/5} \sqrt{x} \sin x\, dx$
22. $\int_0^{1/4} \sin\sqrt{x}\, dx$
23. $\int_0^{1/2} x^2 \cos\sqrt{x}\, dx$
24. $\int_0^{1/2} \tan^{-1} x^3\, dx$

PROBLEMS 10.6

1. Use the identity
$$\frac{\pi}{4} = \tan^{-1} 1 = 7854 \tan^{-1}\left(\frac{1}{10{,}000}\right) - \tan^{-1}\left(\frac{1}{545{,}261}\right)$$
to compute the value of π to twenty-decimal accuracy.

2. The *complete elliptic integral of the first kind* is defined to be
$$K(k) = \int_0^{\pi/2} \frac{dx}{\sqrt{1 - k^2 \sin^2 x}}, \qquad 0 < k < 1$$
Prove that
$$\frac{2}{\pi} K(k) = 1 + \left(\frac{1}{2}\right)^2 k^2 + \left(\frac{1 \cdot 3}{2 \cdot 4}\right)^2 k^4 + \left(\frac{1 \cdot 3 \cdot 5}{2 \cdot 4 \cdot 6}\right)^2 k^6 + \cdots$$

3. Prove that
$$\sum_{n=1}^{\infty} \frac{1}{(n+2)n!} = \frac{1}{2}$$
(Hint: Consider the integral $\int_0^1 xe^x\, dx$.)

4. Determine
$$\lim_{n \to \infty} \sum_{k=1}^{n} \ln\left[\frac{\tan^{-1}(k+1)}{\tan^{-1} k}\right]$$

10.7 SIMPSON'S RULE FOR NUMERICAL INTEGRATION

There are many methods of approximate or numerical integration. Here we discuss just one such method, Simpson's rule. It is the most widely used and is particularly well suited to electronic computer work.

The basis of all numerical integration methods is conceptually easy. We replace a function we cannot integrate by a function we *can* integrate, which approximates the difficult function closely. More precisely, suppose that we need the value of a definite integral $\int_a^b f(x)\,dx$. We subdivide the interval $\{x : a \leq x \leq b\}$ into n subintervals. For each subinterval $\{x : x_{i-1} \leq x \leq x_i\}$, we select some function $g_i(x)$ that approximates $f(x)$ well on this interval. Then we assert that

$$\int_a^b f(x)\,dx \approx \sum_{i=1}^{n} \int_{x_{i-1}}^{x_i} g_i(x)\,dx$$

In Simpson's method each approximating function is quadratic, that is,

$$g_i(x) = a_i + b_i x + c_i x^2, \quad i = 1, 2, \cdots, n$$

A special case suffices to give us the entire theory. We approximate the function $f(x)$ over the interval $\{x : -h \leq x \leq h\}$ by choosing a quadratic function $g(x) = a + bx + cx^2$ such that

$$g(-h) = f(-h), \quad g(0) = f(0), \quad g(h) = f(h)$$

That is, the quadratic function is chosen to agree with $f(x)$ at the endpoints and at the midpoint of the interval. The graphs of $f(x)$ and $g(x)$ over the interval $\{x : -h \leq x \leq h\}$ might look like Fig. 10.1.

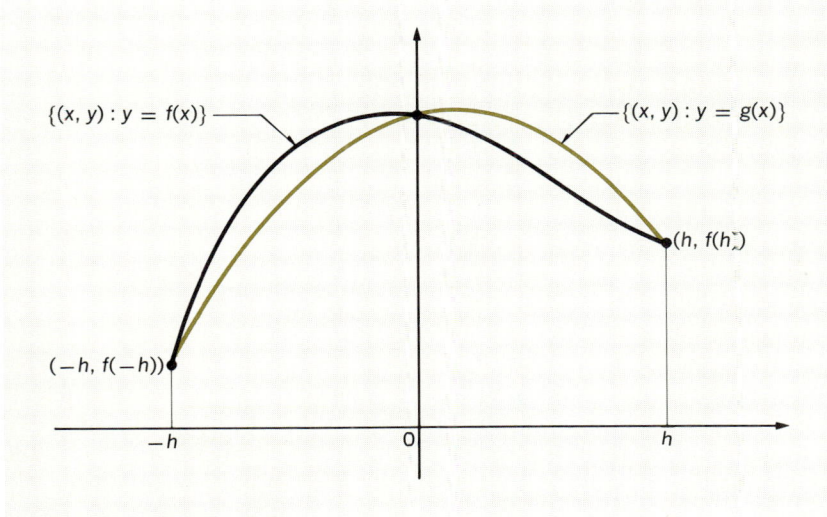

FIGURE 10.1

The three conditions on the function g suffice to determine the coefficients a, b, and c. We arrive at the three simultaneous linear equations in the variables a, b, and c;

$$a + b(-h) + c(-h)^2 = a - bh + ch^2 = f(-h)$$
$$a + b(0) + c(0)^2 = a = f(0)$$
$$a + b(h) + c(h)^2 = a + bh + ch^2 = f(h)$$

It follows that $a = f(0)$, and hence we may write

$$-bh + ch^2 = f(-h) - f(0)$$
$$bh + ch^2 = f(h) - f(0)$$

Adding these together, we get

$$2ch^2 = f(-h) - 2f(0) + f(h)$$

We need go no further in solving these equations, because we do not want the quadratic function g; we want its integral, so that we can write

$$\int_{-h}^{h} f(x)\, dx \approx \int_{-h}^{h} g(x)\, dx$$

But we can find the integral of g as follows

$$\int_{-h}^{h} (a + bx + cx^2)\, dx = ax + \frac{1}{2}bx^2 + \frac{1}{3}cx^3 \Big]_{-h}^{h}$$

$$= ah + \frac{1}{2}bh^2 + \frac{1}{3}ch^3 - \left(-ah + \frac{1}{2}bh^2 - \frac{1}{3}ch^3\right)$$

$$= 2ah + \frac{2}{3}ch^3$$

$$= \frac{h}{3}(6a + 2ch^2)$$

Now, from our algebra just given, we know that $a = f(0)$ and $2ch^2 = f(-h) - 2f(0) + f(h)$. Since $6a + 2ch^2 = f(-h) + 4f(0) + f(h)$, we have the approximate integration formula:

$$\int_{-h}^{h} f(x)\, dx \approx \frac{h}{3}[f(-h) + 4f(0) + f(h)]$$

Now let $\{x : x_i \leq x \leq x_i + 2h\}$ be any interval of length $2h$. It is easy to see that we also have the approximate formula

$$\int_{x_i}^{x_i+2h} f(x)\, dx = \frac{h}{3}[f(x_i) + 4f(x_i + h) + f(x_i + 2h)]$$

This formula is the basis of Simpson's method. Its application is as follows: To obtain an approximation of the definite integral $\int_a^b f(x)\, dx$, we divide the interval $\{x : a \leq x \leq b\}$ into *an even number of subintervals of equal length*. Let $2k$ be the number of subintervals, each of length $(b - a)/2k = h$. Consider these subintervals in pairs and apply the formula above to write

10.7 SIMPSON'S RULE FOR NUMERICAL INTEGRATION

$$\int_a^{a+2h} f(x)\, dx \approx \frac{h}{3}[f(x_0) + 4f(x_0 + h) + f(x_0 + 2h)]$$

$$\int_{a+2h}^{a+4h} f(x)\, dx \approx \frac{h}{3}[f(x_0 + 2h) + 4f(x_0 + 3h) + f(x_0 + 4h)]$$

and so forth. We simplify notation by defining the values

$$y_i = f(a + ih), \quad i = 0, 1, 2, \cdots, 2k$$

Then Simpson's method can be written as

$$\int_a^b f(x)\, dx \approx \frac{b-a}{6k}(y_0 + 4y_1 + 2y_2 + 4y_3 + \cdots$$
$$+ 2y_{2k-2} + 4y_{2k-1} + y_{2k})$$

■ *Example 1.* We apply Simpson's method to the integral

$$\int_1^2 \frac{dx}{x} = \ln 2$$

Let $2k = 10$, so that $h = (2-1)/10 = \frac{1}{10}$. Since $f(x) = 1/x$, we have $y_i = \frac{1}{1 + i/10}$. Thus our approximate value of $\ln 2$ is

$$\frac{1}{30}\left(1 + \frac{4}{1.1} + \frac{2}{1.2} + \frac{4}{1.3} + \frac{2}{1.4} + \frac{4}{1.5} + \frac{2}{1.6} + \frac{4}{1.7}\right.$$
$$\left. - \frac{2}{1.8} + \frac{4}{1.9} + \frac{1}{2}\right) \cong 0.69315 \cdots$$

which is accurate to five decimals

The error in Simpson's rule. We know that Simpson's method provides an approximation, but how good is the approximation? We shall prove that

$$\left|\int_a^b f(x)\, dx - \frac{b-a}{6k}(y_0 + 4y_1 + 2y_2 + \cdots + 4y_{2k-1} + y_{2k})\right|$$
$$\leq \frac{b-a}{180}\left(\frac{b-a}{2k}\right)^4 M$$

where M is the maximum value of $|f^{(4)}(x)|$ over the interval $\{x: a \leq x \leq b\}$. Before doing so, we note that the size of the error can be controlled. By taking $2k$ sufficiently large, we can make the error as small as desired. This sort of *error analysis* is extremely important in numerical work.

Define the constant K by the equation

$$\int_{-h}^h f(t)\, dt = \frac{h}{3}[f(-h) + 4f(0) + f(h)] + Kh^5$$

Then let

$$\phi(x) = \int_{-x}^{x} f(t)\, dt - \frac{x}{3}[f(-x) + 4f(0) + f(x)] + Kx^5$$

Then $\phi(h) = 0$ by the definition of the constant K and obviously we have $\phi(0) = 0$. Hence by Rolle's theorem there is a point x_1, $0 < x_1 < h$, such that $\phi'(x_1) = 0$. But one easily computes that

$$\phi'(x) = \frac{2}{3}f(x) + \frac{2}{3}f(-x) - \frac{4}{3}f(0)$$
$$- \frac{x}{3}[-f'(-x) + f'(x)] - 5Kx^4$$

Again $\phi'(0) = 0$ and we know that $\phi'(x_1) = 0$. Then Rolle's theorem provides a point x_2, $0 < x_2 < x_1$, such that $\phi''(x_2) = 0$. Again it is easily shown that

$$\phi''(x) = \frac{1}{3}f'(x) - \frac{1}{3}f'(-x) - \frac{x}{3}[f''(-x) + f''(x)] - 20Kx^3$$

Once more, $\phi''(0) = 0$ and $\phi''(x_2) = 0$ implies that there is a point x_3, $0 < x_3 < x_2$, for which $\phi'''(x_3) = 0$. But now

$$\phi'''(x) = -\frac{x}{3}[f'''(x) - f'''(-x)] - 60Kx^2$$

Now we apply the mean value theorem. There is a point x_0, $-x < x_0 < x_1$ such that $f'''(x) - f'''(-x) = (x - (-x))f^{(4)}(x_0) = 2xf^{(4)}(x_0)$. Therefore

$$\phi'''(x) = -\frac{x}{3}[2xf^{(4)}(x_0)] - 60Kx^2$$
$$= -x^2\left(\frac{2}{3}f^{(4)}(x_0) + 60K\right)$$

In particular, then

$$\phi'''(x_3) = 0 = -x_3^2[\tfrac{2}{3}f^{(4)}(x_0) + 60K]$$

Because $x_3 > 0$, we must have $\tfrac{2}{3}f^{(4)}(x_0) + 60K = 0$ or
$K = -\tfrac{1}{90}f^{(4)}(x_0)$

We have proved that there exists a point x_0, $-h < x_0 < h$, such that

$$\int_{-h}^{h} f(t)\, dt = \frac{h}{3}[f(-h) + 4f(0) + f(h)] - \frac{h^5}{90}f^{(4)}(x_0)$$

10.7 SIMPSON'S RULE FOR NUMERICAL INTEGRATION

It follows immediately that, when we subdivide the interval $\{x : a \leq x \leq b\}$ into subintervals, each of length $(b - a)/2k = h$, there exists a point x_i in the interval $\{x : a + (2i - 2)h \leq x \leq 2ih\}$ such that

$$\int_{a+(2i-2)h}^{a+2ih} f(t)\, dt = \frac{h}{3}[f(a - (2i - 2)h)$$

$$+ 4f(a + (2i - 1)h) + f(a + 2ih)] - \frac{h^5}{90} f^{(4)}(x_i)$$

Then we have

$$\int_a^b f(t)\, dt = \sum_{i=1}^k \int_{a+(2i-2)h}^{a-2ih} f(t)\, dt$$

$$= \sum_{i=1}^k \frac{h}{3}[f(a + (2i - 2)h) + 4f(a + (2i - 1)h)$$

$$+ f(a + 2ih)] - \frac{h^5}{90} \sum_{i=1}^k f^{(4)}(x_i)$$

Of course, the first summation on the right-hand side of this equation is just our approximate value $\frac{h}{3}(y_0 + 4y_1 + 2y_2 + \cdots + 4y_{2k-1} + y_{2k})$.

The error is

$$E_s = -\frac{h^5}{90} \sum_{i=1}^k f^{(4)}(z_i)$$

Now if M is the maximum value of $|f^{(4)}(x)|$ on $\{x : a \leq x \leq b\}$, we have

$$|E_s| \leq \frac{h^5}{90} \sum_{i=1}^k M = \frac{h^5}{90} \cdot kM = \frac{2kh^5}{180} M$$

But by definition, $2kh = b - a$, so

$$\frac{2kh^5}{180} M = \frac{(b-a)h^4}{180} M$$

and since $h = (b - a)/2k$, we finally have

$$|E_s| \leq \frac{b-a}{180} \left(\frac{b-a}{2k}\right)^4 M$$

■ **Example 2.** The error in using Simpson's method with $2k = 10$ to approximate the integral $\int_1^2 dx/x$ does not exceed

$$\frac{b-a}{180}\left(\frac{b-a}{2k}\right)^4 M = \frac{1}{180}\left(\frac{1}{10}\right)^4 M$$

where M is the maximum value of $D_x^4(1/x) = 24x^{-5}$ on the interval $\{x : 1 \leq x \leq 2\}$. It is easily seen that $M = 24$. Hence the error E_s satisfies

$$|E_s| \leq \frac{1}{180}\left(\frac{1}{10}\right)^4 (24) = \frac{4}{300{,}000} \cong 0.000013$$

In Example 1 we actually obtained better accuracy than this estimate of the error promises.

■ **Example 3.** Again consider the same integral $\int_1^2 dx/x$. For any choice of the number $2k$, the error E_s satisfies

$$|E_s| \leq \frac{1}{180}\left(\frac{1}{2k}\right)^4 (24) = \frac{1}{120k^4}$$

Now suppose we want the approximate to have at least ten-decimal accuracy, that is, we want $|E_s| \leq 0.00000000005 = 5 \times 10^{-11}$. We can be assured of this accuracy if we choose k so that

$$\frac{1}{120k^4} \leq 5 \times 10^{-11}$$

or

$$\frac{1}{120 \times 5 \times 10^{-11}} \leq k^4$$

or

$$\frac{10^{11}}{600} = \frac{10^9}{6} \leq k^4$$

Thus we want

$$k = \sqrt[4]{k^4} \geq \sqrt[4]{\frac{10^9}{6}} = \sqrt[4]{\frac{10^8 \times 10}{6}}$$
$$= 10^2 \sqrt[4]{10/6} = 100 \sqrt[4]{10/6} \cong 113$$

In practice, we would undoubtedly conclude that $k = 100$ is big enough. Then the interval $\{x : 1 \leq x \leq 2\}$ would be subdivided into 200 parts each of length 0.005. We would have to compute 201 values of $f(x) = 1/x$, of course. Therefore an electronic computer would be useful. Indeed, on a modern machine the computation would take less time than printing out the result.

■ **Example 4.** We use Simpson's method with $2k = 10$ to approximate

$$\int_0^1 \frac{dx}{1 + x^2} = \tan^{-1} 1 = \frac{\pi}{4}$$

Again $b - a = 1 - 0 = 1$ and $h = \frac{1}{10}$. The function values $y_i = f(i/10) = 1/(1 + i^2/100)$. Thus our approximate value is

10.7 SIMPSON'S RULE FOR NUMERICAL INTEGRATION

$$\frac{1}{30}\left(1 + \frac{4}{1.01} + \frac{2}{1.04} + \frac{4}{1.09} + \frac{2}{1.16} + \frac{4}{1.25} + \frac{2}{1.36}\right.$$
$$\left. + \frac{4}{1.48} + \frac{2}{1.64} + \frac{4}{1.81} + \frac{1}{2}\right)$$

Using a table of reciprocals, we obtained the value

$$\tfrac{1}{30}(23.56194) = 0.785398$$

Then $\pi \cong 4 \times 0.785398 = 3.141592$, which is accurate to six significant figures.

■ **Example 5.** Simpson's rule also finds application in "integrating" a tabulated function. Suppose that the following table is that of the temperature:

6 A.M.	7 A.M.	8 A.M.	9 A.M.	10 A.M.	11 A.M.	12	1 P.M.	2 P.M.	3 P.M.	4 P.M.	5 P.M.	6 P.M.
49	51	54	58	63	71	80	87	88	90	87	82	79

The actual average temperature is given by the integral

$$\frac{1}{12} \int_{6\text{A.M.}}^{6\text{P.M.}} T(t)\, dt$$

where $T(t)$ is the temperature as a function of time. Using Simpson's rule with $h = 1$, the integral is approximately

$$\tfrac{1}{3}[49 + 4(51) + 2(54) + 4(58) + 2(63) + 4(71)$$
$$+ 2(80) + 4(87) + 2(88) + 4(90) + 2(87) + 4(82) + 79]$$
$$= \tfrac{1}{3}(2628) = 876$$

Therefore the average temperature is $\tfrac{1}{12}(876) = 73$.

In each of Exercises 1 to 10, use Simpson's rule with the value of h given to obtain an approximate value for the integral. Carry out the calculations to three-decimal accuracy. Check your result where possible.

EXERCISES 10.7

1. $\int_2^{10} \dfrac{dx}{1+x}$; $h = 1$

2. $\int_0^1 x^2\, dx$; $h = \dfrac{1}{4}$

3. $\int_0^1 x^4\, dx$; $h = \dfrac{1}{4}$

4. $\int_0^1 (1 + x^2)\, dx$; $h = \dfrac{1}{4}$

5. $\int_0^\pi \dfrac{dx}{2 + \sin x}$; $h = \dfrac{\pi}{4}$

6. $\int_2^8 \dfrac{2x}{\ln x}$; $h = 1$

7. $\int_0^1 \sqrt{1 + x^4}\, dx$; $h = \dfrac{1}{10}$

8. $\int_1^{3/2} \cos(\ln x)\, dx$; $h = \dfrac{1}{8}$

9. $\int_0^{\pi/2} \sqrt{1 - \dfrac{1}{2}\sin^2 x}\, dx$; $h = \dfrac{\pi}{3}$

10. $\int_0^{\pi/3} \sqrt{2 \sec^2 x - 1} \sec x \, dx$; $h = \dfrac{\pi}{12}$

For Exercises 11 to 20, estimate the error in each value obtained in Exercises 1 to 10.

PROBLEMS 10.7

1. Simpson's rule is but one of many formulas for approximate integration. Verify the following formulas.

   (a) The *trapezoid rule*
   $$\int_a^{a+h} f(x) \, dx = \frac{h}{2}[f(a) + f(a+h)] - \frac{h^3}{12} f^{11}(x_0), \quad a < x_0 < a + h$$

   (b) *Weddle's rule*
   $$\int_a^{a+6h} f(x) \, dx = \frac{3h}{10}[f(a) + 5f(a+h) + f(a+2h) + 6f(a+3h)$$
   $$+ f(a+4h) + 5f(a+5h) + f(a+6h)]$$
   $$- \frac{h^7}{140} f^{(6)}(x_0), \quad a < x_0 < a + 6h$$

2. Determine the area $\{(x, y) : 0 \leq x \leq a, \ 0 \leq y \leq (a^{2/3} - x^{2/3})^{3/2}\}$ by first showing the area to be proportional to a^2 and then approximating the constant of proportionality.

3. Determine h so that Simpson's rule will yield five-decimal accuracy in approximating $\int_0^1 \sin(e^x) \, dx$

4. Prove that
   $$\int_0^{\pi/2} e^{-x^2} \sin x \, dx < 1$$

5. (a) Prove that there exist constants c_1 and c_2 such that
   $$\int_a^b p(x) \, dx = \frac{b-a}{3}\left[p(c_1) + p\left(\frac{a+b}{2}\right) + p(c_2)\right]$$
   for every polynomial function $p(x)$ of degree ≤ 3.
   (b) Evaluate c_1 and c_2 in terms of a and b.

6. Prove that there exist constants c_1 and c_2 such that, for every polynomial function $p(x)$ of degree ≤ 3,
   $$\int_0^\infty e^{-x} p(x) \, dx = \frac{1}{4}[c_1 p(c_2) + c_2 p(c_1)]$$

11
THE DIFFERENTIAL
CALCULUS
OF FUNCTIONS
OF SEVERAL
VARIABLES

504 | THE DIFFERENTIAL CALCULUS OF FUNCTIONS OF SEVERAL VARIABLES

This chapter is analogous to Chapters 2, 3, and 4. We study functions of two variables for the most part, using three-dimensional coordinate space to picture the graph $\{(x, y, z) : z = f(x, y)\}$ of a function $f : \mathbf{R} \times \mathbf{R} \to \mathbf{R}$. This representation leads us to use analogous geometric language in discussions of functions of n real variables.

We rely strongly upon the fundamental results in the differential calculus of functions of one variable, in particular, the mean value theorem, in our proofs here. Most of the theorems here are natural generalizations of theorems in the single variable case. Hence much of the motivation is already at hand.

11.1 LINES IN SPACE

Our purpose here is to give an analytic description of the line through two given points in three-dimensional coordinate space. The vector methods used are not only efficient but they foreshadow important applications in geometry. Notice that our development is deliberately free of the use of "distance" and "angle." As has been stressed before, such geometric quantities need not be meaningful when we use the coordinate space as a setting for the graph of a function. We do use the parallelism of lines, however. Parallelism makes good sense in any context because it depends only upon the ratios of "distances" measured along the coordinate axes separately, and such distances are defined.

Consider two fixed points (x_0, y_0, z_0), (x_1, y_1, z_1) in three-dimensional coordinate space. We shall deal with the corresponding position vectors $\mathbf{q}_0 = x_0\mathbf{u} + y_0\mathbf{v} + z_0\mathbf{w}$ and $\mathbf{q}_1 = x_1\mathbf{u} + y_1\mathbf{v} + z_1\mathbf{w}$. By definition of the addition of vectors by the parallelogram rule, the vector $\mathbf{q}_1 - \mathbf{q}_0$ is parallel to the line through the tips of $\mathbf{q}_0$ and $\mathbf{q}_1$. (See Fig. 11.1.) This vector $\mathbf{q}_1 - \mathbf{q}_0$ is called a *direction vector* of the line and its components, the numbers $x_1 - x_0, y_1 - y_0$, and $z_1 - z_0$, are a set of *direction numbers* for the line.

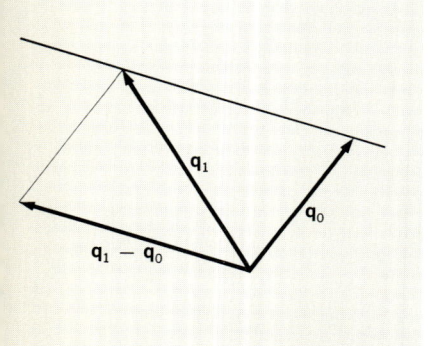

FIGURE 11.1

Let t be any non-zero real number. The position vector $t(\mathbf{q}_1 - \mathbf{q}_0)$ is also parallel to the line through the tips of $\mathbf{q}_0$ and $\mathbf{q}_1$; hence the position vector

$$\mathbf{p} = \mathbf{q}_0 + t(\mathbf{q}_1 - \mathbf{q}_0)$$

has its tip on this line. This follows from the definition of vector addition again. We have drawn a typical situation in Fig. 11.2. Note that we have pictured a case in which $t > 1$. The reader should locate typical vectors $\mathbf{p}$ for the cases $0 < t < 1$ and $t < 0$ on this figure.

Next let $\mathbf{p}$ be the position vector of any point on the line through the tips of $\mathbf{q}_0$ and $\mathbf{q}_1$. From the tip of $\mathbf{p}$ to the line containing the direction vector $\mathbf{q}_1 - \mathbf{q}_0$, draw the line parallel to the vector $\mathbf{q}_0$. This construction is shown in Fig. 11.3 as seen in the plane containing the vectors $\mathbf{q}_0$ and $\mathbf{q}_1$. We thus determine a unique multiple $t(\mathbf{q}_1 - \mathbf{q}_0)$ of the direction vector, and from the construction we have $\mathbf{p} = \mathbf{q}_0 + t(\mathbf{q}_1 - \mathbf{q}_0)$. We establish the following result.

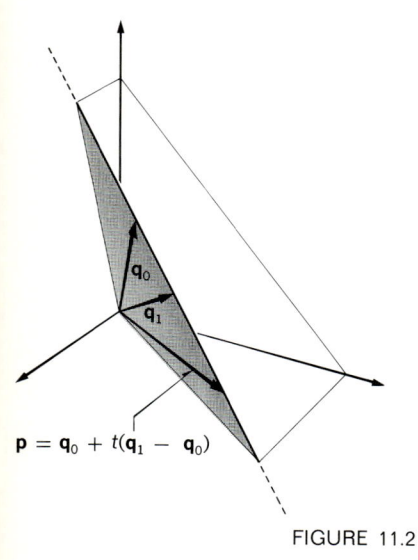

FIGURE 11.2

11.1 LINES IN SPACE

■ **THEOREM 11.1.** Let q_0 and q_1 be fixed position vectors and let L be the line through the tips of q_0 and q_1. Then the position vector p has its tip on the line L if and only if there is a real number $t \in \mathbf{R}$ such that $p = q_0 + t(q_1 - q_0)$.

Note that the direction vector $q_1 - q_0$ is not the only one that can be used in the characterizing equation in Theorem 11.1. If $d = a\mathbf{u} + b\mathbf{v} + c\mathbf{w}$ is any (nonzero) position vector collinear with $q_1 - q_0$, then $p = q_0 + td$, for t any real number, is a vector with its tip on the line L. This clearly implies that if (c, b, c) is a set of direction numbers for L, so is any nonzero multiple (ka, kb, kc) again a set of direction numbers for L.

Next we write the position vector p of a point (x, y, z) on the line L as $p = x\mathbf{u} + y\mathbf{v} + z\mathbf{w}$ where $\mathbf{u}, \mathbf{v}$ and $\mathbf{w}$ are the *unit basis vectors*. (See Section 1.10.) The equation of Theorem 11.1 now becomes

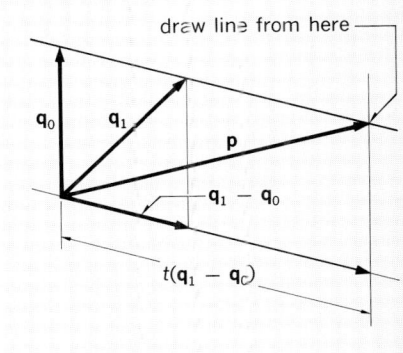

FIGURE 11.3

$$x\mathbf{u} + y\mathbf{v} + z\mathbf{w} = x_0\mathbf{u} + y_0\mathbf{v} + z_0\mathbf{w}$$
$$+ \{(x_1 - x_0)\mathbf{u} + (y_1 - y_0)\mathbf{v} + (z_1 - z_0)\mathbf{w}\}$$
$$= \{x_0 + t(x_1 - x_0)\}\mathbf{u} + \{y_0 + t(y_1 - y_0)\}\mathbf{v}$$
$$+ \{z_0 + t(z_1 - z_0)\}\mathbf{w}$$

Now two vectors are equal if and only if their corresponding components are equal. This fact gives the coordinate equations

$$x = x_0 + t(x_1 - x_0)$$
$$y = y_0 + t(y_1 - y_0)$$
$$z = z_0 + t(z_1 - z_0)$$

These are called *parametric equations* of the line through the points (x_0, y_0, z_0) and (x_1, y_1, z_1).

■ **THEOREM 11.2.** Let (x_0, y_0, z_0) be a fixed point and $a\mathbf{u} + b\mathbf{v} + c\mathbf{w}$ be a fixed position vector. Then the line through (x_0, y_0, z_0) parallel to the vector $a\mathbf{u} + b\mathbf{v} + c\mathbf{w}$ has parametric equations

$$x = x_0 + at, \ y = y_0 + bt, \ z = z_0 + ct$$

PROOF: The position vector

$$x_0\mathbf{u} + y_0\mathbf{v} + z_0\mathbf{w} + t(a\mathbf{u} + b\mathbf{v} + c\mathbf{w})$$

has its tip on a line parallel to $a\mathbf{u} + b\mathbf{v} + c\mathbf{w}$ just by definition of vector addition. That this line passes through (x_0, y_0, z_0) follows by setting $t = 0$. ■

EXERCISES 11.1

1. Find parametric equations of the line through P_0 and P_1 in each of the following cases:
   (a) $P_0 = (3, 1, 2), P_1 = (2, 5, 1)$
   (b) $P_0 = (1, 2, -3), P_1 = (4, 1, -1)$
   (c) $P_0 = (1, -1, 2), P_1 = (0, 1, 4)$
   (d) $P_0 = (0, 2, 2), P_1 = (3, 2, 3)$

2. Find parametric equations of the line through the point P_0 and parallel to the vector $\mathbf{d}$ in each of the following cases.
   (a) $P_0 = (3, -1, 2)$, $\mathbf{d} = -\mathbf{u} + 2\mathbf{v} - \mathbf{w}$
   (b) $P_0 = (1, 2, -3)$, $\mathbf{d} = 3\mathbf{u} - \mathbf{v} + 2\mathbf{w}$
   (c) $P_0 = (-1, 2, -2)$, $\mathbf{d} = 2\mathbf{u} + \mathbf{v} - 3\mathbf{w}$
   (d) $P_0 = (-2, -1, 1)$, $\mathbf{d} = -4\mathbf{u} - 2\mathbf{v} + \mathbf{w}$

3. Find parametric equations of the line through the point P and parallel to the line through P_0 and P_1.
   (a) $P = (1, 0, -2)$, $P_0 = (-2, 3, 1)$, $P_1 = (-1, -3, 4)$
   (b) $P = (1, -1, 0)$, $P_0 = (5, 4, -3)$, $P_1 = (4, 4, 3)$
   (c) $P = (5, 3, -4)$, $P_0 = (1, 3, 0)$, $P_1 = (1, 2, -1)$
   (d) $P = (-2, -1, 3)$, $P_0 = (-3, -2, 1)$, $P_1 = (-3, -1, -4)$

4. Find the point of intersection of the lines described (if they actually intersect, of course.)
   (a) The line through $(3, 1, 2)$ and $(5, 4, 1)$ and the line through $(0, 14, 11)$ and $(3, 8, 5)$.
   (b) The line through $(1, 0, c)$ and $(0, 1, d)$ and the line through $(0, 0, a)$ parallel to the vector $\mathbf{e} = \mathbf{u} + 2\mathbf{v} + (c + 2d)\mathbf{w}$.

5. Determine where the line through P_0 and P_1 meets each of the coordinate planes if
   (a) $P_0 = (3, -1, 1)$ $P_1 = (2, 1, 1)$
   (b) $P_0 = (-1, 2, 2)$ $P_1 = (1, 2, -1)$
   (c) $P_0 = (-3, -1, 1)$ $P_1 = (-2, -1, 3)$
   (d) $P_0 = (4, 0, -3)$ $P_1 = (3, -1, -2)$

PROBLEMS 11.1

1. Let the vector $a\mathbf{u} + b\mathbf{v} + c\mathbf{w}$ have all nonzero components. Show that the line having this direction vector and passing through the point (x_0, y_0, z_0) may be described as the set

$$\left\{(x, y, z) : \frac{x - x_0}{a} = \frac{y - y_0}{b} = \frac{z - z_0}{c}\right\}$$

We say that the equation of the line is in *symmetric form*.

2. Prove that the three points (x_0, y_0, z_0), (x_1, y_1, z_1), (x_2, y_2, z_2) are collinear if and only if the determinant

$$\Delta = \begin{vmatrix} x_0 & y_0 & z_0 \\ x_1 & y_1 & z_1 \\ x_2 & y_2 & z_2 \end{vmatrix}$$

equals zero.

11.2 LINEAR FUNCTIONS AND PLANES

We consider the *linear function*

$$z = f(x, y) = Ax + By + D$$

of the two real variables x and y. This formula provides a real number z for each two-dimensional numerical vector $(x, y) \in \mathbf{R} \times \mathbf{R}$. In view of

11.2 LINEAR FUNCTIONS AND PLANES

Section 2.4 we may consider this as the "rule" for a function f whose domain is $\mathbf{R} \times \mathbf{R}$ and whose codomain is $\mathbf{R}$. We shall show that the graph of this function

$$\{(x, y, z) : z = Ax + By + D\}$$

is a plane in three-dimensional coordinate space.

Let L and L_2 be two intersecting lines in space. The *plane* determined by L and L_2 consists of all those points that lie on some line which is parallel to L_2 and intersects L_1. We use this as the definition of a plane. Figure 11.4 illustrates the role of the lines L_1 and L_2 in defining a plane.

Now we prove that the graph

$$\{(x, y, z) : z = Ax + By + D\}$$

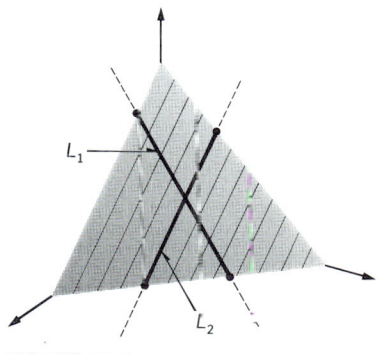

FIGURE 11.4

of a linear function is a plane. Consider the three points on this graph

$$(0, 0, D), (1, 0, A + D), (0, 1, B + D)$$

Let L_1 be the line through $(0, 0, D)$ and $(1, 0, A + D)$. This line has direction numbers $(1, 0, A)$; therefore we may write the parametric equations of L_1 in the form

$$x = s, y = 0, z = D + As$$

Next we let L_2 be the line through $(0, 0, D)$ and $(0, 1, B + D)$ which has direction numbers $(0, 1, B)$. Thus the parametric equations of L_2 have the form

$$x = 0, y = t, z = D + Bt$$

If $(x_0, y_0, Ax_0 + By_0 + D)$ is any point on the graph

$$\{(x, y, z) : z = Ax + By + D\}$$

then the line through this point and parallel to L_2 has parametric equations

$$x = x_0, y = y_0 + u, z = Ax_0 + By_0 + D + Bu$$

By taking $u = -y_0$, we see that the point $(x_0, 0, Ax_0 + D)$ is on this line. But this point is also on L, as we can see by taking $s = x_0$ in the parametric equations of L_1. This proves that every point of the graph

$$\{(x, y, z) : z = Ax + By + D\}$$

lies on a line parallel to L_2 and intersecting L_1. Hence by our definition this graph lies in the plane determined by L_1 and L_2.

In the case where the domain of the linear function $t(x, y) = Ax + By + D$ is the entire set $\mathbf{R} \times \mathbf{R}$, the converse of the last sentence is also true. We leave it as a problem for the reader to complete the proof of the following result.

■ **THEOREM 11.3.** If the domain of the linear function $t(x,y) = Ax + By + D$ is $\mathbf{R} \times \mathbf{R}$, then the graph $\{(x,y,z) : z = f(x,y)\}$ is a plane in $\mathbf{R} \times \mathbf{R} \times \mathbf{R}$.

More generally, we can prove that the set

$$\{(x, y, z) : Ax + By + Cz + D = 0; A^2 + B^2 + C^2 \neq 0\}$$

is a plane in space. (Recall that the condition $A^2 + B^2 + C^2 \neq 0$ merely says that not all three coefficients A, B, and C can be zero.) If $C \neq 0$, then we may solve the equation $Ax + By + Cz + D = 0$ for z in terms of x and y. Then we shall have a case covered by Theorem 11.3 above. The remaining cases are treated as follows. If $C = 0$, assume that $A \neq 0$. Now the equation $Ax + By + D = 0$ is that of a line in the xy coordinate plane. Let this line be L_1. Now in the xz coordinate plane, the equation $Ax + D = 0$ is that of a line parallel to the z axis. Let this line be L_2 and note that L_2 has direction numbers $(0, 0, 1)$. If (x_0, y_0, z_0) is any point on the graph $\{(x, y, z) : Ax + By + D = 0\}$, then this point obviously lies on a vertical line having direction numbers $(0, 0, 1)$ and passing through the point $(x_0, y_0, 0)$, which is on L_1. This completes the proof in this case, and the other cases are analogous.

In summary, note that the plane

$$\{(x, y, z) : Ax + By + D = 0\}$$

is parallel to the z axis. Identical arguments easily show that the plane $\{(x, y, z) : Ax + Cz + D = 0\}$ is parallel to the y axis, while $\{(x, y, z) : By + Cz + D = 0\}$ is parallel to the x axis. (See Fig. 11.5.)

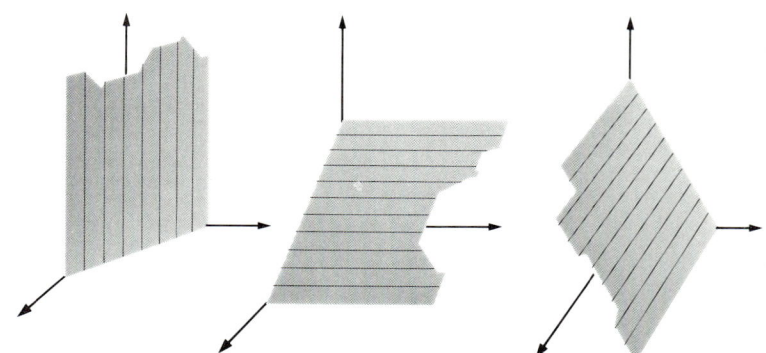

FIGURE 11.5

We shall need an analytic description of the plane that passes through three noncollinear points (x_0, y_0, z_0), (x_1, y_1, z_1) and (x_2, y_2, z_2). A set of parametric equations is given here, and another derivation is called for in Problem 11.2.1. Let L_1 be the line through the points (x_0, y_0, z_0) and (x_1, y_1, z_1); let L_2 be the line through the points (x_0, y_0, z_0) and (x_2, y_2, z_2).

11.2 LINEAR FUNCTIONS AND PLANES

Having assumed that the three points are not collinear, we know that L_1 and L_2 are distinct lines.

Now let (x, y, z) be a point on the plane determined by L_1 and L_2. The line through (x, y, z) parallel to L_2 meets L_1 in some point that has coordinates

$$(x_0 + (x_1 - x_0)s, y_0 + (y_1 - y_0)s, z_0 + (z_1 - z_0)s)$$

for some real number s. (See Fig 11.5.) The line through the point here on L_1, which is parallel to L_2, has parametric equations

$$x = \{x_0 + (x_1 - x_0)s\} + (x_2 - x_0)t = (1 - s - t)x_0 + sx_1 + tx_2$$
$$y = \{y_0 + (y_1 - y_0)s\} + (y_2 - y_0)t = (1 - s - t)y_0 + sy_1 + ty_2$$
$$z = \{z_0 + (z_1 - z_0)s\} + (z_2 - z_0)t = (1 - s - t)z_0 + sz_1 + tz_2$$

Therefore each point (x, y, z) on the plane determines a unique pair (s, t) of parameters. The converse may be proved easily by the reader, thus completing a proof of the following result.

■ **THEOREM 11.4.** The plane through the three noncollinear points (x_0, y_0, z_0), (x_1, y_1, z_1), and (x_2, y_2, z_2) has parametric equations

$$x = (1 - s - t)x_0 + sx_1 + tx_2$$
$$y = (1 - s - t)y_0 + sy_1 + ty_2$$
$$z = (1 - s - t)z_0 + sz_1 + tz_2$$

The equations in Theorem 11.4 take on a more concise form when written as one vector equation. If $\mathbf{p}_i = x_i\mathbf{u} + y_i\mathbf{v} + z_i\mathbf{w}$, $i = 0, 1, 2$, are the position vectors of the three fixed points, then the position vector $\mathbf{p}$ of a point on the plane satisfies the vector equation

$$\mathbf{p} = \mathbf{p}_0 + s(\mathbf{p}_1 - \mathbf{p}_0) + t(\mathbf{p}_2 - \mathbf{p}_0)$$

A *linear function of n real variables* has the form

$$f(x_1, x_2, \cdots, x_n) = A_1 x_1 + \cdots + A_n x_n = \sum_{i=1}^{n} A_i x_i$$

Its graph is the set of $(n + 1)$-dimensional numerical vectors

$$\left\{ (x_1, \cdots, x_n, x_{n+1}) : x_{n+1} = \sum_{i=1}^{n} A_i x_i \right\}$$

The general linear equation in $n + 1$ variables is

$$\sum_{i=1}^{n+1} A_i x_i = C$$

and the set

$$\left\{ (x_1, x_2, \cdots, x_n, x_{n+1}) : \sum_{i=1}^{n+1} A_i x_i = C \right\}$$

is called a *hyperplane in* $(n + 1)$-*dimensional coordinate space*. The "geometry" of such sets can be discussed analytically even though we cannot envision higher dimensional spaces.

The following result will be of use later.

■ **THEOREM 11.5.** Let $\{(x, y, z) : z - z_0 = A(x - x_0) + B(y - y_0)\}$ be a plane and $\{(x, y, z) : x = x_0 + at, y = y_0 + bt, z = z_0 + ct\}$ be a line, both through the point (x_0, y_0, z_0). Then the line lies in the plane if and only if $C = Aa + Bb$.

PROOF: A point $(x_0 + at, y_0 + bt, z_0 + ct)$, $t \neq 0$, satisfies the equation of the plane,

$$z_0 + ct - z_0 = A(x_0 + at - x_0) + B(y_0 + bt - y_0)$$

or $ct = Aat + Bbt$ if and only if $C = Aa + Bb$. ■

EXERCISES 11.2

1. Sketch the graph of each of the linear functions:
    (a) $f(x, y) = x + y + 1$ (b) $f(x, y) = 2x - y - 1$
    (c) $f(x, y) = -x + 3y + 2$ (d) $f(x, y) = -x - 2y + 4$

2. Sketch the following planes:
    (a) $\{(x, y, z) : x + y - z = 1\}$ (b) $\{(x, y, z) : 2x - y + z = 2\}$
    (c) $\{(x, y, z) : -x + y - 2z = 4\}$ (d) $\{(x, y, z) : y + 3z = 6\}$

3. Sketch the plane having parametric equations:
    $x = (1 - s - t) + 2s - t$
    $y = 3(1 - s - t) - s + 2t$
    $z = -(1 - s - t) + s + t$

4. For each triple of points, find parametric equations for the plane containing these points.
    (a) $(1, 1, 2)$, $(3, 0, 1)$, $(-1, -1, 3)$
    (b) $(2, 0, 0)$, $(0, 4, 3)$, $(-1, 1, 1)$
    (c) $(-1, 2, -1)$, $(-1, -1, 2)$, $(1, 1, 1)$
    (d) $(4, 1, -2)$, $(3, -2, 0)$, $(1, 2, 3)$

5. Show that each pair of lines intersects and then determine parametric equations for the plane determined by each pair.
    (a) $\{x, y, z) : x = -1 - 2t, y = 4 + 2t, z = 2 + 3t\}$
    $\{(x, y, z) : x = 4 + 3t, y = 4 + 2t, z = t\}$
    (b) $\{(x, y, z) : x = -5 + 2t, y = -5 + 3t, z = -1 + t\}$
    $\{(x, y, z) : x = 2 + 3t, y = -t, z = -1 - 2t\}$

PROBLEMS 11.2

1. Prove that the equation of the plane through the three noncollinear points (x_0, y_0, z_0), (x_1, y_1, z_1), (x_2, y_2, z_2) can be written in the determinant form

$$\begin{vmatrix} x & y & z & 1 \\ x_0 & y_0 & z_0 & 1 \\ x_1 & y_1 & z_1 & 1 \\ x_2 & y_2 & z_2 & 1 \end{vmatrix} = 0$$

2. If a plane meets the three coordinate axes in the points $(a, 0, 0), (0, b, 0), (0, 0, c)$ where $abc \neq 0$, then show that its equation can be written in the *intercept form*

$$\frac{x}{a} + \frac{y}{b} + \frac{z}{c} = 1$$

3. Given a line and a point not on the line, determine the equation of the plane containing the given line and the given point.

4. Use the result of Problem 3 to obtain an analytic description of the plane containing the given line and points in each of the following cases:
   (a) $\{(x,y,z) : x = t, y = 1 + t, z = 3 + 2t\}$; $(3, -1, 1)$
   (b) $\{(x,y,z) : x = 2 - 3t, y = 4 + t, z = -2 + t\}$; $(0, 0, 1)$
   (c) $\{(x,y,z) : x = -1 + t, y = 2 - t, z = 1 - 2t\}$; $(1, 1, 2)$
   (d) $\{(x,y,z) : x = 5 + 3t, y = -t, z = -5 - 2t\}$; $(2, 1, -1)$

11.3 REGIONS

Our eventual purpose is to give a unified treatment of functions of n variables, where n is any natural number. To attain this goal with precision requires some new concepts which are introduced here. Our definitions are stated for the set $\mathbf{R}^n$ of all n-dimensional numerical vectors. We think of $\mathbf{R}^1 = \mathbf{R}$ as a number line, of $\mathbf{R}^2 = \mathbf{R} \times \mathbf{R}$ as a coordinate plane, and so forth, and hence we shall speak of the "points" in these sets.

Let $(a_1, a_2, \cdots, a_n) = \mathbf{a}$ be a fixed point in $\mathbf{R}^n$. A *basis neighborhood* $U(\mathbf{a}; \delta)$ of this point is a set of the form

$$U(\mathbf{a}; \delta) = \{(x_1, x_2, \cdots, x_n) : |x_i - a_i| < \delta, i = 1, 2, \cdots, n\}$$

where δ is a positive real number. For $n = 1$, such a basis neighborhood is an open interval $U(a; \delta) = \{x : |x - a| < \delta\}$. For $n = 2$, a basis neighborhood is pictured as the interior of a square

$$U((a, b); \delta) = \{(x, y) : |x - a| < \delta, |y - b| < \delta\}$$

For $n = 3$, $U((a, b, c); \delta)$ is seen as a cubical region in space. These three cases were shown in Fig. 11.6. The boundary lines are dashed to remind us that the boundary is *not* part of the basis neighborhood.

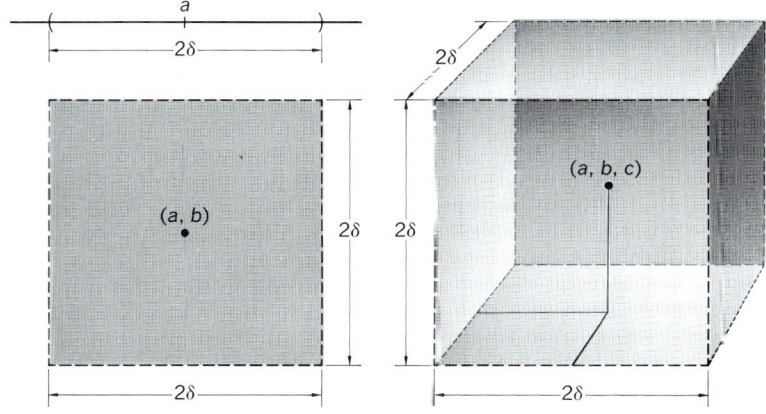

FIGURE 11.6

A subset U of $\mathbf{R}^n$ is *open* if each point of U has a basis neighborhood that lies entirely in U. For technical purposes to be seen later, we also consider the empty set ϕ to be open.

- *Examples of sets that are open*
    1. $\mathbf{R}^n$ itself is open. The proof is trivial.
    2. The rectangular region $R = \{(x,y) : -a < x < a, -b < y < b\}$ is open in $\mathbf{R}^2$. A proof can be given by looking at a fixed point (x_0, y_0) in this region and choosing δ to be the minimum distance from (x_0, y_0) to the boundary of R. Details can be completed by the reader.
    3. The set $\{(x,y) : x^2 + y^2 < 1\}$ is called the *open unit disk* in $\mathbf{R}^2$. In general, the set $\{(x_1, x_2, \cdots, x_n) : \Sigma_{i=1}^{n} x_i^2 < 1\}$ is the open unit disk in $\mathbf{R}^n$. These sets are open in the sense of our definition. The proof of this requires only elementary manipulation of the distance formula but we will not carry it out.
    4. The set $\{(x,y) : xy > 1\}$ is open in $\mathbf{R}^2$. We indicate a geometric argument in Fig. 11.7 where several points of this set are shown in typical basis neighborhoods.

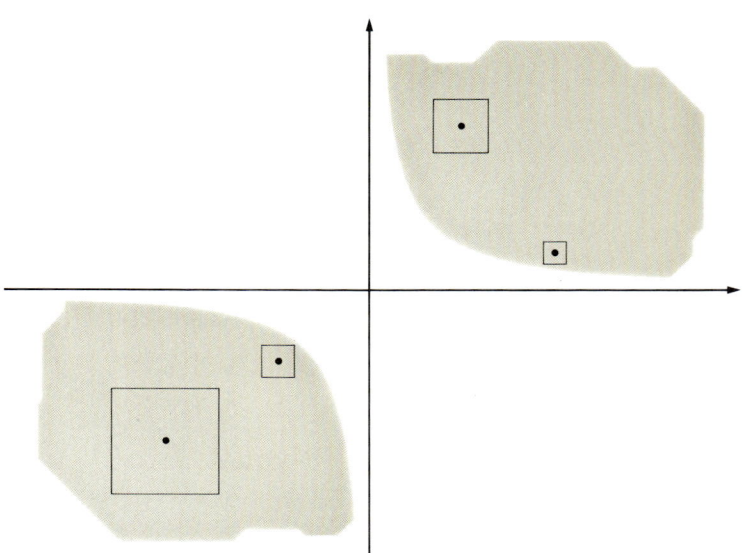

FIGURE 11.7

- *Examples of sets that are not open*
    5. A single point P in $\mathbf{R}^n$ contains no basis neighborhood and hence is not open. Notice, however, that the complement $\mathbf{R}^n - P$ of the point is open.

6. Let L be a line in $\mathbf{R}^n$, $n > 1$. Then L is not an open set but its complement $\mathbf{R}^n - L$ is open.

7. The half-open interval $\{x : a < x \leq b\}$ in $\mathbf{R}^1$ is not open and neither is its complement. The trouble here lies in the fact that no open interval containing the boundary point b lies entirely within the half-open interval. A similar comment applies to the boundary point a of the complement.

8. The *closed unit disk*
$$\left\{(x_1, x_2, \cdots, x_n) : \sum_{i=1}^{n} x_i^2 \leq 1\right\}$$
in $\mathbf{R}^n$ is not open but its complement is. The reader can supply the proof of this statement.

A subset C of $\mathbf{R}^n$ is *closed* if and only if its complement $\mathbf{R}^n - C$ is open. Note, however, that in view of Example 7, there are subsets that are neither closed nor open. Examples 5, 6, and 8 are of closed subsets.

Let U be an open set in $\mathbf{R}^n$. A point $(b_1, b_2, \cdots, b_n)$ is a *boundary point* of U if each basis neighborhood of this point contains both points of U and points of the complement of U. The *boundary* of U consists of all of the boundary points of U. Finally, a *region* in $\mathbf{R}^n$ is an open set or an open set augmented by some part, perhaps all, of its boundary. A point in the region that is not a boundary point is called an *interior point* of the region. If the region is an open set, then each of its points is an interior point.

■ *Examples and counterexamples*

9. The closed upper half-plane $\{(x, y) : y \geq 0\}$ is a region in $\mathbf{R}^2$ whose boundary is the x axis. Any point (x, y) with $y > 0$ is an interior point. On the other hand, if we augment the upper half-plane $\{(x, y) : y > 0\}$ with the negative y axis as shown in Fig. 11.8, we do not have a region. No point of the negative y axis satisfies the definition of being a boundary point of the open set $\{(x, y) : y > 0\}$.

10. The set $\{(x, y) : xy \geq 1\}$ is a region whose boundary is the curve $\{(x, y) : xy = 1\}$. Any point (x, y) with $xy > 1$ is then an interior point. However, if we augment this set with the line segment from $(-1, 1)$ to $(1, 1)$, we no longer have a region. Figure 11.9 shows a basic neighborhood of the origin that fails to intersect the open set $\{(x, y) : xy > 1\}$. Hence the origin is not a boundary point of the open set.

11. The closed unit disk in $\mathbf{R}^n$ is a region whose boundary is the unit $(n - 1)$-dimensional sphere
$$\left\{(x_1, \cdots, x_n) : \sum_{i=1}^{n} x_i^2 = 1\right\}$$
Adding a hyperplane to this set results in a nonregion. We picture a three-dimensional example in Figure 11.10.

FIGURE 11.8

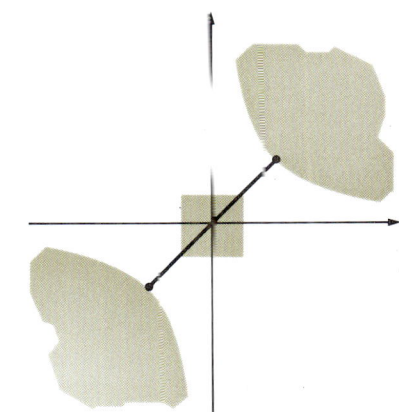

FIGURE 11.9

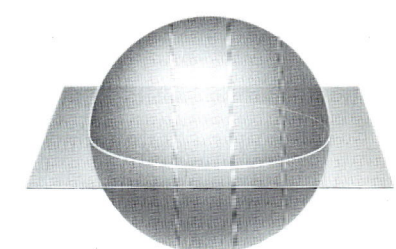

a nonregion

FIGURE 11.10

EXERCISES 11.3

Apply the definition to prove that each subset U in Exercises 1 to 8 is open. This means that, given a point (a, b), you must determine $\delta(a, b) > 0$ such that the basis neighborhood of (a, b) with sides 2δ lies entirely in U.

1. $U = \{(x,y) : y > 0\}$
2. $U = \{(x,y) : x^2 + y^2 < 1\}$
3. $U = \{(x,y) : x^2 + y^2 > 4\}$
4. $U = \{(x,y) : 1 < x^2 + y^2 < 4\}$
5. $U = \{(x,y) : x + 2y - 3 > 0\}$
6. $U = \{(x,y) : 3x - y < 6\}$
7. $U = \{(x,y) : xy > 1\}$
8. $U = \{(x,y) : 3y - 1 < \frac{1}{2}x^2\}$

As Exercises 9 to 16, determine the boundaries of the open sets in Exercises 1 to 8.

PROBLEMS 11.3

1. Prove the following theorems:

   THEOREM A. Consider any collection of open subsets of $\mathbf{R}^n$. The union of the sets in this collection is open. If the collection is finite, then the intersection of the sets in the collection is also open.

   THEOREM B. Consider any collection of closed subsets of $\mathbf{R}^n$. The intersection of the sets in this collection is closed. If the collection is finite, then the union of the sets in the collection is also closed.

2. (a) Find a collection of infinitely many open subsets of $\mathbf{R}^1$ whose intersection is not open.
   (b) Find a collection of infinitely many closed subsets of $\mathbf{R}^1$ whose union is not closed.

3. Let X be any subset of $\mathbf{R}^n$. Our definition applies to define the boundary points of X. Find a subset X which is its own boundary. Are there any sets that have no boundary points?

4. Prove that, if C is a closed subset of $\mathbf{R}^n$, then C contains its boundary.

   (Note: These problems constitute the beginning of a study of the *topology of Euclidean spaces*. The reader may wish to consult a textbook on topology in order to pursue this subject further.)

11.4 FUNCTIONS OF SEVERAL VARIABLES

Let X be a subset of $\mathbf{R}^n$. A real-valued function

$$f : X \to \mathbf{R}$$

whose domain is X will be called a function of n (real) variables. In calculus the domain X is most often a region rather than some arbitrary subset.

As we discussed in Section 2.4, a function f is usually described just by giving a formula for its functional values. Thus we write such equations as

11.4 FUNCTIONS OF SEVERAL VARIABLES

$$f(x,y) = 4 - x^2 - y^2$$
$$g(x,y,z) = x^2 - 3xy + yz - z^2$$

and

$$h(w,x,y,z) = 3w - 2x + y - 5z + 7$$

and refer to them as functions. If we are just given the formula for the values of a function, we are faced with the problem of determining its domain and range.

■ **Example 1.** The formulas for the functions f, g, and h do not contain built-in restrictions on the domains. Therefore, we would assume that f has domain $\mathbf{R}^2$, g has domain $\mathbf{R}^3$, and h has domain $\mathbf{R}^4$.

■ **Example 2.** Suppose we are given

$$f(x,y) = \sqrt{4 - x^2 - y^2}$$

and

$$g(x,y) = \ln(4 - x^2 - y^2)$$

In the formula for f, we recognize that $4 - x^2 - y^2$ must not be negative. Hence the domain of f is the closed subset

$$\{(x,y) : x^2 + y^2 \leq 4\}$$

For the function g we must have a positive number $4 - x^2 - y^2$. Hence the domain of g is the open set

$$\{(x,y) : x^2 + y^2 < 4\}$$

■ **Example 3.** Given the formula

$$f(x,y) = \frac{4}{4 - x^2 - y^2}$$

it is obvious that the domain of f must exclude those values of x and y for which $4 - x^2 - y^2 = 0$. Hence the domain of f is all of $\mathbf{R}^2$, except the circle $\{(x,y) : x^2 + y^2 = 4\}$.

We leave it to the reader to determine the range of each function in the above examples. ■

Given a function $f(x_1, y_2, \cdots, x_n)$ of n variables with the domain X, the subset

$$\{(x_1, \cdots, x_n, x_{n+1}) : (x_1, \cdots, x_n) \in X, x_{n+1} = f(x_1, \cdots x_n)\}$$

of $\mathbf{R}^{n+1}$ is called the *graph* of the function. We cannot draw more than three dimensions and therefore we only draw graphs of functions of two variables. A graph

$$\{(x,y,z) : z = f(x,y)\}$$

is pictured as a surface in three-dimensional coordinate space. From the fact that a function is single-valued, it follows that any line parallel to the z axis meets the graph in at most one point.

We next discuss some techniques for sketching the graph of a function of two variables.

Symmetries. Consider two points (x_1, y_1, z_1) and (x_2, y_2, z_2). These points are

1. symmetrically located across the yz coordinate plane if
$$x_1 = -x_2, y_1 = y_2, z_1 = z_2$$

2. symmetrically located across the xz coordinate plane if
$$x_1 = x_2, y_1 = -y_2, z_1 = z_2$$

3. symmetrically located with respect to the z axis if
$$x_1 = -x_2, y_1 = -y_2, z_1 = z_2$$

Figure 11.11 shows these symmetries.

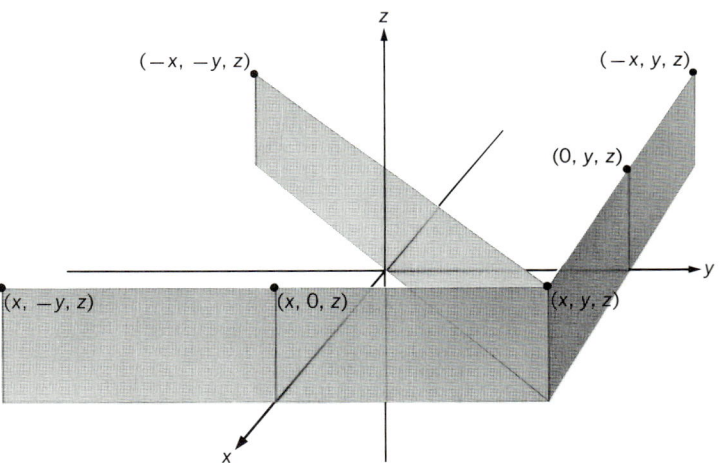

FIGURE 11.11

From the above definitions the following observations are easily made. The graph $\{(x, y, z) : z = f(x, y)\}$ is

1. *symmetric to the yz coordinate plane* if
$$f(-x, y) = f(x, y)$$

2. *symmetric to the xz coordinate plane* if
$$f(x, -y) = f(x, y)$$

3. *symmetric to the z axis* if
$$f(-x, -y) = f(x, y)$$

If the graph is symmetric to both coordinate planes, then it is also symmetric to the z axis.

■ **Example 4.** The graph $\{(x, y, z) : z = 4 - x^2 - y^2\}$ obviously has all three of the symmetries described because
$$4 - (-x)^2 - y^2 = 4 - x^2 - (-y)^2 = 4 - x^2 - y^2$$

■ **Example 5.** The graph $\{(x, y, z) : z = 4 - x^2 - xy\}$ is symmetric to the z axis because $4 - (-x)^2 - (-x)(-y) = 4 - x^2 - xy$. ■

Intercepts. The graph $\{(x, y, z) : z = f(x, y)\}$ has only one intercept on the z axis, namely, the point $(0, 0, f(0, 0))$. Since the x axis is precisely the set $\{(x, y, z) : y = 0\}$, the x *intercepts* of the graph are at those points $(x, 0, 0)$ such that $f(x, 0) = 0$. Similarly, the y *intercepts* are those points $(0, y, 0)$ such that $f(0, y) = 0$.

■ **Example 6.** The graph $\{(x, y, z) : z = 4 - x^2 - y^2\}$ has the single z intercept $(0, 0, 4)$, while the x intercepts are the points $(\pm 2, 0, 0)$ and the y intercepts are $(0, \pm 2, 0)$. ■

Traces. The graph $\{(x, y, z) : z = f(x, y)\}$ may intersect one or more of the coordinate planes. Such an intersection is usually a planar curve, called a *trace* of the graph. The equations of these curves are readily seen to be as follows.

1. The *trace in the xy coordinate plane* is the set
$$\{(x, y) : f(x, y) = 0\}.$$
2. The *trace in the xz coordinate plane* is the set
$$\{(x, z) : z = f(x, 0)\}.$$
3. The *trace in the yz coordinate plane* is the set
$$\{(y, z) : z = f(0, y)\}.$$

■ **Example 7.** The graph $\{(x, y, z) : z = 4 - x^2 - y^2\}$ has traces determined by setting each variable separately equal to zero. Thus, the xy trace is $\{(x, y) : 4 - x^2 - y^2 = 0\}$, which is a circle; the xz trace is $\{(x, z) : z = 4 - x^2\}$. These curves are sketched on the proper coordinate planes in Fig. 11.12. Note that the surface is quite easy to envision when its traces have been drawn. ■

Other sections. At times it is necessary to determine cross sections other than the traces in order to better understand the graph
$$\{(x, y, z) : z = f(x, y)\}.$$

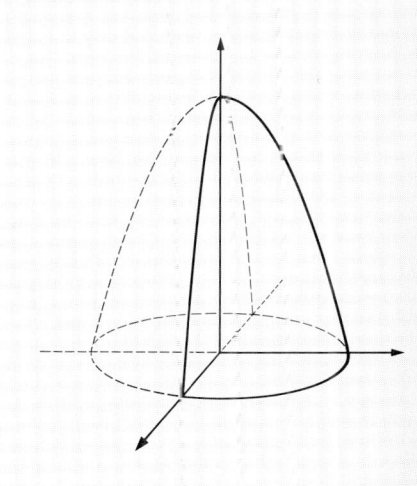

FIGURE 11.12

Cross sections of the surface by planes parallel to the xy coordinate plane have the equations

$$c = f(x, y)$$

where c is a constant. Similarly, the curve with equation $z = f(x, b)$ is the cross section in the plane $\{(x, y, z) : y = b\}$ and the curve with the equation $z = f(a, y)$ is the cross section in the plane $\{(x, y, z) : x = a\}$.

■ *Example 8.* Consider the function

$$f(x, y) = 4y^2 - x^2$$

It is obvious that the origin $(0, 0, 0)$ is the only intercept on any axis. Furthermore, because

$$4(-y)^2 - x^2 = 4y^2 - (-x)^2 = 4(-y)^2 - (-x)^2 = 4y^2 - x^2$$

this function has all three of the symmetries discussed.

The traces of the graph $\{(x, y, z) : z = 4y^2 - x^2\}$ are as follows.

1. The xy trace has the equation $4y^2 - x^2 = 0$ or $y = \pm\frac{1}{2}x$. In short, the xy trace $\{(x, y) : y = \frac{1}{2}x\} \cup \{(x, y) : y = -\frac{1}{2}x\}$ is the union of two lines intersecting at the origin.
2. The xz trace has the equation $z = -x^2$ and hence is the parabola

$$\{(x, z) : z = -x^2\}$$

3. The yz trace is another parabola $\{(y, z) : z = 4y^2\}$.

It is easy to see that cross sections of the graph $\{(x, y, z) : z = 4y^2 - x^2\}$ parallel to either the xz coordinate plane or the yz coordinate plane are also parabolas. The cross sections parallel to the xy coordinate plane are of the form

$$\{(x, y, z) : 4y^2 - x^2 = C\}$$

These curves are called *hyperbolas*. The entire graph is called a *hyperbolic paraboloid* and is illustrated in Fig. 11.14 (page 519).

There is another means of visualizing a function of two variables, the method of level curves. A cross-sectional curve

$$\{(x, y, z) : z = c, f(x, y) = c\}$$

when projected down on the xy coordinate plane as the curve

$$\{(x, y, z) : f(x, y) = c, z = 0\}$$

is called a *level curve*. Drawing several different level curves gives a picture of the behavior of the function f. This is precisely the technique employed in making contour maps and pressure maps, for instance. Figure 11.13 shows the correspondence between a set of level curves and a surface.

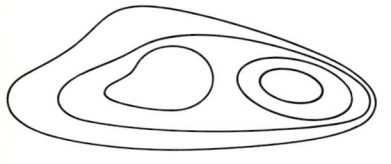

level curves

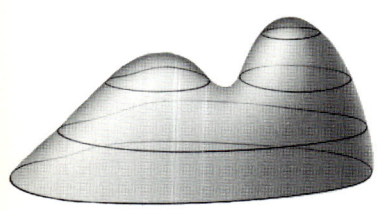

surface

FIGURE 11.13

11.4 FUNCTIONS OF SEVERAL VARIABLES

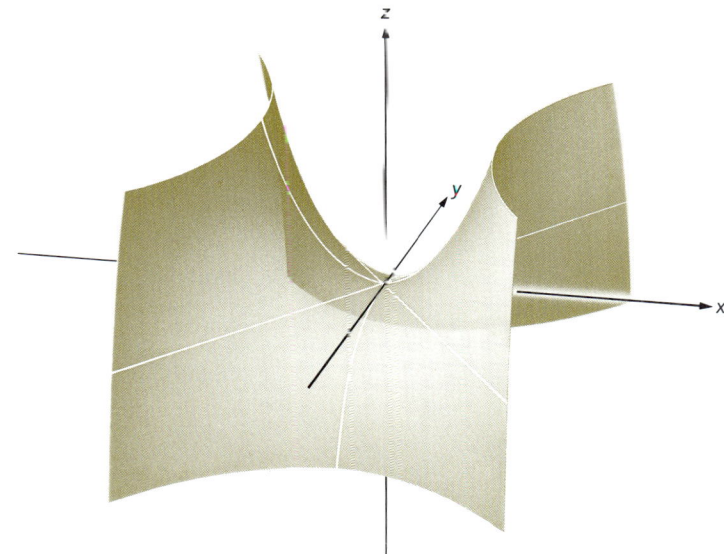

FIGURE 11.14

A graph of the form $\{(x, y, z) : z = f(x)\}$ is a *cylinder* parallel to the y axis. If $(x_0, 0, f(x_0))$ is a point on this graph, then the line $\{(x_0, y, f(x_0))\}$ parallel to the y axis lies on the graph. Similarly, a graph of the form $\{(x, y, z) : z = q(y)\}$ is a cylinder parallel to the x axis. We mention these facts because they are of use shortly.

EXERCISES 11.4

Determine the domain and range of each function in Exercises 1 to 10.

1. $f(x, y) = \dfrac{x - y}{x + y}$
2. $f(x, y) = \sqrt{\dfrac{x - y}{x + y}}$
3. $f(x, y) = \sqrt{xy}$
4. $f(x, y) = \ln xy$
5. $f(x, y) = \tan^{-1}\left(\dfrac{x - y}{x + y}\right)$
6. $f(x, y) = \ln(x - \tan y)$
7. $f(x, y, z) = \dfrac{x}{y} + \dfrac{y}{z} + \dfrac{z}{x}$
8. $f(x, y, z) = \sqrt{xyz}$
9. $f(x, y, z) = x^2 + y \sin^{-1} z$
10. $f(x, y, z) = \tan^{-1}\left(\dfrac{x}{y}\right) + \sec^{-1}\left(\dfrac{y}{z}\right)$

Discuss completely and then sketch the graph of each function in Exercises 11 to 24. Also draw level curves to obtain another picture.

11. $f(x, y) = x + 3y - 1$
12. $f(x, y) = xy$
13. $f(x, y) = (x - y)^2$
14. $f(x, y) = x^2 + y^2 - 2x + 4y - 1$
15. $f(x, y) = \dfrac{x}{x + y}$
16. $f(x, y) = \sqrt{4 - x^2 - y^2}$

17. $f(x,y) = \sqrt{\dfrac{x}{x+y}}$ 18. $f(x,y) = \sin(x^2 + y^2)$

19. $f(x,y) = \sin^{-1}(x^2 + y^2)$ 20. $f(x,y) = \tan^{-1}\left(\dfrac{y}{x}\right)$

21. $f(x,y) = e^{x+y}$ 22. $f(x,y) = x - x^2$

23. $f(x,y) = y^2 - 2$ 24. $f(x,y) = \sin x$

11.5 LIMITS AND CONTINUITY

Let $\mathbf{a} = (a_1, a_2, \cdots, a_n)$ be a fixed point either in the domain of a function or on the boundary of this domain. Then there are points $\mathbf{x} = (x_1, x_2, \cdots, x_n)$ of the domain that are as "close" to $\mathbf{a}$ as desired. We define the statement that $f(\mathbf{x})$ has a limit L as $\mathbf{x}$ approaches the fixed point $\mathbf{a}$; symbolically,

$$\lim_{\mathbf{x} \to \mathbf{a}} f(\mathbf{x}) = L$$

The meaning should be quite clear, of course. We mean that the difference $|f(\mathbf{x}) - L|$ can be made as small as desired by restricting the point $\mathbf{x}$ to lie "near" $\mathbf{a}$. The idea of "near" or "close" depends upon using basis neighborhoods.

■ **DEFINITION.** The function $f(\mathbf{x}) = f(x_1, x_2, \cdots, x_n)$ has the limit L as $\mathbf{x} = (x_1, x_2, \cdots, x_n)$ approaches $\mathbf{a} = (a_1, a_2, \cdots, a_n)$ if, given $\epsilon > 0$, there exists $\delta > 0$ such that

$$|f(x_1, x_2, \cdots, x_n) - L| < \epsilon$$

for each point $(x_1, x_2, \cdots, x_n)$, which is in the domain of f and in the basis neighborhood $U(\mathbf{a}; \delta)$ of $\mathbf{a}$.

■ *Example 1.* We claim that

$$\lim_{(x,y) \to (1,1)} (x^2 + y^2) = 2$$

To prove it, we simply write

$$|x^2 + y^2 - 2| = |x^2 - 1 + y^2 - 1| \leq |x^2 + 1| + |y^2 - 1|$$
$$= |x - 1| \cdot |x + 1| + |y - 1| \cdot |y + 1|$$

This convinces us that we can control the size of $|x^2 + y^2 - 2|$ by restricting the size of both $|x - 1|$ and $|y - 1|$. The reader can carry out the details, choosing $\delta = \tfrac{1}{4}\epsilon$ or 1, whichever is smaller. ■

As applied to functions of several variables, the definition of a limit is more restrictive than it is for functions of one variable. In discussing the one-variable case, it is quite adequate to think intuitively of the variable point x as approaching the fixed point a by sliding along the x axis. Even

11.5 LIMITS AND CONTINUITY

in the two-variable case, however, the variable point (x, y) can approach the fixed point (a, b) from any direction along any curve. Actually, the definition says nothing about a point "approaching" some fixed point. We can recover the intuitive picture by introducing a new idea.

Let g be a function of one variable whose graph $\{(x, y) : y = g(x)\}$ lies in the domain of the function $f(x, y)$, and suppose that $\lim_{x \to a} g(x) = b$. Then the limit

$$\lim_{x \to a} f(x, g(x))$$

if it exists, is called the *limit along the curve* $\{(x, y) : y = g(x)\}$. In particular

$$\lim_{y \to a} f(x, b + m(x - a))$$

is the *limit along the line* $\{(x, y) : y = b + m(x - a)\}$. We point out that, if

$$\lim_{(x,y) \to (a,b)} f(x, y) = L,$$

then the limit of $f(x, y)$ along any curve to the point (a, b) exists and equals L.

■ **Example 2.** Let $f(x, y) = 2xy/(x^2 + y^2)$. The function is defined except at $(0, 0)$. The limit of $f(x, y)$ at $(0, 0)$ along the line $\{(x, y) \: y = x\}$ is

$$\lim_{x \to 0} \frac{2x(x)}{x^2 + (x)^2} = \lim_{x \to 0} 1 = 1$$

However, along the line $y = 2x$, the limit at $(0, 0)$ is

$$\lim_{x \to 0} \frac{2x(2x)}{x^2 + (2x)^2} = \lim_{x \to 0} \frac{4x^2}{5x^2} = \lim_{x \to 0} \frac{4}{5} = \frac{4}{5}$$

Note, however, that along the curve $\{(x, y) : y = x^2\}$ the limit is

$$\lim_{x \to 0} \frac{2x(x^2)}{x^2 + (x^2)^2} = \lim_{x \to 0} \frac{2x^3}{x^2 + x^4} = \lim_{x \to 0} \frac{2x}{1 + x^2} = 0$$

From the remark preceding this example, it follows that there can be no limit of $f(x, y)$ at $(0, 0)$. We return to this example again.

■ **Example 3.** A specific path along which the point (x, y) can be allowed to approach (a, b) is as follows: First let (x, y) approach (x, b) and then allow (x, b) to approach (a, b). This results in the so-called *iterated limit*:

$$\lim_{x \to a} (\lim_{y \to b} f(x, y))$$

The other iterated limit

$$\lim_{y \to b} (\lim_{x \to a} f(x, y))$$

is defined similarly. For instance,

$$\lim_{x \to 0} \left(\lim_{y \to 0} \frac{x-y}{x+y} \right) = \lim_{x \to 0} \left(\frac{x}{x} \right) = 1$$

while

$$\lim_{y \to 0} \left(\lim_{x \to 0} \frac{x-y}{x+y} \right) = \lim_{y \to 0} \left(\frac{-y}{y} \right) = -1$$

Again we point out that, if these iterated limits differ, as in the instance above, then there is no limit at (a, b).

■ **DEFINITION.** Let $f : X \to \mathbf{R}$ be a function of a function of n variables and let $\mathbf{a} = (a_1, a_2, \cdots, a_n)$ be a point in the domain X of f. Then f is *continuous at the point* $\mathbf{a}$ if

$$\lim_{x \to a} f(\mathbf{x}) = f(\mathbf{a})$$

Explicitly, this means that, given any $\epsilon > 0$, we can find $\delta > 0$ (which, in general, will depend upon both ϵ and the point $\mathbf{a}$) such that $|f(\mathbf{x}) - f(\mathbf{a})| < \epsilon$ whenever $\mathbf{x}$ is both in the domain of f and in the basis neighborhood $U(\mathbf{a}; \delta)$ of $\mathbf{a}$. As was emphasized before, the above equation defining continuity means both (1) that the limit of $f(\mathbf{x})$ exists and (2) that the limit equals the functional value. Finally, if f is continuous at each point of its domain X, then we say that f is *continuous* on X.

We should mention that continuity as defined here implicitly includes the cases of one-sided continuity for functions of a single variable. This was done deliberately.

■ ***Example 4.*** The function $f(x, y) = x^2 + y^2$ is continuous at the point $(1, 1)$ because

$$\lim_{(x,y) \to (1,1)} x^2 + y^2 = 2 = 1^2 + 1^2$$

(Example 1). Furthermore, it is easy to prove that this function is continuous on $\mathbf{R}^2$.

■ ***Example 5.*** The function defined by

$$f(x, y) = \frac{2xy}{x^2 + y^2}, (x, y) \neq (0, 0)$$
$$= 0, (x, y) = (0, 0)$$

is not continuous at $(0, 0)$ because it has no limit at $(0, 0)$ (Example 2). It is continuous at any point $(x, y) \neq (0, 0)$; the reader may prove this. ■

The proofs of the following results depend upon limit theorems for functions of several variables. We do not prove these theorems but the reader may wish to do so. The details are practically identical to those found in Chapters 2 and 3.

- **THEOREM 11.6.** Let the functions f and g have a common domain. If both f and g are continuous at the point $\mathbf{a}$, then the functions $f + g, f - g$ and fg are continuous at $\mathbf{a}$. If, in addition, $g(\mathbf{a}) \neq 0$, then f/g is continuous at $\mathbf{a}$.

- **THEOREM 11.7.** Let the functions $g_1(t), g_2(t), \cdots, g_n(t)$ be continuous at t_0 and let $f(x_1, x_2, \cdots, x_n)$ be continuous at the point $(g_1(t_0), g_2(t_0), \cdots, g_n(t_0))$. Then the function $\phi(t) = f(g_1(t), g_2(t), \cdots, g_n(t))$ is continuous at t_0.

- **THEOREM 11.8.** If $f(x_1, x_2, \cdots, x_n)$ is continuous at $\mathbf{a}$ and if $g(t)$ is continuous at $t = f(\mathbf{a})$, then the composite function $(g \circ f)(\mathbf{x}) = g(f(\mathbf{x}))$ is continuous at $\mathbf{a}$.

The reader may state other theorems about compositions of functions of several variables.

EXERCISES 11.5

1. Establish each of the following limits:

   (a) $\displaystyle\lim_{(x,y)\to(2,-1)} \frac{x^2 - y^2}{x - y} = 1$

   (b) $\displaystyle\lim_{(x,y)\to(-2,3)} \frac{xy - y^2}{3 - x} = -3$

   (c) $\displaystyle\lim_{(x,y)\to(0,0)} \sin(x - y) = 0$

   (d) $\displaystyle\lim_{(x,y)\to(1,\frac{1}{2})} \frac{\tan \frac{\pi}{2}(x - y)}{x - y} = 2$

2. Prove that each of the following functions has no limit at the point $(1, -1)$:

   (a) $f(x, y) = \dfrac{x - y}{x + y}$

   (b) $f(x, y) = \dfrac{x + y}{(x - 1)^2}$

   (c) $f(x, y) = x \ln(x + y)$

   (d) $f(x, y) = \dfrac{\sin(x + y)}{(1 + y)^2}$

3. Determine the region on which each of the following functions is continuous

   (a) $f(x) = x\sqrt{4 - x^2 - y^2}$

   (b) $f(x, y) = \dfrac{x}{\sqrt{4 - x^2 - y^2}}$

   (c) $f(x, y) = \cos^{-1}(xy)$

   (d) $f(x, y) = x \ln(xy - 1)$

4. Can the function

   $$f(x, y) = \frac{x^2 - y^2}{x^2 + y^2}, \quad (x, y) \neq (0, 0)$$

   be assigned a value at $(0, 0)$ so as to be continuous everywhere?

PROBLEMS 11.5

1. Which of the following functions is continuous at $(0, 0)$?

   (a) $f(x, y) = \dfrac{x^2 y}{x^3 + y^3}$ if $(x, y) \neq (0, 0)$

              $= 0$ if $(x, y) = (0, 0)$

   (b) $f(x, y) = \dfrac{x^2 y}{x^2 + y^2}$ if $(x, y) \neq (0, 0)$

              $= 0$ if $(x, y) = (0, 0)$

   (c) $f(x, y) = \dfrac{x - y}{|x| + |y|}$ if $(x, y) \neq (0, 0)$

              $= 0$ if $(x, y) = (0, 0)$

   (d) $f(x, y) = \dfrac{xy}{|x| + |y|}$ if $(x, y) \neq (0, 0)$

              $= 0$ if $(x, y) = (0, 0)$

2. Consider the function $f(x, y) = \dfrac{xy^2}{x^2 + y^4}$, $(x, y) \neq (0, 0)$.

   (a) Prove that both iterated limits at $(0, 0)$ exist and are equal.
   (b) Prove that the limit at $(0, 0)$ along any line is zero.
   (c) Prove that $\lim_{(x,y) \to (0,0)} f(x, y)$ fails to exist. (Hint: Consider the curve $\{(x, y) : y = \sqrt{x}\}$.)

3. The function $f(x_1, x_2, \cdots, x_n)$ is said to be *continuous in the first variable*, if for any fixed values $a_2, a_3, \cdots, a_n$, the function of one variable $f(x_1, a_2, a_3, \cdots, a_n)$ is continuous. A similar definition applies to continuity in each of the remaining variables. Find a function $f(x, y)$ that is continuous in each variable separately but is not continuous.

11.6 PARTIAL DERIVATIVES

The partial derivatives of a function of several variables play the same role as does the derivative of a function of one variable. Some complications arise because we deal with higher dimensions. However, these complications are primarily of a technical and not a conceptual nature.

Let $f : X \to \mathbf{R}$ be a function of n variables and let $\mathbf{a} = (a_1, a_2, \cdots, a_n)$ be an interior point of the domain X of f. For each $i = 1, 2, \cdots, n$, the ith *partial derivative* f_i of f is defined at the point $\mathbf{a}$ by

$$f_i(\mathbf{a}) = \lim_{h \to 0} \frac{f(a_1, \cdots, a_{i-1}, a_i + h, \cdots, a_n) - f(a_1, \cdots, a_n)}{h}$$

if this limit exists.

■ **Example 1.** Let $f(x, y) = x^2 - 3xy + y^2$ and $\mathbf{a} = (3, 1)$.

$$f_1(3, 1) = \lim_{h \to 0} \frac{f(3 + h, 1) - f(3, 1)}{h}$$

$$= \lim_{h \to 0} \frac{(3 + h)^2 - 3(3 + h)(1)^2 - (3^2 - 3(3)(1) + 1^2)}{h}$$

$$= \lim_{h \to 0} \frac{3h + h^2}{h} = \lim_{h \to 0} (3 + h) = 3$$

11.6 PARTIAL DERIVATIVES

Also

$$f_2(3, 1) = \lim_{k \to 0} \frac{f(3, 1 + k) - f(3, 1)}{k}$$

$$= \lim_{k \to 0} \frac{3^2 - 3(3)(1 + k)^2 - (3^2 - 3(3)(1) + 1^2)}{k}$$

$$= \lim_{k \to 0} \frac{-7k + k^2}{k} = \lim_{k \to 0} (-7 + k) = -7$$

In general, we compute that the partial derivatives are

$$f_1(x, y) = \lim_{h \to 0} \frac{f(x + h, y) - f(x, y)}{h}$$

$$= \lim_{h \to 0} \frac{(x + h)^2 - 3(x + h)y + y^2 - (x^2 - 3xy + y^2)}{h}$$

$$= \lim_{h \to 0} \frac{2xh - 3yh + h^2}{h} = \lim_{h \to 0} (2x - 3y + h) = 2x - 3y$$

and

$$f_2(x, y) = \lim_{h \to 0} \frac{f(x, y + h) - f(x, y)}{h}$$

$$= \lim_{h \to 0} \frac{x^2 - 3x(y + h) + (y + h)^2 - (x^2 - 3xy + y^2)}{h}$$

$$= \lim_{h \to 0} \frac{-3xh + 2yh + h^2}{h} = \lim_{h \to 0} (-3x + 2y + h)$$

$$= -3x + 2y$$

If we differentiate the function $x^2 - 3yx + y^2$ with respect to x, treating y as a constant, we get $2x - 3y = f_1(x, y)$. Similarly, thinking of x as constant and differentiating with respect to y, we obtain $-3x + 2y = f_2(x, y)$. This is precisely the method of taking partial derivatives. We shall show why this is so for a function of two variables, other cases being analogous.

Let $f(x, y)$ be a function of two variables. For any fixed value y_0 of the second variable, define a function of one variable by setting

$$g(x) = f(x, y_0)$$

Then we compute the value of Dg at $x = x_0$ to be

$$Dg(x_0) = \lim_{h \to 0} \frac{g(x_0 + h) - g(x_0)}{h}$$

$$= \lim_{h \to 0} \frac{f(x_0 + h, y_0) - f(x_0, y_0)}{h} = f_1(x_0, y_0)$$

In this computation, then, we simply treated y_0 as a constant and differentiated with respect to x, as claimed.

■ **Example 2.** Consider the function $f(x,y) = \ln(x^2 + y^2)$, where $(x,y) \neq (0, 0)$. To obtain $f_1(x, y)$, we must multiply $1/(x^2 + y^2)$ by the derivative of $x^2 + y^2$ with respect to x. Since we think of y as a constant, this derivative is just $2x$. Thus we have

$$f_1(x,y) = \frac{2x}{x^2 + y^2}\ \blacksquare$$

As a function of n variables, the partial derivative f_i may also have partial derivatives. We define

$$f_{ij}(x_1, \cdots, x_n)$$
$$= \lim_{h \to 0} \frac{f_i(x_i, \cdots, x_j + h, \cdots, x_n) - f(x_1, \cdots, x_j, \cdots, x_n)}{h}$$

if this limit exists. The functions f_{ij}, $i = 1, 2, \cdots, n$; $j = 1, 2, \cdots, n$ are called the *second partial derivatives* of f. Third and higher order partial derivatives are defined analogously.

■ **Example 3.** Let $f(x,y) = x^3 - 6x^2y + 3y^2$. Then we have

$$f_1(x,y) = 3x^2 - 12xy, f_2(x,y) = -6x^2 + 6y$$

Using the standard method of computing partial derivatives, we get

$$f_{11}(x,y) = 6x - 12y, f_{12}(x,y) = f_{21}(x,y) = -12x, f_{22}(x,y) = 6\ \blacksquare$$

Other notations are often used for partial derivatives, such as the following:

$$f_1(x,y) = D_1 f(x,y) = \frac{\partial}{\partial x} f(x,y)$$

and

$$f_2(x,y) = D_2 f(x,y) = \frac{\partial}{\partial y} f(x,y)$$

Then the second partial derivatives are written as

$$f_{11}(x,y) = D_1\{D_1 f(x,y)\} = D_1^2 f(x,y)$$

or

$$f_{11}(x,y) = \frac{\partial}{\partial x}\left\{\frac{\partial}{\partial x} f(x,y)\right\} = \frac{\partial^2}{\partial x^2} f(x,y)$$

Similarly,

$$f_{12}(x,y) = D_2\{D_1 f(x,y)\} = D_2 D_1 f(x,y)$$

or

$$f_{12}(x,y) = \frac{\partial}{\partial y}\left\{\frac{\partial}{\partial x} f(x,y)\right\} = \frac{\partial^2}{\partial y\,\partial x} f(x,y)$$

11.6 PARTIAL DERIVATIVES

$$f_{21}(x,y) = D_1\{D_2 f(x,y)\} = D_1 D_2 f(x,y)$$
$$= \frac{\partial}{\partial x}\left\{\frac{\partial}{\partial y} f(x,y)\right\} = \frac{\partial^2}{\partial x\, \partial y} f(x,y)$$

and

$$f_{22}(x,y) = D_2\{D_2 f(x,y)\} = D_2{}^2 f(x,y)$$
$$= \frac{\partial}{\partial y}\left\{\frac{\partial}{\partial y} f(x,y)\right\} = \frac{\partial^2}{\partial y^2} f(x,y)$$

The "rounded D" symbol $\frac{\partial}{\partial x} f(x,y)$ can be misleading in certain applications. For instance, what is meant by $\frac{\partial}{\partial x} f(x+y, x-y)$? If we adopt the usual convention that $\partial f/\partial x = f_1$, there should be no difficulty with this notation.

■ **Example 4.** We write the derivatives of $f(x,y) = x^3 y^2 - y^4$ as

$$D_1(x^3 y^2 - y^4) = \frac{\partial}{\partial x}(x^3 y^2 - y^4) = 3x^2 y^2$$

and, for instance,

$$D_1 D_2(x^3 y^2 - y^4) = D_1(2x^3 y - 4y^3) = 6x^2 y$$

or

$$\frac{\partial^2}{\partial x\, \partial y}(x^3 y^2 - y^4) = \frac{\partial}{\partial x}(2x^3 y - 4y^3) = 6x^2 y$$

■ **Example 5.** Let $z = \cos(2x+y) + \sin(2x-y)$. We claim that this function satisfies the equation

$$\frac{\partial^2 z}{\partial x^2} = 4\frac{\partial^2 z}{\partial y^2}$$

or, equivalently,

$$D_1{}^2 z = 4 D_2{}^2 z$$

We first compute

$$\frac{\partial z}{\partial x} = -\sin(2x+y)\frac{\partial}{\partial x}(2x+y) + \cos(2x-y)\frac{\partial}{\partial x}(2x-y)$$
$$= -2\sin(2x+y) + 2\cos(2x-y)$$

and then

$$\frac{\partial^2 z}{\partial y^2} = -4\cos(2x+y) - 4\sin(2x-y)$$

Similarly, the remaining computation is

$$\frac{\partial z}{\partial y} = -\sin(2x+y) \cdot \frac{\partial}{\partial y}(2x+y) + \cos(2x-y) \cdot \frac{\partial}{\partial y}(2x-y)$$
$$= -\sin(2x+y) - \cos(2x-y)$$

and, therefore,

$$\frac{\partial^2 z}{\partial y^2} = -\cos(2x+y) \cdot \frac{\partial}{\partial y}(2x+y) + \sin(2x-y) \cdot \frac{\partial}{\partial y}(2x-y)$$
$$= -\cos(2x+y) - \sin(2x-y)$$

Hence z satisfies the equation. ∎

The reader may have noticed in the examples that

$$f_{12} = f_{21}$$

The following result shows why this equality has been seen.

■ **THEOREM 11.9.** Let f be a function of two variables and suppose that the partial derivatives f_1, f_2, f_{12}, and f_{21} all exist and are continuous on X. Then $f_{12}(x,y) = f_{21}(x,y)$ for any interior point (x,y) of X.

PROOF: The plan of the proof is to show that, in any basis neighborhood $U(x,y;\delta)$, there are points (c_1, d_1) and (c_2, d_2) such that $f_{12}(c_1, d_1) = f_{21}(c_2, d_2)$. Then since δ can be chosen as small as we wish, we use the continuity of f_{12} and f_{21} to prove that $|f_{12}(x,y) - f_{21}(x,y)|$ is as small as is desired. This proves that the two are equal.

The details are straightforward. Let $U(x,y;2\delta)$ be any basis neighborhood of (x,y) lying entirely in X. Then the points $(x+\delta, y), (x, y+\delta)$ and $(x+\delta, y+\delta)$ are also interior points of X. We consider the quantity

$$\Delta = f(x+\delta, y+\delta) - f(x+\delta, y) - f(x, y+\delta) + f(x,y)$$

Defining the function

$$g(s) = f(s, y+\delta) - f(s, y)$$

we obviously have

$$\Delta = g(x+\delta) - g(x)$$

The mean value theorem can be applied to this difference because the derivative

$$\frac{d}{ds}g(s) = f_1(s, y+\delta) - f_1(s, y)$$

is defined for all values of s between x and $x+\delta$. Hence there exists a number c_1, $x < c_1 < x+\delta$, such that

$$\Delta = g(x+\delta) - g(x) = \delta g'(c_1) = \delta\{f_1(c_1, y+\delta) - f_1(c_1, y)\}$$

11.6 PARTIAL DERIVATIVES

Next we define the function
$$h(t) = f_1(c_1, t)$$
so that the previous equation becomes
$$\Delta = \delta\{h(y + \delta) - h(y)\}$$
Again our assumptions are that
$$\frac{d}{dt} h(t) = f_{12}(c_1, t)$$
exist for all t between y and $y + \delta$. Therefore the mean value theorem provides a number d_1 between y and $y + \delta$ such that
$$h(y + \delta) - h(y) = \delta h'(d_1) = \delta f_{12}(c_1, d_1)$$
It follows that
$$\Delta = \delta\{h(y + \delta) - h(y)\} = \delta^2 f_{12}(c_1, d_1)$$
In a completely analogous way we define
$$\phi(t) = f(x + \delta, t) - f(x, t)$$
whence
$$\Delta = \phi(y + \delta) - \phi(y)$$
Because
$$\frac{d}{dt} \phi(t) = f_2(x + \delta, t) - f_2(x, t)$$
exists for all t between y and $y + \delta$, the mean value theorem says there exists a number d_2, $y < d_2 < y + \delta$, such that
$$\Delta = \phi(y + \delta) - \phi(y) = \delta\phi'(d_2) = \delta\{f_2(x + \delta, d_2) - f_2(x, d_2)\}$$
Next we set
$$\psi(s) = f_2(s, d_2)$$
whence
$$\frac{d\psi(s)}{ds} = f_{21}(s, d_2)$$
Once again the mean value theorem comes into play to tell us that there is a number c_2, $x < c_2 < x + \delta$, such that
$$\psi(x + \delta) - \psi(x) = \delta\psi'(c_2) = \delta\{f_{21}(c_2, d_2)\}$$
Thus we have
$$\Delta = \delta\{f_2(x + \delta, d_2) - f_2(x, d_2)\} = \delta^2 f_{21}(c_2, d_2)$$

In these two steps we have shown that

$$\Delta = \delta^2 f_{12}(c_1, d_1) = \delta^2 f_{21}(c_2, d_2)$$

or

$$f_{12}(c_1, d_1) = f_{21}(c_2, d_2)$$

Finally, we use this fact to prove that $f_{12}(x,y) = f_{21}(x,y)$. Given any $\epsilon > 0$, we choose δ from the continuity of f_{12} and f_{21}, so that

$$|f_{12}(x',y') - f_{12}(x,y)| < \frac{\epsilon}{2}$$

$$|f_{21}(x',y') - f_{21}(x,y)| < \frac{\epsilon}{2}$$

whenever (x',y') is in the basis neighborhood $U(x,y;\delta)$. By the argument above, there also exist points (c_1, d_1) and (c_2, d_2) in $U(x,y;\delta)$ such that $f_{12}(c_1, d_1) = f_{21}(c_2, d_2)$. Then we have

$$|f_{12}(x,y) - f_{21}(x,y)|$$
$$= |f_{12}(x,y) - f_{12}(c_1, d_1) + f_{21}(c_2, d_2) - f_{21}(x,y)|$$
$$\leq |f_{12}(x,y) - f_{12}(c_1, d_1)| + |f_{21}(c_2, d_2) - f_{21}(x,y)|$$
$$< \frac{\epsilon}{2} + \frac{\epsilon}{2} = \epsilon$$

Since $f_{12}(x,y)$ and $f_{21}(x,y)$ differ by less than an arbitrary amount, they are equal. ∎

EXERCISES 11.6 Find the first partial derivatives of the functions in Exercises 1 to 40.

1. $f(x,y) = 2x^2 - 3xy$
2. $f(x,y) = 3xy - y^2 + 6x - 3y$
3. $f(x,y) = 2x^2 + 4xy - y^2$
4. $f(x,y) = (2x - y)^3$
5. $f(x,y) = x^2 y + y^3$
6. $f(x,y) = (2x - y)^5$
7. $f(x,y) = \dfrac{x - y}{x + y}$
8. $f(x,y) = \dfrac{x - y^2}{x + y^2}$
9. $f(x,y) = \dfrac{x + y}{xy}$
10. $f(x,y) = \dfrac{2xy}{x^2 + y^2}$
11. $f(x,y) = \sqrt{x^2 + y^2}$
12. $f(x,y) = (x^2 + y^2)^{-3/2}$
13. $f(x,y) = \sqrt{4 + y/x}$
14. $f(x,y) = (\sqrt{x} + \sqrt{y})^2$
15. $f(x,y) = \ln|x + y|$
16. $f(x,y) = \ln\sqrt{x^2 + y^2}$
17. $f(x,y) = x^2 \ln\left|\dfrac{y}{x}\right|$
18. $f(x,y) = e^{-xy}$
19. $f(x,y) = e^{-x^2} - y^2$
20. $f(x,y) = e^{x/y}$
21. $f(x,y) = ye^{x/y}$
22. $f(x,y) = e^{y/x} \ln\left|\dfrac{x}{y}\right|$

11.6 PARTIAL DERIVATIVES

23. $f(x,y) = \sin(x+y)$

24. $f(x,y) = \cos x \sin y$

25. $f(x,y) = \sin 2x \cos 3y$

26. $f(x,y) = x^2 \cos y - 2x \tan y$

27. $f(x,y) = e^x \sin y$

28. $f(x,y) = e^{xy} \sin(x+y)$

29. $f(x,y) = \sin \dfrac{x}{y} + \ln \left| \dfrac{y}{x} \right|$

30. $f(x,y) = x \sec 2x \tan 2y$

31. $f(x,y) = \cos^{-1}\left(\dfrac{y}{x}\right)$

32. $f(x,y) = \tan^{-1}\left(\dfrac{x}{y}\right)$

33. $f(x,y) = x \sin^{-1} y + y \sin^{-1} x$

34. $f(x,y) = \tan^{-1}\left(\dfrac{y-x}{y+x}\right)$

35. $f(x,y,z) = x^2 + y^2 + z^2$

36. $f(x,y,z) = (x^2 + y^2 + z^2)^{1/2}$

37. $f(x,y,z) = x^2 y^2 z^2$

38. $f(x,y,z) = xyz \ln|xyz|$

39. $f(x,y,z) = \cos(xyz)$

40. $f(x,y,z) = \tan^{-1}\sqrt{x^2 y^2 z^2}$

For the functions in Exercises 41 to 50, compute all four second partial derivatives. Check that $f_{12} = f_{21}$ in each case.

41. $f(x,y) = x^2 y - 3xy^2 + y^3$

42. $f(x,y) = (x^2 + xy)^4$

43. $f(x,y) = \dfrac{xy}{x^2 + y^2}$

44. $f(x,y) = \ln \left| \dfrac{y}{x} \right|$

45. $f(x,y) = \ln(x^2 + y^2)^{-1/2}$

46. $f(x,y) = e^{-y/x}$

47. $f(x,y) = \sin(3x - 2y)$

48. $f(x,y) = x \tan^{-1} y$

49. $f(x,y) = \tan^{-1}(x+y)$

50. $f(x,y) = e^x \tan y$

In each of Exercises 51 to 60, a function $z = f(x,y)$ is given, together with a partial differential equation. Show that the function satisfies the equation.

51. $z = x^2 + xy$; $x \dfrac{\partial z}{\partial x} + y \dfrac{\partial z}{\partial y} = 2z$

52. $z = \dfrac{7x^3 - y^3}{x^2}$; $x \dfrac{\partial z}{\partial x} + y \dfrac{\partial z}{\partial y} = z$

53. $z = e^{xy}$; $x \dfrac{\partial z}{\partial x} = y \dfrac{\partial z}{\partial y}$

54. $z = \tan^{-1} \dfrac{y}{x}$; $x \dfrac{\partial z}{\partial x} + y \dfrac{\partial z}{\partial y} = 0$

55. $z = \sin \dfrac{x}{y} + \ln \dfrac{y}{x}$; $x \dfrac{\partial z}{\partial x} + y \dfrac{\partial z}{\partial y} = 0$

56. $z = x \sin\left(\dfrac{x}{y}\right) + y e^{y/x}$; $x \dfrac{\partial z}{\partial x} + y \dfrac{\partial z}{\partial y} = z$

57. $z = \sqrt{x^2 + y^2}$; $\dfrac{\partial^2 z}{\partial x^2} + \dfrac{\partial^2 z}{\partial y^2} = \dfrac{1}{z}$

58. $z = \dfrac{xy}{x+y}$; $x^2 \dfrac{\partial^2 z}{\partial x^2} + 2xy \dfrac{\partial^2 z}{\partial x \partial y} + y^2 \dfrac{\partial^2 z}{\partial y^2} = 0$

59. $z = (2x - 3y)^3 + e^{2x+3y}$; $9\dfrac{\partial^2 z}{\partial x^2} = 4\dfrac{\partial^2 z}{\partial y^2}$

60. $u = (x^2 + y^2 + z^2)^{-1/2}$; $\dfrac{\partial^2 u}{\partial x^2} + \dfrac{\partial^2 u}{\partial y^2} + \dfrac{\partial^2 u}{\partial z^2} = 0$

A function $f(x,y)$ is said to be *harmonic* if it satisfies *Laplace's equation*

$$\dfrac{\partial^2 f}{\partial x^2} + \dfrac{\partial^2 f}{\partial y^2} = 0$$

Prove that each function in Exercises 61 to 70 is harmonic.

61. $f(x,y) = x^2 + xy - y^2 + x - y$

62. $f(x,y) = x^3 - 3xy^2$

63. $f(x,y) = x^4 - 6x^2y^2 + y^4$

64. $f(x,y) = \dfrac{x+y}{x^2+y^2}$

65. $f(x,y) = \ln(x^2 + y^2)$

66. $f(x,y) = e^x \sin y$

67. $f(x,y) = \tan^{-1}\left(\dfrac{2xy}{x^2 - y^2}\right)$

68. $f(x,y) = \ln\sqrt{x^2+y^2} + \tan^{-1}\left(\dfrac{y}{x}\right)$

69. $f(x,y) = e^{x^2-y^2}\cos(2xy)$

70. $f(x,y) = \ln\left(\dfrac{\sqrt{(x-1)^2+y^2}}{\sqrt{(x+1)^2+y^2}}\right)$

PROBLEMS 11.6

1. For the function $f(x,y) = 5x^2y^3 - 6x^3y^2$, show that

$$f_{112} = f_{121} = f_{211} \quad \text{and} \quad f_{122} = f_{212} = f_{221}$$

State conditions similar to those in Theorem 11.9 under which these equations always hold.

2. A function $f(x,y)$ is said to be *homogeneous of order k* if $f(\alpha x, \alpha y) = \alpha^k f(x,y)$ for any constant α. Show that such a function satisfies the partial differential equation

$$xf_1(x,y) + yf_2(x,y) = kf(x,y)$$

3. Prove that

$$\dfrac{\partial^3}{\partial x\, \partial y\, \partial z}\begin{vmatrix} a(x) & b(x) & c(x) \\ d(y) & e(y) & f(y) \\ g(z) & h(z) & k(z) \end{vmatrix} = \begin{vmatrix} a'(x) & b'(x) & c'(x) \\ d'(y) & e'(y) & f'(y) \\ g'(z) & h'(z) & k'(z) \end{vmatrix}$$

4. Show that the functions

$$u = -ky\left(1 - \dfrac{a^2}{x^2+y^2}\right), \quad v = kx\left(1 + \dfrac{a^2}{x^2+y^2}\right)$$

where k and a are constants, satisfy the *Cauchy-Riemann equations*

$$\dfrac{\partial u}{\partial x} = \dfrac{\partial v}{\partial y}, \quad \dfrac{\partial u}{\partial y} = -\dfrac{\partial v}{\partial x}$$

5. If $u(x,y)$ and $v(x,y)$ have continuous second partial derivatives and satisfy the Cauchy-Riemann equations, prove that both u and v satisfy Laplace's equation.

6. The equation

$$\frac{\partial u}{\partial t} = k^2 \frac{\partial^2 u}{\partial x^2}$$

is called the *one-dimensional heat equation*. Prove that both

$$u = e^{-a^2 k^2 t}(A \sin ax + B \cos ax)$$

and

$$v = e^{k^2 t}(Ae^x + Be^{-x})$$

satisfy this equation.

7. The partial differential equation

$$a^2 \frac{\partial^2 u}{\partial x^2} = \frac{\partial^2 u}{\partial t^2}$$

is often called the *vibrating string equation*.
   (a) If $f(x)$ satisfies the ordinary differential equation $a^2 f''(x) = 0$ and $g(t)$ satisfies $g''(t) + k^2 g(t) = 0$, then show that $u(x, t) = f(x)g(t)$ satisfies the vibrating string equation.
   (b) Let f and g be twice differentiable functions of one variable. For any constant a, show that the function

$$w(x, t) = f(x + at) + g(x - at)$$

satisfies the vibrating string equation.

8. Prove that for the function

$$f(x, y) = \frac{x^3 y - xy^3}{x^2 + y^2}, \quad \text{if } (x, y) \neq (0, 0)$$

$$= 0, \quad \text{if } (x, y) = (0, 0)$$

the partial derivatives $f_{12}(0, 0)$ and $f_{21}(0, 0)$ both exist but are not equal.

11.7 TANGENT LINES AND PLANES

There is an easy geometric interpretation of the partial derivatives $f_1(a, b)$ and $f_2(a, b)$. To understand it, consider the functions of one variable defined by

$$g(x) = f(x, b), \quad h(y) = f(a, y)$$

The graphs of these functions may be diagramed as cross sections of the surface, namely, the graph of $g(x)$ is drawn in the plane $\{(x, y, z) : y = b\}$, while the graph of $h(x)$ is in $\{(x, y, z) : x = a\}$. This is shown in Fig. 11.15.

In particular, we have

$$\{(x, y, z) : z = f(x, y)\} \cap \{(x, y, z) : y = b\}$$
$$= \{(x, y, z) : y = b, z = g(x)\}$$

and

$$\{(x, y, z) : z = f(x, y, z)\} \cap \{(x, y, z) : x = a\}$$
$$= \{(x, y, z) : x = a, z = h(y)\}$$

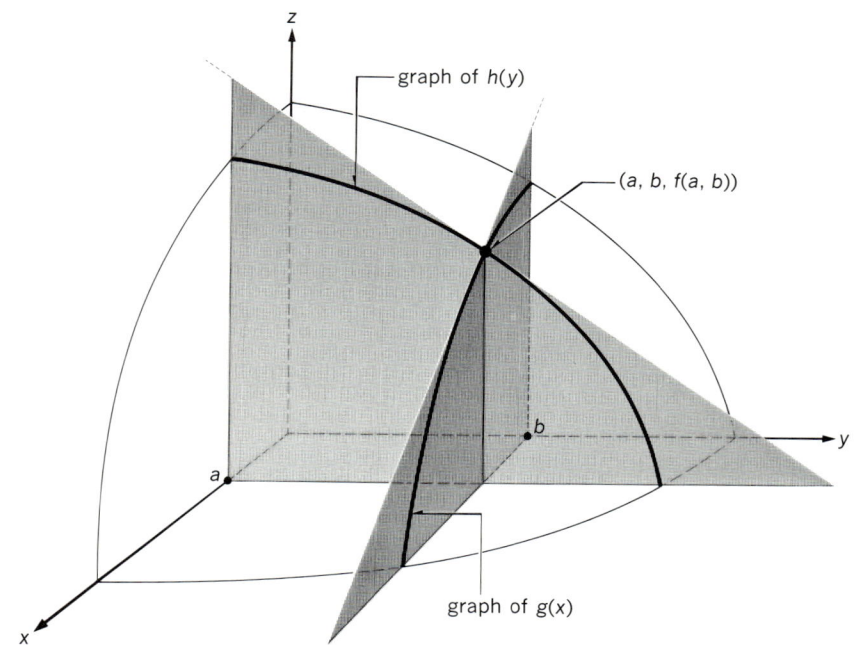

FIGURE 11.15

Now we obviously have

$$Dg(a) = \lim_{x \to a} \frac{g(x) - g(a)}{x - a} = \lim_{x \to a} \frac{f(x, b) - f(a, b)}{x - a} = f_1(a, b)$$

In the same way, it is seen that

$$Dh(b) = f_2(a, b)$$

Therefore, the partial derivatives $f_1(a, b)$ and $f_2(a, b)$ may be interpreted as slopes of cross sections.

The line in the plane $\{(x, y, z) : y = b\}$ which is tangent to the cross section $\{(x, y, z) : y = b, z = f(x, b)\}$ at the point $(a, b, f(a, b))$ has slope $dz/dx = f_1(a, b)$. It is a particular tangent line to the surface $\{(x, y, z) : z = f(x, y)\}$ and it is easy to see that this line has direction numbers $(1, 0, f_1(a, b))$. Similarly, the line in $\{(x, y, z) : x = a\}$, which is tangent to the curve $\{(x, y, z) : x = a, z = f(a, y)\}$, has direction numbers $(0, 1, f_2(a, b))$.

Now consider any line $y - b = m(x - a)$ through (a, b) in the xy plane. The corresponding vertical plane $\{(x, y, z) : y - b = m(x - a)\}$ intersects the surface $\{(x, y, z) : z = f(x, y)\}$ in a planar curve as shown in Fig. 11.16. The line (in this vertical plane) through $(a, b, f(a, b))$ tangent to the planar curve

$$\{(x, y, z) : y - b = m(x - a), z = f(x, b + m(x - a))\}$$

11.7 TANGENT LINES AND PLANES

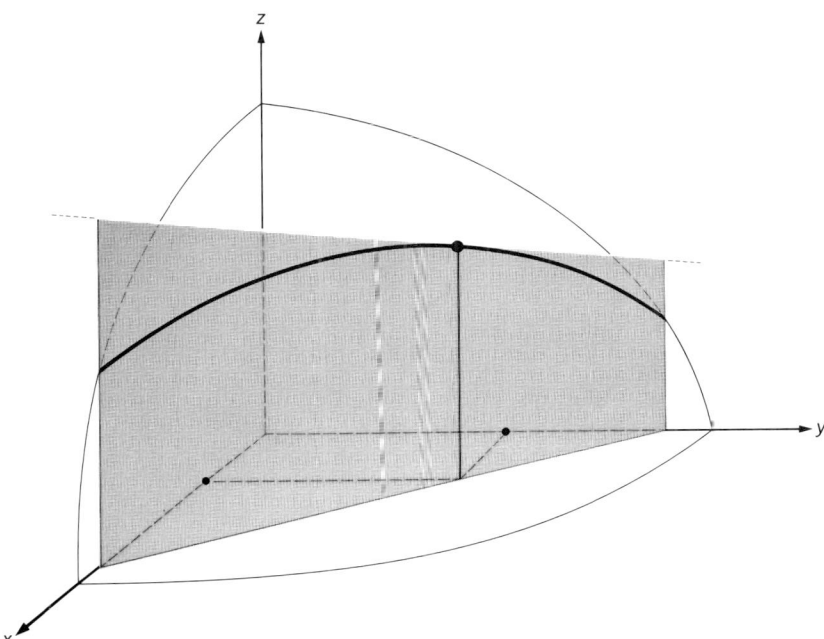

FIGURE 11.16

is also a tangent line to the surface. We compute direction numbers for this line as follows: Consider the two points $(a, b, f(a, b))$ and $(x, m(x - a) + b, f(x, m(x - a) + b))$ in the intersection of the surface and the plane $\{(x, y, z) : y = m(x - a) + b\}$. The direction numbers of the line from $(a, b, f(a, b))$ to the point $(x, m(x - a) + b, f(x, m(x - a) + b))$ are

$$(x - a, m(x - a), f(x, m(x - a) + b) - f(a, b))$$

or, equivalently,

$$\left(1, m, \frac{f(x, m(x - a) + b) - f(a, b)}{x - a}\right)$$

It follows that the direction numbers of the required tangent line may be taken to be

$$\left(1, m, \lim_{x \to a} \frac{f(x, m(x - a) + b) - f(a, b)}{x - a}\right)$$

We next compute this limit.

We suppose that the function $f(x, y)$ has continuous partial derivatives and write

$$f(x, m(x - a) + b) - f(a, b)$$
$$= f(x, m(x - a) + b) - f(x, b) + f(x, b) - f(a, b)$$

Applying the mean value theorem to both of these differences, we see that there are numbers c between a and x, and d between b and $m(x - a) + b$, such that

$$f(x, b) - f(a, b) = (x - a)f_1(c, b)$$
$$f(x, m(x - a) + b) - f(x, b) = \{m(x - a) + b - b\}f_2(x, d)$$
$$= m(x - a)f_2(x, d)$$

Therefore, we may write the equation

$$\frac{f(x, m(x - a) + b) - f(a, b)}{x - a} = mf_2(x, d) + f_1(c, b)$$

It is then obvious that we have

$$\lim_{x \to a} \frac{f(x, m(x - a) + b) - f(a, b)}{x - a} = \lim_{x \to a} [mf_2(x, d) + f(c, b)]$$
$$= mf_2(a, b) + f_1(a, b)$$

the latter equation following from the assumed continuity of f_1 and f_2.

We have shown that the tangent line to the cross-sectional curve $\{(x, y, z) : y = m(x - a) + b, z = f(x, y)\}$ has direction numbers

$$(1, m, f_1(a, b) + mf_2(a, b))$$

This leads to the following definition: Let $f(x, y)$ be a function of two variables with continuous first partial derivatives. Then the plane

$$\Pi = \{(x, y, z) : z - f(a, b) = (x - a)f_1(a, b) + (y - b)f_2(a, b)\}$$

is called the *tangent plane* to the graph of f at the point $(a, b, f(a, b))$. This definition is justified by the fact that every tangent line as defined above lies in this plane. (The reader may apply Theorem 11.5 to prove this fact.)

Let f be a function of two variables, and assume that the first partial derivatives of f exist in some region. Then the numerical vector (f_1, f_2) of first partial derivatives is called the *gradient* of the function f. From the formula for the tangent plane to the graph of f, it is apparent that the tangent plane is horizontal when the gradient vector is zero. We shall see other uses for the gradient later on.

■ **Example.** Let $f(x, y) = x^2 - 3xy + 4y^2 + 2x - y$. Then the gradient vector (f_1, f_2) is $(2x - 3y + 2, -3x + 8y - 1)$. At the point $(2, 1, 5)$ on the surface

$$\{(x, y, z) : z = x^2 - 3xy + 4y^2 + 2x - y\}$$

this gradient vector is $(3, 1)$; the tangent plane at this point has the equation

$$z - 5 = f_1(2, 1)(x - 2) + f_2(2, 1)(y - 1)$$
$$= 3(x - 2) + (y - 1)$$

11.7 TANGENT LINES AND PLANES

or
$$z = 3x + y - 2$$

In addition, the gradient vector (f_1, f_2) will be zero if the equations
$$2x - 3y + 2 = 0$$
$$-3x + 8y - 1 = 0$$
hold simultaneously. Since the simultaneous solution of these equations is $x = -\frac{13}{7}$, $y = -\frac{4}{7}$, it follows that the tangent plane at the point $(-\frac{13}{7}, -\frac{4}{7} - \frac{125}{49})$ is horizontal. ∎

In analogy to the case of a function of two variables, we next define the "linearization" of a function of n variables. We suppose that the first derivatives of the function f all exist at a point $\mathbf{a} = (a_1, a_2, \cdots, a_n)$. Then the linearization of f at $\mathbf{a}$ is the linear function

$$L(x_1, \cdots, x_n) = f(\mathbf{a}) + \sum_{i=1}^{n} f_i(\mathbf{a})(x_i - a_i).$$

At times we use geometric language and speak of this as the equation of a *tangent hyperplane*.

We note also that the *gradient vector* $(f_1, f_2, \cdots, f_n)$ of a function f of n variables is defined when the first derivatives all exist. When use is made later of this vector, it will be clear that the gradient has many important applications.

EXERCISES 11.7

In each of Exercises 1 to 10, a function $f(x, y)$ and a fixed point (a, b) are given. Determine the equation of the tangent plane at the point $(a, b, f(a, b))$ to the graph of f, determine the equations of the tangent lines in the planes $\{(x, y, z) : y = b\}$ and $\{(x, y, z) : x = a\}$ and finally construct a sketch similar to Fig. 11.1.

1. $f(x, y) = x^2 + y;\ (2, -2)$
2. $f(x, y) = x^2 - xy - y^2;\ (2, 1)$
3. $f(x, y) = 4 - x^2 - y^2;\ (1, 1)$
4. $f(x, y) = xy + y;\ (1, 2)$
5. $f(x, y) = \sqrt{x^2 + y^2};\ (3, 3)$
6. $f(x, y) = \sqrt{9 - x^2 - y^2};\ (2, 1)$
7. $f(x, y) = \frac{1}{2}\sqrt{x^2 + y^2 - 4};\ (2, 2)$
8. $f(x, y) = \frac{4}{xy};\ (1, 2)$
9. $f(x, y) = \sqrt{x} + \sqrt{y};\ (1, 1)$
10. $f(x, y) = \ln(x^2 + y^2);\ (0, 1)$

11. Let (x_0, y_0, z_0) be a point on $\{(x, y, z) : z = \sqrt{r^2 - x^2 - y^2}\}$ and show that the tangent plane there is $\{(x, y, z) : x_0 x + y_0 y + z_0 z = r^2\}$.

12. Let (x_0, y_0, z_0) be a point on $\{(x, y, z) : z = kx^2 + y^2\}$ and show that the tangent plane there is $\{(x, y, z) : z_0 z = k(x_0 x + y_0 y)\}$.

11.8 DIFFERENTIABLE FUNCTIONS

Consider a function f of two variables and let (x, y) be an interior point in its domain. Given another point $(x + h, y + k)$ in the interior of the domain, the *increment* in the functional values due to the increments h and k in x and y is defined by

$$\Delta f(x, y; h, k) = f(x + h, y + k) - f(x, y)$$

The function f is said to be *differentiable at the point* (x, y) if this increment can be expressed in the form

$$\Delta f(x, y; h, k) = hf_1(x, y) + kf_2(x, y) + h\epsilon_1 + k\epsilon_2$$

where ϵ_1 and ϵ_2 are functions of x, y, h, and k such that

$$\lim_{(h,k) \to (0,0)} \epsilon_1 = 0 = \lim_{(h,k) \to (0,0)} \epsilon_2$$

Of course, f is differentiable on an open set U if it is differentiable at each point of U. Note that, if f is differentiable, then both the partial derivatives f_1 and f_2 exist. Conversely, if f_1 and f_2 are continuous, then it will be shown (Theorem 11.10) that f is differentiable.

■ **Example 1.** Let $f(x, y) = x^3 - 3xy^2$. Then we have

$$\begin{aligned}\Delta f(x, y; h, k) &= (x + h)^3 - 3(x + h)(y + k)^2 - (x^3 - 3xy^2) \\ &= x^3 + 3x^2h + 3xh^2 + h^3 - 3xy^2 - 6xyk - 3xk^2 \\ &\quad - 3y^2h - 6yhk - 3hk^2 - x^3 + 3xy^2 \\ &= h(3x^2 - 3y^2) + k(-6xy) + h(3xh + h^2) \\ &\quad + k(-3k - 6yh - 3hk)\end{aligned}$$

Here we have

$$\epsilon_1 = 3xh + h^2 \quad \text{and} \quad \epsilon_2 = -3k - 6yh - 3hk$$

both of which approach zero as (h, k) approaches $(0, 0)$. This proves that $f(x, y)$ is differentiable at (x, y). ■

The linear terms in h and k in the increment $\Delta f(x, y; h, k)$ constitute the *total differential*

$$df(x, y; h, k) = hf_1(x, y) + kf_2(x, y)$$

The total differential is used as an approximation to Δf in computing errors, making estimates, and so on. This is analogous to the use of the differential of a function of one variable.

■ **Example 2.** The volume of a cylinder is $V = \pi r^2 h$, a function of the radius r and the height h. We compute the approximate relative error in V by evaluating dV/V (in place of $\Delta V/V$).

Note that

$$dV = 2\pi r h\, \Delta r + \pi r^2\, \Delta h$$

11.8 DIFFERENTIABLE FUNCTIONS

where Δr and Δh are increments in r and h, respectively. Then

$$\frac{dV}{V} = \frac{2\pi r h \, \Delta r}{\pi r^2 h} + \frac{\pi r^2 \, \Delta r}{\pi r^2 h} = 2\left(\frac{\Delta r}{r}\right) + \left(\frac{\Delta h}{h}\right)$$

Thus the relative error in V is twice the relative error in r plus the relative error in h. ∎

If we consider a fixed interior point (a, b) in the domain of the function f, and denote $(a + h, b + h)$ by (x, y), then the equation of differentiability can be rewritten as

$$f(x, y) = f(a, b) + (x - a)f_1(a, b) + (y - b)f_2(a, b) - (x - a)\epsilon_1 + (y - b)\epsilon_2$$

This expresses the function $f(x, y)$ as its linearization

$$f(a, b) + (x - a)f_1(a, b) + (y - b)f_2(a, b)$$

plus the correction terms

$$(x - a)\epsilon_1 + (y - b)\epsilon_2$$

This point of view is often seen in applications.

These concepts generalize directly to functions of more than two variables. The function $f(\mathbf{x}) = f(x_1, x_2, \cdots, x_n)$ is *differentiable at the point* $\mathbf{a} = (a_1, a_2, \cdots, a_n)$ if it can be expressed as

$$f(\mathbf{x}) = f(\mathbf{a}) + \sum_{i=1}^{n}(x_i - a_i)f_i(\mathbf{a}) + \sum_{i=1}^{n}(x_i - a_i)\epsilon_i$$

where $\lim_{x \to a}\epsilon_i = 0$ for each $i = 1, 2, \cdots, n$.

The linear part

$$df(\mathbf{a}; \mathbf{x}) = \sum_{i=1}^{n}(x_i - a_i)f_i(\mathbf{a})$$

is the *total differential* of f and can be used to approximate the increment $\Delta f(\mathbf{a}; \mathbf{x}) = f(\mathbf{x}) - f(\mathbf{a})$.

We conclude this section with one theoretical result.

■ **THEOREM 11.10.** Let the function f of two variables have continuous first partial derivatives in an open set U. Then f is differentiable in U.

PROOF: Let both (x, y) and $(x + h, y + k)$ be points in a basis neighborhood lying entirely in U. For convenience, we define

$$g(x, y; h, k) = f(x + h, y + k) - f(x, y) - hf_1(x, y) - kf_2(x, y)$$

By adding and subtracting $f(x, y + k)$, we may write

$$g(x, y; h, k) = f(x + h, y + k) - f(x, y + k) - hf_1(x, y) \\ + f(x, y + k) - f(x, y) - kf_2(x, y)$$

We apply the mean value theorem to write
$$f(x+h, y+k) - f(x, y+k) = hf_1(c, y+k)$$
for c between x and $x+h$, and
$$f(x, y+k) - f(x, y) = kf_2(x, d)$$
for d between y and $y+k$.

It follows that we may write
$$g(x, y; h, k) = h[f_1(c, y+k) - f_1(x, y)] + k[f_2(x, d) - f_2(x, y)]$$
Clearly, the functions
$$\epsilon_1 = f_1(c, y+k) - f_1(x, y), \qquad \epsilon_2 = f_2(x, d) - f_2(x, y)$$
have the property
$$\lim_{(h,k) \to (0,0)} \epsilon_1 = 0 = \lim_{(h,k) \to (0,0)} \epsilon_2$$
by the assumed continuity of f_1 and f_2.

Thus we may write
$$f(x+h, y+k) - f(x, y) = hf_1(x, y) + kf_2(x, y) + h\epsilon_1 + k\epsilon_2$$
where ϵ_1 and ϵ_2 approach zero as (h, k) approaches $(0, 0)$. This proves that f is differentiable at (x, y). ∎

EXERCISES 11.8

Compute $\Delta f(x, y; h, k)$ and express it as the total differential $df(x, y; h, k)$ plus the correction terms $h\epsilon_1 + k\epsilon_2$, for the functions in Exercises 1 to 4.

1. $f(x, y) = 3x^2 - y^2$
2. $f(x, y) = x^2 y + y^3$
3. $f(x, y) = \dfrac{y}{x+y}$
4. $f(x, y) = \dfrac{xy}{x^2 + y^2}$

For each of the functions in Exercises 5 to 24, compute the total differential.

5. $f(x, y) = \dfrac{Ax + By}{Cx + Dy}$
6. $f(x, y) = \sqrt{4x + 3y}$
7. $f(x, y) = \sin(x + y)$
8. $f(x, y) = \sin x \cos y$
9. $f(x, y) = e^{y/x}$
10. $f(x, y) = \ln\left(\dfrac{y}{x}\right)$
11. $f(x, y) = e^{x^2 - y^2}$
12. $f(x, y) = \ln|x^2 - y^2|$
13. $f(x, y) = x \ln(x^2 + y^2)$
14. $f(x, y) = (x + y) \ln \sqrt{x^2 + y^2}$
15. $f(x, y) = e^x \cos y$
16. $f(x, y) = e^{x/y} \cos\left(\dfrac{y}{x}\right)$
17. $f(x, y) = \tan^{-1}\left(\dfrac{x}{y}\right)$
18. $f(x, y) = \cos^{-1}(x \sin y)$
19. $f(x, y, z) = x^2 + 3xy - 2yz + z^2$
20. $f(x, y, z) = x^2 z - yz^2$

21. $f(x, y, z) = 3x^2y - y^3 + yz^2 - z^5$ 22. $f(x, y, z) = \ln\sqrt{x^2 + y^2 + z^2}$
23. $f(x, y, z) = \tan^{-1}(xyz)$ 24. $f(x, y, z) = \sin(x + y)\sin(y + z)$

Use total differentials to approximate each of Exercises 25 to 28.

25. $\sqrt{(2.9)^2 + (4.1)^2}$ 26. $\cos 44° \sin 29°$
27. $[(2.01)^2 + (1.02)^2 + (1.99)^2]^{-3/2}$ 28. $\sin 31° \cos 29° \tan 46°$

PROBLEMS 11.8

1. Prove that each of the following functions has first partial derivatives at the origin but is not differentiable at the origin.

    (a) $f(x, y) = \sqrt{|xy|}$

    (b) $f(x, y) = \dfrac{x^2y}{x^2 + y^2}$ if $(x, y) \neq (0, 0)$

    $ = 0$ if $(x, y) = (0, 0)$

    (c) $f(x, y, z) = \dfrac{x^2yz}{x^2 + y^2 + z^2}$ if $(x, y, z) \neq (0, 0, 0)$

    $ = 0$ if $(x, y, z) = (0, 0, 0)$

2. Prove that the function

    $$f(x, y) = \frac{x^3 + y^3}{x - y}, \quad x \neq y$$

    $ = 0$ if $(x, y) = (0, 0)$

    is discontinuous at $(0, 0)$ but that $f_{11}(0, 0)$ exists.

3. Let $f(x, y)$ have first partial derivatives and define

    $$g(t) = f(a + bt, c + dt)$$

    Prove that

    $$Dg(t) = bf_1(a + bt, c + dt) + df_2(a + bt, c + dt)$$

    Assume that $f_{12} = f_{21}$; find $D^2g(t)$, too.

4. Establish the customary formulas for the total differentials of sums, products, and quotients.

5. If the total differential of a function is identically zero in an open set, prove that the function is constant.

11.9 THE CHAIN RULE

We shall prove two special cases of the general theorem called the chain rule. Then we state but do not prove the theorem.

Let $f(x, y)$ be a differentiable function of the point (x, y) in the plane. In turn, suppose that the point (x, y) is the location of a particle which is moving. Suppose that the equations of motion of the particle are $x = g(t)$, $y = h(t)$ where g and h are also differentiable. By substituting these values of x and y into the function f, we obtain a function of t, namely

$$F(t) = f(g(t), h(t))$$

It is our purpose to complete the derivative

$$DF(t) = \lim_{\Delta t \to 0} \frac{f(g(t + \Delta t), h(t + \Delta t)) - f(g(t), h(t))}{\Delta t}$$

Because f is differentiable, we write

$$f(g(t + \Delta t), h(t + \Delta t)) - f(g(t), h(t))$$
$$= [g(t + \Delta t) - g(t)][f_1(g(t), h(t)) + \epsilon_1]$$
$$+ [h(t + \Delta t) - h(t)][f_2(g(t), h(t)) + \epsilon_2]$$

where the limit of ϵ_1 and ϵ_2 is zero as $g(t + \Delta t) - g(t)$ and $h(t + \Delta t) - h(t)$ approach zero. It follows that we have the ratio

$$\frac{F(t + \Delta t) - F(t)}{\Delta t} = \left[\frac{g(t + \Delta t) - g(t)}{\Delta t}\right][f_1(g(t), h(t)) + \epsilon_1]$$
$$+ \left[\frac{h(t + \Delta t) - h(t)}{\Delta t}\right][f_2(g(t), h(t)) + \epsilon_2]$$

Now, as Δt approaches zero, we know that both $g(t + \Delta t) - g(t)$ and $h(t + \Delta t) - h(t)$ approach zero. Hence $\lim_{\Delta t \to 0} \epsilon_1 = 0 = \lim_{\Delta t \to 0} \epsilon_2$. We then have

$$DF(t) = Dg(t) \cdot f_1(g(t), h(t)) + Dh(t) \cdot f_2(g(t), h(t))$$

If we write $DF(t) = df/dt$, $Dg(t) = dx/dt$, $Dh(t) = dy/dt$, $f_1 = \partial f/\partial x$, and $f_2 = f/y$, then the previous equation becomes

$$\frac{df}{dt} = \frac{\partial f}{\partial x}\frac{dx}{dt} + \frac{\partial f}{\partial y}\frac{dx}{dt}$$

This formula is quite easy to remember now.

Next, consider the following situation: Again take $f(x, y)$ to be differentiable but now suppose that $x = g(u, v)$ and $y = h(u, v)$ are differentiable functions of two variables. Substituting these values of x and y into f, we obtain a function of u and v:

$$F(u, v) = f(g(u, v), h(u, v))$$

We compute the partial derivative

$$\frac{\partial F}{\partial u} = \lim_{\Delta u \to 0} \frac{F(u + \Delta u, v) - F(u, v)}{\Delta u}$$
$$= \lim_{\Delta u \to 0} \frac{f(g(u + \Delta u, v), h(u + \Delta u, v)) - f(g(u, v), h(u, r))}{\Delta u}$$

We apply the differentiability of f and write

$$F(u + \Delta u, v) - F(u, v)$$
$$= [g(u + \Delta u, v) - g(u, v)][f_1(g(u, v), h(u, v)) + \epsilon_1]$$
$$+ [h(u + \Delta u, v) - h(u, v)][f_2(g(u, v), h(u, v)) + \epsilon_2]$$

11.9 THE CHAIN RULE

where ϵ_1 and ϵ_2 approach zero as $g(u + \Delta u, v) - g(u, v)$ and $h(u + \Delta u, v) - h(u, v)$ approach zero, as both quantities do when Δu approaches zero. It readily follows that

$$\lim_{\Delta u \to 0} \frac{F(u + \Delta u, v) - F(u, v)}{\Delta u}$$

$$= \lim_{\Delta u \to 0} \left[\frac{g(u + \Delta u, v) - g(u, v)}{\Delta u} \right] [f_1(g(u, v), h(u, v)) + \epsilon_1]$$

$$+ \lim_{\Delta u \to 0} \left[\frac{h(u + \Delta u, v) - h(u, v)}{\Delta u} \right] [f_2(g(u, v), h(u, v)) + \epsilon_2]$$

$$= g_1(u, v) \cdot f_1(g(u, v), h(u, v)) + h_1(u, v) \cdot f_2(g(u, v), h(u, v))$$

Again the formula is easier to remember in the following form: Let

$$\frac{\partial F}{\partial u} = \frac{\partial f}{\partial u}, \quad g_1 = \frac{\partial x}{\partial u}, \quad h_1 = \frac{\partial y}{\partial u}, \quad f_1 = \frac{\partial f}{\partial x}, \quad \text{and} \quad f_2 = \frac{\partial f}{\partial y}$$

Then we have

$$\frac{\partial f}{\partial u} = \frac{\partial f}{\partial x}\frac{\partial x}{\partial u} + \frac{\partial f}{\partial y}\frac{\partial y}{\partial u}$$

In exactly the same way we could also derive the formula

$$\frac{\partial f}{\partial v} = \frac{\partial f}{\partial x}\frac{\partial x}{\partial v} + \frac{\partial f}{\partial y}\frac{\partial y}{\partial v}$$

Because the general theorem that follows has precisely the same proof as the special cases, we shall not prove it.

■ **THEOREM 11.11.** Let the function $f(x_1, x_2, \cdots, x_n)$ of n variables have continuous first partial derivatives in an open set X. For each $i = 1, 2, \cdots, n$, let $x_i = \phi_i(u_1, u_2, \cdots, u_m)$ be a differentiable function on an open set U such that, for each point $\mathbf{u}$ of U, the point $(\phi_1(\mathbf{u}), \cdots, \phi_n(\mathbf{u}))$ lies in X. Then we have the partial derivative of f with respect to u_j,

$$\frac{\partial f}{\partial u_j} = \sum_{i=1}^{n} \frac{\partial f}{\partial x_i}\frac{\partial x_i}{\partial u_j} \quad j = 1, 2, \cdots, m$$

where

$$\frac{\partial f}{\partial x_i} = f_i$$

and

$$\frac{\partial x_i}{\partial u_j} = (\phi_i)_j$$

■ **Example 1.** Let $f(x, y) = x^2 - 3xy + y^2$ and suppose that $x = \cos t$ and $y = \sin t$. We have $f_1(x, y) = 2x + 3y$ and $f_2(x, y) = 3x + 2y$.

Therefore, Theorem 11.11 (or the special case discussed first) tells us that
$$\frac{df}{dt} = (2x + 3y)(-\sin t) + (3x + 2y)(\cos t)$$
$$= (2\cos t + 3\sin t)(-\sin t) + (3\cos t + 2\sin t)(\cos t)$$
$$= 3\cos^2 t - 3\sin^2 t = 3\cos 2t$$

This can be verified by direct substitution. We see that
$$f(\cos t, \sin t) = \cos^2 t + 3\sin t \cos t + \sin^2 t$$
$$= 1 + 3\sin t \cos t$$

Hence it follows that
$$\frac{df}{dt} = 3\cos^2 t - 3\sin^2 t$$

■ **Example 2.** Let $f(t)$ be a differentiable function of one variable. If $t = x^2 - y^2$, then define the function
$$u(x, y) = f(x^2 - y^2)$$
By Theorem 11.11, we have
$$\frac{\partial u}{\partial x} = \frac{df}{dt} \cdot \frac{\partial t}{\partial x} = \frac{df}{dt} \cdot 2x$$
and
$$\frac{\partial u}{\partial y} = \frac{df}{dt} \cdot \frac{\partial t}{\partial y} = \frac{df}{dt}(-2y)$$
It follows that the function u satisfies the partial differential equation
$$y \frac{\partial u}{\partial x} + x \frac{\partial u}{\partial y} = 0$$
regardless of the choice of the function f.

■ **Example 3.** Let $f(x, y)$ have continuous first partial derivatives and let $x = r\cos\theta$, $y = r\sin\theta$. Then $\partial x/\partial r = \cos\theta$, $\partial y/\partial r = \sin\theta$, $\partial x/\partial \theta = -r\sin\theta$ and $\partial y/\partial \theta = r\cos\theta$. From Theorem 11.11, we then have
$$\frac{\partial f}{\partial r} = \frac{\partial f}{\partial x}\frac{\partial x}{\partial r} + \frac{\partial f}{\partial y}\frac{\partial y}{\partial r} = \frac{\partial f}{\partial x}\cos\theta + \frac{\partial f}{\partial y}\sin\theta$$
and
$$\frac{\partial f}{\partial \theta} = r\left[-\frac{\partial f}{\partial x}\sin\theta + \frac{\partial f}{\partial y}\cos\theta\right]$$

■ **Example 4.** If $f(x, y) = x^2 - y^2$ and
$$x = u + v - w \quad \text{and} \quad y = u - v + w$$

then
$$\frac{\partial f}{\partial u} = \frac{\partial f}{\partial x}\frac{\partial x}{\partial u} + \frac{\partial f}{\partial y}\frac{\partial y}{\partial u} = \frac{\partial f}{\partial x} + \frac{\partial f}{\partial y}$$
$$\frac{\partial f}{\partial v} = \frac{\partial f}{\partial x}\frac{\partial x}{\partial v} + \frac{\partial f}{\partial y}\frac{\partial y}{\partial v} = \frac{\partial f}{\partial x} - \frac{\partial f}{\partial y}$$
$$\frac{\partial f}{\partial w} = \frac{\partial f}{\partial x}\frac{\partial x}{\partial w} + \frac{\partial f}{\partial y}\frac{\partial y}{\partial w} = -\frac{\partial f}{\partial x} + \frac{\partial f}{\partial y}$$

Therefore,
$$\frac{\partial f}{\partial u} + \frac{\partial f}{\partial v} + \frac{\partial f}{\partial w} = \frac{\partial f}{\partial x} + \frac{\partial f}{\partial y} \quad \blacksquare$$

Implicit differentiation. Let f be a differentiable function of two variables. Suppose that there is a differentiable function $\phi(x)$ such that the equation
$$f(x, \phi(x)) = 0$$
holds for each point in the domain of ϕ. The chain rule then implies that
$$f_1(x, \phi(x))\frac{dx}{dx} + f_2(x, \phi(x))\frac{d\phi}{dx} = 0$$
whence
$$\frac{d\phi(x)}{dx} = -\frac{f_1(x, \phi(x))}{f_2(x, \phi(x))}$$

Most often, we do not (or cannot) solve the equation
$$f(x, y) = 0$$
for y in terms of x. Then we write
$$\frac{dy}{dx} = -\frac{f_1(x, y)}{f_2(x, y)}$$

This formula provides an explicit formula for the slope at the curve $\{(x, y) : f(x, y) = 0\}$.

■ *Example 5.* The equation
$$f(x, y) = x^2 y - xy^2 + 2x - y - 3 = 0$$
defines y as a function of x, although not uniquely if we make no restrictions. In any case, however, the formula above yields
$$\frac{dy}{dx} = -\frac{f_1(x, y)}{f_2(x, y)} = -\frac{2xy - y^2 + 2}{x^2 - 2xy - 1} \quad \blacksquare$$

A function of several variables can also be defined implicitly by an equation.

■ *Example 6.* Consider the equation
$$f(x, y, z) = x^2 + y^2 + z^2 - xyz = 0$$
Suppose that there is a function $\phi(x, y)$ such that
$$x^2 + y^2 + \phi^2(x, y) = xy\phi(x, y)$$
Then we may differentiate, obtaining
$$2x + 2\phi(x, y) \cdot \phi_1(x, y) = y\phi(x, y) + xy\phi_1(x, y)$$
$$2y + 2\phi(x, y) \cdot \phi_2(x, y) = x\phi(x, y) + xy\phi_2(x, y)$$
It follows that
$$\frac{\partial z}{\partial x} = \phi_1(x, y) = \frac{-2x + y\phi(x, y)}{2\phi(x, y) - xy} = \frac{-2x + yz}{2z - xy}$$
$$\frac{\partial z}{\partial y} = \phi_2(x, y) = \frac{-2y + x\phi(x, y)}{2\phi(x, y) - xy} = \frac{-2y + xz}{2z - xy} \quad ■$$

We state a very long theorem which follows, but it is neither proved nor numbered. A proof will be found in advanced courses.

■ **THEOREM.** Let $f(x_1, x_2, \cdots, x_n, z)$ be a continuous function of $n+1$ variables with continuous first partial derivatives. Let $(a_1, \cdots, a_n, b)$ be an interior point of the domain of f such that $f(a_1, \cdots, a_n, b) = 0$ and suppose that $f_{n+1}(a_1, \cdots, a_n, b) \neq 0$. Then there is a basis neighborhood U of $(a_1, \cdots, a_n)$ in $\mathbf{R}^n$ and a basis neighborhood V of b in $\mathbf{R}$ such that, for each point $(x_1, \cdots, x_n) \in U$, the equation $f(x_1, \cdots, x_n, z) = 0$ is satisfied by exactly one value of z in V. Denote this value by
$$z = \phi(x_1, \cdots, x_n)$$
so that $f(x_1, \cdots, x_n, \phi(x_1, \cdots, x_n)) = 0$ is an identity in U. Also note that $b = \phi(a_1, \cdots, a_n)$. Finally, the function ϕ so defined is continuous in U and has first partial derivatives given by the equations
$$f_1 + f_{n+1}\phi_1 = 0$$
$$f_2 + f_{n+1}\phi_2 = 0$$
$$\cdots$$
$$f_n + f_{n+1}\phi_n = 0$$

EXERCISES 11.9 In Exercises 1 to 6, find dw/dt in two ways: (a) by using the chain rule, and (b) by substituting the expressions for xy (and z) in terms of t into the expression for w and then differentiating.

1. $w = x^2 + y^2;\ x = 1 - \dfrac{1}{t}, y = 1 + \dfrac{1}{t}$

2. $w = \dfrac{2xy}{x^2 + y^2};\ x = t^2, y = 2t$

3. $w = e^{x^2+y^2}$; $x = \cos t$, $y = \sin t$

4. $w = \tan^{-1} xy$; $x = t^2$, $y = t^{-3}$

5. $w = x^2 + y^2 + z^2$; $x = e^t \cos t$, $y = e^t \sin t$, $z = e^{-t}$

6. $w = \dfrac{x^2 + y^2 + z^2}{2xy + 2yz}$; $x = t^{-1}$, $y = 2t$, $z = 1 + t$

In Exercises 7 to 12, find the partial derivatives $\partial w/\partial s$ and $\partial w/\partial t$ in two ways: (a) by using the chain rule, and (b) by direct substitution.

7. $w = xy(x^2 + y^2)$; $x = s^2 + t^3$, $y = s^3 + t^2$

8. $w = \sin^{-1}(2x + 3y)$; $x = 2te^s$, $y = s^2 e^t$

9. $w = \cos\left(\dfrac{2y}{x+y}\right)$; $x = st$, $y = e^{st}$

10. $w = (xy + xz + yz)^2$; $x = s^2 + t^2$, $y = st$, $z = (s + t)^2$

11. $w = (x^2 + y^2 + z^2)^{-2}$; $x = s^2 + t^2$, $y = st$, $z = s^2 - t^2$

12. $w = \ln(x^2 + 2y + z^2)$; $x = s + t$, $y = 2st$, $z = s - t$

PROBLEMS 11.9

1. If f is a harmonic function of two variables, then prove that
$$g(x,y) = f\left(\dfrac{x}{x^2 + y^2}, \dfrac{y}{x^2 + y^2}\right)$$
is also harmonic.

2. If f is a differentiable function of one variable and
$$g(x,y) = f(ax + by)$$
find both g_1 and g_2.

3. Given any differentiable function f at one variable, prove that $z = x^2 + xf(xy)$ is a solution of the partial differential equation
$$x \dfrac{\partial z}{\partial x} - y \dfrac{\partial z}{\partial y} = z + x^2$$

4. Given any differentiable function f of one variable, prove that $z = xyf\left(\dfrac{x+y}{xy}\right)$ is a solution of the partial differential equation
$$x^2 \dfrac{\partial z}{\partial x} - y^2 \dfrac{\partial z}{\partial y} = (x - y)z$$

5. If $f(x,y)$ is differentiable and $x = r\cos\theta$, $y = r\sin\theta$, express $(\partial f/\partial x)^2 - (\partial f/\partial y)^2$ in terms of r and θ.

6. If $f(x,y,z)$ is differentiable and if $x = \rho \sin\phi \cos\theta$, $y = \rho \sin\phi \sin\theta$ and $z = \rho \cos\phi$, express $(\partial f/\partial x)^2 + (\partial f/\partial y)^2 + (\partial f/\partial z)^2$ and $\partial^2 f/\partial x^2 + \partial^2 f/\partial y^2 + \partial^2 f/\partial z^2$ in terms of ρ, ϕ, and θ.

7. Let $z = y^{xv}$ and find $\partial z/\partial x$, $\partial z/\partial y$.

8. Let F be a differentiable function of three variables and let f, g, and h be differ-

entiable functions of two variables. Suppose that the six equations

$$F(f(y,z),y,z) = F(x,g(x,z),z) = F(x,y,h(x,y)) = 0$$
$$x = f(y,z), \quad y = g(x,z), \quad z = h(x,y)$$

all hold and prove that

$$f_1(y,z)g_2(x,z)h_1(x,y) = -1$$

9. Let $z = f(x,y)$ satisfy the differential equation

$$a\frac{\partial z}{\partial x} + \frac{\partial z}{\partial y} = 0, \quad a \neq 0$$

Introduce new variables $s = x + ay$, $t = x - ay$ whence

$$z = f(x,y) = f\left(\frac{s+t}{2}, \frac{s-t}{2}\right) = g(s,t)$$

Prove that g is a function of t alone and hence prove that the only solutions of the differential equation are of the form

$$z = g(x - ay)$$

10. Let f and g be differentiable functions of two variables and suppose that the equations

$$u = f(x,y), \quad v = g(x,y)$$

can be solved for x and y in terms of u and v, yielding functions

$$x = h(u,v), \quad y = k(u,v)$$

Assuming that h and k are also differentiable, prove that

$$\begin{vmatrix} f_1 & f_2 \\ g_1 & g_2 \end{vmatrix} \begin{vmatrix} h_1 & h_2 \\ k_1 & k_2 \end{vmatrix} = 1$$

where f and g are evaluated at (x,y), and h and k are evaluated at $(f(x,y), g(x,y))$.

11.10 TAYLOR'S THEOREM IN TWO VARIABLES

Let f be a function of two real variables and let (a, b) be an interior point in the domain of f. Let U be a basis neighborhood of (a, b) lying entirely in the domain and consider any point $(a + h, b + k)$ in U. When f is differentiable, we know that we may write

$$f(a+h, b+k) = f(a,b) + hf_1(a,b) + kf_2(a,b) + h\epsilon_1 + k\epsilon_2,$$

where both ϵ_1 and ϵ_2 approach zero as (h, k) approaches $(0, 0)$.

Taylor's theorem generalizes this equation. Before stating the theorem we introduce a special notation for certain sums of partial derivatives. The symbol

$$(hD_1 + kD_2)^n f$$

denotes the sum

$$h^n D_1^n f + nh^{x-1}kD_1^{n-1}D_2 f + \frac{n(n-1)}{2}h^{n-2}k^2 D_1^{n-2}D_2^2 f + \cdots$$
$$+ \binom{n}{i} h^{n-i}k^i D_1^{n-i}D_2^i f + \cdots + k^n D_2^n f$$

11.10 TAYLOR'S THEOREM IN TWO VARIABLES

where $\binom{n}{i}$ denotes the binomial coefficient, that is,

$$\binom{n}{i} = \frac{n!}{i!(n-i)!}$$

When we write

$$(hD_1 + kD_2)^n f(x, y)$$

we shall mean this sum with each partial derivative evaluated at the point (x, y).

Using the symbol just defined, we can state Taylor's theorem for a function of two variables in a succinct form.

■ **THEOREM 11.12.** Let f be a function of two real variables and let (a, b) be an interior point in the domain of f. Let U be a basis neighborhood of (a, b) lying entirely in the domain and let $(a + h, b + k)$ be any other point in U. If the nth partial derivatives of f are all continuous, then there is a point $(a + h\theta, b + k\theta)$, $0 < \theta < 1$, on the line segment from (a, b) to $(a + h, B + k)$ such that

$$f(a + h, b + k) = f(a, b) + \sum_{j=1}^{n-1} \frac{1}{j!}(hD_1 + kD_2)^j f(a, b) + R_n$$

where the correction term is

$$R_n = \frac{1}{n!}(hD_1 + kD_2)^n f(a + h\theta, b + k\theta)$$

PROOF: This result follows quickly from Taylor's theorem for functions of one variable (Section 5.1). We define the function

$$g(t) = f(a + ht, b + kt)$$

on the interval $\{t : 0 \leq t \leq 1\}$. [Note that $g(0) = f(a, b)$ while $g(1) = f(a + h, b + k)$.] By using the chain rule it is easy to verify that the derivatives of g are

$$Dg(t) = (hD_1 + kD_2)f(a + ht, b + kt)$$
$$D^2g(t) = (hD_1 + kD_2)^2 f(a + ht, b + kt)$$
$$\vdots$$
$$D^j g(t) = (hD_1 + kD_2)^j f(a + ht, b + kt)$$
$$\vdots$$

(We use Theorem 11.9 implicitly in obtaining these formulas.)

Now, by Taylor's theorem for a function of one variable, there exists θ, $0 < \theta < 1$, such that

$$g(1) = g(0) + g'(0) + \frac{1}{2}g''(0) + \cdots$$
$$+ \frac{1}{(n-1)!}g^{(n-1)}(0) + \frac{1}{n!}g^{(n)}(\theta)$$

This is precisely the desired equation,

$$f(a+h, b+k) = f(a,b) + \sum_{j=1}^{n-1} \frac{1}{j!}(hD_1 + kD_2)^j f(a,b)$$
$$+ \frac{1}{n!}(hD_1 + kD_2)^n f(a + h\theta, b + k\theta) \quad \blacksquare$$

■ **Example 1.** Let $f(x,y) = \sin(x+y)$. It is easy to compute that
$$f_1(x,y) = f_2(x,y) = \cos(x+y)$$
$$f_{11}(x,y) = f_{12}(x,y) = f_{22}(x,y) = -\sin(x+y)$$
$$f_{111}(x,y) = f_{ijk}(x,y) = -\cos(x+y)$$
and all fourth partial derivatives equal $\sin(x+y)$. At $(a,b) = (0,0)$ we have $f(0,0) = 0$, $f_1(0,0) = 1$, $f_{11}(0,0) = 0$, $f_{111}(0,0) = -1$. Hence Taylor's theorem provides the equation

$$\sin(h+k) = 0 + (h+k) - \tfrac{1}{2}(h+k)^2 \cdot 0 - \tfrac{1}{6}(h+k)^3$$
$$+ \tfrac{1}{24}(h+k)^4 \sin(h\theta, k\theta), \quad 0 < \theta < 1$$
$$= (h+k) - \tfrac{1}{6}(h+k)^3 + \tfrac{1}{24}(h+k)^4 \sin(h\theta, k\theta),$$
$$0 < \theta < 1 \quad \blacksquare$$

Let us rewrite the formula in Theorem 11.12 as follows. In place of $(a+h, b+k)$, let us use (x,y), so that
$$h = x - a, \quad k = y - b$$
Also we shall set $(a + h\theta, b + k\theta) = (c,d)$. Then the theorem states that there is a point (c,d) on the segment from (a,b) to (x,y) such that

$$f(x,y) = f(a,b) + \sum_{j=1}^{n-1} \frac{1}{j!}[(x-a)D_1 + (y-b)D_2]^j f(a,b) + R_n$$

where
$$R_n = \frac{1}{n!}[(x-a)D_1 + (y-b)D_2]^n f(c,d)$$

Now suppose there is a number δ such that
$$\lim_{n \to \infty} R_n = 0$$
whenever both $|x-a| < \delta$ and $|y-b| < \delta$. Then the *double power series*
$$f(x,y) = f(a,b) + \sum_{n=1}^{\infty} \frac{1}{n!}[(x-a)D_1 + (y-b)D_2]^n f(a,b)$$
is said to converge. We also say that $f(x,y)$ has been expanded about the point (a,b) into a double-power series.

■ **Example 2.** We expand $f(x,y) = x^2y - 3xy^2$ about the point $(1,1)$. It is easy to compute that

$$f_1(x,y) = 2xy - 3y^2 \qquad f_2(x,y) = x^2 - 6xy$$

Also $f_{11}(x,y) = 2y$, $f_{12}(x,y) = 2x - 6y$, $f_{22}(x,y) = -6x$, $f_{112}(x,y) = 2$, $f_{122}(x,y) = -6$, and all other derivatives are identically equal to zero. The formula in Theorem 11.12 is then applied with $f(1,1) = -2$, $f_1(1,1) = -1$, $f_2(1,1) = -5$, $f_{11}(1,1) = 2$, $f_{12}(1,1) = -4$, $f_{22}(1,1) = -6$, $f_{112}(1,1) = 2$, and $f_{122}(1,1) = -6$.

Thus we have

$$\begin{aligned}x^2y - 3xy^2 = &-2 - (x-1) - 5(y-1) \\ &+ (x-1)^2 - 4(x-1)(y-1) - 3(y-1)^2 \\ &+ (x-1)^2(y-1) - 3(x-1)(y-1)^2\end{aligned}$$

▪ **Example 3.** If $f(x,y) = e^{x+y}$, then each partial derivative is also e^{x+y}. Hence we have the value $+1$ for each partial derivative at the point $(0,0)$. It follows that we may write

$$\frac{1}{j!}[xD_1 + yD_2]^j f(0,0) = \frac{1}{j!}(x+y)^j$$

Then the formula in Theorem 11.12 becomes

$$e^{x+y} = 1 + (x+y) - \frac{1}{2}(x+y)^2 + \cdots + \frac{1}{(n-1)!} + R_n$$

where

$$R_n = \frac{1}{n!}(xD_1 + yD_2)^n e^{c+d}$$

for some c between 0 and x and some d between 0 and y. It is obvious that $\lim_{n \to \infty} R_n = 0$ for any fixed point (x,y). Therefore e^{x+y} may be expanded into the double-power series

$$e^{x+y} = 1 + x + y + \tfrac{1}{2}x^2 + xy + \tfrac{1}{2}y^2 + \tfrac{1}{6}x^3 + \tfrac{1}{2}x^2y \\ + \tfrac{1}{2}xy^2 + \tfrac{1}{6}y^3 + \cdots$$

which converges for any point (x,y). Note that this is precisely the series obtained by substituting $x + y$ for z in the series

$$e^z = 1 + z + \tfrac{1}{2}z^2 + \tfrac{1}{6}z^3 + \cdots .$$

In each of Exercises 1 to 5, expand the given polynomial about the given point.   **EXERCISES 11.10**

1. $x^3 + 3x^2y - 2xy^2$; $(0, 1)$
2. $2x^3 - 5xy^2$; $(-1, 1)$
3. $x^4 - y^4$; $(1, -1)$
4. $x^4 + 2x^2 - 2y^4$; $(1, 1)$
5. $x^3y^2 + 2y^5$; $(-1, -1)$

In each of Exercises 6 to 15, expand the given function about the point (0, 0) to include all second-order terms.

6. $f(x,y) = \dfrac{1}{1+x+y}$

7. $f(x,y) = \sqrt{1+x+y}$

8. $f(x,y) = \sqrt{1+x^2+y^2}$

9. $f(x,y) = \sin(x+y)$

10. $f(x,y) = \cos x \sin y$

11. $f(x,y) = e^x \ln(1+y)$

12. $f(x,y) = \tan^{-1}\left(\dfrac{x+y}{1-xy}\right)$

13. $f(x,y) = \sin(xy)$

14. $f(x,y) = \ln(1+xy)$

15. $f(x,y) = e^{xy}$

PROBLEMS 11.10

1. Verify the following approximations:
   (a) $\cos x \cos y \cong 1 - \tfrac{1}{2}(x^2+y^2) + \tfrac{1}{24}(x^4 + 6x^2y^2 + y^4)$
   (b) $e^x \tan^{-1} y \cong y + xy + x^2y - \tfrac{1}{3}y^3 + x^3y$

2. Write out Taylor's formula for a function of three variables. Then use methods comparable to those of this section to prove Taylor's theorem for functions of three variables. Can you generalize your results to functions of four and more variables?

11.11 EXTREME VALUES

Recall that we sought a horizontal tangent line in order to locate an extreme value of a function of one variable. Similarly, we now seek a horizontal tangent plane to locate an extreme value of a function of two variables.

■ *Example 1.* Consider the function $f(x,y) = x^2 + 3y^2 - 2x + 6y$ and its derivatives $f_1(x,y) = 2x - 2$ and $f_2(x,y) = 6y + 6$. At the point $(a, b, f(a,b))$, the tangent plane is

$$\{(x,y,z) : z - f(a,b) = (x-a)(2a-2) + (y-b)(6b+6)\}$$

Obviously, this plane is horizontal when both

$$2a - 2 = 0 \quad \text{and} \quad 6b + 6 = 0$$

Thus at the point $(a, b, f(a,b)) = (1, -1, f(1,-1)) = (1, -1, -4)$ we have a horizontal tangent plane, and we expect this to be an extreme point on the graph of f.

By expanding $f(x,y) = x^2 + 3y^2 - 2x + 6y$ about the point $(1, -1)$, we obtain the equation

$$x^2 + 3y^2 - 2x + 6y = -4 + (x-1)^2 + 3(y+1)^2$$

This rewriting makes it obvious that -4 is the (absolute) minimum value of f. ∎

The theory of extreme values of a function of two variables should begin with the following theorem.

■ **THEOREM 11.13.** Let X be a closed and bounded region in the plane. Then any continuous function $f: X \to R$ attains both a maximum and a minimum value.

We shall not prove this result. It is obviously a generalization of Theorem 3.13. On the other hand, this theorem is only a special case of a very general theorem from topology concerning "compact" domains of a continuous function.

■ **DEFINITION.** Let f be a continuous function of two variables with domain X. Then f has a *relative maximum value* at the point $(a, b) \in X$ if there exists a basis neighborhood U of (a, b) such that $f(x,y) \leq f(a, b)$ for each point $(x, y) \in U \cap X$. Similarly, the function f has a *relative minimum value* at (c, d) if there exists a basis neighborhood V of (c, d) such that $f(x,y) \geq f(c, d)$ for each $(x, y) \in V \cap X$. We speak of either as a *relative extreme value* of f.

■ **THEOREM 11.14.** Suppose that f attains a relative extreme value at an interior point (a, b) of its domain. If both $f_1(a, b)$ and $f_2(a, b)$ exist, then $f_1(a, b) = 0 = f_2(a, b)$. (In brief, the gradient vector vanishes at an extreme value.)

PROOF: We prove the result only for a relative maximum value, the proof for a relative minimum value being analogous.

Suppose that $f(x, y) \leq f(a, b)$ for each point (x, y) in a basis neighborhood U of (a, b). Without loss of generality, we may assume that U lies entirely in the domain of f because we assumed that (a, b) is an interior point. This means that, for sufficiently small values of h, we have

$$f(a + h, b) - f(a, b) \leq 0$$

Thus if $h > 0$, we have

$$\frac{f(a + h, b) - f(a, b)}{h} \leq 0$$

while, if $h < 0$, we have

$$\frac{f(a + h, b) - f(a, b)}{h} \geq 0$$

It surely follows that

$$\lim_{h \to 0+} \frac{f(a+h, b) - f(a, b)}{h} = f_1(a, b) \leq 0$$

and

$$\lim_{h \to 0-} \frac{f(a+h, b) - f(a, b)}{h} = f_1(a, b) \geq 0$$

Therefore $f_1(a, b) = 0$. Similarly, we prove that $f_2(a, b) = 0$. ∎

Any point (a, b) at which both f_1 and f_2 have the value zero is called a *critical point* for the function f. In other words, the gradient vector is zero at a critical point.

■ **Example 2.** Let $f(x, y) = x^2 + xy + y^2 - x + y$. Then the gradient is

$$(f_1, f_2) = (2x + y - 1, x + 2y + 1)$$

This is the zero vector if and only if the equations

$$2x + y - 1 = 0$$
$$x + 2y + 1 = 0$$

hold simultaneously. This happens at the point $(1, -1)$ which is therefore a critical point for f.

If we expand $f(x, y) = x^2 + xy + y^2 - x + y$ in a double-power series about its critical point $(1, -1)$, we obtain

$$\begin{aligned} x^2 + xy + y^2 - x + y \\ &= -1 + (x-1)^2 + (x-1)(y+1) + (y+1)^2 \\ &= -1 + (x-1)^2 + (x-1)(y+1) + \tfrac{1}{4}(y+1)^2 + \tfrac{3}{4}(y+1)^2 \\ &= -1 + [(x-1) + \tfrac{1}{2}(y+1)]^2 + \tfrac{3}{4}(y+1)^2 \\ &= -1 + (x + \tfrac{1}{2}y - \tfrac{1}{2})^2 + \tfrac{3}{4}(y+1)^2 \end{aligned}$$

From this final expression for $f(x, y)$, we see that the point $(1, -1)$ actually provides an absolute minimum value of the function. ∎

The critical point (a, b) for f need not locate an extreme value of f. For instance, it could locate a *minimax value* of the function. In Fig. 11.17 we show the graph of $f(x, y) = y^2 - x^2$. We note that $f_1(x, y) = -2x$ and $f_{11}(x, y) = -2$. Therefore, $f_1(0, 0) = 0$, while $f_{11}(0, 0) = -2$. This tells us that the trace of the graph in the xz coordinate plane has a maximum point at the origin. On the other hand, we have $f_2(x, y) = 2y$, whence $f_{22}(x, y) = 2$. Thus $f_2(0, 0) = 0$ and $f_{22}(0, 0) = 2$, so that the trace in the yz plane has a minimum point at the origin. If $f(a, b)$ is a minimax value of f, then the point $(a, b, f(a, b))$ on the graph of f is called a *saddle point* on the surface.

Our next result is a *second derivative test* for determining the nature of a critical point.

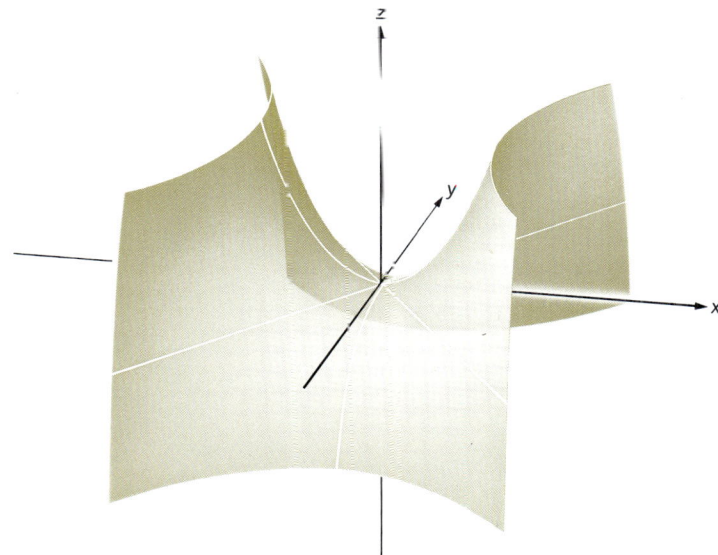

FIGURE 11.17

■ **THEOREM 11.15.** Let f and all of its partial derivatives up to the third order be continuous, and let (a, b) be a critical point in the interior of the domain of f. Then

1. If $f_{11}(a,b)f_{22}(a,b) > f_{12}^2(a,b)$, the value $f(a,b)$ is an extreme value. In particular, $f(a,b)$ is a relative minimum value if $f_{11}(a,b) > 0$ and is a relative maximum value if $f_{11}(a,b) < 0$.
2. If $f_{11}(a,b)f_{22}(a,b) < f_{12}^2(a,b)$, then $f(a,b)$ is a minimax value.
3. If $f_{11}(a,b)f_{22}(a,b) = f_{12}^2(a,b)$, then this test fails.

PROOF: We use Taylor's theorem to expand $f(x,y)$ about (a,b). Since $f_1(a,b) = 0 = f_2(a,b)$, we have

$$f(a+h, b+k) - f(a,b)$$
$$= \tfrac{1}{2}h^2 f_{11}(a,b) + hk f_{12}(a,b) + \tfrac{1}{2}k^2 f_{22}(a,b)$$
$$+ \tfrac{1}{6}(hD_1 + kD_2)^3 f(a+h\theta, b+k\theta)$$

for some θ satisfying $0 < \theta < 1$.

For the sake of brevity, we define

$$A = f_{11}(a,b), \qquad B = f_{12}(a,b), \quad \text{and} \quad C = f_{22}(a,b)$$

Also, we define the numbers

$$r = \sqrt{h^2 + k^2}, \qquad u = \frac{h}{r}, \quad \text{and} \quad v = \frac{k}{r}$$

and note that $u^2 + v^2 = 1$. Then we can rewrite the difference $f(a+h, b+k) - f(a,b)$ in the form

$$\frac{1}{2}Ah^2 + Bhk + \frac{1}{2}Ck^2 + R_2$$
$$= \frac{1}{2}r^2(Au^2 + 2Buv + Cv^2) + \frac{1}{6}r^3(uD_1 + vD_2)^3 f(a + h\theta, b + k\theta)$$
$$= \frac{1}{2}r^2[Au^2 + 2Buv + Cv^2 + \frac{r}{3}(uD_1 + vD_2)^3 f(a + h\theta, b + k\theta)]$$

The third partial derivatives are bounded because they are continuous. Hence there exists $M > 0$ such that

$$|\tfrac{1}{3}(uD_1 + vD_2)^3 f(a + h\theta, b + k\theta)| \leq M$$

for all (u, v) satisfying $u^2 + v^2 = 1$.

If the condition $AC - B^2 > 0$ holds and if $A > 0$, then it follows that the quantity

$$Au^2 + 2Buv + Cv^2, \quad u^2 + v^2 = 1$$

has a positive minimum value, call it m. (The reader should prove this claim.) Select r so small that $m > rM$. Then, we obviously have

$$f(a + h, B + k) - f(a, b) \geq \tfrac{1}{2}r^2(m - rM) \geq 0$$

Hence $f(a, b)$ is a relative minimum value. A similar proof holds if $f_{11}(a, b) = A < 0$. The remaining two cases (2 and 3) we leave for the reader to establish. ∎

■ **Example 3.** The function $f(x, y) = x^3 + y^2 - 6xy + 6x + 3y + 1$ has partial derivatives

$$f_1(x, y) = 3x^2 - 6y + 6, \quad f_2(x, y) = 2y - 6x + 3$$
$$f_{11}(x, y) = 6x, \quad f_{12}(x, y) = 6, \quad f_{22}(x, y) = 2$$

We locate the critical points by solving simultaneously the equations $f_1(x, y) = 0$ and $f_2(x, y) = 0$. In this case these equations are

$$3x^2 - 6y + 6 = 0$$
$$2y - 6x + 3 = 0$$

It is easy to compute that $(1, \tfrac{3}{2})$ and $(5, \tfrac{27}{2})$ are the critical points for f. For the first of these points we have

$$f_{11}(1, \tfrac{3}{2}) = 6(1) = 6 > 0$$

and

$$f_{11}(1, \tfrac{3}{2})f_{22}(1, \tfrac{3}{2}) - f_{12}^2(1, \tfrac{3}{2}) = 6(1) \cdot 2 - (-6)^2 = -24$$

This tells us that $f(1, \tfrac{3}{2}) = \tfrac{3}{4}$ is a minimax value of f. For the other critical point,

$$f_{11}(5, \tfrac{27}{2}) = 6(5) = 30 > 0$$

while

$$f_{11}(5, \tfrac{27}{2})f_{22}(5, \tfrac{27}{2}) - f_{12}{}^2(5, \tfrac{27}{2}) = 6(5) \cdot 2 - (-6)^2 = +24$$

Therefore, by Theorem 11.15 $f(5, \tfrac{27}{2}) = -\tfrac{105}{4}$ is a relative minimum value.

The method of least squares. A useful application of minimizing a function of two variables occurs in the "method of least squares." This is a technique for fitting a line to a set of n points that have been obtained from an experiment, for instance. That is, we are given n points $(x_1, y_1), \cdots, (x_n, y_n)$ and we wish to obtain a line $\{(x, y) : y = mx + b\}$ that in some senses best fits these points. Of course, we do so by choosing values of the slope m and y intercept b.

For any fixed choice of m and b, the vertical distances

$$|y_i - mx_i - b|, \quad i = 1, 2, \cdots, n,$$

between the given points and the corresponding line are called *deviations*. The method of least squares simply minimizes the sum of squares of these deviations. That is, we minimize the nonnegative function

$$f(m, b) = \sum_{i=1}^{n} (y_i - mx_i - b)^2$$

The first partial derivatives are

$$f_1(m, b) = -2 \sum_{i=1}^{n} x_i(y_i - mx_i - b)$$

$$f_2(m, b) = -2 \sum_{i=1}^{n} (y_i - mx_i - b)$$

Thus we obtain the necessary conditions for an extremum value

$$f_1(m, b) = 0 = -2 \sum_{i=1}^{n} (x_i y_i - mx_i^2 - bx_i)$$

and

$$f_2(m, b) = 0 = -2 \sum_{i=1}^{n} (y_i - mx_i - b)$$

or

$$b \sum_{i=1}^{n} x_i + m \sum_{i=1}^{n} x_i^2 = \sum_{i=1}^{n} x_i y_i$$

and

$$nb + m \sum_{i=1}^{n} x_i = \sum_{i=1}^{n} y_i$$

We introduce the average values

$$\bar{x} = \frac{1}{n}\sum_{i=1}^{n} x_i, \qquad \bar{y} = \frac{1}{n}\sum_{i=1}^{n} y_i$$

in order to simplify the solution of these equations. This results in the equations being written as

$$b(n\bar{x}) + m\left(\sum_{i=1}^{n} x_i^2\right) = \sum_{i=1}^{n} x_i y_i$$
$$b + m\bar{x} = \bar{y}$$

The solutions are then obtained as

$$m = \frac{\sum_{i=1}^{n} x_i y_i - n\bar{x}\bar{y}}{\sum_{i=1}^{n} x_i^2 - n\bar{x}^2}$$
$$b = \frac{\bar{y}\sum_{i=1}^{n} x_i^2 - \bar{x}\sum_{i=1}^{n} x_i y_i}{\sum_{i=1}^{n} x_i^2 - n\bar{x}^2}$$

■ *Example 4.* We use least squares to fit a line to the four points $(2, 4.2)$, $(4, 4.9)$, $(6, 5.8)$, and $(8, 7.3)$, which are almost on the line

$$\{(x,y) : y = \tfrac{1}{2}x + 3\}.$$

We have

$$\bar{x} = \tfrac{1}{4}(2 + 4 + 6 + 8) = 5,$$
$$\bar{y} = \tfrac{1}{4}(4.2 + 4.9 + 5.8 + 7.3) = 5.55$$

Then this formula yields

$$m = \frac{2(4.2) + 4(4.9) + 6(5.8) + 8(7.3) - 4(5)(5.55)}{2^2 + 4^2 + 6^2 + 8^2 - 4(5^2)} = 0.51$$

whence

$$b = \bar{y} - m\bar{x} = 5.55 - (0.51)(5) = 3.00$$

The equation of the fitted line is therefore,

$$y = (0.51)x + 3.00 \quad ■$$

For functions of more than two variables the theory of critical points becomes more complicated but does not entail new concepts. If the function $f(w, x, y, z)$ has continuous first partial derivatives, a point (w_0, x_0, y_0, z_0) at which all four derivatives vanish is called a critical point. Taylor's theorem for such functions then provides criteria for maximum, minimum, and minimax values of the function, much as Theorem 11.15. We shall not carry out the details of such an analysis, however.

11.11 EXTREME VALUES

For each function in Exercises 1 to 20, locate and identify the critical points.

EXERCISES 11.11

1. $f(x, y) = x^2 + 3y^2 - 4x + 6y + 1$
2. $f(x, y) = 2x^2 - y^2 + 4x + 4y - 2$
3. $f(x, y) = x^2 + 4xy + 4y^2 + 4x - 8y - 5$
4. $f(x, y) = 14y^2 - 6x^2 - 12x - 28y + 15$
5. $f(x, y) = 9x^2 + 12xy + 7y^2 - 24x - 12y + 3$
6. $f(x, y) = 2x^2 - 6xy + 3y^2 - 6x + 12y - 6$
7. $f(x, y) = y^3 + x^2 - 4xy - 8x + 13y + 2$
8. $f(x, y) = x^3 - 27xy + y^3$
9. $f(x, y) = x^4y^2 + x^3y^3 - 12x^3y^2$
10. $f(x, y) = xy^2(24 - 2x - 3y)$
11. $f(x, y) = xy^2(40 - 2x - 5y)^3$
12. $f(x, y) = xy + \dfrac{1}{x} - \dfrac{64}{y}$
13. $f(x, y) = xye^{-(2x+5y)}$
14. $f(x, y) = e^x \sin y$
15. $f(x, y) = e^{-x} \cos^2 y$
16. $f(x, y) = \sin x + \sin y + \sin(x + y)$
17. $f(x, y, z) = x^2 + 5y^2 + z^2 + 4xy - 2yz - 6x - 8y - 4z$
18. $f(x, y, z) = x^2 + 3y^2 + 2z^2 - 4x + 6y - 8z + 1$
19. $f(x, y, z) = 5x^2 + y^2 + 4z^2 + 2y + 8z + 2$
20. $f(x, y, z) = 5x^2 + 2y^2 + 2z^2 - 2xy - 4yz + 4x + 4y - 4z + 5$

PROBLEMS 11.11

1. Find the minimum distance from the point $(-7, -3)$ to the line
$$\{(x, y) : 3x - 4y = 4\}$$

2. Find the minimum distance from the point $(2, 3, 1)$ to the plane
$$\{(x, y, z) : z = 2x + 2y - 3\}$$

3. What are the least and the greatest distances from the point $(2, 1, -3)$ to the sphere $\{(x, y, z) : x^2 + y^2 + z^2 = 56\}$? Solve this both by calculus and (more simply) by geometric methods.

4. Find the minimum distance between the two lines
$$\{(x, y, z) : x = 1 + s, y = s - 28, z = 1 + 2s\}$$
and
$$\{(x, y, z) : x = -2 - 2t, y = -1 + 2t, z = 5 - t\}$$

5. Let (x_0, y_0, z_0) be a fixed point whose coordinates are all positive. Of all planes that contain this point and have positive intercepts on all three axes, which one cuts off the least volume from the first octant? What is this least volume?

6. Prove that of all triangles having a given perimeter, the equilateral triangle has the largest area.

7. Three points are to be selected on a given circle. How should they be selected so that the corresponding triangle has the longest perimeter?

8. Of all rectangular boxes having a given surface area, show that the cube has the most volume. Conversely, of all rectangular boxes having a given volume, show that the cube has the least surface area.

9. Prove that the rectangular box of largest volume that can be inscribed in a given sphere is a cube.

10. There are S sq ft of material available to make a rectangular box without a top. What dimensions will result in a box of maximum volume?

11. A rectangular box has one vertex at the origin, has three adjacent sides in the coordinate planes and the vertex opposite the origin in the plane

$$\{(x, y, z) : z = -6x - 4y + 108\}$$

If this box has the largest volume of all boxes so located, what are its dimensions?

12. A display case is to have a volume of 99 cu ft, and is to be in the form of a rectangular box with a glass top and a glass front. If glass costs 80 cents per sq ft, if the base material costs 40 cents per sq ft, and if the back and ends cost 30 cents per sq ft, what are the dimensions of the least expensive case and how much will it cost?

13. A silo is to be built in the form of a cylinder with a conical roof. If its total capacity is 1000π cu ft, what dimensions will minimize the surface area to be painted?

14. Let $a_1, a_2, \cdots, a_n$ be n positive numbers. Determine the maximum value of the linear function

$$f(x_1, \cdots, x_n) = \sum_{i=1}^{n} a_i x_i \quad \text{if} \quad \sum_{i=1}^{n} x_i^2 = 1$$

15. Let $U(x, y) = x^{2/3} y^{1/3}$ be the utility to a consumer of quantities x and y of two different commodities whose unit prices are \$7 and \$3, respectively. If the consumer has \$84 to spend, how much should he allot to each commodity in order to maximize his utility? Can you generalize this problem?

16. Two firms are in competition, each having profit depending in part upon the output of the other. Let the outputs of the two firms be O_1 and O_2, while their profits are P_1 and P_2. Suppose these quantities are related by the equations

$$P_1 = 60O_1 - O_1^2 - O_2^2 \qquad P_2 = 80O_2 - 2O_2^2 - \tfrac{1}{2}O_1^2$$

Determine profits if each firm acts independently to maximize its profit. Then determine the maximum value of $P_1 + P_2$ if the firms decide to cooperate.

11.12 EXTREME VALUES WITH CONSTRAINTS

Many problems necessitate finding an extreme value of a function when the variables are restricted by some condition. We already have solved several such problems, of course, but here we introduce the method of *Lagrange multipliers*. Again proofs are deferred.

■ *Example 1.* Suppose we must find the minimum distance from the origin to the curve $\{(x, y) : 16x^2 - 12xy + 25 = 0\}$. The function to be minimized is

$$f(x, y) = \sqrt{x^2 + y^2}$$

while the point (x, y) lies on the curve and hence is constrained by the equation

$$\phi(x, y) = 16x^2 - 12xy + 25 = 0$$

The curve $\{(x, y) : 16x^2 - 12xy + 25 = 0\}$ is seen in Fig. 11.18. The concentric circles shown are just level curves of the function f. As is sug-

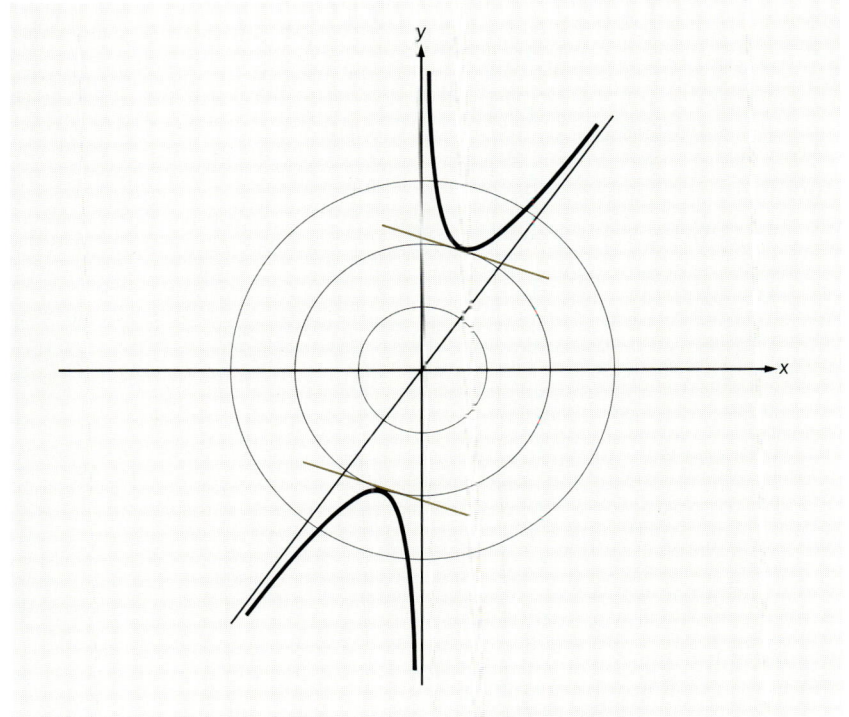

FIGURE 11.18

gested by this figure, we seek those points on the curve whose tangent lines are also tangent to the level curve through those points. From Section 11.9 we know that the level curves have the slope

$$\frac{dy}{dx} = -\frac{f_1(x,y)}{f_2(x,y)} = -\frac{x}{y}$$

The constraint curve has the slope

$$\frac{dy}{dx} = -\frac{\phi_1(x,y)}{\phi_2(x,y)} = -\frac{32x - 12y}{-12x} = \frac{8x - 3y}{3x}$$

And we seek those points for which

$$-\frac{f_1}{f_2} = -\frac{\phi_1}{\phi_2}$$

or, equivalently, such that

$$f_1(x,y) = \lambda \phi_1(x,y)$$

and

$$f_2(x,y) = \lambda \phi_2(x,y)$$

for some constant of proportionality λ. For the present case, we write

$$\frac{x}{\sqrt{x^2 + y^2}} = \lambda(32x - 12y)$$

$$\frac{y}{\sqrt{x^2 + y^2}} = \lambda(-12x)$$

Together with the constraint

$$16x^2 - 12xy + 25 = 0$$

these equations determine the desired point (x,y). (In advanced treatments of this theory, a meaning may be assigned to the optimal value of the constant λ. For our present purposes, however, it is not necessary to evaluate λ.)

From the second equation we have

$$\lambda = -\frac{y}{12x\sqrt{x^2 + y^2}}$$

Substituting this value of λ into the first equation, we obtain

$$\frac{x}{\sqrt{x^2 + y^2}} = -\frac{y}{12x\sqrt{x^2 + y^2}}(32x - 12y)$$

which reduces to

$$(3x - y)(x + 3y) = 0$$

Substituting $y = 3x$ into the constraint equation $16x^2 - 12xy + 25 = 0$, we arrive at $x = \pm\frac{1}{2}\sqrt{5}$, whence $y = \pm\frac{3}{2}\sqrt{5}$. The value $y = -\frac{1}{3}x$ leads to no solutions. Then it follows that the desired minimum distance is $\sqrt{5/4 + 45/4} = \frac{5}{2}\sqrt{2}$. ∎

11.12 EXTREME VALUES WITH CONSTRAINTS

The constant of proportionality λ is a *Lagrange multiplier* and the technique used in Example 1 is the method of *Lagrange multipliers*. The simple case of functions of two variables is covered by the following theorem.

■ **THEOREM 11.16.** Let f and ϕ be differentiable functions of two variables. Subject to the constraint $\phi(x, y) = 0$, suppose that the function f has a relative extreme value at the point (a, b). If we assume that not both $\phi_1(a, b)$ and $\phi_2(a, b)$ are zero, there must be a constant λ such that the equations

$$f_1(a, b) + \lambda \phi_1(a, b) = 0$$
$$f_2(a, b) + \lambda \phi_2(a, b) = 0$$

both hold. (The proof is left as a problem for the reader.)

■ **Example 2.** A box with a square base and no top is to be built to hold V cu ft. We want to be sure that its surface area is as small as possible. If the length of the sides is x, while the height is y, then the area to be minimized is

$$A(x, y) = x^2 + 4xy$$

Subject to the constraint

$$\phi(x, y) = x^2 y - V = 0$$

we form the function, sometimes called the *Lagrangian*,

$$\psi(x, y; \lambda) = A(x, y) + \lambda \phi(x, y)$$
$$= x^2 + 4xy + \lambda(x^2 y - V)$$

Then

$$\psi_1(x, y) = 2x + 4y + 2\lambda xy$$
$$\psi_2(x, y) = 4x + \lambda x^2$$

Setting these equal to zero, we obtain

$$\lambda = -\frac{4}{x}$$

from the second equation, and then

$$2x + 4y + 2\left(-\frac{4}{x}\right)xy = 0$$

or

$$x - 2y = 0$$

Hence the height of the box should be $\frac{1}{2}$ of the length of the one side. ■

We shall give no proofs of the more general theorems that justify the use of Lagrangian multipliers in the following rule: Let $f, \phi_1, \cdots, \phi_k$ be differentiable functions of n variables. To locate the extremum values of f subject to the k constraints,

$$\phi_i(x_1, \cdots, x_n) = 0, \quad i = 1, 2, \cdots, k,$$

we form the new function of $n + k$ variables

$$\psi(x_1, \cdots, x_n; \lambda_1, \cdots, \lambda_k)$$
$$= f(x_1, \cdots, x_n) + \sum_{i=1}^{k} \lambda_i \phi_i(x_1, \cdots, x_n)$$

The $n + k$ first partial derivatives of ψ are set equal to zero, giving us $n + k$ equations in $n + k$ unknowns. Note that each equation

$$\frac{\partial \psi}{\partial \lambda_i} = 0, \quad i = 1, 2, \cdots, k$$

is precisely one of the constraints. We do not require values of the multipliers λ_i, so that we only solve for $(x_1, \cdots, x_n)$.

■ **Example 3.** A display case is to be built to contain 189 cu ft. It is to have a solid base whose cost is $4.50 per sq ft and glass sides and top costing $6 per sq ft. What are the dimensions of the case of lowest cost?

The cost function is

$$6(xy + 2xz + 2yz) + 4.5(xy)$$

while

$$xyz = 189$$

is the constraint. We form

$$10.5xy + 12xz + 12yz + \lambda(xyz - 189)$$

The partial derivatives set equal to zero provide the four equations

$$10.5y + 12z + \lambda yz = 0$$
$$10.5x + 12z + \lambda xz = 0$$
$$12x + 12y + \lambda xy = 0$$
$$xyz - 189 = 0$$

From the first two equations it is easy to conclude that $x = y$; hence the base of the display case should be square. It is also easy to eliminate λ and determine that

$$z = \frac{10.5}{12} y$$

Using the fourth equation, we have

$$y^2\left(\frac{10.5}{12}y\right) = 189$$

or

$$y^3 = 189 \times \frac{12}{10.5} = 216$$

Thus $y = 6 = x$ and $z = \frac{10.5}{12} \times 6 = 5.25$.

■ **Example 4.** Find the maximum and minimum values of z in the curve of intersection of the plane $\{(x, y, z) : x + y + 8z = 8\}$ and the surface $\{(x, y, z) : 2z = \sqrt{16 - x^2 - y^2}\}$. We write the constraints in the form

$$\phi_1(x, y, z) = x + y + 8z - 8 = 0$$
$$\phi_2(x, y, z) = x^2 + y^2 + 4z^2 - 16 = 0$$

It follows that the function ψ is

$$\psi(x, y, z; \lambda_1, \lambda_2)$$
$$= z + \lambda_1(x + y + 8z - 8) + \lambda_2(x^2 + y^2 + 4z^2 - 16)$$

The five equations are

$$\lambda_1 + 2\lambda_2 x = 0$$
$$\lambda_1 + 2\lambda_2 y = 0$$
$$1 + 8\lambda_1 + 8\lambda_2 z = 0$$
$$x + y + 8z - 8 = 0$$
$$x^2 + y^2 + 4z^2 - 16 = 0$$

The first two equations obviously imply that $x = y$. From the fourth equation, we then obtain

$$x + x + 8z - 8 = 0$$

or

$$x = y = 4(1 - z)$$

Substituting these expressions for x and y into the fifth equation, we find

$$16(1 - z)^2 + 16(1 - z)^2 + 4z^2 - 16 = 0$$

or

$$36z^2 - 64z + 16 = 0$$

The desired extreme values of z are therefore

$$z = \frac{8 \pm 2\sqrt{7}}{9} \quad ■$$

The theory concerning extreme values of functions when these values occur on a boundary is a proper subject for advanced calculus. This theory is the essence of *optimization* which finds applications in many disciplines. One optimization problem leads to the theory of *linear programming* in which the extreme values always occur on the boundary of the region involved.

EXERCISES 11.12 In each of Exercises 1 to 18, find the minimum value of the function subject to the given constraints.

1. $f(x,y) = x + y$ if $x^2 + y^2 = 1$
2. $f(x,y) = x^2 + y^2$ if $ax + by + c = 0$
3. $f(x,y) = x^2 + y^2$ if $x^2 y = 2$
4. $f(x,y) = x^2 + y^2$ if $x^3 + y^3 = 6xy + 6$
5. $f(x,y) = x^2 + 24xy + 8y^2$ if $x^2 + y^2 = 25$
6. $f(x,y,z) = x + 2y + 3z$ if $x^2 + y^2 + z^2 = 16$
7. $f(x,y,z) = x^2 + y^2 + z^2$ if $x + 3y - 2z = 4$
8. $f(x,y,z) = x^2 + y^2 + z^2$ if $x^2 + 2xy + y^2 = 100$
9. $f(x,y,z) = 2x^2 + y^2 + z^2$ if $x^2 yz = 1$
10. $f(x,y,z) = x^2 + y^2 + z^2$ if $ax + by + cz = d$
11. $f(x,y,z) = x^2 + y^2 + z^2$ if $x^2 + \dfrac{y^2}{4} + \dfrac{z^2}{9} = 1$
12. $f(x,y,z) = xyz$ if $xy + 2xz + 3yz = 162$
13. $f(x,y,z) = xyz$ if $x^2 + y^2 = 1$ and $x = z$
14. $f(x,y,z) = x^2 + y^2 + z^2$ if $x + 2y + z = 1$ and $2x - y - 3z = 4$
15. $f(x,y,z) = 2x^2 + y^2 + 3z^2$ if $2x + y - 3z = 4$ and $x - y + 2z = 6$
16. $f(w,x,y,z) = w^2 + x^2 + y^2 + z^2$ if $w + x + 3y - z = 2$ and $2w + 2x - y + z = 4$
17. $f(w,x,y,z) = w^2 + x^2 + y^2 + z^2$ if $2w + x + y - z = 2$ and $3w + 2x - y + z = 3$
18. $f(w,x,y,z) = 2w^2 + 2x^2 + y^2 + z^2$ if $-w + x + y + z = 1$, $2w + 2x + y - z = 2$ and $w - x + y - z = -4$

In Exercises 19 to 25, find the extremum distances from the origin to the geometric figure described.

19. The line $\{(x,y) : 4x + 3y = 16\}$
20. The curve $\{(x,y) : x^2 + 3xy - y^2 = 13\}$
21. The surface $\{(x,y,z) : 2z = xy + 4\}$
22. The surface $\{(x,y,z) : 2z = x^2 - y^2 + 2\}$
23. The curve formed by the intersection of the surfaces $\{(x,y,z) : z^2 = 2xy\}$ and $\{(x,y,z) : 3x^2 + 2y + z^2 = 30\}$

24. The curve formed by the intersection of the plane $\{(x,y,z) : z = x+y-1\}$ and the surface $\{(x,y,z) : x^2 + y^2 = z^2\}$

25. The curve formed by the intersection of the surfaces $\{(x,y,z) : x^2 + z^2 = 1\}$ and $\{(x,y,z) : x^2 - y^2 + z^2 - xz + 1 = 0\}$

PROBLEMS 11.12

1. Find the minimum value of $f(x,y,z) = ax^2 + by^2 + cz^2$, $a > 0$, $b > 0$, $c > 0$ if $Ax + By + Cz = D$.

2. Find the minimum value of $f(x,y,z) = x^a y^b z^c$ if $x + y + z = d$ where a, b, c, and d are all positive.

3. Prove that the maximum value of
$$f(x_1, y_2, \ldots, x_n) = x_1 x_2 \cdots x_n$$
when subject to the constraint $x_1 + x_2 + \cdots + x_n = c$ is $(c/n)^n$. Use this fact to prove that
$$\sqrt[n]{x_1 x_2 \cdots x_n} \leq \frac{1}{n}(x_1 + x_2 + \cdots + x_n)$$

4. Let a and b be positive numbers such that $1/a + 1/b = 1$. Show that
$$\frac{x^a}{a} + \frac{y^b}{b} \geq xy$$
for all (x,y) such that $x \geq 0$ and $y \geq 0$.

5. Consider the function
$$f(x,y) = \frac{ax^2 + 2bxy + cy^2}{dx^2 + 2exy + fy^2}, \quad df - e^2 > 0$$
Prove that the maximum value of f equals the larger root of the equation
$$(ac - b^2) - (af - 2be + ca)t + (df - e^2)t^2 = 0$$

6. Prove that the product of three positive numbers whose sum is specified is a maximum if and only if the three numbers are all equal.

7. Find the values of a and b for which the closed curve $\{(x,y) : b^2 x^2 + a^2 y^2 = a^2 b^2\}$ contains the circle $\{(x,y) : (x-1)^2 + y^2 = 1\}$ and has the minimum area.

8. Find the point inside a given triangle for which the sum of the squares of the perpendicular distances from the sides is a minimum.

9. A cross section of a house is a rectangle surmounted by an isosceles triangle. If the area of the cross section is preassigned, what dimensions provide the minimum perimeter?

10. Let P_1, P_2, P_3, and P_4 be the vertices of a quadrilateral. Find the point Q for which the sum of distances from Q to these vertices is a minimum.

11. Given four positive numbers a, b, c, and d, determine which quadrilateral having its edges with these four lengths has a maximum area.

12. Let the sum of the lengths of the 12 edges of a rectangular box be denoted by S. We impose the condition that the sum of the areas of the 6 sides of the box be $\frac{1}{25}S^2$. Consider the cube whose edges have the same length as the shortest edge of the box. Determine the dimensions of the box if its volume less the volume of the cube is to be a maximum.

11.13 EXACT DIFFERENTIALS

The general first order differential equation

$$\frac{dy}{dx} = g(x,y)$$

is often expressed in the form

$$M(x,y)\,dx + N(x,y)\,dy = 0$$

The expression on the left-hand side should remind us of a total differential. Indeed, if there were a function $f(x,y)$ such that

$$f_1(x,y) = M(x,y) \quad \text{and} \quad f_2(x,y) = N(x,y)$$

then the differential equation above would be precisely

$$df(x,y;\,dx,\,dy) = 0$$

Then, in view of Problem 11.8.5, the solution of the differential equation would be

$$f(x,y) = C$$

■ **DEFINITION.** The expression

$$M(x,y)\,dx + N(x,y)\,dy$$

is said to be an *exact differential* if there exists a differentiable function f such that

$$f_1(x,y) = M(x,y) \quad \text{and} \quad f_2(x,y) = N(x,y)$$

and hence

$$M(x,y)\,dx + N(x,y)\,dy = df(x,y;\,dx,\,dy)$$

Confronted with an exact differential $M(x,y)\,dx + N(x,y)\,dy$, the fact that there is a function f such that $M = f_1$ and $N = f_2$ implies that

$$M_2(x,y) = f_{12}(x,y), \quad N_1(x,y) = f_{21}(x,y)$$

Now, if we assume that the second partial derivatives of f are continuous, then Theorem 11.9 concludes that they are equal. This proves one-half of the following result.

■ **THEOREM 11.17.** Let the functions M and N be continuous and have continuous first partial derivatives in R^2. Then the expression $M(x,y)\,dx + N(x,y)\,dy$ is an exact differential if and only if

$$M_2(x,y) = N_1(x,y)$$

for each point (x,y) in R^2.

PROOF: The necessity of this equation has already been established. The fact that the equation is also sufficient requires a new concept if it is to be done with rigor. We shall only give the following sketch of the proof.

Holding y fixed, we integrate $M(x, y)$ with respect to x, obtaining

$$\phi(x, y) = \int M(x, y)\, dx$$

Clearly, we have $\phi_1(x, y) = M(x, y)$. In order that we also have the second desired equation, we form a new function

$$f(x, y) = \phi(x, y) + \psi(y)$$

where $\psi(y)$ is to be determined. Note that $f_1(x, y) = M(x, y)$. We want the equation

$$f_2(x, y) = \phi_2(x, y) + \psi'(y) = N(x, y)$$

or

$$\psi'(y) = N(x, y) - \phi_2(x, y)$$

If the expression $N - \phi_2$ is independent of x, then we can integrate with respect to y to obtain $\psi(y)$ and hence have the desired function f completely determined. We establish this independence by differentiating with respect to x. This results in

$$N_1(x, y) - \phi_{21}(x, y) = N_1(x, y) - \phi_{12}(x, y)$$
$$= N_1(x, y) - M_2(x, y)$$

By assumption $N_1 = M_2$ we obtain the derivative

$$N_1(x, y) - \phi_{21}(x, y) = 0$$

and this completes the argument. The desired function f is given by

$$f(x, y) = \int M(x, y)\, dx + \int \left[N(x, y) - \frac{\partial}{\partial y} \int M(x, y)\, dx \right] dy$$

(In this argument the difficulties involved with "partial" integration have been glossed over. It is precisely at this point that more rigorous details are given in a complete proof.) ∎

■ *Example 1.* The expression

$$(x^2 + 2xy)\, dx + (x^2 - 3y^2)\, dy$$

is an exact differential because

$$\frac{\partial}{\partial y}(x^2 + 2xy) = 2x = \frac{\partial}{\partial x}(x^2 - 3y^2)$$

Furthermore, we have

$$\int M(x, y)\, dx = \int (x^2 + 2xy)\, dx = \frac{1}{3} x^3 + x^2 y$$

and hence
$$\frac{\partial}{\partial y}\int M(x,y)\,dx - \frac{\partial}{\partial y}\left(\frac{1}{3}x^3 + x^2 y\right) = x^2$$

It follows that
$$f(x,y) = \int M(x,y)\,dx + \int \left[N(x,y) - \frac{\partial}{\partial y}\int M(x,y)\,dx\right] dy$$
$$= \frac{1}{3}x^3 + x^2 y + \int [x^2 - 3y^2 - x^2]\,dy$$
$$= \frac{1}{3}x^3 + x^2 y - y^3 + C$$

The differential equation
$$M(x,y)\,dx + N(x,y)\,dy = 0$$
is said to be an *exact first-order differential equation* if
$$M_2(x,y) = N_1(x,y)$$
Its general solution is $f(x,y) = C$, where f is the function produced in the proof of Theorem 11.17.

■ *Example 2.* The differential equation
$$(12x^2 - 6xy + 2y^2)\,dx + (-3x^2 + 4xy + 3y^2)\,dy = 0$$
is exact because
$$\frac{\partial}{\partial y}(12x^2 - 6xy + 24^2) = -6x + 4y = \frac{\partial}{\partial x}(-3x^2 + 4xy + 3y^2)$$

The solution is obtained by first computing
$$\int M(x,y)\,dx = \int (12x^2 - 6xy + 2y^2)\,dx$$
$$= 4x^3 - 3x^2 y + 2xy^2$$
and
$$\frac{\partial}{\partial y}\int M(x,y)\,dx = -3x^2 + 4xy$$

Then
$$N(x,y) - \frac{\partial}{\partial x}\int M(x,y)\,dx$$
$$= -3x^2 + 4xy + 3y^2 - (-3x^2 + 4xy) = 3y^2$$

Therefore

11.13 EXACT DIFFERENTIALS

$$f(x,y) = \int M(x,y)\,dx + \int \left[N(x,y) - \frac{\partial}{\partial y}\int M(x,y)\,dx \right] dy$$

$$= 4x^3 - 3x^2y + 2xy^2 + \int 3y^2\,dy$$

The solution of the equation is then

$$4x^3 - 3x^2y + 2xy^2 + y^3 = C$$

Show that each differential equation in Exercises 1 to 20 is exact and solve.

EXERCISES 11.13

1. $(4x + 2y - 3)\,dx + (2x - 3y + 2)\,dy = 0$
2. $\dfrac{dy}{dx} = \dfrac{x - 3y + 4}{3x - 2y - 1}$
3. $(2x - y^2)\,dx + (y - 2xy)\,dy = 0$
4. $(y^2 + 4x)\,dx + (2xy + 3y^2)\,dy = 0$
5. $\dfrac{dy}{dx} = \dfrac{6x^2y + 2x}{3y - 2x^3}$
6. $\dfrac{(x + y)\,dy + (4x - y)\,dx}{4x^2 + y^2} = 0$
7. $\dfrac{dx - dy}{x - y} + \dfrac{x\,dx + y\,dy}{x^2 + y^2} = 0$
8. $\dfrac{(x - y/x)\,dx + (1 + y)\,dy}{\sqrt{x^2 + y^2}} = 0$
9. $(y + e^x)\,dx + (x + e^y)\,dy = 0$
10. $(ye^x + 2x)\,dx + (e^x + 3)\,dy = 0$
11. $(3x^2 + ye^{xy})\,dx + (xe^{xy} + 2y)\,dy = C$
12. $(y + x \ln y)\,dx + \left(x + \dfrac{x^2}{2y}\right) dy = 0$
13. $\left(e^x + \ln y + \dfrac{y}{x}\right) dx + \left(\dfrac{x}{y} + \ln x + e^y\right) dy = 0$
14. $(1 + y \cos x)\,dx + \sin x\,dy = 0$
15. $(x \sin y - y)\dfrac{dy}{dx} = \cos y$
16. $(2x \sin y + \cos y)\,dx = (x \sin y - x^2 \cos y)\,dy$
17. $(x - 2y)e^{x/y}\,dy = ye^{x/y}\,dx$
18. $(1 + x \cos xy)\,dy + y \cos xy\,dx = 0$
19. $(2x \sin y + e^x \cos y)\,dx = (e^x \sin y - x^2 \cos y)\,dy$
20. $\cos x \cosh y\,(dx + dy) = \sin x \sinh y\,(dx - dy)$

PROBLEMS 11.13
Integrating factors

Consider a first-order differential equation

$$M(x,y)\,dx + N(x,y)\,dy$$

A function $J(x,y)$ such that

$$J(x,y)M(x,y)\,dx + J(x,y)N(x,y)\,dy = 0$$

is exact is called an *integrating factor* for the first equation.

In Problems 1 to 5, show that the given function J is an integrating factor for the equation following it.

1. $J(x,y) = e^x;\ \dfrac{dy}{dx} + y + x = 0$

2. $J(x,y) = e^{x/y};\ (2y - x)\dfrac{dy}{dx} = y.$ (See Exercise 11.13.17)

3. $J(x,y) = \dfrac{1}{xy};\ y\,dx = x\,dy$

4. $J(x,y) = y^{-3};\ (xy + y^4)\,dx + (xy^3 - x^2)\,dy = 0$

5. $J(x,y) = \dfrac{1}{x^2 + y^2};\ (2x^3y - y^3)\,dx + (xy^2 - x^4)\,dy = 0$

6. Find an integrating factor and solve the equations
   (a) $(y + x^2y^2)\,dx = x\,dy$
   (b) $2x^2y\,dx + (x^3 + 2xy)\,dy = 0$

7. Prove that an integrating factor of the form

   $$J(x,y) = x^a y^b$$

   can be chosen for the equation

   $$x^m y^n(Ay\,dx + Bx\,dy) + x^p y^q(Cy\,dx + Dx\,dy) = 0 \qquad \text{if } AD \neq BC$$

8. Show that $e^{\int P(x)\,dx}$ is an integrating factor, in the present sense, for the equation

   $$\dfrac{dy}{dx} + yP(x) = Q(x)$$

12

MULTIPLE INTEGRALS

Just as the concept of the derivative was generalized in Chapter 11, we now extend the meaning of integration to functions of several variables. After a section on "partial integration," we define the double integral, relying initially upon geometric intuition to make the ideas meaningful. Applications of the double integral are followed by an introduction to triple integrals and one application.

12.1 INTEGRATION OF A FUNCTION OF TWO VARIABLES

In Section 11.13 we performed an indefinite integration

$$\int f(x,y)\, dx$$

thinking of y as a constant. This process requires some justification, which we now provide.

Let the function $f(x,y)$ be continuous on the closed rectangular region $R = \{(x,y) : a \leq x \leq b, c \leq y \leq d\}$. For any fixed value y_0 of the second variable, the function $f(x, y_0)$ is continuous on the interval $\{x : a \leq x \leq b\}$. Thus the integral

$$\int_a^b f(x, y_0)\, dx$$

exists. Notice, however, that this integral obviously depends upon the number y_0. In fact, this integral is actually a new function

$$h(y) = \int_b^a f(x, y)\, dx$$

In exactly the same way, we also define the function

$$g(x) = \int_c^d f(x, y)\, dy$$

■ **THEOREM 12.1.** Let the function f have continuous first partial derivatives in the rectangular region $R = \{(x,y) : a \leq x \leq b, c \leq y \leq d\}$. Then the functions

$$g(x) = \int_c^d f(x, y)\, dy, \qquad h(y) = \int_a^b f(x, y)\, dx$$

are both differentiable and

$$g'(x) = \int_c^d f_1(x, y)\, dy, \qquad h'(y) = \int_a^b f_2(x, y)\, dx$$

PROOF: It suffices to prove the formula for $g'(x)$. We want to compute

$$\lim_{h \to 0} \frac{g(x+h) - g(x)}{h} = \lim_{h \to 0} \frac{1}{h} \left[\int_c^d f(x+h, y)\, dy - \int_c^d f(x, y)\, dy \right]$$

$$= \lim_{h \to 0} \frac{1}{h} \int_c^d \left[f(x+h, y) - f(x, y) \right] dy$$

12.1 INTEGRATION OF A FUNCTION OF TWO VARIABLES

Because f is differentiable, the increment in f can be written as

$$f(x + h, y) - f(x, y) = hf_1(x, y) + h\epsilon$$

where $\lim_{h \to 0} \epsilon = 0$. (In this special case, the increment k in the second variable is zero.) Therefore, we may write the equation

$$\begin{aligned}
g'(x) &= \lim_{h \to 0} \frac{1}{h} \int_c^d [f(x + h, y) - f(x, y)] \, dy \\
&= \lim_{h \to 0} \frac{1}{h} \int_c^d [hf_1(x, y) + h\epsilon] \, dy \\
&= \lim_{h \to 0} \int_c^d [f_1(x, y) + \epsilon] \, dy \\
&= \lim_{h \to 0} \int_c^d f_1(x, y) \, dy + \lim_{h \to 0} \epsilon \, dy \\
&= \int_c^d f_1(x, y) + \lim_{h \to 0} \epsilon(d - c)
\end{aligned}$$

Since $\lim_{h \to 0} \epsilon = 0$, the proof is complete. ∎

Applications often involve functions defined on non-rectangular regions. Consider a region of the form

$$R = \{(x, y) : a \leq x \leq b, \gamma(x) \leq y \leq \delta(x)\}$$

bounded by two vertical lines and the graphs of the two functions $\gamma(x)$ and $\delta(x)$. (We assume that $\gamma(x) \leq \delta(x)$, $a \leq x \leq b$, of course.) In Fig. 12.1 we draw such a region. Let the function f be continuous on R. For any fixed x_0, the function $f(x_0, y)$ is continuous on the interval $\{y : \gamma(x_0) \leq y \leq \delta(x_0)\}$. (Refer to Fig. 12.1 again.) Hence the integral

$$\int_{\gamma(x_0)}^{\delta(x_0)} f(x_0, y) \, dy$$

exists. In fact, we obviously have the well-defined function

$$g(x) = \int_{\gamma(x)}^{\delta(x)} f(x, y) \, dy$$

on the interval $\{x : a \leq x \leq b\}$. The following theorem, often called *Leibnitz' rule*, says that g is differentiable when it should be.

■ **THEOREM 12.2.** Let f have continuous first partial derivatives on the region

$$R = \{(x, y) : a \leq x \leq b, \gamma(x) \leq y \leq \delta(x)\}$$

where both γ and δ are differentiable. Then

$$D \int_{\gamma(x)}^{\delta(x)} f(x, y) \, dy = \int_{\gamma(x)}^{\delta(x)} f_1(x, y) \, dy + f(x, \delta(x)) D\delta(x)$$

$$- f(x, \gamma(x)) D\gamma(x)$$

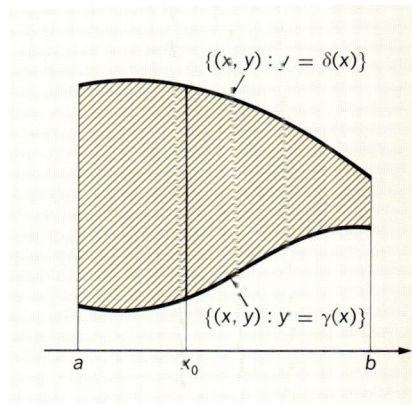

FIGURE 12.1

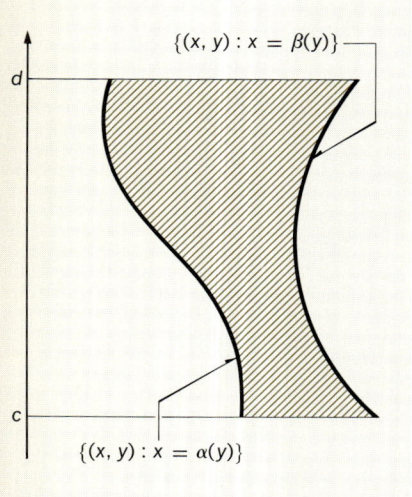

FIGURE 12.2

(By assuming that

$$\int_{\gamma(x)}^{\delta(x)} f(x,y)\,dy = G(x,y)\Big]_{\gamma(x)}^{\delta(x)}$$

and then applying the chain rule, the reader can supply a proof of this result.)

It is obvious that the roles of x and y can be interchanged in the previous discussion. Thus, let f be defined on the region

$$R = \{(x,y) : \alpha(y) \leq x \leq \beta(y),\ c \leq y \leq d\}$$

(See Fig. 12.2). If we then assume the appropriate differentiability conditions, we also have the formula

$$D\int_{\alpha(y)}^{\beta(y)} f(x,y)\,dx = \int_{\alpha(y)}^{\beta(y)} f_2(x,y)\,dx + f(\beta(y),y)\,D\beta(y)\\ -f(\alpha(y),y)\,D\alpha(y)$$

By dividing the region of integration into sections of the two types just described, it is easy to see that Leibnitz' rule for differentiating under an integral applies in more general situations. We do not follow this idea but the reader may wish to do so.

■ **Example 1.** Theorems 12.1 and 12.2 give rise to some clever tricks. Consider the integration formula

$$\int \sin ax\,dx = -\frac{1}{a}\cos ax$$

and think of sin ax as a function of the two variables a and x. Then our theorems imply that

$$D_a \int \sin ax\,dx = \int D_a(\sin ax)\,dx = \int x\cos ax\,dx$$

On the other hand, we have

$$D_a\left(-\frac{1}{a}\cos ax\right) = \frac{1}{a^2}\cos ax + \frac{x}{a}\sin ax$$

This establishes the new integration formula

$$\int x\cos ax\,dx = \frac{1}{a^2}\cos ax + \sin ax \quad ■$$

The differentiable functions,

$$\int_{\gamma(x)}^{\delta(x)} f(x,y)\,dy \quad \text{and} \quad \int_{\alpha(y)}^{\beta(y)} f(x,y)\,dx_1$$

12.1 INTEGRATION OF A FUNCTION OF TWO VARIABLES

discussed in Theorem 12.2 are necessarily continuous. This implies that the following so-called *iterated integrals* exist:

$$\int_a^b \left[\int_{\gamma(x)}^{\delta(x)} f(x,y) \, dy \right] dx, \qquad \int_c^d \left[\int_{\alpha(y)}^{\beta(y)} f(x,y) \, dx \right] dy$$

Such integrals become important to us very soon.

- **Example 2.** Consider the region
$$R = \{(x,y) : 0 \leq x \leq 2, \, 0 \leq y \leq x^2\}$$
in Fig. 12.3. For the function $f(x,y) = x^2 + 2xy$, the iterated integral over R is

$$\int_0^2 \left[\int_0^{x^2} (x^2 + 2xy) \, dy \right] dx = \int_0^2 (x^2 y + xy^2) \Big|_0^{x^2} dx$$

$$= \int_0^2 (x^4 + x^5) \, dx = \frac{256}{15}$$

On the other hand, the region R can also be described as
$$R = \{(x,y) : \sqrt{y} \leq x \leq 2, \, 0 \leq y \leq 4\}$$
Therefore the order of integration can be reversed to give the iterated integral

$$\int_0^4 \left[\int_{\sqrt{y}}^2 (x^2 + 2xy) \, dx \right] dy = \int_0^4 \frac{1}{3} x^3 + x^2 y \Big|_{\sqrt{y}}^2 dy$$

$$= \int_0^4 \left(\frac{8}{3} + 4y - \frac{1}{3} y^{3/2} - y^2 \right) = \frac{256}{15}$$

In view of our next theorem, the equality of the two iterated integrals in Example 2 is to be expected.

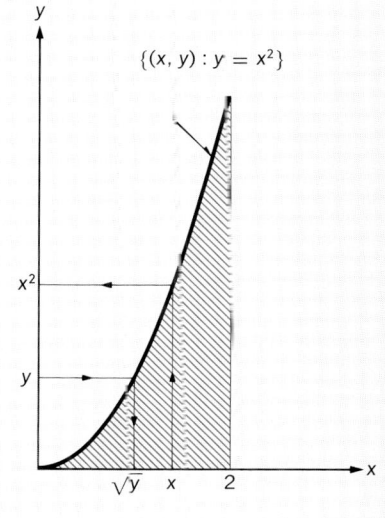

FIGURE 12.3

- **THEOREM 12.3.** Suppose that the region R can be described in the two ways:
$$R = \{(x,y) : a \leq x \leq b, \, \gamma(x) \leq y \leq \delta(x)\}$$
and
$$R = \{(x,y) : \alpha(y) \leq x \leq \beta(y), \, c \leq y \leq d\}$$
If the functions α, β, γ, and δ are differentiable and f is continuous on R, then

$$\int_a^b \left[\int_{\gamma(x)}^{\delta(x)} f(x,y) \, dy \right] dx = \int_c^d \left[\int_{\alpha(y)}^{\beta(y)} f(x,y) \, dx \right] dy$$

We only indicate a method of proof of this difficult theorem in Section 12.3. However, in the case where the region R is rectangular, Theorem 12.3 is quite easy. (See Problem 12.1.13.) Also we might point out that the equality of iterated integrals can be used in more general situations, simply by subdividing the region R appropriately.

EXERCISES 12.1 In Exercises 1 to 10, evaluate the iterated integral

$$\int_a^b \left[\int_{\gamma(x)}^{\delta(x)} f(x,y) \, dy \right] dx$$

for the given function f and the region R. Sketch the region R over which integration is made.

1. $f(x,y) = x + y$; $R = \{(x,y) : 1 \leq x \leq 4, 2 \leq y \leq 4\}$
2. $f(x,y) = 2x + y$; $R = \{(x,y) : 0 \leq x \leq 2, 0 \leq y \leq \sqrt{4 - x^2}\}$
3. $f(x,y) = 3xy^2$; $R = \{(x,y) : 0 \leq x \leq 1, \frac{1}{2}x \leq y \leq \sqrt{x}\}$
4. $f(x,y) = 3xy^2$; $R = \{(x,y) : -1 \leq x \leq 0, 2x \leq y \leq 2x^2\}$
5. $f(x,y) = \sqrt{16 - x^2}$, $R = \{(x,y) : 0 \leq x \leq 4, 0 \leq y \leq x\}$
6. $f(x,y) = y$; $R = \{(x,y) : 0 \leq x \leq 2, 0 \leq y \leq 4 - x^2\}$
7. $f(x,y) = 1 - x^2 - y^2$; $R = \{(x,y) : 0 \leq x \leq 1, 0 \leq y \leq 1 - x\}$
8. $f(x,y) = 2(x + y)$; $R = \left\{ (x,y) : \frac{-\pi}{2} \leq x \leq \frac{\pi}{2}, 0 \leq y \leq \cos x \right\}$
9. $f(x,y) = y\sqrt{9 - x^2 - y^2}$; $R = \{(x,y) : 0 \leq x \leq 1, 0 \leq y \leq x\}$
10. $f(x,y) = \sin(x + y)$; $R = \{(x,y) : 0 \leq x \leq \pi, 0 \leq y \leq \pi - x\}$

Reverse the order of integration for the integrals in Exercises 11 to 16.

11. $\int_0^2 \left[\int_y^2 f(x,y) \, dx \right] dy$
12. $\int_0^2 \left[\int_{y^2}^4 f(x,y) \, dx \right] dy$

13. $\int_0^a \left[\int_0^x f(x,y) \, dy \right] dx$
14. $\int_0^1 \left[\int_{x^2}^x f(x,y) \, dy \right] dx$

15. $\int_1^2 \left[\int_y^{y^2} f(x,y) \, dx \right] dy$
16. $\int_0^2 \left[\int_{-\sqrt{4-x^2}}^{\sqrt{4-x^2}} f(x,y) \, dy \right] dx$

PROBLEMS 12.1 1. Use Theorem 12.2 to deduce the equation

$$\int_0^x \left[\int_0^x e^{a(x-t)} f(t) \, dt \right] dx = \int_0^x (x - t) \, e^{a(x-t)} f(t) \, dt$$

2. First prove that

$$\int_0^\infty e^{-ax} \, dx = \frac{1}{a}$$

and then use Theorem 12.1 and mathematical induction to prove that

$$\int_0^\infty x^n e^{-ax}\, dx = \frac{n!}{a^{n+1}}$$

for each natural number $n \in \mathbf{N}$.

3. Evaluate the integral

$$\int_0^\infty \frac{1}{x}(e^{-ax} - e^{-bx})\, dx$$

4. Prove first that

$$\int_0^\infty \frac{dx}{a^2 + x^2} = \frac{\pi}{2a}, \qquad a > 0$$

Use Theorem 12.1 and mathematical induction to establish the formula

$$\int_0^\infty \frac{dx}{(a^2 + x^2)^n} = \frac{\pi}{2a^{2n-1}} \left[\frac{1 \cdot 3 \cdot 5 \cdots (2n-3)}{2 \cdot 4 \cdot 6 \cdots (2n-2)}\right], \quad a > 0,\ n \in \mathbf{N}$$

Finally, use the substitution $x = a \tan \theta$ to prove that

$$\int_0^{\pi/2} \cos^{2n}\theta\, d\theta = \frac{\pi}{2}\left[\frac{1 \cdot 3 \cdot 5 \cdots (2n-1)}{2 \cdot 4 \cdot 6 \cdots (2n)}\right]$$

5. First prove that

$$\int_0^\infty \frac{dx}{(a^2 + x^2)^{3/2}} = \frac{1}{a^2}$$

Then deduce the formula

$$\int_0^\infty \frac{dx}{(a^2 + x^2)^{n+1/2}} = \frac{1}{a^{2n}}\left[\frac{2 \cdot 4 \cdot 6 \cdots (2n-2)}{3 \cdot 5 \cdot 7 \cdots (2n-1)}\right]$$

Finally, put $x = a \tan \theta$ and prove that

$$\int_0^{\pi/2} \cos^{2n+1}\theta\, d\theta = \frac{2 \cdot 4 \cdot 6 \cdots (2n)}{3 \cdot 5 \cdot 7 \cdots (2n+1)}$$

6. Find two different ways to prove that the integral

$$\int_{a/x}^{b/x} \frac{f(xt)}{t}\, dt$$

does not depend upon x (that is, if $x \neq 0$).

7. In two different ways prove that

$$\int_0^1 x^p (\ln x)^n\, dx = (-1)^n \frac{n!}{(p+1)^{n+1}}, \quad p > 0,\ n \in \mathbf{N}$$

8. If f is periodic of period p, then prove that the integral

$$\int_0^p f(x + t)\, dt$$

does not depend upon x.

9. Given the equation

$$\int_0^\pi \frac{dt}{a+b\cos t} = \frac{\pi}{\sqrt{a^2-b^2}}, \quad (a>b>0)$$

prove that

$$\int_0^\pi \frac{\cos t \, dt}{(a+b\cos t)^3} = \frac{3\pi ab}{2(a^2-b^2)^{5/2}}$$

10. Determine formulas for the integrals

   (a) $\int x^2 \sin ax \, dx$ (b) $\int x^2 \cos ax \, dx$

   (c) $\int x^3 \sin ax \, dx$ (d) $\int x^3 \cos ax \, dx$

11. Establish the formula

$$D^n \int_{-t}^{t} f(x+t)\,dx = 2^n \, D^{n-1} f(2t)$$

12. Prove that

$$\int_0^1 \left[\int_0^1 \frac{x-y}{(x+y)^3}\,dy \right] dx = \frac{1}{2} = - \int_0^1 \left[\int_0^1 \frac{x-y}{(x+y)^3}\,dx \right] dy$$

   Why is this not a contradiction of Theorem 12.3?

13. Let f be continuous on the rectangular region

$$R = \{(x,y) : a \leqq x \leqq b, c \leqq y \leqq d\}$$

   and establish the equation

$$\int_a^b \left[\int_c^d f(x,y)\,dy \right] dx = \int_c^d \left[\int_a^b f(x,y)\,dx \right] dy$$

   (Hint: Define the function

$$G(t,y) = \int_a^t f(x,y)\,dx$$

   and then the function

$$F(t) = \int_c^d G(t,y)\,dy$$

   Use Theorem 12.1 and the fundamental theorem of calculus to evaluate $F(b)$ in two different ways.)

14. Let f and g be continuous on the rectangular region R of Problem 13. Prove the formula

$$\int_a^b \left[\int_c^d [f(x,y)+g(x,y)]\,dy \right] dx = \int_a^b \left[\int_c^d f(x,y)\,dy \right] dx$$
$$+ \int_a^b \left[\int_c^d g(x,y)\,dy \right] dx$$

12.2   VOLUME

We consider *volume* to be a real-valued function V whose domain is a certain class $\mathfrak{M}$ of subsets of three-space $\mathbf{R}^3$. The volume function is assumed to have the following properties:

1. For each member S of $\mathfrak{M}$, $V(S) \geq 0$.
2. If S_1 and S_2 are congruent members of $\mathfrak{M}$, then $V(S_1) = V(S_2)$.
3. If S_1 and S_2 are in $\mathfrak{M}$ and $S_1 \subset S_2$, then $V(S_1) \leq V(S_2)$.
4. If S is in $\mathfrak{M}$, and if S is divided into two pieces S_1 and S_2 by a plane, then both S_1 and S_2 are in $\mathfrak{M}$ and

$$V(S_1) + V(S_2) = V(S)$$

5. Let R be a plane region having area A. Then any right cylinder S over R is in $\mathfrak{M}$ and has a volume equal to the product of its height and the area A of its base.

A discussion of the class $\mathfrak{M}$ would carry us too far from our goals. It is enough to say that $\mathfrak{M}$ is a very extensive class. We now discuss some of its members while showing how an iterated integral is used to define the volume function.

Let f be a nonnegative continuous function defined on the closed region $R = \{(x,y) : a \leq x \leq b, \gamma(x) \leq y \leq \delta(x)\}$ in the xy coordinate plane. Then in space we consider the solid

$$S = \{(x,y,z) : a \leq x \leq b, \gamma(x) \leq y \leq \delta(x), 0 \leq z \leq f(x,y)\}$$

An example is shown in Fig. 12.4. Note that this solid is bounded on the bottom by the xy coordinate plane, is bounded on top by the surface $\{(x,y,z) : z = f(x,y)\}$, and is bounded on the sides by the cylindrical surface constructed over the boundary of the region R.

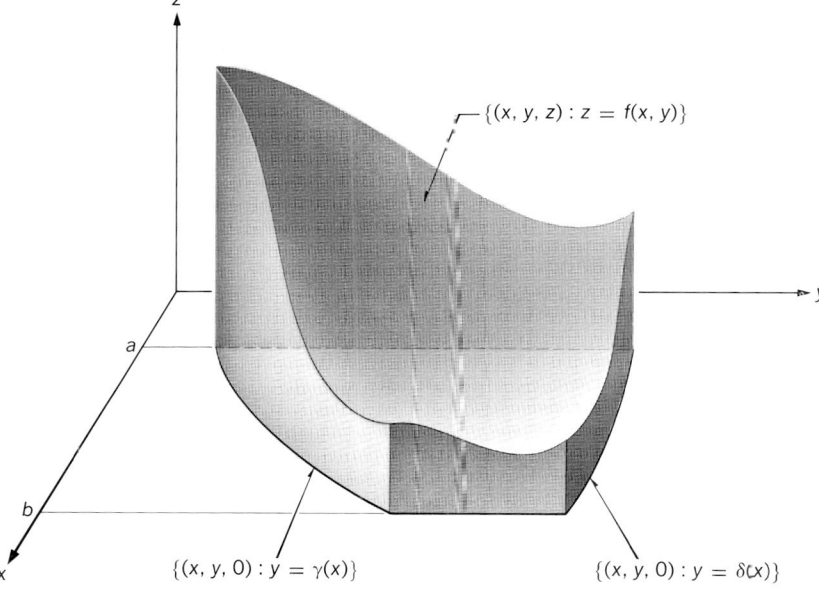

FIGURE 12.4

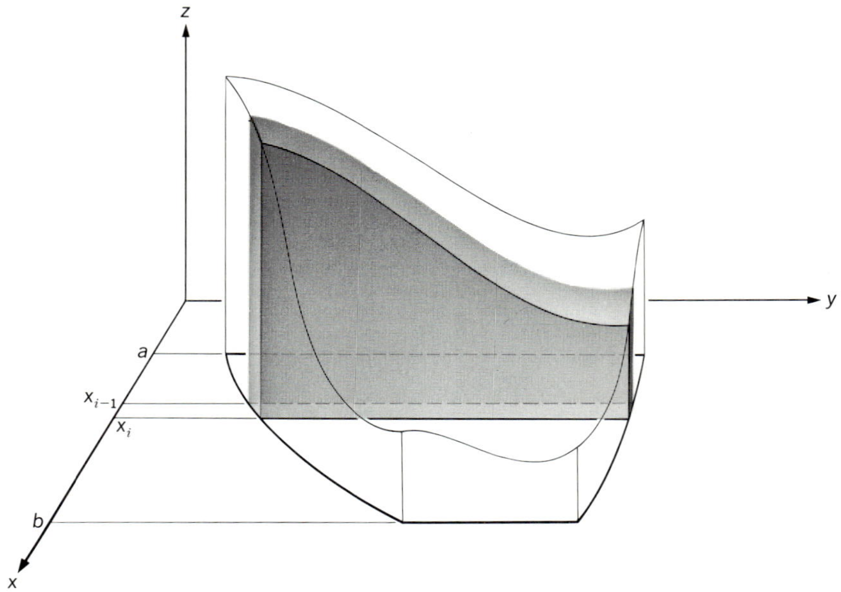

FIGURE 12.5

Imagine slicing this solid into n slabs with the parallel planes

$$\left\{(x, y, z) : x = a + \frac{i}{n}(b - a)\right\}, \quad i = 1, 2, \cdots, n - 1$$

We let V_i denote the volume of the ith slab. In Fig. 12.5 a typical slab is shown. Note that the area of the front face of the ith slab is precisely the integral

$$\int_{\gamma(x_i)}^{\delta(x_i)} f(x_i, y) \, dy$$

where $x_i = a + (i/n)(b - a)$.

Assuming that the functions γ and δ are continuous, it follows that the function

$$G(x) = \int_{\gamma(x)}^{\delta(x)} f(x, y) \, dy$$

is continuous. Let m_i and M_i be the minimum and maximum value of $G(x)$ on the ith subinterval

$$\left\{x : a + \frac{i - 1}{n}(b - a) \leq x \leq a + \frac{i}{n}(b - a)\right\}$$

By our assumption about the volume function, we surely have the inequality

$$m_i\left(\frac{b-a}{n}\right) \leq V_i \leq M_i\left(\frac{b-a}{n}\right)$$

Thus by the mean value theorem there exists a point x_i' in the ith subinterval such that

$$V_i = \left[\int_{\gamma(x_i)}^{\delta(x_i)} f(x_i', y)\, dy\right]\left(\frac{b-a}{n}\right)$$

Having done this for each subinterval, we obtain the Riemann sum

$$V = \sum_{i=1}^{n} \left[\int_{\gamma(x_i')}^{\delta(x_i')} f(x_i', y)\, dy\right]\left(\frac{b-a}{n}\right)$$

This Riemann sum is constant for each value of n; hence the sequence of these sums converges. Obviously, it converges to the iterated integral

$$V(S) = \int_a^b \left[\int_{\gamma(x)}^{\delta(x)} f(x, y)\, dy\right] dx$$

In the same way, the volume of a solid of the form

$$S = \{(x, y, z) : \alpha(y) \leq x \leq \beta(y),\ c \leq y \leq d,\ 0 \leq z \leq f(x, y)\}$$

is defined by the iterated integral

$$\int_c^d \left[\int_{\alpha(y)}^{\beta(y)} f(x, y)\, dx\right] dy$$

A typical example of such a solid is shown in Fig. 12.6.

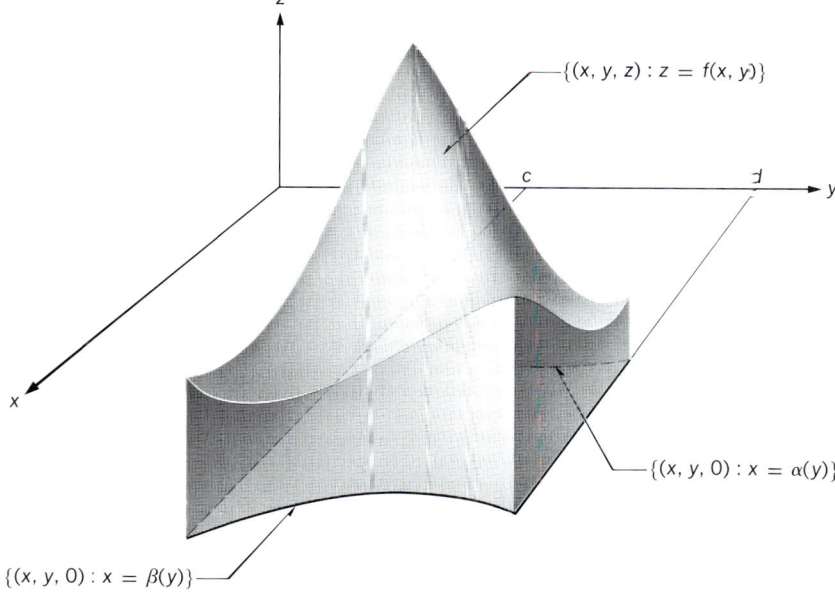

FIGURE 12.6

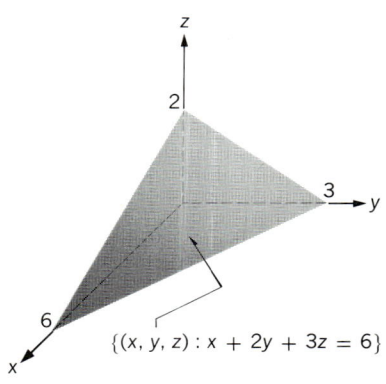

FIGURE 12.7

■ **Example 1.** Consider the tetrahedron cut off the first octant by the plane $\{(x, y, z) : x + 2y + 3z = 6\}$. This is shown in Fig. 12.7.

Clearly, this solid can be expressed in the following two ways
$$R = \{(x, y, z) : 0 \leq x \leq 6, 0 \leq y \leq \tfrac{1}{2}(6 - x),$$
$$0 \leq z \leq \tfrac{1}{3}(6 - x - 2y)\}$$

or

$$R = \{(x, y, z) : 0 \leq x \leq 6 - 2y, 0 \leq y \leq 3,$$
$$0 \leq z \leq \tfrac{1}{3}(6 - x - 2y)\}$$

Therefore the volume is given by either of the iterated integrals

$$\int_0^6 \left[\int_0^{1/2(6-x)} \tfrac{1}{3}(6 - x - 2y)\, dy \right] dx$$

$$= \int_0^3 \left[\int_0^{6-2y} \tfrac{1}{3}(6 - x - 2y)\, dx \right] dy = 6$$

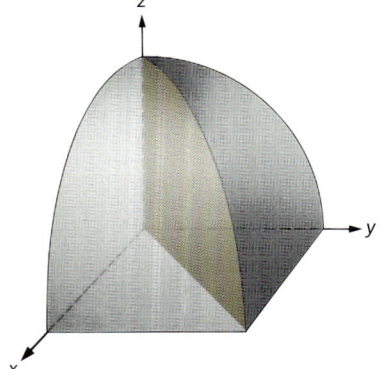

FIGURE 12.8

■ **Example 2.** The volume of the solid in the first octant bounded by the surfaces $\{(x, y, z) : z = \sqrt{4 - x^2}\}$ and $\{(x, y, z) : z = \sqrt{4 - 4^2}\}$ is shown in Fig. 12.8. It is obvious that we can double the volume of the solid

$$\{(x, y, z) : 0 \leq x \leq z, 0 \leq y \leq x, 0 \leq z = \sqrt{4 - x^2}\}$$

in order to obtain the desired volume. Thus

$$V = 2 \int_0^2 \left[\int_0^x \sqrt{4 - x^2}\, dy \right] dx = \frac{16}{3} \quad ■$$

It should be obvious that the region

$$R = \{(x, y) : a \leq x \leq b, \gamma(x) \leq y \leq \delta(x)\}$$

has the area given by the iterated integral

$$A(R) = \int_a^b \left[\int_{\gamma(x)}^{\delta(x)} dy \right] dx = \int_a^b [\delta(x) - \gamma(x)]\, dx$$

Similarly, the area of the region

$$R = \{(x, y) : \alpha(y) \leq x \leq \beta(y), c \leq y \leq d\}$$

is

$$\int_c^d \left[\int_{\alpha(y)}^{\beta(y)} dx \right] dy$$

The reader may illustrate these statements by diagrams. We shall utilize this idea in later applications (Section 12.5).

EXERCISES 12.2

In Exercises 1 to 10, draw a sketch of the solid described and find its volume.

1. $R = \{(x, y, z) : 0 \leq x \leq 1, 0 \leq y \leq 3, 0 \leq z \leq 4 - x - y\}$
2. $R = \{(x, y, z) : 0 \leq x \leq 1, -1 \leq y \leq 2, 0 \leq z \leq 4 - x - y\}$
3. $R = \{(x, y, z) : 1 \leq x \leq 4, 0 \leq y \leq 2, 0 \leq z \leq 2x - y + 4\}$
4. $R = \{(x, y, z) : 0 \leq x \leq 2, 0 \leq y \leq \frac{1}{2}(2 - 3x), 0 \leq z \leq 6 - 3x - 2y\}$
5. $R = \{(x, y, z) : 0 \leq x \leq 4, 0 \leq y \leq \frac{1}{2}x + 1, 0 \leq z \leq 4 + x + y\}$
6. $R = \{(x, y, z) : 0 \leq x \leq 9 - y^2, 0 \leq y \leq 3, 0 \leq z \leq 2x\}$
7. $R = \{(x, y, z) : y \leq x \leq y^2, 1 \leq y \leq 2, 0 \leq z \leq 1 + 2x\}$
8. $R = \{(x, y, z) : 0 \leq x \leq 2, x \leq y \leq 1 + 2x, 0 \leq z \leq 2y(2 - x)\}$
9. $R = \{(x, y, z) : -2 \leq x \leq 0, 0 \leq y \leq \sqrt{2 + x}, 0 \leq z \leq xy(x - y)\}$
10. $R = \{(x, y, z) : 0 \leq x \leq \pi/2, 0 \leq y \leq \pi - 2x, 0 \leq z \leq \cos(x + y)\}$

In Exercises 11 to 25, the bounding surfaces of certain solids are given. Draw a picture of each solid and then find its volume.

11. The coordinate planes and the three planes $\{(x, y, z) : x = 3\}$, $\{(x, y, z) : y = 4\}$ and $\{(x, y, z) : x + y + z = 7\}$.

12. The coordinate planes and the two planes $\{(x, y, z) : 2x + y = 4\}$ and $\{(x, y, z) : 1 + x + 2y = z\}$.

13. The five planes $\{(x, y, z) : z = 0\}$, $\{(x, y, z) : y = 2\}$, $\{(x, y, z) : x = 3y\}$, $\{(x, y, z) : 2x + 3y = 6\}$ and $\{(x, y, z) : 2x - 15y + 12z = 6\}$.

14. The coordinate planes and the plane
$$\left\{(x, y, z) : \frac{x}{a} + \frac{y}{b} + \frac{z}{c} = 1\right\}$$
where a, b, and c are positive constants.

15. The coordinate planes, the two planes $\{(x, y, z) : x = 6\}$ and $\{(x, y, z) : y = 3\}$ and the surface $\{(x, y, z) : z = 2xy\}$.

16. The coordinate planes, the plane $\{(x, y, z) : 2x + 3y = 6\}$ and the surface $\{(x, y, z) : z = x(1 + y)\}$.

17. The coordinate planes, the plane $\{(x, y, z) : x + 2y = 2\}$ and the surface $\{(x, y, z) : z = 4 - x^2 - 4y^2\}$.

18. The plane $\{(x, y, z) : x = y\}$ and the surfaces $\{(x, y, z) : x = y^2\}$ and $\{(x, y, z) : x^2 + z = 1\}$.

19. The plane $\{(x, y, z) : z = 0\}$ and the surface $\{(x, y, z) : z = 9 - x^2 - 3y^2\}$.

20. The four planes $\{(x, y, z) : y = 0\}$, $\{(x, y, z) : z = 0\}$, $\{(x, y, z) : x = 1\}$, $\{(x, y, z) : x = \ln 8\}$ and the surfaces $\{(x, y, z) : y = \ln x\}$ and $\{(x, y, z) : z = e^{x+y}\}$.

21. The sphere $\{(x, y, z) : x^2 + y^2 + z^2 = r^2\}$.

22. The closed surface $\{(x, y, z) : x^2 + y^2 + 4z^2 = 4\}$.

23. The closed surface $\{(x, y, z) : x^{2/3} + y^{2/3} + z^{2/3} = r^{2/3}\}$.

24. The four planes $\{(x, y, z) : x = 1\}$, $\{(x, y, z) : y = 0\}$, $\{(x, y, z) : y = 2\}$, $\{(x, y, z) : z = 0\}$ and the two surfaces $\{(x, y, z) : z = \sqrt{4 - y^2}\}$ and $\{(x, y, z) : x = 2 + y^2 + z^2\}$. (Hint: Look at the solid from a base other than the xy plane.)

25. The solid in the first octant under

$$\left\{(x, y, z) : z = \frac{4}{(x + 1)^2 (y + 1)^2}\right\}$$

(Notice that this solid is infinite in extent but has finite volume.)

12.3 THE DOUBLE INTEGRAL

There will be no need of proofs in this section. Definitions and statements of theorems concerning properties of the double integral are given but primarily as foreshadowing of advanced courses in analysis in which proofs are found. We shall continue using the iterated integral in practical problems.

The chief difficulty in working with the double integral lies in the widely differing shapes of the regions over which integration takes place. Let R be a closed and bounded region in the plane. Because R is bounded, there is a rectangle.

$$Q = \{(x, y) : a \leq x \leq b, c \leq y \leq d\}$$

which contains R. Also, because R is closed, the rectangle Q can be chosen so that R intersects each boundary edge of Q as shown in Fig. 12.9. (These facts have not been established but the reader may find it instructive to do so.)

In order to start the definition of the double integral, we shall consider the subdivisions of the rectangle Q into n^2 subrectangles by the lines

$$\left\{(x, y) : x = a + \frac{i}{n}(b - a)\right\}, \left\{(x, y) : y = c + \frac{i}{n}(d - c)\right\},$$

$$i = 1, 2, \cdots, n - 1$$

for each natural number n.

Next we define two subcollections, $\mathcal{S}_1(n)$ and $\mathcal{S}_2(n)$, of the n^2 subrectangles in the nth subdivisions of Q. A subrectangle is in the collection $\mathcal{S}_1(n)$ if and only if the region R entirely contains the subrectangle. A subrectangle is in the collection $\mathcal{S}_2(n)$ if and only if the subrectangle and the region R intersect. Clearly, as sets of rectangles, these two collections must satisfy the set inclusion

$$\mathcal{S}_1(n) \subset \mathcal{S}_2(n)$$

An example with $n = 8$ is shown in Fig. 12.10.

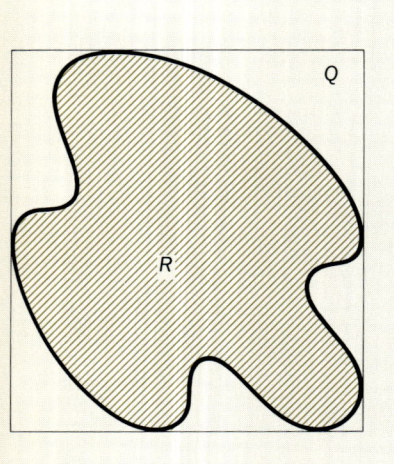

FIGURE 12.9

Number the rectangles in the collection $S_1(n)$ in any order, obtaining the finite sequence

$$Q_1, Q_2, \cdots, Q_{k(n)}$$

Do the same for the rectangles in $S_2(n)$ to obtain the sequence

$$Q'_1, Q'_2, \cdots, Q'_{l(n)}$$

By definition each Q_i is also some Q'_j, and we obviously have the set inclusion

$$\bigcup_{i=1}^{k(n)} Q_i \subset R \subset \bigcup_{j=1}^{l(n)} Q'_j$$

We may now start defining the double integral.

Let f be a bounded function defined on the region R. For a given natural number n, and for each $i = 1, 2, \cdots, k(n)$, define

$$m_i(n) = \text{greatest lower bound of } f(x,y), \quad (x,y) \in Q_i$$

Then we form a *lower sum*

$$s_n = \sum_{i=1}^{k(n)} m_i(n) \, \Delta A$$

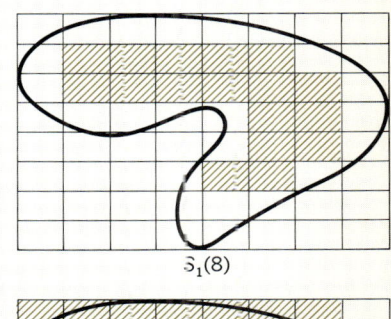

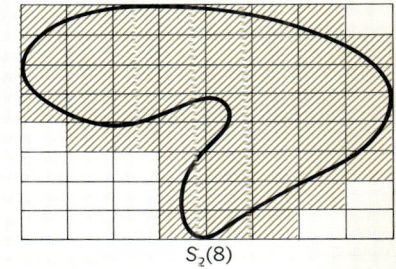

FIGURE 12.10

where $\Delta A = [(b-a)/n][(d-c)/n]$ is the common area of the subrectangles in the nth subdivision of Q. In exactly the same way, for each $j = 1, 2, \cdots, l(n)$, we let

$$M_j(n) = \text{least upper bound of } f(x,y), (x,y) Q'_j \text{ and form an } upper \text{ } sum$$

$$S_n = \sum_{j=1}^{l(n)} M_j(n) \, \Delta A$$

It is obvious that these sums satisfy the inequality

$$s_n \leqq S_n$$

for each $n \in \mathbf{N}$. Furthermore, let us define

$$m = \text{greatest lower bound of } f(x,y), \quad (x,y) \in R$$
$$M = \text{least upper bound of } f(x,y), \quad (x,y) \in R$$

If A denotes the area of the region R, then we also have the inequalities

$$S_n \leqq M \cdot A, \quad m \cdot A \leqq S_n$$

for each natural number n. Therefore the *lower sequence*

$$s_1, s_2, \cdots, s_n, \cdots$$

is bounded above, while the *upper sequence*

$$S_1, S_2, \cdots, S_n, \cdots$$

is bounded below. By showing that the lower sequence is monotone increasing we could apply Theorem 2.10 to establish convergence. We shall not give a general existence proof, however.

▪ **DEFINITION.** The bounded function f is said to be *integrable over the region R*, provided that both the lower and upper sequences converge and

$$\lim_{h \to \infty} s_n = \lim_{h \to \infty} S_n$$

The common limit is called the *double integral* of f over R and is denoted by either of the symbols

$$\iint_R f \quad \text{or} \quad \iint_R f(x, y) \, dA$$

Some of the following results are quite easy to prove and their proofs are requested in problems. Others are quite difficult and are omitted.

▪ **THEOREM 12.4.** Let the functions f and g be integrable on the closed and bounded region R. If α and β are any two real numbers, then $\alpha f + \beta g$ is integrable over R and

$$\iint_R (\alpha f + \beta g) = \alpha \iint_R f + \beta \iint_R g$$

▪ **THEOREM 12.5.** Let f be bounded and integrable on the closed and bounded region R, and let m and M be the greatest lower bound and least upper bound, respectively, of $f(x, y)$ on R. If R has area A, then

$$m \cdot A \leqq \iint_R f \leqq M \cdot A$$

▪ **THEOREM 12.6.** Let the closed and bounded region R be the union of two closed subregions R_1 and R_2 such that R_1 and R_2 have at most boundary points in common. Then for any function f integrable on all three regions, we have

$$\iint_R f = \iint_R f + \iint_R f$$

(Of course, Theorem 12.6 is the two-dimensional analogue of Theorem 6.11. It is much more tedious to prove, however. Only a special case is required in Problem 12.3.5).

▪ **THEOREM 12.7.** Let f be integrable on the closed and bounded region R. If $f(x, y) \geqq 0$ for each point $(x, y) \in R$, then

$$\iint_R f \geq 0$$

- **COROLLARY 12.8.** Let f and g be integrable on the closed and bounded region R. If $f(x,y) \leq g(x,y)$ for each point $(x,y) \in R$, then

$$\iint_R f \leq \iint_R g$$

Consider again the collection $S_2(n)$ of rectangles described earlier. In each rectangle in $S_2(n)$, there are points of the region R. Choose any such point, say (x_j, y_j). It easily follows that we have the inequality

$$m_j(n) \leq f(x_j, y_j) \leq M_j(n)$$

where $m_j(n)$ and $M_j(n)$ are, respectively, the greatest lower bound and least upper bound of $f(x,y)$ on the intersection of R with the jth rectangle in $S_2(n)$, we have the inequality

$$s_n \leq \sum_{j=1}^{l(n)} m_j(n) \, \Delta A \leq \sum_{j=1}^{l(n)} f(x_j, y_j) \, \Delta A$$
$$\leq \sum_{j=1}^{l(n)} M_j(n) \, \Delta A = S_n$$

If f is integrable on R, then Theorem 2.11 proves that

$$\lim_{h \to \infty} \sum_{j=1}^{l(n)} f(x_j, y_j) \, \Delta A = \iint_R f$$

A sum of the form $\sum_{j=1}^{n} f(x_j, y_j) \, \Delta A$ is called a *Riemann sum*, of course. Our discussion shows that we can use such a sum in order to help set up double integrals, for instance.

One last result here justifies our use of iterated integrals.

- **THEOREM 12.9.** Let $R = \{(x,y) : a \leq x \leq b, \gamma(x) \leq y \leq \delta(x)\}$, where γ and δ are continuous functions of x. Then any continuous function f is integrable over R and

$$\iint_R f = \int_a^b \left[\int_{\gamma}^{\delta} f(x,y) \, dy \right] dx$$

Similarly, if $R = \{(x,y) : \alpha(y) \leq x \leq \beta(y), c \leq y \leq d\}$ where α and β are continuous, then the continuous function f is integrable over R and

$$\iint_R f = \int_c^d \left[\int_{\alpha(y)}^{\beta(y)} f(x,y) \, dx \right] dy$$

As a corollary of Theorem 12.9, we have an immediate proof of the equality of iterated integrals in the cases in which the order of integration can be reversed (Theorem 12.3).

PROBLEMS 12.3
Some theory

1. Investigate the existence of the double integral of a continuous function over a closed rectangular region. What properties of such functions are needed to ensure existence?

2. Consider any bounded function over a closed bounded region. Prove that the following subsequences of the lower and upper sequences actually converge:

$$s_2, s_4, s_8, \cdots, s_{2k}, \cdots$$
$$S_5, S_{10}, S_{20}, \cdots, S_{5 \cdot 2k}, \cdots$$

3. Prove Theorem 12.4.

4. Prove Theorem 12.5.

5. Prove Theorem 12.6 in the case where R, R_1, and R_2 are rectangular regions. (Even this simple case is not easy.)

6. Prove Theorem 12.7 and its corollary.

7. Prove Theorem 12.9 in the case where R is a rectangular region. (Hint: Define the functions

$$F(t, y) = \int_a^t f(x, y)\, dx; \quad g(t) = \int_c^d F(t, y)\, dy$$

Compute $g(b)$ in two different ways to obtain the proof.)

8. Let the nonnegative function f be integrable over the two regions R_1 and R_2. If $R_1 \subset R_2$, prove that

$$\iint_{R_1} f \leqq \iint_{R_2} f$$

12.4 POLAR COORDINATES

When dealing with a geometric problem, there is another natural way to locate a point in the plane in addition to the method of rectangular coordinates. Let (x, y) be a point in a coordinatized geometric plane. Then its position vector $\mathbf{p} = x\mathbf{i} + y\mathbf{j}$ forms some angle θ with the positive x axis. The angle θ together with the norm $r = \sqrt{x^2 + y^2}$ of $\mathbf{p}$ surely serve to locate the point. (See Fig. 12.11.)

The above observation leads us to the following definition: Let (r, θ) be any ordered pair of real numbers. These numbers are *polar coordinates* of the point (in a geometric plane) whose position vector is

$$\mathbf{p} = (r \cos \theta)\mathbf{i} + (r \sin \theta)\mathbf{j}$$

Intuitively, we locate the point (r, θ) by standing at the origin, facing in the direction determined by the angle θ, and then moving away from the origin a distance r. (Note that the distance between points in the plane is necessary here.) If the number r is negative, then we face in the given direction and back up.

The relationships

$$x = r \cos \theta \qquad r^2 = x^2 + y^2$$
$$y = r \sin \theta \qquad \tan \theta = \frac{y}{x}$$

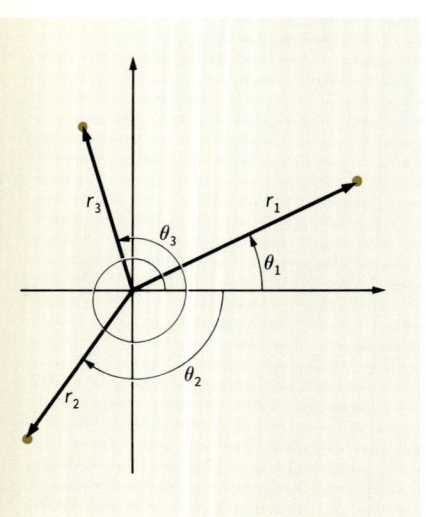

FIGURE 12.11

exist between the polar coordinates (r, θ) and the rectangular coordinates (x, y) of a given point. We notice that there is just one point located by a given pair (r, θ) of polar coordinates. On the other hand, however, there are infinitely many pairs of polar coordinates for any given point. In particular, the point (r, θ) is also located by either

$$(r, \theta + 2n\pi)$$
$$(-r, \theta + (2n + 1)\pi)$$

where n is any integer. The reader may convince himself easily of this fact.

A plane curve such as

$$\{(r, \theta) : r = f(\theta)\}, \{(r, \theta) : \theta = g(r)\}$$

or

$$\{(r, \theta) : F(r, \theta) = 0\}$$

is called a *polar curve*. It is often easier to handle such a curve in its polar form than it would be to change to rectangular coordinates. The problems at the end of this section provide examples of "polar geometry." We only consider one example here.

■ *Example 1.* The polar curve

$$\{(r, \theta) : r = 2 + 4 \cos \theta\}$$

is shown in Fig. 12.12. Note that it is easy to determine the points $(6, 0)$, $(4, \pi/3)$, $(2, \pi/2)$, and $(0, 2\pi/3)$ on this curve. Note, too, that a point moving counterclockwise along the curve from the point $(6, 0)$ reaches the origin when $\theta = 2\pi/3$. Then for any angle θ between $2\pi/3$ and $4\pi/3$, the coordinate r is negative. This means, for instance, we shall find the point $(-2, \pi)$ by facing in the negative x direction and backing up two units. (This locates the point $(2, 0)$ in rectangular coordinates.) Other simple techniques suffice to yield the diagram shown in Fig. 12.12.

In rectangular coordinate form, the equation $r = 2 + 4 \cos \theta$ becomes

$$\pm \sqrt{x^2 + y^2} = 2 + 4\left(\frac{x}{\pm \sqrt{x^2 + y^2}}\right)$$

(which comes from the previous formulas). When we square to remove the radical in this equation, we obtain

$$(x^2 + y^2 - 4x)^2 = 4(x^2 + y^2)$$

Hence the curve in Fig. 12.12 may also be described as

$$\{(x, y) : (x^2 + y^2 - 4x)^2 = 4(x^2 + y^2)\}$$

which is quite difficult to discuss. ■

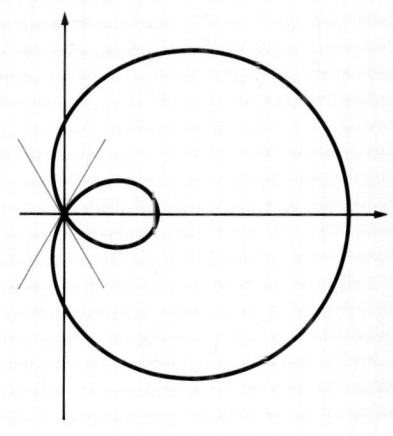

FIGURE 12.12

Consider two types of regions defined in terms of polar coordinates,
$$R_1 = \{(r, \theta) : a(\theta) \leq r \leq b(\theta), \alpha \leq \theta \leq \beta\},$$
where a and b are continuous functions of θ, and
$$R_2 = \{(r, \theta) : a \leq r \leq b, \alpha(r) \leq \theta \leq \beta(r)\},$$
where α and β are continuous functions of r. The illustrations in Fig. 12.13 are typical of the two types.

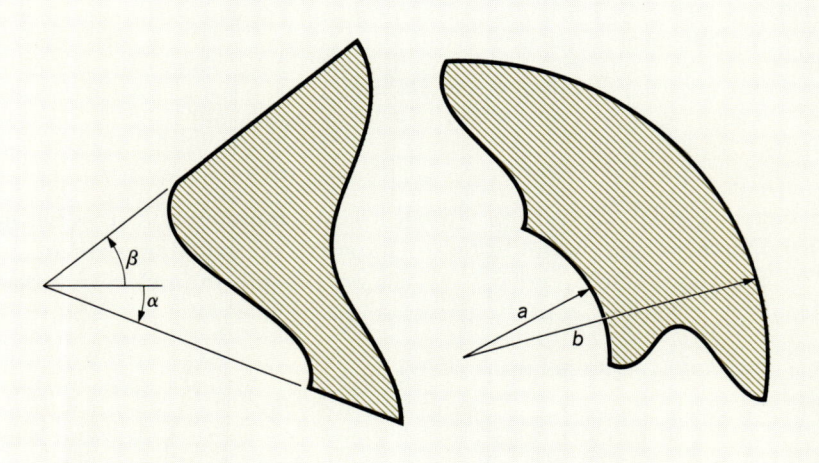

FIGURE 12.13

We recall that the area of a circular sector as shown in Fig. 12.14 is given by the formula
$$A = \tfrac{1}{2} r^2 \theta$$
where the central angle θ is measured in radians. It follows that the region shown in Fig. 12.15, which is the difference of two circular sectors, has the area
$$A = \tfrac{1}{2}(r_2^2 - r_1^2)\theta = \bar{r}(r_2 - r_1)\theta$$
where $\bar{r}$ is the average $\tfrac{1}{2}(r_1 + r_2)$ of the radii. We shall use this formula in the following.

Consider a region
$$R = \{(r, \theta) : a(\theta) \leq r \leq b(\theta), \alpha \leq \theta \leq \beta\}$$
Subdividing R as shown in Fig. 12.16, we readily arrive at an approximation of its area in the form
$$A(R) \cong \sum_{i=1}^{n} \bar{r}_i \, \Delta r \, \Delta \theta$$

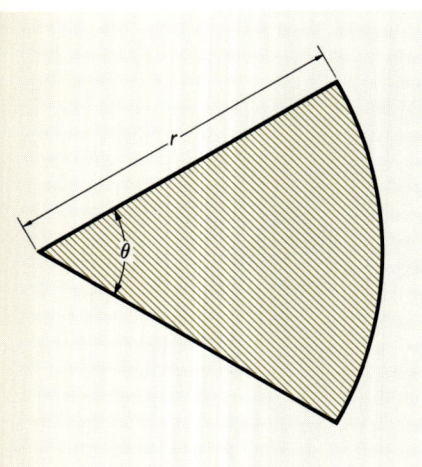

FIGURE 12.14

It follows immediately that the area of R is given by the iterated integral

$$\int_\alpha^\beta \left[\int_a^b r \, dr \right] d\theta$$

More generally, let $f(r, \theta)$ be a nonnegative continuous function defined on the region R. Then the volume of the solid

$$S = \{(r, \theta, z) : (r, \theta) \in R, 0 \leq z \leq f(r, \theta)\}$$

is defined to be

$$V(S) = \int_\alpha^\beta \left[\int_a^b f(r, \theta) \, r \, dr \right] d\theta$$

If the function f defined on the region R is expressed in terms of rectangular coordinates, then we write $f(r \cos \theta, r \sin \theta)$ in place of $f(x, y)$ when we set up the previous integral. Such an interplay between polar and rectangular coordinates is often useful in dealing with double integrals.

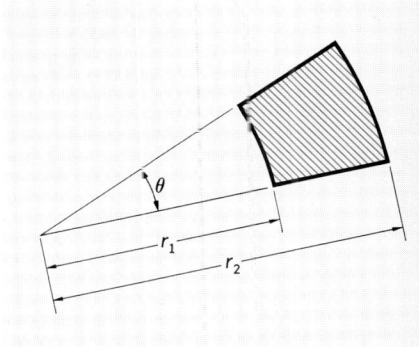

FIGURE 12.15

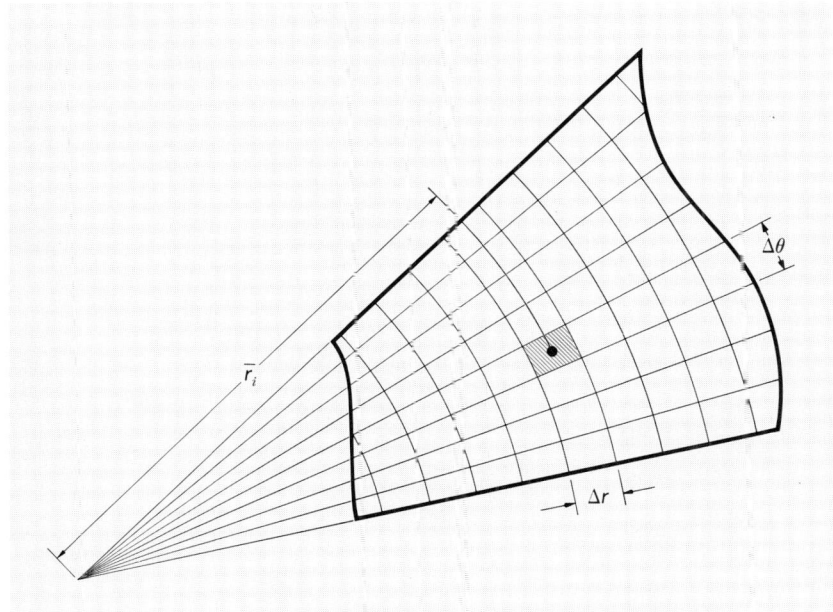

FIGURE 12.16

■ *Example 2.* Consider the function $f(x, y) = \sqrt{x^2 + y^2}$ defined on the square region $R = \{(x, y) : 0 \leq x \leq a, 0 \leq y \leq a\}$. From Fig. 12.17 we see that this square contains the circular sector $R_1 = \{(r, \theta) : 0 \leq r \leq a, 0 \leq \theta \leq \pi/2\}$ and is contained in the sector $R_2 = \{(r, \theta) : 0 \leq r \leq a\sqrt{2}, 0 \leq \theta \leq \pi/2\}$. Therefore from Problem

12.3.8 we have the inequality

$$\iint_{R_1} f \leq \iint_R f \leq \iint_{R_2} f$$

On the regions R_1 and R_2 we can convert to polar coordinates, writing

$$f(x,y) = \sqrt{x^2 + y^2} = r$$

Thus we obtain

$$\iint_{R_1} f = \int_0^{\pi/2} \left[\int_0^a f(r\,dr) \right] d\theta = \frac{1}{3} a^3 \int_0^{\pi/2} d\theta = \frac{\pi a^3}{6}$$

Similarly,

$$\iint_{R_2} f = \int_0^{\pi/2} \left[\int_0^{a\sqrt{2}} r^2\,dr \right] d\theta = \frac{\pi a^3 \sqrt{2}}{3}$$

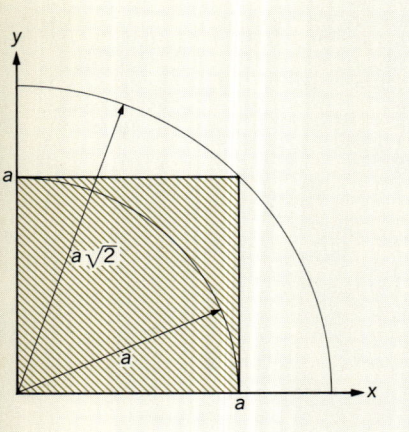

FIGURE 12.17

Therefore we have the inequality

$$\frac{\pi a^3}{6} \leq \int_0^a \left[\int_0^a \sqrt{x^2 + y^2}\,dy \right] dx \leq \frac{a^3 \sqrt{2}}{3}$$

A more important example is the following example.

■ **Example 3.** Define the integral

$$I = \int_0^a e^{-x^2}\,dx$$

which cannot be evaluated in terms of elementary functions. We play a useful trick. Write

$$I^2 = \left(\int_0^a e^{-x^2} \right)^2 = \left(\int_0^a e^{-x^2}\,dx \right)\left(\int_0^a e^{-y^2}\,dy \right)$$

$$= \int_0^a \left[\int_0^a e^{-x^2} e^{-y^2}\,dy \right] dx$$

$$= \iint_R e^{-(x^2+y^2)}\,dA$$

where R is the square region $\{(x,y) : 0 \leq x \leq a, 0 \leq y \leq a\}$. Using the same circular sectors R_1 and R_2 as in Example 2, we have the inequality

$$\iint_{R_1} f \leq I^2 \leq \iint_{R_2} f$$

Now,

$$\iint_{R_1} f = \iint_{R_1} e^{-(x^2+y^2)}\,dA = \int_0^{\pi/2} \left[\int_0^a e^{-r^2}(r\,dr) \right] d\theta$$

$$= \frac{\pi}{4}(1 - e^{-a^2})$$

12.4 POLAR COORDINATES

while, similarly,

$$\iint_{R_2} f = \int_0^{\pi/2} \left[\int_0^{a\sqrt{2}} e^{-r^2}(r\,dr) \right] d\theta = \frac{\pi}{4}(1 - e^{-2a^2})$$

It follows then that

$$\frac{\sqrt{\pi}}{2} \sqrt{1 - e^{-a^2}} \leq I \leq \frac{\sqrt{\pi}}{2} \sqrt{1 - e^{-2a^2}}$$

From the inequality just established it is easy to deduce the value of the improper integral:

$$\int_0^\infty e^{-x^2}\,dx = \lim_{a \to \infty} \int_0^a e^{-x^2} = \frac{1}{2}\sqrt{\pi}$$

This fact is used in probability theory. In particular, the *error function* is defined to be

$$\operatorname{erf}(x) = \frac{2}{\sqrt{\pi}} \int_0^x e^{-t^2}\,dt$$

The multiplicative constant $2/\sqrt{\pi}$ serves to produce the limit

$$\lim_{x \to \infty} \operatorname{erf}(x) = 1$$

EXERCISES 12.4

In Exercises 1 to 14, draw the region described and use an iterated integral to determine its area.

1. $\{(r, \theta) : 0 \leq r \leq 1 + \cos\theta,\ 0 \leq \theta \leq 2\pi\}$
2. $\{(r, \theta) : 0 \leq r \leq \cos 3\theta,\ 0 \leq \theta \leq \pi/6\}$
3. $\{(r, \theta) : 0 \leq r \leq 3 \sin 2\theta,\ 0 \leq \theta \leq \pi/2\}$
4. $\{(r, \theta) : 0 \leq r \leq 1,\ 0 \leq \theta \leq \pi\}$
5. $\{(r, \theta) : 0 \leq r \leq 3\sqrt{\cos 2\theta},\ 0 \leq \theta \leq \pi/6\}$
6. The region enclosed by one loop of the curve
$$\{(r, \theta) : r = 6 \sin 3\theta\}$$
7. The region enclosed by the curve
$$\{(r, \theta) : r = a(1 + \cos\theta)\}$$
8. The region inside the smaller loop of the curve
$$\{(r, \theta) : r = 4 - 8 \sin\theta\}$$
9. The region $\{(r, \theta) : 0 \leq r \leq 2(1 - \cos\theta)\}$
10. The region inside the circle $\{(r, \theta) : r = a\}$ and outside of the curve
$$\{(r, \theta) : r^2 = a^2 \cos 2\theta\}$$
11. The region inside the curve $\{(r, \theta) : r = a(1 + \cos\theta)\}$ and outside of the circle $\{(r, \theta) : r = a\}$

12. The region inside the circle $\{(r, \theta) : r = 5a \sin \theta\}$ and above the line
$$\{(r, \theta) : r \sin \theta = a\}$$

13. The region outside of the curve $\{(r, \theta) : r = 6 + 6 \sin \theta\}$ and inside the circle $\{(r, \theta) : r = 9\}$

14. The region bounded by the curve $\{(r, \theta) : r^2 = 2a^2 \cos 2\theta\}$

In Exercises 15 to 22, draw a careful sketch of the region of integration for the given integral. Use this drawing to change the integral to polar coordinate form. Evaluate the integral

15. $\int_0^1 \left[\int_0^{\sqrt{1-x^2}} (x^2 + y^2)^{3/2} \, dy \right] dx$

16. $\int_0^2 \left[\int_0^{\sqrt{4-y^2}} e^{x^2+y^2} \, dx \right] dy$

17. $\int_{-4}^4 \left[\int_{-\sqrt{16-x^2}}^{\sqrt{16-x^2}} e^{-(x^2+y^2)} \, dy \right] dx$

18. $\int_0^1 \left[\int_x^1 \frac{x \, dy}{\sqrt{x^2 + y^2}} \right] dx$

19. $\int_0^1 \left[\int_x^{\sqrt{2-x^2}} y^2 \, dy \right] dx$

20. $\int_0^1 \left[\int_{2y}^2 y\sqrt{x^2 + y^2} \, dx \right] dy$

21. $\int_0^1 \left[\int_x^1 x(x^2 + y^2)^{1/2} \, dy \right] dx$

22. $\iint_R e^{-(1+x^2+y^2)} \, dA$ where
$$R = \{(x, y) : 1 \leq x^2 + y^2 \leq 4\}$$

Transform to polar coordinates in order to determine the area of the region given in each of Exercises 23 to 30.

23. $\{(x, y) : 1 \leq \sqrt{x^2 + y^2} \leq 2\}$

24. $\{(x, y) : 1 \leq \sqrt{x^2 + y^2} \leq 4, 0 \leq y \leq x\}$

25. The region bounded by the curve $\{(x, y) : y^2 = 4ax\}$ and the vertical line $\{(x, y) : x = a\}$

26. The region bounded by the curve $\{(x, y) : x = a^2/y^2\}$ and the lines $\{(x, y) : x = a\}$, $\{(x, y) : y = a\}$

27. The region inside the circle $\{(x, y) : x^2 + y^2 = 8x\}$ and to the right of the vertical line $\{(x, y) : x = 2\}$

28. The region inside of the circle $\{(x, y) : x^2 + y^2 = 6x\}$ and outside the circle $\{(x, y) : x^2 + y^2 = 9\}$

29. The region enclosed by the curve $\{(x, y) : (x^2 + y^2)^3 = 16x^2\}$

30. The region bounded by the curves $\{(x, y) : y^2 = 8x\}$ and $\{(x, y) : y^2 = 8x - 16\}$

Compute the volume of the solids described in Exercises 31 to 40.

31. $\{(x, y, z) : 0 \leq z \leq 4 - x^2 - y^2\}$

32. $\{(x, y, z) : 0 \leq x^2 + y^2 \leq a^2, 0 \leq z \leq \frac{1}{2}(x^2 + y^2)\}$

33. $\{(x, y, z) : 0 \leq x^2 + y^2 \leq a^2, -\sqrt{x^2 + y^2} \leq z \leq \sqrt{x^2 + y^2}\}$

34. $\left\{ (x, y, z) : 0 \leq x \leq a, 0 \leq y \leq x, 0 \leq z \leq \dfrac{x^2 + y^2}{2a} \right\}$

35. $\{(x, y, z) : 0 \leq x^2 + y^2 \leq a^2, 0 \leq z \leq y - x\}$

36. $\{(x, y, z) : 0 \leq x \leq a, 0 \leq y \leq x, 0 \leq z \leq a^2 - y^2\}$

37. $\left\{(x, y, z) : 0 \leq x^2 + y^2 \leq 2ax, 0 \leq z \leq \dfrac{x^2 + y^2}{a}\right\}$

38. The solid bounded by the surface $\{(x, y, z) : z = \sqrt{x^2 + y^2}\}$ and the plane $\{(x, y, z) : z = a\}$

39. The solid in the first octant which is inside the cylinder $\{(x, y, z) : x^2 + y^2 = a^2\}$ and is outside the cylinder $\{(x, y, z) : z = ax^2\}$

40. The solid cut out of the solid cylinder $\{(x, y, z) : x^2 + y^2 \leq 4\}$ by the sphere $\{(x, y, z) : x^2 + y^2 + z = 16\}$

PROBLEMS 12.4

1. Evaluate the iterated integral

$$\int_0^{\pi/2} \left[\int_0^{\sec\theta} \frac{r\,dr}{1 + r^2 \sin^2\theta}\right] d\theta$$

by transforming it to rectangular coordinate form.

2. Evaluate the double integral

$$\iint_R (1 + x^2 + y^2)^2 \, dA$$

where R is the triangular region with vertices $(0, 0)$, $(0, 2)$, and $(2\sqrt{3}, 2)$.

3. Use an appropriate change of variable to prove that

$$\int_0^\infty e^{-kx^2}\, dx = \frac{1}{2}\sqrt{\frac{\pi}{k}}$$

Then use Theorem 12.1 and mathematical induction to deduce that

$$\int_0^\infty x^{2n} e^{-kx^2}\, dx = \frac{\sqrt{\pi}}{k}\left[\frac{1 \cdot 3 \cdot 5 \cdots (2n-1)}{2^{n+1} k^n}\right], \quad n \in \mathbb{N}$$

4. Prove that

$$\int_0^1 \frac{dx}{\sqrt{|\ln x|}} = \sqrt{\pi} \qquad (\text{Hint: Set } x = e^{-u^2}.)$$

5. Evaluate the improper integrals

(a) $\displaystyle\int_0^\infty x^{-1/2} e^{-k^2 x}\, dk$ (b) $\displaystyle\int_0^\infty x^{1/2} e^{-k^2 x}\, dx$

(c) $\displaystyle\int_0^1 \left(\ln\frac{1}{x}\right)^{1/2} dx$

6. Prove that there exists a constant c such that, for every polynomial function $p(x)$ of degree ≤ 5,

$$\int_{-\infty}^\infty e^{-x^2} p(x)\, dx = \frac{\sqrt{\pi}}{6}[p(-c) + 4p(0) + p(c)]$$

12.5 FIRST MOMENTS AND THE CENTER OF MASS

If a particle of mass m is located at a directed distance d from a line (or a plane), then the particle is said to have *first moment*

$$M = dm$$

with respect to the line (or the plane). If several particles having masses $m_1, m_2, \cdots, m_n$ are located at the points $x_1, x_2, \cdots, x_n$ on the x axis, then the system of particles has the first moment

$$M = \sum_{i=1}^{n} x_i m_i$$

with respect to the y axis. In other words, we assume that *the first moment of a sum is the sum of first moments.*

Now suppose that the entire mass

$$m = \sum_{i=1}^{n} m_i$$

of the system of particles were concentrated into one particle and placed at the point $\bar{x}$ on the x axis. The first moment of this new particle with respect to the y axis is

$$\bar{x}\left(\sum_{i=1}^{n} m_i\right)$$

This first moment will equal that of the entire system if

$$\bar{x}\left(\sum_{i=1}^{n} m_i\right) = \sum_{i=1}^{n} x_i m_i$$

or

$$\bar{x} = \frac{\sum_{i=1}^{n} x_i m_i}{\sum_{i=1}^{n} m_i} = \frac{M}{m}$$

The point $\bar{x}$ is known as the *center of mass* of the system.

Next, suppose that the particles are located at the points (x_1, y_1), $(x_2, y_2), \cdots, (x_n, y_n)$ in the xy plane. The first moment of the system with respect to the y axis is still the same, namely,

$$M_y = \sum_{i=1}^{n} x_i m_i$$

In the same way we have the first moment with respect to the x axis,

$$M_x = \sum_{i=1}^{n} y_i m_i$$

Then the center of mass of the system is located at the point

$$(\bar{x}, \bar{y}) = \left(\frac{\sum x_i m_i}{\sum m_i}, \frac{\sum y_i m_i}{\sum m_i}\right) = \left(\frac{M_y}{m}, \frac{M_x}{m}\right)$$

12.5 FIRST MOMENTS AND THE CENTER OF MASS

Analogously, let the particles be placed at the points (x_1, y_1, z_1), $(x_2, y_2, z_2), \ldots, (x_n, y_n, z_n)$ in space. Then the first moments of the system with respect to the three coordinate planes are

$$M_{yz} = \sum_{i=1}^{n} x_i m_i, \quad M_{xz} = \sum_{i=1}^{n} y_i m_i, \quad M_{xy} = \sum_{i=1}^{n} z_i m_i$$

The center of mass is then

$$(\bar{x}, \bar{y}, \bar{z}) = \left(\frac{M_{yz}}{m}, \frac{M_{xz}}{m}, \frac{M_{xy}}{m} \right)$$

This situation will be utilized in Section 12.10.

In general, we assume that a physical object has matter continuously distributed in space. (This idealization ignores molecular theory, of course.) At a given point in this object, its *density* is defined as follows: Consider a small element ΔS containing the given point. Let this element have volume ΔV and mass Δm. Then the ratio $\Delta m / \Delta V$ is the average density of the element. Then the *density at the point* is defined to be

$$\delta = \lim_{\Delta \to 0} \frac{\Delta m}{\Delta V}$$

We consider a thin sheet of material regarding it as having constant but negligible thickness. Such an object is called a *lamina* and is treated as if it were two-dimensional. Suppose that the region R in the xy plane is covered by a lamina whose density δ is a continuous function of the point $(x, y) \in R$. Let Q be the usual rectangle containing R and consider the nth subdivision of Q into subrectangles. For each subrectangle Q_j of $S_2(n)$ (see Section 12.3), choose a point (x_j, y_j) in $Q_j \cap R$. Then the mass of the piece of the lamina of area ΔA_j is approximately $\delta(x_j, y_j) \Delta A_j$. Hence the Riemann sum

$$\sum_{j=1}^{l(n)} \delta(x_j, y_j) \Delta A_j$$

obviously approximates the mass of the lamina. It follows that the mass of the lamina is defined by the double integral

$$\iint_R \delta(x, y) \, dA$$

Similarly, the piece of the lamina over the region $Q_j \cap R$ has first moments approximated by

$$x_j \Delta m_j \cong x_j \delta(x_j, y_j) \Delta A_j$$
$$y_j \Delta m_j \cong y_j \delta(x_j, y_j) \Delta A_j$$

By considering the corresponding Riemann sums and passing to limits, we see that the first moments of the lamina are defined by the double integrals

$$M_y = \iint_R x\delta(x,y)\, dA \qquad M_x = \iint_R y\delta(x,y)\, dA$$

Finally, the center of mass of the lamina is at the point

$$(\bar{x}, \bar{y}) = \left(\frac{M_y}{m}, \frac{M_x}{m}\right)$$

- **Example 1.** Suppose a lamina covers the region

$$R = \{(x,y) : 0 \leq x \leq 1,\ x^2 \leq y \leq 2x - x^2\}$$

and has the density $\delta(x,y) = x + 2$. Then we have its mass:

$$m = \iint_R (x+2)\, dA = \int_0^1 \left[\int_{x^2}^{2x-x^2} (x+2)\, dy\right] dx$$

and its moments

$$M_y = \iint_R x(x+2)\, dA = \int_0^1 \left[\int_{x^2}^{2x-x^2} x(x+2)\, dy\right] dx$$

and

$$M_x = \iint_R y(x+2)\, dA = \int_0^1 \left[\int_{x^2}^{2x-x^2} y(x+2)\, dy\right] dx$$

It is a routine exercise to compute that $m = \frac{5}{6}$, $M_y = \frac{13}{30}$, $M_x = \frac{13}{30}$. Therefore, the center of mass is at the point $(\bar{x}, \bar{y}) = (\frac{13}{25}, \frac{13}{25})$.

- **Example 2.** Let the lamina cover the region

$$R = \{(x,y) : -a \leq x \leq a,\ 0 \leq y \leq \sqrt{a^2 - x^2}\}$$

and have the density $\delta(x,y) = x + y$. Its mass is given by

$$m = \iint_R (x+y)\, dA = \int_{-a}^{a} \left[\int_0^{\sqrt{a^2-x^2}} (x+y)\, dy\right] dx$$

and its first moments are

$$M_y = \iint_R x(x+y)\, dA = \int_{-a}^{a} \left[\int_0^{\sqrt{a^2-x^2}} x(x+y)\, dy\right] dx$$

and

$$M_x = \iint_R y(x+y)\, dA = \int_{-a}^{a} \left[\int_0^{\sqrt{a^2-x^2}} y(x+y)\, dy\right] dx$$

It is considerably easier to evaluate the above double integrals if we transform to polar coordinates. The reader should verify that we obtain

$$m = \int_0^\pi \left[\int_0^a r(\cos\theta + \sin\theta)\, dr\right] d\theta = \frac{2}{3}a^3$$

$$M_y = \int_0^\pi \left[\int_0^a r^2(\cos^2\theta + \sin\theta\cos\theta)\, r\, dr \right] d\theta = \frac{\pi a^4}{8}$$

and

$$M_x = \int_0^\pi \left[\int_0^a r^2(\sin\theta\cos\theta + \sin^2\theta)\, r\, dr \right] d\theta = \frac{\pi a^4}{8}$$

Thus the center of mass of this semicircular lamina is $[(3\pi a/16), (3\pi a/16)]$.

EXERCISES 12.5

In each of Exercises 1 to 10, find the center of mass of the lamina over the given region having the given density δ. (The number k is a positive constant.)

1. $R = \{(x,y) : 0 \le x \le 2, 0 \le y \le 2 - x\};\ \delta = k$
2. $R = \{(x,y) : 0 \le x \le 2, 0 \le y \le 2 - x\};\ \delta = kx$
3. $R = \{(x,y) : 0 \le x \le 2, 0 \le y \le 2 - x\};\ \delta = k(k^2 + y^2)$
4. $R = \{(x,y) : 0 \le x \le 1, 0 \le y \le 1 - x^2\};\ \delta = ky$
5. $R = \{(x,y) : 0 \le x \le 4, 0 \le y \le \sqrt{x}\};\ \delta = k(x + y)$
6. $R = \{(x,y) : y^2 \le x \le y + 2, -1 \le y \le 2\};\ \delta = kx$
7. $R = \{(x,y) : 0 \le x \le 1, x^2 \le y \le \sqrt{x}\};\ \delta = ky$
8. $R = \{(x,y) : \sqrt{1 + y^2} \le x \le 3, -\sqrt{8} \le y \le \sqrt{8}\};\ \delta = kx$
9. $R = \{(x,y) : 0 \le x \le \pi, 0 \le y \le \sin x\};\ \delta = ky$
10. $R = \{(x,y) : 1 \le x \le e, 0 \le y \le \ln x\};\ \delta = ky$
11. Find the mass of the lamina over the triangle with vertices $(0, 0)$, $(5, 0)$, and $(3, 6)$ if its density is $\delta = ky$.
12. Find the mass of the lamina over the triangle in Exercise 11 if its density is $\delta = k(x + 2y)$.
13. Find the center of mass of the lamina over the square with vertices $(0, 0)$, $(a, 0)$, (a, a), $(0, a)$ if its density is $\delta = ky$,
14. Find the center of mass of the lamina over the square in Exercise 13 if its density is $\delta = k(x^2 + y^2)$.
15. Find the center of mass of the lamina having density $\delta = kxy$ and lying over the region $R = \{(x,y) : 0 \le x \le a, 0 \le y \le (a^{2/3} - x^{2/3})^{3/2}\}$.

PROBLEMS 12.5

1. Develop the double-integral formulas in polar coordinates for the mass and first moments of a plane lamina.

12.6 MOMENTS OF INERTIA

A particle of mass m at the perpendicular distance r from the line L is said to have the *moment of inertia* or *second moment*

$$I = r^2 m$$

with respect to L. A system of particles with masses $m_1, m_2, \cdots, m_n$ at perpendicular distances, respectively, $r_1, r_2, \cdots, r_n$ from the axis L has the moment of inertia

$$I = \sum_{i=1}^{n} r_i^2 m_i$$

with respect to L. Note again that we have the moment of a sum equal to the sum of the moments. In particular, if the system of particles is in the xy plane, then we speak of the *moment of inertia about the x axis*

$$I_x = \sum_{i=1}^{n} y_i^2 m_i$$

the *moment of inertia about the y axis*

$$I_y = \sum_{i=1}^{n} x_i^2 m_i$$

and the *moment of inertia about the z axis*, or the *polar moment of inertia*,

$$I_z = \sum_{i=1}^{n} (x_i^2 + y_i)^2 n_i$$

If the total mass $m = \sum_{i=1}^{n} m_i$ of the system were to be placed at the perpendicular distance r from the axis L, then this single particle would have the same moment of inertia as does the system if

$$r^2 \sum_{i=1}^{n} m_i = \sum_{i=1}^{n} r_i^2 m_i$$

or $r^2 m = I$.

The number r defined by this equation is called the *radius of gyration* of the system.

Now consider a lamina of density δ over the region R in the xy plane. In some subdivision of the region R, look at a small piece of the lamina having area ΔA_j. If we choose any point (x_j, y_j) in this small piece, an approximation to a moment of inertia of this piece is

$$r_j^2 \delta(x_j, y_j) \, \Delta A_j$$

where r_j is the distance from the point (x_j, y_j) to the axis L under consideration. Summing and passing to the limit leads immediately to the following definitions.

The moments of inertia of the lamina of density $\delta(x, y)$ over the region R are given by the double integrals

$$I_y = \iint_R x^2 \delta(x, y) \, dA$$

$$I_x = \iint_R y^2 \delta(x, y) \, dA$$

12.6 MOMENTS OF INERTIA

and

$$I_z = \iint_R (x^2 + y^2)\delta(x,y)\,dA$$

■ **Example 1.** Consider the lamina over the region bounded by the line $\{(x,y) : y = x + 2\}$ and the curve $\{(x,y) : y = x^2\}$. If the density is $\delta = ky$ where k is a positive constant, then the moment of inertia with respect to the x axis is

$$I_x = \iint_R y^2 \cdot ky \cdot dA$$

From Fig 12.18 we readily see that this double integral equals the iterated integral

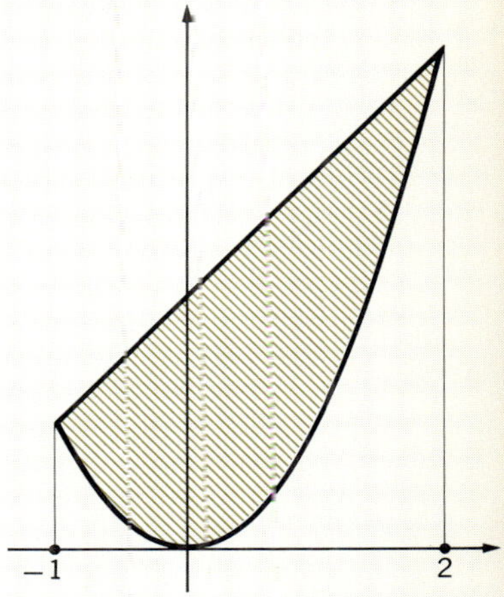

$$\int_{-1}^{2}\left[\int_{x^2}^{x+2} ky^3\,dy\right]dx$$

$$= \int_{-1}^{2}\frac{k}{4}y^4\bigg]_{x^2}^{x+2}dx = \frac{k}{4}\int_{-1}^{2}[(x+2)^4 - x^8]\,dx$$

FIGURE 12.18

$$= \frac{k}{4}\left[\frac{1}{5}(x+2)^5 - \frac{1}{9}x^9\right]_{-1}^{2} = \frac{369}{10}k$$

The mass of this lamina is simply

$$m = \int_{-1}^{2}\left[\int_{x^2}^{x+2} ky\,dy\right]dx$$

$$= \int_{-1}^{2}\frac{k}{2}y^2\bigg]_{x^2}^{x+2}dx = \frac{k}{2}\int_{-1}^{2}[(x+2)^2 - x^4]\,dx$$

$$= \frac{k}{2}\left[\frac{1}{3}(x+2)^3 - \frac{1}{5}x^5\right]_{-1}^{2} = \frac{36}{5}k$$

It follows that the radius of gyration about the x-axis is given by

$$r^2 = \frac{I}{m} = \frac{369k/10}{36k/5} = \frac{41}{8}$$

or $r = \frac{1}{4}\sqrt{82}$.

■ **Example 2.** Consider a circular lamina of constant density. We find its radius of gyration about the axis through the center and perpendicular to its plane. (Imagine the axle of a wheel.)

Let the lamina cover the region $R = \{(x,y) : x^2 + y^2 \leq a^2\}$ or $R = \{(r,\theta) : 0 \leq r \leq a,\ 0 \leq \theta \leq 2\pi\}$. In polar coordinate form, the integrals giving the mass and the polar moment of inertia are

$$m = \int_0^{2\pi}\left[\int_0^a k\cdot r\,dr\right]d\theta = \int_0^{2\pi}\frac{k}{2}r^2\bigg]_0^a d\theta = k\pi a^2$$

and
$$I_z = \int_0^{2\pi} \left[\int_0^a r^2 \cdot k \cdot r \, dr \right] d\theta = \int_0^{2\pi} \frac{k}{4} r^4 \Big|_0^a d\theta = \frac{1}{2} k\pi a^4,$$

where k is the constant density. Thus the radius of gyration is given by

$$r^2 = \frac{I_z}{m} = \frac{1}{2} k\pi a^4 x \frac{1}{k\pi a^2} = \frac{1}{2} a^2$$

or $r = a/\sqrt{2}$.

The moment of inertia plays an important role in the theory of rotating bodies and in certain problems about the deflection of a beam. We leave such applications to other courses.

Higher moments of a system of particles can be defined, too. For instance, the nth *moment* is

$$M^n = \sum_{i=1}^{n} r_i^n m_i$$

Such higher moments find a few applications in mechanics and are actually vital in statistical theory. Let us illustrate this latter point. Suppose we had the results of n measurements of a certain quantity such as the heights of male freshmen at a university. Say that there are n_1 instances of the value y_1, n_2 instances of the value y_2, and so on up to n_k instances of the value y_k, where $n_1 + n_2 + \cdots + n_k = n$. We speak of the "first moment" of this set of measurements.

$$\sum_{i=1}^{k} n_i y_i$$

Note that the measurement y_i has "mass" n_i. Then the "center of mass,"

$$\frac{\sum_{i=1}^{k} n_i y_i}{\sum_{i=1}^{k} n_i} = \frac{1}{n} \sum_{i=1}^{k} n_i y_i$$

is precisely the average value $\bar{y}$ of the measurements.

The "second moment" of this set of measurements,

$$\sum_{i=1}^{k} n_i y_i^2$$

is used to obtain a measure of the *dispersion* of the data. For instance, if the numbers y_i do not differ very much from the average value $\bar{y}$, then this second moment will be close to the number $n\bar{y}^2$. We can use the difference of the two to indicate the dispersion. Indeed, the higher moments are also used for similar purposes. Thus the large mass of data is reduced to a few constants that are representative of the data. Again this subject is left to appropriate statistics courses.

12.7 THE TRIPLE INTEGRAL

EXERCISES 12.6

A lamina lies over the region described in each of Exercises 1 to 10. Its density is given (the number k is a positive constant). Determine both the moment of inertia and the radius of gyration about the given axis.

1. $R = \{(x, y) : 0 \leq x \leq a, 0 \leq y \leq b\}$; $\delta = k$; x axis
2. $R = \{(x, y) : 0 \leq x \leq a, 0 \leq y \leq b\}$; $\delta = ky$; x axis
3. $R = \{(x, y) : 0 \leq x \leq a, 0 \leq y \leq b - \frac{b}{a}x\}$; $\delta = k$; x axis
4. $R = \{(x, y) : 0 \leq x \leq a, 0 \leq y \leq b - \frac{b}{a}x\}$; $\delta = ky$; x axis
5. R = the circular region $\{(x, y) : x^2 + y^2 \leq a^2\}$; $\delta = k$; y axis
6. R = the circular region bounded by $\{(x, y) : x^2 + y^2 = 4x\}$; $\delta = k$; x axis
7. R = the circular region bounded by $\{(x, y) : x^2 + y^2 = 4x - 3\}$; $\delta = k$; x axis
8. $R = \{(x, y) : 1 \leq x \leq 4, \frac{4}{x} \leq y \leq 5 - x\}$; $\delta = k$; z axis
9. $R = \{(x, y) : 0 \leq x \leq 2, -x \leq y \leq x - x^2\}$; $\delta = k(x + y)$; y axis
10. $R = \{(r, \theta) : r = \sin 3\theta; 0 \leq \theta \leq \pi/3\}$; $\delta = kr$; z axis

PROBLEMS 12.6

1. Prove that $I_z = I_x + I_y$.
2. Consider a constant density rectangular lamina of area ab. Find its radius of gyration about (a) one edge, and (b) its diagonal.
3. Prove the following *theorem of parallel axes*: The moment of inertia of a plane lamina about an axis in its plane is equal to the moment of inertia about a parallel axis through the center of mass plus the product of the mass times the square of the distance between the two axes. (Note that this proves that the moment of inertia of the lamina about a line through its center of mass is a minimum.)
4. Consider a plane lamina of density $\delta(x, y)$ over the region R. Prove that the moment of inertia about the line $\{(x, y) : y \cos \theta = x \sin \theta\}$ for any fixed angle θ is

$$I_x \cos^2 \theta - I_{xy} \sin 2\theta + I_y \sin^2 \theta$$

where I_{xy} is called the *product of inertia* and is

$$I_{xy} = \iint_R xy\,\delta(x, y)\,dA$$

5. The counterweight on the flywheel of a steam engine is a lamina with the shape shown in Fig. 12.19. Find the moment of inertia about the axis of the wheel.

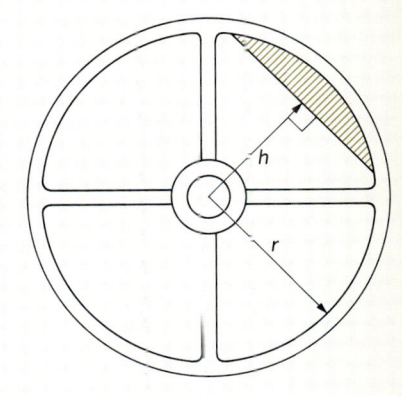

FIGURE 12.19

12.7 THE TRIPLE INTEGRAL

No new ideas are needed to define the triple integral. Here again the major difficulty lies in the treatment of the region in space over which the integral is defined.

Let us start by considering the rectangular parallelopiped
$$Q = \{(x, y, z) : a \leq x \leq b,\ c \leq y \leq d,\ e \leq z \leq f\}$$
We subdivide this solid into n^3 little boxes of similar shape by the planes
$$\left\{(x, y, z) : x = a + \frac{i}{n}(b - a)\right\}$$
$$\left\{(x, y, z) : y = c + \frac{i}{n}(d - c)\right\}$$
and
$$\left\{(x, y, z) : z = e + \frac{i}{n}(f - e)\right\}$$
where $i = 1, 2, \cdots, n - 1$. Order these boxes in any convenient way,
$$Q_1, Q_2, \cdots, Q_j, \cdots, Q_{n^3}$$
and note that each little box has the volume
$$\Delta_n V = \left(\frac{b-a}{n}\right)\left(\frac{d-c}{n}\right)\left(\frac{f-e}{n}\right)$$

Now assume that there is a continuous function f defined on Q. In each Q_j, choose a point (x_j, y_j, z_j) and form the Riemann sum
$$\sum_{j=1}^{n^3} f(x_j, y_j, z_j)\,\Delta_n V$$
The *triple integral* of f over Q is defined to be
$$\iiint_Q f(x, y, z)\,dV = \lim_{n \to \infty} \sum_{j=1}^{n^3} f(x_j, y_j, z_j)\,\Delta_n V$$
It should be obvious that such a triple integral must be evaluated in terms of a thrice-iterated integral or
$$\iiint_Q f(x, y, z)\,dV = \int_a^b \left[\int_c^d \left(\int_e^f f(x, y, z)\,dz\right) dy\right] dx$$

■ **Example 1.** Consider the function $f(x, y, z) = xyz$ over the box $\{(x, y, z) : 0 \leq x \leq 2,\ 1 \leq y \leq 3,\ 0 \leq z \leq 4\}$. Its triple integral is

$$\int_0^2 \left[\int_1^3 \left(\int_0^4 xyz\,dz\right) dy\right] dx = \int_0^2 \left[\int_1^3 \frac{1}{2}xyz^2 \Big|_0^4 dy\right] dx$$
$$= \int_0^2 \left[\int_1^3 8xy\,dy\right] dx$$
$$= \int_0^2 4xy^2 \Big|_1^3 dx$$
$$= \int_0^2 32x\,dx = 16x^2 \Big|_0^2 = 64 \ \blacksquare$$

12.7 THE TRIPLE INTEGRAL

If R is a region in space that is not a rectangular parallelopiped, then we use the same technique as was done for the double integral. Let $Q = \{(x, y, z) : a \leq x \leq b, c \leq y \leq d, e \leq z \leq f\}$ be the smallest such box that contains R. For each natural number n, subdivide Q as before into n^3 sub-boxes. Define the subset $\mathcal{S}(n)$ of the set of all n^3 sub-boxes by agreeing that a sub-box is in $\mathcal{S}(n)$ if it intersects R. Order the sub-boxes in $\mathcal{S}(n)$ in any convenient way,

$$Q_1, Q_2, \ldots, Q_j, \ldots, Q_{k(n)}.$$

Let f be a function continuous on R. For each $Q_j \in \mathcal{S}(n)$, choose a point (x_j, y_j, z_j) in $R \cap Q_j$. Then form the Riemann sum

$$\sum_{j=1}^{k(n)} f(x_j, y_j, z_j) \Delta_n V$$

Then the triple integral of f over R is defined to be

$$\iiint_R f(x, y, z) \, dV = \lim_{h \to \infty} \sum_{j=1}^{k(n)} f(x_j, y_j, z_j) \Delta_n V$$

We shall evaluate such a triple integral only as a thrice-iterated integral, of course. Therefore we next define the kind of regions in space that are used as regions of integration. With appropriate renaming of the axes when necessary, all our regions will be of the form

$$R = \{(x, y, z) : a \leq x \leq b, \gamma(x) \leq y \leq \delta(x), F(x, y) \leq z \leq G(x, y)\}$$

Note that this region is bounded by the planes

$$\{(x, y, z) : x = a\} \quad \text{and} \quad \{(x, y, z) : x = b\}$$

the vertical cylindrical surfaces

$$\{(x, y, z) : y = \gamma(x)\} \quad \text{and} \quad \{(x, y, z) : y = \delta(x)\}$$

and, on the top and bottom of the region, by the two surfaces

$$\{(x, y, z) : z = F(x, y)\} \quad \text{and} \quad \{(x, y, z) : z = G(x, y)\}$$

We assume that all of the functions involved are continuous. A typical such region is shown in Fig. 12.20.

Now if $f(x, y, z)$ is a continuous function on R, then the triple integral is evaluated as the thrice-iterated integral

$$\iiint_R f = \int_a^b \left[\int_{\gamma(x)}^{\delta(x)} \left(\int_{F(x,y)}^{G(x,y)} f(x, y, z) \, dz \right) dy \right] dx$$

In our first application of the triple integral we take the integrand function f to be identically equal to 1 on the region R. Then the value of the triple integral is the volume of the region.

■ **Example 2.** Consider the region bounded by the xy coordinate plane, the two planes $\{(x, y, z) : 2x + 3y = 6\}$, $\{(x, y, z) : 2x + 3z = 6\}$

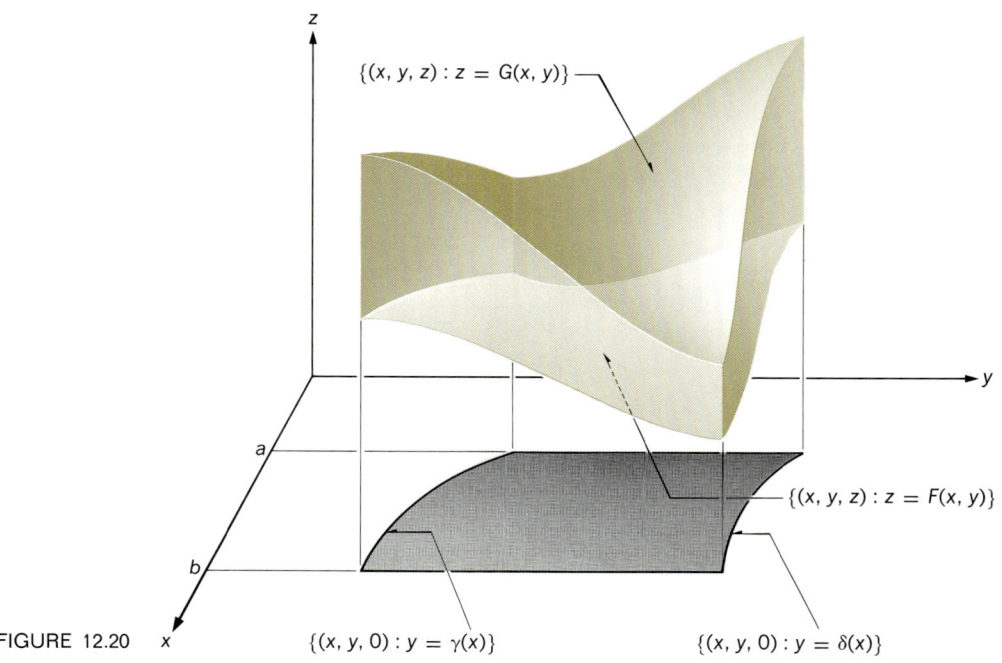

FIGURE 12.20

and the cylinder $\{(x, y, z) : y = 3 + 2x - x^2\}$. This region is shown in Fig. 12.21. By our agreement, the volume of this region is the triple integral

$$\int_0^3 \left[\int_{1/3(6-2x)}^{3+2x-x^2} \left(\int_0^{1/3(6-2x)} dz \right) dy \right] dx$$

$$= \int_0^3 \left[\int_{1/3(6-2x)}^{3+2x-x^2} \frac{1}{3}(6 - 2x) \, dy \right] dx$$

$$= \int_0^3 \frac{1}{3}(6 - 2x) \left[3 + 2x - x^2 - \frac{1}{3}(6 - 2x) \right] dx$$

$$= \frac{1}{9} \int_0^3 (18 + 42x - 34x^2 + 6x^3) \, dx$$

$$= \frac{53}{2}$$

■ **Example 3.** Let R be the tetrahedron cut off of the first octant by the plane

$$\left\{ (x, y, z) : \frac{x}{a} + \frac{y}{b} + \frac{z}{c} = 1 \right\}$$

where a, b, and c are positive constants. From Fig. 12.22 it is easy to see that the volume is given by the triple integral

12.7 THE TRIPLE INTEGRAL

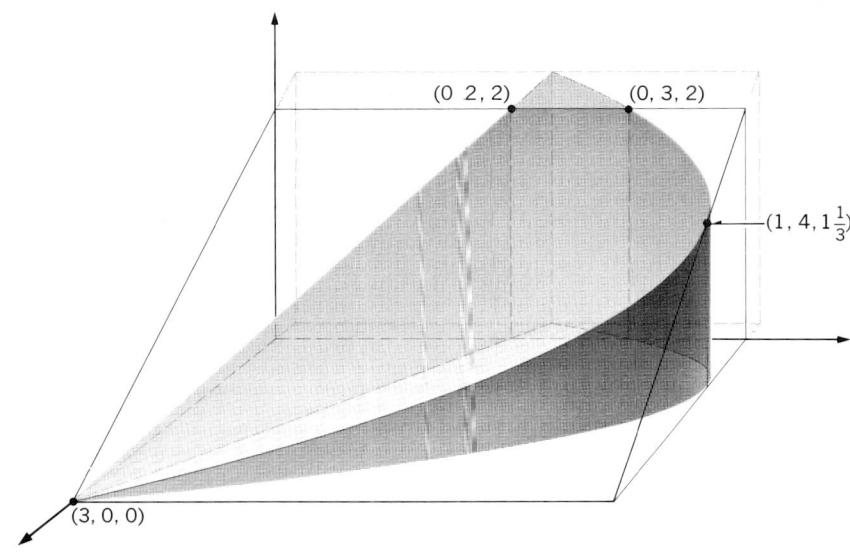

FIGURE 12.21

$$\int_0^a \left[\int_0^{b(1-x/a)} \left(\int_0^{c(1-x/a-y/b)} dz \right) dy \right] dx$$
$$= \int_0^a \left[\int_0^{b(1-x/a)} c\left(1 - \frac{x}{a} - \frac{y}{b}\right) dy \right] dx$$
$$= \int_0^a \frac{bc}{2}\left(1 - \frac{x}{a}\right)^2 dx$$
$$= \frac{abc}{6}$$

Evaluate each thrice-iterated integral in Exercises 1 to 10.

EXERCISES 12.7

1. $\int_0^1 \left[\int_0^1 \left(\int_0^1 xyz\, dz \right) dy \right] dx$

2. $\int_0^1 \left[\int_0^1 \left(\int_{y^2}^y (x+y)\, dz \right) dy \right] dx$

3. $\int_0^1 \left[\int_0^x \left(\int_0^{x+y} xy\, dz \right) dy \right] dx$

4. $\int_0^4 \left[\int_{2x}^{4x} \left(\int_0^y x\, dz \right) dy \right] dx$

5. $\int_0^1 \left[\int_{x^2}^{\sqrt{x}} \left(\int_0^{x+y} 2yz\, dz \right) dy \right] dx$

6. $\int_0^a \left[\int_0^{\sqrt{a^2-x^2}} \left(\int_0^y y\, dz \right) dy \right] dx$

7. $\int_0^2 \left[\int_x^{1+x} \left(\int_0^y z\, dz \right) dy \right] dx$

8. $\int_0^2 \left[\int_0^{\sqrt{4-x^2}} \left(\int_0^{2-y} x\, dz \right) dy \right] dx$

9. $\int_1^2 \left[\int_0^{2-x} \left(\int_0^{2-x-y} z\, dz \right) dy \right] dx$

10. $\int_0^3 \left[\int_0^{\sqrt{9-x^2}} \left(\int_0^{\sqrt{9-x^2y^2}} (x+y+z)\, dz \right) dy \right] dx$

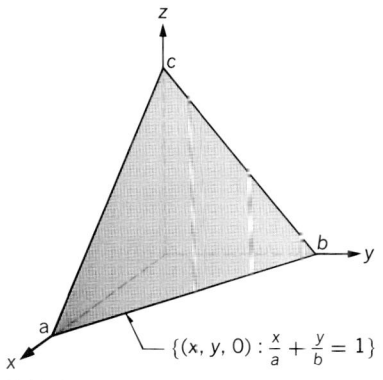

FIGURE 12.22

Determine the volume of the given region in space described in each of exercises 11 to 20.

11. $R = \{(x, y, z) : 0 \leq x \leq 2, 0 \leq y \leq 3 - x, 0 \leq z \leq 4 - x - y\}$
12. $R = \{(x, y, z) : 0 \leq x \leq 4, 0 \leq y \leq 4x - x^2, 0 \leq z \leq x^2 + y^2\}$
13. $R = \{(x, y, z) : 0 \leq x \leq 1, x^2 \leq y \leq \sqrt{x}, 0 \leq z \leq 2xy\}$
14. R is bounded by the coordinate planes, the plane $\{(x, y, z) : x + y = 2\}$ and the surface $\{(x, y, z) : z = \sqrt{16 - x^2}\}$.
15. R is bounded by the three planes $\{(x, y, z) : x = 0\}$, $\{(x, y, z) : y = 4\}$ $\{(x, y, z) : y = x\}$ and the cylinders $\{(x, y, z) : z = \sqrt{y}\}$ and $\{(x, y, z) : z = 6\sqrt{y}\}$.
16. R is bounded by the coordinate planes, the cylinder $\{(x, y, z) : \sqrt{x} + \sqrt{y} = 4\}$ and the surface $\{(x, y, z) : z = 2xy\}$.
17. R is bounded by the surfaces $\{(x, y, z) : z = x^2 + y^2\}$ and
$$\{(x, y, z) : z = 9 - x^2 - y^2\}.$$
18. R is bounded by the sphere $\{(x, y, z) : x^2 + y^2 + z^2 = a^2\}$.
19. R is bounded by the closed surface
$$\left\{(x, y, z) : \frac{x^2}{a^2} + \frac{y^2}{b^2} + \frac{z^2}{c^2} = 1\right\}$$
20. R is the region inside the sphere $\{(x, y, z) : x^2 + y^2 + z^2 = a^2\}$ and outside of the surface $\{(x, y, z) : 2x^2 + 2y^2 + z^2 = a^2\}$.

In each of Exercises 21 to 30, evaluate $\iiint_R f$ for the given region R and given function f.

21. $R = \{(x, y, z) : 0 \leq x \leq 1, 0 \leq y \leq x, y \leq z \leq 1 + y\}; f(x, y, z) = x$
22. $R = \{(x, y, z) : 0 \leq x \leq 2, 1 \leq y \leq 1 + x, 0 \leq z \leq xy\}; f(x, y, z) = \dfrac{1}{xy}$
23. $R = \{(x, y, z) : -1 \leq x \leq 1, 0 \leq y \leq 1, x^2y \leq z \leq 1\}; f(x, y, z) = xy$
24. $R = \{(x, y, z) : -2 \leq x \leq 2, 0 \leq y \leq 4 - x^2, -\sqrt{y} \leq z \leq \sqrt{y}\}; f(x, y, z) = x^2\sqrt{y}$
25. $R = \{(x, y, z) : 0 \leq x \leq 1, 0 \leq y \leq \sqrt{x}, xy \leq z \leq 1\}; f(x, y, z) = 2z$
26. R is bounded by the coordinate planes and the plane
$$\left\{(x, y, z) : \frac{x}{a} + \frac{y}{b} + \frac{z}{c} = 1\right\}$$
$a > 0, b > 0, c > 0; f(x, y, z) = x + y + z$.
27. R is bounded by the plane $\{(x, y, z) : z = 2\}$ and the surface $\{(x, y, z) : z = \frac{1}{2}(x^2 + y^2)\}; f(x, y, z) = x^2 + y^2 + z^2$.
28. R is bounded by the planes $\{(x, y, z) : z = 0\}$ and $\{(x, y, z) : z = 3\}$ and the cylinder $\{(x, y, z) : x^2 + y^2 = 25\}; f(x, y, z) = xy + xz + yz$.
29. R is bounded by the planes $\{(x, y, z) : y = 0\}$, $\{(x, y, z) : 2y + z = 4\}$ and the cylinders $\{(x, y, z) : z = x^2\}$, $\{(x, y, z) : z = 4 - x^2\}; f(x, y, z) = 2z$.

30. R is bounded by the surfaces $\{(x, y, z) : x^2 + y^2 = 4x\}$ and $\{(x, y, z) : z^2 = 8x\}$; $f(x, y, z) = 1 + 2z$.

12.8 CYLINDRICAL COORDINATES

The idea of a cylindrical coordinate system is very easy. A point in space is located by using polar coordinates on the xy plane and the usual z coordinate in the third dimension. (Note that this implies that distance on the xy plane must be meaningful!) The location of a point (r, θ, z) is illustrated in Fig. 12.23.

As in Section 12.4, we have the following relationships between the rectangular coordinates (x, y, z) and the cylindrical coordinates (r, θ, z) of a given point:

$$x = r \cos \theta \qquad r^2 = x^2 + y^2$$
$$y = r \sin \theta \qquad \tan \theta = \frac{y}{x}$$
$$z = z$$

It is easy to see that the "constant r coordinate surface"

$$\{(r, \theta, z) : r = a\}$$

is precisely the right circular cylinder $\{(x, y, z) : x^2 + y^2 = a^2\}$ of radius a with the z axis as its axis of symmetry. (This fact explains the name "cylindrical coordinates.") The other two "constant coordinate" surfaces are the planes

$$\{(r, \theta, z) : \theta = \alpha\} \quad \text{and} \quad \{(r, \theta, z) : z = c\}$$

All three are shown in Fig. 12.24.

We shall only consider regions in space that are described in polar coordinates in the following two ways:

$$S = \{(r, \theta, z) : a(\theta) \leq r \leq b(\theta), \alpha \leq \theta \leq \beta; f(r, \theta) \leq z \leq g(r, \theta)\}$$

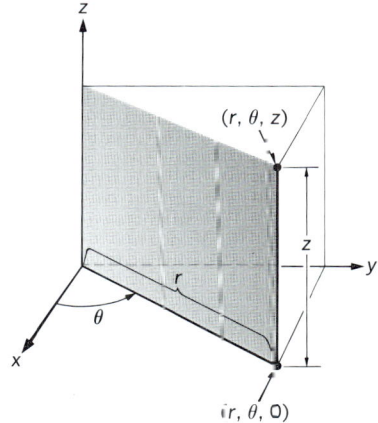

FIGURE 12.23

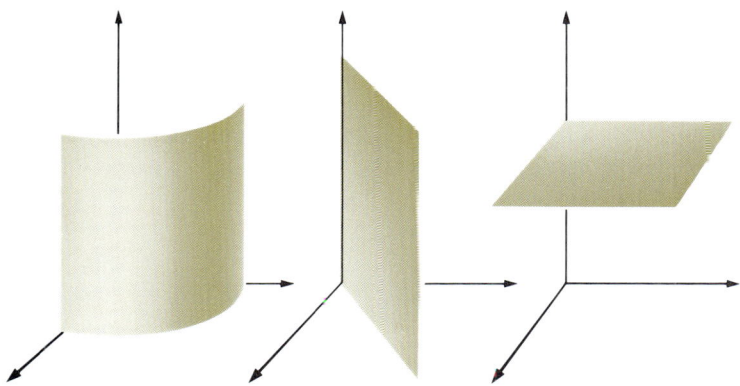

FIGURE 12.24

or
$$S = \{(r, \theta, z) : a \leq r \leq b, \alpha(r) \leq \theta \leq \beta(r), f(r, \theta) \leq z \leq g(r, \theta)\}$$

In particular, a region of the first type is bounded by the two vertical planes $\{(r, \theta, z) : \theta = \alpha\}$ and $\{(r, \theta, z) : \theta = \beta\}$, the two vertical cylinders $\{(r, \theta, z) : r = \alpha(\theta)\}$ and $\{(r, \theta, z) : r = \beta(\theta)\}$ and on the top and bottom by the surfaces $\{(r, \theta, z) : z = f(r, \theta)\}$ and $\{(r, \theta, z) : z = g(r, \theta)\}$, respectively. Such a region in space is shown in Fig. 12.25. Similar remarks apply to the second type.

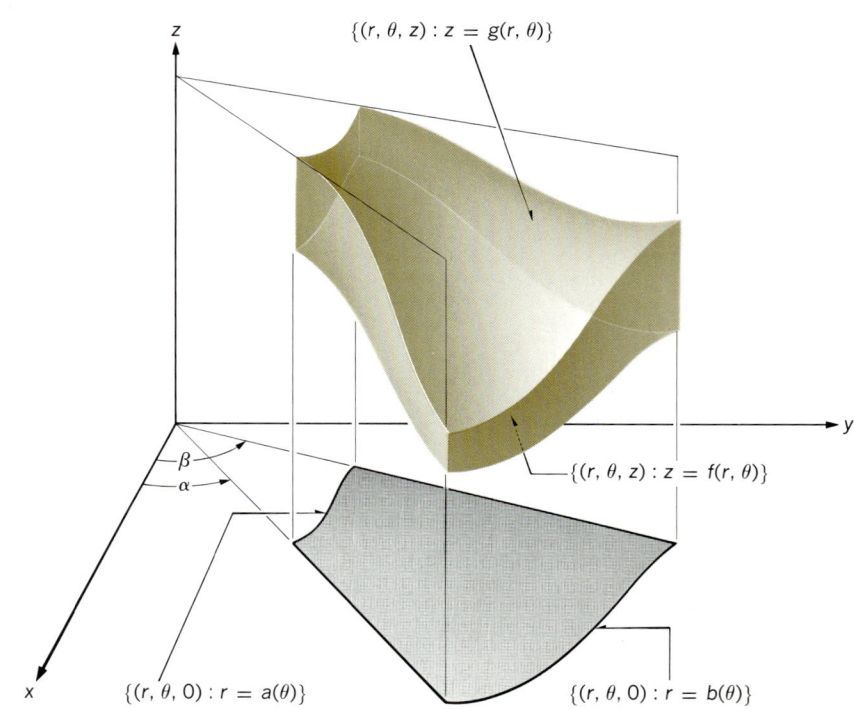

FIGURE 12.25

Let $f(r, \theta, z)$ be defined on
$$S = \{(r, \theta, z) : a(\theta) \leq r \leq b(\theta), \alpha \leq \theta \leq \beta, f(r, \theta) \leq z \leq g(r, \theta)\}$$

If S is subdivided by "constant cylindrical coordinate" surfaces, then a typical element of volume looks like Fig. 12.26, with the volume approximated by
$$\Delta V \cong r \, \Delta r \, \Delta \theta \, \Delta z$$

From here the now-familiar procedure yields the thrice-iterated integral
$$\iiint_S F(r, \theta, z) \, dV = \int_\alpha^\beta \left[\int_{a(\theta)}^{b(\theta)} \left(\int_{f(r,\theta)}^{g(r,\theta)} F(r, \theta, z) r \, dz \right) dr \right] d\theta$$

12.8 CYLINDRICAL COORDINATES

■ **Example 1.** Consider the right circular cylindrical solid
$$S = \{(r, \theta, z) : 0 \leq r \leq a, 0 \leq \theta \leq 2\pi, 0 \leq z \leq h\}$$
of radius a and height h. Its volume is given by the triple integral
$$\int_0^{2\pi} \left[\int_0^a \left(\int_0^h r \, dz \right) dr \right] d\theta = \int_0^{2\pi} \left[\int_0^a hr \, dr \right] d\theta$$
$$= \int_0^{2\pi} \frac{1}{2} a^2 h \, d\theta = \pi a^2 h$$
a formula we know already. ■

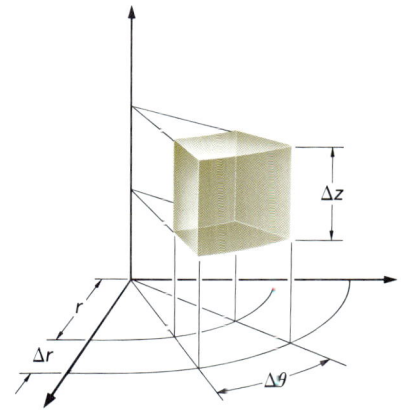

FIGURE 12.26

Cylindrical coordinates are most useful in those triple integrals where the region of integration is bounded on its sides by a closed vertical cylinder of the form $\{(r, \theta, z) : r = a(\theta)\}$ where $a(\theta)$ is a nonnegative function of θ. Then the triple integral of a function $F(x, y, z)$ is given by a thrice-iterated integral of the form
$$\int_0^a \left[\int_0^{a(\theta)} \left(\int_{f(r,\theta)}^{g(r,\theta)} F(r \cos \theta, r \sin \theta, z) r \, dz \right) dr \right] d\theta$$

■ **Example 2.** Consider the region S bounded by the vertical cylinder $\{(x, y, z) : x^2 + y^2 = 4y\}$ and the planes $\{(x, y, z) : z = 0\}$, $\{(x, y, z) : z = 1 + x\}$. The equation $x^2 + y^2 = 4y$ in polar coordinates is $r^2 = 4r \sin \theta$ or just $r = 4 \sin \theta$. This circle is described completely as θ increases from 0 to π radians. Recalling that $1 + x = 1 + r \cos \theta$ in polar coordinates, we may describe the region S as
$$\{(r, \theta, z) : 0 \leq r \leq 4 \sin \theta, 0 \leq \theta \leq \pi, 0 \leq z \leq 1 + r \cos \theta\}$$
Then, for instance, the volume of S is the integral
$$\int_0^\pi \left[\int_0^{4 \sin \theta} \left(\int_0^{1+r \cos \theta} r \, dz \right) dr \right] d\theta$$
We let the reader verify that this integral has the value 4π.

EXERCISES 12.8

For each of the following exercises, compute the volume of the region described. Sketch the region as accurately as you can.

1. $S = \{(r, \theta, z) : 0 \leq r \leq 4, 0 \leq \theta \leq \pi/2, 0 \leq z \leq r\}$
2. $S = \{(r, \theta, z) : 0 \leq r \leq \cos \theta, 0 \leq \theta \leq \pi/2, 0 \leq z \leq r \sin \theta\}$
3. $S = \{(r, \theta, z) : 0 \leq r \leq 2, 0 \leq \theta \leq 2\pi, r \leq z \leq \sqrt{8 - r^2}\}$
4. $S = \{(r, \theta, z) : 0 \leq r \leq a \sin \theta, 0 \leq \theta \leq \pi, \frac{b}{a} \sqrt{a^2 - r^2} \leq z \leq \sqrt{a^2 - r^2}\}$, $b < a$.

5. $S = \{(r, \theta, z) : 0 \leq r \leq 2, 0 \leq \theta \leq \frac{\pi r}{2}, 0 \leq z \leq r^2\}$

6. S is bounded by the cylinder $\{(x, y, z) : x^2 + y^2 = 2ax\}$ and the sphere $\{(x, y, z) : x^2 + y^2 + z^2 = 4a^2\}$.

7. S is bounded by the plane $\{(x, y, z) : z = y\}$ and the surface
$$\{(x, y, z) : z = x^2 + y^2\}.$$

8. S is the region bounded above by $\{(x, y, z) : z = \sqrt{8 - r^2}\}$ and below by $\{(x, y, z) : z = \tfrac{1}{2}r^2\}$.

9. S is bounded by the sphere $\{(x, y, z) : x^2 + y^2 + z^2 = a^2\}$ and by the cylinder $\{(x, y, z) : x^2 + y^2 = b^2, b < a\}$.

10. S is bounded by the closed surface
$$\left\{(x, y, z) : \frac{x^2}{a^2} + \frac{y^2}{b^2} + \frac{z^2}{c^2} = 1\right\}.$$

PROBLEMS 12.8

1. Find the volume inside of the torus generated by revolving a circle of radius a about an axis in its plane at the distance $b > a$ from its center.

2. Prove the following theorem of Pappus: Let R be a plane region and let L be a line in the plane of R that meets R at most in the boundary of R. Rotate R about L to form a solid S of revolution. Then the volume of S equals the product of the area of R and the circumference of the circle described by the centroid of R during the rotation.

12.9 SPHERICAL COORDINATES

As already pointed out, it is convenient to introduce cylindrical coordinates when there is an axis of symmetry in the physical problem. When there is a center of symmetry, it is often more convenient to use a *spherical coordinate system*.

Let (x, y, z) be a point in a coordinatized geometric space. We shall locate the same point with three other numbers (ρ, ϕ, θ) as follows: The first coordinate ρ is just the norm of the position vector of the point (x, y, z), that is,

$$\rho = \sqrt{x^2 + y^2 + z^2}$$

(Note that this implies that distance is meaningful in our problem.) The other two spherical coordinates are angles and are shown in Fig. 12.27. Thus ϕ is the angle from the z axis down to the position vector of the point, while θ is the angle from the positive x axis to the vertical plane containing this position vector.

It is easy to deduce the following relationships between the rectangular coordinates (x, y, z), the cylindrical coordinated (r, θ, z), and the spherical coordinates (ρ, ϕ, θ) of a given point:

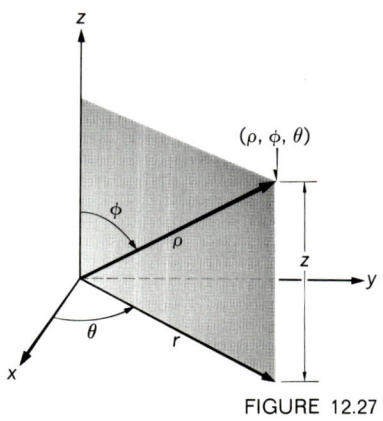

FIGURE 12.27

12.9 SPHERICAL COORDINATES

$r = \rho \sin \phi$ $x = r \cos \theta$ $x = \rho \sin \phi \cos \theta$
$z = \rho \cos \phi$ $y = r \sin \theta$ $y = \rho \sin \phi \sin \theta$
$\theta = \theta$ $z = z$ $z = \rho \cos \phi$

We note that the "constant ρ-coordinate surface" $\{(\rho, \phi, \theta) : \rho = a\}$ is a sphere of radius a about the origin (and hence the name "spherical coordinates"). The "constant ϕ-coordinate surface" $\{(\rho, \phi, \theta) : \phi = \alpha\}$ is a right circular cone symmetric about the z axis. The "constant θ-coordinate surface" $\{(\rho, \phi, \theta) : \theta = \beta\}$ is again a plane.

Now suppose we subdivide a region S in space by means of constant spherical coordinate surfaces. In Fig. 12.28 we show a typical small piece of the subdivided region. Then, considering the diagrams in Fig. 12.29, we compute that this small piece of the region has a volume approximately given by

$$\Delta V \cong (\Delta \rho) \cdot (\rho \Delta \phi) \cdot (\rho \sin \phi \Delta \theta)$$

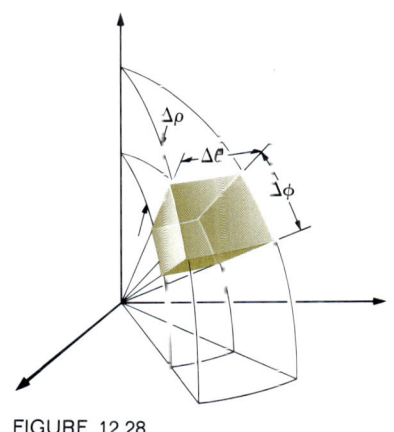

FIGURE 12.28

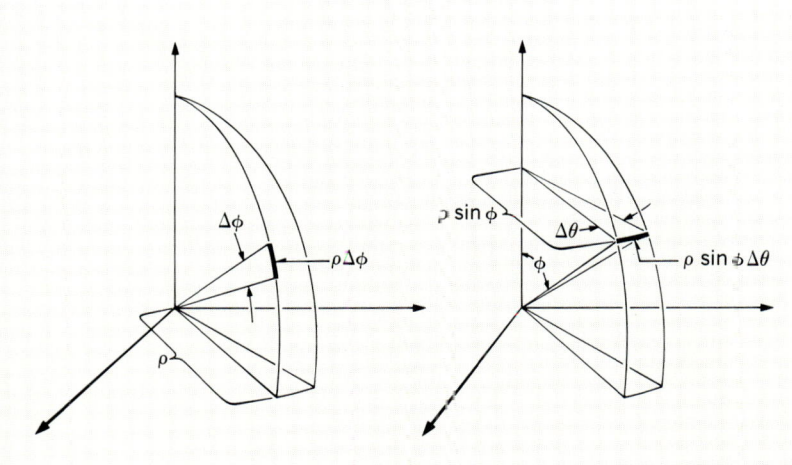

FIGURE 12.29

Because $\Delta V \cong \rho^2 \sin \phi \, \Delta \rho \, \Delta \phi \, \Delta \theta$, we follow the customary procedure to arrive at the triple integral of a function $f(\rho, \phi, \theta)$ defined on the region S

$$\iiint_S f(\rho, \phi, \theta) \, dV = \iiint_S f(\rho, \phi, \theta) \rho^2 \sin \phi \, d\rho \, d\phi \, d\theta$$

■ **Example 1.** We determine the volume of the region below the sphere $\{(x, y, z) : x^2 + y^2 + z^2 = a^2\}$ and above the conical surface $\{(x, y, z) : z = \sqrt{x^2 + y^2}\}$. The reader can verify that these are the surfaces $\{(\rho, \phi, \theta) : \rho = a\}$ and $\{(\rho, \phi, \theta) : \phi = \pi/4\}$, respectively. Thus the region S can be described easily in spherical coordinates as

$$S = \left\{ (\rho, \phi, \theta) : 0 \leq \rho \leq a, 0 \leq \phi \leq \frac{\pi}{4}, 0 \leq \theta \leq 2\pi \right\}$$

Therefore, its volume is the triple integral

$$\iiint_S dr = \int_0^{2\pi} \left[\int_0^{\pi/4} \left(\int_0^a \rho^2 \sin \phi \, d\rho \right) d\phi \right] d\theta$$
$$= \int_0^{2\pi} \left[\int_0^{\pi/4} \frac{1}{3} a^3 \sin \phi \, d\phi \right] d\theta$$
$$= \int_0^{2\pi} \frac{1}{3} a^3 \left(1 - \frac{\sqrt{2}}{2} \right) d\theta$$
$$= \frac{2\pi a^3}{3} \left(1 - \frac{\sqrt{2}}{2} \right)$$

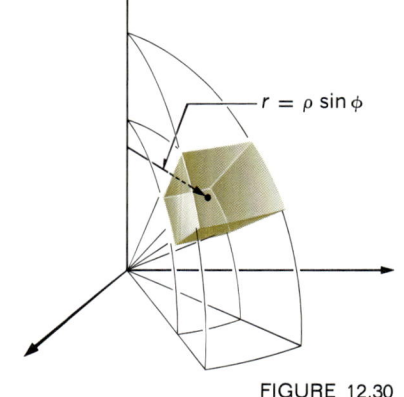

$r = \rho \sin \phi$

FIGURE 12.30

■ *Example 2.* Consider a hemispherical solid of constant density. We shall determine the radius of gyration about its vertical axis of symmetry. It is convenient to use the spherical coordinate system in which this solid occupies the region

$$S = \left\{ (\rho, \phi, \theta) : 0 \leq \rho \leq a, 0 \leq \phi \leq \frac{\pi}{2}, 0 \leq \theta \leq 2\pi \right\}$$

From Fig. 12.30 we easily read off the approximation

$$\Delta I_z \simeq r^2 (\delta \, \Delta \gamma) = (\rho^2 \sin^2 \phi)(\delta \rho^2 \sin \phi \, \Delta \rho \, \Delta \phi \, \Delta \theta)$$

It folows that the polar moment of inertia of this solid is given by the iterated integral

$$\delta \int_0^{2\pi} \left[\int_0^{\pi/2} \left(\int_0^a \rho^4 \sin^3 \phi \, d\rho \right) d\phi \right] d\theta$$
$$= \frac{1}{5} \delta a^5 \int_0^{2\pi} \left[\int_0^{\pi/2} \sin^3 \phi \, d\phi \right] d\theta = \frac{4\pi \delta a^5}{15}$$

Since the volume of this solid is $2\pi a^3/3$ (one-half of a sphere), its mass is $2\pi \delta a^3/3$, and the desired radius of gyration satisfies the equation

$$r^2 = \frac{I_z}{m} = \frac{4\pi \delta a^5}{15} \times \frac{3}{2\pi \delta a^3} = \frac{4a^2}{5}$$

or

$$r = \frac{2}{\sqrt{5}} a$$

EXERCISES 12.9 Sketch the region in space given in each of Exercises 1 to 10 and find its volume. Try to describe the region in terms of common solids.

12.9 SPHERICAL COORDINATES

1. $S = \left\{ (\rho, \phi, \theta) : 0 \leq \rho \leq a, 0 \leq \phi \leq \frac{\pi}{6}, 0 \leq \theta \leq 2\pi \right\}$
2. $S = \left\{ (\rho, \phi, \theta) : 0 \leq \rho \leq a, \frac{\pi}{6} \leq \phi \leq \frac{\pi}{2}, 0 \leq \theta \leq \pi \right\}$
3. $S = \left\{ (\rho, \phi, \theta) : 0 \leq \rho \leq 2a \cos \phi, 0 \leq \phi \leq \frac{\pi}{2}, 0 \leq \theta \leq 2\pi \right\}$
4. $S = \left\{ (\rho, \phi, \theta) : 0 \leq \rho \leq 2a \cos \phi, 0 \leq \phi \leq \frac{\pi}{4}, 0 \leq \theta \leq \frac{\pi}{2} \right\}$
5. $S = \left\{ (\rho, \phi, \theta) : 0 \leq \rho \leq a \sec \phi, 0 \leq \phi \leq \frac{\pi}{6}, 0 \leq \theta \leq 2\pi \right\}$
6. $S = \left\{ (\rho, \phi, \theta) : a \sec \phi \leq \rho \leq 2a \sec \phi, 0 \leq \phi \leq \frac{\pi}{4}, 0 \leq \theta \leq 2\pi \right\}$
7. $S = \left\{ (\rho, \phi, \theta) : 2 \sec \phi \leq \rho \leq 4, 0 \leq \phi \leq \frac{\pi}{3}, 0 \leq \theta \leq 2\pi \right\}$
8. $S = \left\{ (\rho, \phi, \theta) : 0 \leq \rho \leq a, \tan^{-1}(\sqrt{3} \csc \theta) \leq \phi \leq \frac{\pi}{2}, 0 \leq \theta \leq \pi \right\}$
9. $S = \{ (\rho, \phi, \theta) : 0 \leq \rho \leq a \sin \phi, 0 \leq \phi \leq \pi, 0 \leq \theta \leq 2\pi \}$
10. $S = \left\{ (\rho, \phi, \theta) : 0 \leq \rho \leq a \sin^{1/3} \phi, 0 \leq \phi \leq \frac{\pi}{2}, 0 \leq \theta \leq \frac{\pi}{2} \right\}$

For Exercises 11 to 20, consider solids of constant density $\delta = k$ occupying the regions of Exercises 1 to 10. Find the radius of gyration of each solid with respect to the z axis.

PROBLEMS 12.9

1. Two concentric spheres have radii a and b, $a < b$. A plane at a distance $c < a$ from their common center divides the region between the two spheres into two pieces. Find the volume of the smaller piece.

2. Two planes at distances a and b, $a < b$, from the center of a sphere of radius $c > b$ cut out a slab from the solid ball bounded by the sphere. Determine the volume of this slab.

3. A hemispherical shell with constant density $\delta = k$ has an inner radius a and an outer radius b. Locate the center of mass and find the radius of gyration about its axis of symmetry.

4. A solid of constant density $\delta = k$ is bounded by two internally tangent spheres of radii a and b, $a < b$. Locate the center of mass and find the radius of gyration about its axis of symmetry.

5. Newton's law of gravitation states that two particles attract each other with a force acting along the line joining the particles and proportional to the product of their masses and inversely proportional to the distance between them. Prove that a homogeneous solid ball exerts a gravitational force proportional to M/ρ^2 at a point at distance ρ from the center of the ball, where M is the mass of the ball.

12.10 HETEROGENEOUS SOLIDS

We considered heterogeneous laminas in Section 12.5. Here we use triple integration to solve certain problems involving heterogeneous solids.

Let a solid body occupy the region S in space and let the density at a point in the solid be a function δ of position. If the density function is constant over S, then we say that the solid is *homogeneous;* if the density is not constant, we say that the solid is *heterogeneous.*

The *mass* of a solid occupying the region S and having density function δ is given by the triple integral

$$m = \iiint_S \delta(x, y, z) \, dV$$

The *first moments* of this solid with respect to the three coordinate planes are

$$M_{yz} = \iiint_S x\delta(x, y, z) \, dV,$$

$$M_{xz} = \iiint_S y\delta(x, y, z) \, dV,$$

and

$$M_{xy} = \iiint_S z\delta(x, y, z) \, dV$$

Hence the *center of mass* of the solid is

$$(\bar{x}, \bar{y}, \bar{z}) = \left(\frac{M_{yz}}{m}, \frac{M_{xz}}{m}, \frac{M_{xy}}{m} \right)$$

The *moment of inertia* of the solid about the axis L is given by the triple integral

$$I_L = \iiint_S r^2 \, \delta(x, y, z) \, dV$$

where r is the perpendicular distance from the point (x, y, z) in the solid to the line L. Thus, for instance, the moment of inertia I_z about the z axis is

$$I_z = \iiint_S (x^2 + y^2) \, \delta(x, y, z) \, dV$$

The *radius of gyration* of the solid about the axis L is then

$$r_L = \sqrt{\frac{I_L}{m}}$$

■ **Example 1.** Suppose the solid tetrahedron cut off of the first octant by the plane $\{(x, y, z) : x + y + z = 6\}$ has the density function

12.10 HETEROGENEOUS SOLIDS

$\delta(x, y, z) = kz$. (In other words, the density at a point is proportional to the distance of the point from the xy plane.) Then the mass of this solid is

$$\int_0^6 \left[\int_0^{6-x} \left(\int_0^{6-x-y} kz \, dz \right) dy \right] dx$$

$$= \int_0^6 \left[\int_0^{6-x} \frac{1}{2} k(6 - x - y)^2 \, dy \right] dx$$

$$= \int_0^6 \frac{k}{6} (6 - x)^3 \, dx$$

$$= -\frac{k}{24} (6 - x)^4 \Big]_0^6 = 54k$$

The first moments of this solid with respect to the three coordinate planes are worked out next.

$$M_{yz} = \int_0^6 \left[\int_0^{6-x} \left(\int_0^{6-x-y} x \cdot kz \, dz \right) dy \right] dx$$

$$= \int_0^6 \left[\int_0^{6-x} \frac{1}{2} kx(6 - x - y)^2 \, dy \right] dx$$

$$= \frac{1}{6} k \left[108x^2 - 36x^3 + \frac{9}{2} x^4 - \frac{1}{5} x^5 \right]_0^6$$

$$= \frac{324k}{5}$$

$$M_{xz} = \int_0^6 \left[\int_0^{6-x} \left(\int_0^{6-x-y} y \cdot kz \, dz \right) dy \right] dx$$

$$= \int_0^6 \left[\int_0^{6-x} \frac{1}{2} ky(6 - x - y)^2 \, dy \right] dx$$

$$= \int_0^6 \frac{1}{24} k(6 - x)^4 \, dx$$

$$= -\frac{1}{120} k(6 - x)^5 \Big]_0^6 = \frac{324}{5} k$$

$$M_{xy} = \int_0^6 \left[\int_0^{6-x} \left(\int_0^{6-x-y} z \cdot kz \, dz \right) dy \right] dx$$

$$= \int_0^6 \left[\int_0^{6-x} \frac{1}{3} k(6 - x - y)^3 \, dy \right] dx$$

$$= \int_0^6 \frac{1}{12} k(6 - x)^4 \, dx$$

$$= \frac{1}{60} k(6 - x)^5 \Big]_0^6 = \frac{648}{5} k$$

It follows that the center of mass of this solid is at the point

$$\left(\frac{324k/5}{54k}, \frac{324k/5}{54k}, \frac{648k/5}{54k} \right) = \left(\frac{6}{5}, \frac{6}{5}, \frac{12}{5} \right)$$

■ *Example 2.* If the solid under consideration has an axis of symmetry, we customarily introduce cylindrical coordinates. Suppose that a right circular conical solid has a base radius a and height h. Let its density at a point be proportional to the distance of that point from its base. We shall determine the radius of gyration of this solid about its axis of symmetry.

Introduce cylindrical coordinates, taking the axis of symmetry to be the z axis and the base of the solid as the $r\theta$ plane. The conical surface of this solid now has the equation $r/a + z/h = 1$, whence the solid occupies the region

$$S = \left\{(r, \theta, z) : 0 \leq r \leq a, 0 \leq \theta \leq 2\pi, 0 \leq z \leq \frac{h}{a}(a - r)\right\}$$

Then the density function is $\delta(x, y, z) = kz$. It follows that the mass is

$$m = \int_0^{2\pi} \left[\int_0^a \left(\int_0^{h/a(a-r)} kz \cdot r \, dz\right) dr\right] d\theta$$

$$= \int_0^{2\pi} \left[\int_0^a \frac{1}{2} \frac{h^2}{a^2} (a - r)^2 \, r \, dr\right] d\theta$$

$$= \frac{kh^2}{2a^2} \int_0^{2\pi} \left[\int_0^a (a^2 r - 2ar^2 + r^3) \, dr\right] d\theta$$

$$= \frac{kh^2}{2a^2} \cdot \frac{a^4}{12} \int_0^{2\pi} d\theta = \frac{ka^2 h^2 \pi}{12}$$

Similarly, the moment of inertia about the axis of symmetry is

$$I_z = \int_0^{2\pi} \left[\int_0^a \left(\int_0^{h/a(a-r)} r^2 \cdot kz \cdot r \, dz\right) dr\right] d\theta$$

$$= \int_0^{2\pi} \left[\int_0^a \frac{kh^2}{2a^2} (a - r)^2 r^3 \, dr\right] d\theta$$

$$= \frac{kh^2}{2a^2} \int_0^{2\pi} \left[\int_0^a (a^2 r^3 - 2ar^4 + r^5) \, dr\right] d\theta$$

$$= \frac{kh^2}{2a^2} \cdot \frac{a^6}{60} \int_0^{2\pi} d\theta = \frac{ka^4 h^2 \pi}{60}$$

Finally, the radius of gyration about the z axis is given by

$$r_z^2 = \frac{I_z}{m} = \frac{ka^4 h^2 \pi}{60} \times \frac{12}{ka^2 h^2 \pi} = \frac{a^2}{5}, \text{ or } r_z = a/\sqrt{5}$$

■ *Example 3.* If the solid under discussion has a center of symmetry, then the use of spherical coordinates often simplifies the calculations. We determine the moment of inertia about one of the diameters of a spherical shell with inner radius a and outer radius b, assuming that the density is constant, $\delta(x, y, z) = k$.

By selecting spherical coordinates with the origin at the common center of the bounding spheres, the region occupied by the shell is

$$S = \{(\rho, \phi, \theta) : a \leq \rho \leq b, 0 \leq \phi \leq \pi, 0 \leq \theta \leq 2\pi\}$$

Hence the mass of the shell is

$$\int_0^{2\pi} \left[\int_0^{\pi} \left(\int_a^b k \cdot \rho^2 \sin\phi \, d\rho \right) d\phi \right] d\theta = \frac{4\pi k}{3}(b^3 - a^3)$$

If we choose the z axis as the line about which the moment of inertia is taken, it is easy to see that the square of the distance from the point (ρ, ϕ, θ) to the z axis is precisely $\rho^2 \sin^2 \phi = r^2$. Thus the moment of inertia is

$$\int_0^{2\pi} \left[\int_0^{\pi} \left(\int_a^b \rho^2 \sin^2\phi \cdot k\rho^2 \sin\phi \, d\rho \right) d\phi \right] d\theta = \frac{8\pi k}{15}(b^5 - a^5)$$

The radius of gyration is then

$$r_z = \sqrt{\frac{2(b^5 - a^5)}{5(b^3 - a^3)}}$$

(The reader may verify the values given above.)

EXERCISES 12.10

In each of Exercises 1 to 10, the region S occupied by a solid and the density function δ of that solid are given. Determine the center of gravity of the solid. (If you can use symmetry to simplify the work, so much the better.)

1. $S = \{(x, y, z) : -a \leq x \leq a, -b \leq y \leq b, -c \leq z \leq c\}; \delta = kx^2$
2. $S = \{(x, y, z) : 0 \leq x \leq a, 0 \leq y \leq a, 0 \leq z \leq a\}; \delta = kxyz$
3. $S = \{(x, y, z) : -\sqrt{y - y^2} \leq x \leq \sqrt{y - y^2}, 0 \leq y \leq 1, x^2 + y^2 \leq z \leq y\}; \delta = k$
4. $S = \{(\rho, \phi, \theta) : 0 \leq \rho \leq 4, 0 \leq \phi \leq \frac{\pi}{3}, 0 \leq \theta \leq 2\pi\}; \delta = k\rho \cos\phi$
5. $S = \{(x, y, z) : x^2 + y^2 + z^2 \leq a^2\}; \delta = k(x^2 + y^2)^2$
6. S is bounded by the coordinate planes and the plane $\{(x, y, z) : 3x + 4y + 6z = 24\}; \delta = k$.
7. S is bounded by the cylinders $\{(x, y, z) : x^2 + y^2 = 4\}$ and $\{(x, y, z) : x^2 + z^2 = 4\}$ and above the xy plane; $\delta = kz$.
8. S is bounded by the planes $\{(x, y, z) : z = y\}$ and $\{(x, y, z) : z = -y\}$, and by the cylinder $\{(x, y, z) : x^2 = 4 - 2y\}; \delta = k$.
9. S is bounded by the cone $\{(x, y, z) \; z^2 = x^2 + y^2\}$ and the cylinder $\{(x, y, z) : x^2 + y^2 = 6x\}; \delta = \frac{1}{2}(x^2 + y^2)^{1/2}$.
10. S is the region outside of the cone $\{(x, y, z) : z^2 = x^2 + y^2\}$ and inside of the sphere $\{(x, y, z) : x^2 + y^2 + z^2 = 16\}; \delta = kz^2$.

In each of Exercises 11 to 18, determine the moment of inertia of the given solid about the given axis.

11. A homogeneous cylindrical shell of inner radius a and outer radius b and height h; about its axis of symmetry.

12. A homogeneous hemispherical solid; about its axis of symmetry.

13. A homogeneous cube; about one of its edges.

14. A homogeneous cube; about a line through the center of two opposite faces.

15. A homogeneous right circular cylinder of base radius a and height h; about a generating line of its curved surface.

16. A homogeneous right circular cylinder of base radius a and height h; about an axis through the center of symmetry of the cylinder and perpendicular to its axis of symmetry.

17. A homogeneous hemispherical solid; about a diameter of its base.

18. A homogeneous solid torus; about its axis of symmetry.

PROBLEMS 12.10

1. Find the moment of inertia of a homogeneous cube about one of its diagonals.

2. A hemispherical solid occupies the region

$$S = \left\{ (\rho, \phi, \theta) : 0 \leq \rho \leq a, 0 \leq \phi \leq \frac{\pi}{2}, 0 \leq \theta \leq 2\pi \right\}$$

Suppose its density function has the form $\delta = k\rho^{-m}$. For what values of m does the mass actually exist? (When is the mass finite?) Determine I_z when the mass exists.

3. Let a solid with density δ occupy the region S in space and let L denote the line with the vector equation $\mathbf{p} = t(a\mathbf{i} + b\mathbf{j} + c\mathbf{k})$. Prove that the moment of inertia I_L of the solid about the line L is given by

$$I_L = a^2 I_x + b^2 I_y + c^2 I_z - 2 \iiint_S (bcyz + acxz + abxy) \delta \, dV$$

4. State and prove a parallel axis theorem for heterogeneous solids.

13

ANALYTIC
AND VECTOR
GEOMETRY

Little emphasis was placed upon geometry in the preceding chapters. Hence, for students of engineering and physical sciences particularly, this chapter provides some geometric applications of the calculus. Many of these will be repeated and expanded in subsequent course work. Similarly, mention of electromagnetic theory, mechanics, and so forth, is little more than a foreshadowing of material to be found in advanced courses in the student's area of specialization.

13.1 FREE VECTORS

When used for geometric purposes, a coordinate space R^n is assigned a property which as yet we have not fully exploited. Whether parallel to a coordinate axis or not, *each line segment has a length* given by the distance formula (see Section 1.11):

$$d[(x_1, y_1, z_1), (x_2, y_2, z_2)] = \sqrt{(x_1 - x_2)^2 + (y_1 - y_2)^2 + (z_1 - z_2)^2}$$

In particular, each position vector has a length and the unit basis vectors all have the same unit length. A standard notation denotes the three unit basis vectors by

$$\mathbf{i}, \mathbf{j}, \mathbf{k}$$

where $\mathbf{i}$ is the position vector of the point $(1, 0, 0)$, $\mathbf{j}$ is the position vector of $(0, 1, 0)$ and $\mathbf{k}$ is the position vector of $(0, 0, 1)$. Then each position vector $\mathbf{p}$ can be written in one and only one way as a linear combination of the form

$$\mathbf{p} = x\mathbf{i} + y\mathbf{j} + z\mathbf{k}$$

The length or *norm* of this position vector $\mathbf{p}$ is denoted by the symbol

$$\|\mathbf{p}\|$$

This length is the distance from $(0, 0, 0)$ to (x, y, z); hence we have

$$\|\mathbf{p}\| = \|x\mathbf{i} + y\mathbf{j} + z\mathbf{k}\| = \sqrt{x^2 + y^2 + z^2}$$

Note that the norm is a real-valued function whose domain is the set of position vectors. The reader can prove the first theorem by easy applications of the above definitions.

■ **THEOREM 13.1.** The norm function satisfies

N1: $\|\mathbf{p}\| \geq 0$ for each position vector $\mathbf{p}$.
N2: $\|\mathbf{p}\| = 0$ if and only if $\mathbf{p} = 0$.
N3: $\|r\mathbf{p}\| = |r| \cdot \|\mathbf{p}\|$ for any position vector $\mathbf{p}$ and any real number r.
N4: $\|\mathbf{p} + \mathbf{q}\| \leq \|\mathbf{p}\| + \|\mathbf{q}\|$ for any two position vectors. (The triangle inequality.)

The reader probably has a reliable intuitive grasp of the meaning of "line segment." For instance, we realize that a line segment is determined

13.1 FREE VECTORS

by its endpoints and each pair of distinct points determines a unique line segment. In accurate reasoning, however, we need a precise definition such as the following: Let **p** and **q** be the position vectors of two distinct points. We define the *directed line segment* from the tip of **p** to the tip of **q** to be the ordered pair

$$(\mathbf{p}, \mathbf{q})$$

of position vectors. We call **p** the *initial point* and **q** the *terminal point* of the directed line segment.

Just as is the case for ordered pairs of real numbers, two directed line segments $(\mathbf{p}_1, \mathbf{q}_1)$ and $(\mathbf{p}_2, \mathbf{q}_2)$ are *equal* if and only if $\mathbf{p}_1 = \mathbf{p}_2$ and $\mathbf{q}_1 = \mathbf{q}_2$. This means that they coincide.

Associated with the directed line segment $(\mathbf{p}, \mathbf{q})$, there is the position vector $\mathbf{q} - \mathbf{p}$. We use this vector in several ways. First, we define the *length* of the directed line segment to be

$$\|\mathbf{q} - \mathbf{p}\|$$

Then we define two directed line segments $(\mathbf{p}_1, \mathbf{q}_1)$ and $(\mathbf{p}_2, \mathbf{q}_2)$ to be *parallel* if there is a real number r such that

$$\mathbf{q}_2 - \mathbf{p}_2 = r(\mathbf{q}_1 - \mathbf{p}_1)$$

(Fig. 13.1 explains this definition.) Note that the zero vector **0** is defined to be parallel to every vector.

Two directed line segments are said to be *equivalent* if they have the same associated position vector. Thus $(\mathbf{p}_1, \mathbf{q}_1)$ is equivalent to $(\mathbf{p}_2, \mathbf{q}_2)$ if and only if

$$\mathbf{q}_1 - \mathbf{p}_1 = \mathbf{q}_2 - \mathbf{p}_2$$

Note that this implies that the two directed line segments have the same length and are parallel. This equality implies even more, of course.

Geometrically, we can describe the equivalence of $(\mathbf{p}_1, \mathbf{q}_1)$ and $(\mathbf{p}_2, \mathbf{q}_2)$ as follows: The midpoints of the directed line segments $(\mathbf{p}_1, \mathbf{q}_2)$ and $(\mathbf{p}_2, \mathbf{q}_1)$ must coincide. This condition means that $\frac{1}{2}(\mathbf{p}_1 + \mathbf{q}_2) = \frac{1}{2}(\mathbf{p}_2 + \mathbf{q}_1)$. (See Fig. 13.2.) We leave the proof of this characterization to the reader as the Problem 13.1.6.

The reader can prove the following result as an exercise.

■ **THEOREM 13.2.** The equivalence of directed line segments is a true equivalence relation in the sense that it has the three properties:

1. A directed line segment is equivalent to itself. (We say that the relation is *reflexive*.)
2. Let S_1 and S_2 be directed line segments. If S_1 is equivalent to S_2, then S_2 is equivalent to S_1. (We say that the relation is *symmetric*.)
3. Let S_1, S_2, and S_3 be directed line segments. If S_1 is equivalent to S_2 and S_2 is equivalent to S_3, then S_1 is equivalent to S_3. (The relation is *transitive*.)

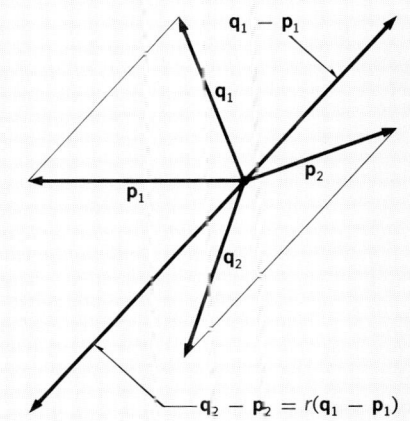

FIGURE 13.1

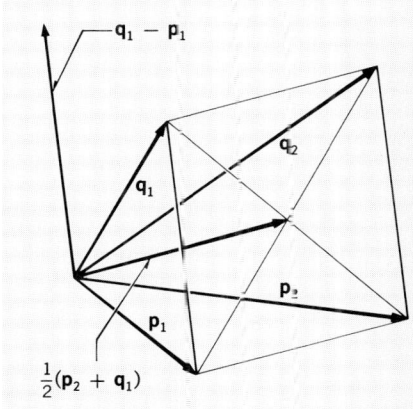

FIGURE 13.2

A position vector is a particular directed line segment if we think of the origin as its initial point. Let (a, b, c) be a point in space. The class of all directed line segments equivalent to the position vector of (a, b, c) is defined to be the *free vector* represented by the given position vector. We now use the notation

$$a\mathbf{i} + b\mathbf{j} + c\mathbf{k}$$

for this free vector (not only for the position vector).

Each directed line segment determines (or represents) a unique free vector. In particular, if the initial and terminal points of the segment have position vectors **p** and **q**, respectively, then the segment determines the free vector represented by $\mathbf{q} - \mathbf{p}$. In terms of the coordinates (x_1, y_1, z_1) of the initial point and (x_2, y_2, z_2) of the terminal point, the free vector is

$$(x_2 - x_1)\mathbf{i} + (y_2 - y_1)\mathbf{j} + (z_2 - z_1)\mathbf{k}$$

Given a free vector

$$\mathbf{v} = a\mathbf{i} + b\mathbf{j} + c\mathbf{k} \neq \mathbf{0}$$

we define the *direction angles* of **v** to be the smallest nonnegative angles α, β, and γ measured from the positive x axis, y axis and z axis, respectively, to the position vector of the point (a, b, c). (This position vector represents **v**.) In Fig. 13.3 we picture the direction angles of a vector whose components are all positive.

It is merely a matter of spacial visualization to see that each direction angle lies in the interval $\{\theta : 0 \leqq \theta \leqq \pi\}$.

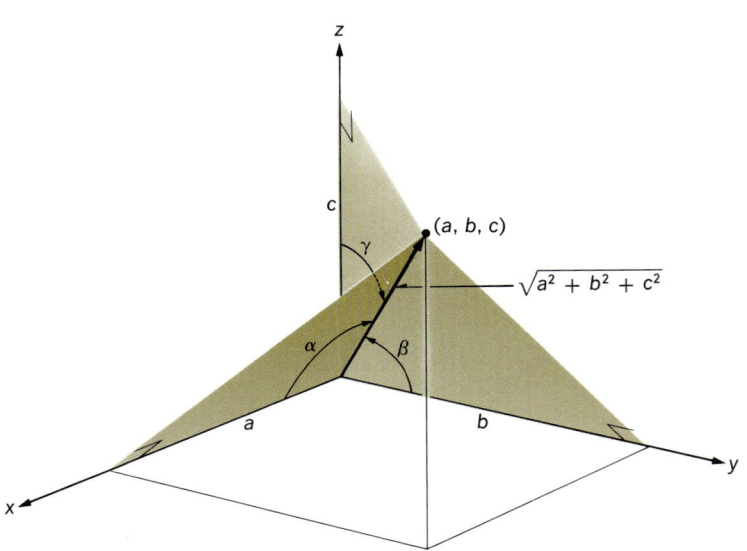

FIGURE 13.3

13.1 FREE VECTORS

The cosines of the direction angles are the *direction cosines*. For the vector $\mathbf{v} = a\mathbf{i} + b\mathbf{j} + c\mathbf{k}$ it is easy to verify (see Fig. 13.3) that

$$\cos \alpha = \frac{a}{\sqrt{a^2 + b^2 + c^2}}, \quad \cos \beta = \frac{b}{\sqrt{a^2 + b^2 + c^2}},$$

$$\cos \gamma = \frac{c}{\sqrt{a^2 + b^2 + c^2}}$$

We immediately find the equation

$$\cos^2 \alpha + \cos^2 \beta + \cos^2 \gamma = 1$$

which tells us that the direction angles are not independent. Knowing any two, we can obtain the value of the third.

Since

$$\sqrt{a^2 + b^2 + c^2} = \|a\mathbf{i} + b\mathbf{j} + c\mathbf{k}\| = \|\mathbf{v}\|$$

the equations giving the values of the direction cosines of $\mathbf{v}$ can be re-written as

$$a = \|\mathbf{v}\| \cos \alpha, \quad b = \|\mathbf{v}\| \cos \beta, \quad c = \|\mathbf{v}\| \cos \gamma$$

Therefore, the vector $\mathbf{v}$ can be expressed as

$$\mathbf{v} = \|\mathbf{v}\|(\mathbf{i} \cos \alpha + \mathbf{j} \cos \beta + \mathbf{k} \cos \gamma)$$

The unit vector

$$\frac{\mathbf{v}}{\|\mathbf{v}\|}$$

is called the *direction vector* of $\mathbf{v}$. Since we obviously have

$$\frac{\mathbf{v}}{\|\mathbf{v}\|} = \mathbf{i} \cos \alpha + \mathbf{j} \cos \beta + \mathbf{k} \cos \gamma$$

it follows that the easy way to determine the direction cosines of the vector $\mathbf{v}$ is to compute its direction vector $\mathbf{v}/\|\mathbf{v}\|$ and look at its components.

The vector equation $\mathbf{q}_1 - \mathbf{p}_1 = \mathbf{q}_2 - \mathbf{p}_2$ defining the equivalence of the directed line segments $(\mathbf{p}_1, \mathbf{q}_1)$ and $(\mathbf{p}_2, \mathbf{q}_2)$ implies that the two segments have the same length and are parallel. Note, however, that the directed line segment $(\mathbf{q}_1, \mathbf{p}_1)$ also has the same length as, and is parallel to, the directed line segment $(\mathbf{p}_1, \mathbf{q}_1)$. We distinguish between the two by introducing the concept of *sense* or *orientation* of a segment.

Two parallel directed line segments $(\mathbf{p}_1, \mathbf{q}_1)$ and $(\mathbf{p}_2, \mathbf{q}_2)$ have the same *sense* if

$$\mathbf{q}_2 - \mathbf{p}_2 = a^2(\mathbf{q}_1 - \mathbf{p}_1), \quad a \neq 0$$

Equivalently, the associated position vectors $\mathbf{q}_1 - \mathbf{p}_1$ and $\mathbf{q}_2 - \mathbf{p}_2$ must have the same direction vector, that is,

$$\frac{\mathbf{q}_1 - \mathbf{p}_1}{\|\mathbf{q}_1 - \mathbf{p}_1\|} = \frac{\mathbf{q}_2 - \mathbf{p}_2}{\|\mathbf{q}_2 - \mathbf{p}_2\|}$$

In these terms we can restate the equivalence of two directed line segments by saying that they are parallel, have the same length, and have the same direction.

A few remarks on the use of direction angles in the plane are in order. In two dimensions we should perhaps use two direction angles α and β as pictured in Fig. 13.4. Note that α is the angle of inclination of the vector. We do not need the angle β, however, because $\beta = \pi/2 - \alpha$. Nevertheless we still have the fundamental identity relating direction cosines, namely,

$$\cos^2 \alpha + \cos^2 \beta = 1$$

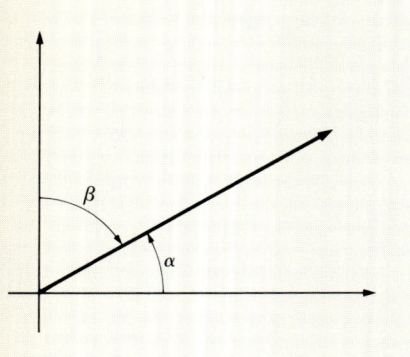

FIGURE 13.4

This follows because $\cos \beta = \cos(\pi/2 - \alpha) = \sin \alpha$. Hence the equation above is merely the identity $\cos^2 \alpha + \sin^2 \alpha = 1$.

EXERCISES 13.1

1. Verify the triangle inequality for each vector pair:
   (a) $2\mathbf{i} + 2\mathbf{j} - \mathbf{k}, \mathbf{i} - 2\mathbf{j} + 2\mathbf{k}$
   (b) $3\mathbf{i} - 4\mathbf{j} - 5\mathbf{k}, 2\mathbf{i} + \mathbf{j} + \mathbf{k}$

2. Find direction cosines for each vector:
   (a) $\mathbf{i} + 2\mathbf{j} + 2\mathbf{k}$ (b) $\mathbf{i} + \mathbf{j} + \mathbf{k}$
   (c) $2\mathbf{i} - 3\mathbf{j} + 3\mathbf{k}$ (d) $6\mathbf{i} + 2\mathbf{j} - 3\mathbf{k}$
   (e) $8\mathbf{i} + 5\mathbf{j} - 6\mathbf{k}$ (f) $-3\mathbf{i} - 4\mathbf{j} + 12\mathbf{k}$

3. Determine the unit vector in the direction of each vector in Exercise 2.

4. Find the direction vector for each of the following directed line segments:
   (a) $((2, -1, 4), (3, 1, 2))$
   (b) $((4, 3, -1), (2, 4, 1))$
   (c) $((2, -3, -4), (-3, 2, 6))$
   (d) $((4, 0, -6), (0, 3, 6))$

5. (a) If the direction angle $\alpha = \pi/4$ and $\gamma = \pi/3$, what is β?
   (b) If $\beta = \pi/3$ and $\gamma = 3\pi/4$, what is α?

6. (a) If $\cos \alpha = \frac{3}{10}$ and $\cos \beta = \frac{2}{5}$, what is γ?
   (b) If $\cos \alpha = \frac{1}{10}$ and $\cos \gamma = \frac{7}{10}$, what is β?

PROBLEMS 13.1

1. If all three direction angles of a vector are equal, what is the angle?

2. Prove that, at most, one of the direction angles of a position vector can be less than $\pi/4$.

3. Determine the set of all points equidistant from the two fixed points (x_1, y_1, z_1) and (x_2, y_2, z_2).

4. If the three real numbers a, b, and c are not all equal, show that the points (a, b, c), (b, c, a) and (c, a, b) are the vertices of an equilateral triangle.

5. A cube 24 units on each edge has its center at the origin and its faces parallel to the coordinate planes. What are its vertices?

6. Prove that the directed line segments $(\mathbf{p}_1, \mathbf{q}_1)$ and $(\mathbf{p}_2, \mathbf{q}_2)$ are equivalent if and only if the midpoints of the directed line segments $(\mathbf{p}_1, \mathbf{q}_2)$ and $(\mathbf{p}_2, \mathbf{q}_1)$ coincide. Why is it not enough to assume that the directed line segments $(\mathbf{p}_1, \mathbf{p}_2)$ and $(\mathbf{q}_1, \mathbf{q}_2)$ are parallel?

13.2 VECTOR ALGEBRA

The addition of free vectors and the multiplication of a free vector by a real number coincide with the corresponding operations on position vectors. For instance, free vectors are added "component-wise." Therefore, there is no need to discuss these operations. The purpose of this section is to introduce two useful operations, the dot product and the cross product.

Consider two free vectors $\mathbf{v}_1$ and $\mathbf{v}_2$. Choose directed line segments that represent these vectors and have the same initial point. Suppose these directed line segments are $(\mathbf{p}_1, \mathbf{q}_1)$ and $(\mathbf{p}_1, \mathbf{q}_2)$. In the plane determined by the endpoints of $\mathbf{p}_1$, $\mathbf{q}_1$, and $\mathbf{q}_2$, we measure the angle θ from $(\mathbf{p}_1, \mathbf{q}_1)$ to $(\mathbf{p}_1, \mathbf{q}_2)$. See Fig. 13.5. This is defined to be the angle from $\mathbf{v}_1$ to $\mathbf{v}_2$. Note that this angle must satisfy the inequality $0 \leq \theta \leq \pi$.

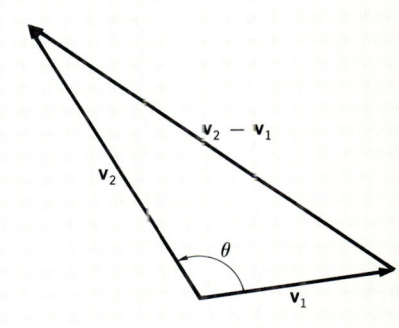

FIGURE 13.5

Now we apply the law of cosines (Section 9.1) to the triangle determined by $\mathbf{v}_1$ and $\mathbf{v}_2$ as in Fig. 13.5. We obtain

$$\|\mathbf{v}_2 - \mathbf{v}_1\|^2 = \|\mathbf{v}_1\|^2 + \|\mathbf{v}_2\|^2 - 2\|\mathbf{v}_1\| \cdot \|\mathbf{v}_2\| \cos \theta$$

where θ is the angle between $\mathbf{v}_1$ and $\mathbf{v}_2$. Solving for $\cos \theta$, we get

$$\cos \theta = \frac{\|\mathbf{v}_1\|^2 + \|\mathbf{v}_2\|^2 - \|\mathbf{v}_2 - \mathbf{v}_1\|^2}{2\|\mathbf{v}_1\| \cdot \|\mathbf{v}_2\|}$$

Next we express $\cos \theta$ in terms of the components of the vectors $\mathbf{v}_1$ and $\mathbf{v}_2$. Suppose $\mathbf{v}_1 = a_1\mathbf{i} + b_1\mathbf{j} + c_1\mathbf{k}$, $\mathbf{v}_2 = a_2\mathbf{i} + b_2\mathbf{j} + c_2\mathbf{k}$. Then we have

$$\cos \theta = \frac{(a_1^2 + b_1^2 + c_1^2) + (a_2^2 + b_2^2 + c_2^2) - (a_2 - a_1)^2 - (b_2 - b_1)^2 - (c_2 - c_1)^2}{2\sqrt{a_1^2 + b_1^2 + c_1^2}\sqrt{a_2^2 + b_2^2 + c_2^2}}$$

or

$$\cos \theta = \frac{a_1 a_2 + b_1 b_2 + c_1 c_2}{\sqrt{a_1^2 + b_1^2 + c_1^2}\sqrt{a_2^2 + b_2^2 + c_2^2}}$$

The numerator of the above fraction is defined to be the *dot product*

$$\mathbf{v}_1 \cdot \mathbf{v}_2$$

of the vectors $\mathbf{v}_1$ and $\mathbf{v}_2$. We have three equivalent expressions for this dot product:

$$\begin{aligned}
\mathbf{v}_1 \cdot \mathbf{v}_2 &= (a_1\mathbf{i} + b_1\mathbf{j} + c_1\mathbf{k}) \cdot (a_2\mathbf{i} + b_2\mathbf{j} + c_2\mathbf{k}) \\
&= a_1 a_2 + b_1 b_2 + c_1 c_2 \\
&= \tfrac{1}{2}[\|\mathbf{v}_1\|^2 + \|\mathbf{v}_2\|^2 - \|\mathbf{v}_2 - \mathbf{v}_1\|^2] \\
&= \|\mathbf{v}_1\| \cdot \|\mathbf{v}_2\| \cos \theta
\end{aligned}$$

The second and third expressions arise easily from manipulating the equation expressing the law of cosines.

The equation
$$\mathbf{v}_1 \cdot \mathbf{v}_2 = \tfrac{1}{2}[\|\mathbf{v}_1\|^2 + \|\mathbf{v}_2\|^2 - \|\mathbf{v}_2 - \mathbf{v}_1\|^2]$$
expresses the dot product in terms of the norm function. Therefore the concept of "angle" can be made to depend upon that of "distance."

We may also point out that the formula
$$\mathbf{v}_1 \cdot \mathbf{v}_2 = \|\mathbf{v}_1\| \cdot \|\mathbf{v}_2\| \cos \theta$$
expresses the dot product as the length of $\mathbf{v}_1$ times the length of the projection of $\mathbf{v}_2$ onto $\mathbf{v}_1$. (See Fig. 13.6.) In this formulation we see that $\mathbf{v}_1 \cdot \mathbf{v}_2$ is defined geometrically and is actually independent of the coordinate system.

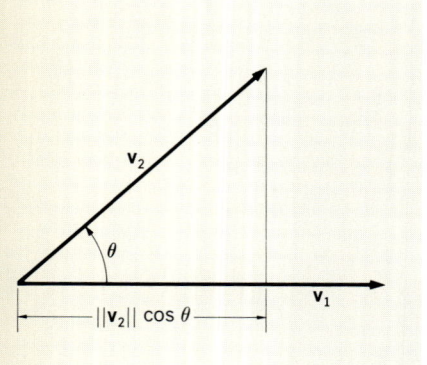

FIGURE 13.6

■ **THEOREM 13.3.** Let $\mathbf{u}$, $\mathbf{v}$, and $\mathbf{w}$ be arbitrary free vectors and let r be a real number. Then

(1) $\mathbf{u} \cdot \mathbf{v} = \mathbf{v} \cdot \mathbf{u}$
(2) $\mathbf{u} \cdot (\mathbf{v} + \mathbf{w}) = \mathbf{u} \cdot \mathbf{v} + \mathbf{u} \cdot \mathbf{w}$
(3) $(r\mathbf{u}) \cdot \mathbf{v} = r(\mathbf{u} \cdot \mathbf{v}) = \mathbf{u} \cdot (r\mathbf{v})$
(4) $\mathbf{u} \cdot \mathbf{u} = \|\mathbf{u}\|^2$

The proof of this theorem requires no more than the easy use of the definitions; hence the proof is left to the reader.

The equation $\mathbf{u} \cdot \mathbf{u} = \|\mathbf{u}\|^2$ tells us that "distance" can be made to depend upon the dot product. The interdependence of "angle" and "distance" is a basic fact in theoretical geometry.

Two vectors $\mathbf{v}_1$ and $\mathbf{v}_2$ are said to be *orthogonal* if the angle from $\mathbf{v}_1$ to $\mathbf{v}_2$ is $\pi/2$ radians. Also, for technical reasons we assume that the zero vector $\mathbf{0}$ is orthogonal to every other vector.

■ **THEOREM 13.4.** The vectors $\mathbf{u}$ and $\mathbf{v}$ are orthogonal if and only if

$$\mathbf{u} \cdot \mathbf{v} = 0$$

PROOF: The equation
$$\mathbf{u} \cdot \mathbf{v} = \|\mathbf{u}\| \cdot \|\mathbf{v}\| \cos \theta$$
reveals that $\mathbf{u} \cdot \mathbf{v} = 0$ if $\theta = \pi/2$, because $\cos \pi/2 = 0$, or if one vector is $\mathbf{0}$ because then one norm would be zero. On the other hand, if $\|\mathbf{u}\| \neq 0 \neq \|\mathbf{v}\|$, whereas $\mathbf{u} \cdot \mathbf{v} = 0$, then we have $\cos \theta = 0$, which implies that $\theta = \pi/2$ because $0 \leqq \theta \leqq \pi$. ■

■ *Example 1.* Let $\mathbf{u} = 2\mathbf{i} + 2\mathbf{j} + \mathbf{k}$ and $\mathbf{v} = \mathbf{i} + 4\mathbf{j} - \mathbf{k}$. Then we have

$$\cos\theta = \frac{\mathbf{u}\cdot\mathbf{v}}{\|\mathbf{u}\|\cdot\|\mathbf{v}\|} = \frac{(2)(1)+(2)(4)+(1)(-1)}{\sqrt{2^2+2^2+1^2}\sqrt{1^2+4^2+(-1)^2}}$$

$$= \frac{9}{\sqrt{9}\sqrt{18}} = \frac{1}{\sqrt{2}}$$

It follows that $\theta = \pi/4$.

- **Example 2.** We prove that the points $P_1 = (-1, 3, 1), P_2 = (5, 5, 4)$, $P_3 = (8, 2, 0)$, and $P_4 = (2, 0, -3)$ are the vertices of a rectangle in space. To do so, we consider the vectors determined by the six directed line segments as follows:

$$\begin{aligned}P_1P_2 &\to 6\mathbf{i}+2\mathbf{j}+3\mathbf{k} & P_1P_3 &\to 9\mathbf{i}-\mathbf{j}-\mathbf{k}\\ P_1P_4 &\to 3\mathbf{i}-3\mathbf{j}-4\mathbf{k} & P_2P_3 &\to 3\mathbf{i}-3\mathbf{j}-4\mathbf{k}\\ P_2P_4 &\to -3\mathbf{i}-5\mathbf{j}-7\mathbf{k} & P_3P_4 &\to -6\mathbf{i}-2\mathbf{j}-3\mathbf{k}\end{aligned}$$

We note that the segments P_1P_2 and P_3P_4 are parallel and so are the segments P_1P_4 and P_2P_3. This proves that the points are vertices of a parallelogram. Then, because

$$(6\mathbf{i}+2\mathbf{j}+3\mathbf{k})\cdot(3\mathbf{i}-3\mathbf{j}-4\mathbf{k}) = 18-6-12 = 0$$

we see that adjacent sides P_1P_2 and P_1P_4 are orthogonal. ■

In order to motivate the cross product, we consider two nonzero vectors $\mathbf{v}_1 = a_1\mathbf{i}+b_1\mathbf{j}+c_1\mathbf{k}$ and $\mathbf{v}_2 = a_2\mathbf{i}+b_2\mathbf{j}+c_2\mathbf{k}$. Suppose we want a vector $\mathbf{w} = x\mathbf{i}+y\mathbf{j}+z\mathbf{k}$ which is orthogonal both to $\mathbf{v}_1$ and to $\mathbf{v}_2$. From Theorem 13.4, the components x, y, z of $\mathbf{w}$ must satisfy simultaneously the two equations

$$\begin{aligned}\mathbf{w}\cdot\mathbf{v}_1 &= a_1x+b_1y+c_1z = 0\\ \mathbf{w}\cdot\mathbf{v}_2 &= a_2x+b_2y+c_2z = 0\end{aligned}$$

Assuming that the vectors $\mathbf{v}_1$ and $\mathbf{v}_2$ are not parallel, the components a_1, b_1, c_1 and a_2, b_2, c_2 are not proportional. Hence as a representative case, let us assume that $a_1b_2 \neq a_2b_1$. Then we can solve these two equations for x and y in terms of z, obtaining

$$x = \frac{b_1c_2-b_2c_1}{a_1b_2-a_2b_1}z, \qquad y = -\frac{a_1c_2-a_2c_1}{a_1b_2-a_2b_1}z$$

It easily follows that any vector of the form

$$r[(b_1c_2-b_2c_1)\mathbf{i}-(a_1c_2-a_2c_1)\mathbf{j}+(a_1b_2-a_2b_1)\mathbf{k}]$$

is orthogonal both to $\mathbf{v}_1$ and to $\mathbf{v}_2$.

The *cross product* of the vectors $\mathbf{v}_1 = a_1\mathbf{i}+b_1\mathbf{j}+c_1\mathbf{k}$ and $\mathbf{v}_2 = a_2\mathbf{i}+b_2\mathbf{j}+c_2\mathbf{k}$ is defined to be the *vector*

$$\mathbf{v}_1\times\mathbf{v}_2 = (b_1c_2-b_2c_1)\mathbf{i}-(a_1c_2-a_2c_1)\mathbf{j}+(a_1b_2-a_2b_1)\mathbf{k}$$

This is also called the *vector product* and defines a new algebraic operation whose basic properties are given in the following theorems.

■ **THEOREM 13.5.** Let **u**, **v**, and **w** be free vectors. Then

(1) $\mathbf{u} \times \mathbf{u} = \mathbf{0}, \mathbf{0} \times \mathbf{u} = \mathbf{0} = \mathbf{u} \times \mathbf{0}$
(2) $\mathbf{u} \times \mathbf{v} = -(\mathbf{v} \times \mathbf{u})$
(3) $(r\mathbf{u}) \times \mathbf{v} = r(\mathbf{u} \times \mathbf{v}) = \mathbf{u} \times (r\mathbf{v})$ for any real number r
(4) $\mathbf{u} \times (\mathbf{v} + \mathbf{w}) = \mathbf{u} \times \mathbf{v} + \mathbf{u} \times \mathbf{w}$

PROOF: (1) It is obvious that $\mathbf{0} \times \mathbf{u} = \mathbf{0} = \mathbf{u} \times \mathbf{0}$. Furthermore, if $\mathbf{u} = a\mathbf{i} + b\mathbf{j} + c\mathbf{k}$, then

$$\mathbf{u} \times \mathbf{u} = (bc - bc)\mathbf{i} - (ac - ac)\mathbf{j} + (ab - ab)\mathbf{k} = \mathbf{0}$$

(2) Let $\mathbf{u} = a\mathbf{i} + b\mathbf{j} + c\mathbf{k}$ and $\mathbf{v} = d\mathbf{i} + e\mathbf{j} + f\mathbf{k}$. Then $\mathbf{u} \times \mathbf{v} = (bf - ce)\mathbf{i} - (af - cd)\mathbf{j} + (ae - bd)\mathbf{k}$ and

$$\mathbf{v} \times \mathbf{u} = (ec - fb)\mathbf{i} - (dc - fa)\mathbf{j} + (db - ea)\mathbf{k} = -(\mathbf{u} \times \mathbf{v})$$

We leave parts (3) and (4) as exercises. ■

■ **THEOREM 13.6.** For any two free vectors **u** and **v**,

$$\|\mathbf{u} \times \mathbf{v}\|^2 = \|\mathbf{u}\|^2 \|\mathbf{v}\|^2 - (\mathbf{u} \cdot \mathbf{v})^2$$

PROOF: Let $\mathbf{u} = a_1\mathbf{i} + b_1\mathbf{j} + c_1\mathbf{k}$ and $\mathbf{v} = a_2\mathbf{i} + b_2\mathbf{j} + c_2\mathbf{k}$. Then

$$\|\mathbf{u} \times \mathbf{v}\|^2 = (b_1c_2 - b_2c_1)^2 + (a_1c_2 - a_2c_1)^2 + (a_1b_2 - a_2b_1)^2$$
$$= b_1^2c_2^2 - 2b_1b_2c_1c_2 + b_2^2c_1^2$$
$$+ a_1^2b_2^2 - 2a_1a_2b_1b_2 + a_2^2b_1^2$$

On the other hand, we have

$$\|\mathbf{u}\|^2 \|\mathbf{v}\|^2 - (\mathbf{u} \cdot \mathbf{v})^2$$
$$= (a_1^2 + b_1^2 + c_1^2)(a_2^2 + b_2^2 + c_2^2) - (a_1a_2 + b_1b_2 + c_1c_2)^2$$
$$= a_1^2a_2^2 + a_1^2b_2^2 + a_1^2c_2^2 + a_2^2b_1^2 + b_1^2b_2^2 + b_1^2c_2^2$$
$$+ a_2^2c_1^2 + b_2^2c_1^2 + c_1^2c_2^2 - a_1^2a_2^2 - b_1^2b_2^2$$
$$- c_1^2c_2^2 - 2a_1a_2b_1b_2 - 2a_1a_2c_1c_2 - 2b_1b_2c_1c_2$$
$$= a_1^2b_2^2 + a_1^2c_2^2 + a_2^2b_1^2 + b_1^2c_2^2 + a_2^2c_1^2 + b_2^2c_1^2$$
$$- 2a_1a_2b_1b_2 - 2a_1a_2c_1c_2 - 2b_1b_2c_1c_2$$

The desired equality is proved by comparing the two expressions. ■

■ COROLLARY 13.7. For any two free vectors **u** and **v**,

$$\|\mathbf{u} \times \mathbf{v}\| = \|\mathbf{u}\| \cdot \|\mathbf{v}\| \sin \theta$$

where θ is the angle between **u** and **v**.

PROOF: From Theorem 13.6 we have

$$\|\mathbf{u} \times \mathbf{v}\|^2 \|\mathbf{u}\|^2 \|\mathbf{v}\|^2 - (\mathbf{u} \cdot \mathbf{v})^2$$

Replacing $\mathbf{u} \cdot \mathbf{v}$ with $\|\mathbf{u}\| \cdot \|\mathbf{v}\| \cos \theta$, we obtain

$$\|\mathbf{u} \times \mathbf{v}\|^2 = \|\mathbf{u}\|^2 \|\mathbf{v}\|^2 - \|\mathbf{u}\|^2 \|\mathbf{v}\|^2 \cos^2 \theta$$
$$= \|\mathbf{u}\|^2 \|\mathbf{v}\|^2 (1 - \cos^2 \theta)$$
$$= \|\mathbf{u}\|^2 \|\mathbf{v}\|^2 \sin^2 \theta$$

Because the angle θ satisfies $0 \leq \theta \leq \Pi$, we have $\sin \theta \geq 0$, whence

$$\|\mathbf{u} \times \mathbf{v}\| = \|\mathbf{u}\| \cdot \|\mathbf{v}\| \sin \theta \quad \blacksquare$$

Two free vectors are *parallel* if the angle from one to the other is either 0 or π radians. We also assume that the zero vector $\mathbf{0}$ is parallel to every vector.

■ **COROLLARY 13.8.** Two vectors $\mathbf{u}$ and $\mathbf{v}$ are parallel if and only if

$$\mathbf{u} \times \mathbf{v} = \mathbf{0}$$

PROOF: If one of the vectors is the zero vector $\mathbf{0}$, then $\mathbf{u} \times \mathbf{v} = \mathbf{0}$ by definition. If $\theta = 0$ or $\theta = \pi$, then $\sin \theta = 0$, whence $\|\mathbf{u} \times \mathbf{v}\| = 0$ by Corollary 13.7. It follows that $\mathbf{u} \times \mathbf{v} = \mathbf{0}$. Conversely, if $\mathbf{u} \times \mathbf{v} = \mathbf{0}$, while $\|\mathbf{u}\| \neq 0$ and $\|\mathbf{v}\| \neq 0$, then from Corollary 13.7, it follows that $\sin \theta = 0$. Thus $\theta = 0$ or $\theta = \pi$. ■

The formula

$$\|\mathbf{u} \times \mathbf{v}\| = \|\mathbf{u}\| \cdot \|\mathbf{v}\| \sin \theta$$

of Corollary 13.7 shows that the norm of the cross product $\mathbf{u} \times \mathbf{v}$ equals the area of the parallelogram having $\mathbf{u}$ and $\mathbf{v}$ as adjacent sides. Figure 13.7 illustrates why this is true. Thus $\|\mathbf{u}\|$ is the base of this parallelogram, while $\|\mathbf{v}\| \sin \theta$ is its height. We already know that $\mathbf{u} \times \mathbf{v}$ is orthogonal both to $\mathbf{u}$ and to $\mathbf{v}$. Therefore, when these vectors are represented by directed line segments having a common initial point, the vector $\mathbf{u} \times \mathbf{v}$ is orthogonal to the plane determined by the line segments. It is an easy matter to verify that the direction of $\mathbf{u} \times \mathbf{v}$ is determined by the *right-hand rule*. If one wraps the fingers of his right hand from $\mathbf{u}$ toward $\mathbf{v}$, then his right thumb points in the direction of $\mathbf{u} \times \mathbf{v}$. Expressed in different terms, a screw with right-hand threads advances in the direction of $\mathbf{u} \times \mathbf{v}$ when rotated from $\mathbf{u}$ to $\mathbf{v}$.

■ **Example 3.** Consider the three points $(3, 2, 2)$, $(1, 1, 4)$ and $(2, 6, 0)$. The plane containing these points contains the two line segments from $(1, 1, 4)$ to $(3, 2, 2)$ and from $(1, 1, 4)$ to $(2, 6, 0)$. These directed line segments represent the vectors

$$\mathbf{u} = (3 - 1)\mathbf{i} + (2 - 1)\mathbf{j} + (2 - 4)\mathbf{k} = 2\mathbf{i} + \mathbf{j} - 2\mathbf{k}$$

and

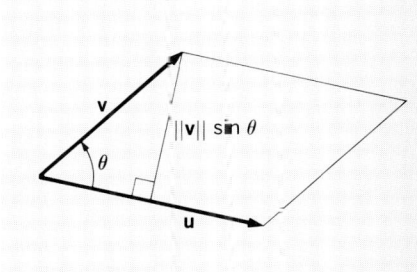

FIGURE 13.7

$$\mathbf{v} = (2-1)\mathbf{i} + (6-1)\mathbf{j} + (0-4)\mathbf{k} = \mathbf{i} + 5\mathbf{j} - 4\mathbf{k}$$

Since the cross product $\mathbf{u} \times \mathbf{v}$ is orthogonal both to $\mathbf{u}$ and to $\mathbf{v}$, it must be orthogonal to the plane containing representatives of $\mathbf{u}$ and $\mathbf{v}$. Now

$$\mathbf{u} \times \mathbf{v} = [(1)(-4) - (5)(-2)]\mathbf{i} - [(2)(-4) - (1)(-2)]\mathbf{j} + [(2)(5) - (1)(1)]\mathbf{k} = 6\mathbf{i} + 6\mathbf{j} + 9\mathbf{k}$$

Then a unit vector orthogonal to the plane through the three given points is

$$\frac{\mathbf{u} \times \mathbf{v}}{\|\mathbf{u} \times \mathbf{v}\|} = \frac{6\mathbf{i} + 6\mathbf{j} + 9\mathbf{k}}{\sqrt{36 + 36 + 81}} = \frac{2\mathbf{i}}{\sqrt{17}} + \frac{2\mathbf{j}}{\sqrt{17}} + \frac{3\mathbf{k}}{\sqrt{17}}$$

■ *Example 4.* The triangle with vertices $(4, -2, 1)$, $(3, 1, -1)$ and $(2, 0, 4)$ has adjacent sides P_1P_2 and P_1P_3 representing vectors

$$\mathbf{u} = (3-4)\mathbf{i} + (1-(-2))\mathbf{j} + (-1-1)\mathbf{k} = -\mathbf{i} + 3\mathbf{j} - 2\mathbf{k}$$

and

$$\mathbf{v} = (2-4)\mathbf{i} + (0-(-2))\mathbf{j} + (4-1)\mathbf{k} = -2\mathbf{i} + 2\mathbf{j} + 3\mathbf{k}$$

The triangle has an area precisely one-half of the area of the parallelogram whose adjacent sides are P_1P_2 and P_1P_3. Hence the area of the triangle is

$$\tfrac{1}{2}\|\mathbf{u} \times \mathbf{v}\| = \tfrac{1}{2}\|(-\mathbf{i} + 3\mathbf{j} - 2\mathbf{k}) \times (-2\mathbf{i} + 2\mathbf{j} + 3\mathbf{k})\|$$
$$= \tfrac{1}{2}\|13\mathbf{i} + 7\mathbf{j} + 4\mathbf{k}\| = \tfrac{1}{2}\sqrt{234}$$

EXERCISES 13.2

1. Let $\mathbf{u} = 3\mathbf{i} - 2\mathbf{j} - \mathbf{k}$, $\mathbf{v} = \mathbf{i} + 2\mathbf{j} + 5\mathbf{k}$ and $\mathbf{w} = -2\mathbf{i} + \mathbf{k}$. Then determine the following quantities:
   (a) $\mathbf{u} \cdot \mathbf{v}$
   (b) $\mathbf{u} \cdot \mathbf{w}$
   (c) $\mathbf{v} \cdot \mathbf{w}$
   (d) $\mathbf{u} \times \mathbf{v}$
   (e) $\mathbf{u} \times \mathbf{w}$
   (f) $\mathbf{v} \times \mathbf{w}$
   (g) $\mathbf{u} \cdot (\mathbf{v} \times \mathbf{w})$
   (h) $(\mathbf{u} \times \mathbf{v}) \cdot \mathbf{w}$
   (i) $\mathbf{u} \times \mathbf{u} + \mathbf{v} \times \mathbf{v} + \mathbf{w} \times \mathbf{w}$
   (j) $\mathbf{u} \times (\mathbf{v} \times \mathbf{w})$
   (k) $(\mathbf{u} \times \mathbf{v}) \times \mathbf{w}$
   (l) $(\mathbf{u} \times \mathbf{v}) \cdot (\mathbf{v} \times \mathbf{w})$

2. Find unit vectors orthogonal to both $\mathbf{u}$ and $\mathbf{v}$ in each instance:
   (a) $\mathbf{u} = \mathbf{i} + 2\mathbf{j} + 2\mathbf{k}$, $\mathbf{v} = -\mathbf{i} + 3\mathbf{j} - \mathbf{k}$
   (b) $\mathbf{u} = 3\mathbf{j} - \mathbf{j} - 3\mathbf{k}$, $\mathbf{v} = 2\mathbf{j} + 4\mathbf{k}$
   (c) $\mathbf{u} = 2\mathbf{i} + 3\mathbf{j}$, $\mathbf{v} = 3\mathbf{j} + 2\mathbf{k}$

3. Find a vector orthogonal to the plane through the given three points in each of the following cases:
   (a) $(0, 0, 0)$, $(1, 3, 1)$, $(-1, -1, 2)$
   (b) $(3, 6, 1)$, $(2, 4, -1)$, $(1, 5, 3)$
   (c) $(0, -1, 2)$, $(3, 3, -10)$, $(1, 1, 4)$
   (d) $(a, 0, 0)$, $(0, b, 0)$, $(0, 0, c)$, a, b, and c positive

4. In each instance in Exercise 3, consider the three given points as the vertices of a triangle and find its area.

5. Find the interior angles of the triangle whose vertices are $(1, -2, 1)$, $(3, 0, 2)$, $(4, 4, 1)$. Express the angles as inverse cosines.

PROBLEMS 13.2

6. Choose constants x and y such that the vector
$$x(\mathbf{i} - 2\mathbf{j} + 3\mathbf{k}) + y(2\mathbf{i} + \mathbf{j} - 2\mathbf{k})$$
is orthogonal to $-\mathbf{i} + 4\mathbf{j} - \mathbf{k}$.

1. Let $\mathbf{u} = a_1\mathbf{i} + b_1\mathbf{j} + c_1\mathbf{k}$ and $\mathbf{v} = a_2\mathbf{i} + b_2\mathbf{j} + c_2\mathbf{k}$ and show that
$$\mathbf{u} \times \mathbf{v} = \begin{vmatrix} \mathbf{i} & \mathbf{j} & \mathbf{k} \\ a_1 & b_1 & c_1 \\ a_2 & b_2 & c_2 \end{vmatrix}$$

2. Let $\mathbf{u}, \mathbf{v},$ and $\mathbf{w}$ be the position vectors of the vertices of a triangle. Show that the area of the triangle is
$$\tfrac{1}{2}\|(\mathbf{v} - \mathbf{u}) \times (\mathbf{w} - \mathbf{u})\|$$

3. Prove that the three position vectors $\mathbf{u}, \mathbf{v},$ and $\mathbf{w}$ are coplanar if and only if there are constants x, y, z such that
$$x\mathbf{u} + y\mathbf{v} + z\mathbf{w} = 0$$

4. Let $\mathbf{u}$ and $\mathbf{v}$ be fixed vectors. Express $\mathbf{u}$ as the sum of two vectors, one parallel to $\mathbf{v}$ and the other orthogonal to $\mathbf{v}$.

5. If the four position vectors $\mathbf{v}_1, \mathbf{v}_2, \mathbf{v}_3,$ and $\mathbf{v}_4$ are coplanar, show that $(\mathbf{v}_1 \times \mathbf{v}_2) \times (\mathbf{v}_3 \times \mathbf{v}_4) = 0$. Is the converse true?

6. Let $\mathbf{u}, \mathbf{v}, \mathbf{w}$ be the position vectors of three noncollinear points. Prove that $(\mathbf{u} \times \mathbf{v}) + (\mathbf{v} \times \mathbf{w}) + (\mathbf{w} \times \mathbf{u})$ is orthogonal to the plane through the three points.

7. Let $\mathbf{u}$ and $\mathbf{v}$ be arbitrary vectors. Prove that the vector
$$\frac{\|\mathbf{v}\|\mathbf{u} + \|\mathbf{u}\|\mathbf{v}}{\|\mathbf{u}\| + \|\mathbf{v}\|}$$
bisects the angle between $\mathbf{u}$ and $\mathbf{v}$.

8. Given two vectors $\mathbf{u}$ and $\mathbf{v}$, prove that the vectors
$$\|\mathbf{u}\|\mathbf{v} + \|\mathbf{v}\|\mathbf{u}, \quad \|\mathbf{v}\|\mathbf{u} - \|\mathbf{u}\|\mathbf{v}$$
are orthogonal.

9. If the three vectors $\mathbf{u}, \mathbf{v}, \mathbf{w}$ satisfy the equation
$$\mathbf{u} + \mathbf{v} + \mathbf{w} = 0$$
show that
$$\mathbf{u} \times \mathbf{v} = \mathbf{v} \times \mathbf{w} = \mathbf{w} \times \mathbf{u}$$

10. Let $\mathbf{u} = a_1\mathbf{i} + b_1\mathbf{j} + c_1\mathbf{k}$, $\mathbf{v} = a_2\mathbf{i} + b_2\mathbf{j} + c_2\mathbf{k}$ and $\mathbf{w} = a_3\mathbf{i} + b_3\mathbf{j} + c_3\mathbf{k}$. Prove that
$$\mathbf{u} \cdot (\mathbf{v} \times \mathbf{w}) = \begin{vmatrix} a_1 & b_1 & c_1 \\ a_2 & b_2 & c_2 \\ a_3 & b_3 & c_3 \end{vmatrix}$$
Interpret this number geometrically as the volume of a parallelopipod.

11. (a) Find the angle between a diagonal of a cube and an edge meeting the diagonal.
    (b) Find the angle between a diagonal of a cube and the diagonal of a face, the two meeting at a vertex.
    (c) Find the angle between the diagonals of adjacent faces of a cube, the two meeting at a vertex.

12. Use vector methods to find the distance between the line through $(1, 0, 2)$ and $(2, -1, 4)$ and the line through $(-1, 2, -1)$ and $(4, 4, -3)$.

13. At the center of a tetrahedron erect four vectors, one perpendicular to each face, having norm equal to the area of that face and directed from the center toward the face. Show that the sum of these vectors is zero.

14. Let **u**, **v**, and **w** be arbitrary vectors and prove the following formulas
    (a) $\mathbf{u} \cdot (\mathbf{v} \times \mathbf{w}) = (\mathbf{u} \times \mathbf{v}) \cdot \mathbf{w}$
    (b) $(\mathbf{u} + \mathbf{v}) \times (\mathbf{u} - \mathbf{v}) = 2(\mathbf{v} \times \mathbf{u})$
    (c) $\mathbf{u} \times (\mathbf{v} \times \mathbf{w}) = (\mathbf{u} \cdot \mathbf{w})\mathbf{v} - (\mathbf{u} \cdot \mathbf{v})\mathbf{w}$
    (d) $\mathbf{u} \times (\mathbf{v} \times \mathbf{w}) + \mathbf{v} \times (\mathbf{w} \times \mathbf{u}) + \mathbf{w} \times (\mathbf{u} \times \mathbf{v}) = 0$
    (e) $(\mathbf{u} \times \mathbf{v}) \cdot [(\mathbf{v} \times \mathbf{w}) \times (\mathbf{w} \times \mathbf{u})] = [\mathbf{u} \cdot (\mathbf{v} \times \mathbf{w})]^2$

13.3 LINES AND PLANES

This section relates the algebra of vectors to previous knowledge about lines and planes in space.

Consider the line L that passes through the point (x_0, y_0, z_0) and is parallel to the vector $\mathbf{v} = a\mathbf{i} + b\mathbf{j} + c\mathbf{k}$. If (x, y, z) is any other point on L, then the segment from (x_0, y_0, z_0) to (x, y, z) represents a vector $\mathbf{u}$ parallel to $\mathbf{v}$. By Corollary 13.8, this parallelism is equivalent to the condition $\mathbf{u} \times \mathbf{v} = 0$. We use this fact to determine equations of the line L.

The vector $\mathbf{u}$ from (x_0, y_0, z_0) to (x, y, z) is

$$\mathbf{u} = (x - x_0)\mathbf{i} + (y - y_0)\mathbf{j} + (z - z_0)\mathbf{k}$$

Then we have

$$\mathbf{u} \times \mathbf{v} = [(y - y_0)c - (z - z_0)b]\mathbf{i} - [(x - x_0)c - (z - z_0)a]\mathbf{j} + [(x - x_0)b - (y - y_0)a]\mathbf{k}$$

Since $\mathbf{u} \times \mathbf{v} = 0$, we have the three component equations

$$(y - y_0)c - (z - z_0)b = 0$$
$$(x - x_0)c - (z - z_0)a = 0$$
$$(x - x_0)b - (y - y_0)a = 0$$

If no component of the vector **v** is zero, then these equations can be written in the well-known form

$$\frac{x - x_0}{a} = \frac{y - y_0}{b} = \frac{z - z_0}{c}$$

The definition of parallelism of vectors **u** and **v** states that we can write the equation

$$\mathbf{u} = t\mathbf{v}$$

13.3 LINES AND PLANES

for some real number t. The corresponding component equations,

$$x - x_0 = at$$
$$y - y_0 = bt$$
$$z - z_0 = ct$$

are parametric equations for the line L.

Next, let (x_0, y_0, z_0) be a fixed point in space and let $\mathbf{N} = A\mathbf{i} + B\mathbf{j} + C\mathbf{k}$ be a nonzero fixed vector. Let $\mathbf{N}$ be represented by a directed line segment with an initial point (x_0, y_0, z_0). Consider the unique plane containing (x_0, y_0, z_0) and orthogonal to $\mathbf{N}$. If (x, y, z) is any point in this plane, then the segment from (x_0, y_0, z_0) to (x, y, z) clearly represents a vector

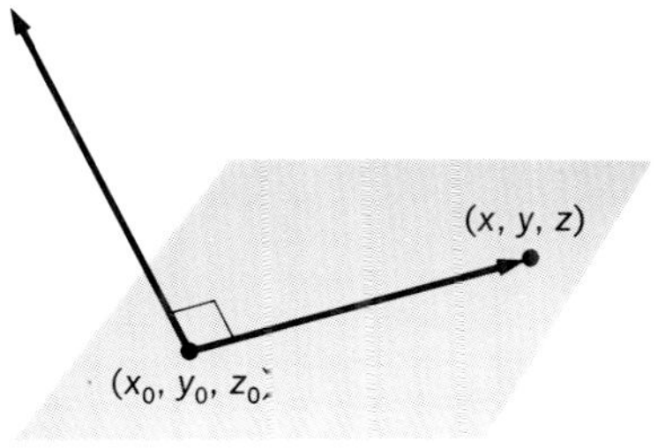

$$\mathbf{u} = (x - x_0)\mathbf{i} + (y - y_0)\mathbf{j} + (z - z_0)\mathbf{k}$$

which is orthogonal to $\mathbf{N}$. See Fig. 13.8. By Theorem 13.4 we have the equation

$$\mathbf{N} \cdot \mathbf{u} = 0$$

FIGURE 13.8

In component form this equation is

$$\mathbf{N} \cdot \mathbf{u} = A(x - x_0) + B(y - y_0) + C(z - z_0) = 0$$

We recognize this as the equation of a plane. This discussion proves the first half of the next result.

■ **THEOREM 13.9.** The plane through the point (x_0, y_0, z_0) and orthogonal to the vector $\mathbf{N} = A\mathbf{i} + B\mathbf{j} + C\mathbf{k}$ has the form

$$\{(x, y, z) : Ax + By + Cz + D = 0\}$$

Conversely, any plane having this form is orthogonal to the vector $\mathbf{N}$.

PROOF: We need only prove the second part, but this follows quickly from the fact that the planes with equations

$$Ax + By + Cz + D_1 = 0$$

and

$$Ax + By + Cz + D_2 = 0$$

are necessarily parallel. ■

In dealing with the plane $\{(x, y, z) : Ax + By + Cz + D = 0\}$ we make much use of its *normal vector* $\mathbf{N} = A\mathbf{i} + B\mathbf{j} + C\mathbf{k}$. The following definition is an example of this usage:

The angle between two planes is defined to be the angle between their normal vectors. Hence the planes are parallel if their normal vectors are parallel and are perpendicular if their normal vectors are orthogonal.

■ *Example 1.* The plane through $(2, -1, 4)$ orthogonal to the vector $\mathbf{N} = -\mathbf{i} + 3\mathbf{j} + 2\mathbf{k}$ has the equation

$$(-1)(x - 2) + 3(y + 1) + 2(z - 4) = 0$$

or

$$-x + 3y + 2z - 3 = 0$$

■ *Example 2.* The plane through the three points $P_1(3, 2, -1)$, $P_2(5, 6, 1)$, and $P_3(4, 1, -2)$ obviously contains the segments P_1P_2 and P_1P_3 representing the vectors

$$\mathbf{u} = (5 - 3)\mathbf{i} + (6 - 2)\mathbf{j} + (1 + 1)\mathbf{k} = 2\mathbf{i} + 4\mathbf{j} + 2\mathbf{k}$$

and

$$\mathbf{v} = (4 - 3)\mathbf{i} + (1 - 2)\mathbf{j} + (-2 + 1)\mathbf{k} = \mathbf{i} - \mathbf{j} - \mathbf{k}$$

It follows that the plane must be orthogonal to $\mathbf{u} \times \mathbf{v}$ which can then be used as a normal vector for the plane. Now we have

$$\mathbf{u} \times \mathbf{v} = [(4)(-1) - (-1)(2)]\mathbf{i} - [(2)(-1) - (1)(2)]\mathbf{j} + [(2)(-1) - (1)(4)]\mathbf{k} = -2\mathbf{i} + 4\mathbf{j} - 6\mathbf{k}$$

Using the point $(3, 2, -1)$, the equation of the plane is

$$-2(x - 3) + 4(y - 2) - 6(z + 1) = 0$$

or

$$-2x + 4y - 6z - 8 = 0$$

or

$$x - 2y + 3z + 4 = 0$$

■ *Example 3.* We use vector methods to determine the (undirected) distance from the point $(3, -2, 1)$ to the plane $\{(x, y, z) : 4x + 5y - 2z = 6\}$. First note that the vector $\mathbf{N} = 4\mathbf{i} + 5\mathbf{j} - 2\mathbf{k}$ is orthogonal to the plane. Now choose any point on the plane, say $(0, 0, -3)$, for instance. Then the segment from $(3, -2, 1)$ to $(0, 0, -3)$ represents the vector

$$\mathbf{u} = (0 - 3)\mathbf{i} + (0 + 2)\mathbf{j} + (-3 - 1)\mathbf{k} = -3\mathbf{i} + 2\mathbf{j} - 4\mathbf{k}$$

In Fig. 13.9 we give a schematic picture of these two vectors. From this figure we see that the desired perpendicular distance d is given by

$$d = \|\mathbf{u}\| \cdot |\cos \theta|$$

Because $\cos \theta$ is given by

$$\cos \theta = \frac{\mathbf{u} \cdot \mathbf{N}}{\|\mathbf{u}\| \cdot \|\mathbf{N}\|}$$

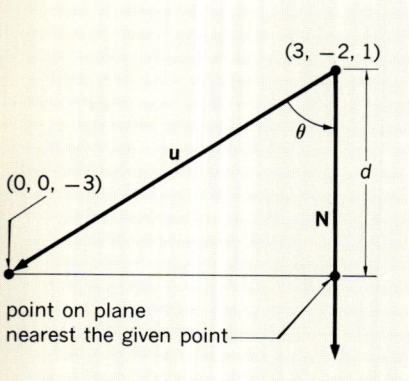

FIGURE 13.9

13.3 LINES AND PLANES

we obtain

$$d = \frac{|\mathbf{u} \cdot \mathbf{N}|}{\|\mathbf{N}\|} = \frac{|(-3\mathbf{i} + 2\mathbf{j} - 4\mathbf{k}) \cdot (4\mathbf{i} + 5\mathbf{j} - 2\mathbf{k})|}{\sqrt{4^2 + 5^2 + (-2)^2}}$$

$$= \frac{|-12 + 10 + 8|}{\sqrt{45}} = \frac{6}{3\sqrt{5}} = \frac{2\sqrt{5}}{5}$$

EXERCISES 13.3

1. Find both parametric equations and equations in symmetric form for the line through the given point with the given direction vector:
   (a) $(1, 1, 0)$; $\mathbf{N} = 3\mathbf{i} - \mathbf{j} + 2\mathbf{k}$
   (b) $(3, -1, 2)$; $\mathbf{N} = -\mathbf{i} + 3\mathbf{j} - 4\mathbf{k}$
   (c) $(-2, 1, -3)$; $\mathbf{N} = 2\mathbf{i} + 3\mathbf{j} - 2\mathbf{k}$
   (d) $(-1, -1, 2)$; $\mathbf{N} = -3\mathbf{i} - 4\mathbf{j} - 5\mathbf{k}$

2. Find both parametric equations and equations in symmetric form for the line through the two given points in each case:
   (a) $(3, 1, 2), (4, 2, 3)$
   (b) $(3, 1, 2), (-1, 3, 6)$
   (c) $(4, -1, -3), (0, 3, 1)$
   (d) $(-2, -3, -1), (4, 5, -1)$

3. In each of the following cases, determine whether the two lines are perpendicular, parallel, or neither.
   (a) $L_1 = \{(x, y, z) : x = 1 + 2t, y = -1 + 3t, z = -2 + 5t\}$
       $L_2 = \{(x, y, z) : x = 3 + t, y = 2 + t, z = 3 - t\}$
   (b) $L_1 = \{(x, y, z) : x = 4 + 3t, y = -1 + 2t, z = -4 - t\}$
       $L_2 = \{(x, y, z) : x = 1 - 6t, y = -3 - 4t, z = -5 + 2t\}$
   (c) $L_1 = \left\{(x, y, z) : \dfrac{x+1}{2} = \dfrac{y}{-3} = \dfrac{z-4}{1}\right\}$
       $L_2 = \left\{(x, y, z) : \dfrac{x+3}{6} = \dfrac{y-3}{-9} = \dfrac{z-3}{3}\right\}$
   (d) $L_1 = \left\{(x, y, z) : \dfrac{x-2}{4} = \dfrac{y-1}{3} = \dfrac{z-1}{-1}\right\}$
       $L_2 = \left\{(x, y, z) : \dfrac{x-2}{8} = \dfrac{y+1}{6} = \dfrac{z-1}{-3}\right\}$

4. Find the points of intersection of each of the following lines with the coordinate planes:
   (a) $\{(x, y, z) : x = 3 + 4t, y = -2 + t, z = 1 - 2t\}$
   (b) $\left\{(x, y, z) : \dfrac{x+2}{-3} = \dfrac{y+1}{2} = \dfrac{z-4}{4}\right\}$

5. In each of the following cases find the equation of the plane through the given point, having the given normal vector:
   (a) $(2, 2, -3)$; $\mathbf{N} = -\mathbf{i} + 3\mathbf{j} - 2\mathbf{k}$
   (b) $(-1, 3, 5)$; $\mathbf{N} = 2\mathbf{i} - 2\mathbf{j} - \mathbf{k}$
   (c) $(4, 4, -6)$; $\mathbf{N} = \mathbf{i} + \mathbf{j} + \mathbf{k}$
   (d) $(3, -6, 12)$; $\mathbf{N} = 4\mathbf{i} + 2\mathbf{j}$

6. In each case find the plane through the given three points:
   (a) $(0, 0, 0), (1, 3, 2), (4, 1, -2)$

(b) $(1, 1, -1)$, $(2, 3, 1)$, $(5, 5, 5)$
(c) $(6, 8, 12)$, $(4, -4, -4)$, $(7, 5, 1)$
(d) $(-1, -2, 4)$, $(3, -3, -2)$, $(6, 4, 4)$

7. (a) Find the line through $(3, -1, -2)$ and orthogonal to the plane $\{(x, y, z) : 2x + 4y - z = 6\}$.
(b) Find the plane through $(-2, -3, 4)$ which is parallel to the plane $\{(x, y, z) : 4x + 5y + 2z + 7 = 0\}$.
(c) Find the plane containing the point $(2, -3, -1)$ and the line
$$\left\{(x, y, z) : \frac{x+1}{-1} = \frac{y-2}{2} = \frac{z+4}{3}\right\}$$
(d) Find the line parallel to each of the planes $\{(x, y, z) : x + y = 6\}$, $\{(x, y, z) : 3x - 2y - 4z = 12\}$ and containing the point $(4, -2, -3)$.

8. (a) Find the distance (undirected) from the point $(-2, -4, -5)$ to the plane $\{(x, y, z) : 3x - 5y + 4z = 30\}$.
(b) Find the smallest angle between the line with parametric equation $x = 1 + 4t$, $y = -2 - 2t$, $z = 4t$ and a line in the plane $\{(x, y, z) : -3x + 4y + z = 6\}$. Express the angle as an inverse cosine.

9. Find the line through the point $(-1, 2, 1)$ and perpendicular to the line $x/2 = y/1 = z/3$.

10. Let A, B, and C be three points in space and let $\mathbf{u}$ and $\mathbf{v}$ be the vectors represented by the segments AB and AC. Show that A, B, and C are collinear if and only if $\mathbf{u} \times \mathbf{v} = 0$.

PROBLEMS 13.3

1. Prove that the lines
$$\left\{(x, y, z) : \frac{x+1}{1} = \frac{y+8}{3} = \frac{z-3}{-2}\right\}$$
and
$$\left\{(x, y, z) : \frac{x+1}{-2} = \frac{y+3}{-1} = \frac{z-3}{4}\right\}$$
intersect and find the equation of the plane containing these lines.

2. If the planes $\{(x, y, z) : A_1 x + B_1 y + C_1 z + D_1 = 0\}$ and $\{(x, y, z) : A_2 x + B_2 y + C_2 z + D_2 = 0\}$ intersect, prove that for any pair $(k_1, k_2) \neq (0, 0)$, the equation
$$k_1(A_1 x + B_1 y + C_1 z + D_1 = 0\} + k_2(A_2 x + B_2 y + C_2 z + D_2| = 0$$
is that of a plane containing the line of intersection of the two given planes.

3. If a line makes equal angles with three different lines in a given plane, prove that the line must be orthogonal to the plane.

4. Determine the equation of the plane containing the point (x_0, y_0, z_0) and the line $\{(x, y, z) : x = x_1 + at, y = y_1 + bt, z = z_1 + ct\}$.

5. Prove that the (undirected) distance from the point (x_0, y_0, z_0) to the plane $\{(x, y, z) : Ax + By + Cz + D = 0\}$ is
$$\frac{|Ax_0 + By_0 + Cz_0 + D|}{\sqrt{A^2 + B^2 + C^2}}$$

6. Consider the cube having vertices $(0, 0, 0)$, $(a, 0, 0)$, $(0, a, 0)$, $(a, a, 0)$, $(0, 0, a)$, $(a, 0, a)$, $(0, a, a)$, and (a, a, a). Find the plane that intersects the surface of this cube in a regular hexagon.

7. Let A, B, C, and D denote four points in space and let $\mathbf{u}$, $\mathbf{v}$, $\mathbf{w}$ be the vectors represented by the segments AB, BC, and AD. Prove that the four points are coplanar if and only if

$$(\mathbf{u} \times \mathbf{v}) \cdot \mathbf{w} = 0$$

13.4 SCALAR FIELDS AND LEVEL SETS

In many areas of application the practitioner speaks of a function of several variables as a *scalar field*. A function of two variables is considered to be a function of the position vector $\mathbf{p} = x\mathbf{i} + y\mathbf{j}$ and, similarly, a function of three variables is a function of the position vector $\mathbf{p} = x\mathbf{i} + y\mathbf{j} + z\mathbf{k}$. We write the symbol $f(\mathbf{p})$ in either case, thus permitting a simultaneous discussion of both two- and three-dimensional cases. In fact, we can use the same symbol in discussing a function of n variables where n is any natural number.

The temperature at a point $\mathbf{p}$ in some solid is a function $T(\mathbf{p})$. This is a typical scalar field. Similarly, the atmospheric pressure at a point on the surface of the earth serves as a two-dimensional example. (Note that the adjective "scalar" serves to distinguish such functions from a vector-valued function, or a *vector field*. A typical vector field is the wind velocity as a function of position on the earth's surface.)

In studying a scalar field f it is useful to know the *level sets* of the function. These are sets of the form

$$L(c) = \{\mathbf{p} : f(\mathbf{p}) = c\}$$

where c is a constant. In two dimensions, such a set is a *level curve*, whereas in three dimensions $L(c)$ is a *level surface*. As a simple example, note that the line $\{(x,y) : Ax + By = C\}$ is a level curve of the linear function $f(x,y) = Ax + By$. Similarly, the circle $\{(x,y) : x^2 + y^2 = 4\}$ is the level curve $L(4)$ of the function $f(x,y) = x^2 + y^2$.

A scalar field is often best understood by sketching some of its level sets. In short, we must draw such curves and surfaces as

$$C = \{(x,y) : f(x,y) = c\}$$

and

$$S = \{(x,y,z) : f(x,y,z) = c\}$$

The techniques used in sketching such level sets are merely easy extensions of those used in drawing graphs of functions

$$\{(x,y) : y = f(x)\} \quad \text{or} \quad \{(x,y,z) : z = f(x,y)\}$$

Indeed, such curves and surfaces are special cases of the level sets under discussion.

Intercepts. The intercepts of a curve or surface are the points where it intersects the coordinate axes. These are found by solving the following equations:

$$
\begin{array}{lll}
x \text{ intercepts} & f(x, 0) = c & \text{or} \quad f(x, 0, 0) = c \\
y \text{ intercepts} & f(0, y) = c & \text{or} \quad f(0, y, 0) = c \\
z \text{ intercepts} & & \phantom{\text{or} \quad} f(0, 0, z) = c
\end{array}
$$

■ **Example 1.** The curve $\{(x, y) : x^2 + 2y^2 - 2x + y = 3\}$ may be considered as a level curve of the function

$$f(x, y) = x^2 + 2y^2 - 2x + y$$

The x intercepts satisfy the equation

$$x^2 - 2x = 3$$

whence $x = 3$ or $x = -1$. Therefore, the x intercepts are the points $(3, 0)$ and $(-1, 0)$. Similarly, the y intercepts are obtained by solving the equation

$$2y^2 + y = 3$$

or

$$(2y + 3)(y - 1) = 0$$

The y intercepts are the points $(0, -\frac{3}{2})$ and $(0, 1)$.

■ **Example 2.** The surface $\{(x, y, z) : x^2 + y^2 + z^2 - 4x + 2y + 4 = 0\}$ has x intercepts determined by the equations

$$x^2 - 4x + 4 = (x - 2)^2 = 0$$

Hence the only x intercept is $(2, 0, 0)$. The equations that locate the intercepts on the other two axes are

$$y^2 + 2y + 4 = 0, \qquad z^2 + 4 = 0$$

Neither of these equations has a real solution, so there are no intercepts on the y and z axes.

Traces. Just as for the graph of a function of two variables, the traces of a surface $\{(x, y, z) : f(x, y, z) = 0\}$ are those curves in which it intersects the coordinate planes. These curves have equations as follows.

$$
\begin{array}{l}
xy \text{ trace} : f(x, y, 0) = 0 \\
xz \text{ trace} : f(x, 0, z) = 0 \\
yz \text{ trace} : f(0, y, z) = 0
\end{array}
$$

■ **Example 3.** The surface $\{(x, y, z) : x^2 + y^2 + 4z^2 - 4 = 0\}$ has the following traces:

xy trace $= \{(x, y, 0) : x^2 + y^2 - 4 = 0\}$
xz trace $= \{(x, 0, z) : x^2 + 4z^2 - 4 = 0\}$
yz trace $= \{(0, y, z) : y^2 + 4z^2 - 4 = 0\}$

We recognize the first as a circle. The other two are ellipses, as will be seen later.

Cross sections. The traces of a surface are special cases of plane cross sections. If the need arises, we can consider cross sections by planes parallel to the xy plane, $\{(x, y, c) : f(x, y, c) = 0\}$, by planes parallel to the xz plane, $\{(x, b, z) : f(x, b, z) = 0\}$ and by planes parallel to the yz plane, $\{(a, y, z) : f(a, y, z) = 0\}$.

■ **Example 4.** The surface $\{(x, y, z) : x^2 + y^2 = 2z^2\}$ has cross sections parallel to the xy plane that are circles. For the intersection of the plane $\{(x, y, z) : z = c\}$ with this surface has equation

$$x^2 + y^2 = 2c^2$$

which is that of a circle of radius $\sqrt{2}|c|$.

Symmetries. We consider only three symmetries of a curve. These are (1) symmetry with respect to the x axis (2) symmetry with respect to the y axis, and (3) symmetry with respect to the origin. If the curve is $\{(x, y) : f(x, y) = 0\}$, then we have

1. x axis symmetry if $f(x, -y) = f(x, y)$
2. y axis symmetry if $f(-x, y) = f(x, y)$
3. origin symmetry if $f(-x, -y) = f(x, y)$

The corresponding symmetries of a surface are seven in number. These are as follows:

1. x axis symmetry if $f(x, -y, -z) = f(x, y, z)$
2. y axis symmetry if $f(-x, y, -z) = f(x, y, z)$
3. z axis symmetry if $f(-x, -y, z) = f(x, y, z)$
4. xy plane symmetry if $f(x, y, -z) = f(x, y, z)$
5. xz plane symmetry if $f(x, -y, z) = f(x, y, z)$
6. yz plane symmetry if $f(-x, y, z) = f(x, y, z)$
7. origin symmetry if $f(-x, -y, -z) = f(x, y, z)$

■ **Example 5.** The surface $\{(x, y, z) : x^2 + y^2 = 2z^2\}$ has all seven symmetries because each variable appears only to even powers. On the other hand, the surface $\{(x, y, z) : xyz = 4\}$ is symmetric to each axis but is not symmetric to the coordinate planes.

Extent. It is obviously helpful to know if the curve or surface under consideration is restricted to certain regions. Techniques for determining such restrictions are easy enough to state but they can be quite difficult to put into practice.

If we are studying the curve $\{(x,y) : f(x,y) = 0\}$, we solve the equation $f(x,y) = 0$ for y in terms of x. This results in several functions

$$y = g_1(x), y = g_2(x), \cdots, y = g_m(x)$$

Then we solve the same equation for x in terms of y, obtaining

$$x = h_1(y), x = h_2(y), \cdots, x = h_n(y)$$

A restriction upon one of these functions is obviously a restriction upon the curve itself.

The same procedure can be attempted when studying the surface $\{(x,y,z) : f(x,y,z) = 0\}$. We solve the equation $f(x,y,z) = 0$ for each variable in terms of the other two. This yields three sets of functions of two variables

$$z = g_1(x,y), z = g_2(x,y), \cdots, z = g_m(x,y)$$
$$y = h_1(x,z), y = h_2(x,z), \cdots, y = h_n(x,z)$$
$$x = k_1(y,z), x = k_2(y,z), \cdots, x = k_p(y,z)$$

A study of these functions will determine restrictions upon the surface.

■ *Example 6.* Consider the surface

$$\{(x,y,z) : x^2 + 3y^2 + z^2 - 4x + 3y + 4z - 5 = 0\}$$

We solve the equation for z in terms of x and y,

$$z^2 + 4z + (x^2 + 3y^2 - 4x + 3y - 5) = 0$$

We obtain

$$z = \frac{-4 \pm \sqrt{16 - 4(x^2 + 3y^2 - 4x + 3y - 5)}}{2}$$
$$= -2 \pm \sqrt{4 - (x^2 + 3y^2 - 4x + 3y - 5)}$$

It follows that we have the restriction

$$4 - (x^2 + 3y - 4x + 3y - 5) \geqq 0$$

By completing the squares, this can be reduced:

$$\tfrac{53}{4} - (x-2)^2 - 3(y + \tfrac{1}{2}) \geqq 0$$

or

$$(x-2)^2 + 3(y + \tfrac{1}{2})^2 \leqq \tfrac{53}{4}$$

This inequality defines a closed elliptical disk centered at $(2, -\tfrac{1}{2})$.

EXERCISES 13.4 In each of Exercises 1 to 20, discuss and sketch the curve whose equation is given.

1. $y^2 = 4x^3$
2. $x^2 = 4(y-2)^3$

3. $9x^2 + 16y^2 = 144$
4. $9x^2 - 16y^2 = 144$
5. $4x^2 + 9y^2 = 8x$
6. $9x^2 - 4y^2 = 36x$
7. $xy - 2x + 2y + 5 = 0$
8. $x^2 - x + y + 1 = 0$
9. $x^2 - xy - 3x - y + 2 = 0$
10. $x^3 + y^3 = 8$
11. $xy^2 = 4$
12. $x^2y = 2x^2 - 8$
13. $x^2y^2 = 4$
14. $y^2 = x^2(4 - x)$
15. $x(y^2 - 4) = 4y$
16. $x^2(y^2 - 4) = 4y$
17. $x^2y^2 = x + 4$
18. $x^2y^2 - x^2 - y^2 + 4 = 0$
19. $3y^2 = (x - 1)(x - 3)(x + 6)$
20. $x^2y^2(x + 4) = (x - 4)^3$

In each of Exercises 21 to 40, discuss and sketch the surface whose equation is given.

21. $x^2 + y^2 + z^2 - 2z = 0$
22. $x^2 + y^2 + z^2 - 4x - 4y = 0$
23. $x^2 + 4y^2 + 4z^2 = 16$
24. $9x^2 + 4y^2 + 9z^2 = 36$
25. $x^2 + y^2 - 2x = 0$
26. $x^2 + y^2 = 4z$
27. $3x + 4y + z^2 = 36$
28. $4x + y^2 + z^2 = 4$
29. $x^2 + 2y^2 + z^2 - 2x + 4y = 0$
30. $x^2 - 2xy + y^2 - 4z^2 = 4$
31. $xy + 4 = z^2$
32. $xz = y^2(z - 1)$
33. $x^2z = 2xy^2 + 8$
34. $z^2 = x^2(4 - y^2)$
35. $x(y^2 + 4) = 4z$
36. $x^2(y^2 - 4) = 4z$
37. $xyz = 6$
38. $x^2yz = 8$
39. $x^2z^2 = x^2y^2 - 16$
40. $xyz(x + 1) = x^2 + y^2$

13.5 DIRECTIONAL DERIVATIVES. THE GRADIENT

We often need the rate of change of a scalar field in some given direction from a point. This requires the new kind of derivative defined here.

Let f be a continuously differentiable scalar field and let $\mathbf{p}_0$ be the position vector of an interior point of the domain of f. Let $\mathbf{u}$ be any unit vector. We determine the rate of change of f at the point $\mathbf{p}_0$ in the direction of $\mathbf{u}$ as follows:

The line through the tip of $\mathbf{p}_0$ with direction vector $\mathbf{u}$ has the vector equation $\mathbf{p} = \mathbf{p}_0 + t\mathbf{u}$. Choose a point on this line near the tip of $\mathbf{p}_0$, say the point with position vector $\mathbf{p}_0 + \Delta t \mathbf{u}$ for a small value of Δt. Form the increment

$$\Delta f = f(\mathbf{p}_0 + \Delta t \mathbf{u}) - f(\mathbf{p}_0)$$

Then the ratio $\Delta f/\Delta t$ is the average rate of change of f in the desired direction and the limit

$$\lim_{\Delta t \to 0} \frac{\Delta f}{\Delta t} = \frac{df}{du}$$

if it exists, is the *directional derivative of f at $\mathbf{p}_0$ in the direction of $\mathbf{u}$.*

Consider the three-dimensional case in coordinates. We let $\mathbf{p}_0 = x_0\mathbf{i} + y_0\mathbf{j} + z_0\mathbf{k}$ and $\mathbf{u} = \mathbf{i}\cos\alpha + \mathbf{j}\cos\beta + \mathbf{k}\cos\gamma$, where $\cos^2\alpha + \cos^2\beta + \cos^2\gamma = 1$. By an easy extension of the results of Section 11.7, we can write the increment Δf in the form

$$\Delta f = f_1(x_0, y_0, z_0)\,\Delta x + f_2(x_0, y_0, z_0)\,\Delta y + f_3(x_0, y_0, z_0)\,\Delta z + R$$

where the remainder term R can be written as

$$R = \epsilon_1\,\Delta x + \epsilon_2\,\Delta y + \epsilon_3\,\Delta z$$

each function ϵ_1, ϵ_2, and ϵ_3 having the limit zero as $(\Delta x, \Delta y, \Delta z)$ approaches $(0, 0, 0)$.

In coordinate form the position vector $\mathbf{p}$ above is

$$\mathbf{p} = \mathbf{p}_0 + \Delta t\,\mathbf{u} = (x_0 + \Delta t\cos\alpha)\mathbf{i} + (y_0 + \Delta t\cos\beta)\mathbf{j} + (z_0 + \Delta t\cos\gamma)\mathbf{k}$$

The increments Δx, Δy, and Δz are thus

$$\Delta x = \Delta t\cos\alpha,\quad \Delta y = \Delta t\cos\beta,\quad \Delta z = \Delta t\cos\gamma$$

It follows that we have

$$\frac{\Delta f}{\Delta t} = f_1(x_0, y_0, z_0)\cos\alpha + f_2(x_0, y_0, z_0)\cos\beta$$
$$+ f_3(x_0, y_0, z_0)\cos\gamma + \epsilon_1\cos\alpha + \epsilon_2\cos\beta + \epsilon_3\cos\gamma$$

Then because ϵ_1, ϵ_2, and ϵ_3 each has the limit zero, we have

$$\lim_{\Delta t \to 0}\frac{\Delta f}{\Delta t} = \frac{df}{d\mathbf{u}} = f_1(x_0, y_0, z_0)\cos\alpha + f_2(x_0, y_0, z_0)\cos\beta$$
$$+ f_2(x_0, y_0, z_0)\cos\gamma$$

where $\cos\alpha$, $\cos\beta$, and $\cos\gamma$ are the components of the unit vector $\mathbf{u}$.

We already have seen certain directional derivatives, of course. The reader can easily prove that the following equations hold:

$$\frac{df}{d\mathbf{i}} = f_1(x_0, y_0, z_0),\quad \frac{df}{d\mathbf{j}} = f_2(x_0, y_0, z_0),\quad \frac{df}{d\mathbf{k}} = f_3(x_0, y_0, z_0)$$

■ *Example 1.* Consider $f(x, y, z) = x^2 + 3xy - y^2 + 2z^2$. We have $f_1(x, y, z) = 2x + 3y$, $f_2(x, y, z) = 3x - 2y$, $f_3(x, y, z) = 4z$. At the point $(1, -1, \tfrac{1}{2})$,

$$f_1(1, -1, \tfrac{1}{2}) = -1,\ f_2(1, -1, \tfrac{1}{2}) = 5,\ f_3(1, -1, \tfrac{1}{2}) = 2$$

Now if $\mathbf{u}$ is the unit vector

$$\mathbf{u} = \tfrac{2}{3}\mathbf{i} - \tfrac{2}{3}\mathbf{j} + \tfrac{1}{3}\mathbf{k}$$

then

$$\frac{df}{d\mathbf{u}} = (-1)\left(\frac{2}{3}\right) + (5)\left(-\frac{2}{3}\right) + (2)\left(\frac{1}{3}\right) = -\frac{10}{3}$$

■ **Example 2.** Let $f(x,y) = xe^y + ye^x$. Then $f_1(x,y) = e^y + ye^x$ and $f_2(x,y) = xe^y + e^x$. Evaluated at the point $(0, 0)$, these yield $f_1(0, 0) = 1$, $f_2(0, 0) = 1$. Hence the directional derivative in the direction of the unit vector $\mathbf{u} = \mathbf{i} \cos \alpha + \mathbf{j} \cos \beta = \mathbf{i} \cos \alpha + \mathbf{j} \sin \alpha$ is

$$\frac{df}{d\mathbf{u}} = f_1(0, 0) \cos \alpha + f_2(0, 0) \cos \beta = \cos \alpha + \sin \alpha$$

It is then easy to maximize $df/d\mathbf{u}$ at $(0, 0)$ by choosing $\alpha = \pi/4$. (This is just a simple problem in maximizing a function of one variable. ■

The *gradient vector* ∇f of the scalar field f is defined by

$$\nabla f = \mathbf{i} f_1(x, y, z) + \mathbf{j} f_2(x, y, z) + \mathbf{k} f_3(x, y, z)$$

It is merely an observation now to point out that the directional derivative of f at (x, y, z) in the direction of the unit vector $\mathbf{u}$ is simply the dot product

$$\frac{df}{d\mathbf{u}} = \nabla f \cdot \mathbf{u}$$

This formulation of $df/d\mathbf{u}$ leads to a better understanding of the gradient vector. We recall that

$$\nabla f \cdot \mathbf{u} = ||\nabla f|| \cdot ||\mathbf{u}|| \cos \theta$$
$$= ||\nabla f|| \cos \theta$$

(Because $\mathbf{u}$ is a unit vector, we have $||\mathbf{u}|| = 1$.) Obviously, the directional derivative $\nabla f \cdot \mathbf{u}$ is a maximum when $\cos \theta = 1$ and thus when the angle θ from ∇f to $\mathbf{u}$ is zero radians. It follows that the maximum rate of change of f takes place in the direction of the gradient vector. (This explains the name "gradient," of course.) Also the magnitude of this maximum rate of change of f is $||\nabla f||$.

■ **Example 3.** Let $f(x, y, z) = x^2 e^y + 2xye^z + 2xze^y$. Then

$$f_1(x, y, z) = 2xe^y + 2ye^z + 2ze^y$$
$$f_2(x, y, z) = x^2 e^y + 2xe^z + 2xze^y$$
$$f_3(x, y, z) = 2xye^z + 2xe^y$$

Therefore, at the point $(0, 1, 1)$, the gradient vector has components $f_1(0, 1, 1) = 4e$, $f_2(0, 1, 1) = 0$, $f_3(0, 1, 1) = 0$. In any direction from $(0, 1, 1)$ we have the directional derivative

$$\frac{df}{d\mathbf{u}} = \nabla f \cdot \mathbf{u}$$
$$= 4e \cos \alpha$$

where $\cos \alpha$ is the x component of the unit vector $\mathbf{u}$. By choosing $\alpha = 0$, $\beta = \pi/2 = \gamma$ we have $\mathbf{u} = \mathbf{i}$ and this obviously maximizes the directional derivative. Note that $\nabla f = 4e\mathbf{i}$ at $(0, 1, 1)$, so that the direction of the maximum rate of change of f is given by ∇f. ∎

Geometrically, the gradient vector ∇f is normal to level surfaces of the scalar field f. More precisely, if (x_0, y_0, z_0) is a point on some level surface $\{(x, y, z); f(x, y, z) = c\}$, then the gradient

$$\nabla f(x_0, y_0, z_0) = \mathbf{i} f_1(x_0, y_0, z_0) + \mathbf{j} f_2(x_0, y_0, z_0) + \mathbf{k} f_3(x_0, y_0, z_0)$$

is normal to the surface at this point. It follows that the tangent plane to the surface can be expressed by the equation

$$(x - x_0) f_1(x_0, y_0, z_0) + (y - y_0) f_2(x_0, y_0, z_0) + (z - z_0) f_3(x_0, y_0, z_0) = 0$$

To see this, note that, if $\mathbf{p}_0$ is the position vector of (x_0, y_0, z_0) and $\mathbf{p}$ is the position vector of a point in the tangent plane, then the vector $\mathbf{p} - \mathbf{p}_0$ in the plane must be orthogonal to the normal vector ∇f. The dot product $(\mathbf{p} - \mathbf{p}_0) \cdot \nabla f = 0$, when written out in coordinate form, is precisely the equation above.

■ **Example 4.** The surface

$$\left\{ (x, y, z) : \frac{x^2}{4} + \frac{y^2}{9} + \frac{z^2}{16} = 1 \right\}$$

is a level set of the scalar field

$$f(x, y, z) = \frac{x^2}{4} + \frac{y^2}{9} + \frac{z^2}{16}$$

At a point (x_0, y_0, z_0) on this surface the normal vector Δf is

$$\frac{2x_0}{4} \mathbf{i} + \frac{2y_0}{9} \mathbf{j} + \frac{2z_0}{16} \mathbf{k}$$

Hence the equation for tangent plane at (x_0, y_0, z_0) to this surface is

$$\frac{2x_0}{4}(x - x_0) + \frac{2y_0}{9}(y - y_0) + \frac{2z_0}{16}(z - z_0) = 0$$

or

$$\frac{x_0 x}{4} + \frac{y_0 y}{9} + \frac{z_0 z}{16} - \left(\frac{x_0^2}{4} + \frac{y_0^2}{9} + \frac{z_0^2}{16} \right) = 0$$

or

$$\frac{x_0 x}{4} + \frac{y_0 y}{9} + \frac{z_0 z}{16} = 1$$

because the point (x_0, y_0, z_0) has coordinates satisfying the equation

$$\frac{x^2}{4} + \frac{4^2}{9} + \frac{z^2}{16} = 1$$

13.5 DIRECTIONAL DERIVATIVES. THE GRADIENT

Thus the tangent plane is

$$\left\{(x, y, z) : \frac{x_0 x}{4} + \frac{y_0 y}{9} + \frac{z_0 z}{16} = 1\right\}.$$

EXERCISES 13.5

In each of Exercises 1 to 8, find the directional derivative at the given point and in the given direction α. Then determine the angle at which the directional derivative is a maximum at the point.

1. $f(x, y) = x^2 + y^2$; $(4, 1)$; $\alpha = \frac{\pi}{4}$

2. $f(x, y) = x^2 - 4y^2$; $(4, 2)$; $\alpha = \frac{\pi}{3}$

3. $f(x, y) = xe^{2y}$; $(4, 0)$; $\alpha = \frac{2\pi}{3}$

4. $f(x, y) = y^2 \tan x$; $\left(\frac{\pi}{4}, 2\right)$; $\alpha = -\frac{\pi}{6}$

5. $f(x, y) = \tan^{-1} \frac{y}{x}$; $\left(1, \frac{\pi}{4}\right)$; $\alpha = \cos^{-1}\left(\frac{4}{5}\right)$

6. $f(x, y) = x^2 \ln xy$; $(1, 1)$; $\alpha = 0$

7. $f(x, y) = \cos xy + \cos x + \cos y$; $\left(0, \frac{\pi}{2}\right)$; $\alpha = \frac{5\pi}{6}$

8. $f(x, y) = e^x \cos y + e^y \cos x$; $(0, 0)$; $\alpha = -\frac{3\pi}{4}$

In each of Exercises 9 to 16, find the directional derivative at the given point in the direction of the given vector. (Some vectors are unit vectors and some are not.)

9. $f(x, y, z) = x^2 + y^2 + z^2$; $(1, 2, 1)$; $\frac{1}{3}\mathbf{i} - \frac{2}{3}\mathbf{j} + \frac{2}{3}\mathbf{k}$

10. $f(x, y, z) = x^2 + y^2 - 4z^2$; $(2, -1, -1)$; $\frac{1}{2}\mathbf{i} + \frac{\sqrt{2}}{2}\mathbf{j} - \frac{1}{2}\mathbf{k}$

11. $f(x, y, z) = x^2 + y^2 + 4xy - 4xz$; $(1, -1, 1)$; $2\mathbf{i} + 4\mathbf{j} - 4\mathbf{k}$

12. $f(x, y, z) = xy^2 + xye^z + yze^x$; $(0, 4, 0)$; $\frac{6}{7}\mathbf{i} - \frac{2}{7}\mathbf{j} - \frac{3}{7}\mathbf{k}$

13. $f(x, y, z) = \ln(x^2 + y^2 + z^2)$; $(2, 2, -1)$; $2\mathbf{i} + \mathbf{j} + 2\mathbf{k}$

14. $f(x, y, z) = \sin xz + \cos yz$; $\left(2, 3, \frac{\pi}{6}\right)$; $\mathbf{i} + \mathbf{j} + \mathbf{k}$

15. $f(x, y, z) = e^x \sin^2 y - z^2$; $\left(0, \frac{\pi}{2}, 2\right)$; $-3\mathbf{i} - 4\mathbf{j} + 5\mathbf{k}$

16. $f(x, y, z) = \tan xyz$; $\left(\frac{\pi}{4}, 1, 1\right)$; $4\mathbf{i} + 2\mathbf{j} - 3\mathbf{k}$

In each of Exercises 17 to 24, find the normal line and the tangent plane to the surface at the given point.

17. $\{(x, y, z) : x^2 + y^2 + z^2 = 16\}$; $(2, 2, 2\sqrt{2})$

18. $\{(x,y,z) : x^2 + 4y^2 + 4z^2 = 16\}$; $(0, 2, 0)$
19. $\{(x,y,z) : 2x^2 - 3xy + 4y^2 = z^2\}$; $(1, -1, 3)$
20. $\{(x,y,z) : x^2 + y^2 = z + 4\}$; $(2, 1, 1)$
21. $\{(x,y,z) : x^2 + y^2 = 8\}$; $(2, 2, 3)$
22. $\{(x,y,z) : yz = x \cos z\}$; $\left(1, 0, \dfrac{\pi}{2}\right)$
23. $\{(x,y,z) : x^2 y + y^2 z + z^2 x + 4 = 0\}$; $(2, -1, 0)$
24. $\{(x,y,z) : x^{2/3} + y^{2/3} + z^{2/3} = 29\}$; $(8, 27, 64)$

PROBLEMS 13.5

1. If $\nabla f = 0$ in some region, what can you say about f?

2. Let f and g be two scalar fields. Compare the directional derivative of f in the direction of ∇g with the directional derivative of g in the direction of ∇f.

3. Find the set of all points on the surface
$$\{(x,y,z) : x^2 + y^2 + 4z^2 - 4yz = 16\}$$
at which the normal line is parallel to the plane $\{(x,y,z) : x + y = 0\}$.

4. Show that every line normal to a sphere passes through the center of the sphere.

5. Let f be a continuously differentiable scalar field and suppose the domain of f contains the entire segment between points (x_0, y_0, z_0) and (x_1, y_1, z_1). If $f(x_0, y_0, z_0) = f(x_1, y_1, z_1)$, prove there is a point (x_2, y_2, z_2) on the segment at which ∇f is orthogonal to the segment.

6. Let $f(x, y, z)$ be differentiable and let $g(w)$ be differentiable. Define the composite function $g \circ f = F$ and show that ∇f and ∇F are parallel wherever neither is zero.

7. Let $\mathbf{u} = \mathbf{i} \cos \theta + \mathbf{j} \sin \theta$ and denote the directional derivative $df/d\mathbf{u}$ of a function $f(x, y)$ by $D_\theta f$. Prove that
   (a) $D_\theta f = -D_{\theta + \pi} f$
   (b) $(D_\theta f)^2 + (D_{\theta + \pi/2} f)^2 = f_x^2 + f_y^2$

8. Find a vector tangent to the curve of intersection of the surfaces $\{(x,y,z) : x^2 + y^2 + 4z^2 = 20\}$ and $\{(x,y,z) : x^2 + y^2 = z^2\}$ at the point $(\sqrt{3}, 1, 2)$.

9. Suppose we need a solution of the equation $f(x, y, z) = 0$. A first approximation (x_1, y_1, z_1) yielding $f(x_1, y_1, z_1) = w_1 \neq 0$ can be improved as follows: Assuming that $w_1 > 0$, we move away from the point (x_1, y_1, z_1) in the direction of "steepest descent," namely, $\nabla f(x_1, y_1, z_1)$. Thus a second approximation of the form
$$x_2 = x_1 - t f_1(x_1, y_1, z_1)$$
$$y_2 = y_1 - t f_2(x_1, y_1, z_1)$$
$$z_2 = z_1 - t f_3(x_1, y_1, z_1)$$
is used. We move along this line far enough to cause the linearization of f at (x_1, y_1, z_1) to reduce in value from w_1 to 0. Find the value of t that should be used. (Further iterations provide better approximations and the technique is called the *method of steepest descent*.)

13.6 CONICS

The plane curves known as *conics* or *conic sections* constitute a well-known family of curves. We use an analytic description here but shall indicate how such curves are defined geometrically in the next section.

Choose a fixed line $\{(x,y) : ax + by + c = 0\}$ and a fixed point (x_0, y_0) not on this line. These are to be the *directrix* and the *focus*, respectively, of a *conic section of eccentricity* η which is defined as follows: A point (x, y) is on the curve if the distance

$$d_1 = \sqrt{(x - x_0)^2 + (y - y_0)^2}$$

from (x, y) to the focus (x_0, y_0) and the perpendicular distance (see Problem 13.3.5)

$$d_2 = \frac{|ax + by + c|}{\sqrt{a^2 + b^2}}$$

from (x, y) to the directrix form the (constant) ratio

$$\frac{d_1}{d_2} = \eta$$

Figure 13.10 illustrates the situation.

From this definition the equation of the conic section, $d_1 = \eta d_2$, is seen to be

$$\sqrt{(x - x_0)^2 + (y - y_0)^2} = \frac{\eta}{\sqrt{a^2 + b^2}} |ax + by + c|$$

By squaring and rearranging we see that the conic of eccentricity η having the focus (x_0, y_0) and the directrix $\{(x, y) : ax + by + c = 0\}$ is

$$\{(x, y) : (a^2 + b^2)[(x - x_0)^2 + (y - y_0)^2] = \eta^2(ax + by + c)^2\}$$

We give this formulation merely to show that any conic has an equation of the form

$$Ax^2 + 2Bxy + Cy^2 + 2Dx + 2Ey + F = 0$$

This is a *quadratic equation in two variables* and the conics are also called *quadratic curves*.

Next we choose a coordinate system that simplifies the above analytic description of a conic. Given the focus and directrix of a conic, we select the y axis of a new coordinate system to be the line through the focus and perpendicular to the directrix. Choose the positive direction on the y axis to be that from directrix to focus. Suppose that the distance from the directrix to the focus is the positive number c. We now select the origin on the y axis (and hence also locate the x axis) so that the coordinates of the focus are

$$\left(0, \frac{c\eta}{1 + \eta}\right)$$

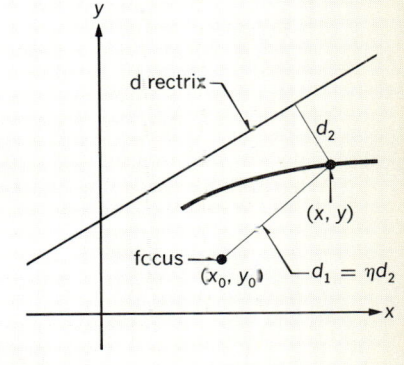

FIGURE 13.10

It follows that the equation of the directrix is

$$y = \frac{c\eta}{1+\eta} - c = -\frac{c}{1+\eta}$$

This choice of the origin divides the segment of the y axis between the focus and the directrix into two segments whose lengths have ratio η. (See Fig. 13.11.)

Using the new coordinate system, the distance d_1 from the focus to a point (x,y) on the curve is

$$d_1 = \sqrt{x^2 + [y - c\eta/(1+\eta)]^2}$$

The perpendicular distance from (x,y) to the directrix is

$$d_2 = \left| y + \frac{c}{1+\eta} \right|$$

Therefore, the defining equation $d_1 = \eta d_2$ now becomes

$$\sqrt{x^2 + [y - c\eta/(1+\eta)]^2} = \eta \left| y + \frac{c}{1+\eta} \right|$$

Squaring and simplifying this equation, we obtain

$$x^2 + (1-\eta^2)y^2 - 2c\eta y = 0$$

Therefore, with respect to the specially chosen coordinate system, the conic of eccentricity η and the distance c between the directrix and focus is the curve

$$\{(x,y) : x^2 + (1-\eta^2)y^2 = 2c\eta y\}$$

There are three distinct types of conics, depending upon the eccentricity η. If $\eta = 1$, the conic is called a *parabola*. If $\eta < 1$, the conic is an *ellipse* and if $\eta > 1$, it is a *hyperbola*. We shall consider these separately in the following discussion. Note, however, that whatever the choice of η, the conic section (in our special coordinate system) is symmetric to the y axis and passes through the origin. These simplifications governed the choice of the coordinate system. Also, this procedure exhibits the three different types as members of the same family of curves.

The parabola. Setting $\eta = 1$ in the equation of a conic, we see that the resulting parabola is

$$\{(x,y) : x^2 = 2cy\}$$

This is a level curve of the two-dimensional scalar field

$$f(x,y) = x^2 - 2cy$$

Since $f_1(x,y) = 2x$ and $f_2(x,y) = -2c$, the tangent line at the point (x_0, y_0) on the curve is

$$\{(x,y) : 2x_0(x - x_0) - 2c(y - y_0) = 0\}$$

The equation of this line can be written as

$$x_0 x - cy = x_0^2 - cy_0$$

Because (x_0, y_0) is on the curve, its coordinates satisfy the equation $x^2 = 2cy$. Thus $x_0^2 = 2cy_0$ and our equation is

$$x_0 x - cy = cy_0$$

We shall use this in drawing a diagram of the parabola shortly.

Note that the parabola $\{(x,y) : x^2 = 2cy\}$ is symmetric to the y axis and that $(0, 0)$ is the only intercept on either axis. Other points on the curve are $(\pm c, c/2)$ and $(\pm 2c, 2c)$. The curves in Fig. 13.12 are three instances of the parabola in our standard position.

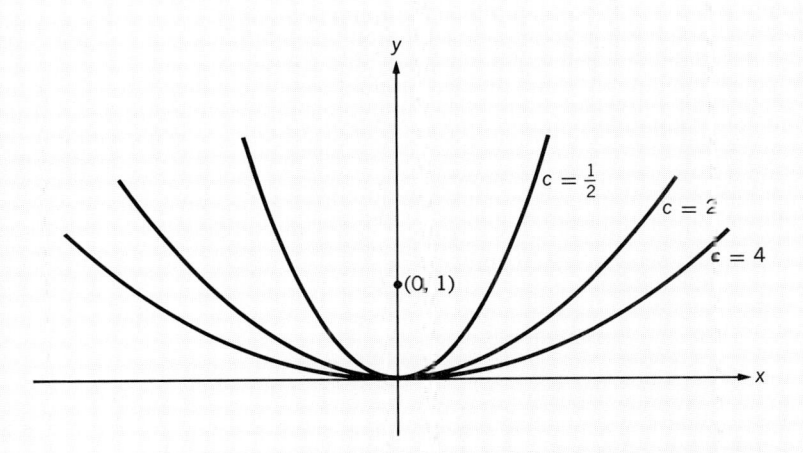

FIGURE 13.12

The ellipse. A conic of eccentricity η where $0 < \eta < 1$ is an *ellipse*. We study this curve by rearranging its equation

$$x^2 + (1 - \eta^2) y^2 - 2c\eta y = 0$$

By completing the square in y, we obtain

$$x^2 + (1 - \eta^2)\left(y - \frac{c\eta}{1 - \eta^2}\right)^2 = \frac{c^2 \eta^2}{1 - \eta^2}$$

A further simplification is made by introducing new constants

$$a = \frac{c\eta}{1 - \eta}, \quad b = \frac{c\eta}{\sqrt{1 - \eta^2}}$$

Note that $1 - \eta^2 < 1$, so that $\sqrt{1 - \eta^2} > 1 - \eta^2$. Therefore, $a > b$. Also we have $1 - \eta^2 = b^2/a^2$. Using these relations, the equation becomes

$$x^2 + \frac{b^2}{a^2}(y - a)^2 = b^2$$

or

$$\frac{x^2}{b^2} + \frac{(y - a)^2}{a^2} = 1$$

We use this form of the equation in order to study the ellipse.

Applying standard techniques, we find that the ellipse

$$\left\{(x, y) : \frac{x^2}{b^2} + \frac{(y - a)^2}{a^2} = 1\right\}$$

has intercepts $(0, 0)$ and $(0, 2a)$. It is symmetric both to the y axis and to the horizontal line $\{(x, y) : y = a\}$. Hence the curve is symmetric to its *center* $(0, a)$. Because it has a center of symmetry, the ellipse is called a *central conic*.

By solving the equation for y in terms of x and for x in terms of y, we get

$$y = a \pm a\sqrt{1 - x^2/b^2}, \quad x = \pm b\sqrt{1 - [(y - a)^2/a^2]}$$

It follows that we must have $x^2 \leq b^2$ and $(y - a)^2 \leq a^2$. Thus the ellipse is contained in the rectangle

$$\{(x, y) : -b \leq x \leq b, \ 0 \leq y \leq 2a\}$$

A bounded curve such as the ellipse has no asymptote.

The ellipse is a level curve of the scalar field

$$f(x, y) = \frac{x^2}{b^2} + \frac{(y - a)^2}{a^2}$$

We have $f_1(x, y) = 2x/b^2$ and $f_2(x, y) = 2(y - a)/a^2$. Thus the tangent line to the ellipse at a point (x_0, y_0) has the equation

$$f_1(x_0, y_0)(x - x_0) + f_2(x_0, y_0)(y - y_0) = 0$$

or

$$\frac{2x_0}{b^2}(x - x_0) + \frac{2(y_0 - a)}{a^2}(y - y_0) = 0$$

or

$$\frac{x_0 x}{b^2} + \frac{y_0 - a}{a^2} y = \frac{y_0}{a}$$

The focus $[0, c\eta/(1 + \eta)]$ can be expressed in terms of the constants a and b as the point $(0, a - \sqrt{a^2 - b^2})$. Similarly, the equation $y =$

$-c/(1+\eta)$ of the directrix becomes

$$y = a - \frac{a^2}{\sqrt{a^2 - b^2}}$$

Utilizing the symmetry of the ellipse with respect to the horizontal line $\{(x,y) : y = a\}$, we point out that the point $(0, a + \sqrt{a^2 - b^2})$ and the line $\{(x,y) : y = a + a^2/\sqrt{a^2 - b^2}\}$ also constitute a focus and a directrix for the same ellipse.

Note that the points $(0, 0)$, $(0, 2a)$, $(\pm b, a)$ and $(\pm b^2/a, a \pm \sqrt{a^2 - b^2})$ are all on the given ellipse. Knowing the tangent lines at these points, we draw the ellipse in standard position. In Fig. 13.13, for example, we show the ellipse

$$\left\{(x,y) : \frac{x^2}{9} + \frac{(y-5)^2}{25} = 1\right\}$$

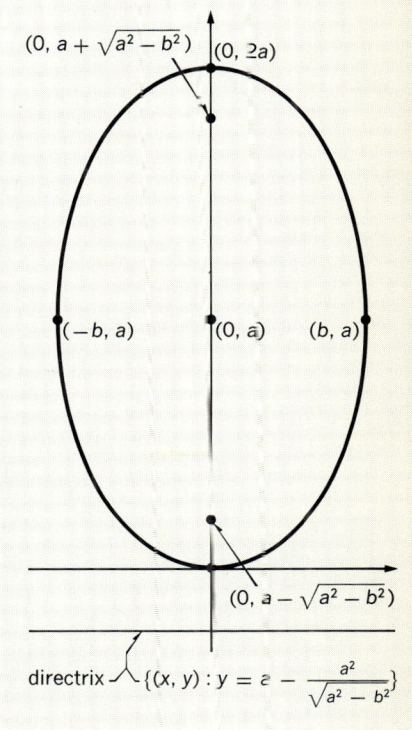

FIGURE 13.13

The hyperbola. A conic of eccentricity $\eta > 1$ is a hyperbola. As in the discussion of the ellipse above, we consider only the equation of the hyperbola in standard position, namely,

$$x^2 + (1 - \eta^2)\left(y - \frac{c\eta}{1 - \eta^2}\right)^2 = \frac{c^2\eta^2}{1 - \eta^2}$$

However, since $\eta > 1$ implies that $1 - \eta^2 < 0$, we now choose the new positive constants

$$a = \frac{c\eta}{\eta^2 - 1}, \quad b = \frac{c\eta}{\sqrt{\eta^2 - 1}}$$

It is easy to see that $\eta^2 - 1 = b^2/a^2$, whence we have

$$\eta = \sqrt{\frac{a^2 + b^2}{a^2}}$$

and

$$c = \frac{b^2}{\sqrt{a^2 + b^2}}$$

In terms of the constants a and b the equation of the hyperbola in standard position becomes

$$\frac{(y+a)^2}{a^2} - \frac{x^2}{b^2} = 1$$

Note first that the only intercepts are the points $(0, 0)$ and $(0, 2a)$. The curve is symmetric both to the y axis and to the horizontal line $\{(x,y) : y = -a\}$. Therefore, the hyperbola is symmetric to its *center* $(0, -a)$.

The focus $[0, c\eta/(1 + \eta)]$ can be given in terms of the constants a and b as

$$(0, \sqrt{a^2 + b^2} - a)$$

and the equation of the directrix $y = -c/(1 + \eta)$ becomes

$$y = \frac{a^2}{\sqrt{a^2 + b^2}} - a$$

The symmetry of the hyperbola about the line $\{(x,y) : y = -a\}$ implies that the point $(0, -\sqrt{a^2 + b^2} - a)$ and the horizontal line

$$\left\{(x,y) : y = \frac{a^2}{\sqrt{a^2 + b^2}} - a\right\}$$

constitute another focus and direction for the same hyperbola.

The hyperbola in our standard position is a level curve of the scalar field

$$f(x,y) = \frac{(y + a)^2}{a^2} - \frac{x^2}{b^2}$$

Hence $f_1(x,y) = -2x/b^2$ and $f_2(x,y) = 2(y + a)/a^2$. The tangent line at a point (x_0, y_0) is therefore

$$\left\{(x,y) : -\frac{2x_0}{b^2}(x - x_0) + \frac{2(y_0 + a)}{a^2}(y - y_0) = 0\right\}$$

By using the fact that

$$\frac{(y_0 + a)^2}{a^2} - \frac{x_0^2}{b^2} = 1$$

the equation of the tangent line reduces to

$$\left(\frac{y_0 + a}{a^2}\right)y - \left(\frac{x_0}{b^2}\right)x + \frac{y_0}{a} = 0$$

The hyperbola is asymptotic to the following two lines through its center:

$$\left\{(x,y) : y = -a + \frac{a}{b}x\right\}, \left\{(x,y) : y = -a - \frac{a}{b}x\right\}$$

The proof that these lines are asymptotes of the hyperbola consists in establishing the limit

$$\lim_{x \to \infty}\left[(-a + a\sqrt{1 + x^2/b^2}) - \left(-a + \frac{a}{b}x\right)\right] = 0$$

Since this limit can be rewritten as

$$\lim_{x \to \infty} \frac{a}{\sqrt{1 + x^2/b^2} + x/b},$$

it is obvious that its value is zero. We have shown that the vertical distance

13.6 CONICS

$$(-a + a\sqrt{1 + x^2/b^2}) - \left(-a + \frac{a}{b}x\right)$$

between the hyperbola, and the line approaches zero as the value of x increases indefinitely. Symmetry completes the argument.

Finally, we note that the points

$$(0, 0),\ (0, -2a),\ (\pm b^2/a,\ -a \pm \sqrt{a^2 + b^2})$$

are on the hyperbola. All of these facts were used in drawing the curve

$$\left\{(x, y) : \frac{(y + 3)^2}{9} - \frac{x^2}{16} = 1\right\}$$

shown in Fig. 13.14.

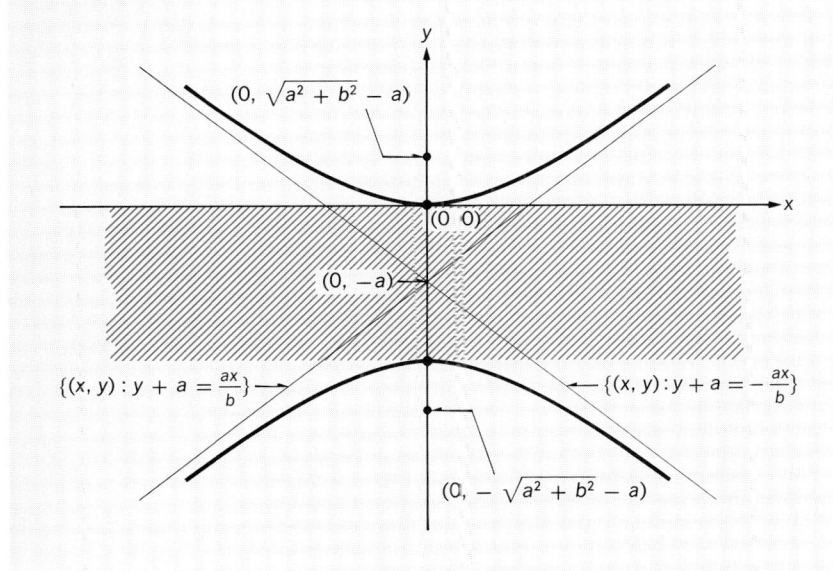

FIGURE 13.14

There are many fascinating properties of the conics which we have no space to develop. Some of these are introduced in the following problems. Note that there is no loss of generality in drawing a conic in standard position when investigating its properties via analysis of its equation. However, if two conics are being studied simultaneously, only one can be assumed to be in standard position.

EXERCISES 13.6

In each of Exercises 1 to 8, identify the conic whose equation is given. Determine its focus and directrix, its center (if it is a central conic), and its asymptotes (if any). Draw the curve together with at least one tangent line.

1. $\{(x, y) : 4y = x^2\}$

2. $\{(x,y) : y = 4x^2\}$

3. $\left\{(x,y) : \dfrac{x^2}{4} + \dfrac{(y-3)^2}{9} = 1\right\}$

4. $\{(x,y) : 4x^2 + y^2 - 4y = 0\}$

5. $\{(x,y) : 25x^2 + 16y^2 - 160y = 0\}$

6. $\left\{(x,y) : \dfrac{(y+4)^2}{16} - \dfrac{x^2}{9} = 1\right\}$

7. $\{(x,y) : 4y^2 + 8y = x^2\}$

8. $\{(x,y) : 4x^2 - 25y^2 - 100y = 0\}$

In each of Exercises 9 to 16, use the definition of a conic to obtain the equation of the particular curve described. Then draw the curve.

9. Parabola with focus $(0, 4)$ and directrix $\{(x,y) : y = -4\}$

10. Parabola with focus $(2, 0)$ and directrix $\{(x,y) : x = -2\}$

11. Parabola with focus $(1, 1)$ and directrix $\{(x,y) : x + y = 8\}$

12. Center at $(0, 0)$, focus $(0, 4)$, directrix $\{(x,y) : y = -8\}$

13. Focus $(-2, 3)$, $\eta = \tfrac{1}{2}$, directrix $\{(x,y) : x = 1\}$

14. Focus $(2, -1)$, $\eta = \tfrac{2}{3}$, directrix $\{(x,y) : y = 4\}$

15. Focus $(0, 4)$, $\eta = \tfrac{3}{2}$, directrix the x axis.

16. Foci $(\pm 3, 2)$, $\eta = 2$, directrix $\{(x,y) : x = 6\}$

PROBLEMS 13.6

1. If $\{(x,y) : ax + by + c = 0\}$ is the directrix of a parabola, show that the parabola has the form $\{(x,y) : (bx - ay)^2 + Ax + By + C = 0\}$.

2. Prove that each curve of the form $\{(x,y) : 4ay = 4a^2 - x^2\}$, $a > 0$, is a parabola and that all such curves have the same focus. (We say this is *a confocal family* of parabolas.)

3. Prove that the parabolas $\{(x,y) : x^2 + 4ay = 4a^2\}$ and $\{(x,y) : x^2 - 4by = 4b^2\}$ are orthogonal at each point of intersection.

4. Let P be a point on a parabola and let F be the focus of the parabola. Prove that the normal line at P bisects the angle between the *focal radius* FP and the ray from P orthogonal to directrix and pointing away from the directrix. (This so-called *reflection property* of the parabola finds many applications in parabolic reflectors, for instance.)

5. The region bounded by a parabola and a line is called a *parabolic segment*. The *base* of a parabolic segment is the segment cut off the line by the parabola. The *height* of the parabolic segment is the distance between the line and a parallel line that is tangent to the parabola. Prove that a parabolic segment of base b and height h has area $\tfrac{2}{3}bh$.

6. Let P be a point on an ellipse whose foci are F_1 and F_2. Prove that the sum of the lengths of the focal radii F_1P and F_2P is a constant independent of the point P.

13.6 CONICS

7. Let P, F_1, and F_2 be as in Problem 6. Prove that the normal line to the ellipse at P bisects the angle formed by the focal radii F_1P and F_2P. (A *reflection property* again.)

8. Determine the area of an ellipse.

9. Rotate an ellipse about its longer axis and determine the volume of the solid bounded by the *ellipsoid* so formed.

10. Prove that the longest chord of an ellipse contains both foci of the ellipse.

11. (a) Show that every rectangle inscribed in an ellipse has its sides parallel to the axes of the ellipse.
    (b) What is the area of the largest rectangle that can be inscribed in an ellipse?

12. Prove that $\{(x, y) : 2xy = a^2\}$ is a hyperbola with foci (a, a) and $(-a, -a)$. What line is its directrix?

13. Prove that the graph of the equation
$$\frac{x^2}{6 - A} + \frac{y^2}{15 - A} = 1$$
is a central conic with foci $(0, \pm 3)$ for any $A < 15$.

14. Prove that the hyperbola has the "same" reflection property as does the ellipse. (See Problem 7.)

15. Prove that the tangent line to a hyperbola and the two asymptotes of the hyperbola form a triangle whose area is independent of the point of tangency.

16. If a circle has the right focal chord of a parabola as a diameter, prove that the circle is tangent to the directrix of the parabola.

    A *focal chord* of a conic is a line segment through a focus with endpoints on the conic. A *right focal chord* is that focal chord orthogonal to the axis of symmetry, that is, parallel to the directrix.

17. Suppose the right focal chord of a parabola has length b. Prove that the focal chord that makes an angle of θ radians with the axis of symmetry has length $b \csc^2 \theta$.

18. Let C_1 and C_2 be the endpoints of the right focal chord of a parabola and let P be the point on the parabola nearest its directrix. (P is sometimes called the *vertex* of the parabola.) Prove that PC_1 and PC_2 form the angle $2 \tan^{-1} 2$.

19. Let C_1 and C_2 be the endpoints of any focal chord of a parabola and let P be the vertex of the parabola. Let the line through C_1 and P meet the directrix of the parabola in the point Q. Prove that the line through Q and C_2 is orthogonal to the directrix.

20. Draw the tangent line at any point P on a parabola. Prove that the line through the focus and orthogonal to this tangent line, the tangent at P, and the tangent line at the vertex of the parabola intersect in one point.

21. Draw the tangent line at any point P on a conic, and let F be a focus of the conic. Prove that the line through F orthogonal to FP meets the tangent line in a point on the directrix.

22. If a circle intersects a conic in four points, prove that the common chords of the two curves form equal angles with the axis of the conic.

23. Prove that the line from the focus of a central conic to the intersection of any two tangent lines bisects the angle formed by the focal radii of the points of tangency.

24. A quadrilateral is circumscribed about an ellipse. Show that the line joining the midpoints of the diagonals of the quadrilateral passes through the center of the ellipse.

25. Prove that any two parallelograms inscribed between a hyperbola and one of its asymptotes have equal areas.

13.7 QUADRICS

A level surface of a quadratic scalar field of the form $f(x, y, z) = Ax^2 + By^2 + Cz^2 + 2(Dxy + Exz + Fyz + Gx + Hy + Iz)$ is called a *quadric* or a *quadric surface*. We shall mention only a few special cases here. A general study of such surfaces requires some tools we have not introduced yet (see Chapter 14).

Ellipsoids. A quadric of the form

$$\left\{(x, y, z) : \frac{(x - x_0)^2}{a^2} + \frac{(y - y_0)^2}{b^2} + \frac{(z - z_0)^2}{c^2} = 1\right\}$$

is an *ellipsoid*. It is easy to see that this surface lies entirely within the rectangular parallelopiped,

$$\{(x, y, z) : |x - x_0| \leq a, |y - y_0| \leq b_1, |z - z_0| \leq c\}$$

Hence the surface is bounded. The points $(x_0 \pm a, y_0, z_0)$, $(x_0, y_0 \pm b, z_0)$ and $(x_0, y_0, z_0 \pm C)$ are on the surface as well as on the boundary of the parallelopiped.

An ellipsoid has several symmetries. It is obviously symmetric to the three planes $\{(x, y, z) : x = x_0\}$, $\{(x, y, z) : y = y_0\}$ and $\{(x, y, z) : z = z_0\}$. This implies that it is symmetric with respect to its *center* (x_0, y_0, z_0) as well as to the three axes formed by intersecting the planes of symmetry in pairs.

■ *Example 1.* Consider the ellipsoid centered at the origin

$$\left\{(x, y, z) : \frac{x^2}{4} + \frac{y^2}{4} + \frac{z^2}{1} = 1\right\}$$

The intercepts on the axes are $(\pm z, 0, 0)$, $(0, \pm 2, 0)$, $(0, 0, \pm 1)$. The trace in the xy plane is the circle $\{(x, y) : x^2 + y^2 = 4\}$; the trace in the xz plane is the ellipse $\{(x, z) : x^2 + 4z^2 = 4\}$; and in the yz plane the trace is the ellipse $\{(y, z) : y^2 + 4z^2 = 4\}$.

Being a level surface of the scalar field

$$f(x, y, z) = \frac{x^2}{4} + \frac{y^2}{4} + z^2$$

which has the gradient vector

$$\nabla f = \tfrac{1}{2}x\mathbf{i} + \tfrac{1}{2}y\mathbf{j} + 2z\mathbf{k}$$

the tangent plane to the surface at a point (x_0, y_0, z_0) has the equation

$$\frac{x_0}{2}(x - x_0) + \frac{y_0}{2}(y - y_0) + 2z_0(z - z_0) = 0$$

This equation can be reduced to

$$\frac{x_0 x}{4} + \frac{y_0 y}{4} + z_0 z = 1$$

by making use of the fact that

$$\frac{x_0^2}{4} + \frac{y_0^2}{4} + z_0^2 = 1$$

This ellipsoid lies in the parallelopiped

$$\{(x, y, z) : -2 \leq x \leq 2, -2 \leq y \leq 2, -1 \leq z \leq 1\}$$

and it is readily seen that each side of this parallelopiped is tangent to the ellipsoid. This surface is shown in Fig. 13.15.

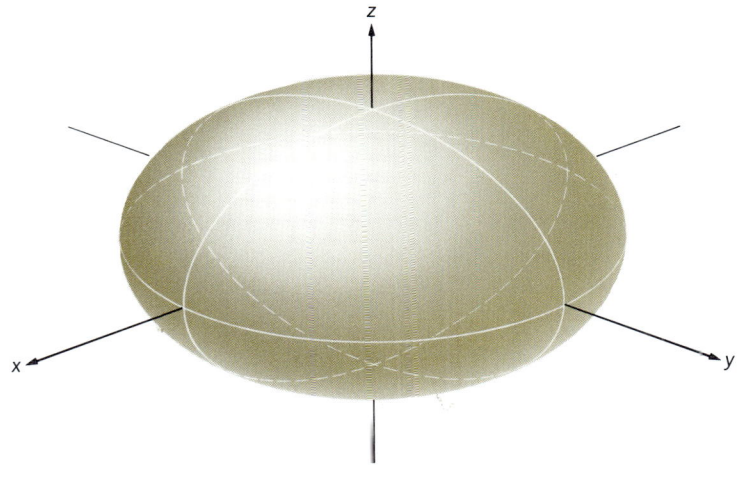

FIGURE 13.15

This ellipsoid is formed by rotating the ellipse $\{(y, z) : y^2 + 4z^2 = 4\}$, its trace in the yz plane, about the z axis. Thus the surface is a *surface of revolution*.

A sphere

$$\{(x, y, z) : (x - x_0)^2 + (y - y_0)^2 + (z - z_0)^2 = r^2\}$$

is a special case of an ellipsoid, of course.

Cones. A quadric of the form

$$\left\{(x, y, z) : \frac{(x - x_0)^2}{a^2} + \frac{(y - y_0)^2}{b^2} = (z - z_0)^2\right\}$$

is a *cone with a vertical axis of symmetry.* We shall consider no other cones. Note that this surface is symmetric to the vertical line through its center of symmetry (x_0, y_0, z_0).

■ *Example 2.* The cone

$$\left\{(x, y, z) : \frac{x^2}{4} + y^2 = z^2\right\}$$

is symmetric to the z axis. Its only intercept on any axis is the origin. The xy trace is also just the origin. The xz trace has the equation $x^2/4 = z^2$ or $z = \pm\frac{1}{2}x$, whence this trace is a pair of lines intersecting at the origin. Similarly, the yz trace has the equation $y^2 = z^2$ or $z = \pm y$, which gives another pair of lines. Finally, a horizontal cross section is an ellipse

$$\left\{(x, y, c) : \frac{x^2}{4} + y^2 = c^2\right\}$$

This *elliptical cone* is shown in Fig. 13.16.

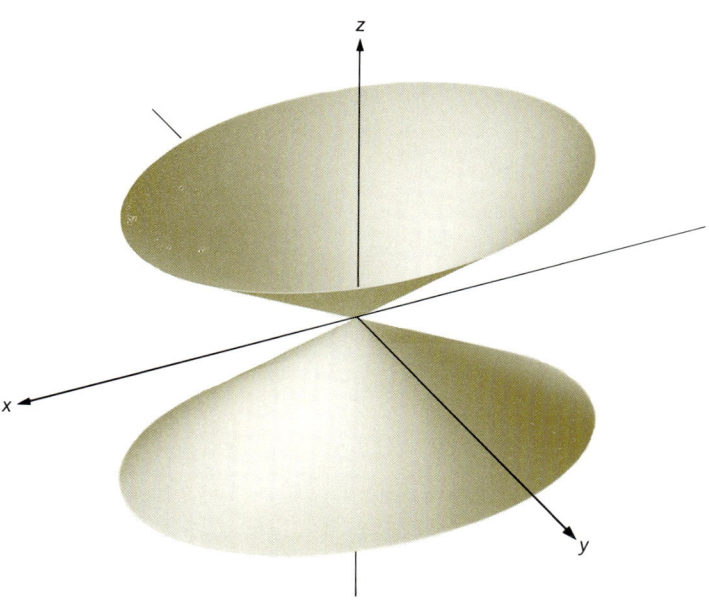

FIGURE 13.16

A cone such as

$$\left\{(x, y, z) : \frac{(x - x_0)^2}{a^2} + \frac{(y - y_0)^2}{a^2} = (z - z_0)^2\right\}$$

is called a *circular cone* because each horizontal cross section is a circle. A circular cone is a surface of revolution.

Hyperboloids. A quadric of the form

$$\left\{(x, y, z) : \frac{(x - x_0)^2}{a^2} + \frac{(y - y_0)^2}{b^2} - \frac{(z - z_0)^2}{c^2} = 1\right\}$$

is called a *hyperboloid of one sheet.* We illustrate such a surface.

■ **Example 3.** Consider the surface

$$\left\{(x, y, z) : \frac{x^2}{4} + \frac{y^2}{9} - \frac{z^2}{4} = 1\right\}$$

It is easy to see that this surface has all of the symmetries we have discussed. The point $(\pm 2, 0, 0)$ and $(0, \pm 3, 0)$ are the only intercepts. (There are no intercepts on the z axis.) The xy trace is the ellipse

$$\left\{(x, y) : \frac{x^2}{4} + \frac{y^2}{9} = 1\right\}$$

while the xz trace and the yz trace are the hyperbolas

$$\left\{(x, z) : \frac{x^2}{4} - \frac{z^2}{4} = 1\right\}, \left\{(y, z) : \frac{y^2}{9} - \frac{z^2}{4} = 1\right\}$$

This elliptical hyperboloid of one sheet is shown in Fig. 13.17.

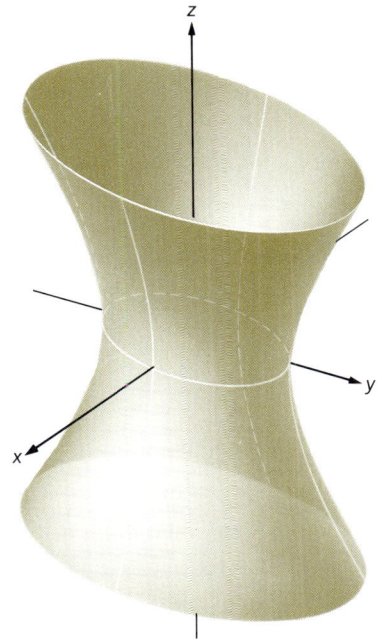

FIGURE 13.17

In a similar way, a quadric of the form

$$\left\{(x,y,z) : \frac{(x-x_0)^2}{a^2} + \frac{(y-y_0)^2}{b^2} + 1 = \frac{(z-z_0)^2}{c^2}\right\}$$

is a *hyperboloid of two sheets*. We can see from the equation that

$$\frac{(z-z_0)^2}{c^a} \geqq 1$$

must hold, whence the entire horizontal slab

$$\{(x,y,z) : z_0 - c < z < z_0 + c\}$$

contains no points on this surface. However, the surface is symmetric to the horizontal plane $\{(x,y,z) : z = z_0\}$ and hence it occurs in two pieces or two sheets. The special case

$$\left\{(x,y,z) : \frac{x^2}{1} + \frac{y^2}{1} + 1 = \frac{z^2}{1}\right\}$$

is illustrated in Fig. 13.18.

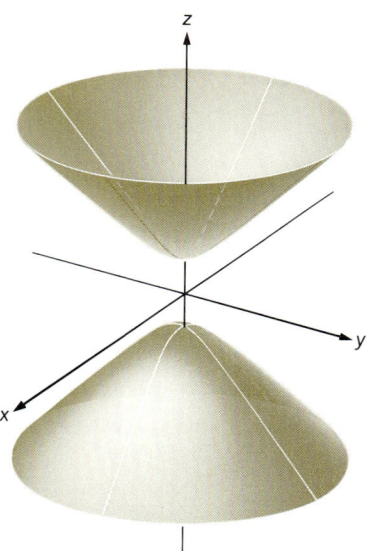

FIGURE 13.18

Paraboloids. A quadric of the form

$$\left\{(x,y,z) : z - z_0 = \frac{(x-x_0)^2}{a^2} + \frac{(y-y_0)^2}{b^2}\right\}$$

is an *elliptical paraboloid* with a vertical axis of symmetry. This surface lies above the horizontal plane $\{(x,y,z) : z = z_0\}$. Its horizontal cross sections are the ellipses

$$\frac{(x-x_0)^2}{a^2} + \frac{(y-y_0)^2}{b^2} = c - z_0, \quad c \geqq z_0$$

■ *Example 4.* Consider the special case

$$\{(x,y,z) : x^2 + y^2 = 2z\}$$

Note that each horizontal cross section is a circle in this case. The xz trace and the yz trace are the parabolas $\{(x, z) : x^2 = 2z\}$ and $\{(y, z) : y^2 = 2z\}$. This circular paraboloid is again a surface of revolution and is shown in Fig. 13.19.

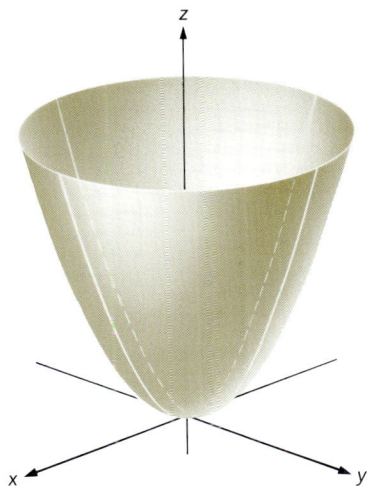

FIGURE 13.19

The quadric

$$\left\{(x,y,z) : z - z_0 = \frac{(x-x_0)^2}{a^2} - \frac{(y-y_0)^2}{b^2}\right\}$$

is a *hyperbolic paraboloid*, with a vertical axis of symmetry.

■ *Example 5.* We have seen the surface

$$\{(x,y,z) : z = x^2 - y^2\}$$

before. (Also see Fig. 13.20.) Note that its traces are (1) the intersecting lines with equations $y = \pm x$ in the xy plane, (2) the parabola $\{(x, z) : z = x^2\}$ in the xz plane, and (3) the parabola $\{(y, z) : z = -y^2\}$ in the yz plane. Other horizontal cross sections than the xy trace are hyperbolas of the form

$$\{(x, y, c) : x^2 - y^2 = c\}$$

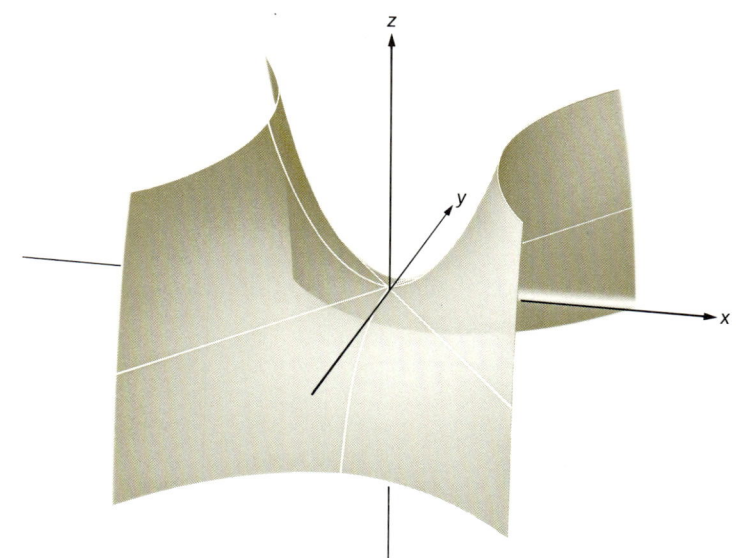

FIGURE 13.20

EXERCISES 13.7 Sketch the surface whose equation is given.

1. $x^2 = 4z$
2. $x^2 + y^2 = 4$
3. $4x^2 + y^2 = 4$
4. $x^2 + y^2 = 4x$
5. $\dfrac{x^2}{16} + \dfrac{y^2}{4} = 1$
6. $\dfrac{x^2}{4} + \dfrac{y^2}{4} + z^2 = 1$
7. $x^2 + 4y^2 = z^2$
8. $x^2 = y^2 + 4z^2$
9. $z = 4x^2 + 4y^2$
10. $z = 4x^2 - 4y^2$
11. $z = 4 - x^2 - y^2$
12. $y = x^2 + 4z^2$
13. $\dfrac{x^2}{9} - \dfrac{y^2}{4} + \dfrac{z^2}{4} = 1$
14. $\dfrac{x^2}{9} - \dfrac{y^2}{4} - \dfrac{z^2}{4} = 1$
15. $\dfrac{x^2}{9} + \dfrac{y^2}{4} - \dfrac{z^2}{4} + 1 = 0$
16. $\dfrac{x^2}{9} - \dfrac{y^2}{4} - \dfrac{z^2}{4} + 1 = 0$

By completing squares, identify the surface whose equation is given.

17. $x^2 + y^2 - 4y = z$
18. $z^2 = x^2 + 4y^2 - 4x + 8y$
19. $x^2 - y^2 - z^2 + 4x - 6z = 0$
20. $x^2 + 4y^2 - 4z^2 - 2x + 8y - 16z = 0$

PROBLEMS 13.7 1. Let F_1 and F_2 be two fixed points in space. Determine the set of all points P such that the sum of the squares of the distances from P to F_1 and to F_2 is constant.

2. Let F_1 and F_2 be two fixed points in space. Determine the set of all points P such that the sum of the distances from P to F_1 and to F_2 is a constant.

3. Let F_1 be a fixed point and let Π be a fixed plane. Determine the set of all points P such that the distance from P to F_1 equals the distance from P to Π.

4. Let the line λ and the plane Π be orthogonal. Find the set of all points P such that the distance from P to λ and the distance from P to Π have a fixed ratio.

5. Let the cone $\{(x,y,z) : x^2+y^2=z^2\}$ be cut with the plane $\{(x,y,z) : z=mx+1\}$. Determine completely the nature of the curve of intersection in the three cases (a) $m = 1$, (b) $0 < m < 1$, and (c) $m > 1$.

6. Determine the volume of an ellipsoid

7. A surface S is *ruled* if each point P on S lies on a line entirely contained in the surface. For any pair (a, b), prove that the line $\{(x,y,z) : x = a + t, y = b + t, z = b^2 - a^2 + 2(b - a)t\}$ lies in the surface $\{(x,y,z) : z = y^2 - x^2\}$ thus proving that this surface is ruled.

8. Prove that a hyperboloid of one sheet is ruled.

13.8 VECTOR FUNCTIONS. PARAMETRIC EQUATIONS

An ordered triple of real-valued functions

$$(x(t), y(t), z(t))$$

defined on an interval $\{t : a \leq t \leq b\}$ provides a transformation of this interval onto the curve in space

$$C = \{(x, y, z) : x = x(t), y = y(t), z = z(t); \; a \leq t \leq b\}$$

The triple of functions is called a *parametrization* of C and the equations $x = x(t), y = y(t), z = z(t)$ are *parametric equations* of the curve. We shall see that there are infinitely many different but related parametrizations of a given curve.

■ **Example 1.** The equations

$$x = x_0 + at, \; y = y_0 + bt, \; z = z_0 + ct$$

are parametric equations of the line through the point (x_0, y_0, z_0) and having the direction vector $\mathbf{v} = a\mathbf{i} + b\mathbf{j} + c\mathbf{k}$. Given any constant $\alpha \neq 0$, however, the vector $\alpha\mathbf{v}$ also serves as a direction vector for this line. Hence the equations

$$x = x_0 + \alpha at, \; y = y_0 + \alpha bt, \; z = z_0 + \alpha ct$$

are parametric equations of the same line.

■ **Example 2.** The parametric equations

$$x = a \cos t, \; y = a \sin t, \; z = t$$

describe a curve known as a *helix*. Because $x^2 + y^2 = a^2$ for each point

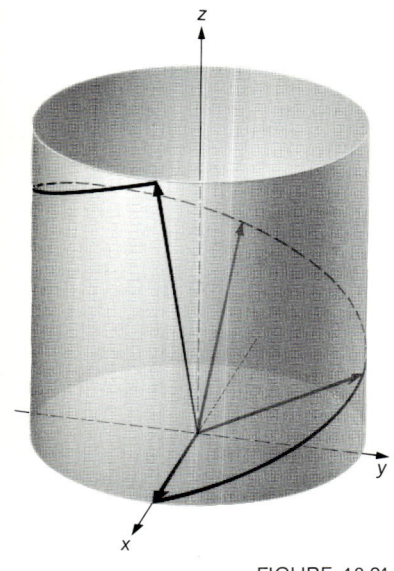

FIGURE 13.21

(x, y, z) on this curve, it is easier to envision the helix if we use cylindrical coordinates. The reader can verify easily that the helix above is also described as

$$\{(r, \theta, z) : r = a, \theta = t, z = t\}$$

The curve lies on the vertical cylinder $\{(r, \theta, z) : r = a\}$ and is shown in Fig. 13.21. ∎

A parametrization $x(t), y(t), z(t), a \leq t \leq b$, of a curve also defines a *vector function*

$$\mathbf{p}(t) = x(t)\mathbf{i} + y(t)\mathbf{j} + z(t)\mathbf{k}$$

where $\mathbf{p}(t)$ is the position vector of the point $(x(t), y(t), z(t))$ on the curve. In this context, $x(t), y(t)$ and $z(t)$ are called *component functions* of the vector function. The vector function is said to be continuous if each of its component functions is continuous. Similarly, the vector function is differentiable if its component functions are differentiable.

We think of a vector function of one real variable as the *vector law of motion* of a particle moving in space, having the position vector $\mathbf{p}(t)$ at time t. This enables us to envision the vector function more readily.

∎ *Example 3.* Consider the vector equation of motion

$$\mathbf{p}(t) = (\cos t)\mathbf{i} + (1 + \cos 2t)\mathbf{j} + 0\mathbf{k}$$

of a particle moving in the xy plane. From the trigonometric identity $\cos 2t = 2 \cos^2 t - 1$, we see that

$$y = 1 + \cos 2t = 1 + (2 \cos^2 t - 1) = 2 \cos^2 t = 2x^2$$

Hence each point $(\cos t, 1 + \cos 2t)$ on the path of the particle lies on the parabola $\{(x, y) : y = 2x^2\}$. However, the inequalities

$$-1 \leq \cos t \leq 1, 0 \leq 1 + \cos 2t \leq 2$$

which hold for every value of t, tell us that the entire parametric curve lies within the rectangle

$$\{(x, y) : -1 \leq x \leq 1, 0 \leq y \leq 2\}$$

Therefore, the parametric curve is just that parabolic arc in Fig. 13.22.

It is not enough to know that the particle moves along this parabolic arc, however. We should know *how* it moves. One easy means to determine this is to plot some points. Consider the following table. Measuring

t	0	$\frac{\pi}{4}$	$\frac{\pi}{2}$	$\frac{3\pi}{4}$	π	$\frac{5\pi}{4}$	$\frac{3\pi}{2}$	$\frac{7\pi}{4}$	2π
x	1	$\frac{1}{2}\sqrt{2}$	0	$-\frac{1}{2}\sqrt{2}$	-1	$-\frac{1}{2}\sqrt{2}$	0	$\frac{1}{2}\sqrt{2}$	1
y	2	1	0	1	2	1	0	1	2

13.8 VECTOR FUNCTIONS. PARAMETRIC EQUATIONS

time t in seconds, the particle starts ($t = 0$) at the point $(1, 2)$. It drops down through $(\frac{1}{2}\sqrt{2}, 1)$ and passes through $(0, 0)$, moving from right to left, at $t = \pi/2$ sec. When $t = \pi$, it is at the point $(-1, 2)$ and starts back along the arc. It passes through $(0, 0)$ again at $t = 3\pi/2$, arriving at its starting point when $t = 2\pi$. We now recognize that the motion is cyclic, that the particle traces out the parabolic arc once in each direction every 2π seconds.

- **Example 4.** The vector function

$$\mathbf{p}(t) = a[(\sin t)\mathbf{i} + (\cos t)\mathbf{j} + (\sin t)\mathbf{k}]$$

may be taken as the law of motion of a particle. The motion is clearly cyclic, starting at $(0, a, 0)$ when $t = 0$ and returning to this point 2π sec later. The reader may verify that the parametric curve traced out by the moving particle is precisely the intersection of the plane $\{(x, y, z) : x = z\}$ with the cylinder $\{(x, y, z) : x^2 + y^2 = a^2\}$. ∎

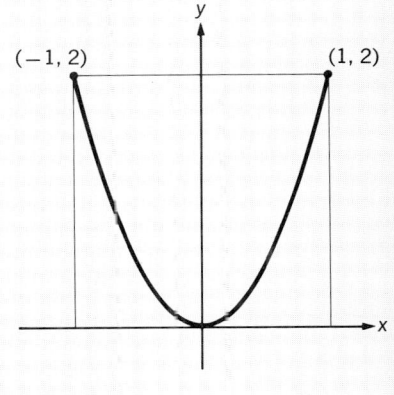

FIGURE 13.22

In precise analogy to the situation just described, an ordered triple of real-valued functions

$$(x(r, s), y(r, s), z(r, s))$$

defined on a plane region R provides a transformation of R onto a surface in space

$$S = \{(x, y, z) : x = x(r, s), y = y(r, s), z = z(r, s), (r, s) \in R\}$$

The triple is a *parametrization* of the surface S and the equations

$$x = x(r, s), y = y(r, s), z = z(r, s)$$

are *parametric equations* of S.

A parametrization also defines the vector function

$$\mathbf{p}(r, s) = x(r, s)\mathbf{i} + y(r, s)\mathbf{j} + z(r, s)\mathbf{k}$$

Such a vector function is continuous, differentiable, and so forth, according as its component functions each has the corresponding property.

- **Example 5.** The plane $\{(x, y, z) : 2x + 3y - z = 6\}$ can be parametrized by setting

$$x = r, y = s, z = 2r + 3s - 6$$

Indeed, the graph of the function $f(x, y)$ is precisely the surface parametrized by the equations

$$x = x, y = y, z = f(x, y)$$

- **Example 6.** The vector function

$$\mathbf{p}(r, s) = r^2\mathbf{i} + rs\mathbf{j} + s^2\mathbf{k}, 0 \leq r \leq 1, 0 \leq s \leq 1$$

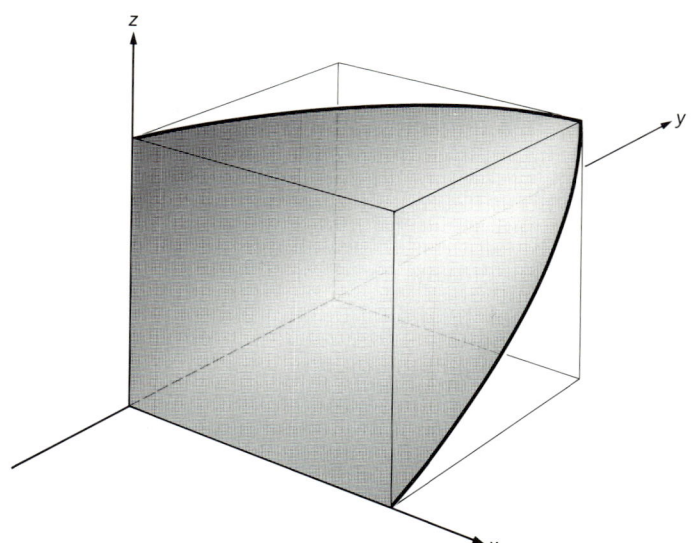

FIGURE 13.23

describes the surface shown in Fig. 13.23. Note that a point $(r, 0)$ has the value $\mathbf{p}(r, 0) = r^2\mathbf{i}$, and similarly, $\mathbf{p}(0, s) = s^2\mathbf{k}$. We also see that $\mathbf{p}(r, 1) = r^2\mathbf{i} + r\mathbf{j} + \mathbf{k}$ and $\mathbf{p}(1, s) = \mathbf{i} + s\mathbf{j} + s^2\mathbf{k}$. Finally, $\mathbf{p}(r, r) = r^2(\mathbf{i} + \mathbf{j} + \mathbf{k})$. This last tells us that the surface contains the segment from the origin to $(1, 1, 1)$. The rest of the surface was obtained by looking at level curves.

EXERCISES 13.8 In each of Exercises 1 to 20, parametric equations and an interval are given. Sketch the corresponding parametric curve in the plane or in space. Also eliminate the parameter to put the equations of the curve into rectangular coordinate form.

1. $x = 2t - 1, y = -1 + 3t; -1 \leq t \leq 1$
2. $x = 2 + 3t, y = 1 - 2t, z = 2 + t; -1 \leq t \leq 1$
3. $x = 3t - 1, y = 1 - 2t^2; 0 \leq t \leq 2$
4. $x = t + 2, y = 4 - t, z = t^2; 0 \leq t \leq 2$
5. $x = t^2, y = t^3; -2 \leq t \leq 0$
6. $x = t, y = t^2, z = t^3; 0 \leq t \leq 2$
7. $x = 4 \sin t, y = -1 + 2 \sin t; 0 \leq t \leq 2\pi$
8. $x = 4 \sin t, y = -1 + 2 \sin t, z = \tfrac{1}{2}t; 0 \leq t \leq 2\pi$
9. $x = 1 + \dfrac{1}{t}, y = t - \dfrac{1}{t}; \dfrac{1}{2} \leq t \leq 4$
10. $x = 1 + \dfrac{1}{t}, y = 2 - \dfrac{1}{t}, z = t + \dfrac{1}{t}; \dfrac{1}{2} \leq t \leq 2$
11. $x = \tan t, y = \sec t; 0 \leq t \leq \dfrac{\pi}{4}$

12. $x = 4 \sec t, y = 3 \tan t, z = t; 0 \leq t < \dfrac{\pi}{2}$

13. $x = \sin t, y = \cos 2t, 0 \leq t \leq \pi$

14. $x = 4 + t, y = 2 \sin t, z = \cos 2t; 0 \leq t \leq \dfrac{\pi}{2}$

15. $x = t^3 - 3t, y = \tan t; 0 \leq t \leq \dfrac{\pi}{4}$

16. $x = t^3 - 3t, y = t + 1, z = \tan t; 0 \leq t \leq \dfrac{\pi}{4}$

17. $x = e^t, y = -1 + 3e^t; -\ln 2 \leq t \leq 0$

18. $x = te^t, y = t \ln t, z = t; 1 \leq t \leq e$

19. $x = 4 \sin^3 t, y = 4 \cos^3 t; 0 \leq t \leq 2\pi$

20. $x = e^{-t} \cos 2t, y = e^{-t} \sin 2t; z = te^{-t}; 0 \leq t \leq 2\pi$

In each of Exercises 21 to 30, a vector function $\mathbf{p}(t)$ is given. Describe the curve traced out by the tip of the vector

21. $\mathbf{p}(t) = \mathbf{i} + 3t\mathbf{j}$

22. $\mathbf{p}(t) = \mathbf{i} + 3t\mathbf{j} + 2\mathbf{k}$

23. $\mathbf{p}(t) = (1 - t + t^2)\mathbf{i} + (2 + t)\mathbf{j}$

24. $\mathbf{p}(t) = (1 - t + t^2)\mathbf{i} + (2 + t)\mathbf{j} + (4 - t)\mathbf{k}$

25. $\mathbf{p}(t) = (4 \cos t)\mathbf{i} + (3 \sin t)\mathbf{j}$

26. $\mathbf{p}(t) = (2 \cos t)\mathbf{i} + (3 \sin t)\mathbf{j} + t\mathbf{k}$

27. $\mathbf{p}(t) = e^t[\mathbf{i} \cos t + \mathbf{j} \sin t]$

28. $\mathbf{p}(t) = e^{-t}[\mathbf{i} \cos t + \mathbf{j} \sin t + \mathbf{k}]$

29. $\mathbf{p}(t) = (\cos t + t \sin t)\mathbf{i} + (\sin t - t \cos t)\mathbf{j}$

30. $\mathbf{p}(t) = t[\mathbf{i} \cos t + \mathbf{j} \sin t + \mathbf{k} \sin 2t]$

In each of Exercises 31 to 40, a vector function of two variables is given. Study and sketch the surface described.

31. $\mathbf{p}(r, s) = (1 + r)\mathbf{i} + (1 - s)\mathbf{j} + (3 + r - s)\mathbf{k}$

32. $\mathbf{p}(r, s) = r\mathbf{i} + s\mathbf{j} + \mathbf{k}(r^2 + s^2)$

33. $\mathbf{p}(r, s) = (1 + r)\mathbf{i} + (1 - s)\mathbf{j} + (r^2 + s^2)\mathbf{k}$

34. $\mathbf{p}(r, s) = r^2\mathbf{i} + 2s\mathbf{j} + rs\mathbf{k}$

35. $\mathbf{p}(r, s) = (r + s)\mathbf{i} + (r - s)\mathbf{j} + (r^2 + s^2)\mathbf{k}$

36. $\mathbf{p}(r, s) = (r \cos s)\mathbf{i} + (r \sin s)\mathbf{j} + r\mathbf{k}$

37. $\mathbf{p}(r, s) = (r \cos s)\mathbf{i} + (r \sin s)\mathbf{j} + r^2\mathbf{k}$

38. $\mathbf{p}(r, s) = e^{-r}[\mathbf{i} \cos s + \mathbf{j} \sin s + \mathbf{k}]$

39. $\mathbf{p}(r, s) = (\cos r \sin s)\mathbf{i} + (\sin r \sin s)\mathbf{j} + (\cos s)\mathbf{k}; -\pi < r \leq \pi, 0 \leq s \leq \pi$

40. $\mathbf{p}(r, s) = [(2 + \cos r) \cos s]\mathbf{i} + [(2 + \cos r) \sin s]\mathbf{j} + (\sin r)\mathbf{k};$
$-\pi < r \leq \pi, 0 \leq s < 2\pi$

PROBLEMS 13.8

1. Prove that the curve described by the vector function
$$\mathbf{p}(t) = (t \cos 3t)\mathbf{i} + (t \sin 3t)\mathbf{j} + (9 - t^2)\mathbf{k}$$
lies in the paraboloid $\{(x, y, z) : z = 9 - x^2 - y^2\}$. Describe the curve in words.

2. Prove that the curve described by the vector function
$$\mathbf{p}(t) = (2 \cos t)(\mathbf{i} + \mathbf{j}) + (2 \sin t)\mathbf{k}$$
lies on both circular cylinders $\{(x, y, z) : x^2 + z^2 = 4\}$ and $\{(x, y, z) : y^2 + z^2 = 4\}$. What is this curve?

3. Prove that a curve of the form
$$\mathbf{p}(t) = at^4\mathbf{i} + bt^3\mathbf{j} + ct^6\mathbf{k}$$
passes through each point on the surface $\{(x, y, z) : z^2 = x^3 + 2y^4\}$ and lies entirely in the surface.

4. Show that all three vector functions
$$\mathbf{p}(r, s) = r\mathbf{i} + s\mathbf{j} + (r^2 + s^2)\mathbf{k}$$
$$\mathbf{p}(r, s) = (r - 2)\mathbf{i} + (S + 1)\mathbf{j} + (r^2 + s^2 - 4r + 2s + 5)\mathbf{k}$$
$$\mathbf{p}(r, s) = (r \cos 3s)\mathbf{i} + (r \sin 3s)\mathbf{j} + r^2\mathbf{k}$$
describe the same paraboloid.

13.9 DIFFERENTIATION OF VECTOR FUNCTIONS

Consider the vector function
$$\mathbf{p}(t) = x(t)\mathbf{i} + y(t)\mathbf{j} + z(t)\mathbf{k}$$

Let t be fixed and let h be an increment. The difference vector
$$\mathbf{p}(t + h) - \mathbf{p}(t)$$
is represented by the line segment from the point $(x(t), y(t), z(t))$ to the point $(x(t + h), y(t + h), z(t + h))$. (See Fig. 13.24). Note that this "secant" vector and its multiple
$$\frac{\mathbf{p}(t + h) - \mathbf{p}(t)}{h}$$
are nearly tangent to the curve described by the position vector $\mathbf{p}(t)$ at the tip of $\mathbf{p}(t)$.

We define the *vector derivative* of $\mathbf{p}(t)$ to be
$$D\mathbf{p}(t) = \lim_{h \to 0} \frac{\mathbf{p}(t + h) - \mathbf{p}(t)}{h}$$
provided that this limit exists. In addition, we define the vector $D\mathbf{p}(t)$ to be tangent to the curve at the tip of $\mathbf{p}(t)$. One easily computes this limit to be

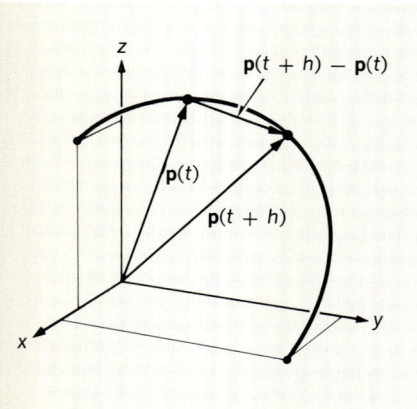

FIGURE 13.24

13.9 DIFFERENTIATION OF VECTOR FUNCTIONS

$$\lim_{h \to 0} \left[\frac{x(t+h) - x(t)}{h} \mathbf{i} + \frac{y(t+h) - y(t)}{h} \mathbf{j} + \frac{z(t+h) - z(t)}{h} \mathbf{k} \right]$$

$$= \left[\lim_{h \to 0} \frac{x(t+h) - x(t)}{h} \right] \mathbf{i} + \left[\lim_{h \to 0} \frac{y(t+h) - y(t)}{h} \right] \mathbf{j} + \left[\lim_{h \to 0} \frac{z(t+h) - z(t)}{h} \right] \mathbf{k}$$

Hence

$$D\mathbf{p}(t) = Dx(t)\mathbf{i} + Dy(t)\mathbf{j} + Dz(t)\mathbf{k}$$

provided that the component functions are differentiable.

The vector function

$$\mathbf{p}(t) = x(t)\mathbf{i} + y(t)\mathbf{j} + z(t)\mathbf{k}$$

is said to be *differentiable*, or *continuously differentiable* or *twice differentiable*, and so forth, according as its component functions $x(t)$, $y(t)$, and $z(t)$ all have the corresponding property. The behavior of differentiation in conjunction with the operations of vector algebra is given next.

■ **THEOREM 13.10.** Let $\mathbf{p}(t)$ and $\mathbf{q}(t)$ be differentiable vector functions and $f(t)$ be a differentiable (scalar) function. Then

1. $\mathbf{p}(t) + \mathbf{q}(t)$ is a differentiable vector function and

   $$D[\mathbf{p}(t) + \mathbf{q}(t)] = D\mathbf{p}(t) + D\mathbf{q}(t)$$

2. $f(t)\mathbf{p}(t)$ is a differentiable vector function and

   $$D[f(t)\mathbf{p}(t)] = Df(t) \cdot \mathbf{p}(t) + f(t) \cdot D\mathbf{p}(t)$$

3. The dot product $\mathbf{p}(t) \cdot \mathbf{q}(t)$ is a differentiable (scalar) function and

   $$D[\mathbf{p}(t) \cdot \mathbf{q}(t)] = D\mathbf{p}(t) \cdot \mathbf{q}(t) + \mathbf{p}(t) \cdot D\mathbf{q}(t)$$

4. The cross product $\mathbf{p}(t) \times \mathbf{q}(t)$ is a differentiable vector function and

   $$D[\mathbf{p}(t) \times \mathbf{q}(t)] = D\mathbf{p}(t) \times \mathbf{q}(t) + \mathbf{p}(t) \times D\mathbf{q}(t)$$

PROOF: We prove only part three. The remaining parts are equally straightforward. Let

$$\mathbf{p}(t) = x_1(t)\mathbf{i} + y_1(t)\mathbf{j} + z_1(t)\mathbf{k}$$

and

$$\mathbf{q}(t) = x_2(t)\mathbf{i} + y_2(t)\mathbf{j} + z_2(t)\mathbf{k}$$

Then

$$\mathbf{p}(t) \cdot \mathbf{q}(t) = x_1(t)x_2(t) + y_1(t)y_2(t) + z_1(t)z_2(t)$$

and its derivative is

$$(Dx_1)x_2 + x_1 Dx_2 + (Dy_1)y_2 + y_1 Dy_2 + (Dz_1)z_2 + z_1 Dz_2$$
$$= (Dx_1)x_2 + (Dy_1)y_2 + (Dz_1)z_2 + x_1 Dx_2 + y_1 Dy_2 + z_1 Dz_2$$
$$= D\mathbf{p}(t) \cdot \mathbf{q}(t) + \mathbf{p}(t) \cdot D\mathbf{q}(t) \quad \blacksquare$$

■ **COROLLARY 13.11.** *If the vector function* $\mathbf{p}(t)$ *has constant norm, the* $\mathbf{p}(t)$ *and* $D\mathbf{p}(t)$ *are orthogonal.*

PROOF: If $\|\mathbf{p}(t)\| = c$, then $\|\mathbf{p}(t)\|^2 = \mathbf{p}(t) \cdot \mathbf{p}(t) = c^2$. By part 3 of Theorem 13.10, we have

$$2D\mathbf{p}(t) \cdot \mathbf{p}(t) = 0 \quad \blacksquare$$

Consider the vector function $\mathbf{p}(t)$ as the law of motion of a particle. Returning to the process of differentiation, we find that the secant vector

$$\mathbf{p}(t + h) - \mathbf{p}(t)$$

is very nearly the same length as the arc from $\mathbf{p}(t)$ to $\mathbf{p}(t + h)$ when h is small. We naturally define

$$\frac{\mathbf{p}(t + h) - \mathbf{p}(t)}{h}$$

to be the average vector velocity of the moving particle during the time interval from t to $t + h$. Hence we now interpret the derivative vector as the instantaneous *vector velocity*. We define

$$\mathbf{v}(t) = D\mathbf{p}(t) = Dx(t)\mathbf{i} + Dy(t)\mathbf{j} + Dz(t)\mathbf{k}$$

The norm of the vector velocity is called the *speed* of the particle,

$$\|\mathbf{v}(t)\| = \sqrt{[Dx(t)]^2 + [Dy(t)]^2 + [Dz(t)]^2}$$

The functions $Dx(t)$, $Dy(t)$, and $Dz(t)$ are the *components of velocity*.

In precise analogy the *vector acceleration* of the moving particle is defined to be

$$\mathbf{a}(t) = D^2\mathbf{p}(t) = D^2x(t)\mathbf{i} + D^2y(t)\mathbf{j} + D^2z(t)\mathbf{k}$$

■ **Example 1.** Consider the law of motion

$$\mathbf{p}(t) = (-1 + 2t + t^2)\mathbf{i} + (2 - t + t^2)\mathbf{j} + (-1 + 2t)\mathbf{k}$$

The vector velocity and vector acceleration are

$$\mathbf{v}(t) = (2 + 2t)\mathbf{i} + (-1 + 2t)\mathbf{j} + 2\mathbf{k}$$

and

$$\mathbf{a}(t) = 2\mathbf{i} + 2\mathbf{j}$$

The last equation tells us that the acceleration is constant and has the magnitude $\|\mathbf{a}(t)\| = 2\sqrt{2}$ directed along a line with direction numbers $(2, 2, 0)$. The three vectors $\mathbf{p}(1)$, $\mathbf{v}(1)$, and $\mathbf{a}(1)$ are shown in Fig. 13.25.

■ **Example 2.** A particle moves in the *xy* plane with law of motion

$$\mathbf{p}(t) = (a \cos \omega t)\mathbf{i} + (a \sin \omega t)\mathbf{j}$$

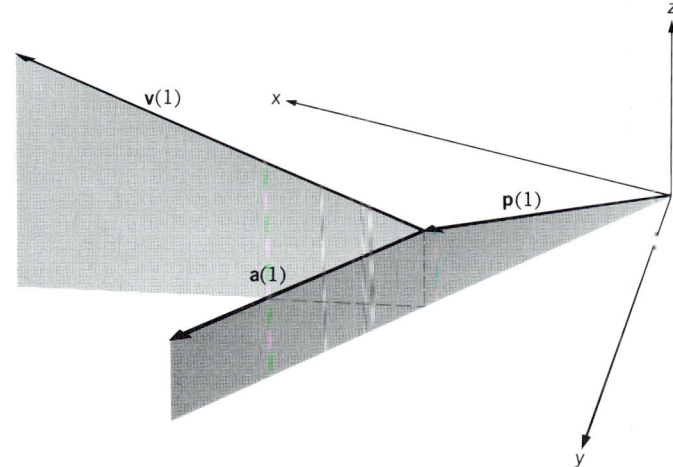

FIGURE 13.25

where $a > 0$ and ω are constants. The particle obviously moves in a circle of radius a about the origin. The velocity and acceleration vectors are

$$\mathbf{v}(t) = (-a\omega \sin \omega t)\mathbf{i} + (a\omega \cos \omega t)\mathbf{j}$$
$$\mathbf{a}(t) = (-a\omega^2 \cos \omega t)\mathbf{i} + (-a\omega^2 \sin \omega t)\mathbf{j} = -\omega^2 \mathbf{p}(t)$$

Because $\|\mathbf{p}(t)\| = a$, Corollary 13.11 concludes that $\mathbf{p}(t)$ and $\mathbf{v}(t)$ are orthogonal. Then, of course, $\mathbf{v}(t)$ and $\mathbf{a}(t)$ are also orthogonal. The speed of the particle is

$$\|\mathbf{v}(t)\| = \sqrt{(-a\omega \sin \omega t)^2 + (a\omega \cos \omega t)^2} = a|\omega|$$

This shows that the speed of a particle in circular motion equals the radius of the circle times the angular velocity $|\omega|$. In passing we remark that either of the equations

$$\mathbf{v}(t) \cdot \mathbf{p}(t) = 0$$

or

$$\mathbf{a}(t) = -\omega^2 \mathbf{p}(t)$$

characterizes uniform circular motion in the plane. ∎

Consider next a vector function of two variables,

$$\mathbf{p}(r, s) = \mathbf{i}x(r, s) + \mathbf{j}y(r, s) + \mathbf{k}z(r, s)$$

We define the partial derivatives of $\mathbf{p}$ to be

$$D_1\mathbf{p}(r, s) = \mathbf{i}x_1(r, s) + \mathbf{j}y_1(r, s) + \mathbf{k}z_1(r, s)$$
$$D_2\mathbf{p}(r, s) = \mathbf{i}x_2(r, s) + \mathbf{j}y_2(r, s) - \mathbf{k}z_2(r, s)$$

Such a vector function $\mathbf{p}$ is *smooth* in a region in the rs plane if it is continuously differentiable in the region and if the vector equation

$$D_1\mathbf{p}(r, s) \times D_2\mathbf{p}(r, s) \neq \mathbf{0}$$

holds at each point. This nonzero vector

$$\mathbf{N}(r, s) = D_1\mathbf{p}(r, s) \times D_2\mathbf{p}(r, s)$$

is normal to the surface described by $\mathbf{p}$ at each point.

If $f(x, y)$ is a differentiable function of two variables, then the vector function

$$\mathbf{p}(x, y) = x\mathbf{i} + y\mathbf{j} + \mathbf{k}f(x, y)$$

has partial derivatives

$$D_1\mathbf{p}(x, y) = \mathbf{i} + \mathbf{k}f_1(x, y)$$
$$D_2\mathbf{p}(x, y) = \mathbf{j} + \mathbf{k}f_2(x, y)$$

It follows that the normal vector $\mathbf{N}$ to the surface is

$$\mathbf{N} = (\mathbf{i} + \mathbf{k}f_1) \times (\mathbf{j} + \mathbf{k}f_2)$$
$$= -\mathbf{i}f_1(x, y) - \mathbf{j}f_2(s, y) + \mathbf{k}$$

This agreement with our earlier results tends to verify the present definitions. We shall not carry out a study of such surfaces, however. The subject is properly part of *differential geometry*. In the next two sections we study curves in space; hence a partial investigation of the geometry of vector functions is made here.

EXERCISES 13.9 In each of Exercises 1 to 15, a vector law of motion is given. Determine the velocity and acceleration vectors in each case. Sketch an arc of the path of motion containing the tip of $\mathbf{p}(t)$ at the given fixed time. Draw both $\mathbf{v}(t)$ and $\mathbf{a}(t)$ at the fixed time.

1. $\mathbf{p}(t) = (1 - 2t)\mathbf{i} + (t - t^2)\mathbf{j}; \ t = 2$
2. $\mathbf{p}(t) = t^2\mathbf{i} + \frac{2}{3}(1 + t)^{3/2}\mathbf{j}; \ t = 3$
3. $\mathbf{p}(t) = (4 \cos \pi t)\mathbf{i} - (2 \sin \pi t)\mathbf{j}; \ t = \frac{1}{2}$
4. $\mathbf{p}(t) = e^t\mathbf{i} + e^{-t}\mathbf{j}; \ t = 0$
5. $\mathbf{p}(t) = e^{-t}(\mathbf{i} \cos \pi t + \mathbf{j} \sin \pi t); \ t = \frac{1}{2}$
6. $\mathbf{p}(t) = (2 - t)\mathbf{i} + (1 + t)\mathbf{j} + 2t\mathbf{k}; \ t = 1$
7. $\mathbf{p}(t) = (1 + t)\mathbf{i} + (1 - t)\mathbf{j} + (1 - t^2)\mathbf{k}; \ t = 1$
8. $\mathbf{p}(t) = t\mathbf{i} + 2t\mathbf{j} + \frac{2}{3}t^{3/2}\mathbf{k}; \ t = 1$
9. $\mathbf{p}(t) = 6t\mathbf{i} + 3\sqrt{2}t^2\mathbf{j} + 2t^3\mathbf{k}; \ t = 1$
10. $\mathbf{p}(t) = (\cos t)\mathbf{i} + (\sin t)\mathbf{j} + \frac{1}{4}t\mathbf{k}; \ t = \frac{\pi}{2}$
11. $\mathbf{p}(t) = (t - \frac{1}{3}t^3)\mathbf{i} + t^2\mathbf{j} + (t + \frac{1}{3}t^3)\mathbf{k}; \ t = 0$
12. $\mathbf{p}(t) = (t + t^2)\mathbf{i} + (t - 3t^2)\mathbf{j} + te^{-t}\mathbf{k}; \ t = -1$
13. $\mathbf{p}(t) = e^t(\mathbf{i} + \mathbf{j} + \mathbf{k} \sin t); \ t = 0$

14. $\mathbf{p}(t) = e^t\mathbf{i} + \sqrt{2}t\mathbf{j} + e^{-t}\mathbf{k}$; $t = 1$

15. $\mathbf{p}(t) = (4\sin t)\mathbf{i} + (2t - \sin t)\mathbf{j} + (\cos 2t)\mathbf{k}$; $t = \dfrac{\pi}{2}$

In each of Exercises 16 to 25, a vector function of two variables is given. Determine where each function is smooth. Determine the normal line and tangent plane at the given point.

16. $\mathbf{p}(r, s) = r\mathbf{i} + s\mathbf{j} + (4 - r^2 - s^2)\mathbf{k}$; $(r, s) = (1, 1)$
17. $\mathbf{p}(r, s) = r^2\mathbf{i} + 25\mathbf{j} + rs\mathbf{k}$; $(r, s) = (1, 1)$
18. $\mathbf{p}(r, s) = (r^2 + 2rs)\mathbf{i} + (r - s)\mathbf{j} + (r^3 + s^3)\mathbf{k}$; $(r, s) = (1, -1)$
19. $\mathbf{p}(r, s) = \dfrac{1}{r^2 + s^2}(r\mathbf{i} + s\mathbf{j} + \mathbf{k})$; $(r, s) = (1, 1)$
20. $\mathbf{p}(r, s) = (r\cos s)\mathbf{i} + (r\sin s)\mathbf{j} + (r + s)\mathbf{k}$; $(r, s) = \left(\dfrac{\pi}{2}, \dfrac{\pi}{2}\right)$
21. $\mathbf{p}(r, s) = e^{-r}(\mathbf{i}\cos s + \mathbf{j}\sin s + s\mathbf{k})$
22. $\mathbf{p}(r, s) = \mathbf{i}\sin rs + \mathbf{j}\cos rs + \mathbf{k}\sin s$; $(r, s) = (\tfrac{1}{2}, \pi)$
23. $\mathbf{p}(r, s) = \sin s(\mathbf{i}\cos r + \mathbf{j}\sin r) + \mathbf{k}\cos s$; $(r, s) = \left(\dfrac{\pi}{4}, \dfrac{\pi}{4}\right)$
24. $\mathbf{p}(r, s) = (r + s)\mathbf{i} + \ln(r + s)\mathbf{j} + e^{r+s}\mathbf{k}$; $(r, s) = (1, 0)$
25. $\mathbf{p}(r, s) = (2 + \cos r)(\mathbf{i}\cos s + \mathbf{j}\sin s) + \mathbf{k}\sin r$; $(r, s) = (0, 0)$

PROBLEMS 13.9

1. A projectile fired from the origin at time $t = 0$ has an initial velocity vector $(v_0\cos\alpha)\mathbf{i} + (v_0\sin\alpha)\mathbf{j}$. If the x component of its acceleration is zero, while the y compartment is the constant $-g$, show that its vector equation of motion is

$$\mathbf{p}(t) = (v_0 t\cos\alpha)\mathbf{i} + (-\tfrac{1}{2}gt^2 + v_0 t\sin\alpha)\mathbf{j}$$

Show further that the projectile recrosses the x axis at the point $x = (v_0^2 \sin 2\alpha)/g$. (This distance is the *range* of the projectile.) Maximize the range by the choice of α.

2. Consider the same projectile as in Problem 1. Let L be a line through $(0, 0)$ with the angle of inclination $\beta \neq 0$. Determine the distance from $(0, 0)$ along L at which the projectile crosses L again. Maximize this "slant range" by the choice of α.

3. Let $\mathbf{p}_1$, $\mathbf{p}_2$, and $\mathbf{p}_3$ be vector functions of two variables. Assuming the appropriate differentiability conditions, express the partial derivatives of
   (a) $\mathbf{p}_1 \cdot \mathbf{p}_2$
   (b) $\mathbf{p}_1 \times \mathbf{p}_2$
   (c) $\mathbf{p}_1 \cdot (\mathbf{p}_2 \times \mathbf{p}_3)$
   in terms of the partial derivatives of $\mathbf{p}_1$, $\mathbf{p}_2$, and $\mathbf{p}_3$.

13.10 ARC LENGTH

Consider the parametric curve
$$C = \{(x, y, z) : x = x(t), y = y(t), z = z(t)\}$$
where $a \leq t \leq b$. Given any natural number n, we set $\Delta t = (b - a)/n$ and define $t_i = a + i\Delta t$, $i = 0, 1, 2, \cdots, n$. Approximate the curve C

with the polygonal curve having consecutive vertices
$$(x(t_i), y(t_i), z(t_i)) \qquad i = 0, 1, 2, \cdots, n$$
This polygonal curve has the length
$$\sum_{i=1}^{n} \sqrt{[x(t_i) - x(t_{i-1})]^2 + [y(t_i) - y(t_{i-1})]^2 + [z(t_i) - z(t_{i-1})]^2}$$

Assuming that the functions $x(t), y(t), z(t)$ are differentiable, we can apply the mean value theorem to show that there exist three points t_i', t_i'', and t_i''' in the ith subinterval $\{t : t_{i-1} \leqq t \leqq t_i\}$ such that
$$x(t_i) - x(t_{i-1}) = (t_i - t_{i-1}) Dx(t_i') = \Delta t \, Dx(t_i')$$
$$y(t_i) - y(t_{i-1}) = \Delta t \, Dy(t_i'')$$
$$z(t_i) - z(t_{i-1}) = \Delta t \, Dz(t_i''')$$

Substituting these in the above sum, we find that the length of the approximating polygonal curve is
$$\sum_{i=1}^{n} \sqrt{[Dx(t_i')]^2 + [Dy(t_i'')]^2 + [Dz(t_i''')]^2} \, \Delta t$$

These are not precisely Riemann sums for the function
$$\sqrt{[Dx(t)]^2 + [Dy(t)]^2 + [Dz(t)]^2}$$
because three different points t_i', t_i'', and t_i''' are used in each subinterval. However, if $Dx(t)$, $Dy(t)$, and $Dz(t)$ are continuous, we can show that these sums converge (Problem 13.10.1). In such cases we define the *arc length* of the parametric curve to be the integral
$$\int_a^b \sqrt{[Dx(t)]^2 + [Dy(t)]^2 + [Dx(t)]^2} \, dt$$

■ **Example 1.** Consider the parametric equations
$$x = t \cos t, \, y = -t \sin t, \, z = t\sqrt{3}; \, 0 \leqq t \leqq 1$$
Clearly, we have
$$Dx(t) = \cos t - t \sin t, \, Dy(t) = -\sin t - t \cos t, \, Dz(t) = \sqrt{3}$$
The sum of the squares of these derivatives is
$$(\cos t - t \sin t)^2 + (-\sin t - t \cos t)^2 + (\sqrt{3})^2$$
$$= \cos^2 t - 2t \cos t \sin t + t^2 \sin^2 t + \sin^2 t$$
$$\qquad + 2t \sin t \cos t + t^2 \cos^2 t + 3$$
$$= 4 + t^2$$
Thus the arc-length integral is
$$\int_0^1 \sqrt{4 + t^2} \, dt = \tfrac{1}{2} t \sqrt{4 + t^2} + 2 \ln \tfrac{1}{2}(t + \sqrt{4 + t^2}) \Big]_0^1$$
$$= \tfrac{1}{2}\sqrt{5} + 2 \ln(\tfrac{1}{2} + \tfrac{1}{2}\sqrt{5}) \quad \blacksquare$$

13.10 ARC LENGTH

We introduced arc length here in order to be able to use the distance along a curve as a parameter for describing the curve. Let (x_0, y_0, z_0) be a fixed point on a curve C. Choose a positive direction along the curve. Now if (x, y, z) is any other point on C, then surely its position vector $\mathbf{p} = x\mathbf{i} + y\mathbf{j} + z\mathbf{k}$ is a vector function of the arc length s from (x_0, y_0, z_0) to (x, y, z). This gives us a parametrization of C, which is very useful. In particular the derivative

$$D\mathbf{p}(s) = Dx(s)\mathbf{i} + Dy(s)\mathbf{j} + Dz(s)\mathbf{k}$$

has an important geometric significance.

Let us return to the "secant" vector

$$\mathbf{p}(s + \Delta s) - \mathbf{p}(s)$$

The length of this vector, when the increment Δs in arc length is small, is very nearly equal to Δs. This is true because the length of the vector is precisely the chord length on the secant line. (See Fig. 13.26.) It follows that the vector

$$\frac{\mathbf{p}(s + \Delta s) - \mathbf{p}(s)}{\Delta s}$$

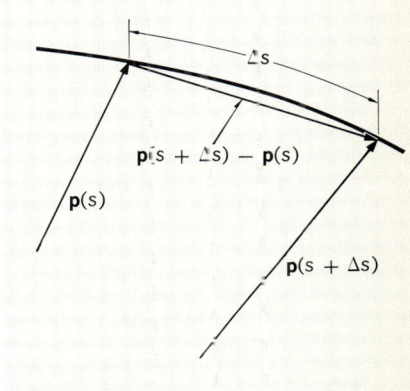

FIGURE 13.26

has a norm approximately equal to one. Now as Δs approaches zero this vector approaches a *unit tangent vector*

$$\frac{d\mathbf{p}}{ds} = D\mathbf{p}(s) = \mathbf{T}$$

■ **Example 2.** Let

$$\mathbf{p}(t) = (-1 + 2t + t^2)\mathbf{i} + (2 - t + t^2)\mathbf{j} + (-1 + 2t)\mathbf{k}$$

Then the unit tangent vector is

$$\mathbf{T} = \frac{d\mathbf{p}}{ds} = \frac{d\mathbf{p}}{dt} \cdot \frac{dt}{ds} = [(2 + 2t)\mathbf{i} + (-1 + 2t)\mathbf{j} + 2\mathbf{k}]\frac{dt}{ds}$$

To determine the factor dt/ds, we use the fact that $\mathbf{T}$ is a unit vector so that $\mathbf{T} \cdot \mathbf{T} = 1$. This equation becomes

$$[(2 + 2t)^2 + (-1 + 2t)^2 + 4]\left(\frac{dt}{ds}\right)^2 = 1$$

whence we have

$$\frac{dt}{ds} = \pm \frac{1}{\sqrt{(2 + 2t)^2 + (-1 + 2t)^2 + 4}} = \pm \frac{1}{\sqrt{9 + 4t + 8t^2}}$$

We choose dt/ds to be positive which is equivalent to saying that t is an increasing function of s and hence that s is an increasing function of t. Thus we have

$$\mathbf{T} = \frac{d\mathbf{p}}{ds} = \frac{(2 + 2t)\mathbf{i} + (-1 - 2t)\mathbf{j} + 2\mathbf{k}}{\sqrt{9 + 4t + 8t^2}}$$

Note that
$$T = \frac{d\mathbf{p}}{ds} = \frac{D\mathbf{p}(t)}{||D\mathbf{p}(t)||}$$

Because **T** is a unit vector, Corollary 13.11 applies to tell us that
$$\frac{d\mathbf{T}}{ds} \cdot \mathbf{T} = 0$$

or $d\mathbf{T}/ds$ is orthogonal to the unit tangent vector **T**. Of course,
$$\frac{d\mathbf{T}}{ds} = \frac{d^2x}{ds^2}\mathbf{i} + \frac{d^2y}{ds^2}\mathbf{j} + \frac{d^2z}{ds^2}\mathbf{k}$$

is not necessarily a unit vector. Assuming that $||d\mathbf{T}/ds||$ is never zero, we define the *normal vector* to the curve
$$N = \frac{d\mathbf{T}}{ds} \bigg/ \left|\left|\frac{d\mathbf{T}}{ds}\right|\right|$$

Finally, the unit vector
$$\mathbf{B} = \mathbf{T} \times \mathbf{N}$$

is called the *binormal vector*. The three unit vectors **T**, **N**, and **B** constitute the vectors associated with the curve C. This triple of vectors is much used in the subject of *differential geometry* in the study of space curves (as discussed briefly in the next section).

■ *Example 3.* Given
$$\mathbf{T} = \frac{(2 + 2t)\mathbf{i} + (-1 + 2t)\mathbf{j} + 2\mathbf{k}}{\sqrt{a + 4t + 8t^2}}$$

from Example 2, we compute
$$\frac{d\mathbf{T}}{ds} = \frac{d\mathbf{T}}{dt} \times \frac{dt}{ds}$$
$$= \frac{d}{dt}\left[\frac{(2 + 2t)}{\sqrt{9 + 4t + 8t^2}}\mathbf{i} + \frac{(-1 + 2t)}{\sqrt{9 + 4t + 8t^2}}\mathbf{j}\right.$$
$$\left. + \frac{2}{\sqrt{9 + 4t + 8t^2}}\mathbf{k}\right] \cdot \frac{1}{\sqrt{9 + 4t + 8t^2}}$$
$$= \left[\frac{(14 - 12t)}{(9 + 4t + 8t^2)^{3/2}}\mathbf{i} + \frac{(20 + 12t)}{(9 + 4t + 8t^2)^{3/2}}\mathbf{j}\right.$$
$$\left. + \frac{(-4 - 16t)}{(9 + 4t + 8t^2)^{3/2}}\mathbf{k}\right] \cdot \frac{1}{\sqrt{9 + 4t + 8t^2}}$$
$$= \frac{1}{(9 + 4t + 8t^2)^2}[(14 - 12t)\mathbf{i} + (20 + 12t)\mathbf{j} + (-4 - 16t)\mathbf{k}]$$

13.10 ARC LENGTH

The norm of $d\mathbf{T}/ds$ is precisely

$$\frac{\sqrt{(14-12t)^2 + (20+12t)^2 + (-4-16t)^2}}{(9-4t+8t^2)^2}$$

$$= \frac{2\sqrt{17}\sqrt{9+4t+8t^2}}{(9+4t+8t^2)^2} = \frac{2\sqrt{17}}{(9+4t+8t^2)^{3/2}}$$

It follows that

$$\mathbf{N} = \frac{d\mathbf{T}}{ds} \Big/ \left\|\frac{d\mathbf{T}}{ds}\right\|$$

$$= \frac{1}{(9+4t+8t^2)^2}[(14-12t)\mathbf{i} + (20+12t)\mathbf{j}$$

$$+ (-4-16t)\mathbf{k}] \times \frac{(9+4t+8t^2)^{3/2}}{2\sqrt{17}}$$

$$= \frac{1}{(9+4t+8t^2)^{1/2}}\left[\frac{7-6t}{\sqrt{17}}\mathbf{i} + \frac{10+6t}{\sqrt{17}}\mathbf{j} - \frac{2+8t}{\sqrt{17}}\mathbf{k}\right]$$

We leave it to the reader to compute the binormal vector.

■ **Example 4.** Consider the helix described by

$$\mathbf{p}(t) = (2\cos t)\mathbf{i} + (2\sin t)\mathbf{j} + t\mathbf{k}$$

We have

$$\frac{d\mathbf{p}}{dt} = (-2\sin t)\mathbf{i} + (2\cos t)\mathbf{j} + \mathbf{k}$$

whence

$$\mathbf{T} = \frac{d\mathbf{p}}{dt} \Big/ \left\|\frac{d\mathbf{p}}{dt}\right\| = \frac{-2\sin t}{\sqrt{5}}\mathbf{i} + \frac{2\cos t}{\sqrt{5}}\mathbf{j} + \frac{1}{\sqrt{5}}\mathbf{k}$$

Then

$$\frac{d\mathbf{T}}{ds} = \frac{d\mathbf{T}}{dt} \times \frac{dt}{ds} = \frac{d\mathbf{T}}{dt} \times \frac{1}{\sqrt{5}} = \frac{-\cos t}{\sqrt{5}}\mathbf{i} + \frac{-2\sin t}{\sqrt{5}}\mathbf{j}$$

We see that

$$\left\|\frac{d\mathbf{T}}{ds}\right\| = \sqrt{\frac{4\cos^2 t}{5} + \frac{4\sin^2 t}{5}} = \frac{2}{\sqrt{5}}$$

Therefore, the unit normal vector $\mathbf{N}$ is given by

$$\mathbf{N} = \frac{d\mathbf{T}}{ds} \Big/ \left\|\frac{d\mathbf{T}}{ds}\right\| = \left(\frac{-2\cos t}{\sqrt{5}}\mathbf{i} + \frac{-2\sin t}{\sqrt{5}}\mathbf{j}\right) \Big/ \frac{2}{\sqrt{5}}$$

$$= -\cos t\,\mathbf{i} - \sin t\,\mathbf{j}$$

This vector is directed oppositely to the projection of $\mathbf{p}(t)$ onto the xy plane. An example is pictured in Fig. 13.27. The calculation of the binormal vector is straightforward. It is

$$\mathbf{B} = \mathbf{T} \times \mathbf{N} = \left(\frac{-2\sin t\,\mathbf{i}}{\sqrt{5}} + \frac{2\cos t\,\mathbf{j}}{\sqrt{5}} + \frac{\mathbf{k}}{\sqrt{5}}\right) \times (-\cos t\,\mathbf{i} - \sin t\,\mathbf{j})$$

$$= \frac{\sin t}{\sqrt{5}}\mathbf{i} - \frac{\cos t}{\sqrt{5}}\mathbf{j} + \frac{2}{\sqrt{5}}\mathbf{k}$$

EXERCISES 13.10

In each of Exercises 1 to 10, a vector function of one variable and an interval are given. Determine the arc length of the corresponding curve. Then find the three unit vectors $\mathbf{T}$, $\mathbf{N}$, and $\mathbf{B}$.

1. $\mathbf{p}(t) = (1 + 2t)\mathbf{i} + (-3 + 4t)\mathbf{j} + (4 - 5t)\mathbf{k}$; $\{t : 0 \leq t \leq 1\}$
2. $\mathbf{p}(t) = (2 - 3t)\mathbf{i} + (2 + 4t)\mathbf{j} + \frac{5}{2}t^2\mathbf{k}$; $\{t : 0 \leq t \leq \frac{3}{4}\}$
3. $\mathbf{p}(t) = t\mathbf{i} + \frac{1}{\sqrt{2}}t^2\mathbf{j} + \frac{1}{3}t^3\mathbf{k}$; $\{t : 0 \leq t \leq 1\}$
4. $\mathbf{p}(t) = t\mathbf{i} + \frac{3}{2}t^2\mathbf{j} + \frac{3}{2}t^3\mathbf{k}$; $\{t : 0 \leq t \leq 3\}$
5. $\mathbf{p}(t) = (\sin 2t)\mathbf{i} + (\cos 2t)\mathbf{j} + 2t^{3/2}\mathbf{k}$; $\{t : 0 \leq t \leq 1\}$
6. $\mathbf{p}(t) = a\mathbf{i}\cos t + a\mathbf{j}\sin t + t\mathbf{k}$; $\{t : 0 \leq t \leq 2\pi\}$
7. $\mathbf{p}(t) = e^t(\mathbf{i}\cos t + \mathbf{j}\sin t + \mathbf{k})$; $\{t : 0 \leq t \leq \pi\}$
8. $\mathbf{p}(t) = t(\mathbf{i}\cos t + \mathbf{j}\sin t + \mathbf{k})$; $\{t : 0 \leq t \leq \frac{\pi}{2}\}$
9. $\mathbf{p}(t) = t^2\mathbf{i} + (t + \frac{1}{3}t^3)\mathbf{j} + (t - \frac{1}{3}t^3)\mathbf{k}$; $\{t : 0 \leq t \leq 1\}$
10. $\mathbf{p}(t) = t\mathbf{i} + \mathbf{j}t\sec t + \mathbf{k}t(\sec t + \tan t)$; $\{t : 0 \leq t \leq \frac{\pi}{4}\}$

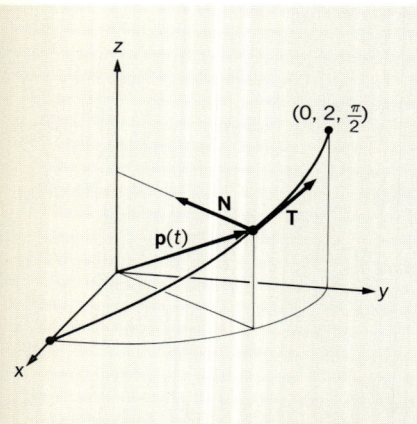

FIGURE 13.27

PROBLEMS 13.10

1. Use lower and upper sequences as in Chapter 6 to prove that the arc-length integral exists over any interval on which the parametric curve is continually differentiable.

2. Prove that the curve described by the vector function $\mathbf{p}(t)$ is planar if and only if
$$D\mathbf{p}(t) \cdot (D^2\mathbf{p}(t) \times D^3\mathbf{p}(t)) = 0$$
for each t.

3. If all of the tangent lines to a given curve pass through a fixed point, show that the curve is a straight line.

4. Find the curve on the cylinder $\{(x, y) : y^2 + z^2 = a^2\}$ that passes through the points $(0, a, 0)$ and $(b, a\cos t, a\sin t)$ and has shortest arc length.

In vector form Newton's third law is $\mathbf{F} = m\mathbf{a}$, where $\mathbf{F}$ is the vector force applied to the body of mass m and $\mathbf{a}$ is the resulting acceleration vector. Let s be the arc length along the path of motion.

**PROBLEMS 13.10
On plane motion**

1. Prove that the velocity vector $\mathbf{v}(t)$ has the form
$$\mathbf{v}(t) = \frac{ds}{dt}\mathbf{T}(t) = v(t)\mathbf{T}(t)$$
and the acceleration vector $\mathbf{a}(t)$ can be written as
$$\mathbf{a}(t) = \frac{d^2s}{dt^2}\mathbf{T}(t) + \left(\frac{ds}{dt}\right)^2 \mathbf{N}(t)$$

2. A man swings a pail in a vertical circle. Assuming that his arm, the handle of the pail, and the pail itself yield a circle of radius $3\frac{1}{2}$ ft, what is the minimum number of revolutions per minute that will keep water in the pail from spilling out?

3. A man swings a pail in a horizontal circle. If the circle is 4 ft in radius, and he moves the pail 45 rpm, will the pail remain full?

4. A two-ton automobile negotiates a curve at a constant 30 mph. If the radius of the curve is 88 ft, what cornering force are the tires producing? Prove that the same curve at 60 mph requires four times the cornering force.

5. To remain in a circular orbit, the normal acceleration acting on a satellite must exactly counteract the gravitational force. If an orbit 220 miles high is planned, what speed is necessary? (Assume that $g = 32$ still holds up there.) What is the orbital period?

6. Starting at the point (0, 10), a bead slides down a straight wire. At what slope should the wire be set in order that the bead reach the vertical line $x = 10$ in the shortest length of time and what is this minimum time? (Assume no friction.)

Let the curve C be described by the twice differentiable vector function $\mathbf{p}(t)$. The unit tangent vector

13.11 CURVATURE

$$\mathbf{T} = \frac{d\mathbf{p}}{ds}$$

changes direction slowly if C is nearly straight and $\mathbf{T}$ changes direction rapidly if C is sharply curved. In Fig. 13.28 we show why the average vector rate of change

$$\frac{\mathbf{T}(s + \Delta s) - \mathbf{T}(s)}{\Delta s}$$

is a measure of the sharpness of the curve.

This discussion leads us to define the *curvature* of C at the point $\mathbf{p}(t)$ to be the norm of $d\mathbf{T}/ds$. Thus we define

$$\kappa(t) = \left\|\frac{d\mathbf{T}}{ds}\right\|$$

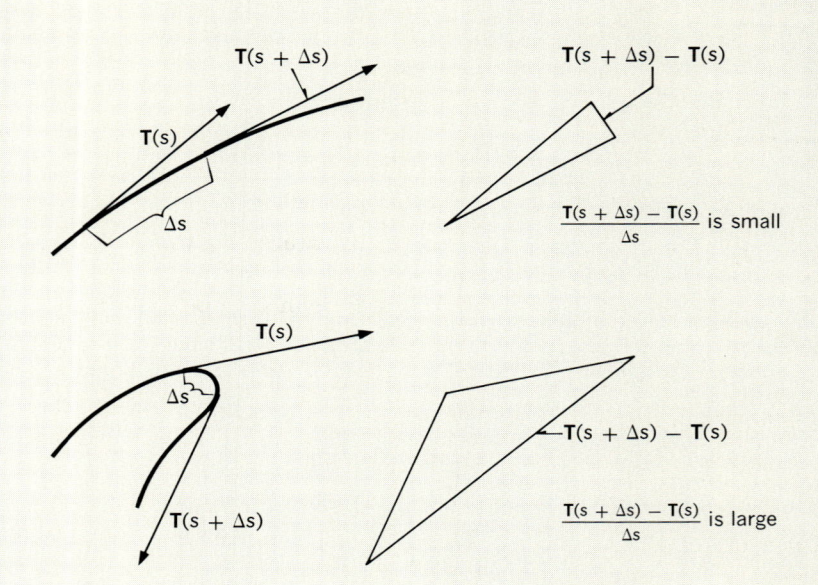

FIGURE 13.28

If $\kappa(t) \neq 0$, then we define its reciprocal

$$\rho(t) = \frac{1}{\kappa(t)}$$

to be the radius of curvature of C at $\mathbf{p}(t)$.

■ **Example 1.** Consider the plane circle described by

$$\mathbf{p}(t) = a \cos \omega t \mathbf{i} + a \sin \omega t \mathbf{j}; \quad a > 0, \omega > 0$$

We easily compute

$$\mathbf{T} = \frac{d\mathbf{p}}{ds} = \frac{d\mathbf{p}}{dt} \times \frac{dt}{ds} = \frac{-a\omega \sin \omega t \mathbf{i} + a\omega \cos \omega t \mathbf{j}}{\sqrt{a^2\omega^2 \sin^2 \omega t + a^2\omega^2 \cos^2 \omega t}}$$
$$= -\sin \omega t \mathbf{i} + \cos \omega t \mathbf{j}$$

Note that

$$\frac{dt}{ds} = \frac{1}{a\omega}$$

Next, we have

$$\frac{d\mathbf{T}}{ds} = \frac{d\mathbf{T}}{dt} \times \frac{dt}{ds} = (-\omega \cos \omega t \mathbf{i} - \omega \sin t \mathbf{j})\left(\frac{1}{a\omega}\right)$$
$$= -\frac{1}{a}(\cos \omega t \mathbf{i} + \sin \omega t \mathbf{j})$$

13.11 CURVATURE

Thus the curvature of the circle is

$$\kappa(t) = \left|\frac{-1}{a}\right| = \frac{1}{a}$$

and its radius of curvature is

$$\rho(t) = \frac{1}{\kappa(t)} = a$$

This agreement between the radius of curvature and the radius of the circle tends to justify the definitions.

Recall that the unit normal vector is defined to be

$$\mathbf{N} = \frac{d\mathbf{T}}{ds} \bigg/ \left\|\frac{d\mathbf{T}}{ds}\right\|$$

It follows that

$$\mathbf{N} = \frac{1}{\kappa(t)}\frac{d\mathbf{T}}{ds} = \rho(t)\frac{d\mathbf{T}}{ds}$$

The vector

$$\mathbf{r}(t) = \rho(t)\mathbf{N} = \rho^2(t)\frac{d\mathbf{T}}{ds}$$

is called the *vector radius of curvature* of the curve C at $\mathbf{p}(t)$. The point whose position vector is

$$\mathbf{q}(t) = \mathbf{p}(t) + \mathbf{r}(t)$$

is called the *center of curvature* of C at $\mathbf{p}(t)$. The plane through the point $\mathbf{p}(t)$ and having the binormal vector $\mathbf{B}$ as its normal vector contains the point $\mathbf{q}(t)$. This is the plane of the vectors $\mathbf{T}$ and $\mathbf{N}$ and is called the *osculating plane*. The circle in this plane whose center is at $\mathbf{q}(t)$ and whose radius is $\rho(t)$ is called the *osculating circle*.

■ **Example 2.** For the plane curve described by

$$\mathbf{p}(t) = (1 - t^2)\mathbf{i} + (2t + t^2)\mathbf{j}$$

we have

$$\mathbf{T} = \frac{d\mathbf{P}}{dt} \cdot \frac{dt}{ds} = \frac{d\mathbf{P}}{dt} \bigg/ \left\|\frac{d\mathbf{P}}{dt}\right\|$$

$$= [-2t\mathbf{i} + (2 + 2t)\mathbf{j}]/2\sqrt{1 + 2t + 2t^2}$$

$$= \frac{-t}{\sqrt{1 + 2t + 2t^2}}\mathbf{i} + \frac{1 + t}{\sqrt{1 + 2t + 2t^2}}\mathbf{j}$$

Noting that

$$\frac{dt}{ds} = \frac{1}{2\sqrt{1 + 2t + 2t^2}}$$

we obtain

$$\frac{d\mathbf{T}}{ds} = \frac{d\mathbf{T}}{dt} \cdot \frac{dt}{ds} = \frac{1}{2\sqrt{1+2t+2t^2}} \frac{d\mathbf{T}}{dt}$$

$$= \frac{-(1+t)}{2(1+2t+2t^2)^2}\mathbf{i} + \frac{-t}{2(1+2t+2t^2)^2}\mathbf{j}$$

It follows that the curvature is

$$\kappa(t) = \left\|\frac{d\mathbf{T}}{ds}\right\| = \frac{1}{2(1+2t+2t^2)^{3/2}}$$

and the radius of curvature is

$$\rho(t) = 2(1+2t+2t^2)^{3/2}$$

The vector radius of curvature is

$$\mathbf{r}(t) = \rho^2(t)\frac{d\mathbf{T}}{ds}$$

$$= 4(1+2t+2t^2)^3\left[\frac{-(1+t)\mathbf{i} - t\mathbf{j}}{2(1+2t+2t^2)^2}\right]$$

$$= -2(1+2t+2t^2)[(1+t)\mathbf{i} + t\mathbf{j}]$$

In particular, when $t = 0$, we have

$$\mathbf{p}(0) = \mathbf{i} \quad \text{and} \quad \mathbf{r}(0) = -2\mathbf{i}$$

The center of curvature is the tip of the vector

$$\mathbf{q}(0) = \mathbf{p}(0) + \mathbf{r}(0) = -\mathbf{i}$$

Similarly, when $t = -\tfrac{1}{2}$, we have

$$\mathbf{p}(-\tfrac{1}{2}) = \tfrac{3}{4}(\mathbf{i} - \mathbf{j}) \quad \text{and} \quad \mathbf{r}(-\tfrac{1}{2}) = \tfrac{1}{2}(-\mathbf{i} + \mathbf{j})$$

whence the center of curvature is the tip of

$$\mathbf{q}(-\tfrac{1}{2}) = \tfrac{3}{4}(\mathbf{i} - \mathbf{j}) + \tfrac{1}{2}(-\mathbf{i} + \mathbf{j}) = \tfrac{1}{4}(\mathbf{i} - \mathbf{j})$$

The corresponding osculating circles are shown in Fig. 13.29.

■ *Example 3.* Consider the "twisted cubic" curve described by

$$\mathbf{p}(t) = 6t\mathbf{i} + 3t^2\mathbf{j} + t^3\mathbf{k}$$

We have

$$\mathbf{T} = \frac{d\mathbf{p}}{dt}\bigg/\left\|\frac{d\mathbf{p}}{dt}\right\| = \frac{6\mathbf{i} + 6t\mathbf{j} + 3t^2\mathbf{k}}{\sqrt{36 + 36t^2 + 9t^4}} = \frac{2\mathbf{i} + 2t\mathbf{j} + t^2\mathbf{k}}{2+t^2}$$

Then we obtain

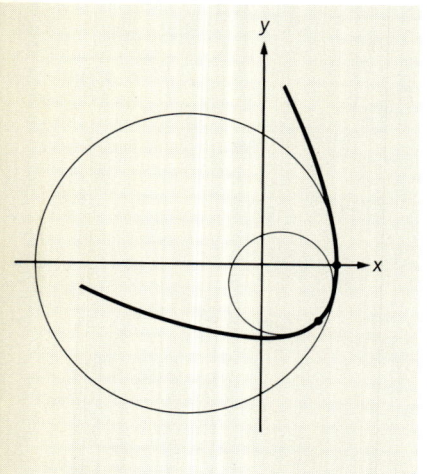

FIGURE 13.29

13.11 CURVATURE

$$\frac{d\mathbf{T}}{ds} = \frac{d\mathbf{T}}{dt} \cdot \frac{dt}{ds} = \frac{1}{3(2+t^2)} \frac{d}{dt} \frac{2\mathbf{i} + 2t\mathbf{j} + t^2\mathbf{k}}{2+t^2}$$

$$= \frac{-4t\mathbf{i} + (4 - 2t)\mathbf{j} + 4t\mathbf{k}}{3(2+t^2)^3}$$

and hence

$$\kappa(t) = \left\|\frac{d\mathbf{T}}{ds}\right\| \frac{2\sqrt{4 - 4t + 9t^2}}{3(2+t^2)}$$

The radius of curvature is

$$\rho(t) = \frac{3(2+t^2)}{2\sqrt{4 - 4t + 9t^2}}$$

and the vector radius of curvature is

$$\mathbf{r}(t) = \rho^2(t)\frac{d\mathbf{T}}{ds} = \frac{3(2+t^2)^3}{2(4 - 4t + 9t^2)}[-2t\mathbf{i} + (2-t)\mathbf{j} + 2t\mathbf{k}]$$

In particular, when $t = 1$, we have

$$\mathbf{p}(1) = 6\mathbf{i} + 3\mathbf{j} + \mathbf{k}$$

while

$$\mathbf{r}(1) = \frac{3(2+1)^3}{2(4-4+9)}[-2\mathbf{i} + \mathbf{j} + 2\mathbf{k}] = -9\mathbf{i} + \frac{9}{2}\mathbf{j} + 9\mathbf{k}$$

The center of curvature is, therefore, the point $(-3, \frac{15}{2}, 10)$. ∎

We have the formula

$$\frac{d\mathbf{T}}{ds} = \kappa(t)\mathbf{N}$$

and shall now compute the derivatives $d\mathbf{N}/ds$ and $d\mathbf{B}/ds$. Knowing that $\mathbf{B}$ is a unit vector, we apply Corollary 13.11 to conclude that

$$\frac{d\mathbf{B}}{ds} \cdot \mathbf{B} = 0$$

Because $\mathbf{T} \cdot \mathbf{B} = 0$, we differentiate to obtain

$$\frac{d}{ds}(\mathbf{T} \cdot \mathbf{B}) = \frac{d\mathbf{T}}{ds} \cdot \mathbf{B} + \mathbf{T} \cdot \frac{d\mathbf{B}}{ds}$$

$$= \kappa(t)\mathbf{N} \cdot \mathbf{B} + \mathbf{T} \cdot \frac{d\mathbf{B}}{ds} = 0$$

Now the fact that $\mathbf{N} \cdot \mathbf{B} = 0$ tells us that

$$\mathbf{T} \cdot \frac{d\mathbf{B}}{ds} = 0$$

This proves that the vector $d\mathbf{B}/ds$ is orthogonal to both $\mathbf{B}$ and $\mathbf{T}$. Therefore, $d\mathbf{B}/ds$ is parallel to $\mathbf{N}$ and there is a scalar $\tau(t)$ such that

$$\frac{d\mathbf{B}}{ds} = -\tau(t)\mathbf{N}$$

This function τ is called the *torsion* of the curve. Lastly, we compute that

$$\frac{d\mathbf{N}}{ds} = \frac{d}{ds}(\mathbf{B} \times \mathbf{T}) = \frac{d\mathbf{B}}{ds} \times \mathbf{T} + \mathbf{B} \times \frac{d\mathbf{T}}{ds}$$
$$= -\tau(t)\mathbf{N} \times \mathbf{T} + \mathbf{B} \times \kappa(t)\mathbf{N}$$
$$= \tau(t)\mathbf{B} - \kappa(t)\mathbf{T}$$

The three equations

$$\frac{d\mathbf{T}}{ds} = \kappa(t)\mathbf{N}, \quad \frac{d\mathbf{N}}{ds} = \tau(t)\mathbf{B} - \kappa(t)\mathbf{T}, \quad \frac{d\mathbf{B}}{ds} = -\tau(t)\mathbf{N}$$

are called the *Frenet formulas,* which are frequently used in the differential geometry of space curves.

EXERCISES 13.11 In each of Exercises 1 to 10, a vector function describing a planar curve is given. Determine the curvature, the vector radius of curvature and the center of curvature in each case.

1. $\mathbf{p}(t) = 4t\mathbf{i} + (2t^2 - 1)\mathbf{j}$
2. $\mathbf{p}(t) = (2t - t^2)\mathbf{i} + (t - t^3)\mathbf{j}$
3. $\mathbf{p}(t) = \left(\frac{1}{1+t}\right)\mathbf{i} + \left(\frac{1}{1-t}\right)\mathbf{j}$
4. $\mathbf{p}(t) = (a \cos t)\mathbf{i} + (b \sin t)\mathbf{j}$
5. $\mathbf{p}(t) = e^{-t}\mathbf{i} + e^{t}\mathbf{j}$
6. $\mathbf{p}(t) = (\sin t)\mathbf{i} + (\cos^2 t)\mathbf{j}$
7. $\mathbf{p}(t) = e^{-t}(\mathbf{i} \cos t + \mathbf{j} \sin t)$
8. $\mathbf{p}(t) = (t - \sin t)\mathbf{i} + (1 - \cos t)\mathbf{j}$
9. $\mathbf{p}(t) = a(\cos t + t \sin t)\mathbf{i} + a(\sin t - t \cos t)\mathbf{j}$
10. $\mathbf{p}(t) = a(\mathbf{i} \cos^3 t + \mathbf{j} \sin^3 t)$

In each of Exercises 11 to 20, a vector function describing a space curve is given. Determine the curvature, the vector radius of curvature, the center of curvature, the osculating plane, and the torsion in each instance.

11. $\mathbf{p}(t) = (1 + 2t)\mathbf{i} + (2 - t)\mathbf{j} + (-1 + 3t)\mathbf{k}$
12. $\mathbf{p}(t) = t\mathbf{i} + t^2\mathbf{j} + (1 + t^2)\mathbf{k}$
13. $\mathbf{p}(t) = t\mathbf{i} + 2t\mathbf{j} + \frac{2}{3}t^{3/2}\mathbf{k}$
14. $\mathbf{p}(t) = t\mathbf{i} + t^2\mathbf{j} + t^3\mathbf{k}$
15. $\mathbf{p}(t) = \mathbf{i} \cos t + \mathbf{j} \sin t + t\mathbf{k}$

13.11 CURVATURE

16. $\mathbf{p}(t) = \mathbf{i} \sin 2t + \mathbf{j} \cos 2t + 2t^{3/2}\mathbf{k}$
17. $\mathbf{p}(t) = t(\mathbf{i} \cos t + \mathbf{j} \sin t + \mathbf{k})$
18. $\mathbf{p}(t) = e^t(\mathbf{i} \cos t + \mathbf{j} \sin t + \mathbf{k})$
19. $\mathbf{p}(t) = \mathbf{i} \sin t + \mathbf{j} \cos t + \mathbf{k} \ln \cos t$
20. $\mathbf{p}(t) = t\mathbf{i} + \mathbf{j} \ln \sec t + \mathbf{k} \ln(\sec t + \tan t)$

PROBLEMS 13.11

1. For the planar graphs $\{(x,y) : y = f(x)\}$ and $\{(r, \theta) : r = f(\theta)\}$ show that the curvatures are given by

$$k = \frac{|f''(x)|}{[1 + (f'(x))^2]^{3/2}} \quad \text{and} \quad k = \frac{\left|r^2 + 2\left(\frac{dr}{d\theta}\right)^2 - r\frac{d^2r}{d\theta^2}\right|}{\left[r^2 + \left(\frac{dr}{d\theta}\right)^2\right]^{3/2}}$$

2. Apply the formulas in Problem 1 above to the graphs of the following equations:
   (a) $y = 2x^2 - x^4$
   (b) $y = x^{3/2}$
   (c) $y = e^{-x}$
   (d) $y = \sin x$
   (e) $y = \ln \sec x$
   (f) $y = \tan^{-1} x$
   (g) $r = 4\theta$
   (h) $r = e^{\theta}$
   (i) $r = 2 \cos \theta$
   (j) $r = \dfrac{1}{2 \sin \theta - 3 \cos \theta}$

3. Prove that curvature is independent of the parametrization of the curve by replacing the parameter t with a new parameter u via a function $t = f(u)$ where $Df(u) > 0$.

4. Prove that a curve is planar if and only if its torsion is identically zero.

5. The quantity $\|d\mathbf{N}/ds\|$ is called the *screw curvature* of the curve. Express it in terms of the curvature and the torsion.

6. Prove that the vector function

$$\mathbf{p}(t) = (t - \tfrac{1}{3}t^3)\mathbf{i} + t^2\mathbf{j} + (t + \tfrac{1}{3}t^3)\mathbf{k}$$

describes a curve with the property that its curvature and torsion are equal at each point.

7. Determine the point of maximum curvature of a conic.

8. Consider the circle

$$\{(x,y) : x^2 + y^2 - 6x - 2y + 10 = 2\sqrt{2}(-x + y + 2)\}$$

Construct a curve that passes through the point $(3, 1)$ on this circle and has the same curvature as the circle at $(3, 1)$ and, also, passes through the origin with the same curvature at $(0, 0)$ as the x axis.

9. Let k be a fixed constant. Given the vector function $\mathbf{p}(t)$ with the unit normal vector $\mathbf{N}(t)$, the vector function

$$\mathbf{q}(t) = \mathbf{p}(t) + k\mathbf{N}(t)$$

describes a *parallel curve*. Determine the curvature of a parallel curve.

10. The *evolute* of a given planar curve is defined to be the set of all of its centers of curvature. Find parametric equations of the evolute of the curve called a cycloid

described by
$$\mathbf{p}(t) = a(t - \sin t)\mathbf{i} + a(1 - \cos t)\mathbf{j}$$

11. The tractrix is the curve in the plane with the property that the axes cut off a segment of length 1 from each tangent line. Find parametric equations for this curve. Prove that its curvature is proportional to the length of that segment of the normal line from the curve to the x axis.

12. For a planar curve, let α denote the angle between the unit basis vector i and the unit tangent vector T. Prove that the curvature can be given as
$$k = \left|\frac{d\alpha}{ds}\right|$$

13.12 VELOCITY AND ACCELERATION IN POLAR COORDINATES

We write the position vector $\mathbf{p} = x\mathbf{i} + y\mathbf{j}$ in the polar coordinate form
$$\mathbf{p} = (r \cos \theta)\mathbf{i} + (r \sin \theta)\mathbf{j} = r\mathbf{u}_\theta$$
where $\mathbf{u}_\theta$ is the unit vector
$$\mathbf{u}_\theta = \mathbf{i} \cos \theta + \mathbf{j} \sin \theta$$
in the direction θ. Rotating $\mathbf{u}_\theta$ through the positive angle $\pi/2$ radians, we obtain the orthogonal unit vector
$$\mathbf{u}_\theta^\perp = -\mathbf{i} \sin \theta + \mathbf{j} \cos \theta$$
From these defining equations, we see that
$$\frac{d\mathbf{u}_\theta}{d\theta} = -\mathbf{i} \sin \theta + \mathbf{j} \cos \theta = \mathbf{u}_\theta^\perp$$
$$\frac{d\mathbf{u}_\theta^\perp}{d\theta} = -\mathbf{i} \cos \theta - \mathbf{j} \sin \theta = -\mathbf{u}_\theta$$

Now consider a particle moving along a curve C described by the two-dimensional vector function $\mathbf{p}(t)$. Its velocity vector
$$\mathbf{v} = \frac{d\mathbf{p}}{dt}$$
can be expressed as a linear combination of $\mathbf{u}_\theta$ and $\mathbf{u}_\theta^\perp$. If we do write
$$\mathbf{v} = v_r \cdot \mathbf{u}_\theta + v_\theta \cdot \mathbf{u}_\theta^\perp$$
then the scalar functions v_r and v_θ are called the *radial* and *transverse components*, respectively, of the velocity. We now determine v_r and v_θ.

We write $\mathbf{p} = r \cdot \mathbf{u}_\theta$, whence the velocity vector is
$$\mathbf{v} = \frac{d}{dt}(r\mathbf{u}_\theta) = \frac{dr}{dt} \cdot \mathbf{u}_\theta + r\frac{d\mathbf{u}_\theta}{dt}$$

Now we have

13.12 VELOCITY AND ACCELERATION IN POLAR COORDINATES

$$\frac{d\mathbf{u}_\theta}{dt} = \frac{d\mathbf{u}_\theta}{d\theta} \cdot \frac{d\theta}{dt} = \frac{d\theta}{dt} \cdot \mathbf{u}_\theta^\perp$$

$$\mathbf{v} = \frac{dr}{dt}\mathbf{u}_\theta + \left(r\frac{d\theta}{dt}\right)\mathbf{u}_\theta^\perp$$

This identifies v_r and v_θ as

$$v_r = \frac{dr}{dt} \quad \text{and} \quad v_\theta = r\frac{d\theta}{dt}$$

We obtain the acceleration vector in the same way

$$\mathbf{a} = \frac{d\mathbf{v}}{dt} = \frac{d}{dt}\left[\frac{dr}{dt}\mathbf{u}_\theta + \left(r\frac{d\theta}{dt}\right)\mathbf{u}_\theta^\perp\right]$$

$$= \frac{d^2r}{dt^2}\mathbf{u}_\theta + \frac{dr}{dt}\frac{d\mathbf{u}_\theta}{dt} + \frac{d}{dt}\left(r\frac{d\theta}{dt}\right)\mathbf{u}_\theta^\perp + \left(r\frac{d\theta}{dt}\right)\frac{d\mathbf{u}_\theta^\perp}{dt}$$

$$= \frac{d^2r}{dt^2}\mathbf{u}_\theta + \frac{dr}{dt}\left(\frac{d\mathbf{u}_\theta}{d\theta} \cdot \frac{d\theta}{dt}\right) + \left(\frac{dr}{dt}\frac{d\theta}{dt} + r\frac{d^2\theta}{dt^2}\right)\mathbf{u}_\theta^\perp$$

$$+ \left(r\frac{d\theta}{dt}\right)\left(\frac{d\mathbf{u}_\theta^\perp}{d\theta} \cdot \frac{d\theta}{dt}\right)$$

$$= \left[\frac{d^2r}{dt^2} - r\left(\frac{d\theta}{dt}\right)^2\right]\mathbf{u}_\theta - \left[2\frac{dr}{dt}\frac{d\theta}{dt} + r\frac{d^2\theta}{dt^2}\right]\mathbf{u}_\theta^\perp$$

Thus the *radial component of acceleration* is

$$a_r = \frac{d^2r}{dt^2} - r\left(\frac{d\theta}{dt}\right)^2$$

and the *transverse component of acceleration* is

$$a_\theta = 2\frac{dr}{dt}\frac{d\theta}{dt} + r\frac{d^2\theta}{dt^2} = \frac{1}{r}\frac{d}{dt}\left(r^2\frac{d\theta}{dt}\right)$$

An important case occurs when the moving particle is acted upon by a *central force* directed along its position vector. In such a case, there is no transverse component of acceleration; hence we have

$$a_\theta = \frac{1}{r}\frac{d\theta}{dt}\left(r^2\frac{d}{dt}\right) = 0 \quad \text{or} \quad \frac{d}{dt}\left(r^2\frac{d\theta}{dt}\right) = 0$$

It follows that

$$r^2\frac{d\theta}{dt} = c, \text{ a constant}$$

We use this equation in the theory of planetary motion below.

Kepler's first law of planetary motion. Kepler's first law states "A planet moves in an elliptical orbit with the sun at one focus of the ellipse." We reproduce Newton's derivation of this law, which Kepler deduced empirically. It might be said that Newton invented the calculus for the purpose of this derivation.

We assume that the orbit of a satellite is a planar curve, and we choose a polar coordinate system with the origin at the center of the primary gravitating body. Suppose that a satellite of mass m is centered at the point (r, θ). It is attracted to its primary by a gravitational force satisfying the *inverse square law* (Newton's assumption)

$$F = \frac{cm}{r^2}$$

where c is a constant. Lastly, we assume Newton's *first law of motion*, namely,

$$\mathbf{F} = m\mathbf{a}$$

The two variables r and θ must satisfy the differential equations

$$ma_r = m\left(\frac{d^2r}{dt^2} - r\left(\frac{d\theta}{dt}\right)^2\right) = -\frac{cm}{r^2}$$

and

$$ma_\theta = \frac{m}{r}\frac{d}{dt}\left(r^2\frac{d\theta}{dt}\right) = 0$$

The minus sign in the first equation only means that the force vector is directly opposite to the position vector. These equations reduce to

$$\frac{d^2r}{dt^2} - r\left(\frac{d\theta}{dt}\right)^2 = -\frac{c}{r^2}$$

$$\frac{d}{dt}\left(r^2\frac{d\theta}{dt}\right) = 0$$

The second equation implies that

$$r^2\frac{d\theta}{dt} = k, \text{ a constant}$$

whence

$$\frac{d\theta}{dt} = \frac{k}{r^2}$$

To continue we employ a trick, introducing the new variable

$$w = \frac{1}{r}$$

Then we have

$$\frac{dr}{dt} = \frac{d}{dt}\frac{1}{w} = -\frac{1}{w^2}\frac{dw}{dt} = -\frac{1}{w^2}\frac{dw}{d\theta}\frac{d\theta}{dt}$$

13.12 VELOCITY AND ACCELERATION IN POLAR COORDINATES

Now because $d\theta/dt = k/r^2 = kw^2$, we can write

$$\frac{dr}{dt} = -\frac{1}{w^2}\frac{dw}{d\theta}\frac{d\theta}{dt} = -\frac{1}{w^2}\frac{dw}{d\theta}\cdot kw^2 = -k\frac{dw}{d\theta}$$

It follows then that we have

$$\frac{d^2r}{dt^2} = \frac{d}{dt}\left(-k\frac{dw}{d\theta}\right) = \frac{d}{d\theta}\left(-k\frac{dw}{d\theta}\right)\frac{d\theta}{dt} = -k\frac{d^2w}{dt^2}(kw^2)$$

or

$$\frac{d^2r}{dt^2} = -k^2w^2\frac{d^2w}{d\theta^2}$$

This change of variable reduces the first differential equation above to

$$\frac{d^2r}{dt^2} - r\left(\frac{d\theta}{dt}\right)^2 = -k^2w^2\frac{d^2w}{d\theta^2} - \frac{1}{w}(kw^2)^2 = -cw^2$$

or

$$k^2w^2\frac{d^2w}{d\theta^2} + k^2w^3 = cw^2$$

or

$$\frac{d^2w}{d\theta^2} + w = \frac{c}{k^2}$$

Another change of variable suffices to complete the solution. We set $z = w - c/k^2$. Then we have $dz/d\theta = dw/d\theta$ and $d^2z/d\theta^2 = d^2w/d\theta^2$. The differential equation then becomes

$$\frac{d^2w}{d\theta^2} + w - \frac{c}{k^2} = \frac{d^2z}{d\theta^2} + z = 0$$

The general solution of this equation was given in Section 9.6 as

$$z = C\cos(\theta - \theta_0)$$

where C and θ_0 are arbitrary constants. Defining the number η by the equation $C = \eta c/k^2$, we write

$$w = z + \frac{c}{k^2} = \frac{c}{k^2}[1 + \eta \cos(\theta - \theta_0)]$$

Finally, setting $r = 1/w$, we obtain

$$r = \frac{k^2/c}{1 + \eta\cos(\theta - \theta_0)}$$

We leave it to the reader to prove that this is the polar form of the equation of a conic section with its focus at the origin. The orbit of a planet is a closed curve and hence is an ellipse. This completes Newton's derivation.

EXERCISES 13.12

In each of Exercises 1 to 8, a law of motion in polar form is given. Find the radial and transverse components of both velocity and acceleration in each case.

1. $\mathbf{p}(\theta) = (2a \cos \theta)\mathbf{u}_\theta;\ \dfrac{d\theta}{dt} = 2$

2. $\mathbf{p}(\theta) = a(1 - \cos \theta)\mathbf{u}_\theta;\ \dfrac{d\theta}{dt} = \dfrac{1}{2}$

3. $\mathbf{p}(\theta) = a(1 + \sin \theta)\mathbf{u}_\theta;\ \theta = t$

4. $\mathbf{p}(\theta) = (2 \sin 2\theta)\mathbf{u}_\theta;\ \theta = \tfrac{1}{2}t^2$

5. $\mathbf{p}(\theta) = (a \cos 4\theta)\mathbf{u}_\theta;\ \theta = \tfrac{1}{2}t$

6. $\mathbf{p}(\theta) = \left(\dfrac{4}{1 - \cos \theta}\right)\mathbf{u}_\theta;\ \dfrac{d\theta}{dt} = -1$

7. $\mathbf{p}(\theta) = \left(\dfrac{2}{1 - 2\cos \theta}\right)\mathbf{u}_\theta;\ \dfrac{d\theta}{dt} = \dfrac{1}{t}$

8. $\mathbf{p}(\theta) = \left(\dfrac{c\eta}{1 - \eta \cos \theta}\right)\mathbf{u}_\theta;\ \dfrac{d\theta}{dt}$ arbitrary

PROBLEMS 13.12

1. A particle moves with constant angular velocity with respect to a point toward which the acceleration vector is always directed. What is its path?

2. A particle moves on an ellipse $\{(r, \theta) : r(1 - \eta \cos \theta) = c\eta\}$ with the acceleration always directed toward the origin. Prove that the magnitude of the acceleration at the point (r, θ) is inversely proportional to r^2.

3. A particle moves along the spiral $\{(r, \theta) : r\theta = k\}$ in the counterclockwise direction with constant speed s. Find both the radial and transverse components of $\mathbf{v}$ and $\mathbf{a}$.

4. A particle moves according to the polar law of motion

$$\mathbf{p}(t) = \cosh wt\ \mathbf{u}_\theta$$

where w is a constant. Discuss this motion.

14

LINEAR ALGEBRA

This final chapter is an important foundation for the future study of a wide variety of subjects. Mathematical subjects such as abstract algebra, functional analysis, and statistics, to name only a few, rely heavily upon linear algebra, both for computational techniques and for motivation. Matrix methods are used in such diverse applications as the theory of molecular and crystal structures, game theory and linear programming, flutter problems in aerodynamics, and factor analysis as used in psychology and sociology.

We begin a study of linear algebra with the practical problem of solving a system of linear equations. The entire algebra of matrices can be based upon this problem. Then, after some computational methods have been developed, we turn briefly to the topics of linear mappings, quadratic forms, and geometry.

14.1 SYSTEMS OF LINEAR EQUATIONS

We inquire here into the solution of systems of linear equations such as

(1)
$$\begin{aligned} x - 3y + 2z &= 4 \\ 3x + y - 3z &= 4 \\ 2x + 2y - 2z &= 3 \end{aligned}$$

(2)
$$\begin{aligned} x + 4y + z - w &= 4 \\ 3x - y - 2z + w &= 6 \end{aligned}$$

and

(3)
$$\begin{aligned} x + 2y &= 6 \\ 3x - 2y &= 2 \\ -x + 5y &= 8 \end{aligned}$$

We see that these three systems are of three different natures. The solution of the first system entails finding the point at which the three planes $\{(x,y,z) : x - 3y + 2z = 4\}$, $\{(x,y,z) : 3x + y - 2z = 6\}$ and $\{(x,y,z) : 2x + 2y - 2z = 3\}$ in $\mathbf{R}^3$ all intersect. The reader can verify that this is the point $(\frac{7}{3}, -\frac{1}{3}, \frac{1}{3})$. The third set of equations has a solution only if the three lines $\{(x,y) : x + 2y = 6\}$, $\{(x,y) : 3x - 2y = 2\}$ and $\{(x,y) : -x + 5y = 8\}$ in $\mathbf{R}^2$ all intersect in the same point. [They *do* meet at $(2, 2)$.] The second set of two equations in four unknowns has an infinity of solutions. The reader can verify that any ordered quadruple of the form

$$(x, y, -10 + 4x + 3y, -14 + 5x + 7y)$$

satisfies both equations of the second set. It may be said that the solutions constitute a two-dimensional plane in $\mathbf{R}^4$, but this does not help envision the solution, of course.

The first system (1) above is a system of three linear equations in

14.1 SYSTEMS OF LINEAR EQUATIONS

three variables. We shall call (1) a 3×3 system (this is read as "three by three"). Similarly, system (2) is a 2×4 system and system (3) is a 3×2 system. For an $m \times n$ system (that is, a system of m equations in n variables, we use a double subscript notation for the coefficients. Thus we write such a system in the form:

$$a_{11}x_1 + a_{12}x_2 + \cdots + a_{1n}x_n = c_1$$
$$a_{21}x_1 + a_{22}x_2 + \cdots + a_{2n}x_n = c_2$$
$$\vdots$$
$$a_{m1}x_1 + a_{m2}x_2 + \cdots + a_{mn}x_n = c_m$$

The number a_{ij} is the coefficient of the jth variable x_j in the ith equation. Using the summation symbol, the ith equation is

$$\sum_{j=1}^{n} a_{ij}x_j = c_i$$

The entire $m \times n$ system can then be written as

$$\sum_{j=1}^{n} a_{ij}x_j = c_i, \quad i = 1, 2, \cdots, m$$

The rectangular array of coefficients

$$A = \begin{bmatrix} a_{11} & a_{12} & \cdots & a_{1n} \\ a_{21} & a_{22} & \cdots & a_{2n} \\ & \vdots & & \\ a_{m1} & a_{m2} & \cdots & a_{mn} \end{bmatrix}$$

is called the *matrix of coefficients* of the $m \times n$ system. The matrix A is called an $m \times n$ matrix because it has m rows and n columns. An n-dimensional numerical vector is a $1 \times n$ matrix; we shall call it a *row vector*. Analogously, an $m \times 1$ matrix is a *column vector*.

We shall have use for the *augmented matrix*

$$[A \ \mathbf{c}]$$

of the $m \times n$ system. This is the $m \times (n + 1)$ matrix obtained by adjoining the column vector $\mathbf{c}$ of constant terms to the matrix A of coefficients. Therefore, if we have

$$A = \begin{bmatrix} a_{11} & a_{12} & \cdots & a_{1n} \\ a_{21} & a_{22} & \cdots & a_{2n} \\ & \vdots & & \\ a_{m1} & a_{m2} & \cdots & a_{mn} \end{bmatrix} \text{ and } \mathbf{c} = \begin{bmatrix} c_1 \\ c_2 \\ \vdots \\ c_m \end{bmatrix}$$

then

$$[A\mathbf{c}] = \begin{bmatrix} a_{11} & a_{12} & \cdots & a_{1n}c_1 \\ a_{21} & a_{22} & \cdots & a_{2n}c_2 \\ & & \vdots & \\ a_{m1} & a_{m2} & \cdots & a_{mn}c_m \end{bmatrix}$$

The augmented matrix $[A\ \mathbf{c}]$, is our first use of a *partitioned matrix,* a matrix subdivided into smaller matrices. For another instance, let the rows of the matrix A above be

$$\mathbf{r}_1 = (a_{11}, a_{12}, \cdots, a_{1n})$$
$$\mathbf{r}_2 = (a_{21}, a_{22}, \cdots, a_{2n})$$
$$\vdots$$
$$\mathbf{r}_m = (a_{m1}, a_{m2}, \cdots, a_{mn})$$

Then A can be written as the partitioned matrix

$$A = \begin{bmatrix} \mathbf{r}_1 \\ \mathbf{r}_2 \\ \vdots \\ \mathbf{r}_m \end{bmatrix}$$

Similarly, if the columns of A are

$$\mathbf{c}_1 = \begin{bmatrix} a_{11} \\ a_{21} \\ \vdots \\ a_{m1} \end{bmatrix}, \mathbf{c}_2 = \begin{bmatrix} a_{12} \\ a_{22} \\ \vdots \\ a_{m2} \end{bmatrix}, \cdots, \mathbf{c}_n = \begin{bmatrix} a_{1n} \\ a_{2n} \\ \vdots \\ a_{mn} \end{bmatrix}$$

then we can write A as the partitioned matrix

$$A = [\mathbf{c}_1 \mathbf{c}_2 \cdots \mathbf{c}_n]$$

For the moment we shall use matrices only as a convenient means of keeping track of the numbers involved in a system of equations. Very soon, however, matrices become important of themselves.

■ **DEFINITION.** The set

$$\left\{ (x_1, x_2, \cdots, x_n) : \sum_{j=i}^{n} a_{ij}x_j = c_i;\ i = 1, 2, \cdots, m \right\} \subset \mathbf{R}^n$$

of all *n*-dimensional numerical vectors $(x_1, \cdots, x_n)$ that satisfy all m equations

$$\sum_{j=1}^{n} a_{ij}x_j = c_i,\ i = 1, 2, \cdots, m$$

is called the *solution set* of the $m \times n$ system of linear equations.

14.1 SYSTEMS OF LINEAR EQUATIONS

Much of our study will center about the nature of the solution set of a system of linear equations. Because the individual sets

$$\{(x_1, x_2, \cdots, x_n) : a_{i1}x_1 + \cdots + a_{in}x_n = c_i\}$$

are hyperplanes in $\mathbf{R}^n$, the solution set is an intersection of such hyperplanes. This however, does not suffice even as a description and it does not aid in actually solving the system. Furthermore, note that a system need not have a solution at all, that is, the solution set of a system can be empty.

■ **DEFINITION.** Two systems of linear equations in n variables are said to be *equivalent* if they have the same solution set. (Note that the two systems need not have the same number of equations.)

■ *Example 1.* The two systems

(1) $\quad \begin{aligned} x - 3y + z &= 5 \\ 2x + 4y + 3x &= -2 \end{aligned}$

and

(2) $\quad \begin{aligned} 3x + y + 4z &= 3 \\ 5x - 5y + 6z &= 13 \\ 7x - y + 9z &= 11 \end{aligned}$

are equivalent. A proof of this fact is given by showing that each equation in system (2) is a linear combination of the equations in system (1). We indicate this as follows:

$$[3x + y + 4z = 3]$$
$$= [x - 3y + z = 5] + [2x + 4y + 3z = -2]$$
$$[5x - 5y + 6z = 13]$$
$$= 3[x - 3y + z = 5] + [2x + 4y + 3z = -2]$$
$$[7x - y + 9z = 11]$$
$$= 3[x - 3y + z = 5] + 2[2x + 4y + 3z = -2]$$

■ **DEFINITION.** The following are called *elementary operations* on a system of equations:

1. The positions of two equations in the system are interchanged.
2. One equation is multiplied by a nonzero constant.
3. One equation is replaced by the sum of itself and a multiple of any other equation.

The same operations applied to the rows of a matrix are called *elementary row operations* on the matrix. These are:

1. Two rows of the matrix are interchanged.
2. One row of the matrix is multiplied by a nonzero constant.

3. One row is replaced by the sum of itself and a multiple of any other row.

Of course, we think of the rows of the matrix as vectors when we multiply them by constants and add them together.

■ **THEOREM 14.1.** If one system of equations can be obtained from another by a sequence of elementary operations, then the two systems are equivalent.

It is obvious that elementary operations of type 1 and type 2 do not alter the solution set of a system. The reader can supply the easy proof that the solution set is not changed by an elementary operation of type 3. Then the proof of Theorem 14.1 is completed by a mathematical induction. Note that it is important that we have a *sequence* of elementary operations. The solution set *can* be altered by doing two elementary operations of type 3 simultaneously, for instance, by replacing two equations both by their sum. Consider the system

$$2x - 3y = -1$$
$$4x + y = 5$$

whose solution set is the single vector (1, 1). If each equation is replaced by their sum, we have a system of two identical equations:

$$6x - 2y = 4$$
$$6x - 2y = 4$$

This system has the line $\{(x,y) : 3x - y = 2\}$ as its solution set. This line contains the point (1, 1) but contains much more. Thus our misuse of the elementary operations has enlarged the solution set.

■ **DEFINITION.** If the matrix A can be obtained from the matrix B by a sequence of elementary row operations, then A is *row equivalent* to B.

This definition obviously stems from Theorem 14.1. We note that row equivalence is a true equivalence relation in the sense that it is reflexive, symmetric, and transitive.

■ **THEOREM 14.2.** If the matrices A and B are row-equivalent, then the rows of B can be expressed as linear combinations of the rows of A. (We think of these rows as vectors, of course.)

PROOF: The matrix B can be obtained by a sequence of elementary row operations on A. Thus the rows of B are produced by a finite number of operations involving the sum of two rows of A. It is obvious then that each row of B is precisely a sum of multiples of the rows of A. ■

EXERCISES 14.1   In each of Exercises 1 to 8, write out the matrix of coefficients and the augmented matrix for the given system.

14.1 SYSTEMS OF LINEAR EQUATIONS

1. $3x - 4y = 5$

2. $2x + y = 3$
   $3x - 2y = 1$

3. $x - 2y + z = 1$
   $3x + 3z = 2$
   $2x - y = 4$

4. $3x + 4y = 6$
   $-x + 2y = 4$
   $x - 4y = -2$

5. $w + 3x - y = 3$
   $-w + 2x = 4$
   $x + 2y = 6$

6. $-w + 2x + y - 4x = 0$
   $2w - x - y + 2z = 2$

7. $x_1 + 2x_2 - 3x_3 + x_4 - x_5 = 0$
   $x_2 + 2x_3 - 2x_4 + 3x_5 = 1$
   $-x_1 - x_2 + 4x_3 + x_5 = 2$

8. $x_1 + 2x_4 = 1$
   $2x_2 + x_3 = -1$
   $-x_3 + x_4 = 0$
   $x_2 - 2x_5 = 3$
   $2x_1 - x_4 = 2$

In each of Exercises 9 to 15, show that the given vector is in the solution set of the accompanying system.

9. $(4, 3)$; $3x - 2y = 6$

10. $(4, 3)$; $3x - 2y = 6$
    $x + 2y = 10$

11. $(4, 3)$; $3x - 2y = 6$
    $x + 2y = 10$
    $-x + 4y = 8$

12. $(\frac{23}{12}, \frac{5}{6}, -\frac{5}{4})$; $x - 2y + z = -1$
    $3x + 3z = 2$
    $2x - y = 3$

13. $(3 + 10c, c, -1 - 7c)$; $x - 3y + z = 2$
    $2x + y + 3z = 3$
    $3x + 5y + 5z = 4$

14. $(\frac{3}{10} - \frac{7}{5}c, \frac{3}{2}, \frac{14}{5} + \frac{3}{5}c, c)$; $w + 3x - y + 2z = 2$
    $-w + x + y - 2z = 4$
    $2w - 2x + 3y + z = 6$

15. $(c, \frac{3}{2} - \frac{1}{2}d, \frac{5}{2} + c + \frac{5}{2}d, d)$; $w + 3x - y + 4z = 2$
    $-w + x + y - 2z = 4$

PROBLEMS 14.1

1. Prove that the 2×2 system

   $ax + by = e$
   $cx + dy = f$

   where $(e, f) \neq (0, 0)$ has a unique solution if and only if $ad \neq bc$.

2. Prove that the 2×2 system

   $ax + by = 0$
   $cx + dy = 0$

   has a nonzero solution if and only if $ad = bc$.

3. Determine all 2×2 systems

$$ax + by = 0$$
$$cx + dy = 0$$

having the nonzero vector $(2, -3)$ as a solution.

4. Prove that any 2×2 matrix is row-equivalent to one and only one of the following matrices

$$\begin{bmatrix} 1 & 0 \\ 0 & 1 \end{bmatrix} \begin{bmatrix} 1 & c \\ 0 & 0 \end{bmatrix} \begin{bmatrix} 0 & 1 \\ 1 & 1 \end{bmatrix} \begin{bmatrix} 0 & 0 \\ 1 & 1 \end{bmatrix}$$

5. A homogeneous equation $ax + by + cz = 0$ has as its graph in $\mathbf{R}^3$ a plane through the origin. Use this fact to give a geometric description of the solution set of a system of m homogeneous equations in three variables.

14.2 REDUCTION TO ROW ECHELON FORM

A standard technique in solving a system of equations uses elementary operations to produce an equivalent system that is easy to solve. The method is an organized "elimination of variables"; the basis of the method is the so-called *Gaussian pivot operation*. This is a sequence of elementary operations which may be described as follows: Let a_{kl} be a fixed nonzero coefficient in the $m \times n$ system

$$\sum_{j=1}^{n} a_{ij}x_j = c_i, \quad i = 1, 2, \cdots, m$$

The first elementary operation in the pivot operation consists of multiplying the kth equation by the constant $1/a_{kl}$. This yields a new kth equation in which the lth variable x_l has the coefficient 1. Then we do elementary operations of type 3 to "eliminate the variable x_l" from the remaining $m - 1$ equations. In particular, the ith equation is replaced by the sum of itself and $(-a_{il})$ times the kth equation. The new ith equation then has the coefficient $a_{il} + (-a_{il})(1) = 0$ for the variable x_l. It follows that we can "eliminate" x_l from all but the kth equation, and in the kth equation x_l has the coefficient 1.

■ *Example 1.* Consider the system

$$x + 3y - z = 6$$
$$2x - 2y + 6z = 5$$
$$-x + 2y + z = 4$$

The coefficient $a_{22} = -2$, being nonzero, can serve as the pivot. The first step is to multiply the second equation by $1/-2$. This gives us the equivalent system

$$x + 3y - z = 6$$
$$-x + y - 3z = -\tfrac{5}{2}$$
$$-x + 2y + z = 4$$

14.2 REDUCTION TO ROW ECHELON FORM

In the first equation we have $a_{12} = 3$. Hence the first equation is replaced by

$$[x + 3y - z = 6] + (-3)[-x + y - 3z = -\tfrac{5}{2}]$$
$$= [4x + 0 \cdot y + 8z = \tfrac{27}{2}]$$

This yields the system

$$\begin{aligned} 4x \phantom{{}+y} + 8z &= \tfrac{27}{2} \\ -x + y - 3z &= -\tfrac{5}{2} \\ -x + 2y + z &= 4 \end{aligned}$$

Because $a_{32} = 2$, the third equation is replaced by

$$[-x + 2y + z = 4] + (-2)[-x + y - 3z = -\tfrac{5}{2}]$$
$$= [x + 0 \cdot y + 7z = 9]$$

The completed pivot operation has resulted in the system

$$\begin{aligned} 4x \phantom{{}+y} + 8z &= \tfrac{27}{2} \\ -x + y - 3z &= -\tfrac{5}{2} \\ x \phantom{{}+2y} + 7z &= 9 \end{aligned}$$

We need not continue writing the variables as we perform these pivot operations. In practice we simply apply elementary row operations to the augmented matrix of a system in order to define the pivot operation. After any sequence of pivot operations has been performed on the matrix, it is very easy to write out the corresponding system of equations.

■ *Example 2.* We reduce the system

$$\begin{aligned} 2w + 3x + 2y - 4z &= 12 \\ -w + 2x - y - 2z &= 8 \\ 3w - x + y - z &= 5 \end{aligned}$$

by operating on its augmented matrix

$$\begin{bmatrix} 2 & 3 & 2 & -4 & 12 \\ -1 & 2 & -1 & -2 & 8 \\ 3 & -1 & 1 & -1 & 5 \end{bmatrix}$$

We first pivot on the coefficient $a_{11} = 2$. The first row is thus multiplied by $1/+2$, resulting in the matrix

$$\begin{bmatrix} 1 & \tfrac{3}{2} & 1 & -2 & 6 \\ -1 & 2 & -1 & -2 & 8 \\ 3 & -1 & 1 & -1 & 5 \end{bmatrix}$$

Because $a_{21} = -1$, the second row is replaced by itself plus $(-(-1))$ times the first row or

$$(-1, 2, -1, -2, 8) + (1, \tfrac{3}{2}, 1, -2, 6) = (0, \tfrac{7}{2}, 0, -4, 14)$$

The coefficient $a_{31} = 3$ tells us to replace the third row by

$$(3, -1, 1, -1, 5) + (-3)(1, \tfrac{3}{2}, 1, -2, 6) = (0, -\tfrac{11}{2}, 2, 5, -13)$$

Our first pivot operation has resulted in the equivalent matrix

$$\begin{bmatrix} 1 & \tfrac{3}{2} & 1 & -2 & 6 \\ 0 & \tfrac{7}{2} & 0 & -4 & 14 \\ 0 & -\tfrac{11}{2} & 2 & 5 & -13 \end{bmatrix}$$

Next we pivot on the new coefficient $a_{22} = \tfrac{7}{2}$. The first step yields

$$\begin{bmatrix} 1 & \tfrac{3}{2} & 1 & -2 & 6 \\ 0 & 1 & 0 & -\tfrac{8}{7} & 4 \\ 0 & -\tfrac{11}{2} & 2 & 5 & -13 \end{bmatrix}$$

The coefficient a_{12} is $\tfrac{3}{2}$; hence the first row is replaced by

$$(1, \tfrac{3}{2}, 1, -2, 6) + (-\tfrac{3}{2})(0, 1, 0, -\tfrac{8}{7}, 4) = (1, 0, 1, -\tfrac{2}{7}, 0)$$

Because a_{32} is $-\tfrac{11}{2}$, the new third row is

$$(0, -\tfrac{11}{2}, 2, 5, -13) + (\tfrac{11}{2})(0, 1, 0, -\tfrac{8}{7}, 4) = (0, 0, 2, -\tfrac{9}{7}, 9)$$

The second pivot has yielded

$$\begin{bmatrix} 1 & 0 & 1 & -\tfrac{2}{7} & 0 \\ 0 & 1 & 0 & -\tfrac{8}{7} & 4 \\ 0 & 0 & 2 & -\tfrac{9}{7} & 9 \end{bmatrix}$$

The last pivot will be the coefficient $a_{33} = 2$. The first step in the pivot operation is to multiply the third row by $\tfrac{1}{2}$, yielding

$$\begin{bmatrix} 1 & 0 & 1 & -\tfrac{2}{7} & 0 \\ 0 & 1 & 0 & -\tfrac{8}{7} & 4 \\ 0 & 0 & 1 & -\tfrac{9}{14} & \tfrac{9}{2} \end{bmatrix}$$

Because $a_{32} = 0$, we leave the second row alone; because a_{31} is 1, the new first row is

$$(1, 0, 1, \tfrac{2}{7}, 0) + (-1)(0, 0, 1, -\tfrac{9}{14}, \tfrac{9}{2})$$
$$= (1, 0, 0, \tfrac{5}{14}, -\tfrac{9}{2})$$

The equivalent matrix is

$$\begin{bmatrix} 1 & 0 & 0 & \tfrac{5}{14} & -\tfrac{9}{2} \\ 0 & 1 & 0 & -\tfrac{8}{7} & 4 \\ 0 & 0 & 1 & -\tfrac{9}{14} & \tfrac{9}{2} \end{bmatrix}$$

14.2 REDUCTION TO ROW ECHELON FORM

which corresponds to the system of equations

$$\begin{aligned} w + \tfrac{5}{14}z &= -\tfrac{9}{2} \\ x - \tfrac{8}{7}z &= 4 \\ y - \tfrac{9}{14}z &= \tfrac{9}{2} \end{aligned}$$

From these equations, it is easy to see that the solution set of the system consists of all four-dimensional numerical vectors of the form

$$(-\tfrac{9}{2} - \tfrac{5}{14}z,\ 4 + \tfrac{8}{7}z,\ \tfrac{9}{2} + \tfrac{9}{14}z,\ z)\ \blacksquare$$

As illustrated in Example 2 the augmented matrix is the entity with which we work. Our definitions also will be stated in terms of matrices. Then we shall utilize the new terminology when speaking of systems of equations.

■ **DEFINITION.** A matrix is said to be in *row echelon form* when it satisfies the following conditions.

1. The nonzero rows precede the zero rows.
2. The first nonzero number in each row is 1.
3. This initial 1 is the only nonzero number in its column.
4. In every row except the first, the initial 1 is farther to the right than is the initial 1 in the preceding row.

■ **Example 3.** The following matrices are in row echelon form:

$$\begin{bmatrix} 1 & 0 & 0 & \tfrac{5}{14} & -\tfrac{9}{2} \\ 0 & 1 & 0 & -\tfrac{8}{7} & 4 \\ 0 & 0 & 1 & -\tfrac{9}{14} & \tfrac{9}{2} \end{bmatrix} \qquad \begin{bmatrix} 1 & 2 & 0 & 0 & 2 & 0 \\ 0 & 0 & 1 & -1 & 1 & 0 \\ 0 & 0 & 0 & 0 & 0 & 1 \end{bmatrix}$$

$$\begin{bmatrix} 0 & 1 & 0 & 0 & 2 & 0 & 1 \\ 0 & 0 & 1 & 1 & 0 & 0 & -1 \\ 0 & 0 & 0 & 0 & 0 & 1 & 3 \\ 0 & 0 & 0 & 0 & 0 & 0 & 0 \end{bmatrix}$$

For any matrix A, there is at least one sequence of elementary row operations that reduces A to row echelon form. In fact, there is an effective procedure that can be programmed for a digital computer. We describe the process in the proof of the next result.

■ **THEOREM 14.3.** Any matrix can be reduced to row echelon form by a sequence of elementary row operations. Furthermore, the row echelon form does not depend upon the particular sequence of elementary row operations used to reduce the matrix.

PROOF: Consider the matrix A as the partitioned matrix

$$A = [\mathbf{c}_1 \mathbf{c}_2 \cdots \mathbf{c}_n]$$

where the column vectors of A are the $\mathbf{c}_j, j = 1, 2, \cdots, n$. Let p_1 be the index of the first nonzero column vector and let a_{ip_1} be the first nonzero component of $\mathbf{c}_{p_1}$. If $i = 1$, we pivot on the number a_{1p_1}. If $i > 1$, we first interchange the ith row and then pivot on a'_{1p_1} ($= a_{ip_1}$). The resulting matrix in partitioned form looks like this:

$$\begin{array}{c|c} \begin{matrix} 0 \cdots 0 \ 1 \\ \hline 0 \cdots 0 \ 0 \\ 0 \cdots 0 \ 0 \\ \uparrow \end{matrix} & \begin{matrix} a_{1p_1+1} \cdots a_{1n} \\ \hline a_{2p_1+1} \cdots a_{2n} \\ a_{mp_1+1} \cdots a_{mn} \end{matrix} \\ P_1 \text{ column} & \end{array}$$

Next we look at the $(m-1) \times (n - p_1)$ matrix in the lower right-hand corner

$$\begin{bmatrix} a_{2p_1+1} & \cdots & a_{2n} \\ a_{mp_1+1} & \cdots & a_{mn} \end{bmatrix}$$

and repeat the procedure. We let p_2 be the index (with respect to the column numbering in A) of the first nonzero column of this matrix. If need be, we interchange rows of A to put the first nonzero number in this column into the first row of this matrix (the second row of A). Then we pivot on it. The reduction to row echelon form follows by an obvious induction argument. We shall establish the second part of the theorem, the uniqueness of the row echelon form, in Section 14.7. ∎

The numbers $p_1, p_2, \cdots, p_r$ selected in the above proof are called the *echelon indices* of the matrix A. As we show in Section 14.7 when we complete the proof of Theorem 14.3, these echelon indices are invariants of the row equivalence class of the matrix A. Some applications of matrices make use of these invariants.

■ **Example 4.** Consider the matrix

$$\begin{bmatrix} 3 & 9 & 9 & 6 & -12 & 30 \\ 2 & 10 & 10 & 4 & -16 & 40 \\ -2 & -4 & -4 & -4 & 8 & -22 \\ 1 & 2 & 2 & 2 & 1 & -4 \end{bmatrix}$$

Because a_{11} is nonzero, our first step is to pivot on a_{11}, obtaining the matrix

$$\begin{array}{c|ccccc} 1 & 3 & 3 & 2 & -4 & 10 \\ \hline 0 & 4 & 4 & 0 & -8 & 20 \\ 0 & 2 & 2 & 0 & 0 & -2 \\ 0 & -1 & -1 & 0 & 5 & -14 \end{array}$$

14.2 REDUCTION TO ROW ECHELON FORM

The first nonzero column of the 3×5 submatrix just outlined is the first, the second in the bigger matrix. Hence $p_2 = 2$ and again we simply pivot on $a_{22} = r$. This results in the matrix

$$\begin{bmatrix} 1 & 0 & 0 & 2 & 2 & -5 \\ 0 & 1 & 1 & 0 & -2 & 5 \\ 0 & 0 & 0 & 0 & 4 & -12 \\ 0 & 0 & 0 & 0 & 3 & -9 \end{bmatrix}$$

The first nonzero column of the 2×4 submatrix outlined is the third, the fifth in the larger matrix. Hence $p_3 = 5$. Again we need not interchange rows—we simply pivot on $a_{35} = 4$. This yields the matrix

$$\begin{bmatrix} 1 & 0 & 0 & 2 & 0 & 1 \\ 0 & 1 & 1 & 0 & 0 & -1 \\ 0 & 0 & 0 & 0 & 1 & -3 \\ 0 & 0 & 0 & 0 & 0 & 0 \end{bmatrix}$$

The outlined 1×3 submatrix has no nonzero columns, so that the procedure is finished; the matrix is in row echelon form. We see that the echelon indices are

$$p_1 = 1, p_2 = 2, p_3 = 5 \quad \blacksquare$$

There are two important consequences of Theorem 14.3. The first provides a procedure for determining when two $m \times n$ matrices are row-equivalent.

■ **COROLLORY 14.4.** Two $m \times n$ matrices are row-equivalent if and only if they have the same row echelon form.

■ **COROLLARY 14.5.** The number of nonzero rows in the row echelon form of a matrix depends only upon the row equivalence class of the matrix.

This second consequence of Theorem 14.3 tells us that the following concept is well-defined: Let A be an $m \times n$ matrix. The *rank* of A is the number of nonzero rows in the row echelon form of A. In particular, the rank of A obviously cannot exceed the number m of rows in A.

The rank of a matrix has important applications. Indeed, the major part of this chapter is devoted to the study of those concepts dependent upon the rank of a matrix. The next result is an important instance of the use of "rank." We shall delay a proof until Section 14.7. (See Corollary 14.27.)

■ **THEOREM 14.6.** A system of linear equations is consistent if and only if the rank of the augmented matrix equals the rank of the matrix of coefficients. If the system is consistent, then the solution is unique if the rank of the matrix of coefficients equals the number of variables; the solution set is infinite if the rank of the matrix of coefficients is less than the number of variables.

We restate Theorem 14.6 in symbols. Consider the system of m equations in n variables,

$$\sum_{j=1}^{n} a_{ij} x_j = c_i, \quad i = 1, 2, \cdots, m$$

The system is inconsistent (has an empty solution set) if

$$\operatorname{rank} [A \ \mathbf{c}] > \operatorname{rank} A$$

The system has a unique solution (a one-point solution set) if

$$\operatorname{rank} [A \ \mathbf{c}] = \operatorname{rank} A = n$$

The system has an infinite solution set if

$$\operatorname{rank} [A \ \mathbf{c}] = \operatorname{rank} A < n$$

Consider our $m \times n$ system. The rank of its coefficient matrix is really the number of "essential" equations in the system. The remaining equations can be expressed as linear combinations of the essential ones. With this in mind, Theorem 14.6 says that the system is inconsistent if there are too many conditions imposed upon the variables. If the number of essential conditions (the rank) equals the number of variables, then we should expect a unique solution. In the final case, there are fewer essential conditions than variables, so that there are many solutions.

■ ***Example 5.*** We apply Theorem 14.6 to the system

$$\begin{aligned} 3x + 2y - z &= 6 \\ 2x - y + 2z &= 4 \\ 5x + y + z &= 8 \end{aligned}$$

The coefficient matrix

$$A = \begin{bmatrix} 3 & 2 & -1 \\ 2 & -1 & 2 \\ 5 & 1 & 1 \end{bmatrix}$$

has row echelon form

$$\begin{bmatrix} 1 & 0 & 0 \\ 0 & 1 & 0 \\ 0 & 0 & 0 \end{bmatrix}$$

14.2 REDUCTION TO ROW ECHELON FORM

Hence rank $A = 2$. The augmented matrix

$$[A \; \mathbf{c}] = \begin{bmatrix} 3 & 2 & -1 & 6 \\ 2 & -1 & 2 & 4 \\ 5 & 1 & 1 & 8 \end{bmatrix}$$

has row echelon form

$$\begin{bmatrix} 1 & 0 & 0 & 0 \\ 0 & 1 & 0 & 0 \\ 0 & 0 & 0 & 1 \end{bmatrix}$$

Hence rank $[A \; \mathbf{c}] = 3$. By Theorem 14.6, this system is inconsistent. We verify this conclusion by noting that by subtracting the first and second equations from the third we obtain the contradictory statement $0 = -2$.

■ **Example 6.** Consider the 4×4 system

$$\begin{aligned} 2w - 2x - 2y + 6z &= -8 \\ 3w - 5x - 7y + 9z &= -22 \\ -3w + x + y - 5z &= 0 \\ 2w \phantom{{}-5x} + 5y + 9z &= -20 \end{aligned}$$

The augmented matrix $[A \; \mathbf{c}]$ is

$$\begin{bmatrix} 2 & -2 & -2 & 6 & -8 \\ 3 & -5 & -7 & 9 & -22 \\ -3 & 1 & 1 & -5 & 0 \\ 2 & 0 & 5 & 9 & -20 \end{bmatrix}$$

Reducing it to row echelon form results in the matrix

$$\begin{bmatrix} 1 & 0 & 0 & 0 & 2 \\ 0 & 1 & 0 & 0 & -1 \\ 0 & 0 & 1 & 0 & 3 \\ 0 & 0 & 0 & 1 & -2 \end{bmatrix}$$

Note that we also have reduced the matrix of coefficients to row echelon form at the same time. (Just omit the column of constant terms.) Hence rank $[A \; \mathbf{c}] = $ rank $A = $ the number of variables $= 4$. Theorem 14.6 tells us that the system has a unique solution, but we can see this from the system corresponding to this row echelon form. These are simply

$$w = 2, \; x = -1, \; y = 3, \; z = -2$$

In other words, the 4×4 system has the unique solution
$$(2, -1, 3, -2)$$

In Example 6 we have used the fact that reducing the augmented matrix $[A \ \mathbf{c}]$ to row echelon form automatically reduces the matrix A to row echelon form. (See Problem 14.2.4.) Note that this also proves another fact, namely, that either

$$\text{rank } [A \ \mathbf{c}] = \text{rank } A$$

or

$$\text{rank } [A \ \mathbf{c}] = 1 + \text{rank } A$$

■ **Example 7.** Consider the 3×4 system of equations
$$w + 3x + 2y - z = 6$$
$$2w + 2x - y + 2z = 4$$
$$4x + 5y - 4z = 8$$

The augmented matrix is
$$\begin{bmatrix} 1 & 3 & 2 & -1 & 6 \\ 2 & 2 & -1 & 2 & 4 \\ 0 & 4 & 5 & -4 & 8 \end{bmatrix}$$

and has the row echelon form
$$\begin{bmatrix} 1 & 0 & -\frac{7}{4} & 2 & 0 \\ 0 & 1 & \frac{5}{4} & -1 & 2 \\ 0 & 0 & 0 & 0 & 0 \end{bmatrix}$$

The corresponding system of equation, equivalent to the given one, is simply
$$w - \tfrac{7}{4}y + 2z = 0$$
$$x + \tfrac{5}{4}y - z = 2$$

Hence any four-dimensional vector of the form
$$(w, x, y, z) = (\tfrac{7}{4}y - 2z, 2 - \tfrac{5}{4}y + z, y, z)$$

satisfies the equations. Obviously there are infinitely many such vectors. Since rank $[A \ \mathbf{c}] = \text{rank } A = 2$ which is less than the number of variables, we have here an instance of part (3) of Theorem 14.6.

EXERCISES 14.2 Reduce each matrix in Exercises 1 to 10 to row echelon form.

1. $\begin{bmatrix} 1 & 1 & 2 \\ 2 & 3 & 4 \end{bmatrix}$
2. $\begin{bmatrix} 2 & -1 & 4 \\ 6 & -3 & 3 \end{bmatrix}$

3. $\begin{bmatrix} 1 & -1 & 1 & 6 \\ 2 & -5 & 2 & 10 \end{bmatrix}$

4. $\begin{bmatrix} 1 & 1 & -3 \\ 2 & 3 & 1 \\ 1 & 5 & 5 \end{bmatrix}$

5. $\begin{bmatrix} 2 & 1 & 1 & -5 \\ 1 & 1 & 1 & -4 \\ 1 & -1 & -1 & 2 \end{bmatrix}$

6. $\begin{bmatrix} 1 & -3 & 4 & 5 \\ 1 & -1 & 1 & 2 \\ 2 & -1 & 1 & -3 \end{bmatrix}$

7. $\begin{bmatrix} 1 & 1 & 1 & -1 & -2 \\ -1 & 1 & 2 & 1 & 1 \\ 1 & 3 & 4 & -1 & -4 \end{bmatrix}$

8. $\begin{bmatrix} 9 & 12 & -8 \\ 0 & 6 & 1 \\ -3 & 8 & -2 \\ 12 & 62 & -3 \end{bmatrix}$

9. $\begin{bmatrix} 1 & 2 & 3 & 4 \\ 5 & 6 & 7 & 8 \\ 9 & 10 & 11 & 12 \\ 13 & 14 & 15 & 16 \end{bmatrix}$

10. $\begin{bmatrix} 2 & 9 & 3 & -1 & 18 \\ 2 & 0 & -3 & 0 & -6 \\ 3 & 13 & -1 & 2 & 5 \\ 0 & 0 & 3 & -2 & 12 \end{bmatrix}$

In each of Exercises 11 to 20, a pair of matrices is given. Show that they are row-equivalent.

11. $\begin{bmatrix} 3 & 4 \\ 2 & -1 \end{bmatrix} \begin{bmatrix} 1 & 5 \\ 0 & 1 \end{bmatrix}$

12. $\begin{bmatrix} 3 & -4 \\ 4 & 1 \end{bmatrix} \begin{bmatrix} 1 & 0 \\ 3 & 1 \end{bmatrix}$

13. $\begin{bmatrix} 3 & 6 \\ 4 & 8 \end{bmatrix} \begin{bmatrix} 1 & 2 \\ 0 & 0 \end{bmatrix}$

14. $\begin{bmatrix} 4 & 2 \\ 2 & 5 \end{bmatrix} \begin{bmatrix} 2 & -3 \\ 0 & 8 \end{bmatrix}$

15. $\begin{bmatrix} 6 & 4 & 4 \\ 5 & 3 & 7 \end{bmatrix} \begin{bmatrix} 1 & 1 & -3 \\ 2 & 0 & 12 \end{bmatrix}$

16. $\begin{bmatrix} 12 & 4 & 6 \\ 9 & 3 & 0 \end{bmatrix} \begin{bmatrix} 0 & 0 & \frac{3}{2} \\ 3 & 1 & 0 \end{bmatrix}$

17. $\begin{bmatrix} 1 & 3 & 1 \\ 2 & -1 & 0 \\ 2 & 4 & -3 \end{bmatrix} \begin{bmatrix} 1 & 1 & -4 \\ 0 & -3 & 8 \\ 0 & 2 & 5 \end{bmatrix}$

18. $\begin{bmatrix} 4 & 6 & 3 \\ 1 & -3 & 4 \\ 5 & 21 & -6 \end{bmatrix} \begin{bmatrix} 1 & -3 & 6 \\ 6 & 0 & 11 \\ 0 & 18 & -13 \end{bmatrix}$

19. $\begin{bmatrix} 1 & 2 & -3 & 2 \\ 3 & 1 & 1 & 6 \\ -3 & 2 & -2 & 4 \end{bmatrix} \begin{bmatrix} 1 & 0 & 0 & 0 \\ 0 & 1 & 0 & 4 \\ 0 & 0 & 1 & 2 \end{bmatrix}$

20. $\begin{bmatrix} 1 & -1 & 2 & 2 & 3 \\ 1 & 1 & -1 & -2 & 4 \\ 2 & -1 & 4 & -3 & 5 \end{bmatrix} \begin{bmatrix} 1 & 0 & 0 & \frac{5}{3} & 4 \\ 0 & 1 & 0 & -7 & -1 \\ 0 & 0 & 1 & -\frac{10}{3} & -1 \end{bmatrix}$

Reduce each system of equations in Exercises 21 to 30 to row echelon form, and use this simplification to determine the solution set.

21. $\begin{aligned} x + y &= 2 \\ 2x + 3y &= 4 \end{aligned}$
22. $\begin{aligned} x - y + z &= 6 \\ 2x - 5y + 2z &= 10 \end{aligned}$

23. $\begin{aligned} x + 2y - 3z &= 3 \\ 2x - y - z &= 1 \\ x + y - 2z &= 2 \end{aligned}$
24. $\begin{aligned} x + 2y - 3z &= 2 \\ 2x - y - z &= 3 \\ x + y - 2z &= 1 \end{aligned}$

25. $\begin{aligned} x + y + 5z &= 4 \\ 3x - z &= 2 \\ 5x + 4y + 3z &= 50 \end{aligned}$
26. $\begin{aligned} x + y - z &= 1 \\ -x + 2y + 3z &= 13 \\ 2x - y - z &= 0 \end{aligned}$

27. $\begin{aligned} w - x + y + 2z &= 0 \\ 2w + x - y - z &= 0 \\ w + 2x - 2y + z &= 0 \end{aligned}$
28. $\begin{aligned} 2w - 3x + y - 2z &= 3 \\ w + 2x - 3y + z &= 4 \\ w - 12x + 11y - 7z &= -6 \end{aligned}$

29. $\begin{aligned} w - x + y + z &= -2 \\ w + x - y + 2z &= 1 \\ 3w - x + y + 2z &= 2 \end{aligned}$
30. $\begin{aligned} w + y &= 1 \\ 2x + z &= 1 \\ 3y + z &= -2 \\ x + 2y &= 5 \end{aligned}$

For the systems of equations in Exercises 31 to 35, compute the ranks of both the coefficient matrix and the augmented matrix. Then apply Theorem 14.6 to determine the nature of the solution set.

31. $\begin{aligned} 3x - y + 2z &= 3 \\ x + 2y + 3z &= 8 \\ -x + y - 3z &= 4 \end{aligned}$
32. $\begin{aligned} 3x - y + 2z &= 3 \\ x + 2y + 3z &= 8 \\ -2x + 3y + z &= 5 \end{aligned}$

33. $\begin{aligned} x + y + z &= 2 \\ 2x - 3y + 7z &= -6 \\ 3x - 2y + 3z &= -9 \end{aligned}$
34. $\begin{aligned} x + y + z &= 2 \\ 2x - 3y + 7z &= -6 \\ 3x - 2y + 8z &= -4 \end{aligned}$

35. $\begin{aligned} w + x + y &= 2 \\ w + x + z &= -5 \\ w + y + z &= 0 \\ x + y + z &= -3 \end{aligned}$

PROBLEMS 14.2

1. What geometric interpretation can be placed upon the solution set of an $m \times 3$ system, $m = 1, 2, 3, 4, \cdots$, when the coefficient matrix and the augmented matrix have equal rank (a) 3, (b) 2, and (c) 1?

2. If $m < n$, show that the homogeneous $m \times n$ system

$$\sum_{j=1}^{n} a_{ij}x_j = 0, \quad i = 1, 2, \cdots, m$$

has a nonzero solution.

3. Each of the following systems contains the parameter α as a coefficient. For what values of α will the system fail to have a unique solution? Does the system have any solution when it fails to have a unique solution?

(a) $\alpha x + y - 2z = 3$
$-\alpha x - 2y + z = 2$
$x + 2y - z = 1$

(b) $\alpha x - 3y - z = 2$
$x + \alpha y + z = -1$
$x - y + \alpha z = \frac{3}{2}$

4. Prove that reducing an augmented matrix $[A\mathbf{c}]$ to row echelon form also reduces the matrix A to row echelon form.

14.3 THE ALGEBRA OF MATRICES

Let $\mathbf{R}^{m \times n}$ denote the set of all $m \times n$ matrices of real numbers. We adopt this symbol to help distinguish between this set and the set $\mathbf{R}^{mn}$ of mn-dimensional numerical vectors. Although the two can be, and sometimes are, identified, we shall not do so here. However, just as in $\mathbf{R}^{mn}$, we define operations on members of $\mathbf{R}^{m \times n}$ "component-wise."

Let $A = (a_{ij})$ and $B = (b_{ij})$ be $m \times n$ matrices and let r be a real number. Then we make the following definitions.

1. $A = B$ if and only if $a_{ij} = b_{ij}$ for each pair (i, j), $i = 1, \cdots, m; j = 1, \cdots, n$.
2. The product rA of the matrix A by the real number r is defined to be

$$rA = r(a_{ij}) = (ra_{ij})$$

3. The sum of the matrices A and B is defined to be

$$A + B = (a_{ij}) + (b_{ij}) = (a_{ij} + b_{ij})$$

■ **Example 1.**

$$3\begin{bmatrix} 4 & 2 & 1 \\ 1 & 0 & 5 \end{bmatrix} + 2\begin{bmatrix} -1 & 0 & 3 \\ 1 & 1 & 4 \end{bmatrix}$$

$$= \begin{bmatrix} 4 & 6 & 3 \\ 3 & 0 & 15 \end{bmatrix} + \begin{bmatrix} -2 & 0 & 6 \\ 2 & 2 & 8 \end{bmatrix}$$

$$= \begin{bmatrix} 4-2 & 6+0 & 3+6 \\ 3+2 & 0+2 & 15+8 \end{bmatrix} = \begin{bmatrix} 2 & 6 & 9 \\ 5 & 2 & 23 \end{bmatrix}$$

The reader can supply a proof for the following result, which tells us all we need to know about the operations defined.

■ **THEOREM 14.7.** The set $\mathbf{R}^{m \times n}$ is a vector space under the two operations of the addition of matrices and multiplication by real numbers. (See Section 1.8 for the definition of a vector space.)

There is also a product of matrices, but not within the set $\mathbf{R}^{m \times n}$ (unless $m = n$). In fact, if A and B are matrices, then we define a product AB only if the number of columns of A equals the number of rows of B. For instance, if A is an $m \times n$ matrix and B is an $n \times p$ matrix, then the product AB is defined and is an $m \times p$ matrix.

We motivate the definition of matrix multiplication by means of an example involving systems of equations. Consider

$$3x + 2y - z = 6$$
$$2x + 4y + 3z = -2$$
$$-x + y + 4z = -4$$

Suppose that x, y, and z can be expressed in terms of two other variables u and v as the linear functions

$$x = 5u - 6v$$
$$y = 7u + 8v$$
$$z = 9u - 11v$$

Then the original system becomes

$$3(5u - 6v) + 2(7u + 8v) - (9u - 11v) = 6$$
$$2(5u - 6v) + 4(7u + 8v) + 3(9u - 11v) = -2$$
$$-(5u - 6v) + (7u + 8v) + 4(9u - 11v) = -4$$

or

$$20u + 9v = 6$$
$$65u - 13v = -2$$
$$38u - 30v = -4$$

Letting the matrix of coefficients of the original equations be

$$A = \begin{bmatrix} 3 & 2 & -1 \\ 2 & 4 & 3 \\ -1 & 1 & 4 \end{bmatrix}$$

and the matrix of coefficients of the substitutions be

$$B = \begin{bmatrix} 5 & -6 \\ 7 & 8 \\ 9 & -11 \end{bmatrix}$$

we want a "multiplication" AB so that $AB = C$ where C is the matrix of coefficients of the new set of equations,

$$\begin{bmatrix} 20 & 9 \\ 65 & -13 \\ 38 & -30 \end{bmatrix}$$

14.3 THE ALGEBRA OF MATRICES

To obtain the "product"

$$\begin{bmatrix} 3 & 2 & -1 \\ 2 & 4 & 3 \\ -1 & 1 & 4 \end{bmatrix} \begin{bmatrix} 5 & -6 \\ 7 & 8 \\ 9 & -11 \end{bmatrix} = \begin{bmatrix} 20 & 9 \\ 65 & -13 \\ 38 & -30 \end{bmatrix}$$

we simply put c_{ij} equal to the dot product of the ith row of A and the jth column of B, considering row and column as vectors. For instance,

$$c_{21} = \begin{bmatrix} 2 & 4 & 3 \end{bmatrix} \cdot \begin{bmatrix} 5 \\ 7 \\ 9 \end{bmatrix} = (2)(5) + (4)(7) + (3)(9) = 65$$

and

$$c_{32} = \begin{bmatrix} -1 & 1 & 4 \end{bmatrix} \cdot \begin{bmatrix} -6 \\ 8 \\ -11 \end{bmatrix} = (-1)(-6) + (1)(8) + (4)(-11) = -30$$

The reader may verify that the other four numbers in C are produced by the same procedure.

■ **DEFINITION.** Let A be an $m \times n$ matrix with n-dimensional row vectors $\mathbf{a}_1, \mathbf{a}_2, \cdots, \mathbf{a}_m$ and let B be an $n \times p$ matrix with n-dimensional column vectors $\mathbf{b}^1, \mathbf{b}^2, \cdots, \mathbf{b}^p$. Then the product matrix $C = AB$ is defined to be the $m \times p$ matrix whose numbers are given by

$$c_{ij} = \mathbf{a}_i \cdot \mathbf{b}^j, \quad i = 1, 2, \cdots, m; \, j = 1, 2, \cdots, p$$

■ **Example 2.**

$$\begin{bmatrix} 3 & 2 \\ 6 & 3 \end{bmatrix} \begin{bmatrix} -2 & 4 \\ 1 & 1 \end{bmatrix} = \begin{bmatrix} (3)(-2) + (2)(1) & (3)(4) + (2)(1) \\ (6)(-2) + (3)(1) & (6)(4) + (3)(1) \end{bmatrix}$$
$$= \begin{bmatrix} -4 & 14 \\ -9 & 27 \end{bmatrix}$$

and

$$\begin{bmatrix} 3 & 2 & -2 \\ 6 & 3 & 1 \\ -2 & 1 & 3 \end{bmatrix} \begin{bmatrix} 4 \\ 1 \\ 0 \end{bmatrix} = \begin{bmatrix} (3)(4) + (2)(1) + (-2)(0) \\ (6)(4) + (3)(1) + (1)(0) \\ (-2)(4) + (1)(1) + (3)(0) \end{bmatrix}$$
$$= \begin{bmatrix} 14 \\ 27 \\ -7 \end{bmatrix} \quad ■$$

One of the many ways in which we use matrix multiplication is to simplify notation. For instance, consider the system of equations

$$\sum_{j=1}^{n} a_{ij} x_j = c_i, \qquad i = 1, 2, \cdots, m$$

As usual, let $A = (a_{ij})$ denote the $m \times n$ matrix of coefficients. Now let x and c denote the column vectors

$$x = \begin{bmatrix} x_1 \\ x_2 \\ \vdots \\ x_n \end{bmatrix} \quad \text{and} \quad c = \begin{bmatrix} c_1 \\ c_2 \\ \vdots \\ c_m \end{bmatrix}$$

Then the product Ax is that of an $m \times n$ matrix times an $n \times 1$ matrix and hence is an $m \times 1$ matrix. Thus the matrix equation

$$Ax = c$$

expresses the system of equations. We shall find this notation of use in later sections.

It is easy to show that matrix multiplication does not satisfy the commutative law. If A is an $m \times n$ matrix and B is an $n \times p$ matrix, then AB is defined. However, BA would only be defined if $m = p$, and even in this case the matrices AB and BA need not even be the same size. (Note that AB will be an $m \times m$ matrix, whereas BA is an $n \times n$ matrix.) Furthermore, even if A and B are "square," that is, if $m = n = p$, the two products AB and BA need not be equal. For instance, let

$$A = \begin{bmatrix} 1 & 3 \\ 2 & -1 \end{bmatrix} \quad \text{and} \quad B = \begin{bmatrix} 2 & -2 \\ 0 & 1 \end{bmatrix}$$

Then it is easily verified that

$$AB = \begin{bmatrix} 2 & 1 \\ 4 & -5 \end{bmatrix} \quad \text{and} \quad BA = \begin{bmatrix} -2 & 8 \\ 2 & -1 \end{bmatrix}$$

The reader may prove the next result, which says that the associative and distributive rules hold for matrix multiplication.

■ **THEOREM 14.8.** Whenever the products and sums of matrices in the following equations are defined, the equations hold:

1. $(AB)C = A(BC)$
2. $(A + B)C = AC + BC$
3. $A(B + C) = AB + AC$

Two distributive laws are needed here because multiplication is not commutative. The order of multiplying must be retained in manipulat-

14.3 THE ALGEBRA OF MATRICES

ing matrices. For example, in the following expansion, we are careful to retain the orders.

$$(A - B)^2 = (A - B)(A - B) = (A - B)A + (A - B)(-B)$$
$$= AA - BA + (A)(-B) + (-B)(-B)$$
$$= A^2 - BA - AB + B^2$$

When it becomes necessary, we shall use phrases such as "multiply the matrix B on the left by the matrix A" or "multiply A on the right by B." (Both of these operations result in the same product AB.)

The *Kronecker delta* is defined by the equations

$$\delta_{ij} = 0 \text{ if } i \neq j$$
$$\delta_{ij} = 1$$

The $n \times n$ matrix $I_n = (\delta_{ij})$ is called an *identity matrix*. The name stems from the obvious fact that if A is an $m \times n$ matrix, then

$$I_m A = A \quad \text{and} \quad A I_n = A$$

Note that I_n has the number 1 along the *main diagonal* and zeros elsewhere.

$$I_n = \begin{bmatrix} 1 & 0 & 0 & 0 & \cdots \\ 0 & 1 & 0 & 0 & \cdots \\ 0 & 0 & 1 & 0 & \cdots \\ 0 & 0 & 0 & 1 & \cdots \\ \cdots & \cdots & \cdots & \cdots & \cdots \end{bmatrix}$$

■ **THEOREM 14.9.** Let E denote the result of performing one of the elementary row operations upon the matrix I_m. Then the result of performing the same elementary row operation upon an $m \times n$ matrix A is the product EA.

PROOF: We shall carry out the proof for an elementary row operation of type 3 and leave the other two cases to the reader. Suppose we add λ times the qth row to the pth row of the matrix I_m. The resulting matrix is simply I_m with λ replacing the 0 in the pth row, qth column position.

$$E = \begin{matrix} 1 \cdots 0 \cdots 0 \cdots 0 \\ 0 \cdots 1 \cdots \lambda \cdots 0 \quad \leftarrow p\text{th row} \\ 0 \cdots 0 \cdots 1 \cdots 0 \quad \leftarrow q\text{th row} \\ 0 \cdots 0 \cdots 0 \cdots 1 \end{matrix}$$

$$p\text{th column} \uparrow \quad \uparrow q\text{th column}$$

Now the number in the ith row and jth column of a product EA is

$$\sum_{k=1}^{m} e_{ik} a_{kj} = a_{ij} \quad \text{if } i \neq p$$
$$= a_{pj} + \lambda a_{qj} \quad \text{if } i = p$$

This is precisely the matrix obtained by adding λ times the qth row of A to the pth row of A. ∎

A matrix E corresponding to an elementary row operation is called an *elementary matrix*. Note that, if the elementary matrices $E_1, E_2, \ldots, E_s$ correspond to a sequence of elementary row operations which reduce a matrix A to its row echelon form, then the product

$$E_s \cdots E_2 E_1 A$$

is the row echelon form of A. We use this fact in the next section.

EXERCISES 14.3 In each of Exercises 1 to 10, compute the indicated product.

1. $[3 \ -1]\begin{bmatrix} 2 \\ 6 \end{bmatrix}$

2. $[3 \ -1]\begin{bmatrix} 2 & 1 \\ 6 & 2 \end{bmatrix}$

3. $\begin{bmatrix} 2 & 0 \\ 0 & -3 \end{bmatrix}\begin{bmatrix} 4 & 1 \\ 1 & 4 \end{bmatrix}$

4. $\begin{bmatrix} 2 & 1 \\ 0 & 3 \end{bmatrix}\begin{bmatrix} -1 & 1 \\ 0 & 2 \end{bmatrix}$

5. $\begin{bmatrix} 2 & 1 \\ 1 & 0 \end{bmatrix}\left(\begin{bmatrix} 1 & 1 \\ 0 & -1 \end{bmatrix} + \begin{bmatrix} 2 & -1 \\ 2 & 2 \end{bmatrix}\right)$

6. $\begin{bmatrix} 2 & -1 \\ 1 & 1 \end{bmatrix}\left(\begin{bmatrix} 3 & 1 & 2 \\ 0 & 2 & -1 \end{bmatrix} + \begin{bmatrix} 2 & -2 & 1 \\ 1 & 1 & 3 \end{bmatrix}\right)$

7. $\begin{bmatrix} 2 & 1 & -2 \\ 0 & 2 & -1 \\ 0 & 0 & 3 \end{bmatrix}\begin{bmatrix} 1 & 2 & -1 \\ 0 & 3 & 2 \\ 0 & 0 & -1 \end{bmatrix}$

8. $\begin{bmatrix} 2 & 0 & 0 \\ 3 & 1 & 0 \\ 2 & 2 & 3 \end{bmatrix}\begin{bmatrix} 4 & 0 & 0 \\ -1 & 2 & 0 \\ -2 & -1 & -1 \end{bmatrix}$

9. $\begin{bmatrix} 2 & 1 & 1 & -1 \\ 0 & 3 & 1 & 1 \\ 0 & 0 & 2 & 1 \\ 1 & 0 & 0 & 2 \end{bmatrix}\begin{bmatrix} 1 & 2 & -2 \\ 3 & 1 & -4 \\ 0 & 1 & 2 \\ -2 & 0 & 1 \end{bmatrix}$

10. $\begin{bmatrix} 2 & -1 & 3 & 2 \\ 1 & 4 & -2 & 1 \end{bmatrix}\begin{bmatrix} 2 & 1 \\ -1 & 4 \\ 3 & -2 \\ 2 & 1 \end{bmatrix}$

14.3 THE ALGEBRA OF MATRICES

For Exercises 11 to 20, let

$$A = \begin{bmatrix} 3 & -1 & 2 \\ 0 & 1 & 1 \\ 4 & 0 & -1 \end{bmatrix} \quad B = \begin{bmatrix} 1 & -2 & 0 \\ 3 & 4 & 1 \\ -1 & 0 & 2 \end{bmatrix}$$

$$C = \begin{bmatrix} 0 & 2 & 1 \\ -2 & 0 & -1 \\ 1 & 1 & -2 \end{bmatrix} \quad D = \begin{bmatrix} 0 & 1 & 3 \\ 0 & 0 & 2 \\ 0 & 0 & 0 \end{bmatrix}$$

Then compute the indicated matrix

11. $A(B + C)$
12. $AB + AC$
13. $(A + B)C$
14. $AC + BC$
15. $AD + DA$
16. $AB - CD$
17. $(A - B)(C - D)$
18. $(AB - C)D$
19. $(2A - 3BC)(C + 2D)$
20. $C^3 - (AD)^2$

PROBLEMS 14.3

1. Prove Theorem 14.8.
2. Express the scalar multiple cA of a constant c times an $m \times n$ matrix A as the product of two matrices.
3. Prove that, for any scalars a and b and matrices A and B (such that AB is defined), we have
   (a) $(aA)B = A(aB) = a(AB)$
   (b) $(aA)(bB) = ab(AB)$
4. (a) If A is a 2×2 matrix such that $AB = BA$ for every 2×2 matrix B, then show that $A = cI$ for some real number c.
   (b) If A is a 2×2 matrix such that $AB = BA$ for every diagonal 2×2 matrix B, then show that A is diagonal.
5. Prove that
$$\begin{pmatrix} 1 & 1 \\ 0 & 1 \end{pmatrix}^n = \begin{pmatrix} 1 & n \\ 0 & 1 \end{pmatrix}$$
6. Find all 2×2 matrices that commute with the matrix
$$\begin{bmatrix} \cos \theta & \sin \theta \\ -\sin \theta & \cos \theta \end{bmatrix}$$
7. Let X be a 2×2 matrix of the form
$$X = \begin{pmatrix} x & y \\ -y & x \end{pmatrix}$$
and solve the matrix equation
$$aX^2 + bX + cI = 0$$

8. Find all 2×2 matrices X such that $X^2 = I$.

9. A matrix A such that $A^2 = I$ is said to be *involutory*. Prove that A is involutory if and only if
$$(I - A)(I + A) = 0$$

10. Find 2×2 matrices $A \neq 0$, $B \neq 0$, such that $A^2 B^2 = 0$.

11. A square matrix is *upper triangular* if all numbers below the main diagonal are zero and is *lower triangular* if all numbers above the main diagonal are zero. We say that the matrix is *strictly upper* (or lower) *triangular* if it also has all zeros on its main diagonal.
    (a) Prove that the product of two triangular matrices of the same type is a triangular matrix of the same type.
    (b) Prove that the product of a triangular matrix and a strictly triangular matrix of the same type is a strictly triangular matrix.

12. If A is a strictly rectangular $n \times n$ matrix, prove that $A^n = 0$.

13. The matrix A is said to be *idempotent* if $A^2 = A$. Find all diagonal $n \times n$ matrices that are idempotent. Find a nondiagonal idempotent 2×2 matrix.

14. Express the dot product of two vectors in $\mathbf{R}^n$ as the product of two matrices.

15. The *trace* $\operatorname{tr}(A)$ of an $n \times n$ matrix is defined to be the sum of the numbers on the main diagonal. Prove
$$\text{(a)} \quad \operatorname{tr}(A + B) = \operatorname{tr}(A) + \operatorname{tr}(B) \qquad \text{(b)} \quad \operatorname{tr}(AB) = \operatorname{tr}(BA)$$
Prove as a corollary that $AB - BA$ can never equal I.

14.4 SQUARE MATRICES

An $n \times n$ matrix is said to be *square*. Our interest in square matrices stems from the fact that any two $n \times n$ matrices can be multiplied to yield another $n \times n$ matrix. In technical terms, the set $\mathbf{R}^{n \times n}$ is closed under matrix multiplication. Adding the operation of matrix multiplication to the vector space structure of $\mathbf{R}^{n \times n}$ results in a much richer algebraic theory. We explore a little of this theory here.

Matrix multiplication satisfies the associative law and the two distributive laws but fails to satisfy the commutative law. We have pointed out that this noncommutative operation must be handled with some care. Further complicating the algebra of square matrices is the existence of zero divisors. A square matrix A is said to be a *zero divisor* if there exists a nonzero square matrix B such that either
$$AB = 0 \quad \text{or} \quad BA = 0$$

■ **Example 1.** The matrix
$$A = \begin{bmatrix} 1 & 0 \\ 1 & 0 \end{bmatrix}$$
is a zero divisor because the nonzero matrix

14.4 SQUARE MATRICES

$$B = \begin{bmatrix} 0 & 0 \\ 1 & 1 \end{bmatrix}$$

has the property that

$$AB = \begin{bmatrix} 1 & 0 \\ 1 & 0 \end{bmatrix}\begin{bmatrix} 0 & 0 \\ 1 & 1 \end{bmatrix} = \begin{bmatrix} 0 & 0 \\ 0 & 0 \end{bmatrix}$$

Note, however, that

$$BA = \begin{bmatrix} 0 & 0 \\ 1 & 1 \end{bmatrix}\begin{bmatrix} 1 & 0 \\ 1 & 0 \end{bmatrix} = \begin{bmatrix} 0 & 0 \\ 2 & 0 \end{bmatrix}$$

is not the zero matrix. ∎

The existence of zero divisors implies that the *cancellation law* of multiplication fails to hold, in general. This means that the equation

$$AB = AC$$

does not necessarily imply that $B = C$, even when A is a nonzero matrix. As an example, note that

$$\begin{bmatrix} 1 & 0 \\ 0 & 0 \end{bmatrix}\begin{bmatrix} 2 & 3 \\ 1 & 2 \end{bmatrix} = \begin{bmatrix} 2 & 3 \\ 0 & 0 \end{bmatrix}$$

and that

$$\begin{bmatrix} 1 & 0 \\ 0 & 0 \end{bmatrix}\begin{bmatrix} 2 & 3 \\ -2 & 1 \end{bmatrix} = \begin{bmatrix} 2 & 3 \\ 0 & 0 \end{bmatrix}$$

but that

$$\begin{bmatrix} 2 & 3 \\ 1 & 2 \end{bmatrix} \neq \begin{bmatrix} 2 & 3 \\ -2 & 1 \end{bmatrix}$$

The lack of the cancellation law adds another problem to the algebra of matrices, of course. However, we can define a class of square matrices in which the cancellation laws hold and for which algebra is almost normal.

■ **DEFINITION.** Let A be an $n \times n$ square matrix. Then the matrix B is an inverse of A if the matrix equations

$$AB = I, \quad BA = I$$

both hold. The matrix A is said to be *nonsingular* if it has an inverse; otherwise, A is *singular*.

■ **THEOREM 14.10.** The inverse of a nonsingular matrix is unique.

PROOF: Let A be a nonsingular matrix and suppose that the matrices B and C are inverses of A, that is, we have

$$AB = I, BA = I, AC = I, CA = I$$

Multiplying the equation $AB = I$ on the left by C, we get

$$C(AB) = CI = C$$

By the associative law, we know that

$$C(AB) = (CA)B$$

and by assumption $CA = I$. Hence we have

$$C(AB) = (CA)B = IB = B = C \blacksquare$$

The unique inverse of a nonsingular matrix A is denoted by

$$A^{-1}$$

Thus by definition we have the equations

$$AA^{-1} = I = A^{-1}A$$

■ *Example 2.* The matrix

$$A = \begin{bmatrix} 2 & 3 \\ 1 & 2 \end{bmatrix}$$

is nonsingular, simply because we can prove that it has the inverse

$$A^{-1} = \begin{bmatrix} 2 & -3 \\ -1 & 2 \end{bmatrix}$$

We check this by multiplication:

$$\begin{bmatrix} 2 & 3 \\ 1 & 2 \end{bmatrix}\begin{bmatrix} 2 & -3 \\ -1 & 2 \end{bmatrix} = \begin{bmatrix} 4-3 & -6+6 \\ 2-2 & -3+4 \end{bmatrix} = \begin{bmatrix} 1 & 0 \\ 0 & 1 \end{bmatrix}$$

and

$$\begin{bmatrix} 2 & -3 \\ -1 & 2 \end{bmatrix}\begin{bmatrix} 2 & 3 \\ 1 & 2 \end{bmatrix} = \begin{bmatrix} 4-3 & 6-6 \\ -2+2 & -3+4 \end{bmatrix} = \begin{bmatrix} 1 & 0 \\ 0 & 1 \end{bmatrix}$$

■ **THEOREM 14.11.** If the $n \times n$ matrices A and B are nonsingular, then their product AB is nonsingular and

$$(AB)^{-1} = B^{-1}A^{-1}$$

PROOF: We need only multiply:

$$(B^{-1}A^{-1})(AB) = B^{-1}(A^{-1}A)B$$
$$= B^{-1}IB$$
$$= B^{-1}B$$
$$= I$$

and

$$(AB)(B^{-1}A^{-1}) = A(BB^{-1})A^{-1}$$
$$= AIA^{-1}$$
$$= AA^{-1}$$
$$= I \ \blacksquare$$

This result says that the set of all nonsingular $n \times n$ matrices is closed under matrix multiplication. Furthermore, the next theorem tells us that there are no zero divisors in this set.

■ **THEOREM 14.12.** Let A be nonsingular. If the product AB is zero, then $B = 0$ and if $CA = 0$, then $C = 0$.
  PROOF: Suppose that $AB = 0$. Then we have

$$A^{-1}(AB) = (A^{-1}A)B = IB = B = 0$$

A similar argument proves the second part. ■

Our next result relates the rank of an $n \times n$ matrix and its singularity.

■ **THEOREM 14.13.** Let A be an $n \times n$ matrix. Then the following four statements are equivalent:

1. A is nonsingular.
2. Rank $A = n$.
3. The row echelon form of A is I.
4. A is a product of elementary matrices.

   PROOF: Let A be nonsingular and let its row echelon form be

$$B = E_s \cdots E_2 E_1 A$$

where $E_1, \cdots, E_s$ are the elementary matrices corresponding to a sequence of elementary row operations that reduce A to B. It is easy to show that each E_i is nonsingular (see Problem 14.4.1). From Theorem 14.11, B is nonsingular, and hence has an inverse. Because $BB^{-1} = I$, we know that the I in the nth row and nth column of I may also be written as the sum

$$\sum_{j=1}^{n} b_{nj} b_{jn}^{-1} = 1$$

where the numbers b_{nj} form the last row of B and b_{jn}^{-1} form the last column of B^{-1}. The fact that this sum equals 1 assures us that not all of the b_{nj} are zero. That is, the last row of the row echelon form B is not zero and hence there are no nonzero rows of B. It follows that rank $A = n$.

Next suppose that the rank of A equals n. It follows that every row of the row echelon form of A has an initial 1. However the initial one is the only nonzero number in its column. Since there are n columns (and n ones), the row echelon form of A must be I. Then there is a product of elementary matrices

$$E_s \cdots E_2 E_1 = R$$

such that

$$RA = I$$

The matrix R is nonsingular (Problem 14.4.1 and Theorem 14.11) and hence has an inverse. Therefore, we have

$$A = IA = (R^{-1}R)A = R^{-1}(RA) = R^{-1}I = R^{-1}$$

and it follows that $A^{-1} = R$. Having shown that $A = R^{-1}$, we need only know that R^{-1} is a product of elementary matrices. However, this too follows from Problem 14.4.1. The proof that part 4 implies part 1 is immediate. ∎

Theorem 14.13 provides a strong clue for computing the inverse of a matrix if the inverse exists. We need only reduce the matrix to row echelon form and keep track of the elementary matrices which correspond to the elementary row operations. If the row echelon form is I and the elementary matrices are $E_1, \cdots, E_s$, then $A^{-1} = E_s \cdots E_1$. The necessary bookkeeping in this procedure can be done automatically and very neatly by the following method. Suppose that A is an $n \times n$ matrix. We consider the $n \times 2n$ matrix designated by $[AI]$ and defined by

$$(AI) = \begin{bmatrix} a_{11} & a_{12} & \cdots & a_{1n} & 1 & 0 & 0 & \cdots & 0 \\ a_{21} & a_{22} & \cdots & a_{2n} & 0 & 1 & 0 & \cdots & 0 \\ \cdots & \cdots & \cdots & \cdots & \cdots & \cdots & \cdots & \cdots & \cdots \\ a_{n1} & a_{n2} & \cdots & a_{nn} & 0 & 0 & 0 & \cdots & 1 \end{bmatrix}$$

Now perform the row operations on $[AI]$, which reduce A to I. Then we automatically obtain the $n \times 2n$ matrix

$$[IA^{-1}]$$

(See Problem 14.4.2.)

14.4 SQUARE MATRICES

- **Example 3.** The matrix

$$A = \begin{bmatrix} 2 & 3 \\ 1 & 2 \end{bmatrix}$$

was shown in Example 2 to have the inverse

$$A^{-1} = \begin{bmatrix} 2 & -3 \\ -1 & 2 \end{bmatrix}$$

We obtain A^{-1} by the technique described above.

$$[AI] = \begin{bmatrix} 2 & 3 & 1 & 0 \\ 1 & 2 & 0 & 1 \end{bmatrix}$$

Using $a_{11} = 2$, we find that

$$\begin{bmatrix} 1 & \frac{3}{2} & \frac{1}{2} & 0 \\ 0 & \frac{1}{2} & -\frac{1}{2} & 1 \end{bmatrix}$$

The substitution of $a_{22} = \frac{1}{2}$ yields

$$\begin{bmatrix} 1 & 0 & 2 & -3 \\ 0 & 1 & -1 & 2 \end{bmatrix}$$

from which we see A^{-1} as stated.

EXERCISES 14.4

Find the inverse of each matrix in Exercises 1 to 10 (where the inverse exists) and verify your calculations by computing AA^{-1}.

1. $\begin{bmatrix} 2 & 1 \\ -1 & 2 \end{bmatrix}$

2. $\begin{bmatrix} 3 & 1 \\ 0 & 2 \end{bmatrix}$

3. $\begin{bmatrix} 1 & 4 & 3 \\ 2 & 0 & 1 \\ 1 & 3 & 0 \end{bmatrix}$

4. $\begin{bmatrix} 1 & -1 & 2 \\ 3 & 1 & -4 \\ 2 & -1 & 3 \end{bmatrix}$

5. $\begin{bmatrix} 1 & 2 & -1 \\ 0 & 1 & 3 \\ 0 & 0 & 2 \end{bmatrix}$

6. $\begin{bmatrix} 2 & 0 & 0 \\ 1 & -1 & 0 \\ 3 & 1 & -2 \end{bmatrix}$

7. $\begin{bmatrix} 1 & 1 & 3 \\ 1 & -1 & 2 \\ 3 & 1 & 1 \end{bmatrix}$

8. $\begin{bmatrix} 1 & 1 & 3 \\ 1 & -1 & 2 \\ 1 & 7 & 6 \end{bmatrix}$

9. $\begin{bmatrix} 1 & 0 & 0 & 1 \\ 0 & 1 & 0 & 1 \\ 1 & 0 & 1 & 0 \\ 1 & 0 & 1 & 1 \end{bmatrix}$ 10. $\begin{bmatrix} 0 & 0 & 0 & 1 & 0 \\ 0 & 1 & 0 & 0 & 0 \\ 1 & 0 & 0 & 0 & 0 \\ 0 & 0 & 1 & 0 & 0 \\ 0 & 0 & 0 & 0 & 1 \end{bmatrix}$

For Exercises 11 to 15, let

$$A = \begin{bmatrix} 3 & -1 & 2 \\ 1 & -1 & 4 \\ -2 & 3 & 2 \end{bmatrix} \qquad B = \begin{bmatrix} 1 & -2 & -3 \\ 0 & 2 & -1 \\ 1 & 4 & 0 \end{bmatrix}$$

and compute the given expression:

11. A^{-1} 12. B^{-1}
13. $(AB)^{-1}$ 14. $(AB^2)^{-1}$
15. $(A^2B^{-1})^{-1}$

PROBLEMS 14.4

1. Prove that each elementary matrix is nonsingular and has an elementary matrix as its inverse.

2. Prove that, if the $n \times 2n$ partitioned matrix $[AI]$, where A is nonsingular, is reduced by elementary row operations to the $n \times 2n$ partitioned matrix $[IB]$, then $B = A^{-1}$.

3. Show that a nonsingular triangular matrix has a triangular inverse.

4. Are there inverses for the following "skew-symmetric" matrices? Generalize your findings to higher degrees.

$$\begin{bmatrix} 0 & 1 & 1 \\ -1 & 0 & 1 \\ -1 & -1 & 0 \end{bmatrix} \qquad \begin{bmatrix} 0 & 1 & 1 & 1 \\ -1 & 0 & 1 & 1 \\ -1 & -1 & 0 & 1 \\ -1 & -1 & -1 & 0 \end{bmatrix}$$

5. Let A be a nilpotent $n \times n$ matrix (that is $A^k = 0$ for some natural number k). Prove that $I + A$ is nonsingular by finding its inverse. (Hint: $(I + A)(I - A) = I - A^2$, and so forth.)

6. Prove that

$$\begin{bmatrix} \cosh x & \sinh x \\ \sinh x & \cosh x \end{bmatrix}^n = \begin{bmatrix} \cosh nx & \sinh nx \\ \sinh nx & \cosh nx \end{bmatrix}$$

7. Let A and B be nonsingular. Is $A + B$ nonsingular? Is $A + B$ nonsingular if it is not zero?

8. Let
$$A = \begin{bmatrix} 1 & 2 & -1 \\ 3 & 0 & 1 \\ 2 & -1 & 2 \end{bmatrix}$$
and prove that
$$A^{-1} = \tfrac{1}{4}[I + 3A - A^2]$$

14.5 DETERMINANTS

Although we make little use of determinants in our study of matrix theory, a knowledge of determinants can be useful in many topics. In this section we list and discuss the basic properties of determinants. However, we leave the proofs of the theorems to the reader.

We associate with a square matrix A a real number called the *determinant* of A denoted by

$$\det A$$

This association provides a function $\det: \mathbf{R}^{n \times n} \to \mathbf{R}$ which we shall define by induction on the *degree* n of the $n \times n$ matrix. A 1×1 matrix $A = (a_{11})$ is simply a real number and we define

$$\det(a_{11}) = a_{11}$$

This gives us the beginning of the inductive definition.

Let A be an $n \times n$ matrix, $n > 1$. The $(n-1) \times (n-1)$ matrix formed by omitting the ith row and jth column of A is called the *minor* of the number a_{ij}. We shall denote the minor of a_{ij} by M_{ij}. Now the inductive definition of the determinant function is completed by the equation

$$\det A = \sum_{j=1}^{n} (-1)^{1+j} a_{ij} \det M_{ij}$$

■ **Example 1.** For a 2×2 matrix we have

$$\det \begin{bmatrix} a_{11} & a_{12} \\ a_{21} & a_{22} \end{bmatrix} = a_{11} \det M_{11} - a_{12} \det M_{12}$$
$$= a_{11} \det(a_{22}) - a_{12} \det(a_{21})$$
$$= a_{11} a_{22} - a_{12} a_{21}$$

Similarly, for a 3×3 matrix the definition provides that

$$\det \begin{bmatrix} a_{11} & a_{12} & a_{13} \\ a_{21} & a_{22} & a_{23} \\ a_{31} & a_{32} & a_{33} \end{bmatrix} = a_{11} \det M_{11} - a_{12} \det M_{12} + a_{13} \det M_{13}$$

$$= a_{11} \det \begin{bmatrix} a_{22} & a_{23} \\ a_{32} & a_{33} \end{bmatrix} - a_{12} \det \begin{bmatrix} a_{21} & a_{23} \\ a_{31} & a_{33} \end{bmatrix} + a_{13} \det \begin{bmatrix} a_{21} & a_{22} \\ a_{31} & a_{32} \end{bmatrix}$$
$$= a_{11}(a_{22}a_{33} - a_{23}a_{32}) - a_{12}(a_{21}a_{33} - a_{23}a_{31}) + a_{13}(a_{21}a_{32} - a_{22}a_{31})$$
$$= a_{11}a_{22}a_{33} - a_{11}a_{23}a_{32} - a_{12}a_{21}a_{33} + a_{12}a_{23}a_{31} + a_{13}a_{21}a_{32} - a_{13}a_{22}a_{31} \blacksquare$$

Let a_{ij} be a number in the $n \times n$ matrix A. Then the number
$$A_{ij} = (-1)^{i+j} \det M_{ij}$$

is called the *cofactor* of a_{ij}. Our inductive definition of the determinant is then written as
$$\det a = \sum_{j=1}^{n} a_{1j} A_{1j}$$

This definition places undue emphasis upon the first row of the matrix. The next result, which we leave unproved, states that we can "expand" $\det A$ in terms of the cofactors of any row or of any column.

■ **THEOREM 14.14.** The determinant of the $n \times n$ matrix A equals the sum of the products of the numbers in any row times their cofactors.

In symbols, we have
$$\det A = \sum_{j=1}^{n} a_{ij} A_{ij} = \sum_{j=i}^{n} (-1)^{i+j} a_{ij} \det M_{ij}$$

for any fixed $i = 1, 2, \cdots, n$. Similarly, $\det A$ equals the sum of the products of the numbers in any column times their cofactors, that is, for any fixed $j = 1, 2, \cdots, n$, we have
$$\det A = \sum_{i=1}^{n} a_{ij} A_{ij} = \sum_{i=1}^{n} (-1)^{i+j} a_{ij} \det M_{ij}$$

■ *Example 2.* We compute the determinant of a 3×3 matrix first by expanding it in terms of the cofactors of its second row and then by expanding it in terms of the cofactors of its third column.

$$\det \begin{bmatrix} 2 & 1 & -1 \\ 0 & 2 & 3 \\ -2 & 1 & 2 \end{bmatrix} = -(0) \det \begin{bmatrix} 1 & -1 \\ 1 & 2 \end{bmatrix} + (2) \det \begin{bmatrix} 2 & -1 \\ -2 & 2 \end{bmatrix}$$

$$-(3) \det \begin{bmatrix} 2 & 1 \\ -2 & 1 \end{bmatrix} = -0(2+1) + 2(4-2) - 3(2+2) = -8$$

$$\det\begin{bmatrix} 2 & 1 & -1 \\ 0 & 2 & 3 \\ -2 & 1 & 2 \end{bmatrix} = +(-1)\det\begin{bmatrix} 0 & 2 \\ -2 & 1 \end{bmatrix} - (3)\det\begin{bmatrix} 2 & 1 \\ -2 & 1 \end{bmatrix}$$

$$+ (2)\det\begin{bmatrix} 2 & 1 \\ 0 & 2 \end{bmatrix} = -1(0+4) - 3(2+2) + 2(4-0) = -8$$

■ **THEOREM 14.15.** (a) If the matrix B is obtained from the square matrix A by an elementary row (or column) operation of type 1, then $\det B = -\det A$.

(b) If the matrix B is obtained from the square matrix A by an elementary row (or column) operation of type 2, then

$$\det B = \lambda \det A$$

where λ is the constant multiplier.

(c) If the matrix B is obtained from the square matrix A by an elementary row (or column) operation of type 3, then

$$\det B = \det A$$

The basic properties of determinants given in Theorem 14.15 can be used to aid in an evaluation of a determinant (see Example 3.) However, even modestly large matrices present severe computation problems, problems of sheer magnitude. For instance, a 10×10 determinant requires over 6,000,000 arithmetic operations in its evaluation. Hence determinants of large order are used primarily for theoretical purposes.

■ *Example 3.* Consider the 4×4 determinant

$$\det\begin{bmatrix} 3 & 2 & 4 & 1 \\ -2 & 4 & 6 & -4 \\ 1 & 2 & 10 & -5 \\ 2 & -4 & -4 & 3 \end{bmatrix}$$

By part (b) of Theorem 14.15, the constant 2, which is a factor of the second row, can be brought out as a factor. Hence the determinant above equals

$$2 \det\begin{bmatrix} 3 & 2 & 4 & 1 \\ -1 & 2 & 3 & -2 \\ 1 & 2 & 10 & -5 \\ 2 & -4 & -4 & 3 \end{bmatrix}$$

Next we subtract the first row from the second row and from the third row, and then add twice the first row to the fourth row. These elementary row operations result in a matrix with equal determinant [part (c) of Theorem 14.15]; hence the original determinant has been simplified to

$$2 \det \begin{bmatrix} 3 & 2 & 4 & 1 \\ -4 & 0 & -1 & -3 \\ -2 & 0 & 6 & -6 \\ 8 & 0 & 4 & 5 \end{bmatrix}$$

The expansion of this determinant in terms of cofactors of the second column has all terms equal to zero except the first. Thus we have the equation

$$\det \begin{bmatrix} 3 & 2 & 4 & 1 \\ -4 & 0 & -1 & -3 \\ -2 & 0 & 6 & -6 \\ 8 & 0 & 4 & 5 \end{bmatrix} = -(2) \det \begin{bmatrix} -4 & -1 & -3 \\ -2 & 6 & -6 \\ 8 & 4 & 5 \end{bmatrix}$$

The reader may verify that the 3×3 determinant has the value -10. Hence the original 4×4 determinant has the value 40.

■ **THEOREM 14.16.** Let A and B be square matrices. Then

$$\det AB = \det A \cdot \det B$$

■ *Example 4.* Consider the matrices

$$A = \begin{bmatrix} 4 & 1 \\ 3 & 2 \end{bmatrix} \quad \text{and} \quad B = \begin{bmatrix} -1 & 3 \\ 2 & -4 \end{bmatrix}$$

Then $\det A = 5$ and $\det B = -2$. We also have

$$AB = \begin{bmatrix} 4 & 1 \\ 3 & 2 \end{bmatrix} \begin{bmatrix} -1 & 3 \\ 2 & -4 \end{bmatrix}$$

$$= \begin{bmatrix} (4)(-1) + (1)(2) & (4)(3) + (1)(-4) \\ (3)(-1) + (2)(2) & (3)(3) + (2)(-4) \end{bmatrix}$$

$$= \begin{bmatrix} -2 & 8 \\ 1 & 1 \end{bmatrix}$$

Hence $\det AB = -10 = \det A \cdot \det B$. ■

Let A be an $m \times n$ matrix. The *transpose* A^T of A is the $n \times m$ matrix whose kth row is the kth column of A. In symbols, if $A = (a_{ij})$. As examples, we have

$$\begin{bmatrix} 1 & 2 & 3 \\ 4 & 5 & 6 \end{bmatrix}^T = \begin{bmatrix} 1 & 4 \\ 2 & 5 \\ 3 & 6 \end{bmatrix}$$

and

$$\begin{bmatrix} 1 & 2 & 3 \\ 4 & 5 & 6 \\ 7 & 8 & 9 \end{bmatrix}^T = \begin{bmatrix} 1 & 4 & 7 \\ 2 & 5 & 8 \\ 3 & 6 & 9 \end{bmatrix}$$

■ **THEOREM 14.17.** Let A be a square matrix. Then

$$\det A^T = \det A$$

Let A be a square matrix. The transpose of the matrix of cofactors of the numbers in A is called the *adjoint* of A, denoted by adj A. Thus if a_{ij} is the number in the ith row and jth column of A, then its cofactor $A_{ij} = (-1)^{i+j} \det M_{ij}$ is the number in the jth row and ith column of adj A. In symbols,

$$\text{adj } A = [A_{ji}]$$

■ **Example 5.** If we have a 2×2 matrix

$$A = \begin{bmatrix} 1 & 2 \\ 3 & 4 \end{bmatrix}$$

then $A_{11} = 4$, $A_{12} = -3$, $A_{21} = -2$ and $A_{22} = 1$. Thus

$$\text{adj } A = \begin{bmatrix} A_{11} & A_{21} \\ A_{12} & A_{22} \end{bmatrix} = \begin{bmatrix} 4 & -2 \\ -3 & 1 \end{bmatrix}$$

■ **Example 6.** Let

$$A = \begin{bmatrix} a_{11} & a_{12} & a_{13} \\ a_{21} & a_{22} & a_{23} \\ a_{31} & a_{32} & a_{33} \end{bmatrix}$$

Then

$$\text{adj } A = \begin{bmatrix} +\det M_{11} & -\det M_{21} & +\det M_{31} \\ -\det M_{12} & +\det M_{22} & -\det M_{32} \\ +\det M_{13} & -\det M_{23} & +\det M_{33} \end{bmatrix}$$

$$= \begin{bmatrix} a_{22}a_{33} - a_{23}a_{32} & -(a_{12}a_{33} - a_{13}a_{32}) & a_{12}a_{23} - a_{13}a_{22} \\ -(a_{21}a_{33} - a_{23}a_{31}) & a_{11}a_{33} - a_{13}a_{31} & -(a_{11}a_{23} - a_{13}a_{21}) \\ a_{21}a_{32} - a_{22}a_{31} & -(a_{11}a_{32} - a_{12}a_{31}) & a_{11}a_{22} - a_{12}a_{21} \end{bmatrix} \blacksquare$$

- **THEOREM 14.18.** Let A be a square matrix. Then we have
$$A(\operatorname{adj} A) = (\det A)I = (\operatorname{adj} A)A$$

- *Example 7.* In Example 5 we computed that
$$\operatorname{adj} \begin{bmatrix} 1 & 2 \\ 3 & 4 \end{bmatrix} = \begin{bmatrix} 4 & -2 \\ -3 & 1 \end{bmatrix}$$

Then
$$\begin{bmatrix} 1 & 2 \\ 3 & 4 \end{bmatrix} \begin{bmatrix} 4 & -2 \\ -3 & 1 \end{bmatrix} = \begin{bmatrix} (1)(4) + (2)(-3) & (1)(-2) + (2)(1) \\ (3)(4) + (4)(-3) & (3)(-2) + (4)(1) \end{bmatrix}$$
$$= \begin{bmatrix} -2 & 0 \\ 0 & -2 \end{bmatrix}$$

Since we have
$$\det \begin{bmatrix} 1 & 2 \\ 3 & 4 \end{bmatrix} = (1)(4) - (2)(3) = -2$$

the equation
$$\begin{bmatrix} 1 & 2 \\ 3 & 4 \end{bmatrix} \cdot \operatorname{adj} \begin{bmatrix} 1 & 2 \\ 3 & 4 \end{bmatrix} = \left(\det \begin{bmatrix} 1 & 2 \\ 3 & 4 \end{bmatrix}\right) \begin{bmatrix} 1 & 0 \\ 0 & 1 \end{bmatrix} = \begin{bmatrix} -2 & 0 \\ 0 & -2 \end{bmatrix}$$

has been verified.

- **THEOREM 14.19.** The square matrix A is nonsingular if and only if $\det A \neq 0$. The inverse of the nonsingular matrix A is given by
$$A^{-1} = \frac{1}{\det A} \operatorname{adj} A$$

- *Example 8.* Consider the 3×3 matrix
$$A = \begin{bmatrix} 1 & 2 & 3 \\ 2 & 3 & 1 \\ 3 & 1 & 2 \end{bmatrix}$$

Then

$$\operatorname{adj} A = \begin{bmatrix} +\det\begin{bmatrix}3 & 1\\1 & 2\end{bmatrix} & -\det\begin{bmatrix}2 & 3\\1 & 2\end{bmatrix} & +\det\begin{bmatrix}2 & 3\\3 & 1\end{bmatrix} \\ -\det\begin{bmatrix}2 & 1\\3 & 2\end{bmatrix} & +\det\begin{bmatrix}1 & 3\\3 & 2\end{bmatrix} & -\det\begin{bmatrix}1 & 3\\2 & 1\end{bmatrix} \\ +\det\begin{bmatrix}2 & 3\\3 & 1\end{bmatrix} & -\det\begin{bmatrix}1 & 2\\3 & 1\end{bmatrix} & +\det\begin{bmatrix}1 & 2\\2 & 3\end{bmatrix} \end{bmatrix}$$

$$= \begin{bmatrix} 5 & -1 & -7 \\ -1 & -7 & 5 \\ -7 & 5 & -1 \end{bmatrix}$$

It is an easy matter to verify the multiplication

$$\begin{bmatrix} 5 & -1 & -7 \\ -1 & -7 & 5 \\ -7 & 5 & -1 \end{bmatrix} \begin{bmatrix} 1 & 2 & 3 \\ 2 & 3 & 1 \\ 3 & 1 & 2 \end{bmatrix} = \begin{bmatrix} -18 & 0 & 0 \\ 0 & -18 & 0 \\ 0 & 0 & -18 \end{bmatrix}$$

Hence A^{-1} is given by

$$A^{-1} = \begin{bmatrix} -\frac{5}{18} & \frac{1}{18} & \frac{7}{18} \\ \frac{1}{18} & \frac{7}{18} & -\frac{5}{18} \\ \frac{7}{18} & -\frac{5}{18} & \frac{1}{18} \end{bmatrix}$$

Using elementary row operations and Theorem 14.15, evaluate each determinant in Exercises 1 to 10.

EXERCISES 14.5

1. $\det\begin{bmatrix}3 & 1\\2 & -3\end{bmatrix}$

2. $\det\begin{bmatrix}4 & 2\\-3 & 1\end{bmatrix}$

3. $\det\begin{bmatrix}1 & 3 & 1\\0 & 1 & 2\\0 & 0 & 3\end{bmatrix}$

4. $\det\begin{bmatrix}1 & 0 & 1\\1 & 1 & 0\\0 & 1 & 1\end{bmatrix}$

5. $\det\begin{bmatrix}1 & -1 & -1\\1 & -1 & 1\\-1 & 1 & -1\end{bmatrix}$

6. $\det\begin{bmatrix}6 & 2 & -3\\-2 & 1 & 2\\4 & 1 & -2\end{bmatrix}$

7. $\det\begin{bmatrix}1 & 0 & 0 & 3\\3 & 0 & 0 & 3\\0 & 1 & 1 & 0\\1 & 0 & 0 & 1\end{bmatrix}$

8. $\det\begin{bmatrix}1 & 3 & 0 & -1\\1 & -1 & 0 & 1\\0 & 1 & -1 & 0\\4 & 2 & 0 & 0\end{bmatrix}$

9. $\det \begin{bmatrix} 1 & 0 & 0 & 0 & 2 \\ -1 & 1 & 0 & 1 & 3 \\ 0 & 1 & 2 & 1 & 0 \\ 1 & 0 & 1 & 1 & 0 \\ 0 & 1 & 0 & 1 & 2 \end{bmatrix}$

10. $\det \begin{bmatrix} 1 & 3 & -2 & 2 & -1 & 1 \\ 1 & -3 & -1 & 1 & 1 & 2 \\ 0 & 0 & 3 & 6 & 1 & -1 \\ 0 & 0 & 0 & 1 & 2 & 2 \\ 0 & 0 & 0 & 0 & 2 & 1 \\ 0 & 0 & 0 & 0 & 3 & -1 \end{bmatrix}$

In Exercises 11 to 16, let A and B be as follows and compute the given determinants

$$A = \begin{bmatrix} 1 & 2 & 0 \\ 3 & -1 & 1 \\ -1 & 2 & 3 \end{bmatrix} \quad B = \begin{bmatrix} 4 & -1 & -2 \\ 1 & 3 & 2 \\ 0 & 2 & -1 \end{bmatrix}$$

11. $\det A$
12. $\det B$
13. $\det(AB)$
14. $\det(BA)$
15. $\det(\operatorname{adj} A)$
16. $\det[A(\operatorname{adj} B)]$
17. $\det(ABA)$
18. $\det(AB^{-1}B)$
19. $\det(A^2 B^2)$
20. $\det(AB^2 A)$

PROBLEMS 14.5

1. The square matrix A is said to be *nilpotent* if $A^p = 0$ for some natural number p. Prove that a nilpotent matrix is singular.

2. Generalize Problem 1 by proving that any zero divisor must be singular.

3. A square matrix A is said to be *skew-symmetric* if $A^T = -A$. For such a matrix of order n prove that

$$\det A = (-1)^n \det A$$

and hence that skew-symmetric matrices of odd order are singular.

4. Let A be the square matrix of order n having

$$a_{ij} = 1 \text{ if } i \neq j, \quad a_{ii} = 0$$

Prove that

$$\det A = (-1)^n (1 - n)$$

5. Show that the equation of the circle passing through three noncollinear points (x_1, y_1), (x_2, y_2) and (x_3, y_3) is

$$\det \begin{bmatrix} x^2 + y^2 & x & y & 1 \\ x_1^2 + y_1^2 & x_1 & y_1 & 1 \\ x_2^2 + y_2^2 & x_2 & y_2 & 1 \\ x_3^2 + y_3^2 & x_3 & y_3 & 1 \end{bmatrix} = 0$$

6. Prove that, for the $n \times n$ matrix A, we have $\det(\operatorname{adj} A) = (\det A)^{n-1}$, $\det A^{-1} = (\det A)^{n-2}$ if $\det A \neq 0$.

14.5 DETERMINANTS

7. Prove that
$$\det \begin{bmatrix} a & 1 & 1 & 1 \\ 1 & a & 1 & 1 \\ 1 & 1 & a & 1 \\ 1 & 1 & 1 & a \end{bmatrix} = (a+3)(a-1)^3$$

8. Let A be the $n \times n$ matrix
$$\begin{bmatrix} a+x & a & a & \cdots & a \\ a & a+x & a & \cdots & a \\ \cdots & \cdots & \cdots & \cdots & \cdots \\ a & a & a & \cdots & a+x \end{bmatrix}$$
and prove that
$$\det A = x^{n-1}(na + x)$$

9. Prove that
$$\det \begin{bmatrix} a-b-c & 2a & 2a \\ 2b & b-c-a & 2b \\ 2c & 2c & c-b-a \end{bmatrix} = (a+b+c)^3$$

10. Determine the value of the *Vandermonde determinant*
$$\det \begin{bmatrix} 1 & 1 & 1 & \cdots & 1 \\ x_1 & x_2 & x_3 & \cdots & x_n \\ x_1^2 & x_2^2 & x_3^2 & \cdots & x_n^2 \\ \cdots & \cdots & \cdots & \cdots & \cdots \\ x_1^{n-1} & x_2^{n-1} & x_3^{n-1} & \cdots & x_n^{n-1} \end{bmatrix}$$

11. If A is nonsingular, show that $(A^T)^{-1} = (A^{-1})^T$.

12. Establish the equalities:
    (a) $\text{adj}(AB) = (\text{adj } B)(\text{adj } A)$
    (b) $\text{adj}(A^T) = (\text{adj } A)^T$

13. A square matrix A is said to be *symmetric* if $A^T = A$. Prove that the inverse of a symmetric matrix is also symmetric.

14. Let f_1 and f_2 be two solutions of the linear differential equation
$$y'' + ay' + by = 0$$
and define the *Wronskian* of f_1 and f_2,
$$W = \det \begin{bmatrix} f_1 & f_2 \\ f_1' & f_2' \end{bmatrix}$$
Prove that
$$W(x) = W_0 e^{-ax}$$
Generalize this result to higher order differential equations.

15. Let φ_{ij}, $i = 1, \cdots, n, j = 1, \cdots, n$, be differentiable functions of one variable. Define the function of n variables

$$f(x_1, x_2, \cdots, x_n) = \det \begin{bmatrix} \varphi_{11}(x_1) & \varphi_{12}(x_1) & \cdots & \varphi_{1n}(x_1) \\ \varphi_{21}(x_2) & \varphi_{22}(x_2) & \cdots & \varphi_{2n}(x_2) \\ \cdots & \cdots & \cdots & \cdots \\ \varphi_{n1}(x_n) & \varphi_{n2}(x_n) & \cdots & \varphi_{nn}(x_n) \end{bmatrix}$$

and determine its first partial derivatives.

14.6 VECTOR SPACES AND SUBSPACES

Consider a homogeneous $m \times n$ system of linear equations

$$A\mathbf{x} = \mathbf{0}$$

Such a system obviously has the solution $\mathbf{x} = \mathbf{0}$ and hence is a consistent system; its solution set is not empty. Of course, its solution set is a subset of $\mathbf{R}^n$. Now suppose that $\mathbf{x}_1$ is a solution of this system and that α is any real number. Then we have

$$A(\alpha \mathbf{x}_1) = \alpha A \mathbf{x}_1 = \alpha \mathbf{0} = \mathbf{0}$$

whence the vector $\alpha \mathbf{x}_1$ is also a solution. Furthermore, given a second solution $\mathbf{x}_2$, we have

$$A(\mathbf{x}_1 + \mathbf{x}_2) = A\mathbf{x}_1 + A\mathbf{x}_2 = \mathbf{0} + \mathbf{0} = \mathbf{0}$$

This proves that $\mathbf{x}_1 + \mathbf{x}_2$ is also a solution. We have shown that the set of solutions of the system $A\mathbf{x} = \mathbf{0}$ is a nonempty subset of $\mathbf{R}^n$ and is closed under scalar multiplication and vector addition. These considerations lead to the concept of a "vector subspace" of a "vector space."

■ **DEFINITION.** A (real) *vector space* is a set $\mathcal{V}$ endowed with two operations, addition and scalar multiplication by real numbers, which satisfy the following properties (see Section 1.8)

V1. If $\mathbf{u}, \mathbf{v} \in \mathcal{V}$ then $\mathbf{u} + \mathbf{v} \in \mathcal{V}$ (closure under addition)
V2. If $\mathbf{u}, \mathbf{v}, \mathbf{w} \in \mathcal{V}$, then

$$\mathbf{u} + (\mathbf{v} + \mathbf{w}) = (\mathbf{u} + \mathbf{v}) + \mathbf{w} \text{ (associative law)}$$

V3. There is an element $\mathbf{0} \in \mathcal{V}$ such that

$$\mathbf{0} + \mathbf{u} = \mathbf{u}$$

for every $\mathbf{u} \in \mathcal{V}$ (the existence of an additive identity).
V4. For each $\mathbf{u} \in \mathcal{V}$, there is an element $-\mathbf{u} \in \mathcal{V}$ such that

$$\mathbf{u} + (-\mathbf{u}) = \mathbf{0} \text{ (existence of additive inverses)}$$

V5. If $\mathbf{u}, \mathbf{v} \in \mathcal{V}$, then

$$\mathbf{u} + \mathbf{v} = \mathbf{v} + \mathbf{u} \text{ (commutative law)}$$

V6. If $\alpha \in \mathbf{R}$ and $\mathbf{u} \in \mathcal{V}$, then $\alpha\mathbf{u} \in \mathcal{V}$ (closure under scalar multiplication)

V7. For every $\mathbf{u} \in \mathcal{V}$, $1\mathbf{u} = \mathbf{u}$

V8. If $r, s \in \mathbf{R}$ and $\mathbf{u} \in \mathcal{V}$, then
$$(r + s)\mathbf{u} = r\mathbf{u} + s\mathbf{u}$$

V9. If $r \in \mathbf{R}$ and $\mathbf{u}, \mathbf{v} \in \mathcal{V}$, then
$$r(\mathbf{u} + \mathbf{v}) = r\mathbf{u} + r\mathbf{v}$$

V10. If $r, s \in \mathbf{R}$ and $\mathbf{u} \in \mathcal{V}$, then
$$(rs)\mathbf{u} = r(s\mathbf{u})$$

■ **Example 1.** (a) Any coordinate space $\mathbf{R}^n$ is a vector space.
(b) The set of convergent infinite sequences is a vector space.
(c) The set $P[x]$ of all polynomial functions is a vector space.
(d) The set $\mathbf{R}^{m \times n}$ of all $m \times n$ matrices is a vector space under the usual addition of matrices and multiplication by real numbers.
(e) The set $C[a, b]$ of all continuous functions in the closed interval $\{x : a \leq x \leq b\}$ is a vector space.
(f) The set of all solutions of a given homogeneous linear differential equation
$$D^n y + a_1(x) D_y^{n-1} + \cdots + a_n(x) y = 0$$
is a vector space.

The reader may provide the verification of these statements.

■ **DEFINITION.** A nonempty subset $\mathcal{U}$ of a vector space $\mathcal{V}$ is said to be a *vector subspace* of $\mathcal{V}$ if $\mathcal{U}$ is closed under addition and scalar multiplication. Note that, because $\mathcal{U}$ is a subset of $\mathcal{V}$, all of the axioms V1 to V10 except V1 and V6 are automatically satisfied. Thus by adding the two closure axioms as conditions on the subset $\mathcal{U}$, we force $\mathcal{U}$ to be a vector space in its own right.

■ **Example 2.** (a) The opening paragraph of this section proves that the set of solutions of a homogeneous $m \times n$ system of linear equations is a vector subspace of $\mathbf{R}^n$.
(b) The set of all infinite sequences that converge to zero is a vector subspace of the set of all convergent infinite sequences.
(c) In $P[x]$, the subset $P_n[x]$ consisting of all polynomial functions of degree $\leq n$ is a vector subspace
(d) The set of all $n \times n$ matrices satisfying the condition
$$a_{ij} = 0 \quad \text{if } i \neq j$$
(the so-called *diagonal* matrices) is a vector subspace of the space of all $n \times n$ matrices.
(e) The subset of $C[a, b]$ consisting of all functions f having $f(c) = 0$

is a vector subspace of $C[a, b]$. (The reader should provide proofs of these statements.) ∎

Let $\mathbf{v}_1, \cdots, \mathbf{v}_k$ be elements of the vector space $\mathcal{V}$. We say that the sequence $(\mathbf{v}_1, \cdots, \mathbf{v}_k)$ is (linearly) *dependent* if there are real numbers $\alpha_1, \cdots, \alpha_k$, not all zero, such that

$$\alpha_1 \mathbf{v}_1 + \cdots + \alpha_k \mathbf{v}_k = \mathbf{0}$$

If the sequence $(\mathbf{v}_1, \cdots, \mathbf{v}_R)$ is not dependent, then it is (linearly) *independent*. Explicitly, the sequence $(\mathbf{v}_1, \cdots, \mathbf{v}_k)$ is independent if the equation $\alpha_1 \mathbf{v}_1 + \cdots + \alpha_k \mathbf{v}_k = \mathbf{0}$ implies $\alpha_1 = \cdots = \alpha_k = 0$.

∎ *Example 3.* (a) The unit basis vectors $\mathbf{u}_1, \mathbf{u}_2, \cdots, \mathbf{u}_n$ in $\mathbf{R}^n$ form an independent sequence.
(b) The vectors $\mathbf{v}_1 = (0, 1, 3)$, $\mathbf{v}_2 = (1, 4, 2)$, $\mathbf{v}_3 = (4, 13, -1)$ form a dependent sequence because

$$-3\mathbf{v}_1 + 4\mathbf{v}_2 - \mathbf{v}_3 = \mathbf{0}$$

(c) In the vector space $P_n[x]$ of all polynomial functions of degree $\leq n$, the sequence $(1, x, x^2, \cdots, x^n)$ is independent. However, any sequence containing more than $n + 1$ polynomial functions is dependent.
(d) In the vector space $C[0, 2\Pi]$ any sequence

$$(\cos x, \cos 2x, \cdots, \cos nx)$$

is independent. [Note that (c) and (d) require proofs. These are left for the reader to supply.]

The proof of the next theorem is an easy exercise in the use of definitions and will be omitted.

∎ **THEOREM 14.20.** Let S be any subset of a vector space $\mathcal{V}$. The collection of all finite linear combinations,

$$\alpha_1 s_1 + \cdots + \alpha_n s_n, \; \alpha_i \in \mathbf{R}$$

of vectors in S is a vector subspace of $\mathcal{V}$.

Let S be a subset of the vector space $\mathcal{V}$. The vector subspace of all finite linear combinations of vectors in S is denoted by $\langle s \rangle$. We say that S *spans* $\langle S \rangle$ or *generates* $\langle S \rangle$.

A sequence $B = (\mathbf{v}_1, \cdots, \mathbf{v}_k, \cdots)$ (which can be infinite) is a *basis* for $\mathcal{V}$ if (1) each finite subsequence of $\mathcal{V}$ is independent and (2) $\langle B \rangle = \mathcal{V}$.

∎ *Example 4.* The sequence of unit vectors $(\mathbf{u}_1, \cdots, \mathbf{u}_n)$ in $\mathbf{R}^n$ is a basis for $\mathbf{R}^n$. However, there are many bases for $\mathbf{R}^n$. For instance, we claim that the vectors $\mathbf{v}_1 = (1, 2, 1)$, $\mathbf{v}_2 = (-1, 0, 3)$, and $\mathbf{v}_3 = (2, -1, -4)$ form a basis for $\mathbf{R}^3$. The proof of this consists merely in solving the vector equation

$$x\mathbf{v}_1 + y\mathbf{v}_2 + z\mathbf{v}_3 = \mathbf{v}$$

14.6 VECTOR SPACES AND SUBSPACES

where $\mathbf{v}$ is a vector in $\mathbf{R}^3$. In particular we set $\mathbf{v} = \mathbf{0}$ and show that $x\mathbf{v}_1 + y\mathbf{v}_2 + z\mathbf{v}_3 = \mathbf{0}$ implies that $x = y = z = 0$, thus proving that $\{\mathbf{v}_1, \mathbf{v}_2, \mathbf{v}_3\}$ is an independent set. Then, setting $\mathbf{v} = (a, b, c)$, a solution of the last equation tells us that $\langle \mathbf{v}_1, \mathbf{v}_2, \mathbf{v}_3 \rangle = \mathbf{R}^3$ and completes the proof. Now we have

$$x\mathbf{v}_1 + y\mathbf{v}_2 + z\mathbf{v}_3 = (x, 2x, x) + (-y, 0, 3y) + (2z, -z, -4z)$$
$$= (x - y + 2z, 2x - z, x + 3y - 4z)$$

and hence the vector equations provide the 3×3 systems

$$\begin{array}{ll} x - y + 2z = 0 & x - y + 2z = a \\ 2x - z = 0 & 2x - z = b \\ x + 3y - 4z = 0 & x + 3y - 4z = c \end{array}$$

The reader may verify that the only solution of the homogeneous system is $x = y = z = 0$ and that the unique solution of the nonhomogeneous system is

$$(\tfrac{3}{8}a + \tfrac{1}{4}b + \tfrac{1}{8}c, \tfrac{7}{8}a - \tfrac{3}{4}b + \tfrac{5}{8}c, \tfrac{3}{4}a - \tfrac{1}{2}b + \tfrac{1}{4}c)$$

Therefore $\{\mathbf{v}_1, \mathbf{v}_2, \mathbf{v}_3\}$ is a basis for $\mathbf{R}^3$.

■ **LEMMA 14.21.** Let $(\mathbf{u}_1, \cdots, \mathbf{u}_n)$ be a basis of a vector space $\mathcal{V}$. If the sequence $(\mathbf{v}_1, \cdots, \mathbf{v}_k)$ is independent, then $k \leq n$.

PROOF: We know that we can write

$$\mathbf{v}_1 = \alpha_{11}\mathbf{u}_1 + \cdots + \alpha_{1n}\mathbf{u}_n$$

By renumbering, if need be, we assume that $\alpha_{1n} \neq 0$. Then we have

$$\mathbf{u}_n = \frac{1}{\alpha_{1n}}\mathbf{v}_1 - \frac{\alpha_{11}}{\alpha_{1n}}\mathbf{u}_1 - \cdots - \frac{\alpha_{1n-1}}{\alpha_{1n}}\mathbf{u}_{n-1}$$

It follows that $\mathcal{V}$ is spanned by the sequence $(\mathbf{v}_1, \mathbf{u}_1, \cdots, \mathbf{u}_{n-1})$. Hence there are constants such that

$$\mathbf{v}_2 = \beta_{21}\mathbf{v}_1 + \alpha_{21}\mathbf{u}_1 + \cdots + \alpha_{2n-1}\mathbf{u}_{n-1}$$

Again renumbering, if necessary, we may assume that $\alpha_{2n-1} \neq 0$. Hence we have

$$\mathbf{u}_{n-1} = \frac{1}{\alpha_{2n-1}}\mathbf{v}_2 - \frac{\beta_{11}}{\alpha_{2n-1}}\mathbf{v}_1 - \frac{\alpha_{21}}{\alpha_{2n-1}}\mathbf{u}_1 - \cdots - \frac{\alpha_{2n-2}}{\alpha_{2n-1}}\mathbf{u}_{n-2}$$

Therefore $(\mathbf{v}_2, \mathbf{v}_1, \mathbf{u}_1, \cdots, \mathbf{u}_{n-2})$ spans $\mathcal{V}$.

If k were larger than n, then we could repeat the above procedure a total of $(n - 2)$ more times to produce a sequence

$$\{\mathbf{v}_n, \mathbf{v}_{n-1}, \cdots, \mathbf{v}_2, \mathbf{v}_1\}$$

which spans $\mathcal{V}$. Then the vectors $\mathbf{v}_{n-1}, \cdots, \mathbf{v}_k$ are linear combinations of $\mathbf{v}_1, \cdots, \mathbf{v}_n$, which contradicts the assumption that $(\mathbf{v}_1, \cdots, \mathbf{v}_k)$ is a linearly independent sequence. Thus $k \leq n$. ■

■ **THEOREM 14.22.** If $(\mathbf{u}_1, \cdots, \mathbf{u}_n)$ is a basis for $\mathcal{V}$, then every basis for $\mathcal{V}$ has precisely n vectors.

PROOF: Let $(\mathbf{v}_1, \cdots, \mathbf{v}_k)$ be a basis for $\mathcal{V}$. By the Lemma 14.21, since this set is independent, we must have $k \leq n$. Then since $(\mathbf{u}_1, \cdots, \mathbf{u}_n)$ is an independent set and $(\mathbf{v}_1, \cdots, \mathbf{v}_k)$ is a basis, Lemma 14.21 tells us that $n \leq k$. Therefore $k = n$. ■

If the number of vectors in a basis for the vector space $\mathcal{V}$ is n, then we say that $\mathcal{V}$ *has dimension n* and that $\mathcal{V}$ is *finite dimensional*. If no finite number of vectors span $\mathcal{V}$ then we say that $\mathcal{V}$ is *infinite dimensional*. Aside from the examples (1) the set of all convergent sequences, (2) the set $P[x]$ of all polynomial functions on $\mathbf{R}$ and (3) the set $C[a, b]$ of all continuous functions on the closed interval $\{x : a \leq x \leq b\}$, we shall discuss only finite dimensional vector spaces.

■ **THEOREM 14.23.** Each vector subspace of $\mathbf{R}^n$ has a basis.

PROOF: If the vector subspace consists of only the zero vector, then we agree that it has the empty set ϕ as a basis. Now let P be any nontrivial vector subspace of $\mathbf{R}^n$. Any one vector in P constitutes an independent sequence and any $n + 1$ vectors in P constitute a dependent sequence. Hence for some number r satisfying $1 \leq r \leq n$, P contains an independent sequence $(\mathbf{v}_1, \cdots, \mathbf{v}_r)$ such that any larger sequence $(\mathbf{v}_1, \cdots, \mathbf{v}_r, \mathbf{v})$ in P is dependent. It follows that $(\mathbf{v}_1, \cdots, \mathbf{v}_r)$ is a basis. ■

■ **THEOREM 14.24.** If $P = \langle \mathbf{v}_1, \cdots, \mathbf{v}_k \rangle$ is the vector subspace of $\mathbf{R}^n$ spanned by the sequence $(\mathbf{v}_1, \cdots, \mathbf{v}_k)$, then a basis for P can be chosen from among the vectors $\mathbf{v}_1, \cdots, \mathbf{v}_k$.

PROOF: For some r satisfying $1 \leq r \leq k$, we may assume that, by renumbering the vectors $\mathbf{v}_i$, if need be, the sequence $(\mathbf{v}_i, \cdots, \mathbf{v}_r)$ is independent, whereas any larger sequence $(\mathbf{v}_1, \cdots, \mathbf{v}_r, \mathbf{v}_j)$ is dependent. It easily follows that $\langle \mathbf{v}_1, \cdots, \mathbf{v}_r \rangle = P$ and being independent, the sequence $(\mathbf{v}_1, \cdots, \mathbf{v}_r)$ is a basis. ■

It should be obvious that a vector subspace of $\mathbf{R}^n$ is the analogue of a line through the origin in the plane and a line or a plane through the origin in space. Our next definition provides the algebraic analogue of lines and planes not passing through the origin.

■ **DEFINITION.** Let S be a vector subspace of a vector space $\mathcal{V}$ and let $\mathbf{v}_0$ be a fixed vector in $\mathcal{V}$. Then the subset V of $\mathcal{V}$ defined by

$$\{V = \mathbf{v} : \mathbf{v} = \mathbf{v}_0 + \mathbf{s}, \mathbf{s} \in S\}$$

is a *linear manifold* in $\mathcal{V}$. The *dimension* of the linear manifold V is the dimension of the vector subspace S.

Our interest in linear manifolds lies in the fact that the set of all solutions of an $m \times n$ system of linear equations is a linear manifold in $\mathbf{R}^n$. The dimension of this linear manifold depends upon the rank of the $m \times n$ matrix of coefficients of the system of equations. We take up these matters in the next section.

EXERCISES 14.6

For each of Exercises 1 to 5, determine if the given vector lies in the accompanying subspace of $\mathbf{R}^n$.

1. $\mathbf{v} = (3, -4, 6)$; $S = \langle (1, 2, -2), (4, 3, 2) \rangle \subset \mathbf{R}^3$
2. $\mathbf{v} = (1, 3, -2)$; $S = \langle (1, 1, 1), (0, 1, -1) \rangle \subset \mathbf{R}^3$
3. $\mathbf{v} = (5, -4, -2)$; $S = \langle (2, -1, 1), (1, 0, 2) \rangle \subset \mathbf{R}^3$
4. $\mathbf{v} = (-1, 4, 4, 1)$; $S = \langle (1, 1, -1, 0), (1, 0, -1, 1), (0, 0, 1, 2) \rangle \subset \mathbf{R}^4$
5. $\mathbf{v} = \langle (3, 1, -2, 2)$; $S = (1, 3, 1, 0), (4, 1, 0, 1), (1, -1, 2, 0) \rangle \subset \mathbf{R}^4$
6. Find a nonzero vector in the intersection of the two subspaces $\langle (1, 2, -1), (3, 2, 0) \rangle$ and $\langle (2, -1, -1), (1, 0, 4) \rangle$ of $\mathbf{R}^3$.
7. Prove that
$$\langle (1, 3, -2), (2, 1, 4) \rangle = \langle (0, 5, -8), (1, -12, 22) \rangle$$
8. Find a basis for $\mathbf{R}^3$ which contains $(1, 0, 2)$ and $(3, 1, 0)$.
9. Find a basis for $\mathbf{R}^4$ which contains $(1, 0, 1, 0), (1, 1, 0, 1)$, and $(0, 1, 1, 0)$.
10. Choose a subset of the set
$$\{(1, 0, 1), (1, -1, 0), (1, 2, 0), (3, 1, 1)\}$$
which forms a basis for $\mathbf{R}^3$ (if such a subset exists).
11. Find a linear combination of the vectors
$$\mathbf{v}_1 = (4, 1, 0, 2), \mathbf{v}_2 = (1, 3, 1, 6), \mathbf{v}_3 = (1, 0, -1, 2), \mathbf{v}_4 = (1, 5, -1, 16)$$
that equals the zero vector.
12. Is the sequence of vectors in $\mathbf{R}^4$
$$\{(0, 1, 1, 2), (3, 1, 5, 2), (-2, 1, 0, 1), (1, 0, 3, -1)\}$$
independent?

PROBLEMS 14.6

1. Verify that each set given in Example 1 is a vector space.
2. Prove the statements in parts (c) and (d) of Example 3.
3. Let S and T be vector subspaces of any vector space V. Prove that $S \cap T$ is again a vector subspace of V.
4. Let S and T be vector subspaces of $\mathbf{R}^n$. If $S \subset T$, then $\dim S \leq \dim T$. Furthermore, if $S \subset T$ and $\dim S = \dim T$, then $S = T$.

5. (a) If the sequence $\{v_1, \cdots, v_r\}$ is independent, then every subset is also independent.
   (b) If some subset of $\{v_1, \cdots, v_r\}$ is dependent, then the entire sequence is dependent.

6. The sequence $\{v_1, \cdots, v_n\}$ is independent in a vector space $\mathcal{V}$ if and only if for each $v \in \langle v_1, \cdots, v_n \rangle$ there are *unique* constants $c_1, \cdots, c_n$ such that $v = c_1 v_1 + \cdots + c_n v_n$.

7. If v_1, v_2, v_3 span $\mathbf{R}^2$, then some pair spans $\mathbf{R}^2$.

8. A set of vectors spanning $\mathbf{R}^n$ must contain at least n vectors.

9. If S and T are subspaces of a vector space $\mathcal{V}$, then the set $S + T$ is defined to be all sums $s + t$, where $s \in S$ and $t \in T$. Prove that $S + T$ is again a vector subspace of $\mathcal{V}$.

10. For any two subspaces of R^n, prove that
$$\dim(S + T) = \dim S + \dim T - \dim(S \cap T)$$

11. If S and T are subspaces of a vector space $\mathcal{V}$ and if $S \cap T = \{0\}$, then the sum of S and T is said to be *direct*. If $S + T$ is a direct sum, show that each vector in $S + T$ has a unique expression $s + t$, $s \in S$, $t \in T$.

12. Prove that $\{1, x, x^2, \cdots, x^n\}$ is a basis for $P_n[x]$.

13. Let a, b, and c be distinct real numbers and prove that the three functions $\sin(x + a)$, $\sin(x + b)$ and $\sin(x + c)$ are dependent in $C[0, \Pi]$.

14. A *hyperplane* in $\mathbf{R}^n$ is a linear manifold of dimension $n - 1$. Prove then that a linear manifold of dimension r is the intersection of exactly $n - r$ hyperplanes.

15. Prove that each pair of k-dimensional vector subspaces of $\mathbf{R}^{2k-1}$ has a nonzero intersection.

16. If each a_{ii} is not zero in an upper triangular $n \times n$ matrix, show that the row vectors of the matrix span $\mathbf{R}^n$.

17. Let $r_1, \cdots, r_n$ be vectors in $\mathbf{R}^n$. Prove that they form an independent sequence if and only if the matrix having these as row vectors has nonzero determinant.

18. Let $\mathcal{V} = C[0, 1]$ and state which of the following subsets are actually vector subspaces:
   (a) $\{f : f(\frac{1}{2}) = 0\}$
   (b) $\{f : f(0) = f(1)\}$
   (c) $\{f : f(0) + f(1) = 0\}$
   (d) $\{f : f(\frac{1}{2}) = \frac{1}{2}(f(0) + f(1))\}$

14.7 THE SOLUTION SET OF A SYSTEM OF LINEAR EQUATIONS

This section is devoted to a theoretical study of the set of solutions of an $m \times n$ system. We shall need a few new facts in order to facilitate our work.

Let $A = (a_{ij})$ be an $m \times n$ matrix. The *row vectors* of A are the n-dimensional numerical vectors

$$r_1 = (a_{11}, a_{12}, \cdots, a_{1n})$$
$$r_2 = (a_{21}, a_{22}, \cdots, a_{2n})$$
$$\cdots\cdots\cdots\cdots\cdots\cdots\cdots$$
$$r_m = (a_{m1}, a_{m2}, \cdots, a_{mn})$$

14.7 THE SOLUTION SET OF A SYSTEM OF LINEAR EQUATIONS

and the *column vectors* of A are the m-dimensional numerical vectors

$$c_1 = \begin{bmatrix} a_{11} \\ a_{21} \\ \vdots \\ a_{m1} \end{bmatrix}, \quad c_2 = \begin{bmatrix} a_{12} \\ a_{22} \\ \vdots \\ a_{m2} \end{bmatrix}, \quad \cdots, \quad c_n = \begin{bmatrix} a_{1n} \\ a_{2n} \\ \vdots \\ a_{mn} \end{bmatrix}$$

The *row space* of A is the subspace $\langle r_1, \cdots, r_m \rangle$ of $\mathbf{R}^n$ which is spanned by the row vectors; the *column space* of A is the subspace $\langle c_1, \cdots, c_n \rangle$ of $\mathbf{R}^m$ which is spanned by the column vectors. The dimension of the row space of A is called the *row rank* of A and the dimension of the column space of A is the *column rank* of a.

■ **THEOREM 14.25.** For any matrix A, the rank (in terms of the row echelon form), the row rank, and the column rank are all equal.

We omit the proof of this result in order to conserve space. Note that the equality of rank and row rank is very easy to prove. The equality of row rank and column rank requires a fairly long argument.

■ **THEOREM 14.26.** A nonhomogeneous system of linear equations

$$A\mathbf{x} = \mathbf{b}$$

has a solution if and only if either

(1) $\mathbf{b}$ is in the column space $\langle c_1, \cdots, c_n \text{ of } A \rangle$, or
(2) the dimension of $\langle c_1, \cdots, c_n, \mathbf{b} \rangle$ equals the column rank of A.

PROOF: If $\mathbf{b}$ is in the column space $\langle c_1, \cdots, c_n \rangle$, then there are n constants $a_1, \cdots, a_n$ such that

$$\mathbf{b} = a_1 c_1 + \cdots + a_n c_n$$

It is immediately obvious that $(a_1, \cdots, a_n)$ is a solution of the system $A\mathbf{x} = \mathbf{b}$. The converse is obvious.

Next we show that condition (2) is equivalent to condition (1). First, if $\mathbf{b}$ lies in $\langle c_1, \cdots, c_n \rangle$, then condition (2) follows from the fact that $\langle c_1, \cdots, c_n, \mathbf{b} \rangle = \langle c_1, \cdots, c_n \rangle$. Conversely, suppose that condition (2) is satisfied. We obviously have the set inclusion

$$\langle c_1, \cdots, c_n \rangle \subset \langle c_1, \cdots, c_n, \mathbf{b} \rangle$$

in any case, if the two have the same dimension, then they are equal (see Problem 14.6.4). ■

As an obvious consequence of Theorem 14.26, we have the proof of Theorem 14.6.

■ **COROLLARY 14.27.** A nonhomogeneous system $A\mathbf{x} = \mathbf{b}$ has a solution if and only if the rank of its coefficient matrix A is equal to the rank of its augmented matrix $(A, \mathbf{b})$.

Next we establish some detailed information about the set of solutions of a system of linear equations. To do so, we return to the case of a homogeneous system $A\mathbf{x} = \mathbf{0}$. We already know that the set of solutions of this homogeneous system is a vector subspace of $\mathbf{R}^n$. We pursue this fact. Suppose that we renumber the column vectors of the coefficient matrix A so that, for some integer r, the set $\{\mathbf{c}_1, \cdots, \mathbf{c}_r\}$ of the first r column vectors of A forms a basis for the column space of A. (See Theorem 14.24.) Thus for each i, $r + 1 \leq i \leq n$, there exist constants $\alpha_{1i}, \cdots, \alpha_{ri}$ such that

$$\mathbf{c}_i = \alpha_{1i}\mathbf{c}_1 + \cdots + \alpha_{ri}\mathbf{c}_r$$

Hence the n-dimensional vector

$$\mathbf{b}_i = (\alpha_{1i}, \cdots, \alpha_{ri}, 0, \cdots, 0, -1, 0, \cdots, 0)$$

where -1 is in the ith position, is a solution of the system $A\mathbf{x} = \mathbf{0}$.

■ **LEMMA 14.28.** The set $\{\mathbf{b}_{r+1}, \cdots, \mathbf{b}_n\}$ described above is a basis for the solution space of the system $A\mathbf{x} = \mathbf{0}$.

PROOF: Suppose there exist constants $\beta_{r+1}, \cdots, \beta_n$ such that

$$\beta_{r+1}\mathbf{b}_{r+1} + \cdots + \beta_n\mathbf{b}_n = \mathbf{0}$$

On the left the ith component, $r + 1 \leq i \leq n$, is precisely $-\beta_i$, whereas the ith component on the right is zero. Therefore, this vector equation implies that $\beta_{r+1} = \cdots = \beta_n = 0$. Hence the set $\{\mathbf{b}_{r+1}, \cdots, \mathbf{b}_n\}$ is linearly independent.

Let $\mathbf{a} = (\alpha_1, \cdots, \alpha_n)$ be any solution of the homogeneous system $A\mathbf{x} = \mathbf{0}$. We obviously have

$$\mathbf{a} + \alpha_{r+1}\mathbf{b}_{r+1} + \cdots + \alpha_n\mathbf{b}_n = (\gamma_1, \cdots, \gamma_r, 0, \cdots, 0)$$

for some real numbers $\gamma_1, \cdots, \gamma_r$. Now, because the set of solutions of $A\mathbf{x} = \mathbf{0}$ is a vector space, this new vector

$$(\gamma_1, \cdots, \gamma_r, 0, \cdots, 0)$$

is also a solution. It easily follows that

$$\gamma_1\mathbf{c}_1 + \cdots + \gamma_r\mathbf{c}_r + 0\mathbf{c}_{r+1} + \cdots + 0\mathbf{c}_n = \mathbf{0}$$

However, the fact that the set $\{\mathbf{c}_1, \cdots, \mathbf{c}_r\}$ is linearly independent implies that $\gamma_1 = \cdots = \gamma_r = 0$. It follows that the vector $\mathbf{a} = \alpha_1, \cdots, \alpha_n$ is a linear combination of $\mathbf{b}_{r+1}, \cdots, \mathbf{b}_n$, that is, the set $\{\mathbf{b}_{r+1}, \cdots, \mathbf{b}_n\}$ spans the solution space. ■

■ **THEOREM 14.29.** The solution space of a homogeneous system of linear equations has dimension $n - r$, where r is the rank of the coefficient matrix. A homogeneous square system has a nonzero solution if and only if its coefficient matrix is singular (that is, has rank $< n$.)

The above result follows directly from Lemma 14.28, of course.

14.7 THE SOLUTION SET OF A SYSTEM OF LINEAR EQUATIONS

■ **THEOREM 14.30.** Let $\mathbf{a} = (a_1, \cdots, a_n)$ be a fixed solution vector of the nonhomogeneous $m \times n$ system of linear equations

$$A\mathbf{x} = \mathbf{b}$$

If $\mathbf{x} = (x_1, \cdots, x_n)$ is any solution vector of the associated homogeneous system $A\mathbf{x} = \mathbf{0}$, then $\mathbf{a} + \mathbf{x}$ is another solution vector of $A\mathbf{x} = \mathbf{b}$. Furthermore, each solution vector of $\mathbf{x}$ of $A\mathbf{x} = \mathbf{0}$.

PROOF: The first part of the theorem follows from the fact that

$$A(\mathbf{a} + \mathbf{x}) = A\mathbf{a} + A\mathbf{x} = \mathbf{b} + \mathbf{0} = \mathbf{b}$$

For the second part, let $\mathbf{y}$ be a solution vector of the nonhomogeneous system $A\mathbf{x} = \mathbf{b}$. By setting $\mathbf{x} = \mathbf{y} - \mathbf{a}$, we have

$$A\mathbf{x} = A(\mathbf{y} - \mathbf{a}) = A\mathbf{y} - A\mathbf{a} = \mathbf{b} - \mathbf{b} = \mathbf{0}$$

Hence $\mathbf{x}$ is a solution of $A\mathbf{x} = \mathbf{0}$ and clearly $\mathbf{y} = \mathbf{a} + \mathbf{x}$. ■

The following theorem summarizes our knowledge of the solution set of a system of linear equations. A proof can be constructed very easily from our earlier results.

■ **THEOREM 14.31.** The set of solutions of a consistent system of linear equations $A\mathbf{x} = \mathbf{b}$ is a linear manifold of dimension $n - r$ in $\mathbf{R}^n$, where n is the number of variables and r is the rank of the coefficient matrix A.

Consider the $n \times n$ system $A\mathbf{x} = \mathbf{b}$. If the matrix A is nonsingular, we can use its inverse A^{-1} as follows: Because

$$A^{-1}(A\mathbf{x}) = A^{-1}\mathbf{b}$$

and

$$A^{-1}(A\mathbf{x}) = (A^{-1}A)\mathbf{x} = I\mathbf{x} = \mathbf{x}$$

we see that the unique solution of the system is

$$\mathbf{x} = A^{-1}\mathbf{b}$$

Recalling that its inverse is given by

$$A^{-1} = \frac{1}{\det A} \operatorname{adj} A$$

we have

$$\mathbf{x} = \left(\frac{1}{\det A}\right)(\operatorname{adj} A)\mathbf{b}$$

It is then only a matter of checking definitions to prove the next theorem, usually called *Cramer's Rule*.

■ **THEOREM 14.32.** If the $n \times n$ system of linear equations $A\mathbf{x} = \mathbf{b}$ has a nonsingular matrix of coefficients, then its solution vector has components

$$x_i = \frac{D_i}{D}, \quad i = 1, 2, \cdots, n$$

where $D = \det A$ and D_i is the determinant of the matrix obtained by replacing the ith column vector of A by the column vector $\mathbf{b}$ of constant terms.

■ *Example.* Consider the 3×3 system

$$\begin{aligned} 2x - 3y - z &= 4 \\ 4x - 7y - 3z &= 2 \\ -2x + 5y + z &= 5 \end{aligned}$$

The coefficient matrix has the determinant

$$\det \begin{bmatrix} 2 & -3 & -1 \\ 4 & -7 & -3 \\ -2 & 5 & 1 \end{bmatrix} = 4$$

The numbers D_1, D_2, and D_3 of Cramer's rule are

$$D_1 = \det \begin{bmatrix} 4 & -3 & -1 \\ 2 & -7 & -3 \\ 5 & 5 & 1 \end{bmatrix} = 38, \quad D_2 = \det \begin{bmatrix} 2 & 4 & -1 \\ 4 & 2 & -3 \\ -2 & 5 & 1 \end{bmatrix} = 28$$

$$D_3 = \det \begin{bmatrix} 2 & -3 & 4 \\ 7 & -7 & 2 \\ -2 & 5 & 5 \end{bmatrix} = 42$$

Therefore, by Theorem 14.32, the components of the solution vector are

$$x_1 = \frac{D_1}{D} = \frac{38}{4} = \frac{19}{2}, \quad x_2 = \frac{D_2}{D} = \frac{28}{4} = 7$$

$$x_3 = \frac{D_3}{D} = \frac{42}{4} = \frac{21}{2}$$

At this point we complete the proof of Theorem 14.3 by showing that the row echelon form of an $m \times n$ matrix is unique. First suppose that C is a nonsingular $m \times m$ matrix and A is an $m \times n$ matrix. Define the $m \times n$ matrix $B = CA$. We claim that the row space of $CA = B$ is precisely the row space of A. This follows because the rows of CA are linear combinations of those of A, while the rows of $A = C^{-1}B$ are linear combinations of those of B.

Suppose that a matrix A has two row echelon forms B and B' obtained by different sequences of elementary row operations. Let $p_1, \cdots, p_r$

be the echelon indices of B (that is, the indices of those columns containing the initial one in each row) and let $p'_1, \cdots, p'_s$ be the echelon indices of B'. We first show that $r = s$ (this number is the rank of A) and that $p'_i = p_i$ for each $i = 1, \cdots, r$.

Let the nonzero rows of E be $\mathbf{b}_1, \cdots, \mathbf{b}_r$ and those of B' be $\mathbf{b}'_1, \cdots, \mathbf{b}'_s$. By our opening remarks, it follows that each row vector $\mathbf{b}'_i$ is a linear combination of the row vectors $\mathbf{b}_j$ and conversely. In particular, let

$$\mathbf{b}'_1 = \alpha_{11}\mathbf{b}_1 + \cdots + \alpha_{1r}\mathbf{b}_r$$

By definition of the row echelon form, the first $p_1 - 1$ columns of B are zero vectors. Hence the first $p_1 - 1$ numbers in $\mathbf{b}_1$ are zero. It follows that $p'_1 \geq p_1$. By reversing the roles of B and B' in the same argument, we show that $p_1 \geq p'_1$, whence $p'_1 = p_1$.

Next consider the equation

$$\mathbf{b}'_2 = \alpha_{21}\mathbf{b}_1 + \cdots + \alpha_{2r}\mathbf{b}_r$$

expressing $\mathbf{b}'_2$ as a linear combination of the $\mathbf{b}_j$. The first $p'_2 - 1$ numbers in $\mathbf{b}'_2$ are zero by definition of the row echelon form. Because $p'_2 > p'_1 = p_1$, it follows that α_{21} must be zero or else $\alpha_{21}\mathbf{b}_1$ would contribute the nonzero number α_{21} to the $p'_1 = p_1$ place of $\mathbf{b}'_2$ and, of course, nothing in the sum can cancel this number. Thus the stage is set for the next step in our induction argument. We simply repeat the same argument to show that $p'_2 = p_2$. This sets up an obvious inductive proof which concludes that $p'_1 = p_1, p'_2 = p_2, \cdots, p'_r = p_r$, and hence that $s \geq r$. By reversing the roles of B and B', we show similarly that $r \geq s$; hence $r = s$.

To complete the proof, we must show that $\mathbf{b}'_i = \mathbf{b}_i$ for each $i = 1, 2, \cdots, r$. Consider the equation,

$$\mathbf{b}'_i = \alpha_{i1}\mathbf{b}_1 + \cdots + \alpha_{ir}\mathbf{b}_r$$

expressing $\mathbf{b}'_i$ as a linear combination of the vectors $\mathbf{b}_j$. Concentrate upon the first nonzero component of $\mathbf{b}'_i$. This is a 1 in the p_ith column by definition. Hence each coefficient $\alpha_{i1}, \alpha_{i2}, \cdots, \alpha_{ii-1}$ must be zero while $\alpha_{i1} = 1$. Furthermore, the component of $\mathbf{b}'_i$ in the p_{i+1} column must be zero again by definition of the row echelon form. The same remark applies to the components of $\mathbf{b}'_i$ in all columns p_j except p_i. It follows that $\alpha_{ij} = \delta_{ij}$ and hence that $\mathbf{b}'_i = \mathbf{b}_i$.

EXERCISES 14.7

For each of Exercises 1 to 6, find a basis for the solution space of the given homogeneous system.

1. $x - y + z = 0$
   $x + y + 3z = 0$
   $x - 3y - z = 0$

2. $x + y - z = 0$
   $2x + 2y + 3z = 0$

3. $w - 2x + y = 0$
   $w + y + z = 0$

4. $2w - 2x - y + z = 0$
   $w - x - 3y + z = 0$
   $-w + 2x + y - 3z = 0$

5. $\begin{aligned} w + x - y - z &= 0 \\ 4w + 2x - 3y + z &= 0 \\ 2x - y - 7z &= 0 \\ w + 5x - 3y - 11z &= 0 \end{aligned}$

6. $\begin{aligned} x_1 + x_2 + x_3 + x_4 + x_5 &= 0 \\ 3x_2 - x_3 + x_4 + 3x_5 &= 0 \\ x_1 + 2x_3 + 4x_5 &= 0 \\ x_1 - x_2 + 3x_4 &= 0 \\ x_1 - x_2 + x_3 - x_4 + x_5 &= 0 \end{aligned}$

In each of Exercises 7 to 15, find a basis for the solution space of the associated homogeneous system and use it to determine the linear manifold of solutions of the given system.

7. $\begin{aligned} 2x - y &= 3 \\ x + y &= 6 \end{aligned}$

8. $\begin{aligned} x + 2y - z &= 2 \\ 2x - y + 3z &= 4 \end{aligned}$

9. $\begin{aligned} x - y + z &= 1 \\ x + y + 3z &= 5 \end{aligned}$

10. $\begin{aligned} w - 2x + y &= 0 \\ w + y + z &= 2 \end{aligned}$

11. $\begin{aligned} w + x - 3y + z &= 0 \\ 2w + 2x - y + z &= 4 \\ -w + 2x + y - 3z &= -1 \end{aligned}$

12. $\begin{aligned} w + x - y - z &= 0 \\ 2x - y - 7z &= -6 \\ w + 5x - 3y - 11z &= -8 \\ 4w + 2x - 3y + z &= 4 \end{aligned}$

13. $\begin{aligned} x_1 + 2x_2 - x_3 + 2x_4 + x_5 &= 5 \\ -x_1 + x_2 - 2x_3 - x_4 + 4x_5 &= 1 \\ 2x_1 + x_3 + x_4 - 3x_5 &= 1 \end{aligned}$

14. $\begin{aligned} x_1 + x_2 + x_3 + x_4 + x_5 &= 1 \\ 3x_2 - x_3 + x_4 + 3x_5 &= -2 \\ x_1 + 2x_3 + 4x_5 &= 7 \\ x_1 - x_2 + 3x_4 &= -1 \\ x_1 - x_2 + x_3 - x_4 + x_5 &= 5 \end{aligned}$

15. $\begin{aligned} x_1 + 2x_2 + 3x_3 + 4x_4 + 5x_5 + 6x_6 &= 0 \\ x_1 + x_2 + x_6 &= 1 \\ x_2 + x_3 + x_6 &= 1 \\ x_3 + x_4 + x_6 &= 1 \\ x_4 + x_5 + x_6 &= 1 \end{aligned}$

For each of Exercises 16 to 20, find a homogeneous linear system whose solution space is as specified.

16. $\langle (1, 1, 1), (0, 1, -1) \rangle$

17. $\langle (1, 2, 0), (3, 1, 0) \rangle$

18. $\langle (2, 1, -3), (1, -1, 0), (1, 3, -4) \rangle$

19. $\langle (1, 0, 1, 1), (1, -1, -1, 0) \rangle$

20. $\langle (1, 2, -1, 1), (0, 1, 2, 1), (1, -1, 0, 2) \rangle$

21. Find a nonhomogeneous system whose general solution vector is of the form

$$\begin{bmatrix} 2 \\ 1 \\ 3 \end{bmatrix} + s \begin{bmatrix} 1 \\ 1 \\ 0 \end{bmatrix} + t \begin{bmatrix} 0 \\ -1 \\ 2 \end{bmatrix}$$

22. Find a nonhomogeneous system whose general solution vector is of the form

$$\begin{bmatrix} 1 \\ 3 \\ 0 \\ -1 \end{bmatrix} + s \begin{bmatrix} 2 \\ 0 \\ 0 \\ 1 \end{bmatrix} + t \begin{bmatrix} 0 \\ 1 \\ -2 \\ 2 \end{bmatrix}$$

In Exercises 23 to 25, use Cramer's rule to solve the given system (each has a unique solution).

23. $\begin{aligned} 3x - y + 2z &= 4 \\ x + 2y + 3z &= 6 \\ -2x + 4y - z &= 1 \end{aligned}$

24. $\begin{aligned} x + 2y - 4z &= 11 \\ 3x - y - 3z &= 15 \\ 4x - 2y + 5z &= 2 \end{aligned}$

25. $\begin{aligned} w + 2x - y + 4z &= -5 \\ 2w - x - y - z &= 6 \\ w - 2x + 5y - 2z &= 17 \\ 3w + 3x + 2y + 5z &= 5 \end{aligned}$

PROBLEMS 14.7

1. Let the matrix product AB be defined. Show that the column vectors of AB are in the column space of A and the row vectors of AB are in the row space of B. (The matrices here need not be square.)

2. If A is a nonsingular matrix and AB is defined, then B and AB have the same row space.

3. Determine all column vectors $\mathbf{b}$ for which the system $A\mathbf{x} = \mathbf{b}$ can be solved if

$$A = \begin{bmatrix} 3 & 1 & -1 & 2 \\ -2 & -1 & 2 & 3 \\ 2 & -2 & 1 & -1 \end{bmatrix}$$

4. Let A be a square matrix and suppose there exists a particular column vector $\mathbf{b}_0$ for which the system $A\mathbf{x} = \mathbf{b}_0$ has a unique solution. Prove that the system $A\mathbf{x} = \mathbf{b}$ has a unique solution for every column vector $\mathbf{b}$.

5. If $\alpha + \beta + \gamma = 0$, prove that the following 3×3 system has nonzero solutions:

$$\begin{aligned} x + y \cos \gamma + z \cos \beta &= 0 \\ x \cos \gamma + y + z \cos \alpha &= 0 \\ x \cos \beta + y \cos \alpha + z &= 0 \end{aligned}$$

14.8 LINEAR MAPPINGS

A function $f: \mathcal{V} \to \mathcal{W}$ of one vector space into another will be called a *mapping* or a *transformation*. We are primarily interested in mappings on coordinate spaces and their vector subspaces. In particular, consider a mapping

$$f: \mathbf{R}^n \to \mathbf{R}^m$$

Let $\mathbf{u}_1, \cdots, \mathbf{u}_n$ be the unit basis vectors in $\mathbf{R}^n$ and $\mathbf{v}_1, \cdots, \mathbf{v}_m$ be the unit basis vectors in $\mathbf{R}^m$. A vector in $\mathbf{R}^n$ will be denoted variously by $\mathbf{x}$, by $(x_1, x_2, \cdots, x_n)$ or by $x_1 \mathbf{u}_1 + \cdots + x_n \mathbf{u}_n$ and a vector in $\mathbf{R}^m$ will be denoted by **y**, by $(y_1, y_2, \cdots, y_m)$ or by $y_1\mathbf{v}_1 + \cdots + y_m\mathbf{v}_m$. Thus, as a vector-valued function of vectors, we write the function f as

$$f(\mathbf{x}) = \mathbf{y}$$

or

$$f(x_1\mathbf{u}_1 + \cdots + x_n\mathbf{u}_n) = y_1\mathbf{v}_1 + \cdots + y_m\mathbf{v}_m$$

The components of the vector-valued function f are real-valued functions f_i of n variables. Thus we have

$$y_i = f_i(x_1, \cdots, x_n)$$

for each $i = 1, 2, \cdots, m$.

We have already considered several kinds of mappings. For instance, a function of several variables is just a mapping $f: \mathbf{R}^n \to \mathbf{R}^1$. A mapping $f: \mathbf{R}^1 \to \mathbf{R}^3$ is a parametric curve in space and a mapping $g: \mathbf{R}^2 \to \mathbf{R}^3$ is a parametric surface. The transformation from polar to rectangular coordinates,

$$(r, \theta) \to (r\cos\theta, r\sin\theta) = (x, y)$$

is a mapping of $\mathbf{R}^2$ into $\mathbf{R}^2$.

We consider only a special class of mappings here. A mapping $L: \mathbf{R}^n \to \mathbf{R}^m$ is said to be *linear* if each of its component functions L_i is a homogeneous linear function. That is, there exist constants a_{ij}, $i = 1, 2, \cdots, m$; $j = 1, 2, \cdots, n$, such that

$$\begin{aligned} y_1 &= L_1(x_1, \cdots, x_n) = a_{11}x_1 + a_{12}x_2 + \cdots + a_{1n}x_n \\ y_2 &= L_2(x_1, \cdots, x_n) = a_{21}x_1 + a_{22}x_2 + \cdots + a_{2n}x_n \\ &\cdots\cdots\cdots\cdots\cdots\cdots\cdots\cdots\cdots\cdots\cdots\cdots\cdots\cdots\cdots \\ y_m &= L_m(x_1, \cdots, x_n) = a_{m1}x_1 + a_{m2}x_2 + \cdots + a_{mn}x_n \end{aligned}$$

The $m \times n$ matrix of coefficients

$$A = \begin{bmatrix} a_{11} & a_{12} & \cdots & a_{1n} \\ a_{21} & a_{22} & \cdots & a_{2n} \\ \cdots & \cdots & \cdots & \cdots \\ a_{m1} & a_{m2} & \cdots & a_{mn} \end{bmatrix}$$

is called the *matrix of the linear mapping*.

If we consider $x \in \mathbf{R}^n$ and $y \in \mathbf{R}^m$ as column vectors,

$$x = \begin{bmatrix} x_1 \\ x_2 \\ \vdots \\ x_n \end{bmatrix}, \quad y = \begin{bmatrix} y_1 \\ y_2 \\ \vdots \\ y_m \end{bmatrix}$$

then we obviously may write the linear mapping in the form

$$\mathbf{y} = L(\mathbf{x}) = A\mathbf{x}$$

where $A\mathbf{x}$ is the product of the matrix A times the vector $\mathbf{x}$. This same equation can be used to define a linear mapping L corresponding to any given $m \times n$ matrix. Thus we have a one-to-one correspondence between $m \times n$ matrices and linear mappings of $\mathbf{R}^n$ into $\mathbf{R}^m$. The correspondence is actually more nearly an identity in that all properties of matrices are reflected in linear mappings, and vice versa. It is logically equivalent to start with linear mappings instead of with matrices, and many authors do so.

■ *Example 1.* Consider a small factory that produces tents and awnings. One tent takes 36 sq yd of canvas, 40 ft of rope and 45 man-hours of labor; one awning requires 6 sq yd of canvas, 20 feet of rope and 14 man-hours of labor. Thus, in order to manufacture x_1 tents and x_2 awnings, the factory needs

$$36x_1 + 6x_2 = y_1 \text{ sq yd of canvas,}$$
$$40x_1 + 20x_2 = y_2 \text{ ft of rope,}$$

and

$$45x_1 + 14x_2 = y_3 \text{ man-hours of labor}$$

The linear mapping described by these equations transforms a desired output (x_1, x_2) into the required input (y_1, y_2, y_3).

■ *Example 2.* Consider $\mathbf{R}^n$ as a Euclidean space. The linear mapping defined by the vector equation

$$L(\mathbf{p}) = \alpha\mathbf{p}, \quad \alpha > 0$$

carries $\mathbf{R}^n$ onto itself. The component functions here are merely

$$y_i = \alpha x_i$$

This geometric transformation is called a *contraction* if $0 < \alpha < 1$, the *identity mapping* if $\alpha = 1$, and a *dilation* if $\alpha > 1$.

■ **THEOREM 14.33.** The mapping $L : \mathbf{R}^n \to \mathbf{R}^m$ is linear if and only if it satisfies the two properties:

(a) $L(\alpha\mathbf{x}) = \alpha L(\mathbf{x})$ (homogeneity) and
(b) $L(\mathbf{x}_1 + \mathbf{x}_2) = L(\mathbf{x}_1) + L(\mathbf{x}_2)$
   for all vectors $\mathbf{x}, \mathbf{x}_1, \mathbf{x}_2$ in $\mathbf{R}^n$ and all real numbers α.

PROOF: Let L be linear. Then

$$L(\mathbf{x}) = A\mathbf{x}$$

where A is the matrix of L. The properties of matrix multiplication easily imply the conditions

$$A(\alpha \mathbf{x}) = \alpha A\mathbf{x} \quad \text{and} \quad A(\mathbf{x}_1 + \mathbf{x}_2) = A\mathbf{x}_1 + A\mathbf{x}_2$$

On the other hand, suppose that L is both homogeneous and additive. Then each unit basis vector $\mathbf{u}_j$, $j = 1, 2, \cdots, n$, in $\mathbf{R}^n$ has a fixed image in $\mathbf{R}^m$,

$$L(\mathbf{u}_j) = a_{1j}\mathbf{v}_1 + \cdots + a_{mj}\mathbf{v}_m$$

where a_{ij}, $i = 1, 2, \cdots, m$, are constants. By the assumed conditions of homogeneity and additivity, it follows that

$$\begin{aligned} L(\mathbf{x}) &= L(x_1\mathbf{u}_1 + \cdots + x_n\mathbf{u}_n) \\ &= x_1(a_{11}\mathbf{v}_1 + \cdots + a_{m1}\mathbf{v}_m) \\ &\quad + \cdots + x_n(a_{1n}\mathbf{v}_1 + \cdots + a_{mn}\mathbf{v}_m) \\ &= (a_{11}x_1 + \cdots + a_{1n}x_n)\mathbf{v}_1 \\ &\quad + \cdots + (a_{m1}x_1 + \cdots + a_{mn}x_n)\mathbf{v}_m \end{aligned}$$

Hence L is linear. ∎

The properties of homogeneity and additivity that characterize linear mappings from one coordinate space to another are adopted as the definition in more general settings. Thus a mapping $f : \mathcal{V} \to \mathcal{W}$ of a vector space $\mathcal{V}$ into a vector space $\mathcal{W}$ is *linear* if

$$f(\alpha \mathbf{u} + \beta \mathbf{v}) = \alpha f(\mathbf{u}) + \beta f(\mathbf{v})$$

for each pair of vectors $\mathbf{u}, \mathbf{v}$ in $\mathcal{V}$ and each pair α, β of real numbers. (Note that the equation here combines homogeneity and additivity into one. We separate the two because they often occur quite independently of each other.)

∎ *Example 3.* Let $\mathcal{V}$ be the vector space $C[a, b]$ of all continuous functions on the closed interval $\{x : a \leqq x \leqq b\}$. Then the mapping L of $C[a, b]$ into itself defined by

$$L(f)(x) = \int_a^x f$$

is linear. ∎

∎ *Example 4.* Consider the vector space $P[x]$ of all polynomial functions on $\mathbf{R}$. It is obvious that differentiation is a linear mapping of $P[x]$ into itself. ∎

Whenever the codomain of a set of functions is a vector space, we can develop an algebra of mappings. Indeed, the set of all mappings from any domain X into the vector space $\mathcal{W}$ forms a vector space. The sum of two mappings $f: X \to \mathcal{W}$ and $g: X \to \mathcal{W}$ is defined by functional addition,

$$(f + g)(x) = f(x) + g(x)$$

and the product of a real number times a mapping is defined by

$$(\alpha f)(x) = \alpha f(x)$$

■ **THEOREM 14.34.** Let L, L_1, and L_2 be linear mappings from a vector space $\mathcal{V}$ to a vector space $\mathcal{W}$ and let α be a real number. Then both $L_1 + L_2$ and αL are linear.

PROOF: By definition, we have

$$\begin{aligned}(L_1 + L_2)(\alpha \mathbf{v}) &= L_1(\alpha \mathbf{v}) + L_2(\alpha \mathbf{v}) \\ &= \alpha L_1(\mathbf{v}) + \alpha L_2(\mathbf{v}), \text{ because } L_1 \text{ and } L_2 \text{ are linear} \\ &= \alpha[L_1(\mathbf{v}) + L_2(\mathbf{v})] \\ &= \alpha(L_1 + L_2)(\mathbf{v})\end{aligned}$$

This proves that $L_1 + L_2$ is homogeneous.

Similarly, we have

$$\begin{aligned}(L_1 + L_2)(\mathbf{v}_1 + \mathbf{v}_2) &= L_1(\mathbf{v}_1 + \mathbf{v}_2) + L_2(\mathbf{v}_1 + \mathbf{v}_2) \\ &= L_1(\mathbf{v}_1) - L_1(\mathbf{v}_2) + L_2(\mathbf{v}_1) + L_2(\mathbf{v}_2) \\ &= L_1(\mathbf{v}_1) - L_2(\mathbf{v}_1) + L_1(\mathbf{v}_2) + L_2(\mathbf{v}_2) \\ &= (L_1 + L_2)(\mathbf{v}_1) + (L_1 + L_2)(\mathbf{v}_2)\end{aligned}$$

and hence $L_1 + L_2$ is additive. The proof that αL is linear is analogous. ■

■ **THEOREM 14.35.** The composition of linear mappings is again a linear mapping.

PROOF: Let $K: \mathcal{U} \to \mathcal{V}$ and $L: \mathcal{V} \to \mathcal{W}$ be linear mappings. Then we have

$$\begin{aligned}(L \circ K)(\alpha \mathbf{u}) &= L[K(\alpha \mathbf{u})], \text{ by definition of } L \circ K \\ &= L[\alpha K(\mathbf{u})], \text{ because } K \text{ is linear} \\ &= \alpha L[K(\mathbf{u})], \text{ because } L \text{ is linear} \\ &= \alpha(L \circ K)(\mathbf{u})\end{aligned}$$

Therefore $L \circ K$ is homogeneous.

Also we may write

$$\begin{aligned}(L \circ K)(\mathbf{u}_1 + \mathbf{u}_2) &= L[K(\mathbf{u}_1 + \mathbf{u}_2)] \\ &= L[K(\mathbf{u}_1) + K(\mathbf{u}_2)] \\ &= L[K(\mathbf{u}_1)] + L[K(\mathbf{u}_2)] \\ &= (L \circ K)(\mathbf{u}_1) + (L \circ K)(\mathbf{u}_2)\end{aligned}$$

Hence $L \circ K$ is additive. ■

These two theorems provide a very interesting algebra of linear mappings, particularly of a vector space $\mathcal{V}$ onto itself. This point of view is exploited in advanced mathematics. Also the algebra of linear mappings on coordinate spaces coincides precisely with the algebra of matrices. In particular, we show how the composition of linear mappings actually defines the multiplication of matrices. Consider linear mappings

$$K: \mathbf{R}^p \to \mathbf{R}^n, \, L: \mathbf{R}^n \to \mathbf{R}^m$$

and their matrices A and B. Note that the matrix A of K is an $n \times p$ matrix and that the matrix B of L is an $m \times n$ matrix. We claim that the matrix of the composition $L \circ K$ is the product matrix BA. One thing is clear. The $m \times p$ matrix BA is the matrix of *some* linear mapping of $\mathbf{R}^p$ into $\mathbf{R}^m$. To be specific, let $\mathbf{x}$ be a vector in $\mathbf{R}^p$ and suppose $\mathbf{x}$ is written in column form. Then we may write

$$\begin{aligned}(L \circ K)(\mathbf{x}) &= L[K(\mathbf{x})] \\ &= L(A\mathbf{x}) \\ &= B(A\mathbf{x}) \\ &= (BA)\mathbf{x}\end{aligned}$$

Note that we have made free use of the representation of a linear mapping as the product of its matrix times a column vector.

Let $L: \mathcal{V} \to \mathcal{W}$ be a linear mapping. The set of all vectors $\mathbf{v}$ in $\mathcal{V}$ which are mapped by L onto the zero vector of $\mathcal{W}$ is called the *kernel* (or the *null space*) of the mapping L. We denote this subset of $\mathcal{V}$ by $\ker(L)$. Thus we have

$$\ker(L) = \{\mathbf{v} : \mathbf{v} \in \mathcal{V}, L(\mathbf{v}) = \mathbf{0}\}$$

The *image* or *range* of L is the set of all vectors $\mathbf{w}$ in $\mathcal{W}$ such that $\mathbf{w} = L(\mathbf{v})$ for some $\mathbf{v}$ in $\mathcal{V}$. This subset of $\mathcal{W}$ will be denoted in $\text{im}(L)$. The reader should prove the following easy result.

■ **THEOREM 14.36.** Let $L: \mathcal{V} \to \mathcal{W}$ be a linear mapping of a vector space $\mathcal{V}$ into a vector space $\mathcal{W}$. Then $\ker(L)$ is a vector subspace of $\mathcal{V}$ and $\text{im}(L)$ is a vector subspace of $\mathcal{W}$.

■ *Example 5.* Consider the vector space $P[x]$ of all polynomial functions on $\mathbf{R}$ and let $D: P[x] \to P[x]$ denote the linear mapping defined by differentiation. It is clear that $\ker(D)$ consists of all constant functions, whereas $\text{im}(D) = P[x]$. To see the latter, we need only note that the polynomial function with values

$$a_0 + a_1 x + \cdots + a_n x^u$$

is the image under differentiation of any polynomial of the form

$$c + a_0 x + \frac{1}{2} a_1 x^2 + \cdots + \frac{1}{n+1} a_n x^n$$

■ **Example 6.** Let A be an $m \times n$ matrix and define the linear mapping $L: \mathbf{R}^n \to \mathbf{R}^m$ by

$$L(\mathbf{x}) = A\mathbf{x}$$

The kernel of L is then the set of vectors $\mathbf{x}$ such that $A\mathbf{x} = 0$. In other words, $\ker(L)$ is the solution space of the homogeneous $m \times n$ system of equations $A\mathbf{x} = 0$. The image of L consists of all vectors $\mathbf{y}$ in $\mathbf{R}^m$ such that $\mathbf{y} = A\mathbf{x}$ for some $\mathbf{x}$ in $\mathbf{R}^n$. Clearly then, such a vector $\mathbf{y}$ is a linear combination of the column vectors of A. Hence $\text{im}(L)$ is simply the column space of the matrix A. ■

A linear mapping $L: \mathcal{V} \to \mathcal{V}$ of a vector space $\mathcal{V}$ into itself is said to be *nonsingular* if $\ker(L) = \{\mathbf{0}\}$ and is *singular* if $\ker(L)$ contains a nonzero vector. Note that the linear mapping $D: P[x] \to P[x]$ of Example 5 is singular despite the fact that $\text{im}(D) = P[x]$.

A linear mapping $L: \mathbf{R}^n \to \mathbf{R}^m$ of a coordinate space $\mathbf{R}^n$ into a coordinate space $\mathbf{R}^m$ has the *rank* equal to the dimension of $\text{im}(L)$ (as a subspace of $\mathbf{R}^m$, of course.)

■ **THEOREM 14.37.** Let $L: \mathbf{R}^n \to \mathbf{R}^m$ be a linear mapping. Then the dimension of $\ker(L)$ plus the rank of L equals n.

(The proof of this result involves no more than a restatement of the chief result concerning the solution of an $m \times n$ system.)

■ **COROLLARY 14.38.** The linear mapping $L: \mathbf{R}^n \to \mathbf{R}^n$ is nonsingular if and only if its rank is n.

The following definitions are included as a foreshadowing of very important developments in subsequent courses in mathematics. The remarks following the definitions are to be interpreted in a similar way.

Let $L: \mathcal{V} \to \mathcal{V}$ be a linear mapping of a vector space $\mathcal{V}$ into itself. A vector $\mathbf{v}$ in $\mathcal{V}$ is a *fixed point* of L if $L(\mathbf{v}) = \mathbf{v}$. We remark that the set of all such fixed points of L constitutes a vector subspace of $\mathcal{V}$. A subspace $\mathcal{S}$ of $\mathcal{V}$ is said to be *pointwise invariant* under L if each vector in $\mathcal{S}$ is a fixed point of L. More generally, a subspace $\mathcal{T}$ of $\mathcal{V}$ is *invariant relative to L,* or simply *L-invariant,* if $L(\mathbf{t})$ lies in $\mathcal{T}$ for every vector $\mathbf{t}$ in $\mathcal{T}$.

■ **Example 7.** Consider the linear mapping $L: \mathbf{R}^3 \to \mathbf{R}^3$ defined by

$$L(x, y, z) = (-x, -y, -z)$$

It is obvious that $L(\mathbf{v}) = \mathbf{v}$ if and only if $\mathbf{v} = \mathbf{0} = (0, 0, 0)$. Hence the only fixed point of L is $\mathbf{0}$. However, L has many invariant subspaces. The reader may verify that any line or plane through the origin is L-invariant. That is, every vector subspace of $\mathbf{R}^3$ is L-invariant.

■ *Example 8.* Consider the linear mapping $L : \mathbf{R}^3 \to \mathbf{R}^3$ defined by

$$L(x, y, z) = (2x + y, y - 2z, x + y + z)$$

The reader can easily see that any vector of the form $(x, -x, 0)$ is a fixed point of L. Furthermore, any fixed point of L must have this form. (Solve the vector equation $(2x + y, y - 2z, x + y + z) = (x, y, z)$.) Hence the only nontrivial pointwise invariant subspace under L is the line with the vector equation $(x, y, z) = t(1, -1, 0)$.

Mappings such as those in Examples 7 and 8 also leave certain geometric properties invariant. For instance, the linear mapping in Example 7 leaves invariant the length of any vector, and hence the volume of any solid. The study of such invariant properties lies of the very foundation of geometry. Indeed, invariance is an extremely important concept throughout mathematics. We shall see some geometric applications of invariant properties in the next two sections.

EXERCISES 14.8 In each of Exercises 1 to 10, a mapping on $\mathbf{R}^3$ is defined. State whether or not the given mapping is linear. If it is linear, write out its matrix and determine its kernal and image.

1. $L(x, y, z) = (x + y, x + z) \in \mathbf{R}^2$
2. $L(x, y, z) = (x + 1, y + z) \in \mathbf{R}^2$
3. $L(x, y, z) = (0, z) \in \mathbf{R}^2$
4. $L(x, y, z) = (x - y, y - x, z - y) \in \mathbf{R}^3$
5. $L(x, y, z) = (x, y, 0) \in \mathbf{R}^3$
6. $L(x, y, z) = (3x - 2y + z, x - 3y + 2z, 2x + 3y - 2z) \in \mathbf{R}^3$
7. $L(x, y, z) = (z, -x + 2, -y) \in \mathbf{R}^3$
8. $L(x, y, z) = (x, -y, -z, 0) \in \mathbf{R}^4$
9. $L(x, y, z) = (x, y, x + z, x + y) \in \mathbf{R}^4$
10. $L(x, y, z) = (x + y, z - x, y, x + y + z) \in \mathbf{R}^4$

In each of Exercises 11 to 15, a mapping on $\mathbf{R}^2$ is described by giving the images of the basis unit vectors $\mathbf{u}_1$ and $\mathbf{u}_2$. Write out the matrix of each mapping (assuming it to be linear).

11. $L(\mathbf{u}_1) = (0, -1)$, $L(\mathbf{u}_2) = (2, 1)$
12. $L(\mathbf{u}_1) = (1, -3)$, $L(\mathbf{u}_2) = (2, 3)$
13. $L(\mathbf{u}_1) = (1, 0)$, $L(\mathbf{u}_2) = (a, 1)$
14. $L(\mathbf{u}_1) = (2, 1)$, $L(\mathbf{u}_2) = (-1, -\frac{1}{2})$
15. $L(\mathbf{u}_1) = (3, -2)$, $L(\mathbf{u}_2) = (2, -3)$

In each of Exercises 16 to 20, a mapping on $\mathbf{R}^3$ is described by giving the images of the basis unit vectors $\mathbf{u}_1$, $\mathbf{u}_2$, and $\mathbf{u}_3$. Write out the matrix of each mapping (assuming it to be linear); find its rank, its kernel, and its image.

16. $L(\mathbf{u}_1) = (1, 3, 1)$, $L(\mathbf{u}_2) = (1, 0, -1)$, $L(\mathbf{u}_3) = (3, 3, -1)$
17. $L(\mathbf{u}_1) = (1, 0, -1)$, $L(\mathbf{u}_2) = (1, 2, 1)$, $L(\mathbf{u}_3) = (2, 2, 0)$
18. $L(\mathbf{u}_1) = (1, 1, -1)$, $L(\mathbf{u}_2) = (1, -1, 1)$, $L(\mathbf{u}_3) = (2, 1, -1)$
19. $L(\mathbf{u}_1) = (1, 3, 2)$, $L(\mathbf{u}_2) = (2, 2, 0)$, $L(\mathbf{u}_3) = (3, 1, 2)$
20. $L(\mathbf{u}_1) = (1, 0, 1, 2)$, $L(\mathbf{u}_2) = (0, 1, -1, 1)$, $L(\mathbf{u}_3) = (2, -3, 5, 1)$

PROBLEMS 14.8

1. Find linear mappings L and K of $\mathbf{R}^3$ into $\mathbf{R}^3$ satisfying the following conditions (separately).
   (a) $L^n = 0$ for some natural number n ($L \neq 0$, of course)
   (b) L, K, and $L \circ K$ are not zero but $K \circ L$ is zero
   (c) $L \neq L$ but $L \circ L = K \circ K$
   (d) $L \circ K \neq K \circ L$ but $L \circ K \circ K = K \circ K \circ L$

2. Let the linear mapping $L : \mathbf{R}^n \to \mathbf{R}^n$ be defined by
$$L(\mathbf{u}_i) = \mathbf{u}_{i+1}, \qquad \text{if } i < n,\ L(\mathbf{u}_n) = \mathbf{u}_1$$
Find the matrix of this mapping and show that $L^n =$ the identity mapping.

3. Find a necessary and sufficient condition on the constant c so that the linear mapping L defined on $\mathbf{R}^3$ by
$$L(x) = \begin{bmatrix} 1 & 3 & c \\ -c & 2 & 1 \\ 2 & -1 & 0 \end{bmatrix} \begin{bmatrix} x \\ y \\ z \end{bmatrix}$$
has a one-dimensional pointwise invariant subspace.

4. What form of matrix is associated with $L : \mathbf{R}^n \to \mathbf{R}^n$ if
$$L(x_1, x_2, \cdots, x_n) = (a_1 x_1, a_2 x_2, \cdots, a_n x_n)?$$

5. Let $\mathbf{v}_0$ be a fixed vector in some vector space V. Prove that the transformation $T(\mathbf{v}) = \mathbf{v} + \mathbf{v}_0$ is not linear. (This is called a *translation*.)

6. Prove that the gradient is a linear mapping of the vector space of differentiable scalar fields on $\mathbf{R}^3$ into the vector space of vector fields on $\mathbf{R}^3$.

7. Let $\{\mathbf{v}_1, \cdots, \mathbf{v}_n\}$ be any set of vectors spanning the domain of the linear mapping $L : V \to W$. Prove that $\{L(\mathbf{v}_1), \cdots, L(\mathbf{v}_n)\}$ spans $\text{im}(L)$.

8. Let V be the direct sum of two of its vector subspaces. That is, $V = S + T$ and $S \cap T = 0$. Define the mapping $L : S + T \to S$ by $L(\mathbf{s} + \mathbf{t}) = \mathbf{s}$ and show that L is linear. (L is called the *projection* of $S + T$ onto S.)

9. Let $S + T$ be a direct sum and let $L : S + T \to S$ and $K : S + T \to T$ be projections. Describe $L + K$, $L \circ K$, $K \circ L$, $L \circ L$ and $K \circ K$ as mappings of $S + T$ into itself.

10. For any linear mapping L of the vector space V into itself show that V is the direct sum of $\ker(L)$ and $\text{im}(L)$.

11. Let $L : V \to V$ and $K : V \to W$ be linear mappings. Show that $\ker(K \circ L)$ contains $\ker(L)$ and $\text{im}(K \circ L)$ lies in $\text{im}(K)$.

12. (a) Find a linear mapping $L: \mathbf{R}^2 \to \mathbf{R}^2$ that transforms the circle $\{(x_1, x_2): x_1^2 + x_2^2 = 1\}$ into the ellipse
$$\left\{(y_1, y_2): \frac{y_1^2}{a^2} + \frac{y_2^2}{b^2} = 1\right\}$$

(b) If $L: \mathbf{R}^2 \to \mathbf{R}^2$ is a linear mapping, prove that there is a constant k, depending only upon L, such that the area of $L(\Gamma)$ is k times the area of Γ for every geometric figure Γ. Use this in conjunction with part (a) to find the area of an ellipse.

13. How can we tell whether a linear mapping $L: \mathbf{R}^3 \to \mathbf{R}^3$ increases or decreases the volume of geometric figures?

14. Let Q be a fixed nonsingular $n \times n$ matrix. Define a mapping $L: \mathbf{R}^{n \times n} \to \mathbf{R}^{n \times n}$ by
$$L(A) = Q^{-1}AQ$$
Show that L is linear.

15. Prove that the sequence $(1, x, 1 - 2x^2, 3x - 4x^3, 1 - 8x^2 + 8x^4)$ is a basis for $P_4[x]$.

14.9 TRANSFORMATIONS OF COORDINATES IN $\mathbf{R}^n$

There are obviously many different sets of basis vectors for the vector space $\mathbf{R}^n$. We consider now the problem of changing from one basis to another.

If $\{\mathbf{v}_1, \cdots, \mathbf{v}_n\}$ is a basis for $\mathbf{R}^n$ and $\mathbf{x}$ is a vector in $\mathbf{R}^n$, then there is a unique ordered n-tuple $(y_1, \cdots, y_n)$ of real numbers such that

$$\mathbf{x} = y_1 \mathbf{v}_1 + \cdots + y_n \mathbf{v}_n$$

The numbers y_i are the *coordinates of* $\mathbf{x}$ *relative to the given basis* $\{\mathbf{v}_i\}$.

Suppose that $\{\mathbf{w}_1, \cdots, \mathbf{w}_n\}$ is another basis for $\mathbf{R}^n$. Then each vector $\mathbf{w}_j, j = 1, \cdots, n$, is a linear combination

$$\mathbf{w}_j = a_{1j} \mathbf{v}_1 + \cdots + a_{nj} \mathbf{v}_n$$

of the basis vectors $\mathbf{v}_i$. This provides us with an $n \times n$ matrix $A = (a_{ij})$ which is necessarily nonsingular (see Problem 14.9.1).

Let the vector $\mathbf{x}$ have coordinates $(y_1, \cdots, y_n)$ relative to the basis $\{\mathbf{v}_j\}$, and let $\mathbf{x}$ have coordinates $(x_1, \cdots, x_n)$ relative to $\{\mathbf{w}_j\}$. The identity $\mathbf{x} = \mathbf{x}$ then can be written as

$$\begin{aligned} y_1 \mathbf{v}_1 + \cdots + y_n \mathbf{v}_n &= x_2(a_{11}\mathbf{v}_1 + \cdots + a_{n1}\mathbf{v}_n) \\ &\quad + \cdots + x_n(a_{1n}\mathbf{v}_1 + \cdots + a_{nn}\mathbf{v}_n) \\ &= (a_{11}x_1 + \cdots + a_{1n}x_n)\mathbf{v}_1 \\ &\quad + \cdots + (a_{n1}x_1 + \cdots + a_{nn}x_n)\mathbf{v}_n \end{aligned}$$

Equating components, we have the equations

$$y_1 = a_{11}x_1 + \cdots + a_{1n}x_n$$
$$\vdots$$
$$y_n = a_{n1}x_1 + \cdots + a_{nn}x_n$$

Thus if we let the n-tuple $(x_1, \cdots, x_n)$ be the numerical vector $\mathbf{w}$, while $(y_1, \cdots, y_n)$ is $\mathbf{v}$, then these equations can be written in matrix form as

$$A\mathbf{w} = \mathbf{v}$$

The change from the basis $\{\mathbf{w}_j\}$ to the basis $\{\mathbf{v}_i\}$ is a *transformation of coordinates* in $\mathbf{R}^n$. The matrix A is the *matrix of the transformation*. If we think of the transformation as a linear mapping, then A is its matrix.

■ **Example 1.** Consider the quadratic function with the values

$$f(x,y) = 3x^2 + 4xy + 6y^2$$

We shall select a new coordinate system with basis vectors

$$\mathbf{v}_1 = a_{11}\mathbf{u}_1 + a_{21}\mathbf{u}_2$$
$$\mathbf{v}_2 = a_{12}\mathbf{u}_1 + a_{22}\mathbf{u}_2$$

where $\mathbf{u}_1 = (1, 0)$ and $\mathbf{u}_2 = (0, 1)$ as usual. We do so in such a way that, when the vector $(x, y) = x\mathbf{u}_1 + y\mathbf{u}_2$ is replaced by $(x', y') = x'\mathbf{v}_1 + y'\mathbf{v}_2$, the form of our function will be simpler. The vector equation

$$\begin{aligned} x\mathbf{u}_1 + y\mathbf{u}_2 &= x'\mathbf{v}_1 + y'\mathbf{v}_2 \\ &= x'(a_{11}\mathbf{u}_1 + a_{21}\mathbf{u}_2) + y'(a_{12}\mathbf{u}_1 + a_{22}\mathbf{u}_2) \\ &= (a_{11}x' + a_{12}y')\mathbf{u}_1 + (a_{21}x' + a_{22}y')\mathbf{u}_2 \end{aligned}$$

provides the component equations

$$x = a_{11}x' + a_{12}y'$$
$$y = a_{21}x' + a_{22}y'$$

Then in terms of the new variables x' and y', the quadratic function has values

$$\begin{aligned} 3(a_{11}x' + a_{12}y')^2 &+ 4(a_{11}x' + a_{12}y')(a_{21}x' + a_{22}y') + 6(a_{21}x' + a_{22}y')^2 \\ &= (3a_{11}^2 + 4a_{11}a_{21} + 6a_{21}^2)x'^2 \\ &\quad + (6a_{11}a_{12} + 4a_{11}a_{22} + 4a_{12}a_{21} + 2a_{21}a_{22})x'y' \\ &\quad + (3a_{12}^2 + 4a_{12}a_{22} + 6a_{22}^2)y'^2 \end{aligned}$$

The matrix

$$A = \begin{bmatrix} a_{11} & a_{12} \\ a_{21} & a_{22} \end{bmatrix}$$

can be selected in many ways to simplify the description of the function. For instance, we may choose

$$A = \begin{bmatrix} 1 & 1 \\ \frac{1}{4} & -1 \end{bmatrix}$$

and then the quadratic function will have the value

$$f(x', y') = \tfrac{75}{8} x'^2 + 5y'^2$$

in terms of the new variables x' and y'. The expression tells us that the level curves of the function are ellipses. ■

Geometric theorems often require a transformation of coordinates. We shall need the following definition; note that it makes use of both distance and orthogonality in $\mathbf{R}^n$. A basis $\{v_1, \cdots, v_n\}$ for $\mathbf{R}^n$ is said to be *orthonormal* if the dot products satisfy

$$\mathbf{v}_i \cdot \mathbf{v}_j = 0 \text{ if } i \neq j$$
$$= 1 \text{ if } i = j$$

Using the Kronecker delta, we may write simply

$$\mathbf{v}_i \cdot \mathbf{v}_j = \delta_{ij}$$

These equations tell us (1) that each vector $\mathbf{v}_i$ in the basis is a unit vector, and (2) that any two distinct vectors in the basis are orthogonal. For instance, the standard basis $\{\mathbf{u}_1, \cdots, \mathbf{u}_n\}$ for $\mathbf{R}^n$ is orthonormal. The vectors $\mathbf{v}_1 = (1/\sqrt{2}, 1/\sqrt{2})$, $\mathbf{v}_1 = (-1/\sqrt{2}, 1/\sqrt{2})$ constitute an orthonormal basis for $\mathbf{R}^2$. The vectors $\mathbf{v}_1 = (\frac{2}{3}, \frac{2}{3}, \frac{1}{3})$, $\mathbf{v}_2 = (-\frac{2}{3}, \frac{1}{3}, \frac{2}{3})$, $\mathbf{v}_3 = (\frac{1}{3}, -\frac{2}{3}, \frac{2}{3})$ form an orthonormal basis for $\mathbf{R}^3$.

Any given basis of a vector subspace of $\mathbf{R}^n$ can be altered into an orthonormal basis. The process of obtaining such a basis is called the *Gram-Schmidt orthonormalization*.

■ **THEOREM 14.39.** Let $\{\mathbf{v}_1, \cdots, \mathbf{v}_k\}$, $1 \leq k \leq n$, be a basis for a vector subspace S of $\mathbf{R}^n$. Then there is an orthonormal basis for S.

PROOF: Knowing that $\mathbf{v}_1 \neq \mathbf{0}$, we define the unit vector

$$\mathbf{w}_1 = \frac{\mathbf{v}_1}{||\mathbf{v}_1||}$$

Note that $\mathbf{v}_1$ and $\mathbf{w}_1$ span the same subspace of $\mathbf{R}^n$. In other words, we have $\langle \mathbf{v}_1 \rangle = \langle \mathbf{w}_1 \rangle$.

Next we define the vector

$$\mathbf{x}_2 = \mathbf{v}_2 - (\mathbf{w}_1 \cdot \mathbf{v}_2)\mathbf{w}_1$$

Clearly, $\mathbf{x}_2 \neq \mathbf{0}$ or else $\mathbf{v}_1$ and $\mathbf{v}_2$ are not linearly independent. Futhermore, we have

$$\mathbf{w}_1 \cdot \mathbf{x}_2 = \mathbf{w}_1 \cdot \mathbf{v}_1 - (\mathbf{w}_1 \cdot \mathbf{v}_2)\mathbf{w}_1 \cdot \mathbf{w}_1$$
$$= \mathbf{w}_1 \cdot \mathbf{v}_2 - (\mathbf{w}_1 \cdot \mathbf{v}_2), \text{ because } \mathbf{w}_1 \text{ is a unit vector}$$
$$= 0$$

Thus $\mathbf{w}_1$ and $\mathbf{x}_2$ are orthogonal. Note that $\langle \mathbf{w}_1, \mathbf{w}_2 \rangle = \langle \mathbf{v}_1, \mathbf{v}_2 \rangle$ because each vector in the pair $\mathbf{w}_1, \mathbf{w}_2$ can be expressed as a linear combination of $\mathbf{v}_1$ and $\mathbf{v}_2$, and conversely.

Now suppose that the system $\{\mathbf{w}_1, \cdots, \mathbf{w}_i\}$ of mutually orthogonal unit vectors has been selected so that we have the equal subspaces $\langle \mathbf{v}_1, \cdots, \mathbf{v}_i \rangle = \langle \mathbf{w}_1, \cdots, \mathbf{w}_i \rangle$. If $i = k$, we are finished. If $i < k$, define

$$\mathbf{x}_{i+1} = \mathbf{v}_{i+1} - (\mathbf{w}_1 \cdot \mathbf{v}_{i+1})\mathbf{w}_1 - \cdots - (\mathbf{w}_i \cdot \mathbf{v}_{i+1})\mathbf{w}_i$$

Again $\mathbf{x}_{i+1} \neq \mathbf{0}$ because $\mathbf{v}_{i+1}$ is not in the space $\langle \mathbf{w}_1, \cdots, \mathbf{w}_i \rangle$. Furthermore, for any $\mathbf{w}_j$, $1 \leq j \leq i$, we have

$$\mathbf{w}_j \cdot \mathbf{x}_{i+1} = \mathbf{w}_j \cdot \mathbf{v}_{i+1} - (\mathbf{w}_1 \cdot \mathbf{v}_{i+1})\mathbf{w}_1 \cdot \mathbf{w}_j - \cdots \\ - (\mathbf{w}_i \cdot \mathbf{v}_{i+1})\mathbf{w}_i \mathbf{w}_j$$

However, the only two terms on the right that are not zero are the first term $\mathbf{w}_j \cdot \mathbf{v}_{i+1}$ and the term $-(\mathbf{w}_j \cdot \mathbf{v}_{i+1})\mathbf{w}_j \cdot \mathbf{w}_j = -\mathbf{w}_j \cdot \mathbf{v}_{i+1}$; these two cancel. Therefore, the vector $\mathbf{x}_{i+1}$ is orthogonal to each vector $\mathbf{w}_1, \cdots, \mathbf{w}_i$ and so is the next unit vector

$$\mathbf{w}_{i+1} = \frac{\mathbf{x}_{i+1}}{\|\mathbf{x}_{i+1}\|}$$

Again we obviously have $\langle \mathbf{v}_1, \cdots, \mathbf{v}_{i+1} \rangle = \langle \mathbf{w}_1, \cdots, \mathbf{w}_{i+1} \rangle$ and this completes a finite induction proof. ∎

■ **Example 2.** The vectors $\mathbf{v}_1 = (1, 1, 0, 0)$, $\mathbf{v}_2 = (0, 1, 1, 1)$ and $\mathbf{v}_3 = (1, 0, -1, 1)$ span a three-dimensional vector subspace of $\mathbf{R}^4$. We start the Gram-Schmidt orthonormalization by setting

$$\mathbf{w}_1 = \frac{\mathbf{v}_1}{\|\mathbf{v}_1\|} = \left(\frac{1}{\sqrt{2}}, \frac{1}{\sqrt{2}}, 0, 0\right)$$

Next, we set

$$\mathbf{x}_2 = \mathbf{v}_2 - (\mathbf{w}_1 \cdot \mathbf{v}_2)\mathbf{w}_1$$
$$= (0, 1, 1, 0) - \left(\frac{1}{\sqrt{2}}\right)\left(\frac{1}{\sqrt{2}}, \frac{1}{\sqrt{2}}, 0, 0\right)$$
$$= (0, 1, 1, 0) - \left(\frac{1}{2}, \frac{1}{2}, 0, 0\right)$$
$$= \left(-\frac{1}{2}, \frac{1}{2}, 1, 0\right)$$

It follows that we have

$$\mathbf{w}_2 = \frac{\mathbf{x}_2}{\|\mathbf{x}_2\|} = \frac{2}{\sqrt{6}}\left(-\frac{1}{2}, \frac{1}{2}, 1, 0\right)$$
$$= \left(-\frac{1}{\sqrt{6}}, \frac{1}{\sqrt{6}}, \frac{2}{\sqrt{6}}, 0\right)$$

Finally, we have

$$\mathbf{x}_3 = \mathbf{v}_3 - (\mathbf{w}_1 \cdot \mathbf{v}_3)\mathbf{w}_1 - (\mathbf{w}_2 \cdot \mathbf{v}_3)\mathbf{w}_2$$
$$= (1, 0, -1, 1) - \left(\frac{1}{\sqrt{2}}\right)\left(\frac{1}{\sqrt{2}}, \frac{1}{\sqrt{2}}, 0, 0\right)$$
$$- \left(-\frac{3}{\sqrt{6}}\right)\left(-\frac{1}{\sqrt{6}}, \frac{1}{\sqrt{6}}, \frac{2}{\sqrt{6}}, 0\right)$$

$$= (1, 0, -1, 1) - \left(\frac{1}{2}, \frac{1}{2}, 0, 0\right) + \left(-\frac{1}{2}, \frac{1}{2}, 1, 0\right)$$
$$= (0, 0, 0, 1)$$

Because $\mathbf{x}_3$ *is* a unit vector, we set $\mathbf{w}_3 = \mathbf{x}_3 = (0, 0, 0, 1)$ and we are finished. Then an orthonormal basis is

$$\left\{ \left(\frac{1}{\sqrt{2}}, \frac{1}{\sqrt{2}}, 0, 0\right), \left(\frac{-1}{\sqrt{6}}, \frac{1}{\sqrt{6}}, \frac{2}{\sqrt{6}}, 0\right), (0, 0, 0, 1) \right\} \blacksquare$$

Suppose that $\{\mathbf{v}_1, \cdots, \mathbf{v}_n\}$ and $\{\mathbf{w}_1, \cdots, \mathbf{w}_n\}$ are two orthonormal bases for $\mathbf{R}^n$. Let the matrix A transform the basis $\{\mathbf{v}_i\}$ into the basis $\{\mathbf{w}_j\}$, that is, let

$$\mathbf{v}_i = a_{i1}\mathbf{w}_1 + \cdots + a_{in}\mathbf{w}_n, \quad i = 1, 2, \cdots, n$$

Writing out the defining equations of an orthonormal system

$$\mathbf{v}_i \cdot \mathbf{v}_j = \delta_{ij}$$

in terms of the basis $\{\mathbf{w}_j\}$, we obtain

$$(a_{i1}\mathbf{w}_1 + \cdots + a_{in}\mathbf{w}_n) \cdot (a_{j1}\mathbf{w}_1 + \cdots + a_{jn}\mathbf{w}_n) = \delta_{ij}$$

Utilizing the fact that $\mathbf{w}_i \cdot \mathbf{w}_j = \delta_{ij}$ also, this dot product reduces to

$$a_{i1}a_{j1} + \cdots + a_{in}a_{jn} = \delta_{ij}$$

This is precisely the equation

$$\mathbf{r}_i \cdot \mathbf{r}_j = \delta_{ij}$$

where $\mathbf{r}_i$ and $\mathbf{r}_j$ are row vectors of the matrix A. Therefore, the row vectors of A are mutually orthogonal unit vectors. This fact can be expressed by the simple matrix equation

$$AA^T = I$$

Note that the number in the ith row and the jth column of the matrix AA^T is the dot product of the ith row vector $\mathbf{r}_i$ of A and the jth column vector of A^T. Since the latter vector is the same as the jth row vector of A, the numbers in AA^T are precisely $\mathbf{r}_i \cdot \mathbf{r}_j$. Of course, the numbers in I are δ_{ij}. Hence the equation $AA^T = I$ is equivalent to the n^2 equations $\mathbf{r}_i \cdot \mathbf{r}_j = \delta_{ij}$.

The matrix A is said to be *orthogonal* if $AA^T = I$. Note that this implies the equation $A^{-1} = A^T$.

■ **THEOREM 14.40.** Let $\{\mathbf{v}_i\}$ be an orthonormal basis for $\mathbf{R}^n$. If A is the matrix transforming $\{\mathbf{v}_i\}$ into the basis $\{\mathbf{w}_j\}$, then $\{\mathbf{w}_j\}$ is orthonormal if and only if A is orthogonal.

PROOF: We have shown above that, if $\{\mathbf{w}_j\}$ is orthonormal, then A is orthogonal. Conversely, if A is orthogonal, then the matrix that transforms the basis $\{\mathbf{w}_j\}$ into $\{\mathbf{v}_i\}$ is $A^{-1} = A^T$. It follows that

14.9 TRANSFORMATIONS OF COORDINATES IN R^n

$$w_j = a_{1j}v_1 + \cdots + a_{nj}v_n, \quad j = 1, 2, \ldots, n$$

Then because $v_i \cdot v_j = \delta_{ij}$, we have the equations

$$w_i \cdot w_j = a_{1i}a_{1j} + \cdots + a_{ni}a_{nj} = \delta_{ij}$$

The second equality is true because A is orthogonal. Therefore, $\{w_j\}$ is orthonormal. ∎

Let A be an orthogonal matrix, that is, $AA^T = I$. Then we have the corresponding equation in determinants

$$(\det A)(\det A^T) = \det I = 1$$

Knowing that $\det A^T = \det A$ for any matrix, we obtain

$$(\det A)^2 = 1$$

for the orthogonal matrix A. Thus either $\det A = 1$ or $\det A = -1$. We say that the orthogonal matrix A is *proper* if $\det A = 1$ and is *improper* if $\det A = -1$.

A transformation of coordinates in R^n from one orthonormal basis to another is said to be a *rotation* if the orthogonal matrix of the transformation is proper. We shall return to this kind of transformation shortly.

■ **Example 3.** Consider the standard orthonormal basis $\{u_1, u_2, u_3\}$ for R^3 and the matrix

$$A = \begin{bmatrix} \frac{2}{3} & \frac{2}{3} & \frac{1}{3} \\ -\frac{2}{3} & \frac{1}{3} & \frac{2}{3} \\ \frac{1}{3} & -\frac{2}{3} & \frac{2}{3} \end{bmatrix}$$

It is easy to verify that $AA^T = I$ and that $\det A = 1$ and hence that A is a proper orthogonal matrix. Therefore, the transformation of $\{u_1, u_2, u_3\}$ into the orthonormal basis

$$v_1 = \tfrac{2}{3}u_1 + \tfrac{2}{3}u_2 + \tfrac{1}{3}u_3 = (\tfrac{2}{3}, \tfrac{2}{3}, \tfrac{1}{3})$$
$$v_2 = -\tfrac{2}{3}u_1 + \tfrac{1}{3}u_2 + \tfrac{2}{3}u_3 = (-\tfrac{2}{3}, \tfrac{1}{3}, \tfrac{2}{3})$$
$$v_3 = \tfrac{1}{3}u_1 - \tfrac{2}{3}u_2 + \tfrac{2}{3}u_3 = (\tfrac{1}{3}, -\tfrac{2}{3}, \tfrac{2}{3})$$

is a rotation. We shall identify this rotation more closely in the next section.

EXERCISES 14.9

1. Find a transformation of coordinates that carries the vector $(2, 1)$ into $(1, 2)$ and the vector $(2, -1)$ into $(1, -2)$.

2. Find a transformation of coordinates in R^3 that carries the vector $(1, 0, -1)$ into $(0, 1, 1)$, the vector $(1, -1, 1)$ into $(1, 0, 1)$ and the vector $(0, 2, 1)$ into $(0, 1, -1)$.

Show that each of the matrices in Exercises 3 to 10 is orthogonal.

3. $\begin{bmatrix} \frac{4}{5} & \frac{3}{5} \\ -\frac{3}{5} & \frac{4}{5} \end{bmatrix}$
4. $\begin{bmatrix} \frac{12}{13} & -\frac{5}{13} \\ \frac{5}{13} & \frac{12}{13} \end{bmatrix}$

5. $\begin{bmatrix} 0 & 1 & 0 \\ 1 & 0 & 0 \\ 0 & 0 & -1 \end{bmatrix}$
6. $\begin{bmatrix} 0 & 0 & 1 \\ 0 & -1 & 0 \\ -1 & 0 & 0 \end{bmatrix}$

7. $\begin{bmatrix} 1 & 0 & 0 \\ 0 & \frac{4}{5} & -\frac{3}{5} \\ 0 & \frac{3}{5} & \frac{4}{5} \end{bmatrix}$
8. $\begin{bmatrix} 1 & 0 & 0 & 0 \\ 0 & 0 & 1 & 0 \\ 0 & 1 & 0 & 0 \\ 0 & 0 & 0 & 1 \end{bmatrix}$

9. $\begin{bmatrix} 1 & 0 & 0 & 0 \\ 0 & \frac{2}{3} & \frac{2}{3} & \frac{1}{3} \\ 0 & -\frac{2}{3} & \frac{1}{3} & \frac{2}{3} \\ 0 & \frac{1}{3} & -\frac{2}{3} & \frac{2}{3} \end{bmatrix}$
10. $\begin{bmatrix} 1/\sqrt{2} & 0 & 1/\sqrt{2} & 0 \\ 0 & 1/\sqrt{2} & 0 & 1/\sqrt{2} \\ \frac{1}{2} & \frac{1}{2} & -\frac{1}{2} & -\frac{1}{2} \\ \frac{1}{2} & -\frac{1}{2} & -\frac{1}{2} & \frac{1}{2} \end{bmatrix}$

In each of Exercises 11 to 20, apply the Gram-Schmidt process to obtain an orthonormal basis for the given vector subspace of $\mathbf{R}^n$.

11. $\langle (2, 1), (-1, 1) \rangle \subset \mathbf{R}^2$
12. $\langle (0, -2), (1, -3) \rangle \subset \mathbf{R}^2$
13. $\langle (1, 3, 2), (1, 0, -1) \rangle \subset \mathbf{R}^3$
14. $\langle (-1, 1, 2), (1, 1, 3), (1, 1, 2) \rangle \subset \mathbf{R}^3$
15. $\langle (2, 1, -1), (3, 1, 2), (0, 1, -7) \rangle \subset \mathbf{R}^3$
16. $\langle (1, 0, 1, -1), (2, 1, 0, 0) \rangle \subset \mathbf{R}^4$
17. $\langle (1, 0, 1, -1), (2, 1, 0, 0), (1, 3, 0, 2) \rangle \subset \mathbf{R}^4$
18. $\langle (1, 1, 1, 1), (1, -1, 1, -1), (-1, 1, 1, -1) \rangle \subset \mathbf{R}^4$
19. $\langle (2, 1, -1, 0), (0, 2, 1, -1), (1, 0, 3, -1), (0, 0, 2, 3) \rangle \subset \mathbf{R}^4$
20. $\langle (2, 1, -1, 0), (0, 1, 1, -1), (1, 0, 3, -1), (0, -1, 1, 0) \rangle \subset \mathbf{R}^4$

PROBLEMS 14.9

1. Let $L: \mathbf{R}^2 \to \mathbf{R}^2$ be a linear mapping. Prove that, if L is proper, then its matrix is of the form

$$\begin{bmatrix} \cos \theta & -\sin \theta \\ \sin \theta & \cos \theta \end{bmatrix}$$

whereas, if L is improper, then its matrix has the form

$$\begin{bmatrix} \cos \theta & \sin \theta \\ \sin \theta & -\cos \theta \end{bmatrix}$$

14.10 SYMMETRIC MATRICES AND DIAGONALIZATION

2. Prove that an orthogonal linear mapping of $\mathbf{R}^n$ onto itself preserves all lengths.

3. Let A be a proper orthogonal matrix. Prove that each element in A equals its own cofactor.

4. Let $L: \mathbf{R}^3 \to \mathbf{R}^3$ be a rotation. Prove that there exists an orthonormal basis for $\mathbf{R}^n$ with respect to which basis the matrix of L has the form
$$\begin{bmatrix} 1 & 0 & 0 \\ 0 & \cos\theta & -\sin\theta \\ 0 & \sin\theta & \cos\theta \end{bmatrix}$$

5. Let A and B be $n \times n$ orthogonal matrices and show that AB is also orthogonal.

PROBLEMS 14.9 Gram-Schmidt Orthonormalization

A vector space $\mathcal{V}$ is said to be an *inner product space* if there is an *inner product* (dot product) $\mathbf{u} \cdot \mathbf{v}$ defined in $\mathcal{V}$ and satisfying

(1) $\mathbf{v} \cdot \mathbf{v} > 0$ if $\mathbf{v} \neq 0$
(2) $\mathbf{v}_1 \cdot \mathbf{v}_2 = \mathbf{v}_2 \cdot \mathbf{v}_1$
(3) $\mathbf{v} \cdot (a\mathbf{v}_1 + b\mathbf{v}_2) = a\mathbf{v} \cdot \mathbf{v}_1 + b\mathbf{v} \cdot \mathbf{v}_2$
(4) $(a\mathbf{v}_1 + b\mathbf{v}_2) \cdot \mathbf{v} = a\mathbf{v}_1 \cdot \mathbf{v} + b\mathbf{v}_2 \cdot \mathbf{v}$

Of course, $\mathbf{R}^n$ is an inner product space with

$$\mathbf{x} \cdot \mathbf{y} = (x_1, \cdots, x_n) \cdot (y_1, \cdots, y_n) = x_1 y_1 + \cdots + x_n y_n$$

1. Prove that the vector space $C[a, b]$ is an inner product space if we define

$$f \cdot g = \int_a^b fg$$

2. Prove that $\{\sin t, \sin 2t, \cdots, \sin nt, \cdots\}$ is an orthogonal set of vectors in $C[0, 2\pi]$.

3. Show that the Gram-Schmidt process can be applied in any inner product space to produce an orthonormal basis, from a given basis.

4. Given the basis $\{1, x, x^2, \cdots, x^n, \cdots\}$ for the vector space of all polynomials in $C[-1, 1]$ (the polynomials *do* form a vector subspace), use the Gram-Schmidt process to obtain an orthonormal basis. The polynomials obtained are called *Legendre* polynomials.

5. Show how the Gram-Schmidt process can be used to determine whether or not a set $\{\mathbf{v}_1, \cdots, \mathbf{v}_n\}$ of vectors in an inner product space $\mathcal{V}$ is linearly independent.

14.10 SYMMETRIC MATRICES AND DIAGONALIZATION

The square matrix B is *similar* to the square matrix A if there is a nonsingular matrix S such that $B = SAS^{-1}$. Note that, if B is similar to A, then $A = S^{-1}BS$ and hence A is similar to B. By taking $S = I$, we see that A is similar to itself. If $B = SAS^{-1}$ and $C = TBT^{-1}$, then we have

$$C = TBT^{-1} + T(SAS^{-1})T^1 = (TS)A(S^{-1}T^{-1})$$
$$= (TS)A(TS)^{-1}$$

This proves that the relation of similarity is reflexive, symmetric, and transitive and hence is an equivalence relation.

■ **Example 1.** The 3×3 matrices

$$A = \begin{bmatrix} 3 & 0 & 0 \\ 0 & 2 & 0 \\ 0 & 0 & -1 \end{bmatrix} \quad \text{and} \quad B = \begin{bmatrix} -67 & 38 & -44 \\ -9 & 8 & -6 \\ 96 & -52 & 63 \end{bmatrix}$$

are similar. It is just a matter of computation to prove that $B = SAS^{-1}$ where

$$S = \begin{bmatrix} -3 & 2 & -2 \\ 9 & -5 & 6 \\ 11 & -6 & 7 \end{bmatrix}. \blacksquare$$

Let $L : \mathbf{R}^n \to \mathbf{R}^n$ be a linear mapping of $\mathbf{R}^n$ into itself and let A be its matrix (with respect to the standard basis). Many of the geometric effects of L can be studied by considering the matrix A. As an important example, suppose we want to find a one-dimensional, L-invariant subspace of $\mathbf{R}^n$. Now a one-dimensional subspace will be L-invariant if and only if it is generated by a vector $\mathbf{v}$ which is mapped by L onto some nonzero multiple $\alpha\mathbf{v}$ of itself. Therefore, we want to find a nonzero vector $\mathbf{v}$ in $\mathbf{R}^n$ with the property that

$$L(\mathbf{v}) = A\mathbf{v} = \alpha\mathbf{v}$$

where α is some nonzero real number. This equation can then be written in the form

$$(A - \alpha I)\mathbf{v} = \mathbf{0}$$

Now this homogeneous $n \times n$ system of equations has a nonzero solution if and only if the coefficient matrix $A - \alpha I$ is singular. This in turn implies that the equation

$$\det(A - \alpha I) = 0$$

must hold. The equation above is obviously a polynomial equation (in α) of degree n. For each real root of this equation a nonzero solution vector $\mathbf{v}$ of the homogeneous system can be found and, of course, each such vector $\mathbf{v}$ generates a one-dimensional, L-invariant subspace of $\mathbf{R}^n$.

■ **DEFINITION.** Let $L : \mathbf{R}^n \to \mathbf{R}^n$ be linear. A generator $\mathbf{v} \neq \mathbf{0}$ of a one-dimensional, L-invariant subspace of $\mathbf{R}^n$ is called a *characteristic vector* of L, and if $L(\mathbf{v}) = \alpha\mathbf{v}$ for some real number α, then α is a *characteristic root* of L. Equivalently, let A be the matrix of L. Then $\mathbf{v}$ is a characteristic vector of either A or L if $A\mathbf{v} = \alpha\mathbf{v}$ for some real number α and α is a characteristic root of A. The characteristic roots of the matrix A are the roots of the nth degree polynomial equation

$$\det(A - \alpha I) = 0$$

which is called the *characteristic equation* of A.

■ **Example 2.** Consider the orthogonal matrix A from Example 2 of the previous section. It is

$$A = \begin{bmatrix} \frac{2}{3} & \frac{2}{3} & \frac{1}{3} \\ -\frac{2}{3} & \frac{1}{3} & \frac{2}{3} \\ \frac{1}{3} & -\frac{2}{3} & \frac{2}{3} \end{bmatrix}$$

Its characteristic equation is

$$\det(A - \alpha I) = \det \begin{bmatrix} \frac{2}{3} - \alpha & \frac{2}{3} & \frac{1}{3} \\ -\frac{2}{3} & \frac{1}{3} - \alpha & \frac{2}{3} \\ \frac{1}{3} & -\frac{2}{3} & \frac{2}{3} - \alpha \end{bmatrix}$$

$$= 1 - \tfrac{5}{3}\alpha + \tfrac{5}{3}\alpha^2 - \alpha^3 = 0$$

One easily sees that $\alpha = 1$ is a real root and then finds it to be the only real root. This characteristic root provides the system of homogeneous equations

$$-\tfrac{1}{3}x + \tfrac{2}{3}y + \tfrac{1}{3}z = 0$$
$$-\tfrac{2}{3}x - \tfrac{2}{3}y + \tfrac{2}{3}z = 0$$
$$\tfrac{1}{3}x - \tfrac{2}{3}y - \tfrac{1}{3}z = 0$$

which we solve to find that all characteristic vectors are of the form $\mathbf{v} = (x, 0, x)$. It follows that the line through the origin and, say, the point $(1, 0, 1)$ is the subspace that is invariant under the rotation defined by A. Hence this rotation is about the line just described.

■ **THEOREM 14.41.** Similar matrices have equal determinants, the same characteristic equations, and the same characteristic roots.

PROOF: If $B = SAS^{-1}$ where S is nonsingular, then $BS = SA$. Therefore, $\det B \cdot \det S = \det S \cdot \det A$; because $\det S \neq 0$, we have $\det B = \det A$. From this it also follows that

$$\det(A - \alpha I) = \det S(A - \alpha I)S^{-1} = \det(SAS^{-1} - \alpha SIS^{-1})$$
$$= \det(B - \alpha I)$$

This shows that A and B have the same characteristic polynomial and the remainder of the theorem is immediately obvious. ■

A *diagonal matrix* is a square matrix in which each element a_{ij}, $i \neq j$, is zero. Thus a diagonal matrix looks like this:

$$D = \begin{bmatrix} a_{11} & 0 & 0 & \cdots & 0 \\ 0 & a_{22} & 0 & \cdots & 0 \\ 0 & 0 & a_{33} & \cdots & 0 \\ \multicolumn{5}{c}{\dotfill} \\ 0 & 0 & 0 & \cdots & a_{nn} \end{bmatrix}$$

The reader should verify that the characteristic roots of a diagonal matrix are precisely the numbers on its main diagonal.

■ **THEOREM 14.42.** If the matrix A is similar to a diagonal matrix

$$D = \begin{bmatrix} \alpha_1 & 0 & \cdots & 0 \\ 0 & \alpha_2 & \cdots & 0 \\ \multicolumn{4}{c}{\dotfill} \\ 0 & 0 & \cdots & \alpha_n \end{bmatrix}$$

then $\alpha_1, \alpha_2, \cdots, \alpha_n$ are the characteristic roots of A.

PROOF: This follows from the preceding comments and Theorem 14.41. ■

A matrix A is *symmetric* if $A^T = A$. This means that the ith row vector and ith column vector of A coincide. We shall show that a symmetric matrix is necessarily similar to a diagonal matrix, and then we use this result to classify completely conics and quadrics. First, however, we need a preliminary result.

■ **THEOREM 14.43.** The characteristic roots of a symmetric matrix are all real.

PROOF: Suppose that the complex number $\alpha + \beta i$ is a root of the characteristic equation of the symmetric matrix A. Because complex roots of a polynomial equation with real coefficients occur in conjugate pairs, the complex number $\alpha - \beta i$ is also a characteristic root. Therefore the matrices

$$A - (\alpha + \beta i)I, \quad A - (\alpha - \beta i)I$$

are singular and so is their product

$$A - (\alpha + \beta i)I, \quad A - (\alpha - \beta i)I$$
$$= A^2 - 2\alpha A + (\alpha^2 + \beta^2)I$$
$$= (A - \alpha I)^2 + \beta^2 I$$

It follows that there is a nonzero column vector **x** such that

$$[(A - \alpha I)^2 - \beta^2 I]\mathbf{x} = \mathbf{0}$$

Therefore we can write the dot product
$$\mathbf{x}^T[(A - \alpha I)^2 + \beta^2 I\,]\mathbf{x} = 0$$
or
$$\mathbf{x}^T(A - \alpha I)^2\mathbf{x} + \beta^2 \mathbf{x}^T\mathbf{x} = 0$$

Setting $\mathbf{y} = (A - \alpha I)\mathbf{x}$, we have $\mathbf{y}^T = \mathbf{x}^T(A - \alpha I)^T = \mathbf{x}^T(A - \alpha I)$ because obviously $A - \alpha I$ is also symmetric. Thus we have the equation
$$\mathbf{y}^T \cdot \mathbf{y} + \beta^2 \mathbf{x}^T \mathbf{x} = 0$$

Knowing that $\mathbf{y}^T\mathbf{y} = \|\mathbf{y}\|^2 \geqq 0$ and that $\mathbf{x}^T\mathbf{x} = \|\mathbf{x}\|^2 > 0$, it must follow that $\beta^2 = 0$. This contradicts the assumption that the characteristic root is complex. ∎

■ **THEOREM 14.44.** Let A be a symmetric matrix. Then there is an orthogonal matrix Q such that $Q^{-1}AQ$ is a diagonal matrix.

PROOF: Let $\alpha_1, \alpha_2, \cdots, \alpha_n$ be the (real) characteristic roots of A. Corresponding to α_1, there is a unit characteristic vector $\mathbf{c}_1$. Consider a matrix C whose first column vector is $\mathbf{c}_1$ and whose column space is $\mathbf{R}^n$. (This can always be constructed, of course.) By the Gram-Schmidt process we alter the remaining columns of C to obtain an orthogonal matrix Q_1 which still has $\mathbf{c}_1$ as its first column vector, of course. Now the first column of the matrix AQ_1 is $A\mathbf{c}_1 = \alpha_1 \mathbf{c}_1$ (because $\mathbf{c}_1$ is a characteristic vector belonging to α_1.) Hence the first column of $Q_1^{-1}AQ_1$ is $\alpha_1 Q_1^{-1}\mathbf{c}_1$. This is precisely the first column of $\alpha_1 Q_1^{-1} Q_1$ which is

$$\begin{bmatrix} \alpha_1 \\ 0 \\ \vdots \\ 0 \end{bmatrix}$$

Furthermore $Q_1^{-1}AQ_1$ is symmetric because we have
$$(Q_1^{-1}AQ_1)^T = (Q_1^T A Q_1) = Q_1^T A^T Q_1 = Q_1^{-1}AQ_1$$

It follows that
$$Q_1^{-1}AQ_1 = \begin{bmatrix} \alpha_1 & 0 & 0 & \cdots & 0 \\ 0 & b_{11} & b_{12} & \cdots & b_{1n-1} \\ 0 & b_{21} & b_{22} & \cdots & b_{2n-1} \\ \vdots & \cdots & \cdots & \cdots & \cdots \\ 0 & b_{n-11} & b_{n-12} & \cdots & b_{n-1n-1} \end{bmatrix}$$

where the $(n-1) \times (n-1)$ matrix $B = (b_{ij})$ is symmetric and has characteristic roots $\alpha_2, \alpha_3, \cdots, \alpha_n$. The proof proceeds now by an obvious finite induction. ∎

From Theorem 14.44 it is evident that the matrix Q which diagonalizes the symmetric matrix A has as its column vectors n mutually orthogonal unit characteristic vectors of A. Thus the problem of diagonalizing a real symmetric matrix is precisely that of finding n mutually orthogonal characteristic vectors. Because we shall make use of this procedure in the next section, we quote the following result.

■ **THEOREM 14.45.** If the real number α is a k-fold root of the characteristic equation of a symmetric matrix A, then A has exactly k mutually orthogonal characteristic vectors belonging to α.

■ *Example 3.* Consider the symmetric matrix

$$A = \begin{bmatrix} -3 & 12 & 18 \\ 12 & 9 & -6 \\ 18 & -6 & -6 \end{bmatrix}$$

Its characteristic equation $\det(A - \alpha I) = 0$ turns out to be

$$-4374 + 567\alpha - \alpha^3 = 0$$

The characteristic roots are $\alpha_1 = 9$, $\alpha_2 = 18$, $\alpha_3 = -27$. Since each of these is a one-fold root, Theorem 14.45 promises us just one characteristic vector for each. The 3×3 system

$$(A - 9I)\mathbf{x} = \mathbf{0}$$

is written out as

$$\begin{bmatrix} -3 & -9 & 9 & -9 & -6 & -9 \\ & 12 & & 12 & & 18 \\ & 18 & & -6 & -6 & -9 \end{bmatrix} \begin{bmatrix} x_1 \\ x_2 \\ x_3 \end{bmatrix} = \begin{bmatrix} 0 \\ 0 \\ 0 \end{bmatrix}$$

or

$$\begin{aligned} -12x_1 + 12x_2 + 18x_3 &= 0 \\ 12x_1 - 6x_3 &= 0 \\ 18x_1 - 6x_2 - 15x_3 &= 0 \end{aligned}$$

It is easy to see that any vector of the form $(x_1, -2x_1, 2x_1)$ is a solution vector of this system and hence is a characteristic vector belonging to the root $\alpha_1 = 9$. A unit vector of this form is $(\frac{1}{3}, -\frac{2}{3}, \frac{2}{3})$. The reader may verify that a characteristic unit vector belonging to $\alpha_2 = 18$ is $(\frac{2}{3}, \frac{2}{3}, \frac{1}{3})$ and that a characteristic unit vector belonging to $\alpha_3 = -27$ is $(-\frac{2}{3}, \frac{1}{3}, \frac{2}{3})$. We use these mutually orthogonal unit vectors as the column vectors of an orthogonal matrix

14.10 SYMMETRIC MATRICES AND DIAGONALIZATION

$$Q = \begin{bmatrix} \frac{1}{3} & \frac{2}{3} & -\frac{2}{3} \\ -\frac{2}{3} & \frac{2}{3} & \frac{1}{3} \\ \frac{2}{3} & \frac{1}{3} & \frac{2}{3} \end{bmatrix}$$

Then the matrix $Q^{-1}AQ$ is

$$\begin{bmatrix} \frac{1}{3} & -\frac{2}{3} & \frac{2}{3} \\ \frac{2}{3} & \frac{2}{3} & \frac{1}{3} \\ -\frac{2}{3} & \frac{1}{3} & \frac{2}{3} \end{bmatrix} \begin{bmatrix} -3 & 12 & 18 \\ 12 & 9 & -6 \\ 18 & -6 & -6 \end{bmatrix} \begin{bmatrix} \frac{1}{3} & \frac{2}{3} & -\frac{2}{3} \\ -\frac{2}{3} & \frac{2}{3} & \frac{1}{3} \\ \frac{2}{3} & \frac{1}{3} & \frac{2}{3} \end{bmatrix}$$

$$= \begin{bmatrix} \frac{1}{3} & -\frac{2}{3} & \frac{2}{3} \\ \frac{2}{3} & \frac{2}{3} & \frac{1}{3} \\ -\frac{2}{3} & \frac{1}{3} & \frac{2}{3} \end{bmatrix} \begin{bmatrix} 3 & 12 & 18 \\ -6 & 12 & -9 \\ 6 & 6 & -18 \end{bmatrix} = \begin{bmatrix} 9 & 0 & 0 \\ 0 & 18 & 0 \\ 0 & 0 & -27 \end{bmatrix}$$

Find the characteristic roots and vectors of each matrix in Exercises 1 to 10.

EXERCISES 14.10

1. $\begin{bmatrix} 2 & 1 \\ 0 & -1 \end{bmatrix}$

2. $\begin{bmatrix} 0 & 1 \\ -1 & 6 \end{bmatrix}$

3. $\begin{bmatrix} 2 & 3 \\ 1 & 4 \end{bmatrix}$

4. $\begin{bmatrix} 2 & -2 \\ -1 & -3 \end{bmatrix}$

5. $\begin{bmatrix} 1 & 0 & 0 \\ 0 & 2 & 0 \\ 0 & 0 & -3 \end{bmatrix}$

6. $\begin{bmatrix} 1 & 0 & 0 \\ 2 & 2 & 0 \\ 3 & 1 & -3 \end{bmatrix}$

7. $\begin{bmatrix} 0 & 1 & 1 \\ 1 & 0 & 0 \\ 1 & 0 & 0 \end{bmatrix}$

8. $\begin{bmatrix} \frac{6}{7} & \frac{2}{7} & -\frac{3}{7} \\ \frac{3}{7} & -\frac{6}{7} & \frac{2}{7} \\ \frac{2}{7} & \frac{3}{7} & \frac{6}{7} \end{bmatrix}$

9. $\begin{bmatrix} 1 & 0 & 0 & 2 \\ 0 & -1 & 0 & -3 \\ 0 & 0 & 2 & 0 \\ 0 & 0 & 0 & 3 \end{bmatrix}$

10. $\begin{bmatrix} 1 & 1 & -1 & -1 \\ 0 & 0 & 1 & 1 \\ 0 & 0 & -1 & 0 \\ 0 & 1 & -1 & -2 \end{bmatrix}$

For each matrix in Exercises 11 to 20, find a nonsingular matrix Q that diagonalizes the given matrix.

11. $\begin{bmatrix} 1 & \sqrt{3} \\ \sqrt{3} & -1 \end{bmatrix}$

12. $\begin{bmatrix} 2 & 1 \\ 1 & -1 \end{bmatrix}$

13. $\begin{bmatrix} 3 & 2 \\ 2 & -1 \end{bmatrix}$

14. $\begin{bmatrix} 3 & \sqrt{2} \\ \sqrt{2} & 3 \end{bmatrix}$

15. $\begin{bmatrix} 3 & -2 \\ -2 & 1 \end{bmatrix}$ 16. $\begin{bmatrix} 2 & 1 \\ 0 & -1 \end{bmatrix}$

17. $\begin{bmatrix} 1 & 0 & 1 \\ 0 & 1 & -1 \\ 1 & -1 & 1 \end{bmatrix}$ 18. $\begin{bmatrix} 0 & 1 & 0 \\ 1 & 0 & 0 \\ 0 & 0 & 0 \end{bmatrix}$

19. $\begin{bmatrix} 1 & 0 & 0 \\ 2 & -1 & 0 \\ 3 & 2 & 2 \end{bmatrix}$ 20. $\begin{bmatrix} 0 & -1 & -2 \\ -1 & 0 & 1 \\ -2 & 1 & 0 \end{bmatrix}$

21. The numbers 4 and -4 are characteristic roots of the following matrix. Find the other two.

$$\begin{bmatrix} -1 & 2 & 1 & 2 \\ 2 & 3 & 2 & -3 \\ 1 & 2 & -1 & 2 \\ 2 & -3 & 2 & 3 \end{bmatrix}$$

22. Prove that the matrix below is not similar to a diagonal one.

$$\begin{bmatrix} 1 & 2 & -3 \\ 0 & -1 & 6 \\ 0 & -1 & 4 \end{bmatrix}$$

PROBLEMS 14.10

1. Let $P(x) = a_0 + a_1 x + \cdots + a_n x^n$ be a polynomial. Given a square matrix A, we form the new matrix

$$P(A) = a_0 I + a_1 A + \cdots + a_n A^n$$

If α is a characteristic root of A, prove that $P(\alpha)$ is then a characteristic root of $P(A)$.

2. The Cayley-Hamilton theorem states that a square matrix satisfies its own characteristic equation. That is, if matrix A has characteristic equation $P(\alpha) = 0$, then $P(A) = 0$ (the zero matrix). Show this to be true for the following matrices.

(a) $\begin{bmatrix} 3 & 2 \\ -1 & 1 \end{bmatrix}$ (b) $\begin{bmatrix} 1 & 3 & 1 \\ 2 & -1 & 0 \\ 0 & 1 & 4 \end{bmatrix}$ (c) $\begin{bmatrix} a & b \\ c & d \end{bmatrix}$

3. Consider the 2×2 matrix

$$A = \begin{bmatrix} a & b \\ c & d \end{bmatrix}$$

Show that A is similar to a diagonal matrix under any one of the following conditions
(a) $(a - d)^2 + bc > 0$
(b) $b = c$
(c) $(a - d)^2 + bc = 0$ only if $A = \frac{1}{2}(a + d)I$.

4. For any matrix prove that both $A^T A$ and AA^T are symmetric.

5. Prove that, if A is either symmetric or skew-symmetric, then $A^T A = AA^T$ and A^2 is symmetric.

6. If A and B are symmetric and if $AB = BA$, prove that AB is symmetric.

7. Prove that zero is a characteristic root of a matrix A if and only if A is singular.

8. Let the characteristic roots $\alpha_1, \ldots, \alpha_n$ of the $n \times n$ matrix A be distinct. Prove that $\alpha_1^p, \ldots, \alpha_n^p$ are characteristic roots of A^p.

9. If α is a characteristic root of the nonsingular matrix A, is $1/\alpha$ a characteristic root of A^{-1}?

10. Prove that ± 1 are the only possible real characteristic roots of an orthogonal matrix.

11. Let all of the characteristic roots of a matrix A be real. Prove that there exists an orthogonal matrix 0 such that $0^{-1}A0$ is triangular.

12. (a) Prove that a (nonzero) matrix is nilpotent if and only if each of its characteristic roots is zero.
    (b) Prove that a nilpotent matrix is not similar to a diagonal matrix.

13. Let A be an idempotent matrix (that is, $A^2 = A$). Prove that each characteristic root is either zero or one.

14. Let A be a nonsingular matrix and prove that the characteristic vectors of A and A^{-1} coincide.

15. Let A have distinct characteristic roots and let B also have distinct characteristic roots. Prove that $AB = BA$ if and only if A and B have the same characteristic roots.

16. Let α be a characteristic root of the matrix A. Prove that $c + \alpha$ is a characteristic root of $cI + A$.

17. Prove that the characteristic roots of a triangular matrix are precisely its main diagonal elements.

18. A linear mapping $L : \mathbf{R}^n \to \mathbf{R}^n$ is said to be *symmetric* if, for each pair of vectors $\mathbf{v}, \mathbf{w}$ in $\mathbf{R}^n$,

$$L(\mathbf{v}) \cdot \mathbf{w} = \mathbf{v} \cdot L(\mathbf{w})$$

Prove that the matrix of a symmetric linear mapping is symmetric.

14.11 QUADRATIC FORMS

A *quadratic form* on $\mathbf{R}^n$ is a real-valued function $Q : \mathbf{R}^n \to \mathbf{R}$ defined by the matrix expression

$$Q(\mathbf{x}) = \mathbf{x}^T A \mathbf{x}$$

where the vector $\mathbf{x}$ is considered to be a column vector (whence $\mathbf{x}^T$ is a row vector) and where A is an $n \times n$ matrix. In particular, if $\mathbf{x}^T = (x_1, x_2, \ldots, x_n)$ and $A = (a_{ij})$, then

$$Q(\mathbf{x}) = \sum_{i=1}^{n} \sum_{j=1}^{n} a_{ij} x_i x_j$$

■ **Example 1.** Consider the quadratic form Q on $\mathbf{R}^2$ defined by

$$Q(\mathbf{x}) = (x_1, x_2) \begin{bmatrix} 3 & -2 \\ 1 & 4 \end{bmatrix} \begin{bmatrix} x_1 \\ x_2 \end{bmatrix}$$

$$= (x_1, x_2) \begin{bmatrix} 3x_1 - 2x_2 \\ x_1 + 4x_2 \end{bmatrix}$$

$$= x_1(3x_1 - 2x_2) + x_2(x_1 + 4x_2)$$
$$= 3x_1^2 - x_1 x_2 + 4x_2^2$$

We remark that any matrix of the form

$$A = \begin{bmatrix} 3 & -c \\ c - 1 & 4 \end{bmatrix}$$

defines the same quadratic form. We simply compute

$$(x_1, x_2) \begin{bmatrix} 3 & -c \\ c - 1 & 4 \end{bmatrix} \begin{bmatrix} x_1 \\ x_2 \end{bmatrix}$$

$$= (x_1, x_2) \begin{bmatrix} 3x_1 & -cx_2 \\ (c-1)x_1 + 4x_2 \end{bmatrix}$$

$$= 3x_1^2 - cx_1 x_2 + (c - 1)x_1 x_2 + 4x_2^2$$
$$= 3x_1^2 - x_1 x_2 + 4x_2^2$$

In particular we can select the number c in such a way that the defining matrix is symmetric. We just want $c - 1 = -c$ or $c = -\frac{1}{2}$. Thus the symmetric matrix

$$A = \begin{bmatrix} 3 & -\frac{1}{2} \\ -\frac{1}{2} & 4 \end{bmatrix}$$

defines the given quadratic form. ■

The situation in the above example generalizes completely, namely, that *any quadratic form can be defined by a symmetric matrix*. To see this, suppose that a quadratic form Q is defined first in terms of an arbitrary matrix $B = (b_{ij})$ so that

$$Q(\mathbf{x}) = \mathbf{x}^T B \mathbf{x} = \sum_{i=1}^{n} \sum_{j=1}^{n} b_{ij} x_i x_j$$

For a fixed pair i, j, there are only two terms in this double sum that involve the product $x_i x_j$, namely, $b_{ij} x_i x_j$ and $b_{ji} x_i x_j$. Thus, in the quadratic form, the coefficient of $x_i x_j$ is $b_{ij} + b_{ji}$. If we define the new matrix A with

$$a_{ii} = b_{ii}$$
$$a_{ij} = a_{ji} = \tfrac{1}{2}(b_{ij} + b_{ji}) \text{ if } i \neq j$$

then we obviously have a symmetric matrix such that

$$\mathbf{x}^T B \mathbf{x} = \mathbf{x}^T A \mathbf{x}$$

for each vector $\mathbf{x}$ in $\mathbf{R}^n$. Therefore, there is no loss of generality in assuming that a quadratic form is defined by a symmetric matrix.

- **Example 2.** The matrix

$$B = \begin{bmatrix} 3 & 2 & 4 \\ -1 & 1 & 6 \\ 2 & -2 & 3 \end{bmatrix}$$

defines the quadratic form

$$Q(\mathbf{x}) = 3x_1^2 + x_2^2 + 3x_3^2 + x_1x_2 + 6x_1x_3 + 4x_2x_3$$

(The reader should verify this statement.) The symmetric matrix A, which defines this same quadratic form, has the following numbers:

$$a_{11} = b_{11} = 3,\ a_{22} = b_{22} = 1,\ a_{33} = b_{33} = 3$$
$$a_{12} = a_{21} = \tfrac{1}{2}(b_{12} + b_{21}) = \tfrac{1}{2}(2 - 1) = \tfrac{1}{2}$$
$$a_{13} = a_{31} = \tfrac{1}{2}(b_{13} + b_{31}) = \tfrac{1}{2}(4 + 2) = 3$$
$$a_{23} = a_{32} = \tfrac{1}{2}(b_{23} + b_{32}) = \tfrac{1}{2}(6 - 2) = 2$$

Therefore the symmetric matrix defining this quadratic form is

$$A = \begin{bmatrix} 3 & \tfrac{1}{2} & 3 \\ \tfrac{1}{2} & 1 & 2 \\ 3 & 2 & 3 \end{bmatrix} \blacksquare$$

Two quadratic forms on $\mathbf{R}^n$,

$$\mathbf{x}^T A \mathbf{x} \quad \text{and} \quad \mathbf{y}^T B \mathbf{y}$$

are said to be *equivalent* if there is a transformation of coordinates

$$\mathbf{x} = S\mathbf{y}$$

where S is a nonsingular matrix, which transforms the first form into the second. That is, we have

$$\mathbf{x}^T A \mathbf{x} = (S\mathbf{y})^T A (S\mathbf{y}) = (\mathbf{y}^T S^T) A (S\mathbf{y}) = \mathbf{y}^T (S^T A S) \mathbf{y} = \mathbf{y}^T B \mathbf{y}$$

It follows that the two forms are equivalent if and only if there exists a nonsingular matrix S such that $B = S^T A S$.

We digress a moment to give a definition suggested by the matrix condition just derived: Two square matrices A and B are *congruent* if there exists a nonsingular matrix S such that

$$B = S^T A S$$

It is easy to prove that congruence is a true equivalence relation on the set of $n \times n$ matrices (see Problem 14.11.1). We also leave the proof of the next result as an exercise.

- **THEOREM 14.46.** If the matrix B is congruent to the symmetric matrix A, the B is also symmetric.

Theorem 14.44 tells us that a symmetric matrix is congruent to a diagonal matrix. In fact, we know more, by another definition. Two quadratic forms on $\mathbf{R}^n$

$$\mathbf{x}^T A \mathbf{x} \quad \text{and} \quad \mathbf{y}^T B \mathbf{y}$$

are said to be *orthogonally equivalent* if there exists an orthogonal transformation of coordinates

$$\mathbf{x} = 0\mathbf{y}$$

where 0 is an orthogonal matrix, which transforms the first form into the second. As with the equivalence of two quadratic forms, we easily show that two forms $\mathbf{x}^T A \mathbf{x}$ and $\mathbf{y}^T B \mathbf{y}$ are orthogonally equivalent if and only if there exists an orthogonal matrix 0 such that

$$B = 0^T A 0$$

Because 0 is orthogonal, we have $0^T = 0^{-1}$. Therefore, B is both congruent to A and similar to A in this case.

- **THEOREM 14.47.** The relations of equivalence and orthogonal equivalence of quadratic forms on $\mathbf{R}^n$ are true equivalence relations.

Now we can conclude from Theorem 14.44 that every quadratic form $\mathbf{x}^T A \mathbf{x}$ is orthogonally equivalent to a diagonal form. This is stated in the following result.

- **THEOREM 14.48.** Every quadratic form $\mathbf{x}^T A \mathbf{x}$ is orthogonally equivalent to a quadratic form

$$\alpha_1 x_1^2 + \alpha_2 x_2^2 + \cdots + \alpha_n x_n^2$$

where $\alpha_1, \cdots, \alpha_n$ are the characteristic roots of the matrix A.

- **THEOREM 14.49.** Every quadratic form $\mathbf{x}^T A \mathbf{x}$ is equivalent to a form

$$x_1^2 + \cdots + x_k^2 - x_{k+1}^2 - \cdots - x_r^2$$

where r is the rank of A and k is the number of positive characteristic roots of A.

PROOF: If A has rank r, then it has exactly r characteristic roots different from zero. Let these be $\alpha_1, \cdots, \alpha_r$ and assume they have been numbered so that $\alpha_1, \cdots, \alpha_k$ are positive and $\alpha_{k+1}, \cdots, \alpha_r$ are negative. We know that there is an orthogonal matrix 0 such that $0^{-1}A0 = 0^T A 0 = D$ where D is the diagonal matrix having the characteristic roots of A on its main diagonal. We denote this matrix schematically by

$$D = \begin{bmatrix} \alpha_1 & & & & & & \\ & \alpha_2 & & & & & \\ & & \ddots & & & 0 & \\ & & & \alpha_r & & & \\ & 0 & & & 0 & & \\ & & & & & \ddots & \\ & & & & & & 0 \end{bmatrix}$$

Now define the diagonal matrix S:

$$S = \begin{bmatrix} 1/\sqrt{\alpha_1} & & & & & & & & \\ & 1/\sqrt{\alpha_2} & & & & & & & \\ & & \ddots & & & & & & \\ & & & 1/\sqrt{\alpha_k} & & & 0 & & \\ & & & & 1/\sqrt{-\alpha_{k+1}} & & & & \\ & & 0 & & & \ddots & & & \\ & & & & & & 1/\sqrt{-\alpha_r} & & \\ & & & & & & & 1 & \\ & & & & & & & & \ddots \\ & & & & & & & & & 1 \end{bmatrix}$$

It is easily computed that $S^T DS = SDS$ is the diagonal matrix having first k ones, then $r - k$ minus ones and then $n - r$ zeros along its main diagonal. It follows that the matrix $(0S)^{-1}$ transforms the quadratic form $\mathbf{x}^T A \mathbf{x}$ into the desired form. ∎

The quadratic form on $\mathbf{R}^n$,

$$x_1^2 + \cdots + x_k^2 - x_{k+1}^2 - \cdots - x_r^2$$

is called the *canonical form* of any quadratic form to which it is equivalent. It is obvious from our results that two quadratic forms are equivalent if and only if they have the same canonical form.

Consider a quadratic form $\mathbf{x}^T A \mathbf{x}$ on $\mathbf{R}^n$. The rank r of the matrix A is also called the *rank* of the quadratic form. The *index* of the quadratic form is the number k of positive characteristic roots of the matrix A. (Equivalently, the index of the quadratic form is the number of positive terms in its canonical form.) The difference $k - (r - k) = 2k - r$ between the number of positive characteristic roots of A and the number of negative characteristic roots is called the *signature* of the quadratic form. We point out that these numbers are invariant under transformations of coordinates in $\mathbf{R}^n$, but we shall not prove this.

EXERCISES 14.11 Diagonalize each of the quadratic forms in Exercises 1 to 10.

1. $3x^2 - 6xy + 4y^2$
2. $6x^2 - 2xy - y^2$
3. $2x^2 + 4xy - y^2$
4. $ax^2 + 2bxy + cy^2$
5. $x^2 + 4y^2 - 2z^2 + 2xy - 4xz + 8yz$
6. $3x^2 - 2y^2 - z^2 - 2xy + 10yz$
7. $2xy + 4yz$
8. $2xy - 2xz + 2yz$
9. $-8x^2 + 7y^2 + 10z^2 - 28xy + 16xz - 44yz$
10. $ax^2 + by^2 + cz^2 + 2(dxy + exz + fyz)$

For each of Exercises 11 to 18, determine whether or not the two given quadratic forms are equivalent.

11. $x_1^2 + x_2^2$; $3y_1^2 + 4y_2^2$
12. $x_1^2 - x_2^2$; $2y_1^2 - y_2^2$
13. $x_1^2 + x_2^2$; $y_1^2 - y_2^2$
14. $x_1^2 + 2x_2^2$; $3y_1^2 + 4y_1y_2 + 2y_2^2$
15. $x_1^2 - 2x_2^2 + 3x_3^2$; $y_1^2 + y_2^2 - y_3^2$
16. $x_1^2 - x_2^2 - 2x_3^2$; $y_1^2 - y_2^2 + y_3^2$
17. x_1x_2; $y_1^2 + y_2^2 + y_3^2$
18. $x_2^2 + 3x_3^2 - 6x_1x_2 + 8x_1x_3 + 4x_2x_3$; $7y_2^2 + 11y_3^2 - 6y_1y_2 + 8y_1y_3 - 10y_2y_3$

19. Given the following matrices A and B, find a nonsingular matrix S such that $B = S^T A S$.
$$A = \begin{bmatrix} 1 & -2 \\ -1 & 1 \end{bmatrix}, B = \begin{bmatrix} -1 & -4 \\ -1 & 5 \end{bmatrix}$$

20. Repeat Exercise 19 for the matrices
$$A = \begin{bmatrix} 2 & 1 & -1 \\ 1 & 2 & 2 \\ -1 & -1 & 0 \end{bmatrix}, B = \begin{bmatrix} 0 & -1 & 1 \\ 5 & 10 & -4 \\ 1 & 2 & 0 \end{bmatrix}$$

14.12 A GEOMETRIC APPLICATION

PROBLEMS 14.11

1. Prove that congruence is an equivalence relation on the set of $n \times n$ matrices.

2. Prove Theorem 14.46.

3. If the quadratic forms $\mathbf{x}^T A \mathbf{x}$ and $\mathbf{x}^T B \mathbf{x}$ are identical and if both A and B are symmetric, prove that $A = B$. (This shows that the symmetric representation of a quadratic form is unique.)

4. Prove that the number of nonequivalent quadratic forms in n variables is $\frac{1}{2}(n+1)(n+2)$.

5. Consider the two quadratic forms on $\mathbf{R}^2$,
$$ax_1^2 + 2bx_1x_2 + cx_2^2 \quad \text{and} \quad dy_1^2 + 2ey_1y_2 + fy_2^2$$
Prove that they are equivalent if and only if one of the following conditions holds:

 (a) $ac - b^2 > 0$, $df - e^2 > 0$ and $a + c$, $d + f$ agree in sign
 (b) $ac - b^2 = 0 = df - e^2$ and $a + c$, $d + f$ agree in sign
 (c) $ac - b^2 > 0$, $df - e^2 > 0$

 In addition, if $ac - b^2 = df - e^2$ and $a + c = d + f$, then the two quadratic forms are orthogonally equivalent.

6. Let
$$m = \frac{1}{n}\sum_{i=1}^{n} x_i$$
be the mean of a set of numbers x_i. Write the quadratic form
$$\sigma^2 = \frac{1}{n-1}\sum_{i=1}^{n}(x_i - m)^2$$
in the form $\mathbf{x}^T A \mathbf{x}$, where A is symmetric. Determine the rank of this quadratic form. For what values of the numbers x_i is this quadratic form equal to zero? (In statistics, the quadratic form σ^2 is called the *variance* of the numbers x_i.)

 A quadratic form $\mathbf{x}^T A \mathbf{x}$ is said to be *positive definite* if $\mathbf{x}^T A \mathbf{x} > 0$ for any vector $\mathbf{x} \neq 0$ and is said to be *positive semidefinite* if $\mathbf{x}^T A \mathbf{x} \geq 0$ for all vectors $\mathbf{x}$. The matrix A is positive definite and positive semidefinite according as the form $\mathbf{x}^T A \mathbf{x}$ is one or the other.

7. Prove that the quadratic form $\mathbf{x}^T A \mathbf{x}$ is positive definite if and only if each characteristic root of A is positive, and is positive semidefinite if and only if each characteristic root of A is nonnegative.

8. For any matrix A prove that AA^T is positive semidefinite and that AA^T is positive definite if and only if A is nonsingular.

14.12 A GEOMETRIC APPLICATION

The fact that the quadratic form $\mathbf{x}^T A \mathbf{x}$ on $\mathbf{R}^n$ is orthogonally equivalent to a diagonal form
$$\alpha_1 y_1^2 + \cdots + \alpha_n y_n^2$$
has an appealing geometric interpretation. We first illustrate the idea with an example in the plane.

■ *Example 1.* We consider the curve
$$\{(x_1, x_2) : 8x_1^2 + 24x_1x_2 + 15x_2^2 = 48\}$$
and set out to identify it. Clearly we have here a level curve of the quadratic form
$$(x_1, x_2)\begin{bmatrix} 8 & 12 \\ 12 & 15 \end{bmatrix}\begin{bmatrix} x_1 \\ x_2 \end{bmatrix}$$
The reader can verify that the characteristic equation of the matrix here is
$$\alpha^2 - 23\alpha - 24 = (\alpha - 24)(\alpha + 1) = 0$$
and that a unit characteristic vector belonging to the root $\alpha_1 = 24$ is $(\frac{3}{5}, \frac{4}{5})$ and a unit characteristic vector belonging to $\alpha_2 = -1$ is $(\frac{4}{5}, -\frac{3}{5})$. Using these as the column vectors of an orthogonal matrix
$$0 = \begin{bmatrix} \frac{3}{5} & \frac{4}{5} \\ \frac{4}{5} & -\frac{3}{5} \end{bmatrix}$$
we know that the orthogonal transformation of coordinates
$$\mathbf{x} = 0\mathbf{y}$$
will transform the given quadratic form into
$$24y_1^2 - y_2^2$$

In terms of the new coordinate system, the original curve has the description $\{(y_1, y_2) : 24y_1^2 - y_2^2 = 48\}$ and we now recognize it to be a hyperbola. ■

The level set
$$\{\mathbf{x} : \mathbf{x}^T A \mathbf{x} = c\}$$
of a quadratic form on $\mathbf{R}^n$ is called a *central quadric*. (In $\mathbf{R}^2$, of course, we say "central conic" instead.) Such a "surface" is studied by choosing a system of coordinates in which the equation is most simple. In short, we use an orthogonal transformation of coordinates to diagonalize the quadratic form.

In view of Theorem 14.44 there is an orthogonal transformation of coordinates $\mathbf{x} = 0\mathbf{y}$ carrying the quadratic form $\mathbf{x}^T A \mathbf{x}$ into the diagonal form $\alpha_1 y_1^2 + \alpha_n y_n^2$, where $\alpha_1, \cdots, \alpha_n$ are the characteristic roots of the matrix A. In terms of the coordinates $y_1, \cdots, y_n$ the central quadric is
$$\{y : \alpha_1 y_1^2 + \cdots + \alpha_n y_n^2 = c\}$$
We study a central conic in this form.

Dimension two. The most general central conic in $\mathbf{R}^2$ is
$$\{(x_1, x_2) : ax_1^2 + 2bx_1x_2 + cx_2^2 = d\}$$

There is an orthogonal matrix 0 whose column vectors are unit characteristic vectors of the matrix

$$A = \begin{bmatrix} a & b \\ b & c \end{bmatrix}$$

and the transformation of coordinates $\mathbf{x} = 0\mathbf{y}$ changes the description of the curve to

$$\{(y_1, y_2) : \alpha_1 y_1^2 + \alpha_2 y_2^2 = d\}$$

We already know such curves, of course. This curve will be an ellipse if α_1 and α_2 have the same sign. (It will be an imaginary ellipse if the sign of α_1 and α_2 disagrees with the sign of d.) This is the case where the index of the quadratic form equals its rank.

The central conic $\{(y_1, y_2) : \alpha_1 y_1^2 + \alpha_2 y_2^2 = d\}$ is a hyperbola if α_1 and α_2 disagree in sign. Note that in this case the quadratic form has rank 2 and index 1. The last case is that in which one characteristic root is zero. Then the quadratic form has rank 1 and the central conic is a pair of parallel lines.

Dimension three. Consider the central quadric

$$\{(x_1, x_2, x_3) : ax_1^2 + bx_2^2 + cx_3^2 + 2dx_1x_2 + 2ex_1x_3 + 2fx_2x_3 = g\}$$

Our investigation recognizes five cases.

CASE 1. If the rank of the quadratic form and its index both equal 3, then the central quadric is an ellipsoid. If any two characteristic roots are equal, the quadric is a surface of revolution; if all three roots are equal, the surface is a sphere.

CASE 2. If the rank equals three and the index equals two, then the central quadric is a hyperboloid of one sheet. It is also a surface of revolution if the two positive roots are equal.

CASE 3. If the rank is three and the index is one, then our surface is a hyperboloid of two sheets. If the two negative roots are equal, it is also a surface of revolution.

CASE 4. If the rank of the quadratic form is two, then the quadric is an elliptic cylinder if the index is also two, and is a hyperbolic cylinder if the index is one.

CASE 5. If the rank of the quadratic form is one, then the surface is a pair of parallel planes.

(The reader may verify these conclusions very easily by referring to Section 13.9.)

■ *Example 2.* Consider the central quadric

$$\{(x_1, x_2, x_3) : x_1^2 - 4x_2x_3 = 4\}$$

The quadratic form has the symmetric matrix
$$A = \begin{bmatrix} 1 & 0 & 0 \\ 0 & 0 & -2 \\ 0 & -2 & 0 \end{bmatrix}$$
and it is easy to verify that its characteristic equation is
$$(1 - \alpha)(\alpha^2 - 4) = 0$$
Hence the characteristic roots are $\alpha_1 = 1, \alpha_2 = 2, \alpha_3 = -2$. Hence, even without actually doing the calculation, we know that the quadratic form can be diagonalized to
$$y_1^2 + 2y_2^2 - 2y_3^2$$
In terms of the new orthogonal coordinate system our quadric is
$$\{(y_1, y_2, y_3) : y_1^2 + 2y_2^2 - 2y_3^2 = 4\}$$
which we recognize as a hyperboloid of one sheet (Section 13.9).

■ **Example 3.** We identify the central quadric
$$\{(x_1, x_2, x_3) : 2x_1^2 + 4x_2^2 + 3x_3^2 + 4x_1x_3 - 4x_2x_3 = 3\}$$
The matrix to be studied is
$$A = \begin{bmatrix} 2 & 0 & 2 \\ 0 & 4 & -2 \\ 2 & -2 & 3 \end{bmatrix}$$
Its characteristic equation is
$$-18\alpha + 9\alpha^2 - \alpha^3 = -\alpha(\alpha - 6)(\alpha - 3) = 0$$
Hence the characteristic roots are $\alpha_1 = 6, \alpha_2 = 3, \alpha_3 = 0$. It follows that there is a new coordinate system in which the central quadric is
$$\{(y_1, y_2, y_3) : 6y_1^2 + 3y_2^2 = 3\}$$
This is an elliptic cylinder, an instance of case 4.

EXERCISES 14.12 Identify each central conic in Exercises 1 to 6. Draw diagrams to illustrate the results.

1. $\{(x, y) : xy = 4\}$
2. $\{(x, y) : x^2 - 4xy + 4y^2 = 4\}$
3. $\{(x, y) : x^2 - 6xy + y^2 = 8\}$
4. $\{(x, y) : x^2 + 8xy + 7y^2 = 3\}$

5. $\{(x, y) : 3x^2 - 14\sqrt{3}\,xy + 18y^2 = 10\}$
6. $\{(x, y) : 29x^2 - 24xy + 36y^2 = 65\}$

Identify each central quadric in Exercises 7 to 12.

7. $\{(x, y, z) : xy + xz + yz = 2\}$
8. $\{(x, y, z) : x^2 - 2yz = 4\}$
9. $\{(x, y, z) : 2x^2 + y^2 - z^2 + 4xy = 1\}$
10. $\{(x, y, z) : -x^2 + y^2 + 4xz - 4yz = 8\}$
11. $\{(x, y, z) : 2x^2 + 2y^2 + 5z^2 - 4xy + 2xz - 2yz = 10\}$
12. $\{(x, y, z) : 13x^2 + 13y^2 + 10z^2 + 8xy + 4xz + 4yz = 36\}$

PROBLEMS 14.12

1. Classify the central conic $\{(x, y) : ax^2 + 2bxy + cy^2 = 1\}$ according to whether the determinant of the matrix
$$A = \begin{bmatrix} a & b \\ b & c \end{bmatrix}$$
is positive, negative, or zero.

2. Classify the central quadric
$$\{(x, y, z) : ax^2 + by^2 + cz^2 + 2dxy + 2exz + 2fyz = 1\}$$
according to whether the determinant
$$\det \begin{bmatrix} a & d & e \\ d & b & f \\ e & f & c \end{bmatrix}$$
has value zero, and is positive or negative. (In this instance, we need additional conditions to classify the surfaces completely.)

3. Let $\{\mathbf{x} : \mathbf{x}^T A \mathbf{x} = 1\}$ be a central quadric in $\mathbf{R}^n$. Show that the minimum distance from the origin to this "surface" is $1/\sqrt{\alpha_m}$, where α_m is the largest positive characteristic root of the matrix A.

4. Prove that the "tangent hyperplane" to the central quadric $\{\mathbf{x} : \mathbf{x}^T A \mathbf{x} = 1\}$ at the fixed point $\mathbf{a}$ on the "surface" is $\{\mathbf{x} : \mathbf{a}^T A \mathbf{x} = 1\}$.

TABLES

THE GREEK ALPHABET

Letters		Names	Letters		Names
A	α	alpha	N	ν	nu
B	β	beta	Ξ	ξ	xi
Γ	γ	gamma	O	o	omicron
Δ	δ	delta	Π	π	pi
E	ϵ	epsilon	P	ρ	rho
Z	ζ	zeta	Σ	σ	sigma
H	η	eta	T	τ	tau
Θ	θ	theta	Υ	υ	upsilon
I	ι	iota	Φ	ϕ	phi
K	κ	kappa	X	χ	chi
Λ	λ	lambda	Ψ	ψ	psi
M	μ	mu	Ω	ω	omega

T-3 | TRIGONOMETRIC FUNCTIONS: ARGUMENTS IN DEGREES

Degrees	Radians	Sine	Tangent	Cotangent	Cosine		
0	0	0	0	———	1.0000	1.5708	90
1	.0175	.0175	.0175	57.290	.9998	1.5533	89
2	.0349	.0349	.0349	28.636	.9994	1.5359	88
3	.0524	.0523	.0524	19.081	.9986	1.5134	87
4	.0698	.0698	.0699	14.301	.9976	1.5010	86
5	.0873	.0872	.0875	11.430	.9962	1.4835	85
6	.1047	.1045	.1051	9.5144	.9945	1.4661	84
7	.1222	.1219	.1228	8.1443	.9925	1.4486	83
8	.1396	.1392	.1405	7.1154	.9903	1.4312	82
9	.1571	.1564	.1584	6.3138	.9877	1.4137	81
10	.1745	.1736	.1763	5.6713	.9848	1.3963	80
11	.1920	.1908	.1944	5.1446	.9816	1.3788	79
12	.2094	.2079	.2126	4.7046	.9781	1.3614	78
13	.2269	.2250	.2309	4.3315	.9744	1.3439	77
14	.2443	.2419	.2493	4.0108	.9703	1.3265	76
15	.2618	.2588	.2679	3.7321	.9659	1.3090	75
16	.2793	.2756	.2867	3.4874	.9613	1.2915	74
17	.2967	.2924	.3057	3.2709	.9563	1.2741	73
18	.3142	.3090	.3249	3.0777	.9511	1.2566	72
19	.3316	.3256	.3443	2.9042	.9455	1.2392	71
20	.3491	.3420	.3640	2.7475	.9397	1.2217	70
21	.3665	.3584	.3839	2.6051	.9336	1.2043	69
22	.3840	.3746	.4040	2.4751	.9272	1.1868	68
23	.4014	.3907	.4245	2.3559	.9205	1.1694	67
24	.4189	.4067	.4452	2.2460	.9135	1.1519	66
25	.4363	.4226	.4663	2.1445	.9063	1.1345	65
26	.4538	.4384	.4877	2.0503	.8988	1.1170	64
27	.4712	.4540	.5095	1.9626	.8910	1.0996	63
28	.4887	.4695	.5317	1.8807	.8829	1.0821	62
29	.5061	.4848	.5543	1.8040	.8746	1.0647	61
30	.5236	.5000	.5774	1.7321	.8660	1.0472	60
31	.5411	.5150	.6009	1.6643	.8572	1.0297	59
32	.5585	.5299	.6249	1.6003	.8480	1.0123	58
33	.5760	.5446	.6494	1.5399	.8387	.9948	57
34	.5934	.5592	.6745	1.4826	.8290	.9774	56
35	.6109	.5736	.7002	1.4281	.8192	.9599	55
36	.6283	.5878	.7265	1.3764	.8090	.9425	54
37	.6458	.6018	.7536	1.3270	.7986	.9250	53
38	.6632	.6157	.7813	1.2799	.7880	.9076	52
39	.6807	.6293	.8098	1.2349	.7771	.8901	51
40	.6981	.6428	.8391	1.1918	.7660	.8727	50
41	.7156	.6561	.8693	1.1504	.7547	.8552	49
42	.7330	.6691	.9004	1.1106	.7431	.8378	48
43	.7505	.6820	.9325	1.0724	.7314	.8203	47
44	.7679	.6947	.9657	1.0355	.7193	.8029	46
45	.7854	.7071	1.0000	1.0000	.7071	.7854	45
		Cosine	Cotangent	Tangent	Sine	Radians	Degrees

T-4 | TRIGONOMETRIC FUNCTIONS OF NUMBERS 0–1.6

x	$\sin x$	$\tan x$	$\cot x$	$\cos x$	x	$\sin x$	$\tan x$	$\cot x$	$\cos x$
0.00	.00000	.00000	—	1.00000	**0.40**	.38942	.42279	2.3652	.92106
.01	.01000	.01000	99.997	0.99995	.41	.39861	.43463	2.3008	.91712
.02	.02000	.02000	49.993	.99980	.42	.40776	.44657	2.2393	.91309
.03	.03000	.03001	33.323	.99955	.43	.41687	.45862	2.1804	.90897
.04	.03999	.04002	24.987	.99920	.44	.42594	.47078	2.1241	.90475
.05	.04998	.05004	19.983	.99875	.45	.43497	.48306	2.0702	.90045
.06	.05996	.06007	16.647	.99820	.46	.44395	.49545	2.0184	.89605
.07	.06994	.07011	14.262	.99755	.47	.45289	.50797	1.9686	.89157
.08	.07991	.08017	12.473	.99680	.48	.46178	.52061	1.9208	.88699
.09	.08988	.09024	11.081	.99595	.49	.47063	.53339	1.8748	.88233
0.10	.09983	.10033	9.9666	.99500	**0.50**	.47943	.54630	1.8305	.87758
.11	.10978	.11045	9.0542	.99396	.51	.48818	.55936	1.7878	.87274
.12	.11971	.12058	8.2933	.99281	.52	.49688	.57256	1.7465	.86782
.13	.12963	.13074	7.6489	.99156	.53	.50553	.58592	1.7067	.86281
.14	.13954	.14092	7.0961	.99022	.54	.51414	.59943	1.6683	.85771
.15	.14944	.15114	6.6166	.98877	.55	.52269	.61311	1.6310	.85252
.16	.15932	.16138	6.1966	.98723	.56	.53119	.62695	1.5950	.84726
.17	.16918	.17166	5.8256	.98558	.57	.53963	.64097	1.5601	.84190
.18	.17903	.18197	5.4954	.98384	.58	.54802	.65517	1.5263	.83646
.19	.18886	.19232	5.1997	.98200	.59	.55636	.66956	1.4935	.83094
0.20	.19867	.20271	4.9332	.98007	**0.60**	.56464	.68414	1.4617	.82534
.21	.20846	.21314	4.6917	.97803	.61	.57287	.69892	1.4308	.81965
.22	.21823	.22362	4.4719	.97590	.62	.58104	.71391	1.4007	.81388
.23	.22798	.23414	4.2709	.97367	.63	.58914	.72911	1.3715	.80803
.24	.23770	.24472	4.0864	.97134	.64	.59720	.74454	1.3431	.80210
.25	.24740	.25534	3.9163	.96891	.65	.60519	.76020	1.3154	.79608
.26	.25708	.26602	3.7591	.96639	.66	.61312	.77610	1.2885	.78999
.27	.26673	.27676	3.6133	.96377	.67	.62099	.79225	1.2622	.78382
.28	.27636	.28755	3.4776	.96106	.68	.62879	.80866	1.2366	.77757
.29	.28595	.29841	3.3511	.95824	.69	.63654	.82534	1.2116	.77125
0.30	.29552	.30934	3.2327	.95534	**0.70**	.64422	.84229	1.1872	.76484
.31	.30506	.32033	3.1218	.95233	.71	.65183	.85953	1.1634	.75836
.32	.31457	.33139	3.0176	.94924	.72	.65938	.87707	1.1402	.75181
.33	.32404	.34252	2.9195	.94604	.73	.66687	.89492	1.1174	.74517
.34	.33349	.35374	2.8270	.94275	.74	.67429	.91309	1.0952	.73847
.35	.34290	.36503	2.7395	.93937	.75	.68164	.93160	1.0734	.73169
.36	.35227	.37640	2.6567	.93590	.76	.68892	.95045	1.0521	.72484
.37	.36162	.38786	2.5782	.93233	.77	.69614	.96967	1.0313	.71791
.38	.37092	.39941	2.5037	.92866	.78	.70328	.98926	1.0109	.71091
.39	.38019	.41105	2.4328	.92491	.79	.71035	1.0092	.99084	.70385
x	$\sin x$	$\tan x$	$\cot x$	$\cos x$	x	$\sin x$	$\tan x$	$\cot x$	$\cos x$

TRIGONOMETRIC FUNCTIONS OF NUMBERS 0–1.6 (cont.)

x	sin x	tan x	cot x	cos x	x	sin x	tan x	cot x	cos x
0.80	.71736	1.0296	.97121	.69671	1.20	.93204	2.5722	.38878	.36236
.81	.72429	1.0505	.95197	.68950	1.21	.93562	2.6503	.37731	.35302
.82	.73115	1.0717	.93309	.68222	1.22	.93910	2.7328	.36593	.34365
.83	.73793	1.0934	.91455	.67488	1.23	.94249	2.8198	.35463	.33424
.84	.74464	1.1156	.89635	.66746	1.24	.94578	2.9119	.34341	.32480
.85	.75128	1.1383	.87848	.65998	1.25	.94898	3.0096	.33227	.31532
.86	.75784	1.1616	.86091	.65244	1.26	.95209	3.1133	.32121	.30582
.87	.76433	1.1853	.84365	.64483	1.27	.95510	3.2236	.31021	.29628
.88	.77074	1.2097	.82668	.63715	1.28	.95802	3.3413	.29928	.28672
.89	.77707	1.2346	.80998	.62941	1.29	.96084	3.4672	.28842	.27712
0.90	.78333	1.2602	.79355	.62161	1.30	.96356	3.6021	.27762	.26750
.91	.78950	1.2864	.77738	.61375	1.31	.96618	3.7471	.26687	.25785
.92	.79560	1.3133	.76146	.60582	1.32	.96872	3.9033	.25619	.24818
.93	.80162	1.3409	.74578	.59783	1.33	.97115	4.0723	.24556	.23848
.94	.80756	1.3692	.73034	.58979	1.34	.97348	4.2556	.23498	.22875
.95	.81342	1.3984	.71511	.58168	1.35	.97572	4.4552	.22446	.21901
.96	.81919	1.4284	.70010	.57352	1.36	.97786	4.6734	.21398	.20924
.97	.82489	1.4592	.68531	.56530	1.37	.97991	4.9131	.20354	.19945
.98	.83050	1.4910	.67071	.55702	1.38	.98185	5.1774	.19315	.18964
.99	.83603	1.5237	.65631	.54869	1.39	.98370	5.4707	.18279	.17981
1.00	.84147	1.5574	.64209	.54030	1.40	.98545	5.7979	.17248	.16997
1.01	.84683	1.5922	.62806	.53186	1.41	.98710	6.1654	.16220	.16010
1.02	.85211	1.6281	.61420	.52337	1.42	.98865	6.5811	.15195	.15023
1.03	.85730	1.6652	.60051	.51482	1.43	.99010	7.0555	.14173	.14033
1.04	.86240	1.7036	.58699	.50622	1.44	.99146	7.6018	.13155	.13042
1.05	.86742	1.7433	.57362	.49757	1.45	.99271	8.2381	.12139	.12050
1.06	.87236	1.7844	.56040	.48837	1.46	.99387	8.9886	.11125	.11057
1.07	.87720	1.8270	.54734	.48012	1.47	.99492	9.8874	.10114	.10063
1.08	.88196	1.8712	.53441	.47133	1.48	.99588	10.983	.09105	.09067
1.09	.88663	1.9171	.52162	.46249	1.49	.99674	12.350	.08097	.08071
1.10	.89121	1.9648	.50897	.45360	1.50	.99749	14.101	.07091	.07074
1.11	.89570	2.0143	.49644	.44466	1.51	.99815	16.428	.06087	.06076
1.12	.90010	2.0660	.48404	.43568	1.52	.99871	19.670	.05084	.05077
1.13	.90441	2.1198	.47175	.42666	1.53	.99917	24.498	.04082	.04079
1.14	.90863	2.1759	.45959	.41759	1.54	.99953	32.461	.03081	.03079
1.15	.91276	2.2345	.44753	.40849	1.55	.99978	48.078	.02080	.02079
1.16	.91680	2.2958	.43558	.39934	1.56	.99994	92.621	.01080	.01080
1.17	.92075	2.3600	.42373	.39015	1.57	1.00000	1255.3	.00080	.00080
1.18	.92461	2.4273	.41199	.38092	1.58	.99996	−108.65	−.00920	−.00920
1.19	.92837	2.4979	.40034	.37166	1.59	.99982	−52.067	−.01921	−.01920
					1.60	.99957	−34.233	−.02921	−.02920
x	sin x	tan x	cot x	cos x					

NATURAL LOGARITHMS OF NUMBERS 1.00–10.09

Logarithms of numbers outside this range can be computed by using the functional equation ($\log ab = \log a + \log b$) and the following values of the logarithmic function:
$\log .1 = .6974-3$; $\log .01 = .3948-5$; $\log .001 = .0922-7$; $\log .0001 = .7897-10$;
$\log .00001 = .4871-12$; $\log .000\,001 = .1845-14$.

N	.00	.01	.02	.03	.04	.05	.06	.07	.08	.09
1.0	0.0000	0.0100	0.0198	0.0296	0.0392	0.0488	0.0583	0.0677	0.0770	0.0862
1.1	0.0953	0.1044	0.1133	0.1222	0.1310	0.1398	0.1484	0.1570	0.1655	0.1740
1.2	0.1823	0.1906	0.1989	0.2070	0.2151	0.2231	0.2311	0.2390	0.2469	0.2546
1.3	0.2624	0.2700	0.2776	0.2852	0.2927	0.3001	0.3075	0.3148	0.3221	0.3293
1.4	0.3365	0.3436	0.3507	0.3577	0.3646	0.3716	0.3784	0.3853	0.3920	0.3988
1.5	0.4055	0.4121	0.4187	0.4253	0.4318	0.4383	0.4447	0.4511	0.4574	0.4637
1.6	0.4700	0.4762	0.4824	0.4886	0.4947	0.5008	0.5068	0.5128	0.5188	0.5247
1.7	0.5306	0.5365	0.5423	0.5481	0.5539	0.5596	0.5653	0.5710	0.5766	0.5822
1.8	0.5878	0.5933	0.5988	0.6043	0.6098	0.6152	0.6206	0.6259	0.6313	0.6366
1.9	0.6419	0.6471	0.6523	0.6575	0.6627	0.6678	0.6729	0.6780	0.6831	0.6881
2.0	0.6931	0.6981	0.7031	0.7080	0.7129	0.7178	0.7227	0.7275	0.7324	0.7372
2.1	0.7419	0.7467	0.7514	0.7561	0.7608	0.7655	0.7701	0.7747	0.7793	0.7839
2.2	0.7885	0.7930	0.7975	0.8020	0.8065	0.8109	0.8154	0.8198	0.8242	0.8286
2.3	0.8329	0.8372	0.8416	0.8459	0.8502	0.8544	0.8587	0.8629	0.8671	0.8713
2.4	0.8755	0.8796	0.8838	0.8879	0.8920	0.8961	0.9002	0.9042	0.9083	0.9123
2.5	0.9163	0.9203	0.9243	0.9282	0.9322	0.9361	0.9400	0.9439	0.9478	0.9517
2.6	0.9555	0.9594	0.9632	0.9670	0.9708	0.9746	0.9783	0.9821	0.9858	0.9895
2.7	0.9933	0.9969	1.0006	1.0043	1.0080	1.0116	1.0152	1.0188	1.0225	1.0260
2.8	1.0296	1.0332	1.0367	1.0403	1.0438	1.0473	1.0508	1.0543	1.0578	1.0613
2.9	1.0647	1.0682	1.0716	1.0750	1.0784	1.0818	1.0852	1.0886	1.0919	1.0953
3.0	1.0986	1.1019	1.1053	1.1086	1.1119	1.1151	1.1184	1.1217	1.1249	1.1282
3.1	1.1314	1.1346	1.1378	1.1410	1.1442	1.1474	1.1506	1.1537	1.1569	1.1600
3.2	1.1632	1.1663	1.1694	1.1725	1.1756	1.1787	1.1817	1.1848	1.1878	1.1909
3.3	1.1939	1.1969	1.2000	1.2030	1.2060	1.2090	1.2119	1.2149	1.2179	1.2208
3.4	1.2238	1.2267	1.2296	1.2326	1.2355	1.2384	1.2413	1.2442	1.2470	1.2499
3.5	1.2528	1.2556	1.2585	1.2613	1.2641	1.2669	1.2698	1.2726	1.2754	1.2782
3.6	1.2809	1.2837	1.2865	1.2892	1.2920	1.2947	1.2975	1.3002	1.3029	1.3056
3.7	1.3083	1.3110	1.3137	1.3164	1.3191	1.3218	1.3244	1.3271	1.3297	1.3324
3.8	1.3350	1.3376	1.3403	1.3429	1.3455	1.3481	1.3507	1.3533	1.3558	1.3584
3.9	1.3610	1.3635	1.3661	1.3686	1.3712	1.3737	1.3762	1.3788	1.3813	1.3838
4.0	1.3863	1.3888	1.3913	1.3938	1.3962	1.3987	1.4012	1.4036	1.4061	1.4085
4.1	1.4110	1.4134	1.4159	1.4183	1.4207	1.4231	1.4255	1.4279	1.4303	1.4327
4.2	1.4351	1.4375	1.4398	1.4422	1.4446	1.4469	1.4493	1.4516	1.4540	1.4563
4.3	1.4586	1.4609	1.4633	1.4656	1.4679	1.4702	1.4725	1.4748	1.4770	1.4793
4.4	1.4816	1.4839	1.4861	1.4884	1.4907	1.4929	1.4951	1.4974	1.4996	1.5019
4.5	1.5041	1.5063	1.5085	1.5107	1.5129	1.5151	1.5173	1.5195	1.5217	1.5239
4.6	1.5261	1.5282	1.5304	1.5326	1.5347	1.5369	1.5390	1.5412	1.5433	1.5454
4.7	1.5476	1.5497	1.5518	1.5539	1.5560	1.5581	1.5602	1.5623	1.5644	1.5665
4.8	1.5686	1.5707	1.5728	1.5748	1.5769	1.5790	1.5810	1.5831	1.5851	1.5872
4.9	1.5892	1.5913	1.5933	1.5953	1.5974	1.5994	1.6014	1.6034	1.6054	1.6074
5.0	1.6094	1.6114	1.6134	1.6154	1.6174	1.6194	1.6214	1.6233	1.6253	1.6273
5.1	1.6292	1.6312	1.6332	1.6351	1.6371	1.6390	1.6409	1.6429	1.6448	1.6467
5.2	1.6487	1.6506	1.6525	1.6544	1.6563	1.6582	1.6601	1.6620	1.6639	1.6658
5.3	1.6677	1.6696	1.6715	1.6734	1.6752	1.6771	1.6790	1.6808	1.6827	1.6845
5.4	1.6864	1.6882	1.6901	1.6919	1.6938	1.6956	1.6974	1.6993	1.7011	1.7029
5.5	1.7047	1.7066	1.7084	1.7102	1.7120	1.7138	1.7156	1.7174	1.7192	1.7210
N	.00	.01	.02	.03	.04	.05	.06	.07	.08	.09

NATURAL LOGARITHMS OF NUMBERS 1.00–10.09 (cont.)

N	.00	.01	.02	.03	.04	.05	.06	.07	.08	.09
5.5	1.7047	1.7066	1.7084	1.7102	1.7120	1.7138	1.7156	1.7174	1.7192	1.7210
5.6	1.7228	1.7246	1.7263	1.7281	1.7299	1.7317	1.7334	1.7352	1.7370	1.7387
5.7	1.7405	1.7422	1.7440	1.7457	1.7475	1.7492	1.7509	1.7527	1.7544	1.7561
5.8	1.7579	1.7596	1.7613	1.7630	1.7647	1.7664	1.7681	1.7699	1.7716	1.7733
5.9	1.7750	1.7766	1.7783	1.7800	1.7817	1.7834	1.7851	1.7867	1.7884	1.7901
6.0	1.7918	1.7934	1.7951	1.7967	1.7984	1.8001	1.8017	1.8034	1.8050	1.8066
6.1	1.8083	1.8099	1.8116	1.8132	1.8148	1.8165	1.8181	1.8197	1.8213	1.8229
6.2	1.8245	1.8262	1.8278	1.8294	1.8310	1.8326	1.8342	1.8358	1.8374	1.8390
6.3	1.8405	1.8421	1.8437	1.8453	1.8469	1.8485	1.8500	1.8516	1.8532	1.8547
6.4	1.8563	1.8579	1.8594	1.8610	1.8625	1.8641	1.8656	1.8672	1.8687	1.8703
6.5	1.8718	1.8733	1.8749	1.8764	1.8779	1.8795	1.8810	1.8825	1.8840	1.8856
6.6	1.8871	1.8886	1.8901	1.8916	1.8931	1.8946	1.8961	1.8976	1.8991	1.9006
6.7	1.9021	1.9036	1.9051	1.9066	1.9081	1.9095	1.9110	1.9125	1.9140	1.9155
6.8	1.9169	1.9184	1.9199	1.9213	1.9228	1.9242	1.9257	1.9272	1.9286	1.9301
6.9	1.9315	1.9330	1.9344	1.9359	1.9373	1.9387	1.9402	1.9416	1.9430	1.9445
7.0	1.9459	1.9473	1.9488	1.9502	1.9516	1.9530	1.9544	1.9559	1.9573	1.9587
7.1	1.9601	1.9615	1.9629	1.9643	1.9657	1.9671	1.9685	1.9699	1.9713	1.9727
7.2	1.9741	1.9755	1.9769	1.9782	1.9796	1.9810	1.9824	1.9838	1.9851	1.9865
7.3	1.9879	1.9892	1.9906	1.9920	1.9933	1.9947	1.9961	1.9974	1.9988	2.0001
7.4	2.0015	2.0028	2.0042	2.0055	2.0069	2.0082	2.0096	2.0109	2.0122	2.0136
7.5	2.0149	2.0162	2.0176	2.0189	2.0202	2.0215	2.0229	2.0242	2.0255	2.0268
7.6	2.0281	2.0295	2.0308	2.0321	2.0334	2.0347	2.0360	2.0373	2.0386	2.0399
7.7	2.0412	2.0425	2.0438	2.0451	2.0464	2.0477	2.0490	2.0503	2.0516	2.0528
7.8	2.0541	2.0554	2.0567	2.0580	2.0592	2.0605	2.0618	2.0631	2.0643	2.0656
7.9	2.0669	2.0681	2.0694	2.0707	2.0719	2.0732	2.0744	2.0757	2.0769	2.0782
8.0	2.0794	2.0807	2.0819	2.0832	2.0844	2.0857	2.0869	2.0882	2.0894	2.0906
8.1	2.0919	2.0931	2.0943	2.0956	2.0968	2.0980	2.0992	2.1005	2.1017	2.1029
8.2	2.1041	2.1054	2.1066	2.1078	2.1090	2.1102	2.1114	2.1126	2.1138	2.1150
8.3	2.1163	2.1175	2.1187	2.1199	2.1211	2.1223	2.1235	2.1247	2.1258	2.1270
8.4	2.1282	2.1294	2.1306	2.1318	2.1330	2.1342	2.1353	2.1365	2.1377	2.1389
8.5	2.1401	2.1412	2.1424	2.1436	2.1448	2.1459	2.1471	2.1483	2.1494	2.1506
8.6	2.1518	2.1529	2.1541	2.1552	2.1564	2.1576	2.1587	2.1599	2.1610	2.1622
8.7	2.1633	2.1645	2.1656	2.1668	2.1679	2.1691	2.1702	2.1713	2.1725	2.1736
8.8	2.1748	2.1759	2.1770	2.1782	2.1793	2.1804	2.1815	2.1827	2.1838	2.1849
8.9	2.1861	2.1872	2.1883	2.1894	2.1905	2.1917	2.1928	2.1939	2.1950	2.1961
9.0	2.1972	2.1983	2.1994	2.2006	2.2017	2.2028	2.2039	2.2050	2.2061	2.2072
9.1	2.2083	2.2094	2.2105	2.2116	2.2127	2.2138	2.2148	2.2159	2.2170	2.2181
9.2	2.2192	2.2203	2.2214	2.2225	2.2235	2.2246	2.2257	2.2268	2.2279	2.2289
9.3	2.2300	2.2311	2.2322	2.2332	2.2343	2.2354	2.2364	2.2375	2.2386	2.2396
9.4	2.2407	2.2418	2.2428	2.2439	2.2450	2.2460	2.2471	2.2481	2.2492	2.2502
9.5	2.2513	2.2523	2.2534	2.2544	2.2555	2.2565	2.2576	2.2586	2.2597	2.2607
9.6	2.2618	2.2628	2.2638	2.2649	2.2659	2.2670	2.2680	2.2690	2.2701	2.2711
9.7	2.2721	2.2732	2.2742	2.2752	2.2762	2.2773	2.2783	2.2793	2.2803	2.2814
9.8	2.2824	2.2834	2.2844	2.2854	2.2865	2.2875	2.2885	2.2895	2.2905	2.2915
9.9	2.2925	2.2935	2.2946	2.2956	2.2966	2.2976	2.2986	2.2996	2.3006	2.3016
10.0	2.3026	2.3036	2.3046	2.3056	2.3066	2.3076	2.3086	2.3096	2.3106	2.3115
N	.00	.01	.02	.03	.04	.05	.06	.07	.08	.09

MANTISSAS OF COMMON LOGARITHMS OF NUMBERS 1000-9999

N	0	1	2	3	4	5	6	7	8	9
10	0000	0043	0086	0128	0170	0212	0253	0294	0334	0374
11	0414	0453	0492	0531	0569	0607	0645	0682	0719	0755
12	0792	0828	0864	0899	0934	0969	1004	1038	1072	1106
13	1139	1173	1206	1239	1271	1303	1335	1367	1399	1430
14	1461	1492	1523	1553	1584	1614	1644	1673	1703	1732
15	1761	1790	1818	1847	1875	1903	1931	1959	1987	2014
16	2041	2068	2095	2122	2148	2175	2201	2227	2253	2279
17	2304	2330	2355	2380	2405	2430	2455	2480	2504	2529
18	2553	2577	2601	2525	2648	2672	2695	2718	2742	2765
19	2788	2810	2833	2856	2878	2900	2923	2945	2967	2989
20	3010	3032	3054	3075	3096	3118	3139	3160	3181	3201
21	3222	3243	3263	3284	3304	3324	3345	3365	3385	3404
22	3424	3444	3464	3483	3502	3522	3541	3560	3579	3598
23	3617	3636	3655	3674	3692	3711	3729	3747	3766	3784
24	3802	3820	3838	3856	3874	3892	3909	3927	3945	3962
25	3979	3997	4014	4031	4048	4065	4082	4099	4116	4133
26	4150	4166	4183	4200	4216	4232	4249	4265	4281	4298
27	4314	4330	4346	4362	4378	4393	4409	4425	4440	4456
28	4472	4487	4502	4518	4533	4548	4564	4579	4594	4609
29	4624	4639	4654	4669	4683	4698	4713	4728	4742	4757
30	4771	4786	4800	4814	4829	4843	4857	4871	4886	4900
31	4914	4928	4942	4955	4969	4983	4997	5011	5024	5038
32	5051	5065	5079	5092	5105	5119	5132	5145	5159	5172
33	5185	5198	5211	5224	5237	5250	5263	5276	5289	5302
34	5315	5328	5340	5353	5366	5378	5391	5403	5416	5428
35	5441	5453	5465	5478	5490	5502	5514	5527	5539	5551
36	5563	5575	5587	5599	5611	5623	5635	5647	5658	5670
37	5682	5694	5705	5717	5729	5740	5752	5763	5775	5786
38	5798	5809	5821	5832	5843	5855	5866	5877	5888	5899
39	5911	5922	5933	5944	5955	5966	5977	5988	5999	6010
40	6021	6031	6042	6053	6064	6075	6085	6096	6107	6117
41	6128	6138	6149	6160	6170	6180	6191	6201	6212	6222
42	6232	6243	6253	6263	6274	6284	6294	6304	6314	6325
43	6335	6345	6355	6365	6375	6385	6395	6405	6415	6425
44	6435	6444	6454	6464	6474	6484	6493	6503	6513	6522
45	6532	6542	6551	6561	6571	6580	6590	6599	6609	6618
46	6628	6637	6646	6656	6665	6675	6684	6693	6702	6712
47	6721	6730	6739	6749	6758	6767	6776	6785	6794	6803
48	6812	6821	6830	6839	6848	6857	6866	6875	6884	6893
49	6902	6911	6920	6928	6937	6946	6955	6964	6972	6981
50	6990	6998	7007	7016	7024	7033	7042	7050	7059	7067
51	7076	7084	7093	7101	7110	7118	7126	7135	7143	7152
52	7160	7168	7177	7185	7193	7202	7210	7218	7226	7235
53	7243	7251	7259	7267	7275	7284	7292	7300	7308	7316
54	7324	7332	7340	7348	7356	7364	7372	7380	7388	7396

MANTISSAS OF COMMON LOGARITHMS OF NUMBERS 1000–9999 (cont.)

N	0	1	2	3	4	5	6	7	8	9
55	7404	7412	7419	7427	7435	7443	7451	7459	7466	7474
56	7482	7490	7497	7505	7513	7520	7528	7536	7543	7551
57	7559	7566	7574	7582	7589	7597	7604	7612	7619	7627
58	7634	7642	7649	7657	7664	7672	7679	7686	7694	7701
59	7709	7716	7723	7731	7738	7745	7752	7760	7767	7774
60	7782	7789	7796	7803	7810	7818	7825	7832	7839	7846
61	7853	7860	7868	7875	7882	7889	7896	7903	7910	7917
62	7924	7931	7938	7945	7952	7959	7966	7973	7980	7987
63	7993	8000	8007	8014	8021	8028	8035	8041	8048	8055
64	8062	8069	8075	8082	8089	8096	8102	8109	8116	8122
65	8129	8136	8142	8149	8156	8162	8169	8176	8182	8189
66	8195	8202	8209	8215	8222	8228	8235	8241	8248	8254
67	8261	8267	8274	8280	8287	8293	8299	8306	8312	8319
68	8325	8331	8338	8344	8351	8357	8363	8370	8376	8382
69	8388	8395	8401	8407	8414	8420	8426	8432	8439	8445
70	8451	8457	8463	8470	8476	8482	8488	8494	8500	8506
71	8513	8519	8525	8531	8537	8543	8549	8555	8561	8567
72	8573	8579	8585	8591	8597	8603	8609	8615	8621	8627
73	8633	8639	8645	8651	8657	8663	8669	8675	8681	8686
74	8692	8698	8704	8710	8716	8722	8727	8733	8739	8745
75	8751	8756	8762	8768	8774	8779	8785	8791	8797	8802
76	8808	8814	8820	8825	8831	8837	8842	8848	8854	8859
77	8865	8871	8876	8882	8887	8893	8899	8904	8910	8915
78	8921	8927	8932	8938	8943	8949	8954	8960	8965	8971
79	8976	8982	8987	8993	8998	9004	9009	9015	9020	9025
80	9031	9036	9042	9047	9053	9058	9063	9069	9074	9079
81	9085	9090	9096	9101	9106	9112	9117	9122	9128	9133
82	9138	9143	9149	9154	9159	9165	9170	9175	9180	9186
83	9191	9196	9201	9206	9212	9217	9222	9227	9232	9238
84	9243	9248	9253	9258	9263	9269	9274	9279	9284	9289
85	9294	9299	9304	9309	9315	9320	9325	9330	9335	9340
86	9345	9350	9355	9360	9365	9370	9375	9380	9385	9390
87	9395	9400	9405	9410	9415	9420	9425	9430	9435	9440
88	9445	9450	9455	9460	9465	9469	9474	9479	9484	9489
89	9494	9499	9504	9509	9513	9518	9523	9528	9533	9538
90	9542	9547	9552	9557	9562	9566	9571	9576	9581	9586
91	9590	9595	9600	9605	9609	9614	9619	9624	9628	9633
92	9638	9643	9647	9652	9657	9661	9666	9671	9675	9680
93	9685	9689	9694	9699	9703	9708	9713	9717	9722	9727
94	9731	9736	9741	9745	9750	9754	9759	9763	9768	9773
95	9777	9782	9786	9791	9795	9800	9805	9809	9814	9818
96	9823	9827	9832	9836	9841	9845	9850	9854	9859	9863
97	9868	9872	9877	9881	9886	9890	9894	9899	9903	9908
98	9912	9917	9921	9926	9930	9934	9939	9943	9948	9952
99	9956	9961	9965	9969	9974	9978	9983	9987	9991	9996

T-10 | THE EXPONENTIAL FUNCTION AND ITS RECIPROCAL IN THE RANGE 0-6

x	e^x	e^{-x}	x	e^x	e^{-x}
0.0	1.0000	1.0000	**3.0**	20.086	.04979
0.1	1.1052	.90484	3.1	22.198	.04505
0.2	1.2214	.81873	3.2	24.533	.04076
0.3	1.3499	.74082	3.3	27.113	.03688
0.4	1.4918	.67032	3.4	29.964	.03337
0.5	1.6487	.60653	3.5	33.115	.03020
0.6	1.8221	.54881	3.6	36.598	.02732
0.7	2.0138	.49659	3.7	40.447	.02472
0.8	2.2255	.44933	3.8	44.701	.02237
0.9	2.4596	.40657	3.9	49.402	.02024
1.0	2.7183	.36788	**4.0**	54.598	.01832
1.1	3.0042	.33287	4.1	60.340	.01657
1.2	3.3201	.30119	4.2	66.686	.01500
1.3	3.6693	.27253	4.3	73.700	.01357
1.4	4.0552	.24660	4.4	81.451	.01228
1.5	4.4817	.22313	4.5	90.017	.01111
1.6	4.9530	.20190	4.6	99.484	.01005
1.7	5.4739	.18268	4.7	109.95	.00910
1.8	6.0496	.16530	4.8	121.51	.00823
1.9	6.6859	.14957	4.9	134.29	.00745
2.0	7.3891	.13534	**5.0**	148.41	.00674
2.1	8.1662	.12246	5.1	164.02	.00610
2.2	9.0250	.11080	5.2	181.27	.00552
2.3	9.9742	.10026	5.3	200.34	.00499
2.4	11.023	.09072	5.4	221.41	.00452
2.5	12.182	.08208	5.5	244.69	.00409
2.6	13.464	.07427	5.6	270.43	.00370
2.7	14.880	.06721	5.7	298.87	.00335
2.8	16.445	.06081	5.8	330.30	.00303
2.9	18.174	.05502	5.9	365.04	.00274
3.0	20.086	.04979	**6.0**	403.43	.00248

ANSWERS TO SELECTED EXERCISES AND PROBLEMS

EXERCISES 1.1

2. $(x + y)(x - y) = [x(x - y)] + [y(x - y)]$ F6
$= [x^2 + x(-y)] + [yx + y(-y)]$ F6
$= (x^2 - xy) + (xy - y^2)$ F5
$= [x^2 + (-xy + xy)] - y^2$ F3
$= (x^2 + 0) - y^2$ F4
$= x^2 - y^2$ F2

5. Suppose $x^2 = 0$. If $x \neq 0$, then by F4 there is a number x^{-1} such that $x^{-1}x = 1$. It follows that $x^{-1}x^2 = (x^{-1}x)x = 1 \cdot x = x$. But $x^{-1}x^2 = x^{-1} \cdot 0 = 0$. Therefore $x = 0$, a contradiction.

8. If $a \neq 0$, then by F4 there is a number a^{-1} such that $a^{-1}a = 1$. Then $ab = ac$ implies that $a^{-1}(ab) = a^{-1}(ac)$ whence $(a^{-1}a)b = 1 \cdot b = b = (a^{-1}a)c = 1 \cdot c = c$.

11. For any real number x, we have $x^2 + [-(-x)^2] = x^2 + [-(-x)(-x)] = x^2 + (x)(-x) = x \cdot x + (x)(-x) = x[x + (-x)] = x \cdot 0 = 0$. Therefore $x^2 - (-x)^2 = 0$ or $x^2 = (-x)^2$.

13. No. [Hint: Show that if $1/a + 1/b = 1/(a + b)$, then $a^2 + ab + b^2 = 0$ and conversely.]

15. Suppose that $a + 1 = r$, a rational number. Then $a = r - 1$ and $r - 1$ is rational by F1. This contradicts the assumption that a is not rational. The second half is similar.

PROBLEMS 1.1

2. Only F4 for multiplication.

5. If there were integers u and v such that $2u + 4v = 9$, then $2(u + 2v) = 9$ and it follows that 9 is even. Since 9 is odd, no such integers exist.

8. If $x \neq -1$, then multiply both sides by $(x + 1)^2(x^2 + 2)$. The equation becomes
$$x^2 = (a + c)x^3 + (a + b + 2c + d)x^2 + (2a + c + 2d)x + 2a + 2b + d$$
Equating coefficients of like powers of x yields the four equations
$$a \phantom{{}+b} + c \phantom{{}+2d} = 0$$
$$a + b + 2c + d = 1$$
$$2a \phantom{{}+b} + c + 2d = 0$$
$$2a + 2b \phantom{{}+2c} + d = 0$$
from which we get $a = -\frac{4}{9}$, $b = \frac{1}{3}$, $c = \frac{4}{9}$, $d = \frac{2}{9}$.

9. Suppose that $\log_{10} 3 = p/q$ where p and q are integers. Then $10^{p/q} = 3$. (We may choose $q > 0$ without loss of generality.) It follows that $10^p = 3^q$ where $p \neq 0$, $q > 0$. One readily sees that the left side is even and the right side is odd. This is a contradiction; hence p and q cannot be chosen so that $\log_{10} 3 = p/q$.

EXERCISES 1.2

1. (a) $\sqrt[3]{3}$ (b) $\sqrt{3}$ (c) neither
   (d) $\sqrt{3} + \sqrt{5}$ (e) $\sqrt[3]{23}$ (f) $\sqrt{19} + \sqrt{21}$

3. (a) all $x \geq -\frac{3}{5}$ (b) all $x \leq \frac{1}{2}$
   (c) if $a = 0$, all x
        if $a > 0$, all $x \leq b^2/4a$
        if $a < 0$, all $x \geq b^2/4a$
   (d) all $x \geq 0$ and   all $x \leq -4$.

4. (a) 2   (b) 0   (c) 0, 2
   (d) If $a = b$, it is true for all x; otherwise $x = \frac{1}{2}(a + b)$.

6. (a) $x < 2$ (b) $x > \frac{3}{7}$ (c) $x > -6$
   (d) $-4 < x < 1$ (e) $1.01 < z < 1.54$
   (f) If $c > 0$, $\dfrac{-c\epsilon - b}{a} < x < \dfrac{c\epsilon - b}{a}$
       If $c < 0$, $\dfrac{c\epsilon - b}{a} < x < \dfrac{-c\epsilon - b}{a}$
   (g) $x < -3$ (h) $x < \frac{5}{2}$

8. (a) $x > 2$ (b) $x < -13$ or $x > -\frac{7}{3}$
   (c) $x \leq -4$ or $x \geq 2$ (d) $x \leq -\sqrt{2}, x \geq \sqrt{2}$ or $-\sqrt{\frac{2}{5}} \leq x \leq \sqrt{\frac{2}{5}}$.

PROBLEMS 1.2

3. If x and y are rational, then $\frac{1}{2}(x + y)$ is a rational number between x and y. If not both x and y are rational, look carefully at the first decimal place where they differ and construct the needed rational number. Two separate cases will be required for a complete proof.

4. Since p, q and n are positive, we may write the inequality $nq < nq + 1 \leq (nq + 1)p$. Setting $m = nq + 1$, we have $nq < m$ or $n < m(p/q)$.

6. The inequality $2^n > n^2$ is true for all integers n except 2, 3 and 4. Similarly, $2^n > n^3$ except for $n = 2, 3, \cdots, 9$. The proofs are via induction.

14. There is no loss in generality in assuming $a < b < c$. Consider the three numbers
$$a|g - e| + b|g - f|, \; a|f - e| - c|f - g|, \; b|e - f| + c|e - g|.$$
If the first is the smallest, then take $x = g$, etc.

15. $-19 + 12x - 2x^2 = -1 - 2(x - 3)^2 \leq -1$

16. $x^2 - 4x + 12 = (x - 2)^2 + 8 \geq 8$

PROBLEMS 1.4

4. It is easy to show that
$$\tfrac{1}{5}(k + 1)^5 + \tfrac{1}{3}(k + 1)^3 + \tfrac{7}{15}(k + 1) = (\tfrac{1}{5}k^5 + \tfrac{1}{3}k^3 + \tfrac{7}{15}k) + k^4 + 2k^3 + 3k^2 + 2k + 1$$
Use this in an inductive argument.

8. It is easy to show that, if $(1 + x)^k > 1 + kx$, then
$$(1 + x)^{k+1} = (1 + x)(1 + x)^k > (1 + x)(1 + kx) = 1 + (k + 1)x + kx^2.$$
The inductive step follows immediately.

11. Use Problem 8 above.

12. First prove that for $n > 3$ we have $2^n < n!$. Then use the binomial theorem to show that
$$(1 + 1/n)^n \leq 1 + 1 + 1/2 + 1/6 + \cdots + 1/n! < 3$$

14. Use the fact from Problem 12 above that $(1 + 1/n)^n < 3$ for all n. The induction step is
$$[(k + 1)!]^2 = (k!)^2(k + 1)^2 \geq k^k(k + 1)^2 \quad \text{by assumption.}$$
To prove that $k^k(k + 1)^2 \geq (k + 1)^{k+1}$, use above fact.

17. Write

$$\left(1 + \frac{1}{6n}\right)^{-n} = \left(\frac{6n+1}{6n}\right)^{-n} = \left(\frac{6n}{6n+1}\right)^{n} = \left(1 - \frac{1}{6n+1}\right)^{n}$$

and use Problem 8.

21. This is more than just an induction problem but we start with an induction as follows: We have

$$a_1 a_2 = \left(\frac{a_1 + a_2}{2}\right)^2 - \left(\frac{a_1 - a_2}{2}\right)^2 \leqq \left(\frac{a_1 + a_2}{2}\right)^2$$

with equality holding if and only if $a_1 = a_2$. It follows that

$$(a_1 a_2)(a_3 a_4) \leqq \left(\frac{a_1 + a_2}{2}\right)^2 \left(\frac{a_3 + a_4}{2}\right)^2 = \left[\left(\frac{a_1 + a_2}{2}\right)\left(\frac{a_3 + a_4}{2}\right)\right]^2$$

By the above reasoning, we also have

$$\left(\frac{a_1 + a_2}{2}\right)\left(\frac{a_3 + a_4}{2}\right) \leqq \left(\frac{\frac{a_1 + a_2}{2} + \frac{a_3 + a_4}{2}}{2}\right)^2$$

whence

$$a_1 a_2 a_3 a_4 \leqq \left(\frac{a_1 + a_2 + a_3 + a_4}{4}\right)^4$$

An induction proves that

$$a_1 \cdots a_n \leqq \left(\frac{a_1 + \cdots + a_n}{n}\right)^n$$

for $n = 2^k$.

If n is not a power of 2, choose k so that $n < 2^k$ and define

$$b_1 = a_1, b_2 = a_2, \cdots, b_n = a_n$$

$$b_{n+1} = \cdots = b_{2^k} = \frac{1}{n}(a_1 + \cdots + a_n)$$

From the induction we have

$$b_1 \times \cdots \times b_{2^k} \leqq \left[\frac{b_1 + \cdots + b_{2^k}}{2^k}\right]^{2^k}$$

which easily reduces to the inequality

$$a_1 \times \cdots \times a_n \leqq \left(\frac{a_1 + \cdots + a_n}{n}\right)^n$$

22. This can be done by sheer calculation of the induction step.

EXERCISES 1.5

2.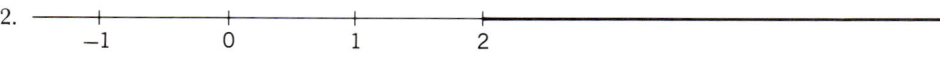

3.

5.

6.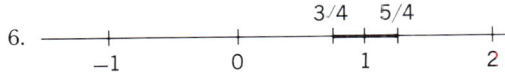

13. 1   14. 0   16. $-39/25$

PROBLEMS 1.5

1. Let y be any element of $\{x : (1-r)a + rb, 0 \leq r \leq 1\}$, i.e. let $y = (1-r)a + rb$ for some value of r satisfying $0 \leq r \leq 1$. Then $y - a = r(b - a)$ which immediately tells us that

$$0 \leq \frac{y-a}{b-a} \leq 1$$

or

$$a \leq y \leq b.$$

The converse is just as easy.

EXERCISES 1.6

2. The set of irrational numbers
$$\{x : x \text{ is irrational and } x < 0\}$$
has zero as its least upper bound.

5. If q is the largest number in X, then $x \leq q$ for each $x \in X$. Thus q is an upper bound. If M is any upper bound, then $x \leq M$ for any $x \in X$. In particular, then, $q \leq M$ whence q is the least upper bound.

6. Since $\frac{2^n - 1}{2^n} = 1 - 1/2^n < 1$ for every n, we see that 1 is an upper bound. Let M be any number less than 1 and we claim that there is an $n \in N$ such that
$$M < 1 - 1/2^n < 1.$$
This follows because, however small $1 - M$ may be, we can find n such that $1/2^n < 1 - M$. It follows that no number less than one is an upper bound.

7. The general formula is
$$\frac{1 + 3 + 9 + \cdots + 3^{n-1}}{3^n} = \frac{1}{2}\left(1 - \frac{1}{3^n}\right)$$
whence $\frac{1}{2}$ is easily shown to be the least upper bound.

8. $\sqrt[3]{6}$

PROBLEMS 1.6

1. It is easy to prove that
$$\frac{1}{1 \cdot 2} + \frac{1}{2 \cdot 3} + \cdots + \frac{1}{n(n+1)} = \frac{n}{n+1}$$
so Exercise 4 above applies.

2. If $x \leq y + \epsilon$ for all $\epsilon > 0$, then x is a lower bound of the set $\{y + \epsilon : \epsilon > 0\}$. Since y is obviously the greatest lower bound of this set, the inequality $x \leq y$ follows. If $x < y + \epsilon$ we can conclude no more because for instance $y < y + \epsilon$ if $\epsilon > 0$.

5. If $k = 2$, then k is the least upper bound. In general, the least upper bound is $\frac{1}{2} + \frac{1}{2}\sqrt{1 + 4k}$.

EXERCISES 1.7

1. $\{x : x < -1\} \cup \{x : x > 2\}$
2. $\{x : -2 < x < \frac{7}{4}\}$
3. $\{x : -1 \leq x \leq 2\}$
4. $\{x : -\frac{8}{3} < x < \frac{8}{3}\}$
5. $\{x : x > 4\} \cup \{x : -\frac{9}{4} < x < \frac{5}{3}\}$
6. $\{x : -2 \leq x \leq -\frac{3}{2}\} \cup \{x : \frac{1}{2} \leq x \leq 3\}$
7. $\{x : -2 < x < 1\} \cup \{x : x > 2\}$
8. $\{x : -4 < x < -1\} \cup \{x : \frac{1}{2} < x < 3\}$
9. $\{x : -2 \leq x \leq 2\}$
10. $\{x : -\frac{5}{2} < x < 3\}$
11. $\{x : \frac{3}{2} < x < 2\}$
12. $\{x : x \leq -\frac{7}{3}\} \cup \{x : \frac{1}{2} \leq x \leq 2\}$
13. $\{x : -4 < x < -\frac{3}{2}\} \cup \{x : x > 2\}$
14. $\{x : x < -2\} \cup \{x : -1 < x \leq 2\}$
15. $\{x : -2 < x < -\frac{2}{3}\} \cup \{x : \frac{5}{2} < x < 4\}$

EXERCISES 1.8

1. (a) $(-2, 3)$ (b) $(8, -8)$ (c) $(0, 1)$
   (d) $(0, 0)$ (e) $(-2, 4)$ (f) $(8, 9)$
2. (a) $x = 1, y = \frac{1}{2}$ (b) $x = 0, y = \frac{1}{2}$
3. (a) $(2, 2, -1)$ (b) $(1, -5, 1)$
   (c) $(\frac{10}{7}, \frac{6}{7}, 0)$ (d) $(0, 1, 0)$
4. $x = -\frac{1}{7}, y = \frac{1}{7}, z = \frac{3}{7}$

PROBLEMS 1.8

4. It is only necessary to check axioms V1, V2, V3, V4 and V6; the others are satisfied automatically.
5. We need only verify F1–F5 for multiplication because addition has been shown (V1–V5) to satisfy F1–F5. Then do F6 separately.
6. If $\mathbf{a} + \mathbf{x} = \mathbf{x}$ for every vector $\mathbf{x}$, then in particular we have $\mathbf{a} + \mathbf{0} = \mathbf{0}$ or $\mathbf{a} = \mathbf{0}$. If $\mathbf{a} + \mathbf{x} = \mathbf{0}$ then $-\mathbf{a} + (\mathbf{a} + \mathbf{x}) = -\mathbf{a}$ or $(-\mathbf{a} + \mathbf{a}) + \mathbf{x} = \mathbf{0} + \mathbf{x} = \mathbf{x} = -\mathbf{a}$.
7. If $\mathbf{a} + \mathbf{x} = \mathbf{b}$, then $-\mathbf{a} + (\mathbf{a} + \mathbf{x}) = -\mathbf{a} + \mathbf{b}$, or $(-\mathbf{a} + \mathbf{a}) + \mathbf{x} = \mathbf{0} + \mathbf{x} = \mathbf{x} = -\mathbf{a} + \mathbf{b}$. Uniqueness follows immediately.

EXERCISES 1.9

1. (a) 1. (c)

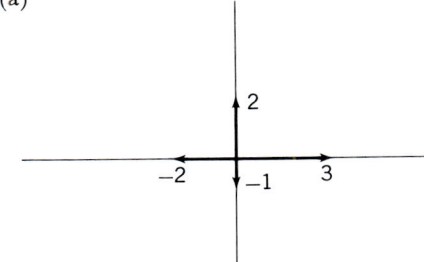 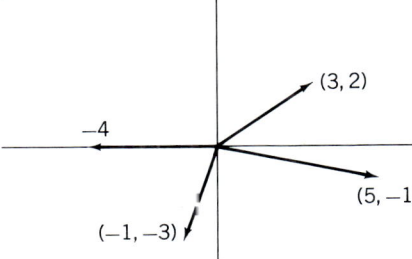

1. (e) 3. (a)

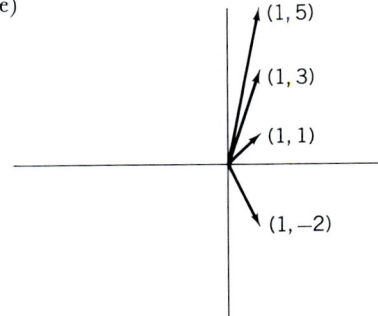 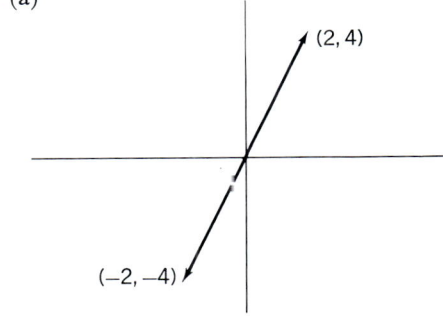

3. (h)

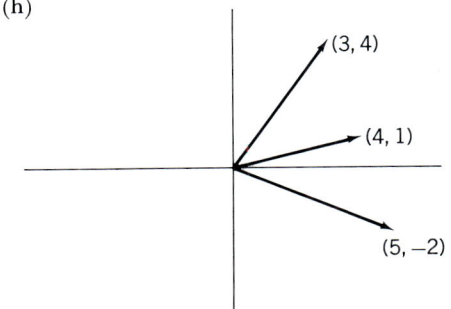

4. (a) $x = \frac{2}{5}, y = \frac{1}{5}$ (b) $x = -\frac{1}{5}, y = \frac{2}{5}$
   (c) $x = -\frac{3}{5}, y = \frac{1}{5}$ (d) $x = \frac{8}{5}, y = -\frac{1}{5}$
   (e) $x = \frac{4}{5}, y = \frac{17}{5}$ (f) $x = (2c - d)/5, y = (c + 2d)/5$

PROBLEMS 1.9

2. Let $a_1 b_2 = a_2 b_1$. If $a_2 \neq 0$, define $r = b_2/a_2$. If $a_2 = 0$, then $b_2 = 0$ and let $r = b_1/a_1$. The rest is easy.

3. Let $x\mathbf{u} + y\mathbf{v} = \mathbf{p}$. It is routine calculation to prove that
$$x\mathbf{u} + y\mathbf{v} = \frac{b_1 y - b_2 x}{a_1 b_2 - a_2 b_1}\mathbf{a} + \frac{a_1 y - a_2 x}{a_1 b_2 - a_2 b_1}\mathbf{b}.$$

4. If the tip of $x\mathbf{u} + y\mathbf{v}$ lies on the line through $(-1, 1)$ and $(1, 3)$, then there is a real number r such that
$$x\mathbf{u} + y\mathbf{v} = (1 - r)(-\mathbf{u} + \mathbf{v}) + r(\mathbf{u} + 3\mathbf{v}) = (-1 + 2r)\mathbf{u} + (1 + 2r)\mathbf{v}$$
Thus $x = -1 + 2r$ and $y = 1 + 2r$. It follows that
$$x + 1 = y - 1 = 2r$$
or $y = x + 2$.

EXERCISES 1.11

1. The equal sides have the following lengths:
   (a) 5   (b) $\sqrt{37}$ (c) $\sqrt{35}$ (d) $\sqrt{14}$

2. The areas are
   (a) 17   (b) $\frac{25}{2}$ (c) $\frac{1}{2}\sqrt{638}$ (d) $\frac{1}{2}\sqrt{870}$

3. The parallelograms have sides of length
   (a) 5 and $3\sqrt{2}$ (b) $\sqrt{21}$ and $2\sqrt{6}$

4. The circle has radius 5.   5. The sphere has radius 6.

6. The center is $(-2, 3)$.   7. The center is $(4, -\frac{1}{2}, \frac{3}{4})$.

8. (a) $(2, 1, 0)$ (b) $(3, 0)$

EXERCISES 2.1

1. $4x - y - 11 = 0$ 2. $x + y - 4 = 0$
3. $3x - 5y - 2 = 0$ 4. $3x + 5y + 4 = 0$
5. $2x + 5y - 10 = 0$ 6. $x - 3y + 3 = 0$
7. $y - 4 = 0$ 8. $y + 1 = 0$
9. $x - 3 = 0$ 10. $2x - y = 0$
11. $(2, 0)$ 12. $(1, 2)$ 13. $(-1, 4)$
14. none   15. $(\frac{25}{21}, \frac{109}{42})$ 16. none
17. $(2, \frac{3}{2})$ 18. $(0, -10)$
19. If $A = 0$, they are the same line. If $A \neq 0$, $(0, \frac{5}{2})$ is the point of intersection.
20. $(1, -1)$
21. $(1, 1)$ is not in this set.

PROBLEMS 2.1

4. You must sell 600 dollars worth of merchandise.
5. 4750 lbs.
6. p is approximately 86 and no. of units is approximately 570.

EXERCISES 2.2

1. (a), (b) and (d) are parallel lines.

2. (a), (b) and (c) are parallelograms.

3. $(7, -4)$ and $(1, -8)$

4. (a) $3x + y - 4 = 0$ (b) $x + y - 4 = 0$
   (c) $x - 2y + 7 = 0$ (d) $3x - 2y - 13 = 0$

PROBLEMS 2.2

1. The desired line has an equation of the form $kAx + kBy + D = 0$. Let $k = 1$ and substitute (a, b) to find D.

2. The first part is very easy, just inspect the equation. The second part is a matter of finding values of k_1 and k_2 which is straightforward.

3. $19x - 68y + 89 = 0$ 4. $t = -2$

EXERCISES 2.3

1. (a) 0   (b) $\frac{2}{5}$ (c) -1 (d) $-\frac{1}{2}$
   (e) 0   (f) none   (g) $-\frac{5}{2}$ (h) 1

2. The slopes are
   (a) 3   (b) $\frac{1}{4}$ (c) $-\frac{3}{2}$ (d) $-\frac{1}{4}$
   (e) $\frac{3}{2}$ (f) $-\frac{2}{7}$ (g) $-b/a$ (h) $-\frac{1}{3}$

3. All four pairs are parallel.

4. (a) 6   (b) none   (c) 3

5. (a), (b), (c) and (e) are collinear sets.

PROBLEMS 2.3

1. Use induction.

2. The two scales agree at $-40°$.

3. It is the change in demand due to a change in the unit price.

EXERCISES 2.4

1. (a) $X = \{x : |x| \neq 2\}$, $Y = \{y : y \leq -\frac{1}{4} \text{ or } y > 0\}$
   (b) $X = \{x : |x| \leq 2\}$, $Y = \{y : 0 \leq y \leq 2\}$
   (c) $X = \{x : |x| \geq 2\}$, $Y = \{y : y \geq 0\}$
   (d) $X = \{x : |x| < 2\}$, $Y = \{y : y \geq \frac{1}{2}\}$
   (e) $X = \{t : t \neq -1\}$, $Y = \{y : y \neq 2\}$
   (f) $X = \{t : t < -1 \text{ or } t \geq 0\}$, $Y = \{y : y \geq 0, y \neq \sqrt{2}\}$
   (g) $X = \{u : |u| < 1\}$, $Y = \{y : y \geq 0\}$
   (h) $X = \{z : z \leq -1 \text{ or } z \geq 2\}$, $Y = \{y : y \geq 0\}$

2. (a) 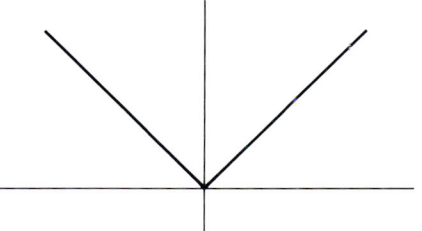   (b)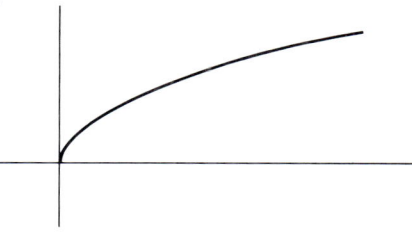

(c)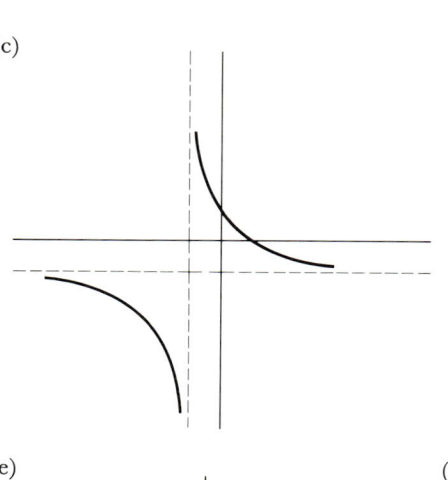

(d)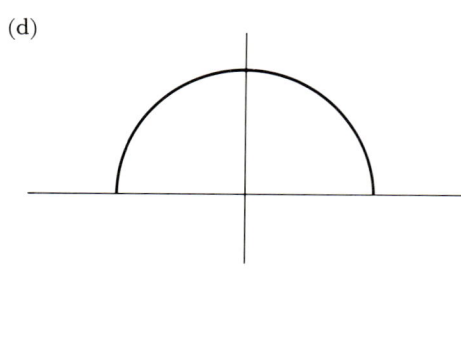

(e)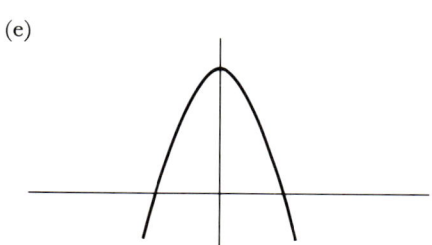

(f)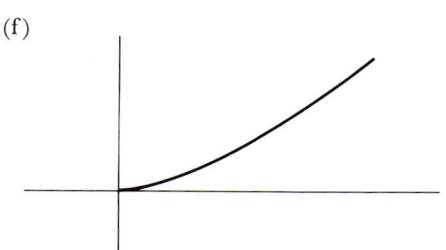

(g)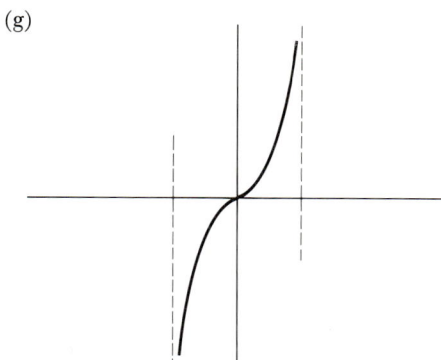

(h)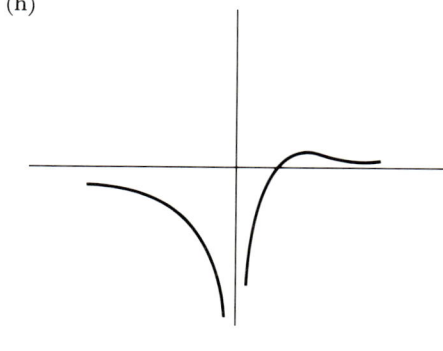

(i)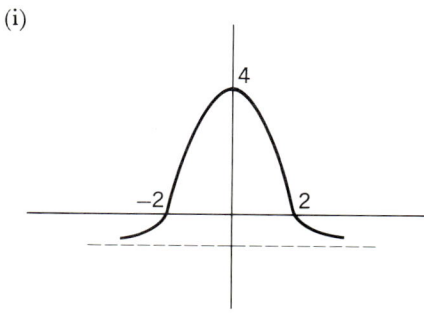

(j)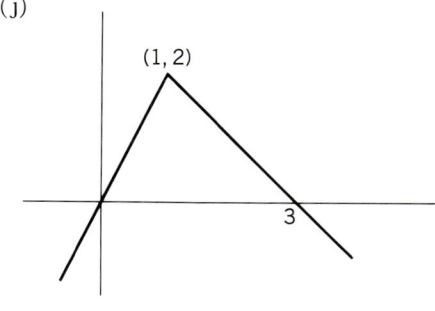

A-11 | SELECTED ANSWERS: PROBLEMS 2.4

4. (a) 3   (b) -1 (c) A
  (d) $2x + h$ (e) $4x - 3 + 2h$ (f) $2Ax + B + Ah$
  (g) $3x^2 + 3xh + h^2$ (h) $\dfrac{-1}{x(x+h)}$
  (i) $-\dfrac{2x+h}{x^2(x+h)^2}$ (j) $\dfrac{1}{(x+1)(x+1+h)}$

PROBLEMS 2.4

1. Vol. of $S = \left(\dfrac{3\pi\sqrt{3}}{2}\right)v$; vol. of $s = \left(\dfrac{\pi\sqrt{3}}{8}\right)v$

2. $f(f(x)) = (x-1)/x$; $f(f(f(x))) = x$, etc.

3. $F(n) = [n/5]$

4. The graph of $g(t)$ is as follows:

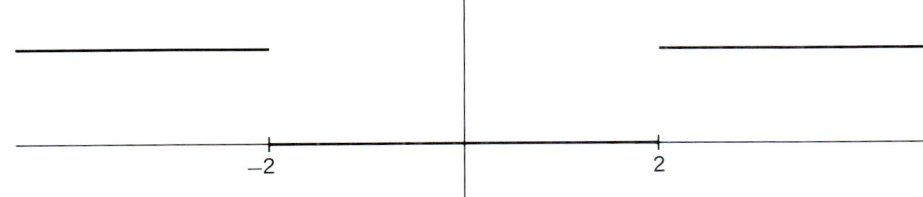

7. $X = \{x : x = 0 \text{ or } \tfrac{1}{2} \leq |x| \leq 1\}$

EXERCISES 2.5

1.
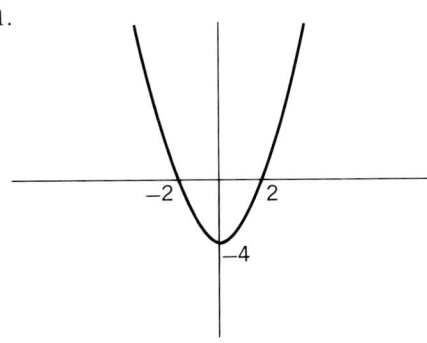

2.
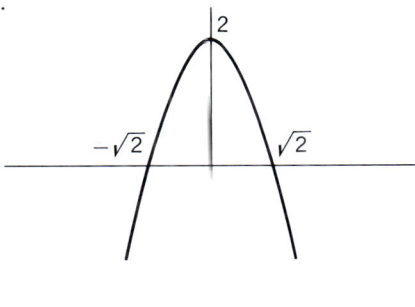

3.
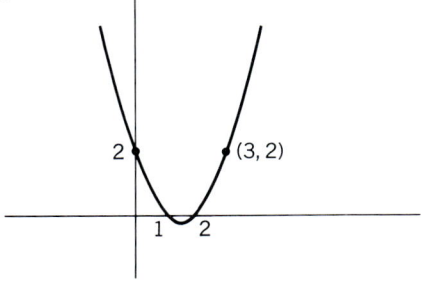

4.
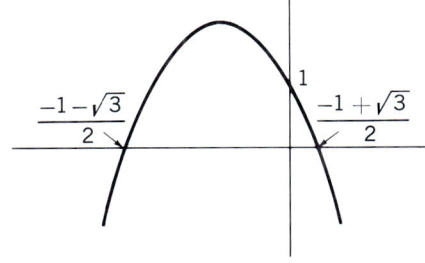

5.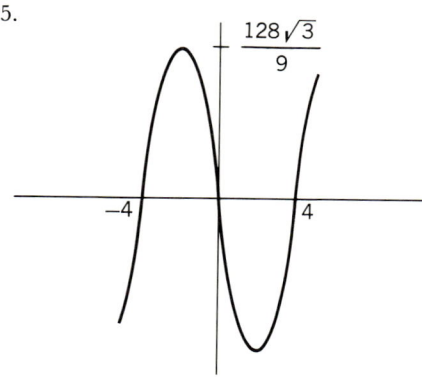

6.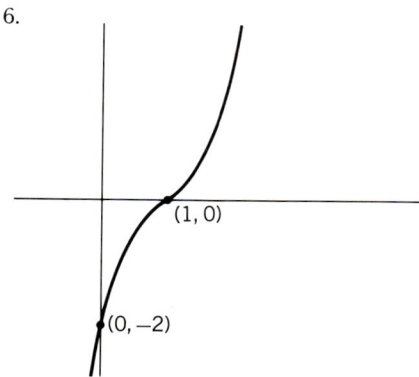

7.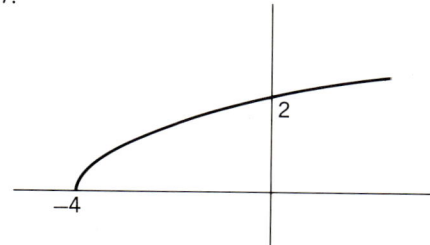

8.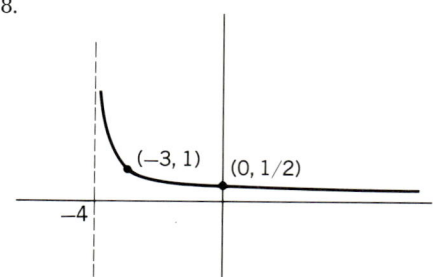

9.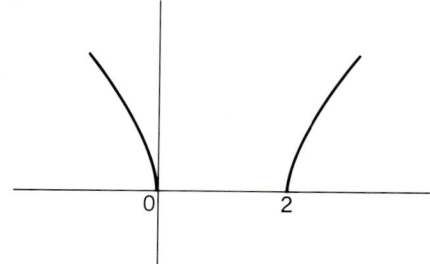

10.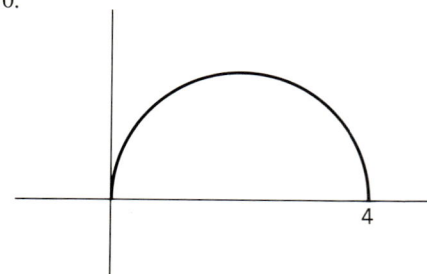

11.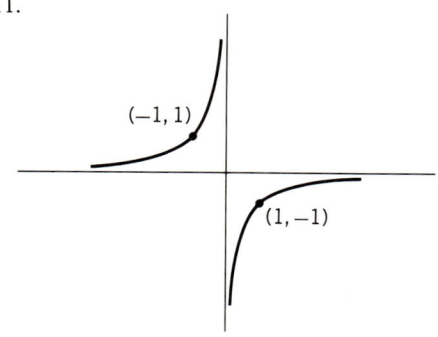

12.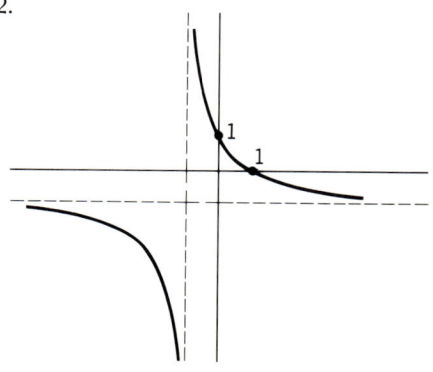

13.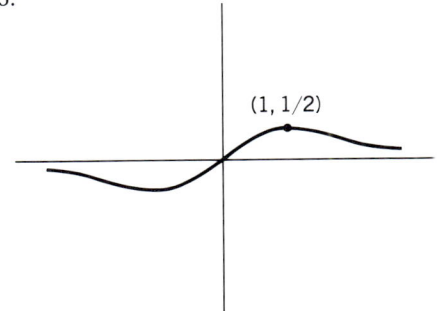

14.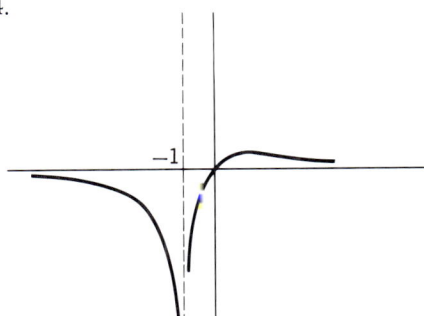

15.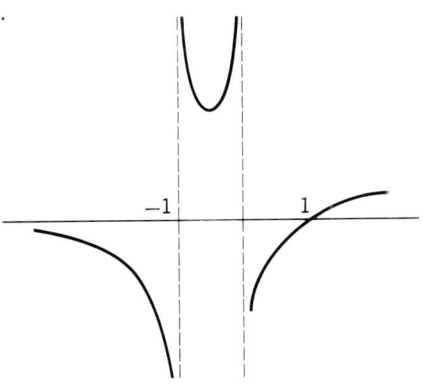

16.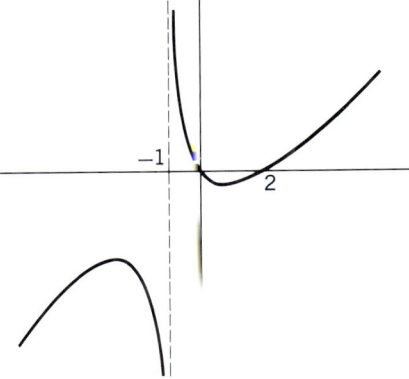

17.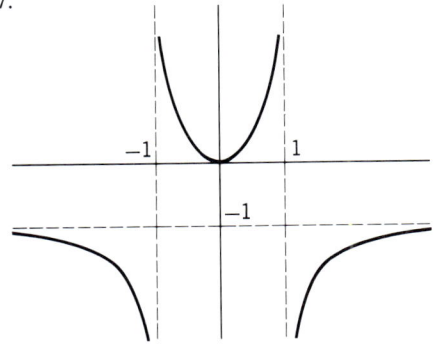

18.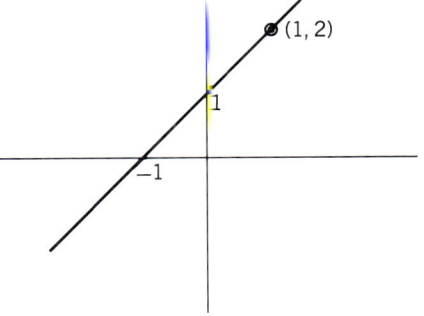

19.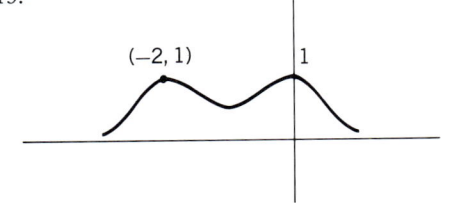

20.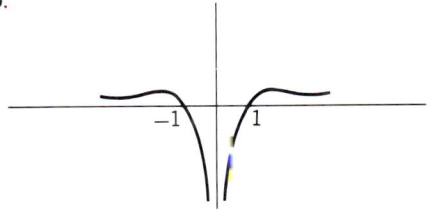

PROBLEMS 2.5

1. The only even and odd function is the zero function.

6. Clearly the function f is uniquely presented as the sum of an even function and an odd function.

EXERCISES 2.6

1. (b) $\lim_{n \to \infty} \dfrac{2n+3}{n} = \lim_{n \to \infty} \left(2 + \dfrac{3}{n}\right) = \lim_{n \to \infty} 2 + \lim_{n \to \infty} \dfrac{3}{n} = 2 + 0$

2. (c) Choose $n_0 > \dfrac{1}{\epsilon^2}$

2. (f) Choose $n_0 > \dfrac{9}{16\epsilon}$

3. (a) $\tfrac{1}{3}$ (b) $-\tfrac{1}{2}$ (c) 0   (d) 0   (e) 2   (f) 0

4. (a) Note that
$$1 + \frac{1}{2} + \frac{1}{4} + \cdots + \frac{1}{2^{n-1}} = 2 - \frac{1}{2^{n-1}}$$

(d) Note that
$$a + ar + ar^2 + \cdots + ar^{n-1} = \frac{a(r^n - 1)}{r - 1}$$

5. (a) $\tfrac{1}{4}$ (b) $= \tfrac{2}{3}$ (c) $= 1$ (d) 1

PROBLEMS 2.6

2. Let $\varphi(n) = \dfrac{1}{(n+1)^2} + \dfrac{1}{(n+2)^2} + \cdots + \dfrac{1}{(n+n)^2}$. In $\varphi(n)$ there are n terms each less than $1/n^2$. Thus we have
$$0 < \varphi(n) < n(1/n^2) = 1/n$$
and $\lim_{n \to \infty} \varphi(n) = 0$ by Theorem 2.11.

3. At 5% annually;   \$1.05
   At 5% semi-annually;   \$1.0506
   At 5% quarterly;   \$1.0510
   At 5% monthly;   \$1.0513
   At 5% weekly;   \$1.0516

4. Compute:
$$\lim_{n \to \infty} (6 + 6(0.9) + 6(0.9)^2 + \cdots + 6(0.9)^n).$$

9. (c) Let $\varphi(n) = (-1)^n$.

10. If the sequence converges, every subsequence converges to the same limit. (Theorem 2.12)

15. The sequence φ is monotone increasing and, for each k,
$$\varphi(2^k) \leq 1 + \frac{1}{2} + \cdots + \frac{1}{2^{k-1}} < 2.$$

16. $\pi = \lim_{n \to \infty} \dfrac{n}{2} \sin \dfrac{2\pi}{n}$.

EXERCISES 2.7

The required rates of change are

1. 2   2. -16 3. $-\frac{3}{2}$ 4. -2
5. 0   6. $\frac{1}{9}$ 7. $-\frac{1}{4}$ 8. 0
9. 2   10. 0   11. 1   12. $\frac{1}{4}$
13. $\frac{1}{4}$ 14. $-\frac{1}{4}$ 15. $\frac{2}{3}\sqrt{3}$ 16. $-\frac{1}{16}$
17. $-\frac{1}{2}$ 18. $\frac{1}{9}\sqrt{3}$ 19. 3   20. $\frac{3}{2}\sqrt{c}$

EXERCISES 2.8

1. $y = 3x - 2$ 2. $y = 4x - 2$
3. $y = 4x - 3$ 4. $y = 2x$
5. $y = -4x - 3$ 6. $y = -x + 3$
7. $y = x$ 8. $y = \frac{1}{4}x + 2$
9. $y = \frac{4}{5}x + \frac{9}{5}$ 10. $y = 1$

11.
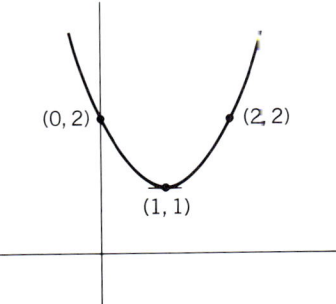

12.
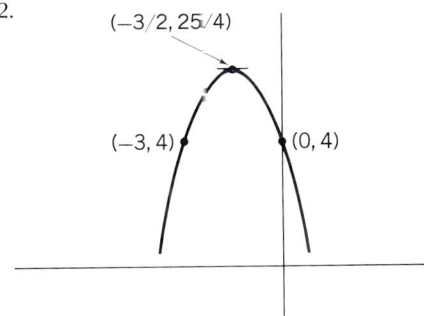

15.
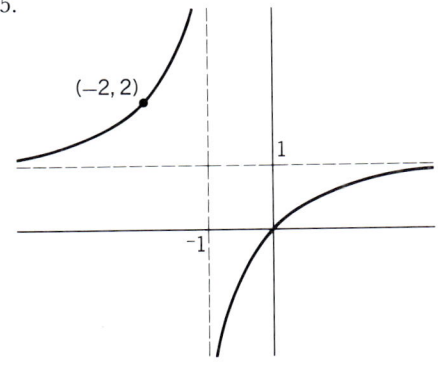

18.
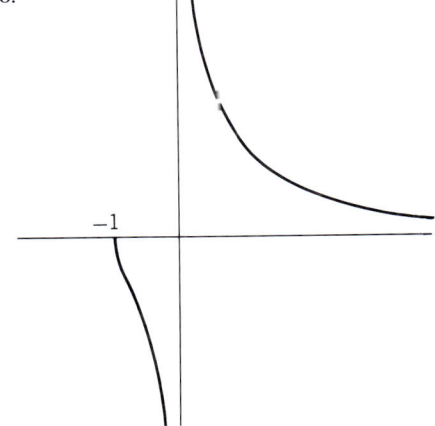

PROBLEMS 2.8

1. $\left(-\dfrac{B}{2A},\; C - \dfrac{B^2}{4A}\right)$ if $A \neq 0$.

2. $(3, 3)$ and $(-3, 33)$

3. $(-1, 2)$

4. Two, one or none according as x_0^2 exceeds, equals or is less than $\dfrac{1}{A}(y_0 - Bx_0 - C)$.

6. $x = 0$

7. At $(2, 2)$, the common tangent line is $\{(x, y) : y = 9x - 16\}$.

8. At (a, a^2), $a \neq 0$, the slopes are $2a$ and $\tfrac{1}{2}a$. The product is 1.

EXERCISES 2.9

Choices of δ can be

1. $\delta = \sqrt{\epsilon}$
4. $\delta = \epsilon$
6. $\delta = \epsilon/4$ if $\epsilon < 4$
8. $\delta = 2a\epsilon$, $\epsilon < \tfrac{1}{2}a$
9. $\delta = \epsilon$ if $\epsilon < \tfrac{1}{4}$
12. $\delta = 3(a + 1)^3 \epsilon$, $\epsilon < a/2$, $a > 0$.

Limits are

13. 4
14. 0
15. none
16. $\tfrac{7}{10}$
17. -1
18. -9
19. $\tfrac{4}{3}a$
20. $\tfrac{2}{3}a$
21. $2/a$
22. $\tfrac{1}{3}a^{-2/3}$
23. $-2/a^3$
24. $\dfrac{1}{2|a|}$

Choices of n can be

25. $n \geqq M^{-1/2}$
26. $n \geqq (M - 3)^{-1/2}$ if $M > 3$
27. $n \geqq M/4$
28. $n \geqq M/2$
29. $n \geqq 1/\epsilon$
30. $n \geqq \sqrt{1 + (2/\epsilon)}$
31. $n \geqq (2/\epsilon) - 1$ if $\epsilon < 2$
32. $n \geqq 4M$ if $M > 1$

Limits are

33. 3
34. $2x - 2$
35. $-2 - 2x$
36. $-6 - 3x^2$
37. $-1/x^2$
38. $\dfrac{1}{2\sqrt{x}}$
39. $-\tfrac{1}{2}x^{-3/2}$
40. $\dfrac{x + 2}{2(x + 1)^{3/2}}$

PROBLEMS 2.9

1. (a) 0 (b) 0 (c) 1 (d) -1

2. Let $f(x) = -1$ if $x < 0$
   $\quad\quad\quad\quad = -1$ if $x \geqq 0$ and x rational
   $\quad\quad\quad\quad = +1$ if $x \geqq 0$ and x irrational

4. At each integer.

PROBLEMS 2.10

1. Assume $\lim_{x \to a} f(x) = L$. Then the condition

$$\text{"}|f(x) - L| < \epsilon \quad \text{whenever} \quad 0 < |x - a| < \delta\text{"}$$

can be re-stated in terms of the function g.

2. (a) Choose $|f(x) - L| < (2nL^{1-1/n})\epsilon$ by controlling $|x - a|$.
   (b) The same as part (a) with changes of sign.

3. Choose $L = \epsilon$. That is, there exists $\delta > 0$ such that

$$|f(x) - L| < L \quad \text{whenever} \quad 0 < |x - a| < \epsilon.$$

The left side of the equivalent inequality

$$-L < f(x) - L < L$$

or $\quad 0 < f(x) < 2L \quad$ completes the proof.

EXERCISES 3.1

1. (a) $f'(x) = 3$; **R**   1. (c) $2x - 2$; **R**   1. (e) $3x^2 + 2$; **R**

1. (h) $f'(x) = -\dfrac{2(x + 1)}{x^3}$; **R** $- \{0\}$

1. (j) $f'(x) = (x + 1)(x^2 + 2x + 2)^{-3/2}$; **R**

4. (b) $f'(-1) = -2 = g'(-1)$

4. (d) $f'(0) = 1 = g'(0)$

PROBLEMS 3.1   *Mathematics*

5. Under the three hypotheses, we may write

$$f'(x) = \lim_{h \to 0} \frac{f(x + h) - f(x)}{h} = \lim_{h \to 0} \frac{f(x) \cdot f(h) - f(x) \cdot f(0)}{h}$$

$$= f(x) \lim_{h \to 0} \frac{f(h) - f(0)}{h} = f(x) f'(0).$$

Economics

1. $C(x)/x = (20{,}000 + 10x + 2x^2)/100x$
   $C'(x) = (2x + 5)/50$
   The average cost is a minimum when $x = 100$.

4. Approximately 164.

Physical Science

2. (a) $\frac{1}{5}\sqrt{15}$ (b) $\sqrt{3}$

3. One of the rates is $\dfrac{9(7t + 5)}{\sqrt{21t^2 + 30t + 25}}$.

EXERCISES 3.2

1. (a) inc. for $x > \frac{3}{2}$; dec. for $x < \frac{3}{2}$

1. (d) inc. for $x < -3$; dec. for $x > -3$
1. (e) inc. for $x < -1$ or $x > 2$; dec. for $-1 < x < 2$
1. (g) inc. for $-1 < x < 1$ or $x > 2$; dec. for $x < -1$ or $1 < x < 2$
1. (i) decreasing on the entire domain

2. (a) rel. min. of $\frac{9}{8}$ at $x = \frac{3}{4}$
   (c) rel. max. of 9 at $x = 3$
   (f) rel. max. of $\frac{16}{3}$ at $x = 2$
       rel. min. of $-\frac{16}{3}$ at $x = -2$
   (h) rel. max. of 4 at $x = -1$
       rel. min. of -4 at $x = 1$
   (j) rel. min. of $\sqrt{3}$ at $x = 0$

3. (d) $f(1) - f(x) = -(1/2x^2)(x^2 - 2x - 1) > 0$ if $1 - \sqrt{2} < x < 1 + \sqrt{2}$, $x \neq 0$

4. (c) $f(-1) - f(x) = -(x + 1)^2(3x + 1)(x - 1) \leq 0$ if $-\frac{4}{3} < x < -\frac{1}{3}$.

5. (c) $a = 0$, $b = 4$ 7. 64

PROBLEMS 3.2

2. The rectangle of maximum area has area r^2.

4. The three numbers are $\frac{1}{9}(140 + 20\sqrt{3})$, $\frac{1}{9}(580 - 20\sqrt{3})$, $\frac{1}{9}(-120 + 40\sqrt{3})$.

6. Shortest path is 500 ft. long.

7. The lot should be 112 ft. × 84 ft.

8. Six feet from the shorter post.

10. The dimensions should be 10 × 10 × 20.

EXERCISES 3.3

1. (a) $\delta = \epsilon/3$ (d) $\delta = \epsilon/2$ if $\epsilon < 2$ (f) $\delta = \epsilon/3$ if $\epsilon < 3$
   (g) $\delta = 2\epsilon$ if $\epsilon < \frac{1}{2}$ (i) $\delta = \epsilon$ if $\epsilon < 1$

3. (b) $x = -2$ (c) $x = -3$, $x = 1$ (e) $x = 0$, $x = 3$
   (i) $|x| = 1$ (The functional value is not real if $|x| > 1$.)

4. $f(1) = -\frac{1}{2}$; No.

PROBLEMS 3.3

1.

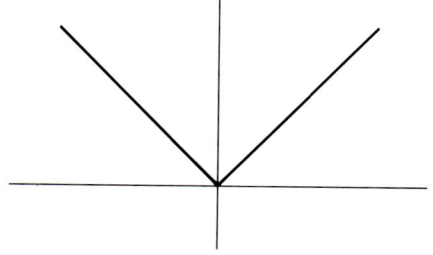

3. $x \neq \pm 2$

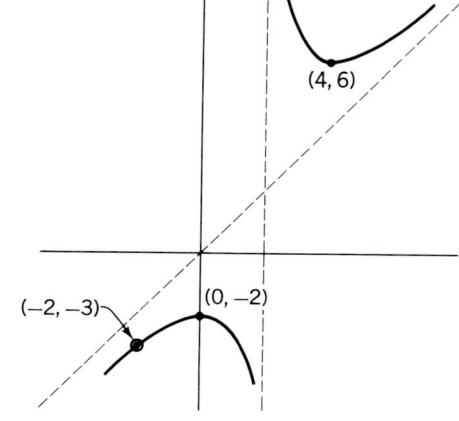

5. Discontinuous at $x = 2$ and $x = 5$

6. If $a > 0$, f is continuous everywhere. If $a < 0$, the function is not defined at $x = 0$.

9. Recall that $[x]$ is the largest integer in x. Let n be an integer and then if $n < x < n + 1$, we have
$$f(x) = 1 - x + 2n$$
Clearly, f is continuous for each such x. However, one easily shows that f is discontinuous at each integer.

EXERCISES 3.4

1. (b) $-2 < r_1 < -1$, $0 < r_2 < 1$ (c) $1 < r < 2$
   (f) 1 is a root; $-10 < r_1 < -9$, two roots between -1 and 0.
   (g) $-\frac{7}{4} < r_1 < -\frac{3}{2}$, $0 < r_2 < 1$ (i) $0 < r < 1$ (j) $-1 < r < 0$

2. (c) $\frac{268}{64} < r < \frac{269}{64}$ (e) $\frac{140}{34} < r < \frac{141}{64}$ (g) $\frac{100}{64} < r < \frac{101}{64}$ (i) $\frac{139}{64} < r < \frac{140}{64}$

PROBLEMS 3.4

1. Assume that the polynomial $p(x) = a_0 + a_1 x + \cdots + a_{2k+1} x^{2k+1}$ has $a_{2k+1} > 0$. Define $M = (|a_0| + \cdots + |a_{2k}|)/a_{2k+1}$. Show that $p(x) < 0$ if $x < -M$ while $p(x) > 0$ if $x > M$.

3. The function $f(y) = y^n$ is continuous for all $y \geq 0$. Given $x > 0$, there is an integer m such that $x < m^n$ whence f takes on the value x somewhere between 0 and m.

4. Let
$$f(x) = \frac{a_1}{b_1^2 + x} + \frac{a_2}{b_2^2 + x} + \frac{a_3}{b_3^2 + x} - 1$$
and consider the right-hand limit at $x = -b_1^2$ and the left-hand limit at $x = -b_2^2$. Similarly for the other roots.

EXERCISES 3.5

1. (c) $X_{f+g} = X_{fg} = \{x : x \geq 0\}$, $X_{f/g} = \{x : x > 0\}$

3. $-3 + 2x$

5. $4 + 3x^2$

8. $-2 - 12x + 6x^2 + 4x^3$

9. $6(3 + 2x)^2$

12. $-1/x^2$

15. $-17/(2 + 3x)^2$

17. $2x/(1 - x^2)^2$

20. $\dfrac{-2 + 4x + 3x^2 + 2x^3 - x^4}{x^2(1 - x)^2}$

22. $(\frac{5}{6}, \frac{169}{12})$ max

23. None

26. $(1, 2)$ max $(-1, -2)$ min

29. $(-2 \pm 2\sqrt{2}, 7 \mp 8\sqrt{2})$ max/min.

41. $(1 \pm \sqrt{\frac{2}{3}}, 2 \pm \frac{7}{3}\sqrt{\frac{2}{3}})$

44. $(0, 0)$, $(-2, 2)$

48. $x = (2a^3 + 1)/(3a^2 - 3)$

50. $x = 2a/(a^2 + 1)$

EXERCISES 3.7

1. (d) Let $h(x) = x^2 - 3x + 2$ and $g(t) = t^4$

2. (b) $X_{g \circ f} = \{x : x \geq -2\}$; range $= \{y : y \geq 0\}$
2. (d) $X_{g \circ f} = \{x : x < -1\} \cup \{x : x \geq 0\}$

3. (b) $f'(x) = 16x^3 - 36x^2 + 18x$
   (d) $f'(x) = -9(3x - 1)^{-4}$

5. (b) $(\frac{1}{3}, 0)$; point of inflection   (d) none   (e) none
   (h) none   (i) $(0, 0)$ min.; $(4, 16)$ max.

PROBLEMS 3.7

2. $1\frac{2}{3}$ ft./hr.   3. $1\frac{2}{3}$ ft./sec.   5. Approx. 33.5 ft./sec.

EXERCISES 3.8

1. (c) $6x, 6$ (d) $6 + 6x, 6$ (e) $6(x - 1)^{-3}, -18(x - 1)^{-4}$
   (g) $\dfrac{1}{x^2(x^2 + 1)^{1/2}}, -\dfrac{3x^2 + 2}{x^3(x^2 + 1)^{3/2}}$ (j) $-\dfrac{1}{4}\dfrac{x + 4}{(x + 1)^{5/2}}, \dfrac{3}{8}\dfrac{x + 6}{(x + 1)^{7/2}}$

2. (b) concave everywhere; none
   (f) convex if $x > 1$, concave if $x < 1$; none
   (i) convex if $x > 0$, concave if $x < 0$; $(0, 0)$

3. (b) $f(x) = -1 - 3x - x^2$ will suffice   (d) $f(x) = -8 + \frac{9}{2}x - x^2$ will suffice

4. (a) $a = -2, b = 3, c = -2$ (c) $a = -3, b = 4, c = -1$

6. $a = \frac{9}{2}, b = \frac{15}{2}, c = -9, d = \frac{1}{2}$

7. $a = 5, b = -\frac{60}{7}, c = \frac{54}{7}, d = -\frac{12}{7}, e = -\frac{3}{7}$

PROBLEMS 3.8

Mathematics

1. (b) $D^3(fg) = (D^3f)g + 3(D^2f)Dg + 3Df(D^3g) + fD^3g$

2. (b) $\dfrac{n!}{(1 - x)^{n+1}}$ (c) $\dfrac{3^n(n + 1)!}{(2 - 3x)^{n+2}}$

5. (c) $f'' = mu^{m-1}v^n u'' + m(m - 1)u^{m-2}v^n(u')^2 + 2mnu^{m-1}v^{n-1}u'v' + n(n - 1)u^m v^{n-2}(v')^2 + nu^m v^{n-1}v''$.

Velocity and Acceleration

1. 40 ft./sec.   2. 150 ft.   3. 296 ft.   4. 56.25 ft.

EXERCISES 3.9

1. (b) $(\frac{3}{2}, \frac{267}{4})$ max.   (c) $(0, 0)$ min.   (f) $(-2, -4)$ max., $(0, 0)$ min.   (j) none

2. (b) $c = -54$ (c) $c = -1$

PROBLEMS 3.9

Mathematics

1. $\frac{32}{81}\pi r^3$ 2. $\frac{8}{3}\pi r^3$

3. (a) 216π (b) 864π (c) $19652\pi/27$

5. (a) $\frac{51}{8}$ (b) $\frac{1}{3}\sqrt{51}$

Physical Science

1. Height equals $\frac{1}{2}$ base.   2. Height equals $\frac{1}{2}$ base.   3. A 2 by 1 rectangle.

4. 4800 sq. ft.   5. $\frac{1}{7}$ miles from the point nearest the town, 1 mile from the water.
6. Height equals $\frac{1}{10}$ diameter.   8. At $a\sqrt[4]{2}$ units.
10. The point dividing the segment into segments with ratio $\sqrt[3]{s/t}$.
11. $\frac{1}{16}(15 - \sqrt{33})L$.   13. Approx. 18.5 ft.

Economics and Management

1. 5   2. $457.50   4. $\sqrt{100d - a/b}$ 5. 7000
6. $2.26   7. 12,000 gal.   8. 16 mph.   9. (a) 32 mph.   (b) 62.7 mph.

PROBLEMS 3.10

1. (a) $|c| > \sqrt{6}$ (b) $c > \frac{9}{16}$ (c) $c = \frac{4}{3}$
2. Use induction to show that
$$f^{(k)}(x) = mn[(m-1)(m-2)\cdots(m-k+1)x^{m-k}(1+mx^n) + km(m-1)\cdots$$
$$(m-k+2)nx^{m-k+1}x^{n-1} + \cdots + (n-1)(n-2)\cdots(n-k+1)x^{n-k}(1+nx^m)]$$
4. Hint: Compute $xf'(x) + x^2 f''(x)$.   5. (a) $1 + \frac{1}{2}x - \frac{3}{8}x^2 + x^3$ 7. Both parts; $p(x) = mx$.
10. Hint: $g(x) = b_1^2 + \cdots + b_n^2 + 2(a_1 b_1 + \cdots + a_n b_n)x + (a_1^2 + \cdots + a_n^2)x^2$.

EXERCISES 3.11

1. $-\dfrac{x+4}{(x-2)^3}, \dfrac{2(x+7)}{(x-2)^4}$ 2. $\dfrac{x^2 - 4x - 12}{(x-2)^2}, \dfrac{32}{(x-2)^3}$

4. $\dfrac{x^2 - 1}{(x^2 + x + 1)^2}, \dfrac{-2(x^3 - 3x - 1)}{(x^2 + x + 1)^3}$ 6. $\dfrac{-4}{x^2\sqrt{4-x^2}}, \dfrac{-4(3x^2 - 8)}{x^3(4-x)^{3/2}}$

7. $\dfrac{1 - 5x}{(x^2+1)^{3/2}}, \dfrac{10x^2 - 3x - 5}{(x^2+1)^{5/2}}$ 9. $f''(x) = \dfrac{-15x^3 + 140x^2 + 488x - 856}{144(x-1)^{5/4}(x+2)^{7/3}}$

PROBLEMS 3.11

1. $r(x) = \dfrac{ax+b}{cx+d}, ad = bc$.

2. $x_{n+1} = x$ if n is odd
   $= \dfrac{1-x}{1+x}$ if n is even

6. $c^2 < ac + bc - ab$ 8. Hint: Use induction on n.

EXERCISES 3.12

1. (a) 30   (b) 1   (c) $-2 + \sqrt{3}$ (d) $\frac{1}{2}(1 - \sqrt{5})$
2. (a) $\frac{3}{2}$ (b) $\frac{1}{3}\sqrt{3}$ (c) $-1 + \sqrt{6}$ (d) $1 - 2\sqrt{2}$
3. (a) $\frac{1}{2}$ (b) $\frac{1}{3}(-2 + \sqrt{13})$ (c) $-1 + \sqrt{2}$ (d) $\frac{7}{6}$

PROBLEMS 3.13

1. (c) $3\frac{1}{9}$ (d) 0.001002
2. Hint: Study the derivative of $f(x) = (x^2 - 4x + 9)/(x^2 + 4x + 9)$.
3. (a) If $f(d) = f(c)$ for some point $d > c$, apply Rolle's Theorem.

4. Hint: This does not require the mean value theorem.

6. $f(x) = 0$ has at most two real roots and $f'(x) = 0$ at most one.

7. Let x_1, x_2 be zeros of $f(x)$. If there is no zero of $g(x)$ between x_1 and x_2, then the function
$$F(x) = f(x)/g(x)$$
satisfies Rolle's Theorem on $\{x : x_1 \leq x \leq x_2\}$. However, the conclusion of Rolle's Theorem is contradictory.

EXERCISES 4.1

1. $-0.3028, 3.303$ 3. ± 3.873 5. 0.6823 6. 1.225

9. 1.769 11. $-1.221, 0.7245$ 13. 0.5823 14. $5.838, 6.155$

PROBLEMS 4.1

2. $x_{n+1} = x_n(2 - x_n r)$ 4. $x_2 = \frac{1393}{985} \cong 1.41421$ 6. Approx. 0.32 ft.

EXERCISES 4.2

3. (a) $v(t) = 12 - 12t + 3t^2$, $a(t) = -12 + 6t$; $v(t) \geq 0$ for all t.
 (d) $v(t) = -125 + 60t + 30t^2 + 20t^3 + 15t^4$; $v(t) < 0$ if $-2.1 < t < 1$.

4. $\frac{1}{4}\sqrt{10}$ seconds.   5. -104 ft./sec.

6. $v(t) = 21 + 24t - 3t^2$, $s(t) = 21t + 12t^2 - t^3$

PROBLEMS 4.2

1. The key to this is the equation
$$a = \frac{dv}{dt} = \frac{dv}{dr} \cdot \frac{dr}{dt} = v\frac{dv}{dr}$$
Using this we obtain
$$v\frac{dv}{dr} = -\frac{gR^2}{r^2}$$
whence it follows that
$$v^2 = \frac{2gR^2}{r} + c.$$

2. Computation alone shows that
$$a(t) = -\frac{gR^2}{r^2(t)}$$

3. The maximum height is $R/(1 - q^2)$ 4. 144 ft.   6. 18 ft./sec.

8. His constant velocity is 12.4 yds/sec.   9. 980.9 cm./sec.^{2}

PROBLEMS 4.3

1. 0.2 ft./min.   2. 0.15 ft./min.   3. $\frac{2}{3\pi}$ in./min.

4. Hint: $V = \frac{1}{3}\pi x^2(3r - x)$, where r is the radius of the sphere and x is the height of the water level.

5. $\frac{51}{10^6}\%$ 6. $-\frac{3}{20}$ amp./sec. 7. 5.6 ft./sec.
8. 1600 ft./sec. 9. $12\frac{4}{7}$ ft./min. 10. $\frac{8}{3}$ mph
12. $35\sqrt{2}$ ft./sec. 13. $\frac{16}{7}\sqrt{7}$ mi./min. 14. 144 mph
15. $\frac{200}{29}\sqrt{29}$ ft./sec. 16. $\frac{88}{3}\sqrt{13}$ ft./sec. 17. Approx. 12:58
18. $\frac{44}{15}\sqrt{177}$ ft./sec. 19. $\frac{280}{243}$ lb./in.² (increasing) 20. 43 hrs.

EXERCISES 4.4

1. (a) $\Delta f(x, h) = (2 + 2x)h + h^2$; $df(x, h) = (2 + 2x)h$

(c) $\Delta f(x, h) = \dfrac{-2xh - h^2}{\sqrt{1 - (x + h)^2} + \sqrt{1 - x^2}}$; $df(x, h) = \dfrac{-xh}{\sqrt{1 - x^2}}$

(f) $\Delta f(x, h) = \dfrac{2xh + h^2}{(1 - x^2)^{1/2}[1 - (x + h)^2]^{1/2}\{[1 - (x + h)^2]^{1/2} + (1 - x^2)^{1/2}\}}$

$df(x, h) = \dfrac{h}{(1 - x^2)^{3/2}}$.

(i) $\Delta f(x, h) = \dfrac{(27x^2 + 95x + 52)h + (27x + 48 + 9h)h^2}{(2 + 3x + 3h)(4 + x + h)^{1/2} + (2 + 3x)(4 + x)^{1/2}}$

$df(x, h) = \dfrac{(26 + 9x)h}{2(4 + x)^{1/2}}$

2. (b) $-\frac{1}{33}, -\frac{1}{36}$ (d) $\frac{7471}{14440}, \frac{17}{32}$ (f) approx. 0.58, $\frac{4}{9}$ (h) approx. 0.041, $\frac{7}{180}$

3. (b) $\frac{1}{50}$ (c) $\frac{1}{12500}$ (d) $\frac{53}{6250}$

4. (a) 10.05 (b) 4.05 (c) 7.014
   (d) 14.97 (e) 2.025 (f) 2.996
   (g) 2.983 (i) 0.227 (j) 0.0169

PROBLEMS 4.4

2. Approx. error $3\pi/16$ in.² 3. $3\pi/8$ in.³ 4. 6.12% 5. 6π in.²
6. $625\pi/16$ lb. 8. $\frac{1}{200}$ 10. Approx. 3.34 in.³; approx. 0.075 in.

PROBLEMS 4.6

1. 35 2. 3000 3. 4459 cans; 12¢ per can
4. If C is the cost per ton of producing low grade steel and P is the selling price per ton, then

$$x = 10 - \sqrt{\dfrac{20P - 15C}{P - C}}$$

5. $\frac{106}{15}$ 6. $\dfrac{x}{g(x)} Dg(x)$ 7. $\dfrac{t}{S(t)} DS(t)$

EXERCISES 4.7

1. -3 2. $-\frac{2}{3}$ 3. -2 4. 5
5. none 6. $\frac{14}{19}$ 7. $\frac{1}{4}$ 8. $-\frac{1}{12}$
9. 1 10. $-\frac{1}{2}$ 11. $\frac{1}{3}$ 12. $\frac{1}{2}$

13. 2   14. 0   15. $\frac{3}{2}$ 16. none

17. 0   18. none   19. 1   20. $\frac{1}{2}$

PROBLEMS 4.7

2. Apply results of Problem 1 to
$$\phi(x) = \begin{vmatrix} f(a) & g(a) & h(a) \\ f(b) & g(b) & h(b) \\ f'(x) & g'(x) & h'(x) \end{vmatrix}$$

EXERCISES 5.1

1. 1.0100   2. 7.0711   3. 9.0554   4. 9.9499

5. 5.9161   6. 8.0623   7. 11.9164   8. 22.6053

9. 4.6875   10. 3.0366   11. 2.9907   12. 2.0235

13. 0.0204   14. 0.0098   15. 0.0044   16. 0.0010

21. $6 - 5(x+1) + (x+1)^2$

23. $238 - 252(x+3) + 100(x+3)^2 - 17(x+3)^3 + (x+3)^4$

25. $\frac{2}{3} - \frac{4}{9}(x - \frac{1}{2}) + \frac{8}{27}(x - \frac{1}{2})^2 - \frac{16}{81}(x - \frac{1}{2})^3$

27. $1 - x^2 + x^4$ 29. $1 + \frac{3}{2}x + \frac{7}{8}x^2 + \frac{11}{16}x^3 + \frac{75}{128}x^4$

PROBLEMS 5.1

3. $x_0 = -\frac{1}{12}\sqrt{66}$ 4. $f - g$ is a polynomial.

EXERCISES 5.2

1. All except (a) and (f) are convergent.

2. Only (a), (b), (c) and (f) are convergent.

3. (b) A geometric series with $r = \sqrt{2} > 1$.
   (c) $\lim_{n \to \infty} a_n \neq 0$
   (d) A geometric series with $r = \frac{11}{10} > 1$.

4. (a) 8   (b) $\frac{1}{6}$ (c) 1   (d) 2   (e) 1

5. Use comparison series with general term as follows:
   (a) $1/6^{n-1}$ (b) $1/2^n$ (c) $1/n$ (d) $1/n^2$ (e) $1/2n$
   (f) $1/(n+1)$ (g) $1/2^{n-1}$ (h) $1/[n(n-1)], (n > 7)$ (i) $1/n^2$ (j) $2/[n(n+1)]$.

6. All except (m) are convergent.

7. All except (a), (b), (e) and (k) are convergent.

8. (a) $1/(n^2+1)$ (b) $1/(3^n+2)$ (c) $1/(3n+1)$ (d) $[(-1)^{n-1}n]/(n^2+1)$ (e) $n^2/2^n$
   (f) $\dfrac{(-1)^{n-1}}{n(n+3)}$ (g) $n(2/3)^{n-1}$ (h) $\dfrac{1}{n}(2/3)^{n-1}$
   (i) $\dfrac{n!}{1 \cdot 4 \cdot 7 \cdots (3n+1)}$ (j) $\dfrac{1 \cdot 3 \cdots (2n+1)}{1 \cdot 5 \cdots (4n+1)}$

PROBLEMS 5.2

1. If $a_n = b_{n+1} - b_n$, then $a_1 + a_2 + \cdots + a_n = b_{n+1} - b_1$ whence we have
$$\lim_{n \to \infty} (a_1 + \cdots + a_n) = \infty$$
If $c_n = \dfrac{1}{b_n} - \dfrac{1}{b_{n+1}}$, then $c_1 + c_2 + \cdots + c_n = \dfrac{1}{b_1} - \dfrac{1}{b_{n+1}}$ whence
$$\lim_{n \to \infty} (c_1 + \cdots + c_n) = \dfrac{1}{b_1}.$$

2. The proof is straightforward epsilontics.

3. Hint: Note that, because each a_i is positive, we have $a_1^2 + \cdots + a_n^2 < (a_1 + \cdots + a_n)^2 < M^2$.

4. First prove there exists n_0 such that if $n > n_0$, then $(a_n + 1/n)^2 \leq 2a_n^2$. Then the series $(a_1 + 1)^2 + (a_2 + \tfrac{1}{2})^2 + \cdots + (a_n + 1/n)^2 + \cdots$ is known to converge by comparison. However, this series is also
$$(a_1^2 + \cdots + a_n^2 + \cdots) + 2(c_1 + \tfrac{1}{2}a_2 + \cdots + (1/n)a_n + \cdots)$$
$$+ (1 + \tfrac{1}{4} + \cdots + 1/n^2 + \cdots)$$
The rest is straightforward.

5. Note first that the comparison test does not apply because it is not assumed that the a_n are positive. The needed hint here is to note that
$$a_1 + a_2 + \cdots + a_n + \cdots = a_1 + \tfrac{1}{2}(a_1 + 2a_2 - a_1) + \tfrac{1}{3}(a_1 + 2a_2 + 3a_3 - a_1 - 2a_2)$$
$$+ \cdots + \tfrac{1}{n}(a_1 + \cdots + na_n - a_1 - \cdots - (n-1)a_{n-1}) + \cdots$$
$$= (1 - \tfrac{1}{2})a_1 + (\tfrac{1}{2} - \tfrac{1}{3})(a_1 + 2a_2) + (\tfrac{1}{3} - \tfrac{1}{4})(a_1 + 2a_2 + 3a_3)$$
$$+ \cdots + \left(\dfrac{1}{n} - \dfrac{1}{n-1}\right)(a_1 + 2a_2 + \cdots + na_n) + \cdots.$$
The rest is routine epsilontics.

6. Hint: $\dfrac{a_n}{(1+a_1)(1+a_2)\cdots(1+a_n)} = \dfrac{1}{(1+a_1)\cdots(1+a_{n-1})} - \dfrac{1}{(1+a_1)\cdots(1+a_n)}$

7. Hint: A comparison with a series with general term c/n^2 can be made.

8. Hint: By induction, show that $2 - \dfrac{1}{\sqrt{n}}\left(1 + \dfrac{1}{\sqrt{2}} + \cdots + \dfrac{1}{\sqrt{n}}\right) \leq \dfrac{1}{\sqrt{n}}$.

9. (a) $\tfrac{1}{2}$ (b) 1   (c) 1

EXERCISES 5.3

1. $\{x : -1 \leq x < 1\}$ 3. $\{x : -1 \leq x \leq 1\}$ 5. $\{x : -3 < x < 6\}$

7. $\mathbb{R}$ 10. $\{x : -1 \leq x \leq 1\}$ 12. $\{0\}$

13. and 14. Depend upon $\lim_{n \to \infty} (1 \pm 1/n)^n$. Setting these limits equal to e and $1/e$, the rest is easy.

16. $\{x : -2 < x < 2\}$ 18. $\{x : -3 < x < 3\}$ 19. $\{x : 1 \leq x < 3\}$

22. $\{x : -2 < x < 6\}$ 24. $\{x : -4 \leq x < -2\}$ 25. $\{x : -7 < x < -1\}$

26. $\{x : -6 \leq x < -3\}$ 28. $\{x : -\tfrac{14}{3} \leq x \leq 2\}$ 30. $\{x : -1 < x < 1\}$

31., 32. and 33. Depend upon knowing

$$\lim_{n\to\infty} \frac{1\cdot 3\cdot 5 \cdots (2n-1)}{2\cdot 4\cdot 6 \cdots (2n)}$$

Call it L and proceed.

34. $\{x: -\frac{1}{2} < x < \frac{1}{2}\}$ 35. **R** 36. $\{x: -1 \leqq x < 1\}$
37. $\{x: -2 - 1/\sqrt{2} < x < -2 + 1/\sqrt{2}\}$ 38. $\{0\}$
39. If $a > b > 0$, then $\{x: -a \leqq x < a\}$. 40. $\{0\}$

PROBLEMS 5.3

2. (a) $\{x: x < -1$ or $x \geqq 1\}$ (b) $\{x: x < -1$ or $x > 1\}$
   (c) $\{x: x < \frac{1}{2}$ or $x > \frac{1}{2}\}$ (d) $\{x: x < -5$ or $x \geqq 5\}$
   (e) $\{x: x \leqq 1$ or $x > 3\}$ (f) $\mathbf{R} - \{0\}$
   (g) Always diverges. (h) $\{x: x < -6$ or $x > -\frac{4}{3}\}$
   (i) $\{x: \frac{11}{4} < x < \frac{17}{2}\}$ (j) $\{x: x < -\frac{9}{7}$ or $x > -1\}$
   (k) $\{x: -1 \leqq x < 1\}$ (l) $\{x: \sqrt{2} < |x| < 2\sqrt{2}\}$

3. Hint: Show by induction that $s_n = nx/(nx+1)$ if $x > 0$.

6. Four terms.

EXERCISES 5.4

1. (a) $17 + 11(x-3) + 2(x-3)^2$ (d) $6 - (x-2)^2$
   (e) $3 + 6(x-1) + 5(x-1)^2 + (x-1)^3$
   (h) $-1 - 10(x-1) - 11(x-1)^2 - 6(x-1)^3 - (x-1)^4$

2. (a) $1 - \frac{1}{2}x - \frac{1}{8}x^2 - \frac{1}{16}x^3 - \frac{5}{128}x^4 - \cdots$
   (c) $1 + \frac{1}{2}x^2 - \frac{1}{8}x^4 + \frac{1}{16}x^6 - \frac{5}{128}x^8 + \cdots$
   (e) $1 + \frac{1}{3}x^3 + \frac{2}{9}x^6 + \frac{14}{81}x^9 + \cdots$
   (f) $\frac{1}{4}\left(1 + \frac{x}{4} + \frac{x^2}{16} + \frac{x^3}{64} + \cdots\right)$
   (h) $x - \frac{1}{2}x^2 + \frac{3}{8}x^3 - \frac{5}{16}x^4 + \frac{35}{128}x^5 - \cdots$

3. (b) $\frac{1}{5} + \frac{2}{25}(x+1) + \frac{3}{250}(x+1)^2 - \frac{1}{625}(x+1)^3 + \cdots$
   (c) $\frac{1}{4}(x+1) + \frac{1}{16}(x+1)^2 + \frac{1}{64}(x+1)^3 + \cdots$
   (e) $1 - \frac{1}{2}(x-1)^2 - \frac{1}{8}(x-1)^4 - \frac{1}{16}(x-1)^6 - \cdots$
   (g) $1 - 2x + x^2 + 2x^3 - 4x^4 + \cdots$
   (i) $-(x-1) + \frac{1}{2}(x-1)^2 - \frac{1}{4}(x-1)^4 + \cdots$

4. (c) $\{x: -5 < x < 3\}$ (d) $\{x: -1 < x < 1\}$

5. (a) 0.97468 (b) 1.09545 (d) 2.02469
   (f) 0.30015 (g) 0.00052 (h) 3.00939

6. (c) $x/\sqrt{1-x^2} = x + \frac{1}{2}x^3 + \frac{3}{8}x^5 + \frac{5}{16}x^7 + \cdots$
   (d) $(1-2x^2)(1-x^2)^{-1/2} = 1 - \frac{3}{2}x^2 - \frac{5}{8}x^4 - \frac{7}{16}x^6 - \frac{45}{128}x^8 - \cdots$

PROBLEMS 5.4

2. Choose n_0 sufficiently large so that $|x|^{n_0} \leqq \frac{1}{2}$. Then compare the given series for $n > n_0$ with $2(1 + x^2 + x^4 + \cdots)$.

4. Write $\sqrt{1+x}/\sqrt{x} = \sqrt{1 + 1/x}$. Let $1/x = y$ and expand $\sqrt{1+y}$ as a binomial series.

5. In each case the induction argument required is quite long. Compute a few terms first.

SELECTED ANSWERS: EXERCISES 6.1

EXERCISES 6.1

1. $\frac{1}{2}$
2. 8
3. $\frac{3}{2}$
4. 2
5. 6
6. $\frac{3}{2}$
7. $\frac{4}{3}$
8. $\frac{4}{3}$
9. $\frac{1}{3}$
11. $\frac{7}{3}$

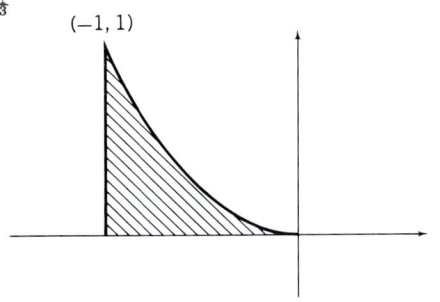
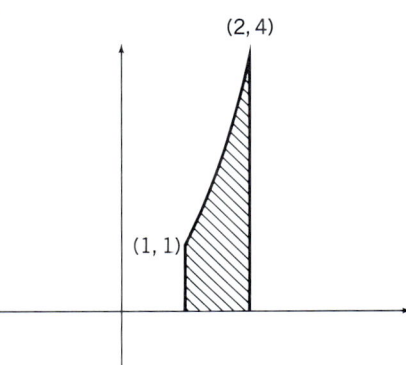

13. $\frac{32}{3}$
15. $\frac{44}{3}$

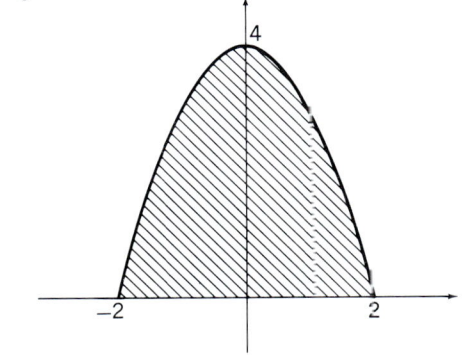
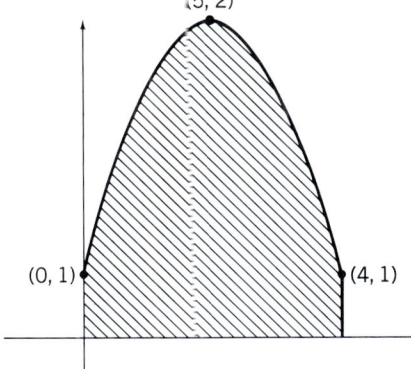

17. $\frac{7}{3} + \frac{3}{2} = \frac{23}{6}$
19. $\frac{1}{3} + \frac{3}{4} = \frac{13}{12}$
20. $1 + 2 = 3$
21. $\frac{1}{2} + \frac{3}{2} = 2$
23. $\frac{1}{3} + \frac{1}{3} = \frac{2}{3}$
24. $\frac{16}{3} + \frac{16}{3} = \frac{32}{3}$

PROBLEMS 6.1

1. Recall the formulas for sums of squares, cubes, etc.
2. It is easy to prove that $S_n - s_n = 1/2n$ whence choose $n > 100$.
3. Each approximating sum is independent of n.

EXERCISES 6.3

2. Simply apply the definition to obtain

$$\sum_{j=2}^{n+1} \varphi(j-2) = \varphi(2-1) + \varphi(3-1) + \cdots + \varphi((n+1)-1)$$
$$= \varphi(1) + \varphi(2) + \cdots + \varphi(n)$$

3. This is called a *telescoping sum*.

4. $\frac{1}{2}n(n + 3)$

5. $n(n + 4)$

6. $\frac{1}{6}n(n + 1)(5 - 2n)$

7. $\frac{1}{6}n(2n^2 + 9n + 13)$

8. $\frac{1}{3}n(n^2 - 1)$

9. $\frac{1}{4}n(n - 1)(n^2 + n + 4)$

10. $\frac{1}{4}(n - 1)n(n + 1)(n + 2)$

11. $\frac{1}{4}n(n + 1)(n + 2)(n + 3)$

12. $\frac{5}{2}$

13. $\sum_{i=1}^{n} (1 + i/n)^2 \cdot \frac{1}{n} = 7/3 + 3/2n + 1/6n^2$

15. $1 - \frac{1}{n + 1}$

16. Write $\frac{1}{i(i + 1)(i + 2)} = \frac{1}{2}\left[\frac{1}{i(i + 1)} - \frac{1}{(i + 1)(i + 2)}\right]$

EXERCISES 6.4

1. 3, 6
2. 4, 100
3. 0, 2
4. $\frac{2}{3}$, 2
5. $\frac{2}{5}$, 2
6. 0, 2
7. 3, 6
8. $\frac{4}{5}$, $\frac{4}{25}\sqrt{41}$
9. $\frac{4}{3}$, 4
10. 4, $3\sqrt{2}$
11. 2.9, 3.1
12. $\frac{501}{75}$, $\frac{167}{25}$
13. 4.9, 8.1
14. 0.397, 0.414
15. 0.596, 0.603
16. 12.415, 12.915
17. 0.517, 0.523
18. 0.321, 0.371
19. 4.872, 4.900
20. 1.093, 1.133

21. $\lim_{n \to \infty} \frac{2(2n + 1)}{n}$

22. $\lim_{n \to \infty} (-2/n)$

23. $\lim_{n \to \infty} \left(\frac{14n^2 + 9n + 1}{6n^2}\right)$

24. $\lim_{n \to \infty} \left(\frac{64n^2 + 96n + 32}{3n^2}\right)$

25. $\lim_{n \to \infty} \left(\frac{n^2 + 6n - 1}{6n^2}\right)$

26. $\lim_{n \to \infty} \left(\frac{20n^2 + 26n + 8}{n^2}\right)$

27. $\lim_{n \to \infty} \left(\frac{17n^2 + 24n + 7}{12n^2}\right)$

28. $\lim_{n \to \infty} \left[\frac{3}{n} \sum_{i=1}^{n} \sqrt{1 + 3i/n}\right]$

29. $\lim_{n \to \infty} \left[\frac{1}{2n} \sum_{i=1}^{n} \sqrt{1 + (i/2n)^2}\right]$ (see Exercise 17)

30. $\lim_{n \to \infty} \left[\frac{1}{n} \sum_{i=1}^{n} \sqrt{1 + (i/n)^3}\right]$ (see Exercise 20)

31. $\int_0^1 x^2$
32. $\int_0^1 x^3$
33. $\int_0^1 (1 + x^2)$
34. $\int_0^1 x^2(1 - x^2)$
35. $\int_0^1 \frac{1}{1 + x}$

PROBLEMS 6.5

1. Each approximating Riemann sum equals $k(b - a)$

3. $\int_a^b (f - c)^2 = (b - a)c^2 - 2c \int_a^b f + \int_a^b f^2$ is a quadratic in c and since $b - a > 0$, this function has minimum value when $c = \frac{1}{b - a} \int_a^b f$. Using this value of c, we have

$$0 \leqq (b - a)\left[\frac{1}{b - a} \int_a^b f\right]^2 - 2\left[\frac{1}{b - a} \int_a^b f\right]^2 + \int_a^b f^2$$

or $$0 \leq -\frac{1}{b-a}\left[\int_a^b f\right]^2 + \int_a^b f^2$$

5. Clearly, we have $\int_a^b f \geq 0$. Let x_0 be a point such that $f(x_0) > 0$. For sufficiently small $\epsilon > 0$, $f(x) > \frac{1}{2}f(x_0)$ for all points in $\{x : x_0 - \epsilon < x < x_0 + \epsilon\}$. Now choose n sufficiently large so that the interval of length $(b-a)/n$ which contains x_0 lies in the 2ϵ interval. Then we have
$$0 < \tfrac{1}{2}f(x_0)\left(\frac{b-a}{n}\right) < \int_a^b f.$$

6. Assuming $g(x) \geq 0$, we have
$$[\min f(x)]g(x) \leq f(x)g(x) \leq [\max f(x)]g(x).$$
It follows that
$$[\min f(x)]\int_a^b g \leq \int_a^b fg \leq [\max f(x)]\int_a^b g.$$
The conclusion follows from the continuity of f.

7. $\int_0^1 \frac{1}{1+x^2}$

EXERCISES 7.1

1. $(x-2)^2$
3. $\sqrt{x}$
5. $x^3 - 3x + 2$
7. $-x^2$
8. $-\sqrt{x}$
9. $7x^2$
11. x^2
12. 99
13. 1
15. $8(x-2)^2$
16. $3(3x-1)^2$
17. $2x\sqrt{x^2(x^2-2)}$
18. $\sqrt{x+1}$
19. $(1-2x)\sqrt{1+x-x^2}$
20. $0, \pm\sqrt{E}$
21. $x + x^3$
23. $x^3 + 3x^2 + 3x$
24. $\frac{1}{3}x^3 + x^2 + x + \frac{1}{3}$
26. $\frac{7}{3}x^3 + \frac{3}{2}x^2$
27. $\frac{2}{3}x^{3/2}$
29. $\frac{2}{3}(x^3 - x^{3/2})$

PROBLEMS 7.1

2. Let a be any fixed number and write
$$\int_{v(x)}^{u(x)} f = \int_{v(x)}^a f + \int_a^{u(x)} f.$$
Then the chain rule and Problem 1 easily complete the proof.

3. The key here is to prove that if $b < c$, then
$$\int_a^c f \geq \int_b^c f.$$
Then write
$$\int_a^x f + \int_a^g f = \int_a^x f + \int_a^{\frac{x+y}{2}} f + \int_{\frac{x+y}{2}}^y f.$$
Assuming $x < y$, we have
$$\int_{\frac{x+y}{2}}^y f \geq \int_x^y f$$
and the rest is easy.

5. $g(T)$ is the operating cost per unit time.

EXERCISES 7.2

1. 15
2. 5
3. 7
4. $\frac{15}{2}$
5. 1
6. $\frac{17}{6}$
7. 39
8. 9
9. $\frac{15}{4}$
10. 0
11. $\frac{1}{5}$
12. $\frac{2}{5}$
13. 0
14. $\frac{1061}{30}$
15. $\frac{1}{2}$
16. $\frac{3}{8}$
17. $\frac{7}{8}$
18. $\frac{50}{27}$
19. $\frac{16}{3}$
20. $\frac{16}{3}$
21. $\frac{52}{3}$
22. 4
23. 4
24. $2(-1+\sqrt{2})$
25. $\frac{384}{7}$
26. $\frac{57}{4}$
27. $\frac{q}{p+q}(2^{1+p/q}-1)$
28. $\frac{q}{b(p+q)}[(a+b)^{1+p/q} - a^{1+p/q}]$
29. $\frac{1}{20}$
30. $-\frac{1}{20}$
31. $\frac{492}{7}$
32. $\frac{q}{2b(p+q)}[(a+4b)^{1+p/q} - a^{1+p/q}]$
33. $\frac{8}{3}$
34. $\frac{2}{3}(-1+2\sqrt{2})$
35. $\frac{1}{2}$
36. $\frac{2}{3}(5\sqrt{5} - 2\sqrt{2})$

PROBLEMS 7.2

1. The sum $\Sigma f(\bar{x}_i)(x_i - x_{i-1}) = \Sigma[F(x_i) - F(x_{i-1})]$ telescopes to $F(b) - F(a)$.
2. Clearly, $D_y x = (1 + 4y^2)^{-1/2}$ whence $D_x y = 1/D_y x = (1 + 4y^2)^{1/2}$. Then the computation of $D_x^2 y$ is straightforward.
3. (a) $s = \frac{1}{2}t^2 + \frac{1}{6}t^3$ (d) $s = \frac{4}{15}t^{5/2} + \frac{4}{3}t^{3/2} - \frac{5}{3}t - \frac{59}{15}$
4. $60\frac{1}{2}\%$ $136\frac{1}{8}\%$.

EXERCISES 7.3

1. $3x + C$
2. $\frac{1}{2}x^2 + C$
4. $\frac{1}{3}x^3 + C$
6. $\frac{1}{2}x^2 + \frac{1}{4}x^4 + C$
7. $\frac{1}{3}x^3 + \frac{1}{5}x^5 + C$
9. $\frac{2}{3}x^{3/2} + C$
11. $3x^{1/3} + C$
12. $\frac{3}{2}x^{4/3} + 9x^{2/3} + C$
14. $\frac{1}{3}x^3 - 2/x + C$
15. $\frac{2}{5}x^{5/2} - 2x^{1/2} + C$
17. $\frac{2}{3}x^{3/2} - x + C$
19. $\frac{1}{3}(1 + x^2)^{3/2} + C$
22. $\frac{1}{8}(1 + 4x^3)^{2/3} + C$
23. $\frac{1}{3}(x^2 + 2x - 1)^{3/2} + C$
24. $\frac{1}{3}(3x^2 + 12x + 4)^{1/2} + C$
25. $y = 2x - \frac{1}{3}x^3 + C$
28. $y = \sqrt{2x^3 + C}$
29. $y = \sqrt[4]{4x^2 + C}$
30. $y = (2x + C)^2$
32. $y = (\frac{1}{3}x^{3/2} + C)^2$
34. $y = -\frac{3}{2} + 1/(C - 4x)$
35. $y = x^2 + x - 1$
36. $y = x^2 - 1/x - 1$
37. $y = (x^2 + 8)^{1/3}$
39. $y = (3x - 2)/(5 - 3x)$
40. $2(y^2 - 1) = x^2 y$
41. $x^3 + 3x^2 + 6x + 6$
42. $\frac{2(c^2 - d)}{d^3} - \frac{2c}{d^2}x + \frac{1}{d}x^2$

PROBLEMS 7.3

2. $y = \frac{W\ell^2 x}{16EI} - \frac{W}{12EI}x^3$

3. On $\{x: -c \leq x \leq c\}$, we have $\sqrt{\tfrac{2}{3}m^2 e^3 + 2b^2 c}$

4. $r = R^{2/3}(R^{1/2} + \tfrac{3}{2}\sqrt{2g}\, t)^{2/3}$

6. From the equation
$$a = \frac{dv}{dt} = \frac{dv}{ds} \cdot \frac{ds}{dt} = v\frac{dv}{ds}$$
we have $a\, ds = v\, dv$ or $\tfrac{1}{2}v^2 = \int a\, ds$.

EXERCISES 7.4

1. $\tfrac{32}{3}$ 2. $\tfrac{1}{12}$ 3. $\tfrac{32}{3}$ 4. 36 5. $\tfrac{1}{6}$
6. $\tfrac{4}{3}$ 7. $\tfrac{4}{3}$ 8. 9 9. 36 10. $\tfrac{243}{2}$
11. $\tfrac{9}{2}$ 12. $\tfrac{128}{15}$ 13. $\tfrac{1}{2}$ 14. $\tfrac{2}{3}$ 15. $\tfrac{7}{12}$
16. $\tfrac{9}{4}$ 17. 8 18. 3 19. 12 20. $\tfrac{115}{12}$

PROBLEMS 7.4

1. $2^{4/3}$ 2. n 3. $n/(n+2)$ (by induction)
4. The (unsigned) area bounded by the vertical lines $\{x: x = a\}$ and $\{x: x = b\}$, by the x axis and by the curve $\{x: y = f(x)\}$.

EXERCISES 7.5

1. $\tfrac{2}{3}(2x+1)^{3/2} + C$
2. $\sqrt{2x+1} + C$
3. $-\tfrac{2}{9}(1-3x)^{3/2} + C$
4. $\tfrac{1}{2}\sqrt{4x+1} + C$
5. $\tfrac{1}{3}(1+x^2)^3 + C$
6. $-\dfrac{1}{2(2+x^2)} + C$
7. $\sqrt{1+x^2} + C$
8. $\tfrac{3}{8}(1+x^2)^{4/3} + C$
9. $-2(4+x^2)^{-1/2} + C$
10. $\tfrac{1}{8}\sqrt{1+8x^2} + C$
11. $-\tfrac{1}{4}(4x-1)^{-1} + C$
12. $\tfrac{1}{3}(2x-x^2)^{3/2} + C$
13. $\sqrt{x^2+4x+1} + C$
14. $\tfrac{3}{4}(x^2+2x-3)^{2/3} + C$
15. $\dfrac{q}{2b(p+q)}(a+bx^2)^{1+p/q} + C$
16. $\tfrac{9}{20}(1+x^{5/3})^{4/3} + C$
17. $-\sqrt{a^2-x^2} + C$
18. $\tfrac{2}{3}(1-1/x)^{3/2} + C$
19. $\tfrac{4}{3}(1+\sqrt{x})^{3/2} + C$
20. $\tfrac{4}{9}(1+x\sqrt{x})^{3/2} + C$
21. $\tfrac{14}{3}$
22. $\tfrac{26}{3}$
23. $\tfrac{13}{6}$
24. 2
25. $\tfrac{1}{6}$
26. $-1+\sqrt{2}$
27. $\tfrac{1}{4}(-3+\sqrt{33})$
28. $\tfrac{8}{3}$
29. $\tfrac{3}{2}(-1+\sqrt{2})$
30. $\tfrac{64}{375}$

EXERCISES 7.6

1. $-\tfrac{2}{3}(x+2)\sqrt{1-x} + C$
2. $\tfrac{1}{21}(3x-1)(x+2)^6 + C$
3. $\tfrac{2}{15}(3x+2)(x-1)^{3/2} + C$
4. $\tfrac{2}{27}(3x-2)\sqrt{3x+1} + C$

5. $\frac{1}{9}(x-1)^2(x-2)^9 - \frac{1}{45}(x-1)(x-2)^{10} + \frac{1}{495}(x-2)^{11} + C$

6. $\frac{1}{3}x^2(x^2+2)^{3/2} - \frac{2}{15}(x^2+2)^{5/2} + C$ 7. $4\sqrt{3} - \frac{8}{3}$

8. $\frac{1696}{105}$ 9. $\frac{16}{3}\sqrt{5} - \frac{16}{15}$ 10. $\frac{4}{15}a^{5/2}$

PROBLEMS 7.6

1. $P(x) = 1 - 3x^2$ 2. $\frac{1}{3960}$

3. $\int f(x)g(x)\,dx = \frac{1}{a+b}[f(x)g'(x) - f'(x)g(x)]$

4. $\frac{3\pi}{16}a^4$ 7. Hint: Let $dv = du$ and $w = \int_0^u f(t)\,dt$.

8. Hint: Work backwards, starting with $\int \frac{dx}{(a^2+x^2)^{n-1}}$.

EXERCISES 7.7

1. 3   2. 1   3. 100   4. 2   5. $\frac{3}{2}$

6. 0   7. diverges   8. 1   9. diverges   10. $2(-1 + \sqrt{2})$

11.–20. All converge.

PROBLEMS 7.7

1. $p < 1$ 2. $\frac{1}{896}$

3. Note that the two are different kinds of improper integral!

4. Hint: Show that the sum approximates an improper integral.

5. As an example, we have

$$\sum_{n=1}^{100} \frac{1}{n^3} + \frac{1}{20402} < \sum_{n=1}^{\infty} \frac{1}{n^3} < \sum_{n=1}^{100} \frac{1}{n^3} + \frac{1}{20000}.$$

EXERCISES 7.8

1. $\frac{256}{3}\pi$ 2. $\frac{256}{3}\pi$ 3. $\frac{128}{3}\pi$ 4. $\frac{320}{3}\pi$ 5. $\frac{512}{15}\pi$

6. $\frac{256}{5}\pi$ 7. $\frac{832}{15}\pi$ 8. $\frac{108}{5}\pi$ 9. $\frac{198}{5}\pi$ 10. $\frac{117}{5}\pi$

11. $\frac{1}{60}$ 12. $\frac{68}{5}\pi$ 13. $\frac{128}{3}\pi$ 14. $\frac{27}{2}\pi$ 15. $\frac{\pi}{3}h(R^2 + Rr + r^2)$

PROBLEMS 7.8

1. $16r^3/3$ (Hint: Appropriately chosen cross-sections are squares.)

2. $2\pi^2 Rr^2$

6. $\frac{2}{3}hr^2$

7. First derive the differential of volume.

EXERCISES 7.9

1. $\frac{14}{3}$ 2. $\frac{2}{27}(10^{3/2} - 1)$ 3. $\frac{9}{2}(2^{2/3} - 1)$ 4. $\frac{339}{16}$

SELECTED ANSWERS: EXERCISES 7.10

5. $\frac{19}{3}$
6. $\frac{123}{32}$
7. $\int_{-2}^{2} \sqrt{1 + 4x^2}\, dx$
8. $7 \int_{0}^{3} \frac{dx}{\sqrt{49 - x^2}}$
9. $\int_{0}^{\frac{1}{2}\sqrt{3}} \frac{1 + x^2}{1 - x^2}\, dx$
10. $\sqrt{8} - \sqrt{3}$
11. $3|a|$
12. $\frac{a^2 + ab + b^2}{a + b}$
13. $4\sqrt{3}$
14. $2\sqrt{3}$
15. and 16. Simply require calculation.

EXERCISES 7.10

1. $\frac{\pi}{27}(145^{3/2} - 1)$
2. $\frac{\pi}{3} a^2 (5^{3/2} - 1)$
3. $\frac{124}{9}\pi$
4. $16\pi(8\sqrt{3} - 9)$
5. $\frac{16911}{1024}\pi$
6. $\frac{24}{5}\pi a^2$

PROBLEMS 7.10

1. The curves rotated will have to be restricted, of course.
2. 3π
4. Of course, one improper integral converges and the other does not.

PROBLEMS 7.11

1. $\frac{275}{8}$ ft.-lbs.
2. $\frac{5}{6}$ ft.-lb.
3. $\frac{125}{6}$ ft.-lbs.
4. $37{,}500\pi$ ft.-lbs.
5. $\frac{32000}{3}\pi$ ft.-lbs.
6. $\frac{125}{24}\pi r^2 (h_1^2 + 4h_1 h_2 + 6h_2^2)$
7. Approx. 8 million ft.-lbs.
9. Depends upon density, of course.
10. $900 \int_{0}^{4} (13 - y)\sqrt{6y - y^2}\, dy$
12. $62{,}500$ ft.-lbs.
14. $422{,}400{,}000$ ft.-lbs.
16. $\frac{1}{2}mgR$
17. mgR
19. $\frac{25}{4}$ ft.-lbs.

PROBLEMS 7.12

1. $16{,}000$ lbs.
2. $48{,}000$
3. $\frac{1}{3}(16{,}000)$ and $\frac{2}{3}(16{,}000)$
4. $8000(5 + 2\sqrt{2})$ ft.-lbs.
5. $90 \int_{0}^{8} (8 - y)\sqrt{8y - y^2}\, dy$
6. $125 \int_{38}^{42} y\sqrt{-y^2 + 80y - 1596}\, dy$
7. The two answers are $\frac{1}{3}(800\sqrt{2})$ and $\frac{1}{3}(3200\sqrt[3]{2})$.

EXERCISES 8.1

1. **R**
2. $\{y : -3 \leq y < \infty\}$
4. **R**
6. $\{y : 1 \leq y < \infty\}$
9. no
10. $\{y : -2\sqrt{5} \leq y \leq 2\sqrt{5}\}$
11. Either $\{x : x \leq -2\}$ or $\{x : x \geq -2\}$.
14. Any one of the four sets $\{x : x \leq -1\}$, $\{x : -1 \leq x < 0\}$, $\{x : 0 < x \leq 1\}$, $\{x : x \geq 1\}$.
16. **R**
17. Same as Exercise 14 with zero included.
19. Any of $\{x : x \leq -1\}$, $\{x : -1 \leq x \leq 1\}$, $\{x : x \geq 1\}$.

22. Slope of the graph of f^{-1} at $(2, 1)$ is $\tfrac{1}{3}$.

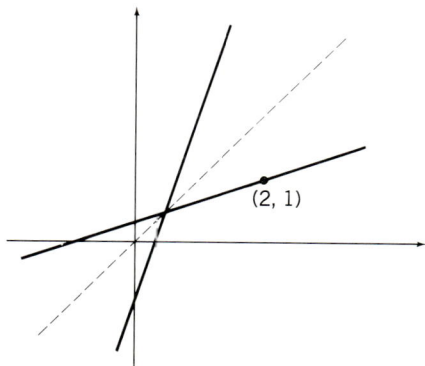

24. Slope of the graph of f^{-1} at $(2, 1)$ is $\tfrac{1}{e}$.

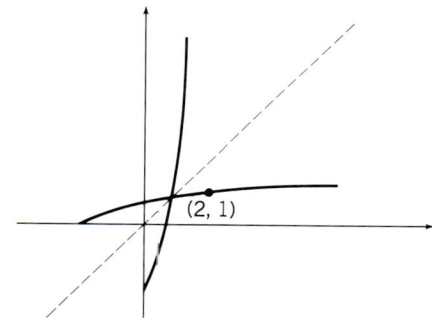

PROBLEMS 8.1

1. $f^{-1}(y) = \dfrac{y}{1 - |y|}$, $-1 < y < 1$.   3. $c = -1$ 4. $b = 1$ 5. $c^2 < 3bd$

EXERCISES 8.2

1. $-\tfrac{1}{2}$ 2. $\dfrac{1}{2\sqrt{x}}$ 3. $\dfrac{1}{\sqrt{9 + 4x}}$

5. $\dfrac{8}{(3 - x)^2}$ 7. $\dfrac{x}{\sqrt{x^2 - 16}}$ 9. $\tfrac{3}{2}\sqrt{x}$

PROBLEMS 8.2

3. The key here is the equation
$$D \int_0^{f(x)} f(t)\, dt \cdot Df^{-1}(x) = f(f^{-1}(x)) Df^{-1}(x)$$

8. The constant is 4.

EXERCISES 8.3

1. a 2. $1 + a^2$ 6. $-\log_3 2$ 7. no root
11. $\tfrac{1}{2}(1 + \sqrt{29})$ 13. 25   15. 3   16. 2, 4

17–20. Hint: Recall that we only define the logarithm of a *positive* number.

22. $f^{-1}(x) = 2 \log_2\left(\frac{x-1}{2}\right)$, $x > 1$

23. $f^{-1}(x) = \log_a(x + \sqrt{x^2 + 1})$

26. $f^{-1}(x) = \frac{1}{2}(3^x + \sqrt{3^x(4 + 3^x)})$

28. $f^{-1}(x) = \sqrt{3 + 2a^x - a^{2x}}$, $x < \log_a 3$

32. $x = (b + c)/(1 + bc)$.

PROBLEMS 8.3

1. Use the intermediate value theorem plus monotonicity. Note that Problems 2 and 3 are equivalent.

4. Take logarithms to the base d twice.

EXERCISES 8.4

2.
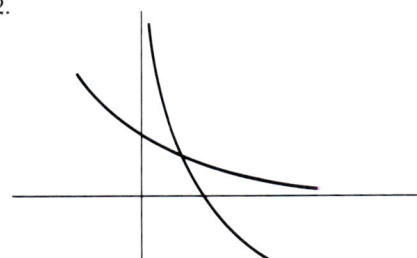

4.
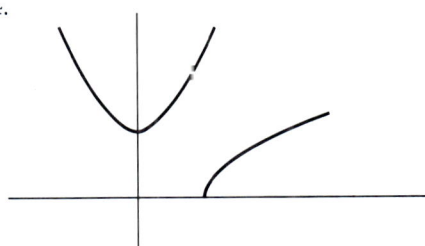

6.
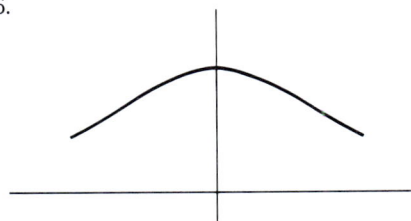

7.
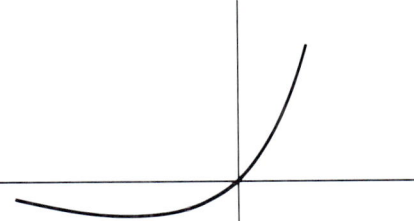

10.
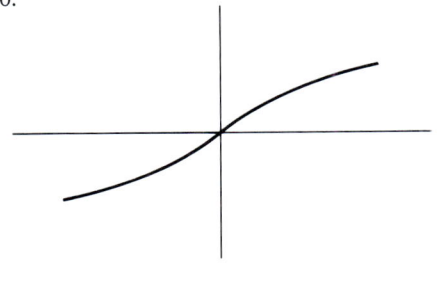

13.
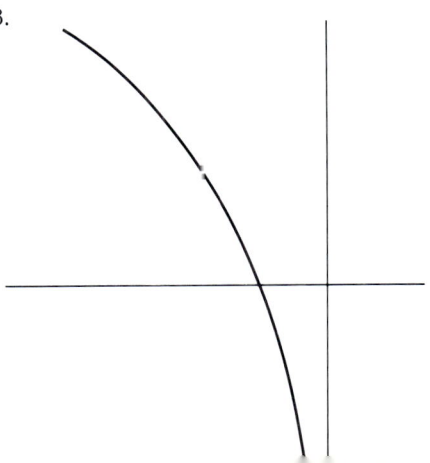

22. (a) 4.3219   (c) -3.6438 (e) 2.6021

PROBLEMS 8.5

4. Simply let $x/n = t$ and apply the definition of e.

5. First note that the sum is a geometric series. Then use L'Hospital's rule to evaluate
$$\lim_{x \to \infty} \frac{\frac{1}{x}e^{1/x}}{e^{1/x} - 1} = \lim_{y \to \infty} \frac{ye^y}{e^y - 1}.$$

6. A special case of Problem 8.   7. Use binomial expansions.

8. It is equivalent to show that $\lim_{n \to \infty} n \ln\left(1 + \frac{a_n x}{n}\right) = 0$.

EXERCISES 8.7

3. $\dfrac{3}{x - 1}$ 6. $\dfrac{2}{x} \ln x$ 8. $\ln x^2$

9. $1 + \ln x$ 12. $\dfrac{a}{x(a + bx)}$ 14. $\dfrac{1}{1 - x^2}$

16. $\dfrac{1}{a - bx}$ 17. $\dfrac{1}{\sqrt{x^2 + a^2}}$ 20. $2\sqrt{x^2 - 4}$

25. $-3^{1-3x} \ln 3$ 27. $(2 - x)e^{2x - x^2/2}$ 29. $\tfrac{1}{2}(e^x - e^{-x})$

32. $(1 + x)e^x$ 35. xe^{ax} 41. $\dfrac{2e^x}{1 - e^{2x}}$

45. $\ln(x^2 + 2) + C$ 47. $x - \ln|x + 1| + C$

49. First divide. Then obtain $\dfrac{1}{3b}x^3 - \dfrac{a}{2b^2}x^2 + \dfrac{a^2}{b^3}x - \dfrac{a^3}{b^4}\ln|a + bx|$

50. $\tfrac{1}{2}(\ln x)^2 + C$ 52. $\dfrac{4^x}{\ln 4} + C$ 54. $2e^{\sqrt{x}} + C$

56. $\ln|e^x - e^{-x}| + C$ 57. $\tfrac{1}{2}x^2 \ln x - \tfrac{1}{4}x^2 + C$

61. $\tfrac{1}{4}(1 - 1/e^8)$ 64. $\tfrac{4}{3}(1 - 1/e^3)$ 66. $\tfrac{1}{2}\ln 13$

69. $\ln\left(\dfrac{\ln 4}{\ln 2}\right)$ 71. $\dfrac{4}{5 \ln 5}$ 72. $\dfrac{15}{2 \ln 2}$

73. Be careful. There are *no* critical points.

74. $(\tfrac{1}{2}, -2 + \ln 2)$ 76. $(e, e), (e^2, \tfrac{1}{2}e^2)$

77. $(-1, -e^{-1}), (e^{-2}, -2e^{-2})$ 79. $(0, 2)$

81. Critical points $(0, 0), (2, 4e^{-2})$. Points of inflection: $x = 2 \pm \sqrt{2}$.

83. There is a critical point for some x_0 between 1 and e.

Problems 85, 86, 87 and 88. The value is zero.

92.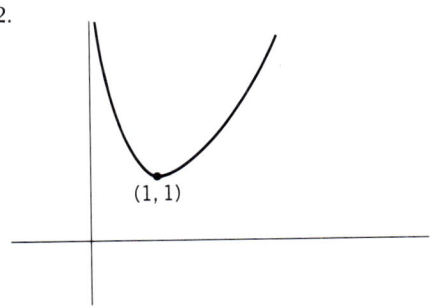

94.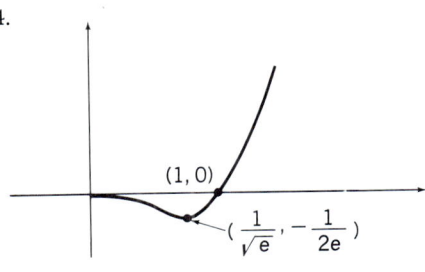

96.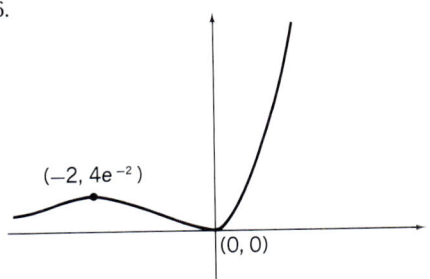

98.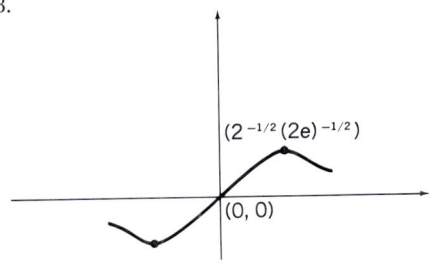

101. 0.567   103. 4.220   106. No root exists.

Problems 108, 109, 110, 111, 112, 115 and 118. The limit is zero.

119. $\frac{1}{3}\ln(x + \sqrt{x^2 - 1})$ 120. $\ln(2x + 2\sqrt{x^2 + 2})$

123. $\frac{1}{2}\ln(1 + e^x)$ 124. $2\ln\frac{1}{3}(2 + e^{xy})$

125. $\ln 13$ 127. $a^2(e - e^{-1})$ 129. $2\pi(4 - 3\ln 3)$

131. (a) $\ln|x/2|$ (b) $\ln|x - 1|$ (c) $\ln|x|$ (d) $\ln|x|$

PROBLEMS 8.7

1. (a) 0.0953   (b) 0.1823   (c) 0.1906   (e) 0.2776
2. (a) 1.1052   (b) 1.2214   (c) 0.9048   (d) 1.3499
3. Use the series from Problem 1.
4. The left inequality is easy. The right inequality can be proved by noting that
$$\frac{1}{5}x^5 + \frac{1}{7}x^7 + \cdots + \frac{1}{2n+1}x^{2n+1} + \cdots < \frac{1}{5}x^5(1 + x^2 + x^4 + \cdots) = \frac{x^5}{5(1-x^2)}$$
Then
$$\frac{x^5}{5(1-x^2)} < \frac{x^4}{4} \quad \text{if} \quad x \leq \frac{1}{2}.$$

5. Hint: Recall that there exists a point c between 0 and x such that
$$e^x = 1 + x + \frac{1}{2}x^2 + \cdots + \frac{1}{n!}x^n + \frac{e^c}{(n+1)!}x^{n+1}.$$

6. Consider the integral $\int_1^{n(k-1)} \frac{dx}{n+x}$ instead of the sum.

7. The inequality $(1 + x/n)^n < e^x$ actually does both sides.

9. Both   10. Yes. No.   11. Area equals $\sqrt{2/e}$ 12. $k = \frac{1}{4}\ln(\frac{46}{147})$

16. The sum $1/2 + 1/3 + \cdots + 1/n$ is a lower sum approximating $\int_1^n dx/x$ and an upper sum for $\int_1^n dx/(x+1)$.

18. $e - 1$ 19. $32/e$

PROBLEMS 8.7 (Economics)

2. k/r if $f(t) = k$ 4. 4%

EXERCISES 8.8

2. $2(6x - 1)^{-2/3}$ 4. $2(x^2 + 1)^{-3/2}$ 5. $-(x + 1)^{-1/2}(x - 1)^{3/2}$

7. $-\dfrac{3x + 7}{6(x + 1)^{2/3}(3x + 1)^{3/2}}$ 9. $\dfrac{5x - 1}{6(x + 1)^{1/2}(x - 1)^{2/3}}$

11. $\dfrac{4x^4 + 5x^2 - 3}{3(x^2 - 1)^{1/2}(x^2 + 1)^{4/3}}$ 14. $x^x(1 + x + x \ln x)$

16. $x^{-1+\ln x} \cdot 2 \ln x$ 19. $(\ln x)^{x-1}[1 + (\ln x)\ln(\ln x)]$

22. $x^{1+x^2}(1 + 2 \ln x)$ 24. $x^x e^{x^x}(1 + \ln x)$

EXERCISES 8.9

2. $3 + 3x + \dfrac{9x^2}{2} + \cdots + \dfrac{(3x)^n}{n!} + \cdots = \sum_{n=0}^{\infty} \dfrac{(3x)^n}{n!}$

4. $x^2 + x^3 + \dfrac{x^4}{2} + \cdots + \dfrac{x^{n+2}}{n!} + \cdots = \sum_{n=0}^{\infty} \dfrac{x^{n+2}}{n!}$

5. $1 + \dfrac{x^2}{2} + \dfrac{x^4}{24} + \cdots + \dfrac{x^{2n}}{(2n)!} + \cdots = \sum_{n=0}^{\infty} \dfrac{x^{2n}}{(2n)!}$

6. $x + \dfrac{x^3}{6} + \dfrac{x^5}{120} + \cdots + \dfrac{x^{2n+1}}{(2n+1)!} + \cdots = \sum_{n=0}^{\infty} \dfrac{x^{2n+1}}{(2n+1)!}$

8. $1 + x - \frac{1}{2}x^2 - \frac{5}{6}x^3$

9. $x + \dfrac{x^2}{2} + \dfrac{x^3}{3} + \cdots + \dfrac{x^n}{n} + \cdots = \sum_{n=1}^{\infty} \dfrac{x^n}{n}$

10. There is no Maclaurin series for $\ln(xe^x)$.

11. $x^2 - \dfrac{x^4}{2} + \dfrac{x^6}{3} - \dfrac{x^8}{4} + \cdots + (-1)^{n-1}\dfrac{x^{2n}}{n} + \cdots$

14. $2x + x^2 - \frac{10}{3}x^3 + \frac{7}{2}x^4 - \cdots$

17. $e + e(x - 1) + \dfrac{e}{2}(x - 1)^2 + \cdots = e\sum_{n=0}^{\infty} \dfrac{(x - 1)^n}{n!}$

20. $e \sum_{n=0}^{\infty} \dfrac{n + 1}{n!}(x - 1)^n$ 22. $\sum_{n=0}^{\infty} (-1)^n \dfrac{x^{2n}}{n!}$

24. $\ln 2 + x - \frac{1}{6}x^3 + \cdots$

25. $(x-1) + \frac{1}{2}(x-1)^2 - \frac{1}{6}(x-1)^3 + \frac{1}{12}(x-1)^4 - \cdots + (-1)^n \frac{(x-1)^n}{(n+1)(n+2)} + \cdots$

PROBLEMS 8.9

1. Note that $\frac{n+1}{n} = \frac{1 + 1/(2n+1)}{1 - 1/(2n+1)}$

2. $1 + \frac{1}{2}x + \frac{1}{6}x^2 + \cdots + x^n/(n+1)! + \cdots$

3. $B_0 = 1$, $B_1 = -\frac{1}{2}$, $B_2 = \frac{1}{6}$, $B_3 = -\frac{1}{30}$, $B_4 = \frac{1}{42}$, $\cdots$

7. The Maclaurin series is identically zero.

9. Use Exercise 8.9.3 and the fact that $\int_0^1 xe^x\, dx = 1$.

11. This is equivalent to the inequality $\frac{1}{n} > \ln\left(\frac{n+1}{n}\right) > \frac{1}{n+1}$

12. Rearrange the partial sums!

EXERCISES 8.10

2. $\frac{1}{2} + ce^{-2x}$
3. $8 + ce^{-x/4}$
6. $xe^{-x} + ce^{-x}$
7. $\frac{1}{3}x - \frac{1}{9} + ce^{-3x}$
8. $\frac{1}{3}x^2 + c/x$
10. $x^4 - c/x^2$
12. $x^4 + cx^3$
15. $\frac{1}{2}(x+1)^4 + c(x+1)^2$
17. $(cx-1)e^x$
20. $x^2(1 + ce^{1/x})$
21. $2e^{-2x}$
23. $-4 + 5e^{x+1}$
25. $2 + (y_0 - 2)e^{x_0 - x}$
26. $(x^3 + 2)/(x+1)$
28. $(x^2 + 4x + 4)/(x^2 + 1)$

PROBLEMS 8.10

1. $2e^x/(2 - e^x)$
2. $2\sqrt[5]{8}$
3. $3 - x + x \ln|x|$
4. $2\sqrt{|x|}$
5. $f(x) = cx^2$ for some constant c.

PROBLEMS 8.11

1. Approx. 45,260
2. 2042 A.D.
3. Approx. 3.9 hrs.
5. Approx. $81\frac{1}{2}\%$
7. 80.4 minutes
10. Approx. 9.7 months
11. 19.2 hrs.
13. 75,001 cm
15. $\lim_{t \to \infty} (C_0 + Ae^{-k^2 t})$

17. This requires two applications of Newton's method.

EXERCISES 9.1

1. $\frac{1}{4}(\sqrt{2} + \sqrt{6})$
3. $\frac{1}{4}(\sqrt{6} - \sqrt{2})$
6. $-\frac{\sqrt{3}}{2}$
8. $\frac{1}{2}\sqrt{2 + \sqrt{2}}$
10. $\frac{1}{2}\sqrt{2 + \sqrt{3}}$
12. $\frac{1}{2}(\cos x - \sqrt{3} \sin x)$
13. $-\cos x$
14. $-\cos x$
15. $\frac{1}{2}(\cos x + \sqrt{3} \sin x)$
17. $\frac{\sqrt{2}}{2}(\cos^2 x - 2 \cos x \sin x - \sin^2 x)$
19. $3 \sin x - 4 \sin^3 x$.

PROBLEMS 9.1

2. For instance, let $a = -1/(2 \sin \frac{1}{2})$, $b = 1$, $c = -\frac{1}{2} + 2\pi$.

3. (c) Let B be an angle with $\cos B = \frac{5}{13}$, $\sin B = \frac{12}{13}$.
   (e) Let B be an angle with $\cos B = -\frac{4}{5}$, $\sin B = \frac{3}{5}$.

6. First prove that no polynomial is periodic.

10. First prove that $f(p) = f(0)$ for each $p \in \mathbf{Z}$. Then prove that
$$f(p/q) = f(1/q + \cdots + 1/q) = f(0)$$
for any rational number p/q, with $0 < p/q < 1$. The rest follows from continuity.

EXERCISES 9.2

1. $6 \cos 2x$
3. $6 \sin 6x$
6. $x \cos x^2$
7. $2(\sin x - \sin 2x)$
10. $2x \cos 2x - \sin 2x$
14. $x \sin x + x^2 \cos x$
16. $2 \cos 2x$
19. $2/(1 + \cos x)$
23. $-(x \sin x + \cos x)/x^2$
26. $(2 \cos 2x)/(\sin 2x)$
29. $1/(\sin x)$
31. $e^x(\cos 2x - 2 \sin 2x)$
34. $(\cos x)e^{\sin x}$
36. $(-2 \sin x) \cdot \sin(\cos x) \cdot \cos(\cos x)$
41. 2
43. 1
45. 2
47. 2
50. $e^{-1/6}$
51. $(\pi/4 + 2k\pi, \sqrt{2})$, $(5\pi/4 + 2k\pi, -\sqrt{2})$
53. $(\pi/3 + 2k\pi, \frac{3}{2}\sqrt{3})$, $(5\pi/3 + 2k\pi, -\frac{3}{2}\sqrt{3})$, $(2k\pi, 0)$
56. $(\pi/3 + 2k\pi, \frac{3}{4}\sqrt{3})$, $(5\pi/3 + 2k\pi, -\frac{3}{4}\sqrt{3})$, $(2k\pi, 0)$

(In Exercises 52, 57 and 58, the trigonometric equations require tables.)

60. $\{(x,y) : y = x - 1\}$
61. $1/(\cos y)$
63. $1/(2 \cos 2y - \sin 2y)$
66. (a) 0 (b) $6(\cos 2x + 1)$ (c) 0
70. (a) 0.51 (c) 1.17 (e) 0.59 (f) 0.26

EXERCISES 9.3

2. Write $f(x)$ as $(\cos x)/(1 + \sin x)$.
5. $Df = -2 \csc^2(2x + 3)$
9. $Df = 0$
12. $Df = 6 \tan^2 2x \sec^2 2x$
15. $4 \sec^2 x \tan x$
17. $(8/x^2) \sec^2(8/x)$
22. $2 \tan x$
23. $(4 \sec^2 x)/(4 \tan x - 1)$
25. $2 \sec x$
27. $e^{\tan x} \sec^2 x$
31. $\frac{1}{2}$
33. $-\frac{1}{3}$
35. 3
38. $e^{1/2}$
45. $-\frac{1}{3} \cos 3x + C$
47. $2 \tan(x/2) + C$
50. $\ln|1 + \sin x| + C$
53. $e^{\sin x} + C$
56. $2\sqrt{3 + \tan x} + C$
59. $-\dfrac{1}{4a} \csc^4 ax + C$
61. $\frac{1}{2}$
63. 1
65. $\frac{1}{3}$
67. 0

In both 69 and 70 the integrals diverge.

72. $\frac{1}{2} \ln 2$
73. $\frac{2}{3}\pi - \sqrt{3}$
75. $2\sqrt{2}$
77. $\pi^2/2$

PROBLEMS 9.3

1. 2
4. No
7. $11\pi/36$
8. $61^{3/2}$

SELECTED ANSWERS: EXERCISES 9.4

10. $4r$ 11. $\frac{1}{2}r\sqrt{2}$ 12. $\dfrac{\mu\omega}{\sqrt{1-\mu^2}}$ 13. $\frac{250}{29}$ ft./sec.

14. Elapsed time $= (a \sec \alpha_1)/v_1 + (b \sec \alpha_2)/v_2$, where α_1 and α_2 satisfy $a \tan \alpha_1 + b \tan \alpha_2 =$ constant.

16. Approx. 2.11 rad. 17. Approx 4.02

18. The key equation is $81x^4 - 54x^3 + 63x^2 - 42x + 7 = 0$.

19. Area equals one half that of the square.

20. Area equals $\frac{1}{2}r^2$ 21. $r^2(1 - \cot x)$

22. There is no largest figure of this form.

23. $2r^2(\sqrt{5} - 1)$ 24. $\pi/4$ 25. Approx. $54°44'$ 26. $12\sqrt{2}$ cu. in.

EXERCISES 9.4

15.
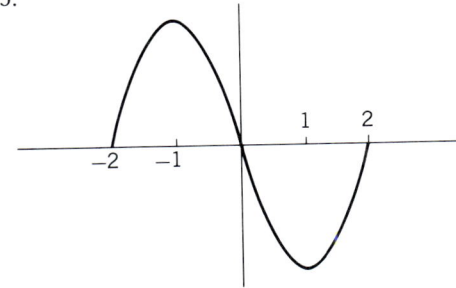

16.
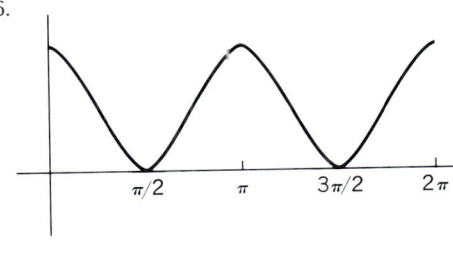

18.
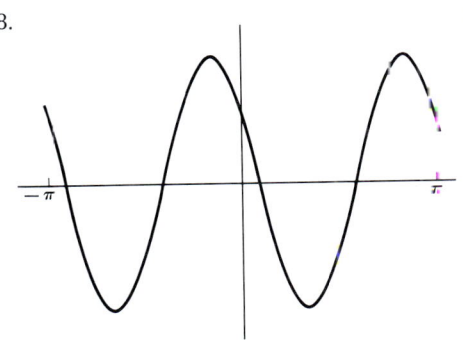

23.
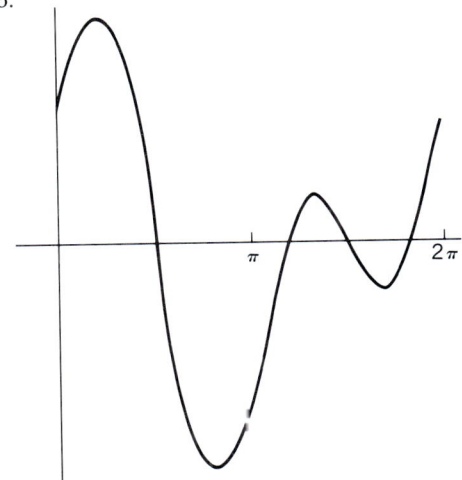

25.

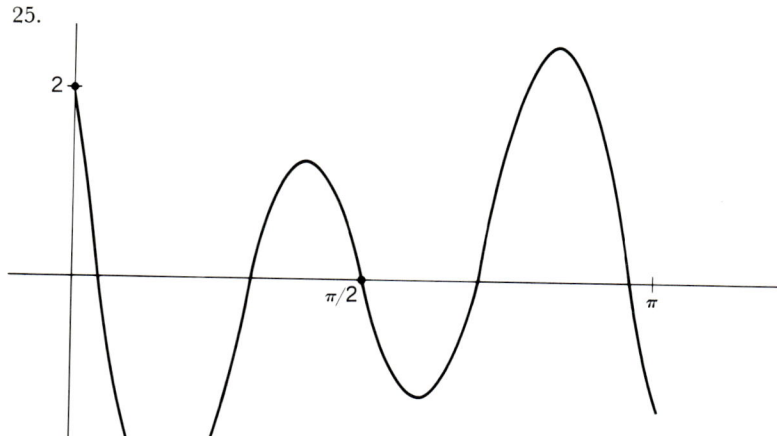

30.

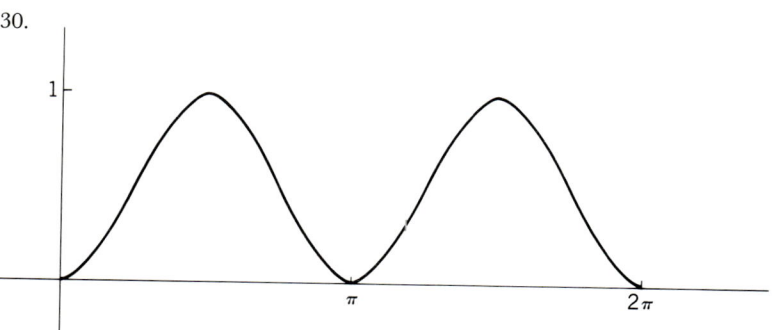

EXERCISES 9.5

11. 0.19867 13. 0.20271 14. 0.13646

16. 0.08716 18. 0.09033 19. -0.03093

PROBLEMS 9.5

1. 0.64279 2. 0.42262 5. $\sin^2 x = \sum_{n=1}^{\infty} (-1)^n \frac{(2x)^{2n}}{2(2n)!}$

6. (a) $\sum_{n=1}^{\infty} (-1)^n \frac{2n}{(n+1)!} x^{2n+1}$

   (b) $x^2 + \frac{1}{3} x^4 + \frac{2}{15} x^6 + \cdots$

   (c) $1 - \frac{1}{3!} x^2 + \frac{1}{5!} x^4 - \cdots$

   (d) $\frac{1}{2} x - \frac{1}{4!} x^3 + \frac{1}{6!} x^5 - \cdots$

SELECTED ANSWERS: PROBLEMS 9.6

PROBLEMS 9.6

1. $\pi/2$ 2. $s = \sqrt{13}\sin 2(t - t_0)$ 3. $2\pi\sqrt{l/k}$ 4. $\tfrac{1}{2}c\sqrt{3}$

5. The differential equation is $\dfrac{d^2\theta}{dt^2} = -\dfrac{g}{l}\sin\theta$

9. $k = \tfrac{1}{2}(-a \pm \sqrt{a^2 - 4b})$ 10. Apply the results of Problem 9

EXERCISES 9.7

1. $\pi/4$ 5. $\tfrac{5}{4}\pi$ 8. $\pi/3$
10. $\pi/6$ 14. $\sqrt{3}/2$ 16. $-\pi/3$
19. $\sqrt{1-t^2}$ 21. $t/\sqrt{1+t^2}$ 23. $(1-t)/2\sqrt{1+t^2}$
41. $1/\sqrt{4-x^2}$ 43. $-1/(1+x^2)$ 47. $1/(1+2x+2x^2)$
50. $1/\sqrt{1-x^2}$ 53. $3/(5+4\cos x)$ 56. $(1/x)\sqrt{x^2-9}$
58. $x/(1-x^8)$ 59. $\sin^{-1} x$ 61. $\tan^{-1} z$
63. $\sec^{-1} x$ 65. $6x^2\tan^{-1} x$ 66. $(\sin^{-1} x)^2$
68. 1   70. $\tfrac{2}{3}$ 71. $\tfrac{1}{2}$ 73. 12
79. $\tfrac{1}{3}$ 81. $\tfrac{1}{2}\sqrt{5}$ 83. $\pm 1/\sqrt{E}$ 85. $-1, \tfrac{1}{6}$
90. $\tfrac{1}{3}\sin^{-1}(3x/2) + C$ 91. $\tfrac{1}{12}\tan^{-1}(3x/4) + C$ 94. $\tfrac{1}{8}\tan^{-1}(x^2/4) + C$
98. $\pi/6$ 101. $\tfrac{1}{3}\sin^{-1}\tfrac{3}{5}$ 105. $\pi/4$

PROBLEMS 9.7

1. (a) $\sin^{-1} x$ (b) $\tan^{-1} x$ (c) $\pi - \cos^{-1} x$ (d) $\cos^{-1}\left(\dfrac{1+x}{2\sqrt{1+x^2}}\right)$

2. (a) $c = 1/\sqrt{a}$ (b) $c = 1/\sqrt{ab}$ (c) $c = 1/\sqrt{a}$ (d) $d = \tfrac{1}{2}\sqrt{a}$

3. $10\sqrt{20}$ ft.   4. 7 rad./min.   5. Approx. $\tfrac{2}{25}$ rad./sec.

6. The distance is a minimum when $\tan\alpha = (\tfrac{4}{5})^{1/3}$.   7. $r = \tfrac{1}{4}c$

8. This requires approximate solutions.   9. Approx. $\tfrac{6}{100}$ shorter.   10. Approx. $20\tfrac{1}{2}$ miles.

EXERCISES 9.8

20. $3\sinh(3x - 1)$ 22. $-3\operatorname{sech} 3x\tanh 3x$ 23. $3\sinh 6x$
24. $2\sinh(4x + 6)$ 26. $2x\cosh 2x + 2x^2\sinh 2x$
28. $-2e^{-2x}$ 30. e^{-2x} 31. e^{2x}
35. $\tanh x - \coth x$ 37. $-\operatorname{csch} x$ 39. $\operatorname{sech} z$
41. $\operatorname{sech} x$ 44. $\tfrac{1}{3}(1 + x^2)$ 46. $\tanh y$
48. 0   49. ∞ 50. $-\tfrac{16}{9}$
51. $\tfrac{1}{3}\sinh 3x + C$ 53. $-\tfrac{1}{2}\coth(2x - 1) + C$ 56. $2\sinh\sqrt{x}$

PROBLEMS 9.8

1. Solve a quadratic equation first.
3. The only method we have is to use power series.
4. $y = (c_1 \sinh bx + c_2 \cosh bx)e^{ax}$
5. (a) 0.3545 (b) 2.18

EXERCISES 9.9

6. $\dfrac{2}{\sqrt{4x^2+1}}$
7. $\dfrac{2}{\sqrt{x^2-2}}$
10. $\dfrac{1}{a^2-x^2}$
12. $\sec x$
15. $-2 \csc 2x$
18. $\cosh^{-1} x$
23. $\cosh^{-1} \dfrac{x}{4} + C$
26. $\sinh^{-1} \dfrac{x}{2} + C$
27. $\tfrac{1}{4} \ln \left| \dfrac{2+x}{2-x} \right| + C$
29. $-\tanh^{-1}(\cos x) + C$
32. $\ln(1 + \sqrt{2})$
36. $\ln(2 + \sqrt{3})$

EXERCISES 10.2

2. $\tfrac{1}{3} \tan^3 x + C$
5. $\tfrac{1}{4} \sec^4 x + C$
6. $\tfrac{1}{5}(\tan 2x)^{5/2} + \tfrac{1}{9}(\tan 2x)^{9/2} + C$
8. $\tfrac{1}{12} \sec^3 4x + C$
10. $-\dfrac{1}{1+\sin x} + C$
13. $-\csc x - \sin x + C$
21. $\tfrac{1}{2} \tan 2x + \tfrac{1}{3} \tan^3 2x + \tfrac{1}{10} \tan^5 2x + C$
24. $\tan^{-1}(\sin x) + C$
26. $x + \cot x - \csc x + C$
29. $\sin x - 4 \sin^3 x + 8 \sin^5 x - \tfrac{32}{7} \sin^7 x + C$
32. $\pi/4 - 2/3$
34. $\tfrac{5}{6}$
37. $\tfrac{1}{2}$
40. $\tfrac{1}{4}$
42. $\tfrac{1}{2}x \sin 2x + \tfrac{1}{4} \cos 2x + C$
43. $2x \sin x + (2 - x^2) \cos x + C$
45. $-x \cot x + \ln \sin x + C$
48. $\tfrac{1}{2}(1 + x^2) \tan^{-1} x - \tfrac{1}{2}x + C$
51. $-\tfrac{1}{5}e^{-x}(2 \cos 2x + \sin 2x) + C$
53. $[(\tfrac{1}{2}x^2 - 1)/4a^2] \sin^{-1} ax + (1/4a)x\sqrt{1 - a^2x^2} + C$
54. $x \operatorname{sech}^{-1} x + \sin^{-1} x + C$
56. $-\dfrac{1}{x} \tan^{-1} x + \ln\left(\dfrac{x}{\sqrt{1+x^2}}\right) + C$

PROBLEMS 10.2

1. (a) $\pi/2$ (b) $\pi/2 - 1$
2. (b) $\tfrac{51}{8} \pi^2$
4. (a) $A = 1/(m+n), B = (n-1)/(m+n)$ (b) $C = 1/(n-1), D = (n-m-2)/(n-1)$
5. (a) 2 (b) Does not converge.
6. (a) $2/\pi$ (b) $\dfrac{2(2 - \sqrt{2})}{3\pi}$
7. Remember the definition of absolute values.
8. Use the "bootstrap" technique of Example 3.
9. (a) $2\sqrt{2} - 2$ (b) $\pi/4$

11. (a) $\int x^n \sin^2 x \, dx = \dfrac{x^{n+1}}{2(n+1)} - \dfrac{x^n}{4}\sin 2x + \dfrac{n}{4}x^{n-1}\sin^2 x - \dfrac{n(n-1)}{4}\int x^{n-2}\sin^2 x \, dx$

14. (a) $\ln(2+\sqrt{3})$ (b) $\ln(1+\sqrt{2})$

EXERCISES 10.3

2. $\sin^{-1}(x/3) + C$

3. $\tfrac{1}{2}x\sqrt{x^2-25} - \tfrac{25}{2}\ln(x+\sqrt{x^2-25}) + C$

5. $\tfrac{1}{27}(9x^2-16)^{3/2}$

8. $-\dfrac{1}{16(4x^2-9)^2} + C$

9. $\tfrac{1}{3}(x^2-9)^{3/2} - 9(x^2-9)^{1/2} + 27\sec^{-1}(x/3) + C$

12. $\dfrac{x}{8(x^2+4)} + \tfrac{1}{16}\tan^{-1}\dfrac{x}{2} + C$

16. $\sqrt{\dfrac{x^2-6}{6x^2}} + C$

17. $-\tfrac{1}{4}x(8-x^2)^{3/2} + x(8-x^2)^{1/2} + 8\sin^{-1}(x/\sqrt{8}) + C$

20. $\tfrac{1}{4}(4+\ln^2 x)^{3/2}\ln x + \tfrac{3}{2}(4+\ln^2 x)^{1/2}\ln x + 6\ln(\ln x + \sqrt{4+\ln^2 x}) + C$

22. $2 + \tfrac{25}{6}\sin^{-1}\tfrac{3}{5}$

25. $-\tfrac{4}{5} + \ln 3$

28. $\tfrac{1}{9}\sqrt{10} - \tfrac{5}{36}$

31. $\tfrac{1}{2}x\sqrt{x^2+a^2} + (a^2/2)\ln(x+\sqrt{x^2+a^2}) + C$

34. $\dfrac{\sqrt{a^2-x^2}}{x} - \sin^{-1}\dfrac{x}{a} + C$

35. $-\dfrac{1}{2}x\sqrt{a^2-x^2} + \dfrac{a^2}{2}\sin^{-1}\dfrac{x}{a} + C$

39. $-(1/x)\sqrt{x^2-a^2} + \ln(x+\sqrt{x^2-a^2}) + C$

41. $\ln(x-2+\sqrt{x^2-4x+20}) + C$

49. $-\sqrt{3-2x-x^2} - \sin^{-1}\left(\dfrac{x+1}{2}\right) + C$

53. $-\sqrt{15+2x-x^2} + C$

54. $\sqrt{x^2+2x-3} - 2\cos^{-1}\dfrac{2}{x+1} + C$

56. $\dfrac{1}{2}\ln(9x^2+24x+25) + \dfrac{1}{9}\tan^{-1}\left(\dfrac{3x+4}{3}\right) + C$

PROBLEMS 10.3

2. Hint: For part (a) let $D = b^2 - 4ac$ and consider the three cases $D < 0$, $D = 0$ and $D > 0$. Each of the integrals (b), (c), (d), (e) and (f) can be reduced to an expression involving the integral (a).

3. Hint: Each integral can be obtained by integration by parts and the "bootstrap" technique. For example, part (a) has the solution

$$-\dfrac{c}{c^2-a^2}\sin(ax+b)\cos(cx+d) + \dfrac{a}{c^2-a^2}\cos(ax+b)\sin(cx+d), \quad a^2 \ne c^2$$

4. Try the substitution $x = \dfrac{a^2}{y}$!

5. (a) $\tfrac{1}{3}(80 - 36\ln 3)$ (b) $18\sqrt{3} - 4\sqrt{7} + 9\ln\left(\dfrac{4+\sqrt{7}}{6+3\sqrt{3}}\right)$

7. (a) $-(1/ax)\sqrt{2ax-x^2} + C$ (c) $-(1/ax)\sqrt{x^2-2ax} + C$

EXERCISES 10.4

2. $3\ln|x-4| - 2\ln|x-2| + C$

4. $\tfrac{1}{2}\ln|x^2-1| + C$

7. $\ln\left|\dfrac{\sqrt{x^2-1}}{x}\right| + C$ 9. $\tfrac{4}{9}\ln\left|\dfrac{x-2}{x+1}\right| - \dfrac{5}{3(x+1)} + C$

11. $-\dfrac{1}{x} - 2\ln|x| + 4\ln|x+1| + C$ 14. $\tfrac{1}{4}\ln\left|\dfrac{x-1}{x+1}\right| + \tfrac{1}{2}\tan^{-1}x + C$

16. $\tfrac{3}{4}\ln(x^4+16) - \tfrac{1}{8}\tan^{-1}(x^2/4) + C$ 20. $\ln|x| - \tfrac{1}{5}\ln|1-x^5| + C$

PROBLEMS 10.4

1. (a) 2 (b) $\pi/32$ (c) $3\pi a^2$ (d) 2π

2. (a) $-\dfrac{\sin^{-1}(x/2)}{2x^2} - \dfrac{\sqrt{4-x^2}}{8x} + C$

   (b) $\tfrac{1}{3}x^3\tan^{-1}(x/3) - \tfrac{1}{2}x^2 + \tfrac{9}{2}\ln(x^2+9) + C$

3. (a) $\dfrac{1}{ad-bc}\ln\left|\dfrac{ax+b}{cx+d}\right| + C$ if $ad \neq bc$.

   (c) $\dfrac{1}{ad-bc}\left[\dfrac{b}{a(ax+b)} + \dfrac{d}{ad-bc}\ln\left|\dfrac{ax+b}{cx+d}\right|\right] + C$ if $ad \neq bc$.

5. $\displaystyle\sum_{k=0}^{\infty}(-1)^k\dfrac{n(n-1)\cdots(n-k+1)}{k!\,n!}\ln|x+k| + C$

6. Hint: Use integration by parts to prove that
$$\int_{-1}^{1}(1-x^2)^n\,dx = \dfrac{2n}{2n+1}\int_{-1}^{1}(1-x^2)^{n-1}\,dx$$

EXERCISES 10.5

1. $\tfrac{2}{15}(3x-10)(x+5)^{3/2} + C$ 4. $\tfrac{2}{105}(15x^2-24x+32)(x+2)^{3/2} + C$

6. $\tfrac{2}{3}(x-6)\sqrt{x+9} + C$ 9. $\sin^{-1}x - \sqrt{1-x^2} + C$

10. $2[1+\sqrt{x}-\ln(1+\sqrt{x})] + C$ 12. $\tfrac{4}{3}(\sqrt{x}-2)\sqrt{1+\sqrt{x}} + C$

16. $3\ln(x^{1/3}+1) + C$ 17. $\ln\left|\dfrac{x}{(1-x^{1/4})^4}\right| + C$

19. $\sqrt{x^2-4} - 2\cos^{-1}(2/x) + C$ 22. $\tfrac{1}{3}(2a^2+6-x^2)\sqrt{a^2+x^2} + C$

24. $\tfrac{3}{8}(x^2+9)(x^2-9)^{1/3} + C$ 27. $\tfrac{1}{2}(\sqrt{x^4-1} - \tan^{-1}\sqrt{x^4-1}) + C$

29. $2(\cos\sqrt{x} + \sqrt{x}\sin\sqrt{x}) + C$ 32. $(x+2)\tan^{-1}\sqrt{x+1} - \sqrt{x+1} + C$

35. $\tfrac{1}{3}\tan^{-1}(\tfrac{1}{3}\tan x) + C$ 37. $\tfrac{1}{2}\ln|\tan(x/2)| - \tfrac{1}{4}\tan^2(x/2) + C$

39. $\tfrac{1}{4}\ln|\sec 2x| + C$ 41. $-\tfrac{1}{2}\ln(1+\cos^2 x) + C$

44. $\tan^{-1}(1+2\tan(x/2)) + C$ 46. $\tfrac{36}{25}$

49. $\tfrac{5648}{35}$ 52. $\tfrac{2}{3}\pi\sqrt{3} - 1$

56. Hint: Consider three cases, $a^2 > b^2$, $a^2 = b^2$ and $a^2 < b^2$.

EXERCISES 10.6

1. $1 - x - \tfrac{1}{3}x^3 - \tfrac{1}{5}x^5 - \tfrac{1}{7}x^7 - \cdots$ 3. $x - \tfrac{1}{3}x^3 + \tfrac{1}{5}x^5 - \tfrac{1}{7}x^7 + \cdots$

5. $\displaystyle\sum_{n=0}^{\infty}(-1)^n\dfrac{1\cdot 3\cdot 5\cdots(2n-1)}{2\cdot 4\cdot 6\cdots 2n}\cdot\dfrac{x^{2n+1}}{2n+1}$ 7. $\displaystyle\sum_{n=0}^{\infty}\dfrac{(-1)^n}{n!}\cdot\dfrac{x^{2n+1}}{2n+1}$

8. $1 + \frac{1}{8}x^4 - \frac{1}{56}x^7 + \frac{1}{160}x^{10} - \cdots$

10. $\sum_{n=1}^{\infty} \frac{(-1)^n}{n!} \cdot \frac{x^n}{r}$

11. 0.493   13. −0.046   15. 2.114   16. 0.492

19. 0.125   20. 0.075   22. 0.081   24. 0.0156

PROBLEMS 10.6

1. 3.14159 26535 89793 23846

2. Expand $(1 - k^2 \sin^2 x)^{-1/2}$ into a binomial series in powers of $k^2 \sin^2 x$ and then integrate term by term.

3. Evaluate the integral two ways.   4. ln 2

EXERCISES 10.7

1. 1.300   3. 0.201   4. $\frac{4}{3}$
5. 0.774   7. 1.089   9. 2.701
11. 0.0044   13. 0.0005   14. 0
15. 0.038   17. 0.00003   19. 0.00025

PROBLEMS 10.7

2. $3\pi a^2/32$ 3. For instance, $h = 0.05$

4. Use Simpson's rule with $h = \pi/4$ and compute the error term.

5. $\frac{1}{4}[2(b + a) \pm (b - a)\sqrt{2}]$ 6. $2 \pm \sqrt{2}$

EXERCISES 11.1

1. (a) $(x, y, z) = (3 + t, 1 - 2t, 2 + t)$ (c) $(x, y, z) = (1 + t, -1 - 2t, 2 - 2t)$
2. (a) $(x, y, z) = (3 - t, -1 + 2t, 2 - t)$ (c) $(x, y, z) = (-1 + 2t, 2 + t, -2 - 3t)$
3. (a) $(x, y, z) = (1 - t, 6t, -2 - 3t)$ (c) $(x, y, z) = (5, 3 + t, -4 + t)$
4. (a) $(5, 4, 1)$ (b) $(\frac{5}{6}, \frac{1}{6}, \frac{5}{6}c + \frac{1}{6}d)$ f $a = \frac{1}{2}c + \frac{5}{6}d$.
5. (a) Parallel to xy plane   (b) Parallel to xz plane
   (c) Parallel to xz plane   (d) $(0, -4, 1), (4, 0, -3), (1, -3, 0)$

EXERCISES 11.2

1. (a)

1. (d)

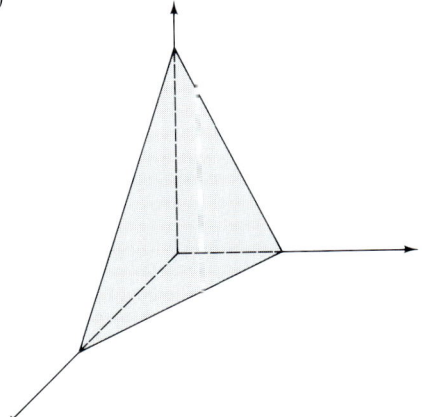

2. (a)

2. (c)

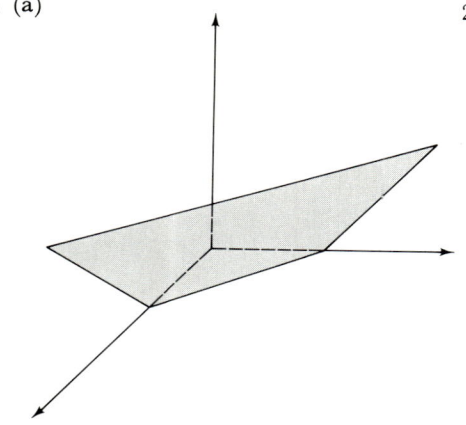
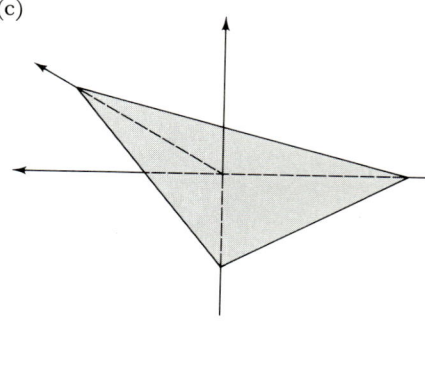

4. (a) $x = 1 - 2s + 2t, y = 1 + s + 2t, z = 2 + s - t$
(d) $x = 4 + s - 2t, y = 1 + 3s - t, z = -2 - 2s - 5t$

5. (a) $x = 1 - 2s + 3t, y = 2 + 2s + 2t, z = -1 + 3s + t$
(b) $x = -1 + 3s + 3t, y = 1 + 2s - t, z = 1 + s - 2t$

PROBLEMS 11.2

For problems 1 and 2 write out parametric equations for the plane and then eliminate the parameters.

3. If the line has parametric equations
$$x = x_0 + at, y = y_0 + bt, z = z_0 + ct$$
and the point is (x_1, y_1, z_1), then equation of the plane is
$$(x - x_0)[b(z_1 - z_0) - c(y_1 - y_0)]$$
$$-(y - y_0)[a(z_1 - z_0) - c(x_1 - x_0)]$$
$$+(z - z_0)[a(y_1 - y_0) - b(x_1 - x_0)] = 0.$$

EXERCISES 11.3

2. $\delta = 1 - \sqrt{x_0^2 + y_0^2}$ 4. Smaller of $\sqrt{x_0^2 + y_0^2} - 1$ and $2 - \sqrt{x_0^2 + y_0^2}$

5. $\delta = \frac{1}{4}(x_0 + 2y_0 - 3)$ 7. Smaller of $\frac{x_0 y_0 - 1}{2|x_0|}$ and $\frac{x_0 y_0 - 1}{2|y_0|}$.

For exercises 9 to 16 replace inequality by equal signs.

PROBLEMS 11.3

1. For Theorem A consider just basis neighborhoods. Theorem B is the set theoretic dual of Theorem A.

2. (a) $\bigcap_{i=1}^{\infty} \{x : -1/i < x < 1 + 1/i\}$ (b) $\bigcup_{i=1}^{\infty} \{x : 1/i \leq x \leq 1 - 1/i\}$

4. A point on the boundary of C cannot be in $R^n - C$.

EXERCISES 11.4

1. $\{(x,y) : x + y \neq 0\}$
3. $\{(x,y) : xy \geqq 0\}$
4. $\{(x,y) : xy > 0\}$
6. $\{(x,y) : x > \tan y\}$
7. $\{(x,y,z) : xyz \neq 0\}$
9. $\{(x,y,z) : |z| \leqq 1\}$

11.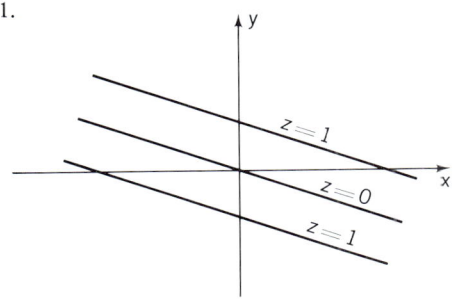

12.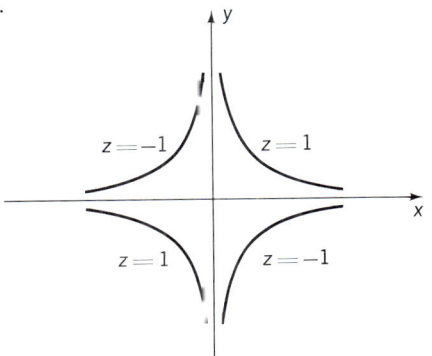

14.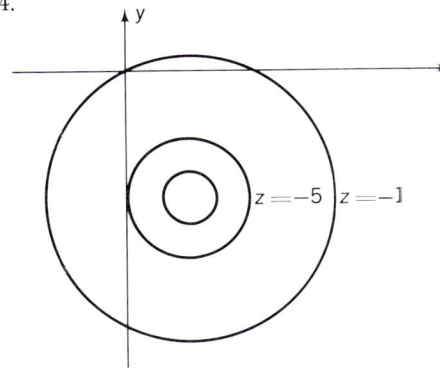

15.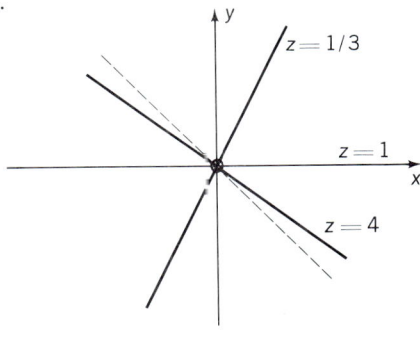

19.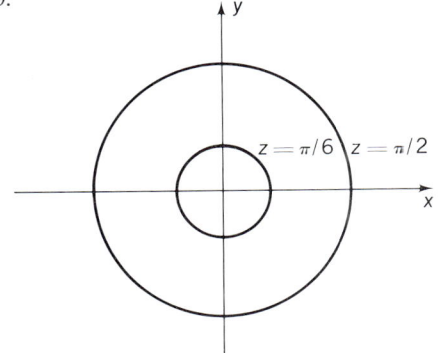

20.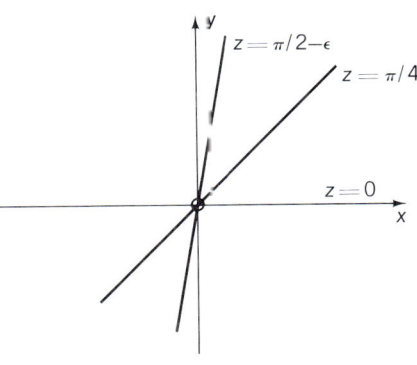

23.

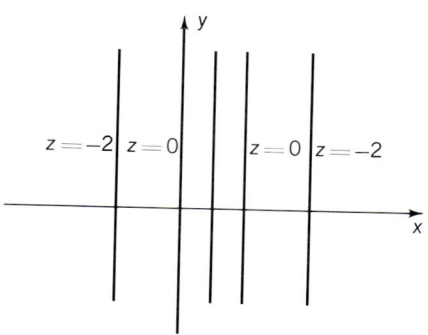
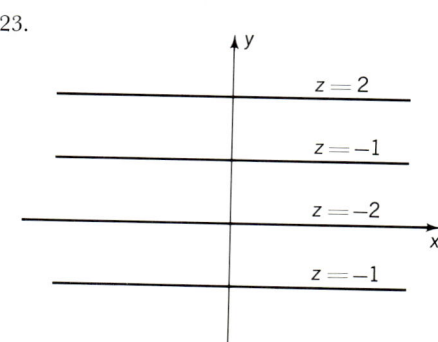

EXERCISES 11.5

1. (a) $\delta = \frac{1}{2}\epsilon$ (d) $\delta = \frac{1}{6}\epsilon$

2. (b) $\lim_{x \to 1} \lim_{y \to -1} \dfrac{x + y}{(x - 1)^2}$ fails to exist.

    (d) $\lim_{y \to -1} \lim_{x \to 1} \dfrac{\sin(x + y)}{(1 + y)^2} = \infty$

3. (a) $\{(x, y) : x^2 + y^2 \leq 4\}$ (d) $\{(x, y) : xy > 1\}$

4. No. $\lim_{x \to 0} \lim_{y \to 0} \dfrac{x^2 - y^2}{x^2 + y^2} \neq \lim_{y \to 0} \lim_{x \to 0} \dfrac{x^2 - y^2}{x^2 + y^2}$.

PROBLEMS 11.5

1. (a) and (c) not continuous.   (b) and (d): choose $\delta = \epsilon$.

3. The function of Problem 2 with $f(0, 0) = 0$.

EXERCISES 11.6

1. $4x - 3y, -3x$ 3. $4x + 4y, 4x - 2y$

6. $10(2x - y)^2, -5(2x - y)^4$ 8. $\dfrac{2y^2}{(x^2 + y^2)^2}, \dfrac{-4xy}{(x^2 + y^2)^2}$

11. $\dfrac{x}{\sqrt{x^2 + y^2}}, \dfrac{y}{\sqrt{x^2 + y^2}}$ 12. $-3x(x^2 + y^2)^{-5/2}, -3y(x^2 + y^2)^{-5/2}$

14. $(\sqrt{x} + \sqrt{y})/\sqrt{x}, (\sqrt{x} + \sqrt{y})/\sqrt{y}$ 17. $-x + 2x \ln|y/x|, x^2/y$

20. $\dfrac{1}{y} e^{x/y}, -\dfrac{x}{y^2} e^{x/y}$ 23. $\cos(x + y), \cos(x + y)$

25. $2 \cos 2x \cos 3y, -3 \sin 2x \sin 3y$ 27. $e^x \sin y, e^x \cos y$

29. $\dfrac{1}{y} \cos \dfrac{x}{y} - \dfrac{1}{y}, -\dfrac{x}{y^2} \cos \dfrac{x}{y} + \dfrac{1}{y}$ 32. $\dfrac{y}{x^2 + y^2}, \dfrac{x}{x^2 + y^2}$

35. $2x, 2y, 2z$ 37. $2xy^2z^2, 2x^2yz^2, 2x^2y^2z$

39. $-yz \sin(xyz), -xz \sin(xyz), -xy \sin(xyz)$

42. $4(x^2 + xy)^2(14x^2 + 14xy + 3y^2), 4(x^2 + xy)^2(x^2 + 13xy), 12x^2(x^2 + xy)^2$

44. $1/x^2, 0, -1/y^2$ 45. $\dfrac{x^2-y^2}{(x^2+y^2)^2}, \dfrac{-2xy}{(x^2+y^2)^2}, \dfrac{y^2-x^2}{(x^2+y^2)^2}$

49. All second partials equal $-\dfrac{2(x+y)}{[1+(x+y)^2]^2}$.

PROBLEMS 11.6

2. Differentiate the defining equation with respect to α and then set $\alpha = 1$. (This requires a chain rule of differentiation.)

3. This can be done very easily, i.e. don't make it difficult.

Problems 4, 5 and 6 are straightforward computation.

8. $f_{12}(0,0) = -1, f_{21}(0,0) = 1$

EXERCISES 11.7

1. $z = 4x + y - 4$ 2. $z = 3(x-1)$
4. $z = 2x + 2y - 2$ 6. $z = -x - \tfrac{1}{2}y + \tfrac{9}{2}$
7. $z = x + y - 3$ 10. $z = 2(y-1)$

EXERCISES 11.8

1. $6xh - 2yk + 3h^2 + k^2$ 3. $\epsilon_1 = \dfrac{(h+k)y}{(x+y)^3 + (x+y)^2(h+k)}$

5. $\dfrac{(AD-BC)(yh-xk)}{(Cx+By)^2}$ 7. $(h+k)\cos(x+y)$

10. $-\dfrac{h}{x} + \dfrac{k}{y}$ 11. $2(hx - ky)e^{x^2-y^2}$

13. $h\ln(x^2+y^2) + \dfrac{2(x^2h+xyk)}{x^2+y^2}$ 15. $e^x(h\cos y - k\sin y)$

17. $\dfrac{yh - xk}{x^2 + y^2}$ 20. $2xzh - z^2k + (x^2 - 2yz)l$

22. $\dfrac{xh + yk + zl}{x^2 + y^2 + z^2}$ 23. $\dfrac{yzh + xzk + xyl}{1 + x^2y^2z^2}$

25. 5.02   26. 0.349   27. 0.0369   28. 0.466

EXERCISES 11.9

1. $-4t^{-3}$ 3. 0   4. $-1/(1+t^2)$ 5. $2(e^{2t} - e^{-2t})$

8. $\dfrac{4te^s + 6se^t}{\sqrt{1-(2x+3y)^2}}, \dfrac{4e^s + 3s^2e^t}{\sqrt{1-(2x+3y)^2}}$

10. $\dfrac{\partial w}{\partial s} = 2(xy + xz + yz)[2xt + (x + 2y + z)(2s + t)]$

12. $\dfrac{\partial w}{\partial s} = \dfrac{2(s+t)}{s^2 + t^2 + 2s^2t^2}$

PROBLEMS 11.9

2. $a\, Df$, $b\, Df$

5. $\left(\dfrac{\partial f}{\partial r}\right)^2 + \dfrac{1}{r^2}\left(\dfrac{\partial f}{\partial \theta}\right)^2$

6. $\left(\dfrac{\partial f}{\partial \rho}\right)^2 + \dfrac{1}{\rho^2}\left(\dfrac{\partial f}{\partial \phi}\right)^2 + \dfrac{1}{\rho^2 \sin^2 \phi}\left(\dfrac{\partial f}{\partial \theta}\right)^2$

7. $\dfrac{\partial z}{\partial x} = (y^{xy})(x^{y-1})y \ln y,\ \dfrac{\partial z}{\partial y} = (y^{xy})(x^y)(\ln x \ln y + 1/y)$

EXERCISES 11.10

1. $-2x + 3x^2 - 4x(y-1) + x^3 + 3x^2(y-1) - 2x(y-1)^2$

2. $-3 + (x-1) - 10(y-1) + 6(x-1)^2 - 10(x-1)(y-1) - 5(y-1)^2 + 2(x-1)^3 - 5(x-1)(y-1)^2$

4. $1 + 8(x-1) - 8(y-1) + 8(x-1)^2 - 12(y-1)^2 + 4(x-1)^3 - 8(y-1)^3 + (x-1)^4 - 2(y-1)^4$

7. $1 + \tfrac{1}{2}x + \tfrac{1}{2}y - \tfrac{1}{8}x^2 - \tfrac{1}{4}xy - \tfrac{1}{8}y^2 + \cdots$

9. $(x+y) - \tfrac{1}{6}(x+y)^3 + \cdots$

11. $y + xy - \tfrac{1}{2}y^2 + \tfrac{1}{2}x^2y - \tfrac{1}{2}xy^2 + \tfrac{1}{3}y^3 + \cdots$

14. $xy - \tfrac{1}{2}x^2y^2 + \cdots$

15. $1 + xy + \tfrac{1}{2}x^2y^2 + \cdots$

EXERCISES 11.11

2. minimax at $(-1, 2)$

3. no critical points

5. minimum at $(2, -1)$

7. min. at $(10, 3)$, minimax at $(\tfrac{10}{3}, -\tfrac{1}{3})$

9. minimum at $(6, 4)$

12. minimum at $(\tfrac{1}{4}, 16)$

14. no critical points

15. test fails

18. minimum at $(2, -1, 2)$

20. minimax at $(-2, -4, -8)$

PROBLEMS 11.11

1. $\tfrac{13}{5}$

2. 2

3. $\sqrt{14},\ 3\sqrt{14}$

4. $\tfrac{16}{17}\sqrt{17}$

5. $\tfrac{9}{2}x_0 y_0 z_0$

7. The points form an equilateral triangle.

10. $\dfrac{\sqrt{3}}{18}s^{3/2}$

11. $6 \times 9 \times 36$

12. $\$59.40$

14. $\sqrt{\sum_{i=1}^{n} a_i^2}$

15. $\$56,\ \$28$

16. $P_1 = 500,\ P_2 = 350;\ P_1 + P_2 = 1866\tfrac{2}{3}$.

EXERCISES 11.12

1. $-\sqrt{2}$

2. $c^2/(a^2 + b^2)$

5. -288

6. $-4\sqrt{14}$

8. 50

11. 9

12. 4374

14. $\tfrac{134}{75}$

16. $\tfrac{58}{30}$

17. $\tfrac{17}{23}$

19. $\tfrac{16}{5}$

20. $2\sqrt{13}$

22. 1

23. $\sqrt{10}$

24. $\sqrt{6 - 3\sqrt{3}}$

25. $\tfrac{1}{2}\sqrt{10}$

PROBLEMS 11.12

1. $\dfrac{abcD^2}{bcA^2 + acB^2 + abC^2}$
2. $a^a b^b c^c \left(\dfrac{d}{a+b+c}\right)^{a+b+c}$

4. Determine the minimum of $\dfrac{x^a}{a} + \dfrac{y^b}{b} - xy$.

9. $y/x = 1 + \tfrac{1}{3}\sqrt{3}$
10. The centroid.
12. $2 \times 2 \times 1$.

EXERCISES 11.13

2. $\tfrac{1}{2}x^2 - 3xy + y^2 + 4x + y = C$
4. $2x^2 + xy^2 + y^3 = C$
5. $2x^3y + x^2 - \tfrac{3}{2}y^2 = C$
7. $\ln(x-y)\sqrt{x^2 + y^2} = C$
9. $xy + e^x + e^y = C$
10. $ye^x + x^2 + 3y = C$
12. $xy + \tfrac{1}{2}x^2 \ln y = C$
14. $x + y \sin x = C$
16. $x^2 \sin y + x \cos y = C$
17. $y^2 e^{x/y} = C$
18. $y + \sin xy = C$
19. $x^2 \sin y + e^x \cos y = C$

PROBLEMS 11.13

6. (a) $x + \tfrac{1}{3}x^3y = Cy$ (b) $x^2y + y^2 = C$

EXERCISES 12.1

1. 33
2. 8
4. $-\tfrac{13}{5}$
6. $\tfrac{128}{15}$
7. $\tfrac{1}{3}$
8. $\pi/2$
10. π
11. $\displaystyle\int_0^2 \int_0^x$
12. $\displaystyle\int_0^4 \int_0^{\sqrt{x}}$
13. $\displaystyle\int_0^a \int_y^a$
15. $\displaystyle\int_1^2 \int_{\sqrt{x}}^x + \int_2^4 \int_{\sqrt{x}}^2$

PROBLEMS 12.1

1. Show that two derivatives with respect to a are equal.
3. $\ln(a/b)$. (Consider an iterated integral of e^{-xy} and interchange the order of integration.)
8. Prove that $D \displaystyle\int_0^p f(x+t)\,dt = 0$
10. (a) $\left(-\dfrac{x^2}{a} + \dfrac{2}{a^3}\right)\cos ax + \dfrac{2x}{a^2}\sin ax$ (c) $\left(\dfrac{6x}{a^3} - \dfrac{x^3}{a}\right)\cos ax + \left(\dfrac{3x^2}{a^2} - \dfrac{6}{a^4}\right)\sin ax$
12. The integrand is not continuous in the region of integration.

EXERCISES 12.2

1. 6
3. 48
4. 6
6. $\tfrac{643}{5}$
7. $\tfrac{57}{10}$
9. $(70 + 32\sqrt{2})/105$
11. 42
12. $\tfrac{52}{3}$
14. $\tfrac{1}{6}a^3c$
16. $\tfrac{9}{2}$
18. $\tfrac{11}{84}$
20. $e - 16 + 8\ln 8$
22. π^2
23. $(\pi^3)/105$
25. 4

EXERCISES 12.4

2. $\pi/24$
4. $\pi/3$
5. $(9\sqrt{3})/8$
7. $\frac{3}{2}\pi a^2$
8. $16\pi - 24\sqrt{3}$
9. 6π
10. $(\pi - 1)a^2$
11. $2a^2(1 + \pi/4)$
13. $18\pi + \frac{81}{2}\sqrt{3}$
14. a^2
15. $\pi/10$
18. $\frac{1}{2}(\sqrt{2} - 1)$
19. $\pi/8 + \frac{1}{4}$
21. $(8\sqrt{3} - 9)/108$
24. $3\pi/8$
26. $a^2 + a - 2a\sqrt{a}$
28. $3\pi - \frac{9}{2}\sqrt{3}$
29. $4\sqrt{2}$
31. 8π
33. $\frac{4}{3}\pi a^3$
35. $\frac{2\sqrt{2}}{3}a^3$
36. $\frac{5}{12}a^4$
37. $\frac{3}{2}\pi a^3$
38. $\frac{\pi}{3}a^3$
39. $\frac{\pi}{16}a^5$

PROBLEMS 12.4

1. $\pi/4$
2. $\frac{246}{5}\sqrt{3}$
5. (b) $\sqrt{\pi}/2k^3$
5. (c) $\sqrt{\pi}/2k^3$
6. $c = \sqrt{\frac{3}{2}}$

EXERCISES 12.5

1. $(\frac{2}{3}, \frac{2}{3})$
2. $(1, \frac{1}{2})$
4. $(\frac{5}{16}, \frac{4}{7})$
6. $(\frac{365}{144}, \frac{685}{144})$
8. $\left(\frac{153}{64} + \frac{3\sqrt{2}}{512}\ln(17 - 12\sqrt{2}), 0\right)$
10. $\left(\frac{e^2 - 1}{4e - 8}, \frac{12 - 4e}{3e - 6}\right)$
11. $30k$
12. $100k$
14. $(\frac{5}{8}a, \frac{5}{8}a)$
15. $(\frac{128}{429}a, \frac{128}{429}a)$

EXERCISES 12.6

The radii of gyration are:

1. $b/\sqrt{3}$
2. $b/\sqrt{2}$
3. $t/\sqrt{6}$
4. $b\sqrt{30}/10$
5. $a/2$
6. 1
7. $\frac{1}{2}$
8. $\sqrt{\frac{45}{15 - 16\ln 2}}$
9. $\sqrt{\frac{8}{7}}$
10. $\frac{4}{5}\sqrt{3}$

PROBLEMS 12.6

2. (a) $b/\sqrt{3}$ (b) $ab/\sqrt{6}$
5. $\left[\frac{r^4 \cos^{-1}(h/r) - \frac{1}{3}(r^2 + 2h^2)\sqrt{r^2 - h^2}}{r^2 \cos^{-1}(h/r) - h\sqrt{r^2 - h^2}}\right]^{1/2}$

EXERCISES 12.7

1. $\frac{1}{8}$
3. $\frac{1}{6}$
4. $\frac{64}{3}$
6. $(\pi/16)a^4$
7. $\frac{8}{3}$
9. $\frac{1}{6}$
10. $\frac{99}{8}\pi$
12. $\frac{9472}{105}$
14. $\frac{4}{3}(2\pi + 9\sqrt{3} - 16)$
15. 64
17. $\frac{81}{4}\pi$
18. $(4\pi/3)a^3$
19. $(4\pi/3)abc$
22. 2
24. $\frac{2048}{105}$
25. $\frac{16}{27}$
26. $(abc/24)(a + b + c)$
28. 0
29. $\frac{832}{105}$
30. $\frac{1024\sqrt{2}}{15}$

EXERCISES 12.8

1. $\frac{32}{3}\pi$
2. $\frac{1}{12}$
3. $\frac{32}{3}\pi$
4. $(\pi/6)a^2(a - b)$
5. $\frac{16}{5}\pi$
6. $\frac{16}{3}\pi a^3$
8. $\frac{1}{3}(16\sqrt{2} - 14)$
9. $(4\pi/3)[a^3 - (a^2 - b^2)^{3/2}]$

PROBLEMS 12.8

1. $2\pi^2 a^2 b$

EXERCISES 12.9

1. $\dfrac{\pi}{3} a^3 (2 - \sqrt{3})$
2. $\dfrac{\pi\sqrt{3}}{6} a^3$
5. $\dfrac{\pi}{9} a^3$
6. $\dfrac{7\pi}{3} a^3$
7. $\dfrac{40\pi}{3}$
8. $\dfrac{\pi}{9} a^3$
9. $\dfrac{\pi^2}{4} a^3$
10. $\dfrac{\pi^2}{24} a^3$
11. $\tfrac{1}{2}\sqrt{1 - \tfrac{2}{5}\sqrt{3}}$
13. $a\sqrt{\tfrac{2}{5}}$
15. $a/\sqrt{10}$
17. $\sqrt{34}/5$

PROBLEMS 12.9

1. $\dfrac{\pi}{3a}(b^3 - a^3)(a - c) + \dfrac{2\pi}{3}(b - a)\left[\dfrac{b^2 c}{a} - \dfrac{1}{2}(a + b)\right]$
2. $\dfrac{\pi c^2}{3b^2}(\varepsilon b^3 - 2a^3 - ab^2)$

3. The radius of gyration is $\left[\dfrac{2(b^5 - a^5)}{5(b^3 - a^3)}\right]^{1/2}$

EXERCISES 12.10

1. $(0, 0, 0)$
3. $(0, \tfrac{1}{4}, \tfrac{9}{20})$
4. $(\tfrac{112}{45}, 0, 0)$
6. $(2, \tfrac{3}{2}, 1)$
8. $(0, \tfrac{8}{7}, 0)$
9. $(4, 0, 0)$
12. $a\sqrt{\tfrac{2}{5}}$
13. $a\sqrt{\tfrac{2}{3}}$
15. $a\sqrt{\tfrac{3}{2}}$
16. $\tfrac{1}{2}\sqrt{a^2 + \tfrac{1}{3}h^2}$
17. $a\sqrt{\tfrac{2}{5}}$
18. $\sqrt{b^2 + \tfrac{3}{4}a^2}$

PROBLEMS 12.10

1. $a\sqrt{\tfrac{3}{2}}$
2. $\dfrac{4\pi k a^{5-m}}{3(5 - m)}$ if $m > 3$.

EXERCISES 13.1

2. (a) $(\tfrac{1}{3}, \tfrac{2}{3}, \tfrac{2}{3})$ (b) $(1/\sqrt{3}, 1/\sqrt{3}, 1/\sqrt{3})$
   (d) $(\tfrac{6}{7}, \tfrac{2}{7}, -\tfrac{3}{7})$ (f) $(-\tfrac{3}{13}, -\tfrac{4}{13}, \tfrac{12}{13})$

3. (b) $\dfrac{1}{\sqrt{3}}(\mathbf{i} + \mathbf{j} + \mathbf{k})$ (c) $\dfrac{1}{\sqrt{22}}(2\mathbf{i} - 3\mathbf{j} + 3\mathbf{k})$
   (d) $\dfrac{1}{7}(6\mathbf{i} + 2\mathbf{j} - 3\mathbf{k})$ (e) $\dfrac{1}{5\sqrt{5}}(8\mathbf{i} + 5\mathbf{j} - 6\mathbf{k})$

4. (a) $\mathbf{i} + 2\mathbf{j} - 2\mathbf{k}$ (d) $-4\mathbf{i} + 3\mathbf{j} + 12\mathbf{k}$
5. (a) $\beta = \pi/6$ (b) $\alpha = \pi/6$
6. (a) $\gamma = \pi/6$ (b) $\beta = \pi/4$

PROBLEMS 13.1

3. $2(x_2 - x_1)x + 2(y_2 - y_1)y + 2(z_2 - z_1)z = x_2^2 + y_2^2 + z_2^2 - x_1^2 - y_1^2 - z_1^2$
5. One vertex is $(12, 12, 12)$.

EXERCISES 13.2

1. (a) -6 (b) -7 (c) 3 (d) $-8\mathbf{i} - 16\mathbf{j} + 8\mathbf{k}$
   (f) $2\mathbf{i} - 11\mathbf{j} + 4\mathbf{k}$ (g) 24 (h) 24 (k) $-16\mathbf{i} - 8\mathbf{j} - 32\mathbf{k}$

2. (b) $\dfrac{1}{\sqrt{46}}(\mathbf{i} - 6\mathbf{j} + 3\mathbf{k})$ (c) $\dfrac{1}{\sqrt{22}}(3\mathbf{i} - 2\mathbf{j} + 3\mathbf{k})$

3. (b) $2\mathbf{i} - 2\mathbf{j} + \mathbf{k}$ (d) $\dfrac{1}{a}\mathbf{i} + \dfrac{1}{b}\mathbf{j} + \dfrac{1}{c}\mathbf{k}$

4. (a) $\tfrac{1}{2}\sqrt{62}$ (c) $13\sqrt{2}$ (d) $\tfrac{1}{2}\sqrt{a^2b^2 + a^2c^2 + b^2c^2}$

5. $\dfrac{\pi}{4}$, $\cos^{-1}\dfrac{3}{\sqrt{10}}$ and $\cos^{-1}\dfrac{2}{\sqrt{5}}$.

PROBLEMS 13.2

4. $\mathbf{u} = \dfrac{\mathbf{u} \cdot \mathbf{v}}{\|\mathbf{v}\|^2}\mathbf{v} + \dfrac{\mathbf{u} \cdot (\mathbf{u} \times \mathbf{v})}{\|\mathbf{u} \times \mathbf{v}\|^2}(\mathbf{u} \times \mathbf{v})$

11. (a) $\cos^{-1}\dfrac{\sqrt{3}}{3}$ (b) $\cos^{-1}\dfrac{\sqrt{6}}{6}$ (c) $\dfrac{\pi}{3}$

12. $\dfrac{10}{3}\sqrt{33}$.

EXERCISES 13.3

1. (b) $(x-3)/-1 = (y+1)/3 = (z-2)/4$
   (d) $(x+1)/3 = (y+1)/4 = (z-2)/5$

2. (a) $x - 3 = y - 1 = z - 2$
   (d) $(x+2)/6 = (y+3)/8$, $z = -1$

3. (b) parallel
   (d) coincident

4. (a) $(0, -\tfrac{11}{4}, \tfrac{5}{2})$, $(11, 0, -3)$, $(5, -\tfrac{3}{2}, 0)$

5. (b) $2x - 2y - z + 13 = 0$
   (c) $x + y + z = 2$

6. (a) $8x - 10y + 11z = 0$
   (d) $36x - 42y + 31z = 172$

7. (b) $4x + 5y + 2z + 15 = 0$
   (d) $(x-4)/-4 = (y+2)/4 = (z+3)/-5$

8. (a) $10/\sqrt{29}$
   (b) $\cos^{-1}(\tfrac{1}{3}\sqrt{170/26})$

9. $(x+1)/4 = (y-2)/-5 = (z-1)/-1$

PROBLEMS 13.3

1. $\{(x, y, z) : 2x + z = 1\}$

4. In determinant form the equation is

$$\begin{vmatrix} x - x_0 & y - y_0 & z - z_0 \\ x_1 - x_0 & y_1 - y_0 & z_1 - z_0 \\ a & b & c \end{vmatrix} = 0$$

6. $x + y + z = \tfrac{3}{2}a$

EXERCISES 13.4

We give the intercepts, the symmetries and the restrictions on the variables in this order:

1. $(0, 0)$; x axis; $x \geq 0$
2. $(0, 2)$; y axis; $y \geq 2$
4. $(\pm 4, 0)$; all symmetries; $|x| \geq 4$
6. $(0, 0)$, $(4, 0)$; x axis; $x \leq 0$ or $x \geq 4$
7. $(\frac{5}{2}, 0)$, $(0, -\frac{5}{2})$; none; $x \neq -2, y \neq 2$
9. $(1, 0)$, $(2, 0)$, $(0, 2)$; none; $x \neq -1, |y + 5| \geq 2\sqrt{6}$
12. $(\pm 2, 0)$; y axis; $x \neq 0, y > 2$
13. None; all symmetries; $xy \neq 0$
15. $(0, 0)$; 0 symmetry; $y \neq \pm 2$
18. $(\pm 2, 0)$, $(0, \pm 2)$; all symmetries; $|x| \geq 2$, $-1 < x < 1$, $|y| \geq 2$, $-1 < y < 1$.
19. $(1, 0)$, $(3, 0)$, $(-6, 0)$, $(0, \pm\sqrt{6})$; x axis; $-6 \leq x \leq 1, x \geq 3$
20. $(4, 0)$; x axis; $x < -4, x \geq 4$.

For the surfaces we give the intercepts; the symmetries; the traces; cross-sections; restrictions on the variables:

21. $(0, 0, 0)$; xz and yz plane; no traces; circles parallel to the xy plane; $x^2 + y^2 \leq 1, 0 \leq z \leq 2$. (A sphere.)
23. $(\pm 4, 0, 0)$, $(0, \pm 2, 0)$, $(0, 0, \pm 2)$; all symmetries; $x^2 + 4y^2 = 16$, $x^2 + 4z^2 = 16$, $y^2 + z^2 = 4$; circles parallel to the xy plane; $|x| \leq 4, |y| \leq 2, |z| \leq 2$. (An ellipsoid.)
25. $(0, 0, 0)$, $(2, 0, 0)$; xz and xy plane; $x^2 + y^2 = 2x$, $(x^2 - 2x = 0, y = 0)$; circles parallel to the xy plane; $0 \leq x \leq 2, |y| \leq 1$. (A circular cylinder.)
28. $(1, 0, 0)$, $(0, \pm 2, 0)$, $(0, 0, \pm 2)$; xz and yz plane; $4x + y^2 = 4$, $4x + z^2 = 4$, $y^2 + z^2 = 4$; circles parallel to yz plane; $x \leq 1$. (A paraboloid.)
30. $(\pm 2, 0, 0)$, $(0, \pm 2, 0)$; xy plane and origin; $x - y = \pm 2$, $x^2 - 4z^2 = 4$, $y^2 - 4z^2 = 4$; cross sections give us no information as yet; $(z - y)^2 \geq 4$.
33. None; xz plane; $x^2 z = 8$, $2xy^2 + 8 = 0$; no useful cross-sections; $x \neq 0$, $(x^2 z - 8)/2x \geq 0$.
35. $(0, 0, 0)$, $(0, \pm 2, 0)$; xz and yz plane; $z = -x^2$, $x = 0 = z$; $4z/(y^2 - 4) \geq 0$.
37. None; symmetric to each axis; none; cross-sections parallel to any coordinate plane are hyperbolas; since $xyz > 0$, the surface appears in only 4 octants.
39. None; all symmetries; $x^2 y^2 = 16$; cross-sections parallel to the yz plane are hyperbolas; $y^2 > z^2, x \neq 0, x^2 y^2 \geq 16$.

EXERCISES 13.5

1. $5\sqrt{2}$; $\tan^{-1} 4$
3. $-\frac{1}{2} + 4\sqrt{3}$; $\alpha = \tan^{-1}(-\frac{1}{8})$
4. $-2 + 4\sqrt{3}$; $\tan^{-1}(-\frac{1}{2})$
6. 1; $-\pi/4$
7. $\frac{1}{2}$; 0
9. $\frac{14}{3}$
11. 2
13. $\frac{8}{27}$
15. $-2\sqrt{2}$
16. $(16 - \pi)/2\sqrt{29}$

We give the equation of the tangent plane:

17. $x + y + z\sqrt{2} = 8$
19. $7x - 11y + 3z = 0$
20. $4x + 2y - z = 9$

22. $\pi y + 2z = \pi$ 23. $4x - 4y - z = 12$ 24. $6x + 4y + 3z = 348$

PROBLEMS 13.5

1. $f \equiv 0$ 2. The two agree.   8. $\mathbf{i} - \mathbf{j}\sqrt{3}$ 9. $\dfrac{\mathbf{p} \cdot \nabla f}{\|\nabla f\|^2}$

EXERCISES 13.6

2. Focus $(0, \frac{1}{8})$; directrix $\{(x,y) : y = -\frac{1}{8}\}$
3. Center $(0, 3)$; foci $(0, 3 \pm \sqrt{5})$; $\{(x,y) : y = 3 \pm 9/\sqrt{5}\}$
5. Center $(0, 5)$; foci $(0, 5 \pm 3)$; $\{(x,y) : y = 5 \pm \frac{25}{3}\}$
7. Center $(0, -1)$; foci $(0, -1 \pm \sqrt{5})$; directrix $\{(x,y) : y = -1 \pm 1/\sqrt{5}\}$; and asymptotes $\{(x,y) : y = -1 \pm \frac{1}{2}x\}$
8. Center $(0, -2)$; foci $(0, -2 \pm \sqrt{29})$; directrix $\{(x,y) : y = -2 \pm 4/\sqrt{29}\}$; and asymptotes $\{(x,y) : y = -2 \pm \frac{2}{5}x\}$
9. $x^2 = 8y$ 11. $(x-y)^2 + 12(x+y) = 60$
12. $x^2/16 + y^2/32 = 1$ 14. $9(x-2)^2 + 5(y+5)^2 = 180$
15. $25(y + \frac{16}{5})^2 - 20x^2 = 576$ 16. $3(x-7)^2 - (y-2)^2 = 12$

PROBLEMS 13.6

6. The sum of the focal radii equals the length of the major axis.
8. πab 9. $\frac{4}{3}\pi ab^2$ 11. (b) $2ab$ 12. $x + y = a$
15. The constant area is ab.

EXERCISES 13.7

1. Parabolic cylinder   3. Elliptical cylinder
6. Ellipsoid of revolution   7. Elliptical cone
9. Circular paraboloid   12. Elliptical paraboloid
17. Circular cone   18. Hyperboloid of one sheet
19. and 20. Hyperboloids of two sheets

PROBLEMS 13.7

1. A sphere   2. An ellipsoid   3. A paraboloid
4. A quadric   5. The conic sections   6. $(4\pi/3)abc$

EXERCISES 13.8

We give the equation in rectangular coordinate form.

1. $3x - 2y = 5$ 2. $(x-2)/3 = (y-1)/-2 = (z-2)/1$
3. $9y = 7 - 4x - 2x^2$ 5. $x^3 = y^2$
6. $y = x^2$, $z = x^3$ 8. $2y = x - 2$, $x = 4\sin 2z$
11. $1 + x^2 = y^2$ 13. $y = 1 - 2x^2$

14. $y = 2\sin(x - 4)$, $z = 1 - \frac{1}{2}y^2$
16. $x = 2 - 3y^2 + y^3$, $z = \tan(y - 1)$
19. $(x/4)^{2/3} + (y/4)^{2/3} = 1$
21. A line parallel to the y axis
24. A parabola in the plane $\{(x, y, z) : y + z = 6\}$
26. An elliptical helix
29. A curve called a cycloid
31. A plane
33. A paraboloid
36. A cone
38. A cone
39. A sphere

PROBLEMS 13.8

3. Given (x_0, y_0, z_0) on the surface, choose $a = x_0$, $b = y_0$, $c = z_0$.
4. The paraboloid has equation $x^2 + y^2 = z$.

EXERCISES 13.9

We give either the velocity or the acceleration vector:
1. $\mathbf{a} = -2\mathbf{j}$
3. $\mathbf{a} = \pi^2 \mathbf{p}$
4. $\mathbf{a} = \mathbf{p}$
6. $\mathbf{a} = 0$
7. $\mathbf{a} = -2\mathbf{k}$
9. $\mathbf{v} = 6(\mathbf{i} + t\sqrt{2}\mathbf{j} + t^2\mathbf{k})$
11. $\mathbf{a} = 2(-t\mathbf{i} + \mathbf{j} + t\mathbf{k})$
12. $\mathbf{a} = 2\mathbf{i} - 6\mathbf{j} + (t - 2)e^{-t}\mathbf{k}$
13. $\mathbf{v} = e^t(\mathbf{i} + \mathbf{j} + (\sin t + \cos t)\mathbf{k})$
14. $\mathbf{a} = e^t\mathbf{i} + e^{-t}\mathbf{k}$

We give the equation of the tangent plane:
16. $x + y - z = 1$
17. $x + y - 2z = 1$
18. $3y - z = 4$
20. $2x - \pi y + \pi z = \pi + \pi^2$
22. $x = 1$
24. none
25. $x = 3$

PROBLEMS 13.9

2. $\tan 2\alpha = \cot \beta$

EXERCISES 13.10

1. $\sqrt{45}$; $\mathbf{T} = \dfrac{1}{3\sqrt{5}}(2\mathbf{i} + 4\mathbf{j} - 5\mathbf{k})$, $\mathbf{N} = 0 = \mathbf{B}$

2. $\frac{75}{32} + 5 \ln \frac{3}{2}$; $\mathbf{T} = \dfrac{1}{5\sqrt{1 + t^2}}(-3\mathbf{i} + 4\mathbf{j} - 5t\mathbf{k})$

$\mathbf{N} = \dfrac{1}{5\sqrt{1 + t^2}}(3t\mathbf{i} - 4t\mathbf{j} + 5\mathbf{k})$

$\mathbf{B} = \frac{4}{5}\mathbf{i} + \frac{3}{5}\mathbf{j}$

4. $\frac{87}{2}$; $\mathbf{T} = \dfrac{2}{2 + 9t^2}(\mathbf{i} + 3t\mathbf{j} + \frac{9}{2}t^2\mathbf{k})$

$\mathbf{N} = \dfrac{2}{2 + 9t^2}(-3t\mathbf{i} + (1 - \frac{9}{2}t^2)\mathbf{j} + 3t\mathbf{k})$

6. $2\pi\sqrt{1 + a^2}$; $\mathbf{T} = \dfrac{1}{\sqrt{1 + a^2}}(-a \sin t\,\mathbf{i} + a\cos t\,\mathbf{j} + \mathbf{k})$

$\mathbf{B} = \dfrac{1}{\sqrt{1 + a^2}}(\sin t\,\mathbf{i} - \cos t\,\mathbf{j} + a\mathbf{k})$

8. $\dfrac{\pi}{8}\sqrt{8+\pi^2} + \ln\left(\dfrac{\pi}{2\sqrt{2}} + \dfrac{1}{2\sqrt{2}}\sqrt{8+\pi^2}\right)$

$$\mathbf{T} = \dfrac{1}{\sqrt{2+t^2}}[(\cos t - t\sin t)\mathbf{i} + (\sin t + t\cos t)\mathbf{j} + \mathbf{k}]$$

9. $\dfrac{4\sqrt{2}}{3}$; $\mathbf{N} = \dfrac{1-t^2}{1+t^2}\mathbf{i} - \dfrac{2t}{1+t^2}\mathbf{k}$

PROBLEMS 13.10

4. A helix

On plane motion

2. Approx. 29 rpm   4. $\tfrac{11}{8}$ g   6. $-1;\ \tfrac{1}{4}\sqrt{5}$ sec.

EXERCISES 13.11

1. $\mathbf{r}(t) = \dfrac{1}{1+t^2}(-4t\mathbf{i} + 4\mathbf{j})$ 3. $\kappa(t) = \sqrt{2}(1-t^2)(1+6t^2+t^4)^{-3/2}$

5. $\kappa(t) = 2(e^{2t} + e^{-2t})^{-3/2}$ 6. $\rho(t) = \tfrac{1}{2}(1 + 4\sin^2 t)^{3/2}$

7. $\kappa(t) = \dfrac{1}{\sqrt{2}} e^t$ 9. $\rho(t) = |at|$

11. A straight line, whence $\mathbf{N} = \mathbf{0} = \mathbf{B}$

13. $\mathbf{T} = \dfrac{1}{\sqrt{t+5}}(\mathbf{i} + 2\mathbf{j} + t^{1/2}\mathbf{k})$

$\mathbf{N} = \sqrt{\dfrac{t}{5(t+5)}}(-\mathbf{i} - 2\mathbf{j} + 5t^{-1/2}\mathbf{k})$

$\mathbf{B} = \dfrac{1}{\sqrt{5}}(2\mathbf{i} - \mathbf{j})$

15. $\mathbf{T} = \dfrac{1}{\sqrt{2}}(-\sin t\,\mathbf{i} + \cos t\,\mathbf{j} + \mathbf{k})$

$\mathbf{N} = -\cos t\,\mathbf{i} - \sin t\,\mathbf{j}$

$\mathbf{B} = \sin t\,\mathbf{i} - \cos t\,\mathbf{j} + \dfrac{1}{\sqrt{2}}\mathbf{k}$

17. $\kappa(t) = \sqrt{\dfrac{8 + 5t^2 + t^4}{(2+t^2)^3}}$ 19. $\kappa(t) = \cos t\sqrt{1 + \cos^2 t}$

PROBLEMS 13.11

2. (b) $6x(4x + 9x)^{-3/2}$ (c) $e^{-x}(1 + e^{-2x})^{-3/2}$ (e) $\cos x$

(g) $\dfrac{2 + \theta^2}{4(1 + \theta^2)^{3/2}}$ (h) $(2e^{2\theta})^{-1/2}$ (i) $\cos^2\theta - \tfrac{1}{2}\sin^2\theta$

5. $\sqrt{\tau^2(t) + \kappa^2(t)}$ 6. $\kappa(t) = \tau(t) = (1+t^2)^{-2}$

7. The vertices.   9. $\sqrt{1 - 2k\kappa(t) + k^2[\kappa^2(t) + \tau^2(t)]}$

11. $\mathbf{p}(\theta) = -a\cos\theta\,\mathbf{i} + a(-\sin\theta + \ln(\sec\theta + \tan\theta))\mathbf{j}$

EXERCISES 13.12

1. $\mathbf{v} = 4a(-\sin\theta\,\mathbf{u}_\theta + \cos\theta\,\mathbf{u}_\theta{}^\perp)$
   $\mathbf{a} = -16a(\cos\theta\,\mathbf{u}_\theta + \sin\theta\,\mathbf{u}_\theta{}^\perp)$

3. $\mathbf{v} = a(\cos\theta\,\mathbf{u}_\theta + (1 + \sin\theta)\mathbf{u}_\theta{}^\perp)$
   $\mathbf{a} = -a(1 + 2\sin\theta)\mathbf{u}_\theta + 2a\cos\theta\,\mathbf{u}_\theta{}^\perp$

5. $\mathbf{v} = -2a\sin 4\theta\,\mathbf{u}_\theta + \tfrac{1}{2}a\cos 4\theta\,\mathbf{u}_\theta{}^\perp$
   $\mathbf{a} = -\tfrac{1}{4}a(16\sin 4\theta + \cos 4\theta)\mathbf{u}_\theta - 2a\sin 4\theta\,\mathbf{u}_\theta{}^\perp$

8. $\mathbf{v} = (1 - \eta\cos\theta)^{-2}(-c\eta^2\sin\theta\,\mathbf{u}_\theta + c\eta(1 - \eta\cos\theta)\mathbf{u}_\theta{}^\perp)\dfrac{d\theta}{dt}$

PROBLEMS 13.12

1. A circle   3. $a_r = -k^3 s\theta^3 \left[\dfrac{2 - \theta^2}{(1 + \theta^2)^2}\right]$

EXERCISES 14.1

2. $\begin{bmatrix} 2 & 1 \\ 3 & -2 \end{bmatrix} \begin{bmatrix} 2 & 1 & 3 \\ 3 & -2 & 1 \end{bmatrix}$ 4. $\begin{bmatrix} 3 & 4 \\ -1 & 2 \\ 1 & -4 \end{bmatrix} \begin{bmatrix} 3 & 4 & 6 \\ -1 & 2 & 4 \\ 1 & -4 & -2 \end{bmatrix}$

6. $\begin{bmatrix} -1 & 2 & 1 & -4 \\ 2 & -1 & -1 & 2 \end{bmatrix} \begin{bmatrix} -1 & 2 & 1 & 4 & 0 \\ 2 & -1 & -1 & 2 & 2 \end{bmatrix}$

8. $\begin{bmatrix} 1 & 0 & 0 & 1 & 0 \\ 0 & 2 & 1 & 0 & 0 \\ 0 & 0 & -1 & 1 & 0 \\ 0 & 1 & 0 & 0 & -2 \\ 2 & 0 & 0 & -1 & 0 \end{bmatrix} \begin{bmatrix} 1 & 0 & 0 & 1 & 0 & 1 \\ 0 & 2 & 1 & 0 & 0 & -1 \\ 0 & 0 & -1 & 1 & 0 & 0 \\ 0 & 1 & 0 & 0 & -2 & 3 \\ 2 & 0 & 0 & -1 & 0 & 2 \end{bmatrix}$

PROBLEMS 14.1

3. $ad = bc$, $3a = 2b$, $3c = 2d$

EXERCISES 14.2

1. $\begin{bmatrix} 1 & 0 & 2 \\ 0 & 1 & 0 \end{bmatrix}$ 2. $\begin{bmatrix} 1 & -\tfrac{1}{2} & 0 \\ 0 & 0 & 1 \end{bmatrix}$ 5. $\begin{bmatrix} 1 & 0 & 0 & -1 \\ 0 & 1 & 0 & -3 \\ 0 & 0 & 0 & 0 \end{bmatrix}$

7. $\begin{bmatrix} 1 & 0 & -\tfrac{1}{2} & -1 & 0 \\ 0 & 1 & \tfrac{3}{2} & 0 & 0 \\ 0 & 0 & 0 & 0 & 1 \end{bmatrix}$ 10. $\begin{bmatrix} 1 & 0 & 0 & -\tfrac{1}{6} & 0 \\ 0 & 1 & 0 & \tfrac{1}{3} & 0 \\ 0 & 0 & 1 & -\tfrac{2}{3} & 0 \\ 0 & 0 & 0 & 0 & 1 \end{bmatrix}$

21. $(2, 0)$ 23. $(1 + z, 1 + z, z)$ 24. Inconsistent.

27. $(0, x, x, 0)$
29. $(\frac{1}{4}, y - \frac{7}{4}, y, -\frac{1}{2})$
30. $(0, 3, 1, -5)$
32. $(2 - z, 3 - z, z)$
34. $(-z, z + 2, z)$
35. $(1, -2, 3, -4)$

PROBLEMS 14.2

3. (a) $\alpha = 1$; no
3. (b) $\alpha = -1$; yes

EXERCISES 14.3

1. 0

3. $\begin{bmatrix} 8 & 2 \\ -3 & -12 \end{bmatrix}$

6. $\begin{bmatrix} 8 & -2 & 6 \\ 4 & 4 & 4 \end{bmatrix}$

8. $\begin{bmatrix} 8 & 0 & 0 \\ 11 & 2 & 0 \\ 0 & 1 & -3 \end{bmatrix}$

9. $\begin{bmatrix} 7 & 6 & -7 \\ 7 & 4 & -9 \\ -2 & 2 & 5 \\ -3 & 2 & 0 \end{bmatrix}$

10. $\begin{bmatrix} 18 & -6 \\ -3 & 22 \end{bmatrix}$

11., 12. $\begin{bmatrix} 2 & -2 & 3 \\ 1 & 5 & 0 \\ 4 & -1 & 4 \end{bmatrix}$

13., 14. $\begin{bmatrix} 8 & 10 & 3 \\ -8 & 8 & -6 \\ 1 & 7 & 1 \end{bmatrix}$

15. $\begin{bmatrix} 12 & 4 & 5 \\ 8 & 0 & 0 \\ 0 & 4 & 12 \end{bmatrix}$

17. $\begin{bmatrix} 0 & 4 & -11 \\ 6 & 3 & 15 \\ -3 & 2 & -4 \end{bmatrix}$

18. $\begin{bmatrix} 0 & -2 & -30 \\ 0 & 4 & 20 \\ 0 & 4 & -6 \end{bmatrix}$

20. $\begin{bmatrix} -6 & -38 & -86 \\ 10 & -10 & -20 \\ 4 & -52 & -164 \end{bmatrix}$

PROBLEMS 14.3

2. $cA = (cI_m)A$
5. Use induction.

6. $\begin{bmatrix} a & b \\ -b & a \end{bmatrix}$

10. $\begin{bmatrix} 1 & 3 \\ 2 & 6 \end{bmatrix} \begin{bmatrix} 6 & 3 \\ -2 & -1 \end{bmatrix}$

13. $\begin{bmatrix} 1 & 1 \\ 0 & 0 \end{bmatrix}$

EXERCISES 14.4

1. $\begin{bmatrix} \frac{1}{3} & -\frac{1}{6} \\ \frac{1}{3} & \frac{1}{3} \end{bmatrix}$

4. $\begin{bmatrix} -\frac{1}{6} & \frac{1}{6} & \frac{1}{3} \\ -\frac{17}{6} & -\frac{1}{6} & \frac{5}{3} \\ -\frac{5}{6} & -\frac{1}{6} & \frac{2}{3} \end{bmatrix}$

6. $\begin{bmatrix} \frac{1}{2} & 0 & 0 \\ \frac{1}{2} & -1 & 0 \\ 1 & -\frac{1}{2} & -\frac{1}{2} \end{bmatrix}$

8. Singular

9. $\begin{bmatrix} 1 & 0 & 1 & -1 \\ 0 & 1 & 1 & -1 \\ -1 & 0 & 0 & 1 \\ 0 & 0 & -1 & 1 \end{bmatrix}$

11. $\frac{1}{30} \begin{bmatrix} 14 & -8 & 2 \\ 10 & -10 & 10 \\ -1 & 7 & 2 \end{bmatrix}$

13. $\begin{bmatrix} -\frac{1}{5} & \frac{2}{5} & -\frac{4}{15} \\ \frac{1}{24} & -\frac{1}{24} & \frac{1}{12} \\ -\frac{1}{4} & \frac{1}{4} & -\frac{1}{6} \end{bmatrix}$

14. $\begin{bmatrix} -\frac{11}{40} & \frac{41}{120} & -\frac{41}{120} \\ \frac{1}{60} & -\frac{1}{30} & \frac{37}{720} \\ -\frac{7}{240} & -\frac{1}{240} & -\frac{1}{40} \end{bmatrix}$

PROBLEMS 14.4

4. Even order such matrices are non-singular
5. $A^{-1} = I - A + A^2 - \cdots + (-1)^{k-1}A^{k-1}$ 7. No to both.

EXERCISES 14.5

1. -11 3. 3 5. 0 6. 2 8. -4
10. 126 12. -33 14. 825 16. -26225 18. -25

PROBLEMS 14.5

2. Use Theorem 14.16.   6. Use Theorems 14.18 and 14.19.   10. $\Pi_{i<j}(x_j - x_i)$
11. Use Theorems 14.17 and 14.19.   13. Use Problem 11.   15. Try a 3×3 case first.

EXERCISES 14.6

2. No.   3. $4(2, -1, 1) - 3(1, 0, 2)$ 4. $4(1, 1, -1, 0) - 5(1, 0, -1, 1) + 3(0, 0, 1, 2)$
6. $(39, -14, 30)$ 8. Add **k**.   9. Add $(0, 1, 0, 0)$
11. $\mathbf{v}_1 - 2\mathbf{v}_2 - 3\mathbf{v}_3 + \mathbf{v}_4$ 12. Yes.

PROBLEMS 14.6   The problems are all routine verifications of definitions.

EXERCISES 14.7

1. $(2, 1, -1)$ 3. $(1, 0, -1, 0), (0, 1, 2, -2)$ 4. $(-6, 4, 1, 5)$ 5. $(1, 1, 2, 0)$
8. $(0, 2, 2) + s(1, -1, -1)$ 10. $(0, 0, 0, 2) + s(2, 1, 0, -2) + t(1, 0, -1, 0)$
11. $(1, 1, 1, 1) + s(-6, 4, 1, 5)$ 13. $(0, 3, 1, 0, 0) + s(1, -6, -5, 3, 0) + t(2, 0, 5, 0, 3)$
14. $(1, -1, 1, -1, 1)$ 17. $x + y + z = 0$
$2x + 2y + z = 0$

19. $w + y - 2z = 0$ 22. $2w + 2x + z = 7$
$w - x + 2y - 3z = 0$ $ 2x + y = 6$
$w + 2x - y = 0$
$w - 2x + 3y - 4z = 0$

23. $(1, 1, 1)$ 24. $(3, 0, -2)$ 25. $(3, -1, 2, -1)$

PROBLEMS 14.7

3. Any **b**   4. A^{-1} must exist.
5. Use trigonometric identities to show that $\alpha + \beta + \gamma = 0$ implies that $1 + 2 \cos \alpha \cos \beta \cos \gamma = \cos^2 \alpha + \cos^2 \beta + \cos^2 \gamma$.

EXERCISES 14.8

1. $\begin{bmatrix} 1 & 1 & 0 \\ 1 & 0 & 1 \end{bmatrix}$ $\ker(L) = \langle (1, -1, -1) \rangle$
$\operatorname{im}(L) = \mathbf{R}^2$

4. $\begin{bmatrix} 1 & -1 & 0 \\ -1 & 1 & 0 \\ 0 & -1 & 1 \end{bmatrix}$ $\ker(L) = \langle(1, 1, 1)\rangle$
$\operatorname{im}(L) = \langle(1, -1, 0), (0, 0, 1)\rangle$

6. $\begin{bmatrix} 3 & -2 & 1 \\ 1 & -3 & 2 \\ 2 & 3 & -2 \end{bmatrix}$ $\ker(L) = \mathbf{0}$
$\operatorname{im}(L) = \mathbf{R}^3$ 7. Not linear.

9. $\begin{bmatrix} 1 & 0 & 0 \\ 0 & 1 & 0 \\ 1 & 0 & 1 \\ 1 & 1 & 0 \end{bmatrix}$ $\ker(L) = \mathbf{0}$
$\operatorname{im}(L) = \langle(1, 0, 1, 1), (0, 1, 0, 1), (0, 0, 1, 0)\rangle$

11. $\begin{bmatrix} 0 & 2 \\ -1 & 1 \end{bmatrix}$ 13. $\begin{bmatrix} 1 & a \\ 0 & 1 \end{bmatrix}$ 15. $\begin{bmatrix} 3 & 2 \\ -2 & -3 \end{bmatrix}$

17. $\begin{bmatrix} 1 & 1 & 2 \\ 0 & 2 & 2 \\ -1 & 1 & 0 \end{bmatrix}$ $\operatorname{rank} A = 2$
$\ker(L) = \langle(1, 1, -1)\rangle$
$\operatorname{im}(L) = \langle(1, 0, -1), (1, 1, 0)\rangle$

19. $\begin{bmatrix} 1 & 2 & 3 \\ 3 & 2 & 1 \\ 2 & 0 & 2 \end{bmatrix}$ $\operatorname{rank} A = 3$
$\ker(L) = \mathbf{0}$
$\operatorname{im}(L) = \mathbf{R}^3$

20. $\begin{bmatrix} 1 & 0 & 2 \\ 0 & 1 & -3 \\ 1 & -1 & 5 \\ 2 & 1 & 1 \end{bmatrix}$ $\operatorname{rank} A = 2$
$\ker(L) = \langle(2, -3, -1)\rangle$
$\operatorname{im}(L) = \langle(1, 0, 1, 2), (0, 1, -1, 1)\rangle$

PROBLEMS 14.8

1. (a) $L(x, y, z) = (y + z, z, 0)$
   (b) $L(x, y, z) = (0, 0, z), K(x, y, z) = (0, 0, x)$
   (c) $L(x, y, z) = (0, 0, x), K(x, y, z) = (y, 0, 0)$

2. $\begin{bmatrix} 0 & 0 & 0 & \cdots & 1 \\ 1 & 0 & 0 & \cdots & 0 \\ 0 & 1 & 0 & \cdots & 0 \\ & & \vdots & & \end{bmatrix}$ 3. $c = 6$ or -1

4. Diagonal   9. $L + K =$ identity, $L \circ K = 0 = K \circ L$, $L \circ L = L$, $K \circ K = K$.

12. $\begin{bmatrix} a & 0 \\ 0 & b \end{bmatrix}$

EXERCISES 14.9

1. $\begin{bmatrix} \frac{1}{2} & 0 \\ 0 & 2 \end{bmatrix}$
2. $\begin{bmatrix} \frac{2}{5} & -\frac{1}{5} & \frac{2}{5} \\ \frac{4}{5} & \frac{3}{5} & -\frac{1}{5} \\ \frac{4}{5} & -\frac{2}{5} & -\frac{1}{5} \end{bmatrix}$

11. $\frac{1}{\sqrt{5}}(2, 1), \frac{1}{\sqrt{5}}(-1, 2)$ 13. $\frac{1}{\sqrt{14}}(1, 3, 2), \frac{1}{\sqrt{42}}(5, 1, -4)$

14. $\frac{1}{\sqrt{6}}(-1, 1, 2), \frac{1}{\sqrt{5}}(2, 0, 1), \frac{1}{\sqrt{30}}(1, 5, -2)$

16. $\frac{1}{\sqrt{3}}(1, 0, 1, -1), \frac{1}{\sqrt{33}}(4, 3, -2, 2)$

18. $(\frac{1}{2}, \frac{1}{2}, \frac{1}{2}, \frac{1}{2}), (\frac{1}{2}, -\frac{1}{2}, \frac{1}{2}, -\frac{1}{2}), (-\frac{1}{2}, \frac{1}{2}, \frac{1}{2}, -\frac{1}{2})$

20. $\frac{1}{\sqrt{6}}(2, 1, -1, 0), \frac{1}{\sqrt{3}}(0, 1, 1, -1), \frac{1}{\sqrt{198}}(8, -7, 9, 2), \frac{1}{\sqrt{99}}(1, -5, -3, -8)$

PROBLEMS 14.9 (Gram-Schmidt Orthonormalization)

4. $\frac{1}{\sqrt{2}}, \sqrt{\frac{3}{2}}x, \frac{1}{8}\sqrt{135}\left(x^2 - \frac{1}{3}\right), \cdots$

EXERCISES 14.10

2. $3 \pm 2\sqrt{2}$
4. $-\frac{1}{2} \pm \frac{1}{2}\sqrt{33}$
5. $1, 2, -3$
7. $0, \pm\sqrt{2}$
9. $1, -1, 2, 3$
10. $1, -1, -1 \pm \sqrt{2}$

11. $\begin{bmatrix} \cos\frac{\pi}{6} & \sin\frac{\pi}{6} \\ \sin\frac{\pi}{6} & -\cos\frac{\pi}{6} \end{bmatrix}$
13. $\begin{bmatrix} \cos\frac{\pi}{8} & \sin\frac{\pi}{8} \\ -\sin\frac{\pi}{8} & \cos\frac{\pi}{8} \end{bmatrix}$
17. $\begin{bmatrix} \frac{1}{\sqrt{2}} & \frac{1}{2} & -\frac{1}{2} \\ \frac{1}{\sqrt{2}} & -\frac{1}{2} & \frac{1}{2} \\ 0 & \frac{1}{\sqrt{2}} & \frac{1}{\sqrt{2}} \end{bmatrix}$

18. $\begin{bmatrix} 0 & \frac{1}{\sqrt{2}} & \frac{1}{\sqrt{2}} \\ 0 & \frac{1}{\sqrt{2}} & -\frac{1}{\sqrt{2}} \\ 1 & 0 & 0 \end{bmatrix}$
20. $\begin{bmatrix} \frac{1}{\sqrt{2}} & \frac{3+2\sqrt{3}}{\sqrt{54+30\sqrt{3}}} & \frac{-3+2\sqrt{3}}{\sqrt{54-30\sqrt{3}}} \\ 0 & \frac{-3-\sqrt{3}}{\sqrt{54+30\sqrt{3}}} & \frac{3-\sqrt{3}}{\sqrt{54-30\sqrt{3}}} \\ \frac{1}{\sqrt{2}} & \frac{-3-2\sqrt{3}}{\sqrt{54+30\sqrt{3}}} & \frac{3-2\sqrt{3}}{\sqrt{54-30\sqrt{3}}} \end{bmatrix}$

21. $6, -2$

PROBLEMS 14.10

4. $(A^T A)^T = A^T (A^T)^T = A^T A$
7. $\det(A - 0I) = \det A = 0$
8. $\det(A^p - \alpha^p I) = \det[(A - \alpha I)(A^{p-1} + \cdots + \alpha^{p-1} I)]$
9. No.
12. Use Problem 8.
16. $\det(cI + A - (c + \alpha)I) = \det(A - \alpha I)$.

EXERCISES 14.11

1. $\frac{1}{36}(137 + 7\sqrt{37})x^2 + \frac{1}{36}(137 - 7\sqrt{37})y^2$
3. $3x^2 - 2y^2$
5. $6x^2 - \frac{3}{2}(1 - \sqrt{5})y^2 - \frac{3}{2}(1 + \sqrt{5})z^2$
7. $\sqrt{5}(y^2 - z^2)$
8. $x^2 + y^2 - 2z^2$
9. $4x^2 - y^2 - 2z^2$
11. Yes
13. No
15. Yes
17. No
19. $\begin{bmatrix} -1 & -1 \\ -1 & -4 \end{bmatrix}$

PROBLEMS 14.11

4. This is the number of different ways to write down s ones followed by t minus ones followed by $n - s - t$ zeros.

5. Recall Problem 14.10.3.
8. Recall Problems 14.10.4 and 14.10.7.

EXERCISES 14.12

1. hyperbola
4. hyperbola
6. ellipse
7. $\alpha = \frac{1}{2}, \frac{1}{2}, -1$
9. $\alpha = (3 \pm \sqrt{17})/2, -1$
10. $\alpha = \pm 3, 0$
11. $\alpha = 5, 4, 0$
12. $\alpha = 18, 9, 9$

PROBLEMS 14.12

Problems 1 and 2 only serve to show that the determinants are not really effective in classification.

INDEX

INDEX

absolute value, 12
acceleration, 184
addition formula (sine and cosine), 402
adjoint matrix, 731
algebraic field, 3
algebraic function, 166
alternating series, 228
amplitude of a sine wave, 423
antiderivative, 293
Archimedean property, 14
arc length, 324, 678
area, 250
area of a surface of revolution, 327
associative laws, 3
asymptote, 73, 656
Atwood's machine, 188 (Problem)
augmented matrix, 697
average unit cost, 111 (Problem)
average value of a function, 257

Bailey root extraction method, 183 (Problem)
basis, 738
basis neighborhood, 511
Bernoulli numbers, 381 (Problem)
Bessel equation, 243 (Problem)
Bessel function, 242 (Problem)
binary splitting, 124
binomial series, 246
binormal vector, 680
bound, upper, 28
bound, lower, 28
boundary, 513
boundary point, 513
bounded sequence, 81
bounded set, 28

canonical form, 777
catenary, 450 (Exercise)
Cauchy-Riemann equations, 532 (Problem)
Cavalieri's theorem, 323 (Problem)
Cayley-Hamilton theorem, 772 (Problem)
center of curvature, 685
center of mass, 598
central conic, 654
central force, 691
chain rule, 137, 541
characteristic equation, 767
characteristic root, 766
characteristic vector, 766
closed interval, 26
closed ray, 26
closed subset, 513

closed unit disk, 513
closure properties, 3
codomain, 62
cofactor, 728
column rank, 743
column space, 743
column vector, 697
common logarithm, 354
commutative laws, 3
commutative ring, 134 (Problem)
comparison test, 225, 316
completeness, 29
component, 34
component functions, 668
component of velocity, 674
component-wise addition, 34
composite function, 135
concave curves, 144
cone, 662
confocal family, 658 (Problem)
congruent matrices, 775
conics; conic sections, 651
consumer demand, 262
continuous function, 119, 522
continuous in one variable, 524 (Problem)
contraction, 751
convergent improper integral, 315
convergent sequence, 78
convergent series, 221
convex curves, 144
coordinate, 25
coordinate line, 25
coordinate plane, 39
coordinate system, 25, 39
cosecant function, 413
cosine function, 399
cotangent function, 413
Cramer's rule, 745
critical point, 113
critical value, 113, 554
cross product, 631
cross-section, 517, 643
curvature, 683
cylinder, 519
cylindrical coordinates, 611

damped vibration, 426
decay law, 386
decreasing function, 113
definite integral, 266
demand function, 61, 201
density of a solid, 599
density property of rational numbers, 14
dependent vectors, 738

derivative, 106
derived function, 106
determinant, 727
diagonal matrix, 737, 767
difference formulas (trigonometric functions), 403, 413
differentiable function, 108, 538, 693
differential, 195
differential equation, 185, 295
differentiation, 106
dilation, 751
dilution problems, 389
dimension, 740
direct sum, 742 (Problem)
directed line segment, 625
direction angles, 626
direction cosines, 627
direction numbers, 504
direction vector, 504, 627
directional derivative, 645
directrix, 651
discontinuity, 120
disjoint sets, 17
distance formula, 25, 45
distributive law, 3
divergent sequence, 82
divergent series, 221
domain, 62
dot product, 629
double angle formulas, 402
double integral, 588
double power series, 550
dummy variable, 291

eccentricity, 651
echelon index, 706
elasticity of demand, 202
elementary matrix, 718
elementary row operations, 699
ellipse, 652
ellipsoid, 660
elliptical cone, 662
elliptical integral, 494 (Problem)
elliptical paraboloid, 664
empty set, 17
equality, 8, 9, 17
equation of a line, 52
equation of motion, 183
equivalent line segments, 625
equivalent systems of equations, 699
error in Simpson's rule, 497
error, percentage and relative, 198
error function, 595
escape velocity, 152 (Problem)

Euler's constant, 382 (Problem)
Euler's formula, 436 (Problem)
even function, 71
evolute, 689 (Problem)
exact differential, 568
exponential function, 350
extended mean value theorem, 173
extent, 643
extreme value, 113, 552
extreme value with constraints, 561

finite dimensional, 740
first moment, 598
first order linear differential equation, 382
fixed point, 755
fluid pressure, 335
focal chord, 659 (Problem)
focal radius, 658 (Problem)
focus, 651
free falling body, 184
free vector, 624
Frenet formulas, 688
frequency, 425
function, 64
functional equation, 356
functional value, 59, 62
fundamental identity (sine and cosine), 399
fundamental theorem of calculus, 288

Gaussian pivot operation, 702
general linear equation, 52
general solution, 298
gradient vector, 536, 537, 647
Gram-Schmidt orthonormalization, 760
graph, 65, 515
greatest lower bound, 28
growth law, 387
gudermannian $gd(x)$, 451 (Problem)

half-angle formulas, 403
half-life, 387
half-open interval, 26
harmonic function, 532 (Exercise)
harmonic series, 222
heat equation, 533 (Problem)
Heaviside's unit function, 69 (Problem)
helix, 667
heterogeneous solid, 618
higher derivative, 143, 526
homogeneous function, 532 (Problem)
homogeneous solid, 613
Hooke's law, 335 (Problem)
hyperbola, 652

hyperbolic function, 445
hyperbolic paraboloid, 518, 665
hyperboloids, 663, 664
hyperplane, 510, 742 (Problem)

idempotent matrix, 720 (Problem)
identity element, 3
identity function, 121
identity matrix, 717
image, 754
image set, 63
implicit differentiation, 545
improper integral, 315
improper orthogonal matrix, 763
increasing function, 113
increment, 193
indefinite integral, 293
independent vectors, 738
indeterminate form, 207
index of a quadratic form, 777
infinite dimensional, 740
infinite sequence, 77
infinite series, 221
initial condition, 295
initial distance, 185
initial point of a segment, 625
initial velocity, 185
inner product space, 765 (Problem)
integers, 2
integrable function, 588
integral test, 317
integrating factor, 383, 572 (Problem)
integration by parts, 309
integration of power series, 489
intercept, 70, 517, 642
intercept form of a linear equation, 55 (Problem), 511 (Problem)
interior, 513
interior point, 513
intermediate value theorem, 125
intersection of sets, 17
interval of convergence, 236
invariance, 755
inverse function, 342
inverse hyperbolic functions, 451
inverse matrix, 721
inverse square law, 692
inverse trigonometric functions, 436
involutory matrix, 720 (Problem)
irrational number, 6
iterated integral, 577
iterated limit, 521

Kepler's first law, 691

kernel, 754
Kronecker delta, 717

Lagrange multiplier, 561
Lagrangian, 563
Laplace's equation, 532 (Exercise)
law of cosines, 401
least squares, method of, 557
least upper bound, 28
left derivative, 110 (Problem)
left limit, 99 (Problem)
Legendre polynomial, 765 (Problem)
Leibnitz' rule, 575
level curve, 518, 641
level set, 641
level surface, 641
l'Hospital's rule, 208
limit, 92, 520
limit along a curve, 521
limiting velocity, 391
linear equations, system of, 696
linear function, 50, 506, 509
linear inequality, 11
linear manifold, 740
linear mapping, 750
logarithm function, 350
logarithmic differentiation, 373
lower bound, 28
lower sequence, 267, 587
lower sum, 277, 587
lower triangular matrix, 720 (Problem)

Maclaurin series, 245
main diagonal, 717
mapping, 749
marginal cost, 111 (Problem), 204
marginal demand, 61 (Problem), 202
marginal profit, 206
marginal revenue, 111 (Problem), 206
market equilibrium, 55 (Problem)
mathematical induction, 17
matrix, 697
matrix of a linear mapping, 750
matrix product, 715
mean value theorem, 170
mean value theorem for integrals, 274
minimax value, 554
Minkowski inequality, 276 (Problem)
minor, 727
moment of inertia, 601, 618
monotone, decreasing, 81, 343
monotone, increasing, 81, 224, 343

natural logarithm function, 359

natural numbers, 2
n-dimensional numerical vector, 34
negative direction, 24
net profit, 111 (Problem)
Newton's first law of motion, 692
Newton's forward interpolation formula, 162
Newton's iterative method, 178
Newton's law of cooling, 391
nilpotent matrix, 734 (Problem)
non-singular matrix, 721
norm, 44, 624
normal vector, 637, 680
null space, 754
nth moment, 604

octant, 42
odd function, 72
Ohm's law, 393
open interval, 26
open ray, 26
open subset, 512
open unit disk, 512
ordered n-tuple, 34
origin, 24
orthogonal matrix, 762
orthogonal vectors, 630
orthogonally equivalent quadratic forms, 776
orthonormal vectors, 760
osculating circle and plane, 685

Pappus' theorem, 614 (Problem)
parabola, 652
parallel axes, theorem of, 605 (Problem)
parallel curve, 689 (Problem)
parallel lines, 56
parallel vectors, 633
parametric equations, 505, 667
parametrization, 667
partial derivative, 524
partial fractions, 479
partial sum of a series, 221
partition of an interval, 276
partitioned matrix, 698
period, 421
periodic function, 400, 421
phase angle, 425
plane, 507
point of inflection, 146
pointwise invariance, 755
polar coordinates, 590
polar curve, 591
polynomial function, 156

position vector, 39, 43
positive, definite and semidefinite, 779 (Problem)
positive direction, 24
power series, 234
power series expansion, 243
present value, 372 (Problem)
prismatoid, 323 (Problem)
product of functions, 128
product of inertia, 605 (Problem)
product of matrices 715
product of sets, 63
proper orthogonal matrix, 763
p-series, 318

quadratic curve, 651
quadratic form, 773
quadratic function, 72
quadric surface, 660
quotient function, 128

radial component, 690, 691
radian, 400
radiocarbon dating, 388
radius of convergence, 234
radius of gyration, 602, 618
range of a function, 63, 754
rank, 707, 777
rate of change, 59, 87, 106, 188
ratio test, 227
rational function, 164
rational number, 2
rationalizing substitutions, 484
ray, open, 26
real number, 4
reduction formula, 458, 470 (Problem)
reflection property, 658 (Problem)
region, 250, 511
related rates, 188
relative extremum, 113, 553
relative maximum value, 113, 553
relative minimum value, 113, 553
remainder in a Taylor series, 244
removable discontinuity, 123 (Problem)
renewal function, 338
Riemann sum, 268, 589
right derivative, 110 (Problem)
right hand rule, 633
right limit, 99 (Problem)
Rolle's theorem, 168
root mean square, 299 (Problem)
root test, 234 (Problem)
rotation, 763
row-echelon form, 705

row-equivalence, 700
row rank, 743
row space, 743
row vector, 697
ruled surface, 667 (Problem)

saddle point, 554
scalar field, 641
scalar multiplication, 35
Schwartz' inequality, 164 (Problem), 275
screw curvature, 689 (Problem)
secant function, 412
second derivative test, 555
second moment, 601
sense of a segment, 627
set, 16
signature of a quadratic form, 778
similar matrices, 765
simple harmonic motion, 433
Simpson's rule, 495
sine function, 399
sine wave, 422
singular matrix, 721
skew-symmetric matrix, 726 (Problem)
slope, 51
slope-intercept form, 53
smooth function, 144, 675
Snell's law of refraction, 418 (Problem)
solid of revolution, 321
solution set, 698
Spearman-Brown function, 69 (Problem)
spherical coordinates, 614
spring law, 331
square matrix, 720
steady state current, 393
steepest descent, method of, 650 (Problem)
strictly monotone function, 343
strictly triangular matrix, 720 (Problem)
subsequence, 83
substitution of variables, 303
sum of functions, 128
sum of numerical vectors, 34
summation notation, 262
supply function, 55 (Problem), 202
surface of revolution, 661
survival function, 338
symmetric form of the equations of a line, 506 (Problem)
symmetric matrix, 735 (Problem)
symmetry, 71, 516, 643

tangency of functions, 108, 149
tangent function, 412

tangent line, 90
tangent plane, 536
Taylor polynomial, 214
Taylor series, 243
Taylor's theorem, 214, 548
telescoping sum, 269
term of a sequence, 77
terminal point of a segment, 625
terminal velocity, 455
tip of a position vector, 39, 43
topology, 514 (Problem)
torsion, 688
total cost, 111 (Problem), 204
total differential, 538
total revenue, 111 (Problem), 205
trace, 517, 642
trace of a matrix, 720 (Problem)
tractrix, 456 (Exercise)
transcendental function, 342
transformation, 749
transformation of coordinates, 759
transitive property, 8
translation, 757 (Problem)
transpose of a matrix, 731
transverse component, 690
trapezoid rule, 502 (Problem)
triangle inequality, 45
trichotomy property, 8
trigonometric substitutions, 471
triple integral, 606

uniformly continuous, 278
union of sets, 17
unit basis vector, 39, 505
unit length, 24
unit point, 24
unit tangent vector, 679
upper bound, 28
upper sequence, 267, 587
upper sum, 277, 587
upper triangular matrix, 720 (Problem)

Vandermonde determinant, 735 (Problem)
variable of integration, 291
variance, 779 (Problem)
vector acceleration, 674
vector derivative, 672
vector field, 641
vector function, 668
vector law of motion, 668
vector product, 632
vector radius of curvature, 685
vector space, 36, 736
vector subspace, 737

vector velocity, 674
velocity, 184
vertex of a parabola, 659 (Problem)
vibrating string equation, 533 (Problem)
volume, 320, 580

Weddle's rule, 502 (Problem)

well-ordering axiom, 18
work, 330
Wronskian, 735 (Problem)

zero divisor, 720
zero vector, 34